Graduate Texts in Mathematics   255

For other titles published in this series, go to
http://www.springer.com/series/136

Roe Goodman
Nolan R. Wallach

Symmetry, Representations,
and Invariants

 Springer

Roe Goodman
Department of Mathematics
Rutgers University
Piscataway, NJ 08854-8019
USA
goodman@math.rutgers.edu

Nolan R. Wallach
Department of Mathematics
University of California, San Diego
La Jolla, CA 92093
USA
nwallach@ucsd.edu

ISBN 978-1-4419-2729-3   e-ISBN 978-0-387-79852-3
DOI 10.1007/978-0-387-79852-3
Springer Dordrecht Heidelberg London New York

Mathematics Subject Classification (2000): 20G05, 14L35, 14M17, 17B10, 20C30, 20G20, 22E10, 22E46, 53B20, 53C35, 57M27

Springer is part of Springer Science+Business Media (www.springer.com)

Contents

Preface

Symmetry, in the title of this book, should be understood as the geometry of Lie (and algebraic) group actions. The basic algebraic and analytic tools in the study of symmetry are representation and invariant theory. These three threads are precisely the topics of this book. The earlier chapters can be studied at several levels. An advanced undergraduate or beginning graduate student can learn the theory for the classical groups using only linear algebra, elementary abstract algebra, and advanced calculus, with further exploration of the key examples and concepts in the numerous exercises following each section. The more sophisticated reader can progress through the first ten chapters with occasional forward references to Chapter 11 for general results about algebraic groups. This allows great flexibility in the use of this book as a course text. The authors have used various chapters in a variety of courses; we suggest ways in which courses can be based on the book later in this preface. Finally, we have taken care to make the main theorems and applications meaningful for the reader who wishes to use the book as a reference to this vast subject.

The authors are gratified that their earlier text, *Representations and Invariants of the Classical Groups* [56], was well received. The present book has the same aim: an entry into the powerful techniques of Lie and algebraic group theory. The parts of the previous book that have withstood the authors' many revisions as they lectured from its material have been retained; these parts appear here after substantial rewriting and reorganization. The first four chapters are, in large part, newly written and offer a more direct and elementary approach to the subject. Several of the later parts of the book are also new. While we continue to look upon the classical groups as both fundamental in their own right and as important examples for the general theory, the results are now stated and proved in their natural generality. These changes justify the more accurate new title for the present book.

We have taken special care to make the book readable at many levels of detail. A reader desiring only the statement of a pertinent result can find it through the table of contents and index, and then read and study it through the examples of its use that are generally given. A more serious reader wishing to delve into a proof of the result can read in detail a more computational proof that uses special properties of the classical groups, or, perhaps in a second reading, the proof in the general case (with occasional forward references to results from later chapters). Usually, there is a third possibility of a proof using analytic methods. Some material in the earlier book, although important in its own right, has been eliminated or replaced. There are new proofs of some of the key results of the theory such as the theorem of the highest weight, the theorem on complete reducibility, the duality theorem, and the Weyl character formula. We hope that our new presentation will make these fundamental tools more accessible.

The last two chapters of the book develop, via a basic introduction to complex algebraic groups, what has come to be called *geometric invariant theory*. This includes the notion of quotient space and the representation-theoretic analysis of the regular functions on a space with an algebraic group action. A full description of the material covered in the book is given later in the preface.

When our earlier text appeared there were few other introductions to the area. The most prominent included the fundamental text of Hermann Weyl, *The Classical Groups: Their Invariants and Representations* [164] and Chevalley's *The Theory of Lie groups I* [33], together with the more recent text *Lie Algebras* by Humphreys [76]. These remarkable volumes should be on the bookshelf of any serious student of the subject. In the interim, several other texts have appeared that cover, for the most part, the material in Chevalley's classic with extensions of his analytic group theory to Lie group theory and that also incorporate much of the material in Humphrey's text. Two books with a more substantial overlap but philosophically very different from ours are those by Knapp [86] and Procesi [123]. There is much for a student to learn from both of these books, which give an exposition of Weyl's methods in invariant theory that is different in emphasis from our book. We have developed the combinatorial aspects of the subject as consequences of the representations and invariants of the classical groups. In Hermann Weyl (and the book of Procesi) the opposite route is followed: the representations and invariants of the classical groups rest on a combinatorial determination of the representations of the symmetric group. Knapp's book is more oriented toward Lie group theory.

Organization

The logical organization of the book is illustrated in the chapter and section dependency chart at the end of the preface. A chapter or section listed in the chart depends on the chapters to which it is connected by a horizontal or rising line. This chart has a central spine; to the right are the more geometric aspects of the subject and on the left the more algebraic aspects. There are several intermediate terminal nodes in this chart (such as Sections 5.6 and 5.7, Chapter 6, and Chapters 9–10) that can serve as goals for courses or self study.

Chapter 1 gives an elementary approach to the classical groups, viewed either as Lie groups or algebraic groups, without using any deep results from differentiable manifold theory or algebraic geometry. Chapter 2 develops the basic structure of the classical groups and their Lie algebras, taking advantage of the defining representations. The complete reducibility of representations of $\mathfrak{sl}(2, \mathbb{C})$ is established by a variant of Cartan's original proof. The key Lie algebra results (Cartan subalgebras and root space decomposition) are then extended to arbitrary semisimple Lie algebras.

Chapter 3 is devoted to Cartan's highest-weight theory and the Weyl group. We give a new algebraic proof of complete reducibility for semisimple Lie algebras following an argument of V. Kac; the only tools needed are the complete reducibility for $\mathfrak{sl}(2,\mathbb{C})$ and the Casimir operator. The general treatment of associative algebras and their representations occurs in Chapter 4, where the key result is the general duality theorem for locally regular representations of a reductive algebraic group. The unifying role of the duality theorem is even more prominent throughout the book than it was in our previous book.

The machinery of Chapters 1–4 is then applied in Chapter 5 to obtain the principal results in classical representations and invariant theory: the first fundamental theorems for the classical groups and the application of invariant theory to representation theory via the duality theorem.

Chapters 6, on spinors, follows the corresponding chapter from our previous book, with some corrections and additional exercises. For the main result in Chapter 7—the Weyl character formula—we give a new algebraic group proof using the radial component of the Casimir operator (replacing the proof via Lie algebra cohomology in the previous book). This proof is a differential operator analogue of Weyl's original proof using compact real forms and the integration formula, which we also present in detail. The treatment of branching laws in Chapter 8 follows the same approach (due to Kostant) as in the previous book.

Chapters 9–10 apply all the machinery developed in previous chapters to analyze the tensor representations of the classical groups. In Chapter 9 we have added a discussion of the Littlewood–Richardson rule (including the role of the $\mathbf{GL}(n,\mathbb{C})$ branching law to reduce the proof to a well-known combinatorial construction). We have removed the partial harmonic decomposition of tensor space under orthogonal and symplectic groups that was treated in Chapter 10 of the previous book, and replaced it with a representation-theoretic treatment of the symmetry properties of curvature tensors for pseudo-Riemannian manifolds.

The general study of algebraic groups over $\mathbb{C}$ and homogeneous spaces begins in Chapter 11 (with the necessary background material from algebraic geometry in Appendix A). In Lie theory the examples are, in many cases, more difficult than the general theorems. As in our previous book, every new concept is detailed with its meaning for each of the classical groups. For example, in Chapter 11 every classical symmetric pair is described and a model is given for the corresponding affine variety, and in Chapter 12 the (complexified) Iwasawa decomposition is worked out explicitly. Also in Chapter 12 a proof of the celebrated Kostant–Rallis theorem for symmetric spaces is given and every implication for the invariant theory of classical groups is explained.

This book can serve for several different courses. An introductory one-term course in Lie groups, algebraic groups, and representation theory with emphasis on the classical groups can be based on Chapters 1–3 (with reference to Appendix D as needed). Chapters 1–3 and 11 (with reference to Appendix A as needed) can be the core of a one-term introductory course on algebraic groups in characteristic zero. For students who have already had an introductory course in Lie algebras and Lie groups, Chapters 3 and 4 together with Chapters 6–10 contain ample material for a second course emphasizing representations, character formulas, and their applications. An alternative (more advanced) second-term course emphasizing the geometric side of the subject can be based on topics from Chapters 3, 4, 11, and 12. A year-long course on representations and classical invariant theory along the lines of Weyl's book would follow Chapters 1–5, 7, 9, and 10. The exercises have been revised and many new ones added (there are now more than 350, most with several parts and detailed hints for solution). Although none of the exercises are used in the proofs of the results in the book, we consider them an essential part of courses based on this book. Working through a significant number of the exercises helps a student learn the general concepts, fine structure, and applications of representation and invariant theory.

Acknowledgments

In the end-of-chapter notes we have attempted to give credits for the results in the book and some idea of the historical development of the subject. We apologize to those whose works we have neglected to cite and for incorrect attributions. We are indebted to many people for finding errors and misprints in the many versions of the material in this book and for suggesting ways to improve the exposition. In particular we would like to thank Ilka Agricola, Laura Barberis, Bachir Bekka, Enriqueta Rodríguez Carrington, Friedrich Knop, Hanspeter Kraft, Peter Landweber, and Tomasz Przebinda. Chapters of the book have been used in many courses, and the interaction with the students was very helpful in arriving at the final version. We thank them all for their patience, comments, and sharp eyes. During the first year that we were writing our previous book (1989–1990), Roger Howe gave a course at Rutgers University on basic invariant theory. We thank him for many interesting conversations on this subject.

The first-named author is grateful for sabbatical leaves from Rutgers University and visiting appointments at the University of California, San Diego; the Université de Metz; Hong Kong University; and the National University of Singapore that were devoted to this book. He thanks the colleagues who arranged programs in which he lectured on this material, including Andrzei Hulanicki and Ewa Damek (University of Wrocław), Bachir Bekka and Jean Ludwig (Université de Metz), Ngai-Ming Mok (Hong Kong University), and Eng-Chye Tan and Chen-Bo Zhu (National University of Singapore). He also is indebted to the mathematics department of the University of California, Berkeley, for its hospitality during several summers of writing the book. The second-named author would like to thank the University of California, San Diego, for sabbatical leaves that were dedicated to this project, and to thank the National Science Foundation for summer support. We thank David Kramer for a superb job of copyediting, and we are especially grateful to our editor Ann Kostant, who has steadfastly supported our efforts on the book from inception to publication.

New Brunswick, New Jersey   *Roe Goodman*
San Diego, California   *Nolan R. Wallach*
February 2009

Organization and Notation

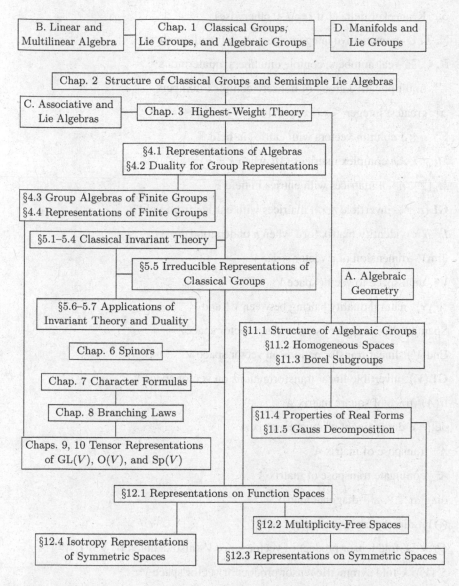

Dependency Chart among Chapters and Sections

"O," said Maggie, pouting, "I dare say I could make it out, if I'd learned what goes before, as you have." "But that's what you just couldn't, Miss Wisdom," said Tom. "For it's all the harder when you know what goes before: for then you've got to say what Definition 3. is and what Axiom V. is."   George Eliot, *The Mill on the Floss*

Some Standard Notation

$\#S$ number of elements in set S (also denoted by Card(S) and $|S|$)

δ_{ij} Kronecker delta (1 if $i = j$, 0 otherwise)

$\mathbb{N}, \mathbb{Z}, \mathbb{Q}$ nonnegative integers, integers, rational numbers

$\mathbb{R}, \mathbb{C}, \mathbb{H}$ real numbers, complex numbers, quaternions

$\mathbb{C}^{\times}$ multiplicative group of nonzero complex numbers

$[x]$ greatest integer $\leq x$ if x is real

$\mathbb{F}^n$ $n \times 1$ column vectors with entries in field $\mathbb{F}$

$M_{k,n}$ $k \times n$ complex matrices (M_n when $k = n$)

$M_n(\mathbb{F})$ $n \times n$ matrices with entries in field $\mathbb{F}$

GL$(n, \mathbb{F})$ invertible $n \times n$ matrices with entries from field $\mathbb{F}$

I_n $n \times n$ identity matrix (or I when n understood)

dim V dimension of a vector space V

V^* dual space to vector space V

$\langle v^*, v \rangle$ natural duality pairing between V^* and V

Span(S) linear span of subset S in a vector space.

End(V) linear transformations on vector space V

GL(V) invertible linear transformations on vector space V

tr(A) trace of square matrix A

det(A) determinant of square matrix A

A^t transpose of matrix A

A^* conjugate transpose of matrix A

diag$[a_1, \ldots, a_n]$ diagonal matrix

$\bigoplus V_i$ direct sum of vector spaces V_i

$\bigotimes^k V$ k-fold tensor product of vector space V (also denoted by $V^{\otimes k}$)

$S^k(V)$ k-fold symmetric tensor product of vector space V

$\bigwedge^k(V)$ k-fold skew-symmetric tensor product of vector space V

$\mathcal{O}[X]$ regular functions on algebraic set X

Other notation is generally defined at its first occurrence and appears in the index of notation at the end of the book.

Chapter 1
Lie Groups and Algebraic Groups

Abstract Hermann Weyl, in his famous book *The Classical Groups, Their Invariants and Representations* [164], coined the name *classical groups* for certain families of matrix groups. In this chapter we introduce these groups and develop the basic ideas of Lie groups, Lie algebras, and linear algebraic groups. We show how to put a Lie group structure on a closed subgroup of the general linear group and determine the Lie algebras of the classical groups. We develop the theory of complex linear algebraic groups far enough to obtain the basic results on their Lie algebras, rational representations, and Jordan–Chevalley decompositions (we defer the deeper results about algebraic groups to Chapter 11). We show that linear algebraic groups are Lie groups, introduce the notion of a *real form* of an algebraic group (considered as a Lie group), and show how the classical groups introduced at the beginning of the chapter appear as real forms of linear algebraic groups.

1.1 The Classical Groups

The *classical groups* are the groups of invertible linear transformations of finite-dimensional vector spaces over the real, complex, and quaternion fields, together with the subgroups that preserve a volume form, a bilinear form, or a sesquilinear form (the forms being nondegenerate and symmetric or skew-symmetric).

1.1.1 General and Special Linear Groups

Let $\mathbb{F}$ denote either the field of real numbers $\mathbb{R}$ or the field of complex numbers $\mathbb{C}$, and let V be a finite-dimensional vector space over $\mathbb{F}$. The set of invertible linear transformations from V to V will be denoted by $\mathbf{GL}(V)$. This set has a group structure under composition of transformations, with identity element the identity transformation $I(x) = x$ for all $x \in V$. The group $\mathbf{GL}(V)$ is the first of the classical

R. Goodman, N.R. Wallach, *Symmetry, Representations, and Invariants,*
Graduate Texts in Mathematics 255, DOI 10.1007/978-0-387-79852-3_1,
© Roe Goodman and Nolan R. Wallach 2009

groups. To study it in more detail, we recall some standard terminology related to linear transformations and their matrices.

Let V and W be finite-dimensional vector spaces over $\mathbb{F}$. Let $\{v_1, \ldots, v_n\}$ and $\{w_1, \ldots, w_m\}$ be bases for V and W, respectively. If $T : V \longrightarrow W$ is a linear map then

$$Tv_j = \sum_{i=1}^{m} a_{ij}w_i \qquad \text{for } j = 1, \ldots, n$$

with $a_{ij} \in \mathbb{F}$. The numbers a_{ij} are called the *matrix coefficients* or *entries* of T with respect to the two bases, and the $m \times n$ array

$$A = \begin{bmatrix} a_{11} & a_{12} & \cdots & a_{1n} \\ a_{21} & a_{22} & \cdots & a_{2n} \\ \vdots & \vdots & \ddots & \vdots \\ a_{m1} & a_{m2} & \cdots & a_{mn} \end{bmatrix}$$

is the *matrix* of T with respect to the two bases. When the elements of V and W are identified with column vectors in $\mathbb{F}^n$ and $\mathbb{F}^m$ using the given bases, then action of T becomes multiplication by the matrix A.

Let $S : W \longrightarrow U$ be another linear transformation, with U an l-dimensional vector space with basis $\{u_1, \ldots, u_l\}$, and let B be the matrix of S with respect to the bases $\{w_1, \ldots, w_m\}$ and $\{u_1, \ldots, u_l\}$. Then the matrix of $S \circ T$ with respect to the bases $\{v_1, \ldots, v_n\}$ and $\{u_1, \ldots, u_l\}$ is given by BA, with the product being the usual product of matrices.

We denote the space of all $n \times n$ matrices over $\mathbb{F}$ by $M_n(\mathbb{F})$, and we denote the $n \times n$ identity matrix by I (or I_n if the size of the matrix needs to be indicated); it has entries $\delta_{ij} = 1$ if $i = j$ and 0 otherwise. Let V be an n-dimensional vector space over $\mathbb{F}$ with basis $\{v_1, \ldots, v_n\}$. If $T : V \longrightarrow V$ is a linear map we write $\mu(T)$ for the matrix of T with respect to this basis. If $T, S \in \mathbf{GL}(V)$ then the preceding observations imply that $\mu(S \circ T) = \mu(S)\mu(T)$. Furthermore, if $T \in \mathbf{GL}(V)$ then $\mu(T \circ T^{-1}) = \mu(T^{-1} \circ T) = \mu(\mathrm{Id}) = I$. The matrix $A \in M_n(\mathbb{F})$ is said to be *invertible* if there is a matrix $B \in M_n(\mathbb{F})$ such that $AB = BA = I$. We note that a linear map $T : V \longrightarrow V$ is in $\mathbf{GL}(V)$ if and only if its matrix $\mu(T)$ is invertible. We also recall that a matrix $A \in M_n(\mathbb{F})$ is invertible if and only if its determinant is nonzero.

We will use the notation $\mathbf{GL}(n, \mathbb{F})$ for the set of $n \times n$ invertible matrices with coefficients in $\mathbb{F}$. Under matrix multiplication $\mathbf{GL}(n, \mathbb{F})$ is a group with the identity matrix as identity element. We note that if V is an n-dimensional vector space over $\mathbb{F}$ with basis $\{v_1, \ldots, v_n\}$, then the map $\mu : \mathbf{GL}(V) \longrightarrow \mathbf{GL}(n, \mathbb{F})$ corresponding to this basis is a group isomorphism. The group $\mathbf{GL}(n, \mathbb{F})$ is called the *general linear group of rank* n.

If $\{w_1, \ldots, w_n\}$ is another basis of V, then there is a matrix $g \in \mathbf{GL}(n, \mathbb{F})$ such that

$$w_j = \sum_{i=1}^{n} g_{ij}v_i \quad \text{and} \quad v_j = \sum_{i=1}^{n} h_{ij}w_i \quad \text{for } j = 1, \ldots, n \,,$$

with $[h_{ij}]$ the inverse matrix to $[g_{ij}]$. Suppose that T is a linear transformation from V to V, that $A = [a_{ij}]$ is the matrix of T with respect to a basis $\{v_1, \ldots, v_n\}$, and that $B = [b_{ij}]$ is the matrix of T with respect to another basis $\{w_1, \ldots, w_n\}$. Then

$$Tw_j = T\left(\sum_i g_{ij}v_i\right) = \sum_i g_{ij}Tv_i$$

$$= \sum_i g_{ij}\left(\sum_k a_{ki}v_k\right) = \sum_l \left(\sum_k \sum_i h_{lk}a_{ki}g_{ij}\right)w_l$$

for $j = 1, \ldots, n$. Thus $B = g^{-1}Ag$ is similar to the matrix A.

Special Linear Group

The special linear group $\mathbf{SL}(n, \mathbb{F})$ is the set of all elements A of $M_n(\mathbb{F})$ such that $\det(A) = 1$. Since $\det(AB) = \det(A)\det(B)$ and $\det(I) = 1$, we see that the special linear group is a subgroup of $\mathbf{GL}(n, \mathbb{F})$.

We note that if V is an n-dimensional vector space over $\mathbb{F}$ with basis $\{v_1, \ldots, v_n\}$ and if $\mu : \mathbf{GL}(V) \longrightarrow \mathbf{GL}(n, \mathbb{F})$ is the map previously defined, then the group

$$\mu^{-1}(\mathbf{SL}(n, \mathbb{F})) = \{T \in \mathbf{GL}(V) : \det(\mu(T)) = 1\}$$

is independent of the choice of basis, by the change of basis formula. We denote this group by $\mathbf{SL}(V)$.

1.1.2 Isometry Groups of Bilinear Forms

Let V be an n-dimensional vector space over $\mathbb{F}$. A bilinear map $B : V \times V \longrightarrow \mathbb{F}$ is called a *bilinear form*. We denote by $\mathbf{O}(V, B)$ (or $\mathbf{O}(B)$ when V is understood) the set of all $g \in \mathbf{GL}(V)$ such that $B(gv, gw) = B(v, w)$ for all $v, w \in V$. We note that $\mathbf{O}(V, B)$ is a subgroup of $\mathbf{GL}(V)$; it is called the *isometry group* of the form B.

Let $\{v_1, \ldots, v_n\}$ be a basis of V and let $\Gamma \in M_n(\mathbb{F})$ be the matrix with $\Gamma_{ij} = B(v_i, v_j)$. If $g \in \mathbf{GL}(V)$ has matrix $A = [a_{ij}]$ relative to this basis, then

$$B(gv_i, gv_j) = \sum_{k,l} a_{ki}a_{lj}B(v_k, v_l) = \sum_{k,l} a_{ki}\Gamma_{kl}a_{lj} .$$

Thus if A^t denotes the transposed matrix $[c_{ij}]$ with $c_{ij} = a_{ji}$, then the condition that $g \in \mathbf{O}(B)$ is that

$$\Gamma = A^t\Gamma A . \tag{1.1}$$

Recall that a bilinear form B is *nondegenerate* if $B(v, w) = 0$ for all w implies that $v = 0$, and likewise $B(v, w) = 0$ for all v implies that $w = 0$. In this case we have $\det\Gamma \neq 0$. Suppose B is nondegenerate. If $T : V \longrightarrow V$ is linear and satisfies

$B(Tv, Tw) = B(v, w)$ for all $v, w \in V$, then $\det(T) \neq 0$ by formula (1.1). Hence $T \in$ $\mathbf{O}(B)$. The next two subsections will discuss the most important special cases of this class of groups.

Orthogonal Groups

We start by introducing the matrix groups; later we will identify these groups with isometry groups of certain classes of bilinear forms. Let $\mathbf{O}(n, \mathbb{F})$ denote the set of all $g \in \mathbf{GL}(n, \mathbb{F})$ such that $gg^t = I$. That is, $g^t = g^{-1}$. We note that $(AB)^t = B^t A^t$ and if $A, B \in \mathbf{GL}(n, \mathbb{F})$ then $(AB)^{-1} = B^{-1}A^{-1}$. It is therefore obvious that $\mathbf{O}(n, \mathbb{F})$ is a subgroup of $\mathbf{GL}(n, \mathbb{F})$. This group is called the *orthogonal group* of $n \times n$ matrices over $\mathbb{F}$. If $\mathbb{F} = \mathbb{R}$ we introduce the *indefinite orthogonal groups*, $\mathrm{O}(p, q)$, with $p + q = n$ and $p, q \in \mathbb{N}$. Let

$$I_{p,q} = \begin{bmatrix} I_p & 0 \\ 0 & -I_q \end{bmatrix}$$

and define

$$\mathbf{O}(p, q) = \{ g \in M_n(\mathbb{R}) : g^t I_{p,q} g = I_{p,q} \}.$$

We note that $\mathbf{O}(n, 0) = \mathbf{O}(0, n) = \mathbf{O}(n, \mathbb{R})$. Also, if

$$s = \begin{bmatrix} 0 & 0 & \cdots & 1 \\ \vdots & \vdots & \ddots & \vdots \\ 0 & 1 & \cdots & 0 \\ 1 & 0 & \cdots & 0 \end{bmatrix}$$

is the matrix with entries 1 on the skew diagonal ($j = n + 1 - i$) and all other entries 0, then $s = s^{-1} = s^t$ and $sI_{p,q}s^{-1} = sI_{p,q}s = sI_{p,q}s = -I_{q,p}$. Thus the map

$$\varphi : \mathbf{O}(p, q) \longrightarrow \mathbf{GL}(n, \mathbb{R})$$

given by $\varphi(g) = sgs$ defines an isomorphism of $\mathbf{O}(p, q)$ onto $\mathbf{O}(q, p)$.

We will now describe these groups in terms of bilinear forms.

Definition 1.1.1. Let V be a vector space over $\mathbb{R}$ and let M be a symmetric bilinear form on V. The form M is *positive definite* if $M(v, v) > 0$ for every $v \in V$ with $v \neq 0$.

Lemma 1.1.2. *Let V be an n-dimensional vector space over $\mathbb{F}$ and let B be a symmetric nondegenerate bilinear form over $\mathbb{F}$.*

1. *If $\mathbb{F} = \mathbb{C}$ then there exists a basis $\{v_1, \ldots, v_n\}$ of V such that $B(v_i, v_j) = \delta_{ij}$.*
2. *If $\mathbb{F} = \mathbb{R}$ then there exist integers $p, q \geq 0$ with $p + q = n$ and a basis $\{v_1, \ldots, v_n\}$ of V such that $B(v_i, v_j) = \varepsilon_i \delta_{ij}$ with $\varepsilon_i = 1$ for $i \leq p$ and $\varepsilon_i = -1$ for $i > p$. Furthermore, if we have another such basis then the corresponding integers (p, q) are the same.*

Remark 1.1.3. The basis for V in part (2) is called a *pseudo-orthonormal basis* relative to B, and the number $p - q$ is called the *signature* of the form (we will also call the pair (p,q) the signature of B). A form is positive definite if and only if its signature is n. In this case a pseudo-orthonormal basis is an orthonormal basis in the usual sense.

Proof. We first observe that if M is a symmetric bilinear form on V such that $M(v,v) = 0$ for all $v \in V$, then $M = 0$. Indeed, using the symmetry and bilinearity we have

$$4M(v,w) = M(v+w, v+w) - M(v-w, v-w) = 0 \quad \text{for all } v, w \in V. \quad (1.2)$$

We now construct a basis $\{w_1, \ldots, w_n\}$ of V such that

$$B(w_i, w_j) = 0 \quad \text{for } i \neq j \text{ and } \quad B(w_i, w_i) \neq 0$$

(such a basis is called an *orthogonal basis* with respect to B). The argument is by induction on n. Since B is nondegenerate, there exists a vector $w_n \in V$ with $B(w_n, w_n) \neq 0$ by (1.2). If $n = 1$ we are done. If $n > 1$, set

$$V' = \{v \in V : B(w_n, v) = 0\}.$$

For $v \in V$ set

$$v' = v - \frac{B(v, w_n)}{B(w_n, w_n)} w_n.$$

Clearly, $v' \in V'$; hence $V = V' + \mathbb{F} w_n$. In particular, this shows that $\dim V' = n - 1$. We assert that the form $B' = B|_{V' \times V'}$ is nondegenerate on V'. Indeed, if $v \in V'$ satisfies $B(v', w) = 0$ for all $w \in V'$, then $B(v', w) = 0$ for all $w \in V$, since $B(v', w_n) = 0$. Hence $v' = 0$, proving nondegeneracy of B'. We may assume by induction that there exists a B'-orthogonal basis $\{w_1, \ldots, w_{n-1}\}$ for V'. Then it is clear that $\{w_1, \ldots, w_n\}$ is a B-orthogonal basis for V.

If $\mathbb{F} = \mathbb{C}$ let $\{w_1, \ldots, w_n\}$ be an orthogonal basis of V with respect to B and let $z_i \in \mathbb{C}$ be a choice of square root of $B(w_i, w_i)$. Setting $v_i = (z_i)^{-1} w_i$, we then obtain the desired normalization $B(v_i, v_j) = \delta_{ij}$.

Now let $\mathbb{F} = \mathbb{R}$. We rearrange the indices (if necessary) so that $B(w_i, w_i) \geq B(w_{i+1}, w_{i+1})$ for $i = 1, \ldots, n-1$. Let $p = 0$ if $B(w_1, w_1) < 0$. Otherwise, let

$$p = \max\{i : B(w_i, w_i) > 0\}.$$

Then $B(w_i, w_i) < 0$ for $i > p$. Take z_i to be a square root of $B(w_i, w_i)$ for $i \leq p$, and take z_i to be a square root of $-B(w_i, w_i)$ for $i > p$. Setting $v_i = (z_i)^{-1} w_i$, we now have $B(v_i, v_j) = \varepsilon_i \delta_{ij}$.

We are left with proving that the integer p is intrinsic to B. Take any basis $\{v_1, \ldots, v_n\}$ such that $B(v_i, v_j) = \varepsilon_i \delta_{ij}$ with $\varepsilon_i = 1$ for $i \leq p$ and $\varepsilon_i = -1$ for $i > p$. Set

$$V_+ = \text{Span}\{v_1, \ldots, v_p\}, \qquad V_- = \text{Span}\{v_{p+1}, \ldots, v_n\}.$$

Then $V = V_+ \oplus V_-$ (direct sum). Let $\pi : V \longrightarrow V_+$ be the projection onto the first factor. We note that $B|_{V_+ \times V_+}$ is positive definite. Let W be any subspace of V such that $B|_{W \times W}$ is positive definite. Suppose that $w \in W$ and $\pi(w) = 0$. Then $w \in V_-$, so it can be written as $w = \sum_{i>p} a_i v_i$. Hence

$$B(w,w) = \sum_{i,j>p} a_i a_j B(v_i, v_j) = -\sum_{i>p} a_i^2 \leq 0 \, .$$

Since $B|_{W \times W}$ has been assumed to be positive definite, it follows that $w = 0$. This implies that $\pi : W \longrightarrow V_+$ is injective, and hence $\dim W \leq \dim V_+ = p$. Thus p is uniquely determined as the maximum dimension of a subspace on which B is positive definite. $\square$

The following result follows immediately from Lemma 1.1.2.

Proposition 1.1.4. *Let B be a nondegenerate symmetric bilinear form on an n-dimensional vector space V over $\mathbb{F}$.*

1. *Let $\mathbb{F} = \mathbb{C}$. If $\{v_1, \dots, v_n\}$ is an orthonormal basis for V with respect to B, then $\mu : \mathbf{O}(V, B) \longrightarrow \mathbf{O}(n, \mathbb{F})$ defines a group isomorphism.*
2. *Let $\mathbb{F} = \mathbb{R}$. If B has signature $(p, n-p)$ and $\{v_1, \dots, v_n\}$ is a pseudo-orthonormal basis of V, then $\mu : \mathbf{O}(V, B) \longrightarrow \mathrm{O}(p, n-p)$ is a group isomorphism.*

Here $\mu(g)$, *for* $g \in \mathbf{GL}(V)$, *is the matrix of g with respect to the given basis.*

The *special orthogonal group* over $\mathbb{F}$ is the subgroup

$$\mathbf{SO}(n, \mathbb{F}) = \mathbf{O}(n, \mathbb{F}) \cap \mathbf{SL}(n, \mathbb{F})$$

of $\mathbf{O}(n, \mathbb{F})$. The *indefinite special orthogonal groups* are the groups

$$\mathbf{SO}(p, q) = \mathbf{O}(p, q) \cap \mathbf{SL}(p+q, \mathbb{R}) \, .$$

Symplectic Group

We set $J = \begin{bmatrix} 0 & I \\ -I & 0 \end{bmatrix}$ with I the $n \times n$ identity matrix. The symplectic group of rank n over $\mathbb{F}$ is defined to be

$$\mathbf{Sp}(n, \mathbb{F}) = \{g \in M_{2n}(\mathbb{F}) : g^t J g = J\}.$$

As in the case of the orthogonal groups one sees without difficulty that $\mathbf{Sp}(n, \mathbb{F})$ is a subgroup of $\mathbf{GL}(2n, \mathbb{F})$.

We will now look at the coordinate-free version of these groups. A bilinear form B is called *skew-symmetric* if $B(v, w) = -B(w, v)$. If B is skew-symmetric and non-degenerate, then $m = \dim V$ must be even, since the matrix of B relative to any basis for V is skew-symmetric and has nonzero determinant.

Lemma 1.1.5. *Let V be a 2n-dimensional vector space over $\mathbb{F}$ and let B be a nondegenerate, skew-symmetric bilinear form on V. Then there exists a basis $\{v_1, \ldots, v_{2n}\}$ for V such that the matrix $[B(v_i, v_j)]$ equals J (call such a basis a B-symplectic basis).*

Proof. Let v be a nonzero element of V. Since B is nondegenerate, there exists $w \in V$ with $B(v, w) \neq 0$. Replacing w with $B(v, w)^{-1}w$, we may assume that $B(v, w) = 1$. Let

$$W = \{x \in V : B(v, x) = 0 \text{ and } B(w, x) = 0\}.$$

For $x \in V$ we set $x' = x - B(v, x)w - B(x, w)v$. Then

$$B(v, x') = B(v, x) - B(v, x)B(v, w) - B(w, x)B(v, v) = 0,$$

since $B(v, w) = 1$ and $B(v, v) = 0$ (by skew symmetry of B). Similarly,

$$B(w, x') = B(w, x) - B(v, x)B(w, w) + B(w, x)B(w, v) = 0,$$

since $B(w, v) = -1$ and $B(w, w) = 0$. Thus $V = U \oplus W$, where U is the span of v and w. It is easily verified that $B|_{U \times U}$ is nondegenerate, and so $U \cap W = \{0\}$. This implies that $\dim W = m - 2$. We leave to the reader to check that $B|_{W \times W}$ also is nondegenerate.

Set $v_n = v$ and $v_{2n} = w$ with v, w as above. Since $B|_{W \times W}$ is nondegenerate, by induction there exists a B-symplectic basis $\{w_1, \ldots, w_{2n-2}\}$ of W. Set $v_i = w_i$ and $v_{n+1-i} = w_{n-i}$ for $i \leq n - 1$. Then $\{v_1, \ldots, v_{2n}\}$ is a B-symplectic basis for V. $\quad\square$

The following result follows immediately from Lemma 1.1.5.

Proposition 1.1.6. *Let V be a 2n-dimensional vector space over $\mathbb{F}$ and let B be a nondegenerate skew-symmetric bilinear form on V. Fix a B-symplectic basis of V and let $\mu(g)$, for $g \in \mathbf{GL}(V)$, be the matrix of g with respect to this basis. Then $\mu : \mathbf{O}(V, B) \longrightarrow \mathbf{Sp}(n, \mathbb{F})$ is a group isomorphism.*

1.1.3 Unitary Groups

Another family of classical subgroups of $\mathbf{GL}(n, \mathbb{C})$ consists of the unitary groups and special unitary groups for definite and indefinite Hermitian forms. If $A \in M_n(\mathbb{C})$ we will use the standard notation $A^* = \overline{A}^t$ for its adjoint matrix, where $\overline{A}$ is the matrix obtained from A by complex conjugating all of the entries. The *unitary group* of rank n is the group

$$\mathbf{U}(n) = \{g \in M_n(\mathbb{C}) : g^*g = I\}.$$

The *special unitary group* is $\mathbf{SU}(n) = \mathbf{U}(n) \cap \mathbf{SL}(n, \mathbb{C})$. Let the matrix $I_{p,q}$ be as in Section 1.1.2. We define the *indefinite unitary group* of signature (p, q) to be

$$\mathbf{U}(p, q) = \{g \in M_n(\mathbb{C}) : g^*I_{p,q}g = I_{p,q}\}.$$

The special indefinite unitary group of signature (p,q) is $\mathbf{SU}(p,q) = \mathbf{U}(p,q) \cap \mathbf{SL}(n,\mathbb{C})$.

We will now obtain a coordinate-free description of these groups. Let V be an n-dimensional vector space over $\mathbb{C}$. An $\mathbb{R}$ bilinear map $B: V \times V \longrightarrow \mathbb{C}$ (where we view V as a vector space over $\mathbb{R}$) is said to be a *Hermitian* form if it satisfies

1. $B(av,w) = \overline{a}B(v,w)$ for all $a \in \mathbb{C}$ and all $v,w \in V$.
2. $B(w,v) = \overline{B(v,w)}$ for all $v,w \in V$.

By the second condition, we see that a Hermitian form is nondegenerate provided $B(v,w) = 0$ for all $w \in V$ implies that $v = 0$. The form is said to be *positive definite* if $B(v,v) > 0$ for all $v \in V$ with $v \neq 0$. (Note that if M is a Hermitian form, then $M(v,v) \in \mathbb{R}$ for all $v \in V$.) We define $\mathbf{U}(V,B)$ (also denoted by $\mathbf{U}(B)$ when V is understood) to be the group of all elements $g \in \mathbf{GL}(V)$ such that $B(gv,gw) = B(v,w)$ for all $v,w \in V$. We call $\mathbf{U}(B)$ the *unitary group* of B.

Lemma 1.1.7. *Let V be an n-dimensional vector space over $\mathbb{C}$ and let B be a nondegenerate Hermitian form on V. Then there exist an integer p, with $n \geq p \geq 0$, and a basis $\{v_1, \ldots, v_n\}$ of V such that $B(v_i,v_j) = \varepsilon_i \delta_{ij}$, with $\varepsilon_i = 1$ for $i \leq p$ and $\varepsilon_i = -1$ for $i > p$. The number p depends only on B and not on the choice of basis.*

The proof of Lemma 1.1.7 is almost identical to that of Lemma 1.1.2 and will be left as an exercise.

If V is an n-dimensional vector space over $\mathbb{C}$ and B is a nondegenerate Hermitian form on V, then a basis as in Lemma 1.1.7 will be called a *pseudo-orthonormal basis* (if $p = n$ then it is an *orthonormal basis* in the usual sense). The pair $(p,n-p)$ will be called the *signature* of B. The following result is proved in exactly the same way as the corresponding result for orthogonal groups.

Proposition 1.1.8. *Let V be a finite-dimensional vector space over $\mathbb{C}$ and let B be a nondegenerate Hermitian form on V of signature (p,q). Fix a pseudo-orthonormal basis of V relative to B and let $\mu(g)$, for $g \in \mathbf{GL}(V)$, be the matrix of g with respect to this basis. Then $\mu : U(V,B) \longrightarrow U(p,q)$ is a group isomorphism.*

1.1.4 Quaternionic Groups

We recall some basic properties of the quaternions. Consider the four-dimensional real vector space $\mathbb{H}$ consisting of the 2×2 complex matrices

$$w = \begin{bmatrix} x & -\overline{y} \\ y & \overline{x} \end{bmatrix} \quad \text{with } x,y \in \mathbb{C}. \tag{1.3}$$

One checks directly that $\mathbb{H}$ is closed under multiplication in $M_2(\mathbb{C})$. If $w \in \mathbb{H}$ then $w^* \in \mathbb{H}$ and

$$w^*w = ww^* = (|x|^2 + |y|^2)I$$

(where w^* denotes the conjugate-transpose matrix). Hence every nonzero element of $\mathbb{H}$ is invertible. Thus $\mathbb{H}$ is a *division algebra* (or skew field) over $\mathbb{R}$. This division algebra is a realization of the quaternions.

The more usual way of introducing the quaternions is to consider the vector space $\mathbb{H}$ over $\mathbb{R}$ with basis $\{\mathbf{1}, \mathbf{i}, \mathbf{j}, \mathbf{k}\}$. Define a multiplication so that $\mathbf{1}$ is the identity and

$$\mathbf{i}^2 = \mathbf{j}^2 = \mathbf{k}^2 = -\mathbf{1},$$

$$\mathbf{ij} = -\mathbf{ji} = \mathbf{k}, \quad \mathbf{ki} = -\mathbf{ik} = \mathbf{j}, \quad \mathbf{jk} = -\mathbf{kj} = \mathbf{i};$$

then extend the multiplication to $\mathbb{H}$ by linearity relative to real scalars. To obtain an isomorphism between this version of $\mathbb{H}$ and the 2×2 complex matrix version, take

$$\mathbf{1} = I, \quad \mathbf{i} = \begin{bmatrix} i & 0 \\ 0 & -i \end{bmatrix}, \quad \mathbf{j} = \begin{bmatrix} 0 & 1 \\ -1 & 0 \end{bmatrix}, \quad \mathbf{k} = \begin{bmatrix} 0 & i \\ i & 0 \end{bmatrix},$$

where i is a fixed choice of $\sqrt{-1}$. The *conjugation* $w \mapsto w^*$ satisfies $(uv)^* = v^* u^*$. In terms of real components, $(a + b\mathbf{i} + c\mathbf{j} + d\mathbf{k})^* = a - b\mathbf{i} - c\mathbf{j} - d\mathbf{k}$ for $a, b, c, d \in \mathbb{R}$. It is useful to write quaternions in complex form as $x + \mathbf{j}y$ with $x, y \in \mathbb{C}$; however, note that the conjugation is then given as

$$(x + \mathbf{j}y)^* = \bar{x} + \bar{y}\mathbf{j} = \bar{x} - \mathbf{j}y.$$

On the $4n$-dimensional real vector space $\mathbb{H}^n$ we define multiplication by $a \in \mathbb{H}$ on the right:

$$(u_1, \ldots, u_n) \cdot a = (u_1 a, \ldots, u_n a).$$

We note that $u \cdot \mathbf{1} = u$ and $u \cdot (ab) = (u \cdot a) \cdot b$. We can therefore think of $\mathbb{H}^n$ as a vector space over $\mathbb{H}$. Viewing elements of $\mathbb{H}^n$ as $n \times 1$ column vectors, we define Au for $u \in \mathbb{H}^n$ and $A \in M_n(\mathbb{H})$ by matrix multiplication. Then $A(u \cdot a) = (Au) \cdot a$ for $a \in \mathbb{H}$; hence A defines a quaternionic linear map. Here matrix multiplication is defined as usual, but one must be careful about the order of multiplication of the entries.

We can make $\mathbb{H}^n$ into a $2n$-dimensional vector space over $\mathbb{C}$ in many ways; for example, we can embed $\mathbb{C}$ into $\mathbb{H}$ as any of the subfields

$$\mathbb{R}\mathbf{1} + \mathbb{R}\mathbf{i}, \quad \mathbb{R}\mathbf{1} + \mathbb{R}\mathbf{j}, \quad \mathbb{R}\mathbf{1} + \mathbb{R}\mathbf{k}. \tag{1.4}$$

Using the first of these embeddings, we write $z = x + \mathbf{j}y \in \mathbb{H}^n$ with $x, y \in \mathbb{C}^n$, and likewise $C = A + \mathbf{j}B \in M_n(\mathbb{H})$ with $A, B \in M_n(\mathbb{C})$. The maps

$$z \mapsto \begin{bmatrix} x \\ y \end{bmatrix} \quad \text{and} \quad C \mapsto \begin{bmatrix} A & -\bar{B} \\ B & \bar{A} \end{bmatrix}$$

identify $\mathbb{H}^n$ with $\mathbb{C}^{2n}$ and $M_n(\mathbb{H})$ with the real subalgebra of $M_{2n}(\mathbb{C})$ consisting of matrices T such that

$$JT = \overline{T}J, \quad \text{where} \quad J = \begin{bmatrix} 0 & I \\ -I & 0 \end{bmatrix}. \tag{1.5}$$

We define $\mathbf{GL}(n, \mathbb{H})$ to be the group of all invertible $n \times n$ matrices over $\mathbb{H}$. Then $\mathbf{GL}(n, \mathbb{H})$ acts on $\mathbb{H}^n$ by complex linear transformations relative to each of the complex structures (1.4). If we use the embedding of $M_n(\mathbb{H})$ into $M_{2n}(\mathbb{C})$ just described, then from (1.5) we see that $\mathbf{GL}(n, \mathbb{H}) = \{g \in \mathbf{GL}(2n, \mathbb{C}) : Jg = \overline{g}J\}$.

Quaternionic Special Linear Group

We leave it to the reader to prove that the determinant of $A \in \mathbf{GL}(n, \mathbb{H})$ as a complex linear transformation with respect to any of the complex structures (1.4) is the same. We can thus define $\mathbf{SL}(n, \mathbb{H})$ to be the elements of determinant one in $\mathbf{GL}(n, \mathbb{H})$ with respect to any of these complex structures. This group is usually denoted by $\mathbf{SU}^*(2n)$.

The Quaternionic Unitary Groups

For $X = [x_{ij}] \in M_n(\mathbb{H})$ we define $X^* = [x_{ji}^*]$ (here we take the quaternionic matrix entries $x_{ij} \in M_2(\mathbb{C})$ given by (1.3)). Let the diagonal matrix $I_{p,q}$ (with $p + q = n$) be as in Section 1.1.2. The *indefinite quaternionic unitary groups* are the groups

$$\mathbf{Sp}(p, q) = \{g \in \mathbf{GL}(p + q, \mathbb{H}) : g^* I_{p,q} g = I_{p,q}\}.$$

We leave it to the reader to prove that this set is a subgroup of $\mathbf{GL}(p + q, \mathbb{H})$.

The group $\mathbf{Sp}(p, q)$ is the isometry group of the nondegenerate *quaternionic Hermitian form*

$$B(w, z) = w^* I_{p,q} z, \quad \text{for } w, z \in \mathbb{H}^n. \tag{1.6}$$

(Note that this form satisfies $B(w, z) = B(z, w)^*$ and $B(w\alpha, z\beta) = \alpha^* B(w, z)\beta$ for $\alpha, \beta \in \mathbb{H}$.) If we write $w = u + \mathbf{j}v$ and $z = x + \mathbf{j}y$ with $u, v, x, y \in \mathbb{C}^n$, and set $K_{p,q} = \text{diag}[I_{p,q} \ I_{p,q}] \in M_{2n}(\mathbb{R})$, then

$$B(w, z) = \begin{bmatrix} u^* & v^* \end{bmatrix} K_{p,q} \begin{bmatrix} x \\ y \end{bmatrix} + \mathbf{j} \begin{bmatrix} u^t & v^t \end{bmatrix} K_{p,q} \begin{bmatrix} -y \\ x \end{bmatrix}.$$

Thus the elements of $\mathbf{Sp}(p, q)$, viewed as linear transformations of $\mathbb{C}^{2n}$, preserve both a Hermitian form of signature $(2p, 2q)$ and a nondegenerate skew-symmetric form.

The Group SO*(2n)

Let J be the $2n \times 2n$ skew-symmetric matrix from Section 1.1.2. Since $J^2 = -I_{2n}$ (the $2n \times 2n$ identity matrix), the map of $\mathbf{GL}(2n, \mathbb{C})$ to itself given by $\theta(g) = -JgJ$ defines an automorphism whose square is the identity. Our last family of classical groups is

$$\mathbf{SO}^*(2n) = \{g \in \mathbf{SO}(2n, \mathbb{C}) : \theta(\overline{g}) = g\}.$$

We identify $\mathbb{C}^{2n}$ with $\mathbb{H}^n$ as a vector space over $\mathbb{C}$ by the map $\begin{bmatrix} a \\ b \end{bmatrix} \mapsto a + \mathbf{j}b$, where $a, b \in \mathbb{C}^n$. The group $\mathbf{SO}^*(2n)$ then becomes the isometry group of the nondegenerate *quaternionic skew-Hermitian form*

$$C(x, y) = x^* \mathbf{j} y, \quad \text{for } x, y \in \mathbb{H}^n. \tag{1.7}$$

This form satisfies $C(x, y) = -C(y, x)^*$ and $C(x\alpha, y\beta) = \alpha^* C(x, y)\beta$ for $\alpha, \beta \in \mathbb{H}$.

We have now completed the list of the classical groups associated with $\mathbb{R}$, $\mathbb{C}$, and $\mathbb{H}$. We will return to this list at the end of the chapter when we consider *real forms* of complex algebraic groups. Later we will define *covering groups*; any group covering one of the groups on this list—for example, a *spin group* in Chapter 7—will also be called a classical group.

1.1.5 Exercises

In these exercises $\mathbb{F}$ denotes either $\mathbb{R}$ or $\mathbb{C}$. See Appendix B.2 for notation and properties of tensor and exterior products of vector spaces.

1. Let $\{v_1, \ldots, v_n\}$ and $\{w_1, \ldots, w_n\}$ be bases for an $\mathbb{F}$ vector space V. Suppose a linear map $T : V \longrightarrow V$ has matrices A and B, respectively, relative to these bases. Show that $\det A = \det B$.
2. Determine the signature of the form $B(x, y) = \sum_{i=1}^{n} x_i y_{n+1-i}$ on $\mathbb{R}^n$.
3. Let V be a vector space over $\mathbb{F}$ and let B be a skew-symmetric or symmetric nondegenerate bilinear form on V. Assume that W is a subspace of V on which B restricts to a nondegenerate form. Prove that the restriction of B to the subspace $W^\perp = \{v \in V : B(v, w) = 0 \text{ for all } w \in W\}$ is nondegenerate.
4. Let V denote the vector space of symmetric 2×2 matrices over $\mathbb{F}$. If $x, y \in V$ define $B(x, y) = \det(x + y) - \det(x) - \det(y)$.
   (a) Show that B is nondegenerate, and that if $\mathbb{F} = \mathbb{R}$ then the signature of the form B is $(1, 2)$.
   (b) If $g \in \mathbf{SL}(2, \mathbb{F})$ define $\varphi(g) \in \mathbf{GL}(V)$ by $\varphi(g)(v) = gvg^t$. Show that the map $\varphi : \mathbf{SL}(2, \mathbb{F}) \longrightarrow \mathbf{SO}(V, B)$ is a group homomorphism with kernel $\{\pm I\}$.
5. The purpose of this exercise is to prove Lemma 1.1.7 by the method of proof of Lemma 1.1.2.

(a) Prove that if M is a Hermitian form such that $M(v,v) = 0$ for all v then $M = 0$. (HINT: Show that $M(v + sw, v + sw) = sM(w,v) + \bar{s}M(v,w)$ for all $s \in \mathbb{C}$; then substitute values for s to see that $M(v,w) = 0$.)

(b) Use the result of part (a) to complete the proof of Lemma 1.1.7. (HINT: Note that $M(v,v) \in \mathbb{R}$ since M is Hermitian.)

6. Let V be a $2n$-dimensional vector space over $\mathbb{F}$. Consider the space $W = \bigwedge^n V$. Fix a basis ω of the one-dimensional vector space $\bigwedge^{2n} V$. Consider the bilinear form $B(u,v)$ on W defined by $u \wedge v = B(u,v)\omega$.
   (a) Show that B is nondegenerate.
   (b) Show that B is skew-symmetric if n is odd and symmetric if n is even.
   (c) Determine the signature of B when n is even and $\mathbb{F} = \mathbb{R}$,

7. *(Notation of the previous exercise)* Let $V = \mathbb{F}^4$ with basis $\{e_1, e_2, e_3, e_4\}$ and let $\omega = e_1 \wedge e_2 \wedge e_3 \wedge e_4$. Define $\varphi(g)(u \wedge v) = gu \wedge gv$ for $g \in \mathbf{SL}(4, \mathbb{F})$ and $u, v \in \mathbb{F}^4$. Show that $\varphi : \mathbf{SL}(4, \mathbb{F}) \longrightarrow \mathbf{SO}(\bigwedge^2 \mathbb{F}^4, B)$ is a group homomorphism with kernel $\{\pm I\}$. (HINT: Use Jordan canonical form to determine the kernel.)

8. *(Notation of the previous exercise)* Let ψ be the restriction of φ to $\mathbf{Sp}(2, \mathbb{F})$. Let $v = e_1 \wedge e_3 + e_2 \wedge e_4$.
   (a) Show that $\psi(g)v = v$ and $B(v,v) = -2$. (HINT: The map $e_i \wedge e_j \mapsto e_{ij} - e_{ji}$ is a linear isomorphism between $\bigwedge^2 \mathbb{F}^4$ and the subspace of skew-symmetric matrices in $M_4(\mathbb{F})$. Show that this map takes v to J and $\varphi(g)$ to the transformation $A \mapsto gAg^t$.)
   (b) Let $W = \{w \in \bigwedge^2 \mathbb{F}^4 : B(v,w) = 0\}$. Show that $B|_{W \times W}$ is nondegenerate and has signature $(3,2)$ when $\mathbb{F} = \mathbb{R}$.
   (c) Set $\rho(g) = \psi(g)|_W$. Show that ρ is a group homomorphism from $\mathbf{Sp}(2, \mathbb{F})$ to $\mathbf{SO}(W, B|_{W \times W})$ with kernel $\{\pm 1\}$. (HINT: Use the previous exercise to determine the kernel.)

9. Let $V = M_2(\mathbb{F})$. For $x, y \in V$ define $B(x,y) = \det(x + y) - \det(x) - \det(y)$.
   (a) Show that B is a symmetric nondegenerate form on V, and calculate the signature of B when $\mathbb{F} = \mathbb{R}$.
   (b) Let $G = \mathbf{SL}(2, \mathbb{F}) \times \mathbf{SL}(2, \mathbb{F})$ and define $\varphi : G \longrightarrow \mathbf{GL}(V)$ by $\varphi(a,b)v = axb^t$ for $a, b \in \mathbf{SL}(2, \mathbb{F})$ and $v \in V$. Show that φ is a group homomorphism and $\varphi(G) \subset \mathbf{SO}(V, B)$. Determine $\mathrm{Ker}(\varphi)$. (HINT: Use Jordan canonical form to determine the kernel.)

10. Identify $\mathbb{H}^n$ with $\mathbb{C}^{2n}$ as a vector space over $\mathbb{C}$ by the map $a + \mathbf{j}b \mapsto \begin{bmatrix} a \\ b \end{bmatrix}$, where $a, b \in \mathbb{C}^n$. Let $T = A + \mathbf{j}B \in M_n(\mathbb{H})$ with $A, B \in M_n(\mathbb{C})$.
    (a) Show that left multiplication by T on $\mathbb{H}^n$ corresponds to multiplication by the matrix $\begin{bmatrix} A & -\bar{B} \\ B & \bar{A} \end{bmatrix} \in M_{2n}(\mathbb{C})$ on $\mathbb{C}^{2n}$.
    (b) Show that multiplication by $\mathbf{i}$ on $M_n(\mathbb{H})$ becomes the transformation

$$\begin{bmatrix} A & -\bar{B} \\ B & -\bar{A} \end{bmatrix} \mapsto \begin{bmatrix} \mathbf{i}A & -\mathbf{i}\bar{B} \\ -\mathbf{i}B & -\mathbf{i}\bar{A} \end{bmatrix}.$$

11. Use the identification of $\mathbb{H}^n$ with $\mathbb{C}^{2n}$ in the previous exercise to view the form $B(x,y)$ in equation (1.6) as an $\mathbb{H}$-valued function on $\mathbb{C}^{2n} \times \mathbb{C}^{2n}$.

(a) Show that $B(x,y) = \overline{B_0(x,y)} + \mathbf{j}B_1(x,y)$, where B_0 is a $\mathbb{C}$-Hermitian form on $\mathbb{C}^{2n}$ of signature $(2p,2q)$ and B_1 is a nondegenerate skew-symmetric $\mathbb{C}$-bilinear form on $\mathbb{C}^{2n}$.

(b) Use part (a) to prove that $\mathbf{Sp}(p,q) = \mathbf{Sp}(\mathbb{C}^{2n}, B_1) \cap \mathbf{U}(\mathbb{C}^{2n}, B_0)$.

12. Use the identification of $\mathbb{H}^n$ with $\mathbb{C}^{2n}$ in the previous exercise to view the form $C(x,y)$ from equation (1.7) as an $\mathbb{H}$-valued function on $\mathbb{C}^{2n} \times \mathbb{C}^{2n}$.

(a) Show that $C(x,y) = \overline{C_0(x,y)} + \mathbf{j}x^t y$ for $x,y \in \mathbb{C}^{2n}$, where $C_0(x,y)$ is a $\mathbb{C}$-Hermitian form on $\mathbb{C}^{2n}$ of signature (n,n).

(b) Use the result of part (a) to prove that $\mathbf{SO}^*(2n) = \mathbf{SO}(2n,\mathbb{C}) \cap \mathbf{U}(\mathbb{C}^{2n}, C_0)$.

13. Why can't we just define $\mathbf{SL}(n,\mathbb{H})$ by taking all $g \in \mathbf{GL}(n,\mathbb{H})$ such that the usual formula for the determinant of g yields 1?

14. Consider the three embeddings of $\mathbb{C}$ in $\mathbb{H}$ given by the subfields (1.4). These give three ways of writing $X \in M_n(\mathbb{H})$ as a $2n \times 2n$ matrix over $\mathbb{C}$. Show that these three matrices have the same determinant.

1.2 The Classical Lie Algebras

Let V be a vector space over $\mathbb{F}$. Let $\mathrm{End}(V)$ denote the algebra (under composition) of $\mathbb{F}$-linear maps of V to V. If $X,Y \in \mathrm{End}(V)$ then we set $[X,Y] = XY - YX$. This defines a new product on $\mathrm{End}(V)$ that satisfies two properties:

(1) $[X,Y] = -[Y,X]$ for all X,Y *(skew symmetry)* .

(2) $[X,[Y,Z]] = [[X,Y],Z] + [Y,[X,Z]]$ for all X,Y,Z *(Jacobi identity)* .

Definition 1.2.1. A vector space $\mathfrak{g}$ over $\mathbb{F}$ together with a bilinear map $X,Y \mapsto [X,Y]$ of $\mathfrak{g} \times \mathfrak{g}$ to $\mathfrak{g}$ is said to be a *Lie algebra* if conditions (1) and (2) are satisfied.

In particular, $\mathrm{End}(V)$ is a Lie algebra under the binary operation $[X,Y] = XY - YX$. Condition (2) is a substitute for the associative rule for multiplication; it says that for fixed X, the linear transformation $Y \mapsto [X,Y]$ is a *derivation* of the (nonassociative) algebra $(\mathfrak{g}, [\cdot,\cdot])$.

If $\mathfrak{g}$ is a Lie algebra and if $\mathfrak{h}$ is a subspace such that $X,Y \in \mathfrak{h}$ implies that $[X,Y]$ in $\mathfrak{h}$, then $\mathfrak{h}$ is a Lie algebra under the restriction of $[\cdot,\cdot]$. We will call $\mathfrak{h}$ a *Lie subalgebra* of $\mathfrak{g}$ (or *subalgebra*, when the Lie algebra context is clear).

Suppose that $\mathfrak{g}$ and $\mathfrak{h}$ are Lie algebras over $\mathbb{F}$. A *Lie algebra homomorphism* of $\mathfrak{g}$ to $\mathfrak{h}$ is an $\mathbb{F}$-linear map $T : \mathfrak{g} \longrightarrow \mathfrak{h}$ such that $T[X,Y] = [TX,TY]$ for all $X,Y \in \mathfrak{g}$. A Lie algebra homomorphism is an *isomorphism* if it is bijective.

1.2.1 General and Special Linear Lie Algebras

If V is a vector space over $\mathbb{F}$, we write $\mathfrak{gl}(V)$ for $\mathrm{End}(V)$ looked upon as a Lie algebra under $[X,Y] = XY - YX$. We write $\mathfrak{gl}(n,\mathbb{F})$ to denote $M_n(\mathbb{F})$ as a Lie algebra under the matrix commutator bracket. If $\dim V = n$ and we fix a basis for V, then the correspondence between linear transformations and their matrices gives a Lie algebra isomorphism $\mathfrak{gl}(V) \cong \mathfrak{gl}(n, \mathbb{R})$. These Lie algebras will be called the *general linear Lie algebras*.

If $A = [a_{ij}] \in M_n(\mathbb{F})$ then its *trace* is $\mathrm{tr}(A) = \sum_i a_{ii}$. We note that $\mathrm{tr}(AB) = \mathrm{tr}(BA)$. This implies that if A is the matrix of $T \in \mathrm{End}(V)$ with respect to some basis, then $\mathrm{tr}(A)$ is independent of the choice of basis. We will write $\mathrm{tr}(T) = \mathrm{tr}(A)$. We define

$$\mathfrak{sl}(V) = \{T \in \mathrm{End}(V) : \mathrm{tr}(T) = 0\} .$$

Since $\mathrm{tr}([S, T]) = 0$ for all $S, T \in \mathrm{End}(V)$, we see that $\mathfrak{sl}(V)$ is a Lie subalgebra of $\mathfrak{gl}(V)$. Choosing a basis for V, we may identify this Lie algebra with

$$\mathfrak{sl}(n, \mathbb{F}) = \{A \in \mathfrak{gl}(n, \mathbb{F}) : \mathrm{tr}(A) = 0\} .$$

These Lie algebras will be called the *special linear Lie algebras*.

1.2.2 Lie Algebras Associated with Bilinear Forms

Let V be a vector space over $\mathbb{F}$ and let $B : V \times V \longrightarrow \mathbb{F}$ be a bilinear map. We define

$$\mathfrak{so}(V, B) = \{X \in \mathrm{End}(V) : B(Xv, w) = -B(v, Xw)\} .$$

Thus $\mathfrak{so}(V, B)$ consists of the linear transformations that are *skew-symmetric* relative to the form B, and is obviously a linear subspace of $\mathfrak{gl}(V)$. If $X, Y \in \mathfrak{so}(V, B)$, then

$$B(XYv, w) = -B(Yv, Xw) = B(v, YXw) .$$

It follows that $B([X, Y]v, w) = -B(v, [X, Y]w)$, and hence $\mathfrak{so}(V, B)$ is a Lie subalgebra of $\mathfrak{gl}(V)$.

Suppose V is finite-dimensional. Fix a basis $\{v_1, \ldots, v_n\}$ for V and let Γ be the $n \times n$ matrix with entries $\Gamma_{ij} = B(v_i, v_j)$. By a calculation analogous to that in Section 1.1.2, we see that $T \in \mathfrak{so}(V, B)$ if and only if its matrix A relative to this basis satisfies

$$A^t \Gamma + \Gamma A = 0 . \tag{1.8}$$

When B is nondegenerate then Γ is invertible, and equation (1.8) can be written as $A^t = -\Gamma A \Gamma^{-1}$. In particular, this implies that $\mathrm{tr}(T) = 0$ for all $T \in \mathfrak{so}(V, B)$.

Orthogonal Lie Algebras

Take $V = \mathbb{F}^n$ and the bilinear form B with matrix $\Gamma = I_n$ relative to the standard basis for $\mathbb{F}^n$. Define

$$\mathfrak{so}(n,\mathbb{F}) = \{X \in M_n(\mathbb{F}) : X^t = -X\}.$$

Since B is nondegenerate, $\mathfrak{so}(n,\mathbb{F})$ is a Lie subalgebra of $\mathfrak{sl}(n,\mathbb{F})$.

When $\mathbb{F} = \mathbb{R}$ we take integers $p, q \geq 0$ such that $p + q = n$ and let B be the bilinear form on $\mathbb{R}^n$ whose matrix relative to the standard basis is $I_{p,q}$ (as in Section 1.1.2). Define

$$\mathfrak{so}(p,q) = \{X \in M_n(\mathbb{R}) : X^t I_{p,q} = -I_{p,q}X\}.$$

Since B is nondegenerate, $\mathfrak{so}(p,q)$ is a Lie subalgebra of $\mathfrak{sl}(n,\mathbb{R})$.

To obtain a basis-free definition of this family of Lie algebras, let B be a non-degenerate symmetric bilinear form on an n-dimensional vector space V over $\mathbb{F}$. Let $\{v_1, \ldots, v_n\}$ be a basis for V that is orthonormal (when $\mathbb{F} = \mathbb{C}$) or pseudo-orthonormal (when $\mathbb{F} = \mathbb{R}$) relative to B (see Lemma 1.1.2). Let $\mu(T)$ be the matrix of $T \in \mathrm{End}(V)$ relative to this basis . When $\mathbb{F} = \mathbb{C}$, then μ defines a Lie algebra iso-morphism of $\mathfrak{so}(V,B)$ onto $\mathfrak{so}(n,\mathbb{C})$. When $\mathbb{F} = \mathbb{R}$ and B has signature (p,q), then μ defines a Lie algebra isomorphism of $\mathfrak{so}(V,B)$ onto $\mathfrak{so}(p,q)$.

Symplectic Lie Algebra

Let J be the $2n \times 2n$ skew-symmetric matrix from Section 1.1.2. We define

$$\mathfrak{sp}(n,\mathbb{F}) = \{X \in M_{2n}(\mathbb{F}) : X^t J = -JX\}.$$

This subspace of $\mathfrak{gl}(n,\mathbb{F})$ is a Lie subalgebra that we call the *symplectic Lie algebra of rank n*.

To obtain a basis-free definition of this family of Lie algebras, let B be a non-degenerate skew-symmetric bilinear form on a $2n$-dimensional vector space V over $\mathbb{F}$. Let $\{v_1, \ldots, v_{2n}\}$ be a B-symplectic basis for V (see Lemma 1.1.5). The map μ that assigns to an endomorphism of V its matrix relative to this basis defines an isomorphism of $\mathfrak{so}(V,B)$ onto $\mathfrak{sp}(n,\mathbb{F})$.

1.2.3 Unitary Lie Algebras

Let $p, q \geq 0$ be integers such that $p + q = n$ and let $I_{p,q}$ be the $n \times n$ matrix from Section 1.1.2. We define

$$\mathfrak{u}(p,q) = \{X \in M_n(\mathbb{C}) : X^* I_{p,q} = -I_{p,q}X\}$$

(notice that this space is a *real* subspace of $M_n(\mathbb{C})$). One checks directly that $\mathfrak{u}(p,q)$ is a Lie subalgebra of $\mathfrak{gl}_n(\mathbb{C})$ (considered as a Lie algebra over $\mathbb{R}$). We define $\mathfrak{su}(p,q) = \mathfrak{u}(p,q) \cap \mathfrak{sl}(n,\mathbb{C})$.

To obtain a basis-free description of this family of Lie algebras, let V be an n-dimensional vector space over $\mathbb{C}$, and let B be a nondegenerate Hermitian form on

V. We define

$$\mathfrak{u}(V,B) = \{T \in \text{End}_{\mathbb{C}}(V) : B(Tv,w) = -B(v,Tw) \text{ for all } v,w \in V\} .$$

We set $\mathfrak{su}(V,B) = \mathfrak{u}(V,B) \cap \mathfrak{sl}(V)$. If B has signature (p,q) and if $\{v_1,\ldots,v_n\}$ is a pseudo-orthogonal basis of V relative to B (see Lemma 1.1.7), then the assignment $T \mapsto \mu(T)$ of T to its matrix relative to this basis defines a Lie algebra isomorphism of $\mathfrak{u}(V,B)$ with $\mathfrak{u}(p,q)$ and of $\mathfrak{su}(V,B)$ with $\mathfrak{su}(p,q)$.

1.2.4 Quaternionic Lie Algebras

Quaternionic General and Special Linear Lie Algebras

We follow the notation of Section 1.1.4. Consider the $n \times n$ matrices over the quaternions with the usual matrix commutator. We will denote this Lie algebra by $\mathfrak{gl}(n,\mathbb{H})$, considered as a Lie algebra over $\mathbb{R}$ (we have not defined Lie algebras over skew fields). We can identify $\mathbb{H}^n$ with $\mathbb{C}^{2n}$ using one of the isomorphic copies of $\mathbb{C}$ ($\mathbb{R}1 + \mathbb{R}\mathbf{i}$, $\mathbb{R}1 + \mathbb{R}\mathbf{j}$, or $\mathbb{R}1 + \mathbb{R}\mathbf{k}$) in $\mathbb{H}$. Define

$$\mathfrak{sl}(n,\mathbb{H}) = \{X \in \mathfrak{gl}(n,\mathbb{H}) : \text{tr}(X) = 0\} .$$

Then $\mathfrak{sl}(n,\mathbb{H})$ is the real Lie algebra that is usually denoted by $\mathfrak{su}^*(2n)$.

Quaternionic Unitary Lie Algebras

For $n = p + q$ with p,q nonnegative integers, we define

$$\mathfrak{sp}(p,q) = \{X \in \mathfrak{gl}(n,\mathbb{H}) : X^* I_{p,q} = -I_{p,q} X\}$$

(the quaternionic adjoint X^* was defined in Section 1.1.4). We leave it as an exercise to check that $\mathfrak{sp}(p,q)$ is a real Lie subalgebra of $\mathfrak{gl}(n,\mathbb{H})$. Let the quaternionic Hermitian form $B(x,y)$ be defined as in (1.6). Then $\mathfrak{sp}(p,q)$ consists of the matrices $X \in M_n(\mathbb{H})$ that satisfy

$$B(Xx,y) = -B(x,X^*y) \quad \text{for all } x,y \in \mathbb{H}^n .$$

The Lie Algebra $\mathfrak{so}^*(2n)$

Let the automorphism θ of $M_{2n}(\mathbb{C})$ be as defined in Section 1.1.4 ($\theta(A) = -JAJ$). Define

$$\mathfrak{so}^*(2n) = \{X \in \mathfrak{so}(2n,\mathbb{C}) : \theta(\overline{X}) = X\} .$$

This real vector subspace of $\mathfrak{so}(2n, \mathbb{C})$ is a real Lie subalgebra of $\mathfrak{so}(2n, \mathbb{C})$ (considered as a Lie algebra over $\mathbb{R}$). Identify $\mathbb{C}^{2n}$ with $\mathbb{H}^n$ as in Section 1.2.4 and let the quaternionic skew-Hermitian form $C(x, y)$ be defined as in (1.7). Then $\mathfrak{so}^*(2n)$ corresponds to the matrices $X \in M_n(\mathbb{H})$ that satisfy

$$C(Xx, y) = -C(x, X^*y) \quad \text{for all } x, y \in \mathbb{H}^n .$$

1.2.5 Lie Algebras Associated with Classical Groups

The Lie algebras $\mathfrak{g}$ described in the preceding sections constitute the list of *classical Lie algebras* over $\mathbb{R}$ and $\mathbb{C}$. These Lie algebras will be a major subject of study throughout the remainder of this book. We will find, however, that the given matrix form of $\mathfrak{g}$ is not always the most convenient; other choices of bases will be needed to determine the structure of $\mathfrak{g}$. This is one of the reasons that we have stressed the intrinsic basis-free characterizations.

Following the standard convention, we have labeled each classical Lie algebra with a fraktur-font version of the name of a corresponding classical group. This passage from a Lie group to a Lie algebra, which is fundamental to Lie theory, arises by *differentiating* the defining equations for the group. In brief, each classical group G is a subgroup of $\mathbf{GL}(V)$ (where V is a real vector space) that is defined by a set $\mathcal{R}$ of algebraic equations. The corresponding Lie subalgebra $\mathfrak{g}$ of $\mathfrak{gl}(V)$ is determined by taking differentiable curves $\sigma : (-\varepsilon, \varepsilon) \to \mathbf{GL}(V)$ such that $\sigma(0) = I$ and $\sigma(t)$ satisfies the equations in $\mathcal{R}$. Then $\sigma'(0) \in \mathfrak{g}$, and all elements of $\mathfrak{g}$ are obtained in this way. This is the reason why $\mathfrak{g}$ is called the *infinitesimal form* of G.

For example, if G is the subgroup $\mathbf{O}(V, B)$ of $\mathbf{GL}(V)$ defined by a bilinear form B, then the curve σ must satisfy $B(\sigma(t)v, \sigma(t)w) = B(v, w)$ for all $v, w \in V$ and $t \in (-\varepsilon, \varepsilon)$. If we differentiate these relations we have

$$0 = \frac{d}{dt} B(\sigma(t)v, \sigma(t)w)\Big|_{t=0} = B(\sigma'(0)v, \sigma(0)w) + B(\sigma(0)v, \sigma'(0)w)$$

for all $v, w \in V$. Since $\sigma(0) = I$, we see that $\sigma'(0) \in \mathfrak{so}(V, B)$, as asserted.

We will return to these ideas in Section 1.3.4 after developing some basic aspects of Lie group theory.

1.2.6 Exercises

1. Prove that the Jacobi identity (2) holds for $\text{End}(V)$.
2. Prove that the inverse of a bijective Lie algebra homomorphism is a Lie algebra homomorphism.
3. Let B be a bilinear form on a finite-dimensional vector space V over $\mathbb{F}$.
   (a) Prove that $\mathfrak{so}(V, B)$ is a Lie subalgebra of $\mathfrak{gl}(V)$.

(b) Suppose that B is nondegenerate. Prove that $\text{tr}(X) = 0$ for all $X \in \mathfrak{so}(V,B)$.
4. Prove that $\mathfrak{u}(p,q)$, $\mathfrak{sp}(p,q)$, and $\mathfrak{so}^*(2n)$ are real Lie algebras.
5. Let $B_0(x,y)$ be the Hermitian form and $B_1(x,y)$ the skew-symmetric form on $\mathbb{C}^{2n}$ in Exercises 1.1.5 #11.
   (a) Show that $\mathfrak{sp}(p,q) = \mathfrak{su}(\mathbb{C}^{2n}, B_0) \cap \mathfrak{sp}(\mathbb{C}^{2n}, B_1)$ when $M_n(\mathbb{H})$ is identified with a real subspace of $M_{2n}(\mathbb{C})$ as in Exercises 1.1.5 #10.
   (b) Use part (a) to show that $\mathfrak{sp}(p,q) \subset \mathfrak{sl}(p+q, \mathbb{H})$.
6. Let $X \in M_n(\mathbb{H})$. For each of the three choices of a copy of $\mathbb{C}$ in $\mathbb{H}$ given by (1.4) write out the corresponding matrix of X as an element of $M_{2n}(\mathbb{C})$. Use this formula to show that the trace of X is independent of the choice.

1.3 Closed Subgroups of $\mathbf{GL}(\mathbf{n}, \mathbb{R})$

In this section we introduce some basic aspects of Lie groups that motivate the later developments in this book. We begin with the definition of a topological group and then emphasize the topological groups that are closed subgroups of $\mathbf{GL}(n, \mathbb{R})$. Our main tool is the *exponential map*, which we treat by explicit matrix calculations.

1.3.1 Topological Groups

Let G be a group with a Hausdorff topology. If the multiplication and inversion maps

$$G \times G \longrightarrow G \quad (g,h \mapsto gh) \quad \text{and} \quad G \longrightarrow G \quad (g \mapsto g^{-1})$$

are continuous, G is called a *topological group* (in this definition, the set $G \times G$ is given the product topology). For example, $\mathbf{GL}(n, \mathbb{F})$ is a topological group when endowed with the topology of the open subset $\{X : \det(X) \neq 0\}$ of $M_n(\mathbb{F})$. The multiplication is continuous and Cramer's rule implies that the inverse is continuous.

If G is a topological group, each element $g \in G$ defines *translation maps*

$$L_g : G \longrightarrow G \quad \text{and} \quad R_g : G \longrightarrow G,$$

given by $L_g(x) = gx$ and $R_g(x) = xg$. The group properties and continuity imply that R_g and L_g are homeomorphisms.

If G is a topological group and H is a subgroup, then H is also a topological group (in the relative topology). If H is also a closed subset of G, then we call it a *topological subgroup* of G. For example, the defining equations of each classical group show that it is a closed subset of $\mathbf{GL}(V)$ for some V, and hence it is a topological subgroup of $\mathbf{GL}(V)$.

A *topological group homomorphism* will mean a group homomorphism that is also a continuous map. A topological group homomorphism is said to be a *topological group isomorphism* if it is bijective and its inverse is also a topological group homomorphism. An isomorphism of a group with itself is called an *automorphism*. For example, if G is a topological group then each element $g \in G$ defines an automorphism $\tau(g)$ by conjugation: $\tau(g)x = gxg^{-1}$. Such automorphisms are called *inner*.

Before we study our main examples we prove two useful general results about topological groups.

Proposition 1.3.1. *If H is an open subgroup of a topological group G, then H is also closed in G.*

Proof. We note that G is a disjoint union of left cosets. If $g \in G$ then the left coset $gH = L_g(H)$ is open, since L_g is a homeomorphism. Hence the union of all the left cosets other than H is open, and so H is closed. $\qquad\square$

Proposition 1.3.2. *Let G be a topological group. Then the identity component of G (that is, the connected component G° that contains the identity element e) is a normal subgroup.*

Proof. Let G° be the identity component of G. If $h \in G^\circ$ then $h \in L_h(G^\circ)$ because $e \in G^\circ$. Since L_h is a homeomorphism and $G^\circ \cap L_h(G^\circ)$ is nonempty, it follows that $L_h(G^\circ) = G^\circ$, showing that G° is closed under multiplication. Since $e \in L_h G^\circ$, we also have $h^{-1} \in G^\circ$, and so G° is a subgroup. If $g \in G$ the inner automorphism $\tau(g)$ is a homeomorphism that fixes e and hence maps G° into G°. $\qquad\square$

1.3.2 Exponential Map

On $M_n(\mathbb{R})$ we define the inner product $\langle X, Y \rangle = \mathrm{tr}(XY^t)$. The corresponding norm

$$\|X\| = \langle X, X \rangle^{\frac{1}{2}} = \Big(\sum_{i,j=1}^{n} x_{ij}^2 \Big)^{1/2}$$

has the following properties (where $X, Y \in M_n(\mathbb{R})$ and $c \in \mathbb{R}$):

1. $\|X + Y\| \le \|X\| + \|Y\|, \quad \|cX\| = |c|\,\|X\|,$
2. $\|XY\| \le \|X\|\,\|Y\|,$
3. $\|X\| = 0$ if and only if $X = 0$.

Properties (1) and (3) follow by identifying $M_n(\mathbb{R})$ as a real vector space with $\mathbb{R}^{n^2}$ using the matrix entries. To verify property (2), observe that

$$\|XY\|^2 = \sum_{i,j} \Big(\sum_{k} x_{ik} y_{kj} \Big)^2 .$$

Now $\big| \sum_k x_{ik} y_{kj} \big|^2 \le \big(\sum_k x_{ik}^2 \big) \big(\sum_k y_{kj}^2 \big)$ by the Cauchy–Schwarz inequality. Hence

$$\|XY\|^2 \leq \sum_{i,j} \left(\sum_k x_{ik}^2 \right) \left(\sum_k y_{kj}^2 \right) = \left(\sum_{i,k} x_{ik}^2 \right) \left(\sum_{k,j} y_{kj}^2 \right) = \|X\|^2 \|Y\|^2 .$$

Taking the square root of both sides completes the proof.

We define matrix-valued analytic functions by substitution in convergent power series. Let $\{a_m\}$ be a sequence of real numbers such that

$$\sum_{m=0}^{\infty} |a_m| r^m < \infty \quad \text{for some } r > 0 .$$

For $A \in M_n(\mathbb{R})$ and $r > 0$ let

$$B_r(A) = \{X \in M_n(\mathbb{R}) : \|X - A\| < r\}$$

(the open ball of radius r around A). If $X \in B_r(0)$ and $k \geq l$ then by properties (1) and (2) of the norm we have

$$\left\| \sum_{0 \leq m \leq k} a_m X^m - \sum_{0 \leq m \leq l} a_m X^m \right\| = \left\| \sum_{l < m \leq k} a_m X^m \right\| \leq \sum_{l < m \leq k} |a_m| \|X^m\|$$

$$\leq \sum_{l < m \leq k} |a_m| \|X\|^m \leq \sum_{m > l}^{\infty} |a_m| r^m .$$

The last series goes to 0 as $l \to \infty$ by the convergence assumption. We define

$$f(X) = \sum_{m=0}^{\infty} a_m X^m \quad \text{for } X \in B_r(0) .$$

The function $X \mapsto f(X)$ is *real analytic* (each entry in the matrix $f(X)$ is a convergent power series in the entries of X when $\|X\| < r$).

Substitution Principle: Any equation involving power series in a complex variable x that holds as an identity of absolutely convergent series when $|x| < r$ also holds as an identity of matrix power series in a matrix variable X, and these series converge absolutely in the matrix norm when $\|X\| < r$.

This follows by rearranging the power series, which is permissible by absolute convergence.

In Lie theory two functions play a special role:

$$\exp(X) = \sum_{m=0}^{\infty} \frac{1}{m!} X^m \quad \text{and} \quad \log(1 + X) = \sum_{m=1}^{\infty} (-1)^{m+1} \frac{1}{m} X^m .$$

The exponential series converges absolutely for all X, and the logarithm series converges absolutely for $\|X\| < 1$. We therefore have two analytic matrix-valued functions,

$$\exp : M_n(\mathbb{R}) \longrightarrow M_n(\mathbb{R}) \quad \text{and} \quad \log : B_1(I) \longrightarrow M_n(\mathbb{R}) .$$

If $X,Y \in M_n(\mathbb{R})$ and $XY = YX$, then each term $(X+Y)^m$ can be expanded by the binomial formula. Rearranging the series for $\exp(X+Y)$ (which is justified by absolute convergence), we obtain the identity

$$\exp(X+Y) = \exp(X)\exp(Y). \tag{1.9}$$

In particular, this implies that $\exp(X)\exp(-X) = \exp(0) = I$. Thus

$$\exp : M_n(\mathbb{R}) \longrightarrow \mathbf{GL}(n,\mathbb{R}).$$

The power series for the exponential and logarithm satisfy the identities

$$\log(\exp(x)) = x \qquad \text{for } |x| < \log 2, \tag{1.10}$$
$$\exp(\log(1+x)) = 1+x \quad \text{for } |x| < 1. \tag{1.11}$$

To verify (1.10), use the chain rule to show that the derivative of $\log(\exp(x))$ is 1; since this function vanishes at $x = 0$, it is x. To verify (1.11), use the chain rule twice to show that the second derivative of $\exp(\log(1+x))$ is zero; thus the function is a polynomial of degree one. This polynomial and its first derivative have the value 1 at $x = 0$; hence it is $x+1$.

We use these identities to show that the matrix logarithm function gives a local inverse to the exponential function.

Lemma 1.3.3. *Suppose $g \in \mathbf{GL}(n,\mathbb{R})$ satisfies $\|g - I\| < \log 2/(1 + \log 2)$. Then $\|\log(g)\| < \log 2$ and $\exp(\log(g)) = g$. Furthermore, if $X \in B_{\log 2}(0)$ and $\exp X = g$, then $X = \log(g)$.*

Proof. Since $\log 2/(1 + \log 2) < 1$, the power series for $\log(g)$ is absolutely convergent and

$$\|\log(g)\| \leq \sum_{m=1}^{\infty} \|g - I\|^m = \frac{\|g - I\|}{1 - \|g - I\|} < \log 2.$$

Since $\|g - I\| < 1$, we can replace z by $g - I$ in identity (1.11) by the substitution principle. Hence $\exp(\log(g)) = g$.

If $X \in B_{\log 2}(0)$ then

$$\|\exp(X) - I\| \leq e^{\|X\|} - 1 < 1.$$

Hence we can replace x by X in identity (1.10) by the substitution principle. If $\exp X = g$, this identity yields $X = \log(g)$. $\qquad\square$

Remark 1.3.4. Lemma 1.3.3 asserts that the exponential map is a bijection from a neighborhood of 0 in $M_n(\mathbb{R})$ onto a neighborhood of I in $\mathbf{GL}(n,\mathbb{R})$. However, if $n \geq 2$ then the map $\exp : M_n(\mathbb{R}) \longrightarrow \{g : \det(g) > 0\} \subset \mathbf{GL}(n,\mathbb{R})$ is neither injective nor surjective, in contrast to the scalar case.

If $X \in M_n(\mathbb{R})$, then the continuous function $\varphi(t) = \exp(tX)$ from $\mathbb{R}$ to $\mathbf{GL}(n,\mathbb{R})$ satisfies $\varphi(0) = I$ and $\varphi(s+t) = \varphi(s)\varphi(t)$ for all $s,t \in \mathbb{R}$, by equation (1.9). Thus given X, we obtain a homomorphism φ from the additive group of real numbers to the group $\mathbf{GL}(n,\mathbb{R})$. We call this homomorphism the *one-parameter group generated by X*. It is a fundamental result in Lie theory that all homomorphisms from $\mathbb{R}$ to $\mathbf{GL}(n,\mathbb{R})$ are obtained in this way.

Theorem 1.3.5. *Let $\varphi : \mathbb{R} \longrightarrow \mathbf{GL}(n,\mathbb{R})$ be a continuous homomorphism from the additive group $\mathbb{R}$ to $\mathbf{GL}(n,\mathbb{R})$. Then there exists a unique $X \in M_n(\mathbb{R})$ such that $\varphi(t) = \exp(tX)$ for all $t \in \mathbb{R}$.*

Proof. The uniqueness of X is immediate, since

$$\frac{d}{dt}\exp(tX)\bigg|_{t=0} = X \,.$$

To prove the existence of X, let $\varepsilon > 0$ and set $\varphi_\varepsilon(t) = \varphi(\varepsilon t)$. Then φ_ε is also a continuous homomorphism of $\mathbb{R}$ into $\mathbf{GL}(n,\mathbb{R})$. Since φ is continuous and $\varphi(0) = I$, from Lemma 1.3.3 we can choose ε such that $\varphi_\varepsilon(t) \in \exp B_r(0)$ for $|t| < 2$, where $r = (1/2)\log 2$. If we can show that $\varphi_\varepsilon(t) = \exp(tX)$ for some $X \in M_n(\mathbb{R})$ and all $t \in \mathbb{R}$, then $\varphi(t) = \exp\big((t/\varepsilon)X\big)$. Thus it suffices to treat the case $\varepsilon = 1$.

Assume now that $\varphi(t) \in \exp B_r(0)$ for $|t| < 2$, with $r = (1/2)\log 2$. Then there exists $X \in B_r(0)$ such that $\varphi(1) = \exp X$. Likewise, there exists $Z \in B_r(0)$ such that $\varphi(1/2) = \exp Z$. But

$$\varphi(1) = \varphi(1/2) \cdot \varphi(1/2) = \exp(Z) \cdot \exp(Z) = \exp(2Z) \,.$$

Since $\|2Z\| < \log 2$ and $\|X\| < \log 2$, Lemma 1.3.3 implies that $Z = (1/2)X$. Since $\varphi(1/4) = \exp(W)$ with $W \in B_r(0)$, we likewise have $W = (1/2)Z = (1/4)X$. Continuing this argument, we conclude that

$$\varphi(1/2^k) = \exp((1/2^k)X) \quad \text{for all integers } k \geq 0 \,.$$

Let $a = (1/2)a_1 + (1/4)a_2 + \cdots + (1/2^k)a_k + \cdots$, with $a_j \in \{0,1\}$, be the dyadic expansion of the real number $0 \leq a < 1$. Then by continuity and the assumption that φ is a group homomorphism we have

$$\begin{aligned}
\varphi(a) &= \lim_{k\to\infty} \varphi\big((1/2)a_1 + (1/4)a_2 + \cdots + (1/2^k)a_k\big) \\
&= \lim_{k\to\infty} \varphi(1/2)^{a_1}\varphi(1/4)^{a_2}\cdots\varphi(1/2^k)^{a_k} \\
&= \lim_{k\to\infty} \big(\exp(1/2)X\big)^{a_1}\cdots\big(\exp(1/2^k)X\big)^{a_k} \\
&= \lim_{k\to\infty} \exp\big\{\big((1/2)a_1 + (1/4)a_2 + \cdots + (1/2^k)a_k\big)X\big\} = \exp(aX) \,.
\end{aligned}$$

Now if $0 \leq a < 1$ then $\varphi(-a) = \varphi(a)^{-1} = \exp(aX)^{-1} = \exp(-aX)$. Finally, given $a \in \mathbb{R}$ choose an integer $k > |a|$. Then

$$\varphi(a) = \varphi(a/k)^k = \big(\exp(a/k)X\big)^k = \exp(aX) \,.$$

This shows that φ is the one-parameter subgroup generated by X. $\qquad\square$

1.3.3 Lie Algebra of a Closed Subgroup of $\mathbf{GL}(\mathbf{n}, \mathbb{R})$

Let G be a closed subgroup of $\mathbf{GL}(n, \mathbb{R})$. We define

$$\mathrm{Lie}(G) = \{X \in M_n(\mathbb{R}) : \exp(tX) \in G \text{ for all } t \in \mathbb{R}\}. \qquad (1.12)$$

Thus by Theorem 1.3.5 each matrix in $\mathrm{Lie}(G)$ corresponds to a unique continuous one-parameter subgroup of G. To show that $\mathrm{Lie}(G)$ is a Lie subalgebra of $\mathfrak{gl}(n, \mathbb{R})$, we need more information about the product $\exp X \exp Y$.

Fix $X, Y \in M_n(\mathbb{R})$. By Lemma 1.3.3 there is an analytic matrix-valued function $Z(s,t)$, defined for (s,t) in a neighborhood of zero in $\mathbb{R}^2$, such that $Z(0,0) = 0$ and

$$\exp(sX)\exp(tY) = \exp(Z(s,t)).$$

It is easy to calculate the linear and quadratic terms in the power series of $Z(s,t)$. Since $Z(s,t) = \log(\exp(sX)\exp(tY))$, the power series for the logarithm and exponential functions give

$$
\begin{aligned}
Z(s,t) &= \big(\exp(sX)\exp(tY) - I\big) - \frac{1}{2}\big(\exp(sX)\exp(tY) - I\big)^2 + \cdots \\
&= \Big(\big(I + sX + \frac{1}{2}s^2X^2\big)\big(I + tY + \frac{1}{2}t^2Y^2\big) - I\Big) - \frac{1}{2}(sX + tY)^2 + \cdots \\
&= \Big(sX + tY + \frac{1}{2}s^2X^2 + stXY + \frac{1}{2}t^2Y^2\Big) \\
&\quad - \frac{1}{2}\big(s^2X^2 + st(XY + YX) + t^2Y^2\big) + \cdots,
\end{aligned}
$$

where $\cdots$ indicates terms that are of total degree three and higher in s, t. The first-degree term is $sX + tY$, as expected (the series terminates after this term when X and Y commute). The quadratic terms involving only X or Y cancel; the only remaining term involving both X and Y is the commutator:

$$Z(s,t) = sX + tY + \frac{st}{2}[X,Y] + \cdots. \qquad (1.13)$$

Rescaling X and Y, we can state formula (1.13) as follows:

Lemma 1.3.6. *There exist $\varepsilon > 0$ and an analytic matrix-valued function $R(X,Y)$ on $B_\varepsilon(0) \times B_\varepsilon(0)$ such that*

$$\exp X \exp Y = \exp\Big(X + Y + \frac{1}{2}[X,Y] + R(X,Y)\Big)$$

when $X, Y \in B_\varepsilon(0)$. *Furthermore,* $\|R(X,Y)\| \le C(\|X\| + \|Y\|)^3$ *for some constant* C *and all* $X, Y \in B_\varepsilon(0)$.

From Lemma 1.3.6 we now obtain the fundamental identities relating the Lie algebra structure of $\mathfrak{gl}(n, \mathbb{R})$ to the group structure of $\mathbf{GL}(n, \mathbb{R})$.

Proposition 1.3.7. *If* $X, Y \in M_n(\mathbb{R})$, *then*

$$\exp(X+Y) = \lim_{k\to\infty} \left(\exp\left(\frac{1}{k}X\right) \exp\left(\frac{1}{k}Y\right) \right)^k, \tag{1.14}$$

$$\exp([X,Y]) = \lim_{k\to\infty} \left(\exp\left(\frac{1}{k}X\right) \exp\left(\frac{1}{k}Y\right) \exp\left(-\frac{1}{k}X\right) \exp\left(-\frac{1}{k}Y\right) \right)^{k^2}. \tag{1.15}$$

Proof. For k a sufficiently large integer, Lemma 1.3.6 implies that

$$\exp\left(\frac{1}{k}X\right) \exp\left(\frac{1}{k}Y\right) = \exp\left(\frac{1}{k}(X+Y) + \mathrm{O}(1/k^2)\right),$$

where $\mathrm{O}(r)$ denotes a matrix function of r whose norm is bounded by Cr for some constant C (depending only on $\|X\| + \|Y\|$) and all small r. Hence

$$\left(\exp\left(\frac{1}{k}X\right) \exp\left(\frac{1}{k}Y\right) \right)^k = \exp k\left(\frac{1}{k}(X+Y) + \mathrm{O}(1/k^2)\right)$$
$$= \exp\left(X + Y + \mathrm{O}(1/k)\right).$$

Letting $k \to \infty$, we obtain formula (1.14).

Likewise, we have

$$\exp\left(\frac{1}{k}X\right) \exp\left(\frac{1}{k}Y\right) \exp\left(-\frac{1}{k}X\right) \exp\left(-\frac{1}{k}Y\right)$$
$$= \exp\left(\frac{1}{k}(X+Y) + \frac{1}{2k^2}[X,Y] + \mathrm{O}(1/k^3)\right)$$
$$\times \exp\left(-\frac{1}{k}(X+Y) + \frac{1}{2k^2}[X,Y] + \mathrm{O}(1/k^3)\right)$$
$$= \exp\left(\frac{1}{k^2}[X,Y] + \mathrm{O}(1/k^3)\right).$$

(Of course, each occurrence of $\mathrm{O}(1/k^3)$ in these formulas stands for a different function.) Thus

$$\left(\exp\left(\frac{1}{k}X\right) \exp\left(\frac{1}{k}Y\right) \exp\left(-\frac{1}{k}X\right) \exp\left(-\frac{1}{k}Y\right) \right)^{k^2}$$
$$= \exp k^2 \left(\frac{1}{k^2}[X,Y] + \mathrm{O}(1/k^3)\right) = \exp\left([X,Y] + \mathrm{O}(1/k)\right).$$

This implies formula (1.15). $\square$

Theorem 1.3.8. *If* G *is a closed subgroup of* $\mathbf{GL}(n, \mathbb{R})$, *then* $\mathrm{Lie}(G)$ *is a Lie subalgebra of* $M_n(\mathbb{R})$.

Proof. If $X \in \mathrm{Lie}(G)$, then $tX \in \mathrm{Lie}(G)$ for all $t \in \mathbb{R}$. If $X, Y \in \mathrm{Lie}(G)$ and $t \in \mathbb{R}$, then

$$\exp\left(t(X+Y)\right) = \lim_{k \to \infty} \left(\exp\left(\frac{t}{k}X\right) \exp\left(\frac{t}{k}Y\right) \right)^k$$

is in G, since G is a closed subgroup. Similarly,

$$\exp(t[X,Y]) = \lim_{k \to \infty} \left(\exp\left(\frac{t}{k}X\right) \exp\left(\frac{t}{k}Y\right) \exp\left(-\frac{t}{k}X\right) \exp\left(-\frac{t}{k}Y\right) \right)^{k^2}$$

is in G. $\square$

If G is a closed subgroup of $\mathbf{GL}(n, \mathbb{R})$, then the elements of G act on the one-parameter subgroups in G by conjugation. Since $gX^k g^{-1} = (gXg^{-1})^k$ for $g \in G$, $X \in \mathrm{Lie}(G)$, and all positive integers k, we have

$$g(\exp tX)g^{-1} = \exp(tgXg^{-1}) \quad \text{for all } t \in \mathbb{R}.$$

The left side of this equation is a one-parameter subgroup of G. Hence $gXg^{-1} \in \mathrm{Lie}(G)$. We define $\mathrm{Ad}(g) \in \mathbf{GL}(\mathrm{Lie}(G))$ by

$$\mathrm{Ad}(g)X = gXg^{-1} \quad \text{for } X \in \mathrm{Lie}(G). \tag{1.16}$$

Clearly $\mathrm{Ad}(g_1 g_2) = \mathrm{Ad}(g_1)\,\mathrm{Ad}(g_2)$ for $g_1, g_2 \in G$, so $g \mapsto \mathrm{Ad}(g)$ is a continuous group homomorphism from G to $\mathbf{GL}(\mathrm{Lie}(G))$. Furthermore, if $X, Y \in \mathrm{Lie}(G)$ and $g \in G$, then the relation $gXYg^{-1} = (gXg^{-1})(gYg^{-1})$ implies that

$$\mathrm{Ad}(g)([X,Y]) = [\mathrm{Ad}(g)X, \mathrm{Ad}(g)Y]. \tag{1.17}$$

Hence $\mathrm{Ad}(g)$ is an *automorphism* of the Lie algebra structure.

Remark 1.3.9. There are several ways of associating a Lie algebra with a closed subgroup of $\mathbf{GL}(n, \mathbb{R})$; in the course of chapter we shall prove that the different Lie algebras are all isomorphic.

1.3.4 Lie Algebras of the Classical Groups

To determine the Lie algebras of the classical groups, we fix the following embeddings of $\mathbf{GL}(n, \mathbb{F})$ as a closed subgroup of $\mathbf{GL}(dn, \mathbb{R})$. Here $\mathbb{F}$ is $\mathbb{C}$ or $\mathbb{H}$ and $d = \dim_{\mathbb{R}} \mathbb{F}$.

We take $\mathbb{C}^n$ to be $\mathbb{R}^{2n}$ and let multiplication by $\sqrt{-1}$ be given by the matrix

$$J = \begin{bmatrix} 0 & I \\ -I & 0 \end{bmatrix},$$

with I the $n \times n$ identity matrix. Then $M_n(\mathbb{C})$ is identified with the matrices in $M_{2n}(\mathbb{R})$ that commute with J, and $\mathbf{GL}(n, \mathbb{C})$ is identified with the invertible matrices in $M_n(\mathbb{C})$. Thus $\mathbf{GL}(n, \mathbb{C})$ is a closed subgroup of $\mathbf{GL}(2n, \mathbb{R})$.

The case of the quaternionic groups is handled similarly. We take $\mathbb{H}^n$ to be $\mathbb{R}^{4n}$ and use the $4n \times 4n$ matrices

$$J_1 = \begin{bmatrix} J & 0 \\ 0 & -J \end{bmatrix}, \quad J_2 = \begin{bmatrix} 0 & I \\ -I & 0 \end{bmatrix}, \quad \text{and} \quad J_3 = \begin{bmatrix} 0 & J \\ J & 0 \end{bmatrix}$$

(with J as above but now I is the $2n \times 2n$ identity matrix) to give multiplication by $\mathbf{i}, \mathbf{j}$, and $\mathbf{k}$, respectively. This gives a model for $\mathbb{H}^n$, since these matrices satisfy the quaternion relations

$$J_p^2 = -I, \quad J_1 J_2 = J_3, \quad J_2 J_3 = J_1, \quad J_3 J_1 = J_2,$$
$$\text{and} \quad J_p J_l = -J_l J_p \quad \text{for } p \neq l.$$

In this model $M_n(\mathbb{H})$ is identified with the matrices in $M_{4n}(\mathbb{R})$ that commute with J_p ($p = 1, 2, 3$), and $\mathbf{GL}(n, \mathbb{H})$ consists of the invertible matrices in $M_n(\mathbb{H})$. Thus $\mathbf{GL}(n, \mathbb{H})$ is a closed subgroup of $\mathbf{GL}(4n, \mathbb{R})$.

Since each classical group G is a closed subgroup of $\mathbf{GL}(n, \mathbb{F})$ with $\mathbb{F}$ either $\mathbb{R}$, $\mathbb{C}$, or $\mathbb{H}$, the embeddings just defined make G a closed subgroup of $\mathbf{GL}(dn, \mathbb{R})$. With these identifications the names of the Lie algebras in Section 1.2 are consistent with the names attached to the groups in Section 1.1; to obtain the Lie algebra corresponding to a classical group, one replaces the initial capital letters in the group name with fraktur letters. We work out the details for a few examples and leave the rest as an exercise.

It is clear from the definition that $\mathrm{Lie}(\mathbf{GL}(n, \mathbb{R})) = M_n(\mathbb{R}) = \mathfrak{gl}(n, \mathbb{R})$. The Lie algebra of $\mathbf{GL}(n, \mathbb{C})$ consists of all $X \in M_{2n}(\mathbb{R})$ such that $J^{-1} \exp(tX) J = \exp(tX)$ for all $t \in \mathbb{R}$. Since $A^{-1} \exp(X) A = \exp(A^{-1} X A)$ for any $A \in \mathbf{GL}(n, \mathbb{R})$, we see that $X \in \mathrm{Lie}(\mathbf{GL}(n, \mathbb{C}))$ if and only if $\exp(tJ^{-1}XJ) = \exp(tX)$ for all $t \in \mathbb{R}$. This relation holds if and only if $J^{-1}XJ = X$, so we conclude that $\mathrm{Lie}(\mathbf{GL}(n, \mathbb{C})) = \mathfrak{gl}(n, \mathbb{C})$. The same argument (using the matrices $\{J_i\}$) shows that $\mathrm{Lie}(\mathbf{GL}(n, \mathbb{H})) = \mathfrak{gl}(n, \mathbb{H})$.

We now look at $\mathbf{SL}(n, \mathbb{R})$. For any $X \in M_n(\mathbb{C})$ there exist $U \in \mathbf{U}(n)$ and an upper-triangular matrix $T = [t_{ij}]$ such that $X = UTU^{-1}$ (this is the *Schur triangular form*). Thus $\exp(X) = U \exp(T) U^{-1}$ and so $\det(\exp(X)) = \det(\exp(T))$. Since $\exp(T)$ is upper triangular with ith diagonal entry $e^{t_{ii}}$, we have $\det(\exp(T)) = e^{\mathrm{tr}(T)}$. But $\mathrm{tr}(T) = \mathrm{tr}(X)$, so we conclude that

$$\det(\exp(X)) = e^{\mathrm{tr}(X)}. \tag{1.18}$$

If $X \in M_n(\mathbb{R})$, then from equation (1.18) we see that the one-parameter subgroup $t \mapsto \exp(tX)$ is in $\mathbf{SL}(n, \mathbb{R})$ if and only if $\mathrm{tr}(X) = 0$. Hence

$$\mathrm{Lie}(\mathbf{SL}(n, \mathbb{R})) = \{X \in M_n(\mathbb{R}) : \mathrm{tr}(X) = 0\} = \mathfrak{sl}(n, \mathbb{R}).$$

For the other classical groups it is convenient to use the following simple result:

Lemma 1.3.10. *Suppose $H \subset G \subset \mathbf{GL}(n,\mathbb{R})$ with H a closed subgroup of G and G a closed subgroup of $\mathbf{GL}(n,\mathbb{R})$. Then H is a closed subgroup of $\mathbf{GL}(n,\mathbb{R})$, and*

$$\mathrm{Lie}(H) = \{X \in \mathrm{Lie}(G) : \exp(tX) \in H \text{ for all } t \in \mathbb{R}\}\,.$$

Proof. It is obvious that H is a closed subgroup of $\mathbf{GL}(n,\mathbb{R})$. If $X \in \mathrm{Lie}(H)$ then $\exp(tX) \in H \subset G$ for all $t \in \mathbb{R}$. Thus $X \in \mathrm{Lie}(G)$. $\qquad\square$

We consider $\mathrm{Lie}(\mathbf{Sp}(n,\mathbb{C}))$. Since $\mathbf{Sp}(n,\mathbb{C}) \subset \mathbf{GL}(2n,\mathbb{C}) \subset \mathbf{GL}(2n,\mathbb{R})$, we can look upon $\mathrm{Lie}(\mathbf{Sp}(n,\mathbb{C}))$ as the set of $X \in M_{2n}(\mathbb{C})$ such that $\exp tX \in \mathbf{Sp}(n,\mathbb{C})$ for all $t \in \mathbb{R}$. This condition can be expressed as

$$\exp(tX^t)J\exp(tX) = J \quad \text{for all } t \in \mathbb{R}\,. \tag{1.19}$$

Differentiating this equation at $t = 0$, we find that $X^tJ + JX = 0$ for all $X \in \mathrm{Lie}(\mathbf{Sp}(n,\mathbb{C}))$. Conversely, if X satisfies this last equation, then $JXJ^{-1} = -X^t$, and so

$$J\exp(tX)J^{-1} = \exp(tJXJ^{-1}) = \exp(-tX^t) \quad \text{for all } t \in \mathbb{R}\,.$$

Hence X satisfies condition (1.19). This proves that $\mathrm{Lie}(\mathbf{Sp}(n,\mathbb{C})) = \mathfrak{sp}(n,\mathbb{C})$.

We do one more family of examples. Let $G = \mathbf{U}(p,q) \subset \mathbf{GL}(p+q,\mathbb{C})$. Then

$$\mathrm{Lie}(G) = \{X \in M_n(\mathbb{C}) : \exp(tX)^*I_{p,q}\exp(tX) = I_{p,q} \text{ for all } t \in \mathbb{R}\}\,.$$

We note that for $t \in \mathbb{R}$,

$$\left(\exp tX\right)^* = \left(I + tX + \frac{1}{2}t^2X^2 + \cdots\right)^* = I + tX^* + \frac{1}{2}t^2(X^*)^2 + \cdots\,.$$

Thus if $X \in \mathrm{Lie}(G)$, then

$$\left(\exp tX\right)^*I_{p,q}\exp tX = I_{p,q}\,.$$

Differentiating this equation with respect to t at $t = 0$, we obtain $X^*I_{p,q} + I_{p,q}X = 0$. This shows that $\mathrm{Lie}(\mathbf{U}(p,q)) \subset \mathfrak{u}(p,q)$. Conversely, if $X \in \mathfrak{u}(p,q)$, then

$$(X^*)^kI_{p,q} = (-1)^kI_{p,q}X^k \quad \text{for all integers } k\,.$$

Using this relation in the power series for the exponential function, we have

$$\exp(tX^*)I_{p,q} = I_{p,q}\exp(-tX)\,.$$

This equation can be written as $\exp(tX)^*I_{p,q}\exp(tX) = I_{p,q}$; hence $\exp(tX)$ is in $\mathbf{U}(p,q)$ for all $t \in \mathbb{R}$. This proves that $\mathrm{Lie}(\mathbf{U}(p,q)) = \mathfrak{u}(p,q)$.

1.3.5 Exponential Coordinates on Closed Subgroups

We will now study in more detail the relationship between the Lie algebra of a closed subgroup H of $\mathbf{GL}(n, \mathbb{R})$ and the group structure of H. We first note that for $X \in \mathrm{Lie}(H)$ the map $t \mapsto \exp tX$ from $\mathbb{R}$ to H has range in the identity component of H. Hence the Lie algebra of H is the same as the Lie algebra of the identity component of H. It is therefore reasonable to confine our attention to connected groups in this discussion.

Theorem 1.3.11. *Let H be a closed subgroup of $\mathbf{GL}(n, \mathbb{R})$. There exist an open neighborhood V of 0 in $\mathrm{Lie}(H)$ and an open neighborhood Ω of I in $\mathbf{GL}(n, \mathbb{R})$ such that $\exp(V) = H \cap \Omega$ and $\exp : V \longrightarrow \exp(V)$ is a homeomorphism onto the open neighborhood $H \cap \Omega$ of I in H.*

Proof. Let $K = \{X \in M_n(\mathbb{R}) : \mathrm{tr}(X^t Y) = 0 \text{ for all } Y \in \mathrm{Lie}(H)\}$ be the orthogonal complement of $\mathrm{Lie}(H)$ in $M_n(\mathbb{R})$ relative to the trace form inner product. Then there is an orthogonal direct sum decomposition

$$M_n(\mathbb{R}) = \mathrm{Lie}(H) \oplus K . \tag{1.20}$$

Using decomposition (1.20), we define an analytic map $\varphi : M_n(\mathbb{R}) \longrightarrow \mathbf{GL}(n, \mathbb{R})$ by $\varphi(X) = \exp(X_1) \exp(X_2)$ when $X = X_1 + X_2$ with $X_1 \in \mathrm{Lie}(H)$ and $X_2 \in K$. We note that $\varphi(0) = I$ and

$$\varphi(tX) = \big(I + tX_1 + \mathrm{O}(t^2)\big)\big(I + tX_2 + \mathrm{O}(t^2)\big) = I + tX + \mathrm{O}(t^2) .$$

Hence the differential of φ at 0 is the identity map. The inverse function theorem implies that there exists $s_1 > 0$ such that $\varphi : B_{s_1}(0) \longrightarrow \mathbf{GL}(n, \mathbb{R})$ has an open image U_1 and the map $\varphi : B_{s_1}(0) \longrightarrow U_1$ is a homeomorphism.

With these preliminaries established we can begin the argument. Suppose, for the sake of obtaining a contradiction, that for every ε such that $0 < \varepsilon \le s_1$, the set $\varphi(B_\varepsilon(0)) \cap H$ contains an element that is not in $\exp(\mathrm{Lie}(H))$. In this case for each integer $k \ge 1/s_1$ there exists an element in $Z_k \in B_{1/k}(0)$ such that $\exp(Z_k) \in H$ and $Z_k \notin \mathrm{Lie}(H)$. We write $Z_k = X_k + Y_k$ with $X_k \in \mathrm{Lie}(H)$ and $0 \ne Y_k \in K$. Then

$$\varphi(Z_k) = \exp(X_k) \exp(Y_k) .$$

Since $\exp(X_k) \in H$, we see that $\exp(Y_k) \in H$. We also observe that $\|Y_k\| \le 1/k$. Let $\varepsilon_k = \|Y_k\|$. Then $0 < \varepsilon_k \le 1/k \le s_1$. For each k there exists a positive integer m_k such that $s_1 \le m_k \varepsilon_k < 2s_1$. Hence

$$s_1 \le \|m_k Y_k\| < 2s_1 . \tag{1.21}$$

Since the sequence $m_k Y_k$ is bounded, we can replace it with a subsequence that converges. We may therefore assume that there exists $Y \in W$ with $\lim_{k \to \infty} m_k Y_k = Y$. Then $\|Y\| \ge s_1 > 0$ by inequalities (1.21), so $Y \ne 0$.

We claim that $\exp(tY) \in H$ for all $t \in \mathbb{R}$. Indeed, given t, we write $tm_k = a_k + b_k$ with $a_k \in \mathbb{Z}$ the integer part of tm_k and $0 \le b_k < 1$. Then $tm_k Y_k = a_k Y_k + b_k Y_k$ and

$$\exp(tm_k Y_k) = \left(\exp Y_k\right)^{a_k} \exp\left(b_k Y_k\right).$$

We have $\left(\exp Y_k\right)^{m_k} \in H$ for all n. Since $\lim_{k \to \infty} Y_k = 0$ and $0 \le b_k < 1$, it follows that $\lim_{k \to \infty} \exp(b_k Y_k) = I$. Hence

$$\exp(tY) = \lim_{k \to \infty} \exp\left(tm_k Y_k\right) = \lim_{k \to \infty} \left(\exp Y_k\right)^{a_k} \in H,$$

since H is closed. But this implies that $Y \in \mathrm{Lie}(H) \cap K = \{0\}$, which is a contradiction, since $Y \ne 0$. This proves that there must exist an $\varepsilon > 0$ such that $\varphi(B_\varepsilon(0)) \cap H \subset \exp(\mathrm{Lie}(H))$. Set $V = B_\varepsilon(0) \cap \mathrm{Lie} H$. Then $\exp V = \varphi(B_\varepsilon(0)) \cap H$ is an open neighborhood of I in H, by definition of the relative topology on H, and the restriction of $\exp$ to V is a homeomorphism onto $\exp V$. $\qquad \square$

A topological group G is a *Lie group* if there is a differentiable manifold structure on G (see Appendix D.1.1) such that the following conditions are satisfied:

(i) The manifold topology on G is the same as the group topology.
(ii) The multiplication map $G \times G \longrightarrow G$ and the inversion map $G \longrightarrow G$ are C^∞.

The group $\mathbf{GL}(n, \mathbb{R})$ is a Lie group, with its manifold structure as an open subset of the vector space $M_n(\mathbb{R})$. The multiplication and inversion maps are real analytic.

Theorem 1.3.12. *Let H be a closed subgroup of $\mathbf{GL}(n, \mathbb{R})$, considered as a topological group with the relative topology. Then H has a Lie group structure that is compatible with its topological group structure.*

Proof. Let $K \subset M_n(\mathbb{R})$ be the orthogonal complement to $\mathrm{Lie}(H)$, as in equation (1.20). The map $X \oplus Y \mapsto \exp(X) \exp(Y)$ from $\mathrm{Lie}(H) \oplus K$ to $\mathbf{GL}(n, \mathbb{R})$ has differential $X \oplus Y \mapsto X + Y$ at 0 by Lemma 1.3.6. Hence by the inverse function theorem, Lemma 1.3.3, and Theorem 1.3.11 there are open neighborhoods of 0

$$U \subset \mathrm{Lie}(H), \quad V \subset K, \quad W \subset M_n(\mathbb{R}),$$

with the following properties:

1. If $\Omega = \{g_1 g_2 g_3 : g_i \in \exp W\}$, then the map $\log : \Omega \longrightarrow M_n(\mathbb{R})$ is a diffeomorphism onto its image. Furthermore, $W = -W$.
2. There are real-analytic maps $\varphi : \Omega \longrightarrow U$ and $\psi : \Omega \longrightarrow V$ such that $g \in \Omega$ can be factored as $g = \exp(\varphi(g)) \exp(\psi(g))$.
3. $H \cap \Omega = \{g \in \Omega : \psi(g) = 0\}$.

Give H the relative topology as a closed subset of $\mathbf{GL}(n, \mathbb{R})$. We will define a C^∞ d-atlas for H (where $d = \dim \mathrm{Lie}(H)$) as follows:

Given $h \in H$, we can write $h \exp U = (h \exp W) \cap H$ by property (3). Hence $U_h = h \exp U$ is an open neighborhood of h in H. Define

$$\Phi_h(h\exp X) = X \quad \text{for } X \in U.$$

Then by property (2) the map $\Phi_h : U_h \longrightarrow U$ is a homeomorphism. Suppose $h_1 \exp X_1 = h_2 \exp X_2$ with $h_i \in H$ and $X_i \in U$. Then $\Phi_{h_2}(\Phi_{h_1}^{-1}(X_1)) = X_2$. Now

$$h_2^{-1} h_1 = \exp(X_2)\exp(-X_1) \in (\exp W)^2,$$

so we see that $\exp X_2 = h_2^{-1} h_1 \exp X_1 \in \Omega$. It follows from properties (1) and (2) that $X_2 = \log(h_2^{-1} h_1 \exp X_1)$ is a C^∞ function of X_1 with values in $\mathrm{Lie}(H)$ (in fact, it is a real-analytic function). Thus $\{(U_h, \Phi_h)\}_{h \in H}$ is a C^∞ d-atlas for H.

It remains to show that the map $h_1, h_2 \mapsto h_1 h_2^{-1}$ is C^∞ from $H \times H$ to H. Let $X_i \in \mathfrak{h}$. Then $h_1 \exp X_1 \exp(-X_2) h_2^{-1} = h_1 h_2^{-1} \exp(\mathrm{Ad}(h_2)X_1)\exp(-\mathrm{Ad}(h_2)X_2)$. Fix h_2 and set

$$U^{(2)} = \{(X_1, X_2) \in U \times U : \exp(\mathrm{Ad}(h_2)X_1)\exp(-\mathrm{Ad}(h_2)X_2) \in \exp W\}.$$

Then $U^{(2)}$ is an open neighborhood of $(0,0)$ in $\mathrm{Lie}(H) \times \mathrm{Lie}(H)$. By property (3),

$$\beta(X_1, X_2) = \log\left(\exp(\mathrm{Ad}(h_2)X_1)\exp(-\mathrm{Ad}(h_2)X_2)\right) \in U$$

for $(X_1, X_2) \in U^{(2)}$. The map $\beta : U^{(2)} \longrightarrow U$ is clearly C^∞ and we can write

$$h_1 \exp X_1 \exp(-X_2) h_2^{-1} = h_1 h_2^{-1} \exp\beta(X_1, X_2)$$

for $(X_1, X_2) \in U^{(2)}$. Thus multiplication and inversion are C^∞ maps on H. $\square$

Remark 1.3.13. An atlas $\mathcal{A} = \{U_\alpha, \Phi_\alpha\}_{\alpha \in I}$ on a C^∞ manifold X is *real analytic* if each transition map $\Phi_\alpha \circ \Phi_\beta^{-1}$ is given by a convergent power series in the local coordinates at each point in its domain. Such an atlas defines a *real-analytic* (class C^ω) manifold structure on X, just as in the C^∞ case, since the composition of real-analytic functions is real analytic. A map between manifolds of class C^ω is *real analytic* if it is given by convergent power series in local real-analytic coordinates. The exponential coordinate atlas on the subgroup H defined in the proof of Theorem 1.3.12 is real analytic, and the group operations on H are real-analytic maps relative to the C^ω manifold structure defined by this atlas.

1.3.6 Differentials of Homomorphisms

Let $G \subset \mathbf{GL}(n, \mathbb{R})$ and $H \subset \mathbf{GL}(m, \mathbb{R})$ be closed subgroups.

Proposition 1.3.14. *Let $\varphi : H \longrightarrow G$ be a continuous homomorphism. There exists a unique Lie algebra homomorphism $d\varphi : \mathrm{Lie}(H) \longrightarrow \mathrm{Lie}(G)$, called the* differential *of φ, such that $\varphi(\exp(X)) = \exp(d\varphi(X))$ for all $X \in \mathrm{Lie}(H)$.*

Proof. If $X \in \mathrm{Lie}(H)$ then $t \mapsto \varphi(\exp tX)$ defines a continuous homomorphism of $\mathbb{R}$ into $\mathbf{GL}(n, \mathbb{R})$. Hence Theorem 1.3.5 implies that there exists $\mu(X) \in M_n(\mathbb{R})$ such that

$$\varphi(\exp(tX)) = \exp(t\mu(X)) \quad \text{for all } t \in R \ .$$

It is clear from the definition that $\mu(tX) = t\mu(X)$ for all $t \in \mathbb{R}$. We will use Proposition 1.3.7 to prove that $\mu : \text{Lie}(H) \longrightarrow \text{Lie}(G)$ is a Lie algebra homomorphism.

If $X, Y \in \text{Lie}(H)$ and $t \in \mathbb{R}$, then by continuity of φ and formula (1.14) we have

$$\begin{aligned}
\varphi\Big(\exp(tX + tY)\Big) &= \lim_{k \to \infty} \varphi\Big(\exp\Big(\frac{t}{k}X\Big)\exp\Big(\frac{t}{k}Y\Big)\Big)^k \\
&= \lim_{k \to \infty} \Big(\exp\Big(\frac{t}{k}\mu(X)\Big)\exp\Big(\frac{t}{k}\mu(Y)\Big)\Big)^k \\
&= \exp\big(t\mu(X) + t\mu(Y)\big) \ .
\end{aligned}$$

Hence the uniqueness assertion in Theorem 1.3.5 gives $\mu(X + Y) = \mu(X) + \mu(Y)$. Likewise, now using formula (1.15), we prove that $\mu([X,Y]) = [\mu(X), \mu(Y)]$. We define $d\varphi(X) = \mu(X)$. $\square$

By Remark 1.3.13, G and H are real-analytic manifolds relative to charts given by exponential coordinates.

Corollary 1.3.15. *The homomorphism φ is real analytic.*

Proof. This follows immediately from the definition of the Lie group structures on G and H using exponential coordinates (as in the proof of Theorem 1.3.12), together with Proposition 1.3.14. $\square$

1.3.7 Lie Algebras and Vector Fields

We call the entries x_{ij} in the matrix $X = [x_{ij}] \in M_n(\mathbb{R})$ the *standard coordinates* on $M_n(\mathbb{R})$. That is, the functions x_{ij} are the components of X with respect to the standard basis $\{e_{ij}\}$ for $M_n(\mathbb{R})$ (where the elementary matrix e_{ij} has exactly one nonzero entry, which is a 1 in the i, j position). If U is an open neighborhood of I in $M_n(\mathbb{R})$ and $f \in C^\infty(U)$, then

$$\frac{\partial}{\partial x_{ij}} f(u) = \frac{d}{dt} f(u + te_i)\Big|_{t=0}$$

is the usual partial derivative relative to the standard coordinates.

If we use the multiplicative structure on $M_n(\mathbb{R})$ and the exponential map instead of the additive structure, then we can define

$$\frac{d}{dt} f\big(u\exp(te_{ij})\big)\Big|_{t=0} = \frac{d}{dt} f\big(u + tue_{ij})\big)\Big|_{t=0} \ ,$$

since $\exp(tA) = I + tA + O(t^2)$ for $A \in M_n(\mathbb{R})$. Now $ue_{ij} = \sum_{k=1}^{n} x_{ki}(u)e_{kj}$. Thus by the chain rule we find that

$$\frac{d}{dt} f\big(u\exp(te_{ij})\big)\Big|_{t=0} = E_{ij}f(u) \quad \text{for } u \in U ,$$

where E_{ij} is the vector field

$$E_{ij} = \sum_{k=1}^{n} x_{ki} \frac{\partial}{\partial x_{kj}} \tag{1.22}$$

on U. In general, if $A = \sum_{i,j=1}^{n} a_{ij} e_{ij} \in M_n(\mathbb{R})$, then we can define a vector field on $M_n(\mathbb{R})$ by $X_A = \sum_{i,j=1}^{n} a_{ij} E_{ij}$. By the chain rule we have

$$\sum_{i,j=1}^{n} a_{ij} E_{ij} f(u) = \frac{d}{dt} f\big(u\exp\big(\textstyle\sum_{i,j} a_{ij} e_{ij}\big)\big)\Big|_{t=0} .$$

Hence

$$X_A f(u) = \frac{d}{dt} f\big(u\exp(tA)\big)\Big|_{t=0} . \tag{1.23}$$

Define the *left translation operator* $L(y)$ by $L(y)f(g) = f(y^{-1}g)$ for f a C^∞ function on $\mathbf{GL}(n,\mathbb{R})$ and $y \in \mathbf{GL}(n,\mathbb{R})$. It is clear from (1.23) that X_A commutes with $L(y)$ for all $y \in \mathbf{GL}(n,\mathbb{R})$. Furthermore, at the identity element we have

$$(X_A)_I = \sum_{i,j} a_{ij} \left(\frac{\partial}{\partial x_{ij}} \right)_I \in T(M_n(\mathbb{R}))_I , \tag{1.24}$$

since $(E_{ij})_I = (\partial/\partial x_{kj})_I$. It is important to observe that equation (1.24) holds only at I; the vector field X_A is a linear combination (with real coefficients) of the *variable-coefficient* vector fields $\{E_{ij}\}$, whereas the *constant-coefficient* vector field $\sum_{i,j} a_{ij} (\partial/\partial x_{ij})$ does not commute with $L(y)$.

The map $A \mapsto X_A$ is obviously linear; we claim that it also satisfies

$$[X_A, X_B] = X_{[A,B]} \tag{1.25}$$

and hence is a Lie algebra homomorphism. Indeed, using linearity in A and B, we see that it suffices to verify formula (1.25) when $A = e_{ij}$ and $B = e_{kl}$. In this case $[e_{ij}, e_{kl}] = \delta_{jk} e_{il} - \delta_{il} e_{kj}$ by matrix multiplication, whereas the commutator of the vector fields is

$$[E_{ij}, E_{kl}] = \sum_{p,q} x_{pi} \Big\{ \frac{\partial}{\partial x_{pj}} (x_{qk}) \Big\} \frac{\partial}{\partial x_{ql}} - \sum_{p,q} x_{qk} \Big\{ \frac{\partial}{\partial x_{ql}} (x_{pi}) \Big\} \frac{\partial}{\partial x_{pj}}$$

$$= \delta_{jk} E_{il} - \delta_{il} E_{kj} .$$

Hence $[e_{ij}, e_{kl}] \mapsto [E_{ij}, E_{kl}]$ as claimed.

Now assume that G is a closed subgroup of $\mathbf{GL}(n,\mathbb{R})$ and let $\mathrm{Lie}(G)$ be defined as in (1.12) using one-parameter subgroups. We know from Corollary 1.3.15 that the inclusion map $\iota_G : G \longrightarrow \mathbf{GL}(n,\mathbb{R})$ is C^∞ (in fact, real analytic).

Lemma 1.3.16. *The differential of the inclusion map satisfies* $(d\iota_G)_I(T(G)_I) = \{(X_A)_I : A \in \mathrm{Lie}(G)\}$.

Proof. For $A \in \mathrm{Lie}(G)$ the one-parameter group $t \mapsto \exp(tA)$ is a C^∞ map from $\mathbb{R}$ to G, by definition of the manifold structure of G (see Theorem 1.3.12). We define the tangent vector $v_A \in T(G)_I$ by

$$v_A f = \frac{d}{dt} f\big(\exp(tA)\big)\Big|_{t=0} \quad \text{for } f \in C^\infty(G) \, .$$

By definition of the differential of a smooth map, we then have $(dt_G)_I(v_A)f = (X_A)_I$. This shows that

$$(dt_G)_I(T(G)_I) \supset \{(X_A)_I : A \in \mathrm{Lie}(G)\} \, . \tag{1.26}$$

Since $\dim \mathrm{Lie}(A) = \dim T(G)_I$, the two spaces in (1.26) are the same. $\qquad\square$

Define the *left translation operator* $L(y)$ on $C^\infty(G)$ by $L(y)f(g) = f(y^{-1}g)$ for $y \in G$ and $f \in C^\infty(G)$. We say that a smooth vector field X on G is *left invariant* if it commutes with the operators $L(y)$ for all $y \in G$:

$$X(L(y)f)(g) = (L(y)Xf)(g) \quad \text{for all } y, g \in G \text{ and } f \in C^\infty(G) \, .$$

If $A \in \mathrm{Lie}(G)$ we set

$$X_A^G f(g) = \frac{d}{dt} f(g\exp(tA))\Big|_{t=0} \quad \text{for } f \in C^\infty(G) \, .$$

Since the map $\mathbb{R} \times G \longrightarrow G$ given by $t, g \mapsto g\exp(tA)$ is smooth, we see that X_A^G is a left-invariant vector field on G. When $G = \mathbf{GL}(n, \mathbb{R})$ then $X_A^G = X_A$ as defined in (1.24).

Proposition 1.3.17. *Every left-invariant regular vector field on G is of the form X_A^G for a unique $A \in \mathrm{Lie}(G)$. Furthermore, if $A, B \in \mathrm{Lie}(G)$ then $[X_A^G, X_B^G] = X_{[A,B]}^G$.*

Proof. Since a left-invariant vector field X is uniquely determined by the tangent vector X_I at I, the first statement follows from Lemma 1.26. Likewise, to prove the commutator formula it suffices to show that

$$[X_A^G, X_B^G]_I = \big(X_{[A,B]}^G\big)_I \quad \text{for all } A, B \in \mathrm{Lie}(G) \, . \tag{1.27}$$

From Theorem 1.3.12 there is a coordinate chart for $\mathbf{GL}(n, \mathbb{R})$ at I whose first $d = \dim G$ coordinates are the linear coordinates given by a basis for $\mathrm{Lie}(G)$. Thus there is a neighborhood Ω of I in $\mathbf{GL}(n, \mathbb{R})$ such that every C^∞ function f on the corresponding neighborhood $U = \Omega \cap G$ of I in G is of the form $\varphi|_U$, with $\varphi \in C^\infty(\Omega)$. If $g \in U$ and $A \in \mathrm{Lie}(G)$, then for $t \in \mathbb{R}$ near zero we have $g\exp tA \in U$. Hence

$$X_A \varphi(g) = \frac{d}{dt}\varphi(g\exp tA)\Big|_{t=0} = \frac{d}{dt}f(g\exp tA)\Big|_{t=0} = X_A^G f(g).$$

Thus $(X_A\varphi)|_U = X_A^G f$. Now take $B \in \mathrm{Lie}(G)$. Then

$$[X_A^G, X_B^G]f = X_A^G X_B^G f - X_B^G X_A^G f = (X_A X_B \varphi - X_B X_A \varphi)\big|_U = ([X_A, X_B]\varphi)\big|_U \, .$$

But by (1.25) we have $[X_A, X_B]\varphi = X_{[A,B]}\varphi$. Hence

$$[X_A^G, X_B^G]f = (X_{[A,B]}\varphi)\big|_U = X_{[A,B]}^G f \,.$$

Since this last equation holds for all $f \in C^\infty(U)$, it proves (1.27). $\qquad\qquad\square$

Let $G \subset \mathbf{GL}(n,\mathbb{R})$ and $H \subset \mathbf{GL}(m,\mathbb{R})$ be closed subgroups. If $\varphi : H \longrightarrow G$ is a continuous homomorphism, we know from Corollary 1.3.15 that φ must be real analytic. We now calculate $d\varphi_I : T(H)_I \longrightarrow T(G)_I$. Using the notation in the proof of Lemma 1.26, we have the linear map $\mathrm{Lie}(H) \longrightarrow T(H)_I$ given by $A \mapsto v_A$ for $A \in \mathrm{Lie}(H)$. If $f \in C^\infty(G)$ then

$$d\varphi_I(v_A)f = \frac{d}{dt}f(\varphi(\exp tA))\Big|_{t=0}\,.$$

By Proposition 1.3.14 there is a Lie algebra homomorphism $d\varphi : \mathrm{Lie}(H) \longrightarrow \mathrm{Lie}(G)$ with $\varphi(\exp(tA)) = \exp(td\varphi(A))$. Thus $d\varphi_I(v_A) = v_{d\varphi(A)}$. Since the vector field X_A^H on H is left invariant, we conclude that

$$d\varphi_h(X_A^H)_h = (X_{d\varphi(A)}^G)_{\varphi(h)} \quad \text{for all } h \in H \,.$$

Thus for a closed subgroup G of $\mathbf{GL}(n,\mathbb{R})$ the matrix algebra version of its Lie algebra and the geometric version of its Lie algebra as the left-invariant vector fields on G are isomorphic under the correspondence $A \mapsto X_A^G$, by Proposition 1.3.17. Furthermore, under this correspondence the differential of a homomorphism given in Proposition 1.3.14 is the same as the differential defined in the general Lie group context (see Appendix D.2.2).

1.3.8 Exercises

1. Show that $\exp : M_n(\mathbb{C}) \longrightarrow \mathbf{GL}(n,\mathbb{C})$ is surjective. (HINT: Use Jordan canonical form.)
2. This exercise shows that $\exp : M_n(\mathbb{R}) \longrightarrow \mathbf{GL}(n,\mathbb{R})_+ = \{g \in M_n(\mathbb{R}) : \det g > 0\}$ is neither injective nor surjective when $n \geq 2$.
   (a) Let $X = \left[\begin{smallmatrix} 0 & 1 \\ -1 & 0 \end{smallmatrix}\right]$. Calculate the matrix form of the one-parameter subgroup $\varphi(t) = \exp(tX)$ and show that the kernel of the homomorphism $t \mapsto \varphi(t)$ is $2\pi\mathbb{Z}$.
   (b) Let $g = \left[\begin{smallmatrix} -1 & 1 \\ 0 & -1 \end{smallmatrix}\right]$. Show that g is not the exponential of any real 2×2 matrix. (HINT: Assume $g = \exp(X)$. Compare the eigenvectors of X and g to conclude that X can have only one eigenvalue. Then use $\mathrm{tr}(X) = 0$ to show that this eigenvalue must be zero.)
3. Complete the proof that the Lie algebras of the classical groups in Section 1.1 are the Lie algebras with the corresponding fraktur names in Section 1.2, following the same technique used for $\mathfrak{sl}(n,\mathbb{R})$, $\mathfrak{sp}(n,\mathbb{F})$, and $\mathfrak{su}(p,q)$ in Section 1.3.4.

4. *(Notation of Exercises 1.1.5, # 4)* Show that φ is continuous and prove that $d\varphi$ is a Lie algebra isomorphism. Use this result to prove that the image of φ is open (and hence also closed) in $\mathbf{SO}(V, B)$.

5. *(Notation of Exercises 1.1.5, #6 and #7)* Show that φ is continuous and prove that $d\varphi$ is a Lie algebra isomorphism. Use this result to prove that the image of φ is open and closed in the corresponding orthogonal group.

6. *(Notation of Exercises 1.1.5, #8 and #9)* Prove that the differentials of ψ and φ are Lie algebra isomorphisms.

7. Let $X, Y \in M_n(\mathbb{R})$. Use Lemma 1.3.6 to prove that there exist an $\varepsilon > 0$ and a constant $C > 0$ such that the following holds for $\|X\| + \|Y\| < \varepsilon$:
   (a) $\exp X \exp Y \exp(-X) = \exp\left(Y + [X, Y] + Q(X, Y)\right)$, with $Q(X, Y) \in M_n(\mathbb{R})$ and $\|Q(X, Y)\| \leq C(\|X\| + \|Y\|)^3$.
   (b) $\exp X \exp Y \exp(-X) \exp(-Y) = \exp\left([X, Y] + P(X, Y)\right)$, with $P(X, Y) \in M_n(\mathbb{R})$ and $\|P(X, Y)\| \leq C(\|X\| + \|Y\|)^3$.

1.4 Linear Algebraic Groups

Since each classical group $G \subset \mathbf{GL}_n(\mathbb{F})$ is defined by algebraic equations, we can also study G using algebraic techniques instead of analysis. We will take the field $\mathbb{F} = \mathbb{C}$ in this setting (it could be any algebraically closed field of characteristic zero). We also require that the equations defining G be polynomials in the *complex* matrix entries (that is, they do not involve complex conjugation).

1.4.1 Definitions and Examples

Definition 1.4.1. A subgroup G of $\mathbf{GL}(n, \mathbb{C})$ is a *linear algebraic group* if there is a set $\mathcal{A}$ of polynomial functions on $M_n(\mathbb{C})$ such that

$$G = \{g \in \mathbf{GL}(n, \mathbb{C}) : f(g) = 0 \text{ for all } f \in \mathcal{A}\}.$$

Here a function f on $M_n(\mathbb{C})$ is a *polynomial function* if

$$f(y) = p(x_{11}(y), x_{12}(y), \ldots, x_{nn}(y)) \quad \text{for all } y \in M_n(\mathbb{C}),$$

where $p \in \mathbb{C}[x_{11}, x_{12}, \ldots, x_{nn}]$ is a polynomial and x_{ij} are the matrix entry functions on $M_n(\mathbb{C})$.

Given a complex vector space V with $\dim V = n$, we fix a basis for V and we let $\mu : \mathbf{GL}(V) \longrightarrow \mathbf{GL}(n, \mathbb{C})$ be the corresponding isomorphism as in Section 1.1.1. We call a subgroup $G \subset \mathbf{GL}(V)$ a *linear algebraic group* if $\mu(G)$ is an algebraic group in the sense of Definition 1.4.1 (this definition is clearly independent of the choice of basis).

Examples

1. The basic example of a linear algebraic group is $\mathbf{GL}(n,\mathbb{C})$ (take the defining set $\mathcal{A}$ of relations to consist of the zero polynomial). In the case $n = 1$ we have $\mathbf{GL}(1,\mathbb{C}) = \mathbb{C} \setminus \{0\} = \mathbb{C}^{\times}$, the multiplicative group of the field $\mathbb{C}$.

2. The special linear group $\mathbf{SL}(n,\mathbb{C})$ is algebraic and defined by one polynomial equation $\det(g) - 1 = 0$.

3. Let $D_n \subset \mathbf{GL}(n,\mathbb{C})$ be the subgroup of diagonal matrices. The defining equations for D_n are $x_{ij}(g) = 0$ for all $i \neq j$, so D_n is an algebraic group.

4. Let $N_n^+ \subset \mathbf{GL}(n,\mathbb{C})$ be the subgroup of upper-triangular matrices with diagonal entries 1. The defining equations in this case are $x_{ii}(g) = 1$ for all i and $x_{ij}(g) = 0$ for all $i > j$. When $n = 2$, the group N_2^+ is isomorphic (as an abstract group) to the additive group of the field $\mathbb{C}$, via the map $z \mapsto \left[\begin{smallmatrix} 1 & z \\ 0 & 1 \end{smallmatrix}\right]$ from $\mathbb{C}$ to N_2^+. We will look upon $\mathbb{C}$ as the linear algebraic group N_2^+.

5. Let $B_n \subset \mathbf{GL}(n,\mathbb{C})$ be the subgroup of upper-triangular matrices. The defining equations for B_n are $x_{ij}(g) = 0$ for all $i > j$, so B_n is an algebraic group.

6. Let $\Gamma \in \mathbf{GL}(n,\mathbb{C})$ and let $B_\Gamma(x,y) = x^t \Gamma y$ for $x, y \in \mathbb{C}^n$. Then B_Γ is a nondegenerate bilinear form on $\mathbb{C}^n$. Let $G_\Gamma = \{g \in \mathbf{GL}(n,\mathbb{C}) : g^t \Gamma g = \Gamma\}$ be the subgroup that preserves this form. Since G_Γ is defined by quadratic equations in the matrix entries, it is an algebraic group. This shows that the orthogonal groups and the symplectic groups are algebraic subgroups of $\mathbf{GL}(n,\mathbb{C})$.

For the orthogonal or symplectic groups (when $\Gamma^t = \pm \Gamma$), there is another description of G_Γ that will be important in connection with *real forms* in this chapter and *symmetric spaces* in Chapters 11 and 12. Define

$$\sigma_\Gamma(g) = \Gamma^{-1}(g^t)^{-1}\Gamma \quad \text{for } g \in \mathbf{GL}(n,\mathbb{C}) .$$

Then $\sigma_\Gamma(gh) = \sigma_\Gamma(g)\sigma_\Gamma(h)$ for $g, h \in \mathbf{GL}(n,\mathbb{C})$, $\sigma_\Gamma(I) = I$, and

$$\sigma_\Gamma(\sigma_\Gamma(g)) = \Gamma^{-1}(\Gamma^t g (\Gamma^t)^{-1})\Gamma = g,$$

since $\Gamma^{-1}\Gamma^t = \pm I$. Such a map σ_S is called an *involutory automorphism* of $\mathbf{GL}(n,\mathbb{C})$. We have $g \in G_\Gamma$ if and only if $\sigma_\Gamma(g) = g$, and hence the group G_Γ is the set of *fixed points* of σ_Γ.

1.4.2 Regular Functions

We now establish some basic properties of linear algebraic groups. We begin with the notion of *regular function*. For the group $\mathbf{GL}(n,\mathbb{C})$, the algebra of *regular functions* is defined as

$$\mathcal{O}[\mathbf{GL}(n,\mathbb{C})] = \mathbb{C}[x_{11}, x_{12}, \ldots, x_{nn}, \det(x)^{-1}] .$$

This is the commutative algebra over $\mathbb{C}$ generated by the matrix entry functions $\{x_{ij}\}$ and the function $\det(x)^{-1}$, with the relation $\det(x) \cdot \det(x)^{-1} = 1$ (where $\det(x)$ is expressed as a polynomial in $\{x_{ij}\}$ as usual).

For any complex vector space V of dimension n, let $\varphi : \mathbf{GL}(V) \longrightarrow \mathbf{GL}(n, \mathbb{C})$ be the group isomorphism defined in terms of a basis for V. The algebra $\mathcal{O}[\mathbf{GL}(V)]$ of *regular functions* on $\mathbf{GL}(V)$ is defined as all functions $f \circ \varphi$, where f is a regular function on $\mathbf{GL}(n, \mathbb{C})$. This definition is clearly independent of the choice of basis for V.

The regular functions on $\mathbf{GL}(V)$ that are linear combinations of the matrix entry functions x_{ij}, relative to some basis for V, can be described in the following basis-free way: Given $B \in \mathrm{End}(V)$, we define a function f_B on $\mathrm{End}(V)$ by

$$f_B(Y) = \mathrm{tr}_V(YB), \quad \text{for } Y \in \mathrm{End}(V). \tag{1.28}$$

For example, when $V = \mathbb{C}^n$ and $B = e_{ij}$, then $f_{e_{ij}}(Y) = x_{ji}(Y)$. Since the map $B \mapsto f_B$ is linear, it follows that each function f_B on $\mathbf{GL}(n, \mathbb{C})$ is a linear combination of the matrix-entry functions and hence is regular. Furthermore, the algebra $\mathcal{O}[\mathbf{GL}(n, \mathbb{C})]$ is generated by $\{f_B : B \in M_n(\mathbb{C})\}$ and $(\det)^{-1}$. Thus for any finite-dimensional vector space V the algebra $\mathcal{O}[\mathbf{GL}(V)]$ is generated by $(\det)^{-1}$ and the functions f_B, for $B \in \mathrm{End}(V)$.

An element $g \in \mathbf{GL}(V)$ acts on $\mathrm{End}(V)$ by left and right multiplication, and we have

$$f_B(gY) = f_{Bg}(Y), \quad f_B(Yg) = f_{gB}(Y) \quad \text{for } B, Y \in \mathrm{End}(V).$$

Thus the functions f_B allow us to transfer properties of the linear action of g on $\mathrm{End}(V)$ to properties of the action of g on functions on $\mathbf{GL}(V)$, as we will see in later sections.

Definition 1.4.2. Let $G \subset \mathbf{GL}(V)$ be an algebraic subgroup. A complex-valued function f on G is *regular* if it is the restriction to G of a regular function on $\mathbf{GL}(V)$.

The set $\mathcal{O}[G]$ of regular functions on G is a commutative algebra over $\mathbb{C}$ under pointwise multiplication. It has a finite set of generators, namely the restrictions to G of $(\det)^{-1}$ and the functions f_B, with B varying over any linear basis for $\mathrm{End}(V)$. Set

$$\mathcal{I}_G = \{f \in \mathcal{O}[\mathbf{GL}(V)] : f(G) = 0\}.$$

This is an ideal in $\mathcal{O}[\mathbf{GL}(V)]$ that we can describe in terms of the algebra $\mathcal{P}(\mathrm{End}(V))$ of polynomials on $\mathrm{End}(V)$ by

$$\mathcal{I}_G = \bigcup_{p \geq 0} \{(\det)^{-p} f : f \in \mathcal{P}(\mathrm{End}(V)), f(G) = 0\}. \tag{1.29}$$

The map $f \mapsto f|_G$ gives an algebra isomorphism

$$\mathcal{O}[G] \cong \mathcal{O}[\mathbf{GL}(V)]/\mathcal{I}_G. \tag{1.30}$$

Let G and H be linear algebraic groups and let $\varphi : G \longrightarrow H$ be a map. For $f \in \mathcal{O}[H]$ define the function $\varphi^*(f)$ on G by $\varphi^*(f)(g) = f(\varphi(g))$. We say that φ is a *regular map* if $\varphi^*(\mathcal{O}[H]) \subset \mathcal{O}[G]$.

Definition 1.4.3. *An* algebraic group homomorphism $\varphi : G \longrightarrow H$ *is a group homomorphism that is a regular map. We say that G and H are* isomorphic *as algebraic groups if there exists an algebraic group homomorphism $\varphi : G \longrightarrow H$ that has a regular inverse.*

Given linear algebraic groups $G \subset \mathbf{GL}(m, \mathbb{C})$ and $H \subset \mathbf{GL}(n, \mathbb{C})$, we make the group-theoretic direct product $K = G \times H$ into an algebraic group by the natural block-diagonal embedding $K \longrightarrow \mathbf{GL}(m + n, \mathbb{C})$ as the block-diagonal matrices

$$k = \begin{bmatrix} g & 0 \\ 0 & h \end{bmatrix} \quad \text{with } g \in G \text{ and } h \in H .$$

Since polynomials in the matrix entries of g and h are polynomials in the matrix entries of k, we see that K is an algebraic subgroup of $\mathbf{GL}(m + n, \mathbb{C})$. The algebra homomorphism carrying $f' \otimes f'' \in \mathcal{O}[G] \otimes \mathcal{O}[H]$ to the function $(g, h) \mapsto f'(g)f''(h)$ on $G \times H$ gives an isomorphism

$$\mathcal{O}[G] \otimes \mathcal{O}[H] \overset{\cong}{\longrightarrow} \mathcal{O}[K]$$

(see Lemma A.1.9). In particular, $G \times G$ is an algebraic group with the algebra of regular functions $\mathcal{O}[G \times G] \cong \mathcal{O}[G] \otimes \mathcal{O}[G]$.

Proposition 1.4.4. *The maps $\mu : G \times G \longrightarrow G$ and $\eta : G \longrightarrow G$ given by multiplication and inversion are regular. If $f \in \mathcal{O}[G]$ then there exist an integer p and $f_i', f_i'' \in \mathcal{O}[G]$ for $i = 1, \ldots, p$ such that*

$$f(gh) = \sum_{i=1}^{p} f_i'(g) f_i''(h) \quad \text{for } g, h \in G . \tag{1.31}$$

Furthermore, for fixed $g \in G$ the maps $x \mapsto L_g(x) = gx$ and $x \mapsto R_g(x) = xg$ on G are regular.

Proof. Cramer's rule says that $\eta(g) = \det(g)^{-1} \mathrm{adj}(g)$, where $\mathrm{adj}(g)$ is the transposed cofactor matrix of g. Since the matrix entries of $\mathrm{adj}(g)$ are polynomials in the matrix entries $x_{ij}(g)$, it is clear from (1.30) that $\eta^* f \in \mathcal{O}[G]$ whenever $f \in \mathcal{O}[G]$.

Let $g, h \in G$. Then

$$x_{ij}(gh) = \sum_r x_{ir}(g) x_{rj}(h) .$$

Hence (1.31) is valid when $f = x_{ij}|_G$. It also holds when $f = (\det)^{-1}|_G$ by the multiplicative property of the determinant. Let $\mathcal{F}$ be the set of $f \in \mathcal{O}[G]$ for which (1.31) is valid. Then $\mathcal{F}$ is a subalgebra of $\mathcal{O}[G]$, and we have just verified that the matrix entry functions and $\det^{-1}$ are in $\mathcal{F}$. Since these functions generate $\mathcal{O}[G]$ as an algebra, it follows that $\mathcal{F} = \mathcal{O}[G]$.

Using the identification $\mathcal{O}[G \times G] = \mathcal{O}[G] \otimes \mathcal{O}[G]$, we can write (1.31) as

$$\mu^*(f) = \sum_i f_i' \otimes f_i'' \ .$$

This shows that μ is a regular map. Furthermore, $L_g^*(f) = \sum_i f_i'(g) f_i''$ and $R_g^*(f) = \sum_i f_i''(g) f_i'$, which proves that L_g and R_g are regular maps. $\square$

Examples

1. Let D_n be the subgroup of diagonal matrices in $\mathbf{GL}(n, \mathbb{C})$. The map

$$(x_1, \ldots, x_n) \mapsto \mathrm{diag}[x_1, \ldots, x_n]$$

from $(\mathbb{C}^\times)^n$ to D_n is obviously an isomorphism of algebraic groups. Since $\mathcal{O}[\mathbb{C}^\times] = \mathbb{C}[x, x^{-1}]$ consists of the Laurent polynomials in one variable, it follows that

$$\mathcal{O}[D_n] \cong \mathbb{C}[x_1, x_1^{-1}, \ldots, x_n, x_n^{-1}]$$

is the algebra of the Laurent polynomials in n variables. We call an algebraic group H that is isomorphic to D_n an *algebraic torus* of *rank n*.

2. Let $N_n^+ \subset \mathbf{GL}(n, \mathbb{C})$ be the subgroup of upper-triangular matrices with unit diagonal. It is easy to show that the functions x_{ij} for $i > j$ and $x_{ii} - 1$ generate $\mathcal{I}_{N_n^+}$, and that

$$\mathcal{O}[N_n^+] \cong \mathbb{C}[x_{12}, x_{13}, \ldots, x_{n-1,n}]$$

is the algebra of polynomials in the $n(n-1)/2$ variables $\{x_{ij} : i < j\}$.

Remark 1.4.5. In the examples of algebraic groups G just given, the determination of generators for the ideal $\mathcal{I}_G$ and the structure of $\mathcal{O}[G]$ is straightforward because $\mathcal{I}_G$ is generated by linear functions of the matrix entries. In general, it is a difficult problem to find generators for $\mathcal{I}_G$ and to determine the structure of the algebra $\mathcal{O}[G]$.

1.4.3 Lie Algebra of an Algebraic Group

The next step in developing the theory of algebraic groups over $\mathbb{C}$ is to associate a Lie algebra of matrices to a linear algebraic group $G \subset \mathbf{GL}(n, \mathbb{C})$. We want the definition to be purely algebraic. Since the exponential function on matrices is not a polynomial, our approach is somewhat different than that in Section 1.3.3. Our strategy is to adapt the vector field point of view in Section 1.3.7 to the setting of linear algebraic groups; the main change is to replace the algebra of smooth functions on G by the algebra $\mathcal{O}[G]$ of regular (rational) functions. The Lie algebra of G will then be defined as the derivations of $\mathcal{O}[G]$ that commute with left translations. The following notion of a derivation (*infinitesimal transformation*) plays an important role in Lie theory.

Definition 1.4.6. Let $\mathcal{A}$ be an algebra (not assumed to be associative) over a field $\mathbb{F}$. Then $\mathrm{Der}(\mathcal{A}) \subset \mathrm{End}(\mathcal{A})$ is the set of all linear transformations $D : \mathcal{A} \longrightarrow \mathcal{A}$ that satisfy $D(ab) = (Da)b + a(Db)$ for all $a, b \in \mathcal{A}$ (call D a *derivation* of $\mathcal{A}$).

We leave it as an exercise to verify that $\mathrm{Der}(\mathcal{A})$ is a Lie subalgebra of $\mathrm{End}(\mathcal{A})$, called the algebra of *derivations* of $\mathcal{A}$.

We begin with the case $G = \mathbf{GL}(n, \mathbb{C})$, which we view as a linear algebraic group with the algebra of regular functions $\mathcal{O}[G] = \mathbb{C}[x_{11}, x_{12}, \ldots, x_{nn}, \det^{-1}]$. A *regular vector field* on G is a complex linear transformation $X : \mathcal{O}[G] \longrightarrow \mathcal{O}[G]$ of the form

$$Xf(g) = \sum_{i,j=1}^{n} c_{ij}(g) \frac{\partial f}{\partial x_{ij}}(g) \tag{1.32}$$

for $f \in \mathcal{O}[G]$ and $g \in G$, where we assume that the coefficients c_{ij} are in $\mathcal{O}[G]$. In addition to being a linear transformation of $\mathcal{O}[G]$, the operator X satisfies

$$X(f_1 f_2)(g) = (Xf_1)(g)f_2(g) + f_1(g)(Xf_2)(g) \tag{1.33}$$

for $f_1, f_2 \in \mathcal{O}[G]$ and $g \in G$, by the product rule for differentiation. Any linear transformation X of $\mathcal{O}[G]$ that satisfies (1.33) is called a *derivation* of the algebra $\mathcal{O}[G]$. If X_1 and X_2 are derivations, then so is the linear transformation $[X_1, X_2]$, and we write $\mathrm{Der}(\mathcal{O}[G])$ for the Lie algebra of all derivations of $\mathcal{O}[G]$.

We will show that every derivation of $\mathcal{O}[G]$ is given by a regular vector field on G. For this purpose it is useful to consider equation (1.33) with g fixed. We say that a complex linear map $v : \mathcal{O}[G] \longrightarrow \mathbb{C}$ is a *tangent vector to G at g* if

$$v(f_1 f_2) = v(f_1)f_2(g) + f_1(g)v(f_2) . \tag{1.34}$$

The set of all tangent vectors at g is a vector subspace of the complex dual vector space $\mathcal{O}[G]^*$, since equation (1.34) is linear in v. We call this vector space the *tangent space* to G at g (in the sense of algebraic groups), and denote it by $T(G)_g$. For any $A = [a_{ij}] \in M_n(\mathbb{C})$ we can define a tangent vector v_A at g by

$$v_A(f) = \sum_{i,j=1}^{n} a_{ij} \frac{\partial f}{\partial x_{ij}}(g) \quad \text{for } f \in \mathcal{O}[G] . \tag{1.35}$$

Lemma 1.4.7. *Let $G = \mathbf{GL}(n, \mathbb{C})$ and let $v \in T(G)_g$. Set $a_{ij} = v(x_{ij})$ and $A = [a_{ij}] \in M_n(\mathbb{C})$. Then $v = v_A$. Hence the map $A \mapsto v_A$ is a linear isomorphism from $M_n(\mathbb{C})$ to $T(G)_g$.*

Proof. By (1.34) we have $v(1) = v(1 \cdot 1) = 2v(1)$. Hence $v(1) = 0$. In particular, if $f = \det^k$ for some positive integer k, then

$$0 = v(f \cdot f^{-1}) = v(f)f(g)^{-1} + f(g)v(f^{-1}) ,$$

and so $v(1/f) = -v(f)/f(g)^2$. Hence v is uniquely determined by its restriction to the polynomial functions on G. Furthermore, $v(f_1 f_2) = 0$ whenever f_1 and f_2 are polynomials on $M_n(\mathbb{C})$ with $f_1(g) = 0$ and $f_2(g) = 0$. Let f be a polynomial function on $M_n(\mathbb{C})$. When v is evaluated on the Taylor polynomial of f centered at g, one obtains zero for the constant term and for all terms of degree greater than one. Also $v(x_{ij} - x_{ij}(g)) = v(x_{ij})$. This implies that $v = v_A$, where $a_{ij} = v(x_{ij})$. $\square$

Corollary 1.4.8. $(G = \mathbf{GL}(n,\mathbb{C}))$ *If* $X \in \mathrm{Der}(\mathcal{O}[G])$ *then* X *is given by* (1.32), *where* $c_{ij} = X(x_{ij})$.

Proof. For fixed $g \in G$, the linear functional $f \mapsto Xf(g)$ is a tangent vector at g. Hence $Xf(g) = v_A(f)$, where $a_{ij} = X(x_{ij})(g)$. Now define $c_{ij}(g) = X(x_{ij})(g)$ for all $g \in G$. Then $c_{ij} \in \mathcal{O}[G]$ by assumption, and equation (1.32) holds. $\square$

We continue to study the group $G = \mathbf{GL}(n,\mathbb{C})$ as an algebraic group. Just as in the Lie group case, we say that a regular vector field X on G is *left invariant* if it commutes with the left translation operators $L(y)$ for all $y \in G$ (where now these operators are understood to act on $\mathcal{O}[G]$).

Let $A \in M_n(\mathbb{C})$. Define a derivation X_A of $\mathcal{O}[G]$ by

$$X_A f(u) = \frac{d}{dt} f\big(u(I + tA)\big)\Big|_{t=0}$$

for $u \in G$ and $f \in \mathcal{O}[G]$, where the derivative is defined algebraically as usual for rational functions of the complex variable t. When $A = e_{ij}$ is an elementary matrix, we write $X_{e_{ij}} = E_{ij}$, as in Section 1.3.7 (but now understood as acting on $\mathcal{O}[G]$). Then the map $A \mapsto X_A$ is complex linear, and when $A = [a_{ij}]$ we have

$$X_A = \sum_{i,j} a_{ij} E_{ij}\,, \quad \text{with} \quad E_{ij} = \sum_{k=1}^{n} x_{ki} \frac{\partial}{\partial x_{kj}}\,,$$

by the same proof as for (1.22). The commutator correspondence (1.25) holds as an equality of regular vector fields on $\mathbf{GL}(n,\mathbb{C})$ (with the same proof). Thus the map $A \mapsto X_A$ is a complex Lie algebra isomorphism from $M_n(n,\mathbb{C})$ onto the Lie algebra of left-invariant regular vector fields on $\mathbf{GL}(n,\mathbb{C})$. Furthermore,

$$X_A f_B(u) = \frac{d}{dt} \mathrm{tr}\big(u(I + tA)B\big)\Big|_{t=0} = \mathrm{tr}(uAB) = f_{AB}(u) \qquad (1.36)$$

for all $A, B \in M_n(\mathbb{C})$, where the trace function f_B is defined by (1.28).

Now let $G \subset \mathbf{GL}(n,\mathbb{C})$ be a linear algebraic group. We define its Lie algebra $\mathfrak{g}$ as a complex Lie subalgebra of $M_n(\mathbb{C})$ as follows: Recall that $\mathcal{I}_G \subset \mathcal{O}[\mathbf{GL}(n,\mathbb{C})]$ is the ideal of regular functions that vanish on G. Define

$$\mathfrak{g} = \{A \in M_n(\mathbb{C}) : X_A f \in \mathcal{I}_G \quad \text{for all } f \in \mathcal{I}_G\}\,. \qquad (1.37)$$

When $G = \mathbf{GL}(n,\mathbb{C})$, we have $\mathcal{I}_G = 0$, so $\mathfrak{g} = M_n(\mathbb{C})$ in this case, in agreement with the previous definition of $\mathrm{Lie}(G)$. An arbitrary algebraic subgroup G of $\mathbf{GL}(n,\mathbb{C})$ is closed, and hence a Lie group by Theorem 1.3.11. After developing some algebraic tools, we shall show (in Section 1.4.4) that $\mathfrak{g} = \mathrm{Lie}(G)$ is the same set of matrices, whether we consider G as an algebraic group or as a Lie group.

Let $A \in \mathfrak{g}$. Then the left-invariant vector field X_A on $\mathbf{GL}(n, \mathbb{C})$ induces a linear transformation of the quotient algebra $\mathcal{O}[G] = \mathcal{O}[\mathbf{GL}(n, \mathbb{C})]/\mathfrak{I}_G$:

$$X_A(f + \mathfrak{I}_G) = X_A(f) + \mathfrak{I}_G$$

(since $X_A(\mathfrak{I}_G) \subset \mathfrak{I}_G$). For simplicity of notation we will also denote this transformation by X_A when the domain is clear. Clearly X_A is a derivation of $\mathcal{O}[G]$ that commutes with left translations by elements of G.

Proposition 1.4.9. *Let G be an algebraic subgroup of $\mathbf{GL}(n, \mathbb{C})$. Then $\mathfrak{g}$ is a Lie subalgebra of $M_n(\mathbb{C})$ (viewed as a Lie algebra over $\mathbb{C}$). Furthermore, the map $A \mapsto X_A$ is an injective complex linear Lie algebra homomorphism from $\mathfrak{g}$ to $\mathrm{Der}(\mathcal{O}[G])$.*

Proof. Since the map $A \mapsto X_A$ is complex linear, it follows that $A + \lambda B \in \mathfrak{g}$ if $A, B \in \mathfrak{g}$ and $\lambda \in \mathbb{C}$. The differential operators $X_A X_B$ and $X_B X_A$ on $\mathcal{O}[\mathbf{GL}(V)]$ leave the subspace $\mathfrak{I}_G$ invariant. Hence $[X_A, X_B]$ also leaves this space invariant. But $[X_A, X_B] = X_{[A,B]}$ by (1.25), so we have $[A, B] \in \mathfrak{g}$.

Suppose $A \in \mathrm{Lie}(G)$ and X_A acts by zero on $\mathcal{O}[G]$. Then $X_A f|_G = 0$ for all $f \in \mathcal{O}[\mathbf{GL}(n, \mathbb{C})]$. Since $I \in G$ and X_A commutes with left translations by $\mathbf{GL}(n, \mathbb{C})$, it follows that $X_A f = 0$ for all regular functions f on $\mathbf{GL}(n, \mathbb{C})$. Hence $A = 0$ by Corollary 1.4.8. $\square$

To calculate $\mathfrak{g}$ it is convenient to use the following property: If $G \subset \mathbf{GL}(n, \mathbb{C})$ and $A \in M_n(\mathbb{C})$, then A is in $\mathfrak{g}$ if and only if

$$X_A f|_G = 0 \quad \text{for all } f \in \mathcal{P}(M_n(\mathbb{C})) \cap \mathfrak{I}_G. \tag{1.38}$$

This is an easy consequence of the definition of $\mathfrak{g}$ and (1.29), and we leave the proof as an exercise. Another basic relation between algebraic groups and their Lie algebras is the following:

> If $G \subset H$ are linear algebraic groups with Lie algebras $\mathfrak{g}$ and $\mathfrak{h}$, respectively, then $\mathfrak{g} \subset \mathfrak{h}$.
$$\tag{1.39}$$

This is clear from the definition of the Lie algebras, since $\mathfrak{I}_H \subset \mathfrak{I}_G$.

Examples

1. Let D_n be the group of invertible diagonal $n \times n$ matrices. Then the Lie algebra $\mathfrak{d}_n$ of D_n (in the sense of algebraic groups) consists of the diagonal matrices in $M_n(\mathbb{C})$. To prove this, take any polynomial f on $M_n(\mathbb{C})$ that vanishes on D_n. Then we can write

$$f = \sum_{i \neq j} x_{ij} f_{ij},$$

where $f_{ij} \in \mathcal{P}(M_n(\mathbb{C}))$ and $1 \leq i, j \leq n$. Hence $A = [a_{ij}] \in \mathfrak{d}_n$ if and only if $X_A x_{ij}|_{D_n} = 0$ for all $i \neq j$. Set $B = A e_{ji}$. Then

$$X_A x_{ij} = f_B = \sum_{p=1}^{n} a_{pj} x_{ip}$$

by (1.36). Thus we see that $X_A x_{ij}$ vanishes on D_n for all $i \neq j$ if and only if $a_{ij} = 0$ for all $i \neq j$.

2. Let N_n^+ be the group of upper-triangular matrices with diagonal entries 1. Then its Lie algebra $\mathfrak{n}_n^+$ consists of the strictly upper-triangular matrices in $M_n(\mathbb{C})$. To prove this, let f be any polynomial on $M_n(\mathbb{C})$ that vanishes on N_n^+. Then we can write

$$f = \sum_{i=1}^{n} (x_{ii} - 1) f_i + \sum_{1 \leq j < i \leq n} x_{ij} f_{ij},$$

where f_i and f_{ij} are polynomials on $M_n(\mathbb{C})$. Hence $A \in \mathfrak{n}_n^+$ if and only if $X_A x_{ij}|_{N_n^+} = 0$ for all $1 \leq j \leq i \leq n$. By the same calculation as in Example 1, we have

$$X_A x_{ij}|_{N_n^+} = a_{ij} + \sum_{p=i+1}^{n} a_{pj} x_{ip}.$$

Hence $A \in \mathfrak{n}_n^+$ if and only if $a_{ij} = 0$ for all $1 \leq j \leq i \leq n$.

3. Let $1 \leq p \leq n$ and let P be the subgroup of $\mathbf{GL}(n, \mathbb{C})$ consisting of all matrices in block upper-triangular form

$$g = \begin{bmatrix} a & b \\ 0 & d \end{bmatrix}, \quad \text{where } a \in \mathbf{GL}(p, \mathbb{C}), d \in \mathbf{GL}(n-p, \mathbb{C}), \text{ and } b \in M_{p,n-p}(\mathbb{C}).$$

The same arguments as in Example 2 show that the ideal $\mathcal{I}_P$ is generated by the matrix entry functions x_{ij} with $p < i \leq n$ and $1 \leq j \leq p$ and that the Lie algebra of P (as an algebraic group) consists of all matrices X in block upper-triangular form

$$X = \begin{bmatrix} A & B \\ 0 & D \end{bmatrix}, \quad \text{where } A \in M_p(\mathbb{C}), \ D \in M_{n-p}(\mathbb{C}), \text{ and } B \in M_{p,n-p}(\mathbb{C}).$$

1.4.4 Algebraic Groups as Lie Groups

We now show that a linear algebraic group over $\mathbb{C}$ is a Lie group and that the Lie algebra defined using continuous one-parameter subgroups coincides with the Lie algebra defined using left-invariant derivations of the algebra of regular functions. For $Z = [z_{pq}] \in M_n(\mathbb{C})$ we write $Z = X + iY$, where i is a fixed choice of $\sqrt{-1}$ and $X, Y \in M_n(\mathbb{R})$. Then the map

$$Z \mapsto \begin{bmatrix} X & Y \\ -Y & X \end{bmatrix}$$

is an isomorphism between $M_n(\mathbb{C})$ considered as an associative algebra over $\mathbb{R}$ and the subalgebra of $M_{2n}(\mathbb{R})$ consisting of matrices A such that $AJ = JA$, where

$$J = \begin{bmatrix} 0 & I \\ -I & 0 \end{bmatrix} \quad \text{with } I = I_n$$

as in Section 1.3.4.

Recall that we associate to a closed subgroup G of $\mathbf{GL}(2n, \mathbb{R})$ the matrix Lie algebra

$$\text{Lie}(G) = \{A \in M_{2n}(\mathbb{R}) : \exp(tA) \in G \text{ for all } t \in \mathbb{R}\}, \tag{1.40}$$

and we give G the Lie group structure using exponential coordinates (Theorem 1.3.12). For example, when $G = \mathbf{GL}(n, \mathbb{C})$, then the Lie algebra of $\mathbf{GL}(n, \mathbb{C})$ (as a real Lie group) is just $M_n(\mathbb{C})$ looked upon as a subspace of $M_{2n}(\mathbb{R})$ as above. This is the same matrix Lie algebra as in the sense of linear algebraic groups, but with scalar multiplication restricted to $\mathbb{R}$. We now prove that the same relation between the Lie algebras holds for every linear algebraic group.

Theorem 1.4.10. *Let G be an algebraic subgroup of $\mathbf{GL}(n, \mathbb{C})$ with Lie algebra $\mathfrak{g}$ as an algebraic group. Then G has the structure of a Lie group whose Lie algebra as a Lie group is $\mathfrak{g}$ looked upon as a Lie algebra over $\mathbb{R}$. If G and H are linear algebraic groups then a homomorphism in the sense of linear algebraic groups is a Lie group homomorphism.*

Proof. By definition, G is a closed subgroup of $\mathbf{GL}(n, \mathbb{C})$ and hence of $\mathbf{GL}(2n, \mathbb{R})$. Thus G has a Lie group structure and a Lie algebra $\text{Lie}(G)$ defined by (1.40), which is a Lie subalgebra of $M_n(\mathbb{C})$ looked upon as a vector space over $\mathbb{R}$.

If $A \in \text{Lie}(G)$ and $f \in \mathfrak{I}_G$ (see Section 1.4.3), then $f(g \exp(tA)) = 0$ for $g \in G$ and all $t \in \mathbb{R}$. Hence

$$0 = \frac{d}{dt} f(g \exp(tA)) \Big|_{t=0} = \frac{d}{dt} f(g(I + tA)) \Big|_{t=0} = X_A f(g)$$

(see Section 1.4.3), so we have $A \in \mathfrak{g}$. Thus $\text{Lie}(G)$ is a subalgebra of $\mathfrak{g}$ (looked upon as a real vector space).

Conversely, given $A \in \mathfrak{g}$, we must show that $\exp tA \in G$ for all $t \in \mathbb{R}$. Since G is algebraic, this is the same as showing that

$$f(\exp tA) = 0 \quad \text{for all } f \in \mathfrak{I}_G \text{ and all } t \in \mathbb{R}. \tag{1.41}$$

Given $f \in \mathfrak{I}_G$, we set $\varphi(t) = f(\exp tA)$ for $t \in \mathbb{C}$. Then $\varphi(t)$ is an analytic function of t, since it is a polynomial in the complex matrix entries $z_{pq}(\exp tA)$ and $1/\det(\exp tA) = \det(\exp -tA)$. Hence by Taylor's theorem

$$\varphi(t) = \sum_{k=0}^{\infty} \varphi^{(k)}(0) \frac{t^k}{k!},$$

with the series converging absolutely for all $t \in \mathbb{C}$. Since $\exp(tA) = I + tA + O(t^2)$, it follows from the definition of the vector field X_A that

$$\varphi^{(k)}(0) = (X_A^k f)(I) \quad \text{for all nonnegative integers } k .$$

But $X_A^k f \in \mathcal{I}_G$, since $A \in \mathfrak{g}$, so we have $(X_A^k f)(I) = 0$. Hence $\varphi(t) = 0$ for all t, which proves (1.41). Thus $\mathfrak{g} \subset \mathrm{Lie}(G)$.

The last assertion of the theorem is clear because polynomials in the matrix entries $\{z_{ij}\}$ and $\det^{-1}$ are C^∞ functions relative to the real Lie group structure. $\qquad\square$

When G is a linear algebraic group, we shall denote the Lie algebra of G either by $\mathfrak{g}$ or by $\mathrm{Lie}(G)$, as a Lie algebra over $\mathbb{C}$. When G is viewed as a real Lie group, then $\mathrm{Lie}(G)$ is viewed as a vector space over $\mathbb{R}$.

1.4.5 Exercises

1. Let $D_n = (\mathbb{C}^\times)^n$ (an algebraic torus of rank n). Suppose D_k is isomorphic to D_n as an algebraic group. Prove that $k = n$. (HINT: The given group isomorphism induces a surjective algebra homomorphism from $\mathcal{O}[D_k]$ onto $\mathcal{O}[D_n]$; clear denominators to obtain a polynomial relation of the form $x_n f(x_1, \ldots, x_k) = g(x_1, \ldots, x_k)$, which implies $n \leq k$.)
2. Let $\mathcal{A}$ be a finite-dimensional associative algebra over $\mathbb{C}$ with unit 1. Let G be the set of all $g \in \mathcal{A}$ such that g is invertible in $\mathcal{A}$. For $z \in \mathcal{A}$ let $L_a \in \mathrm{End}(\mathcal{A})$ be the operator of left multiplication by a. Define $\Phi : G \longrightarrow \mathbf{GL}(\mathcal{A})$ by $\Phi(g) = L_g$.
   (a) Show that $\Phi(G)$ is a linear algebraic subgroup in $\mathbf{GL}(\mathcal{A})$. (HINT: To find a set of algebraic equations for $\Phi(G)$, prove that $T \in \mathrm{End}(\mathcal{A})$ commutes with all the operators of right multiplication by elements of $\mathcal{A}$ if and only if $T = L_a$ for some $a \in \mathcal{A}$.)
   (b) For $a \in \mathcal{A}$, show that there is a left-invariant vector field X_a on G such that

$$X_a f(g) = \frac{d}{dt} f(g(1 + ta)) \Big|_{t=0} \quad \text{for } f \in \mathcal{O}[G] .$$

   (c) Let $\mathcal{A}_{\mathrm{Lie}}$ be the vector space $\mathcal{A}$ with Lie bracket $[a, b] = ab - ba$. Prove that the map $a \mapsto X_a$ is an isomorphism from $\mathcal{A}_{\mathrm{Lie}}$ onto the left-invariant vector fields on G. (HINT: Adapt the arguments used for $\mathbf{GL}(n, \mathbb{C})$ in Section 1.4.3.)
   (d) Let $\{u_\alpha\}$ be a basis for $\mathcal{A}$ (as a vector space), and let $\{u_\alpha^*\}$ be the dual basis. Define the *structure constants* $c_{\alpha\beta\gamma}$ by $u_\alpha u_\beta = \sum_\gamma c_{\alpha\beta\gamma} u_\gamma$. Let $\partial/\partial u_\alpha$ denote the directional derivative in the direction u_α. Prove that

$$X_{u_\beta} = \sum_\gamma \varphi_{\beta\gamma}(\partial/\partial u_\gamma) ,$$

where $\varphi_{\beta\gamma} = \sum_\alpha c_{\alpha\beta\gamma} u_\alpha^*$ is a linear function on $\mathcal{A}$. (HINT: Adapt the argument used for Corollary 1.4.8.)

3. Let $\mathcal{A}$ be a finite-dimensional algebra over $\mathbb{C}$. This means that there is a *multiplication map* $\mu : \mathcal{A} \times \mathcal{A} \longrightarrow \mathcal{A}$ that is bilinear (it is not assumed to be associative). Define the *automorphism group* of $\mathcal{A}$ to be

$$\text{Aut}(\mathcal{A}) = \{g \in \mathbf{GL}(\mathcal{A}) : g\mu(X,Y) = \mu(gX, gY), \text{ for } X, Y \in \mathcal{A}\}.$$

Show that $\text{Aut}(\mathcal{A})$ is an algebraic subgroup of $\mathbf{GL}(\mathcal{A})$.

4. Suppose $G \subset \mathbf{GL}(n, \mathbb{C})$ is a linear algebraic group. Let $z \mapsto \varphi(z)$ be an analytic map from $\{z \in \mathbb{C} : |z| < r\}$ to $M_n(\mathbb{C})$ for some $r > 0$. Assume that $\varphi(0) = I$ and $\varphi(z) \in G$ for all $|z| < r$. Prove that the matrix $A = (d/dz)\varphi(z)|_{z=0}$ is in Lie(G). (HINT: Write $\varphi(z) = I + zA + z^2 F(z)$, where $F(z)$ is an analytic matrix-valued function. Then show that $X_A f(g) = (d/dz)f(g\varphi(z))|_{z=0}$ for all $f \in \mathcal{O}[\mathbf{GL}(n, \mathbb{C})]$.)

5. Let $B_\Gamma(x, y) = x^t \Gamma y$ be a nondegenerate bilinear form on $\mathbb{C}^n$, where $\Gamma \in \mathbf{GL}_n(\mathbb{C})$. Let G_Γ be the isometry group of this form. Define the *Cayley transform* $c(A) = (I + A)(I - A)^{-1}$ for $A \in M_n(\mathbb{C})$ with $\det(I - A) \neq 0$.

   (a) Suppose $A \in M_n(\mathbb{C})$ and $\det(I - A) \neq 0$. Prove that $c(A) \in G_\Gamma$ if and only if $A^t \Gamma + \Gamma A = 0$. (HINT: Use the equation $g^t \Gamma g = \Gamma$ characterizing elements $g \in G_\Gamma$.)

   (b) Give an algebraic proof (without using the exponential map) that Lie(G_Γ) consists of all $A \in M_n(\mathbb{C})$ such that

   $$A^t \Gamma + \Gamma A = 0. \tag{$\star$}$$

   Conclude that the Lie algebra of G_Γ is the same, whether G_Γ be viewed as a Lie group or as a linear algebraic group.
   (HINT: Define $\psi_B(g) = \text{tr}((g^t \Gamma g - \Gamma)B)$ for $g \in \mathbf{GL}(n, \mathbb{C})$ and $B \in M_n(\mathbb{C})$. Show that $X_A \psi_B(I) = \text{tr}((A^t \Gamma + \Gamma A)B)$ for any $A \in M_n(\mathbb{C})$. Since ψ_B vanishes on G_Γ, conclude that every $A \in$ Lie(G_Γ) satisfies $(\star)$. For the converse, take A satisfying $(\star)$, define $\varphi(z) = c(zA)$, and then apply the previous exercise and part (a).)

6. Let V be a finite-dimensional complex vector space with a nondegenerate skew-symmetric bilinear form Ω. Define $\mathbf{GSp}(V, \Omega)$ to be all $g \in \mathbf{GL}(V)$ for which there is a $\lambda \in \mathbb{C}^\times$ (depending on g) such that $\Omega(gx, gy) = \lambda \Omega(x, y)$ for all $x, y \in V$.

   (a) Show that the homomorphism $\mathbb{C}^\times \times \mathbf{Sp}(V, \Omega) \longrightarrow \mathbf{GSp}(V, \Omega)$ given by $(\lambda, g) \mapsto \lambda g$ is surjective. What is its kernel?

   (b) Show that $\mathbf{GSp}(V, \Omega)$ is an algebraic subgroup of $\mathbf{GL}(V)$.

   (c) Find Lie(G). (HINT: Show that (a) implies $\dim_\mathbb{C}$ Lie$(G) = \dim_\mathbb{C} \mathfrak{sp}(\mathbb{C}^{2l}, \Omega) + 1$.)

7. Let $G = \mathbf{GL}(1, \mathbb{C})$ and let $\varphi : G \longrightarrow G$ by $\varphi(z) = \bar{z}$. Show that φ is a group homomorphism that is *not* regular.

8. Let $P \subset \mathbf{GL}(n, \mathbb{C})$ be the subgroup defined in Example 3 of Section 1.4.3.

   (a) Prove that the ideal $\mathcal{J}_P$ is generated by the matrix entry functions x_{ij} with $p < i \leq n$ and $1 \leq j \leq p$.

   (b) Use (a) to prove that Lie(P) consists of all matrices in 2×2 block upper triangular form (with diagonal blocks of sizes $p \times p$ and $(n - p) \times (n - p)$).

9. Let $G \subset \mathbf{GL}(n, \mathbb{C})$. Prove that condition (1.38) characterizes $\mathrm{Lie}(G)$. (HINT: The functions in $\mathcal{I}_G$ are of the form $\det^{-p} f$, where $f \in \mathcal{P}(M_n(\mathbb{C}))$ vanishes on G. Use this to show that if f and $X_A f$ vanish on G then so does $X_A(\det^{-p} f)$.)

10. Let $\mathcal{A}$ be an algebra over a field $\mathbb{F}$, and let D_1, D_2 be derivations of $\mathcal{A}$. Verify that $[D_1, D_2] = D_1 \circ D_2 - D_2 \circ D_1$ is a derivation of $\mathcal{A}$.

1.5 Rational Representations

Now that we have introduced the symmetries associated with the classical groups, we turn to the second main theme of the book: linear actions (*representations*) of an algebraic group G on finite-dimensional vector spaces. Determining all such actions might seem much harder than studying the group directly, but it turns out, thanks to the work of É. Cartan and H. Weyl, that these representations have a very explicit structure that also yields information about G. Linear representations are also the natural setting for studying *invariants* of G, the third theme of the book.

1.5.1 Definitions and Examples

Let G be a linear algebraic group. A *representation* of G is a pair (ρ, V), where V is a complex vector space (not necessarily finite-dimensional), and $\rho : G \longrightarrow \mathbf{GL}(V)$ is a group homomorphism. We say that the representation is *regular* if $\dim V < \infty$ and the functions on G,

$$g \mapsto \langle v^*, \rho(g)v \rangle , \tag{1.42}$$

which we call *matrix coefficients* of ρ, are regular for all $v \in V$ and $v^* \in V^*$ (recall that $\langle v^*, v \rangle$ denotes the natural pairing between a vector space and its dual).

If we fix a basis for V and write out the matrix for $\rho(g)$ in this basis ($d = \dim V$),

$$\rho(g) = \begin{bmatrix} \rho_{11}(g) & \cdots & \rho_{1d}(g) \\ \vdots & \ddots & \vdots \\ \rho_{d1}(g) & \cdots & \rho_{dd}(g) \end{bmatrix},$$

then all the functions $\rho_{ij}(g)$ on G are regular. Furthermore, ρ is a regular homomorphism from G to $\mathbf{GL}(V)$, since the regular functions on $\mathbf{GL}(V)$ are generated by the matrix entry functions and $\det^{-1}$, and we have $(\det \rho(g))^{-1} = \det \rho(g^{-1})$, which is a regular function on G. Regular representations are often called *rational* representations, since each entry $\rho_{ij}(g)$ is a rational function of the matrix entries of g (however, the only denominators that occur are powers of $\det g$, so these functions are defined everywhere on G).

It will be convenient to phrase the definition of regularity as follows: On $\mathrm{End}(V)$ we have the symmetric bilinear form $A, B \mapsto \mathrm{tr}_V(AB)$. This form is nondegenerate, so if $\lambda \in \mathrm{End}(V)^*$ then there exists $A_\lambda \in \mathrm{End}(V)$ such that $\lambda(X) = \mathrm{tr}_V(A_\lambda X)$. For $B \in \mathrm{End}(V)$ define the function f_B^ρ on G by

$$f_B^\rho(g) = \mathrm{tr}_V(\rho(g)B)$$

(when B has rank one, then this function is of the form (1.42)). Then (ρ, V) is regular if and only if f_B^ρ is a regular function on G, for all $B \in \mathrm{End}(V)$. We set

$$E^\rho = \{f_B^\rho : B \in \mathrm{End}(V)\}\,.$$

This is the linear subspace of $\mathcal{O}[G]$ spanned by the functions in the matrix for ρ. It is finite-dimensional and invariant under left and right translations by G. We call it the space of *representative functions* associated with ρ.

Suppose ρ is a representation of G on an infinite-dimensional vector space V. We say that (ρ, V) is *locally regular* if every finite-dimensional subspace $E \subset V$ is contained in a finite-dimensional G-invariant subspace F such that the restriction of ρ to F is a regular representation.

If (ρ, V) is a regular representation and $W \subset V$ is a linear subspace, then we say that W is *G-invariant* if $\rho(g)w \in W$ for all $g \in G$ and $w \in W$. In this case we obtain a representation σ of G on W by restriction of $\rho(g)$. We also obtain a representation τ of G on the quotient space V/W by setting $\tau(g)(v+W) = \rho(g)v + W$. If we take a basis for W and complete it to a basis for V, then the matrix of $\rho(g)$ relative to this basis has the block form

$$\rho(g) = \begin{bmatrix} \sigma(g) & * \\ 0 & \tau(g) \end{bmatrix} \tag{1.43}$$

(with the basis for W listed first). This matrix form shows that the representations (σ, W) and $(\tau, V/W)$ are regular.

If (ρ, V) and (τ, W) are representations of G and $T \in \mathrm{Hom}(V, W)$, we say that T is a *G intertwining map* if

$$\tau(g)T = T\rho(g) \quad \text{for all } g \in G\,.$$

We denote by $\mathrm{Hom}_G(V, W)$ the vector space of all G intertwining maps. The representations ρ and τ are *equivalent* if there exists an invertible G intertwining map. In this case we write $\rho \cong \tau$.

We say that a representation (ρ, V) with $V \neq \{0\}$ is *reducible* if there is a G-invariant subspace $W \subset V$ such that $W \neq \{0\}$ and $W \neq V$. This means that there exists a basis for V such that $\rho(g)$ has the block form (1.43) with all blocks of size at least 1×1. A representation that is not reducible is called *irreducible*.

Consider now the representations L and R of G on the infinite-dimensional vector space $\mathcal{O}[G]$ given by *left* and *right translations*:

$$L(x)f(y) = f(x^{-1}y), \quad R(x)f(y) = f(yx) \text{ for } f \in \mathcal{O}[G]\,.$$

Proposition 1.5.1. *The representations $(L, \mathcal{O}[G])$ and $(R, \mathcal{O}[G])$ are locally regular.*

Proof. For any $f \in \mathcal{O}[G]$, equation (1.31) furnishes functions $f_i', f_i'' \in \mathcal{O}[G]$ such that

$$L(x)f = \sum_{i=1}^{n} f_i'(x^{-1}) f_i'', \quad R(x)f = \sum_{i=1}^{n} f_i''(x) f_i'. \tag{1.44}$$

Thus the subspaces

$$V_L(f) = \mathrm{Span}\{L(x)f : x \in G\} \quad \text{and} \quad V_R(f) = \mathrm{Span}\{R(x)f : x \in G\}$$

are finite-dimensional. By definition, $V_L(f)$ is invariant under left translations, while $V_R(f)$ is invariant under right translations. If $E \subset \mathcal{O}[G]$ is any finite-dimensional subspace, let $f_1, \ldots, f_k$ be a basis for E. Then $F_L = \sum_{i=1}^{k} V_L(f_i)$ is a finite-dimensional subspace invariant under left translations; likewise, the finite-dimensional subspace $F_R = \sum_{i=1}^{k} V_R(f_i)$ is invariant under right translations. From (1.44) we see that the restrictions of the representations L to F_L and R to F_R are regular. $\qquad\square$

We note that $L(x)R(y)f = R(y)L(x)f$ for $f \in \mathcal{O}[G]$. We can thus define a locally regular representation τ of the product group $G \times G$ on $\mathcal{O}[G]$ by $\tau(x, y) = L(x)R(y)$. We recover the left and right translation representations of G by restricting τ to the subgroups $G \times \{1\}$ and $\{1\} \times G$, each of which is isomorphic to G.

We may also embed G into $G \times G$ as $\Delta(G) = \{(x, x) : x \in G\}$ (the *diagonal subgroup*). The restriction of τ to $\Delta(G)$ gives the *conjugation representation* of G on $\mathcal{O}[G]$, which we denote by Int. It acts by

$$\mathrm{Int}(x)f(y) = f(x^{-1}yx) \quad \text{for } f \in \mathcal{O}[G] \text{ and } x \in G.$$

By the observations above, $(\mathrm{Int}, \mathcal{O}[G])$ is a locally regular representation.

1.5.2 Differential of a Rational Representation

Let $G \subset \mathbf{GL}(n, \mathbb{C})$ be a linear algebraic group with Lie algebra $\mathfrak{g} \subset \mathfrak{gl}(n, \mathbb{C})$. Let (π, V) be a rational representation of G. Viewing G and $\mathbf{GL}(V)$ as Lie groups, we can apply Proposition 1.3.14 to obtain a homomorphism (of real Lie algebras)

$$d\pi : \mathfrak{g} \longrightarrow \mathfrak{gl}(V).$$

We call $d\pi$ the *differential* of the representation π. Since $\mathfrak{g}$ is a Lie algebra over $\mathbb{C}$ in this case, we have $\pi(\exp(tA)) = \exp(d\pi(tA))$ for $A \in \mathfrak{g}$ and $t \in \mathbb{C}$. The entries in the matrix $\pi(g)$ (relative to any basis for V) are regular functions on G, so it follows that $t \mapsto \pi(\exp(tA))$ is an analytic (matrix-valued) function of the complex variable t. Thus

$$d\pi(A) = \frac{d}{dt}\pi(\exp(tA))\Big|_{t=0},$$

and the map $A \mapsto d\pi(A)$ is complex linear. Thus $d\pi$ is a homomorphism of complex Lie algebras when G is a linear algebraic group.

This definition of $d\pi$ has made use of the exponential map and the Lie group structure on G. We can also define $d\pi$ in a purely algebraic fashion, as follows: View the elements of $\mathfrak{g}$ as left-invariant vector fields on G by the map $A \mapsto X_A$ and differentiate the entries in the matrix for π using X_A. To express this in a basis-free way, recall that every linear transformation $B \in \mathrm{End}(V)$ defines a linear function f_C on $\mathrm{End}(V)$ by

$$f_C(B) = \mathrm{tr}_V(BC) \quad \text{for } B \in \mathrm{End}(V) \ .$$

The representative function $f_C^{\pi} = f_C \circ \pi$ on G is then a regular function.

Theorem 1.5.2. *The differential of a rational representation (π, V) is the unique linear map $d\pi : \mathfrak{g} \longrightarrow \mathrm{End}(V)$ such that*

$$X_A(f_C \circ \pi)(I) = f_{d\pi(A)C}(I) \quad \text{for all } A \in \mathfrak{g} \text{ and } C \in \mathrm{End}(V) \ . \tag{1.45}$$

Furthermore, for $A \in \mathrm{Lie}(G)$, one has

$$X_A(f \circ \pi) = (X_{d\pi(A)}f) \circ \pi \quad \text{for all } f \in \mathcal{O}[\mathbf{GL}(V)] \ . \tag{1.46}$$

Proof. For fixed $A \in \mathfrak{g}$, the map $C \mapsto X_A(f_C \circ \pi)(I)$ is a linear functional on $\mathrm{End}(V)$. Hence there exists a unique $D \in \mathrm{End}(V)$ such that

$$X_A(f_C \circ \pi)(I) = \mathrm{tr}_V(DC) = f_{DC}(I) \ .$$

But $f_{DC} = X_D f_C$ by equation (1.36). Hence to show that $d\pi(A) = D$, it suffices to prove that equation (1.46) holds. Let $f \in \mathcal{O}[\mathbf{GL}(V)]$ and $g \in G$. Then

$$X_A(f \circ \pi)(g) = \left. \frac{d}{dt} f\big(\pi(g\exp(tA))\big) \right|_{t=0} = \left. \frac{d}{dt} f\big(\pi(g)\exp(td\pi(A))\big) \right|_{t=0}$$
$$= (X_{d\pi(A)}f)(\pi(g))$$

by definition of the vector fields X_A on G and $X_{d\pi(A)}$ on $\mathbf{GL}(V)$. $\qquad\square$

Remark 1.5.3. An algebraic-group proof of Theorem 1.5.2 and the property that $d\pi$ is a Lie algebra homomorphism (taking (1.45) as the definition of $d\pi(A)$) is outlined in Exercises 1.5.4.

Let G and H be linear algebraic groups with Lie algebras $\mathfrak{g}$ and $\mathfrak{h}$, respectively, and let $\pi : G \longrightarrow H$ be a regular homomorphism. If $H \subset \mathbf{GL}(V)$, then we may view (π, V) as a regular representation of G with differential $d\pi : \mathfrak{g} \longrightarrow \mathrm{End}(V)$.

Proposition 1.5.4. *The range of $d\pi$ is contained in $\mathfrak{h}$. Hence $d\pi$ is a Lie algebra homomorphism from $\mathfrak{g}$ to $\mathfrak{h}$. Furthermore, if $K \subset \mathbf{GL}(W)$ is a linear algebraic group and $\rho : H \longrightarrow K$ is a regular homomorphism, then $d(\rho \circ \pi) = d\rho \circ d\pi$. In particular, if $G = K$ and $\rho \circ \pi$ is the identity map, then $d\rho \circ d\pi$ is the identity map. Hence isomorphic linear algebraic groups have isomorphic Lie algebras.*

Proof. We first verify that $d\pi(A) \in \mathfrak{h}$ for all $A \in \mathfrak{g}$. Let $f \in \mathcal{I}_H$ and $h \in H$. Then

$$(X_{d\pi(A)}f)(h) = L(h^{-1})(X_{d\pi(A)}f)(I) = X_{d\pi(A)}(L(h^{-1})f)(I)$$
$$= X_A((L(h^{-1})f) \circ \pi)(I)$$

by (1.46). But $L(h^{-1})f \in \mathfrak{I}_H$, so $(L(h^{-1})f) \circ \pi = 0$, since $\pi(G) \subset H$. Hence we have $X_{d\pi(A)}f(h) = 0$ for all $h \in H$. This shows that $d\pi(A) \in \mathfrak{h}$.

Given regular homomorphisms

$$G \xrightarrow{\pi} H \xrightarrow{\rho} K,$$

we set $\sigma = \rho \circ \pi$ and take $A \in \mathfrak{g}$ and $f \in \mathcal{O}[K]$. Then by (1.46) we have

$$(X_{d\sigma(A)}f) \circ \sigma = X_A((f \circ \rho) \circ \pi) = (X_{d\pi(A)}(f \circ \rho)) \circ \pi = (X_{d\rho(d\pi(A))}f) \circ \sigma.$$

Taking $f = f_C$ for $C \in \operatorname{End}(W)$ and evaluating the functions in this equation at I, we conclude from (1.45) that $d\sigma(A) = d\rho(d\pi(A))$. $\qquad\qquad\square$

Corollary 1.5.5. *Suppose $G \subset H$ are algebraic subgroups of $\mathbf{GL}(n, \mathbb{C})$. If (π, V) is a regular representation of H, then the differential of $\pi|_G$ is $d\pi|_{\mathfrak{g}}$.*

Examples

1. Let $G \subset \mathbf{GL}(V)$ be a linear algebraic group. By definition of $\mathcal{O}[G]$, the representation $\pi(g) = g$ on V is regular. We call (π, V) the *defining* representation of G. It follows directly from the definition that $d\pi(A) = A$ for $A \in \mathfrak{g}$.

2. Let (π, V) be a regular representation. Define the *contragredient* (or *dual*) representation (π^*, V^*) by $\pi^*(g)v^* = v^* \circ \pi(g^{-1})$. Then π^* is obviously regular, since

$$\langle v^*, \pi(g)v \rangle = \langle \pi^*(g^{-1})v^*, v \rangle \quad \text{for } v \in V \text{ and } v^* \in V^*.$$

If $\dim V = d$ (the *degree* of π) and V is identified with $d \times 1$ column vectors by a choice of basis, then V^* is identified with $1 \times d$ row vectors. Viewing $\pi(g)$ as a $d \times d$ matrix using the basis, we have

$$\langle v^*, \pi(g)v \rangle = v^*\pi(g)v \quad \text{(matrix multiplication)}.$$

Thus $\pi^*(g)$ acts by right multiplication on row vectors by the matrix $\pi(g^{-1})$.

The space of representative functions for π^* consists of the functions $g \mapsto f(g^{-1})$, where f is a representative function for π. If $W \subset V$ is a G-invariant subspace, then

$$W^\perp = \{v^* \in V^* : \langle v^*, w \rangle = 0 \quad \text{for all } w \in W\}$$

is a G-invariant subspace of V^*. In particular, if (π, V) is irreducible then so is (π^*, V^*). The natural vector-space isomorphism $(V^*)^* \cong V$ gives an equivalence $(\pi^*)^* \cong \pi$.

To calculate the differential of π^*, let $A \in \mathfrak{g}$, $v \in V$, and $v^* \in V^*$. Then

$$\langle d\pi^*(A)v^*, v \rangle = \frac{d}{dt} \langle \pi^*(\exp tA)v^*, v \rangle \Big|_{t=0} = \frac{d}{dt} \langle v^*, \pi(\exp(-tA))v \rangle \Big|_{t=0}$$
$$= -\langle v^*, d\pi(A)v \rangle.$$

Since this holds for all v and v^*, we conclude that

$$d\pi^*(A) = -d\pi(A)^t \quad \text{for } A \in \mathfrak{g}. \tag{1.47}$$

Caution: The notation $\pi^*(g)$ for the contragredient representation should not be confused with the notation B^* for the conjugate transpose of a matrix B. The pairing $\langle v^*, v \rangle$ between V^* and V is *complex linear* in each argument.

3. Let (π_1, V_1) and (π_2, V_2) be regular representations of G. Define the *direct sum* representation $\pi_1 \oplus \pi_2$ on $V_1 \oplus V_2$ by

$$(\pi_1 \oplus \pi_2)(g)(v_1 \oplus v_2) = \pi_1(g)v_1 \oplus \pi_2(g)v_2 \quad \text{for } g \in G \text{ and } v_i \in V_i.$$

Then $\pi_1 \oplus \pi_2$ is obviously a representation of G. It is regular, since

$$\langle v_1^* \oplus v_2^*, (\pi_1 \oplus \pi_2)(g)(v_1 \oplus v_2) \rangle = \langle v_1^*, \pi_1(g)v_1 \rangle + \langle v_2^*, \pi_2(g)v_2 \rangle$$

for $v_i \in V_i$ and $v_i^* \in V_i^*$. This shows that the space of representative functions for $\pi_1 \oplus \pi_2$ is $E^{\pi_1 \oplus \pi_2} = E^{\pi_1} + E^{\pi_2}$. If $\pi = \pi_1 \oplus \pi_2$, then in matrix form we have

$$\pi(g) = \begin{bmatrix} \pi_1(g) & 0 \\ 0 & \pi_2(g) \end{bmatrix}.$$

Differentiating the matrix entries, we find that

$$d\pi(A) = \begin{bmatrix} d\pi_1(A) & 0 \\ 0 & d\pi_2(A) \end{bmatrix} \quad \text{for } A \in \mathfrak{g}.$$

Thus $d\pi(A) = d\pi_1(A) \oplus d\pi_2(A)$.

4. Let (π_1, V_1) and (π_2, V_2) be regular representations of G. Define the *tensor product* representation $\pi_1 \otimes \pi_2$ on $V_1 \otimes V_2$ by

$$(\pi_1 \otimes \pi_2)(g)(v_1 \otimes v_2) = \pi_1(g)v_1 \otimes \pi_2(g)v_2$$

for $g \in G$ and $v_i \in V$. It is clear that $\pi_1 \otimes \pi_2$ is a representation. It is regular, since

$$\langle v_1^* \otimes v_2^*, (\pi_1 \otimes \pi_2)(g)(v_1 \otimes v_2) \rangle = \langle v_1^*, \pi_1(g)v \rangle \langle v_2^*, \pi_2(g)v_2 \rangle$$

for $v_i \in V$ and $v_i^* \in V_i^*$. In terms of representative functions, we have

$$E^{\pi_1 \otimes \pi_2} = \text{Span}(E^{\pi_1} \cdot E^{\pi_2})$$

(the sums of products of representative functions of π_1 and π_2). Set $\pi = \pi_1 \otimes \pi_2$. Then

$$d\pi(A) = \frac{d}{dt}\Big\{ \exp\big(td\pi_1(A)\big) \otimes \exp\big(td\pi_2(A)\big) \Big\}\Big|_{t=0}$$
$$= d\pi_1(A) \otimes I + I \otimes d\pi_2(A). \tag{1.48}$$

5. Let (π, V) be a regular representation of G and set $\rho = \pi \otimes \pi^*$ on $V \otimes V^*$. Then by Examples 2 and 4 we see that

$$d\rho(A) = d\pi(A) \otimes I - I \otimes d\pi(A)^t. \tag{1.49}$$

However, there is the canonical isomorphism $T : V \otimes V^* \cong \mathrm{End}(V)$, with

$$T(v \otimes v^*)(u) = \langle v^*, u \rangle v.$$

Set $\sigma(g) = T\rho(g)T^{-1}$. If $Y \in \mathrm{End}(V)$ then $T(Y \otimes I) = YT$ and $T(I \otimes Y^t) = TY$. Hence $\sigma(g)(Y) = \pi(g)Y\pi(g)^{-1}$ and

$$d\sigma(A)(Y) = d\pi(A)Y - Yd\pi(A) \quad \text{for } A \in \mathfrak{g}. \tag{1.50}$$

6. Let (π, V) be a regular representation of G and set $\rho = \pi^* \otimes \pi^*$ on $V^* \otimes V^*$. Then by Examples 2 and 4 we see that

$$d\rho(A) = -d\pi(A)^t \otimes I - I \otimes d\pi(A)^t.$$

However, there is a canonical isomorphism between $V^* \otimes V^*$ and the space of bilinear forms on V, where $g \in G$ acts on a bilinear form B by

$$g \cdot B(x,y) = B(\pi(g^{-1})x, \pi(g^{-1})y).$$

If V is identified with column vectors by a choice of a basis and $B(x,y) = y^t \Gamma x$, then $g \cdot \Gamma = \pi(g^{-1})^t \Gamma \pi(g^{-1})$ (matrix multiplication). The action of $A \in \mathfrak{g}$ on B is

$$A \cdot B(x,y) = -B(d\pi(A)x, y) - B(x, d\pi(A)y).$$

We say that a bilinear form B is *invariant* under G if $g \cdot B = B$ for all $g \in G$. Likewise, we say that B is *invariant* under $\mathfrak{g}$ if $A \cdot B = 0$ for all $A \in \mathfrak{g}$. This invariance property can be expressed as

$$B(d\pi(A)x, y) + B(x, d\pi(A)y) = 0 \quad \text{for all } x, y \in V \text{ and } A \in \mathfrak{g}.$$

1.5.3 The Adjoint Representation

Let $G \subset \mathbf{GL}(n, \mathbb{C})$ be an algebraic group with Lie algebra $\mathfrak{g}$. The representation of $\mathbf{GL}(n, \mathbb{C})$ on $M_n(\mathbb{C})$ by similarity $(A \mapsto gAg^{-1})$ is regular (see Example 5 of Section 1.5.2). We now show that the restriction of this representation to G furnishes a regular representation of G. The following lemma is the key point.

Lemma 1.5.6. *Let $A \in \mathfrak{g}$ and $g \in G$. Then $gAg^{-1} \in \mathfrak{g}$.*

Proof. For $A \in M_n(\mathbb{C})$, $g \in \mathbf{GL}(n, \mathbb{C})$, and $t \in \mathbb{C}$ we have

$$g \exp(tA) g^{-1} = \sum_{k=0}^{\infty} \frac{t^k}{k!} (gAg^{-1})^k = \exp(tgAg^{-1}) \,.$$

Now assume $A \in \mathfrak{g}$ and $g \in G$. Since $\mathfrak{g} = \mathrm{Lie}(G)$ by Theorem 1.4.10, we have $\exp(tgAg^{-1}) = g \exp(t\dot{A}) g^{-1} \in G$ for all $t \in \mathbb{C}$. Hence $gAg^{-1} \in \mathfrak{g}$.   □

We define $\mathrm{Ad}(g)A = gAg^{-1}$ for $g \in G$ and $A \in \mathfrak{g}$. Then by Lemma 1.5.6 we have $\mathrm{Ad}(g) : \mathfrak{g} \longrightarrow \mathfrak{g}$. The representation $(\mathrm{Ad}, \mathfrak{g})$ is called the *adjoint representation* of G. For $A, B \in \mathfrak{g}$ we calculate

$$\mathrm{Ad}(g)[A,B] = gABg^{-1} - gBAg^{-1} = gAg^{-1}gBg^{-1} - gBg^{-1}gAg^{-1}$$
$$= [\mathrm{Ad}(g)A, \mathrm{Ad}(g)B] \,.$$

Thus $\mathrm{Ad}(g)$ is a Lie algebra automorphism and $\mathrm{Ad} : G \longrightarrow \mathrm{Aut}(\mathfrak{g})$ (the group of automorphisms of $\mathfrak{g}$).

If $H \subset \mathbf{GL}(n, \mathbb{C})$ is another algebraic group with Lie algebra $\mathfrak{h}$, we denote the adjoint representations of G and H by Ad_G and Ad_H, respectively. Suppose that $G \subset H$. Since $\mathfrak{g} \subset \mathfrak{h}$ by property (1.39), we have

$$\mathrm{Ad}_H(g)A = \mathrm{Ad}_G(g)A \quad \text{for } g \in G \text{ and } A \in \mathfrak{g} \,. \tag{1.51}$$

Theorem 1.5.7. *The differential of the adjoint representation of G is the representation* $\mathrm{ad} : \mathfrak{g} \longrightarrow \mathrm{End}(\mathfrak{g})$ *given by*

$$\mathrm{ad}(A)(B) = [A,B] \quad \text{for } A, B \in \mathfrak{g} \,. \tag{1.52}$$

Furthermore, $\mathrm{ad}(A)$ *is a derivation of* $\mathfrak{g}$*, and hence* $\mathrm{ad}(\mathfrak{g}) \subset \mathrm{Der}(\mathfrak{g})$.

Proof. Equation (1.52) is the special case of equation (1.50) with π the defining representation of G on $\mathbb{C}^n$ and $d\pi(A) = A$. The derivation property follows from the Jacobi identity.   □

Remark 1.5.8. If $G \subset \mathbf{GL}(n, \mathbb{R})$ is any closed subgroup, then $gAg^{-1} \in \mathrm{Lie}(G)$ for all $g \in G$ and $A \in \mathrm{Lie}(G)$ (by the same argument as in Lemma 1.5.6). Thus we can define the adjoint representation Ad of G on the real vector space $\mathrm{Lie}(G)$ as for algebraic groups. Clearly $\mathrm{Ad} : G \longrightarrow \mathrm{Aut}(\mathfrak{g})$ is a homomorphism from G to the group of automorphisms of $\mathrm{Lie}(G)$, and Theorem 1.5.7 holds for $\mathrm{Lie}(G)$.

1.5.4 Exercises

1. Let (π, V) be a rational representation of a linear algebraic group G.

(a) Using equation (1.45) to define $d\pi(A)$, deduce from Proposition 1.4.9 (without using the exponential map) that $d\pi([A,B]) = [d\pi(A), d\pi(B)]$ for $A, B \in \mathfrak{g}$.

(b) Prove (without using the exponential map) that equation (1.45) implies equation (1.46). (HINT: For fixed $g \in G$ consider the linear functional

$$f \mapsto (X_{d\pi(A)}f)(\pi(g)) - X_A(f \circ \pi)(g) \quad \text{for } f \in \mathcal{O}[\mathbf{GL}(V)] \ .$$

This functional vanishes when $f = f_C$. Now apply Lemma 1.4.7.)

2. Give an algebraic proof of formula (1.47) that does not use the exponential map. (HINT: Assume $G \subset \mathbf{GL}(n, \mathbb{C})$, replace $\exp(tA)$ by the rational map $t \mapsto I + tA$ from $\mathbb{C}$ to $\mathbf{GL}(n, \mathbb{C})$, and use Theorem 1.5.2.)

3. Give an algebraic proof of formula (1.48) that does not use the exponential map. (HINT: Use the method of the previous exercise.)

4. (a) Let $A \in M_n(\mathbb{C})$ and $g \in \mathbf{GL}(n, \mathbb{C})$. Give an algebraic proof (without using the exponential map) that $R(g)X_A R(g^{-1}) = X_{gAg^{-1}}$.

(b) Use the result of (a) to give an algebraic proof of Lemma 1.5.6. (HINT: If $f \in \mathcal{I}_G$ then $R(g)f$ and $X_A f$ are also in $\mathcal{I}_G$.)

5. Define $\varphi(A) = \begin{bmatrix} \det(A)^{-1} & 0 \\ 0 & A \end{bmatrix}$ for $A \in \mathbf{GL}(n, \mathbb{C})$. Show that $A \mapsto \varphi(A)$ is an injective regular homomorphism from $\mathbf{GL}(n, \mathbb{C})$ to $\mathbf{SL}(n+1, \mathbb{C})$, and that $d\varphi(X) = \begin{bmatrix} -\mathrm{tr}(X) & 0 \\ 0 & X \end{bmatrix}$ for $X \in \mathfrak{gl}(n, \mathbb{C})$.

1.6 Jordan Decomposition

In the theory of Lie groups the additive group $\mathbb{R}$ and the (connected) multiplicative group $\mathbb{R}_{>0}$ of positive real numbers are isomorphic under the map $x \mapsto \exp x$. By contrast, in the theory of algebraic groups the additive group $\mathbb{C}$ and the multiplicative group $\mathbb{C}^\times$ are not isomorphic. In this section we find all regular representations of these two groups and obtain the algebraic-group version of the Jordan canonical form of a matrix.

1.6.1 Rational Representations of $\mathbb{C}$

Recall that we have given the additive group $\mathbb{C}$ the structure of a linear algebraic group by embedding it into $\mathbf{SL}(2, \mathbb{C})$ with the homomorphism

$$z \mapsto \varphi(z) = \begin{bmatrix} 1 & z \\ 0 & 1 \end{bmatrix} = I + z e_{12} \ .$$

The regular functions on $\mathbb{C}$ are the polynomials in z, and the Lie algebra of $\mathbb{C}$ is spanned by the matrix e_{12}, which satisfies $(e_{12})^2 = 0$. Thus $\varphi(z) = \exp(z e_{12})$. We now determine all the regular representations of $\mathbb{C}$.

A matrix $A \in M_n(\mathbb{C})$ is called *nilpotent* if $A^k = 0$ for some positive integer k. A nilpotent matrix has trace zero, since zero is its only eigenvalue. A matrix $u \in M_n(\mathbb{C})$ is called *unipotent* if $u - I$ is nilpotent. Note that a unipotent transformation is nonsingular and has determinant 1, since 1 is its only eigenvalue.

Let $A \in M_n(\mathbb{C})$ be nilpotent. Then $A^n = 0$ and for $t \in \mathbb{C}$ we have

$$\exp tA = I + Y, \quad \text{where} \quad Y = tA + \frac{t^2}{2!}A^2 + \cdots + \frac{t^{n-1}}{(n-1)!}A^{n-1}$$

is also nilpotent. Hence the matrix $\exp tA$ is unipotent and $t \mapsto \exp(tA)$ is a regular homomorphism from the additive group $\mathbb{C}$ to $\mathbf{GL}(n, \mathbb{C})$.

Conversely, if $u = I + Y \in \mathbf{GL}(n, \mathbb{C})$ is unipotent, then $Y^n = 0$ and we define

$$\log u = \sum_{k=1}^{n-1} (-1)^{k+1} \frac{1}{k} Y^k .$$

By the substitution principle (Section 1.3.2), we have $\exp(\log(I + A)) = I + A$. Thus the exponential function is a bijective polynomial map from the nilpotent elements in $M_n(\mathbb{C})$ onto the unipotent elements in $\mathbf{GL}(n, \mathbb{C})$, with inverse $u \mapsto \log u$.

Lemma 1.6.1 (Taylor's formula). *Suppose $A \in M_n(\mathbb{C})$ is nilpotent and f is a regular function on $\mathbf{GL}(n, \mathbb{C})$. Then there exists an integer k such that $(X_A)^k f = 0$ and*

$$f(\exp A) = \sum_{m=0}^{k-1} \frac{1}{m!} (X_A^m f)(I) . \tag{1.53}$$

Proof. Since $\det(\exp zA) = 1$, the function $z \mapsto \varphi(z) = f(\exp zA)$ is a polynomial in $z \in \mathbb{C}$. Hence there exists a positive integer k such that $(d/dz)^k \varphi(z) = 0$. Furthermore,

$$\varphi^{(m)}(0) = (X_A^m f)(I) . \tag{1.54}$$

Equation (1.53) now follows from (1.54) by evaluating $\varphi(1)$ using the Taylor expansion centered at 0. $\square$

Theorem 1.6.2. *Let $G \subset \mathbf{GL}(n, \mathbb{C})$ be a linear algebraic group with Lie algebra $\mathfrak{g}$. If $A \in M_n(\mathbb{C})$ is nilpotent, then $A \in \mathfrak{g}$ if and only if $\exp A \in G$. Furthermore, if $A \in \mathfrak{g}$ is a nilpotent matrix and (ρ, V) is a regular representation of G, then $d\rho(A)$ is a nilpotent transformation on V, and*

$$\rho(\exp A) = \exp d\rho(A) . \tag{1.55}$$

Proof. Take $f \in \mathfrak{I}_G$. If $A \in \mathfrak{g}$, then $X_A^m f \in \mathfrak{I}_G$ for all integers $m \geq 0$. Hence $(X_A)^m f(I) = 0$ for all m, and so by Taylor's formula (1.53) we have $f(\exp A) = 0$. Thus $\exp A \in G$. Conversely, if $\exp A \in G$, then the function $z \mapsto \varphi(z) = f(\exp zA)$ on $\mathbb{C}$ vanishes when z is an integer, so it must vanish for all z, since it is a polynomial. Hence $X_A f(I) = 0$ for all $f \in \mathfrak{I}_G$. By the left invariance of X_A we then have $X_A f(g) = 0$ for all $g \in G$. Thus $A \in \mathfrak{g}$, proving the first assertion.

To prove the second assertion, apply Lemma 1.6.1 to the finite-dimensional space of functions f_B^ρ, where $B \in \text{End}(V)$. This gives a positive integer k such that

$$0 = X_A^k f_B^\rho(I) = \text{tr}_V(d\rho(A)^k B) \quad \text{for all } B \in \text{End}(V) .$$

Hence $(d\rho(A))^k = 0$. Applying Taylor's formula to the function f_B^ρ, we obtain

$$\text{tr}_V(B\rho(\exp A)) = \sum_{m=0}^{k-1} \frac{1}{m!} X_A^m f_B^\rho(I) = \sum_{m=0}^{k-1} \frac{1}{m!} \text{tr}_V(d\rho(A)^m B)$$
$$= \text{tr}_V(B \exp d\rho(A)) .$$

This holds for all B, so we obtain (1.55). $\qquad\qquad\qquad\qquad\qquad\qquad\qquad$ $\square$

Corollary 1.6.3. *If (π, V) is a regular representation of the additive group $\mathbb{C}$, then there exists a unique nilpotent $A \in \text{End}(V)$ such that $\pi(z) = \exp(zA)$ for all $z \in \mathbb{C}$.*

1.6.2 Rational Representations of $\mathbb{C}^\times$

The regular representations of $\mathbb{C}^\times = \mathbf{GL}(1, \mathbb{C})$ have the following form:

Lemma 1.6.4. *Let $(\varphi, \mathbb{C}^n)$ be a regular representation of $\mathbb{C}^\times$. For $p \in \mathbb{Z}$ define $E_p = \{v \in \mathbb{C}^n : \varphi(z)v = z^p v \text{ for all } z \in \mathbb{C}^\times\}$. Then*

$$\mathbb{C}^n = \bigoplus_{p \in \mathbb{Z}} E_p , \qquad\qquad (1.56)$$

and hence $\varphi(z)$ is a semisimple transformation. Conversely, given a direct sum decomposition (1.56) of $\mathbb{C}^n$, define $\varphi(z)v = z^p v$ for $z \in \mathbb{C}^\times$ and $v \in E_p$. Then φ is a regular representation of $\mathbb{C}^\times$ on $\mathbb{C}^n$ that is determined (up to equivalence) by the set of integers $\{\dim E_p : p \in \mathbb{Z}\}$.

Proof. Since $\mathcal{O}[\mathbb{C}^\times] = \mathbb{C}[z, z^{-1}]$, the entries in the matrix $\varphi(z)$ are Laurent polynomials. Thus there is an expansion

$$\varphi(z) = \sum_{p \in \mathbb{Z}} z^p T_p , \qquad\qquad (1.57)$$

where the coefficients T_p are in $M_n(\mathbb{C})$ and only a finite number of them are nonzero. Since $\varphi(z)\varphi(w) = \varphi(zw)$, we have

$$\sum_{p,q \in \mathbb{Z}} z^p w^q T_p T_q = \sum_{r \in \mathbb{Z}} z^r w^r T_r .$$

Equating coefficients of $z^p w^q$ yields the relations

$$T_p T_q = 0 \quad \text{for } p \neq q, \quad T_p^2 = T_p . \qquad\qquad (1.58)$$

Furthermore, since $\varphi(1) = I_n$, one has $\sum_{p \in \mathbb{Z}} T_p = I_n$. Thus the family of matrices $\{T_p : p \in \mathbb{Z}\}$ consists of mutually commuting projections and gives a *resolution of the identity* on $\mathbb{C}^n$. If $v \in \mathbb{C}^n$ and $T_p v = v$, then

$$\varphi(z)v = \sum_{q \in \mathbb{Z}} z^q T_q T_p v = z^p v$$

by (1.58), so $\mathrm{Range}(T_p) \subset E_p$. The opposite inclusion is obvious from the uniqueness of the expansion (1.57). Thus $E_p = \mathrm{Range}(T_p)$, which proves (1.56).

Conversely, given a decomposition (1.56), we let T_p be the projection onto E_p defined by this decomposition, and we define $\varphi(z)$ by (1.57). Then φ is clearly a regular homomorphism from $\mathbb{C}^\times$ into $\mathbf{GL}(n, \mathbb{C})$.   □

1.6.3 Jordan–Chevalley Decomposition

A matrix $A \in M_n(\mathbb{C})$ has a unique *additive* Jordan decomposition $A = S + N$ with S semisimple, N nilpotent, and $SN = NS$. Likewise, $g \in \mathbf{GL}(n, \mathbb{C})$ has a unique *multiplicative* Jordan decomposition $g = su$ with s semisimple, u unipotent, and $su = us$ (see Sections B.1.2 and B.1.3).

Theorem 1.6.5. *Let $G \subset \mathbf{GL}(n, \mathbb{C})$ be an algebraic group with Lie algebra $\mathfrak{g}$. If $A \in \mathfrak{g}$ and $A = S + N$ is its additive Jordan decomposition, then $S, N \in \mathfrak{g}$. Furthermore, if $g \in G$ and $g = su$ is its multiplicative Jordan decomposition, then $s, u \in G$.*

Proof. For k a nonnegative integer let $\mathcal{P}^{(k)}(M_n(\mathbb{C}))$ be the space of homogeneous polynomials of degree k in the matrix entry functions $\{x_{ij} : 1 \leq i, j \leq n\}$. This space is invariant under the right translations $R(g)$ for $g \in \mathbf{GL}(n, \mathbb{C})$ and the vector fields X_A for $A \in M_n(\mathbb{C})$, by the formula for matrix multiplication and from (1.22). Set

$$W_m = \sum_{k,r=0}^{m} (\det)^{-r} \mathcal{P}^{(k)}(M_n(\mathbb{C})) . \qquad (1.59)$$

The space W_m is finite-dimensional and invariant under $R(g)$ and X_A because $R(g)$ preserves products of functions, X_A is a derivation, and powers of the determinant transform by

$$R(g)(\det)^{-r} = (\det g)^{-r}(\det)^{-r} \quad \text{and} \quad X_A(\det)^{-r} = -r\,\mathrm{tr}(A)(\det)^{-r} .$$

Furthermore, $\mathcal{O}[\mathbf{GL}(n, \mathbb{C})] = \bigcup_{m \geq 0} W_m$.

Suppose $S \in M_n(\mathbb{C})$ is semisimple. We claim that the restriction of X_S to W_m is a semisimple operator for all nonnegative integers m. To verify this, we may assume $S = \mathrm{diag}[\lambda_1, \ldots, \lambda_n]$. Then the action of X_S on the generators of $\mathcal{O}[\mathbf{GL}(n, \mathbb{C})]$ is

$$X_S f_{e_{ij}} = f_{S e_{ij}} = \lambda_i f_{e_{ij}} , \quad X_S(\det)^{-1} = -\,\mathrm{tr}(S)(\det)^{-1} .$$

Since X_S is a derivation, it follows that any product of the functions $f_{e_{ij}}$ and $\det^{-r}$ is an eigenvector for X_S. Because such products span W_m, we see that W_m has a basis consisting of eigenvectors for X_S.

Given a semisimple element $s \in \mathbf{GL}(n, \mathbb{C})$, we use a similar argument to show that the restriction of $R(s)$ to W_m is a semisimple operator. Namely, we may assume that $s = \mathrm{diag}[\sigma_1, \ldots, \sigma_n]$ with $\sigma_i \neq 0$. Then the action of $R(s)$ on the generators of $\mathcal{O}[\mathbf{GL}(n, \mathbb{C})]$ is

$$R(s)f_{e_{ij}} = f_{se_{ij}} = \sigma_i f_{e_{ij}}, \quad R(s)(\det)^{-1} = \det(s)^{-1}(\det)^{-1}.$$

Since $R(s)(f_1 f_2) = (R(s)f_1)(R(s)f_2)$ for $f_1, f_2 \in \mathcal{O}[\mathbf{GL}(n, \mathbb{C})]$, it follows that any product of the functions $f_{e_{ij}}$ and $\det^{-r}$ is an eigenvector for $R(s)$. Because such products span W_m, we see that W_m has a basis consisting of eigenvectors for $R(s)$.

Let $N \in M_n(\mathbb{C})$ be nilpotent and let $u \in \mathbf{GL}(n, \mathbb{C})$ be unipotent. Then by Theorem 1.6.2 the vector field X_N acts nilpotently on W_m and the operator $R(u)$ is unipotent on W_m.

The multiplicative Jordan decomposition $g = su$ for $g \in \mathbf{GL}(n, \mathbb{C})$ gives the decomposition $R(g) = R(s)R(u)$, with commuting factors. From the argument above and the uniqueness of the Jordan decomposition we conclude that the restrictions of $R(s)$ and $R(u)$ to W_m provide the semisimple and unipotent factors for the restriction of $R(g)$. Starting with the additive Jordan decomposition $A = S + N$ in $M_n(\mathbb{C})$, we likewise see that the restrictions of X_S and X_N to W_m furnish the semisimple and nilpotent parts of the restriction of X_A.

With these properties of the Jordan decompositions established, we can complete the proof as follows. Given $f \in \mathcal{I}_G$, choose m large enough that $f \in W_m$. The Jordan decompositions of $R(g)$ and X_A on W_m are

$$R(g)|_{W_m} = (R(s)|_{W_m})(R(u)|_{W_m}), \qquad X_A|_{W_m} = X_S|_{W_m} + X_N|_{W_m}.$$

Hence there exist polynomials $\varphi(z)$ and $\psi(z)$ such that $R(s)f = \varphi(R(g))f$ and $X_S f = \psi(X_A)$ for all $f \in W_m$. Thus $R(s)f$ and $X_S f$ are in $\mathcal{I}_G$, which implies that $s \in G$ and $S \in \mathfrak{g}$. $\qquad \square$

Theorem 1.6.6. *Let G be an algebraic group with Lie algebra $\mathfrak{g}$. Suppose (ρ, V) is a regular representation of G.*

1. *If $A \in \mathfrak{g}$ and $A = S + N$ is its additive Jordan decomposition, then $d\rho(S)$ is semisimple, $d\rho(N)$ is nilpotent, and $d\rho(A) = d\rho(S) + d\rho(N)$ is the additive Jordan decomposition of $d\rho(A)$ in $\mathrm{End}(V)$.*
2. *If $g \in G$ and $g = su$ is its multiplicative Jordan decomposition in G, then $\rho(s)$ is semisimple, $\rho(u)$ is unipotent, and $\rho(g) = \rho(s)\rho(u)$ is the multiplicative Jordan decomposition of $\rho(g)$ in $\mathbf{GL}(V)$.*

Proof. We know from Theorem 1.6.2 that $d\rho(N)$ is nilpotent and $\rho(u)$ is unipotent, and since $d\rho$ is a Lie algebra homomorphism, we have

$$[d\rho(N), d\rho(S)] = d\rho([N, S]) = 0.$$

Likewise, $\rho(u)\rho(s) = \rho(us) = \rho(su) = \rho(s)\rho(u)$. Thus by the uniqueness of the Jordan decomposition, it suffices to prove that $d\rho(S)$ and $\rho(s)$ are semisimple. Let

$$E^\rho = \{f_B^\rho : B \in \mathrm{End}(V)\} \subset \mathcal{O}[G]$$

be the space of representative functions for ρ. Assume that $G \subset \mathbf{GL}(n,\mathbb{C})$ as an algebraic subgroup. Let $W_m \subset \mathcal{O}[\mathbf{GL}(n,\mathbb{C})]$ be the space introduced in the proof of Theorem 1.6.5, and choose an integer m such that $E^\rho \subset W_m|_G$. We have shown in Theorem 1.6.5 that $R(s)|_{W_m}$ is semisimple. Hence $R(s)$ acts semisimply on E^ρ. Thus there is a polynomial $\varphi(z)$ with distinct roots such that

$$\varphi(R(s))E^\rho = 0. \tag{1.60}$$

However, we have

$$R(s)^k f_B^\rho = f_{\rho(s)^k B}^\rho \quad \text{for all positive integers } k.$$

By the linearity of the trace and (1.60) we conclude that $\mathrm{tr}(\varphi(\rho(s))B) = 0$ for all $B \in \mathrm{End}(V)$. Hence $\varphi(\rho(s)) = 0$, which implies that $\rho(s)$ is semisimple. The same proof applies to $d\rho(S)$. $\qquad\square$

From Theorems 1.6.5 and 1.6.6 we see that every element g of G has a *semisimple component* g_s and a *unipotent component* g_u such that $g = g_s g_u$. Furthermore, this factorization is independent of the choice of defining representation $G \subset \mathbf{GL}(V)$. Likewise, every element $Y \in \mathfrak{g}$ has a unique *semisimple component* Y_s and a unique *nilpotent component* Y_n such that $Y = Y_s + Y_n$.

We denote the set of all semisimple elements of G by G_s and the set of all unipotent elements by G_u. Likewise, we denote the set of all semisimple elements of $\mathfrak{g}$ by $\mathfrak{g}_s$ and the set of all nilpotent elements by $\mathfrak{g}_n$. Suppose $G \subset \mathbf{GL}(n,\mathbb{C})$ as an algebraic subgroup. Since $T \in M_n(\mathbb{C})$ is nilpotent if and only if $T^n = 0$, we have

$$\mathfrak{g}_n = \mathfrak{g} \cap \{T \in M_n(\mathbb{C}) : T^n = 0\},$$
$$G_u = G \cap \{g \in \mathbf{GL}(n,\mathbb{C}) : (I - g)^n = 0\}.$$

Thus $\mathfrak{g}_n$ is an algebraic subset of $M_n(\mathbb{C})$ and G_u is an algebraic subset of $\mathbf{GL}(n,\mathbb{C})$.

Corollary 1.6.7. *Suppose G and H are algebraic groups with Lie algebras $\mathfrak{g}$ and $\mathfrak{h}$. Let $\rho : G \longrightarrow H$ be a regular homomorphism such that $d\rho : \mathfrak{g} \longrightarrow \mathfrak{h}$ is surjective. Then $\rho(G_u) = H_u$.*

Proof. By Theorem 1.6.2 the map $N \mapsto \exp(N)$ from $\mathfrak{g}_n$ to G_u is a bijection, and by Theorem 1.6.6 we have $H_u = \exp(\mathfrak{h}_n) = \exp(d\rho(\mathfrak{g}_n)) = \rho(G_u)$. $\qquad\square$

1.6.4 Exercises

1. Let $H, X \in M_n(\mathbb{C})$ be such that $[H, X] = 2X$. Show that X is nilpotent. (HINT: Show that $[H, X^k] = 2kX^k$. Then consider the eigenvalues of ad H on $M_n(\mathbb{C})$.)

2. Show that if $X \in M_n(\mathbb{C})$ is nilpotent then there exists $H \in M_n(\mathbb{C})$ such that $[H, X] = 2X$. (HINT: Use the Jordan canonical form to write $X = gJg^{-1}$ with $g \in \mathbf{GL}(n, \mathbb{C})$ and $J = \mathrm{diag}[J_1, \ldots, J_k]$ with each J_i either 0 or a $p_i \times p_i$ matrix of the form

$$
\begin{bmatrix}
0 & 1 & 0 & \cdots & 0 \\
0 & 0 & 1 & \cdots & 0 \\
\vdots & \vdots & \vdots & \ddots & \vdots \\
0 & 0 & 0 & \cdots & 1 \\
0 & 0 & 0 & \cdots & 0
\end{bmatrix}.
$$

Show that there exists $H_i \in M_{p_i}(\mathbb{C})$ such that $[H_i, J_i] = 2J_i$, and then take $H = g\,\mathrm{diag}[H_1, \ldots, H_k]g^{-1}$.)

3. Show that if $0 \neq X \in M_n(\mathbb{C})$ is nilpotent, then there exist $H, Y \in M_n(\mathbb{C})$ such that $[X, Y] = H$ and $[H, X] = 2X$, $[H, Y] = -2Y$. Conclude that $\mathbb{C}X + \mathbb{C}Y + \mathbb{C}H$ is a Lie subalgebra of $M_n(\mathbb{C})$ isomorphic to $\mathfrak{sl}(2, \mathbb{C})$.

4. Suppose V and W are finite-dimensional vector spaces over $\mathbb{C}$. Let $x \in \mathbf{GL}(V)$ and $y \in \mathbf{GL}(W)$ have multiplicative Jordan decompositions $x = x_s x_u$ and $y = y_s y_u$. Prove that the multiplicative Jordan decomposition of $x \otimes y$ in $\mathbf{GL}(V \otimes W)$ is $x \otimes y = (x_s \otimes y_s)(x_u \otimes y_u)$.

5. Suppose $\mathcal{A}$ is a finite-dimensional algebra over $\mathbb{C}$ (not necessarily associative). For example, $\mathcal{A}$ could be a Lie algebra. Let $g \in \mathrm{Aut}(\mathcal{A})$ have multiplicative Jordan decomposition $g = g_s g_u$ in $\mathbf{GL}(\mathcal{A})$. Show that g_s and g_u are also in $\mathrm{Aut}(\mathcal{A})$.

6. Suppose $g \in \mathbf{GL}(n, \mathbb{C})$ satisfies $g^k = I$ for some positive integer k. Prove that g is semisimple.

7. Let $G = \mathbf{SL}(2, \mathbb{C})$.
   (a) Show that $\{g \in G : \mathrm{tr}(g)^2 \neq 4\} \subset G_s$. (HINT: Show that the elements in this set have distinct eigenvalues.)
   (b) Let $u(t) = \begin{bmatrix} 1 & t \\ 0 & 1 \end{bmatrix}$ and $v(t) = \begin{bmatrix} 1 & 0 \\ t & 1 \end{bmatrix}$ for $t \in \mathbb{C}$. Show that $u(r)v(t) \in G_s$ whenever $rt(4 + rt) \neq 0$ and that $u(r)v(t)u(r) \in G_s$ whenever $rt(2 + rt) \neq 0$.
   (c) Show that G_s and G_u are not subgroups of G.

8. Let $G = \{\exp(tA) : t \in \mathbb{C}\}$, where $A = \begin{bmatrix} 1 & 1 \\ 0 & 1 \end{bmatrix}$.
   (a) Show that G is a closed subgroup of $\mathbf{GL}(2, \mathbb{C})$. (HINT: Calculate the matrix entries of $\exp(tA)$.)
   (b) Show that G is not an algebraic subgroup of $\mathbf{GL}(2, \mathbb{C})$. (HINT: If G were an algebraic group, then G would contain the semisimple and unipotent components of $\exp(tA)$. Show that this is a contradiction.)
   (c) Find the smallest algebraic subgroup $H \subset \mathbf{GL}(2, \mathbb{C})$ such that $G \subset H$. (HINT: Use the calculations from (b).)

1.7 Real Forms of Complex Algebraic Groups

In this section we introduce the notion of a *real form* of a complex linear algebraic group. We list them for the classical groups (these Lie groups already appeared in Section 1.1); in each case among the many real forms there is a unique compact form.

1.7.1 Real Forms and Complex Conjugations

We begin with a definition that refers to subgroups of $\mathbf{GL}(n,\mathbb{C})$. We will obtain a more general notion of a *real form* later in this section.

Definition 1.7.1. Let $G \subset \mathbf{GL}(n,\mathbb{C})$ be an algebraic subgroup. Then G is *defined over* $\mathbb{R}$ if the ideal $\mathfrak{I}_G$ is generated by $\mathfrak{I}_{\mathbb{R},G} = \{f \in \mathfrak{I}_G : f(\mathbf{GL}(n,\mathbb{R})) \subset \mathbb{R}\}$. If G is defined over $\mathbb{R}$, then we set $G_\mathbb{R} = G \cap \mathbf{GL}(n,\mathbb{R})$ and call $G_\mathbb{R}$ the *group of* $\mathbb{R}$-*rational points* of G.

Examples

1. The group $G = \mathbf{GL}(n,\mathbb{C})$ is defined over $\mathbb{R}$ (since $\mathfrak{I}_G = 0$), and $G_\mathbb{R} = \mathbf{GL}(n,\mathbb{R})$.

2. The group $G = B_n$ of $n \times n$ invertible upper-triangular matrices is defined over $\mathbb{R}$, since $\mathfrak{I}_G$ is generated by the matrix-entry functions $\{x_{ij} : n \geq i > j \geq 1\}$, which are real valued on $\mathbf{GL}(n,\mathbb{R})$. In this case $G_\mathbb{R}$ is the group of $n \times n$ real invertible upper-triangular matrices.

For $g \in \mathbf{GL}(n,\mathbb{C})$ we set $\sigma(g) = \bar{g}$ (complex conjugation of matrix entries). Then σ is an involutive automorphism of $\mathbf{GL}(n,\mathbb{C})$ as a real Lie group (σ^2 is the identity) and $d\sigma(A) = \bar{A}$ for $A \in M_n(\mathbb{C})$. If $f \in \mathcal{O}[\mathbf{GL}(n,\mathbb{C})]$ then we set

$$\bar{f}(g) = \overline{f(\sigma(g))}.$$

Here the overline on the right denotes complex conjugation. Since f is the product of $\det^{-k}$ (for some nonnegative integer k) and a polynomial φ in the matrix-entry functions, we obtain the function $\bar{f}$ by conjugating the coefficients of φ. We can write $f = f_1 + if_2$, where $f_1 = (f + \bar{f})/2$, $f_2 = (f - \bar{f})/(2i)$, and $i = \sqrt{-1}$. The functions f_1 and f_2 are real-valued on $\mathbf{GL}(n,\mathbb{R})$, and $\bar{f} = f_1 - if_2$. Thus $f(\mathbf{GL}(n,\mathbb{R})) \subset \mathbb{R}$ if and only if $\bar{f} = f$.

Lemma 1.7.2. *Let $G \subset \mathbf{GL}(n,\mathbb{C})$ be an algebraic subgroup. Then G is defined over $\mathbb{R}$ if and only if $\mathfrak{I}_G$ is invariant under $f \mapsto \bar{f}$.*

Proof. Assume that G is defined over $\mathbb{R}$. If $f_1 \in \mathfrak{I}_{\mathbb{R},G}$ then $f_1 = \bar{f}_1$. Hence $f_1(\sigma(g)) = \overline{f_1(g)} = 0$ for $g \in G$. Since $\mathfrak{I}_{\mathbb{R},G}$ is assumed to generate $\mathfrak{I}_G$, it follows that $\sigma(g) \in G$ for all $g \in G$. Thus for any $f \in \mathfrak{I}_G$ we have $\bar{f}(g) = 0$, and hence $\bar{f} \in \mathfrak{I}_G$.

Conversely, if $\mathcal{I}_G$ is invariant under $f \mapsto \bar{f}$, then every $f \in \mathcal{I}_G$ is of the form $f_1 + i f_2$ as above, where $f_j \in \mathcal{I}_{\mathbb{R},G}$. Thus $\mathcal{I}_{\mathbb{R},G}$ generates $\mathcal{I}_G$, and so G is defined over $\mathbb{R}$. $\square$

Assume that $G \subset \mathbf{GL}(n,\mathbb{C})$ is an algebraic group defined over $\mathbb{R}$. Let $\mathfrak{g} \subset M_n(\mathbb{C})$ be the Lie algebra of G. Since $\mathcal{I}_{\mathbb{R},G}$ generates $\mathcal{I}_G$ and σ^2 is the identity map, this implies that $\sigma(G) = G$. Hence σ defines a Lie group automorphism of G and $d\sigma(A) = \bar{A} \in \mathfrak{g}$ for all $A \in \mathfrak{g}$. By definition, $G_{\mathbb{R}} = \{g \in G : \sigma(g) = g\}$. Hence $G_{\mathbb{R}}$ is a Lie subgroup of G and

$$\mathrm{Lie}(G_{\mathbb{R}}) = \{A \in \mathfrak{g} : \bar{A} = A\}.$$

If $A \in \mathfrak{g}$ then $A = A_1 + iA_2$, where $A_1 = (A + \bar{A})/2$ and $A_2 = (A - \bar{A})/2i$ are in $\mathrm{Lie}(G_{\mathbb{R}})$. Thus

$$\mathfrak{g} = \mathrm{Lie}(G_{\mathbb{R}}) \oplus i\,\mathrm{Lie}(G_{\mathbb{R}}) \tag{1.61}$$

as a real vector space, so $\dim_{\mathbb{R}} \mathrm{Lie}(G_{\mathbb{R}}) = \dim_{\mathbb{C}} \mathfrak{g}$. Therefore the dimension of the Lie group $G_{\mathbb{R}}$ is the same as the dimension of G as a linear algebraic group over $\mathbb{C}$ (see Appendix A.1.6).

Remark 1.7.3. If a linear algebraic group G is defined over $\mathbb{R}$, then there is a set $\mathcal{A}$ of polynomials with real coefficients such that G consists of the common zeros of these polynomials in $\mathbf{GL}(n,\mathbb{C})$. The converse assertion is more subtle, however, since it is possible that $\mathcal{A}$ does not generate the ideal $\mathcal{I}_G$, as required by Definition 1.7.1. For example, the group B_n of upper-triangular $n \times n$ matrices is the zero set of the polynomials $\{x_{ij}^2 : n \geq i > j \geq 1\}$; these polynomials are real on $\mathbf{GL}(n,\mathbb{R})$ but do not generate $\mathcal{I}_{B_n}$ (of course, in this case we already know that B_n is defined over $\mathbb{R}$).

By generalizing the notion of complex conjugation we now obtain a useful criterion (not involving a specific matrix form of G) for G to be isomorphic to a linear algebraic group defined over $\mathbb{R}$. This will also furnish the general notion of a *real form* of G.

Definition 1.7.4. Let G be a linear algebraic group and let τ be an automorphism of G as a real Lie group such that τ^2 is the identity. For $f \in \mathcal{O}[G]$ define f^τ by

$$f^\tau(g) = \overline{f(\tau(g))}$$

(with the overline denoting complex conjugation). Then τ is a *complex conjugation* on G if $f^\tau \in \mathcal{O}[G]$ for all $f \in \mathcal{O}[G]$.

When $G \subset \mathbf{GL}(n,\mathbb{C})$ is defined over $\mathbb{R}$, then the map $\sigma(g) = \bar{g}$ introduced previously is a complex conjugation. In Section 1.7.2 we shall give examples of complex conjugations when G is a classical group.

Theorem 1.7.5. *Let G be a linear algebraic group and let τ be a complex conjugation on G. Then there exists a linear algebraic group $H \subset \mathbf{GL}(n,\mathbb{C})$ defined over $\mathbb{R}$ and an isomorphism $\rho : G \longrightarrow H$ such that $\rho(\tau(g)) = \sigma(\rho(g))$, where σ is the conjugation of $\mathbf{GL}(n,\mathbb{C})$ given by complex conjugation of matrix entries.*

Proof. Fix a finite set $\{1, f_1, \ldots, f_m\}$ of regular functions on G that generate $\mathcal{O}[G]$ as an algebra over $\mathbb{C}$ (for example, the restrictions to G of the matrix entry functions and $\det^{-1}$ given by the defining representation of G). Set $C(f) = f^\tau$ for $f \in \mathcal{O}[G]$ and let

$$V = \mathrm{Span}_\mathbb{C}\{R(g)f_k, R(g)Cf_k : g \in G \text{ and } k = 1, \ldots, m\}.$$

Then V is invariant under G and C, since $CR(g) = R(\tau(g))C$. Let $\rho(g) = R(g)|_V$. It follows from Proposition 1.5.1 that V is finite-dimensional and (ρ, V) is a regular representation of G.

We note that if $g, g' \in G$ and $f_k(g) = f_k(g')$ for all k, then $f(g) = f(g')$ for all $f \in \mathcal{O}[G]$, since the set $\{1, f_1, \ldots, f_m\}$ generates $\mathcal{O}[G]$. Letting f run over the restrictions to G of the matrix entry functions (relative to some matrix form of G), we conclude that $g = g'$. Thus if $\rho(g)f = f$ for all $f \in V$, then $g = I$, proving that $\mathrm{Ker}(\rho) = \{I\}$.

Since C^2 is the identity map, we can decompose $V = V_+ \oplus V_-$ as a vector space over $\mathbb{R}$, where

$$V_+ = \{f \in V : C(f) = f\}, \quad V_- = \{f \in V : C(f) = -f\}.$$

Because $C(if) = -iC(f)$ we have $V_- = iV_+$. Choose a basis (over $\mathbb{R}$) of V_+, say $\{v_1, \ldots, v_n\}$. Then $\{v_1, \ldots, v_n\}$ is also a basis of V over $\mathbb{C}$. If we use this basis to identify V with $\mathbb{C}^n$ then C becomes complex conjugation. To simplify the notation we will also write $\rho(g)$ for the matrix of $\rho(g)$ relative to this basis.

We now have an injective regular homomorphism $\rho : G \longrightarrow \mathbf{GL}(n, \mathbb{C})$ such that $\rho(\tau(g)) = \sigma(\rho(g))$, where σ denotes complex conjugation of matrix entries. In Chapter 11 (Theorem 11.1.5) we will prove that the image of a linear algebraic group under a regular homomorphism is always a linear algebraic group (i.e., a closed subgroup in the *Zariski topology*). Assuming this result (whose proof does not depend on the current argument), we conclude that $H = \rho(G)$ is an algebraic subgroup of $\mathbf{GL}(n, \mathbb{C})$. Furthermore, if $\delta \in V^*$ is the linear functional $f \mapsto f(I)$, then

$$f(g) = R(g)f(I) = \langle \delta, R(g)f \rangle. \tag{1.62}$$

Hence $\rho^*(\mathcal{O}[H]) = \mathcal{O}[G]$, since by (1.62) the functions $f_1, \ldots, f_m$ are matrix entries of (ρ, V). This proves that ρ^{-1} is a regular map.

Finally, let $f \in \mathcal{I}_H$. Then for $h = \rho(g) \in H$ we have

$$\bar{f}(h) = \overline{f(\sigma(\rho(g)))} = \overline{f(\rho(\tau(g)))} = 0.$$

Hence $\bar{f} \in \mathcal{I}_H$, so from Lemma 1.7.2 we conclude that H is defined over $\mathbb{R}$. $\square$

Definition 1.7.6. Let G be a linear algebraic group. A subgroup K of G is called a *real form* of G if there exists a complex conjugation τ on G such that

$$K = \{g \in G : \tau(g) = g\}.$$

Let K be a real form of G. Then K is a closed subgroup of G, and from Theorem 1.7.5 and (1.61) we see that the dimension of K as a real Lie group is equal to the dimension of G as a complex linear algebraic group, and

$$\mathfrak{g} = \mathrm{Lie}(K) \oplus \mathrm{i}\,\mathrm{Lie}(K) \tag{1.63}$$

as a real vector space.

One of the motivations for introducing real forms is that we can study the representations of G using the real form and its Lie algebra. Let G be a linear algebraic group, and let G° be the connected component of the identity of G (as a real Lie group). Let K be a real form of G and set $\mathfrak{k} = \mathrm{Lie}(K)$.

Proposition 1.7.7. *Suppose* (ρ, V) *is a regular representation of* G. *Then a subspace* $W \subset V$ *is invariant under* $d\rho(\mathfrak{k})$ *if and only if it is invariant under* G°. *In particular,* V *is irreducible under* $\mathfrak{k}$ *if and only if it is irreducible under* G°.

Proof. Assume W is invariant under $\mathfrak{k}$. Since the map $X \mapsto d\rho(X)$ from $\mathfrak{g}$ to $\mathrm{End}(V)$ is complex linear, it follows from (1.63) that W is invariant under $\mathfrak{g}$. Let $W^\perp \subset V^*$ be the annihilator of W. Then $\langle w^*, (d\rho(X))^k w \rangle = 0$ for $w \in W$, $w^* \in W^\perp$, $X \in \mathfrak{g}$, and all integers k. Hence

$$\langle w^*, \rho(\exp X) w \rangle = \langle w^*, \exp(d\rho(X)) w \rangle = \sum_{k=0}^\infty \frac{1}{k!} \langle w^*, d\rho(X)^k w \rangle = 0\,,$$

so $\rho(\exp X) W \subset W$. Since G° is generated by $\exp(\mathfrak{g})$, this proves that W is invariant under G°. To prove the converse we reverse this argument, replacing X by tX and differentiating at $t = 0$. $\qquad\square$

1.7.2 Real Forms of the Classical Groups

We now describe the complex conjugations and real forms of the complex classical groups. We label the groups and their real forms using É. Cartan's classification. For each complex group there is one real form that is compact.

1. (Type AI) Let $G = \mathbf{GL}(n, \mathbb{C})$ (resp. $\mathbf{SL}(n, \mathbb{C})$) and define $\tau(g) = \bar{g}$ for $g \in G$. Then $f^\tau = \bar{f}$ for $f \in \mathbb{C}[G]$, and so τ is a complex conjugation on G. The associated real form is $\mathbf{GL}(n, \mathbb{R})$ (resp. $\mathbf{SL}(n, \mathbb{R})$).

2. (Type AII) Let $G = \mathbf{GL}(2n, \mathbb{C})$ (resp. $\mathbf{SL}(2n, \mathbb{C})$) and let J be the $2n \times 2n$ skew-symmetric matrix from Section 1.1.2. Define $\tau(g) = J\bar{g}J^t$ for $g \in G$. Since $J^2 = -I$, we see that τ^2 is the identity. Also if f is a regular function on G then $f^\tau(g) = \bar{f}(JgJ^t)$, and so f^τ is also a regular function on G. Hence τ is a complex conjugation on G. The equation $\tau(g) = g$ can be written as $Jg = \bar{g}J$. Hence the associated real form of G is the group $\mathbf{GL}(n, \mathbb{H})$ (resp. $\mathbf{SL}(n, \mathbb{H})$) from Section 1.1.4, where we view $\mathbb{H}^n$ as a $2n$-dimensional vector space over $\mathbb{C}$.

3. (Type AIII) Let $G = \mathbf{GL}(n, \mathbb{C})$ (resp. $\mathbf{SL}(n, \mathbb{C})$) and let $p, q \in \mathbb{N}$ be such that $p + q = n$. Let $I_{p,q} = \mathrm{diag}[I_p, -I_q]$ as in Section 1.1.2 and define $\tau(g) = I_{p,q}(g^*)^{-1}I_{p,q}$ for $g \in G$. Since $I_{p,q}^2 = I_n$, we see that τ^2 is the identity. Also if f is a regular function on G then $f^\tau(g) = \bar{f}(I_{p,q}(g^t)^{-1}I_{p,q})$, and so f^τ is also a regular function on G.

Hence τ is a complex conjugation on G. The equation $\tau(g) = g$ can be written as $g^* I_{p,q} g = I_{p,q}$, so the indefinite unitary group $\mathbf{U}(p,q)$ (resp. $\mathbf{SU}(p,q)$) defined in Section 1.1.3 is the real form of G defined by τ. The *unitary group* $\mathbf{U}(n,0) = \mathbf{U}(n)$ (resp. $\mathbf{SU}(n)$) is a compact real form of G.

4. (Type BDI) Let G be $\mathbf{O}(n,\mathbb{C}) = \{g \in \mathbf{GL}(n,\mathbb{C}) : gg^t = 1\}$ (resp. $\mathbf{SO}(n,\mathbb{C})$) and let $p,q \in \mathbb{N}$ be such that $p+q = n$. Let the matrix $I_{p,q}$ be as in Type AIII. Define $\tau(g) = I_{p,q} \overline{g} I_{p,q}$ for $g \in G$. Since $(g^t)^{-1} = g$ for $g \in G$, τ is the restriction to G of the complex conjugation in Example 3. We leave it as an exercise to show that the corresponding real form is isomorphic to the group $\mathbf{O}(p,q)$ (resp. $\mathbf{SO}(p,q)$) defined in Section 1.1.2. When $p = n$ we obtain the compact real form $\mathbf{O}(n)$ (resp. $\mathbf{SO}(n)$).

5. (Type DIII) Let G be $\mathbf{SO}(2n,\mathbb{C})$ and let J be the $2n \times 2n$ skew-symmetric matrix as in Type AII. Define $\tau(g) = J\overline{g}J^t$ for $g \in G$. Just as in Type AII, we see that τ is a complex conjugation of G. The corresponding real form is the group $\mathbf{SO}^*(2n)$ defined in Section 1.1.4 (see Exercises 1.1.5, #12).

6. (Type CI) Let G be $\mathbf{Sp}(n,\mathbb{C}) \subset \mathbf{SL}(2n,\mathbb{C})$. The equation defining G is $g^t J g = J$, where J is the skew-symmetric matrix in Type AII. Since J is real, we may define $\tau(g) = \overline{g}$ for $g \in G$ and obtain a complex conjugation on G. The associated real form is $\mathbf{Sp}(n,\mathbb{R})$.

7. (Type CII) Let $p,q \in \mathbb{N}$ be such that $p+q = n$ and let $K_{p,q} = \text{diag}[I_{p,q}, I_{p,q}] \in M_{2n}(\mathbb{R})$ as in Section 1.1.4. Let Ω be the nondegenerate skew form on $\mathbb{C}^{2n}$ with matrix

$$K_{p,q}J = \begin{bmatrix} 0 & I_{p,q} \\ -I_{p,q} & 0 \end{bmatrix},$$

with J as in Type CI. Let $G = \mathbf{Sp}(\mathbb{C}^{2n}, \Omega)$ and define $\tau(g) = K_{p,q}(g^*)^{-1}K_{p,q}$ for $g \in G$. We leave it as an exercise to prove that τ is a complex conjugation of G. The corresponding real form is the group $\mathbf{Sp}(p,q)$ defined in Section 1.1.4. When $p = n$ we use the notation $\mathbf{Sp}(n) = \mathbf{Sp}(n,0)$. Since $K_{n,0} = I_{2n}$, it follows that $\mathbf{Sp}(n) = \mathbf{SU}(2n) \cap \mathbf{Sp}(n,\mathbb{C})$. Hence $\mathbf{Sp}(n)$ is a compact real form of $\mathbf{Sp}(n,\mathbb{C})$.

Summary

We have shown that the classical groups (with the condition $\det(g) = 1$ included for conciseness) can be viewed either as

- the complex linear algebraic groups $\mathbf{SL}(n,\mathbb{C})$, $\mathbf{SO}(n,\mathbb{C})$, and $\mathbf{Sp}(n,\mathbb{C})$ together with their real forms, or alternatively as
- the special linear groups over the fields $\mathbb{R}$, $\mathbb{C}$, and $\mathbb{H}$, together with the special isometry groups of nondegenerate forms (symmetric or skew symmetric, Hermitian or skew Hermitian) over these fields.

Thus we have the following families of classical groups:

Special linear groups: $\mathbf{SL}(n,\mathbb{R})$, $\mathbf{SL}(n,\mathbb{C})$, and $\mathbf{SL}(n,\mathbb{H})$. Of these, only $\mathbf{SL}(n,\mathbb{C})$ is an algebraic group over $\mathbb{C}$, whereas the other two are real forms of $\mathbf{SL}(n,\mathbb{C})$ (resp. $\mathbf{SL}(2n,\mathbb{C})$).

Automorphism groups of forms: On a real vector space, a Hermitian (resp. skew-Hermitian) form is the same as a symmetric (resp. skew-symmetric) form. On a complex vector space skew-Hermitian forms become Hermitian after multiplication by i, and vice versa. On a quaternionic vector space there are no nonzero bilinear forms at all (by the noncommutativity of quaternionic multiplication), so the form must be either Hermitian or skew-Hermitian. Taking these restrictions into account, we see that the possibilities for unimodular isometry groups are those in Table 1.1.

Table 1.1 Isometry Groups of Forms.

Group	Field	Form
$\mathbf{SO}(p,q)$	$\mathbb{R}$	Symmetric
$\mathbf{SO}(n,\mathbb{C})$	$\mathbb{C}$	Symmetric
$\mathbf{Sp}(n,\mathbb{R})$	$\mathbb{R}$	Skew-symmetric
$\mathbf{Sp}(n,\mathbb{C})$	$\mathbb{C}$	Skew-symmetric
$\mathbf{SU}(p,q)$	$\mathbb{C}$	Hermitian
$\mathbf{Sp}(p,q)$	$\mathbb{H}$	Hermitian
$\mathbf{SO}^*(2n)$	$\mathbb{H}$	Skew-Hermitian

Note that the group $\mathbf{SU}(p,q)$ is *not* an algebraic group over $\mathbb{C}$, even though the field is $\mathbb{C}$, since its defining equations involve complex conjugation. Likewise, the groups for the field $\mathbb{H}$ are not algebraic groups over $\mathbb{C}$, even though they can be viewed as complex matrix groups.

1.7.3 Exercises

1. On $G = \mathbb{C}^\times$ define the conjugation $\tau(z) = \bar{z}^{-1}$. Let $V \subset \mathcal{O}[G]$ be the subspace with basis $f_1(z) = z$ and $f_2(z) = z^{-1}$. Define $Cf(z) = \overline{f(\tau(z))}$ and $\rho(z)f(w) = f(wz)$ for $f \in V$ and $z \in G$, as in Theorem 1.7.5.
   (a) Find a basis $\{v_1, v_2\}$ for the real subspace $V_+ = \{f \in V : Cf = f\}$ so that in this basis, $\rho(z) = \begin{bmatrix} (z+z^{-1})/2 & (z-z^{-1})/2\mathrm{i} \\ -(z-z^{-1})/2\mathrm{i} & (z+z^{-1})/2 \end{bmatrix}$ for $z \in \mathbb{C}^\times$.
   (b) Let $K = \{z \in G : \tau(z) = z\}$. Use (a) to show that $G \cong \mathbf{SO}(2,\mathbb{C})$ as an algebraic group and $K \cong \mathbf{SO}(2)$ as a Lie group.
2. Show that $\mathbf{Sp}(1)$ is isomorphic to $\mathbf{SU}(2)$ as a Lie group. (HINT: Consider the adjoint representation of $\mathbf{Sp}(1)$.)
3. Let ψ be the real linear transformation of $\mathbb{C}^{2n}$ defined by

$$\psi[z_1,\ldots,z_n,z_{n+1},\ldots,z_{2n}] = [\bar{z}_{n+1},\ldots,\bar{z}_{2n},-\bar{z}_1,\ldots,-\bar{z}_n] \ .$$

Define $\mathbf{SU}^*(2n) = \{g \in \mathbf{SL}(2n,\mathbb{C}) : g\psi = \psi g\}$. Show that $\mathbf{SU}^*(2n)$ is isomorphic to $\mathbf{SL}(n,\mathbb{H})$ as a Lie group.

4. Let $G = \mathbf{Sp}(\mathbb{C}^{2n},\Omega)$ be the group for the real form of Type CII. Show that the map $g \mapsto (g^*)^{-1}$ defines an involutory automorphism of G as a real Lie group, and that $g \mapsto \tau(g) = K_{p,q}(g^*)^{-1}K_{p,q}$ defines a complex conjugation of G.

5. Let $G = \mathbf{O}(n,\mathbb{C})$ and let $\tau(g) = I_{p,q}\bar{g}I_{p,q}$ be the complex conjugation of Type BDI. Let $H = \{g \in G : \tau(g) = g\}$ be the associated real form. Define $J_{p,q} = \mathrm{diag}[I_p, iI_q]$ and set $\gamma(g) = J_{p,q}^{-1}gJ_{p,q}$ for $g \in G$.

   (a) Prove that $\gamma(\tau(g)) = \overline{\gamma(g)}$ for $g \in G$. Hence $\gamma(H) \subset \mathbf{GL}(n,\mathbb{R})$. (HINT: Note that $J_{p,q}^2 = I_{p,q}$ and $J_{p,q}^{-1} = \bar{J}_{p,q}$.)

   (b) Prove that $\gamma(g)^t I_{p,q}\gamma(g) = I_{p,q}$ for $g \in G$. Together with the result from part (a) this shows that $\gamma(H) = \mathbf{O}(p,q)$.

1.8 Notes

Section 1.3. For a more complete introduction to Lie groups through matrix groups, see Rossmann [127].

Section 1.4. Although Hermann Weyl seemed well aware that there could be a theory of algebraic groups (for example he calculated the ideal of the orthogonal groups in [164]), he studied the classical groups as individuals with many similarities rather than as examples of linear algebraic groups. Chevalley considered algebraic groups to be a natural subclass of the class of Lie groups and devoted Volumes II and III of his *Theory of Lie Groups* [35], [36] to the development of their basic properties. The modern theory of linear algebraic groups has its genesis in the work of Borel [15], [16]; see [17] for a detailed historical account. Additional books on algebraic groups are Humphreys [77], Kraft [92], Springer [136], and Onishchik and Vinberg [118].

Section 1.7. Proposition 1.7.7 is the Lie algebra version of Weyl's *unitary trick*. A detailed discussion of real forms of complex semisimple Lie groups and É. Cartan's classification can be found in Helgason [66]. One can see from [66, Chapter X, Table V] that the real forms of the classical groups contain a substantial portion of the connected simple Lie groups. The remaining simple Lie groups are the real forms of the five *exceptional* simple Lie groups (Cartan's types G2, F4, E6, E7, and E8 of dimension 14, 52, 78, 133, and 248 respectively).

Chapter 2
Structure of Classical Groups

Abstract In this chapter we study the structure of a classical group G and its Lie algebra. We choose a matrix realization of G such that the diagonal subgroup $H \subset G$ is a *maximal torus*; by elementary linear algebra every conjugacy class of semisimple elements intersects H. Using the unipotent elements in G, we show that the groups $\mathbf{GL}(n, \mathbb{C})$, $\mathbf{SL}(n, \mathbb{C})$, $\mathbf{SO}(n, \mathbb{C})$, and $\mathbf{Sp}(n, \mathbb{C})$ are connected (as Lie groups and as algebraic groups). We examine the group $\mathbf{SL}(2, \mathbb{C})$, find its irreducible representations, and show that every regular representation decomposes as the direct sum of irreducible representations. This group and its Lie algebra play a basic role in the structure of the other classical groups and Lie algebras. We decompose the Lie algebra of a classical group under the adjoint action of a maximal torus and find the invariant subspaces (called *root spaces*) and the corresponding characters (called *roots*). The commutation relations of the root spaces are encoded by the set of roots; we use this information to prove that the classical (trace-zero) Lie algebras are simple (or semisimple). In the final section of the chapter we develop some general Lie algebra methods (solvable Lie algebras, Killing form) and show that every semisimple Lie algebra has a root-space decomposition with the same properties as those of the classical Lie algebras.

2.1 Semisimple Elements

A semisimple matrix can be diagonalized, relative to a suitable basis. In this section we show that a maximal set of mutually commuting semisimple elements in a classical group can be simultaneously diagonalized by an element of the group.

R. Goodman, N.R. Wallach, *Symmetry, Representations, and Invariants*,
Graduate Texts in Mathematics 255, DOI 10.1007/978-0-387-79852-3_2,
© Roe Goodman and Nolan R. Wallach 2009

2.1.1 Toral Groups

Recall that an (algebraic) torus is an algebraic group T isomorphic to $(\mathbb{C}^\times)^l$; the integer l is called the *rank* of T. The rank is uniquely determined by the algebraic group structure of T (this follows from Lemma 2.1.2 below or Exercises 1.4.5 #1).

Definition 2.1.1. A *rational character* of a linear algebraic group K is a regular homomorphism $\chi : K \longrightarrow \mathbb{C}^\times$.

The set $\mathfrak{X}(K)$ of rational characters of K has the natural structure of an abelian group with $(\chi_1\chi_2)(k) = \chi_1(k)\chi_2(k)$ for $k \in K$. The identity element of $\mathfrak{X}(K)$ is the *trivial character* $\chi_0(k) = 1$ for all $k \in K$.

Lemma 2.1.2. *Let T be an algebraic torus of rank l. The group $\mathfrak{X}(T)$ is isomorphic to $\mathbb{Z}^l$. Furthermore, $\mathfrak{X}(T)$ is a basis for $\mathcal{O}[T]$ as a vector space over $\mathbb{C}$.*

Proof. We may assume that $T = (\mathbb{C}^\times)^l$ with coordinate functions $x_1, \ldots, x_l$. Thus $\mathcal{O}[T] = \mathbb{C}[x_1, \ldots, x_l, x_1^{-1}, \ldots, x_l^{-1}]$. For $t = [x_1(t), \ldots, x_l(t)] \in T$ and $\lambda = [p_1, \ldots, p_l] \in \mathbb{Z}^l$ we set

$$t^\lambda = \prod_{k=1}^{l} x_k(t)^{p_k} . \tag{2.1}$$

Then $t \mapsto t^\lambda$ is a rational character of T, which we will denote by χ_λ. Since $t^{\lambda+\mu} = t^\lambda t^\mu$ for $\lambda, \mu \in \mathbb{Z}^l$, the map $\lambda \mapsto \chi_\lambda$ is an injective group homomorphism from $\mathbb{Z}^l$ to $\mathfrak{X}(T)$. Clearly, the set of functions $\{\chi_\lambda : \lambda \in \mathbb{Z}^l\}$ is a basis for $\mathcal{O}[T]$ as a vector space over $\mathbb{C}$.

Conversely, let χ be a rational character of T. Then for $k = 1, \ldots, l$ the function

$$z \mapsto \varphi_k(z) = \chi(1, \ldots, z, \ldots, 1) \quad (z \text{ in } k\text{th coordinate})$$

is a one-dimensional regular representation of $\mathbb{C}^\times$. From Lemma 1.6.4 we have $\varphi_k(z) = z^{p_k}$ for some $p_k \in \mathbb{Z}$. Hence

$$\chi(x_1, \ldots, x_l) = \prod_{k=1}^{l} \varphi_k(x_k) = \chi_\lambda(x_1, \ldots, x_l) ,$$

where $\lambda = [p_1, \ldots, p_l]$. Thus every rational character of T is of the form χ_λ for some $\lambda \in \mathbb{Z}^l$. $\square$

Proposition 2.1.3. *Let T be an algebraic torus. Suppose (ρ, V) is a regular representation of T. Then there exists a finite subset $\Psi \subset \mathfrak{X}(T)$ such that*

$$V = \bigoplus_{\chi \in \Psi} V(\chi) , \tag{2.2}$$

where $V(\chi) = \{v \in V : \rho(t)v = \chi(t)v \text{ for all } t \in T\}$ is the χ weight space of T on V. If $g \in \mathrm{End}(V)$ commutes with $\rho(t)$ for all $t \in T$, then $gV(\chi) \subset V(\chi)$.

Proof. Since $(\mathbb{C}^\times)^l \cong \mathbb{C}^\times \times (\mathbb{C}^\times)^{l-1}$, the existence of the decomposition (2.2) follows from Lemma 1.6.4 by induction on l. The last statement is clear from the definition of $V(\chi)$. $\qquad\square$

Lemma 2.1.4. *Let T be an algebraic torus. Then there exists an element $t \in T$ with the following property: If $f \in \mathcal{O}[T]$ and $f(t^n) = 0$ for all $n \in \mathbb{Z}$, then $f = 0$.*

Proof. We may assume $T = (\mathbb{C}^\times)^l$. Choose $t \in T$ such that its coordinates $t_i = x_i(t)$ satisfy

$$t_1^{p_1} \cdots t_l^{p_l} \neq 1 \quad \text{for all } (p_1, \ldots, p_l) \in \mathbb{Z}^l \setminus \{0\}. \tag{2.3}$$

This is always possible; for example we can take $t_1, \ldots, t_l$ to be algebraically independent over the rationals.

Let $f \in \mathcal{O}[T]$ satisfy $f(t^n) = 0$ for all $n \in \mathbb{Z}$. Replacing f by $(x_1 \cdots x_l)^r f$ for a suitably large r, we may assume that

$$f = \sum_{|K| \leq p} a_K x^K$$

for some positive integer p, where the exponents K are in $\mathbb{N}^l$. Since $f(t^n) = 0$ for all $n \in \mathbb{Z}$, the coefficients $\{a_K\}$ satisfy the equations

$$\sum_K a_K (t^K)^n = 0 \quad \text{for all } n \in \mathbb{Z}. \tag{2.4}$$

We claim that the numbers $\{t^K : K \in \mathbb{N}^l\}$ are all distinct. Indeed, if $t^K = t^L$ for some $K, L \in \mathbb{N}^l$ with $K \neq L$, then $t^P = 1$, where $P = K - L \neq 0$, which would violate (2.3). Enumerate the coefficients a_K of f as $b_1, \ldots, b_r$ and the corresponding character values t^K as $y_1, \ldots, y_r$. Then (2.4) implies that

$$\sum_{j=1}^r b_j (y_j)^n = 0 \qquad \text{for } n = 0, 1, \ldots, r-1.$$

We view these equations as a homogeneous linear system for $b_1, \ldots, b_r$. The coefficient matrix is the $r \times r$ *Vandermonde matrix:*

$$V_r(y) = \begin{bmatrix} y_1^{r-1} & y_1^{r-2} & \cdots & y_1 & 1 \\ y_2^{r-1} & y_2^{r-2} & \cdots & y_2 & 1 \\ \vdots & \vdots & \ddots & \vdots & \vdots \\ y_r^{r-1} & y_r^{r-2} & \cdots & y_r & 1 \end{bmatrix}.$$

The determinant of this matrix is the *Vandermonde determinant* $\prod_{1 \leq i < j \leq r} (y_i - y_j)$ (see Exercises 2.1.3). Since $y_i \neq y_j$ for $i \neq j$, the determinant is nonzero, and hence $b_K = 0$ for all K. Thus $f = 0$. $\qquad\square$

2.1.2 Maximal Torus in a Classical Group

If G is a linear algebraic group, then a torus $H \subset G$ is *maximal* if it is not contained in any larger torus in G. When G is one of the classical linear algebraic groups $\mathbf{GL}(n,\mathbb{C})$, $\mathbf{SL}(n,\mathbb{C})$, $\mathbf{Sp}(\mathbb{C}^n,\Omega)$, $\mathbf{SO}(\mathbb{C}^n,B)$ (where Ω is a nondegenerate skew-symmetric bilinear form and B is a nondegenerate symmetric bilinear form) we would like the subgroup H of diagonal matrices in G to be a maximal torus. For this purpose we make the following choices of B and Ω:

We denote by s_l the $l \times l$ matrix

$$s_l = \begin{bmatrix} 0 & 0 & \cdots & 0 & 1 \\ 0 & 0 & \cdots & 1 & 0 \\ \vdots & \vdots & \ddots & \vdots & \vdots \\ 0 & 1 & \cdots & 0 & 0 \\ 1 & 0 & \cdots & 0 & 0 \end{bmatrix} \tag{2.5}$$

with 1 on the skew diagonal and 0 elsewhere. Let $n = 2l$ be even, set

$$J_+ = \begin{bmatrix} 0 & s_l \\ s_l & 0 \end{bmatrix}, \qquad J_- = \begin{bmatrix} 0 & s_l \\ -s_l & 0 \end{bmatrix},$$

and define the bilinear forms

$$B(x,y) = x^t J_+ y, \qquad \Omega(x,y) = x^t J_- y \quad \text{for } x,y \in \mathbb{C}^n . \tag{2.6}$$

The form B is nondegenerate and *symmetric*, and the form Ω is nondegenerate and *skew-symmetric*. From equation (1.8) we calculate that the Lie algebra $\mathfrak{so}(\mathbb{C}^{2l},B)$ of $\mathbf{SO}(\mathbb{C}^{2l},B)$ consists of all matrices

$$A = \begin{bmatrix} a & b \\ c & -s_l a^t s_l \end{bmatrix}, \qquad \begin{cases} a,b,c \in M_l(\mathbb{C}) , \\ b^t = -s_l b s_l , \quad c^t = -s_l c s_l \end{cases} \tag{2.7}$$

(thus b and c are *skew-symmetric* around the skew diagonal). Likewise, the Lie algebra $\mathfrak{sp}(\mathbb{C}^{2l},\Omega)$ of $\mathbf{Sp}(\mathbb{C}^{2l},\Omega)$ consists of all matrices

$$A = \begin{bmatrix} a & b \\ c & -s_l a^t s_l \end{bmatrix}, \qquad \begin{cases} a,b,c \in M_l(\mathbb{C}) , \\ b^t = s_l b s_l , \quad c^t = s_l c s_l \end{cases} \tag{2.8}$$

(b and c are *symmetric* around the skew diagonal).

Finally, we consider the orthogonal group on $\mathbb{C}^n$ when $n = 2l+1$ is odd. We take the symmetric bilinear form

$$B(x,y) = \sum_{i+j=n+1} x_i y_j \quad \text{for } x,y \in \mathbb{C}^n . \tag{2.9}$$

We can write this form as $B(x,y) = x^t S y$, where the $n \times n$ symmetric matrix $S = s_{2l+1}$ has block form

$$S = \begin{bmatrix} 0 & 0 & s_l \\ 0 & 1 & 0 \\ s_l & 0 & 0 \end{bmatrix}.$$

Writing the elements of $M_n(\mathbb{C})$ in the same block form and making a matrix calculation from equation (1.8), we find that the Lie algebra $\mathfrak{so}(\mathbb{C}^{2l+1}, B)$ of $\mathbf{SO}(\mathbb{C}^{2l+1}, B)$ consists of all matrices

$$A = \begin{bmatrix} a & w & b \\ u^t & 0 & -w^t s_l \\ c & -s_l u & -s_l a^t s_l \end{bmatrix}, \qquad \begin{cases} a, b, c \in M_l(\mathbb{C}), \\ b^t = -s_l b s_l, \quad c^t = -s_l c s_l, \\ \text{and} \quad u, w \in \mathbb{C}^l. \end{cases} \qquad (2.10)$$

Suppose now that G is $\mathbf{GL}(n, \mathbb{C})$, $\mathbf{SL}(n, \mathbb{C})$, $\mathbf{Sp}(\mathbb{C}^n, \Omega)$, or $\mathbf{SO}(\mathbb{C}^n, B)$ with Ω and B chosen as above. Let H be the subgroup of diagonal matrices in G; write $\mathfrak{g} = \mathrm{Lie}(G)$ and $\mathfrak{h} = \mathrm{Lie}(H)$. By Example 1 of Section 1.4.3 and (1.39) we know that $\mathfrak{h}$ consists of all diagonal matrices that are in $\mathfrak{g}$. We have the following case-by-case description of H and $\mathfrak{h}$:

1. When $G = \mathbf{SL}(l+1, \mathbb{C})$ (we say that G is of **type A_ℓ**), then

$$H = \{\mathrm{diag}[x_1, \ldots, x_l, (x_1 \cdots x_l)^{-1}] : x_i \in \mathbb{C}^\times\},$$
$$\mathrm{Lie}(H) = \{\mathrm{diag}[a_1, \ldots, a_{l+1}] : a_i \in \mathbb{C}, \quad \textstyle\sum_i a_i = 0\}.$$

2. When $G = \mathbf{Sp}(\mathbb{C}^{2l}, \Omega)$ (we say that G is of **type C_ℓ**) or $G = \mathbf{SO}(\mathbb{C}^{2l}, B)$ (we say that G is of **type D_ℓ**), then by (2.7) and (2.8),

$$H = \{\mathrm{diag}[x_1, \ldots, x_l, x_l^{-1}, \ldots, x_1^{-1}] : x_i \in \mathbb{C}^\times\},$$
$$\mathfrak{h} = \{\mathrm{diag}[a_1, \ldots, a_l, -a_l, \ldots, -a_1] : a_i \in \mathbb{C}\}.$$

3. When $G = \mathbf{SO}(\mathbb{C}^{2l+1}, B)$ (we say that G is of **type B_ℓ**), then by (2.10),

$$H = \{\mathrm{diag}[x_1, \ldots, x_l, 1, x_l^{-1}, \ldots, x_1^{-1}] : x_i \in \mathbb{C}^\times\},$$
$$\mathfrak{h} = \{\mathrm{diag}[a_1, \ldots, a_l, 0, -a_l, \ldots, -a_1] : a_i \in \mathbb{C}\}.$$

In all cases H is isomorphic as an algebraic group to the product of l copies of $\mathbb{C}^\times$, so it is a torus of rank l. The Lie algebra $\mathfrak{h}$ is isomorphic to the vector space $\mathbb{C}^l$ with all Lie brackets zero. Define coordinate functions $x_1, \ldots, x_l$ on H as above. Then $\mathcal{O}[H] = \mathbb{C}[x_1, \ldots, x_l, x_1^{-1}, \ldots, x_l^{-1}]$.

Theorem 2.1.5. *Let G be $\mathbf{GL}(n, \mathbb{C})$, $\mathbf{SL}(n, \mathbb{C})$, $\mathbf{SO}(\mathbb{C}^n, B)$ or $\mathbf{Sp}(\mathbb{C}^{2l}, \Omega)$ in the form given above, where H is the diagonal subgroup in G. Suppose $g \in G$ and $gh = hg$ for all $h \in H$. Then $g \in H$.*

Proof. We have $G \subset \mathbf{GL}(n, \mathbb{C})$. An element $h \in H$ acts on the standard basis $\{e_1, \ldots, e_n\}$ for $\mathbb{C}^n$ by $he_i = \theta_i(h)e_i$. Here the characters θ_i are given as follows in terms of the coordinate functions $x_1, \ldots, x_l$ on H:

1. $G = \mathbf{GL}(l, \mathbb{C})$: $\theta_i = x_i$ for $i = 1, \ldots, l$.

2. $G = \mathbf{SL}(l+1, \mathbb{C})$: $\theta_i = x_i$ for $i = 1, \ldots, l$ and $\theta_{l+1} = (x_1 \cdots x_l)^{-1}$.

3. $G = \mathbf{SO}(\mathbb{C}^{2l}, B)$ or $\mathbf{Sp}(\mathbb{C}^{2l}, \Omega)$: $\theta_i = x_i$ and $\theta_{2l+1-i} = x_i^{-1}$ for $i = 1, \ldots, l$.

4. $G = \mathbf{SO}(\mathbb{C}^{2l+1}, B)$: $\theta_i = x_i$, $\theta_{2l+2-i} = x_i^{-1}$ for $i = 1, \ldots, l$, and $\theta_{l+1} = 1$.

Since the characters $\theta_1, \ldots, \theta_n$ are all distinct, the weight space decomposition (2.2) of $\mathbb{C}^n$ under H is given by the one-dimensional subspaces $\mathbb{C}e_i$. If $gh = hg$ for all $h \in H$, then g preserves the weight spaces and hence is a diagonal matrix. $\square$

Corollary 2.1.6. *Let G and H be as in Theorem* 2.1.5. *Suppose $T \subset G$ is an abelian subgroup (not assumed to be algebraic). If $H \subset T$ then $H = T$. In particular, H is a maximal torus in G.*

The choice of the maximal torus H depended on choosing a particular matrix form of G. We shall prove that if T is any maximal torus in G then there exists an element $\gamma \in G$ such that $T = \gamma H \gamma^{-1}$. We begin by conjugating individual semisimple elements into H.

Theorem 2.1.7. (Notation as in Theorem 2.1.5) *If $g \in G$ is semisimple then there exists $\gamma \in G$ such that $\gamma g \gamma^{-1} \in H$.*

Proof. When G is $\mathbf{GL}(n, \mathbb{C})$ or $\mathbf{SL}(n, \mathbb{C})$, let $\{v_1, \ldots, v_n\}$ be a basis of eigenvectors for g and define $\gamma v_i = e_i$, where $\{e_i\}$ is the standard basis for $\mathbb{C}^n$. Multiplying v_1 by a suitable constant, we can arrange that $\det \gamma = 1$. Then $\gamma \in G$ and $\gamma g \gamma^{-1} \in H$.

If $g \in \mathbf{SL}(n, \mathbb{C})$ is semisimple and preserves a nondegenerate bilinear form ω on $\mathbb{C}^n$, then there is an eigenspace decomposition

$$\mathbb{C}^n = \bigoplus V_\lambda, \quad gv = \lambda v \quad \text{for } v \in V_\lambda . \tag{2.11}$$

Furthermore, $\omega(u, v) = \omega(gu, gv) = \lambda \mu \, \omega(u, v)$ for $u \in V_\lambda$ and $v \in V_\mu$. Hence

$$\omega(V_\lambda, V_\mu) = 0 \quad \text{if } \lambda \mu \neq 1 . \tag{2.12}$$

Since ω is nondegenerate, it follows from (2.11) and (2.12) that

$$\dim V_{1/\mu} = \dim V_\mu . \tag{2.13}$$

Let $\mu_1, \ldots, \mu_k$ be the (distinct) eigenvalues of g that are not ± 1. From (2.13) we see that $k = 2r$ is even and that we can take $\mu_i^{-1} = \mu_{r+i}$ for $i = 1, \ldots, r$.

Recall that a subspace $W \subset \mathbb{C}^n$ is ω *isotropic* if $\omega(u, v) = 0$ for all $u, v \in W$ (see Appendix B.2.1). By (2.12) the subspaces V_{μ_i} and V_{1/μ_i} are ω isotropic and the restriction of ω to $V_{\mu_i} \times V_{1/\mu_i}$ is nondegenerate. Let $W_i = V_{\mu_i} \oplus V_{1/\mu_i}$ for $i = 1, \ldots, r$. Then

(a) the subspaces V_1, V_{-1}, and W_i are mutually orthogonal relative to the form ω, and the restriction of ω to each of these subspaces is nondegenerate;

(b) $\mathbb{C}^n = V_1 \oplus V_{-1} \oplus W_1 \oplus \cdots \oplus W_r$;

(c) $\det g = (-1)^k$, where $k = \dim V_{-1}$.

Now suppose $\omega = \Omega$ is the skew-symmetric form (2.6) and $g \in \mathbf{Sp}(\mathbb{C}^{2l}, \Omega)$. From (a) we see that $\dim V_1$ and $\dim V_{-1}$ are even. By Lemma 1.1.5 we can find canonical symplectic bases in each of the subspaces in decomposition (b); in the case of W_i we may take a basis $v_1, \ldots, v_s$ for V_{μ_i} and an Ω-dual basis $v_{-1}, \ldots, v_{-s}$ for V_{1/μ_i}. Altogether, these bases give a canonical symplectic basis for $\mathbb{C}^{2l}$. We may enumerate it as $v_1, \ldots, v_l, v_{-1}, \ldots, v_{-l}$, so that

$$gv_i = \lambda_i v_i, \quad gv_{-i} = \lambda_i^{-1} v_{-i} \quad \text{for } i = 1, \ldots, l.$$

The linear transformation γ such that $\gamma v_i = e_i$ and $\gamma v_{-i} = e_{2l+1-i}$ for $i = 1, \ldots, l$ is in G due to the choice (2.6) of the matrix for Ω. Furthermore,

$$\gamma g \gamma^{-1} = \mathrm{diag}[\lambda_1, \ldots, \lambda_l, \lambda_l^{-1}, \ldots, \lambda_1^{-1}] \in H.$$

This proves the theorem in the symplectic case.

Now assume that G is the orthogonal group for the form B in (2.6) or (2.9). Since $\det g = 1$, we see from (c) that $\dim V_{-1} = 2q$ is even, and by (2.13) $\dim W_i$ is even. Hence n is odd if and only if $\dim V_1$ is odd. Just as in the symplectic case, we construct canonical B-isotropic bases in each of the subspaces in decomposition (b) (see Section B.2.1); the union of these bases gives an isotropic basis for $\mathbb{C}^n$. When $n = 2l$ and $\dim V_1 = 2r$ we can enumerate this basis so that

$$gv_i = \lambda_i v_i, \quad gv_{-i} = \lambda_i^{-1} v_{-i} \quad \text{for } i = 1, \ldots, l.$$

The linear transformation γ such that $\gamma v_i = e_i$ and $\gamma v_{-i} = e_{n+1-i}$ is in $\mathbf{O}(\mathbb{C}^n, B)$, and we can interchange v_l and v_{-l} if necessary to get $\det \gamma = 1$. Then

$$\gamma g \gamma^{-1} = \mathrm{diag}[\lambda_1, \ldots, \lambda_l, \lambda_l^{-1}, \ldots, \lambda_1^{-1}] \in H.$$

When $n = 2l + 1$ we know that $\lambda = 1$ occurs as an eigenvalue of g, so we can enumerate this basis so that

$$gv_0 = v_0, \quad gv_i = \lambda_i v_i, \quad gv_{-i} = \lambda_i^{-1} v_{-i} \quad \text{for } i = 1, \ldots, l.$$

The linear transformation γ such that $\gamma v_0 = e_{l+1}$, $\gamma v_i = e_i$, and $\gamma v_{-i} = e_{n+1-i}$ is in $\mathbf{O}(\mathbb{C}^n, B)$. Replacing γ by $-\gamma$ if necessary, we have $\gamma \in \mathbf{SO}(\mathbb{C}^n, B)$ and

$$\gamma g \gamma^{-1} = \mathrm{diag}[\lambda_1, \ldots, \lambda_l, 1, \lambda_l^{-1}, \ldots, \lambda_1^{-1}] \in H.$$

This completes the proof of the theorem. $\qquad\square$

Corollary 2.1.8. *If T is any torus in G, then there exists $\gamma \in G$ such that $\gamma T \gamma^{-1} \subset H$. In particular, if T is a maximal torus in G, then $\gamma T \gamma^{-1} = H$.*

Proof. Choose $t \in T$ satisfying the condition of Lemma 2.1.4. By Theorem 2.1.7 there exists $\gamma \in G$ such that $\gamma t \gamma^{-1} \in H$. We want to show that $\gamma x \gamma^{-1} \in H$ for all $x \in T$. To prove this, take any function $\varphi \in \mathcal{I}_H$ and define a regular function f on T by $f(x) = \varphi(\gamma x \gamma^{-1})$. Then $f(t^p) = 0$ for all $p \in \mathbb{Z}$, since $\gamma t^p \gamma^{-1} \in H$. Hence

Lemma 2.1.4 implies that $f(x) = 0$ for all $x \in T$. Since φ was any function in $\mathfrak{I}_H$, we conclude that $\gamma x \gamma^{-1} \in H$. If T is a maximal torus then so is $\gamma T \gamma^{-1}$. Hence $\gamma T \gamma^{-1} = H$ in this case. $\qquad\qquad\square$

From Corollary 2.1.8, we see that the integer $l = \dim H$ does not depend on the choice of a particular maximal torus in G. We call l the *rank* of G.

2.1.3 Exercises

1. Verify that the Lie algebras of the orthogonal and symplectic groups are given in the matrix forms (2.7), (2.8), and (2.10).
2. Let $V_r(y)$ be the Vandermonde matrix, as in Section 2.1.2. Prove that

$$\det V_r(y) = \prod_{1 \le i < j \le r} (y_i - y_j).$$

(HINT: Fix $y_2, \ldots, y_r$ and consider $\det V_r(y)$ as a polynomial in y_1. Show that it has degree $r - 1$ with roots $y_2, \ldots, y_r$ and that the coefficient of y_1^{r-1} is the Vandermonde determinant for $y_2, \ldots, y_r$. Now use induction on r.)
3. Let H be a torus of rank n. Let $\mathcal{X}_*(H)$ be the set of all regular homomorphisms from $\mathbb{C}^\times$ into H. Define a group structure on $\mathcal{X}_*(H)$ by pointwise multiplication: $(\pi_1 \pi_2)(z) = \pi_1(z)\pi_2(z)$ for $\pi_1, \pi_2 \in \mathcal{X}_*(H)$.
   (a) Prove that $\mathcal{X}_*(H)$ is isomorphic to $\mathbb{Z}^n$ as an abstract group. (HINT: Use Lemma 1.6.4.)
   (b) Prove that if $\pi \in \mathcal{X}_*(H)$ and $\chi \in \mathcal{X}(H)$ then there is an integer $\langle \pi, \chi \rangle \in \mathbb{Z}$ such that
$$\chi(\pi(z)) = z^{\langle \pi, \chi \rangle} \qquad \text{for all } z \in \mathbb{C}^\times.$$
   (c) Show that the pairing $\pi, \chi \mapsto \langle \pi, \chi \rangle$ is additive in each variable (relative to the abelian group structures on $\mathcal{X}(H)$ and $\mathcal{X}_*(H)$) and is *nondegenerate* (this means that if $\langle \pi, \chi \rangle = 0$ for all χ then $\pi = 1$, and similarly for χ).
4. Let $G \subset \mathbf{GL}(n, \mathbb{C})$ be a classical group with Lie algebra $\mathfrak{g} \subset \mathfrak{gl}(n, \mathbb{C})$ (for the orthogonal and symplectic groups use the bilinear forms (2.6) and (2.9)). Define $\theta(g) = (g^t)^{-1}$ for $g \in G$.
   (a) Show that θ is a regular automorphism of G and that $d\theta(X) = -X^t$ for $X \in \mathfrak{g}$.
   (b) Define $K = \{g \in G : \theta(g) = g\}$ and let $\mathfrak{k}$ be the Lie algebra of K. Show that $\mathfrak{k} = \{X \in \mathfrak{g} : d\theta(X) = X\}$.
   (c) Define $\mathfrak{p} = \{X \in \mathfrak{g} : d\theta(X) = -X\}$. Show that $\mathrm{Ad}(K)\mathfrak{p} \subset \mathfrak{p}$, $\mathfrak{g} = \mathfrak{k} \oplus \mathfrak{p}$, $[\mathfrak{k}, \mathfrak{p}] \subset \mathfrak{p}$, and $[\mathfrak{p}, \mathfrak{p}] \subset \mathfrak{k}$. (HINT: $d\theta$ is a derivation of $\mathfrak{g}$ and has eigenvalues ± 1.)
   (d) Determine the explicit matrix form of $\mathfrak{k}$ and $\mathfrak{p}$ when $G = \mathbf{Sp}(\mathbb{C}^{2l}, \Omega)$, with Ω given by (2.6). Show that $\mathfrak{k}$ is isomorphic to $\mathfrak{gl}(l, \mathbb{C})$ in this case. (HINT: Write $X \in \mathfrak{g}$ in block form $\begin{bmatrix} A & B \\ C & D \end{bmatrix}$ and show that the map $X \mapsto A + iBs_l$ gives a Lie algebra isomorphism from $\mathfrak{k}$ to $\mathfrak{gl}(l, \mathbb{C})$.)

2.2 Unipotent Elements

Unipotent elements give an algebraic relation between a linear algebraic group and its Lie algebra, since they are exponentials of nilpotent elements and the exponential map is a polynomial function on nilpotent matrices. In this section we exploit this property to prove the connectedness of the classical groups.

2.2.1 Low-Rank Examples

We shall show that the classical groups $\mathbf{SL}(n,\mathbb{C})$, $\mathbf{SO}(n,\mathbb{C})$, and $\mathbf{Sp}(n,\mathbb{C})$ are generated by their unipotent elements. We begin with the basic case $G = \mathbf{SL}(2,\mathbb{C})$. Let $N^+ = \{u(z) : z \in \mathbb{C}\}$ and $N^- = \{v(z) : z \in \mathbb{C}\}$, where

$$u(z) = \begin{bmatrix} 1 & z \\ 0 & 1 \end{bmatrix} \quad \text{and} \quad v(z) = \begin{bmatrix} 1 & 0 \\ z & 1 \end{bmatrix}.$$

The groups N^+ and N^- are isomorphic to the additive group of the field $\mathbb{C}$.

Lemma 2.2.1. *The group* $\mathbf{SL}(2,\mathbb{C})$ *is generated by* $N^+ \cup N^-$.

Proof. Let $g = \begin{bmatrix} a & b \\ c & d \end{bmatrix}$ with $ad - bc = 1$. If $a \neq 0$ we can use elementary row and column operations to factor

$$g = \begin{bmatrix} 1 & 0 \\ a^{-1}c & 1 \end{bmatrix} \begin{bmatrix} a & 0 \\ 0 & a^{-1} \end{bmatrix} \begin{bmatrix} 1 & a^{-1}b \\ 0 & 1 \end{bmatrix}.$$

If $a = 0$ then $c \neq 0$ and we can likewise factor

$$g = \begin{bmatrix} 0 & -1 \\ 1 & 0 \end{bmatrix} \begin{bmatrix} c & 0 \\ 0 & c^{-1} \end{bmatrix} \begin{bmatrix} 1 & c^{-1}d \\ 0 & 1 \end{bmatrix}.$$

Finally, we factor

$$\begin{bmatrix} 0 & -1 \\ 1 & 0 \end{bmatrix} = \begin{bmatrix} 1 & -1 \\ 0 & 1 \end{bmatrix} \begin{bmatrix} 1 & 0 \\ 1 & 1 \end{bmatrix} \begin{bmatrix} 1 & -1 \\ 0 & 1 \end{bmatrix},$$

$$\begin{bmatrix} a & 0 \\ 0 & a^{-1} \end{bmatrix} = \begin{bmatrix} 1 & -a \\ 0 & 1 \end{bmatrix} \begin{bmatrix} 1 & 0 \\ (a^{-1} - 1) & 1 \end{bmatrix} \begin{bmatrix} 1 & 1 \\ 0 & 1 \end{bmatrix} \begin{bmatrix} 1 & 0 \\ (a-1) & 1 \end{bmatrix},$$

to complete the proof.   □

The orthogonal and symplectic groups of low rank are closely related to $\mathbf{GL}(1,\mathbb{C})$ and $\mathbf{SL}(2,\mathbb{C})$, as follows. Define a skew-symmetric bilinear form Ω on $\mathbb{C}^2$ by

$$\Omega(v,w) = \det[v, w],$$

where $[v, w] \in M_2(\mathbb{C})$ has columns v, w. We have $\det[e_1, e_1] = \det[e_2, e_2] = 0$ and $\det[e_1, e_2] = 1$, showing that the form Ω is nondegenerate. Since the determinant function is multiplicative, the form Ω satisfies

$$\Omega(gv, gw) = (\det g)\Omega(v, w) \quad \text{for } g \in \mathbf{GL}(2, \mathbb{C}) .$$

Hence g preserves Ω if and only if $\det g = 1$. This proves that $\mathbf{Sp}(\mathbb{C}^2, \Omega) = \mathbf{SL}(2, \mathbb{C})$.

Next, consider the group $\mathbf{SO}(\mathbb{C}^2, B)$ with B the bilinear form with matrix s_2 in (2.5). We calculate that

$$g^t s_2 g = \begin{bmatrix} 2ac & ad+bc \\ ad+bc & 2bd \end{bmatrix} \quad \text{for} \quad g = \begin{bmatrix} a & b \\ c & d \end{bmatrix} \in \mathbf{SL}(2, \mathbb{C}) .$$

Since $ad - bc = 1$, it follows that $ad + bc = 2ad - 1$. Hence $g^t s_2 g = s_2$ if and only if $ad = 1$ and $b = c = 0$. Thus $\mathbf{SO}(\mathbb{C}^2, B)$ consists of the matrices

$$\begin{bmatrix} a & 0 \\ 0 & a^{-1} \end{bmatrix} \quad \text{for} \quad a \in \mathbb{C}^\times.$$

This furnishes an isomorphism $\mathbf{SO}(\mathbb{C}^2, B) \cong \mathbf{GL}(1, \mathbb{C})$.

Now consider the group $G = \mathbf{SO}(\mathbb{C}^3, B)$, where B is the bilinear form on $\mathbb{C}^3$ with matrix s_3 as in (2.5). From Section 2.1.2 we know that the subgroup

$$H = \{\mathrm{diag}[x, 1, x^{-1}] : x \in \mathbb{C}^\times\}$$

of diagonal matrices in G is a maximal torus. Set $\widetilde{G} = \mathbf{SL}(2, \mathbb{C})$ and let

$$\widetilde{H} = \{\mathrm{diag}[x, x^{-1}] : x \in \mathbb{C}^\times\}$$

be the subgroup of diagonal matrices in $\widetilde{G}$.

We now define a homomorphism $\rho : \widetilde{G} \longrightarrow G$ that maps $\widetilde{H}$ onto H. Set

$$V = \{X \in M_2(\mathbb{C}) : \mathrm{tr}(X) = 0\}$$

and let $\widetilde{G}$ act on V by $\rho(g)X = gXg^{-1}$ (this is the adjoint representation of $\widetilde{G}$). The symmetric bilinear form

$$\omega(X, Y) = \frac{1}{2}\mathrm{tr}(XY)$$

is obviously invariant under $\rho(\widetilde{G})$, since $\mathrm{tr}(XY) = \mathrm{tr}(YX)$ for all $X, Y \in M_n(\mathbb{C})$. The basis

$$v_0 = \begin{bmatrix} 1 & 0 \\ 0 & -1 \end{bmatrix}, \quad v_1 = \begin{bmatrix} 0 & \sqrt{2} \\ 0 & 0 \end{bmatrix}, \quad v_{-1} = \begin{bmatrix} 0 & 0 \\ \sqrt{2} & 0 \end{bmatrix}$$

for V is ω isotropic. We identify V with $\mathbb{C}^3$ via the map $v_1 \mapsto e_1, v_0 \mapsto e_2$, and $v_{-1} \mapsto e_3$. Then ω becomes B. From Corollary 1.6.3 we know that any element of the subgroup N^+ or N^- in Lemma 2.2.1 is carried by the homomorphism ρ to a unipotent matrix. Hence by Lemma 2.2.1 we conclude that $\det(\rho(g)) = 1$ for all $g \in \widetilde{G}$. Hence $\rho(\widetilde{G}) \subset G$ by Lemma 2.2.1. If $h = \mathrm{diag}[x, x^{-1}] \in \widetilde{H}$, then $\rho(h)$ has the matrix $\mathrm{diag}[x^2, 1, x^{-2}]$, relative to the ordered basis $\{v_1, v_0, v_{-1}\}$ for V. Thus $\rho(\widetilde{H}) = H$.

Finally, we consider $G = \mathbf{SO}(\mathbb{C}^4, B)$, where B is the symmetric bilinear form on $\mathbb{C}^4$ with matrix s_4 as in (2.5). From Section 2.1.2 we know that the subgroup

$$H = \{\mathrm{diag}[x_1, x_2, x_2^{-1}, x_1^{-1}] : x_1, x_2 \in \mathbb{C}^\times\}$$

of diagonal matrices in G is a maximal torus. Set $\widetilde{G} = \mathbf{SL}(2, \mathbb{C}) \times \mathbf{SL}(2, \mathbb{C})$ and let $\widetilde{H}$ be the product of the diagonal subgroups of the factors of $\widetilde{G}$. We now define a homomorphism $\pi : \widetilde{G} \longrightarrow G$ that maps $\widetilde{H}$ onto H, as follows. Set $V = M_2(\mathbb{C})$ and let $\widetilde{G}$ act on V by $\pi(a, b)X = aXb^{-1}$. From the quadratic form $Q(X) = 2\det X$ on V we obtain the symmetric bilinear form $\beta(X, Y) = \det(X + Y) - \det X - \det Y$. Set

$$v_1 = e_{11}, \quad v_2 = e_{12}, \quad v_3 = -e_{21}, \text{ and } \quad v_4 = e_{22}.$$

Clearly $\beta(\pi(g)X, \pi(g)Y) = \beta(X, Y)$ for $g \in \widetilde{G}$. The vectors v_j are β-isotropic and $\beta(v_1, v_4) = \beta(v_2, v_3) = 1$. If we identify V with $\mathbb{C}^4$ via the basis $\{v_1, v_2, v_3, v_4\}$, then β becomes the form B.

Let $g \in \widetilde{G}$ be of the form (I, b) or (b, I), where b is either in the subgroup N^+ or in the subgroup N^- of Lemma 2.2.1. From Corollary 1.6.3 we know that $\pi(g)$ is a unipotent matrix, and so from Lemma 2.2.1 we conclude that $\det(\pi(g)) = 1$ for all $g \in \widetilde{G}$. Hence $\pi(\widetilde{G}) \subset \mathbf{SO}(\mathbb{C}^4, B)$. Given $h = (\mathrm{diag}[x_1, x_1^{-1}], \mathrm{diag}[x_2, x_2^{-1}]) \in \widetilde{H}$, we have

$$\pi(h) = \mathrm{diag}[x_1 x_2^{-1}, x_1 x_2, x_1^{-1} x_2^{-1}, x_1^{-1} x_2].$$

Since the map $(x_1, x_2) \mapsto (x_1 x_2^{-1}, x_1 x_2)$ from $(\mathbb{C}^\times)^2$ to $(\mathbb{C}^\times)^2$ is surjective, we have shown that $\pi(\widetilde{H}) = H$.

2.2.2 Unipotent Generation of Classical Groups

The differential of a regular representation of an algebraic group G gives a representation of $\mathrm{Lie}(G)$. On the nilpotent elements in $\mathrm{Lie}(G)$ the exponential map is algebraic and maps them to unipotent elements in G. This gives an algebraic link from Lie algebra representations to group representations, provided the unipotent elements generate G. We now prove that this is the case for the following families of classical groups.

Theorem 2.2.2. *Suppse that G is $\mathbf{SL}(l+1, \mathbb{C})$, $\mathbf{SO}(2l+1, \mathbb{C})$, or $\mathbf{Sp}(l, \mathbb{C})$ with $l \geq 1$, or that G is $\mathbf{SO}(2l, \mathbb{C})$ with $l \geq 2$. Then G is generated by its unipotent elements.*

Proof. We have $G \subset \mathbf{GL}(n,\mathbb{C})$ (where $n = l+1, 2l$, or $2l+1$). Let G' be the subgroup generated by the unipotent elements of G. Since the conjugate of a unipotent element is unipotent, we see that G' is a normal subgroup of G. In the orthogonal or symplectic case we take the matrix form of G as in Theorem 2.1.5 so that the subgroup H of diagonal matrices is a maximal torus in G. To prove the theorem, it suffices by Theorems 1.6.5 and 2.1.7 to show that $H \subset G'$.

Type A: When $G = \mathbf{SL}(2,\mathbb{C})$, we have $G' = G$ by Lemma 2.2.1. For $G = \mathbf{SL}(n,\mathbb{C})$ with $n \geq 3$ and $h = \mathrm{diag}[x_1,\ldots,x_n] \in H$ we factor $h = h'h''$, where

$$h' = \mathrm{diag}[x_1, x_1^{-1}, 1, \ldots, 1], \quad h'' = \mathrm{diag}[1, x_1 x_2, x_3, \ldots, x_n].$$

Let $G_1 \cong \mathbf{SL}(2,\mathbb{C})$ be the subgroup of matrices in block form $\mathrm{diag}[a, I_{n-2}]$ with $a \in \mathbf{SL}(2,\mathbb{C})$, and let $G_2 \cong \mathbf{SL}(n-1,\mathbb{C})$ be the subgroup of matrices in block form $\mathrm{diag}[1, b]$ with $b \in \mathbf{SL}(n-1,\mathbb{C})$. Then $h' \in G_1$ and $h'' \in G_2$. By induction on n, we may assume that h' and h'' are products of unipotent elements. Hence h is also, so we conclude that $G = G'$.

Type C: Let Ω be the symplectic form (2.6). From Section 2.2.1 we know that $\mathbf{Sp}(\mathbb{C}^2, \Omega) = \mathbf{SL}(2,\mathbb{C})$. Hence from Lemma 2.2.1 we conclude that $\mathbf{Sp}(\mathbb{C}^2, \Omega)$ is generated by its unipotent elements. For $G = \mathbf{Sp}(\mathbb{C}^{2l}, \Omega)$ with $l > 1$ and $h = \mathrm{diag}[x_1,\ldots,x_l,x_1^{-1},\ldots,x_1^{-1}] \in H$, we factor $h = h'h''$, where

$$h' = \mathrm{diag}[x_1, 1, \ldots, 1, x_1^{-1}], \quad h'' = \mathrm{diag}[1, x_2, \ldots, x_l, x_l^{-1}, \ldots, x_2^{-1}, 1].$$

We split $\mathbb{C}^{2l} = V_1 \oplus V_2$, where $V_1 = \mathrm{Span}\{e_1, e_{2l}\}$ and $V_2 = \mathrm{Span}\{e_2,\ldots,e_{2l-1}\}$. The restrictions of the symplectic form Ω to V_1 and to V_2 are nondegenerate. Define

$$G_1 = \{g \in G : gV_1 = V_1 \text{ and } g = I \text{ on } V_2\},$$
$$G_2 = \{g \in G : g = I \text{ on } V_1 \text{ and } gV_2 = V_2\}.$$

Then $G_1 \cong \mathbf{Sp}(1,\mathbb{C})$, while $G_2 \cong \mathbf{Sp}(l-1,\mathbb{C})$, and we have $h' \in G_1$ and $h'' \in G_2$. By induction on l, we may assume that h' and h'' are products of unipotent elements. Hence h is also, so we conclude that $G = G'$.

Types B and D: Let B be the symmetric form (2.9) on $\mathbb{C}^n$. Suppose first that $G = \mathbf{SO}(\mathbb{C}^3, B)$. Let $\widetilde{G} = \mathbf{SL}(2,\mathbb{C})$. In Section 2.2.1 we constructed a regular homomorphism $\rho : \widetilde{G} \longrightarrow \mathbf{SO}(\mathbb{C}^3, B)$ that maps the diagonal subgroup $\widetilde{H} \subset \widetilde{G}$ onto the diagonal subgroup $H \subset G$. Since every element of $\widetilde{H}$ is a product of unipotent elements, the same is true for H. Hence $G = \mathbf{SO}(3,\mathbb{C})$ is generated by its unipotent elements.

Now let $G = \mathbf{SO}(\mathbb{C}^4, B)$ and set $\widetilde{G} = \mathbf{SL}(2,\mathbb{C}) \times \mathbf{SL}(2,\mathbb{C})$. Let H be the diagonal subgroup of G and let $\widetilde{H}$ be the product of the diagonal subgroups of the factors of $\widetilde{G}$. In Section 2.2.1 we constructed a regular homomorphism $\pi : \widetilde{G} \longrightarrow \mathbf{SO}(\mathbb{C}^4, B)$ that maps $\widetilde{H}$ onto H. Hence the argument just given for $\mathbf{SO}(3,\mathbb{C})$ applies in this case, and we conclude that $\mathbf{SO}(4,\mathbb{C})$ is generated by its unipotent elements.

Finally, we consider the groups $G = \mathbf{SO}(\mathbb{C}^n, B)$ with $n \geq 5$. Embed $\mathbf{SO}(\mathbb{C}^{2l}, B)$ into $\mathbf{SO}(\mathbb{C}^{2l+1}, B)$ by the regular homomorphism

$$\begin{bmatrix} a & b \\ c & d \end{bmatrix} \mapsto \begin{bmatrix} a & 0 & b \\ 0 & 1 & 0 \\ c & 0 & d \end{bmatrix}. \tag{2.14}$$

The diagonal subgroup of $\mathbf{SO}(\mathbb{C}^{2l}, B)$ is isomorphic to the diagonal subgroup of $\mathbf{SO}(\mathbb{C}^{2l+1}, B)$ via this embedding, so it suffices to prove that every diagonal element in $\mathbf{SO}(\mathbb{C}^n, B)$ is a product of unipotent elements when n is even. We just proved this to be the case when $n = 4$, so we may assume $n = 2l \geq 6$. For

$$h = \mathrm{diag}[x_1, \ldots, x_l, x_l^{-1}, \ldots, x_1^{-1}] \in H$$

we factor $h = h'h''$, where

$$h' = \mathrm{diag}[x_1, x_2, 1, \ldots, 1, x_2^{-1}, x_1^{-1}],$$
$$h'' = \mathrm{diag}[1, 1, x_3, \ldots, x_l, x_l^{-1}, \ldots, x_3^{-1}, 1, 1].$$

We split $\mathbb{C}^n = V_1 \oplus V_2$, where

$$V_1 = \mathrm{Span}\{e_1, e_2, e_{n-1}, e_n\}, \qquad V_2 = \mathrm{Span}\{e_3, \ldots, e_{n-2}\}.$$

The restriction of the symmetric form B to V_i is nondegenerate. If we set

$$G_1 = \{g \in G : gV_1 = V_1 \text{ and } g = I \text{ on } V_2\},$$

then $h \in G_1 \cong \mathbf{SO}(4, \mathbb{C})$. Let $W_1 = \mathrm{Span}\{e_1, e_n\}$ and $W_2 = \mathrm{Span}\{e_2, \ldots, e_{n-1}\}$. Set

$$G_2 = \{g \in G : g = I \text{ on } W_1 \text{ and } gW_2 = W_2\}.$$

We have $G_2 \cong \mathbf{SO}(2l - 2, \mathbb{C})$ and $h'' \in G_2$. Since $2l - 2 \geq 4$, we may assume by induction that h' and h'' are products of unipotent elements. Hence h is also a product of unipotent elements, proving that $G = G'$. $\qquad\square$

2.2.3 Connected Groups

Definition 2.2.3. A linear algebraic group G is *connected* (in the sense of algebraic groups) if the ring $\mathcal{O}[G]$ has no zero divisors.

Examples

1. The rings $\mathbb{C}[t]$ and $\mathbb{C}[t, t^{-1}]$ obviously have no zero divisors; hence the additive group $\mathbb{C}$ and the multiplicative group $\mathbb{C}^\times$ are connected. Likewise, the torus D_n of diagonal matrices and the group N_n^+ of upper-triangular unipotent matrices are connected (see Examples 1 and 2 of Section 1.4.2).

2. If G and H are connected linear algebraic groups, then the group $G \times H$ is connected, since $\mathcal{O}[G \times H] \cong \mathcal{O}[G] \otimes \mathcal{O}[H]$.

3. If G is a connected linear algebraic group and there is a surjective regular homomorphism $\rho : G \longrightarrow H$, then H is connected, since $\rho^* : \mathcal{O}[H] \longrightarrow \mathcal{O}[G]$ is injective.

Theorem 2.2.4. *Let G be a linear algebraic group that is generated by unipotent elements. Then G is connected as an algebraic group and as a Lie group.*

Proof. Suppose $f_1, f_2 \in \mathcal{O}[G]$, $f_1 \neq 0$, and $f_1 f_2 = 0$. We must show that $f_2 = 0$. Translating f_1 and f_2 by an element of G if necessary, we may assume that $f_1(I) \neq 0$. Let $g \in G$. Since g is a product of unipotent elements, Theorem 1.6.2 implies that there exist nilpotent elements $X_1, \ldots, X_r$ in $\mathfrak{g}$ such that $g = \exp(X_1) \cdots \exp(X_r)$. Define $\varphi(t) = \exp(tX_1) \cdots \exp(tX_r)$ for $t \in \mathbb{C}$. The entries in the matrix $\varphi(t)$ are polynomials in t, and $\varphi(1) = g$. Since X_j is nilpotent, we have $\det(\varphi(t)) = 1$ for all t. Hence the functions $p_1(t) = f_1(\varphi(t))$ and $p_2(t) = f_2(\varphi(t))$ are polynomials in t. Since $p_1(0) \neq 0$ while $p_1(t)p_2(t) = 0$ for all t, it follows that $p_2(t) = 0$ for all t. In particular, $f_2(g) = 0$. This holds for all $g \in G$, so $f_2 = 0$, proving that G is connected as a linear algebraic group. This argument also shows that G is arcwise connected, and hence connected, as a Lie group.   □

Theorem 2.2.5. *The groups $\mathbf{GL}(n, \mathbb{C})$, $\mathbf{SL}(n, \mathbb{C})$, $\mathbf{SO}(n, \mathbb{C})$, and $\mathbf{Sp}(n, \mathbb{C})$ are connected (as linear algebraic groups and Lie groups) for all $n \geq 1$.*

Proof. The homomorphism $\lambda, g \mapsto \lambda g$ from $\mathbb{C}^\times \times \mathbf{SL}(n, \mathbb{C})$ to $\mathbf{GL}(n, \mathbb{C})$ is surjective. Hence the connectedness of $\mathbf{GL}(n, \mathbb{C})$ will follow from the connectedness of $\mathbb{C}^\times$ and $\mathbf{SL}(n, \mathbb{C})$, as in Examples 2 and 3 above. The groups $\mathbf{SL}(1, \mathbb{C})$ and $\mathbf{SO}(1, \mathbb{C})$ are trivial, and we showed in Section 2.2.1 that $\mathbf{SO}(2, \mathbb{C})$ is isomorphic to $\mathbb{C}^\times$, hence connected. For the remaining cases use Theorems 2.2.2 and 2.2.4.   □

Remark 2.2.6. The regular homomorphisms $\rho : \mathbf{SL}(2, \mathbb{C}) \longrightarrow \mathbf{SO}(3, \mathbb{C})$ and $\pi : \mathbf{SL}(2, \mathbb{C}) \times \mathbf{SL}(2, \mathbb{C}) \longrightarrow \mathbf{SO}(4, \mathbb{C})$ constructed in Section 2.2.1 have kernels $\pm I$; hence $d\rho$ and $d\pi$ are bijective by dimensional considerations. Since $\mathbf{SO}(n, \mathbb{C})$ is connected, it follows that these homomorphisms are surjective. After we introduce the spin groups in Chapter 6, we will see that $\mathbf{SL}(2, \mathbb{C}) \cong \mathbf{Spin}(3, \mathbb{C})$ and $\mathbf{SL}(2, \mathbb{C}) \times \mathbf{SL}(2, \mathbb{C}) \cong \mathbf{Spin}(4, \mathbb{C})$.

We shall study regular representations of a linear algebraic group in terms of the associated representations of its Lie algebra. The following theorem will be a basic tool.

Theorem 2.2.7. *Suppose G is a linear algebraic group with Lie algebra $\mathfrak{g}$. Let (π, V) be a regular representation of G and $W \subset V$ a subspace.*

1. If $\pi(g)W \subset W$ for all $g \in G$ then $d\pi(A)W \subset W$ for all $A \in \mathfrak{g}$.
2. Assume that G is generated by unipotent elements. If $d\pi(X)W \subset W$ for all $X \in \mathfrak{g}$ then $\pi(g)W \subset W$ for all $g \in G$. Hence V is irreducible under the action of G if and only if it is irreducible under the action of $\mathfrak{g}$.

Proof. This follows by the same argument as in Proposition 1.7.7, using the exponentials of nilpotent elements of $\mathfrak{g}$ to generate G in part (2). □

Remark 2.2.8. In Chapter 11 we shall show that the algebraic notion of connectedness can be expressed in terms of the *Zariski topology*, and that a connected linear algebraic group is also connected relative to its topology as a Lie group (Theorem 11.2.9). Since a connected Lie group is generated by $\{\exp X : X \in \mathfrak{g}\}$, this will imply part (2) of Theorem 2.2.7 without assuming unipotent generation of G.

2.2.4 Exercises

1. (*Cayley Parameters*) Let G be $\mathbf{SO}(n,\mathbb{C})$ or $\mathbf{Sp}(n,\mathbb{C})$ and let $\mathfrak{g} = \mathrm{Lie}(G)$. Define $\mathcal{V}_G = \{g \in G : \det(I+g) \neq 0\}$ and $\mathcal{V}_{\mathfrak{g}} = \{X \in \mathfrak{g} : \det(I-X) \neq 0\}$. For $X \in \mathcal{V}_{\mathfrak{g}}$ define the Cayley transform $c(X) = (I+X)(I-X)^{-1}$. (Recall that $c(X) \in G$ by Exercises 1.4.5 #5.)
   (a) Show that c is a bijection from $\mathcal{V}_{\mathfrak{g}}$ onto $\mathcal{V}_G$.
   (b) Show that $\mathcal{V}_{\mathfrak{g}}$ is invariant under the adjoint action of G on $\mathfrak{g}$, and show that $gc(X)g^{-1} = c(gXg^{-1})$ for $g \in G$ and $X \in \mathcal{V}_{\mathfrak{g}}$.
   (c) Suppose that $f \in \mathcal{O}[G]$ and f vanishes on $\mathcal{V}_G$. Prove that $f = 0$. (HINT: Consider the function $g \mapsto f(g)\det(I+g)$ and use Theorem 2.2.5.)
2. Let $\rho : \mathbf{SL}(2,\mathbb{C}) \longrightarrow \mathbf{SO}(\mathbb{C}^3, B)$ as in Section 2.2.1. Let H (resp. $\widetilde{H}$) be the diagonal subgroup in $\mathbf{SO}(\mathbb{C}^3, B)$ (resp. $\mathbf{SL}(2,\mathbb{C})$). Let $\rho^* : \mathfrak{X}(H) \longrightarrow \mathfrak{X}(\widetilde{H})$ be the homomorphism of the character groups given by $\chi \mapsto \chi \circ \rho$. Determine the image of ρ^*. (HINT: $\mathfrak{X}(H)$ and $\mathfrak{X}(\widetilde{H})$ are isomorphic to the additive group $\mathbb{Z}$, and the image of ρ^* can be identified with a subgroup of $\mathbb{Z}$.)
3. Let $\pi : \mathbf{SL}(2,\mathbb{C}) \times \mathbf{SL}(2,\mathbb{C}) \longrightarrow \mathbf{SO}(\mathbb{C}^4, B)$ as in Section 2.2.1. Repeat the calculations of the previous exercise in this case. (HINT: Now $\mathfrak{X}(H)$ and $\mathfrak{X}(\widetilde{H})$ are isomorphic to the additive group $\mathbb{Z}^2$, and the image of π^* can be identified with a lattice in $\mathbb{Z}^2$.)

2.3 Regular Representations of $\mathbf{SL}(2,\mathbb{C})$

The group $G = \mathbf{SL}(2,\mathbb{C})$ and its Lie algebra $\mathfrak{g} = \mathfrak{sl}(2,\mathbb{C})$ play central roles in determining the structure of the classical groups and their representations. To find all the regular representations of G, we begin by finding all the irreducible finite-dimensional representations of $\mathfrak{g}$. Then we show that every such representation is the differential of an irreducible regular representation of G, thereby obtaining all irreducible regular representations of G. Next we show that an every finite-dimensional representation of $\mathfrak{g}$ decomposes as a direct sum of irreducible representations (the *complete reducibility* property), and conclude that every regular representation of G is completely reducible.

2.3.1 Irreducible Representations of $\mathfrak{sl}(2,\mathbb{C})$

Recall that a *representation* of a complex Lie algebra $\mathfrak{g}$ on a complex vector space V is a linear map $\pi : \mathfrak{g} \longrightarrow \mathrm{End}(V)$ such that

$$\pi([A,B]) = \pi(A)\pi(B) - \pi(B)\pi(A) \quad \text{for all } A,B \in \mathfrak{g} .$$

Here the Lie bracket $[A,B]$ on the left is calculated in $\mathfrak{g}$, whereas the product on the right is composition of linear transformations. We shall call V a $\mathfrak{g}$-*module* and write $\pi(A)v$ simply as Av when $v \in V$, provided that the representation π is understood from the context. Thus, even if $\mathfrak{g}$ is a Lie subalgebra of $M_n(\mathbb{C})$, an expression such as $A^k v$, for a nonnegative integer k, means $\pi(A)^k v$.

Let $\mathfrak{g} = \mathfrak{sl}(2,\mathbb{C})$. The matrices $x = \begin{bmatrix} 0 & 1 \\ 0 & 0 \end{bmatrix}$, $y = \begin{bmatrix} 0 & 0 \\ 1 & 0 \end{bmatrix}$, $h = \begin{bmatrix} 1 & 0 \\ 0 & -1 \end{bmatrix}$ are a basis for $\mathfrak{g}$ and satisfy the commutation relations

$$[h,x] = 2x , \quad [h,y] = -2y , \quad [x,y] = h . \tag{2.15}$$

Any triple $\{x,y,h\}$ of nonzero elements in a Lie algebra satisfying (2.15) will be called a TDS (three-dimensional simple) triple.

Lemma 2.3.1. *Let V be a $\mathfrak{g}$-module (possibly infinite-dimensional) and let $v_0 \in V$ be such that $x v_0 = 0$ and $h v_0 = \lambda v_0$ for some $\lambda \in \mathbb{C}$. Set $v_j = y^j v_0$ for $j \in \mathbb{N}$ and $v_j = 0$ for $j < 0$. Then $y v_j = v_{j+1}$, $h v_j = (\lambda - 2j)v_j$, and*

$$x v_j = j(\lambda - j + 1)v_{j-1} \quad \text{for } j \in \mathbb{N} . \tag{2.16}$$

Proof. The equation for $y v_j$ follows by definition, and the equation for $h v_j$ follows from the commutation relation (proved by induction on j)

$$h y^j v = y^j h v - 2 j v \quad \text{for all } v \in V \text{ and } j \in \mathbb{N} . \tag{2.17}$$

From (2.17) and the relation $x y v = y x v + h v$ one proves by induction on j that

$$x y^j v = j y^{j-1}(h - j + 1)v + y^j x v \quad \text{for all } v \in V \text{ and } j \in \mathbb{N} . \tag{2.18}$$

Taking $v = v_0$ and using $x v_0 = 0$, we obtain equation (2.16). $\square$

Let V be a finite-dimensional $\mathfrak{g}$-module. We decompose V into generalized eigenspaces for the action of h:

$$V = \bigoplus_{\lambda \in \mathbb{C}} V(\lambda), \quad \text{where } V(\lambda) = \bigcup_{k \geq 1} \mathrm{Ker}(h - \lambda)^k .$$

If $v \in V(\lambda)$ then $(h - \lambda)^k v = 0$ for some $k \geq 1$. As linear transformations on V,

$$x(h - \lambda) = (h - \lambda - 2)x \quad \text{and} \quad y(h - \lambda) = (h - \lambda + 2)x .$$

Hence $(h - \lambda - 2)^k x v = x(h - \lambda)^k v = 0$ and $(h - \lambda + 2)^k y v = y(h - \lambda)^k v = 0$. Thus

$$xV(\lambda) \subset V(\lambda+2) \quad \text{and} \quad yV(\lambda) \subset V(\lambda-2) \quad \text{for all } \lambda \in \mathbb{C}. \qquad (2.19)$$

If $V(\lambda) \neq 0$ then λ is called a *weight* of V with *weight space* $V(\lambda)$.

Lemma 2.3.2. *Suppose V is a finite-dimensional $\mathfrak{g}$-module and $0 \neq v_0 \in V$ satisfies $hv_0 = \lambda v_0$ and $xv_0 = 0$. Let k be the smallest nonnegative integer such that $y^k v_0 \neq 0$ and $y^{k+1} v_0 = 0$. Then $\lambda = k$ and the space $W = \mathrm{Span}\{v_0, yv_0, \ldots, y^k v_0\}$ is a $(k+1)$-dimensional $\mathfrak{g}$-module.*

Proof. Such an integer k exists by (2.19), since V is finite-dimensional and the weight spaces are linearly independent. Lemma 2.3.1 implies that W is invariant under x, y, and h. Furthermore, $v_0, yv_0, \ldots, y^k v_0$ are eigenvectors for h with respective eigenvalues $\lambda, \lambda - 2, \ldots, \lambda - 2k$. Hence these vectors are a basis for W. By (2.16),

$$0 = xy^{k+1}v_0 = (k+1)(\lambda-k)y^k v_0.$$

Since $y^k v_0 \neq 0$, it follows that $\lambda = k$. $\qquad\square$

We can describe the action of $\mathfrak{g}$ on the subspace W in Lemma 2.3.2 in matrix form as follows: For $k \in \mathbb{N}$ define the $(k+1) \times (k+1)$ matrices

$$X_k = \begin{bmatrix} 0 & k & 0 & 0 & \cdots & 0 \\ 0 & 0 & 2(k-1) & 0 & \cdots & 0 \\ 0 & 0 & 0 & 3(k-2) & \cdots & 0 \\ \vdots & \vdots & & \vdots & \ddots & \vdots \\ 0 & 0 & 0 & 0 & \cdots & k \\ 0 & 0 & 0 & 0 & \cdots & 0 \end{bmatrix}, \quad Y_k = \begin{bmatrix} 0 & 0 & \cdots & 0 & 0 \\ 1 & 0 & \cdots & 0 & 0 \\ 0 & 1 & \cdots & 0 & 0 \\ \vdots & \vdots & \ddots & \vdots & \vdots \\ 0 & 0 & \cdots & 1 & 0 \end{bmatrix},$$

and $H_k = \mathrm{diag}[k, k-2, \ldots, 2-k, -k]$. A direct check yields

$$[X_k, Y_k] = H_k, \quad [H_k, X_k] = 2X_k, \quad \text{and} \quad [H_k, Y_k] = -2Y_k.$$

With all of this in place we can classify the irreducible finite-dimensional modules for $\mathfrak{g}$.

Proposition 2.3.3. *Let $k \geq 0$ be an integer. The representation $(\rho_k, F^{(k)})$ of $\mathfrak{g}$ on $\mathbb{C}^{k+1}$ defined by*

$$\rho_k(x) = X_k, \quad \rho_k(h) = H_k, \quad \text{and} \quad \rho_k(y) = Y_k$$

is irreducible. Furthermore, if (σ, W) is an irreducible representation of $\mathfrak{g}$ with $\dim W = k+1 > 0$, then (σ, W) is equivalent to $(\rho_k, F^{(k)})$. In particular, W is equivalent to W^* as a $\mathfrak{g}$-module.

Proof. Suppose that $W \subset F^{(k)}$ is a nonzero invariant subspace. Since $xW(\lambda) \subset W(\lambda+2)$, there must be λ with $W(\lambda) \neq 0$ and $xW(\lambda) = 0$. But from the echelon form of X_k we see that $\mathrm{Ker}(X_k) = \mathbb{C}e_1$. Hence $\lambda = k$ and $W(k) = \mathbb{C}e_1$. Since $Y_k e_j = e_{j+1}$ for $1 \leq j \leq k$, it follows that $W = F^{(k)}$.

Let (σ, W) be any irreducible representation of $\mathfrak{g}$ with $\dim W = k+1 > 0$. There exists an eigenvalue λ of h such that $xW(\lambda) = 0$ and $0 \neq w_0 \in W(\lambda)$ such that $hw_0 = \lambda w_0$. By Lemma 2.3.2 we know that λ is a nonnegative integer, and the space spanned by the set $\{w_0, yw_0, y^2 w_0, \ldots\}$ is invariant under $\mathfrak{g}$ and has dimension $\lambda + 1$. But this space is all of W, since σ is irreducible. Hence $\lambda = k$, and by Lemma 2.3.1 the matrices of the actions of x, y, h with respect to the ordered basis $\{w_0, yw_0, \ldots, y^k w_0\}$ are X_k, Y_k, and H_k, respectively. Since W^* is an irreducible $\mathfrak{g}$-module of the same dimension as W, it must be equivalent to W. $\square$

Corollary 2.3.4. *The weights of a finite-dimensional $\mathfrak{g}$-module V are integers.*

Proof. There are $\mathfrak{g}$-invariant subspaces $0 = V_0 \subset V_1 \subset \cdots \subset V_k = V$ such that the quotient modules $W_j = V_j / V_{j-1}$ are irreducible for $j = 1, \ldots, k-1$. The weights are the eigenvalues of h on V, and this set is the union of the sets of eigenvalues of h on the modules W_j. Hence all weights are integers by Proposition 2.3.3. $\square$

2.3.2 Irreducible Regular Representations of $\mathbf{SL}(2, \mathbb{C})$

We now turn to the construction of irreducible regular representations of $\mathbf{SL}(2, \mathbb{C})$. Let the subgroups N^+ of upper-triangular unipotent matrices and N^- of lower-triangular unipotent matrices be as in Section 2.2.1. Set $d(a) = \mathrm{diag}[a, a^{-1}]$ for $a \in \mathbb{C}^\times$.

Proposition 2.3.5. *For every integer $k \geq 0$ there is a unique (up to equivalence) irreducible regular representation (π, V) of $\mathbf{SL}(2, \mathbb{C})$ of dimension $k+1$ whose differential is the representation ρ_k in Proposition 2.3.3. It has the following properties:*

1. *The semisimple operator $\pi(d(a))$ has eigenvalues $a^k, a^{k-2}, \ldots, a^{-k+2}, a^{-k}$.*
2. *$\pi(d(a))$ acts on by the scalar a^k on the one-dimensional space V^{N^+} of N^+-fixed vectors.*
3. *$\pi(d(a))$ acts on by the scalar a^{-k} on the one-dimensional space V^{N^-} of N^--fixed vectors.*

Proof. Let $\mathcal{P}(\mathbb{C}^2)$ be the polynomial functions on $\mathbb{C}^2$ and let $V = \mathcal{P}^k(\mathbb{C}^2)$ be the space of polynomials that are homogeneous of degree k. Here it is convenient to identify elements of $\mathbb{C}^2$ with row vectors $x = [x_1, x_2]$ and have $G = \mathbf{SL}(2, \mathbb{C})$ act by multiplication on the right. We then can define a representation of G on V by $\pi(g)\varphi(x) = \varphi(xg)$ for $\varphi \in V$. Thus

$$\pi(g)\varphi(x_1, x_2) = \varphi(ax_1 + cx_2, bx_1 + dx_2) \quad \text{when } g = \begin{bmatrix} a & b \\ c & d \end{bmatrix}.$$

In particular, the one-parameter subgroups $d(a)$, $u(z)$, and $v(z)$ act by

$$\pi(d(a))\varphi(x_1,x_2) = \varphi(ax_1, a^{-1}x_2),$$
$$\pi(u(z))\varphi(x_1,x_2) = \varphi(x_1, x_2 + zx_1),$$
$$\pi(v(z))\varphi(x_1,x_2) = \varphi(x_1 + zx_2, x_2).$$

As a basis for V we take the monomials

$$\mathbf{v}_j(x_1,x_2) = \frac{k!}{(k-j)!} x_1^{k-j} x_2^j \quad \text{for } j = 0, 1, \ldots, k.$$

From the formulas above for the action of $\pi(d(a))$ we see that these functions are eigenvectors for $\pi(d(a))$:

$$\pi(d(a))\mathbf{v}_j = a^{k-2j}\mathbf{v}_j.$$

Also, V^{N^+} is the space of polynomials depending only on x_1, so it consists of multiples of $\mathbf{v}_0$, whereas V^{N^-} is the space of polynomials depending only on x_2, so it consists of multiples of $\mathbf{v}_k$.

We now calculate the representation $d\pi$ of $\mathfrak{g}$. Since $u(z) = \exp(zx)$ and $v(z) = \exp(zy)$, we have $\pi(u(z)) = \exp(zd\pi(x))$ and $\pi(\dot{v}(z)) = \exp(zd\pi(y))$ by Theorem 1.6.2. Taking the z derivative, we obtain

$$d\pi(x)\varphi(x_1,x_2) = \left.\frac{\partial}{\partial z}\varphi(x_1, x_2 + zx_1)\right|_{z=0} = x_1\frac{\partial}{\partial x_2}\varphi(x_1,x_2),$$

$$d\pi(y)\varphi(x_1,x_2) = \left.\frac{\partial}{\partial z}\varphi(x_1 + zx_2, x_2)\right|_{z=0} = x_2\frac{\partial}{\partial x_1}\varphi(x_1,x_2).$$

Since $d\pi(h) = d\pi(x)d\pi(y) - d\pi(y)d\pi(x)$, we also have

$$d\pi(h)\varphi(x_1,x_2) = \left(x_1\frac{\partial}{\partial x_1} - x_2\frac{\partial}{\partial x_2}\right)\varphi(x_1,x_2).$$

On the basis vectors $\mathbf{v}_j$ we thus have

$$d\pi(h)\mathbf{v}_j = \frac{k!}{(k-j)!}\left(x_1\frac{\partial}{\partial x_1} - x_2\frac{\partial}{\partial x_2}\right)(x_1^{k-j}x_2^j) = (k-2j)\mathbf{v}_j,$$

$$d\pi(x)\mathbf{v}_j = \frac{k!}{(k-j)!}\left(x_1\frac{\partial}{\partial x_2}\right)(x_1^{k-j}x_2^j) = j(k-j+1)\mathbf{v}_{j-1},$$

$$d\pi(x)\mathbf{v}_j = \frac{k!}{(k-j)!}\left(x_2\frac{\partial}{\partial x_1}\right)(x_1^{k-j}x_2^j) = \mathbf{v}_{j+1}.$$

It follows from Proposition 2.3.3 that $d\pi \cong \rho_k$ is an irreducible representation of $\mathfrak{g}$, and all irreducible representations of $\mathfrak{g}$ are obtained this way. Theorem 2.2.7 now implies that π is an irreducible representation of G. Furthermore, π is uniquely determined by $d\pi$, since $\pi(u)$, for u unipotent, is uniquely determined by $d\pi(u)$ (Theorem 1.6.2) and G is generated by unipotent elements (Lemma 2.2.1). $\qquad\square$

2.3.3 Complete Reducibility of $\mathbf{SL}(2, \mathbb{C})$

Now that we have determined the irreducible regular representations of $\mathbf{SL}(2, \mathbb{C})$, we turn to the problem of finding all the regular representations. We first solve this problem for finite-dimensional representations of $\mathfrak{g} = \mathfrak{sl}(2, \mathbb{C})$.

Theorem 2.3.6. *Let V be a finite-dimensional $\mathfrak{g}$-module with $\dim V > 0$. Then there exist integers $k_1, \ldots, k_r$ (not necessarily distinct) such that V is equivalent to $F^{(k_1)} \bigoplus F^{(k_2)} \bigoplus \cdots \bigoplus F^{(k_r)}$.*

The key step in the proof of Theorem 2.3.6 is the following result:

Lemma 2.3.7. *Suppose W is a $\mathfrak{g}$-module with a submodule Z such that Z is equivalent to $F^{(k)}$ and W/Z is equivalent to $F^{(l)}$. Then W is equivalent to $F^{(k)} \bigoplus F^{(l)}$.*

Proof. Suppose first that $k \neq l$. The lemma is true for W if and only if it is true for W^*. The modules $F^{(k)}$ are self-dual, and replacing W by W^* interchanges the submodule and quotient module. Hence we may assume that $k < l$. By putting h in upper-triangular matrix form, we see that the set of eigenvalues of h on W (ignoring multiplicities) is

$$\{k, k-2, \ldots, -k+2, -k\} \cup \{l, l-2, \ldots, -l+2, -l\} \, .$$

Thus there exists $0 \neq u_0 \in W$ such that $hu_0 = lu_0$ and $xu_0 = 0$. Since $k < l$, the vector u_0 is not in Z, so the vectors $u_j = y^j u_0$ are not in Z for $j = 0, 1, \ldots, l$ (since $xu_j = j(l - j + 1)u_{j-1}$). By Proposition 2.3.3 these vectors span an irreducible $\mathfrak{g}$-module isomorphic to $F^{(l)}$ that has zero intersection with Z. Since $\dim W = k + l + 2$, this module is a complement to Z in W.

Now assume that $k = l$. Then $\dim W(l) = 2$, while $\dim Z(l) = 1$. Thus there exist nonzero vectors $z_0 \in Z(l)$ and $w_0 \in W(l)$ with $w_0 \notin Z$ and

$$hw_0 = lw_0 + az_0 \quad \text{for some } a \in \mathbb{C} \, .$$

Set $z_j = y^j z_0$ and $w_j = y^j w_0$. Using (2.17) we calculate that

$$
\begin{aligned}
hw_j = hy^j w_0 &= -2jy^j w_0 + y^j hw_0 \\
&= -2jw_j + y^j(lw_0 + az_0) = (l - 2j)w_j + az_j \, .
\end{aligned}
$$

Since $W(l+2) = 0$, we have $xz_0 = 0$ and $xw_0 = 0$. Thus equation (2.18) gives $xz_j = j(l - j + 1)z_{j-1}$ and

$$
\begin{aligned}
xw_j = jy^{j-1}(h - j + 1)w_0 &= j(l - j + 1)y^{j-1}w_0 + ajy^{j-1}z_0 \\
&= j(l - j + 1)w_{j-1} + ajz_{j-1} \, .
\end{aligned}
$$

It follows by induction on j that $\{z_j, w_j\}$ is linearly independent for $j = 0, 1, \ldots, l$. Since the weight spaces $W(l), \ldots, W(-l)$ are linearly independent, we conclude that

$$\{z_0, z_1, \ldots, z_l, w_0, w_1, \ldots, w_l\}$$

is a basis for W. Let X_l, Y_l, and H_l be the matrices in Section 2.3.1. Then relative to this basis the matrices for h, y, and x are

$$H = \begin{bmatrix} H_l & aI \\ 0 & H_l \end{bmatrix}, \quad Y = \begin{bmatrix} Y_l & 0 \\ 0 & Y_l \end{bmatrix}, \quad X = \begin{bmatrix} X_l & A \\ 0 & X_l \end{bmatrix},$$

respectively, where $A = \mathrm{diag}[0, a, 2a, \ldots, la]$. But

$$H = [X, Y] = \begin{bmatrix} H_l & [A, Y_l] \\ 0 & H_l \end{bmatrix}.$$

This implies that $[A, Y_l] = aI$. Hence $0 = \mathrm{tr}(aI) = (l+1)a$, so we have $a = 0$. The matrices H, Y, and X show that W is equivalent to the direct sum of two copies of $F^{(l)}$. $\square$

Proof of Theorem 2.3.6. If $\dim V = 1$ the result is true with $r = 1$ and $k_1 = 0$. Assume that the theorem is true for all $\mathfrak{g}$-modules of dimension less than m, and let V be an m-dimensional $\mathfrak{g}$-module.

The eigenvalues of h on V are integers by Corollary 2.3.4. Let k_1 be the biggest eigenvalue. Then $k_1 \geq 0$ and $V(l) = 0$ for $l > k_1$, so we have an injective module homomorphism of $F^{(k_1)}$ into V by Lemma 2.3.1. Let Z be the image of $F^{(k_1)}$. If $Z = V$ we are done. Otherwise, since $\dim V/Z < \dim V$, we can apply the inductive hypothesis to conclude that V/Z is equivalent to $F^{(k_2)} \oplus \cdots \oplus F^{(k_r)}$. Let

$$T : V \longrightarrow F^{(k_2)} \oplus \cdots \oplus F^{(k_r)}$$

be a surjective intertwining operator with kernel Z. For each $i = 2, \ldots, r$ choose $v_i \in V(k_i)$ such that

$$\mathbb{C}Tv_i = 0 \oplus \cdots \oplus F^{(k_i)}(k_i) \oplus \cdots \oplus 0.$$

Let $W_i = Z + \mathrm{Span}\{v_i, yv_i, \ldots, y^{k_i}v_i\}$ and $T_i = T|_{W_i}$. Then W_i is invariant under $\mathfrak{g}$ and $T_i : W_i \longrightarrow F^{(k_i)}$ is a surjective intertwining operator with kernel Z. Lemma 2.3.7 implies that $W_i = Z \oplus U_i$ and T_i defines an equivalence between U_i and $F^{(k_i)}$. Now set $U = U_2 + \cdots + U_r$. Then

$$T(U) = T(U_2) + \cdots + T(U_r) = F^{(k_2)} \oplus \cdots \oplus F^{(k_r)}.$$

Thus $T|_U$ is surjective. Since $\dim U \leq \dim U_2 + \cdots + \dim U_r = \dim T(U)$, it follows that $T|_U$ is bijective. Hence $V = Z \oplus U$, completing the induction. $\square$

Corollary 2.3.8. *Let (ρ, V) be a finite-dimensional representation of $\mathfrak{sl}(2,\mathbb{C})$. There exists a regular representation (π, W) of $\mathbf{SL}(2,\mathbb{C})$ such that $(d\pi, W)$ is equivalent to (ρ, V). Furthermore, every regular representation of $\mathbf{SL}(2,\mathbb{C})$ is a direct sum of irreducible subrepresentations.*

Proof. By Theorem 2.3.6 we may assume that $V = F^{(k_1)} \oplus F^{(k_2)} \oplus \cdots \oplus F^{(k_r)}$. Each of the summands is the differential of a representation of $\mathbf{SL}(2,\mathbb{C})$ by Proposition 2.3.5. $\square$

2.3.4 Exercises

1. Let $e_{ij} \in M_3(\mathbb{C})$ be the usual elementary matrices. Set $x = e_{13}$, $y = e_{31}$, and $h = e_{11} - e_{33}$.
   (a) Verify that $\{x, y, h\}$ is a TDS triple in $\mathfrak{sl}(3, \mathbb{C})$.
   (b) Let $\mathfrak{g} = \mathbb{C}x + \mathbb{C}y + \mathbb{C}h \cong \mathfrak{sl}(2, \mathbb{C})$ and let $U = M_3(\mathbb{C})$. Define a representation ρ of $\mathfrak{g}$ on U by $\rho(A)X = [A, X]$ for $A \in \mathfrak{g}$ and $X \in M_3(\mathbb{C})$. Show that $\rho(h)$ is diagonalizable, with eigenvalues ± 2 (multiplicity 1), ± 1 (multiplicity 2), and 0 (multiplicity 3). Find all $u \in U$ such that $\rho(h)u = \lambda u$ and $\rho(x)u = 0$, where $\lambda = 0, 1, 2$.
   (c) Let $F^{(k)}$ be the irreducible $(k+1)$-dimensional representation of $\mathfrak{g}$. Show that

$$U \cong F^{(2)} \oplus F^{(1)} \oplus F^{(1)} \oplus F^{(0)} \oplus F^{(0)}$$

   as a $\mathfrak{g}$-module. (HINT: Use the results of (b) and Theorem 2.3.6.)
2. Let k be a nonnegative integer and let W_k be the polynomials in $\mathbb{C}[x]$ of degree at most k. If $f \in W_k$ set

$$\sigma_k(g)f(x) = (cx+a)^k f\left(\frac{dx+b}{cx+a}\right) \quad \text{for } g = \begin{bmatrix} a & b \\ c & d \end{bmatrix} \in \mathbf{SL}(2,\mathbb{C}).$$

   Show that $\sigma_k(g)W_k = W_k$ and that (σ_k, W_k) defines a representation of $\mathbf{SL}(2, \mathbb{C})$ equivalent to the irreducible $(k+1)$-dimensional representation. (HINT: Find an intertwining operator between this representation and the representation used in the proof of Proposition 2.3.5.)
3. Find the irreducible regular representations of $\mathbf{SO}(3, \mathbb{C})$. (HINT: Use the homomorphism $\rho : \mathbf{SL}(2, \mathbb{C}) \longrightarrow \mathbf{SO}(3, \mathbb{C})$ from Section 2.2.1.)
4. Let $V = \mathbb{C}[x]$. Define operators E and F on V by

$$E\varphi(x) = -\frac{1}{2}\frac{d^2\varphi(x)}{dx^2}, \qquad F\varphi(x) = \frac{1}{2}x^2\varphi(x) \quad \text{for } \varphi \in V.$$

   Set $H = [E, F]$.
   (a) Show that $H = -x(d/dx) - 1/2$ and that $\{E, F, H\}$ is a TDS triple.
   (b) Find the space $V^E = \{\varphi \in V : E\varphi = 0\}$.
   (c) Let $V_{\text{even}} \subset V$ be the space of *even* polynomials and $V_{\text{odd}} \subset V$ the space of *odd* polynomials. Let $\mathfrak{g} \subset \text{End}(V)$ be the Lie algebra spanned by E, F, H. Show that each of these spaces is invariant and irreducible under $\mathfrak{g}$. (HINT: Use (b) and Lemma 2.3.1.)

(d) Show that $V = V_{\text{even}} \oplus V_{\text{odd}}$ and that V_{even} is not equivalent to V_{odd} as a module for $\mathfrak{g}$. (HINT: Show that the operator H is diagonalizable on V_{even} and V_{odd} and find its eigenvalues.)

5. Let $X \in M_n(\mathbb{C})$ be a nilpotent and nonzero. By Exercise 1.6.4 #3 there exist $H, Y \in M_n(\mathbb{C})$ such that $\{X, Y, H\}$ is a TDS triple. Let $\mathfrak{g} = \text{Span}\{H, X, Y\} \cong \mathfrak{sl}(2, \mathbb{C})$ and consider $V = \mathbb{C}^n$ as a representation π of $\mathfrak{g}$ by left multiplication of matrices on column vectors.
   (a) Show that π is irreducible if and only if the Jordan canonical form of X consists of a single block.
   (b) In the decomposition of V into irreducible subspaces given by Theorem 2.3.6, let m_j be the number of times the representation $F^{(j)}$ occurs. Show that m_j is the number of Jordan blocks of size $j + 1$ in the Jordan canonical form of X.
   (c) Show that π is determined (up to isomorphism) by the eigenvalues (with multiplicities) of H on $\text{Ker}(X)$.
6. Let (ρ, W) be a finite-dimensional representation of $\mathfrak{sl}(2, \mathbb{C})$. For $k \in \mathbb{Z}$ set $f(k) = \dim\{w \in W : \rho(h)w = kw\}$.
   (a) Show that $f(k) = f(-k)$.
   (b) Let $g_{\text{even}}(k) = f(2k)$ and $g_{\text{odd}}(k) = f(2k + 1)$. Show that g_{even} and g_{odd} are unimodal functions from $\mathbb{Z}$ to $\mathbb{N}$. Here a function ϕ is called *unimodal* if there exists k_0 such that $\phi(a) \leq \phi(b)$ for all $a < b \leq k_0$ and $\phi(a) \geq \phi(b)$ for all $k_0 \leq a < b$. (HINT: Decompose W into a direct sum of irreducible subspaces and use Proposition 2.3.3.)

2.4 The Adjoint Representation

We now use the maximal torus in a classical group to decompose the Lie algebra of the group into eigenspaces, traditionally called *root spaces*, under the adjoint representation.

2.4.1 Roots with Respect to a Maximal Torus

Throughout this section G will denote a connected classical group of rank l. Thus G is $\mathbf{GL}(l, \mathbb{C})$, $\mathbf{SL}(l + 1, \mathbb{C})$, $\mathbf{Sp}(\mathbb{C}^{2l}, \Omega)$, $\mathbf{SO}(\mathbb{C}^{2l}, B)$, or $\mathbf{SO}(\mathbb{C}^{2l+1}, B)$, where we take as Ω and B the bilinear forms (2.6) and (2.9). We set $\mathfrak{g} = \text{Lie}(G)$. The subgroup H of diagonal matrices in G is a maximal torus of rank l, and we denote its Lie algebra by $\mathfrak{h}$. In this section we will study the regular representation π of H on the vector space $\mathfrak{g}$ given by $\pi(h)X = hXh^{-1}$ for $h \in H$ and $X \in \mathfrak{g}$.

Let $x_1, \ldots, x_l$ be the coordinate functions on H used in the proof of Theorem 2.1.5. Using these coordinates we obtain an isomorphism between the group $\mathcal{X}(H)$ of rational characters of H and the additive group $\mathbb{Z}^l$ (see Lemma 2.1.2). Under this isomorphism, $\lambda = [\lambda_1, \ldots, \lambda_l] \in \mathbb{Z}^l$ corresponds to the character $h \mapsto h^\lambda$, where

$$h^\lambda = \prod_{k=1}^{l} x_k(h)^{\lambda_k}, \quad \text{for } h \in H. \tag{2.20}$$

For $\lambda, \mu \in \mathbb{Z}^l$ and $h \in H$ we have $h^\lambda h^\mu = h^{\lambda+\mu}$.

For making calculations it is convenient to fix the following bases for $\mathfrak{h}^*$:

(a) Let $G = \mathbf{GL}(l, \mathbb{C})$. Define $\langle \varepsilon_i, A \rangle = a_i$ for $A = \mathrm{diag}[a_1, \ldots, a_l] \in \mathfrak{h}$. Then $\{\varepsilon_1, \ldots, \varepsilon_l\}$ is a basis for $\mathfrak{h}^*$.

(b) Let $G = \mathbf{SL}(l+1, \mathbb{C})$. Then $\mathfrak{h}$ consists of all diagonal matrices of trace zero. With an abuse of notation we will continue to denote the restrictions to $\mathfrak{h}$ of the linear functionals in (a) by ε_i. The elements of $\mathfrak{h}^*$ can then be written uniquely as $\sum_{i=1}^{l+1} \lambda_i \varepsilon_i$ with $\lambda_i \in \mathbb{C}$ and $\sum_{i=1}^{l+1} \lambda_i = 0$. A basis for $\mathfrak{h}^*$ is furnished by the functionals

$$\varepsilon_i - \frac{1}{l+1}(\varepsilon_1 + \cdots + \varepsilon_{l+1}) \quad \text{for } i = 1, \ldots, l.$$

(c) Let G be $\mathbf{Sp}(\mathbb{C}^{2l}, \Omega)$ or $\mathbf{SO}(\mathbb{C}^{2l}, B)$. For $i = 1, \ldots, l$ define $\langle \varepsilon_i, A \rangle = a_i$, where $A = \mathrm{diag}[a_1, \ldots, a_l, -a_l, \ldots, -a_1] \in \mathfrak{h}$. Then $\{\varepsilon_1, \ldots, \varepsilon_l\}$ is a basis for $\mathfrak{h}^*$.

(d) Let $G = \mathbf{SO}(\mathbb{C}^{2l+1}, B)$. For $A = \mathrm{diag}[a_1, \ldots, a_l, 0, -a_l, \ldots, -a_1] \in \mathfrak{h}$ and $i = 1, \ldots, l$ define $\langle \varepsilon_i, A \rangle = a_i$. Then $\{\varepsilon_1, \ldots, \varepsilon_l\}$ is a basis for $\mathfrak{h}^*$.

We define $P(G) = \{d\theta : \theta \in \mathfrak{X}(H)\} \subset \mathfrak{h}^*$. With the functionals ε_i defined as above, we have

$$P(G) = \bigoplus_{k=1}^{l} \mathbb{Z}\varepsilon_k. \tag{2.21}$$

Indeed, given $\lambda = \lambda_1 \varepsilon_1 + \cdots + \lambda_l \varepsilon_l$ with $\lambda_i \in \mathbb{Z}$, let e^λ denote the rational character of H determined by $[\lambda_1, \ldots, \lambda_l] \in \mathbb{Z}^l$ as in (2.20). Every element of $\mathfrak{X}(H)$ is of this form, and we claim that $de^\lambda(A) = \langle \lambda, A \rangle$ for $A \in \mathfrak{h}$. To prove this, recall from Section 1.4.3 that $A \in \mathfrak{h}$ acts by the vector field

$$X_A = \sum_{i=1}^{l} \langle \varepsilon_i, A \rangle x_i \frac{\partial}{\partial x_i}$$

on $\mathbb{C}[x_1, x_1^{-1}, \ldots, x_l, x_l^{-1}]$. By definition of the differential of a representation we have

$$de^\lambda(A) = X_A(x_1^{\lambda_1} \cdots x_l^{\lambda_l})(1) = \sum_{i=1}^{l} \lambda_i \langle \varepsilon_i, A \rangle = \langle \lambda, A \rangle$$

as claimed. This proves (2.21). The map $\lambda \mapsto e^\lambda$ is thus an isomorphism between the additive group $P(G)$ and the character group $\mathfrak{X}(H)$, by Lemma 2.1.2. From (2.21) we see that $P(G)$ is a *lattice* (free abelian subgroup of rank l) in $\mathfrak{h}^*$, which is called the *weight lattice* of G (the notation $P(G)$ is justified, since all maximal tori are conjugate in G).

We now study the adjoint action of H and $\mathfrak{h}$ on $\mathfrak{g}$. For $\alpha \in P(G)$ let

$$\mathfrak{g}_\alpha = \{X \in \mathfrak{g} : hXh^{-1} = h^\alpha X \text{ for all } h \in H\}$$
$$= \{X \in \mathfrak{g} : [A,X] = \langle \alpha, A \rangle X \text{ for all } A \in \mathfrak{h}\} \,.$$

(The equivalence of these two formulas for $\mathfrak{g}_\alpha$ is clear from the discussion above.) For $\alpha = 0$ we have $\mathfrak{g}_0 = \mathfrak{h}$, by the same argument as in the proof of Theorem 2.1.5. If $\alpha \neq 0$ and $\mathfrak{g}_\alpha \neq 0$ then α is called a *root* of H on $\mathfrak{g}$ and $\mathfrak{g}_\alpha$ is called a *root space*. If α is a root then a nonzero element of $\mathfrak{g}_\alpha$ is called a *root vector* for α. We call the set Φ of roots the *root system* of $\mathfrak{g}$. Its definition requires fixing a choice of maximal torus, so we write $\Phi = \Phi(\mathfrak{g},\mathfrak{h})$ when we want to make this choice explicit. Applying Proposition 2.1.3, we have the *root space decomposition*

$$\mathfrak{g} = \mathfrak{h} \oplus \bigoplus_{\alpha \in \Phi} \mathfrak{g}_\alpha \,. \tag{2.22}$$

Theorem 2.4.1. *Let $G \subset \mathbf{GL}(n,\mathbb{C})$ be a connected classical group, and let $H \subset G$ be a maximal torus with Lie algebra $\mathfrak{h}$. Let $\Phi \subset \mathfrak{h}^*$ be the root system of $\mathfrak{g}$.*

1. $\dim \mathfrak{g}_\alpha = 1$ for all $\alpha \in \Phi$.
2. If $\alpha \in \Phi$ and $c\alpha \in \Phi$ for some $c \in \mathbb{C}$ then $c = \pm 1$.
3. The symmetric bilinear form $(X,Y) = \mathrm{tr}_{\mathbb{C}^n}(XY)$ on $\mathfrak{g}$ is invariant:

$$([X,Y],Z) = -(Y,[X,Z]) \quad \text{for } X,Y,Z \in \mathfrak{g} \,.$$

4. Let $\alpha,\beta \in \Phi$ and $\alpha \neq -\beta$. Then $(\mathfrak{h},\mathfrak{g}_\alpha) = 0$ and $(\mathfrak{g}_\alpha,\mathfrak{g}_\beta) = 0$.
5. The form (X,Y) on $\mathfrak{g}$ is nondegenerate.

Proof of (1): We shall calculate the roots and root vectors for each type of classical group. We take the Lie algebras in the matrix form of Section 2.1.2. In this realization the algebras are invariant under the transpose. For $A \in \mathfrak{h}$ and $X \in \mathfrak{g}$ we have $[A,X]^t = -[A,X^t]$. Hence if X is a root vector for the root α, then X^t is a root vector for $-\alpha$.

Type A: Let G be $\mathbf{GL}(n,\mathbb{C})$ or $\mathbf{SL}(n,\mathbb{C})$. For $A = \mathrm{diag}[a_1,\dots,a_n] \in \mathfrak{h}$ we have

$$[A, e_{ij}] = (a_i - a_j)e_{ij} = \langle \varepsilon_i - \varepsilon_j, A \rangle e_{ij} \,.$$

Since the set $\{e_{ij} : 1 \leq i,j \leq n, i \neq j\}$ is a basis of $\mathfrak{g}$ modulo $\mathfrak{h}$, the roots are

$$\{\pm(\varepsilon_i - \varepsilon_j) : 1 \leq i < j \leq n\} \,,$$

each with multiplicity 1. The root space $\mathfrak{g}_\lambda$ is $\mathbb{C}e_{ij}$ for $\lambda = \varepsilon_i - \varepsilon_j$.

Type C: Let $G = \mathbf{Sp}(\mathbb{C}^{2l},\Omega)$. Label the basis for $\mathbb{C}^{2l}$ as $e_{\pm 1},\dots,e_{\pm l}$, where $e_{-i} = e_{2l+1-i}$. Let $e_{i,j}$ be the matrix that takes the basis vector e_j to e_i and annihilates e_k for $k \neq j$ (here i and j range over $\pm 1,\dots,\pm l$). Set $X_{\varepsilon_i - \varepsilon_j} = e_{i,j} - e_{-j,-i}$ for $1 \leq i,j \leq l$, $i \neq j$. Then $X_{\varepsilon_i - \varepsilon_j} \in \mathfrak{g}$ and

$$[A, X_{\varepsilon_i - \varepsilon_j}] = \langle \varepsilon_i - \varepsilon_j, A \rangle X_{\varepsilon_i - \varepsilon_j} \,, \tag{2.23}$$

for $A \in \mathfrak{h}$. Hence $\varepsilon_i - \varepsilon_j$ is a root. These roots are associated with the embedding $\mathfrak{gl}(l,\mathbb{C}) \longrightarrow \mathfrak{g}$ given by $Y \mapsto \begin{bmatrix} Y & 0 \\ 0 & -s_l Y^t s_l \end{bmatrix}$ for $Y \in \mathfrak{gl}(l,\mathbb{C})$, where s_l is defined in (2.5). Set $X_{\varepsilon_i+\varepsilon_j} = e_{i,-j} + e_{j,-i}$, $X_{-\varepsilon_i-\varepsilon_j} = e_{-j,i} + e_{-i,j}$ for $1 \le i < j \le l$, and set $X_{2\varepsilon_i} = e_{i,-i}$ for $1 \le i \le l$. These matrices are in $\mathfrak{g}$, and

$$[A, X_{\pm(\varepsilon_i+\varepsilon_j)}] = \pm \langle \varepsilon_i + \varepsilon_j, A \rangle X_{\pm(\varepsilon_i+\varepsilon_j)}$$

for $A \in \mathfrak{h}$. Hence $\pm(\varepsilon_i + \varepsilon_j)$ are roots for $1 \le i \le j \le l$. From the block matrix form (2.8) of $\mathfrak{g}$ we see that

$$\{X_{\pm(\varepsilon_i-\varepsilon_j)}, X_{\pm(\varepsilon_i+\varepsilon_j)} : 1 \le i < j \le l\} \cup \{X_{\pm 2\varepsilon_i} : 1 \le i \le l\}$$

is a basis for $\mathfrak{g}$ modulo $\mathfrak{h}$. This shows that the roots have multiplicity one and are

$$\pm(\varepsilon_i - \varepsilon_j) \text{ and } \pm(\varepsilon_i + \varepsilon_j) \text{ for } 1 \le i < j \le l, \quad \pm 2\varepsilon_k \text{ for } 1 \le k \le l.$$

Type D: Let $G = \mathbf{SO}(\mathbb{C}^{2l}, B)$. Label the basis for $\mathbb{C}^{2l}$ and define $X_{\varepsilon_i-\varepsilon_j}$ as in the case of $\mathbf{Sp}(\mathbb{C}^{2l}, \Omega)$. Then $X_{\varepsilon_i-\varepsilon_j} \in \mathfrak{g}$ and (2.23) holds for $A \in \mathfrak{h}$, so $\varepsilon_i - \varepsilon_j$ is a root. These roots arise from the same embedding $\mathfrak{gl}(l,\mathbb{C}) \longrightarrow \mathfrak{g}$ as in the symplectic case. Set $X_{\varepsilon_i+\varepsilon_j} = e_{i,-j} - e_{j,-i}$ and $X_{-\varepsilon_i-\varepsilon_j} = e_{-j,i} - e_{-i,j}$ for $1 \le i < j \le l$. Then $X_{\pm(\varepsilon_i+\varepsilon_j)} \in \mathfrak{g}$ and

$$[A, X_{\pm(\varepsilon_i+\varepsilon_j)}] = \pm \langle \varepsilon_i + \varepsilon_j, A \rangle X_{\pm(\varepsilon_i+\varepsilon_j)}$$

for $A \in \mathfrak{h}$. Thus $\pm(\varepsilon_i + \varepsilon_j)$ is a root. From the block matrix form (2.7) for $\mathfrak{g}$ we see that

$$\{X_{\pm(\varepsilon_i-\varepsilon_j)} : 1 \le i < j \le l\} \cup \{X_{\pm(\varepsilon_i+\varepsilon_j)} : 1 \le i < j \le l\}$$

is a basis for $\mathfrak{g}$ modulo $\mathfrak{h}$. This shows that the roots have multiplicity one and are

$$\pm(\varepsilon_i - \varepsilon_j) \text{ and } \pm(\varepsilon_i + \varepsilon_j) \text{ for } 1 \le i < j \le l.$$

Type B: Let $G = \mathbf{SO}(\mathbb{C}^{2l+1}, B)$. We embed $\mathbf{SO}(\mathbb{C}^{2l}, B)$ into G by equation (2.14). Since $H \subset \mathbf{SO}(\mathbb{C}^{2l}, B) \subset G$ via this embedding, the roots $\pm \varepsilon_i \pm \varepsilon_j$ of $\mathrm{ad}(\mathfrak{h})$ on $\mathfrak{so}(\mathbb{C}^{2l}, B)$ also occur for the adjoint action of $\mathfrak{h}$ on $\mathfrak{g}$. We label the basis for $\mathbb{C}^{2l+1}$ as $\{e_{-l}, \ldots, e_{-1}, e_0, e_1, \ldots, e_l\}$, where $e_0 = e_{l+1}$ and $e_{-i} = e_{2l+2-i}$. Let $e_{i,j}$ be the matrix that takes the basis vector e_j to e_i and annihilates e_k for $k \ne j$ (here i and j range over $0, \pm 1, \ldots, \pm l$). Then the corresponding root vectors from type D are

$$X_{\varepsilon_i-\varepsilon_j} = e_{i,j} - e_{-j,-i}, \quad X_{\varepsilon_j-\varepsilon_i} = e_{j,i} - e_{-i,-j},$$

$$X_{\varepsilon_i+\varepsilon_j} = e_{i,-j} - e_{j,-i}, \quad X_{-\varepsilon_i-\varepsilon_j} = e_{-j,i} - e_{-i,j},$$

for $1 \le i < j \le l$. Define

$$X_{\varepsilon_i} = e_{i,0} - e_{0,-i}, \quad X_{-\varepsilon_i} = e_{0,i} - e_{-i,0},$$

for $1 \le i \le l$. Then $X_{\pm \varepsilon_i} \in \mathfrak{g}$ and $[A, X_{\pm \varepsilon_i}] = \pm \langle \varepsilon_i, A \rangle X_{\varepsilon_i}$ for $A \in \mathfrak{h}$. From the block matrix form (2.10) for $\mathfrak{g}$ we see that $\{X_{\pm \varepsilon_i} : 1 \le i \le l\}$ is a basis for $\mathfrak{g}$ modulo $\mathfrak{so}(\mathbb{C}^{2l}, B)$. Hence the results above for $\mathfrak{so}(\mathbb{C}^{2l}, B)$ imply that the roots of $\mathfrak{so}(\mathbb{C}^{2l+1}, B)$ have multiplicity one and are

$$\pm (\varepsilon_i - \varepsilon_j) \text{ and } \pm (\varepsilon_i + \varepsilon_j) \text{ for } 1 \le i < j \le l, \quad \pm \varepsilon_k \text{ for } 1 \le k \le l.$$

Proof of (2):   This is clear from the calculations above.

Proof of (3):   Let $X, Y, Z \in \mathfrak{g}$. Since $\text{tr}(AB) = \text{tr}(BA)$, we have

$$([X, Y], Z) = \text{tr}(XYZ - YXZ) = \text{tr}(YZX - YXZ)$$
$$= -\text{tr}(Y[X, Z]) = -(Y, [X, Z]).$$

Proof of (4):   Let $X \in \mathfrak{g}_\alpha, Y \in \mathfrak{g}_\beta$, and $A \in \mathfrak{h}$. Then

$$0 = ([A, X], Y) + (X, [A, Y]) = \langle \alpha + \beta, A \rangle (X, Y).$$

Since $\alpha + \beta \ne 0$ we can take A such that $\langle \alpha + \beta, A \rangle \ne 0$. Hence $(X, Y) = 0$ in this case. The same argument, but with $Y \in \mathfrak{h}$, shows that $(\mathfrak{h}, \mathfrak{g}_\alpha) = 0$.

Proof of (5):   By (4), we only need to show that the restrictions of the trace form to $\mathfrak{h} \times \mathfrak{h}$ and to $\mathfrak{g}_\alpha \times \mathfrak{g}_{-\alpha}$ are nondegenerate for all $\alpha \in \Phi$. Suppose $X, Y \in \mathfrak{h}$. Then

$$\text{tr}(XY) = \begin{cases} \sum_{i=1}^n \varepsilon_i(X) \varepsilon_i(Y) & \text{if } G = \mathbf{GL}(n, \mathbb{C}) \text{ or } G = \mathbf{SL}(n, \mathbb{C}), \\ 2 \sum_{i=1}^l \varepsilon_i(X) \varepsilon_i(Y) & \text{otherwise.} \end{cases} \tag{2.24}$$

From this it is clear that the restriction of the trace form to $\mathfrak{h} \times \mathfrak{h}$ is nondegenerate.

For $\alpha \in \Phi$ we define $X_\alpha \in \mathfrak{g}_\alpha$ for types A, B, C, and D in terms of the elementary matrices $e_{i,j}$ as above. Then $X_\alpha X_{-\alpha}$ is given as follows (the case of $\mathbf{GL}(n, \mathbb{C})$ is the same as type A):

Type A: $X_{\varepsilon_i - \varepsilon_j} X_{\varepsilon_j - \varepsilon_i} = e_{i,i}$ for $1 \le i < j \le l+1$.

Type B: $X_{\varepsilon_i - \varepsilon_j} X_{\varepsilon_j - \varepsilon_i} = e_{i,i} + e_{-j,-j}$ and $X_{\varepsilon_i + \varepsilon_j} X_{-\varepsilon_j - \varepsilon_i} = e_{i,i} + e_{j,j}$ for $1 \le i < j \le l$. Also $X_{\varepsilon_i} X_{-\varepsilon_i} = e_{i,i} + e_{0,0}$ for $1 \le i \le l$.

Type C: $X_{\varepsilon_i - \varepsilon_j} X_{\varepsilon_j - \varepsilon_i} = e_{i,i} + e_{-j,-j}$ for $1 \le i < j \le l$ and $X_{\varepsilon_i + \varepsilon_j} X_{-\varepsilon_j - \varepsilon_i} = e_{i,i} + e_{j,j}$ for $1 \le i \le j \le l$.

Type D: $X_{\varepsilon_i - \varepsilon_j} X_{\varepsilon_j - \varepsilon_i} = e_{i,i} + e_{-j,-j}$ and $X_{\varepsilon_i + \varepsilon_j} X_{-\varepsilon_j - \varepsilon_i} = e_{i,i} + e_{j,j}$ for $1 \le i < j \le l$.

From these formulas it is evident that $\text{tr}(X_\alpha X_{-\alpha}) \ne 0$ for all $\alpha \in \Phi$. $\square$

2.4.2 Commutation Relations of Root Spaces

We continue the notation of the previous section ($G \subset \mathbf{GL}(n, \mathbb{C})$ a connected classical group). Now that we have decomposed the Lie algebra $\mathfrak{g}$ of G into root spaces under the action of a maximal torus, the next step is to find the commutation relations among the root spaces.

We first observe that

$$[\mathfrak{g}_\alpha, \mathfrak{g}_\beta] \subset \mathfrak{g}_{\alpha+\beta} \quad \text{for } \alpha, \beta \in \mathfrak{h}^* . \tag{2.25}$$

Indeed, let $A \in \mathfrak{h}$. Then

$$[A, [X, Y]] = [[A, X], Y] + [X, [A, Y]] = \langle \alpha + \beta, A \rangle [X, Y]$$

for $X \in \mathfrak{g}_\alpha$ and $Y \in \mathfrak{g}_\beta$. Hence $[X, Y] \in \mathfrak{g}_{\alpha+\beta}$. In particular, if $\alpha + \beta$ is not a root, then $\mathfrak{g}_{\alpha+\beta} = 0$, so X and Y commute in this case. We also see from (2.25) that

$$[\mathfrak{g}_\alpha, \mathfrak{g}_{-\alpha}] \subset \mathfrak{g}_0 = \mathfrak{h} .$$

When α, β, and $\alpha + \beta$ are all roots, then it turns out that $[\mathfrak{g}_\alpha, \mathfrak{g}_\beta] \neq 0$, and hence the inclusion in (2.25) is an equality (recall that $\dim \mathfrak{g}_\alpha = 1$ for all $\alpha \in \Phi$). One way to prove this is to calculate all possible commutators for each type of classical group. Instead of doing this, we shall follow a more conceptual approach using the representation theory of $\mathfrak{sl}(2, \mathbb{C})$ and the invariant bilinear form on $\mathfrak{g}$ from Theorem 2.4.1.

We begin by showing that for each root α, the subalgebra of $\mathfrak{g}$ generated by $\mathfrak{g}_\alpha$ and $\mathfrak{g}_{-\alpha}$ is isomorphic to $\mathfrak{sl}(2, \mathbb{C})$.

Lemma 2.4.2. (Notation as in Theorem 2.4.1) *For each $\alpha \in \Phi$ there exist $e_\alpha \in \mathfrak{g}_\alpha$ and $f_\alpha \in \mathfrak{g}_{-\alpha}$ such that the element $h_\alpha = [e_\alpha, f_\alpha] \in \mathfrak{h}$ satisfies $\langle \alpha, h_\alpha \rangle = 2$. Hence*

$$[h_\alpha, e_\alpha] = 2e_\alpha , \quad [h_\alpha, f_\alpha] = -2f_\alpha ,$$

so that $\{e_\alpha, f_\alpha, h_\alpha\}$ is a TDS triple.

Proof. By Theorem 2.4.1 we can pick $X \in \mathfrak{g}_\alpha$ and $Y \in \mathfrak{g}_{-\alpha}$ such that $(X, Y) \neq 0$. Set $A = [X, Y] \in \mathfrak{h}$. Then

$$[A, X] = \langle \alpha, A \rangle X , \quad [A, Y] = -\langle \alpha, A \rangle Y . \tag{2.26}$$

We claim that $A \neq 0$. To prove this take any $B \in \mathfrak{h}$ such that $\langle \alpha, B \rangle \neq 0$. Then

$$(A, B) = ([X, Y], B) = (Y, [B, X]) = \langle \alpha, B \rangle (Y, X) \neq 0 . \tag{2.27}$$

We now prove that $\langle \alpha, A \rangle \neq 0$. Since $A \in \mathfrak{h}$, it is a semisimple matrix. For $\lambda \in \mathbb{C}$ let

$$V_\lambda = \{v \in \mathbb{C}^n : Av = \lambda v\}$$

be the λ eigenspace of A. Assume for the sake of contradiction that $\langle \alpha, A \rangle = 0$. Then from (2.26) we see that X and Y would commute with A, and hence V_λ would be invariant under X and Y. But this would imply that

$$\lambda \dim V_\lambda = \mathrm{tr}_{V_\lambda}(A) = \mathrm{tr}_{V_\lambda}([X, Y]|_{V_\lambda}) = 0 .$$

Hence $V_\lambda = 0$ for all $\lambda \neq 0$, making $A = 0$, which is a contradiction.

Now that we know $\langle \alpha, A \rangle \neq 0$, we can rescale X, Y, and A, as follows: Set $e_\alpha = sX$, $f_\alpha = tY$, and $h_\alpha = stA$, where $s, t \in \mathbb{C}^\times$. Then

$$[h_\alpha, e_\alpha] = st\langle \alpha, A \rangle e_\alpha , \quad [h_\alpha, f_\alpha] = -st\langle \alpha, A \rangle f_\alpha ,$$
$$[e_\alpha, f_\alpha] = st[X, Y] = h_\alpha .$$

Thus any choice of s, t such that $st\langle \alpha, A \rangle = 2$ gives $\langle \alpha, h_\alpha \rangle = 2$ and the desired TDS triple. □

For future calculations it will be useful to have explicit choices of e_α and f_α for each pair of roots $\pm \alpha \in \Phi$. If $\{e_\alpha, f_\alpha, h_\alpha\}$ is a TDS triple that satisfies the conditions in Lemma 2.4.2 for a root α, then $\{f_\alpha, e_\alpha, -h_\alpha\}$ satisfies the conditions for $-\alpha$. So we may take $e_{-\alpha} = f_\alpha$ and $f_{-\alpha} = e_\alpha$ once we have chosen e_α and f_α. We shall follow the notation of Section 2.4.1.

Type A:
Let $\alpha = \varepsilon_i - \varepsilon_j$ with $1 \leq i < j \leq l+1$. Set $e_\alpha = e_{ij}$ and $f_\alpha = e_{ji}$. Then $h_\alpha = e_{ii} - e_{jj}$.

Type B:
(a) For $\alpha = \varepsilon_i - \varepsilon_j$ with $1 \leq i < j \leq l$ set $e_\alpha = e_{i,j} - e_{-j,-i}$ and $f_\alpha = e_{j,i} - e_{-i,-j}$.
Then $h_\alpha = e_{i,i} - e_{j,j} + e_{-j,-j} - e_{-i,-i}$.
(b) For $\alpha = \varepsilon_i + \varepsilon_j$ with $1 \leq i < j \leq l$ set $e_\alpha = e_{i,-j} - e_{j,-i}$ and $f_\alpha = e_{-j,i} - e_{-i,j}$.
Then $h_\alpha = e_{i,i} + e_{j,j} - e_{-j,-j} - e_{-i,-i}$.
(c) For $\alpha = \varepsilon_i$ with $1 \leq i \leq l$ set $e_\alpha = e_{i,0} - e_{0,-i}$ and $f_\alpha = 2e_{0,i} - 2e_{-i,0}$.
Then $h_\alpha = 2e_{i,i} - 2e_{-i,-i}$.

Type C:
(a) For $\alpha = \varepsilon_i - \varepsilon_j$ with $1 \leq i < j \leq l$ set $e_\alpha = e_{i,j} - e_{-j,-i}$ and $f_\alpha = e_{j,i} - e_{-i,-j}$.
Then $h_\alpha = e_{i,i} - e_{j,j} + e_{-j,-j} - e_{-i,-i}$.
(b) For $\alpha = \varepsilon_i + \varepsilon_j$ with $1 \leq i < j \leq l$ set $e_\alpha = e_{i,-j} + e_{j,-i}$ and $f_\alpha = e_{-j,i} - e_{-i,j}$.
Then $h_\alpha = e_{i,i} + e_{j,j} - e_{-j,-j} - e_{-i,-i}$.
(c) For $\alpha = 2\varepsilon_i$ with $1 \leq i \leq l$ set $e_\alpha = e_{i,-i}$ and $f_\alpha = e_{-i,i}$.
Then $h_\alpha = e_{i,i} - e_{-i,-i}$.

Type D:
(a) For $\alpha = \varepsilon_i - \varepsilon_j$ with $1 \leq i < j \leq l$ set $e_\alpha = e_{i,j} - e_{-j,-i}$ and $f_\alpha = e_{j,i} - e_{-i,-j}$.
Then $h_\alpha = e_{i,i} - e_{j,j} + e_{-j,-j} - e_{-i,-i}$.
(b) For $\alpha = \varepsilon_i + \varepsilon_j$ with $1 \leq i < j \leq l$ set $e_\alpha = e_{i,-j} - e_{j,-i}$ and $f_\alpha = e_{-j,i} - e_{-i,j}$.
Then $h_\alpha = e_{i,i} + e_{j,j} - e_{-j,-j} - e_{-i,-i}$.

In all cases it is evident that $\langle \alpha, h_\alpha \rangle = 2$, so e_α, f_α satisfy the conditions of Lemma 2.4.2.

We call h_α the *coroot* to α. Since the space $[\mathfrak{g}_\alpha, \mathfrak{g}_{-\alpha}]$ has dimension one, h_α is uniquely determined by the properties $h_\alpha \in [\mathfrak{g}_\alpha, \mathfrak{g}_{-\alpha}]$ and $\langle \alpha, h_\alpha \rangle = 2$. For $X, Y \in \mathfrak{g}$ let the bilinear form (X, Y) be defined as in Theorem 2.4.1. This form is nondegenerate on $\mathfrak{h} \times \mathfrak{h}$; hence we may use it to identify $\mathfrak{h}$ with $\mathfrak{h}^*$. Then (2.27) implies that

h_α is proportional to α. Furthermore, $(h_\alpha, h_\alpha) = \langle \alpha, h_\alpha \rangle (e_\alpha, f_\alpha) \neq 0$. Hence with $\mathfrak{h}$ identified with $\mathfrak{h}^*$ we have

$$\alpha = \frac{2}{(h_\alpha, h_\alpha)} h_\alpha \,. \tag{2.28}$$

We will also use the notation $\check{\alpha}$ for the coroot h_α.

For $\alpha \in \Phi$ we denote by $\mathfrak{s}(\alpha)$ the algebra spanned by $\{e_\alpha, f_\alpha, h_\alpha\}$. It is isomorphic to $\mathfrak{sl}(2, \mathbb{C})$ under the map $e \mapsto e_\alpha, f \mapsto f_\alpha, h \mapsto h_\alpha$. The algebra $\mathfrak{g}$ becomes a module for $\mathfrak{s}(\alpha)$ by restricting the adjoint representation of $\mathfrak{g}$ to $\mathfrak{s}(\alpha)$. We can thus apply the results on the representations of $\mathfrak{sl}(2, \mathbb{C})$ that we obtained in Section 2.3.3 to study commutation relations in $\mathfrak{g}$.

Let $\alpha, \beta \in \Phi$ with $\alpha \neq \pm\beta$. We observe that $\beta + k\alpha \neq 0$, by Theorem 2.4.1 (2). Hence for every $k \in \mathbb{Z}$,

$$\dim \mathfrak{g}_{\beta + k\alpha} = \begin{cases} 1 \text{ if } \beta + k\alpha \in \Phi \,, \\ 0 \text{ otherwise.} \end{cases}$$

Let

$$R(\alpha, \beta) = \{\beta + k\alpha : k \in \mathbb{Z}\} \cap \Phi \,,$$

which we call the α *root string* through β. The number of elements of a root string is called the *length* of the string. Define

$$V_{\alpha, \beta} = \sum_{\gamma \in R(\alpha, \beta)} \mathfrak{g}_\gamma \,.$$

Lemma 2.4.3. *For every* $\alpha, \beta \in \Phi$ *with* $\alpha \neq \pm\beta$, *the space* $V_{\alpha, \beta}$ *is invariant and irreducible under* $\mathrm{ad}(\mathfrak{s}(\alpha))$.

Proof. From (2.25) we have $[\mathfrak{g}_\alpha, \mathfrak{g}_{\beta + k\alpha}] \subset \mathfrak{g}_{\beta + (k+1)\alpha}$ and $[\mathfrak{g}_{-\alpha}, \mathfrak{g}_{\beta + k\alpha}] \subset \mathfrak{g}_{\beta + (k-1)\alpha}$, so we see that $V_{\alpha, \beta}$ is invariant under $\mathrm{ad}(\mathfrak{s}(\alpha))$. Denote by π the representation of $\mathfrak{s}(\alpha)$ on $V_{\alpha, \beta}$.

If $\gamma = \beta + k\alpha \in \Phi$, then $\pi(h_\alpha)$ acts on the one-dimensional space $\mathfrak{g}_\gamma$ by the scalar

$$\langle \gamma, h_\alpha \rangle = \langle \beta, h_\alpha \rangle + k\langle \alpha, h_\alpha \rangle = \langle \beta, h_\alpha \rangle + 2k \,.$$

Thus by (2.29) we see that the eigenvalues of $\pi(h_\alpha)$ are integers and are either all even or all odd. Furthermore, each eigenvalue occurs with multiplicity one.

Suppose for the sake of contradiction that $V_{\alpha, \beta}$ is not irreducible under $\mathfrak{s}(\alpha)$. Then by Theorem 2.3.6, $V_{\alpha, \beta}$ contains nonzero irreducible invariant subspaces U and W with $W \cap U = \{0\}$. By Proposition 2.3.3 the eigenvalues of h_α on W are n, $n - 2, \ldots, -n + 2, -n$ and the eigenvalues of h_α on U are $m, m - 2, \ldots, -m + 2$, $-m$, where m and n are nonnegative integers. The eigenvalues of h_α on W and on U are subsets of the set of eigenvalues of $\pi(h_\alpha)$, so it follows that m and n are both even or both odd. But this implies that the eigenvalue $\min(m, n)$ of $\pi(h_\alpha)$ has multiplicity greater than one, which is a contradiction. $\qquad\square$

Corollary 2.4.4. *If* $\alpha, \beta \in \Phi$ *and* $\alpha + \beta \in \Phi$, *then* $[\mathfrak{g}_\alpha, \mathfrak{g}_\beta] = \mathfrak{g}_{\alpha + \beta}$.

Proof. Since $\alpha + \beta \in \Phi$, we have $\alpha \neq \pm\beta$. Thus $V_{\alpha,\beta}$ is irreducible under $\mathfrak{s}_\alpha$ and contains $\mathfrak{g}_{\alpha+\beta}$. Hence by (2.16) the operator $E = \pi(e_\alpha)$ maps $\mathfrak{g}_\beta$ onto $\mathfrak{g}_{\alpha+\beta}$. $\qquad\square$

Corollary 2.4.5. *Let $\alpha, \beta \in \Phi$ with $\beta \neq \pm\alpha$. Let p be the largest integer $j \geq 0$ such that $\beta + j\alpha \in \Phi$ and let q be the largest integer $k \geq 0$ such that $\beta - k\alpha \in \Phi$. Then*

$$\langle \beta, h_\alpha \rangle = q - p \in \mathbb{Z},$$

and $\beta + r\alpha \in \Phi$ for all integers r with $-q \leq r \leq p$. In particular, $\beta - \langle \beta, h_\alpha \rangle \alpha \in \Phi$.

Proof. The largest eigenvalue of $\pi(h_\alpha)$ is the positive integer $n = \langle \beta, h_\alpha \rangle + 2p$. Since π is irreducible, Proposition 2.3.3 implies that the eigenspaces of $\pi(h_\alpha)$ are $\mathfrak{g}_{\beta+r\alpha}$ for $r = p, p-1, \ldots, -q+1, -q$. Hence the α-string through β is $\beta + r\alpha$ with $r = p, p-1, \ldots, -q+1, -q$. Furthermore, the smallest eigenvalue of $\pi(h)$ is $-n = \langle \beta, h_\alpha \rangle - 2q$. This gives the relation

$$-\langle \beta, h_\alpha \rangle - 2p = \langle \beta, h_\alpha \rangle - 2q.$$

Hence $\langle \beta, h_\alpha \rangle = q - p$. Since $p \geq 0$ and $q \geq 0$, we see that $-q \leq -\langle \beta, h_\alpha \rangle \leq p$. Thus $\beta - \langle \beta, h_\alpha \rangle \alpha \in \Phi$. $\qquad\square$

Remark 2.4.6. From the case-by-case calculations for types **A–D** made above we see that

$$\langle \beta, h_\alpha \rangle \in \{0, \pm 1, \pm 2\} \quad \text{for all } \alpha, \beta \in \Phi. \tag{2.29}$$

2.4.3 *Structure of Classical Root Systems*

In the previous section we saw that the commutation relations in the Lie algebra of a classical group are controlled by the root system. We now study the root systems in more detail. Let Φ be the root system for a classical Lie algebra $\mathfrak{g}$ of type A_l, B_l, C_l, or D_l (with $l \geq 3$ for D_l). Then Φ spans $\mathfrak{h}^*$ (this is clear from the descriptions in Section 2.4.1). Thus we can choose (in many ways) a set of roots that is a basis for $\mathfrak{h}^*$. An optimal choice of basis is the following:

Definition 2.4.7. A subset $\Delta = \{\alpha_1, \ldots, \alpha_l\} \subset \Phi$ is a set of *simple roots* if every $\gamma \in \Phi$ can be written uniquely as

$$\gamma = n_1\alpha_1 + \cdots + n_l\alpha_l, \text{ with } n_1, \ldots, n_l \text{ integers all of the same sign.} \tag{2.30}$$

Notice that the requirement of uniqueness in expression (2.30), together with the fact that Φ spans $\mathfrak{h}^*$, implies that Δ is a basis for $\mathfrak{h}^*$. Furthermore, if Δ is a set of simple roots, then it partitions Φ into two disjoint subsets

$$\Phi = \Phi^+ \cup (-\Phi^+),$$

where Φ^+ consists of all the roots for which the coefficients n_i in (2.30) are non-negative. We call $\gamma \in \Phi^+$ a *positive root*, relative to Δ.

We shall show, with a case-by-case analysis, that Φ has a set of simple roots. We first prove that if $\Delta = \{\alpha_1, \ldots, \alpha_l\}$ is a set of simple roots and $i \neq j$, then

$$\langle \alpha_i, h_{\alpha_j} \rangle \in \{0, -1, -2\}.$$

Indeed, we have already observed that $\langle \alpha, h_\beta \rangle \in \{0, \pm 1, \pm 2\}$ for all roots α, β. Let $H_i = h_{\alpha_i}$ be the coroot to α_i and define

$$C_{ij} = \langle \alpha_j, H_i \rangle. \tag{2.31}$$

Set $\gamma = \alpha_j - C_{ij}\alpha_i$. By Corollary 2.4.5 we have $\gamma \in \Phi$. If $C_{ij} > 0$ this expansion of γ would contradict (2.30). Hence $C_{ij} \leq 0$ for all $i \neq j$.

Remark 2.4.8. The integers C_{ij} in (2.31) are called the *Cartan integers*, and the $l \times l$ matrix $C = [C_{ij}]$ is called the *Cartan matrix* for the set Δ. Note that the diagonal entries of C are $\langle \alpha_i, H_i \rangle = 2$.

If Δ is a set of simple roots and $\beta = n_1\alpha_1 + \cdots + n_l\alpha_l$ is a root, then we define the *height* of β (relative to Δ) as

$$\mathrm{ht}(\beta) = n_1 + \cdots + n_l.$$

The positive roots are then the roots β with $\mathrm{ht}(\beta) > 0$. A root β is called the *highest root* of Φ, relative to a set Δ of simple roots, if

$$\mathrm{ht}(\beta) > \mathrm{ht}(\gamma) \quad \text{for all roots } \gamma \neq \beta.$$

If such a root exists, it is clearly unique.

We now give a set of simple roots and the associated Cartan matrix and positive roots for each classical root system, and we show that there is a highest root, denoted by $\widetilde{\alpha}$ (in type D_l we assume $l \geq 3$). We write the coroots H_i in terms of the elementary diagonal matrices $E_i = e_{i,i}$, as in Section 2.4.1. The Cartan matrix is very sparse, and it can be efficiently encoded in terms of a *Dynkin diagram*. This is a graph with a node for each root $\alpha_i \in \Delta$. The nodes corresponding to α_i and α_j are joined by $C_{ij}C_{ji}$ lines for $i \neq j$. Furthermore, if the two roots are of different lengths (relative to the inner product for which $\{\varepsilon_i\}$ is an orthonormal basis), then an inequality sign is placed on the lines to indicate which root is longer. We give the Dynkin diagrams and indicate the root corresponding to each node in each case. Above the node for α_i we put the coefficient of α_i in the highest root.

Type A $(G = \mathbf{SL}(l+1, \mathbb{C}))$: Let $\alpha_i = \varepsilon_i - \varepsilon_{i+1}$ and $\Delta = \{\alpha_1, \ldots, \alpha_l\}$. Since

$$\varepsilon_i - \varepsilon_j = \alpha_i + \cdots + \alpha_{j-1} \quad \text{for } 1 \leq i < j \leq l+1,$$

we see that Δ is a set of simple roots. The associated set of positive roots is

$$\Phi^+ = \{\varepsilon_i - \varepsilon_j : 1 \leq i < j \leq l+1\} \tag{2.32}$$

and the highest root is $\tilde{\alpha} = \varepsilon_1 - \varepsilon_{l+1} = \alpha_1 + \cdots + \alpha_l$ with $\mathrm{ht}(\tilde{\alpha}) = l$. Here $H_i = E_i - E_{i+1}$. Thus the Cartan matrix has $C_{ij} = -1$ if $|i - j| = 1$ and $C_{ij} = 0$ if $|i - j| > 1$. The Dynkin diagram is shown in Figure 2.1.

Fig. 2.1 Dynkin diagram of
type A_l.

Type B ($G = \mathbf{SO}(2l+1, \mathbb{C})$): Let $\alpha_i = \varepsilon_i - \varepsilon_{i+1}$ for $1 \le i \le l-1$ and $\alpha_l = \varepsilon_l$. Take $\Delta = \{\alpha_1, \ldots, \alpha_l\}$. For $1 \le i < j \le l$, we can write $\varepsilon_i - \varepsilon_j = \alpha_i + \cdots + \alpha_{j-1}$ as in type A, whereas

$$
\begin{aligned}
\varepsilon_i + \varepsilon_j &= (\varepsilon_i - \varepsilon_l) + (\varepsilon_j - \varepsilon_l) + 2\varepsilon_l \\
&= \alpha_i + \cdots + \alpha_{l-1} + \alpha_j + \cdots + \alpha_{l-1} + 2\alpha_l \\
&= \alpha_i + \cdots + \alpha_{j-1} + 2\alpha_j + \cdots + 2\alpha_l \, .
\end{aligned}
$$

For $1 \le i \le l$ we have $\varepsilon_i = (\varepsilon_i - \varepsilon_l) + \varepsilon_l = \alpha_i + \cdots + \alpha_l$. These formulas show that Δ is a set of simple roots. The associated set of positive roots is

$$\Phi^+ = \{\varepsilon_i - \varepsilon_j, \varepsilon_i + \varepsilon_j : 1 \le i < j \le l\} \cup \{\varepsilon_i : 1 \le i \le l\}. \qquad (2.33)$$

The highest root is $\tilde{\alpha} = \varepsilon_1 + \varepsilon_2 = \alpha_1 + 2\alpha_2 + \cdots + 2\alpha_l$ with $\mathrm{ht}(\tilde{\alpha}) = 2l - 1$. The simple coroots are

$$H_i = E_i - E_{i+1} + E_{-i-1} - E_{-i} \quad \text{for } 1 \le i \le l-1 \, ,$$

and $H_l = 2E_l - 2E_{-l}$, where we are using the same enumeration of the basis for $\mathbb{C}^{2l+1}$ as in Section 2.4.1. Thus the Cartan matrix has $C_{ij} = -1$ if $|i - j| = 1$ and $i, j \le l-1$, whereas $C_{l-1,l} = -2$ and $C_{l,l-1} = -1$. All other nondiagonal entries are zero. The Dynkin diagram is shown in Figure 2.2 for $l \ge 2$.

Fig. 2.2 Dynkin diagram of
type B_l.

Type C ($G = \mathbf{Sp}(l, \mathbb{C})$): Let $\alpha_i = \varepsilon_i - \varepsilon_{i+1}$ for $1 \le i \le l-1$ and $\alpha_l = 2\varepsilon_l$. Take $\Delta = \{\alpha_1, \ldots, \alpha_l\}$. For $1 \le i < j \le l$ we can write $\varepsilon_i - \varepsilon_j = \alpha_i + \cdots + \alpha_{j-1}$ and $\varepsilon_i + \varepsilon_l = \alpha_i + \cdots + \alpha_l$, whereas for $1 \le i < j \le l-1$ we have

$$
\begin{aligned}
\varepsilon_i + \varepsilon_j &= (\varepsilon_i - \varepsilon_l) + (\varepsilon_j - \varepsilon_l) + 2\varepsilon_l \\
&= \alpha_i + \cdots + \alpha_{l-1} + \alpha_j + \cdots + \alpha_{l-1} + \alpha_l \\
&= \alpha_i + \cdots + \alpha_{j-1} + 2\alpha_j + \cdots + 2\alpha_{l-1} + \alpha_l \, .
\end{aligned}
$$

For $1 \le i < l$ we have $2\varepsilon_i = 2(\varepsilon_i - \varepsilon_l) + 2\varepsilon_l = 2\alpha_i + \cdots + 2\alpha_{l-1} + \alpha_l$. These formulas show that Δ is a set of simple roots. The associated set of positive roots is

$$\Phi^+ = \{\varepsilon_i - \varepsilon_j, \varepsilon_i + \varepsilon_j : 1 \le i < j \le l\} \cup \{2\varepsilon_i : 1 \le i \le l\}. \tag{2.34}$$

The highest root is $\widetilde{\alpha} = 2\varepsilon_1 = 2\alpha_1 + \cdots + 2\alpha_{l-1} + \alpha_l$ with $\mathrm{ht}(\widetilde{\alpha}) = 2l - 1$. The simple coroots are

$$H_i = E_i - E_{i+1} + E_{-i-1} - E_{-i} \quad \text{for } 1 \le i \le l-1,$$

and $H_l = E_l - E_{-l}$, where we are using the same enumeration of the basis for $\mathbb{C}^{2l+1}$ as in Section 2.4.1. The Cartan matrix has $C_{ij} = -1$ if $|i - j| = 1$ and $i, j \le l-1$, whereas now $C_{l-1,l} = -1$ and $C_{l,l-1} = -2$. All other nondiagonal entries are zero. Notice that this is the transpose of the Cartan matrix of type B. If $l \ge 2$ the Dynkin diagram is shown in Figure 2.3. It can be obtained from the Dynkin diagram of type B_l by reversing the arrow on the double bond and reversing the coefficients of the highest root. In particular, the diagrams B_2 and C_2 are identical. (This low-rank coincidence was already noted in Exercises 1.1.5 #8; it is examined further in Exercises 2.4.5 #6.)

Fig. 2.3 Dynkin diagram of type C_l.

Type D $(G = \mathbf{SO}(2l, \mathbb{C})$ with $l \ge 3)$: Let $\alpha_i = \varepsilon_i - \varepsilon_{i+1}$ for $1 \le i \le l-1$ and $\alpha_l = \varepsilon_{l-1} + \varepsilon_l$. For $1 \le i < j \le l$ we can write $\varepsilon_i - \varepsilon_j = \alpha_i + \cdots + \alpha_{j-1}$ as in type A, whereas for $1 \le i < l-1$ we have

$$\varepsilon_i + \varepsilon_{l-1} = \alpha_i + \cdots + \alpha_l, \qquad \varepsilon_i + \varepsilon_l = \alpha_i + \cdots + \alpha_{l-2} + \alpha_l.$$

For $1 \le i < j \le l-2$ we have

$$\begin{aligned}
\varepsilon_i + \varepsilon_j &= (\varepsilon_i - \varepsilon_{l-1}) + (\varepsilon_j - \varepsilon_l) + (\varepsilon_{l-1} + \varepsilon_l) \\
&= \alpha_i + \cdots + \alpha_{l-2} + \alpha_j + \cdots + \alpha_{l-1} + \alpha_l \\
&= \alpha_i + \cdots + \alpha_{j-1} + 2\alpha_j + \cdots + 2\alpha_{l-2} + \alpha_{l-1} + \alpha_l.
\end{aligned}$$

These formulas show that Δ is a set of simple roots. The associated set of positive roots is

$$\Phi^+ = \{\varepsilon_i - \varepsilon_j, \varepsilon_i + \varepsilon_j : 1 \le i < j \le l\}. \tag{2.35}$$

The highest root is $\widetilde{\alpha} = \varepsilon_1 + \varepsilon_2 = \alpha_1 + 2\alpha_2 + \cdots + 2\alpha_{l-2} + \alpha_{l-1} + \alpha_l$ with $\mathrm{ht}(\widetilde{\alpha}) = 2l - 3$. The simple coroots are

$$H_i = E_i - E_{i+1} + E_{-i-1} - E_{-i} \quad \text{for } 1 \le i \le l-1,$$

and $H_l = E_{l-1} + E_l - E_{-l} - E_{-l+1}$, with the same enumeration of the basis for $\mathbb{C}^{2l}$ as in type C. Thus the Cartan matrix has $C_{ij} = -1$ if $|i - j| = 1$ and $i, j \le l-1$, whereas $C_{l-2,l} = C_{l,l-2} = -1$. All other nondiagonal entries are zero. The Dynkin diagram is shown in Figure 2.4. Notice that when $l = 2$ the diagram is not connected (it is the diagram for $\mathfrak{sl}(2, \mathbb{C}) \oplus \mathfrak{sl}(2, \mathbb{C})$; see Remark 2.2.6). When $l = 3$ the diagram is the same as the diagram for type A_3. This low-rank coincidence was already noted in Exercises 1.1.5 #7; it is examined further in Exercises 2.4.5 #5.

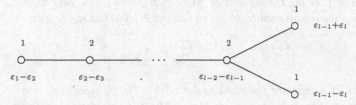

Fig. 2.4 Dynkin diagram of type D_l.

Remark 2.4.9. The Dynkin diagrams of the four types of classical groups are distinct except in the cases $A_1 = B_1 = C_1$, $B_2 = C_2$, and $A_3 = D_3$. In these cases there are corresponding Lie algebra isomorphisms; see Section 2.2.1 for the rank-one simple algebras and see Exercises 2.4.5 for the isomorphisms $\mathfrak{so}(\mathbb{C}^5) \cong \mathfrak{sp}(\mathbb{C}^4)$ and $\mathfrak{sl}(\mathbb{C}^4) \cong \mathfrak{so}(\mathbb{C}^6)$. We will show in Chapter 3 that all systems of simple roots are conjugate by the Weyl group; hence the Dynkin diagram is uniquely defined by the root system and does not depend on the choice of a simple set of roots. Thus the Dynkin diagram completely determines the Lie algebra up to isomorphism.

For a root system of types A or D, in which all the roots have squared length two (relative to the trace form inner product on $\mathfrak{h}$), the identification of $\mathfrak{h}$ with $\mathfrak{h}^*$ takes roots to coroots. For root systems of type B or C, in which the roots have two lengths, the roots of type B_l are identified with the coroots of type C_l and vice versa (e.g., ε_i is identified with the coroot to $2\varepsilon_i$ and vice versa). This allows us to transfer results known for roots to analogous results for coroots. For example, if $\alpha \in \Phi^+$ then

$$H_\alpha = m_1 H_1 + \cdots + m_l H_l, \tag{2.36}$$

where m_i is a nonnegative integer for $i = 1, \ldots, l$.

Lemma 2.4.10. *Let Φ be the root system for a classical Lie algebra $\mathfrak{g}$ of rank l and type $A, B, C,$ or D (in the case of type D assume that $l \geq 3$). Let the system of simple roots $\Delta \subset \Phi$ be chosen as above. Let Φ^+ be the positive roots and let $\tilde{\alpha}$ be the highest root relative to Δ. Then the following properties hold:*

1. *If $\alpha, \beta \in \Phi^+$ and $\alpha + \beta \in \Phi$, then $\alpha + \beta \in \Phi^+$.*
2. *If $\beta \in \Phi^+$ and β is not a simple root, then there exist $\gamma, \delta \in \Phi^+$ such that $\beta = \gamma + \delta$.*
3. *$\tilde{\alpha} = n_1 \alpha_1 + \cdots + n_l \alpha_l$ with $n_i \geq 1$ for $i = 1, \ldots, l$.*
4. *For any $\beta \in \Phi^+$ with $\beta \neq \tilde{\alpha}$ there exists $\alpha \in \Phi^+$ such that $\alpha + \beta \in \Phi^+$.*
5. *If $\alpha \in \Phi^+$ and $\alpha \neq \tilde{\alpha}$, then there exist $1 \leq i_1, i_2, \ldots, i_r \leq l$ such that $\alpha = \tilde{\alpha} - \alpha_{i_1} - \cdots - \alpha_{i_r}$ and $\tilde{\alpha} - \alpha_{i_1} - \cdots - \alpha_{i_j} \in \Phi$ for all $1 \leq j \leq r$.*

Proof. Property (1) is clear from the definition of a system of simple roots. Properties (2)–(5) follow on a case-by-case basis from the calculations made above. We leave the details as an exercise. □

We can now state the second structure theorem for $\mathfrak{g}$.

Theorem 2.4.11. *Let* $\mathfrak{g}$ *be the Lie algebra of* $\mathbf{SL}(l+1, \mathbb{C})$, $\mathbf{Sp}(\mathbb{C}^{2l}, \Omega)$, *or* $\mathbf{SO}(\mathbb{C}^{2l+1}, B)$ *with* $l \geq 1$, *or the Lie algebra of* $\mathbf{SO}(\mathbb{C}^{2l}, B)$ *with* $l \geq 3$. *Take the set of simple roots* Δ *and the positive roots* Φ^+ *as in Lemma 2.4.10. The subspaces*

$$\mathfrak{n}^+ = \bigoplus_{\alpha \in \Phi^+} \mathfrak{g}_\alpha, \quad \mathfrak{n}^- = \bigoplus_{\alpha \in \Phi^+} \mathfrak{g}_{-\alpha}$$

are Lie subalgebras of $\mathfrak{g}$ *that are invariant under* $\mathrm{ad}(\mathfrak{h})$. *The subspace* $\mathfrak{n}^+ + \mathfrak{n}^-$ *generates* $\mathfrak{g}$ *as a Lie algebra. In particular,* $\mathfrak{g} = [\mathfrak{g}, \mathfrak{g}]$. *There is a vector space direct sum decomposition*

$$\mathfrak{g} = \mathfrak{n}^- + \mathfrak{h} + \mathfrak{n}^+ . \tag{2.37}$$

Furthermore, the iterated Lie brackets of the root spaces $\mathfrak{g}_{\alpha_1}, \ldots, \mathfrak{g}_{\alpha_l}$ *span* $\mathfrak{n}^+$, *and the iterated Lie brackets of the root spaces* $\mathfrak{g}_{-\alpha_1}, \ldots, \mathfrak{g}_{-\alpha_l}$ *span* $\mathfrak{n}^-$.

Proof. The fact that $\mathfrak{n}^+$ and $\mathfrak{n}^-$ are subalgebras follows from property (1) in Lemma 2.4.10. Equation (2.37) is clear from Theorem 2.4.1 and the decomposition

$$\Phi = \Phi^+ \cup (-\Phi^+) .$$

For $\alpha \in \Phi$ let $h_\alpha \in \mathfrak{h}$ be the coroot. From the calculations above it is clear that $\mathfrak{h} = \mathrm{Span}\{h_\alpha : \alpha \in \Phi\}$. Since $h_\alpha \in [\mathfrak{g}_\alpha, \mathfrak{g}_{-\alpha}]$ by Lemma 2.4.2, we conclude from (2.37) that $\mathfrak{n}^+ + \mathfrak{n}^-$ generates $\mathfrak{g}$ as a Lie algebra.

To verify that $\mathfrak{n}^+$ is generated by the simple root spaces, we use induction on the height of $\beta \in \Phi^+$ (the simple roots being the roots of height 1). If β is not simple, then $\beta = \gamma + \delta$ for some $\gamma, \delta \in \Phi^+$ (Lemma 2.4.10 (2)). But we know that $[\mathfrak{g}_\gamma, \mathfrak{g}_\delta] = \mathfrak{g}_\beta$ from Corollary 2.4.4. Since the heights of γ and δ are less than the height of β, the induction continues. The same argument applies to $\mathfrak{n}^-$. $\qquad\square$

Remark 2.4.12. When $\mathfrak{g}$ is taken in the matrix form of Section 2.4.1, then $\mathfrak{n}^+$ consists of all strictly upper-triangular matrices in $\mathfrak{g}$, and $\mathfrak{n}^-$ consists of all strictly lower-triangular matrices in $\mathfrak{g}$. Furthermore, $\mathfrak{g}$ is invariant under the map $\theta(X) = -X^t$ (negative transpose). This map is an automorphism of $\mathfrak{g}$ with $\theta^2 = \mathrm{Identity}$. Since $\theta(H) = -H$ for $H \in \mathfrak{h}$, it follows that $\theta(\mathfrak{g}_\alpha) = \mathfrak{g}_{-\alpha}$. Indeed, if $[H, X] = \alpha(H)X$ then

$$[H, \theta(X)] = \theta([-H, X]) = -\alpha(H)\theta(X) .$$

In particular, $\theta(\mathfrak{n}^+) = \mathfrak{n}^-$.

2.4.4 Irreducibility of the Adjoint Representation

Now that we have the root space decompositions of the Lie algebras of the classical groups, we can prove the following fundamental result:

Theorem 2.4.13. *Let G be one of the groups $\mathbf{SL}(\mathbb{C}^{l+1})$, $\mathbf{Sp}(\mathbb{C}^{2l})$, $\mathbf{SO}(\mathbb{C}^{2l+1})$ with $l \geq 1$, or $\mathbf{SO}(\mathbb{C}^{2l})$ with $l \geq 3$. Then the adjoint representation of G is irreducible.*

Proof. By Theorems 2.2.2 and 2.2.7 it will suffice to show that $\mathrm{ad}(\mathfrak{g})$ acts irreducibly on $\mathfrak{g} = \mathrm{Lie}(G)$. Let Φ, Φ^+, Δ, and $\widetilde{\alpha}$ be as in Lemma 2.4.10.

Suppose U is a nonzero $\mathrm{ad}(\mathfrak{g})$-invariant subspace of $\mathfrak{g}$. We shall prove that $U = \mathfrak{g}$. Since $[\mathfrak{h}, U] \subset U$ and each root space $\mathfrak{g}_\alpha$ has dimension one, we have a decomposition

$$U = (U \cap \mathfrak{h}) \oplus \left(\bigoplus_{\alpha \in S} \mathfrak{g}_\alpha \right),$$

where $S = \{\alpha \in \Phi : \mathfrak{g}_\alpha \subset U\}$. We claim that

(1) S is nonempty.

Indeed, if $U \subset \mathfrak{h}$, then we would have $[U, \mathfrak{g}_\alpha] \subset U \cap \mathfrak{g}_\alpha = 0$ for all $\alpha \in \Phi$. Hence $\alpha(U) = 0$ for all roots α, which would imply $U = 0$, since the roots span $\mathfrak{h}^*$, a contradiction. This proves (1). Next we prove

(2) $U \cap \mathfrak{h} \neq 0$.

To see this, take $\alpha \in S$. Then by Lemma 2.4.2 we have $h_\alpha = -[f_\alpha, e_\alpha] \in U \cap \mathfrak{h}$. Now let $\alpha \in \Phi$. Then we have the following:

(3)   If $\alpha(U \cap \mathfrak{h}) \neq 0$ then $\mathfrak{g}_\alpha \subset U$.

Indeed, $[U \cap \mathfrak{h}, \mathfrak{g}_\alpha] = \mathfrak{g}_\alpha$ in this case.

From (3) we see that if $\alpha \in S$ then $-\alpha \in S$. Set $S^+ = S \cap \Phi^+$. If $\alpha \in S^+$ and $\alpha \neq \widetilde{\alpha}$, then by Lemma 2.4.10 (3) there exists $\gamma \in \Phi^+$ such that $\alpha + \gamma \in \Phi$. Since $[\mathfrak{g}_\alpha, \mathfrak{g}_\gamma] = \mathfrak{g}_{\alpha+\gamma}$ by Corollary 2.4.4, we see that $\mathfrak{g}_{\alpha+\gamma} \subset U$. Hence $\alpha + \gamma \in S^+$ and has a height greater than that of α. Thus if $\beta \in S^+$ has maximum height among the elements of S^+, then $\beta = \widetilde{\alpha}$. This proves that $\widetilde{\alpha} \in S^+$. We can now prove

(4) $S = \Phi$.

By (3) it suffices to show that $S^+ = \Phi^+$. Given $\alpha \in \Phi^+$ choose $i_1, \ldots, i_r$ as in Lemma 2.4.10 (5) and set

$$\beta_j = \widetilde{\alpha} - \alpha_{i_1} - \cdots - \alpha_{i_j} \quad \text{for } j = 1, \ldots, r.$$

Write $F_i = f_{\alpha_i}$ for the element in Lemma 2.4.2. Then by induction on j and Corollary 2.4.4 we have

$$\mathfrak{g}_{\beta_j} = \mathrm{ad}(F_{i_j}) \cdots \mathrm{ad}(F_{i_1}) \mathfrak{g}_{\widetilde{\alpha}} \subset U \quad \text{for } j = 1, \ldots, r.$$

Taking $j = r$, we conclude that $\mathfrak{g}_\alpha \subset U$, which proves (4). Hence $U \cap \mathfrak{h} = \mathfrak{h}$, since $\mathfrak{h} \subset [\mathfrak{n}^+, \mathfrak{n}^-]$. This shows that $U = \mathfrak{g}$. $\square$

Remark 2.4.14. For any Lie algebra $\mathfrak{g}$, the subspaces of $\mathfrak{g}$ that are invariant under $\mathrm{ad}(\mathfrak{g})$ are the *ideals* of $\mathfrak{g}$. A Lie algebra is called *simple* if it is not abelian and it has no proper ideals. (By this definition the one-dimensional Lie algebra is not simple, even though it has no proper ideals.) The classical Lie algebras occurring in Theorem 2.4.13 are thus simple. Note that their Dynkin diagrams are connected graphs.

Remark 2.4.15. A Lie algebra is called *semisimple* if it is a direct sum of simple Lie algebras. The low-dimensional orthogonal Lie algebras excluded from Theorem 2.4.11 and Theorem 2.4.13 are $\mathfrak{so}(4,\mathbb{C}) \cong \mathfrak{sl}(2,\mathbb{C}) \oplus \mathfrak{sl}(2,\mathbb{C})$, which is semisimple (with a disconnected Dynkin diagram), and $\mathfrak{so}(2,\mathbb{C}) \cong \mathfrak{gl}(1,\mathbb{C})$, which is abelian (and has no roots).

2.4.5 Exercises

1. For each type of classical group write out the coroots in terms of the ε_i (after the identification of $\mathfrak{h}$ with $\mathfrak{h}^*$ as in Section 2.4.1). Show that for types A and D the roots and coroots are the same. Show that for type B the coroots are the same as the roots for C and vice versa.

2. Let G be a classical group. Let Φ be the root system for G, $\alpha_1, \ldots, \alpha_l$ the simple roots, and Φ^+ the positive roots as in Lemma 2.4.10. Verify that the calculations in Section 2.4.3 can be expressed as follows:

   (a) For G of type A_l, $\Phi^+ \setminus \Delta$ consists of the roots

   $$\alpha_i + \cdots + \alpha_j \quad \text{for } 1 \leq i < j \leq l.$$

   (b) For G of type B_l with $l \geq 2$, $\Phi^+ \setminus \Delta$ consists of the roots

   $$\alpha_i + \cdots + \alpha_j \quad \text{for } 1 \leq i < j \leq l,$$
   $$\alpha_i + \cdots + \alpha_{j-1} + 2\alpha_j + \cdots + 2\alpha_l \quad \text{for } 1 \leq i < j \leq l.$$

   (c) For G of type C_l with $l \geq 2$, $\Phi^+ \setminus \Delta$ consists of the roots

   $$\alpha_i + \cdots + \alpha_j \quad \text{for } 1 \leq i < j \leq l,$$
   $$\alpha_i + \cdots + \alpha_{j-1} + 2\alpha_j + \cdots + 2\alpha_{l-1} + \alpha_l \quad \text{for } 1 \leq i < j < l,$$
   $$2\alpha_i + \cdots + 2\alpha_{l-1} + \alpha_l \quad \text{for } 1 \leq i < l.$$

   (d) For G of type D_l with $l \geq 3$, $\Phi^+ \setminus \Delta$ consists of the roots

   $$\alpha_i + \cdots + \alpha_j \quad \text{for } 1 \leq i < j < l,$$
   $$\alpha_i + \cdots + \alpha_l \quad \text{for } 1 \leq i < l-1,$$
   $$\alpha_i + \cdots + \alpha_{l-2} + \alpha_l \quad \text{for } 1 \leq i < l-1,$$
   $$\alpha_i + \cdots + \alpha_{j-1} + 2\alpha_j + \cdots + 2\alpha_{l-2} + \alpha_{l-1} + \alpha_l \quad \text{for } 1 \leq i < j < l-1.$$

   Now use (a)–(d) to prove assertions (2)–(5) in Lemma 2.4.10.

3. (Assumptions and notation as in Lemma 2.4.10.) Let $S \subset \Delta$ be any subset that corresponds to a *connected* subgraph of the Dynkin diagram of Δ. Use the previous exercise to verify that $\sum_{\alpha \in S} \alpha$ is a root.

4. (Assumptions and notation as in Lemma 2.4.2 and Lemma 2.4.10.) Let $1 \le i, j \le l$ with $i \ne j$ and let C_{ij} be the Cartan integers.
   (a) Show that the α_j root string through α_i is $\alpha_i, \ldots, \alpha_i - C_{ji}\alpha_j$. (HINT: Use the fact that $\alpha_i - \alpha_j$ is not a root and the proof of Corollary 2.4.5.)
   (b) Show that $[e_{\alpha_j}, e_{-\alpha_i}] = 0$ and

$$\mathrm{ad}(e_{\alpha_j})^k(e_{\alpha_i}) \ne 0 \quad \text{for } k = 0, \ldots, -C_{ji},$$
$$\mathrm{ad}(e_{\alpha_j})^k(e_{\alpha_i}) = 0 \quad \text{for } k = -C_{ji} + 1.$$

(HINT: Use (a) and Corollary 2.4.4.)

5. Consider the representation ρ of $\mathbf{SL}(4, \mathbb{C})$ on $\bigwedge^2 \mathbb{C}^4$, where $\rho(g)(v_1 \wedge v_2) = gv_1 \wedge gv_2$ for $g \in \mathbf{SL}(4, \mathbb{C})$ and $v_1, v_2 \in \mathbb{C}^4$. Let $\Omega = e_1 \wedge e_2 \wedge e_3 \wedge e_4$ and let B be the nondegenerate symmetric bilinear form such that $a \wedge b = B(a, b)\Omega$ for $a, b \in \bigwedge^2 \mathbb{C}^4$, as in Exercises 1.1.5 #6 and #7.
   (a) Let $g \in \mathbf{SL}(4, \mathbb{C})$, $X \in \mathfrak{sl}(4, \mathbb{C})$, and $a, b \in \bigwedge^2 \mathbb{C}^4$. Show that

$$B(\rho(g)a, \rho(g)b) = B(a, b) \quad \text{and} \quad B(d\rho(X)a, b) + B(a, d\rho(X)b) = 0.$$

(b) Use $d\rho$ to obtain a Lie algebra isomorphism $\mathfrak{sl}(4, \mathbb{C}) \cong \mathfrak{so}(\bigwedge^2 \mathbb{C}^4, B)$. (HINT: $\mathfrak{sl}(4, \mathbb{C})$ is a simple Lie algebra.)
   (c) Show that $\rho : \mathbf{SL}(4, \mathbb{C}) \longrightarrow \mathbf{SO}(\bigwedge^2 \mathbb{C}^4, B)$ is surjective, and $\mathrm{Ker}(\rho) = \{\pm I\}$. (HINT: For the surjectivity, use (b) and Theorem 2.2.2. To determine $\mathrm{Ker}(\rho)$, use (b) to show that $\mathrm{Ad}(g) = I$ for all $g \in \mathrm{Ker}(\rho)$, and then use Theorem 2.1.5.)

6. Let B be the symmetric bilinear form on $\bigwedge^2 \mathbb{C}^4$ and ρ the representation of $\mathbf{SL}(4, \mathbb{C})$ on $\bigwedge^2 \mathbb{C}$ as in the previous exercise. Let $\omega = e_1 \wedge e_4 + e_2 \wedge e_3$. Identify $\mathbb{C}^4$ with $(\mathbb{C}^4)^*$ by the inner product $(x, y) = x^t y$, so that ω can also be viewed as a skew-symmetric bilinear form on $\mathbb{C}^4$. Define

$$\mathcal{L} = \{a \in \bigwedge^2 \mathbb{C}^4 : B(a, \omega) = 0\}.$$

Then $\rho(g)\mathcal{L} \subset \mathcal{L}$ for all $g \in \mathbf{Sp}(\mathbb{C}^4, \omega)$ and $\bigwedge^2 \mathbb{C}^4 = \mathbb{C}\omega \oplus \mathcal{L}$. Furthermore, if β is the restriction of the bilinear form B to $\mathcal{L} \times \mathcal{L}$, then β is nondegenerate (see Exercises 1.1.5 #8).
   (a) Let $\varphi(g)$ be the restriction of $\rho(g)$ to the subspace $\mathcal{L}$, for $g \in \mathbf{Sp}(\mathbb{C}^4, \omega)$. Use $d\varphi$ to obtain a Lie algebra isomorphism $\mathfrak{sp}(\mathbb{C}^4, \omega) \cong \mathfrak{so}(\mathbb{C}^5, \beta)$. (HINT: $\mathfrak{sp}(\mathbb{C}^4, \omega)$ is a simple Lie algebra.)
   (b) Show that $\varphi : \mathbf{Sp}(\mathbb{C}^4, \omega) \longrightarrow \mathbf{SO}(\mathcal{L}, \beta)$ is surjective and $\mathrm{Ker}(\varphi) = \{\pm I\}$. (HINT: For the surjectivity, use Theorem 2.2.2. To determine $\mathrm{Ker}(\varphi)$, use (a) to show that $\mathrm{Ad}(g) = I$ for all $g \in \mathrm{Ker}(\varphi)$, and then use Theorem 2.1.5.)

2.5 Semisimple Lie Algebras

We will show that the structural features of the Lie algebras of the classical groups studied in Section 2.4 carry over to the class of *semisimple* Lie algebras. This requires some preliminary general results on Lie algebras. These results will be used again in Chapters 11 and 12, but the remainder of the current chapter may be omitted by the reader interested only in the classical groups (in fact, it turns out that there are only five *exceptional* simple Lie algebras, traditionally labeled E_6, E_7, E_8, F_4, and G_2, that are not Lie algebras of classical groups).

2.5.1 Solvable Lie Algebras

We begin with a Lie-algebraic condition for nilpotence of a linear transformation.

Lemma 2.5.1. *Let V be a finite-dimensional complex vector space and let $A \in \mathrm{End}(V)$. Suppose there exist $X_i, Y_i \in \mathrm{End}(V)$ such that $A = \sum_{i=1}^k [X_i, Y_i]$ and $[A, X_i] = 0$ for all i. Then A is nilpotent.*

Proof. Let Σ be the spectrum of A, and let $\{P_\lambda\}_{\lambda \in \Sigma}$ be the resolution of the identity for A (see Lemma B.1.1). Then $P_\lambda X_i = X_i P_\lambda = P_\lambda X_i P_\lambda$ for all i, so

$$P_\lambda [X_i, Y_i] P_\lambda = P_\lambda X_i P_\lambda Y_i P_\lambda - P_\lambda Y_i P_\lambda X_i P_\lambda = [P_\lambda X_i P_\lambda, P_\lambda Y_i P_\lambda] \, .$$

Hence $\mathrm{tr}(P_\lambda [X_i, Y_i] P_\lambda) = 0$ for all i, so we obtain $\mathrm{tr}(P_\lambda A) = 0$ for all $\lambda \in \Sigma$. However, $\mathrm{tr}(P_\lambda A) = \lambda \dim V_\lambda$, where

$$V_\lambda = \{v \in V : (A - \lambda)^k v = 0 \quad \text{for some } k\} \, .$$

It follows that $V_\lambda = 0$ for all $\lambda \neq 0$, so that A is nilpotent.   □

Definition 2.5.2. A finite-dimensional representation (π, V) of a Lie algebra $\mathfrak{g}$ is *completely reducible* if every $\mathfrak{g}$-invariant subspace $W \subset V$ has a $\mathfrak{g}$-invariant complementary subspace U. Thus $W \cap U = \{0\}$ and $V = W \oplus U$.

Theorem 2.5.3. *Let V be a finite-dimensional complex vector space. Suppose $\mathfrak{g}$ is a Lie subalgebra of $\mathrm{End}(V)$ such that V is completely reducible as a representation of $\mathfrak{g}$. Let $\mathfrak{z} = \{X \in \mathfrak{g} : [X, Y] = 0 \text{ for all } Y \in \mathfrak{g}\}$ be the center of $\mathfrak{g}$. Then*

1. every $A \in \mathfrak{z}$ is a semisimple linear transformation;
2. $[\mathfrak{g}, \mathfrak{g}] \cap \mathfrak{z} = 0$;
3. $\mathfrak{g}/\mathfrak{z}$ has no nonzero abelian ideal.

Proof. Complete reducibility implies that $V = \bigoplus_i V_i$, where each V_i is invariant and irreducible under the action of $\mathfrak{g}$. If $Z \in \mathfrak{z}$ then the restriction of Z to V_i commutes with the action of $\mathfrak{g}$, hence is a scalar by Schur's lemma (Lemma 4.1.4). This proves (1). Then (2) follows from (1) and Lemma 2.5.1.

To prove (3), let $\mathfrak{a} \subset \mathfrak{g}/\mathfrak{z}$ be an abelian ideal. Then $\mathfrak{a} = \mathfrak{h}/\mathfrak{z}$, where $\mathfrak{h}$ is an ideal in $\mathfrak{g}$ such that $[\mathfrak{h},\mathfrak{h}] \subset \mathfrak{z}$. But by (2) this implies that $[\mathfrak{h},\mathfrak{h}] = 0$, so $\mathfrak{h}$ is an abelian ideal in $\mathfrak{g}$. Let $\mathcal{B}$ be the associative subalgebra of $\mathrm{End}(V)$ generated by $[\mathfrak{h},\mathfrak{g}]$. By Lemma 2.5.1 we know that the elements of $[\mathfrak{h},\mathfrak{g}]$ are nilpotent endomorphisms of V. Since $[\mathfrak{h},\mathfrak{g}] \subset \mathfrak{h}$ is abelian, it follows that the elements of $\mathcal{B}$ are nilpotent endomorphisms. If we can prove that $\mathcal{B} = 0$, then $\mathfrak{h} \subset \mathfrak{z}$ and hence $\mathfrak{a} = 0$, establishing (3).

We now turn to the proof that $\mathcal{B} = 0$. Let $\mathcal{A}$ be the associative subalgebra of $\mathrm{End}(V)$ generated by $\mathfrak{g}$. We claim that

$$\mathcal{A}\mathcal{B} \subset \mathcal{B}\mathcal{A} + \mathcal{B}. \tag{2.38}$$

Indeed, for $X, Y \in \mathfrak{g}$ and $Z \in \mathfrak{h}$ we have $[X, [Y, Z]] \in [\mathfrak{g}, \mathfrak{h}]$ by the Jacobi identity, since $\mathfrak{h}$ is an ideal. Hence

$$X[Y, Z] = [Y, Z]X + [X, [Y, Z]] \in \mathcal{B}\mathcal{A} + \mathcal{B}. \tag{2.39}$$

Let $b \in \mathcal{B}$ and suppose that $Xb \in \mathcal{B}\mathcal{A} + \mathcal{B}$. Then by (2.39) we have

$$X[Y, Z]b = [Y, Z]Xb + [X, [Y, Z]]b \in [Y, Z]\mathcal{B}\mathcal{A} + \mathcal{B} \subset \mathcal{B}\mathcal{A} + \mathcal{B}.$$

Now (2.38) follows from this last relation by induction on the degree (in terms of the generators from $\mathfrak{g}$ and $[\mathfrak{h}, \mathfrak{g}]$) of the elements in $\mathcal{A}$ and $\mathcal{B}$.

We next show that

$$(\mathcal{A}\mathcal{B})^k \subset \mathcal{B}^k\mathcal{A} + \mathcal{B}^k \tag{2.40}$$

for every positive integer k. This is true for $k = 1$ by (2.38). Assuming that it holds for k, we use (2.38) to get the inclusions

$$(\mathcal{A}\mathcal{B})^{k+1} = (\mathcal{A}\mathcal{B})^k(\mathcal{A}\mathcal{B}) \subset (\mathcal{B}^k\mathcal{A} + \mathcal{B}^k)(\mathcal{A}\mathcal{B}) \subset \mathcal{B}^k\mathcal{A}\mathcal{B}$$
$$\subset \mathcal{B}^k(\mathcal{B}\mathcal{A} + \mathcal{B}) \subset \mathcal{B}^{k+1}\mathcal{A} + \mathcal{B}^{k+1}.$$

Hence (2.40) holds for all k.

We now complete the proof as follows. Since $\mathcal{B}^k = 0$ for k sufficiently large, the same is true for $(\mathcal{A}\mathcal{B})^k$ by (2.40). Suppose $(\mathcal{A}\mathcal{B})^{k+1} = 0$ for some $k \geq 1$. Set $\mathcal{C} = (\mathcal{A}\mathcal{B})^k$. Then $\mathcal{C}^2 = 0$. Set $W = \mathcal{C}V$. Since $\mathcal{A}\mathcal{C} \subset \mathcal{C}$, the subspace W is $\mathcal{A}$-invariant. Hence by complete reducibility of V relative to the action of $\mathfrak{g}$, there is an $\mathcal{A}$-invariant complementary subspace U such that $V = W \oplus U$. Now $\mathcal{C}W = \mathcal{C}^2V = 0$ and $\mathcal{C}U \subset \mathcal{C}V = W$. But $\mathcal{C}U \subset U$ also, so $\mathcal{C}U \subset U \cap W = \{0\}$. Hence $\mathcal{C}V = 0$. Thus $\mathcal{C} = 0$. It follows (by downward induction on k) that $\mathcal{A}\mathcal{B} = 0$. Since $I \in \mathcal{A}$, we conclude that $\mathcal{B} = 0$. $\qquad\square$

For a Lie algebra $\mathfrak{g}$ we define the *derived algebra* $\mathcal{D}(\mathfrak{g}) = [\mathfrak{g}, \mathfrak{g}]$ and we set $\mathcal{D}^{k+1}(\mathfrak{g}) = \mathcal{D}(\mathcal{D}^k(\mathfrak{g}))$ for $k = 1, 2, \ldots$. One shows by induction on k that $\mathcal{D}^k(\mathfrak{g})$ is invariant under all derivations of $\mathfrak{g}$. In particular, $\mathcal{D}^k(\mathfrak{g})$ is an ideal in $\mathfrak{g}$ for each k, and $\mathcal{D}^k(\mathfrak{g})/cD^{k+1}(\mathfrak{g})$ is abelian.

Definition 2.5.4. $\mathfrak{g}$ is *solvable* if there exists an integer $k \geq 1$ such that $\mathcal{D}^k\mathfrak{g} = 0$.

It is clear from the definition that a Lie subalgebra of a solvable Lie algebra is also solvable. Also, if $\pi : \mathfrak{g} \longrightarrow \mathfrak{h}$ is a surjective Lie algebra homomorphism, then

$$\pi(\mathcal{D}^k(\mathfrak{g})) = \mathcal{D}^k(\mathfrak{h}) \,.$$

Hence the solvability of $\mathfrak{g}$ implies the solvability of $\mathfrak{h}$. Furthermore, if $\mathfrak{g}$ is a nonzero solvable Lie algebra and we choose k such that $\mathcal{D}^k(\mathfrak{g}) \neq 0$ and $\mathcal{D}^{k+1}(\mathfrak{g}) = 0$, then $\mathcal{D}^k(\mathfrak{g})$ is an abelian ideal in $\mathfrak{g}$ that is invariant under all derivations of $\mathfrak{g}$. ·

Remark 2.5.5. The archetypical example of a solvable Lie algebra is the $n \times n$ upper-triangular matrices $\mathfrak{b}_n$. Indeed, we have $\mathcal{D}(\mathfrak{b}_n) = \mathfrak{n}_n^+$, the Lie algebra of $n \times n$ upper-triangular matrices with zeros on the main diagonal. If $\mathfrak{n}_{n,r}^+$ is the Lie subalgebra of $\mathfrak{n}_n^+$ consisting of matrices $X = [x_{ij}]$ such that $x_{ij} = 0$ for $j - i \leq r - 1$, then $\mathfrak{n}_n^+ = \mathfrak{n}_{n,1}^+$ and $[\mathfrak{n}_n^+, \mathfrak{n}_{n,r}^+] \subset \mathfrak{n}_{n,r+1}^+$ for $r = 1, 2, \ldots$. Hence $\mathcal{D}^k(\mathfrak{b}_n) \subset \mathfrak{n}_{n,k}^+$, and so $\mathcal{D}^k(\mathfrak{b}_n) = 0$ for $k > n$.

Corollary 2.5.6. *Suppose $\mathfrak{g} \subset \mathrm{End}(V)$ is a solvable Lie algebra and that V is completely reducible as a $\mathfrak{g}$-module. Then $\mathfrak{g}$ is abelian. In particular, if V is an irreducible $\mathfrak{g}$-module, then $\dim V = 1$.*

Proof. Let $\mathfrak{z}$ be the center of $\mathfrak{g}$. If $\mathfrak{z} \neq \mathfrak{g}$, then $\mathfrak{g}/\mathfrak{z}$ would be a nonzero solvable Lie algebra and hence would contain a nonzero abelian ideal. But this would contradict part (3) of Theorem 2.5.3, so we must have $\mathfrak{z} = \mathfrak{g}$. Given that $\mathfrak{g}$ is abelian and V is completely reducible, we can find a basis for V consisting of simultaneous eigenvectors for all the transformations $X \in \mathfrak{g}$; thus V is the direct sum of invariant one-dimensional subspaces. This implies the last statement of the corollary.   □

We can now obtain Cartan's trace-form criterion for solvability of a Lie algebra.

Theorem 2.5.7. *Let V be a finite-dimensional complex vector space. Let $\mathfrak{g} \subset \mathrm{End}(V)$ be a Lie subalgebra such that $\mathrm{tr}(XY) = 0$ for all $X, Y \in \mathfrak{g}$. Then $\mathfrak{g}$ is solvable.*

Proof. We use induction on $\dim \mathfrak{g}$. A one-dimensional Lie algebra is solvable. Also, if $[\mathfrak{g}, \mathfrak{g}]$ is solvable, then so is $\mathfrak{g}$, since $\mathcal{D}^{k+1}(\mathfrak{g}) = \mathcal{D}^k([\mathfrak{g}, \mathfrak{g}])$. Thus by induction we need to consider only the case $\mathfrak{g} = [\mathfrak{g}, \mathfrak{g}]$.

Take any maximal proper Lie subalgebra $\mathfrak{h} \subset \mathfrak{g}$. Then $\mathfrak{h}$ is solvable, by induction. Hence the natural representation of $\mathfrak{h}$ on $\mathfrak{g}/\mathfrak{h}$ has a one-dimensional invariant subspace, by Corollary 2.5.6. This means that there exist $0 \neq Y \in \mathfrak{g}$ and $\mu \in \mathfrak{h}^*$ such that

$$[X, Y] \equiv \mu(X)Y \qquad (\mathrm{mod}\ \mathfrak{h})$$

for all $X \in \mathfrak{h}$. But this commutation relation implies that $\mathbb{C}Y + \mathfrak{h}$ is a Lie subalgebra of $\mathfrak{g}$. Since $\mathfrak{h}$ was chosen as a maximal subalgebra, we must have $\mathbb{C}Y + \mathfrak{h} = \mathfrak{g}$. Furthermore, $\mu \neq 0$ because we are assuming $\mathfrak{g} = [\mathfrak{g}, \mathfrak{g}]$.

Given the structure of $\mathfrak{g}$ as above, we next determine the structure of an arbitrary irreducible $\mathfrak{g}$-module (π, W). By Corollary 2.5.6 again, there exist $w_0 \in W$ and $\sigma \in \mathfrak{h}^*$ such that

$$\pi(X)w_0 = \sigma(X)w_0 \quad \text{for all } X \in \mathfrak{h} \,.$$

Set $w_k = \pi(Y)^k w_0$ and $W_k = \mathbb{C}w_k + \cdots + \mathbb{C}w_0$. We claim that for $X \in \mathfrak{h}$,

$$\pi(X)w_k \equiv (\sigma(X) + k\mu(X))w_k \qquad (\mathrm{mod}\ W_{k-1}) \qquad (2.41)$$

(where $W_{-1} = \{0\}$). Indeed, this is true for $k = 0$ by definition. If it holds for k then $\pi(\mathfrak{h})W_k \subset W_k$ and

$$\begin{aligned}
\pi(X)w_{k+1} &= \pi(X)\pi(Y)w_k = \pi(Y)\pi(X)w_k + \pi([X,Y])w_k \\
&\equiv (\sigma(X) + (k+1)\mu(X))w_{k+1} \qquad (\mathrm{mod}\ W_k) .
\end{aligned}$$

Thus (2.41) holds for all k. Let m be the smallest integer such that $W_m = W_{m+1}$. Then W_m is invariant under $\mathfrak{g}$, and hence $W_m = W$ by irreducibility. Thus $\dim W = m + 1$ and

$$\mathrm{tr}(\pi(X)) = \sum_{k=0}^m \sigma(X) + k\mu(X) = (m+1)\left(\sigma(X) + \frac{m}{2}\mu(X)\right)$$

for all $X \in \mathfrak{h}$. However, $\mathfrak{g} = [\mathfrak{g}, \mathfrak{g}]$, so $\mathrm{tr}(\pi(X)) = 0$. Thus

$$\sigma(X) = -\frac{m}{2}\mu(X) \quad \text{for all } X \in \mathfrak{h} .$$

From (2.41) again we get

$$\mathrm{tr}(\pi(X)^2) = \sum_{k=0}^m \left(k - \frac{m}{2}\right)^2 \mu(X)^2 \quad \text{for all } X \in \mathfrak{h} . \qquad (2.42)$$

We finally apply these results to the given representation of $\mathfrak{g}$ on V. Take a composition series $\{0\} = V_0 \subset V_1 \subset \cdots \subset V_r = V$, where each subspace V_j is invariant under $\mathfrak{g}$ and $W_i = V_i/V_{i-1}$ is an irreducible $\mathfrak{g}$-module. Write $\dim W_i = m_i + 1$. Then (2.42) implies that

$$\mathrm{tr}_V(X^2) = \mu(X)^2 \sum_{i=1}^r \sum_{k=0}^{m_i} \left(k - \frac{1}{2}m_i\right)^2$$

for all $X \in \mathfrak{h}$. But by assumption, $\mathrm{tr}_V(X^2) = 0$ and there exists $X \in \mathfrak{h}$ with $\mu(X) \neq 0$. This forces $m_i = 0$ for $i = 1, \ldots, r$. Hence $\dim W_i = 1$ for each i. Since $\mathfrak{g} = [\mathfrak{g}, \mathfrak{g}]$, this implies that $\mathfrak{g}V_i \subset V_{i-1}$. If we take a basis for V consisting of a nonzero vector from each W_i, then the matrices for $\mathfrak{g}$ relative to this basis are strictly upper triangular. Hence $\mathfrak{g}$ is solvable, by Remark 2.5.5. $\qquad \square$

Recall that a finite-dimensional Lie algebra is *simple* if it is not abelian and has no proper ideals.

Corollary 2.5.8. *Let $\mathfrak{g}$ be a Lie subalgebra of $\mathrm{End}(V)$ that has no nonzero abelian ideals. Then the bilinear form $\mathrm{tr}(XY)$ on $\mathfrak{g}$ is nondegenerate, and $\mathfrak{g} = \mathfrak{g}_1 \oplus \cdots \oplus \mathfrak{g}_r$ (Lie algebra direct sum), where each $\mathfrak{g}_i$ is a simple Lie algebra.*

Proof. Let $\mathfrak{r} = \{X \in \mathfrak{g} : \mathrm{tr}(XY) = 0 \quad \text{for all } Y \in \mathfrak{g}\}$ be the radical of the trace form. Then $\mathfrak{r}$ is an ideal in $\mathfrak{g}$, and by Cartan's criterion $\mathfrak{r}$ is a solvable Lie algebra. Suppose

$\mathfrak{r} \neq 0$. Then $\mathfrak{r}$ contains a nonzero abelian ideal $\mathfrak{a}$ that is invariant under all derivations of $\mathfrak{r}$. Hence $\mathfrak{a}$ is an abelian ideal in $\mathfrak{g}$, which is a contradiction. Thus the trace form is nondegenerate.

To prove the second assertion, let $\mathfrak{g}_1 \subset \mathfrak{g}$ be an irreducible subspace for the adjoint representation of $\mathfrak{g}$ and define

$$\mathfrak{g}_1^\perp = \{X \in \mathfrak{g} : \operatorname{tr}(XY) = 0 \quad \text{for all } Y \in \mathfrak{g}_1\}\ .$$

Then $\mathfrak{g}_1^\perp$ is an ideal in $\mathfrak{g}$, and $\mathfrak{g}_1 \cap \mathfrak{g}_1^\perp$ is solvable by Cartan's criterion. Hence $\mathfrak{g}_1 \cap \mathfrak{g}_1^\perp = 0$ by the same argument as before. Thus $[\mathfrak{g}_1, \mathfrak{g}_1^\perp] = 0$, so we have the decomposition

$$\mathfrak{g} = \mathfrak{g}_1 \oplus \mathfrak{g}_1^\perp \quad \text{(direct sum of Lie algebras)}\ .$$

In particular, $\mathfrak{g}_1$ is irreducible as an $\operatorname{ad}\mathfrak{g}_1$-module. It cannot be abelian, so it is a simple Lie algebra. Now use induction on $\dim \mathfrak{g}$. $\square$

Corollary 2.5.9. *Let V be a finite-dimensional complex vector space. Suppose $\mathfrak{g}$ is a Lie subalgebra of* $\operatorname{End}(V)$ *such that V is completely reducible as a representation of $\mathfrak{g}$. Let $\mathfrak{z} = \{X \in \mathfrak{g} : [X,Y] = 0 \text{ for all } Y \in \mathfrak{g}\}$ be the center of $\mathfrak{g}$. Then the derived Lie algebra $[\mathfrak{g},\mathfrak{g}]$ is semisimple, and $\mathfrak{g} = [\mathfrak{g},\mathfrak{g}] \oplus \mathfrak{z}$.*

Proof. Theorem 2.5.3 implies that $\mathfrak{g}/\mathfrak{z}$ has no nonzero abelian ideals; hence $\mathfrak{g}/\mathfrak{z}$ is semisimple (Corollary 2.5.8). Since $\mathfrak{g}/\mathfrak{z}$ is a direct sum of simple algebras, it satisfies $[\mathfrak{g}/\mathfrak{z}, \mathfrak{g}/\mathfrak{z}] = \mathfrak{g}/\mathfrak{z}$. Let $p : \mathfrak{g} \longrightarrow \mathfrak{g}/\mathfrak{z}$ be the natural surjection. If $u, v \in \mathfrak{g}$ then $p([u,v]) = [p(u), p(v)]$. Since p is surjective, it follows that $\mathfrak{g}/\mathfrak{z}$ is spanned by the elements $p([u,v])$ for $u, v \in \mathfrak{g}$. Thus $p([\mathfrak{g},\mathfrak{g}]) = \mathfrak{g}/\mathfrak{z}$. Now Theorem 2.5.3 (2) implies that the restriction of p to $[\mathfrak{g},\mathfrak{g}]$ gives a Lie algebra isomorphism with $\mathfrak{g}/\mathfrak{z}$ and that $\dim([\mathfrak{g},\mathfrak{g}]) + \dim \mathfrak{z} = \dim \mathfrak{g}$. Hence $\mathfrak{g} = [\mathfrak{g},\mathfrak{g}] \oplus \mathfrak{z}$. $\square$

Let $\mathfrak{g}$ be a finite-dimensional complex Lie algebra.

Definition 2.5.10. *The Killing form of $\mathfrak{g}$ is the bilinear form $B(X,Y) = \operatorname{tr}(\operatorname{ad}X \operatorname{ad}Y)$ for $X, Y \in \mathfrak{g}$.*

Recall that $\mathfrak{g}$ is *semisimple* if it is the direct sum of simple Lie algebras. We now obtain *Cartan's criterion* for semisimplicity.

Theorem 2.5.11. *The Lie algebra $\mathfrak{g}$ is semisimple if and only if its Killing form is nondegenerate.*

Proof. Assume that $\mathfrak{g}$ is semisimple. Since the adjoint representation of a simple Lie algebra is faithful, the same is true for a semisimple Lie algebra. Hence a semisimple Lie algebra $\mathfrak{g}$ is isomorphic to a Lie subalgebra of $\operatorname{End}(\mathfrak{g})$. Let

$$\mathfrak{g} = \mathfrak{g}_1 \oplus \cdots \oplus \mathfrak{g}_r$$

(Lie algebra direct sum), where each $\mathfrak{g}_i$ is a simple Lie algebra. If $\mathfrak{m}$ is an abelian ideal in $\mathfrak{g}$, then $\mathfrak{m} \cap \mathfrak{g}_i$ is an abelian ideal in $\mathfrak{g}_i$, for each i, and hence is zero. Thus $\mathfrak{m} = 0$. Hence B is nondegenerate by Corollary 2.5.8.

Conversely, suppose the Killing form is nondegenerate. Then the adjoint representation is faithful. To show that $\mathfrak{g}$ is semisimple, it suffices by Corollary 2.5.8 to show that $\mathfrak{g}$ has no nonzero abelian ideals.

Suppose $\mathfrak{a}$ is an ideal in $\mathfrak{g}$, $X \in \mathfrak{a}$, and $Y \in \mathfrak{g}$. Then $\operatorname{ad}X \operatorname{ad}Y$ maps $\mathfrak{g}$ into $\mathfrak{a}$ and leaves $\mathfrak{a}$ invariant. Hence

$$B(X,Y) = \operatorname{tr}(\operatorname{ad}X|_{\mathfrak{a}} \operatorname{ad}Y|_{\mathfrak{a}}) . \tag{2.43}$$

If $\mathfrak{a}$ is an abelian ideal, then $\operatorname{ad}X|_{\mathfrak{a}} = 0$. Since B is nondegenerate, (2.43) implies that $X = 0$. Thus $\mathfrak{a} = 0$. □

Corollary 2.5.12. *Suppose $\mathfrak{g}$ is a semisimple Lie algebra and $D \in \operatorname{Der}(\mathfrak{g})$. Then there exists $X \in \mathfrak{g}$ such that $D = \operatorname{ad}X$.*

Proof. The derivation property $D([Y,Z]) = [D(Y),Z] + [Y,D(Z)]$ can be expressed as the commutation relation

$$[D, \operatorname{ad}Y] = \operatorname{ad}D(Y) \quad \text{for all } Y \in \mathfrak{g} . \tag{2.44}$$

Consider the linear functional $Y \mapsto \operatorname{tr}(D \operatorname{ad}Y)$ on $\mathfrak{g}$. Since the Killing form is nondegenerate, there exists $X \in \mathfrak{g}$ such that $\operatorname{tr}(D \operatorname{ad}Y) = B(X,Y)$ for all $Y \in \mathfrak{g}$. Take $Y, Z \in \mathfrak{g}$ and use the invariance of B to obtain

$$\begin{aligned} B(\operatorname{ad}X(Y),Z) &= B(X,[Y,Z]) = \operatorname{tr}(D \operatorname{ad}[Y,Z]) = \operatorname{tr}(D[\operatorname{ad}Y, \operatorname{ad}Z]) \\ &= \operatorname{tr}(D \operatorname{ad}Y \operatorname{ad}Z) - \operatorname{tr}(D \operatorname{ad}Z \operatorname{ad}Y) = \operatorname{tr}([D, \operatorname{ad}Y] \operatorname{ad}Z) . \end{aligned}$$

Hence (2.44) and the nondegeneracy of B give $\operatorname{ad}X = D$. □

For the next result we need the following formula, valid for any elements Y, Z in a Lie algebra $\mathfrak{g}$, any $D \in \operatorname{Der}(\mathfrak{g})$, and any scalars λ, μ:

$$\left(D - (\lambda + \mu)\right)^k [Y,Z] = \sum_r \binom{k}{r} [(D-\lambda)^r Y, (D-\mu)^{k-r} Z] . \tag{2.45}$$

(The proof is by induction on k using the derivation property and the inclusion–exclusion identity for binomial coefficients.)

Corollary 2.5.13. *Let $\mathfrak{g}$ be a semisimple Lie algebra. If $X \in \mathfrak{g}$ and $\operatorname{ad}X = S + N$ is the additive Jordan decomposition in $\operatorname{End}(\mathfrak{g})$ (with S semisimple, N nilpotent, and $[S,N] = 0$), then there exist $X_s, X_n \in \mathfrak{g}$ such that $\operatorname{ad}X_s = S$ and $\operatorname{ad}X_n = N$.*

Proof. Let $\lambda \in \mathbb{C}$ and set

$$\mathfrak{g}_\lambda(X) = \bigcup_{k \geq 1} \operatorname{Ker}(\operatorname{ad}X - \lambda)^k$$

(the generalized λ eigenspace of $\operatorname{ad}X$). The Jordan decomposition of $\operatorname{ad}X$ then gives a direct-sum decomposition

$$\mathfrak{g} = \bigoplus_\lambda \mathfrak{g}_\lambda(X) \,,$$

and S acts by λ on $\mathfrak{g}_\lambda(X)$. Taking $D = \operatorname{ad} X, Y \in \mathfrak{g}_\lambda(X), Z \in \mathfrak{g}_\mu(X)$, and k sufficiently large in (2.45), we see that

$$[\mathfrak{g}_\lambda(X), \mathfrak{g}_\mu(X)] \subset \mathfrak{g}_{\lambda+\mu}(X) \,. \tag{2.46}$$

Hence S is a derivation of $\mathfrak{g}$. By Corollary 2.5.12 there exists $X_s \in \mathfrak{g}$ such that $\operatorname{ad} X_s = S$. Set $X_n = X - X_s$. $\square$

2.5.2 Root Space Decomposition

In this section we shall show that every semisimple Lie algebra has a *root space decomposition* with the properties that we established in Section 2.4 for the Lie algebras of the classical groups. We begin with the following Lie algebra generalization of a familiar property of nilpotent linear transformations:

Theorem 2.5.14 (Engel). *Let V be a nonzero finite-dimensional vector space and let $\mathfrak{g} \subset \operatorname{End}(V)$ be a Lie algebra. Assume that every $X \in \mathfrak{g}$ is a nilpotent linear transformation. Then there exists a nonzero vector $v_0 \in V$ such that $X v_0 = 0$ for all $X \in \mathfrak{g}$.*

Proof. For $X \in \operatorname{End}(V)$ write L_X and R_X for the linear transformations of $\operatorname{End}(V)$ given by left and right multiplication by X, respectively. Then $\operatorname{ad} X = L_X - R_X$ and L_X commutes with R_X. Hence

$$(\operatorname{ad} X)^k = \sum_j \binom{k}{j} (-1)^{k-j} \left(L_X\right)^j \left(R_X\right)^{k-j}$$

by the binomial expansion. If X is nilpotent on V then $X^n = 0$, where $n = \dim V$. Thus $\left(L_X\right)^j \left(R_X\right)^{2n-j} = 0$ if $0 \le j \le 2n$. Hence $(\operatorname{ad} X)^{2n} = 0$, so $\operatorname{ad} X$ is nilpotent on $\operatorname{End}(V)$.

We prove the theorem by induction on $\dim \mathfrak{g}$ (when $\dim \mathfrak{g} = 1$ the theorem is clearly true). Take a proper subalgebra $\mathfrak{h} \subset \mathfrak{g}$ of maximal dimension. Then $\mathfrak{h}$ acts on $\mathfrak{g}/\mathfrak{h}$ by the adjoint representation. This action is by nilpotent linear transformations, so the induction hypothesis implies that there exists $Y \notin \mathfrak{h}$ such that

$$[X,Y] \equiv 0 \bmod \mathfrak{h} \quad \text{for all } X \in \mathfrak{h} \,.$$

Thus $\mathbb{C}Y + \mathfrak{h}$ is a Lie subalgebra of $\mathfrak{g}$, since $[Y, \mathfrak{h}] \subset \mathfrak{h}$. But $\mathfrak{h}$ was chosen maximal, so we must have $\mathfrak{g} = \mathbb{C}Y + \mathfrak{h}$. Set

$$W = \{v \in V : Xv = 0 \text{ for all } X \in \mathfrak{h}\} \,.$$

By the induction hypothesis we know that $W \neq 0$. If $v \in W$ then

$$XYv = YXv + [X,Y]v = 0$$

for all $X \in \mathfrak{h}$, since $[X,Y] \in \mathfrak{h}$. Thus W is invariant under Y, so there exists a nonzero vector $v_0 \in W$ such that $Yv_0 = 0$. It follows that $\mathfrak{g}v_0 = 0$. $\qquad\square$

Corollary 2.5.15. *There exists a basis for V in which the elements of $\mathfrak{g}$ are represented by strictly upper-triangular matrices.*

Proof. This follows by repeated application of Theorem 2.5.14, replacing V by $V/\mathbb{C}v_0$ at each step. $\qquad\square$

Corollary 2.5.16. *Suppose $\mathfrak{g}$ is a semisimple Lie algebra. Then there exists a nonzero element $X \in \mathfrak{g}$ such that $\mathrm{ad}X$ is semisimple.*

Proof. We argue by contradiction. If $\mathfrak{g}$ contained no nonzero elements X with $\mathrm{ad}X$ semisimple, then Corollary 2.5.13 would imply that $\mathrm{ad}X$ is nilpotent for all $X \in \mathfrak{g}$. Hence Corollary 2.5.15 would furnish a basis for $\mathfrak{g}$ such that $\mathrm{ad}X$ is strictly upper triangular. But then $\mathrm{ad}X\,\mathrm{ad}Y$ would also be strictly upper triangular for all $X,Y \in \mathfrak{g}$, and hence the Killing form would be zero, contradicting Theorem 2.5.11. $\qquad\square$

For the rest of this section we let $\mathfrak{g}$ be a semisimple Lie algebra. We call a subalgebra $\mathfrak{h} \subset \mathfrak{g}$ a *toral subalgebra* if $\mathrm{ad}X$ is semisimple for all $X \in \mathfrak{h}$. Corollary 2.5.16 implies the existence of nonzero toral subalgebras.

Lemma 2.5.17. *Let $\mathfrak{h}$ be a toral subalgebra. Then $[\mathfrak{h},\mathfrak{h}] = 0$.*

Proof. Let $X \in \mathfrak{h}$. Then $\mathfrak{h}$ is an invariant subspace for the semisimple transformation $\mathrm{ad}X$. If $[X,\mathfrak{h}] \neq 0$ then there would exist an eigenvalue $\lambda \neq 0$ and an eigenvector $Y \in \mathfrak{h}$ such that $[X,Y] = \lambda Y$. But then

$$(\mathrm{ad}Y)(X) = -\lambda Y \neq 0, \quad (\mathrm{ad}Y)^2(X) = 0,$$

which would imply that $\mathrm{ad}Y$ is not a semisimple transformation. Hence we must have $[X,\mathfrak{h}] = 0$ for all $X \in \mathfrak{h}$. $\qquad\square$

We shall call a toral subalgebra $\mathfrak{h} \subset \mathfrak{g}$ a *Cartan subalgebra* if it has maximal dimension among all toral subalgebras of $\mathfrak{g}$. From Corollary 2.5.16 and Lemma 2.5.17 we see that $\mathfrak{g}$ contains nonzero Cartan subalgebras and that Cartan subalgebras are abelian. We fix a choice of a Cartan subalgebra $\mathfrak{h}$. For $\lambda \in \mathfrak{h}^*$ let

$$\mathfrak{g}_\lambda = \{Y \in \mathfrak{g} : [X,Y] = \langle \lambda, X \rangle Y \text{ for all } X \in \mathfrak{h}\}.$$

In particular, $\mathfrak{g}_0 = \{Y \in \mathfrak{g} : [X,Y] = 0 \text{ for all } X \in \mathfrak{h}\}$ is the *centralizer* of $\mathfrak{h}$ in $\mathfrak{g}$. Let $\Phi \subset \mathfrak{g}^* \setminus \{0\}$ be the set of λ such that $\mathfrak{g}_\lambda \neq 0$. We call Φ the set of *roots* of $\mathfrak{h}$ on $\mathfrak{g}$. Since the mutually commuting linear transformations $\mathrm{ad}X$ are semisimple (for $X \in \mathfrak{h}$), there is a *root space decomposition*

$$\mathfrak{g} = \mathfrak{g}_0 \oplus \bigoplus_{\lambda \in \Phi} \mathfrak{g}_\lambda \,.$$

Let B denote the Killing form of $\mathfrak{g}$. By the same arguments used for the classical groups in Sections 2.4.1 and 2.4.2 (but now using B instead of the trace form on the defining representation of a classical group), it follows that

1. $[\mathfrak{g}_\lambda, \mathfrak{g}_\mu] \subset \mathfrak{g}_{\lambda+\mu}$;
2. $B(\mathfrak{g}_\lambda, \mathfrak{g}_\mu) = 0$ if $\lambda + \mu \neq 0$;
3. the restriction of B to $\mathfrak{g}_0 \times \mathfrak{g}_0$ is nondegenerate;
4. if $\lambda \in \Phi$ then $-\lambda \in \Phi$ and the restriction of B to $\mathfrak{g}_\lambda \times \mathfrak{g}_{-\lambda}$ is nondegenerate.

New arguments are needed to prove the following key result:

Proposition 2.5.18. *A Cartan algebra is its own centralizer in $\mathfrak{g}$; thus $\mathfrak{h} = \mathfrak{g}_0$.*

Proof. Since $\mathfrak{h}$ is abelian, we have $\mathfrak{h} \subset \mathfrak{g}_0$. Let $X \in \mathfrak{g}_0$ and let $X = X_s + X_n$ be the Jordan decomposition of X given by Corollary 2.5.13.

(i) X_s and X_n are in $\mathfrak{g}_0$.

Indeed, since $[X, \mathfrak{h}] = 0$ and the adjoint representation of $\mathfrak{g}$ is faithful, we have $[X_s, \mathfrak{h}] = 0$. Hence $X_s \in \mathfrak{h}$ by the maximality of $\mathfrak{h}$, which implies that $X_n = X - X_s$ is also in $\mathfrak{h}$.

(ii)   The restriction of B to $\mathfrak{h} \times \mathfrak{h}$ is nondegenerate.

To prove this, let $0 \neq h \in \mathfrak{h}$. Then by property (3) there exists $X \in \mathfrak{g}_0$ such that $B(h, X) \neq 0$. Since $X_n \in \mathfrak{g}_0$ by (i), we have $[h, X_n] = 0$ and hence $\operatorname{ad} h \operatorname{ad} X_n$ is nilpotent on $\mathfrak{g}$. Thus $B(h, X_n) = 0$ and so $B(h, X_s) \neq 0$. Since $X_s \in \mathfrak{h}$, this proves (ii).

(iii) $[\mathfrak{g}_0, \mathfrak{g}_0] = 0$.

For the proof of (iii), we observe that if $X \in \mathfrak{g}_0$, then $\operatorname{ad} X_s$ acts by zero on $\mathfrak{g}_0$, since $X_s \in \mathfrak{h}$. Hence $\operatorname{ad} X|_{\mathfrak{g}_0} = \operatorname{ad} X_n|_{\mathfrak{g}_0}$ is nilpotent. Suppose for the sake of contradiction that $[\mathfrak{g}_0, \mathfrak{g}_0] \neq 0$ and consider the adjoint action of $\mathfrak{g}_0$ on the invariant subspace $[\mathfrak{g}_0, \mathfrak{g}_0]$. By Theorem 2.5.14 there would exist $0 \neq Z \in [\mathfrak{g}_0, \mathfrak{g}_0]$ such that $[\mathfrak{g}_0, Z] = 0$. Then $[\mathfrak{g}_0, Z_n] = 0$ and hence $\operatorname{ad} Y \operatorname{ad} Z_n$ is nilpotent for all $Y \in \mathfrak{g}_0$. This implies that $B(Y, Z_n) = 0$ for all $Y \in \mathfrak{g}_0$, so we conclude from (3) that $Z_n = 0$. Thus $Z = Z_s$ must be in $\mathfrak{h}$. Now

$$B(h, [X, Y]) = B([h, X], Y) = 0 \quad \text{for all } h \in \mathfrak{h} \text{ and } X, Y \in \mathfrak{g}_0 .$$

Hence $\mathfrak{h} \cap [\mathfrak{g}_0, \mathfrak{g}_0] = 0$ by (ii), and so $Z = 0$, giving a contradiction.

It is now easy to complete the proof of the proposition. If $X, Y \in \mathfrak{g}_0$ then $\operatorname{ad} X_n \operatorname{ad} Y$ is nilpotent, since $\mathfrak{g}_0$ is abelian by (iii). Hence $B(X_n, Y) = 0$, and so $X_n = 0$ by (3). Thus $X = X_s \in \mathfrak{h}$. $\square$

Corollary 2.5.19. *Let $\mathfrak{g}$ be a semisimple Lie algebra and $\mathfrak{h}$ a Cartan subalgebra. Then*

$$\mathfrak{g} = \mathfrak{h} \oplus \bigoplus_{\lambda \in \Phi} \mathfrak{g}_\lambda . \tag{2.47}$$

Hence if $Y \in \mathfrak{g}$ and $[Y, \mathfrak{h}] \subset \mathfrak{h}$, then $Y \in \mathfrak{h}$. In particular, $\mathfrak{h}$ is a maximal abelian subalgebra of $\mathfrak{g}$.

Since the form B is nondegenerate on $\mathfrak{h} \times \mathfrak{h}$, it defines a bilinear form on $\mathfrak{h}^*$ that we denote by (α, β).

Theorem 2.5.20. *The roots and root spaces satisfy the following properties:*

1. Φ *spans* $\mathfrak{h}^*$.
2. *If* $\alpha \in \Phi$ *then* $\dim [\mathfrak{g}_\alpha, \mathfrak{g}_{-\alpha}] = 1$ *and there is a unique element* $h_\alpha \in [\mathfrak{g}_\alpha, \mathfrak{g}_{-\alpha}]$ *such that* $\langle \alpha, h_\alpha \rangle = 2$ *(call* h_α *the coroot to* α*).*
3. *If* $\alpha \in \Phi$ *and* $c \in \mathbb{C}$ *then* $c\alpha \in \Phi$ *if and only if* $c = \pm 1$. *Also* $\dim \mathfrak{g}_\alpha = 1$.
4. *Let* $\alpha, \beta \in \Phi$ *with* $\beta \neq \pm \alpha$. *Let* p *be the largest integer* $j \geq 0$ *with* $\beta + j\alpha \in \Phi$ *and let* q *be the largest integer* $k \geq 0$ *with* $\beta - k\alpha \in \Phi$. *Then*

$$\langle \beta, h_\alpha \rangle = q - p \in \mathbb{Z} \tag{2.48}$$

and $\beta + r\alpha \in \Phi$ *for all integers* r *with* $-q \leq r \leq p$. *Hence* $\beta - \langle \beta, h_\alpha \rangle \alpha \in \Phi$.
5. *If* $\alpha, \beta \in \Phi$ *and* $\alpha + \beta \in \Phi$, *then* $[\mathfrak{g}_\alpha, \mathfrak{g}_\beta] = \mathfrak{g}_{\alpha+\beta}$.

Proof. (1): If $h \in \mathfrak{h}$ and $\langle \alpha, h \rangle = 0$ for all $\alpha \in \Phi$, then $[h, \mathfrak{g}_\alpha] = 0$ and hence $[h, \mathfrak{g}] = 0$. The center of $\mathfrak{g}$ is trivial, since $\mathfrak{g}$ has no abelian ideals, so $h = 0$. Thus Φ spans $\mathfrak{h}^*$.

(2): Let $X \in \mathfrak{g}_\alpha$ and $Y \in \mathfrak{g}_{-\alpha}$. Then $[X, Y] \in \mathfrak{g}_0 = \mathfrak{h}$ and for $h \in \mathfrak{h}$ we have

$$B(h, [X, Y]) = B([h, X], Y) = \langle \alpha, h \rangle B(X, Y).$$

Thus $[X, Y]$ corresponds to $B(X, Y)\alpha$ under the isomorphism $\mathfrak{h} \cong \mathfrak{h}^*$ given by the form B. Since B is nondegenerate on $\mathfrak{g}_\alpha \times \mathfrak{g}_{-\alpha}$, it follows that $\dim [\mathfrak{g}_\alpha, \mathfrak{g}_{-\alpha}] = 1$.

Suppose $B(X, Y) \neq 0$ and set $H = [X, Y]$. Then $0 \neq H \in \mathfrak{h}$. If $\langle \alpha, H \rangle = 0$ then H would commute with X and Y, and hence $\mathrm{ad}\, H$ would be nilpotent by Lemma 2.5.1, which is a contradiction. Hence $\langle \alpha, H \rangle \neq 0$ and we can rescale X and Y to obtain elements $e_\alpha \in \mathfrak{g}_\alpha$ and $f_\alpha \in \mathfrak{g}_{-\alpha}$ such that $\langle \alpha, h_\alpha \rangle = 2$, where $h_\alpha = [e_\alpha, f_\alpha]$.

(3): Let $\mathfrak{s}(\alpha) = \mathrm{Span}\{e_\alpha, f_\alpha, h_\alpha\} \cong \mathfrak{sl}(2, \mathbb{C})$ and set

$$M_\alpha = \mathbb{C}h_\alpha + \sum_{c \neq 0} \mathfrak{g}_{c\alpha}.$$

Since $[e_\alpha, \mathfrak{g}_{c\alpha}] \subset \mathfrak{g}_{(c+1)\alpha}$, $[f_\alpha, \mathfrak{g}_{c\alpha}] \subset \mathfrak{g}_{(c-1)\alpha}$, and $[e_\alpha, \mathfrak{g}_{-\alpha}] = [f_\alpha, \mathfrak{g}_\alpha] = \mathbb{C}h_\alpha$, we see that M_α is invariant under the adjoint action of $\mathfrak{s}(\alpha)$.

The eigenvalues of $\mathrm{ad}\, h_\alpha$ on M_α are $2c$ with multiplicity $\dim \mathfrak{g}_{c\alpha}$ and 0 with multiplicity one. By the complete reducibility of representations of $\mathfrak{sl}(2, \mathbb{C})$ (Theorem 2.3.6) and the classification of irreducible representations (Proposition 2.3.3) these eigenvalues must be integers. Hence $c\alpha \in \Phi$ implies that $2c$ is an integer. The eigenvalues in any irreducible representation are all even or all odd. Hence $c\alpha$ is not a root for any integer c with $|c| > 1$, since $\mathfrak{s}(\alpha)$ contains the zero eigenspace in M_α. This also proves that the only irreducible component of M_α with even eigenvalues is $\mathfrak{s}(\alpha)$, and it occurs with multiplicity one.

Suppose $(p + 1/2)\alpha \in \Phi$ for some positive integer p. Then $\mathrm{ad}\, h_\alpha$ would have eigenvalues $2p + 1, 2p - 1, \ldots, 3, 1$ on M_α, and hence $(1/2)\alpha$ would be a root. But then α could not be a root, by the argument just given, which is a contradiction. Thus we conclude that $M_\alpha = \mathbb{C}h_\alpha + \mathbb{C}e_\alpha + \mathbb{C}f_\alpha$. Hence $\dim \mathfrak{g}_\alpha = 1$.

(4): The notion of α root string through β from Section 2.4.2 carries over verbatim, as does Lemma 2.4.3. Hence the argument in Corollary 2.4.5 applies.

(5): This follows from the same argument as Corollary 2.4.4.   □

2.5.3 Geometry of Root Systems

Let $\mathfrak{g}$ be a semisimple Lie algebra. Fix a Cartan subalgebra $\mathfrak{h}$ and let Φ be the set of roots of $\mathfrak{h}$ on $\mathfrak{g}$. For $\alpha \in \Phi$ there is a TDS triple $\{e_\alpha, f_\alpha, h_\alpha\}$ with $\langle \alpha, h_\alpha \rangle = 2$. Define

$$\check{\alpha} = n_\alpha \alpha, \quad \text{where } n_\alpha = B(e_\alpha, f_\alpha) \in \mathbb{Z} \setminus \{0\}. \tag{2.49}$$

Then $h_\alpha \longleftrightarrow \check{\alpha}$ under the isomorphism $\mathfrak{h} \cong \mathfrak{h}^*$ given by the Killing form B (see the proof of Theorem 2.5.20 (2)), and we shall call $\check{\alpha}$ the *coroot* to α.

By complete reducibility of representations of $\mathfrak{sl}(2, \mathbb{C})$ we know that $\mathfrak{g}$ decomposes into the direct sum of irreducible representations under the adjoint action of $\mathfrak{s}(\alpha) = \mathrm{Span}\{e_\alpha, f_\alpha, h_\alpha\}$. From Proposition 2.3.3 and Theorem 2.3.6 we see that e_α and f_α act by integer matrices relative to a suitable basis for any finite-dimensional representation of $\mathfrak{sl}(2, \mathbb{C})$. Hence the trace of $\mathrm{ad}(e_\alpha)\,\mathrm{ad}(f_\alpha)$ is an integer.

Since $\mathrm{Span}\,\Phi = \mathfrak{h}^*$ we can choose a basis $\{\alpha_1, \ldots, \alpha_l\}$ for $\mathfrak{h}^*$ consisting of roots. Setting $H_i = h_{\alpha_i}$, we see from (2.49) that $\{H_1, \ldots, H_l\}$ is a basis for $\mathfrak{h}$. Let

$$\mathfrak{h}_\mathbb{Q} = \mathrm{Span}_\mathbb{Q}\{H_1, \ldots, H_l\}, \quad \mathfrak{h}_\mathbb{Q}^* = \mathrm{Span}_\mathbb{Q}\{\alpha_1, \ldots, \alpha_l\},$$

where $\mathbb{Q}$ denotes the field of rational numbers.

Lemma 2.5.21. *Each root $\alpha \in \Phi$ is in $\mathfrak{h}_\mathbb{Q}^*$, and the element h_α is in $\mathfrak{h}_\mathbb{Q}$. Let $a, b \in \mathfrak{h}_\mathbb{Q}$. Then $B(a, b) \in \mathbb{Q}$ and $B(a, a) > 0$ if $a \neq 0$.*

Proof. Set $a_{ij} = \langle \alpha_j, H_i \rangle$ and let $A = [a_{ij}]$ be the corresponding $l \times l$ matrix. The entries of A are integers by Theorem 2.5.20 (4), and the columns of A are linearly independent. Hence A is invertible. For $\alpha \in \Phi$ we can write $\alpha = \sum_i c_i \alpha_i$ for unique coefficients $c_i \in \mathbb{C}$. These coefficients satisfy the system of equations

$$\sum_j a_{ij} c_j = \langle \alpha, H_i \rangle \quad \text{for } i = 1, \ldots, l.$$

Since the right side of this system consists of integers, it follows that $c_j \in \mathbb{Q}$ and hence $\alpha \in \mathfrak{h}_\mathbb{Q}^*$. From (2.49) we then see that $h_\alpha \in \mathfrak{h}_\mathbb{Q}$ also.

Given $a, b \in \mathfrak{h}_\mathbb{Q}$, we can write $a = \sum_i c_i H_i$ and $b = \sum_j d_j H_j$ with $c_i, d_j \in \mathbb{Q}$. Thus

$$B(a, b) = \mathrm{tr}(\mathrm{ad}(a)\,\mathrm{ad}(b)) = \sum_{i,j} c_i d_j \,\mathrm{tr}(\mathrm{ad}(H_i)\,\mathrm{ad}(H_j)).$$

By Theorem 2.5.20 (3) we have

$$\text{tr}(\text{ad}(H_i)\,\text{ad}(H_j)) = \sum_{\alpha \in \Phi} \langle \alpha, H_i \rangle \langle \alpha, H_j \rangle \,.$$

This is an integer by (2.48), so $B(a,b) \in \mathbb{Q}$. Likewise,

$$B(a,a) = \text{tr}(\text{ad}(a)^2) = \sum_{\alpha \in \Phi} \langle \alpha, a \rangle^2 \,,$$

and we have just proved that $\langle \alpha, a \rangle \in \mathbb{Q}$. If $a \neq 0$ then there exists $\alpha \in \Phi$ such that $\langle \alpha, a \rangle \neq 0$, because the center of $\mathfrak{g}$ is trivial. Hence $B(a,a) > 0$. $\qquad\square$

Corollary 2.5.22. *Let $\mathfrak{h}_{\mathbb{R}}$ be the real span of $\{h_\alpha : \alpha \in \Phi\}$ and let $\mathfrak{h}_{\mathbb{R}}^*$ be the real span of the roots. Then the Killing form is real-valued and positive definite on $\mathfrak{h}_{\mathbb{R}}$. Furthermore, $\mathfrak{h}_{\mathbb{R}} \cong \mathfrak{h}_{\mathbb{R}}^*$ under the Killing-form duality.*

Proof. This follows immediately from (2.49) and Lemma 2.5.21. $\qquad\square$

Let $E = \mathfrak{h}_{\mathbb{R}}^*$ with the bilinear form $(\cdot, \cdot)$ defined by the dual of the Killing form. By Corollary 2.5.22, E is an l-dimensional real Euclidean vector space. We have $\Phi \subset E$, and the coroots are related to the roots by

$$\check{\alpha} = \frac{2}{(\alpha, \alpha)} \alpha \quad \text{for } \alpha \in \Phi$$

by (2.49). Let $\check{\Phi} = \{\check{\alpha} : \alpha \in \Phi\}$ be the set of coroots. Then $(\beta, \check{\alpha}) \in \mathbb{Z}$ for all $\alpha, \beta \in \Phi$ by (2.48).

An element $h \in E$ is called *regular* if $(\alpha, h) \neq 0$ for all $\alpha \in \Phi$. Since the set

$$\bigcup_{\alpha \in \Phi} \{h \in E : (\alpha, h) = 0\}$$

is a finite union of hyperplanes, regular elements exist. Fix a regular element h_0 and define

$$\Phi^+ = \{\alpha \in \Phi : (\alpha, h_0) > 0\} \,.$$

Then $\Phi = \Phi^+ \cup (-\Phi^+)$. We call the elements of Φ^+ the *positive roots*. A positive root α is called *indecomposable* if there do not exist $\beta, \gamma \in \Phi^+$ such that $\alpha = \beta + \gamma$ (these definitions depend on the choice of h_0, of course).

Proposition 2.5.23. *Let Δ be the set of indecomposable positive roots.*

1. *Δ is a basis for the vector space E.*
2. *Every positive root is a linear combination of the elements of Δ with nonnegative integer coefficients.*
3. *If $\beta \in \Phi^+ \setminus \Delta$ then there exists $\alpha \in \Delta$ such that $\beta - \alpha \in \Phi^+$.*
4. *If $\alpha, \beta \in \Delta$ then the α root string through β is*

$$\beta, \beta + \alpha, \ldots, \beta + p\alpha, \quad \text{where } p = -(\beta, \check{\alpha}) \geq 0 \,. \tag{2.50}$$

Proof. The key to the proof is the following property of root systems:

($\star$)   If $\alpha, \beta \in \Phi$ and $(\alpha, \beta) > 0$ then $\beta - \alpha \in \Phi$.

This property holds by Theorem 2.5.20 (4): $\beta - (\beta, \check{\alpha})\alpha \in \Phi$ and $(\beta, \check{\alpha}) \geq 1$, since $(\alpha, \beta) > 0$; hence $\beta - \alpha \in \Phi$. It follows from ($\star$) that

$$(\alpha, \beta) \leq 0 \quad \text{for all } \alpha, \beta \in \Delta \text{ with } \alpha \neq \beta . \tag{2.51}$$

Indeed, if $(\alpha, \beta) > 0$ then ($\star$) would imply that $\beta - \alpha \in \Phi$. If $\beta - \alpha \in \Phi^+$ then $\alpha = \beta + (\beta - \alpha)$, contradicting the indecomposability of α. Likewise, $\alpha - \beta \in \Phi^+$ would contradict the indecomposability of β. We now use these results to prove the assertions of the proposition.

(1): Any real linear relation among the elements of Δ can be written as

$$\sum_{\alpha \in \Delta_1} c_\alpha \alpha = \sum_{\beta \in \Delta_2} d_\beta \beta , \tag{2.52}$$

where Δ_1 and Δ_2 are disjoint subsets of Δ and the coefficients c_α and d_β are nonnegative. Denote the sum in (2.52) by γ. Then by (2.51) we have

$$0 \leq (\gamma, \gamma) = \sum_{\alpha \in \Delta_1} \sum_{\beta \in \Delta_2} c_\alpha d_\beta (\alpha, \beta) \leq 0 .$$

Hence $\gamma = 0$, and so we have

$$0 = (\gamma, h_0) = \sum_{\alpha \in \Delta_1} c_\alpha (\alpha, h_0) = \sum_{\beta \in \Delta_2} d_\beta (\beta, h_0) .$$

Since $(\alpha, h_0) > 0$ and $(\beta, h_0) > 0$, it follows that $c_\alpha = d_\beta = 0$ for all α, β.

(2): The set $M = \{(\alpha, h_0) : \alpha \in \Phi^+\}$ of positive real numbers is finite and totally ordered. If m_0 is the smallest number in M, then any $\alpha \in \Phi^+$ with $(\alpha, h_0) = m_0$ is indecomposable; hence $\alpha \in \Delta$. Given $\beta \in \Phi^+ \setminus \Delta$, then $m = (\beta, h_0) > m_0$ and $\beta = \gamma + \delta$ for some $\gamma, \delta \in \Phi^+$. Since $(\gamma, h_0) < m$ and $(\delta, h_0) < m$, we may assume by induction on m that γ and δ are nonnegative integral combinations of elements of Δ, and hence so is β.

(3): Let $\beta \in \Phi^+ \setminus \Delta$. There must exist $\alpha \in \Delta$ such that $(\alpha, \beta) > 0$, since otherwise the set $\Delta \cup \{\beta\}$ would be linearly independent by the argument at the beginning of the proof. This is impossible, since Δ is a basis for E by (1) and (2). Thus $\gamma = \beta - \alpha \in \Phi$ by ($\star$). Since $\beta \neq \alpha$, there is some $\delta \in \Delta$ that occurs with positive coefficient in γ. Hence $\gamma \in \Phi^+$.

(4): Since $\beta - \alpha$ cannot be a root, the α-string through β begins at β. Now apply Theorem 2.5.20 (4). $\square$

We call the elements of Δ the *simple roots* (relative to the choice of Φ^+). Fix an enumeration $\alpha_1, \ldots, \alpha_l$ of Δ and write $E_i = e_{\alpha_i}$, $F_i = f_{\alpha_i}$, and $H_i = h_{\alpha_i}$ for the elements of the TDS triple associated with α_i. Define the *Cartan integers*

$C_{ij} = \langle \alpha_j, H_i \rangle$ and the $l \times l$ *Cartan matrix* $C = [C_{ij}]$ as in Section 2.4.3. Note that $C_{ii} = 2$ and $C_{ij} \le 0$ for $i \ne j$.

Theorem 2.5.24. *The simple root vectors* $\{E_1, \ldots, E_l, F_1, \ldots, F_l\}$ *generate* $\mathfrak{g}$. *They satisfy the relations* $[E_i, F_j] = 0$ *for* $i \ne j$ *and* $[H_i, H_j] = 0$, *where* $H_i = [E_i, F_i]$. *They also satisfy the following relations determined by the Cartan matrix:*

$$[H_i, E_j] = C_{ij} E_j \,, \quad [H_i, F_j] = -C_{ij} F_j \,; \tag{2.53}$$

$$\mathrm{ad}(E_i)^{-C_{ij}+1} E_j = 0 \ \ for \ i \ne j \,; \tag{2.54}$$

$$\mathrm{ad}(F_i)^{-C_{ij}+1} F_j = 0 \ \ for \ i \ne j \,. \tag{2.55}$$

Proof. Let $\mathfrak{g}'$ be the Lie subalgebra generated by the E_i and F_j. Since $\{H_1, \ldots, H_l\}$ is a basis for $\mathfrak{h}$, we have $\mathfrak{h} \subset \mathfrak{g}'$. We show that $\mathfrak{g}_\beta \in \mathfrak{g}'$ for all $\beta \in \Phi^+$ by induction on the height of β, exactly as in the proof of Theorem 2.4.11. The same argument with β replaced by $-\beta$ shows that $\mathfrak{g}_{-\beta} \subset \mathfrak{g}'$. Hence $\mathfrak{g}' = \mathfrak{g}$.

The commutation relations in the theorem follow from the definition of the Cartan integers and Proposition 2.5.23 (4). $\qquad\square$

The proof of Theorem 2.5.24 also gives the following generalization of Theorem 2.4.11:

Corollary 2.5.25. *Define* $\mathfrak{n}^+ = \sum_{\alpha \subset \Phi^+} \mathfrak{g}_\alpha$ *and* $\mathfrak{n}^- = \sum_{\alpha \in \Phi^+} \mathfrak{g}_{-\alpha}$. *Then* $\mathfrak{n}^+$ *and* $\mathfrak{n}^-$ *are Lie subalgebras of* $\mathfrak{g}$ *that are invariant under* $\mathrm{ad}\,\mathfrak{h}$, *and* $\mathfrak{g} = \mathfrak{n}^- + \mathfrak{h} + \mathfrak{n}^+$. *Furthermore,* $\mathfrak{n}^+$ *is generated by* $\{E_1, \ldots, E_l\}$ *and* $\mathfrak{n}^-$ *is generated by* $\{F_1, \ldots, F_l\}$.

Remark 2.5.26. We define the *height* of a root (relative to the system of positive roots) just as for the Lie algebras of the classical groups: $\mathrm{ht}\left(\sum_i c_i \alpha_i\right) = \sum_i c_i$ (the coefficients c_i are integers all of the same sign). Then

$$\mathfrak{n}^- = \sum_{\mathrm{ht}(\alpha) < 0} \mathfrak{g}_\alpha \quad \text{and} \quad \mathfrak{n}^+ = \sum_{\mathrm{ht}(\alpha) > 0} \mathfrak{g}_\alpha \,.$$

Let $\mathfrak{b} = \mathfrak{h} + \mathfrak{n}^+$. Then $\mathfrak{b}$ is a maximal solvable subalgebra of $\mathfrak{g}$ that we call a *Borel subalgebra*.

We call the set Δ of simple roots *decomposable* if it can be partitioned into nonempty disjoint subsets $\Delta_1 \cup \Delta_2$, with $\Delta_1 \perp \Delta_2$ relative to the inner product on E. Otherwise, we call Δ *indecomposable*.

Theorem 2.5.27. *The semisimple Lie algebra* $\mathfrak{g}$ *is simple if and only if* Δ *is indecomposable.*

Proof. Assume that $\Delta = \Delta_1 \cup \Delta_2$ is decomposable. Let $\alpha \in \Delta_1$ and $\beta \in \Delta_2$. Then $p = 0$ in (2.50), since $(\alpha, \beta) = 0$. Hence $\beta + \alpha$ is not a root, and we already know that $\beta - \alpha$ is not a root. Thus

$$[\mathfrak{g}_{\pm\alpha}, \mathfrak{g}_{\pm\beta}] = 0 \quad \text{for all } \alpha \in \Delta_1 \text{ and } \beta \in \Delta_2 \,. \tag{2.56}$$

Let $\mathfrak{m}$ be the Lie subalgebra of $\mathfrak{g}$ generated by the root spaces $\mathfrak{g}_{\pm\alpha}$ with α ranging over Δ_1. It is clear from (2.56) and Theorem 2.5.24 that $\mathfrak{m}$ is a proper ideal in $\mathfrak{g}$. Hence $\mathfrak{g}$ is not simple.

Conversely, suppose $\mathfrak{g}$ is not simple. Then $\mathfrak{g} = \mathfrak{g}_1 \oplus \cdots \oplus \mathfrak{g}_r$, where each $\mathfrak{g}_i$ is a simple Lie algebra. The Cartan subalgebra $\mathfrak{h}$ must decompose as $\mathfrak{h} = \mathfrak{h}_1 \oplus \cdots \oplus \mathfrak{h}_r$, and by maximality of $\mathfrak{h}$ we see that $\mathfrak{h}_i$ is a Cartan subalgebra in $\mathfrak{g}_i$. It is clear from the definition of the Killing form that the roots of $\mathfrak{g}_i$ are orthogonal to the roots of $\mathfrak{g}_j$ for $i \neq j$. Since Δ is a basis for $\mathfrak{h}^*$, it must contain a basis for each $\mathfrak{h}_i^*$. Hence Δ is decomposable. $\square$

2.5.4 Conjugacy of Cartan Subalgebras

Our results about the semisimple Lie algebra $\mathfrak{g}$ have been based on the choice of a particular Cartan subalgebra $\mathfrak{h} \subset \mathfrak{g}$. We now show that this choice is irrelevant, generalizing Corollary 2.1.8.

If $X \in \mathfrak{g}$ is nilpotent, then $\mathrm{ad}\,X$ is a nilpotent derivation of $\mathfrak{g}$, and $\exp(\mathrm{ad}\,X)$ is a Lie algebra automorphism of $\mathfrak{g}$, called an *elementary automorphism*. It satisfies

$$\mathrm{ad}\big(\exp(\mathrm{ad}\,X)Y\big) = \exp(\mathrm{ad}\,X)\,\mathrm{ad}\,Y\exp(-\mathrm{ad}\,X) \quad \text{for } Y \in \mathfrak{g} \qquad (2.57)$$

by Proposition 1.3.14. Let $\mathrm{Int}(\mathfrak{g})$ be the subgroup of $\mathrm{Aut}(\mathfrak{g})$ generated by the elementary automorphisms.

Theorem 2.5.28. *Let $\mathfrak{g}$ be a semisimple Lie algebra over $\mathbb{C}$ and let $\mathfrak{h}_1$ and $\mathfrak{h}_2$ be Cartan subalgebras of $\mathfrak{g}$. Then there exists an automorphism $\varphi \in \mathrm{Int}(\mathfrak{g})$ such that $\varphi(\mathfrak{h}_1) = \mathfrak{h}_2$.*

To prove this theorem, we need some preliminary results. Let $\mathfrak{g} = \mathfrak{n}^- + \mathfrak{h} + \mathfrak{n}^+$ be the triangular decomposition of $\mathfrak{g}$ from Corollary 2.5.25 and let $\mathfrak{b} = \mathfrak{h} + \mathfrak{n}^+$ be the corresponding Borel subalgebra. We shall call an element $H \in \mathfrak{h}$ *regular* if $\alpha(H) \neq 0$ for all roots α. From the root space decomposition of $\mathfrak{g}$ under $\mathrm{ad}\,\mathfrak{h}$, we see that this condition is the same as $\dim \mathrm{Ker}(\mathrm{ad}\,H) = \dim \mathfrak{h}$.

Lemma 2.5.29. *Suppose $Z \in \mathfrak{b}$ is semisimple. Write $Z = H + Y$, where $H \in \mathfrak{h}$ and $Y \in \mathfrak{n}^+$. Then $\dim \mathrm{Ker}(\mathrm{ad}\,Z) = \dim \mathrm{Ker}(\mathrm{ad}\,H) \geq \dim \mathfrak{h}$, with equality if and only if H is regular.*

Proof. Enumerate the positive roots in order of nondecreasing height as $\{\beta_1, \ldots, \beta_n\}$ and take an ordered basis for $\mathfrak{g}$ as

$$\{X_{-\beta_n}, \ldots, X_{-\beta_1}, H_1, \ldots, H_l, X_{\beta_1}, \ldots, X_{\beta_n}\}\ .$$

Here $X_\alpha \in \mathfrak{g}_\alpha$ and $\{H_1, \ldots, H_l\}$ is any basis for $\mathfrak{h}$. Then the matrix for $\mathrm{ad}\,Z$ relative to this basis is upper triangular and has the same diagonal as $\mathrm{ad}\,H$, namely

$$[-\beta_n(H), \ldots, -\beta_1(H), \underbrace{0, \ldots, 0}_{l}, \beta_1(H), \ldots, \beta_n(H)] \; .$$

Since $\operatorname{ad} Z$ is semisimple, these diagonal entries are its eigenvalues, repeated according to multiplicity. Hence

$$\dim \operatorname{Ker}(\operatorname{ad} Z) = \dim \mathfrak{h} + 2\operatorname{Card}\{\alpha \in \Phi^+ : \alpha(H) = 0\} \; .$$

This implies the statement of the lemma. $\square$

Lemma 2.5.30. *Let $H \in \mathfrak{h}$ be regular. Define $f(X) = \exp(\operatorname{ad} X)H - H$ for $X \in \mathfrak{n}^+$. Then f is a polynomial map of $\mathfrak{n}^+$ onto $\mathfrak{n}^+$.*

Proof. Write the elements of $\mathfrak{n}^+$ as $X = \sum_{\alpha \in \Phi^+} X_\alpha$ with $X_\alpha \in \mathfrak{g}_\alpha$. Then

$$f(X) = \sum_{k \geq 1} \frac{1}{k!} (\operatorname{ad} X)^k H = -\sum_{\alpha \in \Phi^+} \alpha(H) X_\alpha + \sum_{k \geq 2} p_k(X) \; ,$$

where $p_k(X)$ is a homogeneous polynomial map of degree k on $\mathfrak{h}$. Note that $p_k(X) = 0$ for all sufficiently large k by the nilpotence of $\operatorname{ad} X$. From this formula it is clear that f maps a neighborhood of zero in $\mathfrak{n}^+$ bijectively onto some neighborhood U of zero in $\mathfrak{n}^+$.

To show that f is globally surjective, we introduce a one-parameter group of *grading automorphisms* of $\mathfrak{g}$ as follows: Set

$$\mathfrak{g}_0 = \mathfrak{h} \; , \quad \mathfrak{g}_n = \sum_{\operatorname{ht}(\beta)=n} \mathfrak{g}_\beta \quad \text{for } n \neq 0 \; .$$

This makes $\mathfrak{g}$ a *graded Lie algebra*: $[\mathfrak{g}_k, \mathfrak{g}_n] \subset \mathfrak{g}_{k+n}$ and $\mathfrak{g} = \bigoplus_{n \in \mathbb{Z}} \mathfrak{g}_n$. For $t \in \mathbb{C}^\times$ and $X_n \in \mathfrak{g}_n$ define

$$\delta_t \left(\sum_n X_n \right) = \sum_n t^n X_n \; .$$

The graded commutation relations imply that $\delta_t \in \operatorname{Aut}(\mathfrak{g})$. Thus $t \mapsto \delta_t$ is a regular homomorphism from $\mathbb{C}^\times$ to $\operatorname{Aut}(\mathfrak{g})$ (clearly $\delta_s \delta_t = \delta_{st}$). Since $\delta_t H = H$ for $H \in \mathfrak{h}$, we have $\delta_t f(X) = f(\delta_t X)$. Now let $Y \in \mathfrak{n}^+$. Since $\lim_{t \to 0} \delta_t Y = 0$, we can choose t sufficiently small that $\delta_t Y \in U$. Then there exists $X \in \mathfrak{n}^+$ such that $\delta_t Y = f(X)$, and hence $Y = \delta_{t^{-1}} f(X) = f(\delta_{t^{-1}} X)$. $\square$

Corollary 2.5.31. *Suppose $Z \in \mathfrak{b}$ is semisimple and $\dim \operatorname{Ker}(\operatorname{ad} Z) = \dim \mathfrak{h}$. Then there exist $X \in \mathfrak{n}^+$ and a regular element $H \in \mathfrak{h}$ such that $\exp(\operatorname{ad} X)H = Z$.*

Proof. Write $Z = H + Y$ with $H \in \mathfrak{h}$ and $Y \in \mathfrak{n}^+$. By Lemma 2.5.29, H is regular, so by Lemma 2.5.30 there exists $X \in \mathfrak{n}^+$ with $\exp(\operatorname{ad} X)H = H + Y = Z$. $\square$

We now come to the key result relating two Borel subalgebras.

Lemma 2.5.32. *Suppose $\mathfrak{b}_i = \mathfrak{h}_i + \mathfrak{n}_i$ are Borel subalgebras of $\mathfrak{g}$, for $i = 1, 2$. Then*

$$\mathfrak{b}_1 = \mathfrak{b}_1 \cap \mathfrak{b}_2 + \mathfrak{n}_1 \; . \tag{2.58}$$

Proof. The right side of (2.58) is contained in the left side, so it suffices to show that both sides have the same dimension. For any subspace $V \subset \mathfrak{g}$ let $V^\perp$ be the orthogonal of V relative to the Killing form on $\mathfrak{g}$. Then $\dim V^\perp = \dim \mathfrak{g} - \dim V$, since the Killing form is nondegenerate. It is easy to see from the root space decomposition that $\mathfrak{n}_i \subset (\mathfrak{b}_i)^\perp$. Since $\dim \mathfrak{n}_i = \dim \mathfrak{g} - \dim \mathfrak{b}_i$, it follows that $(\mathfrak{b}_i)^\perp = \mathfrak{n}_i$. Thus we have

$$(\mathfrak{b}_1 + \mathfrak{b}_2)^\perp = (\mathfrak{b}_1)^\perp \cap (\mathfrak{b}_2)^\perp = \mathfrak{n}_1 \cap \mathfrak{n}_2 . \tag{2.59}$$

But $\mathfrak{n}_2$ contains all the nilpotent elements of $\mathfrak{b}_2$, so $\mathfrak{n}_1 \cap \mathfrak{n}_2 = \mathfrak{n}_1 \cap \mathfrak{b}_2$. Thus (2.59) implies that

$$\dim(\mathfrak{b}_1 + \mathfrak{b}_2) = \dim \mathfrak{g} - \dim(\mathfrak{n}_1 \cap \mathfrak{b}_2) . \tag{2.60}$$

Set $d = \dim(\mathfrak{b}_1 \cap \mathfrak{b}_2 + \mathfrak{n}_1)$. Then by (2.60) we have

$$\begin{aligned}
d &= \dim(\mathfrak{b}_1 \cap \mathfrak{b}_2) + \dim \mathfrak{n}_1 - \dim(\mathfrak{n}_1 \cap \mathfrak{b}_2) \\
&= \dim(\mathfrak{b}_1 \cap \mathfrak{b}_2) + \dim(\mathfrak{b}_1 + \mathfrak{b}_2) + \dim \mathfrak{n}_1 - \dim \mathfrak{g} \\
&= \dim \mathfrak{b}_1 + \dim \mathfrak{b}_2 + \dim \mathfrak{n}_1 - \dim \mathfrak{g} .
\end{aligned}$$

Since $\dim \mathfrak{b}_1 + \dim \mathfrak{n}_1 = \dim \mathfrak{g}$, we have shown that $d = \dim \mathfrak{b}_2$. Clearly $d \leq \dim \mathfrak{b}_1$, so this proves that $\dim \mathfrak{b}_2 \leq \dim \mathfrak{b}_1$. Reversing the roles of $\mathfrak{b}_1$ and $\mathfrak{b}_2$, we conclude that $\dim \mathfrak{b}_1 = \dim \mathfrak{b}_2 = d$, and hence (2.58) holds. $\qquad\square$

Proof of Theorem 2.5.28.   We may assume that $\dim \mathfrak{h}_1 \leq \dim \mathfrak{h}_2$. Choose systems of positive roots relative to $\mathfrak{h}_1$ and $\mathfrak{h}_2$ and let $\mathfrak{b}_i = \mathfrak{h}_i + \mathfrak{n}_i$ be the corresponding Borel subalgebras, for $i = 1, 2$. Let H_1 be a regular element in $\mathfrak{h}_1$. By Lemma 2.5.32 there exist $Z \in \mathfrak{b}_1 \cap \mathfrak{b}_2$ and $Y_1 \in \mathfrak{n}_1$ such that $H_1 = Z + Y_1$. Then by Lemma 2.5.30 there exists $X_1 \in \mathfrak{n}_1$ with $\exp(\operatorname{ad} X_1) H_1 = Z$. In particular, Z is a semisimple element of $\mathfrak{g}$ and by Lemma 2.5.29 we have

$$\dim \operatorname{Ker}(\operatorname{ad} Z) = \dim \operatorname{Ker}(\operatorname{ad} H_1) = \dim \mathfrak{h}_1 .$$

But $Z \in \mathfrak{b}_2$, so Lemma 2.5.29 gives $\dim \operatorname{Ker}(\operatorname{ad} Z) \geq \dim \mathfrak{h}_2$. This proves that $\dim \mathfrak{h}_1 = \dim \mathfrak{h}_2$. Now apply Corollary 2.5.31: there exists $X_2 \in \mathfrak{n}_2$ such that

$$\exp(\operatorname{ad} X_2) Z = H_2 \in \mathfrak{h}_2 .$$

Since $\dim \operatorname{Ker}(\operatorname{ad} H_2) = \dim \operatorname{Ker}(\operatorname{ad} Z) = \dim \mathfrak{h}_2$, we see that H_2 is regular. Hence $\mathfrak{h}_2 = \operatorname{Ker}(\operatorname{ad} H_2)$. Thus the automorphism $\varphi = \exp(\operatorname{ad} X_2) \exp(\operatorname{ad} X_1) \in \operatorname{Int} \mathfrak{g}$ maps $\mathfrak{h}_1$ onto $\mathfrak{h}_2$. $\qquad\square$

Remark 2.5.33. Let $Z \in \mathfrak{g}$ be a semisimple element. We say that Z is *regular* if $\dim \operatorname{Ker}(\operatorname{ad} Z)$ has the smallest possible value among all elements of $\mathfrak{g}$. From Theorem 2.5.28 we see that this minimal dimension is the *rank* of $\mathfrak{g}$. Furthermore, if Z is regular then $\operatorname{Ker}(\operatorname{ad} Z)$ is a Cartan subalgebra of $\mathfrak{g}$ and all Cartan subalgebras are obtained this way.

2.5.5 Exercises

1. Let $\mathfrak{g}$ be a finite-dimensional Lie algebra and let B be the Killing form of $\mathfrak{g}$. Show that $B([X,Y],Z) = B(X,[Y,Z])$ for all $X,Y,Z \in \mathfrak{g}$.
2. Let $\mathfrak{g} = \mathbb{C}X + \mathbb{C}Y$ be the two-dimensional Lie algebra with commutation relations $[X,Y] = Y$. Calculate the Killing form of $\mathfrak{g}$.
3. Suppose $\mathfrak{g}$ is a simple Lie algebra and $\omega(X,Y)$ is an invariant symmetric bilinear form on $\mathfrak{g}$. Show that ω is a multiple of the Killing form B of $\mathfrak{g}$. (HINT: Use the nondegeneracy of B to write $\omega(X,Y) = B(TX,Y)$ for some $T \in \mathrm{End}(\mathfrak{g})$. Then show that the eigenspaces of T are invariant under $\mathrm{ad}\,\mathfrak{g}$.)
4. Let $\mathfrak{g} = \mathfrak{sl}(n,\mathbb{C})$. Show that the Killing form B of $\mathfrak{g}$ is $2n\,\mathrm{tr}_{\mathbb{C}^n}(XY)$. (HINT: Calculate $B(H,H)$ for $H = \mathrm{diag}[1,-1,0,\ldots,0]$ and then use the previous exercise.)
5. Let $\mathfrak{g}$ be a finite-dimensional Lie algebra and let $\mathfrak{h} \subset \mathfrak{g}$ be an ideal. Prove that the Killing form of $\mathfrak{h}$ is the restriction to $\mathfrak{h}$ of the Killing form of $\mathfrak{g}$.
6. Prove formula (2.45).
7. Let D be a derivation of a finite-dimensional Lie algebra $\mathfrak{g}$. Prove that $\exp(tD)$ is an automorphism of $\mathfrak{g}$ for all scalars t. (HINT: Let $X,Y \in \mathfrak{g}$ and consider the curves $\varphi(t) = \exp(tD)[X,Y]$ and $\psi(t) = [\exp(tD)X,\exp(tD)Y]$ in $\mathfrak{g}$. Show that $\varphi(t)$ and $\psi(t)$ satisfy the same differential equation and $\varphi(0) = \psi(0)$.)

2.6 Notes

Section 2.1.2. The proof of the conjugacy of maximal tori for the classical groups given here takes advantage of a special property of the defining representation of a classical group, namely that it is *multiplicity-free* for the maximal torus. In Chapter 11 we will prove the conjugacy of maximal tori in any connected linear algebraic group using the general structural results developed there.

Section 2.2.2. A linear algebraic group $G \subset \mathbf{GL}(n,\mathbb{C})$ is connected if and only if the defining ideal for G in $\mathbb{C}[G]$ is *prime*. Weyl [164, Chapter X, Supplement B] gives a direct argument for this in the case of the symplectic and orthogonal groups.

Sections 2.4.1 and 2.5.2. The *roots* of a semisimple Lie algebra were introduced by Killing as the roots of the characteristic polynomial $\det(\mathrm{ad}(x) - \lambda I)$, for $x \in \mathfrak{g}$ (by the Jordan decomposition, one may assume that x is semisimple and hence that $x \in \mathfrak{h}$). See the Note Historique in Bourbaki [12] and Hawkins [63] for details.

Section 2.3.3. See Borel [17, Chapter II] for the history of the proof of complete reducibility for representations of $\mathbf{SL}(2,\mathbb{C})$. The proof given here is based on arguments first used by Cartan [26].

Sections 2.4.3 and 2.5.3. Using the set of roots to study the structure of $\mathfrak{g}$ is a fundamental technique going back to Killing and Cartan. The most thorough axiomatic treatment of root systems is in Bourbaki [12]; for recent developments see Humphreys [78] and Kane [83]. The notion of a set of simple roots and the associated Dynkin diagram was introduced in Dynkin [44], which gives a self-contained development of the structure of semisimple Lie algebras.

Section 2.5.1. In this section we follow the exposition in Hochschild [68]. The proof of Theorem 2.5.3 is from Hochschild [68, Theorem XI.1.2], and the proof of Theorem 2.5.7 is from Hochschild [68, Theorem XI.1.6].

Chapter 3
Highest-Weight Theory

Abstract In this chapter we study the regular representations of a classical group G by the same method used for the adjoint representation. When G is a connected classical group, an irreducible regular G-module decomposes into a direct sum of weight spaces relative to the action of a maximal torus of G. The *theorem of the highest weight* asserts that among the weights that occur in the decomposition, there is a unique *maximal* element, relative to a partial order coming from a choice of positive roots for G. We prove that every dominant integral weight of a semisimple Lie algebra $\mathfrak{g}$ is the highest weight of an irreducible finite-dimensional representation of $\mathfrak{g}$. When $\mathfrak{g}$ is the Lie algebra of a classical group G, the corresponding regular representations of G are constructed in Chapters 5 and 6 and studied in greater detail in Chapters 9 and 10. A crucial property of a classical group is the *complete reducibility* of its regular representations. We give two (independent) proofs of this: one algebraic using the *Casimir operator*, and one analytic using Weyl's *unitarian trick*.

3.1 Roots and Weights

The restriction to a maximal torus of a regular representation of a classical group decomposes into a direct sum of *weight spaces* that are permuted by the action of the *Weyl group* of G. This finite group plays an important role in the representation theory of G, and we determine its structure for each type of classical group. The Weyl group acts faithfully on the dual of the Lie algebra of the maximal torus as the group generated by reflections in root hyperplanes. We then study this reflection group for an arbitrary semisimple Lie algebra.

R. Goodman, N.R. Wallach, *Symmetry, Representations, and Invariants,*
Graduate Texts in Mathematics 255, DOI 10.1007/978-0-387-79852-3_3,
© Roe Goodman and Nolan R. Wallach 2009

3.1.1 Weyl Group

Let G be a connected classical group and let H be a maximal torus in G. Define the *normalizer* of H in G to be

$$\mathrm{Norm}_G(H) = \{g \in G : ghg^{-1} \in H \text{ for all } h \in H\},$$

and define the *Weyl group* $W_G = \mathrm{Norm}_G(H)/H$. Since all maximal tori of G are conjugate, the group W_G is uniquely defined (as an abstract group) by G, and it acts by conjugation as automorphisms of H. We shall see in later chapters that many aspects of the representation theory and invariant theory for G can be reduced to questions about functions on H that are invariant under W_G. The success of this approach rests on the fact that W_G is a finite group of known structure (either the symmetric group or a somewhat larger group). We proceed with the details.

Since H is abelian, there is a natural homomorphism $\varphi : W_G \longrightarrow \mathrm{Aut}(H)$ given by $\varphi(sH)h = shs^{-1}$ for $s \in \mathrm{Norm}_G(H)$. This homomorphism gives an action of W_G on the character group $\mathfrak{X}(H)$, where for $\theta \in \mathfrak{X}(H)$ the character $s \cdot \theta$ is defined by

$$s \cdot \theta(h) = \theta(s^{-1}hs), \quad \text{for } h \in H.$$

(Note that the right side of this equation depends only on the coset sH.) Writing $\theta = e^\lambda$ for $\lambda \in P(G)$ as in Section 2.4.1, we can describe this action as

$$s \cdot e^\lambda = e^{s \cdot \lambda}, \quad \text{where} \quad \langle s \cdot \lambda, x \rangle = \langle \lambda, \mathrm{Ad}(s)^{-1}x \rangle \quad \text{for } x \in \mathfrak{h}.$$

This defines a representation of W_G on $\mathfrak{h}^*$.

Theorem 3.1.1. *W_G is a finite group and the representation of W_G on $\mathfrak{h}^*$ is faithful.*

Proof. Let $s \in \mathrm{Norm}_G(H)$. Suppose $s \cdot \theta = \theta$ for all $\theta \in \mathfrak{X}(H)$. Then $s^{-1}hs = h$ for all $h \in H$, and hence $s \in H$ by Theorem 2.1.5. This proves that the representation of W_G on $\mathfrak{h}^*$ is faithful.

To prove the finiteness of W_G, we shall assume that $G \subset \mathbf{GL}(n, \mathbb{C})$ is in the matrix form of Section 2.4.1, so that H is the group of diagonal matrices in G. In the proof of Theorem 2.1.5 we noted that $h \in H$ acts on the standard basis for $\mathbb{C}^n$ by

$$he_i = \theta_i(h)e_i \text{ for } i = 1, \dots, n,$$

where $\theta_i \in \mathfrak{X}(H)$ and $\theta_i \neq \theta_j$ for $i \neq j$. Let $s \in \mathrm{Norm}_G(H)$. Then

$$hse_i = s(s^{-1}hs)e_i = s\theta_i(s^{-1}hs)e_i = (s \cdot \theta_i)(h)se_i. \tag{3.1}$$

Hence se_i is an eigenvector for h with eigenvalue $(s \cdot \theta_i)(h)$. Since the characters $s \cdot \theta_1, \dots, s \cdot \theta_n$ are all distinct, this implies that there is a permutation $\sigma \in \mathfrak{S}_n$ and there are scalars $\lambda_i \in \mathbb{C}^\times$ such that

$$se_i = \lambda_i e_{\sigma(i)} \quad \text{for } i = 1, \dots, n. \tag{3.2}$$

Since $she_i = \lambda_i \theta_i(h) e_{\sigma(i)}$ for $h \in H$, the permutation σ depends only on the coset sH. If $t \in \text{Norm}_G(H)$ and $te_i = \mu_i e_{\tau(i)}$ with $\tau \in \mathfrak{S}_n$ and $\mu_i \in \mathbb{C}^\times$, then

$$ tse_i = \lambda_i te_{\sigma(i)} = \mu_{\sigma(i)} \lambda_i e_{\tau\sigma(i)} \,. $$

Hence the map $s \mapsto \sigma$ is a homomorphism from $\text{Norm}_G(H)$ into $\mathfrak{S}_n$ that is constant on the cosets of H. If σ is the identity permutation, then s commutes with H and therefore $s \in H$ by Theorem 2.1.5. Thus we have defined an injective homomorphism from W_G into $\mathfrak{S}_n$, so W_G is a finite group. $\qquad\qquad\square$

We now describe W_G for each type of classical group. We will use the embedding of W_G into $\mathfrak{S}_n$ employed in the proof of Theorem 3.1.1. For $\sigma \in \mathfrak{S}_n$ let $s_\sigma \in \textbf{GL}(n, \mathbb{C})$ be the matrix such that $s_\sigma e_i = e_{\sigma(i)}$ for $i = 1, \ldots, n$. This is the usual representation of $\mathfrak{S}_n$ on $\mathbb{C}^n$ as *permutation matrices*.

Suppose $G = \textbf{GL}(n, \mathbb{C})$. Then H is the group of all $n \times n$ diagonal matrices. Clearly $s_\sigma \in \text{Norm}_G(H)$ for every $\sigma \in \mathfrak{S}_n$. From the proof of Theorem 3.1.1 we know that every coset in W_G is of the form $s_\sigma H$ for some $\sigma \in \mathfrak{S}_n$. Hence $W_G \cong \mathfrak{S}_n$. The action of $\sigma \in \mathfrak{S}_n$ on the diagonal coordinate functions $x_1, \ldots, x_n$ for H is $\sigma \cdot x_i = x_{\sigma^{-1}(i)}$.

Let $G = \textbf{SL}(n, \mathbb{C})$. Now H consists of all diagonal matrices of determinant 1. Given $\sigma \in \mathfrak{S}_n$, we may pick $\lambda_i \in \mathbb{C}^\times$ such that the transformation s defined by (3.2) has determinant 1 and hence is in $\text{Norm}_G(H)$. To prove this, recall that every permutation is a product of cyclic permutations, and every cyclic permutation is a product of transpositions (for example, the cycle $(1, 2, \ldots, k)$ is equal to $(1, k) \cdots (1, 3)(1, 2)$). Consequently, it is enough to verify this when σ is the transposition $i \leftrightarrow j$. In this case we take $\lambda_j = -1$ and $\lambda_k = 1$ for $k \neq j$. Since $\det(s_\sigma) = -1$, we obtain $\det s = 1$. Thus the homomorphism $W_G \longrightarrow \mathfrak{S}_n$ constructed in the proof of Theorem 3.1.1 is surjective. Hence $W_G \cong \mathfrak{S}_n$. Notice, however, that this isomorphism arises by choosing elements of $\text{Norm}_G(H)$ whose adjoint action on $\mathfrak{h}$ is given by permutation matrices; the group of all permutation matrices is not a subgroup of G.

Next, consider the case $G = \textbf{Sp}(\mathbb{C}^{2l}, \Omega)$, with Ω as in (2.6). Let $s_l \in \textbf{GL}(l, \mathbb{C})$ be the matrix for the permutation $(1, l)(2, l-1)(3, l-2)\cdots$, as in equation (2.5). For $\sigma \in \mathfrak{S}_l$ let $s_\sigma \in \textbf{GL}(l, \mathbb{C})$ be the corresponding permutation matrix. Clearly $s_\sigma^t = s_\sigma^{-1}$, so if we define

$$ \pi(\sigma) = \begin{bmatrix} s_\sigma & 0 \\ 0 & s_l s_\sigma s_l \end{bmatrix}, $$

then $\pi(\sigma) \in G$ and hence $\pi(\sigma) \in \text{Norm}_G(H)$. Obviously $\pi(\sigma) \in H$ if and only if $\sigma = 1$, so we obtain an injective homomorphism $\bar{\pi} : \mathfrak{S}_l \longrightarrow W_G$.

To find other elements of W_G, consider the transpositions $(i, 2l + 1 - i)$ in $\mathfrak{S}_{2l}$, where $1 \leq i \leq l$. Set $e_{-i} = e_{2l+1-i}$, where $\{e_i\}$ is the standard basis for $\mathbb{C}^{2l}$. Define $\tau_i \in \textbf{GL}(2l, \mathbb{C})$ by

$$ \tau_i e_i = e_{-i}, \quad \tau_i e_{-i} = -e_i, \quad \tau_i e_k = e_k \text{ for } k \neq i, -i \,. $$

Since $\{\tau_i e_j : j = \pm 1, \ldots, \pm l\}$ is an Ω-symplectic basis for $\mathbb{C}^{2l}$, we have $\tau_i \in$ $\mathbf{Sp}(\mathbb{C}^{2l}, \Omega)$ by Lemma 1.1.5. Clearly $\tau_i \in \mathrm{Norm}_G(H)$ and $\tau_i^2 \in H$. Furthermore, $\tau_i \tau_j = \tau_j \tau_i$ if $1 \le i, j \le l$. Given $F \subset \{1, \ldots, l\}$, define

$$\tau_F = \prod_{i \in F} \tau_i \in \mathrm{Norm}_G(H) .$$

Then the H-cosets of the elements $\{\tau_F\}$ form an abelian subgroup $T_l \cong (\mathbb{Z}/2\mathbb{Z})^l$ of W_G. The action of τ_F on the coordinate functions $x_1, \ldots, x_l$ for H is $x_i \mapsto x_i^{-1}$ for $i \in F$ and $x_j \mapsto x_j$ for $j \notin F$. This makes it evident that

$$\pi(\sigma) \tau_F \pi(\sigma)^{-1} = \tau_{\sigma F} \quad \text{for } F \subset \{1, \ldots, l\} \text{ and } \sigma \in \mathfrak{S}_l . \tag{3.3}$$

Clearly, $T_l \cap \bar{\pi}(\mathfrak{S}_l) H = \{1\}$.

Lemma 3.1.2. *For $G = \mathbf{Sp}(\mathbb{C}^{2l}, \Omega)$, the subgroup $T_l \subset W_G$ is normal, and W_G is the semidirect product of T_l and $\bar{\pi}(\mathfrak{S}_l)$. The action of W_G on the coordinate functions in $\mathcal{O}[H]$ is by $x_i \mapsto (x_{\sigma(i)})^{\pm 1}$ $(i = 1, \ldots, l)$, for every permutation σ and choice ± 1 of exponents.*

Proof. Recall that a group K is a *semidirect product* of subgroups L and M if M is a normal subgroup of K, $L \cap M = 1$, and $K = L \cdot M$.

By (3.3) we see that it suffices to prove that $W_G = T_l \pi(\mathfrak{S}_l)$. Suppose $s \in \mathrm{Norm}_G(H)$. Then there exists $\sigma \in \mathfrak{S}_{2l}$ such that s is given by (3.2). Define

$$F = \{i : i \le l \text{ and } \sigma(i) \ge l+1\} ,$$

and let $\mu \in \mathfrak{S}_{2l}$ be the product of the transpositions interchanging $\sigma(i)$ with $\sigma(i) - l$ for $i \in F$. Then $\mu \sigma$ stabilizes the set $\{1, \ldots, l\}$. Let $\nu \in \mathfrak{S}_l$ be the corresponding permutation of this set. Then $\pi(\nu)^{-1} \tau_F s e_i = \pm \lambda_i e_i$ for $i = 1, \ldots, l$. Thus $s \in \tau_F \pi(\nu) H$. $\square$

Now consider the case $G = \mathbf{SO}(\mathbb{C}^{2l+1}, B)$, with the symmetric form B as in (2.9). For $\sigma \in \mathfrak{S}_l$ define

$$\varphi(\sigma) = \begin{bmatrix} s_\sigma & 0 & 0 \\ 0 & 1 & 0 \\ 0 & 0 & s_l s_\sigma s_l \end{bmatrix} .$$

Then $\varphi(\sigma) \in G$ and hence $\varphi(\sigma) \in \mathrm{Norm}_G(H)$. Obviously, $\varphi(\sigma) \in H$ if and only if $\sigma = 1$, so we get an injective homomorphism $\bar{\varphi} : \mathfrak{S}_l \longrightarrow W_G$.

We construct other elements of W_G in this case by the same method as for the symplectic group. Set

$$e_{-i} = e_{2l+2-i} \quad \text{for } i = 1, \ldots, l+1 .$$

For each transposition $(i, 2l+2-i)$ in $\mathfrak{S}_{2l+1}$, where $1 \le i \le l$, we define $\gamma_i \in$ $\mathbf{GL}(2l+1, \mathbb{C})$ by

$$\gamma_i e_i = e_{-i}, \quad \gamma_i e_{-i} = e_i, \quad \gamma_i e_0 = -e_0,$$

and $\quad \gamma_i e_k = e_k \quad$ for $k \neq i, 0, -i$.

Then $\gamma_i \in \mathbf{O}(B, \mathbb{C})$ and by making γ_i act on e_{l+1} by -1 we obtain $\det \gamma_i = 1$. Hence $\gamma_i \in \mathrm{Norm}_G(H)$. Furthermore, $\gamma_i^2 \in H$ and $\gamma_i \gamma_j = \gamma_j \gamma_i$ if $1 \leq i, j \leq l$. Given $F \subset \{1, \ldots, l\}$, we define

$$\gamma_F = \prod_{i \in F} \gamma_i \in \mathrm{Norm}_G(H).$$

Then the H-cosets of the elements $\{\gamma_F\}$ form an abelian subgroup $T_l \cong (\mathbb{Z}/2\mathbb{Z})^l$ of W_G. The action of γ_F on the coordinate functions $x_1, \ldots, x_l$ for $\mathcal{O}[H]$ is the same as that of τ_F for the symplectic group.

Lemma 3.1.3. *Let $G = \mathbf{SO}(\mathbb{C}^{2l+1}, B)$. The subgroup $T_l \subset W_G$ is normal, and W_G is the semidirect product of T_l and $\bar{\varphi}(\mathfrak{S}_l)$. The action of W_G on the coordinate functions in $\mathcal{O}[H]$ is by $x_i \mapsto (x_{\sigma(i)})^{\pm 1}$ $(i = 1, \ldots, l)$, for every permutation σ and choice ± 1 of exponents.*

Proof. Suppose $s \in \mathrm{Norm}_G(H)$. Then there exists $\sigma \in \mathfrak{S}_{2l+1}$ such that s is given by (3.2), with $n = 2l + 1$. The action of s as an automorphism of H is

$$s \cdot \mathrm{diag}[a_1, \ldots, a_n] s^{-1} = \mathrm{diag}[a_{\sigma^{-1}(1)}, \ldots, a_{\sigma^{-1}(n)}].$$

Since $a_{l+1} = 1$ for elements of H, we must have $\sigma(l+1) = l+1$. Now use the same argument as in the proof of Lemma 3.1.2. $\qquad\square$

Finally, we consider the case $G = \mathbf{SO}(\mathbb{C}^{2l}, B)$, with B as in (2.6). For $\sigma \in \mathfrak{S}_l$ define $\pi(\sigma)$ as in the symplectic case. Then $\pi(\sigma) \in \mathrm{Norm}_G(H)$. Obviously, $\pi(\sigma) \in H$ if and only if $\sigma = 1$, so we have an injective homomorphism $\bar{\pi} : \mathfrak{S}_l \longrightarrow W_G$. The automorphism of H induced by $\sigma \in \mathfrak{S}_l$ is the same as for the symplectic group.

We have slightly less freedom in constructing other elements of W_G than in the case of $\mathbf{SO}(\mathbb{C}^{2l+1}, B)$. Set

$$e_{-i} = e_{2l+1-i} \quad \text{for } i = 1, \ldots, l.$$

For each transposition $(i, 2l + 1 - i)$ in $\mathfrak{S}_{2l}$, where $1 \leq i \leq l$, we define $\beta_i \in \mathbf{GL}(2l, \mathbb{C})$ by

$$\beta_i e_i = e_{-i}, \qquad \beta_i e_{-i} = e_i,$$

$$\beta_i e_k = e_k \quad \text{for } k \neq i, -i.$$

Then $\beta_i \in \mathbf{O}(\mathbb{C}^{2l}, B)$ and clearly $\beta_i H \beta_i^{-1} = H$. However, $\det \beta_i = -1$, so $\beta_i \notin \mathbf{SO}(\mathbb{C}^{2l}, B)$. Nonetheless, we still have $\beta_i^2 \in H$ and $\beta_i \beta_j = \beta_j \beta_i$ if $1 \leq i, j \leq l$. Given $F \subset \{1, \ldots, l\}$, we define

$$\beta_F = \prod_{i \in F} \beta_i.$$

If $\mathrm{Card}(F)$ is *even*, then $\det \beta_F = 1$ and hence $\beta_F \in \mathrm{Norm}_G(H)$. Thus the H cosets of the elements $\{\beta_F : \mathrm{Card}(F) \text{ even }\}$ form an abelian subgroup R_l of W_G.

Lemma 3.1.4. *Let* $G = \mathbf{SO}(\mathbb{C}^{2l}, B)$. *The subgroup* $R_l \subset W_G$ *is normal, and* W_G *is the semidirect product of* R_l *and* $\bar{\pi}(\mathfrak{S}_l)$. *The action of* W_G *on the coordinate functions in* $\mathcal{O}[H]$ *is by* $x_i \mapsto \left(x_{\sigma(i)}\right)^{\pm 1}$ $(i = 1, \ldots, l)$, *for every permutation* σ *and choice* ± 1 *of exponents with an* even *number of negative exponents.*

Proof. By the same argument as in the proof of Lemma 3.1.2 we see that the normalizer of H in $\mathbf{O}(\mathbb{C}^{2l}, B)$ is given by the H cosets of the elements $\beta_F \pi(\sigma)$ as σ ranges over $\mathfrak{S}_l$ and F ranges over all subsets of $\{1, \ldots, l\}$. Since $\pi(\sigma) \in \mathrm{Norm}_G(H)$, we have $\beta_F \pi(\sigma) \in \mathrm{Norm}_G(H)$ if and only if $\mathrm{Card}(F)$ is even. $\qquad \square$

3.1.2 Root Reflections

In this section we will give a geometric definition of the Weyl group in terms of reflections in root hyperplanes. This definition is phrased entirely in terms of the system of roots of a Cartan subalgebra. At the end of the section we use it to define the Weyl group for any semisimple Lie algebra.

Let G be $\mathbf{Sp}(n, \mathbb{C})$, $\mathbf{SL}(n, \mathbb{C})$ $(n \geq 2)$, or $\mathbf{SO}(n, \mathbb{C})$ $(n \geq 3)$. Let $\Phi \subset \mathfrak{h}^*$ be the roots and Δ the simple roots of $\mathfrak{g}$ relative to a choice Φ^+ of positive roots. For each $\alpha \in \Phi$ let $h_\alpha \in [\mathfrak{g}_\alpha, \mathfrak{g}_{-\alpha}]$ be the coroot to α (see Section 2.4.3). We define the *root reflection* $s_\alpha : \mathfrak{h}^* \longrightarrow \mathfrak{h}^*$ by

$$s_\alpha(\beta) = \beta - \langle \beta, h_\alpha \rangle \alpha \quad \text{for } \beta \in \mathfrak{h}^* . \tag{3.4}$$

The operator s_α satisfies

$$s_\alpha(\beta) = \begin{cases} -\beta & \text{if } \beta \in \mathbb{C}\alpha , \\ \beta & \text{if } \langle \beta, h_\alpha \rangle = 0 . \end{cases} \tag{3.5}$$

Thus $s_\alpha^2 = I$, and it is clear that s_α is uniquely determined by (3.5). It can be described geometrically as the *reflection through the hyperplane* $(h_\alpha)^\perp$.

Remark 3.1.5. Write $\mathfrak{h}_\mathbb{R}$ for the real linear span of the coroots. Since the roots take integer values on the coroots, the real dual space $\mathfrak{h}_\mathbb{R}^*$ is the real linear span of the roots. The calculations of Section 2.4.1 show that the trace form of the defining representation for $\mathfrak{g}$ is positive definite on $\mathfrak{h}_\mathbb{R}$. Using this form to identify $\mathfrak{h}_\mathbb{R}$ with $\mathfrak{h}_\mathbb{R}^*$, we obtain a positive definite inner product (α, β) on $\mathfrak{h}_\mathbb{R}^*$. From (2.28) we see that the formula for s_α can be written as

$$s_\alpha(\beta) = \beta - \frac{2(\beta, \alpha)}{(\alpha, \alpha)} \alpha . \tag{3.6}$$

By Corollary 2.4.5 we know that s_α permutes the set Φ. From (3.5) we see that it acts as an orthogonal transformation on $\mathfrak{h}_\mathbb{R}^*$; in particular, it permutes the short roots and the long roots when G is of type B or C (the cases with roots of two lengths).

The action of the Weyl group on $\mathfrak{h}^*$ can be expressed in terms of these reflection operators, as follows:

Lemma 3.1.6. *Let $W = \mathrm{Norm}_G(H)/H$ be the Weyl group of G. Identify W with a subgroup of $\mathbf{GL}(\mathfrak{h}^*)$ by the natural action of W on $\mathcal{X}(H)$.*

1. For every $\alpha \in \Phi$ there exists $w \in W$ such that w acts on $\mathfrak{h}^*$ by the reflection s_α .
2. $W \cdot \Delta = \Phi$.
3. W is generated by the reflections $\{s_\alpha : \alpha \in \Delta\}$.
4. If $w \in W$ and $w\Phi^+ = \Phi^+$ then $w = 1$.
5. There exists a unique element $w_0 \in W$ such that $w_0\Phi^+ = -\Phi^+$.

Proof. We proceed case by case using the enumeration of Δ from Section 2.4.3 and the description of W from Section 3.1.1. In all cases we use the characterization (3.5) of s_α.

Type A $(G = \mathbf{SL}(l+1,\mathbb{C}))$: Here $W \cong \mathfrak{S}_{l+1}$ acts on $\mathfrak{h}^*$ by permutations of $\varepsilon_1,\dots,\varepsilon_{l+1}$. Let $\alpha = \varepsilon_i - \varepsilon_j$. Then

$$\langle \varepsilon_k, h_\alpha \rangle = \begin{cases} 1 & \text{if } k = i, \\ -1 & \text{if } k = j, \\ 0 & \text{otherwise.} \end{cases}$$

This shows that $s_\alpha\varepsilon_k = \varepsilon_{\sigma(k)}$, where σ is the transposition $(i,j) \in \mathfrak{S}_{l+1}$. Hence $s_\alpha \in W$, which proves (1). Property (2) is clear, since the transposition $\sigma = (i+1, j)$ carries the simple root $\varepsilon_i - \varepsilon_{i+1}$ to $\varepsilon_i - \varepsilon_j$ for any $j \neq i$. Property (3) follows from the fact that any permutation is the product of transpositions $(i, i+1)$. If the permutation σ preserves Φ^+ then $\sigma(i) < \sigma(j)$ for all $i < j$ and hence σ is the identity, which proves (4). To prove (5), let $\sigma \in \mathfrak{S}_{l+1}$ act by $\sigma(1) = l+1, \sigma(2) = l, \dots, \sigma(l+1) = 1$ and let $w_0 \in W$ correspond to σ. Then $w_0\alpha_i = \varepsilon_{\sigma(i)} - \varepsilon_{\sigma(i+1)} = -\alpha_{l+1-i}$ for $i = 1,\dots,l$. Hence w_0 is the desired element. It is unique, by (4).

The root systems of types B_l, C_l, and D_l each contain subsystems of type A_{l-1} and the corresponding Weyl groups contain $\mathfrak{S}_l$. Furthermore, the set of simple roots for these systems is obtained by adjoining one root to the simple roots for A_{l-1}. So we need consider only the roots and Weyl group elements not in the A_{l-1} subsystem in these remaining cases. We use the same notation for elements of W in these cases as in Section 3.1.1.

Type B $(G = \mathbf{SO}(\mathbb{C}^{2l+1}, B))$: For $\alpha = \varepsilon_i$ we have $s_\alpha\varepsilon_i = -\varepsilon_i$, whereas $s_\alpha\varepsilon_j = \varepsilon_j$ if $i \neq j$. So s_α gives the action of γ_i on $\mathfrak{h}^*$. When $\alpha = \varepsilon_i + \varepsilon_j$ with $i \neq j$, then $s_\alpha\varepsilon_i = -\varepsilon_j$ and s_α fixes ε_k for $k \neq i, j$. Hence s_α has the same action on $\mathfrak{h}^*$ as $\gamma_i\gamma_j\varphi(\sigma)$, where $\sigma = (ij)$, proving (1). Since γ_j transforms $\varepsilon_i - \varepsilon_j$ into $\varepsilon_i + \varepsilon_j$ and the transposition $\sigma = (il)$ interchanges ε_i and ε_l, we obtain (2). We know from Lemma 3.1.3 that W contains elements that act on $\mathfrak{h}^*$ by $s_{\alpha_1},\dots,s_{\alpha_{l-1}}$ and generate a subgroup of W isomorphic to $\mathfrak{S}_l$. Combining this with the relation $\gamma_i = \varphi(\sigma)\gamma_l\varphi(\sigma)$, where $\sigma = (il)$, we obtain (3). Hence W acts by orthogonal transformations of $\mathfrak{h}^*_\mathbb{R}$. If w preserves Φ^+ then w must permute the set of short positive roots, and so $w\varepsilon_i = \varepsilon_{\sigma(i)}$

for some $\sigma \in \mathfrak{S}_l$. Arguing as in the case of type A_l we then conclude that $\sigma = 1$. Let w_0 be the product of the cosets $\gamma_i H$ for $i = 1, \ldots, l$. Then w_0 acts by $-I$ on $\mathfrak{h}^*$, so it is the desired element.

Type C $(G = \mathbf{Sp}(\mathbb{C}^{2l}, \Omega))$: Since $s_{2\varepsilon_i} = s_{\varepsilon_i}$, we can use Lemma 3.1.2 and the same argument as for type B, replacing γ_i by τ_i. In this case an element $w \in W$ that preserves Φ^+ will permute the set $\{2\varepsilon_i\}$ of *long* positive roots. Again w_0 acts by $-I$.

Type D $(G = \mathbf{SO}(\mathbb{C}^{2l}, B))$: For $\alpha = \varepsilon_i + \varepsilon_j$ the reflection s_α has the same action on $\mathfrak{h}^*$ as $\beta_i \beta_j \pi(\sigma)$, where $\sigma = (ij)$. This proves (1). We know from Lemma 3.1.4 that W contains elements that act on $\mathfrak{h}^*$ by $s_{\alpha_1}, \ldots, s_{\alpha_{l-1}}$ and generate a subgroup of W isomorphic to $\mathfrak{S}_l$. Since we can move the simple root $\alpha_l = \varepsilon_{l-1} + \varepsilon_l$ to α by a permutation σ, we obtain (2).

This same permutation action conjugates the reflection s_{α_l} to s_α, so we get (3). If w preserves Φ^+ then for $1 \leq i < j \leq l$ we have

$$w(\varepsilon_i + \varepsilon_j) = \varepsilon_{\sigma(i)} \pm \varepsilon_{\sigma(j)}$$

for some $\sigma \in \mathfrak{S}_l$ with $\sigma(i) < \sigma(j)$. Hence σ is the identity and $w\varepsilon_i = \varepsilon_i$ for $1 \leq i \leq l-1$. Since w can only change the sign of an even number of the ε_i, it follows that w fixes ε_l, which proves (4). If l is even, let w_0 be the product of all the cosets $\beta_i H$ for $1 \leq i \leq l$. Then w_0 acts by $-I$. If l is odd take w_0 to be the product of these cosets for $1 \leq i \leq l-1$. In this case we have $w_0 \alpha_{l-1} = -\alpha_l$, $w_0 \alpha_l = -\alpha_{l-1}$, and $w_0 \alpha_i = -\alpha_i$ for $i = 1, \ldots, l-2$, which shows that w_0 is the desired element. $\square$

Now we consider a semisimple Lie algebra $\mathfrak{g}$ over $\mathbb{C}$ (see Section 2.5); the reader who has omitted Section 2.5 can take $\mathfrak{g}$ to be a classical semisimple Lie algebra in all that follows. Fix a Cartan subalgebra $\mathfrak{h} \subset \mathfrak{g}$ and let Φ be the set of roots of $\mathfrak{h}$ on $\mathfrak{g}$ (the particular choice of $\mathfrak{h}$ is irrelevant, due to Theorem 2.5.28). Choose a set Φ^+ of positive roots and let $\Delta \subset \Phi^+$ be the simple roots (see Section 2.5.3). Enumerate $\Delta = \{\alpha_1, \ldots, \alpha_l\}$. We introduce a basis for $\mathfrak{h}^*$ whose significance will be more evident later.

Definition 3.1.7. The *fundamental weights* (relative to the choice of simple roots Δ) are elements $\{\varpi_1, \ldots, \varpi_l\}$ of $\mathfrak{h}^*$ dual to the coroot basis $\{\check{\alpha}_1, \ldots, \check{\alpha}_l\}$ for $\mathfrak{h}^*$. Thus $(\varpi_i, \check{\alpha}_j) = \delta_{ij}$ for $i, j = 1, \ldots, l$.

For $\alpha \in \Phi$ define the *root reflection* s_α by equation (3.4). Then s_α is the orthogonal reflection in the hyperplane $\alpha^\perp$ and acts as a permutation of the set Φ by Theorem 2.5.20 (4). In particular, the reflections in the simple root hyperplanes transform the fundamental weights by

$$s_{\alpha_i} \varpi_j = \varpi_j - \delta_{ij} \alpha_i . \tag{3.7}$$

Definition 3.1.8. The *Weyl group* of $(\mathfrak{g}, \mathfrak{h})$ is the group $W = W(\mathfrak{g}, \mathfrak{h})$ of orthogonal transformations of $\mathfrak{h}_\mathbb{R}^*$ generated by the root reflections.

We note that W is finite, since $w \in W$ is determined by the corresponding permutation of the finite set Φ. In the case of the Lie algebra of a classical group, Lemma 3.1.6 (3) shows that Definition 3.1.8 is consistent with that of Section 3.1.1.

Theorem 3.1.9. *Let Φ be the roots, $\Delta \subset \Phi$ a set of simple roots, and W the Weyl group for a semisimple Lie algebra $\mathfrak{g}$ and Cartan subalgebra $\mathfrak{h} \subset \mathfrak{g}$. Then all the properties (1)–(5) of Lemma 3.1.6 are satisfied.*

Proof. Property (1) is true by definition of W. Let $W_0 \subset W$ be the subgroup generated by the reflections $\{s_{\alpha_1}, \ldots, s_{\alpha_l}\}$. The following geometric property is basic to all arguments involving root reflections:

$$\text{If } \beta \in \Phi^+ \setminus \{\alpha_k\}, \quad \text{then} \quad s_{\alpha_k}\beta \in \Phi^+ \setminus \{\alpha_k\} . \tag{3.8}$$

(Thus the reflection in a simple root hyperplane sends the simple root to its negative and permutes the other positive roots.) To prove (3.8), let $\beta = \sum_i c_i \alpha_i$. There is an index $j \neq k$ such that $c_j > 0$, whereas in $s_{\alpha_k}\beta$ the coefficients of the simple roots other than α_k are the same as those of β. Hence *all* the coefficients of $s_{\alpha_k}\beta$ must be nonnegative.

Proof of (2): Let $\beta \in \Phi^+ \setminus \Delta$. We shall prove by induction on $\text{ht}(\beta)$ that $\beta \in W_0 \cdot \Delta$. We can write $\beta = \sum_i c_i \alpha_i$ with $c_i \geq 0$ and $c_i \neq 0$ for at least two indices i. Furthermore, there must exist an index k such that $(\beta, \alpha_k) > 0$, since otherwise we would have $(\beta, \beta) = \sum_i c_i(\beta, \alpha_i) \leq 0$, forcing $\beta = 0$. We have $s_{\alpha_k}\beta \in \Phi^+$ by (3.8), and we claim that

$$\text{ht}(s_{\alpha_k}\beta) < \text{ht}(\beta) . \tag{3.9}$$

Indeed, $s_{\alpha_k}\beta = \beta - d_k\alpha_k$, where $d_k = 2(\beta, \alpha_k)/(\alpha_k, \alpha_k) > 0$ by (3.6). Thus in $s_{\alpha_k}\beta$ the coefficient of α_k is $c_k - d_k < c_k$. This proves (3.9).

By induction $s_{\alpha_k}\beta \in W_0 \cdot \Delta$, hence $\beta \in W_0 \cdot \Delta$. Thus we can write $\beta = w\alpha_j$ for some $w \in W_0$. Hence

$$-\beta = w(-\alpha_j) = w(s_{\alpha_j}\alpha_j) \in W_0\Delta .$$

This completes the proof that $\Phi = W_0 \cdot \Delta$, which implies (2).

Proof of (3): Let $\beta \in \Phi$. Then by (2) there exist $w \in W_0$ and an index i such that $\beta = w\alpha_i$. Hence for $\gamma \in \mathfrak{h}_{\mathbb{R}}^*$ we have

$$s_\beta\gamma = \gamma - \frac{2(w\alpha_i, \gamma)}{(w\alpha_i, w\alpha_i)} w\alpha_i = w\left(w^{-1}\gamma - \frac{2(\alpha_i, w^{-1}\gamma)}{(\alpha_i, \alpha_i)} \alpha_i\right)$$

$$= \left(ws_{\alpha_i}w^{-1}\right)\gamma .$$

This calculation shows that $s_\beta = ws_{\alpha_i}w^{-1} \in W_0$, proving (3).

Proof of (4): Let $w \in W$ and suppose $w\Phi^+ = \Phi^+$. Assume for the sake of contradiction that $w \neq 1$. Then by (3) and (3.8) we can write $w = s_1 \cdots s_r$, where s_j is the reflection relative to a simple root α_{i_j} and $r \geq 2$. Among such presentations of w we choose one with the smallest value of r. We have

$$s_1 \cdots s_{r-1}\alpha_{i_r} = -s_1 \cdots s_{r-1}s_r\alpha_{i_r} = -w\alpha_{i_r} \in -\Phi^+ .$$

Since $\alpha_{i_r} \in \Phi^+$, there must exist an index $1 \leq j < r$ such that

$$s_j s_{j+1} \cdots s_{r-1} \alpha_{i_r} \in -\Phi^+ \quad \text{and} \quad s_{j+1} \cdots s_{r-1} \alpha_{i_r} \in \Phi^+ .$$

(If $j = r - 1$ then the product $s_{j+1} \cdots s_{r-1}$ equals 1.) Hence by (3.8) we know that

$$s_{j+1} \cdots s_{r-1} \alpha_{i_r} = \alpha_{i_j} . \tag{3.10}$$

Set $w_1 = s_{j+1} \ldots s_{r-1} \in W$. Then (3.10) implies that $w_1 s_r (w_1)^{-1} = s_j$ (as in the proof of (3)). We now use this relation to write w as a product of $r - 2$ simple reflections:

$$w = s_1 \cdots s_{j-1} \left(w_1 s_r (w_1)^{-1} \right) w_1 s_r = s_1 \cdots s_{j-1} s_{j+1} \cdots s_{r-1}$$

(since $(s_r)^2 = 1$). This contradicts the minimality of r, and hence we conclude that $w = 1$, proving (4).

Proof of (5): Let $h \in \mathfrak{h}_{\mathbb{R}}^*$ be a regular element. Define

$$\rho = \varpi_1 + \cdots + \varpi_l$$

and choose $w \in W$ to maximize $(s(h), \rho)$. We claim that

$$(w(h), \alpha_j) > 0 \quad \text{for } j = 1, \ldots, l . \tag{3.11}$$

To prove this, note that $s_{\alpha_j} \rho = \rho - \alpha_j$ for $j = 1, \ldots, l$ by (3.7). Thus

$$(w(h), \rho) \geq (s_{\alpha_j} w(h), \rho) = (w(h), \rho) - (w(h), \check{\alpha}_j)(\alpha_j, \rho) .$$

Hence $(w(h), \alpha_j) \geq 0$, since $(\alpha_j, \rho) = (\alpha_j, \alpha_j)(\check{\alpha}_j, \rho)/2 = (\alpha_j, \alpha_j)/2 > 0$. But $(w(h), \alpha_j) = (h, w^{-1} \alpha_j) \neq 0$, since h is regular. Thus we have proved (3.11).

In particular, taking $h = -h_0$, where h_0 is a regular element defining Φ^+ (see Section 2.5.3), we obtain an element $w \in W$ such that $(w(h_0), \alpha) < 0$ for all $\alpha \in \Phi^+$. Set $w_0 = w^{-1}$. Then $w_0 \Phi^+ = -\Phi^+$. If $w_1 \in W$ is another element sending Φ^+ to $-\Phi^+$ then $w_0(w_1)^{-1}$ preserves the set Φ^+, and hence must be the identity, by (4). Thus w_0 is unique.   □

Remark 3.1.10. The proof of Theorem 3.1.9 has introduced new arguments that will be used in the next proposition and in later sections. For $\mathfrak{g}$ a classical Lie algebra the proof also furnishes a more geometric and less computational explanation for the validity of Lemma 3.1.6.

Definition 3.1.11. Let $C = \{\mu \in \mathfrak{h}_{\mathbb{R}}^* : (\mu, \alpha_i) \geq 0 \text{ for } i = 1, \ldots, l\}$. Then C is a closed convex cone in the Euclidean space $\mathfrak{h}_{\mathbb{R}}^*$ that is called the *positive Weyl chamber* (relative to the choice Φ^+ of positive roots).

If $\mu = \sum_{i=1}^{l} c_i \varpi_i$, then $c_i = (\mu, \check{\alpha}_i) = 2(\mu, \alpha_i)/(\alpha_i, \alpha_i)$. Hence

$$\mu \in C \quad \text{if and only if } c_i \geq 0 \text{ for } i = 1, \ldots, l . \tag{3.12}$$

We shall also need the *dual cone*

$$C^* = \{\lambda \in \mathfrak{h}_{\mathbb{R}}^* : (\lambda, \varpi_i) \geq 0 \text{ for } i = 1, \ldots, l\} .$$

If $\lambda = \sum_{i=1}^{l} d_i \alpha_i$, then $d_i = 2(\lambda, \varpi_i)/(\alpha_i, \alpha_i)$. Hence

$$\lambda \in C^* \quad \text{if and only if } d_i \geq 0 \text{ for } i = 1, \ldots, l . \tag{3.13}$$

We now prove that the positive Weyl chamber is a *fundamental domain* for the action of W on $\mathfrak{h}_{\mathbb{R}}^*$.

Proposition 3.1.12. *If* $\lambda \in \mathfrak{h}_{\mathbb{R}}^*$ *then there exist* $\mu \in C$ *and* $w \in W$ *such that* $w \cdot \lambda = \mu$. *The element* μ *is unique, and if* λ *is regular then* w *is also unique.*

Proof. (For a case-by-case proof of this result for the classical Lie algebras, see Proposition 3.1.20.) We use the dual cone C^* to define a partial order on $\mathfrak{h}_{\mathbb{R}}^*$: say that $\lambda \preceq \mu$ if $\mu - \lambda \in C^*$. This partial order is compatible with addition and multiplication by positive scalars. For a fixed λ the set

$$\{w \cdot \lambda : w \in W \text{ and } \lambda \preceq w \cdot \lambda\}$$

is nonempty and finite. Let μ be a maximal element in this set (relative to the partial order $\preceq$). Since $s_{\alpha_i} \mu = \mu - (\mu, \check{\alpha}_i)\alpha_i$ and μ is maximal, the inequality $(\mu, \check{\alpha}_i) < 0$ is impossible by (3.13). Thus $\mu \in C$.

To prove uniqueness of μ, we may assume that $\lambda, \mu \in C \setminus \{0\}$ and $w \cdot \lambda = \mu$ for some $w \in W$. We will use the same type of argument as in the proof of Theorem 3.1.9 (4). Write $w = s_1 \cdots s_r$, where s_j is the reflection relative to a simple root α_{i_j}. Assume $w \neq 1$. Then there exists an index k such that $w\alpha_k \in -\Phi^+$ (since otherwise $w\Delta \subset \Phi^+$, which would imply $w\Phi^+ = \Phi^+$, contradicting Theorem 3.1.9 (4)). Thus there exists an index $1 \leq j \leq r$ such that

$$s_j s_{j+1} \cdots s_r \alpha_k \in -\Phi^+ \quad \text{and} \quad s_{j+1} \cdots s_r \alpha_k \in \Phi^+ .$$

(If $j = r$ then the product $s_{j+1} \cdots s_r$ equals 1.) Hence by (3.8) we have $s_{j+1} \cdots s_r \alpha_k = \alpha_{i_j}$. Set $w_1 = s_{j+1} \cdots s_r \in W$. Then $w_1 s_{\alpha_k}(w_1)^{-1} = s_j$. We can use this relation to write w as

$$w = s_1 \cdots s_{j-1}\big(w_1 s_{\alpha_k}(w_1)^{-1}\big)w_1 = s_1 \cdots s_{j-1}s_{j+1} \cdots s_r s_{\alpha_k} .$$

Hence $w_1 = w s_{\alpha_k} = s_1 \cdots s_{j-1}s_{j+1} \cdots s_r$ is a product of $r-1$ simple reflections. Since $\lambda, \mu \in C$ and $w\alpha_k \in -\Phi^+$, we have

$$0 \geq (\mu, w\alpha_k) = (w^{-1}\mu, \alpha_k) = (\lambda, \alpha_k) \geq 0 .$$

Hence $(\lambda, \alpha_k) = 0$; thus $s_{\alpha_k}\lambda = \lambda$ and $w_1\lambda = \mu$. If $w_1 \neq 1$ we can continue this shortening process until we reach the identity. This proves that $\lambda = \mu$ and that w is the product of simple reflections that fix λ. In particular, if λ is regular, then no simple reflection fixes it, so $w = 1$ in this case. $\qquad\square$

3.1.3 Weight Lattice

We now introduce a class of Lie algebras larger than the semisimple algebras that includes the Lie algebras of all the classical groups (for example, $\mathbf{GL}(n,\mathbb{C})$ and $\mathbf{SO}(2,\mathbb{C})$).

Definition 3.1.13. A Lie algebra $\mathfrak{g}$ over $\mathbb{C}$ is *reductive* if $\mathfrak{g} = \mathfrak{z}(\mathfrak{g}) \oplus \mathfrak{g}_1$, where $\mathfrak{z}(\mathfrak{g})$ is the center of $\mathfrak{g}$ and $\mathfrak{g}_1$ is semisimple.

If $\mathfrak{g}$ is reductive then $\mathfrak{g}_1 = \mathfrak{g}'$, where $\mathfrak{g}' = [\mathfrak{g}, \mathfrak{g}]$ is the derived algebra (since a semisimple Lie algebra is equal to its derived algebra and $[\mathfrak{g}, \mathfrak{z}(\mathfrak{g})] = 0$). Thus the decomposition of $\mathfrak{g}$ in Definition 3.1.13 is unique. For example, $\mathfrak{gl}(n,\mathbb{C}) = \mathbb{C}I \oplus \mathfrak{sl}(n,\mathbb{C})$ is reductive.

Remark 3.1.14. If V is completely reducible under the action of a Lie algebra $\mathfrak{g} \subset \mathrm{End}(V)$, then Corollary 2.5.9 shows that $\mathfrak{g}$ is reductive. The converse holds if and only if $\mathfrak{z}(\mathfrak{g})$ acts by semisimply on V.

   Assume that $\mathfrak{g}$ is reductive. Let $\mathfrak{h}_0$ be a Cartan subalgebra of $\mathfrak{g}'$ and set $\mathfrak{h} = \mathfrak{z}(\mathfrak{g}) + \mathfrak{h}_0$. Then the root space decomposition of $\mathfrak{g}'$ under $\mathfrak{h}_0$ furnishes a root space decomposition (2.47) of $\mathfrak{g}$ with $\mathfrak{g}_\lambda \subset \mathfrak{g}'$ and $\lambda = 0$ on $\mathfrak{z}(\mathfrak{g})$ for all $\lambda \in \Phi$. For example, the roots $\varepsilon_i - \varepsilon_j$ of $\mathfrak{sl}(n,\mathbb{C})$ all vanish on $\mathbb{C}I$. For a root $\alpha \in \Phi$ let $h_\alpha \in [\mathfrak{g}_\alpha, \mathfrak{g}_\alpha]$ be the coroot (with the normalization $\langle \alpha, h_\alpha \rangle = 2$).
   We define the *weight lattice* for $\mathfrak{g}$ as

$$P(\mathfrak{g}) = \{\mu \in \mathfrak{h}^* : \langle \mu, h_\alpha \rangle \in \mathbb{Z} \quad \text{for all } \alpha \in \Phi\}\ .$$

For example, the weights of the adjoint representation are $\Phi \cup \{0\}$. Clearly, $P(\mathfrak{g})$ is an additive subgroup of $\mathfrak{h}^*$. We define the *root lattice* $Q(\mathfrak{g})$ to be the additive subgroup of $\mathfrak{h}^*$ generated by Φ. Thus $Q(\mathfrak{g}) \subset P(\mathfrak{g})$.

Remark 3.1.15. Although the definition of $P(\mathfrak{g})$ and $Q(\mathfrak{g})$ requires picking a particular Cartan subalgebra, we know that all Cartan subalgebras of $\mathfrak{g}$ are conjugate by an element of $\mathrm{Int}(\mathfrak{g})$ (Theorem 2.5.28), so in this sense $P(\mathfrak{g})$ and $Q(\mathfrak{g})$ depend only on $\mathfrak{g}$.

Theorem 3.1.16. *Let (π, V) be a finite-dimensional representation of $\mathfrak{g}$. For $\mu \in \mathfrak{h}^*$ set*

$$V(\mu) = \{v \in V : \pi(Y)v = \langle \mu, Y \rangle v \quad \text{for all } Y \in \mathfrak{h}\}\ .$$

Call $\mu \in \mathfrak{h}^*$ a weight of (π, V) if $V(\mu) \neq 0$, and let $\mathfrak{X}(V) \subset \mathfrak{h}^*$ be the set of weights of (π, V). Then $\mathfrak{X}(V) \subset P(\mathfrak{g})$. If $\pi(Z)$ is diagonalizable for all $Z \in \mathfrak{z}(\mathfrak{g})$, then

$$V = \bigoplus_{\mu \in \mathfrak{X}(V)} V(\mu)\ . \tag{3.14}$$

Proof. Take the three-dimensional simple algebra $\mathfrak{s}(\alpha)$ containing h_α, as in Sections 2.4.2 and 2.5.2, and apply Theorem 2.3.6 to the restriction of π to $\mathfrak{s}(\alpha)$ to prove the first assertion. To obtain (3.14), observe that the coroots h_α span $\mathfrak{h}_0$ and the operators $\{\pi(h_\alpha) : \alpha \in \Phi\}$ mutually commute. Thus we only need to show that $\pi(h_\alpha)$ is diagonalizable for each $\alpha \in \Phi$. This follows from Theorem 2.3.6. $\square$

Lemma 3.1.17. $P(\mathfrak{g})$ *and* $Q(\mathfrak{g})$ *are invariant under the Weyl group* W.

Proof. Since W permutes the elements of Φ and acts linearly on $\mathfrak{h}^*$, it leaves the root lattice invariant. Let $\alpha \in \Phi$ and $\mu \in P(\mathfrak{g})$. Then

$$s_\alpha \mu = \mu - \langle \mu, h_\alpha \rangle \alpha \in \mu + Q(\mathfrak{g}) \subset P(\mathfrak{g}) ,$$

since $\langle \mu, h_\alpha \rangle \in \mathbb{Z}$. This shows that $P(\mathfrak{g})$ is invariant under the reflections s_α for all $\alpha \in \Phi$; hence it is invariant under W (for a classical Lie algebra we are using Theorem 3.1.9 (3), whereas for a general reductive Lie algebra this invariance follows from the definition of the Weyl group). $\square$

Fix a set Φ^+ of positive roots. Let $\Delta = \{\alpha_1, \ldots, \alpha_l\} \subset \Phi^+$ be the simple roots in Φ^+. Then

$$Q(\mathfrak{g}) = \mathbb{Z}\alpha_1 + \cdots + \mathbb{Z}\alpha_l$$

is a free abelian group of rank l, since every root is an integer linear combination of the simple roots and the simple roots are linearly independent. Denote by H_i the coroot to α_i and let $\{\varpi_1, \ldots, \varpi_l\} \subset \mathfrak{h}_0^*$ be the fundamental weights (Definition 3.1.7). When $\mathfrak{z}(\mathfrak{g}) \neq 0$ we extend ϖ_i to an element of $\mathfrak{h}^*$ by setting $\varpi_i = 0$ on $\mathfrak{z}(\mathfrak{g})$. We claim that

$$P(\mathfrak{g}) = \mathfrak{z}(\mathfrak{g})^* \oplus \{n_1\varpi_1 + \cdots + n_l\varpi_l : n_i \in \mathbb{Z}\} . \tag{3.15}$$

To prove (3.15), let $\mu \in \mathfrak{h}^*$. Then $\mu = \lambda + \sum_{i=1}^l n_i\varpi_i$, where $n_i = \langle \mu, H_i \rangle$ and $\langle \lambda, H_i \rangle = 0$ for $i = 1, \ldots, l$. If $\mu \in P(\mathfrak{g})$ then $n_i \in \mathbb{Z}$ by definition. To prove the converse, we may assume that $\mathfrak{z}(\mathfrak{g}) = 0$. Let $\mu = \sum_{i=1}^l n_i\varpi_i$ with $n_i \in \mathbb{Z}$. We must show that $\langle \mu, h_\alpha \rangle \in \mathbb{Z}$ for all $\alpha \in \Phi$. For a classical Lie algebra this follows from (2.36). In general, we know from Proposition 3.1.12 that there exist $w \in W$ and an index i such that $\alpha = w \cdot \alpha_i$. Identifying $\mathfrak{h}_\mathbb{R}$ with $\mathfrak{h}_\mathbb{R}^*$ using the Killing form, we can write

$$h_\alpha = \frac{2}{(\alpha, \alpha)}\alpha = \frac{2}{(w\alpha_i, w\alpha_i)}\alpha_i = wH_i .$$

By Theorem 3.1.9, w is a product of simple reflections. Now for $i, j = 1, \ldots, l$ we have

$$s_{\alpha_j} H_i = H_i - \langle \alpha_j, H_i \rangle H_j = H_i - C_{ij} H_j ,$$

where $C_{ij} \in \mathbb{Z}$ is the Cartan integer for the pair (i, j). Hence there are integers m_i such that

$$h_\alpha = m_1 H_1 + \cdots + m_l H_l .$$

Thus $\langle \mu, h_\alpha \rangle = \sum_{i=1}^l m_i n_i \in \mathbb{Z}$. This completes the proof of (3.15).

Remark 3.1.18. When $\mathfrak{g}$ is semisimple, (3.15) shows that $P(\mathfrak{g})$ is a free abelian group of rank l having the fundamental weights as basis. In this case $P(\mathfrak{g})$ and $Q(\mathfrak{g})$ are both *lattices* in the l-dimensional vector space $\mathfrak{h}_\mathbb{R}^*$ in the usual sense of the term.

We now give the fundamental weights for each type of classical group in terms of the weights $\{\varepsilon_i\}$ (see the formulas in Section 2.4.3 giving H_i in terms of the diagonal matrices $E_i = e_{ii}$).

Type A ($\mathfrak{g} = \mathfrak{sl}(l+1, \mathbb{C})$): Since $H_i = E_i - E_{i+1}$, we have

$$\varpi_i = \varepsilon_1 + \cdots + \varepsilon_i - \frac{i}{l+1}(\varepsilon_1 + \cdots + \varepsilon_{l+1}) \quad \text{for } 1 \leq i \leq l.$$

Type B ($\mathfrak{g} = \mathfrak{so}(2l+1, \mathbb{C})$): Now $H_i = E_i - E_{i+1} + E_{-i-1} - E_{-i}$ for $1 \leq i \leq l-1$ and $H_l = 2E_l - 2E_{-l}$, so we have

$$\varpi_i = \varepsilon_1 + \cdots + \varepsilon_i \quad \text{for } 1 \leq i \leq l-1, \quad \text{and} \quad \varpi_l = \frac{1}{2}(\varepsilon_1 + \cdots + \varepsilon_l).$$

Type C ($\mathfrak{g} = \mathfrak{sp}(l, \mathbb{C})$): In this case $H_l = E_l - E_{-l}$ and for $1 \leq i \leq l-1$ we have $H_i = E_i - E_{i+1} + E_{-i-1} - E_{-i}$. Thus

$$\varpi_i = \varepsilon_1 + \cdots + \varepsilon_i \quad \text{for } 1 \leq i \leq l.$$

Type D ($\mathfrak{g} = \mathfrak{so}(2l, \mathbb{C})$ with $l \geq 2$): For $1 \leq i \leq l-1$ we have

$$H_i = E_i - E_{i+1} + E_{-i-1} - E_{-i} \quad \text{and} \quad H_l = E_{l-1} + E_l - E_{-l} - E_{-l+1}.$$

A direct calculation shows that $\varpi_i = \varepsilon_1 + \cdots + \varepsilon_i$ for $1 \leq i \leq l-2$, and

$$\varpi_{l-1} = \frac{1}{2}(\varepsilon_1 + \cdots + \varepsilon_{l-1} - \varepsilon_l), \quad \varpi_l = \frac{1}{2}(\varepsilon_1 + \cdots + \varepsilon_{l-1} + \varepsilon_l).$$

Let G be a connected classical group. Since the functionals ε_i are weights of the defining representation of G, we have $\varepsilon_i \in P(\mathfrak{g})$ for $i = 1, \ldots, l$. Thus $P(G) \subset P(\mathfrak{g})$ by (2.21). For $\mathfrak{g}$ of type A or C all the fundamental weights are in $P(G)$, so $P(G) = P(\mathfrak{g})$ when $G = \mathbf{SL}(n, \mathbb{C})$ or $\mathbf{Sp}(n, \mathbb{C})$. However, for $G = \mathbf{SO}(2l+1, \mathbb{C})$ we have

$$\varpi_i \in P(G) \quad \text{for } 1 \leq i \leq l-1, \quad 2\varpi_l \in P(G),$$

but $\varpi_l \notin P(G)$. For $G = \mathbf{SO}(2l, \mathbb{C})$ we have

$$\varpi_i \in P(G) \quad \text{for } 1 \leq i \leq l-2, \quad m\varpi_{l-1} + n\varpi_l \in P(G) \quad \text{if } m+n \in 2\mathbb{Z},$$

but ϖ_{l-1} and ϖ_l are not in $P(G)$. Therefore

$$P(\mathfrak{g})/P(G) \cong \mathbb{Z}/2\mathbb{Z} \quad \text{when } G = \mathbf{SO}(n, \mathbb{C}).$$

This means that for the orthogonal groups in odd (resp. even) dimensions there is no single-valued character χ on the maximal torus whose differential is ϖ_l (resp. ϖ_{l-1} or ϖ_l). We will resolve this difficulty in Chapter 6 with the construction of the groups $\mathbf{Spin}(n, \mathbb{C})$ and the *spin representations*.

3.1.4 Dominant Weights

We continue the same notation as in Section 3.1.3. Define the *dominant integral weights* (relative to the given choice of positive roots) to be

$$P_{++}(\mathfrak{g}) = \{\lambda \in \mathfrak{h}^* : \langle \lambda, H_i \rangle \in \mathbb{N} \text{ for } i = 1, \ldots, l\},$$

where $\mathbb{N} = \{0, 1, 2, \ldots\}$. From (3.15) we have

$$P_{++}(\mathfrak{g}) = \mathfrak{z}(\mathfrak{g}) + \mathbb{N}\varpi_1 + \cdots + \mathbb{N}\varpi_l.$$

We say that $\mu \in P_{++}(\mathfrak{g})$ is *regular* if $\langle \mu, H_i \rangle > 0$ for $i = 1, \ldots, l$. This is equivalent to

$$\lambda = \zeta + n_1\varpi_1 + \cdots + n_l\varpi_l, \text{ with } \zeta \in \mathfrak{z}(\mathfrak{g})^* \text{ and } n_i \geq 1 \text{ for all } i.$$

When G is a classical group and $\mathfrak{h}$ is the Lie algebra of a maximal torus of G, we define the *dominant weights* for G to be

$$P_{++}(G) = P(G) \cap P_{++}(\mathfrak{g})$$

(see Section 2.4.1). Then $P_{++}(G) = P_{++}(\mathfrak{g})$ when G is $\mathbf{SL}(n, \mathbb{C})$ or $\mathbf{Sp}(n, \mathbb{C})$ by (2.21) and the formulas for $\{\varpi_i\}$ in Section 3.1.3.

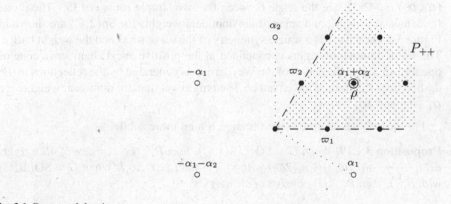

Fig. 3.1 Roots and dominant weights for $\mathbf{SL}(3, \mathbb{C})$.

Examples

Type A$_2$: We have $P(\mathfrak{g}) \subset \mathbb{R}\alpha_1 + \mathbb{R}\alpha_2$. Since $(\alpha_1, \alpha_1) = (\alpha_2, \alpha_2) = 2$ and $(\alpha_1, \alpha_2) = -1$, we see that the two simple roots have the same length, and the angle between them is $120°$. The roots (indicated by $\circ$), the fundamental weights, and some of the dominant weights (indicated by $\bullet$) for $\mathbf{SL}(3, \mathbb{C})$ are shown in Figure 3.1. Notice the hexagonal symmetry of the set of roots and the weight lattice. The set of dominant weights is contained in the positive Weyl chamber (a cone of opening $60°$), and the action of the Weyl group is generated by the reflections in the dashed lines that bound the chamber. The only root that is a dominant weight is $\alpha_1 + \alpha_2 = \varpi_1 + \varpi_2$, and it is regular.

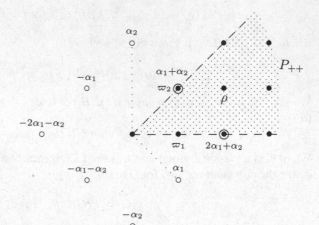

Fig. 3.2 Roots and dominant weights for $\mathbf{Sp}(2,\mathbb{C})$.

Type C_2: In this case $P(\mathfrak{g}) \subset \mathbb{R}\alpha_1 + \mathbb{R}\alpha_2$, but now $(\alpha_1, \alpha_1) = 2$, $(\alpha_2, \alpha_2) = 4$, and $(\alpha_1, \alpha_2) = -2$. Hence the angle between the two simple roots is $135°$. The roots, fundamental weights, and some of the dominant weights for $\mathbf{Sp}(2,\mathbb{C})$ are shown in Figure 3.2. Here there is a square symmetry of the set of roots and the weight lattice. The set of dominant weights is contained in the positive Weyl chamber (a cone of opening $45°$), and the action of the Weyl group is generated by the reflections in the dashed lines that bound the chamber. The only roots that are dominant weights are $\alpha_1 + \alpha_2$ and $2\alpha_1 + \alpha_2$.

For the orthogonal groups the situation is a bit more subtle.

Proposition 3.1.19. *When $G = \mathbf{SO}(2l+1,\mathbb{C})$, then $P_{++}(G)$ consists of all weights $n_1\varpi_1 + \cdots + n_{l-1}\varpi_{l-1} + n_l(2\varpi_l)$ with $n_i \in \mathbb{N}$ for $i = 1,\ldots,l$. When $G = \mathbf{SO}(2l,\mathbb{C})$ with $l \geq 2$, then $P_{++}(G)$ consists of all weights*

$$n_1\varpi_1 + \cdots + n_{l-2}\varpi_{l-2} + n_{l-1}(2\varpi_{l-1}) + n_l(2\varpi_l) + n_{l+1}(\varpi_{l-1} + \varpi_l) \qquad (3.16)$$

with $n_i \in \mathbb{N}$ for $i = 1,\ldots,l+1$.

Proof. In both cases we have $P(G) = \sum_{i=1}^{l} \mathbb{Z}\varepsilon_i$. Thus the first assertion is obvious from the formulas for ϖ_i. Now assume $G = \mathbf{SO}(2l,\mathbb{C})$. Then every weight of the form (3.16) is in $P_{++}(G)$. Conversely, suppose that $\lambda = k_1\varpi_1 + \cdots + k_l\varpi_l \in P_{++}(\mathfrak{g})$. Then the coefficients of ε_{l-1} and ε_l in λ are $(k_l + k_{l-1})/2$ and $(k_l - k_{l-1})/2$, respectively. Thus $\lambda \in P(G)$ if and only if $k_l + k_{l-1} = 2p$ and $k_l - k_{l-1} = 2q$ with $p,q \in \mathbb{Z}$.

Suppose $\lambda \in P_{++}(G)$. Since $k_i \geq 0$ for all i, we have $p \geq 0$ and $-p \leq q \leq p$. Set $n_i = k_i$ for $i = 1,\ldots,l-2$. If $q \geq 0$ set $n_{l-1} = 0$, $n_l = q$, and $n_{l+1} = p - q$. If $q \leq 0$ set

$n_{l-1} = -q$, $n_l = 0$, and $n_{l+1} = p + q$. Then λ is given by (3.16) with all coefficients $n_i \in \mathbb{N}$. □

The roots, fundamental weights, and some of the dominant weights for $\mathfrak{so}(4, \mathbb{C})$ are shown in Figure 3.3. The dominant weights of $\mathbf{SO}(4, \mathbb{C})$ are indicated by $\star$, and the dominant weights of $\mathfrak{so}(4, \mathbb{C})$ that are not weights of $\mathbf{SO}(4, \mathbb{C})$ are indicated by •. In this case the two simple roots are perpendicular and the Dynkin diagram is disconnected. The root system and the weight lattice are the product of two copies of the $\mathfrak{sl}(2, \mathbb{C})$ system (this occurs only in rank two), corresponding to the isomorphism $\mathfrak{so}(4, \mathbb{C}) \cong \mathfrak{sl}(2, \mathbb{C}) \oplus \mathfrak{sl}(2, \mathbb{C})$. The set of dominant weights is contained in the positive Weyl chamber, which is a cone of opening $90°$. The dominant weights for $\mathbf{SO}(4, \mathbb{C})$ are the nonnegative integer combinations of $2\varpi_1$, $2\varpi_2$, and $\varpi_1 + \varpi_2$ in this case.

The root systems of types B_2 and C_2 are isomorphic; to obtain the roots, fundamental weights, and dominant weights for $\mathfrak{so}(5, \mathbb{C})$, we just interchange the subscripts 1 and 2 in Figure 3.2.

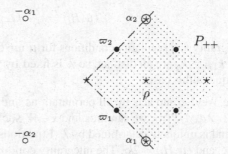

Fig. 3.3 Roots and dominant weights for $\mathfrak{g} = \mathfrak{so}(4, \mathbb{C})$.

The definition of dominant weight depends on a choice of the system Φ^+ of positive roots. We now prove that any weight can be transformed into a unique dominant weight by the action of the Weyl group. This means that the dominant weights give a cross section for the orbits of the Weyl group on the weight lattice.

Proposition 3.1.20. *For every* $\lambda \in P(\mathfrak{g})$ *there is* $\mu \in P_{++}(\mathfrak{g})$ *and* $s \in W$ *such that* $\lambda = s \cdot \mu$. *The weight* μ *is uniquely determined by* λ. *If* μ *is regular, then* s *is uniquely determined by* λ *and hence the orbit* $W \cdot \mu$ *has* $|W|$ *elements.*

For each type of classical group the dominant weights are given in terms of the weights $\{\varepsilon_i\}$ *as follows:*

1. Let $G = \mathbf{GL}(n, \mathbb{C})$ *or* $\mathbf{SL}(n, \mathbb{C})$. *Then* $P_{++}(\mathfrak{g})$ *consists of all weights*

$$\mu = k_1 \varepsilon_1 + \cdots + k_n \varepsilon_n \text{ with } k_1 \geq k_2 \geq \cdots \geq k_n \text{ and } k_i - k_{i+1} \in \mathbb{Z}. \tag{3.17}$$

2. Let $G = \mathbf{SO}(2l + 1, \mathbb{C})$. *Then* $P_{++}(\mathfrak{g})$ *consists of all weights*

$$\mu = k_1 \varepsilon_1 + \cdots + k_l \varepsilon_l \text{ with } k_1 \geq k_2 \geq \cdots \geq k_l \geq 0. \tag{3.18}$$

Here $2k_i$ and $k_i - k_j$ are integers for all i, j.

3. *Let $G = \mathbf{Sp}(l, \mathbb{C})$. Then $P_{++}(\mathfrak{g})$ consists of all μ satisfying (3.18) with k_i integers for all i.*

4. *Let $G = \mathbf{SO}(2l, \mathbb{C})$, $l \geq 2$. Then $P_{++}(\mathfrak{g})$ consists of all weights*

$$\mu = k_1 \varepsilon_1 + \cdots + k_l \varepsilon_l \text{ with } k_1 \geq \cdots \geq k_{l-1} \geq |k_l| . \tag{3.19}$$

Here $2k_i$ and $k_i - k_j$ are integers for all i, j.

The weight μ is regular when all inequalities in (3.17), (3.18), or (3.19) are strict.

Proof. For a general reductive Lie algebra, the first part of the proposition follows from Proposition 3.1.12. We will prove (1)–(4) for the classical Lie algebras by explicit calculation.

(1): The Weyl group is $W = \mathfrak{S}_n$, acting on $\mathfrak{h}^*$ by permuting $\varepsilon_1, \ldots, \varepsilon_n$. If $\lambda \in P(\mathfrak{g})$, then by a suitable permutation s we can make $\lambda = s \cdot \mu$, where $\mu = \sum_{i=1}^{n} k_i \varepsilon_i$ with $k_1 \geq \cdots \geq k_n$. Clearly μ is uniquely determined by λ. Since

$$\langle \mu, H_i \rangle = k_i - k_{i+1} , \tag{3.20}$$

the integrality and regularity conditions for μ are clear. When μ is regular, the coefficients $k_1, \ldots, k_n$ are all distinct, so λ is fixed by no nontrivial permutation. Hence s is unique in this case.

(2): The Weyl group acts by all permutations and sign changes of $\varepsilon_1, \ldots, \varepsilon_l$. Given $\lambda = \sum_{i=1}^{l} \lambda_i \varepsilon_i \in P(\mathfrak{g})$, we can thus find $s \in W$ such that $\lambda = s \cdot \mu$, where μ satisfies (3.18) and is uniquely determined by λ. Equations (3.20) hold for $i = 1, \ldots, l - 1$ in this case, and $\langle \mu, H_l \rangle = 2k_l$. The integrality conditions on k_i and the condition for μ to be regular are now clear. Evidently if μ is regular and $s \in W$ fixes μ then $s = 1$.

Conversely, every $\mu \in P_{++}(\mathfrak{g})$ is of the form

$$\mu = n_1 \varpi_1 + \cdots + n_l \varpi_l, \qquad n_i \in \mathbb{N} .$$

A calculation shows that μ satisfies (3.18) with $k_i = n_i + \cdots + n_{l-1} + (1/2)n_l$.

(3): The Weyl group W is the same as in (2), so the dominant weights are given by (3.18). In this case $\langle \mu, H_l \rangle = k_l$, so the coefficients k_i are all integers in this case. The regularity condition is obtained as in (2).

(4): The Weyl group acts by all permutations of $\varepsilon_1, \ldots, \varepsilon_l$ and all changes of an *even* number of signs. So by the action of W we can always make at least $l - 1$ of the coefficients of λ positive. Then we can permute the ε_i such that $\mu = s \cdot \lambda$ satisfies (3.19). In this case (3.20) holds for $i = 1, \ldots, l - 1$, and $\langle \mu, H_l \rangle = k_{l-1} + k_l$. The integrality condition and the condition for regularity are clear. Just as in (2) we see that if μ is regular then the only element of W that fixes μ is the identity.

Conversely, if $\mu \in P_{++}(\mathfrak{g})$ then μ is given as

$$\mu = n_1 \varpi_1 + \cdots + n_l \varpi_l, \qquad n_i \in \mathbb{N} .$$

Let $k_i = n_i + \cdots + n_{l-2} + (1/2)(n_{l-1} + n_l)$ for $i = 1, \ldots, l-2$. If we set $k_{l-1} = (1/2)(n_l + n_{l-1})$ and $k_l = (1/2)(n_l - n_{l-1})$, then μ satisfies (3.19). $\qquad\square$

There is a particular dominant weight that plays an important role in representation theory and which already appeared in the proof of Theorem 3.1.9. Recall that the choice of positive roots gives a triangular decomposition $\mathfrak{g} = \mathfrak{n}^- + \mathfrak{h} + \mathfrak{n}^+$.

Lemma 3.1.21. *Define $\rho \in \mathfrak{h}^*$ by*

$$\langle \rho, Y \rangle = \frac{1}{2} \operatorname{tr}(\operatorname{ad}(Y)|_{\mathfrak{n}^+}) = \frac{1}{2} \sum_{\alpha \in \Phi^+} \langle \alpha, Y \rangle$$

for $Y \in \mathfrak{h}$. Then $\rho = \varpi_1 + \cdots + \varpi_l$. Hence ρ is a dominant regular weight.

Proof. Let $s_i \in W$ be the reflection in the root α_i. By (3.8) we have

$$s_i(\rho) = -\frac{1}{2}\alpha_i + \frac{1}{2} \sum_{\beta \in \Phi^+ \setminus \{\alpha_i\}} \beta = \rho - \alpha_i .$$

But we also have $s_i(\rho) = \rho - \langle \rho, H_i \rangle \alpha_i$ by the definition of s_i. Hence $\langle \rho, H_i \rangle = 1$ and $\rho = \varpi_1 + \cdots + \varpi_l$. $\qquad\square$

Let $\mathfrak{g}$ be a classical Lie algebra. From the equations $\langle \rho, H_i \rangle = 1$ it is easy to calculate that ρ is given as follows:

$$\rho = \frac{l}{2}\varepsilon_1 + \frac{l-2}{2}\varepsilon_2 + \cdots - \frac{l-2}{2}\varepsilon_l - \frac{l}{2}\varepsilon_{l+1} , \qquad \text{(Type } A_l)$$

$$\rho = \left(l - \frac{1}{2}\right)\varepsilon_1 + \left(l - \frac{3}{2}\right)\varepsilon_2 + \cdots + \frac{3}{2}\varepsilon_{l-1} + \frac{1}{2}\varepsilon_l , \qquad \text{(Type } B_l)$$

$$\rho = l\varepsilon_1 + (l-1)\varepsilon_2 + \cdots + 2\varepsilon_{l-1} + \varepsilon_l , \qquad \text{(Type } C_l)$$

$$\rho = (l-1)\varepsilon_1 + (l-2)\varepsilon_2 + \cdots + 2\varepsilon_{l-2} + \varepsilon_{l-1} . \qquad \text{(Type } D_l)$$

See Figures 3.1, 3.2, and 3.3 for the rank-two examples.

3.1.5 Exercises

1. Let $G \subset \mathbf{GL}(n, \mathbb{C})$ be a classical group in the matrix form of Section 2.4.1 and let Φ be the root system of G. Set $V = \sum_{i=1}^{n} \mathbb{R}\varepsilon_i$. Give V the inner product $(\cdot, \cdot)$ such that $(\varepsilon_i, \varepsilon_j) = \delta_{ij}$.
   (a) Show that (α, α), for $\alpha \in \Phi$, is 1, 2, or 4, and that at most two distinct lengths occur. (The system Φ is called *simply laced* when all roots have the same length, because the Dynkin diagram has no double lines in this case.)
   (b) Let $\alpha, \beta \in \Phi$ with $(\alpha, \alpha) = (\beta, \beta)$. Show that there exists $w \in W_G$ such that $w \cdot \alpha = \beta$. (If $G = \mathbf{SO}(2l, \mathbb{C})$ assume that $l \geq 3$.)

2. Let $\mathfrak{g}$ be a classical Lie algebra.
   (a) Verify the formulas for the fundamental weights in terms of $\{\varepsilon_i\}$.
   (b) Show that the fundamental weights are given in terms of the simple roots as follows:
   Type A_l: For $1 \le i \le l$,

$$\varpi_i = \frac{l+1-i}{l+1}(\alpha_1 + 2\alpha_2 + \cdots + (i-1)\alpha_{i-1})$$
$$+ \frac{i}{l+1}((l-i+1)\alpha_i + (l-i)\alpha_{i+1} + \cdots + \alpha_l).$$

Type B_l: For $i < l$,

$$\varpi_i = \alpha_1 + 2\alpha_2 + \cdots + (i-1)\alpha_{i-1} + i(\alpha_i + \alpha_{i+1} + \cdots + \alpha_l),$$
$$\varpi_l = \frac{1}{2}(\alpha_1 + 2\alpha_2 + \cdots + l\alpha_l).$$

Type C_l: For $1 \le i \le l$,

$$\varpi_i = \alpha_1 + 2\alpha_2 + \cdots + (i-1)\alpha_{i-1} + i\left(\alpha_i + \alpha_{i+1} + \cdots + \alpha_{l-1} + \frac{1}{2}\alpha_l\right).$$

Type D_l: For $i < l-1$,

$$\varpi_i = \alpha_1 + 2\alpha_2 + \cdots + (i-1)\alpha_{i-1} + i(\alpha_i + \alpha_{i+1} + \cdots + \alpha_{l-2})$$
$$+ \frac{i}{2}(\alpha_{l-1} + \alpha_l),$$
$$\varpi_{l-1} = \frac{1}{2}\left(\alpha_1 + 2\alpha_2 + \cdots + (l-2)\alpha_{l-2} + \frac{l}{2}\alpha_{l-1} + \frac{l-2}{2}\alpha_l\right),$$
$$\varpi_l = \frac{1}{2}\left(\alpha_1 + 2\alpha_2 + \cdots + (l-2)\alpha_{l-2} + \frac{l-2}{2}\alpha_{l-1} + \frac{l}{2}\alpha_l\right).$$

3. Use the preceding exercise to verify the following relation between the root lattice and the weight lattice of the Lie algebra of a classical group (the order of the abelian group $P(\mathfrak{g})/Q(\mathfrak{g})$ is called the *index of connection* of the root system).
   Type A_l: $P(\mathfrak{g})/Q(\mathfrak{g}) \cong \mathbb{Z}/(l+1)\mathbb{Z}$;
   Type B_l: $P(\mathfrak{g})/Q(\mathfrak{g}) \cong \mathbb{Z}/2\mathbb{Z}$;
   Type C_l: $P(\mathfrak{g})/Q(\mathfrak{g}) \cong \mathbb{Z}/2\mathbb{Z}$;
   Type D_l $(l \ge 2)$: $P(\mathfrak{g})/Q(\mathfrak{g}) \cong \begin{cases} \mathbb{Z}/4\mathbb{Z} & \text{for } l \text{ odd,} \\ (\mathbb{Z}/2\mathbb{Z}) \times (\mathbb{Z}/2\mathbb{Z}) & \text{for } l \text{ even.} \end{cases}$
   In all cases find representatives for the cosets of $Q(\mathfrak{g})$ in $P(\mathfrak{g})$. (HINT: First verify the following congruences modulo $Q(\mathfrak{g})$: For types A_l and C_l, $\varpi_i \equiv i\varpi_1$ for $i = 2, \ldots, l$. For type D_l, $\varpi_i \equiv i\varpi_1$ for $i = 2, \ldots, l-2$, $\varpi_{l-1} - \varpi_l \equiv \varpi_1$ and $\varpi_{l-1} + \varpi_l \equiv (l-1)\varpi_1$.)

4. (a) Verify the formulas given for the weight ρ in terms of $\{\varepsilon_i\}$.
   (b) Show that ρ is given in terms of the simple roots as follows:
   Type A_l: $2\rho = \sum_{i=1}^{l} i(l-i+1)\alpha_i$;

Type B_l: $2\rho = \sum_{i=1}^{l} i(2l-i)\alpha_i$;

Type C_l: $2\rho = \sum_{i=1}^{l-1} i(2l-i+1)\alpha_i + \frac{l(l+1)}{2}\alpha_l$;

Type D_l: $2\rho = \sum_{i=1}^{l-2} 2\left(il - \frac{i(i+1)}{2}\right)\alpha_i + \frac{l(l-1)}{2}(\alpha_{l-1}+\alpha_l)$.

5. Let Φ be the root system of a semisimple Lie algebra, and let Δ_1 and Δ_2 be two systems of simple roots. Prove that there is an element w of the Weyl group such that $w\Delta_1 = \Delta_2$. (HINT: Let Φ_i be the positive roots relative to Δ_i, for $i = 1, 2$, and argue by induction on $r = |\Phi_1 \cap (-\Phi_2)|$. If $r = 0$ then $\Phi_1 = \Phi_2$ and hence $\Delta_1 = \Delta_2$. If $r > 0$ then $-\Phi_2$ must contain some $\alpha \in \Delta_1$. Using the fact that s_α permutes the set $\Phi_1 \setminus \{\alpha\}$, show that $(s_\alpha \Phi_1) \cap (-\Phi_2)$ has cardinality $r-1$. Hence the induction hypothesis applies to $s_\alpha \Delta_1$ and Δ_2.)

6. Let $G = \mathbf{SL}(3,\mathbb{C})$, let $H \subset G$ be the maximal torus of diagonal matrices, and consider the representation of G on $V = \mathbb{C}^3 \otimes \mathbb{C}^3$.

   (a) Find the set of weights $\mathfrak{X}(V)$ in terms of the functionals ε_1, ε_2, ε_3. For each weight determine its multiplicity, and verify that the multiplicities are invariant under the Weyl group W of G.

   (b) Verify that each Weyl group orbit in $\mathfrak{X}(V)$ contains exactly one dominant weight. Find the *extreme* dominant weights β (those such that $\beta + \alpha \notin \mathfrak{X}(V)$ for any positive root α).

   (c) Write the elements of $\mathfrak{X}(V)$ in terms of the fundamental weights ϖ_1, ϖ_2. Plot the set of weights as in Figure 3.1 (label multiplicities and indicate W-orbits).

   (d) There is a G-invariant decomposition $V = V_+ \oplus V_-$, where V_+ is the symmetric 2-tensors, and V_- are the skew-symmetric 2-tensors. Determine the weights and multiplicities of $V_\pm$. Verify that the weight multiplicities are invariant under W.

7. Let $G = \mathbf{Sp}(\mathbb{C}^4, \Omega)$, where Ω is the skew form (2.6), let $H \subset G$ be the maximal torus of diagonal matrices, and consider the representation of G on $V = \bigwedge^2 \mathbb{C}^4$.

   (a) Find the set of weights $\mathfrak{X}(V)$ in terms of the functionals ε_1, ε_2, and for each weight determine its multiplicity (note that $\varepsilon_3 = -\varepsilon_2$ and $\varepsilon_4 = -\varepsilon_1$ as elements of $\mathfrak{h}^*$). Verify that the multiplicities are invariant under the Weyl group W of G.

   (b) Verify that each Weyl group orbit in $\mathfrak{X}(V)$ contains exactly one dominant weight. Find the *extreme* dominant weights β (those such that $\beta + \alpha \notin \mathfrak{X}(V)$ for any positive root α).

   (c) Write the elements of $\mathfrak{X}(V)$ in terms of the fundamental weights ϖ_1, ϖ_2. Plot the set of weights as in Figure 3.2 (label multiplicities and indicate W-orbits).

3.2 Irreducible Representations

The principal result of this section is the *theorem of the highest weight*. The highest weight of an irreducible representation occurs with multiplicity one and uniquely determines the representation, just as in the case of $\mathfrak{sl}(2,\mathbb{C})$ treated in Chapter 2.

3.2.1 Theorem of the Highest Weight

Throughout this section $\mathfrak{g}$ will be a semisimple Lie algebra. Fix a Cartan subalgebra $\mathfrak{h}$ of $\mathfrak{g}$ and let Φ be the root system of $\mathfrak{g}$ with respect to $\mathfrak{h}$. Choose a set Φ^+ of positive roots in Φ. If $\alpha \in \Phi^+$ then let $\mathfrak{g}_\alpha$ denote the corresponding root space. As in Section 2.5.3, we define $\mathfrak{n}^+ = \bigoplus_{\alpha \in \Phi^+} \mathfrak{g}_\alpha$ and $\mathfrak{n}^- = \bigoplus_{\alpha \in \Phi^+} \mathfrak{g}_{-\alpha}$. We also write

$$\mathfrak{s}(\alpha) = \mathfrak{g}_{-\alpha} + [\mathfrak{g}_\alpha, \mathfrak{g}_{-\alpha}] + \mathfrak{g}_\alpha \, ,$$

as in Section 2.4.2. Then $\mathfrak{s}(\alpha)$ is a Lie subalgebra of $\mathfrak{g}$ that has basis $x \in \mathfrak{g}_\alpha$, $h \in [\mathfrak{g}_\alpha, \mathfrak{g}_{-\alpha}]$, and $y \in \mathfrak{g}_{-\alpha}$ such that $[x,y] = h$, $[h,x] = 2x$, and $[h,y] = -2y$. Recall that the element h, called the *coroot associated with* α, is unique, and is denoted by h_α.

Let (π, V) be a representation of $\mathfrak{g}$ (we do not assume that $\dim V < \infty$). If $\lambda \in \mathfrak{h}^*$ then we set

$$V(\lambda) = \{ v \in V : \pi(h)v = \langle \lambda, h \rangle v \text{ for all } h \in \mathfrak{h} \} \, ,$$

as in the finite-dimensional case. We define

$$\mathfrak{X}(V) = \{ \lambda \in \mathfrak{h}^* : V(\lambda) \neq 0 \} \, .$$

We call an element of $\mathfrak{X}(V)$ a *weight* of the representation π and the space $V(\lambda)$ the λ *weight space* of V. We note that if $\alpha \in \Phi^+$ and $x \in \mathfrak{g}_\alpha$ then $xV(\lambda) \subset V(\lambda + \alpha)$. Furthermore, the weight spaces for different weights are linearly independent.

Put a partial order on $\mathfrak{h}^*$ by defining $\mu \prec \lambda$ if $\mu = \lambda - \beta_1 - \cdots - \beta_r$ for some (not necessarily distinct) $\beta_i \in \Phi^+$ and some integer $r \geq 1$. We call this the *root order*. From (3.13) we see that on pairs λ, μ with $\lambda - \mu \in Q(\mathfrak{g})$ this partial order is the same as the partial order defined by the dual positive Weyl chamber (see the proof of Proposition 3.1.12).

If V is infinite-dimensional, then $\mathfrak{X}(V)$ can be empty. We now introduce a class of $\mathfrak{g}$-modules that have weight-space decompositions with finite-dimensional weight spaces. If $\mathfrak{g}$ is a Lie algebra then $U(\mathfrak{g})$ denotes its universal enveloping algebra; every Lie algebra representation (π, V) of $\mathfrak{g}$ extends uniquely to an associative algebra representation (π, V) of $U(\mathfrak{g})$ (see Section C.2.1). We use module notation and write $\pi(T)v = Tv$ for $T \in U(\mathfrak{g})$ and $v \in V$.

Definition 3.2.1. A $\mathfrak{g}$-module V is a *highest-weight representation* (relative to a fixed set Φ^+ of positive roots) if there are $\lambda \in \mathfrak{h}^*$ and a nonzero vector $v_0 \in V$ such that (1) $\mathfrak{n}^+ v_0 = 0$, (2) $h v_0 = \langle \lambda, h \rangle v_0$ for all $h \in \mathfrak{h}$, and (3) $V = U(\mathfrak{g})v_0$.

Set $\mathfrak{b} = \mathfrak{h} + \mathfrak{n}^+$. A vector v_0 satisfying properties (1) and (2) in Definition 3.2.1 is called a $\mathfrak{b}$-*extreme* vector. A vector v_0 satisfying property (3) in Definition 3.2.1 is called a $\mathfrak{g}$-*cyclic vector* for the representation.

Lemma 3.2.2. *Let V be a highest-weight representation of $\mathfrak{g}$ as in Definition 3.2.1. Then*

$$V = \mathbb{C}v_0 \oplus \bigoplus_{\mu \prec \lambda} V(\mu) \, , \tag{3.21}$$

with $\dim V(\mu) < \infty$ for all μ. In particular, $\dim V(\lambda) = 1$ and λ is the unique max-imal element in $\mathfrak{X}(V)$, relative to the root order; it is called the highest weight of V.

Proof. Since $\mathfrak{g} = \mathfrak{n}^- \oplus \mathfrak{h} \oplus \mathfrak{n}^+$, the Poincaré–Birkhoff–Witt theorem implies that there is a linear bijection $U(\mathfrak{n}^-) \otimes U(\mathfrak{h}) \otimes U(\mathfrak{n}^+) \longrightarrow U(\mathfrak{g})$ such that

$$Y \otimes H \otimes X \mapsto YHX \quad \text{for } Y \in U(\mathfrak{n}^-), H \in U(\mathfrak{h}), \text{ and } X \in U(\mathfrak{n}^+)$$

(see Corollary C.2.3). We have $U(\mathfrak{h})U(\mathfrak{n}^+)v_0 = \mathbb{C}v_0$, so it follows that $V = U(\mathfrak{n}^-)v_0$. Thus V is the linear span of v_0 together with the elements $y_1 y_2 \cdots y_r v_0$, for all $y_j \in \mathfrak{g}_{-\beta_j}$ with $\beta_j \in \Phi^+$ and $r = 1, 2, \ldots$. If $h \in \mathfrak{h}$ then

$$hy_1 y_2 \cdots y_r v_0 = y_1 y_2 \cdots y_r h v_0 + \sum_{i=1}^{r} y_1 \cdots y_{i-1}[h, y_i]y_{i+1} \cdots y_r v_0$$

$$= \Big(\langle \lambda, h \rangle - \sum_{i=1}^{r} \langle \beta_i, h \rangle \Big) y_1 \cdots y_r v_0 .$$

Thus $y_1 y_2 \cdots y_r v_0 \in V(\mu)$, where $\mu = \lambda - \beta_1 - \cdots - \beta_r$ satisfies $\mu \prec \lambda$. This implies (3.21). We have $\dim \mathfrak{g}_{-\beta_j} = 1$, and for a fixed μ there is only a finite number of choices of $\beta_j \in \Phi^+$ such that $\mu = \lambda - \beta_1 - \cdots - \beta_r$. Hence $\dim V(\mu) < \infty$. $\square$

Corollary 3.2.3. *Let (π, V) be a nonzero irreducible finite-dimensional representa-tion of $\mathfrak{g}$. There exists a unique dominant integral $\lambda \in \mathfrak{h}^*$ such that $\dim V(\lambda) = 1$. Furthermore, every $\mu \in \mathfrak{X}(V)$ with $\mu \neq \lambda$ satisfies $\mu \prec \lambda$.*

Proof. By Theorem 3.1.16 we know that $\mathfrak{X}(V) \subset P(\mathfrak{g})$. Let λ be any maximal ele-ment (relative to the root order) in the finite set $\mathfrak{X}(V)$. If $\alpha \in \Phi^+$ and $x \in \mathfrak{g}_\alpha$ then $xV(\lambda) \subset V(\lambda + \alpha)$. But $V(\lambda + \alpha) = 0$ by the maximality of λ. Hence $\pi(\mathfrak{n}^+)V(\lambda) = 0$. Let $0 \neq v_0 \in V(\lambda)$. Then $L = U(\mathfrak{g})v_0 \neq (0)$ is a $\mathfrak{g}$-invariant subspace, and hence $L = V$. Now apply Lemma 3.2.2. The fact that λ is dominant follows from the rep-resentation theory of $\mathfrak{sl}(2, \mathbb{C})$ (Theorem 2.3.6 applied to the subalgebra $\mathfrak{s}(\alpha)$). $\square$

Definition 3.2.4. If (π, V) is a nonzero finite-dimensional irreducible representation of $\mathfrak{g}$ then the element $\lambda \in \mathfrak{X}(V)$ in Corollary 3.2.3 is called the *highest weight of V*. We will denote it by λ_V.

We now use the universal enveloping algebra $U(\mathfrak{g})$ to prove that the irreducible highest-weight representations of $\mathfrak{g}$ (infinite-dimensional, in general) are in one-to-one correspondence with $\mathfrak{h}^*$ via their highest weights. By the universal property of $U(\mathfrak{g})$, representations of $\mathfrak{g}$ are the same as modules for $U(\mathfrak{g})$. We define a $U(\mathfrak{g})$-module structure on $U(\mathfrak{g})^*$ by

$$gf(u) = f(ug) \quad \text{for } g, u \in U(\mathfrak{g}) \text{ and } f \in U(\mathfrak{g})^*. \tag{3.22}$$

Clearly $gf \in U(\mathfrak{g})^*$ and the map $g, f \mapsto gf$ is bilinear. Also for $g, g' \in U(\mathfrak{g})$ we have

$$g(g'f)(u) = g'f(ug) = f(ugg') = (gg')f(u) \,,$$

so definition (3.22) does give a representation of $U(\mathfrak{g})$ on $U(\mathfrak{g})^*$. Here the algebra $U(\mathfrak{g})$ is playing the role previously assigned to the algebraic group G, while the vector space $U(\mathfrak{g})^*$ is serving as the replacement for the space of regular functions on G.

Theorem 3.2.5. *Let $\lambda \in \mathfrak{h}^*$.*

1. *There exists an irreducible highest-weight representation (σ, L^λ) of $\mathfrak{g}$ with highest weight λ.*
2. *Let (π, V) be an irreducible highest-weight representation of $\mathfrak{g}$ with highest weight λ. Then (π, V) is equivalent to (σ, L^λ).*

Proof. To prove (1) we use λ to define a particular element of $U(\mathfrak{g})^*$ as follows. Since $\mathfrak{h}$ is commutative, the algebra $U(\mathfrak{h})$ is isomorphic to the symmetric algebra $S(\mathfrak{h})$, which in turn is isomorphic to the polynomial functions on $\mathfrak{h}^*$. Using these isomorphisms, we define $\lambda(H)$, for $H \in \mathfrak{h}$, to be evaluation at λ. This gives an algebra homomorphism $H \mapsto \lambda(H)$ from $U(\mathfrak{h})$ to $\mathbb{C}$. If $H = h_1 \cdots h_n$ with $h_j \in \mathfrak{h}$, then

$$\lambda(H) = \langle \lambda, h_1 \rangle \cdots \langle \lambda, h_n \rangle \,.$$

For any Lie algebra $\mathfrak{m}$ there is a unique algebra homomorphism $\varepsilon : U(\mathfrak{m}) \longrightarrow \mathbb{C}$ such that

$$\varepsilon(1) = 1 \quad \text{and} \quad \varepsilon(U(\mathfrak{m})\mathfrak{m}) = 0 \,.$$

When $Z \in U(\mathfrak{m})$ is written in terms of a P–B–W basis, $\varepsilon(Z)$ is the coefficient of 1. Combining these homomorphisms (when $\mathfrak{m} = \mathfrak{n}^\pm$) with the linear isomorphism $U(\mathfrak{g}) \cong U(\mathfrak{n}^-) \otimes U(\mathfrak{h}) \otimes U(\mathfrak{n}^+)$ already employed in the proof of Lemma 3.2.2, we can construct a unique element $f_\lambda \in U(\mathfrak{g})^*$ that satisfies

$$f_\lambda(YHX) = \varepsilon(Y)\varepsilon(X)\lambda(H) \quad \text{for } Y \in U(\mathfrak{n}^-), H \in U(\mathfrak{h}), \text{ and } X \in U(\mathfrak{n}^+) \,.$$

If $h \in \mathfrak{h}$, then $YHXh = YHhX + YHZ$ with $Z = [X, h] \in \mathfrak{n}^+ U(\mathfrak{n}^+)$. Since $\varepsilon(Z) = 0$ and $\lambda(Hh) = \langle \lambda, h \rangle \lambda(H)$, we obtain

$$hf_\lambda(YHX) = f_\lambda(YHhX) + f_\lambda(YHZ) = \langle \lambda, h \rangle f_\lambda(YHX) \,. \tag{3.23}$$

Furthermore, if $x \in \mathfrak{n}^+$ then $\varepsilon(x) = 0$, so we also have

$$xf_\lambda(YHX) = f_\lambda(YHXx) = \varepsilon(Y)\lambda(H)\varepsilon(X)\varepsilon(x) = 0 \,. \tag{3.24}$$

Define $L^\lambda = \{gf_\lambda : g \in U(\mathfrak{g})\}$ to be the $U(\mathfrak{g})$-cyclic subspace generated by f_λ relative to the action (3.22) of $U(\mathfrak{g})$ on $U(\mathfrak{g})^*$, and let σ be the representation of $\mathfrak{g}$ on L^λ obtained by restriction of this action. Then (3.23) and (3.24) show that (σ, L^λ) is a highest-weight representation with highest weight λ. Since $HX \in U(\mathfrak{h})U(\mathfrak{n}^+)$ acts on f_λ by the scalar $\lambda(H)\varepsilon(X)$, we have

$$L^\lambda = \{Yf_\lambda : Y \in U(\mathfrak{n}^-)\}. \tag{3.25}$$

We now prove that L^λ is irreducible. Suppose $0 \neq M \subset L^\lambda$ is a $\mathfrak{g}$-invariant subspace. By Lemma 3.2.2 we see that M has a weight-space decomposition under $\mathfrak{h}$:

$$M = \bigoplus_{\mu \preceq \lambda} M \cap L^\lambda(\mu).$$

If $\mu \preceq \lambda$ then there are only finitely many $\nu \in \mathfrak{h}^*$ such that $\mu \preceq \nu \preceq \lambda$. (This is clear by writing $\lambda - \mu$, $\nu - \mu$, and $\lambda - \nu$ as nonnegative integer combinations of the simple roots.) Hence there exists a weight of M, say μ, that is maximal. Take $0 \neq f \in M(\mu)$. Then $\mathfrak{n}^+ f = 0$ by maximality of μ, and hence $Xf = \varepsilon(X)f$ for all $X \in U(\mathfrak{n}^+)$. By (3.25) we know that $f = Y_0 f_\lambda$ for some $Y_0 \in U(\mathfrak{n}^-)$. Thus for $Y \in U(\mathfrak{n}^-), H \in U(\mathfrak{h})$, and $X \in U(\mathfrak{n}^+)$ we can evaluate

$$f(YHX) = Xf(YH) = \varepsilon(X)f(YH) = \varepsilon(X)f_\lambda(YHY_0).$$

But $HY_0 = Y_0 H + [H, Y_0]$, and $[H, Y_0] \in \mathfrak{n}^- U(\mathfrak{n}^-)$. Thus by definition of f_λ we obtain

$$f(YHX) = \varepsilon(X)\varepsilon(Y)\varepsilon(Y_0)\lambda(H) = \varepsilon(Y_0)f_\lambda(YHX).$$

Thus $\varepsilon(Y_0) \neq 0$ and $f = \varepsilon(Y_0)f_\lambda$. This implies that $M = L^\lambda$, and hence L^λ is irreducible, completing the proof of (1).

To prove (2), we note by Lemma 3.2.2 that V has a weight-space decomposition and $\dim(V/\mathfrak{n}^- V) = 1$. Let $p : V \longrightarrow V/\mathfrak{n}^- V$ be the natural projection. Then the restriction of p to $V(\lambda)$ is bijective. Fix nonzero elements $v_0 \in V(\lambda)$ and $u_0 \in V/\mathfrak{n}^- V$ such that $p(v_0) = u_0$. The representation π extends canonically to a representation of $U(\mathfrak{g})$ on V. For $X \in U(\mathfrak{n}^+)$ we have $Xv_0 = \varepsilon(X)v_0$, for $H \in U(\mathfrak{h})$ we have $Hv_0 = \lambda(H)v_0$, while for $Y \in U(\mathfrak{n}^-)$ we have $Yv_0 \equiv \varepsilon(Y)v_0 \mod \mathfrak{n}^- V$. Thus

$$p(YHXv_0) = \varepsilon(X)\lambda(H)p(Yv_0) = \varepsilon(Y)\varepsilon(X)\lambda(H)u_0 = f_\lambda(YHX)u_0. \tag{3.26}$$

For any $g \in U(\mathfrak{g})$ and $v \in V$, the vector $p(gv)$ is a scalar multiple of u_0. If we fix $v \in V$, then we obtain a linear functional $T(v) \in U(\mathfrak{g})^*$ such that

$$p(gv) = T(v)(g)u_0 \quad \text{for all } g \in U(\mathfrak{g})$$

(since the map $g \mapsto p(gv)$ is linear). The map $v \mapsto T(v)$ defines a linear transformation $T : V \longrightarrow U(\mathfrak{g})^*$, and we calculated in (3.26) that $T(v_0) = f_\lambda$. If $x \in \mathfrak{g}$ then

$$T(xv)(g)u_0 = p(gxv) = T(v)(gx)u_0 = (xT(v))(g)u_0.$$

Hence $T(xv) = xT(v)$. Since V is irreducible, we have $V = U(\mathfrak{g})v_0$, and so $T(V) = U(\mathfrak{g})f_\lambda = L^\lambda$. Thus T is a $\mathfrak{g}$-homomorphism from V onto L^λ.

We claim that $\mathrm{Ker}(T) = 0$. Indeed, if $T(v) = 0$ then $p(gv) = 0$ for all $g \in U(\mathfrak{g})$. Hence $U(\mathfrak{g})v \subset \mathfrak{n}^- V$. Since $\mathfrak{n}^- V \neq V$, we see that $U(\mathfrak{g})v$ is a proper $\mathfrak{g}$-invariant subspace of V. But V is irreducible, so $U(\mathfrak{g})v = 0$, and hence $v = 0$. Thus T gives an isomorphism between V and L^λ. $\square$

The vector space L^λ in Theorem 3.2.5 is infinite-dimensional, in general. From Corollary 3.2.3 a necessary condition for L^λ to be finite-dimensional is that λ be dominant integral. We now show that this condition is sufficient.

Theorem 3.2.6. *Let $\lambda \in \mathfrak{h}^*$ be dominant integral. Then the irreducible highest weight representation L^λ is finite-dimensional.*

Proof. We write $\mathfrak{X}(\lambda)$ for the set of weights of L^λ. Since λ is integral, we know from Lemma 3.2.2 that $\mathfrak{X}(\lambda) \subset P(\mathfrak{g}) \subset \mathfrak{h}_{\mathbb{R}}^*$. Fix a highest-weight vector $0 \neq v_0 \in L^\lambda(\lambda)$.

Let $\alpha \in \Delta$ be a simple root. We will show that $\mathfrak{X}(\lambda)$ is invariant under the action of the reflection s_α on $\mathfrak{h}_{\mathbb{R}}^*$. The argument proceeds in several steps. Fix a TDS basis $x \in \mathfrak{g}_\alpha$, $y \in \mathfrak{g}_{-\alpha}$, and $h = [x,y] \in \mathfrak{h}$ for the subalgebra $\mathfrak{s}(\alpha)$. Set $n = \langle \alpha, h \rangle$ and $v_j = y^j v_0$. Then $n \in \mathbb{N}$, since λ is dominant integral.

(i)   If β is a simple root and $z \in \mathfrak{g}_\beta$, then $zv_{n+1} = 0$.

To prove (i), suppose first that $\beta \neq \alpha$. Then $\beta - \alpha$ is not a root, since the coefficients of the simple roots are of opposite signs. Hence $[z,y] = 0$, so in this case $zv_j = y^j z v_0 = 0$. Now suppose that $\beta = \alpha$. Then we may assume that $z = x$. We know that $xv_j = j(n+1-j)v_{j-1}$ by equation (2.16). Thus $xv_{n+1} = 0$.

(ii)   The subspace $U(\mathfrak{s}(\alpha))v_0$ is finite-dimensional.

We have $\mathfrak{n}^+ v_{n+1} = 0$ by (i), since the subspaces $\mathfrak{g}_\beta$ with $\beta \in \Delta$ generate $\mathfrak{n}^+$ by Theorem 2.5.24. Furthermore $v_{n+1} \in L^\lambda(\mu)$ with $\mu = \lambda - (n+1)\alpha$. Hence the $\mathfrak{g}$-submodule $Z = U(\mathfrak{g})v_{n+1}$ is a highest-weight module with highest weight μ. By Lemma 3.2.2 every weight γ of Z satisfies $\gamma \preceq \mu$. Since $\mu \prec \alpha$, it follows that Z is a proper $\mathfrak{g}$-submodule of L^λ. But L^λ is irreducible, so $Z = 0$. The subspace $U(\mathfrak{s}(\alpha))v_0$ is spanned by $\{v_j : j \in \mathbb{N}\}$, since it is a highest-weight $\mathfrak{s}(\alpha)$ module. But $v_j \in Z$ for $j \geq n+1$, so $v_j = 0$ for $j \geq n+1$. Hence $U(\mathfrak{s}(\alpha))v_0 = \mathrm{Span}\{v_0, \ldots, v_n\}$.

(iii)   Let $v \in L^\lambda$ be arbitrary and let $F = U(\mathfrak{s}(\alpha))v$. Then $\dim F < \infty$.

Since $L^\lambda = U(\mathfrak{g})v_0$, there is an integer j such that $v \in U_j(\mathfrak{g})v_0$, where $U_j(\mathfrak{g})$ is the subspace of $U(\mathfrak{g})$ spanned by products of j or fewer elements of $\mathfrak{g}$. By Leibniz's rule $[\mathfrak{g}, U_j(\mathfrak{g})] \subset U_j(\mathfrak{g})$. Hence $F \subset U_j(\mathfrak{g})U(\mathfrak{s}(\alpha))v_0$. Since $U_j(\mathfrak{g})$ and $U(\mathfrak{s}(\alpha))v_0$ are finite-dimensional, this proves (iii).

(iv) $s_\alpha \mathfrak{X}(\lambda) \subset \mathfrak{X}(\lambda)$.

Let $\mu \in \mathfrak{X}(\lambda)$ and let $0 \neq v \in L^\lambda(\mu)$. Since $[\mathfrak{h}, \mathfrak{s}(\alpha)] \subset \mathfrak{s}(\alpha)$, the space $F = U(\mathfrak{s}(\alpha))v$ is invariant under $\mathfrak{h}$ and the weights of $\mathfrak{h}$ on F are of the form $\mu + k\alpha$, where $k \in \mathbb{Z}$. We know that F is finite-dimensional by (iii), so Theorem 2.3.6 implies that F is equivalent to

$$F^{(k_1)} \oplus \cdots \oplus F^{(k_r)} \tag{3.27}$$

as an $\mathfrak{s}(\alpha)$-module. Now $\langle \mu, h \rangle$ must occur as an eigenvalue of h in one of the submodules in (3.27), and hence $-\langle \mu, h \rangle$ also occurs as an eigenvalue by Lemma 2.3.2. It follows that there exists a weight of the form $\mu + k\alpha$ in $\mathfrak{X}(\lambda)$ with

$$\langle \mu + k\alpha, h \rangle = -\langle \mu, h \rangle \, .$$

Thus $2k = -2\langle \mu, h \rangle$. Hence $s_\alpha \mu = \mu + k\alpha \in \mathfrak{X}(\lambda)$, which proves (iv).

We can now complete the proof of the theorem. The Weyl group W is generated by the reflections s_α with $\alpha \in \Delta$ by Theorem 3.1.9. Since (iv) holds for all simple roots α, we conclude that the set $\mathfrak{X}(\lambda)$ is invariant under W. We already know from Lemma 3.2.2 that $\dim L^\lambda(\mu) < \infty$ for all μ. Thus the finite-dimensionality of L^λ is a consequence of the following property:

(v)   The cardinality of $\mathfrak{X}(\lambda)$ is finite.

Indeed, let $\mu \in \mathfrak{X}(\lambda)$. Then $\mu \in \mathfrak{h}_\mathbb{R}^*$. By Proposition 3.1.12 there exists $s \in W$ such that $\xi = s\mu$ is in the positive Weyl chamber C. Since $\xi \in \mathfrak{X}(\lambda)$, we know from Lemma 3.2.2 that $\xi = \lambda - Q$, where $Q = \beta_1 + \cdots + \beta_r$ with $\beta_i \in \Phi^+$. Let $(\cdot, \cdot)$ be the inner product on $\mathfrak{h}_\mathbb{R}^*$ as in Remark 3.1.5. Then

$$\begin{aligned}
(\xi, \xi) = (\lambda - Q, \xi) &= (\lambda, \xi) - (Q, \xi) \\
&\leq (\lambda, \xi) = (\lambda, \lambda - Q) \\
&\leq (\lambda, \lambda) \, .
\end{aligned}$$

Here we have used the inequalities $(\lambda, \beta_i) \geq 0$ and $(\xi, \beta_i) \geq 0$, which hold for elements of C. Since W acts by orthogonal transformations, we have thus shown that

$$(\mu, \mu) \leq (\lambda, \lambda) \, . \tag{3.28}$$

This implies that $\mathfrak{X}(\lambda)$ is contained in the intersection of the ball of radius $\|\lambda\|$ with the weight lattice $P(\mathfrak{g})$. This subset of $P(\mathfrak{g})$ is finite, which proves (v). $\square$

3.2.2 Weights of Irreducible Representations

Let $\mathfrak{g}$ be a semisimple Lie algebra with Cartan subalgebra $\mathfrak{h}$. We shall examine the set of weights of the finite-dimensional irreducible representations of $\mathfrak{g}$ in more detail, using the representation theory of $\mathfrak{sl}(2, \mathbb{C})$. We write $\Phi = \Phi(\mathfrak{g}, \mathfrak{h})$ for the set of roots of $\mathfrak{h}$ on $\mathfrak{g}$ and $Q_+ = Q_+(\mathfrak{g})$ for the semigroup generated by a fixed choice of positive roots Φ^+. Enumerate the simple roots in Φ^+ as $\alpha_1, \ldots, \alpha_l$ and write $H_i = h_{\alpha_i}$ for the corresponding coroots. For each root $\alpha \in \Phi$ we have the associated root reflection s_α; it acts on $Y \in \mathfrak{h}$ by

$$s_\alpha Y = Y - \langle \alpha, Y \rangle h_\alpha \, .$$

The dual action of s_α on $\beta \in \mathfrak{h}^*$ is $s_\alpha \beta = \beta - \langle \beta, h_\alpha \rangle \alpha$. The Weyl group W of $\mathfrak{g}$ is the finite group generated by the root reflections.

The root lattice and weight lattice are contained in the real vector space $\mathfrak{h}_\mathbb{R}^*$ spanned by the roots, and the inner product (α, β) on $\mathfrak{h}_\mathbb{R}^*$ defined by the Killing form is positive definite (see Corollary 2.5.22); for the classical groups this inner product is proportional to the standard inner product with $(\varepsilon_i, \varepsilon_j) = \delta_{ij}$. Let $\|\alpha\|^2 = (\alpha, \alpha)$ be the associated norm. Since the root reflections are real orthogonal transformations, the Weyl group preserves this inner product and norm.

Proposition 3.2.7. *Let (π, V) be a finite-dimensional representation of $\mathfrak{g}$ with weight space decomposition*

$$V = \bigoplus_{\mu \in \mathfrak{X}(V)} V(\mu) .$$

For $\alpha \in \Phi$ let $\{e_\alpha, f_\alpha, h_\alpha\}$ be a TDS triple associated with α. Define $E = \pi(e_\alpha)$, $F = \pi(f_\alpha)$, and $\tau_\alpha = \exp(E)\exp(-F)\exp(E) \in \mathbf{GL}(V)$. Then

1. *$\tau_\alpha \pi(Y) \tau_\alpha^{-1} = \pi(s_\alpha Y)$ for $Y \in \mathfrak{h}$;*
2. *$\tau_\alpha V(\mu) = V(s_\alpha \mu)$ for all $\mu \in \mathfrak{h}^*$;*
3. *$\dim V(\mu) = \dim V(s \cdot \mu)$ for all $s \in W$.*

Hence the weights $\mathfrak{X}(V)$ and the weight multiplicity function $m_V(\mu) = \dim V(\mu)$ are invariant under W.

Proof. (1): From Theorem 2.3.6 and Proposition 2.3.3 we know that E and F are nilpotent transformations. If X is any nilpotent linear transformation on V, then $\mathrm{ad}(X)$ is nilpotent on $\mathrm{End}(V)$ and we have

$$\exp(X)A\exp(-X) = \exp(\mathrm{ad}\,X)A \quad \text{for all } A \in \mathrm{End}(V) . \tag{3.29}$$

(This follows from Lemma 1.6.1 and the fact that the differential of the representation Ad is the representation ad.) For $Y \in \mathfrak{h}$ we have

$$\mathrm{ad}(E)\pi(Y) = -\pi(\mathrm{ad}(Y)e_\alpha) = -\langle \alpha, Y \rangle E .$$

Hence $\mathrm{ad}(E)^2(\pi(Y)) = 0$, and so from (3.29) we obtain

$$\exp(E)\pi(Y)\exp(-E) = \pi(Y) - \langle \alpha, Y \rangle E .$$

In particular, $\exp(E)\pi(h_\alpha)\exp(-E) \doteq \pi(h_\alpha) - 2E$. We also have

$$(\mathrm{ad}\,E)F = \pi(h_\alpha) \quad \text{and} \quad (\mathrm{ad}\,E)^2 F = -\langle \alpha, h_\alpha \rangle E = -2E .$$

Hence $(\mathrm{ad}\,E)^3 F = 0$, and so from (3.29) we obtain

$$\exp(E)F\exp(-E) = F + \pi(h_\alpha) - E .$$

The linear map taking e_α to $-f_\alpha$, f_α to $-e_\alpha$, and h_α to $-h_\alpha$ is an automorphism of $\mathfrak{s}(\alpha)$ (on $\mathfrak{sl}(2, \mathbb{C})$ this is the map $X \mapsto -X^t$). Hence the calculations just made also prove that

$$\exp(-F)\pi(Y)\exp(F) = \pi(Y) - \langle \alpha, Y \rangle F \,,$$
$$\exp(-F)E\exp(F) = E + \pi(h_\alpha) - F \,.$$

Combining these relations we obtain

$$\begin{aligned}
\tau_\alpha \pi(Y) \tau_\alpha^{-1} &= \exp(E)\exp(-F)\{\pi(Y) - \langle \alpha, Y \rangle E\}\exp(F)\exp(-E) \\
&= \exp E\{\pi(Y) - \langle \alpha, Y \rangle E - \langle \alpha, Y \rangle \pi(h_\alpha)\}\exp(-E) \\
&= \pi(Y) - \langle Y, \alpha \rangle \pi(h_\alpha) = \pi(s_\alpha Y) \,.
\end{aligned}$$

(2): Let $v \in V(\mu)$ and $Y \in \mathfrak{h}$. Then by (1) we have $\pi(Y)\tau_\alpha v = \tau_\alpha \pi(s_\alpha Y)v = \langle s_\alpha \mu, Y \rangle \tau_\alpha v$. This shows that $\tau_\alpha v \in V(s_\alpha \mu)$.

(3): This follows from (2), since W is generated by the reflections s_α. $\qquad\square$

Remark 3.2.8. The definition of the linear transformation τ_α comes from the matrix identity $\begin{bmatrix} 1 & 1 \\ 0 & 1 \end{bmatrix} \begin{bmatrix} 1 & 0 \\ -1 & 1 \end{bmatrix} \begin{bmatrix} 1 & 1 \\ 0 & 1 \end{bmatrix} = \begin{bmatrix} 0 & 1 \\ -1 & 0 \end{bmatrix}$ in $\mathbf{SL}(2, \mathbb{C})$, where the element on the right is a representative for the nontrivial Weyl group element.

Lemma 3.2.9. *Let (π, V) be a finite-dimensional representation of $\mathfrak{g}$ and let $\mathfrak{X}(V)$ be the set of weights of V. If $\lambda \in \mathfrak{X}(V)$ then $\lambda - k\alpha \in \mathfrak{X}(V)$ for all roots $\alpha \in \Phi$ and all integers k between 0 and $\langle \lambda, h_\alpha \rangle$, inclusive, where $h_\alpha \in \mathfrak{h}$ is the coroot to α.*

Proof. We may suppose that the integer $m = \langle \lambda, h_\alpha \rangle$ is nonzero. Since $s_\alpha \cdot \lambda = \lambda - m\alpha$, we have

$$\dim V(\lambda) = \dim V(\lambda - m\alpha)$$

by Proposition 3.2.7. Take $0 \neq v \in V(\lambda)$. If $m > 0$ then from Theorem 2.3.6 and Proposition 2.3.3 we have $0 \neq \pi(f_\alpha)^k v \in V(\lambda - k\alpha)$ for $k = 0, 1, \dots, m$. If $m < 0$ then likewise we have $0 \neq \pi(e_\alpha)^{-k}v \in V(\lambda - k\alpha)$ for $k = 0, -1, \dots, m$. This shows that $\lambda - k\alpha \in \mathfrak{X}(V)$. $\qquad\square$

We say that a subset $\Psi \subset P(\mathfrak{g})$ is Φ-*saturated* if for all $\lambda \in \Psi$ and $\alpha \in \Phi$, one has $\lambda - k\alpha \in \Psi$ for all integers k between 0 and $\langle \lambda, h_\alpha \rangle$. In particular, a Φ-saturated set is invariant under the Weyl group W, since W is generated by the reflections $s_\alpha(\lambda) = \lambda - \langle \lambda, h_\alpha \rangle \alpha$. An element $\lambda \in \Psi$ is called Φ-*extreme* if for all $\alpha \in \Phi$, either $\lambda + \alpha \notin \Psi$ or $\lambda - \alpha \notin \Psi$.

From Lemma 3.2.9 the set of weights of a finite-dimensional representation of $\mathfrak{g}$ is Φ-saturated, and the highest weight is Φ-extreme. Using these notions we now show how to construct the complete set of weights of an irreducible representation starting with the highest weight.

Proposition 3.2.10. *Let V be the finite-dimensional irreducible $\mathfrak{g}$-module with highest weight λ.*

1. *$\mathfrak{X}(V)$ is the smallest Φ-saturated subset of $P(\mathfrak{g})$ containing λ.*
2. *The orbit of λ under the Weyl group is the set of Φ-extreme elements of $\mathfrak{X}(V)$.*

Proof. (1): Let $\Psi' \subset \mathfrak{X}(V)$ be the smallest Φ-saturated subset of $P(\mathfrak{g})$ containing λ. If $\Psi'' = \mathfrak{X}(V) \setminus \Psi'$ were nonempty, then it would contain a maximal element μ (relative to the root order). Since μ is not the highest weight, there must exist $\alpha \in \Phi^+$ such that $\mu + \alpha \in \mathfrak{X}(V)$. Let p, q be the largest integers such that $\mu + p\alpha$ and $\mu - q\alpha$ are weights of V. Then $p \geq 1$, $q \geq 0$, and $s_\alpha(\lambda + p\alpha) = \lambda - q\alpha$ by Corollary 2.4.5. Because μ is maximal in Ψ''', we have $\mu + p\alpha \in \Psi'$. However, Ψ', being Φ-saturated, is invariant under W, so $\mu - q\alpha \in \Psi'$ also. Hence $\mu + k\alpha \in \Psi'$ for all integers $k \in [-q, p]$. In particular, taking $k = 0$ we conclude that $\mu \in \Psi'$, which is a contradiction.

(2): Since $W \cdot \Phi = \Phi$, the set of Φ-extreme elements of $\mathfrak{X}(V)$ is invariant under W. Thus it suffices to show that if $\mu \in P_{++}(\mathfrak{g}) \cap \mathfrak{X}(V)$ is Φ-extreme then $\mu = \lambda$. Take a simple root α_i and corresponding reflection s_i. Let the nonnegative integers p, q be as in (1) relative to μ and α_i. Since $s_i(\mu + p\alpha_i) = \mu - q\alpha_i$ we have $q - p = \langle \mu, H_i \rangle$. But since μ is dominant, $\langle \mu, H_i \rangle \geq 0$. Hence $q \geq p$. If $p \geq 1$ then $q \geq 1$ and $\mu \pm \alpha_i \in \mathfrak{X}(V)$. This would contradict the assumption that μ is Φ-extreme. Hence $p = 0$ and $\mu + \alpha_i \notin \mathfrak{X}(V)$ for $i = 1, \ldots, l$. We conclude that μ is a maximal element of $\mathfrak{X}(V)$. But we have already shown in Corollary 3.2.3 that λ is the unique maximal weight. $\qquad\square$

On the set of dominant weights of a representation, the root order $\prec$ has the following inductive property:

Proposition 3.2.11. *Let V be any finite-dimensional representation of $\mathfrak{g}$. Suppose $\mu \in P_{++}(\mathfrak{g})$, $\nu \in \mathfrak{X}(V)$, and $\mu \prec \nu$. Then $\mu \in \mathfrak{X}(V)$.*

Proof. By assumption, $\nu = \mu + \beta$, where $\beta = \sum_{i=1}^l n_i \alpha_i \in Q_+$. We proceed by induction on $\mathrm{ht}(\beta) = \sum n_i$, the result being true if $\beta = 0$. If $\beta \neq 0$ then

$$0 < (\beta, \beta) = \sum n_i (\beta, \alpha_i) \, .$$

Thus there exists an index i such that $n_i \geq 1$ and $(\beta, \alpha_i) > 0$. For this value of i we have $\langle \beta, H_i \rangle \geq 1$. Since $\langle \mu, H_j \rangle \geq 0$ for all j, it follows that $\langle \nu, H_i \rangle \geq 1$. But $\mathfrak{X}(V)$ is Φ-saturated, so $\nu' = \nu - \alpha_i \in \mathfrak{X}(V)$. Set $\beta' = \beta - \alpha_i$. Then $\beta' \in Q_+$, $\mathrm{ht}(\beta') = \mathrm{ht}(\beta) - 1$, and $\mu = \nu' - \beta'$. By induction, $\mu \in \mathfrak{X}(V)$. $\qquad\square$

Corollary 3.2.12. *Let L^λ be the finite-dimensional irreducible $\mathfrak{g}$-module with highest weight λ. Then $\mathfrak{X}(L^\lambda) \cap P_{++}(\mathfrak{g})$ consists of all $\mu \in P_{++}(\mathfrak{g})$ such that $\mu \preceq \lambda$.*

Corollary 3.2.12 and inequality (3.28) give an explicit algorithm for finding the weights of L^λ. Take all $\beta \in Q_+$ such that $\|\lambda - \beta\| \leq \|\lambda\|$ (there are only finitely many) and write $\mu = \lambda - \beta$ in terms of the basis of fundamental weights. If all the coefficients are nonnegative, then μ is a weight of L^λ. This gives all the dominant weights, and the Weyl group orbits of these weights make up the entire set of weights.

Example

Consider the representation of $\mathfrak{sl}(3,\mathbb{C})$ with highest weight $\lambda = \varpi_1 + 2\varpi_2$. The weights of this representation (as determined by the algorithm just described) are shown in Figure 3.4.

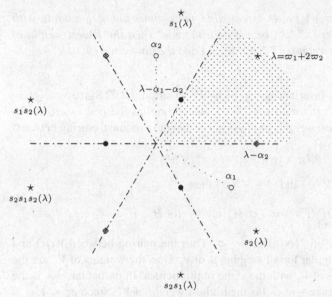

Fig. 3.4 Weights of the representation $L^{\varpi_1+2\varpi_2}$ of $\mathfrak{sl}(3,\mathbb{C})$.

It is clear from the picture that $\lambda - \alpha_2$ and $\lambda - \alpha_1 - \alpha_2$ are the only dominant weights in $\lambda - Q_+$. The highest weight is regular, and its orbit under the Weyl group $W = \mathfrak{S}_3$ (indicated by $\star$) has $|W| = 6$ elements (in Figure 5.1, s_1 and s_2 denote the reflections for the simple roots α_1 and α_2, respectively). Each of the weights $\lambda - \alpha_2$ and $\lambda - \alpha_1 - \alpha_2$ is fixed by one of the simple reflections, and their W-orbits (indicated by $\circledast$ and $\bullet$, respectively) have three elements. The weights $\star$ in the *outer shell* have multiplicity one, by Proposition 3.2.7 and Corollary 3.2.3. The weights marked by $\circledast$ have multiplicity one, whereas the weights marked by $\bullet$ have multiplicity two, so $\dim V^\mu = 15$ (we shall see how to obtain these multiplicities in Section 8.1.2).

3.2.3 Lowest Weights and Dual Representations

We continue the notation and assumptions of Section 3.2.2. If we choose $-\Phi^+$ as the system of positive roots instead of Φ^+, then the subalgebra $\mathfrak{n}^+$ is replaced by $\mathfrak{n}^-$ and $\mathfrak{b}$ is replaced by $\bar{\mathfrak{b}} = \mathfrak{h} + \mathfrak{n}^-$. We call $\bar{\mathfrak{b}}$ the Borel subalgebra *opposite* to

b. From Corollary 3.2.3 we conclude that if V is an irreducible finite-dimensional $\mathfrak{g}$-module, then it has a unique *lowest weight* $\mu \in -P_{++}(\mathfrak{g})$, characterized by the property $\mu \preceq \nu$ for all $\nu \in \mathfrak{X}(V)$. The weight space $V(\mu)$ is one-dimensional; a nonzero vector in this weight space is called a *lowest-weight vector* for V. Let w_0 be the unique element in the Weyl group W such that $w_0 \Phi^+ = -\Phi^+$ (see Lemma 3.1.6 for the classical groups and Theorem 3.1.9 in the general case).

Theorem 3.2.13. *Let (π, V) be an irreducible finite-dimensional $\mathfrak{g}$-module with highest weight λ and let (π^*, V^*) be the dual module. Then the lowest weight of V is $w_0(\lambda)$. The highest weight of V^* is $-w_0(\lambda)$ and the lowest weight of V^* is $-\lambda$.*

Proof. The set $\mathfrak{X}(V)$ is invariant under W by Proposition 3.2.7. Since $w_0 \cdot Q_+ = -Q_+$, we have $w_0(\lambda) \prec \mu$ for all $\mu \in \mathfrak{X}(V)$, which implies the first assertion. To find the highest and lowest weights of V^*, observe that the canonical pairing between V and V^* satisfies

$$\langle \pi(X)v, v^* \rangle = -\langle v, \pi^*(X)v^* \rangle$$

for $X \in \mathfrak{g}$. Hence if $v \in V(\mu)$ and $v^* \in V^*(\sigma)$ then

$$\mu(H)\langle v, v^* \rangle = -\sigma(H)\langle v, v^* \rangle \quad \text{for } H \in \mathfrak{h} \,.$$

This implies that $V(\mu) \perp V^*(\sigma)$ if $\mu \neq -\sigma$. Thus the pairing between $V(\mu)$ and $V^*(-\mu)$ must be nonsingular for all weights μ of V. Thus the weights of V^* are the negatives of the weights of V, with the same multiplicities. In particular, $-\lambda$ is the lowest weight of V^*. Hence $-w_0(\lambda)$ is the highest weight of V^*, since $w_0^2 = 1$. $\quad\square$

3.2.4 Symplectic and Orthogonal Representations

We continue the notation and assumptions of Section 3.2.2. We will also use Schur's lemma (Lemma 4.1.4) from Chapter 4.

Theorem 3.2.14. *Suppose (π, V) is an irreducible finite-dimensional representation of $\mathfrak{g}$ with highest weight λ. There is a nonzero $\mathfrak{g}$-invariant bilinear form on V if and only if $-w_0\lambda = \lambda$. In this case the form is nonsingular and unique up to a scalar multiple. Furthermore, the form is either symmetric or skew-symmetric.*

Proof. We can identify the invariant bilinear forms on V with $\mathrm{Hom}_{\mathfrak{g}}(V, V^*)$ as a $\mathfrak{g}$-module. In this identification a $\mathfrak{g}$-intertwining operator $T : V \longrightarrow V^*$ corresponds to the bilinear form

$$\Omega(x, y) = \langle Tx, y \rangle, \quad \text{for } x, y \in V \,.$$

Hence by Schur's lemma the space of invariant bilinear forms has dimension 1 or 0, depending on whether π and π^* are equivalent or not. But we know from Theorem 3.2.5 that π and π^* are equivalent if and only if $\lambda = -w_0\lambda$.

Suppose Ω is a nonzero invariant bilinear form on V. The radical of Ω is a proper g-invariant subspace, and hence is zero, by the irreducibility of π. Thus Ω is nondegenerate. Define bilinear forms

$$\Omega_\pm(x,y) = \Omega(x,y) \pm \Omega(y,x).$$

Then Ω_+ is symmetric, Ω_- is skew-symmetric, and each is g-invariant. Since the space of invariant forms is one-dimensional, either $\Omega_+ = 0$ or $\Omega_- = 0$. □

Suppose the irreducible representation (π,V) admits a nonzero g-invariant bilinear form Ω. We call π *symplectic* or *orthogonal* depending on whether Ω is skew-symmetric or symmetric, respectively. We now seek a criterion in terms of the highest weight λ of π to distinguish between these two possibilities. We begin with the basic case.

Lemma 3.2.15. *Suppose $G = \mathbf{SL}(2,\mathbb{C})$ and (π,V) is the $(m+1)$-dimensional irreducible representation of G. Then π is symplectic if m is odd, and π is orthogonal if m is even.*

Proof. The element w_0 acts on $\mathfrak{h}^*$ by -1, so every irreducible representation of G is self-dual. Recall from Section 2.3.2 that we can take for V the space of polynomials $f(x_1,x_2)$ that are homogeneous of degree m, with action

$$\pi(g)f(x_1,x_2) = f(ax_1 \dotplus cx_2, bx_1 + dx_2), \quad \text{for } g = \begin{bmatrix} a & b \\ c & d \end{bmatrix}.$$

The function $f(x_1,x_2) = x_1^m$ is the highest-weight vector.

We take the normalized monomials

$$v_k = \binom{m}{k} x_1^{m-k} x_2^k$$

as a basis for V, and we define a bilinear form Ω on V by

$$\Omega(v_k, v_{m-k}) = (-1)^k \binom{m}{k}$$

and $\Omega(v_k, v_p) = 0$ if $p \neq m - k$. Observe that Ω is skew if m is odd and is symmetric if m is even. If $u = ax_1 + bx_2$ and $v = cx_1 + dx_2$, then $u^m, v^m \in V$ and from the binomial expansion we find that

$$\Omega(u^m, v^m) = \det \begin{bmatrix} a & b \\ c & d \end{bmatrix}^m.$$

Hence $\Omega(g \cdot u^m, g \cdot v^m) = (\det g)^m \Omega(u^m, v^m)$ for $g \in \mathbf{GL}(2,\mathbb{C})$. Since $V = \mathrm{Span}\{u^m : u = ax_1 + bx_2\}$, we see that Ω is invariant under $\mathbf{SL}(2,\mathbb{C})$, and hence is also invariant under $\mathfrak{sl}(2,\mathbb{C})$. □

We now return to the general case. Define $h^0 = \sum_{\alpha \in \Phi^+} h_\alpha$, where h_α is the coroot to α.

Lemma 3.2.16. *Let $\alpha_1, \ldots, \alpha_l$ be the simple roots in Φ^+ and let H_i be the coroot to α_i. Then $\langle \alpha_i, h^0 \rangle = 2$ for $i = 1, \ldots, l$. Furthermore, there are integers $c_i \geq 1$ such that*

$$h^0 = \sum_{i=1}^{l} c_i H_i \, .$$

Proof. Let $s_i \in W$ be the reflection in the root α_i and let $\check{\Phi}_i^+$ be the set of positive coroots with H_i deleted. Then s_i preserves $\check{\Phi}_i^+$ and $s_i(H_i) = -H_i$. Hence

$$s_i(h^0) = h^0 - 2H_i$$

(see the proof of Lemma 3.1.21). However, $s_i(h^0) = h^0 - \langle \alpha_i, h^0 \rangle H_i$, so we conclude that $\langle \alpha_i, h^0 \rangle = 2$.

We calculate

$$c_i = \langle \varpi_i, h^0 \rangle = \sum_{\alpha \in \Phi^+} \langle \varpi_i, h_\alpha \rangle \, . \tag{3.30}$$

But $\langle \varpi_i, h_\alpha \rangle \geq 0$ for all $\alpha \in \Phi^+$, and $\langle \varpi_i, H_i \rangle = 1$. Hence $c_i \geq 1$. □

Let $\{E_i, F_i, H_i\}$ be a TDS with $E_i \in \mathfrak{g}_{\alpha_i}$ and $F_i \in \mathfrak{g}_{-\alpha_i}$. Define

$$e^0 = \sum_{i=1}^{l} E_i \quad \text{and} \quad f^0 = \sum_{i=1}^{l} c_i F_i \, ,$$

where c_i is given by (3.30). We claim that $\{e^0, f^0, h^0\}$ is a TDS triple. Clearly,

$$[h^0, e^0] = 2e^0 \quad \text{and} \quad [h^0, f^0] = -2f^0 \, ,$$

since $\langle \alpha_i, h^0 \rangle = 2$ for $i = 1, \ldots, l$. Since $\alpha_i - \alpha_j$ is not a root, we have $[E_i, F_j] = 0$ when $i \neq j$. Hence

$$[e^0, f^0] = \sum_{i=1}^{l} c_i [E_i, F_i] = h^0 \, ,$$

which proves our assertion. Now we apply Lemma 3.2.15 to the subalgebra $\mathfrak{g}^0 = \mathrm{Span}\{e^0, f^0, h^0\}$ of $\mathfrak{g}$ to obtain the following criterion:

Theorem 3.2.17. *Let (π, V) be the irreducible representation of $\mathfrak{g}$ with highest weight $\lambda \neq 0$. Assume that $-w_0 \lambda = \lambda$. Set $m = \langle \lambda, h^0 \rangle$. Then m is a positive integer. If m is odd, then π is symplectic, whereas if m is even, then π is orthogonal.*

Proof. Write $\lambda = m_1 \varpi_1 + \cdots + m_l \varpi_l$ in terms of the fundamental weights and let c_i be the integers in Lemma 3.2.16. Then

$$m = \sum_{i=1}^{l} c_i m_i \, ,$$

from which it is clear that m is a positive integer.

Fix a nonzero vector v_0 of weight λ in V and let $V^{(0)}$ be the $\mathfrak{g}^0$-cyclic subspace generated by v_0. Let σ be the representation of $\mathfrak{g}^0$ on $V^{(0)}$ obtained from the restriction of π. By Proposition 2.3.3 $(\sigma, V^{(0)})$ is an irreducible representation of $\mathfrak{g}^0$ of dimension $m+1$ and $\sigma(h^0)$ has eigenvalues $m, m-2, \ldots, -m+2, m$.

The complete set of eigenvalues of $\pi(h^0)$ on V consists of the numbers $\langle \mu, h^0 \rangle$, where μ is a weight of π. For any such μ we have $\mu \preceq \lambda$, since λ is the highest weight. Furthermore, $-\mu$ is also a weight of π, since $\pi \cong \pi^*$, so we also have $-\mu \preceq \lambda$. Thus if $\mu \neq \pm\lambda$, then $-\lambda \prec \mu \prec \lambda$. Now since $\langle \alpha_i, h^0 \rangle = 2$ for $i = 1, \ldots l$, it follows that $\langle \beta, h^0 \rangle \geq 2$ for all nonzero $\beta \in Q_+$. Hence

$$-m = \langle \lambda, h^0 \rangle < \langle \mu, h^0 \rangle < \langle \lambda, h^0 \rangle = m \qquad (3.31)$$

for all weights μ of π except $\mu = \pm\lambda$. Since the highest weight λ occurs with multiplicity one in V, it follows from (3.31) that the representation $(\sigma, V^{(0)})$ of $\mathfrak{g}^0$ occurs with multiplicity one when V is decomposed into isotypic components under the action of $\pi(\mathfrak{g}^0)$ (see Theorem 2.3.6). In particular, $V(-\lambda) \subset V^{(0)}$ since $\sigma(h^0)$ acts by $-m$ on $V(-\lambda)$.

Let Ω be a nonzero $\mathfrak{g}$-invariant form on V and let Ω^0 be the restriction of Ω to $V^{(0)}$. Since $\Omega|_{V(\lambda) \times V(-\lambda)} \neq 0$ we know that Ω^0 is nonzero. Thus we can apply Lemma 3.2.15 to determine whether Ω is symmetric or skew-symmetric. $\square$

3.2.5 Exercises

In the following exercises $\mathfrak{g}$ is a simple Lie algebra of the specified type.

1. Let $\tilde{\alpha}$ be the highest root; it is dominant since it is the highest weight of the adjoint representation.
   (a) Let Ψ be the set of positive roots that are also dominant weights. Verify that Ψ is given as follows:
   Type A_l: $\tilde{\alpha} = \varpi_1 + \varpi_l$,   Type B_l: ϖ_1 and $\tilde{\alpha} = \varpi_2$,
   Type C_l: ϖ_2 and $\tilde{\alpha} = 2\varpi_1$,   Type D_l: $\tilde{\alpha} = \varpi_2$.
   (b) Show that the set of dominant weights $\mu \preceq \tilde{\alpha}$ consists of $\Psi \cup \{0\}$, where Ψ is given in (a).
   (c) Verify case by case that the set of weights of the adjoint representation of $\mathfrak{g}$ (namely $\Phi \cup \{0\}$) consists of $W \cdot \Psi \cup \{0\}$, as predicted by Corollary 3.2.12.
2. Draw a figure similar to Figure 3.4 for the weight $\lambda = \varpi_1 + 2\varpi_2$ when $\mathfrak{g} = \mathfrak{sp}(2, \mathbb{C})$. Find the dominant weights of L^λ and their orbits under the Weyl group of $\mathfrak{g}$ (see Figure 3.2 in Section 3.1.4).
3. Let $\mathfrak{g}$ be the Lie algebra of a classical group of rank l and let $\varpi_1, \ldots, \varpi_l$ be the fundamental weights. Suppose $\lambda = m_1\varpi_1 + \cdots + m_l\varpi_l$ is the highest weight of an irreducible $\mathfrak{g}$-module V. Let λ^* be the highest weight of the dual module V^*. Use the formula $\lambda^* = -w_0 \cdot \lambda$ and the results of Section 3.1.2 to show that λ^* is given as follows:
   Type A_l: $\lambda^* = m_l\varpi_1 + m_{l-1}\varpi_2 + \cdots + m_2\varpi_{l-1} + m_1\varpi_l$,

Type B_l or C_l: $\lambda^* = \lambda$,

Type D_l: $\lambda^* = m_1 \varpi_1 + \cdots + m_{l-2} \varpi_{l-2} + m_l \varpi_{l-1} + m_{l-1} \varpi_l$ if l is odd, and $\lambda^* = \lambda$ if l is even.

4. Let σ_r be the irreducible representation of $\mathfrak{sl}(n, \mathbb{C})$ with highest weight ϖ_r. In Section 5.5.2 we will show that σ_r is the differential of the natural representation of $\mathbf{SL}(n, \mathbb{C})$ on $\bigwedge^r \mathbb{C}^n$. Assuming this result, show that the duality $\sigma_r^* \cong \sigma_{n-r}$, from the previous exercise can be realized as follows:

(a) Show that $\bigwedge^r \mathbb{C}^n \times \bigwedge^{n-r} \mathbb{C}^n$ has a unique nonsingular bilinear form $\langle \cdot, \cdot \rangle$ such that $u \wedge w = \langle u, w \rangle e_1 \wedge \cdots \wedge e_n$ for $u \in \bigwedge^r \mathbb{C}^n$ and $w \in \bigwedge^{n-r} \mathbb{C}^n$. (HINT: Let I and J be ordered subsets of $\{1, \ldots, n\}$ with $|I| = r$ and $|J| = n - r$. Then $v_I \wedge v_J = \pm e_1 \wedge \cdots \wedge e_n$ if $I \cap J = \emptyset$, and $v_I \wedge v_J = 0$ otherwise.)

(b) Show that for $g \in \mathbf{SL}(n, \mathbb{C})$ one has $\langle \sigma_r(g)u, \sigma_{n-r}(g)w \rangle = \langle u, w \rangle$.

5. Let $G = \mathbf{SO}(n, \mathbb{C})$. Show that $\bigwedge^p \mathbb{C}^n \cong \bigwedge^{n-p} \mathbb{C}^n$ as a G-module for $0 \le p \le n$. (HINT: Use the previous exercise.)

6. Show that the element h^0 in Lemma 3.2.16 is given in terms of the simple coroots as follows:

Type A_l: $h^0 = \sum_{i=1}^{l} i(l+1-i) H_i$,

Type B_l: $h^0 = \sum_{i=1}^{l-1} i(2l+1-i) H_i + \frac{l(l+1)}{2} H_l$,

Type C_l: $h^0 = \sum_{i=1}^{l} i(2l-i) H_i$,

Type D_l: $h^0 = \sum_{i=1}^{l-2} 2(il - \frac{i(i+1)}{2}) H_i + \frac{l(l-1)}{2}(H_{l-1} + H_l)$.

(HINT: Use the results of Exercises 3.1.5, #3 and the fact that h^0 corresponds to 2ρ for the dual root system, where α_i is replaced by H_i.)

7. Let $\mu = m_1 \varpi_1 + \cdots + m_l \varpi_l \in P_{++}(\mathfrak{g})$. Use the previous exercise to show that the integer m in Theorem 3.2.17 is given as follows:

Type A_l: $m = \sum_{i=1}^{l} i(l+1-i) m_i$,

Type B_l: $m = \sum_{i=1}^{l-1} i(2l+1-i) m_i + \frac{l(l+1)}{2} m_l$,

Type C_l: $m = \sum_{i=1}^{l} i(2l-i) m_i$,

Type D_l: $m = \sum_{i=1}^{l-2} 2(il - \frac{i(i+1)}{2}) m_i + \frac{l(l-1)}{2}(m_{l-1} + m_l)$.

8. Use the criterion of Theorem 3.2.17 and the previous exercise to show that the adjoint representation of $\mathfrak{g}$ is orthogonal.

9. Let (π, V) be the irreducible $\mathfrak{g}$-module with highest weight ϖ_i.

(a) Show that π is an orthogonal representation in the following cases:

Type A_l: $i = p$ when $l = 2p - 1$ and p is even,

Type B_l: $i = 1, 2, \ldots, l-1$ for all l, and $i = l$ if $l \equiv 0, 3 \pmod 4$,

Type C_l: $2 \le i \le l$ and i even,

Type D_l: $i = 1, 2, \ldots, l-2$ for all l, and $i = l-1, l$ if $l \equiv 0 \pmod 4$.

(b) Show that π is a symplectic representation in the following cases:

Type A_l: $i = p$ when $l = 2p - 1$ and p is odd,

Type B_l: $i = l$ if $l \equiv 1, 2 \pmod 4$,

Type C_l: $1 \le i \le l$ and i odd,

Type D_l: $i = l-1, l$ if $l \equiv 2 \pmod 4$.

10. Let (π, V) be the irreducible $\mathfrak{g}$-module with highest weight $\rho = \varpi_1 + \cdots + \varpi_l$ (the smallest regular dominant weight).

(a) Show that π is an orthogonal representation in the following cases:

Type A_l: $l \equiv 0,2,3 \pmod 4$,
Types B_l and C_l: $l \equiv 0,3 \pmod 4$,
Type D_l: all $l \geq 3$.
(b) Show that π is a symplectic representation in the following cases:
Type A_l: $l \equiv 1 \pmod 4$,
Types B_l and C_l : $l \equiv 1,2 \pmod 4$.

3.3 Reductivity of Classical Groups

In this section we give two proofs of the complete reducibility of regular representations of a classical group G. The first proof is algebraic and applies to any semisimple Lie algebra; the key tool is the *Casimir operator*, which will also play an important role in Chapter 7. The second proof is analytic and uses integration on a compact real form of G.

3.3.1 Reductive Groups

Let G be a linear algebraic group. We say that a rational representation (ρ, V) of G is *completely reducible* if for every G-invariant subspace $W \subset V$ there exists a G-invariant subspace $U \subset V$ such that $V = W \oplus U$. In matrix terms, this means that any basis $\{w_1, \ldots, w_p\}$ for W can be completed to a basis $\{w_1, \ldots, w_p, u_1, \ldots, u_q\}$ for V such that the subspace $U = \mathrm{Span}\{u_1, \ldots, u_q\}$ is invariant under $\rho(g)$ for all $g \in G$. Thus the matrix of $\rho(g)$ relative to this basis has the block-diagonal form

$$\begin{bmatrix} \sigma(g) & 0 \\ 0 & \tau(g) \end{bmatrix},$$

where $\sigma(g) = \rho(g)|_W$ and $\tau(g) = \rho(g)|_U$.

Definition 3.3.1. A linear algebraic group G is *reductive* if every rational representation (ρ, V) of G is completely reducible.

We shall show that the classical groups are reductive. Both proofs we give require some preparation; we will emphasize the algebraic approach because it involves techniques that we will use later in the book (some of the preliminary material for the analytic proof, which involves integration over a compact real form of G, is in Appendix D).

Lemma 3.3.2. *Let (ρ, V) be a completely reducible rational representation of the algebraic group G. Suppose $W \subset V$ is an invariant subspace. Set $\sigma(x) = \rho(x)|_W$ and $\pi(x)(v + W) = \rho(x)v + W$ for $x \in \mathcal{A}$ and $v \in V$. Then the representations (σ, W) and $(\pi, V/W)$ are completely reducible.*

Proof. Write $V = W \oplus U$ for some invariant subspace U, and let P be the projection onto W with kernel U. If $Y \subset W$ is an invariant subspace, then the subspace $U \oplus Y$ is invariant. Hence there is an invariant subspace $Z \subset V$ such that

$$V = (U \oplus Y) \oplus Z. \tag{3.32}$$

The subspace $P(Z) \subset W$ is invariant, and we claim that

$$W = Y \oplus P(Z). \tag{3.33}$$

We have $\dim W = \dim V/U = \dim Y + \dim Z$ by (3.32). Since $\operatorname{Ker} P = U$, the map $z \mapsto P(z)$ is a bijective intertwining operator from Z to $P(Z)$, so that $\dim Z = \dim P(Z)$. Hence $\dim W = \dim Y + \dim P(Z)$. Also,

$$W = P(V) = P(U) + P(Y) + P(Z) = Y + P(Z).$$

Thus (3.33) holds, which proves the complete reducibility of (σ, W).

Let $M \subset V/W$ be an invariant subspace, and $p : V \longrightarrow V/W$ the canonical quotient map. Define $\widetilde{M} = p^{-1}(M) \subset V$. Then $\widetilde{M}$ is invariant, so there exists an invariant subspace $\widetilde{N}$ with $V = \widetilde{M} \oplus \widetilde{N}$. Set $N = p(\widetilde{N})$. This is an invariant subspace, and $V/W = M \oplus N$. Thus $(\pi, V/W)$ is completely reducible. $\square$

Proposition 3.3.3. *Let (ρ, V) be a rational representation of an algebraic group G. The following are equivalent:*

1. *(ρ, V) is completely reducible.*
2. *$V = V_1 \oplus \cdots \oplus V_d$ with each V_i invariant and irreducible under G.*
3. *$V = V_1 + \cdots + V_d$ as a vector space, where each V_i is an irreducible G-submodule of V.*

Proof. (1) $\Rightarrow$ (2): If $\dim V = 1$ then V is irreducible and (2) trivially holds. Assume that (1) $\Rightarrow$ (2) for all rational G-modules V of dimension less than r. Let V be a module of dimension r. If V is irreducible then (2) trivially holds. Otherwise there are nonzero submodules W and U such that $V = U \oplus W$. These submodules are completely reducible by Lemma 3.3.2, and hence they decompose as the direct sum of irreducible submodules by the induction hypothesis. Thus (2) also holds for V.

(2) $\Rightarrow$ (1): We prove (1) by induction on the number d of irreducible summands in (2). Let $0 \neq W \subset V$ be a submodule. If $d = 1$, then $W = V$ by irreducibility, and we are done. If $d > 1$, let $P_1 : V \longrightarrow V_1$ be the projection operator associated with the direct sum decomposition (2). If $P_1 W = 0$, then

$$W \subset V_2 \oplus \cdots \oplus V_d,$$

so by the induction hypothesis there is a G-invariant complement to W. If $P_1 W \neq 0$ then $P_1 W = V_1$, since it is a G-invariant subspace. Set $W' = \operatorname{Ker}(P_1|_W)$. We have $W' \subset V_2 \oplus \cdots \oplus V_d$, so by the induction hypothesis there exists a G-invariant subspace $U \subset V_2 \oplus \cdots \oplus V_d$ such that

$$V_2 \oplus \cdots \oplus V_d = W' \oplus U .$$

Since $P_1 U = 0$, we have $W \cap U \subset W' \cap U = 0$. Also,

$$\dim W = \dim V_1 + \dim W' ,$$

and therefore

$$\dim W + \dim U = \dim V_1 + \dim W' + \dim U$$
$$= \dim V_1 + \sum_{i \geq 2} \dim V_i = \dim V .$$

Hence $V = U \oplus W$.

Obviously (2) implies (3). Assuming (3), we will prove that

$$V = W_1 \oplus \cdots \oplus W_s$$

with W_j, for $j = 1, \ldots, s$, being irreducible G-submodules. The case $d = 1$ is obvious. Let $d > 1$ and assume that the result is true for any regular G-module that is the sum of fewer than d irreducible submodules. Applying the induction hypothesis to the regular G-module $U = V_1 + \cdots + V_{d-1}$, we can find irreducible G-modules W_i such that

$$U = W_1 \oplus \cdots \oplus W_r .$$

By assumption, $V = U + V_d$. The subspace $U \cap V_d$ is a G-submodule of V_d; hence it is either (0) or V_d by irreducibility. In the first case we have the desired direct-sum decomposition with $s = r + 1$ and $W_s = V_d$; in the second case we have the direct-sum decomposition with $r = s$. $\qquad\square$

Remark. The methods in this proof apply more generally to the context of modules for an algebra over $\mathbb{C}$ treated in Section 4.1.4. In that section we will leave it to the reader to make the appropriate changes in the proof for this wider context.

Corollary 3.3.4. *Suppose (ρ, V) and (σ, W) are completely reducible regular representations of G. Then $(\rho \oplus \sigma, V \oplus W)$ is a completely reducible representation.*

Proof. By Proposition 3.3.3, V and W are direct sums of irreducible invariant subspaces. Thus

$$V = V_1 \oplus \cdots \oplus V_m \quad \text{and} \quad W = W_1 \oplus \cdots \oplus W_n .$$

It follows that $V \oplus W$ satisfies condition (2) in Proposition 3.3.3. $\qquad\square$

Proposition 3.3.5. *Let G and H be linear algebraic groups with $H \subset G$. Assume that H is reductive and has finite index in G. Then G is reductive.*

Proof. Let (ρ, V) be a rational representation of G and suppose $W \subset V$ is a G-invariant subspace. Since H is reductive, there exists a H-invariant subspace Z such that $V = W \oplus Z$. Let $P : V \longrightarrow W$ be the projection along Z. Then

$$\rho(h) P \rho(h)^{-1} = P \quad \text{for } h \in H , \tag{3.34}$$

since Z is invariant under H. Given a coset $\gamma = gH$ of H in G, let $P(\gamma) = \rho(g)P\rho(g)^{-1}$. This is well defined by (3.34). Denote the set of left H cosets in G by Γ and form the average

$$Q = \frac{1}{|\Gamma|} \sum_{\gamma \in \Gamma} P(\gamma)$$

(at this point we use the assumption that Γ is finite). Let $w \in W$. Then $\rho(g)^{-1}w \in W$, since W is G-invariant. Hence $P\rho(g)^{-1}w = \rho(g)^{-1}w$, so we have $\rho(g)P\rho(g)^{-1}w = w$ for every $g \in G$. Thus $Qw = w$. On the other hand, for any $v \in V$ the vector $P\rho(g)^{-1}v$ is in W, so by the invariance of W we know that $\rho(g)P\rho(g)^{-1}v$ is in W. This shows that $\mathrm{Range}(Q) \subset W$ and $Q^2 = Q$. Hence Q is a projection with range W.

If $g \in G$ then

$$\rho(g)Q = \frac{1}{|\Gamma|} \sum_{\gamma \in \Gamma} \rho(g\gamma)P\rho(\gamma)^{-1} = \frac{1}{|\Gamma|} \sum_{\gamma \in \Gamma} \rho(g\gamma)P\rho(g\gamma)^{-1}\rho(g)$$

$$= Q\rho(g)$$

(replace γ by $g\gamma$ in the second summation). Since Q commutes with $\rho(G)$, the subspace $Y = \mathrm{Range}(I_V - Q)$ is a G-invariant complement to W.   □

In particular, every finite group G is a linear algebraic group, so we can take $H = \{1\}$ in Proposition 3.3.5 and obtain *Maschke's theorem*:

Corollary 3.3.6. *Let G be a finite group. Then G is reductive.*

3.3.2 Casimir Operator

We now introduce the key ingredient for the algebraic proof that the classical groups are reductive. Let $\mathfrak{g}$ be a semisimple Lie algebra. Fix a Cartan subalgebra $\mathfrak{h}$ in $\mathfrak{g}$, let Φ be the root system of $\mathfrak{g}$ with respect to $\mathfrak{h}$, and fix a set Φ^+ of positive roots in Φ. Recall that the Killing form B on $\mathfrak{g}$ is nondegenerate by Theorem 2.5.11. The restriction of B to $\mathfrak{h}_{\mathbb{R}}$ is positive definite and gives inner products and norms, denoted by $(\cdot, \cdot)$ and $\| \cdot \|$, on $\mathfrak{h}_{\mathbb{R}}$ and $\mathfrak{h}_{\mathbb{R}}^*$.

Fix a basis $\{X_i\}$ for $\mathfrak{g}$ and let $\{Y_i\}$ be the B-dual basis: $B(X_i, Y_j) = \delta_{ij}$ for all i, j. If (π, V) is a representation of $\mathfrak{g}$ (not necessarily finite-dimensional), we define

$$C_\pi = \sum_i \pi(X_i)\,\pi(Y_i)\,. \tag{3.35}$$

This linear transformation on V is called the *Casimir operator* of the representation.

Lemma 3.3.7. *The Casimir operator is independent of the choice of basis for $\mathfrak{g}$ and commutes with $\pi(\mathfrak{g})$.*

Proof. We can choose a basis $\{Z_i\}$ for $\mathfrak{g}$ such that $B(Z_i, Z_j) = \delta_{ij}$. Write $X_i = \sum_j B(X_i, Z_j) Z_j$ and $Y_i = \sum_k B(Y_i, Z_k) Z_k$ and substitute in the formula for C_π to obtain

$$C_\pi = \sum_{i,j,k} B(X_i, Z_j) B(Y_i, Z_k) \pi(Z_j)\, \pi(Z_k)\, .$$

For fixed j, j, the sum over i on the right side is

$$B(\textstyle\sum_i B(X_i, Z_j) Y_i, Z_k) = B(Z_j, Z_k) = \delta_{jk}\, .$$

Hence $C_\pi = \sum_j \pi(Z_j)^2$, which proves that C_π does not depend on the choice of basis.

Now let $Z \in \mathfrak{g}$. Using the expansion of $[Z, Z_i]$ terms of the B-orthonormal basis $\{Z_j\}$, we can write

$$[Z, Z_i] = \sum_j B([Z, Z_i], Z_j) Z_j = \sum_j B(Z, [Z_i, Z_j]) Z_j\, .$$

Here we have used the $\mathfrak{g}$ invariance of B in the second equation. Since $[A, BC] = [A, B]C + B[A, C]$ for any $A, B, C \in \mathrm{End}(V)$, we can use these expansions to write

$$\begin{aligned} [\pi(Z), C_\pi] &= \sum_i \pi([Z, Z_i])\, \pi(Z_i) + \sum_j \pi(Z_j)\, \pi([Z, Z_j]) \\ &= \sum_{i,j} \big\{ B(Z, [Z_i, Z_j]) + B(Z, [Z_j, Z_i]) \big\}\, \pi(Z_j)^2\, . \end{aligned}$$

However, this last sum is zero by the skew symmetry of the Lie bracket. $\qquad\square$

Lemma 3.3.8. *Let (π, V) be a highest-weight representation of $\mathfrak{g}$ with highest weight λ and let $\rho = (1/2) \sum_{\alpha \in \Phi^+} \alpha$. Then the Casimir operator acts on V as a scalar:*

$$C_\pi v = \big((\lambda + \rho, \lambda + \rho) - (\rho, \rho) \big) v \tag{3.36}$$

for all $v \in V$.

Proof. Let $H_1, \dots, H_l$ be an orthonormal basis of $\mathfrak{h}_\mathbb{R}$ with respect to B. Enumerate $\Phi^+ = \{\alpha_1, \dots, \alpha_d\}$ and for $\alpha \in \Phi^+$ fix $X_{\pm\alpha} \in \mathfrak{g}_{\pm\alpha}$, normalized such that $B(X_\alpha, X_{-\alpha}) = 1$. Then

$$\begin{aligned} &\{H_1, \dots, H_l, X_{\alpha_1}, X_{-\alpha_1}, \dots, X_{\alpha_d}, X_{-\alpha_d}\}\, , \\ &\{H_1, \dots, H_l, X_{-\alpha_1}, X_{\alpha_1}, \dots, X_{-\alpha_d}, X_{\alpha_d}\} \end{aligned} \tag{3.37}$$

are dual bases for $\mathfrak{g}$. For the pair of bases (3.37) the Casimir operator is given by

$$C_\pi = \sum_{i=1}^l \pi(H_i)^2 + \sum_{\alpha \in \Phi^+} \big(\pi(X_\alpha)\pi(X_{-\alpha}) + \pi(X_{-\alpha})\pi(X_\alpha) \big)\, . \tag{3.38}$$

Let $H_\rho \in \mathfrak{h}$ satisfy $B(H_\rho, H) = \langle \rho, H \rangle$ for all $H \in \mathfrak{h}$. We can rewrite formula (3.38) using the commutation relation $\pi(X_\alpha)\pi(X_{-\alpha}) = \pi(H_\alpha) + \pi(X_{-\alpha})\pi(X_\alpha)$ to obtain

$$C_\pi = \sum_{i=1}^{l} \pi(H_i)^2 + 2\pi(H_\rho) + 2 \sum_{\alpha \in \Phi^+} \pi(X_{-\alpha})\pi(X_\alpha) \,. \tag{3.39}$$

Let $v_0 \in V(\lambda)$ be a nonzero highest-weight vector. By formula (3.39),

$$C_\pi v_0 = \Big(\sum_{i=1}^{l} \pi(H_i)^2 + 2\pi(H_\rho) \Big) v_0 = \Big(\sum_{i=1}^{l} \langle \lambda, H_i \rangle^2 + 2\langle \lambda, H_\rho \rangle \Big) v_0$$
$$= ((\lambda, \lambda) + 2(\rho, \lambda)) v_0 \,.$$

Here we have used the fact that $\{H_i\}$ is a B-orthonormal basis for $\mathfrak{h}_\mathbb{R}$ to express $(\lambda, \lambda) = \sum_i \langle \lambda, H_i \rangle^2$. Now write

$$(\lambda, \lambda) + 2(\lambda, \rho) = (\lambda + \rho, \lambda + \rho) - (\rho, \rho)$$

to see that (3.36) holds when $v = v_0$. Since $V = U(\mathfrak{g})v_0$ and C_π commutes with the action of $U(\mathfrak{g})$ by Lemma 3.3.7, it follows that (3.36) also holds for all $v \in V$. $\square$

The following result is our first application of the Casimir operator.

Proposition 3.3.9. *Let V be a finite-dimensional highest-weight $\mathfrak{g}$-module with highest weight λ. Then λ is dominant integral and V is irreducible. Hence V is isomorphic to L^λ.*

Proof. The assumption of finite-dimensionality implies that $\lambda \in P(\mathfrak{g})$ by Theorem 3.1.16. Since λ is the maximal weight of V, Theorem 2.3.6 applied to the subalgebras $\mathfrak{s}(\alpha)$ for $\alpha \in \Phi^+$ shows that λ is dominant (as in the proof of Corollary 3.2.3).

Let L be any nonzero irreducible submodule of V. By Corollary 3.2.3 we know that L is a highest-weight module. Let μ be the highest weight of L. Since $L \subset V$, the Casimir operator C_L on L is the restriction of the Casimir operator C_V on V. Hence Lemma 3.3.8 gives

$$C_V w = ((\lambda + \rho, \lambda + \rho) - (\rho, \rho))w = ((\mu + \rho, \mu + \rho) - (\rho, \rho))w \tag{3.40}$$

for all $w \in L$. Since $L \neq 0$, we conclude from equation (3.40) that

$$\|\lambda + \rho\|^2 = \|\mu + \rho\|^2 \,. \tag{3.41}$$

Suppose, for the sake of contradiction, that $\mu \neq \lambda$. Since $\mu \in \mathfrak{X}(V)$, Lemma 3.2.2 shows that $\mu = \lambda - \beta$ for some $0 \neq \beta \in Q_+$, and by Lemma 3.1.21 we have

$$(\rho, \beta) = \sum_{i=1}^{l} (\varpi_i, \beta) > 0 \,.$$

We also know that $(\mu, \beta) \geq 0$ and $(\lambda, \beta) \geq 0$, since μ and λ are dominant. In particular, $(\mu + \rho, \alpha) > 0$ for all $\alpha \in \Phi^+$, so $\mu + \rho \neq 0$. From these inequalities we obtain

$$\|\mu+\rho\|^2 = (\mu+\rho,\lambda-\beta+\rho)$$
$$= (\mu,\lambda)-(\mu,\beta)+(\mu,\rho)+(\rho,\lambda)-(\rho,\beta)+(\rho,\rho)$$
$$< (\mu,\lambda)+(\mu,\rho)+(\rho,\lambda)+(\rho,\rho)=(\mu+\rho,\lambda+\rho)$$
$$\leq \|\mu+\rho\|\|\lambda+\rho\|,$$

where we have used the Cauchy–Schwarz inequality to obtain the last inequality. We have proved that $\|\mu+\rho\|<\|\lambda+\rho\|$, which contradicts (3.41). Hence $\mu=\lambda$ and $V=L$ is irreducible. $\qquad\square$

Remark 3.3.10. In contrast to the situation of Proposition 3.3.9, a highest-weight module that is infinite-dimensional can be reducible, even when the highest weight is dominant integral (see Exercises 3.3.5 #2).

3.3.3 Algebraic Proof of Complete Reducibility

We now come to the main result of this section.

Theorem 3.3.11. *Let G be a classical group. Then G is reductive.*

For the algebraic proof of this theorem we use the following result: We will say that a finite-dimensional representation (π,V) of a Lie algebra $\mathfrak{g}$ is *completely reducible* if every $\mathfrak{g}$-invariant subspace $W \subset V$ has a $\mathfrak{g}$-invariant complementary subspace in V. Proposition 3.3.3, which holds for finite-dimensional representations of $\mathfrak{g}$ (with the same proof), shows that this definition is equivalent to the assertion that V decomposes into a direct sum of irreducible $\mathfrak{g}$-submodules

Theorem 3.3.12. *Let $\mathfrak{g}$ be a semisimple Lie algebra. Then every finite-dimensional $\mathfrak{g}$-module V is completely reducible.*

Proof of Theorem 3.3.11 from Theorem 3.3.12. From the correspondence between G-invariant subspaces and $\mathfrak{g}$-invariant subspaces (Theorem 2.2.2 and Theorem 2.2.7), we see that Theorem 3.3.12 implies Theorem 3.3.11 when $\mathfrak{g}$ is semisimple. For the remaining classical groups whose Lie algebras are not semisimple, we note that $\mathbf{SO}(2,\mathbb{C}) \cong \mathbb{C}^\times$ is reductive by Lemma 1.6.4. There is a surjective regular homomorphism $\mathbb{C}^\times \times \mathbf{SL}(n,\mathbb{C}) \longrightarrow \mathbf{GL}(n,\mathbb{C})$ given by $(\lambda,g) \mapsto \lambda g$. In Chapter 4 we will prove, without using Theorem 3.3.11, that a product of reductive groups is reductive (Proposition 4.2.6). It is obvious from the definition that if G is reductive and $\varphi : G \longrightarrow H$ is a regular surjective homomorphism, then H is reductive. It follows that $\mathbf{GL}(n,\mathbb{C})$ is reductive. If $n \geq 3$, then $\mathbf{SO}(n,\mathbb{C})$ is a normal subgroup of index two in $\mathbf{O}(n,\mathbb{C})$. Given that $\mathbf{SO}(n,\mathbb{C})$ is reductive, then $\mathbf{O}(n,\mathbb{C})$ is reductive by Proposition 3.3.5. $\qquad\square$

We prepare the proof of Theorem 3.3.12 by the following lemma, which is analogous to Lemma 2.3.7 but with the Casimir operator replacing the diagonal element h in $\mathfrak{sl}(2,\mathbb{C})$.

Lemma 3.3.13. *Let λ and μ be dominant integral weights. Suppose the finite-dimensional $\mathfrak{g}$-module V contains a submodule Z that is equivalent to L^λ. Assume that V/Z is equivalent to L^μ. Then V is equivalent to $L^\lambda \oplus L^\mu$.*

Proof. By assumption, $Z \cong L^\lambda$, so Lemma 3.3.8 implies that the Casimir operator C_Z acts on Z as the scalar $a = (\lambda + \rho, \lambda + \rho) - (\rho, \rho)$. Likewise, the Casimir operator $C_{V/Z}$ acts on V/Z as the scalar $d = (\mu + \rho, \mu + \rho) - (\rho, \rho)$. Thus the matrix of the Casimir operator C_V, relative to an ordered basis for V that begins with a basis for Z, is of the form

$$\begin{bmatrix} aI & B \\ 0 & dI \end{bmatrix}, \tag{3.42}$$

where I denotes identity matrices of the appropriate sizes and B is the matrix for a linear transformation from V/Z to Z. Let $p : V \longrightarrow V/Z$ be the natural projection. Since the action of $\mathfrak{h}$ on V is diagonalizable, we can choose $u_0 \in V(\mu)$ such that $p(u_0) \neq 0$.

Case 1: Suppose there exists $\alpha \in \Phi^+$ such that $xu_0 \neq 0$ for $0 \neq x \in \mathfrak{g}_\alpha$. Then $p(xu_0) = xp(u_0) = 0$, since μ is the highest weight of V/Z. Thus $xu_0 \in Z(\mu + \alpha)$, so we see that $Z(\mu + \alpha) \neq 0$. Hence $\mu + \alpha \preceq \lambda$, since λ is the highest weight of Z. This shows that $\mu \neq \lambda$ in this case. Since λ and μ are dominant and $\mu \prec \lambda$, we have

$$(\mu + \rho, \mu + \rho) - (\rho, \rho) < (\lambda + \rho, \lambda + \rho) - (\rho, \rho),$$

as in the proof of Proposition 3.3.9. This implies that the linear transformation C_V diagonalizes, with Z the eigenspace for the eigenvalue $(\lambda + \rho, \lambda + \rho) - (\rho, \rho)$. Indeed, since we know that $a \neq d$ in the matrix (3.42), we have

$$\begin{bmatrix} I & -(d-a)^{-1}B \\ 0 & I \end{bmatrix} \begin{bmatrix} aI & B \\ 0 & dI \end{bmatrix} \begin{bmatrix} I & (d-a)^{-1}B \\ 0 & I \end{bmatrix} = \begin{bmatrix} aI & 0 \\ 0 & dI \end{bmatrix}.$$

Define $U = \{u \in V : C_V u = du\}$. Then U is invariant under $\mathfrak{g}$, since C_V commutes with $\mathfrak{g}$ on V. Furthermore, $\dim U = \dim(V/Z)$ by the diagonalization of C_V. Hence $p(U) = V/Z$, so we have $U \cong L^\mu$ as a $\mathfrak{g}$-module and $V = Z \oplus U$.

Case 2: Suppose that $\mathfrak{n}^+ u_0 = 0$. In this case Proposition 3.3.9 implies that the highest-weight $\mathfrak{g}$-submodule $U = U(\mathfrak{g})u_0$ of V is equivalent to L^μ. In particular,

$$\dim U = \dim L^\mu = \dim(V/Z).$$

Suppose that $U \cap Z \neq (0)$. Then $U = U \cap Z = Z$ by the irreducibility of U and Z. But then we have $u_0 \in Z$, and so $p(u_0) = 0$, which is a contradiction. Since $\dim U + \dim Z = \dim V$, this completes the proof that $V = Z \oplus U$.   □

Proof of Theorem 3.3.12. We will show that there are dominant integral weights $\lambda_1, \ldots, \lambda_r$ (not necessarily distinct) such that

$$V \cong L^{\lambda_1} \oplus \cdots \oplus L^{\lambda_r}. \tag{3.43}$$

We proceed by induction on $\dim V$, following the argument in Section 2.3.3 for $\mathfrak{sl}(2,\mathbb{C})$. If $\dim V = 1$ we may take $r = 1$ and $\lambda_1 = 0$. Now assume that (3.43) holds when $\dim V < m$ and consider the case $\dim V = m$. Let $Z \subset V$ be a nonzero irreducible $\mathfrak{g}$-submodule. Then $Z \cong L^{\lambda_1}$ by Proposition 3.3.9. If $Z = V$ we are done. Otherwise, $\dim V/Z < m$ and the inductive hypothesis implies that

$$V/Z \cong L^{\lambda_2} \oplus \cdots \oplus L^{\lambda_r}$$

for some dominant integral weights λ_i. Let

$$T : V \longrightarrow L^{\lambda_2} \oplus \cdots \oplus L^{\lambda_r}$$

be a surjective intertwining operator with $\operatorname{Ker} T = Z$. Since the action of $\mathfrak{h}$ on V is diagonalizable, we can choose elements $v_i \in V(\lambda_i)$, for $i = 2, \ldots, r$, such that $T(v_i)$ is a basis of the one-dimensional λ_i weight space in the summand L^{λ_i}. We set

$$V_i = Z + U(\mathfrak{g})v_i, \quad T_i = T|_{V_i} : V_i \longrightarrow L^{\lambda_i}.$$

Then V_i is a $\mathfrak{g}$-submodule of V and T_i is a surjective intertwining operator with kernel Z. Lemma 3.3.13 implies that there exists a $\mathfrak{g}$-submodule U_i of V_i such that $V_i = Z \oplus U_i$, and that T_i defines a $\mathfrak{g}$-module equivalence between U_i and L^{λ_i}. Now set $U = U_2 + \cdots + U_r$. Then

$$T(U) = T(U_2) + \cdots + T(U_r) = L^{\lambda_2} \oplus \cdots \oplus L^{\lambda_r}.$$

Thus $T|_U$ is surjective. Since $\dim U \leq \dim U_2 + \cdots + \dim U_r = \dim T(U)$, it follows that $T|_U$ is bijective. Hence $U = U_2 \oplus \cdots \oplus U_r$ and $V = Z \oplus U$. This completes the induction proof. $\qquad \square$

If V is a $\mathfrak{g}$-module we set $V^{\mathfrak{n}^+} = \{v \in V : X \cdot v = 0 \quad \text{for all } X \in \mathfrak{n}^+\}$.

Corollary 3.3.14. *Let V be a finite-dimensional $\mathfrak{g}$-module. Then V is irreducible if and only if $\dim V^{\mathfrak{n}^+} = 1$.*

Proof. If V is irreducible then $\dim V^{\mathfrak{n}^+} = 1$ by Corollary 3.2.3 and Proposition 3.3.9. Conversely, assume that $\dim V^{\mathfrak{n}^+} = 1$. By Theorem 3.3.12 there is a $\mathfrak{g}$-module decomposition $V = V_1 \oplus \cdots \oplus V_r$ with V_i irreducible. Since $\dim (V_i)^{\mathfrak{n}^+} = 1$, we see that $\dim V^{\mathfrak{n}^+} = r$. Hence $r = 1$ and V is irreducible. $\qquad \square$

3.3.4 The Unitarian Trick

In this section we will give an analytic proof of Theorem 3.3.11. The main tool is the following analytic criterion for reductivity:

Theorem 3.3.15. *Suppose G is a connected algebraic group that has a compact real form. Then G is reductive.*

Before proving the theorem, we recall some properties of real forms. Let K be a compact real form of G. Write $\iota : K \to G$ for the embedding map ($\iota(k) = k$). Let $\mathfrak{k} = d\iota(\text{Lie}(K))$. Then $\mathfrak{k}$ is a real Lie subalgebra of $\mathfrak{g}$ (the Lie algebra of G as a linear algebraic group). Let τ denote the complex conjugation (see Section 1.7.1) that defines K in G. Then $\sigma = d\tau$ is a Lie algebra isomorphism of $\mathfrak{g}$ (as a Lie algebra over $\mathbb{R}$) and $\sigma(iX) = -i\sigma(X)$. If we define $\mathfrak{g}_+ = \{X \in \mathfrak{g} : \sigma(X) = X\}$, then $\mathfrak{g} = \mathfrak{g}_+ + i\mathfrak{g}_+$. This implies that as real vector spaces, $\mathfrak{k}$ and $\mathfrak{g}_+$ have the same dimension. Since $\mathfrak{k} \subset \mathfrak{g}_+$, we obtain

$$\mathfrak{g} = \mathfrak{k} \oplus i\mathfrak{k} . \tag{3.44}$$

Let (ρ, V) be a regular representation of G. On V we put take any Hermitian inner product $(\cdot, \cdot)$. For example, let $\{v_1, \ldots, v_n\}$ be a basis of V and define

$$(z_1 v_1 + \cdots + z_n v_n, w_1 v_1 + \cdots + w_n v_n) = z_1 \overline{w}_1 + \cdots + z_n \overline{w}_n .$$

We now introduce the unitarian trick. Let dk denote normalized invariant measure on K (Section D.2.4). If $v, w \in V$ then we set

$$\langle v | w \rangle = \int_K (\rho(k)v, \rho(k)w) \, dk .$$

We note that the function $k \mapsto (\rho(k)v, \rho(k)w)$ is continuous on K. If $v \in V$ and $v \neq 0$, then $(\rho(k)v, \rho(k)v) > 0$. Thus Lemma D.1.15 implies that $\langle v | v \rangle > 0$. It is obvious from the linearity of integration that $\langle \cdot | \cdot \rangle$ satisfies the other conditions of a Hermitian inner product.

Lemma 3.3.16. *If $u \in K$, $v, w \in V$ then $\langle \rho(u)v | \rho(u)w \rangle = \langle v | w \rangle$.*

Proof. By Lemmas D.2.11 and D.2.12 we have

$$\langle \rho(u)v | \rho(u)w \rangle = \int_K (\rho(k)\rho(u)v, \, \rho(k)\rho(u)w) \, dk$$

$$= \int_K (\rho(ku)v, \, \rho(ku)w) \, dk$$

$$= \int_K (\rho(k)v, \, \rho(k)w) \, dk = \langle v | w \rangle . \qquad \square$$

Completion of proof of Theorem 3.3.15.   Given a subspace W of V, we set

$$W^\perp = \{v \in V : \langle v | w \rangle = 0, w \in W\} .$$

It is easily seen that $\dim W^\perp = \dim V - \dim W$. Also, if $v \in W \cap W^\perp$, then $\langle v | v \rangle = 0$ and so $v = 0$. Thus

$$V = W \oplus W^\perp . \tag{3.45}$$

Now let W be a $\rho(G)$-invariant subspace of V. We will show that $W^\perp$ is also $\rho(G)$-invariant. This will prove that (ρ, V) is completely reducible. Since (ρ, V) is an arbitrary regular representation of G, this will imply that G is reductive.

It is obvious that $\rho(K)W^\perp \subset W^\perp$. Let $X \in \mathfrak{k}$, $v \in W^\perp$, and $w \in W$. Then $\langle \rho(\exp(tX))v \,|\, w \rangle = 0$. Differentiating in t at $t = 0$ and using the relation

$$\rho(\exp(tX)) = \exp(t \, d\rho(X)),$$

we obtain $\langle d\rho(X)v \,|\, w \rangle = 0$. Hence $W^\perp$ is invariant under $\mathfrak{k}$, so by Proposition 1.7.7 it is also invariant under G. □

Corollary 3.3.17. *Let G be a classical group. Then G is reductive.*

Proof. If G is connected, then Section 1.7.2 furnishes a compact real form for G, so we may apply Theorem 3.3.15. From Theorem 2.2.5 we know that the only non-connected classical groups are the groups $\mathbf{O}(n, \mathbb{C})$ for $n \geq 3$. Since $\mathbf{SO}(n, \mathbb{C})$ is a connected normal subgroup of index two in this case, we conclude from Proposition 3.3.5 that $\mathbf{O}(n, \mathbb{C})$ is also reductive. □

Remark 3.3.18. If G is any linear algebraic group whose Lie algebra is semisimple, then G has a compact real form (see Section 11.5.1). Hence G is reductive.

3.3.5 Exercises

1. Let $\mathfrak{g} = \mathfrak{sl}(2, \mathbb{C})$. Fix a TDS basis $\{e, f, h\}$ and let B be the Killing form on $\mathfrak{g}$.
   (a) Show that the B-dual basis is $\{(1/4)f, (1/4)e, (1/8)h\}$. Hence the Casimir operator for a representation π of $\mathfrak{g}$ is

   $$C_\pi = \frac{1}{4}\Big(\pi(e)\pi(f) + \pi(f)\pi(e)\Big) + \frac{1}{8}\pi(h)^2.$$

   (b) Show that the Casimir operator acts in the irreducible $(n+1)$-dimensional representation of $\mathfrak{g}$ by the scalar $n(n+2)/8$. (HINT: Use (a), Lemma 3.3.8, and the fact that $\pi(h)$ has largest eigenvalue n.)
2. Let V be a countable-dimensional vector space with basis $\{v_j : j \in \mathbb{N}\}$ and let $\lambda \in \mathbb{C}$. Define linear transformations E, F, H on V by

   $$Ev_j = j(\lambda - j + 1)v_{j-1}, \quad Fv_j = v_{j+1}, \quad Hv_j = (\lambda - 2j)v_j$$

   for $j \in \mathbb{N}$ (where $v_{-1} = 0$). Let $\mathfrak{g} = \mathfrak{sl}(2, \mathbb{C})$ and fix a TDS basis $\{e, f, h\}$ for $\mathfrak{g}$.
   (a) Show that the correspondence $e \mapsto E$, $f \mapsto F$, and $h \mapsto H$ defines a highest-weight representation π_λ of $\mathfrak{g}$ on V (relative to the given TDS basis).
   (b) Show that the representation π_λ is irreducible if $\lambda \notin \mathbb{N}$.
   (c) Let $\lambda = n$ be a nonnegative integer. Show that the subspace Z spanned by $\{v_j : j \geq n + 1\}$ is an irreducible $\mathfrak{g}$-submodule of V and the quotient V/Z is isomorphic to the $(n+1)$-dimensional irreducible representation $F^{(n)}$ of $\mathfrak{g}$, but there is no $\mathfrak{g}$-submodule of V isomorphic to $F^{(n)}$. Thus π_n is a highest-weight representation that is not irreducible, even though the highest weight is dominant integral.

3. (Notation of Section 3.3.2) Let $\mathfrak{g}$ be a semisimple Lie algebra and V the finite-dimensional $\mathfrak{g}$-module with highest weight λ. Let $\mu \in \mathfrak{X}(V)$ with $\mu \neq \lambda$. Show that $\|\mu + \rho\| < \|\lambda + \rho\|$. (HINT: Modify the proof of Proposition 3.3.9, using inequality (3.28) to replace the dominance assumption on μ.)

4. Let $G_1, \ldots, G_r$ be classical groups. Show that $G = G_1 \times G_2 \times \cdots \times G_r$ is reductive. (HINT: Reduce to the case that G is connected. Then take compact real forms $K_i \subset G_i$ and show that $K_1 \times K_2 \times \cdots \times K_r$ is a compact real form of G.)

3.4 Notes

Sections 3.1.1 and 3.1.2. Although the group generated by the root reflections appears in Cartan [26] and [27], the role of the group $\mathrm{Norm}_G(H)/H$ for the representations and characters of a semisimple Lie group and the generation of this group by simple root reflections was first emphasized in Weyl [162]. Soon thereafter, Coxeter [41] made a thorough study of groups generated by reflections (see Bourbaki [12], Humphreys [78], and Kane [83]). An elementary introduction with lots of pictures is in Grove and Benson [57].

Sections 3.1.3 and 3.1.4. For the examples of rank-3 classical groups and pictures of their root and weight lattices, see Fulton and Harris [52].

Section 3.2.1. The theorem of the highest weight is due to Cartan [27], who proved existence of finite-dimensional irreducible representations with a given dominant integral highest weight by explicit and lengthy constructions. The approach here via the universal enveloping algebra, which applies to arbitrary highest weights and yields irreducible representations that are infinite-dimensional when the weight is not dominant integral, is due to Chevalley [34] and Harish-Chandra [60]. See Borel [17, Chapter VII, §3.6-7] for more historical details and Jacobson [79], Bourbaki [13, Chapitre VIII, §7], or Humphreys [76, §21]. The proof of uniqueness (which does not use complete reducibility) is from Wallach [152].

Section 3.2.4. The approach in this section is from Bourbaki [13, Chapitre VIII, §7.5].

Section 3.3.2. The Casimir operator was introduced by Casimir [30]. It will reappear in Chapter 7 in the algebraic proof of the Weyl character formula.

Section 3.3.3. The proof of Theorem 3.3.12 given here is due to V. Kac. Like all algebraic proofs of this result, it uses the Casimir operator. Using the irreducibility of finite-dimensional highest-weight modules, which is proved in Section 3.3.2 via the Casimir operator, the key step in the proof is to diagonalize the Casimir operator. Other algebraic proofs have been found by Brauer [19], Casimir and van der Waerden [31], and Whitehead [165] and [166].

Section 3.3.4. The translation-invariant measure on a compact classical group was introduced by A. Hurwitz in 1897 and developed further by Schur [131]. It plays a central role in Weyl's work [162] (the unitarian trick and character formula).

Chapter 4
Algebras and Representations

Abstract In this chapter we develop some algebraic tools needed for the general theory of representations and invariants. The central result is a duality theorem for locally regular representations of a reductive algebraic group G. The duality between the irreducible regular representations of G and irreducible representations of the commuting algebra of G plays a fundamental role in classical invariant theory. We study the representations of a finite group through its group algebra and characters, and we construct induced representations and calculate their characters.

4.1 Representations of Associative Algebras

In this section we obtain the basic facts about representations of associative algebras: a general version of Schur's lemma, the Jacobson density theorem, the notion of complete reducibility of representations, the double commutant theorem, and the isotypic decomposition of a locally completely reducible representation of an algebraic group.

4.1.1 Definitions and Examples

We know from the previous chapter that every regular representation (ρ, V) of a reductive linear algebraic group G decomposes into a direct sum of irreducible representations (in particular, this is true when G is a classical group). The same is true for finite-dimensional representations of a semisimple Lie algebra $\mathfrak{g}$. The next task is to determine the extent of uniqueness of such a decomposition and to find explicit projection operators onto irreducible subspaces of V. In the tradition of modern mathematics we will attack these problems by putting them in a more general (abstract) context, which we have already employed, for example, in the proof of the theorem of the highest weight in Section 3.2.1.

R. Goodman, N.R. Wallach, *Symmetry, Representations, and Invariants,*
Graduate Texts in Mathematics 255, DOI 10.1007/978-0-387-79852-3_4,
© Roe Goodman and Nolan R. Wallach 2009

Definition 4.1.1. An *associative algebra* over the complex field $\mathbb{C}$ is a vector space $\mathcal{A}$ over $\mathbb{C}$ together with a bilinear multiplication map

$$\mu : \mathcal{A} \times \mathcal{A} \longrightarrow \mathcal{A}, \quad x, y \mapsto xy = \mu(x, y),$$

such that $(xy)z = x(yz)$. The algebra $\mathcal{A}$ is said to have a *unit element* if there exists $e \in \mathcal{A}$ such that $ae = ea = a$ for all $a \in \mathcal{A}$. If $\mathcal{A}$ has a unit element it is unique and it will usually be denoted by 1.

Examples

1. Let V be a vector space over $\mathbb{C}$ (possibly infinite-dimensional), and let $\mathcal{A} = \text{End}(V)$ be the space of $\mathbb{C}$-linear transformations on V. Then $\mathcal{A}$ is an associative algebra with multiplication the composition of transformations. When $\dim V = n < \infty$, then this algebra has a basis consisting of the *elementary matrices* e_{ij} that multiply by $e_{ij} e_{km} = \delta_{jk} e_{im}$ for $1 \leq i, j \leq n$. This algebra will play a fundamental role in our study of associative algebras and their representations.

2. Let G be a group. We define an associative algebra $\mathcal{A}[G]$, called the *group algebra* of G, as follows: As a vector space, $\mathcal{A}[G]$ is the set of all functions $f : G \longrightarrow \mathbb{C}$ such that the *support* of f (the set where $f(g) \neq 0$) is finite. This space has a basis consisting of the functions $\{\delta_g : g \in G\}$, where

$$\delta_g(x) = \begin{cases} 1 & \text{if } x = g, \\ 0 & \text{otherwise.} \end{cases}$$

Thus an element x of $\mathcal{A}[G]$ has a unique expression as a formal sum $\sum_{g \in G} x(g) \delta_g$ with only a finite number of coefficients $x(g) \neq 0$.

We identify $g \in G$ with the element $\delta_g \in \mathcal{A}[G]$, and we define multiplication on $\mathcal{A}[G]$ as the bilinear extension of group multiplication. Thus, given functions $x, y \in \mathcal{A}[G]$, we define their product $x * y$ by

$$\left(\sum_{g \in G} x(g) \delta_g \right) * \left(\sum_{h \in G} y(h) \delta_h \right) = \sum_{g, h \in G} x(g) y(h) \delta_{gh},$$

with the sum over $g, h \in G$. (We indicate the multiplication by $*$ so it will not be confused with the pointwise multiplication of functions on G.) This product is associative by the associativity of group multiplication. The identity element $e \in G$ becomes the unit element δ_e in $\mathcal{A}[G]$ and G is a subgroup of the group of invertible elements of $\mathcal{A}[G]$. The function $x * y$ is called the *convolution* of the functions x and y; from the definition it is clear that

$$(x * y)(g) = \sum_{hk=g} x(h) y(k) = \sum_{h \in G} x(h) y(h^{-1} g).$$

If $\varphi : G \longrightarrow H$ is a group homomorphism, then we can extend φ uniquely to a linear map $\widetilde{\varphi} : \mathcal{A}[G] \longrightarrow \mathcal{A}[H]$ by the rule

$$\widetilde{\varphi}\left(\sum_{g \in G} x(g)\delta_g\right) = \sum_{g \in G} x(g)\delta_{\varphi(g)} \, .$$

From the definition of multiplication in $\mathcal{A}[G]$ we see that the extended map $\widetilde{\varphi}$ is an associative algebra homomorphism. Furthermore, if $\psi : H \longrightarrow K$ is another group homomorphism, then $\widetilde{\psi \circ \varphi} = \widetilde{\psi} \circ \widetilde{\varphi}$.

An important special case occurs when G is a subgroup of H and φ is the inclusion map. Then $\widetilde{\varphi}$ is injective (since $\{\delta_g\}$ is a basis of $\mathcal{A}[G]$). Thus we can identify $\mathcal{A}[G]$ with the subalgebra of $\mathcal{A}[H]$ consisting of functions supported on G.

3. Let $\mathfrak{g}$ be a Lie algebra over $\mathcal{A}$. Just as in the case of group algebras, there is an associative algebra $U(\mathfrak{g})$ (the *universal enveloping algebra* of $\mathfrak{g}$) and an injective linear map $j : \mathfrak{g} \longrightarrow U(\mathfrak{g})$ such that $j(\mathfrak{g})$ generates $U(\mathfrak{g})$ and

$$j([X,Y]) = j(X)j(Y) - j(Y)j(X)$$

(the multiplication on the right is in $U(\mathfrak{g})$; see Appendix C.2.1 and Theorem C.2.2). Since $U(\mathfrak{g})$ is uniquely determined by $\mathfrak{g}$, up to isomorphism, we will identify $\mathfrak{g}$ with $j(\mathfrak{g})$. If $\mathfrak{h} \subset \mathfrak{g}$ is a Lie subalgebra then the Poincaré–Birkhoff–Witt Theorem C.2.2 allows us to identify $U(\mathfrak{h})$ with the associative subalgebra of $U(\mathfrak{g})$ generated by $\mathfrak{h}$, so we have the same situation as for the group algebra of a subgroup $H \subset G$.

Definition 4.1.2. Let $\mathcal{A}$ be an associative algebra over $\mathcal{C}$. A *representation* of $\mathcal{A}$ is a pair (ρ, V), where V is a vector space over $\mathbb{C}$ and $\rho : \mathcal{A} \longrightarrow \mathrm{End}(V)$ is an associative algebra homomorphism. If $\mathcal{A}$ has an identity element 1, then we require that $\rho(1)$ act as the identity transformation I_V on V.

When the map ρ is understood from the context, we shall call V an $\mathcal{A}$-*module* and write av for $\rho(a)v$. If V, W are both $\mathcal{A}$-modules, then we make the vector space $V \oplus W$ into an $\mathcal{A}$-module by the action $a \cdot (v \oplus w) = av \oplus aw$.

If $U \subset V$ is a linear subspace such that $\rho(a)U \subset U$ for all $a \in \mathcal{A}$, then we say that U is *invariant* under the representation. In this case we can define a representation (ρ_U, U) by the restriction of $\rho(\mathcal{A})$ to U and a representation $(\rho_{V/U}, V/U)$ by the natural quotient action of $\rho(\mathcal{A})$ on V/U. A representation (ρ, V) is *irreducible* if the only invariant subspaces are $\{0\}$ and V.

Define $\mathrm{Ker}(\rho) = \{x \in \mathcal{A} : \rho(x) = 0\}$. This is a two-sided ideal in $\mathcal{A}$, and V is a module for the quotient algebra $\mathcal{A}/\mathrm{Ker}(\rho)$ via the natural quotient map. A representation ρ is *faithful* if $\mathrm{Ker}(\rho) = 0$.

Definition 4.1.3. Let (ρ, V) and (τ, W) be representations of $\mathcal{A}$, and let $\mathrm{Hom}(V, W)$ be the space of $\mathbb{C}$-linear maps from V to W. We denote by $\mathrm{Hom}_{\mathcal{A}}(V, W)$ the set of all $T \in \mathrm{Hom}(V, W)$ such that $T\rho(a) = \tau(a)T$ for all $a \in \mathcal{A}$. Such a map is called an *intertwining operator* between the two representations or a *module homomorphism*.

If $U \subset V$ is an invariant subspace, then the inclusion map $U \longrightarrow V$ and the quotient map $V \longrightarrow V/U$ are intertwining operators. We say that the representations (ρ, V) and (τ, W) are *equivalent* if there exists an invertible operator in $\mathrm{Hom}_{\mathcal{A}}(V, W)$. In this case we write $(\rho, V) \cong (\tau, W)$.

The composition of two intertwining operators, when defined, is again an intertwining operator. In particular, when $V = W$ and $\rho = \tau$, then $\mathrm{Hom}_{\mathcal{A}}(V,V)$ is an associative algebra, which we denote by $\mathrm{End}_{\mathcal{A}}(V)$.

Examples

1. Let $\mathcal{A} = \mathbb{C}[x]$ be the polynomial ring in one indeterminate. Let V be a finite-dimensional vector space, and let $T \in \mathrm{End}(V)$. Define a representation (ρ, V) of $\mathcal{A}$ by $\rho(f) = f(T)$ for $f \in \mathbb{C}[x]$. Then $\mathrm{Ker}(\rho)$ is the ideal in $\mathcal{A}$ generated by the *minimal polynomial* of T. The problem of finding a canonical form for this representation is the same as finding the Jordan canonical form for T (see Section B.1.2).

2. Let G be a group and let $\mathcal{A} = \mathcal{A}[G]$ be the group algebra of G. If (ρ, V) is a representation of $\mathcal{A}$, then the map $g \mapsto \rho(\delta_g)$ is a group homomorphism from G to $\mathbf{GL}(V)$. Conversely, every representation $\rho : G \longrightarrow \mathbf{GL}(V)$ extends uniquely to a representation ρ of $\mathcal{A}[G]$ on V by

$$\rho(f) = \sum_{g \in G} f(g)\rho(g)$$

for $f \in \mathcal{A}[G]$. We shall use the same symbol to denote a representation of a group and its group algebra.

Suppose $W \subset V$ is a linear subspace. If W is invariant under G and $w \in W$, then $\rho(f)w \in W$, since $\rho(g)w \in W$. Conversely, if $\rho(f)W \subset W$ for all $f \in \mathcal{A}[G]$, then $\rho(G)W \subset W$, since we can take $f = \delta_g$ with g arbitrary in G. Furthermore, an operator $R \in \mathrm{End}(V)$ commutes with the action of G if and only if it commutes with $\rho(f)$ for all $f \in \mathcal{A}[G]$.

Two important new constructions are possible in the case of group representations (we already encountered them in Section 1.5.1 when G is a linear algebraic group). The first is the *contragredient* or *dual* representation (ρ^*, V^*), where

$$\langle \rho^*(g)f, v \rangle = \langle f, \rho(g^{-1})v \rangle$$

for $g \in G$, $v \in V$, and $f \in V^*$. The second is the *tensor product* $(\rho \otimes \sigma, V \otimes W)$ of two representations defined by

$$(\rho \otimes \sigma)(g)(v \otimes w) = \rho(g)v \otimes \sigma(g)w.$$

For example, let (ρ, V) and (σ, W) be finite-dimensional representations of G. There is a representation π of G on $\mathrm{Hom}(V, W)$ by $\pi(g)T = \sigma(g)T\rho(g)^{-1}$ for $T \in \mathrm{Hom}(V, W)$. There is a natural linear isomorphism

$$\mathrm{Hom}(V, W) \cong W \otimes V^* \tag{4.1}$$

(see Section B.2.2). Here a tensor of the form $w \otimes v^*$ gives the linear transformation $Tv = \langle v^*, v \rangle w$ from V to W. Since the tensor $\sigma(g)w \otimes \rho^*(g)v^*$ gives the linear transformation

$$v \mapsto \langle \rho^*(g)v^*, v \rangle \sigma(g)w = \langle v^*, \rho(g)^{-1}v \rangle \sigma(g)w = \sigma(g)T\rho(g)^{-1}v,$$

we see that π is equivalent to $\sigma \otimes \rho^*$. In particular, the space $\mathrm{Hom}_G(V,W)$ of G-intertwining maps between V and W corresponds to the space $(W \otimes V^*)^G$ of G-fixed elements in $W \otimes V^*$.

We can iterate the tensor product construction to obtain G-modules $\bigotimes^k V = V^{\otimes k}$ (the k-fold tensor product of V with itself) with $g \in G$ acting by

$$\rho^{\otimes k}(g)(v_1 \otimes \cdots \otimes v_k) = \rho(g)v_1 \otimes \cdots \otimes \rho(g)v_k$$

on decomposable tensors. The subspaces $S^k(V)$ (symmetric tensors) and $\bigwedge^k V$ (skew-symmetric tensors) are G-invariant (see Sections B.2.3 and B.2.4). These modules are called the *symmetric* and *skew-symmetric* powers of ρ.

The contragredient and tensor product constructions for group representations are associated with the *inversion map* $g \mapsto g^{-1}$ and the *diagonal map* $g \mapsto (g,g)$. The properties of these maps can be described axiomatically using the notion of a *Hopf algebra* (see Exercises 4.1.8).

3. Let $\mathfrak{g}$ be a Lie algebra over $\mathbb{C}$, and let (ρ, V) be a representation of $\mathfrak{g}$. The universal mapping property implies that ρ extends uniquely to a representation of $U(\mathfrak{g})$ (see Section C.2.1) and that every representation of $\mathfrak{g}$ comes from a unique representation of $U(\mathfrak{g})$, just as in the case of group algebras. In this case we define the *dual* representation (ρ^*, V^*) by

$$\langle \rho^*(X)f, v \rangle = -\langle f, \rho(X)v \rangle \quad \text{for } X \in \mathfrak{g} \text{ and } f \in V^*.$$

We can also define the *tensor product* $(\rho \otimes \sigma, V \otimes W)$ of two representations by letting $X \in \mathfrak{g}$ act by

$$X \cdot (v \otimes w) = \rho(X)v \otimes w + v \otimes \sigma(X)w.$$

When $\mathfrak{g}$ is the Lie algebra of a linear algebraic group G and ρ, σ are the differentials of regular representations of G, then this action of $\mathfrak{g}$ is the differential of the tensor product of the G representations (see Sections 1.5.2).

These constructions are associated with the maps $X \mapsto -X$ and $X \mapsto X \otimes I + I \otimes X$. As in the case of group algebras, the properties of these maps can be described axiomatically using the notion of a *Hopf algebra* (see Exercises 4.1.8). The k-fold tensor powers of ρ and the symmetric and skew-symmetric powers are defined by analogy with the case of group representations. Here $X \in \mathfrak{g}$ acts by

$$\rho^{\otimes k}(X)(v_1 \otimes \cdots \otimes v_k) = \rho(X)v_1 \otimes \cdots \otimes v_k + v_1 \otimes \rho(X)v_2 \otimes \cdots \otimes v_k$$
$$+ \cdots + v_1 \otimes \cdots \otimes \rho(X)v_k$$

on decomposable tensors. This action extends linearly to all tensors.

4.1.2 Schur's Lemma

We say that a vector space has *countable dimension* if the cardinality of every linear independent set of vectors is countable.

Lemma 4.1.4. *Let (ρ,V) and (τ,W) be irreducible representations of an associative algebra A. Assume that V and W have countable dimension over $\mathbb{C}$. Then*

$$\dim \mathrm{Hom}_A(V,W) = \begin{cases} 1 \text{ if } (\rho,V) \cong (\tau,W), \\ 0 \text{ otherwise.} \end{cases}$$

Proof. Let $T \in \mathrm{Hom}_A(V,W)$. Then $\mathrm{Ker}(T)$ and $\mathrm{Range}(T)$ are invariant subspaces of V and W, respectively. If $T \neq 0$, then $\mathrm{Ker}(T) \neq V$ and $\mathrm{Range}(T) \neq 0$. Hence by the irreducibility of the representations, $\mathrm{Ker}(T) = 0$ and $\mathrm{Range}(T) = W$, so that T is a linear isomorphism. Thus $\mathrm{Hom}_A(V,W) \neq 0$ if and only if $(\rho,V) \cong (\tau,W)$.

Suppose the representations are equivalent. If $S,T \in \mathrm{Hom}_A(V,W)$ are nonzero, then $R = T^{-1}S \in \mathrm{End}_A(V)$. Assume, for the sake of contradiction, that R is not a multiple of the identity operator. Then for all $\lambda \in \mathbb{C}$ we would have $R - \lambda I$ nonzero and hence invertible. We assert that this implies that for any nonzero vector $v \in V$ and distinct scalars $\lambda_1, \ldots, \lambda_m$, the set

$$\{(R - \lambda_1 I)^{-1}v, \ldots, (R - \lambda_m I)^{-1}v\} \tag{4.2}$$

is linearly independent. We note that this would contradict the countable dimensionality of V and the lemma would follow.

Thus it suffices to prove the linear independence of (4.2) under the hypothesis on R. Suppose there is a linear relation

$$\sum_{i=1}^m a_i (R - \lambda_i I)^{-1} v = 0 \,.$$

Multiplying through by $\prod_j (R - \lambda_j I)$, we obtain the relation $f(R)v = 0$, where

$$f(x) = \sum_{i=1}^m a_i \Big\{ \prod_{j \neq i} (x - \lambda_j) \Big\} \,.$$

The polynomial $f(x)$ takes the value $a_i \prod_{j \neq i}(\lambda_i - \lambda_j)$ at $x = \lambda_i$. If $a_i \neq 0$ for some i, then $f(x)$ is a nonzero polynomial and has a factorization

$$f(x) = c(x - \mu_1) \cdots (x - \mu_m) \,,$$

with $c \neq 0$ and $\mu_i \in \mathbb{C}$. But by our assumption on R the operators $R - \mu_i I$ are invertible for each i, and hence $f(R)$ is invertible. This contradicts the relation $f(R)v = 0$. Thus $a_i = 0$ for all i and the set (4.2) is linearly independent.   □

4.1.3 Jacobson Density Theorem

If V is a complex vector space, $v_j \in V$, and $T \in \text{End}(V)$, then we write

$$V^{(n)} = \underbrace{V \oplus \cdots \oplus V}_{n \text{ copies}} \quad \text{and} \quad T^{(n)}[v_1, \ldots, v_n] = [Tv_1, \ldots, Tv_n] \,.$$

The map $T \mapsto T^{(n)}$ is a representation of $\text{End}(V)$ on $V^{(n)}$. If Z is a subspace of V, then we identify $Z^{(n)}$ with the subspace $\{[z_1, \ldots, z_n] : z_j \in Z\}$ of $V^{(n)}$. If $\mathcal{R} \subset \text{End}(V)$ is a subalgebra, then we consider $V^{(n)}$ to be an $\mathcal{R}$-module with $r \in \mathcal{R}$ acting as $r^{(n)}$; we write rv for $r^{(n)}v$ when $v \in V^{(n)}$.

Theorem 4.1.5. *Let V be a countable-dimensional vector space over $\mathbb{C}$. Let $\mathcal{R}$ be a subalgebra of $\text{End}(V)$ that acts irreducibly on V. Assume that for every finite-dimensional subspace W of V there exists $r \in \mathcal{R}$ so that $r|_W = I|_W$. Then $\mathcal{R}[v_1, \ldots, v_n] = V^{(n)}$ whenever $\{v_1, \ldots, v_n\}$ is a linearly independent subset of V.*

Proof. The proof is by induction on n. If $n = 1$ the assertion is the definition of irreducibility. Assume that the theorem holds for n and suppose $\{v_1, \ldots, v_{n+1}\}$ is a linearly independent set in V. Given any elements $x_1, \ldots, x_{n+1}$ in V, we must find $r \in \mathcal{R}$ such that

$$rv_j = x_j \quad \text{for} \quad j = 1, \ldots, n+1 \,. \tag{4.3}$$

The inductive hypothesis implies that there is an element $r_0 \in \mathcal{R}$ such that $r_0 v_j = x_j$ for $j = 1, \ldots, n$. Define $\mathcal{B} = \{r \in \mathcal{R} : r[v_1, \ldots, v_n] = 0\}$. The subspace $\mathcal{B}v_{n+1}$ of V is invariant under $\mathcal{R}$. Suppose $\mathcal{B}v_{n+1} \neq 0$; then $\mathcal{B}v_{n+1} = V$, since $\mathcal{R}$ acts irreducibly on V. Hence there exists $b_0 \in \mathcal{B}$ such that $b_0 v_{n+1} = x_{n+1} - r_0 v_{n+1}$. Since $b_0 v_j = 0$ for $j = 1, \ldots, n$, we see that the element $r = r_0 + b_0$ of $\mathcal{R}$ satisfies (4.3), and we are done in this case.

To complete the inductive step, it thus suffices to show that $\mathcal{B}v_{n+1} \neq 0$. We assume the contrary and show that this leads to a contradiction. Set

$$W = \mathcal{R}[v_1, \ldots, v_n, v_{n+1}] \quad \text{and} \quad U = \{[\underbrace{0, \ldots, 0}_{n}, v] : v \in V\} \,.$$

Then $[v_1, \ldots, v_n, v_{n+1}] \in W$. By the inductive hypothesis $V^{(n+1)} = W + U$. If $r \in \mathcal{R}$ and $[rv_1, \ldots, rv_n, rv_{n+1}] \in U \cap W$, then $rv_j = 0$ for $j = 1, \ldots, n$. Hence $r \in \mathcal{B}$ and consequently $rv_{n+1} = 0$ by the assumption $\mathcal{B}v_{n+1} = 0$. Thus $W \cap U = 0$, so we conclude that

$$V^{(n+1)} \cong W \oplus U \tag{4.4}$$

as an $\mathcal{R}$ module. Let $P : V^{(n+1)} \longrightarrow W$ be the projection corresponding to this direct sum decomposition. Then P commutes with the action of $\mathcal{R}$ and can be written as

$$P[x_1, \ldots, x_{n+1}] = \left[\sum_j P_{1,j} x_j, \ldots, \sum_j P_{n+1,j} x_j \right]$$

with $P_{i,j} \in \text{End}_{\mathcal{R}}(V)$. Thus by Lemma 4.1.4, each operator $P_{i,j}$ equals $q_{i,j}I$ for some scalar $q_{i,j} \in \mathbb{C}$. Hence for any subspace Z of V we have $P(Z^{(n+1)}) \subset Z^{(n+1)}$.

We can now obtain the desired contradiction. Set $Z = \text{Span}\{v_1, \ldots, v_{n+1}\}$ and let $w_1, \ldots, w_{n+1}$ be arbitrary elements of V. Since $\{v_1, \ldots, v_{n+1}\}$ is linearly independent, there is a linear transformation $T : Z \longrightarrow V$ with $Tv_j = w_j$ for $j = 1, \ldots, n+1$. We calculate

$$
\begin{aligned}
T^{(n+1)}P[v_1, \ldots, v_{n+1}] &= \left[\textstyle\sum_j q_{1,j}Tv_j, \ldots, \sum_j q_{n+1,j}v_j\right] \\
&= \left[\textstyle\sum_j q_{1,j}w_j, \ldots, \sum_j q_{n+1,j}w_j\right] \\
&= P[w_1, \ldots, w_{n+1}].
\end{aligned}
$$

On the other hand,

$$
P[v_1, \ldots, v_{n+1}] = [v_1, \ldots, v_{n+1}] \quad \text{and} \quad T^{(n+1)}[v_1, \ldots, v_{n+1}] = [w_1, \ldots, w_{n+1}],
$$

so we conclude that $[w_1, \ldots, w_{n+1}] = P[w_1, \ldots, w_{n+1}]$. Hence $[w_1, \ldots, w_{n+1}] \in W$. Since w_j are any elements of V, this implies that $W = V^{(n+1)}$, which contradicts (4.4). □

Corollary 4.1.6. *If X is a finite-dimensional subspace of V and $f \in \text{Hom}(X, L)$, then there exists $r \in \mathcal{R}$ such that $f = r|_X$.*

Proof. Let $\{v_1, \ldots, v_n\}$ be a basis for X and set $w_j = f(v_j)$ for $j = 1, \ldots, n$. By Theorem 4.1.5 there exists $r \in \mathcal{R}$ such that $rv_j = w_j$ for $j = 1, \ldots, n$. Hence by linearity $r|_X = f$. □

Corollary 4.1.7 (Burnside's Theorem). *If $\mathcal{R}$ acts irreducibly on L and $\dim L < \infty$, then $\mathcal{R} = \text{End}(L)$.*

Thus the image of an associative algebra in a finite-dimensional irreducible representation (ρ, L) is completely determined by $\dim L$ (the *degree* of the representation).

4.1.4 Complete Reducibility

Let (ρ, V) be a finite-dimensional representation of the associative algebra $\mathcal{A}$. When $V = W \oplus U$ with W and U invariant subspaces, then $U \cong V/W$ as an $\mathcal{A}$-module. In general, if $W \subset V$ is an $\mathcal{A}$-invariant subspace, then by extending a basis for W to a basis for V, we obtain a *vector-space isomorphism* $V \cong W \oplus (V/W)$. However, this isomorphism is not necessarily an isomorphism of $\mathcal{A}$-modules.

Definition 4.1.8. A finite-dimensional $\mathcal{A}$-module V is *completely reducible* if for every $\mathcal{A}$-invariant subspace $W \subset V$ there exists a complementary invariant subspace $U \subset V$ such that $V = W \oplus U$.

We proved in Chapter 3 that rational representations of classical groups and finite-dimensional representations of semisimple Lie algebras are completely reducible. For any associative algebra the property of complete reducibility is inherited by subrepresentations and quotient representations.

Lemma 4.1.9. *Let (ρ, V) be completely reducible and suppose $W \subset V$ is an invariant subspace. Set $\sigma(x) = \rho(x)|_W$ and $\pi(x)(v+W) = \rho(x)v + W$ for $x \in \mathcal{A}$ and $v \in V$. Then the representations (σ, W) and $(\pi, V/W)$ are completely reducible.*

Proof. The proof of Lemma 3.3.2 applies verbatim to this context. $\square$

Remark 4.1.10. The converse to Lemma 4.1.9 is not true. For example, let $\mathcal{A}$ be the algebra of matrices of the form $\begin{bmatrix} x & y \\ 0 & x \end{bmatrix}$ with $x, y \in \mathbb{C}$, acting on $V = \mathbb{C}^2$ by left multiplication. The space $W = \mathbb{C}e_1$ is invariant and irreducible. Since V/W is one-dimensional, it is also irreducible. But the matrices in $\mathcal{A}$ have only one distinct eigenvalue and are not diagonal, so there is no invariant complement to W in V. Thus V is not completely reducible as an $\mathcal{A}$-module.

Proposition 4.1.11. *Let (ρ, V) be a finite-dimensional representation of the associative algebra $\mathcal{A}$. The following are equivalent:*

1. *(ρ, V) is completely reducible.*
2. *$V = W_1 \oplus \cdots \oplus W_s$ with each W_i an irreducible $\mathcal{A}$-module.*
3. *$V = V_1 + \cdots + V_d$ as a vector space, where each V_i is an irreducible $\mathcal{A}$-submodule.*

Furthermore, if V satisfies these conditions and if all the V_i in (3) are equivalent to a single irreducible $\mathcal{A}$-module W, then every $\mathcal{A}$-submodule of V is isomorphic to a direct sum of copies of W.

Proof. The equivalence of the three conditions follows by the proof of Proposition 3.3.3. Now assume that V satisfies these conditions and that the V_i are all mutually equivalent as $\mathcal{A}$-modules. Let M be an $\mathcal{A}$-submodule of V. Since V is completely reducible by (1), it follows from Lemma 4.1.9 that M is completely reducible. Hence by (2) we have $M = W_1 \oplus \cdots \oplus W_r$ with W_i an irreducible $\mathcal{A}$-module. Furthermore, there is a complementary $\mathcal{A}$-submodule N such that $V = M \oplus N$. Hence

$$V = W_1 \oplus \cdots \oplus W_r \oplus N .$$

Let $p_i : V \longrightarrow W_i$ be the projection corresponding to this decomposition. By (3) we have $W_i = p_i(V_1) + \cdots + p_i(V_d)$. Thus for each i there exists j such that $p_i(V_j) \neq (0)$. Since W_i and V_j are irreducible and p_i is an $\mathcal{A}$-module map, Schur's lemma implies that $W_i \cong V_j$ as an $\mathcal{A}$-module. Hence $W_i \cong W$ for all i. $\square$

Corollary 4.1.12. *Suppose (ρ, V) and (σ, W) are completely reducible representations of $\mathcal{A}$. Then $(\rho \oplus \sigma, V \oplus W)$ is a completely reducible representation.*

Proof. This follows from the equivalence between conditions (1) and (2) in Proposition 4.1.11. $\square$

4.1.5 Double Commutant Theorem

Let V be a vector space. For any subset $\mathcal{S} \subset \mathrm{End}(V)$ we define

$$\mathrm{Comm}(\mathcal{S}) = \{x \in \mathrm{End}(V) : xs = sx \quad \text{for all} \ \ s \in \mathcal{S}\}$$

and call it the *commutant* of $\mathcal{S}$. We observe that $\mathrm{Comm}(\mathcal{S})$ is an associative algebra with unit I_V.

Theorem 4.1.13 (Double Commutant). *Suppose $\mathcal{A} \subset \mathrm{End}\,V$ is an associative algebra with identity I_V. Set $\mathcal{B} = \mathrm{Comm}(\mathcal{A})$. If V is a completely reducible $\mathcal{A}$-module, then $\mathrm{Comm}(\mathcal{B}) = \mathcal{A}$.*

Proof. By definition we have $\mathcal{A} \subset \mathrm{Comm}(\mathcal{B})$. Let $T \in \mathrm{Comm}(\mathcal{B})$ and fix a basis $\{v_1, \ldots, v_n\}$ for V. It will suffice to find an element $S \in \mathcal{A}$ such that $Sv_i = Tv_i$ for $i = 1, \ldots, n$. Let $w_0 = v_1 \oplus \cdots \oplus v_n \in V^{(n)}$. Since $V^{(n)}$ is a completely reducible $\mathcal{A}$-module by Proposition 4.1.11, the cyclic submodule $M = \mathcal{A} \cdot w_0$ has an $\mathcal{A}$-invariant complement. Thus there is a projection $P : V^{(n)} \longrightarrow M$ that commutes with $\mathcal{A}$. The action of P is given by an $n \times n$ matrix $[p_{ij}]$, where $p_{ij} \in \mathcal{B}$. Since $Pw_0 = w_0$ and $Tp_{ij} = p_{ij}T$, we have

$$P(Tv_1 \oplus \cdots \oplus Tv_n) = Tv_1 \oplus \cdots \oplus Tv_n \in M \ .$$

Hence by definition of M there exists $S \in \mathcal{A}$ such that

$$Sv_1 \oplus \cdots \oplus Sv_n = Tv_1 \oplus \cdots \oplus Tv_n \ .$$

This proves that $T = S$, so $T \in \mathcal{A}$. $\square$

4.1.6 Isotypic Decomposition and Multiplicities

Let $\mathcal{A}$ be an associative algebra with unit 1. If U is a finite-dimensional irreducible $\mathcal{A}$-module, we denote by $[U]$ the equivalence class of all $\mathcal{A}$-modules equivalent to U. Let $\widehat{\mathcal{A}}$ be the set of all equivalence classes of finite-dimensional irreducible $\mathcal{A}$-modules. Suppose that V is an $\mathcal{A}$-module (we do not assume that V is finite-dimensional). For each $\lambda \in \widehat{\mathcal{A}}$ we define the λ-*isotypic subspace*

$$V_{(\lambda)} = \sum_{U \subset V, [U] = \lambda} U \ .$$

Fix a module F^λ in the class λ for each $\lambda \in \widehat{\mathcal{A}}$. There is a tautological linear map

$$S_\lambda : \mathrm{Hom}_{\mathcal{A}}(F^\lambda, V) \otimes F^\lambda \longrightarrow V \ , \qquad S_\lambda(u \otimes w) = u(w) \ . \tag{4.5}$$

Make $\mathrm{Hom}_A(F^\lambda, V) \otimes F^\lambda$ into an A-module with action $x \cdot (u \otimes w) = u \otimes (xw)$ for $x \in A$. Then S_λ is an A-intertwining map. If $0 \neq u \in \mathrm{Hom}_A(F^\lambda, V)$ then Schur's lemma (Lemma 4.1.4) implies that $u(F^\lambda)$ is an irreducible A-submodule of V isomorphic to F^λ. Hence

$$S_\lambda\left(\mathrm{Hom}_A(F^\lambda, V) \otimes F^\lambda\right) \subset V_{(\lambda)} \quad \text{for every } \lambda \in \widehat{A}.$$

Definition 4.1.14. The A-module V is *locally completely reducible* if the cyclic A-submodule Av is finite-dimensional and completely reducible for every $v \in V$.

For example, if G is a reductive linear algebraic group, then by Proposition 1.4.4 $\mathcal{O}[G]$ is a locally completely reducible module for the group algebra $A[G]$ relative to the left or right translation action of G.

Proposition 4.1.15. *Let V be a locally completely reducible A-module. Then the map S_λ gives an A-module isomorphism $\mathrm{Hom}_A(F^\lambda, V) \otimes F^\lambda \cong V_{(\lambda)}$ for each $\lambda \in \widehat{A}$. Furthermore,*

$$V = \bigoplus_{\lambda \in \widehat{A}} V_{(\lambda)} \quad \text{(algebraic direct sum)}. \tag{4.6}$$

Proof. If $U \subset V$ is an A-invariant finite-dimensional irreducible subspace with $[U] = \lambda$, then there exists $u \in \mathrm{Hom}_A(F^\lambda, V)$ such that $\mathrm{Range}(u) = U$. Hence S_λ is surjective.

To show that S_λ is injective, let $u_i \in \mathrm{Hom}_A(F^\lambda, V)$ and $w_i \in F^\lambda$ for $i = 1, \ldots, k$, and suppose that $\sum_i u_i(w_i) = 0$. We may assume that $\{w_1, \ldots, w_k\}$ is linearly independent and that $u_i \neq 0$ for all i. Let $W = u_1(F^\lambda) + \cdots + u_k(F^\lambda)$. Then W is a finite-dimensional A-submodule of $V_{(\lambda)}$; hence by Proposition 4.1.11, $W = W_1 \oplus \cdots \oplus W_m$ with W_j irreducible and $[W_j] = \lambda$. Let $\varphi_j : W \longrightarrow F^\lambda$ be the projection onto the subspace W_j followed by an A-module isomorphism with F^λ. Then $\varphi_j \circ u_i \in \mathrm{End}_A(F^\lambda)$. Thus $\varphi_j \circ u_i = c_{ij}I$ with $c_{ij} \in \mathbb{C}$ (Schur's lemma), and we have

$$0 = \sum_i \varphi_j u_i(w_i) = \sum_i c_{ij} w_i \quad \text{for } j = 1, \ldots, m.$$

Since $\{w_1, \ldots, w_k\}$ is linearly independent, we conclude that $c_{ij} = 0$. Hence the projection of $\mathrm{Range}(u_i)$ onto W_j is zero for $j = 1, \ldots, m$. This implies that $u_i = 0$, proving injectivity of S_λ.

The definition of local complete reducibility implies that V is spanned by the spaces $V_{(\lambda)}$ ($\lambda \in \widehat{A}$). So it remains to prove only that these spaces are linearly independent. Fix distinct classes $\{\lambda_1, \ldots, \lambda_d\} \subset \widehat{A}$ such that $V_{(\lambda_i)} \neq \{0\}$. We will prove by induction on d that the sum $E = V_{(\lambda_1)} + \cdots + V_{(\lambda_d)}$ is direct. If $d = 1$ there is nothing to prove. Let $d > 1$ and assume that the result holds for $d - 1$ summands. Set $U = V_{(\lambda_1)} + \cdots + V_{(\lambda_{d-1})}$. Then $E = U + V_{(\lambda_d)}$ and $U = V_{(\lambda_1)} \oplus \cdots \oplus V_{(\lambda_{d-1})}$ by the induction hypothesis. For $i < d$ let $p_i : U \longrightarrow V_{(\lambda_i)}$ be the projection corresponding to this direct sum decomposition. Suppose, for the sake of contradiction, that there exists a nonzero vector $v \in U \cap V_{(\lambda_d)}$. The A-submodule Av of $V_{(\lambda_d)}$ is then nonzero, finite-dimensional, and completely reducible. Hence by the last part of Proposition 4.1.11 there is a decomposition

$$\mathcal{A}v = W_1 \oplus \cdots \oplus W_r \quad \text{with} \quad [W_i] = \lambda_d . \tag{4.7}$$

On the other hand, since $\mathcal{A}v \subset U$, there must exist an $i < d$ such that $p_i(\mathcal{A}v)$ is nonzero. But Proposition 4.1.11 then implies that $p_i(\mathcal{A}v)$ is a direct sum of irreducible modules of type λ_i. Since $\lambda_i \neq \lambda_d$, this contradicts (4.7), by Schur's lemma. Hence $U \cap V_{(\lambda_d)} = (0)$, and we have $E = V_{(\lambda_1)} \oplus \cdots \oplus V_{(\lambda_d)}$.   □

We call (4.6) the *primary decomposition* of V. We set

$$m_V(\lambda) = \dim \mathrm{Hom}_A(F^\lambda, V) \quad \text{for } \lambda \in \widehat{\mathcal{A}} ,$$

and call $m_V(\lambda)$ the *multiplicity* of ξ in V. The multiplicities may be finite or infinite; likewise for the number of nonzero summands in the primary decomposition. We call the set

$$\mathrm{Spec}(V) = \{\lambda \in \widehat{\mathcal{A}} : m_V(\lambda) \neq 0\}$$

the *spectrum* of the $\mathcal{A}$-module V. The primary decomposition of V gives an isomorphism

$$V \cong \bigoplus_{\lambda \in \mathrm{Spec}(V)} \mathrm{Hom}_A(F^\lambda, V) \otimes F^\lambda , \tag{4.8}$$

with the action of $\mathcal{A}$ only on the second factor in each summand.

Assume that V is completely reducible under $\mathcal{A}$. The primary decomposition has a finite number of summands, since V is finite-dimensional, and the multiplicities are finite. We claim that $m_V(\lambda)$ is also given by

$$m_V(\lambda) = \dim \mathrm{Hom}_A(V, F^\lambda) . \tag{4.9}$$

To prove this, let $m = m_V(\lambda)$. Then $V = W \oplus (F^\lambda)^{(m)}$, where W is the sum of the isotypic subspaces for representations not equivalent to λ. If $T \in \mathrm{Hom}_A(V, F^\lambda)$, then by Schur's lemma $T(W) = 0$ and T is a linear combination of the operators $\{T_1, \ldots, T_m\}$, where

$$T_i(w \oplus v_1 \oplus \cdots \oplus v_m) = v_i \quad \text{for } w \in W \text{ and } v_i \in V .$$

These operators are linearly independent, so they furnish a basis for $\mathrm{Hom}_A(V, F^\lambda)$.

Remark 4.1.16. Let U and V be completely reducible $\mathcal{A}$-modules. Define

$$\langle U, V \rangle = \dim \mathrm{Hom}_A(U, V) .$$

Then from Proposition 4.1.15 we have

$$\langle U, V \rangle = \sum_{\lambda \in \widehat{\mathcal{A}}} m_U(\lambda) m_V(\lambda) . \tag{4.10}$$

It follows that $\langle U,V \rangle = \langle V,U \rangle$ and $\langle U,V \oplus W \rangle = \langle U,V \rangle + \langle U,W \rangle$ for any completely reducible $\mathcal{A}$-modules U, V, and W.

4.1.7 Characters

Let $\mathcal{A}$ be an associative algebra with unit 1. If (ρ,V) is a finite-dimensional representation of $\mathcal{A}$, then the *character* of the representation is the linear functional $\mathrm{ch}\,V$ on $\mathcal{A}$ given by

$$\mathrm{ch}\,V(a) = \mathrm{tr}_V(\rho(a)) \qquad \text{for } a \in \mathcal{A} \ .$$

Proposition 4.1.17. *Characters satisfy* $\mathrm{ch}\,V(ab) = \mathrm{ch}\,V(ba)$ *for all* $a,b \in \mathcal{A}$ *and* $\mathrm{ch}\,V(1) = \dim V$. *Furthermore, if* $U \subset V$ *is a submodule and* $W = V/U$, *then* $\mathrm{ch}\,V = \mathrm{ch}\,U + \mathrm{ch}\,W$.

Proof. The first two properties are obvious from the definition. The third follows by picking a subspace $Z \subset V$ complementary to U. Then the matrix of $\rho(a), a \in \mathcal{A}$, is in block triangular form relative to the decomposition $V = U \oplus Z$, and the diagonal blocks give the action of a on U and on V/U. $\qquad\square$

The use of characters in representation theory is a powerful tool (similar to the use of generating functions in combinatorics). This will become apparent in later chapters. Let us find the extent to which a representation is determined by its character.

Lemma 4.1.18. *Suppose* $(\rho_1,V_1),\ldots,(\rho_r,V_r)$ *are finite-dimensional irreducible representations of* $\mathcal{A}$ *such that* ρ_i *is not equivalent to* ρ_j *when* $i \neq j$. *Then the set* $\{\mathrm{ch}\,V_1,\ldots,\mathrm{ch}\,V_r\}$ *of linear functionals on* $\mathcal{A}$ *is linearly independent.*

Proof. Set $V = V_1 \oplus \cdots \oplus V_r$ and $\rho = \rho_1 \oplus \cdots \oplus \rho_r$. Then (ρ,V) is a completely reducible representation of $\mathcal{A}$ by Proposition 4.1.11. Let $\mathcal{B}$ be the commutant of $\rho(\mathcal{A})$. Since the representations are irreducible and mutually inequivalent, Schur's lemma (Lemma 4.1.4) implies that the elements of $\mathcal{B}$ preserve each subspace V_j and act on it by scalars. Hence by the double commutant theorem (Theorem 4.1.13),

$$\rho(\mathcal{A}) = \mathrm{End}(V_1) \oplus \cdots \oplus \mathrm{End}(V_r) \ .$$

Let $I_i \in \mathrm{End}(V_i)$ be the identity operator on V_i. For each i there exists $Q_i \in \mathcal{A}$ with $\rho(Q_i)|_{V_j} = \delta_{ij}I_j$. We have

$$\mathrm{ch}\,V_j(Q_i) = \delta_{ij}\dim V_i \ .$$

Thus given a linear relation $\sum a_j\,\mathrm{ch}\,V_j = 0$, we may evaluate on Q_i to conclude that $a_i\dim V_i = 0$. Hence $a_i = 0$ for all i. $\qquad\square$

Let (ρ,V) be a finite-dimensional $\mathcal{A}$-module. A *composition series* (or *Jordan–Hölder series*) for V is a sequence of submodules

$$(0) = V_0 \subset V_1 \subset \cdots \subset V_r = V$$

such that $0 \neq W_i = V_i/V_{i-1}$ is irreducible for $i = 1, \ldots, r$. It is clear by induction on $\dim V$ that a composition series always exists. We define the *semisimplification* of V to be the module

$$V_{ss} = \bigoplus_{i=1}^{r} W_i \,.$$

By (3) of Proposition 4.1.17 and the obvious induction, we see that

$$\mathrm{ch}\, V = \sum_{i=1}^{r} \mathrm{ch}(V_i/V_{i-1}) = \mathrm{ch}\, V_{ss} \,. \tag{4.11}$$

Theorem 4.1.19. *Let* (ρ, V) *be a finite-dimensional* A-*module. The irreducible factors in a composition series for* V *are unique up to isomorphism and order of appearance. Furthermore, the module* V_{ss} *is uniquely determined by* $\mathrm{ch}\, V$ *up to isomorphism. In particular, if* V *is completely reducible, then* V *is uniquely determined up to isomorphism by* $\mathrm{ch}\, V$.

Proof. Let (ρ_i, U_i), for $i = 1, \ldots, n$, be the pairwise inequivalent irreducible representations that occur in the composition series for V, with corresponding multiplicities m_i. Then

$$\mathrm{ch}\, V = \sum_{i=1}^{n} m_i \, \mathrm{ch}\, U_i$$

by (4.11). Lemma 4.1.18 implies that the multiplicities m_i are uniquely determined by $\mathrm{ch}\, V$. $\square$

Example

Let $G = \mathbf{SL}(2, \mathbb{C})$ and let (ρ, V) be a regular representation of G. Let

$$d(q) = \mathrm{diag}[q, q^{-1}] \quad \text{for } q \in \mathbb{C}^{\times} \,.$$

If $g \in G$ and $\mathrm{tr}(g)^2 \neq 4$, then $g = h d(q) h^{-1}$ for some $h \in G$ (see Exercises 1.6.4, #7), where the eigenvalues of g are $\{q, q^{-1}\}$. Hence $\mathrm{ch}\, V(g) = \mathrm{ch}\, V(d(q))$. Since the function $g \mapsto \mathrm{ch}\, V(\rho(g))$ is regular and G is connected (Theorem 2.2.5), the character is determined by its restriction to the set $\{g \in G : \mathrm{tr}(g)^2 \neq 4\}$. Hence $\mathrm{ch}\, V$ is uniquely determined by the function $q \mapsto \mathrm{ch}\, V(d(q))$ for $q \in \mathbb{C}^{\times}$. Furthermore, since $d(q)$ is conjugate to $d(q^{-1})$ in G, this function on $\mathbb{C}^{\times}$ is invariant under the symmetry $q \mapsto q^{-1}$ arising from the action of the Weyl group of G on the diagonal matrices.

Let $(\rho_k, F^{(k)})$ be the $(k+1)$-dimensional irreducible representation of $\mathbf{SL}(2, \mathbb{C})$ (see Proposition 2.3.5). Then

$$\mathrm{ch}\, F^{(k)}(d(q)) = q^k + q^{k-2} + \cdots + q^{-k+2} + q^{-k}$$

(note the invariance under $q \mapsto q^{-1}$). For a positive integer n we define

$$[n]_q = q^{n-1} + q^{n-3} + \cdots + q^{-n+3} + q^{-n+1} = \frac{q^n - q^{-n}}{q - q^{-1}}$$

as a rational function of q. Then we can write $\operatorname{ch} F^{(k)}(d(q)) = [k+1]_q$.

Define $[0]_q = 1$, $[n]_q! = \prod_{j=0}^{n}[n-j]_q$ (the q-*factorial*), and

$$\begin{bmatrix} m+n \\ n \end{bmatrix}_q = \frac{[m+n]_q!}{[m]_q! [n]_q!}. \qquad \text{(q-binomial coefficient)}$$

Theorem 4.1.20 (Hermite Reciprocity). *Let $S^j(F^{(k)})$ be the jth symmetric power of $F^{(k)}$. Then for $q \in \mathbb{C}^\times$,*

$$\operatorname{ch} S^j(F^{(k)})(d(q)) = \begin{bmatrix} k+j \\ k \end{bmatrix}_q. \qquad (4.12)$$

In particular, $S^j(F^{(k)}) \cong S^k(F^{(j)})$ as representations of $\mathbf{SL}(2,\mathbb{C})$.

To prove this theorem we need some further character identities. Fix k and write $f_j(q) = \operatorname{ch} S^j(F^{(k)})(d(q))$ for $q \in \mathbb{C}^\times$. Let $\{x_0, \ldots, x_k\}$ be a basis for $F^{(k)}$ such that

$$\rho_k(d(q))x_j = q^{k-2j} x_j.$$

Then the monomials $x_0^{m_0} x_1^{m_1} \cdots x_k^{m_k}$ with $m_0 + \cdots + m_k = j$ give a basis for $S^j(F^{(k)})$, and $d(q)$ acts on such a monomial by the scalar q^r, where

$$r = km_0 + (k-2)m_1 + \cdots + (2-k)m_{k-1} - km_k.$$

Hence

$$f_j(q) = \sum_{m_0, \ldots, m_k} q^{km_0 + (k-2)m_1 + \cdots + (2-k)m_{k-1} - km_k}$$

with the sum over all nonnegative integers $m_0, \ldots, m_k$ such that $m_0 + \cdots + m_k = j$.

We form the *generating function*

$$f(t,q) = \sum_{j=0}^{\infty} t^j f_j(q),$$

which we view as a formal power series in the indeterminate t with coefficients in the ring $\mathbb{C}(q)$ of rational functions of q. If we let $\mathbb{C}^\times$ act by scalar multiplication on $F^{(k)}$, then $t \in \mathbb{C}^\times$ acts by multiplication by t^j on $S_j(F^{(k)})$ and this action commutes with the action of $\mathbf{SL}(2,\mathbb{C})$. Thus we can also view $f(t,q)$ as a *formal character* for the joint action of $\mathbb{C}^\times \times \mathbf{SL}(2,\mathbb{C})$ on the infinite-dimensional graded vector space $S(F^{(k)})$.

Lemma 4.1.21. *The generating function factors as*

$$f(t,q) = \prod_{j=0}^{k}(1 - tq^{k-2j})^{-1} . \tag{4.13}$$

Proof. By definition $(1 - tq^{k-2j})^{-1}$ is the formal power series

$$\sum_{m=0}^{\infty} t^m q^{m(k-2j)} . \tag{4.14}$$

Hence the right side of (4.13) is

$$\sum_{m_0,\dots,m_k} t^{m_0+\cdots+m_k} q^{km_0+(k-2)m_1+\cdots+(2-k)m_{k-1}-km_k} .$$

with the sum over all nonnegative integers $m_0,\dots,m_k$. Thus the coefficient of t^j is $f_j(q)$. $\square$

Since the representation $S^j(F^{(k)})$ is completely reducible, it is determined up to equivalence by its character, by Theorem 4.1.19. Just as for the ordinary binomial coefficients, one has the symmetry

$$\begin{bmatrix} m+n \\ n \end{bmatrix}_q = \begin{bmatrix} m+n \\ m \end{bmatrix}_q .$$

To complete the proof of Theorem 4.1.20, it thus suffices to prove the following result:

Lemma 4.1.22. *One has the formal power series identity*

$$\prod_{j=0}^{k}(1 - tq^{k-2j})^{-1} = \sum_{j=0}^{\infty} t^j \begin{bmatrix} k+j \\ k \end{bmatrix}_q ,$$

where the factors on the left side are defined by (4.14).

Proof. The proof proceeds by induction on k. The case $k = 0$ is the formal power series identity $(1-t)^{-1} = \sum_{j=0}^{\infty} t^j$. Now set

$$H_k(t,q) = \sum_{j=0}^{\infty} t^j \begin{bmatrix} k+j \\ k \end{bmatrix}_q ,$$

and assume that

$$H_k(t,q) = \prod_{j=0}^{k}(1 - tq^{k-2j})^{-1} .$$

It is easy to check that the q-binomial coefficients satisfy the recurrence

$$\begin{bmatrix} k+1+j \\ k+1 \end{bmatrix}_q = \frac{q^{k+1+j} - q^{-k-1-j}}{q^{k+1} - q^{-k-1}} \begin{bmatrix} k+j \\ k \end{bmatrix}_q.$$

Thus

$$H_{k+1}(t,q) = \frac{q^{k+1}}{q^{k+1} - q^{-k-1}} H_k(tq,q) - \frac{q^{-k-1}}{q^{k+1} - q^{-k-1}} H_k(tq^{-1},q).$$

Hence by the induction hypothesis we have

$$H_{k+1}(t,q) = \frac{q^{k+1}}{(q^{k+1} - q^{-k-1}) \prod_{j=0}^{k}(1 - tq^{k+1-2j})}$$

$$- \frac{q^{-k-1}}{(q^{k+1} - q^{-k-1}) \prod_{j=0}^{k}(1 - tq^{k-1-2j})}$$

$$= \left(\frac{q^{k+1}}{1 - tq^{k+1}} - \frac{q^{-k-1}}{1 - tq^{-k-1}} \right) \Big/ \left((q^{k+1} - q^{-k-1}) \prod_{j=1}^{k}(1 - tq^{k+1-2j}) \right)$$

$$= \prod_{j=0}^{k+1}(1 - tq^{k+1-2j})^{-1}. \qquad \square$$

4.1.8 Exercises

1. Let $\mathcal{A}$ be an associative algebra over $\mathbb{C}$ with unit element 1. Then $\mathcal{A} \otimes \mathcal{A}$ is an associative algebra with unit element $1 \otimes 1$, where the multiplication is defined by $(a \otimes b)(c \otimes d) = (ac) \otimes (bc)$ on decomposable tensors and extended to be bilinear. A *bialgebra structure* on $\mathcal{A}$ consists of an algebra homomorphism $\Delta : \mathcal{A} \longrightarrow \mathcal{A} \otimes \mathcal{A}$ (called the *comultiplication*) and an algebra homomorphism $\varepsilon : \mathcal{A} \longrightarrow \mathbb{C}$ (called the *counit*) that satisfy the following:

   (coassociativity) The maps $\Delta \otimes I_{\mathcal{A}}$ and $I_{\mathcal{A}} \otimes \Delta$ from $\mathcal{A}$ to $\mathcal{A} \otimes \mathcal{A} \otimes \mathcal{A}$ coincide: $(\Delta \otimes I_{\mathcal{A}})(\Delta(a)) = (I_{\mathcal{A}} \otimes \Delta)(\Delta(a))$ for all $a \in \mathcal{A}$, where $(\mathcal{A} \otimes \mathcal{A}) \otimes \mathcal{A}$ is identified with $\mathcal{A} \otimes (\mathcal{A} \otimes \mathcal{A})$ as usual and $I_{\mathcal{A}} : \mathcal{A} \longrightarrow \mathcal{A}$ is the identity map.

   (counit) The maps $(I_{\mathcal{A}} \otimes \varepsilon) \circ \Delta$ and $(\varepsilon \otimes I_{\mathcal{A}}) \circ \Delta$ from $\mathcal{A}$ to $\mathcal{A}$ coincide: $(I_{\mathcal{A}} \otimes \varepsilon)(\Delta(a)) = (\varepsilon \otimes I_{\mathcal{A}})(\Delta(a))$ for all $a \in \mathcal{A}$, where we identify $\mathbb{C} \otimes \mathcal{A}$ with $\mathcal{A}$ as usual.

   (a) Let G be a group and let $\mathcal{A} = \mathcal{A}[G]$ with convolution product. Define Δ and ε on the basis elements δ_x for $x \in G$ by $\Delta(\delta_x) = \delta_x \otimes \delta_x$ and $\varepsilon(\delta_x) = 1$, and extend these maps by linearity. Show that Δ and ε satisfy the conditions for a bialgebra structure on $\mathcal{A}$ and that $\langle \Delta(f), g \otimes h \rangle = \langle f, gh \rangle$ for $f, g, h \in \mathcal{A}[G]$. Here we write $\langle \phi, \psi \rangle = \sum_{x \in X} \phi(x)\psi(x)$ for complex-valued functions ϕ, ψ on a set X, and gh denotes the pointwise product of the functions g and h.

(b) Let G be a group and consider $\mathcal{A}[G]$ as the commutative algebra of $\mathbb{C}$-valued functions on G with pointwise multiplication and the constant function 1 as identity element. Identify $\mathcal{A}[G] \otimes \mathcal{A}[G]$ with $\mathcal{A}[G \times G]$ by $\delta_x \otimes \delta_y \longleftrightarrow \delta_{(x,y)}$ for $x, y \in G$. Define Δ by $\Delta(f)(x, y) = f(xy)$ and define $\varepsilon(f) = f(1)$, where $1 \in G$ is the identity element. Show that this defines a bialgebra structure on $\mathcal{A}[G]$ and that $\langle \Delta(f), g \otimes h \rangle = \langle f, g * h \rangle$ for $f, g, h \in \mathcal{A}[G]$, where $\langle \phi, \psi \rangle$ is defined as in (a) and where $g * h$ denotes the convolution product of the functions g and h.

(c) Let G be a linear algebraic group. Consider $\mathcal{A} = \mathcal{O}[G]$ as a (commutative) algebra with pointwise multiplication of functions and the constant function 1 as the identity element. Identify $\mathcal{A} \otimes \mathcal{A}$ with $\mathcal{O}[G \times G]$ as in Proposition 1.4.4 and define Δ and ε by the same formulas as in (b). Show that this defines a bialgebra structure on $\mathcal{A}$.

(d) Let $\mathfrak{g}$ be a Lie algebra over $\mathbb{C}$ and let $U(\mathfrak{g})$ be the universal enveloping algebra of $\mathfrak{g}$. Define $\Delta(X) = X \otimes 1 + 1 \otimes X$ for $X \in \mathfrak{g}$. Show that $\Delta([X, Y]) = \Delta(X)\Delta(Y) - \Delta(Y)\Delta(X)$, and conclude that Δ extends uniquely to an algebra homomorphism $\Delta : U(\mathfrak{g}) \longrightarrow U(\mathfrak{g}) \otimes U(\mathfrak{g})$. Let $\varepsilon : U(\mathfrak{g}) \longrightarrow \mathbb{C}$ be the unique algebra homomorphism such that $\varepsilon(X) = 0$ for all $X \in \mathfrak{g}$, as in Section 3.2.1. Show that Δ and ε define a bialgebra structure on $U(\mathfrak{g})$.

(e) Suppose G is a linear algebraic group with Lie algebra $\mathfrak{g}$. Define a bilinear form on $U(\mathfrak{g}) \times \mathcal{O}[G]$ by $\langle T, f \rangle = T f(I)$ for $T \in U(\mathfrak{g})$ and $f \in \mathcal{O}[G]$, where the action of $U(\mathfrak{g})$ on $\mathcal{O}[G]$ comes from the action of $\mathfrak{g}$ as left-invariant vector fields. Show that $\langle \Delta(T), f \otimes g \rangle = \langle T, fg \rangle$ for all $T \in U(\mathfrak{g})$ and $f, g \in \mathcal{O}[G]$, where Δ is defined as in (d). (This shows that the comultiplication on $U(\mathfrak{g})$ is dual to the pointwise multiplication on $\mathcal{O}[G]$.)

2. Let $\mathcal{A}$ be an associative algebra over $\mathbb{C}$, and suppose Δ and ε give $\mathcal{A}$ the structure of a bialgebra, in the sense of the previous exercise. Let (V, ρ) and (W, σ) be representations of $\mathcal{A}$.

(a) Show that the map $(a, b) \mapsto \rho(a) \otimes \sigma(b)$ extends to a representation of $\mathcal{A} \otimes \mathcal{A}$ on $V \otimes W$, denoted by $\rho \widehat{\otimes} \sigma$.

(b) Define $(\rho \otimes \sigma)(a) = (\rho \widehat{\otimes} \sigma)(\Delta(a))$ for $a \in \mathcal{A}$. Show that $\rho \otimes \sigma$ is a representation of $\mathcal{A}$, called the *tensor product* $\rho \otimes \sigma$ of the representations ρ and σ.

(c) When $\mathcal{A}$ and Δ are given as in (a) or (d) of the previous exercise, verify that the tensor product defined via the map Δ is the same as the tensor product defined in Section 4.1.1.

3. Let $\mathcal{A}$ be a bialgebra, in the sense of the previous exercises with comultiplication map Δ and counit ε. Let $S : \mathcal{A} \longrightarrow \mathcal{A}$ be an *antiautomorphsim* ($S(xy) = S(y)S(x)$ for all $x, y \in \mathcal{A}$). Then S is called an *antipode* if $\mu((S \otimes I_A)(\Delta(a))) = \varepsilon(a)1$ and $\mu((I_A \otimes S)(\Delta(a))) = \varepsilon(a)1$ for all $a \in \mathcal{A}$, where $\mu : \mathcal{A} \otimes \mathcal{A} \longrightarrow \mathcal{A}$ is the multiplication map. A bialgebra with an antipode is called a *Hopf algebra*.

(a) Let G be a group, and let $\mathcal{A} = \mathcal{A}[G]$ with convolution multiplication. Let Δ and ε be defined as in the exercise above, and let $Sf(x) = f(x^{-1})$ for $f \in \mathcal{A}[G]$ and $x \in G$. Show that S is an antipode.

(b) Let G be a group, and let $\mathcal{A} = \mathcal{A}[G]$ with pointwise multiplication. Let Δ and ε be defined as in the exercise above, and let $Sf(x) = f(x^{-1})$ for $f \in \mathcal{A}[G]$ and

$x \in G$. Show that S is an antipode (the same holds when G is a linear algebraic group and $\mathcal{A} = \mathcal{O}[G]$).

(c) Let $\mathfrak{g}$ be a Lie algebra over $\mathbb{C}$. Define the maps Δ and ε on $U(\mathfrak{g})$ as in the exercise above. Let $S(X) = -X$ for X in $\mathfrak{g}$. Show that S extends to an antiautomorphism of $U(\mathfrak{g})$ and satisfies the conditions for an antipode.

4. Let $\mathcal{A}$ be a Hopf algebra over $\mathbb{C}$ with antipode S.

(a) Given a representation (ρ, V) of $\mathcal{A}$, define $\rho^S(x) = \rho(Sx)^*$ for $x \in \mathcal{A}$. Show that (ρ^S, V^*) is a representation of $\mathcal{A}$.

(b) Show that the representation (ρ^S, V^*) is the *dual* representation to (ρ, V) when $\mathcal{A}$ is either $\mathcal{A}[G]$ with convolution multiplication or $U(\mathfrak{g})$ (where $\mathfrak{g}$ is a Lie algebra) and the antipode is defined as in the exercise above.

5. Let $\mathcal{A} = \mathcal{A}[x]$ and let $T \in M_n[\mathbb{C}]$. Define a representation ρ of $\mathcal{A}$ on $\mathbb{C}^n$ by $\rho(x) = T$. When is the representation ρ completely reducible? (HINT: Put T into Jordan canonical form.)

6. Let $\mathcal{A}$ be an associative algebra and let V be a completely reducible finite-dimensional $\mathcal{A}$-module.

(a) Show that V is irreducible if and only if $\dim \mathrm{Hom}_{\mathcal{A}}(V, V) = 1$.

(b) Does (a) hold if V is not completely reducible? (HINT: Consider the algebra of all upper-triangular 2×2 matrices.)

7. Let (ρ, V) and (σ, W) be finite-dimensional representations of a group G and let $g \in G$.

(a) Show that $\mathrm{ch}(V \otimes W)(g) = \mathrm{ch}\,V(g) \cdot \mathrm{ch}\,W(g)$.

(b) Show that $\mathrm{ch}(\bigwedge^2 V)(g) = \frac{1}{2}\Big((\mathrm{ch}\,V(g))^2 - \mathrm{ch}\,V(g^2)\Big)$.

(c) Show that $\mathrm{ch}(S^2(V))(g) = \frac{1}{2}\Big((\mathrm{ch}\,V(g))^2 + \mathrm{ch}\,V(g^2)\Big)$.

(HINT: Let $\{\lambda_i\}$ be the eigenvalues of $\rho(g)$ on V. Then $\{\lambda_i \lambda_j\}_{i<j}$ are the eigenvalues of g on $\bigwedge^2 V$ and $\{\lambda_i \lambda_j\}_{i \le j}$ are the eigenvalues of g on $S^2(V)$.)

The following exercises use the notation in Section 4.1.7.

8. Let (σ, W) be a regular representation of $\mathbf{SL}(2, \mathbb{C})$. For $q \in \mathbb{C}^\times$ let $f(q) = \mathrm{ch}\,W(d(q))$. Write $f(q) = f_{\mathrm{even}}(q) + f_{\mathrm{odd}}(q)$, where $f_{\mathrm{even}}(-q) = f_{\mathrm{even}}(q)$ and $f_{\mathrm{odd}}(-q) = -f_{\mathrm{odd}}(q)$.

(a) Show that $f_{\mathrm{even}}(q) = f_{\mathrm{even}}(q^{-1})$ and $f_{\mathrm{odd}}(q) = f_{\mathrm{odd}}(q^{-1})$.

(b) Let $f_{\mathrm{even}}(q) = \sum_{k \in \mathbb{Z}} a_k q^{2k}$ and $f_{\mathrm{odd}}(q) = \sum_{k \in \mathbb{Z}} b_k q^{2k+1}$. Show that the sequences $\{a_k\}$ and $\{b_k\}$ are unimodal. (HINT: See Exercises 2.3.4 #6.)

9. Let (σ, W) be a regular representation of $\mathbf{SL}(2, \mathbb{C})$ and let $W \cong \bigoplus_{k \ge 0} m_k F^{(k)}$ be the decomposition of W into isotypic components. We say that W is *even* if $m_k = 0$ for all odd integers k, and we say that W is *odd* if $m_k = 0$ for all even integers.

(a) Show W is even if and only if $\mathrm{ch}\,W(d(-q)) = \mathrm{ch}\,W(d(q))$, and odd if and only if $\mathrm{ch}\,W(d(-q)) = -\mathrm{ch}\,W(d(q))$. (HINT: Use Proposition 2.3.5.)

(b) Show that $S^j(F^{(k)})$ is even if jk is even and odd if jk is odd. (HINT: Use the model for $F^{(k)}$ from Section 2.3.2 to show that $-I \in \mathbf{SL}(2, \mathbb{C})$ acts on $F^{(k)}$ by $(-1)^k$ and hence acts by $(-1)^{jk}$ on $S^j(F^{(k)})$.)

10. Set $f(q) = \left[{m+n \atop m}\right]_q$ for $q \in \mathbb{C}^\times$ and positive integers m and n.

    (a) Show that $f(q) = f(q^{-1})$.

    (b) Show that $f(q) = \sum_{k \in \mathbb{Z}} a_k q^{2k+\varepsilon}$, where $\varepsilon = 0$ when mn is even and $\varepsilon = 1$ when mn is odd.

    (c) Show that the sequence $\{a_k\}$ in (b) is unimodal. (HINT: Use the previous exercises and Theorem 4.1.20.)

11. (a) Show (by a computer algebra system or otherwise) that

$$\left[{4+3 \atop 3}\right]_q = q^{12} + q^{10} + 2q^8 + 3q^6 + 4q^4 + 4q^2 + 5 + \cdots$$

    (where $\cdots$ indicates terms in negative powers of q).

    (b) Use (a) to prove that $S^3(V_4) \cong S^4(V_3) \cong V_{12} \oplus V_8 \oplus V_6 \oplus V_4 \oplus V_0$.
    (HINT: Use Proposition 2.3.5 and Theorem 4.1.20.)

12. (a) Show (by a computer algebra system or otherwise) that

$$\left[{5+3 \atop 3}\right]_q = q^{15} + q^{13} + 2q^{11} + 3q^9 + 4q^7 + 5q^5 + 6q^3 + 6q + \cdots$$

    (where $\cdots$ indicates terms in negative powers of q).

    (b) Use (a) to prove that $S^3(V_5) \cong S^5(V_3) \cong V_{15} \oplus V_{11} \oplus V_9 \oplus V_7 \oplus V_5 \oplus V_3$.
    (HINT: Use Proposition 2.3.5 and Theorem 4.1.20.)

13. For $n \in \mathbb{N}$ and $q \in \mathbb{C}$ define $\{n\}_1 = n$ and

$$\{n\}_q = q^{n-1} + q^{n-2} + \cdots + q + 1 = (q^n - 1)/(q-1) \quad \text{for } q \neq 1.$$

    (a) Show that $\{n\}_{q^2} = q^{n-1}[n]_q$.

    (b) Define

$$C_{n+m,m}(q) = \frac{\{m+n\}_q!}{\{m\}_q!\{n\}_q!}.$$

(This is an alternative version of the q-binomial coefficient that also gives the ordinary binomial coefficient when specialized to $q = 1$.) Let p be a prime and let $\mathbb{F}$ be the field with $q = p^n$ elements. Prove that $C_{m+n,m}(q)$ is the number of m-dimensional subspaces in the vector space $\mathbb{F}^{m+n}$. (HINT: The number of nonzero elements of $\mathbb{F}^{m+n}$ is $q^{n+m} - 1$. If $v \in \mathbb{F}^{m+n} - \{0\}$ then the number of elements that are not multiples of v is $q^{n+m} - q$. Continuing in this way we find that the cardinality of the set of all linearly independent m-tuples $\{v_1, ..., v_m\}$ is $(q^{n+m} - 1)(q^{n+m-1} - 1) \cdots (q^{n+1} - 1) = a_{n,m}$. The desired cardinality is thus $a_{n,m}/a_{0,m} = C_{n+m,m}(q)$.)

4.2 Duality for Group Representations

In this section we prove the duality theorem. As first applications we obtain *Schur–Weyl duality* for representations of $\mathbf{GL}(n,\mathbb{C})$ on tensor spaces and the decomposition of the representation of $G \times G$ on $\mathcal{O}[G]$. Further applications of the duality theorem will occur in later chapters.

4.2.1 General Duality Theorem

Assume that $G \subset \mathbf{GL}(n,\mathbb{C})$ is a reductive linear algebraic group. Let $\widehat{G}$ denote the equivalence classes of irreducible regular representations of G and fix a representation (π^λ, F^λ) in the class λ for each $\lambda \in \widehat{G}$. We view representation spaces for G as modules for the group algebra $\mathcal{A}[G]$, as in Section 4.1.1, and identify $\widehat{G}$ with a subset of $\widehat{\mathcal{A}[G]}$.

Let (ρ, L) be a locally regular representation of G with $\dim L$ countable. Then ρ is a locally completely reducible representation of $\mathcal{A}[G]$, and the irreducible $\mathcal{A}[G]$ submodules of L are irreducible regular representations of G (since G is reductive). Thus the nonzero isotypic components $L_{(\lambda)}$ are labeled by $\lambda \in \widehat{G}$. We shall write $\mathrm{Spec}(\rho)$ for the set of representation types that occur in the primary decomposition of L. Then by Proposition 4.1.15 we have

$$L \cong \bigoplus_{\lambda \in \mathrm{Spec}(\rho)} \mathrm{Hom}(F^\lambda, L) \otimes F^\lambda$$

as a G-module, with $g \in G$ acting by $I \otimes \pi^\lambda(g)$ on the summand of type λ. We now focus on the *multiplicity spaces* $\mathrm{Hom}(F^\lambda, L)$ in this decomposition.

Assume that $\mathcal{R} \subset \mathrm{End}(L)$ is a subalgebra that satisfies the following conditions:

(i) $\mathcal{R}$ acts irreducibly on L,
(ii) if $g \in G$ and $T \in \mathcal{R}$ then $\rho(g)T\rho(g)^{-1} \in \mathcal{R}$ (so G acts on $\mathcal{R}$), and
(iii) the representation of G on $\mathcal{R}$ in (ii) is locally regular.

If $\dim L < \infty$, the only choice for $\mathcal{R}$ is $\mathrm{End}(L)$ by Corollary 4.1.7. By contrast, when $\dim L = \infty$ there may exist many such algebras $\mathcal{R}$; we shall see an important example in Section 5.6.1 (the *Weyl algebra* of linear differential operators with polynomial coefficients).

Fix $\mathcal{R}$ satisfying the conditions (i) and (ii) and let

$$\mathcal{R}^G = \{T \in \mathcal{R} : \rho(g)T = T\rho(g) \quad \text{for all } g \in G\}$$

(the commutant of $\rho(G)$ in $\mathcal{R}$). Since G is reductive, we may view L as a locally completely irreducible representation of $\mathcal{A}[G]$. Since elements of $\mathcal{R}^G$ commute with elements of $\mathcal{A}[G]$, we have a representation of the algebra $\mathcal{R}^G \otimes \mathcal{A}[G]$ on L. The *duality theorem* describes the decomposition of this representation.

Let

$$E^\lambda = \mathrm{Hom}_G(F^\lambda, L) \quad \text{for } \lambda \in \widehat{G}.$$

Then E^λ is a module for $\mathcal{R}^G$ in a natural way by left multiplication, since

$$Tu(\pi^\lambda(g)v) = T\rho(g)u(v) = \rho(g)(Tu(v))$$

for $T \in \mathcal{R}^G$, $u \in E^\lambda$, $g \in G$, and $v \in F^\lambda$. Hence as a module for the algebra $\mathcal{R}^G \otimes \mathcal{A}[G]$ the space L decomposes as

$$L \cong \bigoplus_{\lambda \in \mathrm{Spec}(\rho)} E^\lambda \otimes F^\lambda. \tag{4.15}$$

In (4.15) an operator $T \in \mathcal{R}^G$ acts by $T \otimes I$ on the summand of type λ.

Theorem 4.2.1 (Duality). *Each multiplicity space E^λ is an irreducible $\mathcal{R}^G$ module. Furthermore, if $\lambda, \mu \in \mathrm{Spec}(\rho)$ and $E^\lambda \cong E^\mu$ as an $\mathcal{R}^G$ module, then $\lambda = \mu$.*

The duality theorem plays a central role in the representation and invariant theory of the classical groups. Here is an immediate consequence.

Corollary 4.2.2 (Duality Correspondence). *Let σ be the representation of $\mathcal{R}^G$ on L and let $\mathrm{Spec}(\sigma)$ denote the set of equivalence classes of the irreducible representations $\{E^\lambda\}$ of the algebra $\mathcal{R}^G$ that occur in L. Then the following hold:*

1. *The representation (σ, L) is a direct sum of irreducible $\mathcal{R}^G$ modules, and each irreducible submodule E^λ occurs with finite multiplicity $\dim F^\lambda$.*
2. *The map $F^\lambda \longrightarrow E^\lambda$ sets up a bijection between $\mathrm{Spec}(\rho)$ and $\mathrm{Spec}(\sigma)$.*

The proof of the duality theorem will use the following result:

Lemma 4.2.3. *Let $X \subset L$ be a finite-dimensional G-invariant subspace. Then $\mathcal{R}^G\big|_X = \mathrm{Hom}_G(X, L)$.*

Proof. Let $T \in \mathrm{Hom}_G(X, L)$. Then by Corollary 4.1.6 there exists $r \in \mathcal{R}$ such that $r|_X = T$. Since G is reductive, condition (iii) and Proposition 4.1.15 furnish a projection $r \mapsto r^\natural$ from $\mathcal{R}$ to $\mathcal{R}^G$. But the map $\mathcal{R} \longrightarrow \mathrm{Hom}(X, L)$ given by $y \mapsto y|_X$ intertwines the G actions, by the G invariance of X. Hence $T = T^\natural = r^\natural\big|_X$. $\qquad\square$

Proof of Theorem 4.2.1. We first prove that the action of $\mathcal{R}^G$ on $\mathrm{Hom}_G(F^\lambda, L)$ is irreducible. Let $T \in \mathrm{Hom}_G(F^\lambda, L)$ be nonzero. Given another nonzero element $S \in \mathrm{Hom}_G(F^\lambda, L)$ we need to find $r \in \mathcal{R}^G$ such that $rT = S$. Let $X = TF^\lambda$ and $Y = SF^\lambda$. Then by Schur's lemma X and Y are isomorphic G-modules of class λ. Thus Lemma 4.2.3 implies that there exists $u \in \mathcal{R}^G$ such that $u|_X$ implements this isomorphism. Thus $uT : F^\lambda \longrightarrow SF^\lambda$ is a G-module isomorphism. Schur's lemma implies that there exists $c \in \mathbb{C}$ such that $cuT = S$, so we may take $r = cu$.

We now show that if $\lambda \neq \mu$ then $\mathrm{Hom}_G(F^\lambda, L)$ and $\mathrm{Hom}_G(F^\mu, L)$ are inequivalent modules for $\mathcal{R}^G$. Suppose

$$\varphi : \mathrm{Hom}_G(F^\lambda, L) \longrightarrow \mathrm{Hom}_G(F^\mu, L)$$

is an intertwining operator for the action of $\mathcal{R}^G$. Let $T \in \mathrm{Hom}_G(F^\lambda, L)$ be nonzero and set $S = \varphi(T)$. We want to show that $S = 0$. Set $U = TF^\lambda + SF^\mu$. Then since we are assuming $\lambda \neq \mu$, the sum is direct. Let $p : U \longrightarrow SF^\mu$ be the corresponding projection. Then Lemma 4.2.3 implies that there exists $r \in \mathcal{R}^G$ such that $r|_U = p$. Since $pT = 0$, we have $rT = 0$. Hence

$$0 = \varphi(rT) = r\varphi(T) = rS = pS = S,$$

which proves that $\varphi = 0$. $\qquad \square$

In the finite-dimensional case we can combine the duality theorem with the double commutant theorem.

Corollary 4.2.4. *Assume* $\dim L < \infty$. *Set* $\mathcal{A} = \mathrm{Span}\,\rho(G)$ *and* $\mathcal{B} = \mathrm{End}_\mathcal{A}(L)$. *Then* L *is a completely reducible* $\mathcal{B}$*-module. Furthermore, the following hold:*

1. *Suppose that for every* $\lambda \in \mathrm{Spec}(\rho)$ *there is given an operator* $T_\lambda \in \mathrm{End}(F^\lambda)$. *Then there exists* $T \in \mathcal{A}$ *that acts by* $I \otimes T_\lambda$ *on the* λ *summand in the decomposition* (4.15).
2. *Let* $T \in \mathcal{A} \cap \mathcal{B}$ *(the center of* $\mathcal{A}$*). Then* T *is diagonalized by the decomposition* (4.15) *and acts by a scalar* $\widehat{T}(\lambda) \in \mathbb{C}$ *on* $E^\lambda \otimes F^\lambda$. *Conversely, given any complex-valued function* f *on* $\mathrm{Spec}(\rho)$, *there exists* $T \in \mathcal{A} \cap \mathcal{B}$ *such that* $\widehat{T}(\lambda) = f(\lambda)$.

Proof. Since L is the direct sum of $\mathcal{B}$-invariant irreducible subspaces by Theorem 4.2.1, it is a completely reducible $\mathcal{B}$-module by Proposition 4.1.11. We now prove the other assertions.

(1): Let $T \in \mathrm{End}(L)$ be the operator that acts by $I \otimes T_\lambda$ on the λ summand. Then $T \in \mathrm{Comm}(\mathcal{B})$, and hence $T \in \mathcal{A}$ by the double commutant theorem (Theorem 4.1.13).

(2): Each summand in (4.15) is invariant under T, and the action of T on the λ summand is by an operator of the form $R_\lambda \otimes I = I \otimes S_\lambda$ with $R_\lambda \in \mathrm{End}(E^\lambda)$ and $S_\lambda \in \mathrm{End}(F^\lambda)$. Such an operator must be a scalar multiple of the identity operator. The converse follows from (1). $\qquad \square$

4.2.2 Products of Reductive Groups

We now apply the duality theorem to determine the regular representations of the product of two reductive linear algebraic groups H and K. Let $G = H \times K$ be the direct product linear algebraic group. Recall that $\mathcal{O}[G] \cong \mathcal{O}[H] \otimes \mathcal{O}[K]$ under the natural pointwise multiplication map. Let (σ, V) and (τ, W) be regular representations of H and K respectively. The *outer tensor product* is the representation $(\sigma \widehat{\otimes} \tau, V \otimes W)$ of $H \times K$, where

$$(\sigma \widehat{\otimes} \tau)(h,k) = \sigma(h) \otimes \tau(k) \quad \text{for } h \in H \text{ and } k \in K .$$

Notice that when $H = K$, the restriction of the outer tensor product $\sigma \widehat{\otimes} \tau$ to the diagonal subgroup $\{(h,h) : h \in H\}$ of $H \times H$ is the tensor product $\sigma \otimes \tau$.

Proposition 4.2.5. *Suppose (σ, V) and (τ, W) are irreducible. Then the outer tensor product $(\sigma \widehat{\otimes} \tau, V \otimes W)$ is an irreducible representation of $H \times K$, and every irreducible regular representation of $H \times K$ is of this form.*

Proof. We have $\text{End}(V \otimes W) = \text{End}(V) \otimes \text{End}(W) = \text{Span}\{\sigma(H) \otimes \tau(K)\}$ by Corollary 4.1.7. Hence if $0 \neq u \in V \otimes W$, then $\text{Span}\{(\sigma(H) \otimes \tau(K))u\} = V \otimes W$. This shows that $\rho \widehat{\otimes} \sigma$ is irreducible.

Conversely, given an irreducible regular representation (ρ, L) of $H \times K$, set $\tau(k) = \rho(1,k)$ for $k \in K$, and use Theorem 4.2.1 (with $\mathcal{R} = \text{End}(L)$) to decompose L as a K-module:

$$L = \bigoplus_{\lambda \in \text{Spec}(\tau)} E^\lambda \otimes F^\lambda . \tag{4.16}$$

Set $\sigma(h) = \rho(h,1)$ for $h \in H$. Then $\sigma(H) \subset \text{End}_K(L)$, and thus H preserves decomposition (4.16) and acts on the λ summand by $h \mapsto \sigma_\lambda(h) \otimes I$ for some representation σ_λ. We claim that σ_λ is irreducible. To prove this, note that since $\text{End}_K(L)$ acts irreducibly on E^λ by Theorem 4.2.1, we have

$$\text{End}_K(L) \cong \bigoplus_{\lambda \in \text{Spec}(\tau)} \text{End}(E^\lambda) \otimes I . \tag{4.17}$$

But ρ is an irreducible representation, so $\text{End}(L)$ is spanned by the transformations $\rho(h,k) = \sigma(h)\tau(k)$ with $h \in H$ and $k \in K$. Since K is reductive, there is a projection $T \mapsto T^\natural$ from $\text{End}(L)$ onto $\text{End}_K(L)$, and $\tau(k)^\natural$, for $k \in K$, acts by a scalar in each summand in (4.16) by Schur's lemma. Hence $\text{End}(E^\lambda)$ is spanned by $\sigma_\lambda(H)$, proving that σ_λ is irreducible. Thus each summand in (4.16) is an irreducible module for $H \times K$, by the earlier argument. Since ρ is irreducible, there can be only one summand in (4.16). Hence $\rho = \sigma \widehat{\otimes} \tau$. $\square$

Proposition 4.2.6. *If H and K are reductive linear algebraic groups, then $H \times K$ is reductive.*

Proof. Let ρ be a regular representation of $H \times K$. As in the proof of Proposition 4.2.5 we set $\tau(k) = \rho(1,k)$ for $k \in K$, and we use Theorem 4.2.1 (with $\mathcal{R} = \text{End}(L)$) to obtain decomposition (4.16) of L as a K-module. Set $\sigma(h) = \rho(h,1)$ for $h \in H$. Then $\sigma(H) \subset \text{End}_K(L)$, and thus we have a regular representation of H on E^λ for each $\lambda \in \text{Spec}(\tau)$. Since H is reductive, these representations of H decompose as direct sums of irreducible representations. Using these decompositions in (4.16), we obtain a decomposition of L as a direct sum of representations of $H \times K$ that are irreducible by Proposition 4.2.5. $\square$

4.2.3 Isotypic Decomposition of $\mathcal{O}[G]$

Let G be a reductive linear algebraic group. The group $G \times G$ is reductive by Proposition 4.2.6 and it acts on $\mathcal{O}[G]$ by left and right translations. Denote this representation by ρ:

$$\rho(y,z)f(x) = f(y^{-1}xz), \quad \text{for} \ f \in \mathcal{O}[G] \ \text{and} \ x,y,z \in G.$$

For each $\lambda \in \widehat{G}$ fix an irreducible representation (π^λ, F^λ) in the class λ. Denote by λ^* the class of the representation contragredient to λ. We choose the representations π^λ so that the vector space F^{λ^*} equals $(F^\lambda)^*$ and the operator $\pi^{\lambda^*}(g)$ equals $\pi^\lambda(g^{-1})^t$. We write $d_\lambda = \dim V^\lambda$ and call d_λ the *degree* of the representation. Note that $d_\lambda = d_{\lambda^*}$.

Theorem 4.2.7. *For* $\lambda \in \widehat{G}$ *define* $\varphi_\lambda(v^* \otimes v)(g) = \langle v^*, \pi^\lambda(g)v \rangle$ *for* $g \in G$, $v^* \in V^{\lambda^*}$, *and* $v \in V$. *Extend* φ_λ *to a linear map from* $F^{\lambda^*} \otimes F^\lambda$ *to* $\mathcal{O}[G]$. *Then the following hold:*

1. Range(φ_λ) *is independent of the choice of the model* (π^λ, F^λ) *and furnishes an irreducible regular representation of* $G \times G$ *isomorphic to* $F^{\lambda^*} \widehat{\otimes} F^\lambda$.
2. *Under the action of* $G \times G$, *the space* $\mathcal{O}[G]$ *decomposes as*

$$\mathcal{O}[G] = \bigoplus_{\lambda \in \widehat{G}} \varphi_\lambda(F^{\lambda^*} \otimes F^\lambda). \tag{4.18}$$

Proof. Given $v \in F^\lambda$ and $v^* \in F^{\lambda^*}$, we set $f_{v^*,v} = \varphi_\lambda(v^* \otimes v)$. Then for $x,y,z \in G$ we have

$$f_{x \cdot v^*, y \cdot v}(z) = \langle \pi^{\lambda^*}(x)v^*, \pi^\lambda(z)\pi^\lambda(y)v \rangle = f_{v^*,v}(x^{-1}zy).$$

This shows that φ_λ intertwines the action of $G \times G$. Since $F^{\lambda^*} \otimes F^\lambda$ is irreducible as a $G \times G$ module by Proposition 4.2.5, Schur's lemma implies that φ_λ must be injective. It is clear that the range of φ_λ depends only on the equivalence class of (π^λ, F^λ).

Let $\mathcal{O}[G]_{(\lambda)}$ be the λ-isotypic subspace relative to the right action R of G. The calculation above shows that Range$(\varphi_\lambda) \subset \mathcal{O}[G]_{(\lambda)}$, so by Proposition 4.1.15 we need to show only the opposite inclusion. Let $W \subset \mathcal{O}[G]_{(\lambda)}$ be any irreducible subspace for the right action of G. We may then take W as the model for λ in the definition of the map φ_λ. Define $\delta \in W^*$ by $\langle \delta, w \rangle = w(1)$. Then

$$f_{\delta,w}(g) = \langle \delta, R(g)w \rangle = (R(g)w)(1) = w(g).$$

Hence $\varphi_\lambda(\delta \otimes w) = w$, completing the proof. $\square$

Corollary 4.2.8. *In the right-translation representation of* G *on* $\mathcal{O}[G]$ *every irreducible representation of* G *occurs with multiplicity equal to its dimension.*

Remark 4.2.9. The representations of $G \times G$ that occur in $\mathcal{O}[G]$ are the outer tensor products $\lambda^* \widehat{\otimes} \lambda$ for all $\lambda \in \widehat{G}$, and each representation occurs with multiplicity one.

Under the isomorphism $F^{\lambda^*} \otimes F^\lambda \cong \text{End}(F^\lambda)$ that sends the tensor $v^* \otimes v$ to the rank-one operator $u \mapsto \langle v^*, u \rangle v$, the map φ_λ in Theorem 4.2.7 is given by $\varphi_\lambda(T)(g) = \text{tr}(\pi^\lambda(g)T)$ for $T \in \text{End}(F^\lambda)$. In this model for the isotypic components the element $(g, g') \in G \times G$ acts by $T \mapsto \pi^\lambda(g')T\pi^\lambda(g)^{-1}$ on the λ summand.

The *duality principle* in Theorem 4.2.1 asserts that the commutant of G explains the multiplicities in the primary decomposition. For example, the space $\mathcal{O}[G]$ is a direct sum of irreducible representations of G, relative to the right translation action, since G is reductive. Obtaining such a decomposition requires decomposing each isotypic component into irreducible subspaces. If G is not an algebraic torus, then it has irreducible representations of dimension greater than one, and the decomposition of the corresponding isotypic component is not unique. However, when we include the additional symmetries coming from the commuting left translation action by G, then each isotypic component becomes irreducible under the action of $G \times G$.

4.2.4 Schur–Weyl Duality

We now apply the duality theorem to obtain a result that will play a central role in our study of tensor and polynomial invariants for the classical groups. Let ρ be the defining representation of $\mathbf{GL}(n, \mathbb{C})$ on $\mathbb{C}^n$. For all integers $k \geq 0$ we can construct the representation $\rho_k = \rho^{\otimes k}$ on $\bigotimes^k \mathbb{C}^n$. Since

$$\rho_k(g)(v_1 \otimes \cdots \otimes v_k) = gv_1 \otimes \cdots \otimes gv_k$$

for $g \in \mathbf{GL}(n, \mathbb{C})$, we can permute the positions of the vectors in the tensor product without changing the action of G. Let $\mathfrak{S}_k$ be the group of permutations of $\{1, 2, \ldots, k\}$. We define a representation σ_k of $\mathfrak{S}_k$ on $\bigotimes^k \mathbb{C}^n$ by

$$\sigma_k(s)(v_1 \otimes \cdots \otimes v_k) = v_{s^{-1}(1)} \otimes \cdots \otimes v_{s^{-1}(k)}$$

for $s \in \mathfrak{S}_k$. Notice that $\sigma_k(s)$ moves the vector in the ith position in the tensor product to the position $s(i)$. It is clear that $\sigma_k(s)$ commutes with $\rho_k(g)$ for all $s \in \mathfrak{S}_k$ and $g \in \mathbf{GL}(n, \mathbb{C})$. Let $\mathcal{A} = \text{Span}\,\rho_k(\mathbf{GL}(n, \mathbb{C}))$ and $\mathcal{B} = \text{Span}\,\sigma_k(\mathfrak{S}_k)$. Then we have $\mathcal{A} \subset \text{Comm}(\mathcal{B})$.

Theorem 4.2.10 (Schur). *One has* $\text{Comm}(\mathcal{B}) = \mathcal{A}$ *and* $\text{Comm}(\mathcal{A}) = \mathcal{B}$.

Proof. The representations σ_k and ρ_k are completely reducible (by Corollary 3.3.6 and Theorem 3.3.11). From the double commutant theorem (Theorem 4.1.13) it suffices to prove that $\text{Comm}(\mathcal{B}) \subset \mathcal{A}$.

Let $\{e_1, \ldots, e_n\}$ be the standard basis for $\mathbb{C}^n$. For an ordered k-tuple $I = (i_1, \ldots, i_k)$ with $1 \leq i_j \leq n$, define $|I| = k$ and $e_I = e_{i_1} \otimes \cdots \otimes e_{i_k}$. The tensors $\{e_I\}$, with I ranging over the all such k-tuples, give a basis for $\bigotimes^k \mathbb{C}^n$. The group $\mathfrak{S}_k$ permutes this basis by the action $\sigma_k(s)e_I = e_{s \cdot I}$, where for $I = (i_1, \ldots, i_k)$ and $s \in \mathfrak{S}_k$ we define

$$s \cdot (i_1, \ldots, i_k) = (i_{s^{-1}(1)}, \ldots, i_{s^{-1}(k)}) \, .$$

Note that s changes the *positions* (1 to k) of the indices, not their values (1 to n), and we have $(st) \cdot I = s \cdot (t \cdot I)$ for $s, t \in \mathfrak{S}_k$.

Suppose $T \in \mathrm{End}(\bigotimes^k \mathbb{C}^n)$ has matrix $[a_{I,J}]$ relative to the basis $\{e_I\}$:

$$Te_J = \sum_I a_{I,J} e_I \, .$$

We have

$$T(\sigma_k(s)e_J) = T(e_{s \cdot J}) = \sum_I a_{I, s \cdot J} e_I$$

for $s \in \mathfrak{S}_k$, whereas

$$\sigma_k(s)(Te_J) = \sum_I a_{I,J} e_{s \cdot I} = \sum_I a_{s^{-1} \cdot I, J} e_I \, .$$

Thus $T \in \mathrm{Comm}(\mathcal{B})$ if and only if $a_{I, s \cdot J} = a_{s^{-1} \cdot I, J}$ for all multi-indices I, J and all $s \in \mathfrak{S}_k$. Replacing I by $s \cdot I$, we can write this condition as

$$a_{s \cdot I, s \cdot J} = a_{I,J} \quad \text{for all } I, J \text{ and all } s \in \mathfrak{S}_k \, . \tag{4.19}$$

Consider the nondegenerate bilinear form $(X, Y) = \mathrm{tr}(XY)$ on $\mathrm{End}\left(\bigotimes^k \mathbb{C}^n\right)$. We claim that the restriction of this form to $\mathrm{Comm}(\mathcal{B})$ is nondegenerate. Indeed, we have a projection $X \mapsto X^\natural$ of $\mathrm{End}\left(\bigotimes^k \mathbb{C}^n\right)$ onto $\mathrm{Comm}(\mathcal{B})$ given by averaging over $\mathfrak{S}_k$:

$$X^\natural = \frac{1}{k!} \sum_{s \in \mathfrak{S}_k} \sigma_k(s) X \sigma_k(s)^{-1} \, .$$

If $T \in \mathrm{Comm}(\mathcal{B})$ then

$$(X^\natural, T) = \frac{1}{k!} \sum_{s \in \mathfrak{S}_k} \mathrm{tr}(\sigma_k(s) X \sigma_k(s)^{-1} T) = (X, T) \, ,$$

since $\sigma_k(s) T = T \sigma_k(s)$. Thus $(\mathrm{Comm}(\mathcal{B}), T) = 0$ implies that $(X, T) = 0$ for all $X \in \mathrm{End}\left(\bigotimes^k \mathbb{C}^n\right)$, and so $T = 0$. Hence the trace form on $\mathrm{Comm}(\mathcal{B})$ is nondegenerate.

To prove that $\mathrm{Comm}(\mathcal{B}) = \mathcal{A}$, it thus suffices to show that if $T \in \mathrm{Comm}(\mathcal{B})$ is orthogonal to $\mathcal{A}$ then $T = 0$. Now if $g = [g_{ij}] \in \mathbf{GL}(n, \mathbb{C})$, then $\rho_k(g)$ has matrix $g_{I,J} = g_{i_1 j_1} \cdots g_{i_k j_k}$ relative to the basis $\{e_I\}$. Thus we assume that

$$(T, \rho_k(g)) = \sum_{I,J} a_{I,J} g_{j_1 i_1} \cdots g_{j_k i_k} = 0 \tag{4.20}$$

for all $g \in \mathbf{GL}(n, \mathbb{C})$, where $[a_{I,J}]$ is the matrix of T. Define a polynomial function f_T on $M_n(\mathbb{C})$ by

$$f_T(X) = \sum_{I,J} a_{I,J} x_{j_1 i_1} \cdots x_{j_k i_k}$$

for $X = [x_{ij}] \in M_n(\mathbb{C})$. From (4.20) we have $\det(X)f_T(X) = 0$ for all $X \in M_n(\mathbb{C})$. Hence f_T is identically zero, so for all $[x_{ij}] \in M_n(\mathbb{C})$ we have

$$\sum_{I,J} a_{I,J} x_{j_1 i_1} \cdots x_{j_k i_k} = 0 . \qquad (4.21)$$

We now show that (4.19) and (4.21) imply that $a_{I,J} = 0$ for all I, J. We begin by grouping the terms in (4.21) according to distinct monomials in the matrix entries $\{x_{ij}\}$. Introduce the notation $x_{I,J} = x_{i_1 j_1} \cdots x_{i_k j_k}$, and view these monomials as polynomial functions on $M_n(\mathbb{C})$. Let Ξ be the set of all ordered pairs (I, J) of multi-indices with $|I| = |J| = k$. The group $\mathfrak{S}_k$ acts on Ξ by

$$s \cdot (I, J) = (s \cdot I, s \cdot J) .$$

From (4.19) we see that T commutes with $\mathfrak{S}_k$ if and only if the function $(I, J) \mapsto a_{I,J}$ is constant on the orbits of $\mathfrak{S}_k$ in Ξ.

The action of $\mathfrak{S}_k$ on Ξ defines an equivalence relation on Ξ, where $(I, J) \equiv (I', J')$ if $(I', J') = (s \cdot I, s \cdot J)$ for some $s \in \mathfrak{S}_k$. This gives a decomposition of Ξ into disjoint equivalence classes. Choose a set Γ of representatives for the equivalence classes. Then every monomial $x_{I,J}$ with $|I| = |J| = k$ can be written as x_γ for some $\gamma \in \Gamma$. Indeed, since the variables x_{ij} mutually commute, we have

$$x_\gamma = x_{s \cdot \gamma} \quad \text{for all } s \in \mathfrak{S}_k \text{ and } \gamma \in \Gamma .$$

Suppose $x_{I,J} = x_{I',J'}$. Then there must be an integer p such that $x_{i'_1 j'_1} = x_{i_p j_p}$. Call $p = 1'$. Similarly, there must be an integer $q \neq p$ such that $x_{i'_2 j'_2} = x_{i_q j_q}$. Call $q = 2'$. Continuing this way, we obtain a permutation

$$s : (1, 2, \ldots, k) \rightarrow (1', 2', \ldots, k')$$

such that $I = s \cdot I'$ and $J = s \cdot J'$. This proves that γ is uniquely determined by x_γ. For $\gamma \in \Gamma$ let $n_\gamma = |\mathfrak{S}_k \cdot \gamma|$ be the cardinality of the corresponding orbit.

Assume that the coefficients $a_{I,J}$ satisfy (4.19) and (4.21). Since $a_{I,J} = a_\gamma$ for all $(I, J) \in \mathfrak{S}_k \cdot \gamma$, it follows from (4.21) that

$$\sum_{\gamma \in \Gamma} n_\gamma a_\gamma x_\gamma = 0 .$$

Since the set of monomials $\{x_\gamma : \gamma \in \Gamma\}$ is linearly independent, this implies that $a_{I,J} = 0$ for all $(I, J) \in \Xi$. This proves that $T = 0$. Hence $\mathcal{A} = \mathrm{Comm}(\mathcal{B})$. $\square$

From Theorems 4.2.1 and 4.2.10 we obtain a preliminary version of *Schur–Weyl duality*:

Corollary 4.2.11. *There are irreducible, mutually inequivalent $\mathfrak{S}_k$-modules E^λ and irreducible, mutually inequivalent $\mathbf{GL}(n, \mathbb{C})$-modules F^λ such that*

$$\bigotimes^k \mathbb{C}^n \cong \bigoplus_{\lambda \in \mathrm{Spec}(\rho_k)} E^\lambda \otimes F^\lambda$$

as a representation of $\mathfrak{S}_k \times \mathbf{GL}(n, \mathbb{C})$. The representation E^λ uniquely determines F^λ and conversely.

In Chapter 9 we shall determine the explicit form of the irreducible representations and the duality correspondence in Corollary 4.2.11.

4.2.5 Commuting Algebra and Highest-Weight Vectors

Let $\mathfrak{g}$ be a semisimple Lie algebra and let V be a finite-dimensional $\mathfrak{g}$-module. We shall apply the theorem of the highest weight to decompose the commuting algebra $\mathrm{End}_\mathfrak{g}(V)$ as a direct sum of full matrix algebras.

Fix a Cartan subalgebra $\mathfrak{h}$ of $\mathfrak{g}$ and a choice of positive roots of $\mathfrak{h}$, and let $\mathfrak{g} = \mathfrak{n}^- + \mathfrak{h} + \mathfrak{n}^+$ be the associated triangular decomposition of $\mathfrak{g}$, as in Corollary 2.5.25. Set

$$V^{\mathfrak{n}^+} = \{v \in V : X \cdot v = 0 \quad \text{for all } X \in \mathfrak{n}^+\}.$$

Note that if $T \in \mathrm{End}_\mathfrak{g}(V)$ then it preserves $V^{\mathfrak{n}^+}$ and it preserves the weight space decomposition

$$V^{\mathfrak{n}^+} = \bigoplus_{\mu \in \mathcal{S}} V^{\mathfrak{n}^+}(\mu).$$

Here $\mathcal{S} = \{\mu \in P_{++}(\mathfrak{g}) : V^{\mathfrak{n}^+}(\mu) \neq 0\}$. By Theorem 3.2.5 we can label the equivalence classes of irreducible $\mathfrak{g}$-modules by their highest weights. For each $\mu \in \mathcal{S}$ choose an irreducible representation (π^μ, V^μ) with highest weight μ.

Theorem 4.2.12. *The restriction map $\varphi : T \mapsto T|_{V^{\mathfrak{n}^+}}$ for $T \in \mathrm{End}_\mathfrak{g}(V)$ gives an algebra isomorphism*

$$\mathrm{End}_\mathfrak{g}(V) \cong \bigoplus_{\mu \in \mathcal{S}} \mathrm{End}(V^{\mathfrak{n}^+}(\mu)). \tag{4.22}$$

For every $\mu \in \mathcal{S}$ the space $V^{\mathfrak{n}^+}(\mu)$ is an irreducible module for $\mathrm{End}_\mathfrak{g}(V)$. Furthermore, distinct values of μ give inequivalent modules for $\mathrm{End}_\mathfrak{g}(V)$. Under the joint action of $\mathfrak{g}$ and $\mathrm{End}_\mathfrak{g}(V)$ the space V decomposes as

$$V \cong \bigoplus_{\mu \in \mathcal{S}} V^\mu \otimes V^{\mathfrak{n}^+}(\mu). \tag{4.23}$$

Proof. Since every finite-dimensional representation of $\mathfrak{g}$ is completely reducible by Theorem 3.3.12, we can apply Proposition 4.1.15 (viewing V as a $U(\mathfrak{g})$-module) to obtain the primary decomposition

$$V = \bigoplus_{\mu \in P_{++}(\mathfrak{g})} V_{(\mu)}, \qquad \mathrm{End}_{\mathfrak{g}}(V) \cong \bigoplus_{\mu \in P_{++}(\mathfrak{g})} \mathrm{End}_{\mathfrak{g}}(V_{(\mu)}) . \qquad (4.24)$$

Here we write $V_{(\mu)}$ for the isotypic component of V of type V^{μ}. For each $V_{(\mu)} \neq 0$ we choose irreducible submodules $V_{\mu,i} \cong V^{\mu}$ for $i = 1, \dots, d(\mu)$ such that

$$V_{(\mu)} = V_{\mu,1} \oplus \cdots \oplus V_{\mu,d(\mu)} , \qquad (4.25)$$

where $d(\mu) = \mathrm{mult}_V(\pi^{\mu})$. Let $v_{\mu,i} \in V_{\mu,i}$ be a highest-weight vector. Then (4.25) and Corollary 3.3.14 imply that

$$\mathrm{mult}_V(\pi^{\mu}) = \dim V^{\mathfrak{n}^+}(\mu) . \qquad (4.26)$$

Hence the nonzero terms in (4.24) are those with $\mu \in \mathcal{S}$.

Let $T \in \mathrm{End}_{\mathfrak{g}}(V)$ and suppose $\varphi(T) = 0$. Then $T v_{\mu,i} = 0$ for all μ and $i = 1, \dots, d(\mu)$. If $v = x_1 \cdots x_p v_{\mu,i}$ with $x_i \in \mathfrak{g}$, then

$$T v = x_1 \cdots x_p T v_{\mu,i} = 0 .$$

But $v_{\mu,i}$ is a cyclic vector for $V_{\mu,i}$ by Theorem 3.2.5, so $T V_{\mu,i} = 0$. Hence $T V_{(\mu)} = 0$ for all $\mu \in P_{++}(\mathfrak{g})$. Thus $T = 0$, which shows that φ is injective. We also have

$$\dim \mathrm{End}_{\mathfrak{g}}(V_{(\mu)}) = (\mathrm{mult}_V(\pi^{\mu}))^2 = \left(\dim V^{\mathfrak{n}^+}(\mu) \right)^2 = \dim \left(\mathrm{End}\, V^{\mathfrak{n}^+}(\mu) \right)$$

by (4.26). Since φ is injective, it follows that $\varphi(\mathrm{End}_{\mathfrak{g}}(V_{(\mu)})) = \mathrm{End}(V^{\mathfrak{n}^+}(\mu))$ for all $\mu \in P_{++}(\mathfrak{g})$. Hence by (4.24) we see that φ is also surjective. This proves (4.22). The other assertions of the theorem now follow from (4.22) and (4.25). $\square$

4.2.6 Abstract Capelli Theorem

Let G be a reductive linear algebraic group, and let (ρ, L) be a locally regular representation of G with $\dim L$ countable. Recall that ρ is a locally completely reducible representation of $\mathcal{A}[G]$, and the irreducible $\mathcal{A}[G]$ submodules of L are irreducible regular representations of G.

There is a representation $d\rho$ of the Lie algebra $\mathfrak{g} = \mathrm{Lie}(G)$ on L such that on every finite-dimensional G-submodule $V \subset L$ one has $d\rho|_V = d(\rho|_V)$. We extend $d\rho$ to a representation of the universal enveloping algebra $U(\mathfrak{g})$ on L (see Appendix C.2.1). Denote by $Z(\mathfrak{g})$ the *center* of the algebra $U(\mathfrak{g})$ (the elements T such that $TX = XT$ for all $X \in \mathfrak{g}$). Assume that $\mathcal{R} \subset \mathrm{End}(L)$ is a subalgebra that satisfies the conditions (i), (ii), (iii) in Section 4.2.1.

Theorem 4.2.13. *Suppose $\mathcal{R}^G \subset d\rho(U(\mathfrak{g}))$. Then $\mathcal{R}^G \subset d\rho(Z(\mathfrak{g}))$ and $\mathcal{R}^G$ is commutative. Furthermore, in the decomposition (4.15) the irreducible $\mathcal{R}^G$-modules E^{λ} are all one-dimensional. Hence L is multiplicity-free as a G-module.*

Proof. Let τ be the representation of G on $\mathcal{R}$ given by

$$\tau(g)r = \rho(g)r\rho(g^{-1}) \,.$$

Then $\tau(g) \in \mathrm{Aut}(\mathcal{R})$ and the representation τ is locally regular, by conditions (ii) and (iii) of Section 4.2.1. Hence there is a representation $d\tau : \mathfrak{g} \longrightarrow \mathrm{End}(\mathcal{R})$ such that on every finite-dimensional G-submodule $\mathcal{W} \subset \mathcal{R}$ one has $d\tau|_{\mathcal{W}} = d(\tau|_{\mathcal{W}})$. We claim that

$$d\tau(X)T = [d\rho(X), T] \qquad \text{for } X \in \mathfrak{g} \text{ and } T \in \mathcal{R} \,. \tag{4.27}$$

Indeed, given $v \in L$ and $T \in \mathcal{R}$, there are finite-dimensional G-submodules $V_0 \subset V_1 \subset L$ and $\mathcal{W} \subset \mathcal{R}$ such that $v \in V$, $T \in \mathcal{W}$, and $TV_0 \in V_1$. Thus the functions

$$t \mapsto \rho(\exp tX)T\rho(\exp -tX)v \quad \text{and} \quad t \mapsto \rho(\exp tX)T\rho(\exp -tX)$$

are analytic from $\mathbb{C}$ to the finite-dimensional spaces V_1 and $\mathcal{W}$, respectively. By definition of the differential of a representation,

$$\begin{aligned}
(d\tau(X)T)v &= \frac{d}{dt}\rho(\exp tX)T\rho(\exp -tX)v\Big|_{t=0} \\
&= \frac{d}{dt}\rho(\exp tX)Tv\Big|_{t=0} + T\frac{d}{dt}\rho(\exp -tX)v\Big|_{t=0} \\
&= [d\rho(X), T]v \,,
\end{aligned}$$

proving (4.27).

Now suppose $T \in \mathcal{R}^G$. Then $\tau(g)T = T$ for all $g \in G$, so $d\tau(X)T = 0$ for all $X \in \mathfrak{g}$. Hence by (4.27) we have

$$d\rho(X)T = Td\rho(X) \quad \text{for all } X \in \mathfrak{g} \,. \tag{4.28}$$

By assumption, there exists $\widetilde{T} \in U(\mathfrak{g})$ such that $T = d\rho(\widetilde{T})$. One has $\widetilde{T} \in U_k(\mathfrak{g})$ for some integer k. Set $K = U_k(\mathfrak{g}) \cap \mathrm{Ker}(d\rho)$. Then $\mathrm{Ad}(G)K = K$. Since G is reductive and the adjoint representation of G on $U_k(\mathfrak{g})$ is regular, there is an $\mathrm{Ad}(G)$-invariant subspace $M \subset U_k(\mathfrak{g})$ such that

$$U_k(\mathfrak{g}) = K \oplus M \,.$$

Write $\widetilde{T} = T_0 + T_1$, where $T_0 \in K$ and $T_1 \in M$. Then $d\rho(\widetilde{T}) = d\rho(T_1) = T$. From (4.28) we have

$$d\rho(\mathrm{ad}(X)T_1) = [d\rho(X), d\rho(T_1)] = 0 \quad \text{for all } X \in \mathfrak{g} \,.$$

Hence $\mathrm{ad}(X)T_1 \in \mathrm{Ker}(d\rho)$ for all $X \in \mathfrak{g}$. But the subspace M is invariant under $\mathrm{ad}(\mathfrak{g})$, since it is invariant under G. Thus

$$\mathrm{ad}(X)T_1 \in \mathrm{Ker}(d\rho) \cap M = \{0\} \quad \text{for all } X \in \mathfrak{g} \,.$$

This proves that $T_1 \in Z(\mathfrak{g})$. The algebra $d\rho(Z(\mathfrak{g}))$ is commutative, since $d\rho$ is a homomorphism of associative algebras. Hence the subalgebra $\mathcal{R}^G$ is commutative.

To prove that the irreducible $\mathcal{R}^G$-module E^λ has dimension one, let $\mathcal{B} = \mathcal{R}^G|_{E^\lambda}$. Then $\mathcal{B} \subset \mathrm{End}_{\mathcal{B}}(E^\lambda)$, since $\mathcal{B}$ is commutative. Hence by Schur's lemma (Lemma 4.1.4), we have $\dim \mathcal{B} = 1$, and hence $\dim E^\lambda = 1$. Since E^λ uniquely determines λ, it follows that L is multiplicity-free as a G-module. $\square$

4.2.7 Exercises

1. Let $\mathcal{A}$ be an associative algebra with 1 and let $L : \mathcal{A} \longrightarrow \mathrm{End}(\mathcal{A})$ be the *left multiplication representation* $L(a)x = ax$. Suppose $T \in \mathrm{End}(\mathcal{A})$ commutes with $L(\mathcal{A})$. Prove that there is an element $b \in \mathcal{A}$ such that $T(a) = ab$ for all $a \in \mathcal{A}$. (HINT: Consider the action of T on 1.)
2. Let G be a group. Suppose $T \in \mathrm{End}(\mathcal{A}[G])$ commutes with left translations by G. Show that there is a function $\varphi \in \mathcal{A}[G]$ such that $Tf = f * \varphi$ (convolution product) for all $f \in \mathcal{A}[G]$. (HINT: Use the previous exercise.)
3. Let G be a linear algebraic group and (ρ, V) a regular representation of G. Define a representation π of $G \times G$ on $\mathrm{End}(V)$ by $\pi(x,y)T = \rho(x)T\rho(y^{-1})$ for $T \in \mathrm{End}(V)$ and $x, y \in G$.
   (a) Show that the space E^ρ of representative functions (see Section 1.5.1) is invariant under $G \times G$ (acting by left and right translations) and that the map $B \mapsto f_B$ from $\mathrm{End}(V)$ to E^ρ intertwines the actions π and $L \widehat{\otimes} R$ of $G \times G$.
   (b) Suppose ρ is irreducible. Prove that the map $B \mapsto f_B$ from $\mathrm{End}(V)$ to $\mathcal{O}[G]$ is injective. (HINT: Use Corollary 4.1.7.)

4.3 Group Algebras of Finite Groups

In this section apply the general results of the chapter to the case of the group algebra of a finite group, and we obtain the representation-theoretic version of Fourier analysis.

4.3.1 Structure of Group Algebras

Let G be a finite group. Thus $\mathcal{A}[G]$ consists of all complex-valued functions on G. We denote by L and R the left and right translation representations of G on $\mathcal{A}[G]$:

$$L(g)f(x) = f(g^{-1}x), \quad R(g)f(x) = f(xg) .$$

By Corollary 3.3.6 we know that G is a reductive group. Each irreducible representation is finite-dimensional and has a G-invariant positive definite Hermitian inner product (obtained by averaging any inner product over G). Thus we may take each model (π^λ, F^λ) to be unitary for $\lambda \in \widehat{G}$. The space F^{λ^*} can be taken as F^λ with

$$\pi^{\lambda^*}(g) = \overline{\pi^\lambda(g)} \,. \tag{4.29}$$

Here the bar denotes the complex conjugate of the matrix of $\pi^\lambda(g)$ relative to any orthonormal basis for F^λ. Equation (4.29) holds because the transpose inverse of a unitary matrix is the complex conjugate of the matrix.

From Theorem 4.2.7 and Remark 4.2.9 the vector space $\mathcal{A}[G]$ decomposes under $G \times G$ as

$$\mathcal{A}[G] \cong \bigoplus_{\lambda \in \widehat{G}} \operatorname{End}(F^\lambda) \,, \tag{4.30}$$

with $(g, h) \in G \times G$ acting on $T \in \operatorname{End}(F^\lambda)$ by $T \mapsto \pi^\lambda(g) T \pi^\lambda(h)^{-1}$. In particular, since $\dim \mathcal{A}[G] = |G|$ and $\dim \operatorname{End}(F^\lambda) = (d_\lambda)^2$, the isomorphism (4.30) implies that

$$|G| = \sum_{\lambda \in \widehat{G}} (d_\lambda)^2 \,. \tag{4.31}$$

We recall that $\mathcal{A}[G]$ is an associative algebra relative to the convolution product, with identity element δ_1. It has a conjugate-linear antiautomorphism $f \mapsto f^*$ given by

$$f^*(g) = \overline{f(g^{-1})}$$

(the conjugate-linear extension of the inversion map on G to $\mathcal{A}[G]$). If we view the right side of (4.30) as block diagonal matrices (one block for each $\lambda \in \widehat{G}$ and an element of $\operatorname{End} F^\lambda$ in the block indexed by λ), then these matrices also form an associative algebra under matrix multiplication. For $T \in \operatorname{End} F^\lambda$ let T^* denote the adjoint operator relative to the G-invariant inner product on F^λ:

$$(Tu, v) = (u, T^*v) \quad \text{for } u, v \in F^\lambda \,.$$

We define a conjugate-linear antiautomorphism of the algebra $\bigoplus_{\lambda \in \widehat{G}} \operatorname{End}(F^\lambda)$ by using the map $T \mapsto T^*$ on each summand.

We will now define an explicit isomorphism between these two algebras. Given $f \in \mathcal{A}[G]$ and $\lambda \in \widehat{G}$, we define an operator $\mathcal{F}f(\lambda)$ on F^λ by

$$\mathcal{F}f(\lambda) = \sum_{x \in G} f(x) \pi^\lambda(x) \,.$$

In particular, when f is the function δ_g with $g \in G$, then $\mathcal{F}\delta_g(\lambda) = \pi^\lambda(g)$. Hence the map $f \mapsto \mathcal{F}f(\lambda)$ is the canonical extension of the representation π^λ of G to a representation of $\mathcal{A}[G]$. We define the *Fourier transform* $\mathcal{F}f$ of f to be the element of the algebra $\bigoplus_{\lambda \in \widehat{G}} \operatorname{End}(F^\lambda)$ with λ component $\mathcal{F}f(\lambda)$.

Theorem 4.3.1. *The Fourier transform*

$$\mathcal{F} : \mathcal{A}[G] \longrightarrow \bigoplus_{\lambda \in \widehat{G}} \mathrm{End}(F^\lambda)$$

is an algebra isomorphism that preserves the $*$ operation on each algebra. Furthermore, for $f \in \mathcal{A}[G]$ and $g \in G$ one has

$$\mathcal{F}(L(g)f)(\lambda) = \pi^\lambda(g)\mathcal{F}f(\lambda), \qquad \mathcal{F}(R(g)f)(\lambda) = \mathcal{F}f(\lambda)\pi^\lambda(g^{-1}). \qquad (4.32)$$

Proof. Since $\mathcal{F}(\delta_{g_1 g_2}) = \pi^\lambda(g_1 g_2) = \pi^\lambda(g_1)\pi^\lambda(g_2) = \mathcal{F}(\delta_{g_1})\mathcal{F}(\delta_{g_2})$, the map $\mathcal{F}$ transforms convolution of functions on G into multiplication of operators on each space F^λ:

$$\mathcal{F}(f_1 * f_2)(\lambda) = \mathcal{F}f_1(\lambda)\mathcal{F}f_2(\lambda)$$

for $f_1, f_2 \in \mathcal{A}[G]$ and $\lambda \in \widehat{G}$. Also, $\mathcal{F}\delta_1(\lambda) = I_{F^\lambda}$. This shows that $\mathcal{F}$ is an algebra homomorphism. Hence equations (4.32) follow from $L(g)f = \delta_g * f$ and $R(g)f = f * \delta_{g^{-1}}$. The $*$ operation is preserved by $\mathcal{F}$, since $(\delta_g)^* = \delta_{g^{-1}}$ and $(\pi^\lambda(g))^* = \pi^\lambda(g^{-1})$. From Corollary 4.2.4 (1) we see that $\mathcal{F}$ is surjective. Then (4.31) shows that it is bijective. $\qquad \square$

4.3.2 Schur Orthogonality Relations

We begin with a variant of Schur's lemma. Let G be a group and let U and V be finite-dimensional G-modules.

Lemma 4.3.2. *Suppose C is a G-invariant bilinear form on $U \times V$. Then $C = 0$ if U is not equivalent to V^* as a G-module. If $U = V^*$ there is a constant κ such that $C(u,v) = \kappa\langle u,v \rangle$, where $\langle u,v \rangle$ denotes the canonical bilinear pairing of V^* and V.*

Proof. We can write C as $C(u,v) = \langle Tu, v \rangle$, where $T \in \mathrm{Hom}(U, V^*)$. Since the form C and the canonical bilinear pairing of V^* and V are both G invariant, we have

$$\langle g^{-1}Tgu, v \rangle = \langle Tgu, gv \rangle = \langle Tu, v \rangle$$

for all $u \in U$, $v \in V$, and $g \in G$. Hence $g^{-1}Tg = T$, and so $T \in \mathrm{Hom}_G(U, V^*)$. The conclusion now follows from Lemma 4.1.4. $\qquad \square$

Let $\lambda \in \widehat{G}$. For $A \in \mathrm{End}(F^\lambda)$ we define the *representative function* f_A^λ on G by $f_A^\lambda(g) = \mathrm{tr}(\pi^\lambda(g)A)$ for $g \in G$, as in Section 1.5.1.

Lemma 4.3.3 (Schur Orthogonality Relations). *Suppose G is a finite group and $\lambda, \mu \in \widehat{G}$. Let $A \in \mathrm{End}(F^\lambda)$ and $B \in \mathrm{End}(F^\mu)$. Then*

$$\frac{1}{|G|}\sum_{g \in G} f_A^\lambda(g) f_B^\mu(g) = \begin{cases} (1/d_\lambda)\,\mathrm{tr}(AB^t) & \text{if } \mu = \lambda^*, \\ 0 & \text{otherwise.} \end{cases} \qquad (4.33)$$

Proof. Define a bilinear form C on $\text{End}(F^\lambda) \times \text{End}(F^\mu)$ by

$$C(A,B) = \frac{1}{|G|} \sum_{g \in G} f_A^\lambda(g) f_B^\mu(g) \,. \tag{4.34}$$

We have $f_A^\lambda(xgy) = f_{\pi^\lambda(y)A\pi^\lambda(x)}^\lambda(g)$ for $x, y \in G$, with an analogous transformation law for f_B^μ. Replacing g by xgy in (4.34), we see that

$$C(\pi^\lambda(y)A\pi^\lambda(x),\ \pi^\mu(y)B\pi^\mu(x)) = C(A,B) \,. \tag{4.35}$$

Thus C is invariant under $G \times G$. Since $\text{End}(F^\lambda)$ is an irreducible module for $G \times G$ (isomorphic to the outer tensor product module $F^\lambda \widehat{\otimes} F^{\lambda^*}$), Lemma 4.3.2 implies that $C = 0$ if $\mu \neq \lambda^*$.

Suppose now $\mu = \lambda^*$ and write $\pi = \pi^\lambda$, $V = F^\lambda$, $\pi^* = \pi^{\lambda^*}$, and $V^* = F^{\lambda^*}$. The bilinear form $\langle A, B \rangle = \text{tr}_V(AB^t)$, for $A \in \text{End}(V)$ and $B \in \text{End}(V^*)$, is G-invariant and nondegenerate, so by Lemma 4.3.2 there is a constant κ such that $C(A,B) = \kappa \, \text{tr}_V(AB^t)$. To determine κ, we recall that for $A \in \text{End}(V)$ and $B \in \text{End}(V^*)$ we have $\text{tr}_V(A) \, \text{tr}_{V^*}(B) = \text{tr}_{V \otimes V^*}(A \otimes B)$. Thus

$$f_A^\lambda(g) f_B^\mu(g) = \text{tr}_{V \otimes V^*}(\pi(g)A \otimes \pi^*(g)B) \,.$$

Now take $A = I_V$ and $B = I_{V^*}$. Then $\text{tr}_V(AB^t) = d_\lambda$, and hence $\kappa d_\lambda = \text{tr}_{V \otimes V^*}(P)$, where

$$P = \frac{1}{|G|} \sum_{g \in G} \pi(g) \otimes \pi^*(g)$$

is the projection onto the G-invariant subspace of $V \otimes V^*$. But by Lemma 4.3.2 this subspace has dimension one. Hence $\text{tr}_{V \otimes V^*}(P) = 1$, proving that $\kappa = 1/d_\lambda$. $\qquad \square$

4.3.3 Fourier Inversion Formula

With the aid of the Schur orthogonality relations we can now find an explicit inverse to the Fourier transform $\mathcal{F}$ on $\mathcal{A}[G]$.

Theorem 4.3.4 (Fourier Inversion Formula). *Suppose G is a finite group. Let $F = \{F(\lambda)\}_{\lambda \in \widehat{G}}$ be in $\mathcal{F}\mathcal{A}[G]$. Define a function $f \in \mathcal{A}[G]$ by*

$$f(g) = \frac{1}{|G|} \sum_{\lambda \in \widehat{G}} d_\lambda \, \text{tr}\left(\pi^\lambda(g) F(\lambda^*)^t\right) \,. \tag{4.36}$$

Then $\mathcal{F}f(\lambda) = F(\lambda)$ for all $\lambda \in \widehat{G}$.

Proof. The operator $\mathcal{F}f(\lambda)$ is uniquely determined by $\text{tr}(\mathcal{F}f(\lambda)A)$, with A varying over $\text{End}(V^\lambda)$. Replacing each representation by its dual, we write the formula for f as

$$f(g) = \frac{1}{|G|} \sum_{\mu \in \widehat{G}} d_\mu \operatorname{tr}\left(\pi^{\mu^*}(g) F(\mu)^t\right).$$

Then we calculate

$$\operatorname{tr}(\mathcal{F}f(\lambda)A) = \frac{1}{|G|} \sum_{g \in G} \sum_{\mu \in \widehat{G}} d_\mu \operatorname{tr}\left(\pi^\lambda(g)A\right) \operatorname{tr}\left(\pi^{\mu^*}(g) F(\mu)^t\right)$$

$$= \sum_{\mu \in \widehat{G}} d_\mu \left\{ \frac{1}{|G|} \sum_{g \in G} f_A(g) f_{F(\mu)^t}(g) \right\}.$$

Applying the Schur orthogonality relations (4.33), we find that the terms with $\mu \neq \lambda$ vanish and $\operatorname{tr}(\mathcal{F}f(\lambda)A) = \operatorname{tr}(AF(\lambda))$. Since this holds for all A, we conclude that $\mathcal{F}f(\lambda) = F(\lambda)$. $\square$

Corollary 4.3.5 (Plancherel Formula). *Let* $\varphi, \psi \in \mathcal{A}[G]$. *Then*

$$\sum_{g \in G} \varphi(g)\overline{\psi(g)} = \frac{1}{|G|} \sum_{\lambda \in \widehat{G}} d_\lambda \operatorname{tr}\left(\mathcal{F}\varphi(\lambda)\mathcal{F}\psi(\lambda)^*\right). \qquad (4.37)$$

Proof. Let $f = \varphi * (\psi)^*$. Then

$$f(1) = \sum_{g \in G} \varphi(g)\overline{\psi(g)}.$$

We can also express $f(1)$ by the Fourier inversion formula evaluated at $g = 1$:

$$f(1) = \frac{1}{|G|} \sum_{\lambda \in \widehat{G}} d_\lambda \operatorname{tr}(\mathcal{F}f(\lambda^*)^t) = \frac{1}{|G|} \sum_{\lambda \in \widehat{G}} d_\lambda \operatorname{tr}(\mathcal{F}f(\lambda)).$$

Since $\mathcal{F}f(\lambda) = \mathcal{F}\varphi(\lambda)\mathcal{F}\psi(\lambda)^*$, we obtain (4.37). $\square$

Remark 4.3.6. If we use the *normalized* Fourier transform $\Phi(\lambda) = (1/|G|)\mathcal{F}\phi(\lambda)$, then (4.37) becomes

$$\frac{1}{|G|} \sum_{g \in G} \varphi(g)\overline{\psi(g)} = \sum_{\lambda \in \widehat{G}} d_\lambda \operatorname{tr}\left(\Phi(\lambda)\Psi(\lambda)^*\right). \qquad (4.38)$$

The left side of (4.38) is a positive definite Hermitian inner product on $\mathcal{A}[G]$ that is invariant under the operators $L(g)$ and $R(g)$ for $g \in G$, normalized so that the constant function $\varphi(g) = 1$ has norm 1. The Plancherel formula expresses this inner product in terms of the inner products on $\operatorname{End}(E^\lambda)$ given by $\operatorname{tr}(ST^*)$; these inner products are invariant under left and right multiplication by the unitary operators $\pi^\lambda(g)$ for $g \in G$. In this form the Plancherel formula applies to every compact topological group, with $\mathcal{A}(G)$ replaced by $L^2(G)$ and summation over G replaced by integration relative to the normalized invariant measure.

4.3.4 The Algebra of Central Functions

We continue our investigation of the group algebra of a finite group G. Let $\mathcal{A}[G]^G$ be the *center* of $\mathcal{A}[G]$: by definition, $f \in \mathcal{A}[G]^G$ if and only if

$$f * \varphi = \varphi * f \quad \text{for all } \varphi \in \mathcal{A}[G] . \tag{4.39}$$

We call such a function f a *central function* on G. The space of central functions on G is a commutative algebra (under convolution multiplication). In this section we shall write down two different bases for the space $\mathcal{A}[G]^G$ and use the Fourier transform on G to study the relation between them.

We first observe that in (4.39) it suffices to take $\varphi = \delta_x$, with x ranging over G, since these functions span $\mathcal{A}[G]$. Thus f is central if and only if $f(yx) = f(xy)$ for all $x, y \in G$. Replacing y by yx^{-1}, we can express this condition as

$$f(xyx^{-1}) = f(y) \quad \text{for } x, y \in G .$$

Thus we can also describe $\mathcal{A}[G]^G$ as the space of functions f on G that are constant on the conjugacy classes of G. From this observation we obtain the following basis for $\mathcal{A}[G]^G$:

Proposition 4.3.7. *Let* $\mathrm{Conj}(G)$ *be the set of conjugacy classes in* G. *For each* $C \in \mathrm{Conj}(G)$ *let* φ_C *be the characteristic function of* C. *Then the set* $\{\varphi_C\}_{C \in \mathrm{Conj}(G)}$ *is a basis for* $\mathcal{A}[G]^G$, *and every function* $f \in \mathcal{A}[G]^G$ *has the expansion*

$$f = \sum_{C \in \mathrm{Conj}(G)} f(C) \varphi_C$$

In particular,

$$\dim \mathcal{A}[G]^G = |\mathrm{Conj}(G)| . \tag{4.40}$$

We denote the character of a finite-dimensional representation ρ by χ_ρ, viewed as a function on G: $\chi_\rho(g) = \mathrm{tr}(\rho(g))$. Characters are central functions because

$$\mathrm{tr}(\rho(xy)) = \mathrm{tr}(\rho(x)\rho(y)) = \mathrm{tr}(\rho(y)\rho(x)) .$$

We note that

$$\chi_\rho(g^{-1}) = \overline{\chi_\rho(g)} , \tag{4.41}$$

where the bar denotes complex conjugate. Indeed, since $\rho(g)$ can be taken as a unitary matrix relative to a suitable basis, the eigenvalues of $\rho(g)$ have absolute value 1. Hence the eigenvalues of $\rho(g^{-1})$ are the complex conjugates of those of $\rho(g)$, and the trace is the sum of these eigenvalues. We write χ_λ for the character of the irreducible representation π^λ.

We have another representation of $\mathcal{A}[G]^G$ obtained from the Fourier transform. We know that the map $\mathcal{F}$ is an algebra isomorphism from $\mathcal{A}[G]$ (with convolution multiplication) to

$$\mathcal{F}\mathcal{A}[G] = \bigoplus_{\lambda \in \widehat{G}} \mathrm{End}(F^\lambda)$$

by Theorem 4.3.1. Since the center of each ideal $\mathrm{End}(F^\lambda)$ in $\mathcal{F}\mathcal{A}[G]$ consists of scalar multiples of the identity operator, we conclude that f is a central function on G if and only if

$$\mathcal{F}f(\lambda) = c_\lambda I_{F^\lambda} \quad \text{for all } \lambda \in \widehat{G}, \tag{4.42}$$

where $c_\lambda \in \mathbb{C}$. For each $\lambda \in \widehat{G}$ define $E_\lambda \in \mathcal{F}\mathcal{A}[G]$ to be the identity operator on F^λ and zero on F^μ for $\mu \neq \lambda$. The set of operator-valued functions $\{E_\lambda\}_{\lambda \in \widehat{G}}$ is obviously linearly independent, and from (4.42) we see that it is a basis for $\mathcal{F}\mathcal{A}[G]^G$.

Proposition 4.3.8. *The Fourier transform of $f \in \mathcal{A}[G]^G$ has the expansion*

$$\mathcal{F}f = \sum_{\lambda \in \widehat{G}} \mathcal{F}f(\lambda)E_\lambda . \tag{4.43}$$

In particular, $\dim \mathcal{A}[G]^G = |\widehat{G}|$, *and hence*

$$|\widehat{G}| = |\mathrm{Conj}(G)| . \tag{4.44}$$

Example

Suppose $G = \mathfrak{S}_n$, the symmetric group on n letters. Every $g \in G$ can be written uniquely as a product of disjoint cyclic permutations. For example, $(123)(45)$ is the permutation $1 \to 2, 2 \to 3, 3 \to 1, 4 \to 5, 5 \to 4$ in $\mathfrak{S}_5$. Furthermore, g is conjugate to g' if and only if the number of cycles of length $1, 2, \ldots, n$ is the same for g and g'. Thus each conjugacy class C in G corresponds to a *partition* of the integer n as the sum of positive integers:
$$n = k_1 + k_2 + \cdots + k_d ,$$
with $k_1 \geq k_2 \geq \cdots \geq k_d > 0$. The class C consists of all elements with cycle lengths $k_1, k_2, \ldots, k_d$. From (4.44) it follows that $\mathfrak{S}_n$ has $p(n)$ inequivalent irreducible representations, where $p(n)$ is the number of partitions of n. $\qquad\square$

We return to a general finite group G. Under the inverse Fourier transform, the operator E_λ corresponds to convolution by a central function e_λ on G. To determine e_λ, we apply the Fourier inversion formula (4.36):

$$e_\lambda(g) = \mathcal{F}^{-1}E_\lambda(g) = \frac{d_\lambda}{|G|}\chi_\lambda(g^{-1}) . \tag{4.45}$$

Since $\mathcal{F}^{-1}$ is an algebra isomorphism, the family of functions $\{e_\lambda : \lambda \in \widehat{G}\}$ gives a resolution of the identity for the algebra $\mathcal{A}[G]$:

$$e_\lambda * e_\mu = \begin{cases} e_\lambda & \text{for } \lambda = \mu , \\ 0 & \text{otherwise}, \end{cases} \quad \text{and} \quad \sum_{\lambda \in \widehat{G}} e_\lambda = \delta_1 . \tag{4.46}$$

Since $E_\lambda = \mathcal{F}e_\lambda$ and $\chi_\lambda(g^{-1}) = \chi_{\lambda^*}(g)$, we find from (4.45) that

$$\mathcal{F}\chi_{\lambda^*}(\mu) = \begin{cases} (|G|/d_\lambda)I_{F^\lambda} & \text{if } \mu = \lambda, \\ 0 & \text{otherwise}. \end{cases} \tag{4.47}$$

Thus the irreducible characters have Fourier transforms that vanish except on a single irreducible representation. Furthermore, from Proposition 4.3.8 we see that the irreducible characters give a basis for $\mathcal{A}[G]^G$. The explicit form of the expansion of a central function in terms of irreducible characters is as follows:

Theorem 4.3.9. *Let* $\varphi, \psi \in \mathcal{A}[G]^G$ *and* $g \in G$. *Then*

$$\varphi(g) = \sum_{\lambda \in \widehat{G}} \widehat{\varphi}(\lambda)\chi_\lambda(g), \text{ where } \widehat{\varphi}(\lambda) = \frac{1}{|G|}\sum_{g \in G} \varphi(g)\overline{\chi_\lambda(g)}, \text{ and} \tag{4.48}$$

$$\frac{1}{|G|}\sum_{g \in G} \varphi(g)\overline{\psi(g)} = \sum_{\lambda \in \widehat{G}} \widehat{\varphi}(\lambda)\overline{\widehat{\psi}(\lambda)}. \tag{4.49}$$

Proof. Define a positive definite inner product on $\mathcal{A}[G]$ by

$$\langle \varphi \mid \psi \rangle = \frac{1}{|G|}\sum_{g \in G} \varphi(g)\overline{\psi(g)}.$$

Let $\lambda, \mu \in \widehat{G}$. Then $\chi_\lambda(g) = f_A^\lambda(g)$ and $\overline{\chi_\mu(g)} = f_B^{\mu^*}(g)$, where A is the identity operator on F^λ and B is the identity operator on F^{μ^*}. Hence the Schur orthogonality relations imply that

$$\langle \chi_\mu \mid \chi_\lambda \rangle = \begin{cases} 1 & \text{if } \mu = \lambda, \\ 0 & \text{otherwise}. \end{cases}$$

Thus $\{\chi_\lambda\}_{\lambda \in \widehat{G}}$ is an orthonormal basis for $\mathcal{A}[G]^G$, relative to the inner product $\langle \cdot \mid \cdot \rangle$. This implies formulas (4.48) and (4.49). $\qquad\square$

Corollary 4.3.10 (Dual Orthogonality Relations). *Suppose* C_1 *and* C_2 *are conjugacy classes in* G. *Then*

$$\sum_{\lambda \in \widehat{G}} \chi_\lambda(C_1)\overline{\chi_\lambda(C_2)} = \begin{cases} |G|/|C_1| & \text{if } C_1 = C_2, \\ 0 & \text{otherwise}. \end{cases} \tag{4.50}$$

Proof. Let $C \subset G$ be a conjugacy class. Then

$$|G|\widehat{\varphi_C}(\lambda) = |C|\chi_{\lambda^*}(C). \tag{4.51}$$

Taking $C = C_1$ and $C = C_2$ in (4.51) and then using (4.49), we obtain (4.50). $\qquad\square$

Corollary 4.3.11. *Suppose* (ρ, V) *is any finite-dimensional representation of* G. *For* $\lambda \in \widehat{G}$ *let* $m_\rho(\lambda)$ *be the multiplicity of* λ *in* ρ. *Then* $m_\rho(\lambda) = \langle \chi_\rho \mid \chi_\lambda \rangle$ *and*

$$\langle \chi_\rho \mid \chi_\rho \rangle = \sum_{\lambda \in \widehat{G}} m_\rho(\lambda)^2.$$

In particular, $\langle \chi_\rho \mid \chi_\lambda \rangle$ is a positive integer, and ρ is irreducible if and only if $\langle \chi_\rho \mid \chi_\rho \rangle = 1$. The operator

$$P_\lambda = \frac{d_\lambda}{|G|} \sum_{g \in G} \overline{\chi_\lambda(g)} \rho(g) \tag{4.52}$$

is the projection onto the λ-isotypic component of V.

Proof. We have

$$\chi_\rho = \sum_{\lambda \in \widehat{G}} m_\rho(\lambda) \chi_\lambda \,,$$

so the result on multiplicities follows from (4.48) and (4.49).

By Corollary 4.2.4 (2) there exists $f \in \mathcal{A}[G]^G$ such that $\rho(f) = P_\lambda$. To show that

$$f(g) = \frac{d_\lambda}{|G|} \overline{\chi_\lambda(g)} \quad \text{for } g \in G \,,$$

it suffices (by complete reducibility of ρ) to show that $P_\lambda = \rho(f)$ when $\rho = \pi^\mu$ for some $\mu \in \widehat{G}$. In this case $\rho(f) = \delta_{\lambda\mu} I_{F^\mu}$ by (4.47), and the same formula holds for P_λ by definition. $\qquad\square$

Finding an explicit formula for χ_λ or for $\mathcal{F}\varphi_C$ is a difficult problem whenever G is a noncommutative finite group. We shall solve this problem for the symmetric group in Chapter 9 by relating the representations of the symmetric group to representations of the general linear group.

Remark 4.3.12. The sets of functions $\{\chi_\lambda : \lambda \in \widehat{G}\}$ and $\{\varphi_C : C \in \text{Conj}(G)\}$ on G have the same cardinality. However, there is no other simple relationship between them. This is a representation-theoretic version of the *uncertainty principle*: The function φ_C is supported on a single conjugacy class. If $C \neq \{1\}$ this forces $\mathcal{F}\varphi_C$ to be nonzero on at least two irreducible representations. (Let $\varphi = \varphi_C$ and $\psi = \delta_1$; then the left side of (4.49) is zero, while the right side is $\sum_\lambda d_\lambda \widehat{\varphi_C}(\lambda)$.) In the other direction, $\mathcal{F}\chi_\lambda$ is supported on the single irreducible representation λ. If λ is not the trivial representation, this forces χ_λ to be nonzero on at least two nontrivial conjugacy classes. (Since the trivial representation has character 1, the orthogonality of characters yields $\sum_{C \in \text{Conj}(G)} |C| \chi_\lambda(C) = 0$.)

4.3.5 Exercises

1. Let $n > 1$ be an integer, and let $\mathbb{Z}_n = \mathbb{Z}/n\mathbb{Z}$ be the additive group of integers mod n.
   (a) Let $e(k) = e^{2\pi i k/n}$ for $k \in \mathbb{Z}_n$. Show that the characters of $\mathbb{Z}_n$ are the functions $\chi_q(k) = e(kq)$ for $q = 0, 1, \ldots, n-1$.
   (b) For $f \in \mathcal{A}[\mathbb{Z}_n]$, define $\hat{f} \in \mathcal{A}[\mathbb{Z}_n]$ by $\hat{f}(q) = (1/n) \sum_{k=0}^{n-1} f(k) e(-kq)$. Show that $f(k) = \sum_{q=0}^{n-1} \hat{f}(q) e(kq)$, and that

$$\frac{1}{n}\sum_{k=0}^{n-1}|f(k)|^2 = \sum_{q=0}^{n-1}|\hat{f}(q)|^2 \, .$$

2. Let $\mathbb{F}$ be a finite field of characteristic p (so $\mathbb{F}$ has $q = p^n$ elements for some integer n). This exercise and several that follow apply Fourier analysis to the additive group of $\mathbb{F}$. This requires no detailed knowledge of the structure of $\mathbb{F}$ when $\bar{p}$ does not divide n; for the case that p divides n you will need to know more about finite fields to verify part (a) of the exercise. Let $\mathbb{S}^1$ be the multiplicative group of complex numbers of absolute value 1. Let $\chi : \mathbb{F} \to \mathbb{S}^1$ be such that $\chi(x+y) = \chi(x)\chi(y)$ and $\chi(0) = 1$ (i.e., χ is an *additive character* of $\mathbb{F}$). The smallest subfield of $\mathbb{F}$, call it $\mathbb{K}$, is isomorphic to $\mathbb{Z}/p\mathbb{Z}$. Define $e(k) = e^{2\pi i k/p}$ for $k \in \mathbb{Z}/p\mathbb{Z}$. This defines an additive character of $\mathbb{Z}/p\mathbb{Z}$ and hence a character of $\mathbb{K}$. $\mathbb{F}$ is a finite-dimensional vector space over $\mathbb{K}$. If $a \in \mathbb{F}$ define a linear transformation $L_a : \mathbb{F} \to \mathbb{F}$ by $L_a x = ax$. Set $\chi_1(a) = e(\operatorname{tr}(L_a))$.
(a) We say that an additive character χ is *nontrivial* if $\chi(x) \neq 1$ for some $x \in \mathbb{F}$. Let u be a nonzero element of $\mathbb{F}$ and define $\eta(x) = \chi_1(ux)$. Show that η is a nontrivial additive character of $\mathbb{F}$.
(b) Show that if η is an additive character of $\mathbb{F}$ then there exists a unique $u \in \mathbb{F}$ such that $\eta(x) = \chi_1(ux)$ for all $x \in \mathbb{F}$.
(c) Show that if χ is any nontrivial additive character of $\mathbb{F}$ and if η is an additive character of $\mathbb{F}$ then there exists a unique $u \in \mathbb{F}$ such that $\eta(x) = \chi(ux)$ for all $x \in \mathbb{F}$.
3. Let $\mathbb{F}$ be a finite field. Fix a nontrivial additive character χ of $\mathbb{F}$. Show that the Fourier transform on $\mathcal{A}[\mathbb{F}]$ (relative to the additive group structure of $\mathbb{F}$) can be expressed as follows: For $f \in \mathcal{A}[\mathbb{F}]$, define $\hat{f} \in \mathcal{A}[\mathbb{F}]$ by

$$\hat{f}(\xi) = \frac{1}{|\mathbb{F}|}\sum_{x \in \mathbb{F}} f(y)\chi(-x\xi)$$

for $\xi \in \mathbb{F}$. Then the Fourier inversion formula becomes

$$f(x) = \sum_{\xi \in \mathbb{F}} \hat{f}(\xi)\chi(x\xi) \, ,$$

and one has $(1/|\mathbb{F}|)\sum_{x \in \mathbb{F}}|f(x)|^2 = \sum_{\xi \in \mathbb{F}}|\hat{f}(\xi)|^2 \, .$

4.4 Representations of Finite Groups

Constructing irreducible representations of finite nonabelian groups is a difficult problem. As a preliminary step, we construct in this section a more accessible class of representations, the *induced representations*, and calculate their characters.

4.4.1 Induced Representations

Let G be a finite group and H a subgroup of G. Given any finite-dimensional representation (ρ, F) of G, we obtain a representation $(\text{Res}_H^G(\rho), F)$ of H by restriction. Conversely, a representation (μ, E) of H *induces* a representation of G by the following construction: We take the space $\mathbb{C}[G; E]$ of all functions from G to E, and let G act by left translations: $L(g)f(x) = f(g^{-1}x)$. Let $\mathfrak{I}_\mu \subset \mathbb{C}[G; E]$ be the subspace of functions such that

$$f(gh) = \mu(h)^{-1} f(g), \quad \text{for all } h \in H \text{ and } g \in G.$$

The space $\mathfrak{I}_\mu$ is invariant under left translation by G. We define the *induced representation* $\text{Ind}_H^G(\mu)$ of G to be the left translation action of G on the space $\mathfrak{I}_\mu$.

Theorem 4.4.1 (Frobenius Reciprocity). *There is a vector space isomorphism*

$$\text{Hom}_G\left(\rho, \text{Ind}_H^G(\mu)\right) \cong \text{Hom}_H\left(\text{Res}_H^G(\rho), \mu\right). \tag{4.53}$$

Proof. Let $T \in \text{Hom}_G\left(\rho, \text{Ind}_H^G(\mu)\right)$. Then $T : F \longrightarrow \mathfrak{I}_\mu$, so we obtain a map $\hat{T} : F \longrightarrow E$ by evaluation at the identity element: $\hat{T}v = (Tv)(1)$ for $v \in F$. The map $\hat{T}$ then intertwines the action of H on these two spaces, since

$$\hat{T}(\rho(h)v) = (T\rho(h)v)(1) = (L(h)Tv)(1) = (Tv)(h^{-1})$$
$$= \mu(h)Tv(1) = \mu(h)\hat{T}v$$

for $h \in H$ and $v \in E$. Thus $\hat{T} \in \text{Hom}_H(\text{Res}_H^G(\rho), \mu)$.

Conversely, given $S \in \text{Hom}_H(\text{Res}_H^G(\rho), \mu)$, define $\check{S} : F \longrightarrow \mathfrak{I}_\mu$ by

$$(\check{S}v)(g) = S(\rho(g)^{-1}v) \quad \text{for } v \in F \text{ and } g \in G.$$

We check that the function $g \mapsto (\check{S}v)(g)$ has the appropriate transformation property under right translation by $h \in H$:

$$(\check{S}v)(gh) = S(\rho(h)^{-1}\rho(g)^{-1}v) = \mu(h)^{-1}S(\rho(g)^{-1}v) = \mu(h)^{-1}(\check{S}v)(g).$$

Thus $\check{S}v \in \mathfrak{I}_\mu$. Furthermore, for $g, g' \in G$,

$$(L(g)\check{S}v)(g') = (\check{S}v)(g^{-1}g') = S(\rho(g')^{-1}\rho(g)v) = (\check{S}\rho(g)v)(g'),$$

which shows that $\check{S} \in \text{Hom}_G(\rho, \text{Ind}_H^G(\mu))$.

Finally, we verify that the maps $T \mapsto \hat{T}$ and $S \mapsto \check{S}$ are inverses and thus give the isomorphism (4.53). Let $v \in E$. Since $(\check{S}v)(1) = Sv$, we have $(\check{S})^\wedge = S$. In the other direction, $(\hat{T})^\vee v$ is the E-valued function

$$g \mapsto \hat{T}(\rho(g^{-1})v) = (T\rho(g^{-1})v)(1) = (L(g^{-1})Tv)(1) = (Tv)(g).$$

Thus $(\hat{T})^\vee = T$. $\qquad\qquad\qquad\qquad\qquad\qquad\qquad\qquad\qquad\qquad\qquad\qquad\qquad\square$

Relation (4.53) is called *Frobenius reciprocity*. It ranks with Schur's lemma as one of the most useful general results in representation theory. Frobenius reciprocity can be expressed as an equality of multiplicities for induced and restricted representations as follows:

Assume that (ρ, F) is an irreducible representation of G and (μ, E) is an irreducible representation of H. Then $\dim \mathrm{Hom}_H(E, F)$ is the multiplicity of μ in $\mathrm{Res}_H^G(\rho)$. Likewise, $\dim \mathrm{Hom}_G(F, \mathcal{I}_\mu)$ is the multiplicity of ρ in $\mathrm{Ind}_H^G(\mu)$. But we have $\dim \mathrm{Hom}_H(F, E) = \dim \mathrm{Hom}_H(E, F)$ by (4.9), so the equality (4.53) can be expressed as

$$\text{Multiplicity of } \rho \text{ in } \mathrm{Ind}_H^G(\mu) = \text{Multiplicity of } \mu \text{ in } \mathrm{Res}_H^G(\rho) \,. \qquad (4.54)$$

4.4.2 Characters of Induced Representations

Let G be a finite group and let (μ, E) be a finite-dimensional representation of a subgroup H of G. We now obtain a formula for the character of the induced representation $\mathrm{Ind}_H^G(\mu)$ in terms of the character of μ and the action of G on G/H. For $g \in G$ let

$$(G/H)^g = \{x \in G/H : g \cdot x = x\}$$

be the set of the fixed points of g on G/H. Let $z \mapsto \bar{z} = zH$ be the canonical quotient map from G to G/H. We observe that $\bar{z} \in (G/H)^g$ if and only if $z^{-1}gz \in H$. Thus if $\bar{z}$ is a fixed point for g, then the character value $\chi(\bar{z}) = \chi_\mu(z^{-1}gz)$ is defined and depends only on the coset $\bar{z}$, since χ_μ is constant on conjugacy classes.

Theorem 4.4.2. *Let* $\pi = \mathrm{Ind}_H^G(\mu)$. *Then the character of* π *is*

$$\chi_\pi(g) = \sum_{\bar{z} \in (G/H)^g} \chi_\mu(\bar{z}) \,. \qquad (4.55)$$

Proof. The space $\mathcal{I}_\mu$ on which π acts is complicated, although the action of $\pi(g)$ by left translations is straightforward. To calculate the trace of $\pi(g)$ we first construct a linear isomorphism between $\mathcal{I}_\mu$ and the space $\mathbb{C}[G/H; E]$ of all functions on G/H with values in E. We then take a basis for $\mathbb{C}[G/H; E]$ on which the trace of the G action is easily calculated. This will yield the character formula.

We begin by fixing a section $\sigma : G/H \longrightarrow G$. This means that $g \in \sigma(\bar{g})H$ for all $g \in G$. Thus $\sigma(\bar{g})^{-1}g \in H$ for all $g \in G$. Now given $f \in \mathcal{I}_\mu$, we define $\tilde{f}(g) = \mu(\sigma(\bar{g})^{-1}g)f(g)$. Then

$$\tilde{f}(gh) = \mu(\sigma(\bar{g})^{-1}g)\mu(h)f(gh) = \tilde{f}(g)$$

because $f(gh) = \mu(h)^{-1}f(g)$. This shows that $\tilde{f} \in \mathbb{C}[G/H; E]$; we write $\tilde{f}(\bar{g})$ for $g \in G$. The map $f \mapsto \tilde{f}$ is bijective from $\mathcal{I}_\mu$ to $\mathbb{C}[G/H; E]$, since the function $f(g) = \mu(g^{-1}\sigma(\bar{g}))\tilde{f}(\bar{g})$ is in $\mathcal{I}_\mu$ for any $\tilde{f} \in \mathbb{C}[G/H; E]$.

Next we calculate how the action of G is transformed by the map $f \mapsto \widetilde{f}$:

$$(\pi(g)f)^{\sim}(\bar{z}) = \mu(\sigma(\bar{z})^{-1}z)\,(\pi(g)f)(z) = \mu(\sigma(\bar{z})^{-1}z)f(g^{-1}z)$$
$$= \mu(\sigma(\bar{z})^{-1}z)\,\mu(z^{-1}g\sigma(g^{-1}\bar{z}))\,\widetilde{f}(g^{-1}\bar{z})\,.$$

Define a function $M : G \times (G/H) \longrightarrow \mathbf{GL}(E)$ by

$$M(g,\bar{z}) = \mu(\sigma(\bar{z})^{-1}z)\,\mu(z^{-1}g\sigma(g^{-1}\bar{z}))\,,$$

and define the operator $\widetilde{\pi}(g)$ on $\mathbb{C}[G/H;E]$ by

$$\widetilde{\pi}(g)\widetilde{f}(\bar{z}) = M(g,\bar{z})\,\widetilde{f}(g^{-1}\bar{z})\,.$$

Then the calculation above shows that $(\pi(g)f)^{\sim}(\bar{z}) = \widetilde{\pi}(g)\widetilde{f}(\bar{z})$. It follows that $\widetilde{\pi}$ is a representation of G, equivalent to π under the map $f \mapsto \widetilde{f}$. Hence $\chi_\pi(g) = \operatorname{tr}(\widetilde{\pi}(g))$.

We now proceed to calculate the trace of $\widetilde{\pi}(g)$. For $x \in G/H$ and $v \in E$ we denote by $\delta_x \otimes v$ the function from G/H to E that takes the value v at x and is 0 elsewhere. Likewise, for $v^* \in E^*$ we denote by $\delta_x^* \otimes v^*$ the linear functional $f \mapsto \langle f(x), v^*\rangle$ on $\mathbb{C}[G/H;E]$. If $g \in G$ then the function $\widetilde{\pi}(g)(\delta_x \otimes v)$ is zero except at $g^{-1} \cdot x$, and so we have

$$\langle\,\widetilde{\pi}(g)(\delta_x \otimes v),\, \delta_x^* \otimes v^*\rangle = \begin{cases} \langle M(g,x)v,\, v^*\rangle & \text{if } g \cdot x = x, \\ 0 & \text{otherwise}. \end{cases}$$

Letting v run over a basis for E and v^* over the dual basis, we obtain

$$\operatorname{tr}(\widetilde{\pi}(g)) = \sum_{x \in (G/H)^g} \operatorname{tr}(M(g,x))\,. \tag{4.56}$$

For $x = \bar{z} \in (G/H)^g$ we have $z^{-1}gz \in H$, and the formula for $M(g,\bar{z})$ simplifies to

$$M(g,\bar{z}) = \mu(\sigma(\bar{z})^{-1}z)\mu(z^{-1}gz)\mu(\sigma(\bar{z})^{-1}z)^{-1}\,.$$

Hence $\operatorname{tr}(M(g,\bar{z})) = \operatorname{tr}(\mu(z^{-1}gz)) = \chi_\mu(\bar{z})$, so (4.55) follows from (4.56). $\qquad\square$

Corollary 4.4.3 (Fixed-Point Formula). *Let* 1 *be the one-dimensional trivial representation of* H. *The character of the representation* $\pi = \operatorname{Ind}_H^G(1)$ *is*

$$\chi_\pi(g) = \textit{number of fixed points of } g \textit{ on } G/H\,. \tag{4.57}$$

4.4.3 Standard Representation of $\mathfrak{S}_n$

We recall some properties of the symmetric group $\mathfrak{S}_n$ on n letters. There is a non-trivial one-dimensional representation $\operatorname{sgn} : \mathfrak{S}_n \longrightarrow \pm 1$. It can be defined using the representation of $\mathfrak{S}_n$ on the one-dimensional space spanned by the polynomial

$$\Delta(x) = \prod_{i<j}(x_i - x_j)$$

in the variables $x_1, \dots, x_n$. The group $\mathfrak{S}_n$ acts on $\mathbb{C}[x_1, \dots, x_n]$ by permuting the variables, and by definition

$$s \cdot \Delta(x) = \operatorname{sgn}(s)\Delta(x).$$

We embed $\mathfrak{S}_n \subset \mathfrak{S}_{n+1}$ as the subgroup fixing $n+1$. It is clear that $\operatorname{sgn}(s)$, for $s \in \mathfrak{S}_n$, has the same value if we consider s as an element of $\mathfrak{S}_p$ for any $p \geq n$. Thus the sgn character is consistently defined on the whole family of symmetric groups.

A permutation s is *even* if $\operatorname{sgn}(s) = 1$. The *alternating group* $\mathfrak{A}_n$ is the subgroup of $\mathfrak{S}_n$ consisting of all even permutations. When $n \geq 5$ the group $\mathfrak{A}_n$ is *simple* (has no proper normal subgroups).

We construct the *standard* irreducible representation of the symmetric group as follows: Let $\mathfrak{S}_n$ act on $\mathbb{C}^n$ by permutation of the basis vectors $e_1, \dots, e_n$. Taking coordinates relative to this basis, we identify $\mathbb{C}^n$ with $\mathbb{C}[X]$ (the space of complex-valued functions on the set $X = \{1, \dots, n\}$). The action of $\mathfrak{S}_n$ on $\mathbb{C}^n$ then becomes

$$\pi(g)f(x) = f(g^{-1}x). \tag{4.58}$$

The set X, as a homogeneous space for $\mathfrak{S}_n$, is isomorphic to $\mathfrak{S}_n/\mathfrak{S}_{n-1}$. Hence $\pi \cong \operatorname{Ind}_{\mathfrak{S}_{n-1}}^{\mathfrak{S}_n}(1)$. We shall obtain the decomposition of this representation from the following general considerations.

Let a group G act transitively on a set X. Then G acts on $X \times X$ by

$$g \cdot (x,y) = (gx, gy),$$

and the diagonal $\widetilde{X} = \{(x,x) : x \in X\}$ is a single G-orbit. We say that the action of G on X is *doubly transitive* if G has exactly *two* orbits in $X \times X$, namely $\widetilde{X}$ and its complement $\{(x,y) : x \neq y\}$. It is easy to check that double transitivity is equivalent to the following two properties:

1. G acts transitively on X.
2. For $x \in X$, the stability group G_x of x acts transitively on $X \setminus \{x\}$.

For a doubly transitive action, the induced representation decomposes as follows:

Proposition 4.4.4. *Suppose that G is a finite group that acts doubly transitively on a set X and $|X| \geq 2$. Then $\mathbb{C}[X]$ decomposes into two irreducible subspaces under G, namely the constant functions and*

$$V = \left\{ f \in \mathbb{C}[X] : \sum_{x \in X} f(x) = 0 \right\}.$$

Let $\rho(g)$ be the restriction of $\pi(g)$ to V. Then ρ has character

$$\chi_\rho(g) = \#\{x : gx = x\} - 1. \tag{4.59}$$

Proof. Clearly $\mathbb{C}[X] = \mathbb{C}1 \oplus V$ and each summand is invariant under G. To see that V is irreducible under G, let $T \in \mathrm{End}_G(\mathbb{C}[X])$. Then

$$Tf(x) = \sum_{y \in X} K(x,y) f(y)$$

for $f \in \mathbb{C}[X]$, and the G-intertwining property of T is equivalent to

$$K(x,y) = K(gx,gy) \quad \text{for } x, y \in X \text{ and } g \in G \cdot.$$

This means that the function K is constant on the G orbits in $X \times X$. Since there are exactly two orbits, by the double transitivity assumption, we conclude that $\dim \mathrm{End}_G(\mathbb{C}[X]) = 2$. However, the operator

$$Pf(x) = \frac{1}{|X|} \sum_{x \in X} f(x)$$

that projects $\mathbb{C}[X]$ onto $\mathbb{C}1$ is G-invariant and not the identity operator, so we have

$$\mathrm{End}_G(\mathbb{C}[X]) = \mathbb{C}I + \mathbb{C}P.$$

Since π is completely reducible, it follows that $V = \mathrm{Range}(I - P)$ is irreducible under the action of G.

To obtain the character formula, write $\pi = \rho \oplus \iota$, where ι is the trivial representation. This gives $\chi_\pi = \chi_\rho + 1$. If we fix $x \in X$ and set $H = G_x$, then $\pi \cong \mathrm{Ind}_H^G(1)$, so from formula (4.57) we know that $\chi_\pi(g)$ is the number of fixed points of the permutation g. This proves (4.59). $\square$

Corollary 4.4.5. *Let $V \subset \mathbb{C}^n$ be the subspace $\{x : \sum x_i = 0\}$. Then $\mathfrak{S}_n$ acts irreducibly on V for $n \geq 2$, and $\mathfrak{A}_n$ acts irreducibly on V for $n \geq 4$.*

Proof. Clearly $\mathfrak{S}_n$ acts doubly transitively on $X = \{1, 2, \ldots, n\}$. We claim that $\mathfrak{A}_n$ also acts doubly transitively on X when $n \geq 4$. Indeed, the isotropy group of $\{n\}$ in $\mathfrak{A}_n$ is $\mathfrak{A}_{n-1}$, and it is easy to check that $\mathfrak{A}_n$ acts transitively on $\{1, 2, \ldots, n\}$ when $n \geq 3$. Now apply Proposition 4.4.4. $\square$

Let $\rho(g)$ be the restriction to V of the permutation action (4.58) of $\mathfrak{S}_n$ or $\mathfrak{A}_n$. We call (ρ, V) the *standard representation* of these groups.

Example

For the group $\mathfrak{S}_3$ we have three partitions of 3: 3, $2 + 1$, and $1 + 1 + 1$, and the corresponding three conjugacy classes (3-cycles, transpositions, and identity). Thus there are three irreducible representations. These must be the identity, the standard representation, and sgn. As a check, the sum of the squares of their dimensions is $1^2 + 2^2 + 1^2 = 6 = |\mathfrak{S}_3|$, as required by Theorem 4.3.1.

4.4.4 Representations of $\mathfrak{S}_k$ on Tensors

In Section 4.2.4 we defined a representation σ_k of the symmetric group $\mathfrak{S}_k$ on the k-fold tensor product $\bigotimes^k \mathbb{C}^n$ by permutation of tensor positions. Now we will show how an important family of induced representations of $\mathfrak{S}_k$ naturally occurs on subspaces of $\bigotimes^k \mathbb{C}^n$.

Given an ordered k-tuple $I = (i_1, \ldots, i_k)$ of positive integers with $1 \le i_p \le n$, we set $e_I = e_{i_1} \otimes \cdots \otimes e_{i_k}$, and we write $|I| = k$ for the number of entries in I, as in the proof of Theorem 4.2.10. Let H be the diagonal subgroup of $\mathbf{GL}(n, \mathbb{C})$. It acts on $\mathbb{C}^n$ by $te_i = t_i e_i$, for $t = \mathrm{diag}[t_1, \ldots, t_n]$. We parameterize the characters of H by $\mathbb{Z}^n$, where $\lambda = [\lambda_1, \ldots, \lambda_n] \in \mathbb{Z}^n$ gives the character

$$t \mapsto t^\lambda = t_1^{\lambda_1} \cdots t_n^{\lambda_n} \, .$$

The action of H on $\mathbb{C}^n$ extends to a representation on $\bigotimes^k \mathbb{C}^n$ by the representation ρ_k of $\mathbf{GL}(n, \mathbb{C})$ (see Section 4.1.5). For $\lambda \in \mathbb{Z}^n$ let

$$\textstyle\bigotimes^k \mathbb{C}^n(\lambda) = \{ u \in \bigotimes^k \mathbb{C}^n : \rho_k(t)u = t^\lambda u \}$$

be the λ weight space of H. Given a k-tuple I as above, define

$$\mu_j = \#\{ p : i_p = j \}$$

and set $\mu_I = [\mu_1, \ldots, \mu_n] \in \mathbb{Z}^n$. Then $\rho_k(t)e_I = t^{\mu_I} e_I$ for $t \in H$, so e_I is a weight vector for H with weight $\mu_I \in \mathbb{N}^n$. Furthermore,

$$|\mu_I| = \mu_1 + \cdots + \mu_n = |I| = k \, .$$

This shows that

$$\textstyle\bigotimes^k \mathbb{C}^n(\lambda) = \begin{cases} \mathrm{Span}\{ e_I \, ; \, \mu_I = \lambda \} & \text{if } \lambda \in \mathbb{N}^n \text{ and } |\lambda| = k, \\ 0 & \text{otherwise.} \end{cases} \tag{4.60}$$

Since the actions of H and $\mathfrak{S}_k$ on $\bigotimes^k \mathbb{C}^n$ mutually commute, the weight space $\bigotimes^k \mathbb{C}^n(\lambda)$ is a module for $\mathfrak{S}_k$. We will show that it is equivalent to an induced representation. Let $\lambda \in \mathbb{N}^k$ with $|\lambda| = k$. Set

$$u(\lambda) = \underbrace{e_1 \otimes \cdots \otimes e_1}_{\lambda_1 \text{ factors}} \otimes \underbrace{e_2 \otimes \cdots \otimes e_2}_{\lambda_2 \text{ factors}} \otimes \cdots .$$

Then $u(\lambda) \in \bigotimes^k \mathbb{C}^n(\lambda)$. Let $\mathfrak{S}_\lambda$ be the subgroup of $\mathfrak{S}_k$ fixing $u(\lambda)$. We have

$$\mathfrak{S}_\lambda \cong \mathfrak{S}_{\lambda_1} \times \cdots \times \mathfrak{S}_{\lambda_n} \, ,$$

where the first factor permutes positions $1, \ldots, \lambda_1$, the second factor permutes positions $\lambda_1 + 1, \ldots, \lambda_1 + \lambda_2$, and so on.

Proposition 4.4.6. *The restriction of the representation* σ_k *of* $\mathfrak{S}_k$ *to the subspace* $\bigotimes^k \mathbb{C}^n(\lambda)$ *is equivalent to the representation* $\mathrm{Ind}_{\mathfrak{S}_\lambda}^{\mathfrak{S}_k}(1)$ *on* $\mathbb{C}[\mathfrak{S}_k/\mathfrak{S}_\lambda]$. *Furthermore, if* $\lambda' \in \mathbb{N}^n$ *is obtained by permuting the entries of* λ, *then* $\bigotimes^k \mathbb{C}^n(\lambda) \cong \bigotimes^k \mathbb{C}^n(\lambda')$ *as a module for* $\mathfrak{S}_k$.

Proof. Let I be a k-tuple such that $\mu_I = \lambda$. Then there is a permutation s such that $\sigma_k(s)u(\lambda) = e_I$. Thus the map $s \mapsto \sigma_k(s)u(\lambda)$ gives a bijection from $\mathfrak{S}_k/\mathfrak{S}_\lambda$ to a basis for $\bigotimes^k \mathbb{C}^n(\lambda)$ and intertwines the left multiplication action of $\mathfrak{S}_k$ with the representation σ_k. To verify the last statement, we observe that $\mathbf{GL}(n, \mathbb{C})$ contains a subgroup isomorphic to $\mathfrak{S}_n$ (the permutation matrices), which acts on V by permuting the basis vectors $e_1, \ldots, e_n$. The action of this subgroup on $\bigotimes^k \mathbb{C}^n$ commutes with the action of $\sigma_k(\mathfrak{S}_k)$ and permutes the weight spaces for H as needed. $\qquad \square$

We will write $I^\lambda = \bigotimes^k \mathbb{C}^n(\lambda)$, viewed as a module for $\mathfrak{S}_k$. We obtain the character of this module from Corollary 4.4.3.

Corollary 4.4.7. *The representation* I^λ *has the character*

$$\mathrm{ch}_{\mathfrak{S}_k}(I^\lambda)(y) = \#\{\text{fixed points of } y \text{ on } \mathfrak{S}_k/\mathfrak{S}_\lambda\} \qquad (4.61)$$

for $y \in \mathfrak{S}_k$. *In particular, the values of the character are nonnegative integers.*

Remark 4.4.8. A conjugacy class C in $\mathfrak{S}_k$ is determined by the *cycle decomposition* of the elements in the class. We will write $C = C(1^{i_1} 2^{i_2} \cdots k^{i_k})$ to denote the class of elements with i_1 cycles of length 1, i_2 cycles of length 2, and so forth. For example, when $k = 5$ one has $(1)(3)(245) \in C(1^2 3^1)$ and $(12)(345) \in C(2^1 3^1)$. The cycle lengths for a class satisfy the constraint

$$1 \cdot i_1 + 2 \cdot i_2 + \cdots + k \cdot i_k = k,$$

but are otherwise arbitrary. Thus the conjugacy classes in $\mathfrak{S}_k$ correspond to the partitions $k = k_1 + k_2 + \cdots + k_p$ of k. Here we may assume $k_1 \geq k_2 \geq \cdots \geq k_p > 0$. Then the partition is uniquely described by the p-tuple $\beta = [k_1, \ldots, k_p]$, and we say that the partition has p *parts*. If β is a partition of k with i_q parts of size q, we will denote by

$$C_\beta = C(1^{i_1} 2^{i_2} \cdots k^{i_k})$$

the corresponding conjugacy class in $\mathfrak{S}_k$.

4.4.5 Exercises

1. Let $\mathbb{F}$ be a finite field, and let χ be a nontrivial additive character of $\mathbb{F}$. For $u \in \mathbb{F}$ let χ_u be the additive character $x \mapsto \chi(xu)$. For $\varphi \in \mathbb{C}[\mathbb{F}]$, let M_φ be the operator of multiplication by φ on $\mathbb{C}[\mathbb{F}]$: $(M_\varphi f)(x) = \varphi(x)f(x)$ for $f \in \mathbb{C}[\mathbb{F}]$ and $x \in \mathbb{F}$.

(a) Suppose $A \in \text{End}(\mathbb{C}[\mathbb{F}])$ and A commutes with the operators M_{χ_u} for all $u \in \mathbb{F}$. Show that A commutes with M_φ for all $\varphi \in \mathbb{C}[\mathbb{F}]$. (HINT: Apply the Fourier inversion formula to φ.)

(b) Suppose $A \in \text{End}(\mathbb{C}[\mathbb{F}])$ and A commutes with the left translation operators $L(t)f(x) = f(x-t)$ for all $t \in \mathbb{F}$. Prove that there is a function $g \in \mathbb{C}[\mathbb{F}]$ such that $Af = f * g$ (convolution product) for all $f \in \mathbb{C}[\mathbb{F}]$. (HINT: Consider the action of A on delta functions.)

(c) Suppose $A \in \text{End}(\mathbb{C}[\mathbb{F}])$ satisfies the conditions of (a) and (b). Show that A is a multiple of the identity operator.

2. Let $\mathbb{F}$ be a finite field and let G be the group of all matrices of the form

$$g = \begin{bmatrix} 1 & x & z \\ 0 & 1 & y \\ 0 & 0 & 1 \end{bmatrix}, \quad x,y,z \in \mathbb{F}$$

(the *Heisenberg group* over $\mathbb{F}$). Let $\chi : \mathbb{F} \to \mathbb{S}^1$ be an additive character. Let H be the subgroup of all $g \in G$ such that $y = 0$. Define a one-dimensional representation μ_χ of H by $\mu_\chi \begin{bmatrix} 1 & x & z \\ 0 & 1 & 0 \\ 0 & 0 & 1 \end{bmatrix} = \chi(z)$.

(a) Show that $\text{Ind}_H^G(\mu_\chi)$ is irreducible for every nontrivial χ. (HINT: Use a variant of the argument in the proof of Proposition 4.4.4. Define $\Phi : H \times \mathbb{F} \to G$ by $\Phi(h,y) = h \begin{bmatrix} 1 & 0 & 0 \\ 0 & 1 & y \\ 0 & 0 & 1 \end{bmatrix}$. Show that Φ is bijective. For $f \in \text{Ind}_H^G(\mu_x)$ let $T(f)(x) = f(\Phi(1,x))$. Then $T : \text{Ind}_H^G(\mu_x) \to \mathbb{C}[\mathbb{F}]$ is bijective. Set $\alpha(x) = \begin{bmatrix} 1 & x & 0 \\ 0 & 1 & 0 \\ 0 & 0 & 1 \end{bmatrix}$ and $\beta(x) = \Phi(1,x)$ and write $\pi = \pi_{\mu_\chi}$. Show that $T(\pi(\beta(x))f)(t) = T(f)(t-x)$ and $T(\pi(\alpha(x))f)(t) = \chi(-tx)T(f)(t)$. Now use the previous exercise to show that an operator that commutes with $\pi(g)$ for all $g \in G$ is a multiple of the identity.)

(b) Find the conjugacy classes in G.

(c) Prove that $\widehat{G}$ consists of the classes of the representations $\text{Ind}_H^G(\mu_\chi)$ for all nontrivial additive characters χ, together with the classes of the representations of G/K of dimension one, where K is the subgroup $\begin{bmatrix} 1 & 0 & z \\ 0 & 1 & 0 \\ 0 & 0 & 1 \end{bmatrix}$ with $z \in \mathbb{F}$. (HINT: Show that there are exactly $|\mathbb{F}| - 1$ nontrivial additive characters of $\mathbb{F}$ and $|\mathbb{F}|^2$ one-dimensional representations of G/K.)

3. Let $\mathbb{F}$ be a finite field and let G be the group of all matrices $g = \begin{bmatrix} a & b \\ 0 & 1 \end{bmatrix}$ with $a,b \in \mathbb{F}$ and $a \neq 0$. Let G act on $\mathbb{F}$ by $g \cdot x = ax + b$.

(a) Show that G acts doubly transitively on $\mathbb{F}$.

(b) Let $V = \{f \in \mathbb{C}[\mathbb{F}] : \sum_{x \in \mathbb{F}} f(x) = 0\}$. Let G act on V by $\rho(g)f(x) = f(g^{-1} \cdot x)$. Show that (ρ, V) is an irreducible representation.

(c) Find the conjugacy classes in G.

(d) Show that $\widehat{G}$ consists of the class of V and the classes of representations of G/U of dimension one, where $U = \left\{ \begin{bmatrix} 1 & x \\ 0 & 1 \end{bmatrix} : x \in \mathbb{F} \right\}$ is a normal subgroup. (HINT: The group G/U is isomorphic to the multiplicative group of $\mathbb{F}$, so it has $|\mathbb{F}| - 1$ irreducible representations. Combine this with the result from (c).)

(e) If $\mathbb{F} = \mathbb{Z}_3$ show that G is isomorphic to $\mathfrak{S}_3$.

4.5 Notes

Section 4.1. The study of noncommutative associative algebras (formerly called *hypercomplex numbers*) began in the mid-nineteenth century with Hamilton's quaternions and Grassmann's exterior algebra. See the Note Historique in Bourbaki [10] and the bibliographic notes in Weyl [164]. Group representations (for finite groups) were first studied by Frobenius [47]. For historical accounts and further references see Mackey [107] and Hawkins [63].

Sections 4.1.2. The finite-dimensional version of Lemma 4.1.4 appears in Schur's thesis [129]. The version here is due to Dixmier.

Section 4.1.7. Characters of noncommutative groups were introduced by Frobenius [46]. Hermite reciprocity was a major result in nineteenth-century invariant theory; see Howe [69] for applications and generalizations.

Section 4.2.1. The abstract duality theorem (Theorem 4.2.1) is a generalization of a result of Wallach [155, Proposition 1.5]. In the first edition of this book the algebra $\mathcal{R}$ was assumed to be graded, and this was generalized by I. Agricola [1] to apply to nongraded algebras $\mathcal{R}$ such as the ring of differential operators on a smooth algebraic variety. The short proof presented here based on the Jacobson density theorem is new.

Section 4.2.3. Theorem 4.2.7 is an algebraic version of the famous Peter–Weyl theorem for compact topological groups (see Wallach [153, Chapter 2], for example). See Borel [17, Chapter III] for the history and comments on this pivotal result in representation theory. We shall use this decomposition extensively in Chapter 12.

Section 4.2.4. Theorem 4.2.10 appears in Schur's thesis [129]. The duality between representations of the symmetric group and the general linear group is a central theme in Weyl's book [164].

Section 4.2.5. Theorem 4.2.12 is one of the most useful consequences of the theorem of the highest weight. It shows that the highest weight is much more than just a label for an irreducible representation of $\mathfrak{g}$, since the space of $\mathfrak{n}$-invariants of a given weight also furnishes an irreducible representation of the commuting algebra of $\mathfrak{g}$. This result will be fundamental in Chapters 9 and 10.

Section 4.3.2. The orthogonality relations were first obtained by Schur [130].

Sections 4.3.3 and 4.3.4. For a sample of the applications of representation theory of finite groups see Fässler and Stiefel [45], Sternberg [139], and Terras [142]. See Wallach [153, Chapter 2], for example, for the generalization of the Plancherel formula to compact groups as in Remark 4.3.6. See Terras [142] for more on the uncertainty principle in Remark 4.3.12.

Sections 4.4.1 and 4.4.2. The notion of an induced representation and the calculation of its character is due to Frobenius [48].

Chapter 5
Classical Invariant Theory

Abstract For a linear algebraic group G and a regular representation (ρ, V) of G, the *basic problem of invariant theory* is to describe the G-invariant elements $(\bigotimes^k V)^G$ of the k-fold tensor product for all k. If G is a reductive, then a solution to this problem for (ρ^*, V^*) leads to a determination of the polynomial invariants $\mathcal{P}(V)^G$. When $G \subset \mathbf{GL}(W)$ is a classical group and $V = W^k \oplus (W^*)^l$ (k copies of W and l copies of W^*), explicit and elegant solutions to the basic problem of invariant theory, known as the *first fundamental theorem* (FFT) of invariant theory for G, were found by Schur, Weyl, Brauer, and others. The fundamental case is $G = \mathbf{GL}(V)$ acting on V. Following Schur and Weyl, we turn the problem of finding tensor invariants into the problem of finding the operators commuting with the action of $\mathbf{GL}(V)$ on $\bigotimes^k V$, which we solved in Chapter 4. This gives an FFT for $\mathbf{GL}(V)$ in terms of *complete contractions* of vectors and covectors. When G is the orthogonal or symplectic group we first find all polynomial invariants of at most $\dim V$ vectors. We then use this special case to transform the general problem of tensor invariants for an arbitrary number of vectors into an invariant problem for $\mathbf{GL}(V)$ of the type we have already solved.

The FFT furnishes generators for the commutant of the action of a classical group G on the exterior algebra $\bigwedge V$ of the defining representation; the general duality theorem from Chapter 4 gives the G-isotypic decomposition of $\bigwedge V$. This furnishes irreducible representations for each *fundamental weight* of the special linear group and the symplectic group. For $G = \mathbf{SO}(V)$ it gives representations for all G-integral fundamental weights (the *spin representations* for the half-integral fundamental weights of $\mathfrak{so}(V)$ will be constructed in Chapter 6 using the *Clifford algebra*). Irreducible representations with arbitrary dominant integral highest weights are obtained as iterated *Cartan products* of the fundamental representations. In Chapters 9 and 10 we shall return to this construction and obtain a precise description of the tensor subspaces on which the irreducible representations are realized.

Combining the FFT with the general duality theorem we obtain *Howe duality* for the classical groups: When V is a multiple of the basic representation of a classical group G and $\mathbb{D}(V)$ is the algebra of polynomial-coefficient differential operators on V, then the commuting algebra $\mathbb{D}(V)^G$ is generated by a Lie algebra of

R. Goodman, N.R. Wallach, *Symmetry, Representations, and Invariants*,
Graduate Texts in Mathematics 255, DOI 10.1007/978-0-387-79852-3_5,
© Roe Goodman and Nolan R. Wallach 2009

differential operators isomorphic to $\mathfrak{g}'$, where $\mathfrak{g}'$ is another classical Lie algebra (the *Howe dual* to $\mathfrak{g} = \mathrm{Lie}(G)$). The general duality theorem from Chapter 4 then sets up a correspondence between the irreducible regular (finite-dimensional) representations of G occurring in $\mathcal{P}(V)$ and certain irreducible representations of $\mathfrak{g}'$ (generally infinite-dimensional). As a special case, we obtain the classical theory of spherical harmonics. The chapter concludes with several more applications of the FFT.

5.1 Polynomial Invariants for Reductive Groups

For an algebraic group G acting by a regular representation on a vector space V, the *ring of invariants* is the algebra of polynomial functions f on V such that $f(gv) = f(v)$ for all $v \in V$ and $g \in G$. In this section we show that when G is reductive, then the algebra of invariants always has a finite set of generators. We find such a set of generators for the case $G = \mathfrak{S}_n$, acting by permuting the coordinates in $\mathbb{C}^n$.

5.1.1 The Ring of Invariants

Let G be a reductive linear algebraic group. Suppose (π, V) is a regular representation of G. We write $\pi(g)v = gv$ for $g \in G$ and $v \in V$ when the representation π is understood from the context. We define a representation ρ of G on the algebra $\mathcal{P}(V)$ of polynomial functions on V by

$$\rho(g)f(v) = f(g^{-1}v) \quad \text{for } f \in \mathcal{P}(V) .$$

The finite-dimensional spaces

$$\mathcal{P}^k(V) = \{f \in \mathcal{P}(V) : f(zv) = z^k f(v) \quad \text{for } z \in \mathbb{C}^\times\}$$

of homogeneous polynomials of degree k are G-invariant, for $k = 0, 1, \ldots$, and the restriction ρ_k of ρ to $\mathcal{P}^k(V)$ is a regular representation of G. Since G is reductive, the finite-dimensional algebra $\mathbb{C}[\rho_k(G)]$ is semisimple. From Proposition 4.1.15 we have a primary decomposition

$$\mathcal{P}^k(V) = \bigoplus_{\sigma \in \widehat{G}} W_{(\sigma)} \tag{5.1}$$

into G-isotypic subspaces, where $\widehat{G}$ is the set of (equivalence classes of) irreducible regular representations of G and $W_{(\sigma)}$ is the σ-isotypic component of $\mathcal{P}^k(V)$.

Let $f \in \mathcal{P}(V)$. Then $f = f_0 + \cdots + f_d$ with f_j homogeneous of degree j. Decomposing each f_k according to (5.1) and collecting terms with the same σ, we can write

$$f = \sum_{\sigma \in \hat{G}} f_\sigma, \tag{5.2}$$

where $f_\sigma \in W_{(\sigma)}$ is the σ-isotypic component of f. We write $f^\natural$ for the isotypic component of f corresponding to the trivial representation.

We denote the space of G-invariant polynomials on V by $\mathcal{P}(V)^G$. This subalgebra of $\mathcal{P}(V)$ is called the *algebra of G-invariants*. Since multiplication by a G-invariant function φ leaves the isotypic subspaces invariant, we have

$$(\varphi f)^\natural = \varphi f^\natural \quad \text{for } f \in \mathcal{P}(V) \text{ and } \varphi \in \mathcal{P}(V)^G .$$

Thus the projection operator $f \mapsto f^\natural$ is a $\mathcal{P}(V)^G$-module map when $\mathcal{P}(V)$ is viewed as a module for the algebra of invariants.

Theorem 5.1.1. *Suppose G is a reductive linear algebraic group acting by a regular representation on a vector space V. Then the algebra $\mathcal{P}(V)^G$ of G-invariant polynomials on V is finitely generated as a $\mathbb{C}$-algebra.*

Proof. Write $\mathcal{R} = \mathcal{P}(V)$ and $\mathcal{J} = \mathcal{P}(V)^G$. By the Hilbert basis theorem (Theorem A.1.2), every ideal $\mathcal{B} \subset \mathcal{R}$ and every quotient $\mathcal{R}/\mathcal{B}$ is finitely generated as an $\mathcal{R}$ module. To show that $\mathcal{J}$ is finitely generated as an algebra over $\mathbb{C}$, consider the ideal $\mathcal{R}\mathcal{J}_+$, where $\mathcal{J}_+$ is the space of invariant polynomials with zero constant term. This ideal has a finite set of generators (as an $\mathcal{R}$-module), say $\varphi_1, \ldots, \varphi_n$, and we may take φ_i to be a homogeneous G-invariant polynomial of degree $d_i \geq 1$. We claim that $\varphi_1, \ldots, \varphi_n$ generate $\mathcal{J}$ as an algebra over $\mathbb{C}$. Indeed, if $\varphi \in \mathcal{J}$ then there exist $f_1, \ldots, f_n$ in $\mathcal{R}$ such that $\varphi = \sum f_i \varphi_i$. Now apply the operator $\varphi \mapsto \varphi^\natural$ to obtain

$$\varphi = \sum (f_i \varphi_i)^\natural = \sum f_i^\natural \varphi_i .$$

Since $\deg f_i^\natural \leq \deg f_i < \deg \varphi$, we may assume by induction that $f_i^\natural$ is in the algebra generated by $\varphi_1, \ldots, \varphi_n$. Hence so is φ. $\square$

Definition 5.1.2. Let $\{f_1, \ldots, f_n\}$ be a set of generators for $\mathcal{P}(V)^G$ with n as small as possible. Then $\{f_1, \ldots, f_n\}$ is called a set of *basic invariants*.

Theorem 5.1.1 asserts that there always exists a finite set of basic invariants when G is reductive. Since $\mathcal{P}(V)$ and $\mathcal{J} = \mathcal{P}(V)^G$ are graded algebras, relative to the usual degree of a polynomial, there is a set of basic invariants with each f_i homogeneous, say of degree d_i. If we enumerate the f_i so that $d_1 \leq d_2 \leq \cdots$ then the sequence $\{d_i\}$ is uniquely determined (even though the set of basic invariants is not unique). To prove this, define

$$m_k = \dim \mathcal{J}_k/(\mathcal{J}_+^2)_k ,$$

where $\mathcal{J}_k$ is the homogeneous component of degree k of $\mathcal{J}$. We claim that

$$m_k = \mathrm{Card}\{j : d_j = k\} . \tag{5.3}$$

Indeed, if $\varphi \in \mathcal{J}_k$ then

$$\varphi = \sum_I a_I f^I \qquad (\text{sum over } I \text{ with } d_1 i_1 + \cdots + d_n i_n = k),$$

where $f^I = f_1^{i_1} \cdots f_n^{i_n}$ and $a_I \in \mathbb{C}$. Since $f^I \in \mathcal{J}_+^2$ if $i_1 + \cdots + i_n \geq 2$, we can write

$$\varphi = \sum b_i f_i + \psi \qquad (\text{sum over } i \text{ with } d_i = k),$$

where $\psi \in (\mathcal{J}_+^2)_k$ and $b_i \in \mathbb{C}$. This proves (5.3) and shows that the set $\{d_i\}$ is intrinsically determined by $\mathcal{J}$ as a graded algebra.

We can also associate the *Hilbert series*

$$H(t) = \sum_{k=0}^{\infty} \dim(\mathcal{J}_k) t^k$$

to the graded algebra $\mathcal{J}$. When the set of basic invariants is algebraically independent the Hilbert series is convergent for $|t| < 1$ and is given by the rational function

$$H(t) = \prod_{i=1}^{n} (1 - t^{d_i})^{-1},$$

where $d_1, \ldots, d_n$ are the degrees of the basic invariants.

5.1.2 Invariant Polynomials for $\mathfrak{S}_n$

Let the symmetric group $\mathfrak{S}_n$ act on the polynomial ring $\mathbb{C}[x_1, \ldots, x_n]$ by

$$(s \cdot f)(x_1, \ldots, x_n) = f(x_{s(1)}, \ldots, x_{s(n)})$$

for $s \in \mathfrak{S}_n$ and $f \in \mathbb{C}[x_1, \ldots, x_n]$. We can view this as the action of $\mathfrak{S}_n$ on $\mathcal{P}(\mathbb{C}^n)$ arising from the representation of $\mathfrak{S}_n$ on $\mathbb{C}^n$ as permutation matrices, with $x_1, \ldots, x_n$ being the coordinate functions on $\mathbb{C}^n$.

Define the *elementary symmetric functions* $\sigma_1, \ldots, \sigma_n$ by

$$\sigma_i(x_1, \ldots, x_n) = \sum_{1 \leq j_1 < \cdots < j_i \leq n} x_{j_1} \cdots x_{j_i},$$

and set $\sigma_0 = 1$. For example, $\sigma_1(x) = x_1 + \cdots + x_n$ and $\sigma_n(x) = x_1 \cdots x_n$. Clearly, $\sigma_i \in \mathbb{C}[x_1, \ldots, x_n]^{\mathfrak{S}_n}$. Furthermore, we have the identity

$$\prod_{i=1}^{n}(t - x_i) = \sum_{j=0}^{n} t^{n-j}(-1)^j \sigma_j(x). \tag{5.4}$$

Thus the functions $\{\sigma_i(x)\}$ express the coefficients of a monic polynomial in the variable t as symmetric functions of the roots $x_1, \ldots, x_n$ of the polynomial.

Theorem 5.1.3. *The set of functions* $\{\sigma_1, \ldots, \sigma_n\}$ *is algebraically independent and* $\mathbb{C}[x_1, \ldots, x_n]^{\mathfrak{S}_n} = \mathbb{C}[\sigma_1, \ldots, \sigma_n]$. *Hence the elementary symmetric functions are a set of basic invariants for* $\mathfrak{S}_n$.

Proof. We put the *graded lexicographic order* on $\mathbb{N}^n$. Let $I \neq J \in \mathbb{N}^n$. We define $I \overset{\text{grlex}}{>} J$ if either $|I| > |J|$ or else $|I| = |J|$, $i_p = j_p$ for $p < r$, and $i_r > j_r$. This is a total order on $\mathbb{N}^n$, and the set $\{J \in \mathbb{N}^n : I \overset{\text{grlex}}{>} J\}$ is finite for every $I \in \mathbb{N}^n$. Furthermore, this order is compatible with addition:

$$\text{if } I \overset{\text{grlex}}{>} J \text{ and } P \overset{\text{grlex}}{>} Q \text{ then } I + P \overset{\text{grlex}}{>} J + Q .$$

Given $f = \sum_I a_I x^I \in \mathbb{C}[x_1, \ldots, x_n]$, we define the *support* of f to be

$$\mathcal{S}(f) = \{I \in \mathbb{N}^n : a_I \neq 0\} .$$

Assume that f is invariant under $\mathfrak{S}_n$. Then $a_I = a_{s \cdot I}$ for all $s \in \mathfrak{S}_n$, where $s \cdot [i_1, \ldots, i_n] = [i_{s^{-1}(1)}, \ldots, i_{s^{-1}(n)}]$. Thus $\mathcal{S}(f)$ is invariant under $\mathfrak{S}_n$. If $J \in \mathcal{S}(f)$ then the set of indices $\{s \cdot J : s \in \mathfrak{S}_n\}$ contains a unique index $I = [i_1, \ldots, i_n]$ with $i_1 \geq i_2 \geq \cdots \geq i_n$ (we call such an index I *dominant*). If I is the largest index in $\mathcal{S}(f)$ (for the graded lexicographic order), then I must be dominant (since $|s \cdot I| = |I|$ for $s \in \mathfrak{S}_n$). The corresponding term $a_I x^I$ is called the *dominant term* of f. For example, the dominant term in σ_i is $x_1 x_2 \cdots x_i$.

Given a dominant index I, set

$$\dot{\sigma}_I = \sigma_1^{i_1 - i_2} \sigma_2^{i_2 - i_3} \cdots \sigma_n^{i_n} .$$

Then σ_I is a homogeneous polynomial of degree $|I|$ that is invariant under $\mathfrak{S}_n$. We claim that

$$\mathcal{S}(\sigma_I - x^I) \subset \{J \in \mathbb{N}^n : J \overset{\text{grlex}}{<} I\} . \tag{5.5}$$

We prove (5.5) by induction on the graded lexicographic order of I. The smallest dominant index in this order is $I = [1, 0, \ldots, 0]$, and in this case

$$\sigma_I - x^I = \sigma_1 - x_1 = x_2 + \cdots + x_n .$$

Thus (5.5) holds for $I = [1, 0, \ldots, 0]$. Given a dominant index I, we may thus assume that (5.5) holds for all dominant indices less than I. We can write $I = J + M_i$ for some $i \leq n$, where $M_i = [\underbrace{1, \ldots, 1}_{i}, 0, \ldots, 0]$ and J is a dominant index less than I. Thus $\sigma_I = \sigma_i \sigma_J$. Now

$$\sigma_i = \Big(\prod_{p=1}^{i} x_p \Big) + \cdots ,$$

where $\cdots$ indicates a linear combination of monomials x^K with $K \overset{\text{grlex}}{<} M_i$. By induction, $\sigma_J = x^J + \cdots$, where $\cdots$ indicates a linear combination of monomials x^L

with $L \overset{\text{grlex}}{<} J$. Hence $\sigma_i \sigma_J = x^{J+M_i} + \cdots$, where $\cdots$ indicates a linear combination of monomials x^P with $P \overset{\text{grlex}}{<} M_i + J = I$. This proves (5.5).

Suppose $f(x) \in \mathbb{C}[x_1, \ldots, x_n]^{\mathfrak{S}_n}$. Let ax^I be the dominant term in f. Then $g(x) = f(x) - a\sigma^I \in \mathbb{C}[x_1, \ldots, x_n]$. If bx^J is the dominant term in $g(x)$ then (5.5) implies that $J \overset{\text{grlex}}{<} I$. By induction we may assume that $g \in \mathbb{C}[\sigma_1, \ldots, \sigma_n]$, so it follows that $f \in \mathbb{C}[\sigma_1, \ldots, \sigma_n]$.

It remains to prove that the set $\{\sigma_1, \ldots, \sigma_n\} \subset \mathbb{C}[x_1, \ldots, x_n]$ is algebraically independent. This is true for $n = 1$, since $\sigma_1(x_1) = x_1$. Assume that this is true for n. Suppose for the sake of contradiction that the elementary symmetric functions in the variables $x_1, \ldots, x_{n+1}$ satisfy a nontrivial polynomial relation. We can write such a relation as

$$\sum_{j=0}^{p} f_j(\sigma_1, \ldots, \sigma_n) \sigma_{n+1}^j = 0, \tag{5.6}$$

where each f_j is a polynomial in n variables and $f_p \neq 0$. We take the smallest p for which a relation (5.6) holds. Then $f_0 \neq 0$ (otherwise we could divide by the nonzero polynomial σ_{n+1} and obtain a relation with a smaller value of p). Now we substitute $x_{n+1} = 0$ in (5.6); the function σ_{n+1} becomes zero and $\sigma_1, \ldots, \sigma_n$ become the elementary symmetric functions in $x_1, \ldots, x_n$. Hence we obtain a nontrivial relation $f_0(\sigma_1, \ldots, \sigma_n) = 0$, which contradicts the induction hypothesis. $\qquad\square$

Write $\mathcal{P} = \mathbb{C}[x_1, \ldots, x_n]$ and $\mathcal{J} = \mathcal{P}^{\mathfrak{S}_n}$. We now turn to a description of $\mathcal{P}$ as a module for $\mathcal{J}$. Let $\mathcal{P}^k$ be the homogeneous polynomials of degree k. If $f = \sum a_I x^I \in \mathcal{P}$ we write $\partial(f)$ for the constant-coefficient differential operator $\sum_I a_I (\partial/\partial x)^I$, where

$$(\partial/\partial x)^I = \prod_{j=1}^{n} (\partial/\partial x_j)^{i_j} \qquad \text{for } I = (i_1, \ldots, i_n) .$$

The map $f \mapsto \partial(f)$ is an algebra homomorphism from $\mathcal{P}$ to differential operators that is uniquely determined by the property that $\partial(x_i) = \partial/\partial x_i$.

Let $\mathcal{J}_+ \subset \mathcal{P}^{\mathfrak{S}_n}$ be the set of invariant polynomials with constant term zero. We say that a polynomial $h(x) \in \mathcal{P}$ is *harmonic* (relative to $\mathfrak{S}_n$) if $\partial(f)h = 0$ for all $f \in \mathcal{J}_+$. Since $\mathcal{J}$ is generated by the power sums $s_1(x), \ldots, s_n(x)$ (see Exercises 5.1.3), the condition for $h(x)$ to be harmonic relative to $\mathfrak{S}_n$ can also be expressed as

$$\sum_{j=1}^{n} (\partial/\partial x_j)^k h(x) = 0 \qquad \text{for } k = 1, 2, \ldots, n . \tag{5.7}$$

We denote the space of $\mathfrak{S}_n$-harmonic polynomials by $\mathcal{H}$. If g is an $\mathfrak{S}_n$-harmonic polynomial, then each homogeneous component of g is also harmonic. Hence

$$\mathcal{H} = \bigoplus_{k \geq 0} \mathcal{H}^k ,$$

where $\mathcal{H}^k$ are the harmonic polynomials homogeneous of degree k. For example, we see from (5.7) that $\mathcal{H}^0 = \mathbb{C}$ and $\mathcal{H}^1$ consists of all linear forms $a_1 x_1 + \cdots + a_n x_n$ with $\sum_i a_i = 0$.

Theorem 5.1.4. *The multiplication map $f, g \mapsto fg$ extends to a linear isomorphism $\mathcal{J} \otimes \mathcal{H} \longrightarrow \mathcal{P}$. Hence $\mathcal{P}$ is a free $\mathcal{J}$-module on $\dim \mathcal{H}$ generators.*

The proof of this theorem will require some preliminary results. Let $(\mathcal{P}\mathcal{J}_+)^k$ be the homogeneous polynomials of degree k in $\mathcal{P}\mathcal{J}_+$.

Lemma 5.1.5. *One has $\mathcal{P}^k = \mathcal{H}^k \oplus (\mathcal{P}\mathcal{J}_+)^k$ for all k.*

Proof. Define $\langle f \mid g \rangle = \partial(f)g^*(0)$ for $f, g \in \mathcal{P}$, where g^* denotes the polynomial whose coefficients are the complex conjugates of those of g. Since $\langle x^I \mid x^J \rangle = I!$ if $I = J$ and is zero otherwise, the form $\langle f \mid g \rangle$ is Hermitian symmetric and positive definite. Furthermore,

$$\langle f \mid \partial(g)h \rangle = \langle fg^* \mid h \rangle .$$

Hence $h \in \mathcal{H}$ if and only if $\langle fg \mid h \rangle = 0$ for all $f \in \mathcal{P}$ and all $g \in \mathcal{J}_+$. Since $\langle \mathcal{P}^j \mid \mathcal{P}^k \rangle = 0$ if $j \neq k$, it follows that the space $\mathcal{H}^k$ is the orthogonal complement of $(\mathcal{P}\mathcal{J}_+)^k$ in $\mathcal{P}^k$. This implies the lemma. $\qquad\square$

Here is the key result needed in the proof of Theorem 5.1.4.

Lemma 5.1.6. *Let $f_1, \ldots, f_m \in \mathcal{J}$ and suppose $f_1 \notin \sum_{j=2}^m f_j \mathcal{J}$. If $g_1, \ldots, g_m \in \mathcal{P}$ satisfy the relation $\sum_{j=1}^m f_j g_j = 0$ and g_1 is homogeneous, then $g_1 \in \mathcal{P}\mathcal{J}_+$.*

Proof. Suppose $\deg g_1 = 0$. Then $g_1 = c$ is constant. If c were not zero, then we could write $f_1 = -(1/c)\sum_{j=2}^m f_j g_j$, which would be a contradiction. Hence $g_1 = 0$ and the lemma is true in this case.

Now assume that $\deg g_1 = d$ and the lemma is true for all relations with the degree of the first term less than d. Let $s \in \mathfrak{S}_n$ be a transposition $p \leftrightarrow q$. We have the relation

$$0 = \sum_{j=1}^m f_j(x)g_j(x) = \sum_{j=1}^m f_j(x)g_j(s \cdot x) ,$$

since $f_j \in \mathcal{J}$. Hence

$$\sum_{j=1}^m f_j(x)(g_j(x) - g_j(s \cdot x)) = 0 .$$

Now $g_j(x) - g_j(s \cdot x) = 0$ when $x_p = x_q$, so for $j = 1, \ldots, m$ there are polynomials $h_j(x)$ such that

$$g_j(x) - g_j(s \cdot x) = (x_p - x_q)h_j(x) .$$

Furthermore, $h_1(x)$ is homogeneous of degree $d - 1$. This gives the relation

$$(x_p - x_q)\sum_{j=1}^m f_j(x)h_j(x) = 0 .$$

Thus $\sum_{j=1}^{m} f_j h_j = 0$, so by the induction hypothesis $h_1(x) \in \mathcal{P}\mathcal{J}_+$. Hence

$$g_1 - \rho(s)g_1 \in \mathcal{P}\mathcal{J}_+ \tag{5.8}$$

for every transposition $s \in \mathfrak{S}_n$. Let $s \in \mathfrak{S}_n$ be arbitrary. Then there are transpositions $s_1, \ldots, s_l$ such that $s = s_1 \cdots s_l$. By writing $g_1 - \rho(s)g_1$ as a telescoping sum

$$g_1 - \rho(s)g_1 = g_1 - \rho(s_1)g_1 + \rho(s_1)(g_1 - \rho(s_2)g_1)$$
$$+ \cdots + \rho(s_1 \cdots s_{l-1})(g_1 - \rho(s_l)g_1)$$

and using the invariance of $\mathcal{P}\mathcal{J}_+$ under $\rho(\mathfrak{S}_n)$, we conclude that (5.8) holds for every element of $\mathfrak{S}_n$. Averaging this relation over $\mathfrak{S}_n$, we obtain

$$g_1 \equiv \frac{1}{n!} \sum_{s \in \mathfrak{S}_n} \rho(s)g_1 \quad \text{modulo } \mathcal{P}\mathcal{J}_+ . \tag{5.9}$$

Since the right side is in $\mathcal{J}_+$ (because g_1 is homogeneous of degree ≥ 1), this proves the lemma.   □

Proof of Theorem 5.1.4. We first prove that the map $\mathcal{H} \otimes \mathcal{J} \longrightarrow \mathcal{P}$ given by multiplication of functions is surjective. Let $f \in \mathcal{P}^k$. Since $1 \in \mathcal{H}$, we may assume by induction that $\mathcal{H} \cdot (\mathcal{P}\mathcal{J}_+)$ contains all polynomials of degree less than the degree of f. By Lemma 5.1.5 we may write $f = h + \sum_i f_i g_i$ with $h \in \mathcal{H}^k$ and $f_i \in \mathcal{P}$, $g_i \in \mathcal{J}_+$. Since g_i has zero constant term, the degree of f_i is less than the degree of f. By induction each f_i is in the space $\mathcal{H} \cdot (\mathcal{P}\mathcal{J}_+)$ and hence so is f. This proves that the map from $\mathcal{H} \otimes \mathcal{J}$ to $\mathcal{P}$ is surjective.

It remains to prove that the map is injective. Suppose there is a nontrivial relation

$$\sum_{j=1}^{m} f_j g_j = 0 , \tag{5.10}$$

where $0 \neq f_j \in \mathcal{J}$ and $g_j \in \mathcal{H}$. Since $\mathcal{H}$ is spanned by homogeneous elements, we may assume that each g_j is homogeneous and $\{g_1, \ldots, g_m\}$ is linearly independent. We may also assume that m is minimal among all such relations. Then there exists an index j such that $f_j \notin \sum_{i \neq j} f_i \mathcal{J}$ (since otherwise we could replace relation (5.10) by a relation with $m - 1$ summands). By renumbering, we can assume that f_1 has this property. But now Lemma 5.1.6 implies that $g_1 \in \mathcal{P}\mathcal{J}_+$. Hence $g_1 = 0$ by Lemma 5.1.5 (1), which contradicts the linear independence of $\{g_1, \ldots, g_m\}$. This proves the injectivity of the map.   □

Corollary 5.1.7. *The series $p_{\mathcal{H}}(t) = \sum_{j \geq 0} (\dim \mathcal{H}^j) t^j$ is a polynomial and has the factorization*

$$p_{\mathcal{H}}(t) = \prod_{k=1}^{n} (1 + t + \cdots + t^{k-1}) . \tag{5.11}$$

In particular, $\dim \mathcal{H} = p_{\mathcal{H}}(1) = n! = |\mathfrak{S}_n|$.

Proof. Define $f(t) = \sum_{j=0}^{\infty} (\dim \mathcal{J}_j) t^j$ and $g(t) = \sum_{j=0}^{\infty} (\dim \mathcal{P}^j) t^j$. Then for $|t| < 1$,

$$f(t) = \prod_{k=1}^{n}(1-t^k)^{-1} \quad \text{and} \quad g(t) = \prod_{k=1}^{n}(1-t)^{-1}$$

(see Exercises 5.1.3). The isomorphism $\mathcal{H} \otimes \mathcal{J} \cong \mathcal{P}$ as graded vector spaces implies that $p_{\mathcal{H}}(t)f(t) = g(t)$. Hence

$$p_{\mathcal{H}}(t) = \prod_{k=1}^{n}\left(\frac{1-t^k}{1-t}\right).$$

We obtain (5.11) by carrying out the division in each factor in this product. $\square$

From Corollary 5.1.7 we see that $p_{\mathcal{H}}(t) = t^{n(n-1)/2} + \cdots$. Hence the $\mathfrak{S}_n$-harmonic polynomials have degree at most $n(n-1)/2$, and there is a unique (up to a constant multiple) polynomial in $\mathcal{H}$ of degree $n(n-1)/2$. We now find this polynomial and show that it generates $\mathcal{H}$ as a $\mathcal{P}$-module, where $f \in \mathcal{P}$ acts on $\mathcal{H}$ by the differential operator $\partial(f)$.

Theorem 5.1.8. *The space $\mathcal{H}$ is spanned by the polynomial*

$$\Delta(x) = \prod_{1 \le i < j \le n}(x_i - x_j)$$

and its partial derivatives of all orders.

As a preliminary to proving this theorem, we observe that $\rho(s)\Delta(x) = \mathrm{sgn}(s)\Delta(x)$ for all $s \in \mathfrak{S}_n$, so $\Delta(x)$ is *skew invariant* under $\mathfrak{S}_n$. Furthermore, if $g(x)$ is any skew-invariant polynomial, then $g(x)$ is divisible by $x_i - x_j$ for every $i \ne j$, since $g(x)$ vanishes on the hyperplane $x_i = x_j$ in $\mathbb{C}^n$. Hence $\Delta(x)$ divides $g(x)$. Now to prove that $\Delta(x)$ is harmonic, take $f \in \mathcal{J}_+$. Then $g(x) = \partial(f)\Delta(x)$ is skew invariant, since f is invariant. Hence $g(x)$ is divisible by $\Delta(x)$. But $g(x)$ has degree less than the degree of $\Delta(x)$, so $g = 0$.

Let $\mathcal{E} = \mathrm{Span}\{\partial(g)\Delta(x) : g \in \mathcal{P}\}$. Since $\Delta(x) \in \mathcal{H}$, we have

$$\partial(f)\partial(g)\Delta(x) = \partial(g)\partial(f)\Delta(x) = 0 \quad \text{for all } f \in \mathcal{J}_+ \text{ and } g \in \mathcal{P}.$$

Hence $\mathcal{E} \subset \mathcal{H}$. We now need the following result:

Lemma 5.1.9. *Suppose $g \in \mathcal{P}^m$ and $\partial(g)\Delta = 0$. Then $g \in (\mathcal{P}\mathcal{J}_+)^m$.*

Proof. Since $\mathcal{H}$ is finite-dimensional, we have $\mathcal{P}^m = (\mathcal{P}\mathcal{J}_+)^m$ for m sufficiently large, by Lemma 5.1.5. Hence the lemma is true in this case. We assume by induction that the lemma holds for polynomials of degree greater than m. Take $1 \le i < j \le n$ and set $f(x) = x_i - x_j$. Then $fg \in \mathcal{P}^{m+1}$ and $\partial(fg)\Delta = \partial(f)\partial(g)\Delta = 0$. Hence by the induction hypothesis

$$fg = \sum_k u_k v_k \quad \text{with } u_k \in \mathcal{P} \text{ and } v_k \in \mathcal{J}_+.$$

Let $s \in \mathfrak{S}_n$ be the transposition $i \leftrightarrow j$. Then

$$fg - \rho(s)(fg) = \sum_k (u_k - \rho(s)u_k)v_k \,.$$

Since $u_k(x) - \rho(s)u_k(x) = 0$ when $x_i = x_j$, we can write $u_k - \rho(s)u_k = fw_k$ for some polynomial w_k. Also, $\rho(s)f = -f$, so

$$fg - \rho(s)(fg) = fg - (\rho(s)f)(\rho(s)g) = f(g + \rho(s)g) \,.$$

Thus $f \cdot (g + \rho(s)g) = f \cdot \sum_k w_k v_k$. Dividing by f, we obtain

$$g - \mathrm{sgn}(s)\rho(s)g \in \mathcal{P}\mathcal{J}_+ \tag{5.12}$$

for every transposition s. By a telescoping sum argument (see the proof of Lemma 5.1.6) we conclude that (5.12) holds for all $s \in \mathfrak{S}_n$. Now set

$$h = \frac{1}{|\mathfrak{S}_n|} \sum_{s \in \mathfrak{S}_n} \mathrm{sgn}(s)\rho(s)g \in \mathcal{P}^m \,.$$

Then (5.12) shows that $g - h \in \mathcal{P}\mathcal{J}_+$. Since $\rho(s)h = \mathrm{sgn}(s)h$ for all $s \in \mathfrak{S}_n$, we can write $h = \varphi\Delta$ for some homogeneous polynomial $\varphi \in \mathcal{J}$, as already noted. If φ has positive degree, then $g \in \varphi\Delta + \mathcal{P}\mathcal{J}_+ \subset \mathcal{P}\mathcal{J}_+$ as desired. If $\varphi = c \in \mathbb{C}$, then $c\partial(\Delta)\Delta = \partial(g)\Delta = 0$. But

$$\partial(\Delta)\Delta = \partial(\Delta)\Delta(0) = \langle \Delta \mid \Delta \rangle \neq 0 \,.$$

Hence $c = 0$ and $g \in \mathcal{P}\mathcal{J}_+$. $\square$

Proof of Theorem 5.1.8. Consider the map $\mathcal{H} \longrightarrow \mathcal{E}$ given by $h \mapsto \partial(h)\Delta$. If $\partial(h)\Delta = 0$, then since Δ is homogeneous we have $\partial(h_m)\Delta = 0$ for all m, where h_m is the homogeneous component of h of degree m. Hence $h_m \in \mathcal{H}^m \cap (\mathcal{P}\mathcal{J}_+)^m = 0$ (Lemma 5.1.5). This proves that $\dim \mathcal{H} \leq \dim \mathcal{E}$. Since $\mathcal{E} \subset \mathcal{H}$, we have $\mathcal{E} = \mathcal{H}$. $\square$

5.1.3 Exercises

1. A *binary form of degree n* is a homogeneous polynomial

$$a_0 x^n + a_1 x^{n-1}y + \cdots + a_{n-1}xy^{n-1} + a_n y^n$$

of degree n in two variables, with coefficients $a_i \in \mathbb{C}$. Consider the space $F^{(2)}$ of binary forms $f(x,y) = a_0 x^2 + 2a_1 xy + a_2 y^2$ of degree 2 and the representation ρ of $G = \mathbf{SL}(2,\mathbb{C})$ on $\mathcal{P}(F^{(2)})$.
(a) Show that the *discriminant* $D(f) = a_1^2 - a_0 a_2$ is a G-invariant polynomial on $F^{(2)}$. (HINT: Write $f(x,y) = [x,y]A\begin{bmatrix} x \\ y \end{bmatrix}$, where $A = \begin{bmatrix} a_0 & a_1 \\ a_1 & a_2 \end{bmatrix}$. Show that $D(f) = \det(A)$ and that for $g \in G$ the quadratic form $g \cdot f$ has the matrix gAg^t.)

(b) Show that $\mathcal{P}^k\big(F^{(2)}\big) \cong F^{(2k)} \oplus F^{(2k-4)} \oplus F^{(2k-8)} \oplus \ldots$; hence $\dim \mathcal{P}^k\big(F^{(2)}\big)^G$ is zero if k is odd and is one if k is even. (HINT: Use Hermite reciprocity (Theorem 4.1.20) to show that $\mathcal{P}^k\big(F^{(2)}\big)$ has character

$$\frac{(q^{k+2} - q^{-k-2})(q^{k+1} - q^{-k-1})}{(q - q^{-1})(q^2 - q^{-2})}$$
$$= q^{2k} + q^{2k-2} + 2q^{2k-4} + 2q^{2k-6} + 3q^{2k-8} + \cdots .$$

Then use Proposition 2.3.5 and the complete reducibility of G.)

(c) Show that $\mathcal{P}\big(F^{(2)}\big)^G = \mathbb{C}[D]$ is a polynomial ring in one variable. (HINT: We have $\dim \mathcal{P}^{2m}\big(F^{(2)}\big)^G \geq 1$ by (a). Now use (b).)

2. Let $G = \mathbf{O}(n, \mathbb{C}) = \{g \in M_n(\mathbb{C}) : g^t g = I\}$ and write r^2 for the polynomial $x_1^2 + \cdots + x_n^2$.

   (a) Show that $\mathcal{P}(\mathbb{C}^n)^G = \mathbb{C}[r^2]$. (HINT: $\{v \in \mathbb{C}^n : r^2(v) \neq 0\} = G \cdot e_1$. Thus if $f \in \mathcal{P}(\mathbb{C}^n)^G$ then f is completely determined by the polynomial $f(x, 0, \ldots, 0)$ in one variable.)

   (b) Prove that if $n \geq 2$ then the same result holds when $\mathbf{O}(n, \mathbb{C})$ is replaced by $\mathbf{SO}(n, \mathbb{C})$.

3. Let $G = \mathbf{Sp}(\mathbb{C}^{2n})$. Show that $\mathcal{P}(\mathbb{C}^{2n})^G = \mathbb{C} \cdot 1$. (HINT: $G \cdot e_1 = \mathbb{C}^{2n} - \{0\}$.)

4. For $x = [x_1, \ldots, x_n]$ let $s_k(x) = \sum_{i=1}^n x_i^k$ be the kth *power sum*. Prove that the set of functions $\{s_1(x), \ldots, s_n(x)\}$ is algebraically independent and $\mathbb{C}[x_1, \ldots, x_n]^{\mathfrak{S}_n} = \mathbb{C}[s_1, \ldots, s_n]$. (HINT: Show that the Jacobian determinant is given by

$$\frac{\partial(s_1, \ldots, s_n)}{\partial(x_1, \ldots, x_n)} = n! \prod_{1 \leq i < j \leq n} (x_j - x_i)$$

by reducing it to a Vandermonde determinant. Conclude that $\{s_1, \ldots, s_n\}$ is algebraically independent. Now use Theorem 5.1.3.)

5. Let $\mathbb{Z}_2^n = \{[\varepsilon_1, \ldots, \varepsilon_n] : \varepsilon_i = \pm 1\}$ and let $\mathfrak{S}_n$ act on $\mathbb{Z}_2^n$ by permuting the entries. Let $\mathfrak{B}_n = \mathfrak{S}_n \ltimes \mathbb{Z}_2^n$ be the semidirect product group (the Weyl group of type BC). There is a representation of $\mathfrak{B}_n$ on $\mathbb{C}^n$ where $\mathfrak{S}_n$ acts as permutations of coordinates and $\gamma \in \mathbb{Z}_2^n$ acts by $\gamma \cdot x = [\varepsilon_1 x_1, \ldots, \varepsilon_n x_n]$.

   (a) Prove that $\{s_2, s_4, \ldots, s_{2n}\}$ is a set of basic invariants for the action of $\mathfrak{B}_n$ on $\mathcal{P}(\mathbb{C}^n)$, where $s_k(x)$ is the kth power sum as in the previous exercise. (HINT: Show that $\mathbb{C}[x_1, \ldots, x_n]^{\mathbb{Z}_2^n} = \mathbb{C}[x_1^2, \ldots, x_n^2]$. Then set $y_k = x_k^2$ and consider the action of $\mathfrak{S}_n$ on $\mathbb{C}[y_1, \ldots, y_n]$.)

   (b) Prove that the set $\{s_2, s_4, \ldots, s_{2n}\}$ is algebraically independent. (HINT: Show that the Jacobian determinant is given by

$$\frac{\partial(s_2, s_4, \ldots, s_{2n})}{\partial(x_1, \ldots, x_n)} = 2^n n! (x_1 \cdots x_n) \prod_{1 \leq i < j \leq n} (x_j^2 - x_i^2)$$

by reducing it to a Vandermonde determinant.)

6. (*Notation as in previous exercise*) Let $(\mathbb{Z}_2^n)_{\text{even}} \subset \mathbb{Z}_2^n$ be the kernel of the homo-morphism $[\varepsilon_1, \dots, \varepsilon_n] \mapsto \varepsilon_1 \cdots \varepsilon_n$. Let $\mathfrak{D}_n = \mathfrak{S}_n \ltimes (\mathbb{Z}_2^n)_{\text{even}} \subset \mathfrak{B}_n$ be the semidirect product group (the Weyl group of type D).

(a) Prove that $\{\varphi, s_2, s_4, \dots, s_{2n-2}\}$ is a set of basic invariants for the action of $\mathfrak{D}_n$ on $\mathcal{P}(\mathbb{C}^n)$, where $s_k(x)$ is the kth power sum and $\varphi = x_1 \cdots x_n$. (HINT: Show that the $(\mathbb{Z}_2^n)_{\text{even}}$ invariant polynomials are $\mathbb{C}[x_1^2, \dots, x_n^2] \oplus \varphi \mathbb{C}[x_1^2, \dots, x_n^2]$. Conclude that $\mathbb{C}[x_1, \dots, x_n]^{\mathfrak{D}_n} = \mathbb{C}[\varphi, s_2, s_4, \dots, s_{2n}]$. Then use the relation $\varphi^2 = \sigma_{2n}$ to show that s_{2n} is a polynomial in $\varphi, s_2, s_4, \dots, s_{2n-2}$.)

(b) Prove that $\{\varphi, s_2, s_4, \dots, s_{2n-2}\}$ is algebraically independent. (HINT: Show that the Jacobian determinant is given by

$$\frac{\partial(\varphi, s_2, s_4, \dots, s_{2n-2})}{\partial(x_1, \dots, x_n)} = 2^{n-1}(n-1)! \prod_{1 \le i < j \le n} (x_j^2 - x_i^2)$$

by reducing it to a Vandermonde determinant.)

7. Let G be a reductive linear algebraic group G acting on a vector space V by a regular representation.

(a) Suppose there exists an algebraically independent set of homogeneous basic invariants $\{f_1, \dots, f_n\}$. Prove that the Hilbert series of $\mathcal{P}(V)^G$ is

$$H(t) = \prod_{i=1}^{n} (1 - t^{d_i})^{-1}, \quad \text{where } d_i = \deg f_i \ .$$

(b) Let $G = \mathfrak{S}_n, \mathfrak{B}_n$, or $\mathfrak{D}_n$ acting on $V = \mathbb{C}^n$ as in the previous exercises. Calculate the Hilbert series of $\mathcal{P}(V)^G$.

8. Let $\mathcal{H}$ be the $\mathfrak{S}_3$-harmonic polynomials in $\mathbb{C}[x_1, x_2, x_3]$.

(a) Show that $\dim \mathcal{H}^2 = \dim \mathcal{H}^1 = 2$. (HINT: Calculate the polynomial $p_{\mathcal{H}}(t)$.)

(b) Find a basis for $\mathcal{H}$.

9. Let the group $\mathfrak{B} = \mathfrak{B}_n = \mathfrak{S}_n \ltimes \mathbb{Z}_2^n$ be as in exercise #5 and let $\mathcal{P} = \mathbb{C}[x_1, \dots, x_n]$. Set $\mathcal{J}_{\mathfrak{B}} = \mathcal{P}^{\mathfrak{B}}$ and let $(\mathcal{J}_{\mathfrak{B}})_+$ be the $\mathfrak{B}$-invariant polynomials with constant term 0. We say that a polynomial $h(x)$ is $\mathfrak{B}$-*harmonic* if $\partial(f)h(x) = 0$ for all $f \in (\mathcal{J}_{\mathfrak{B}})_+$. Let $\mathcal{H}_{\mathfrak{B}}$ be the space of all $\mathfrak{B}$-harmonic polynomials and $\mathcal{H}_{\mathfrak{B}}^k$ the subspace of homogeneous polynomials of degree k.

(a) Prove that the multiplication map $f, g \mapsto fg$ extends to a linear isomorphism $\mathcal{J}_{\mathfrak{B}} \otimes \mathcal{H}_{\mathfrak{B}} \longrightarrow \mathcal{P}$. (HINT: The proof of Theorem 5.1.4 uses only Lemmas 5.1.5 and 5.1.6. The proof of Lemma 5.1.5 applies without change. In the proof of Lemma 5.1.6 for this case use the fact that $\mathfrak{B}$ is generated by transpositions and the transformation τ, where $\tau(x_1) = -x_1$ and $\tau(x_j) = x_j$ for $j > 1$. Note that if $g \in \mathcal{P}$ then $g(\tau \cdot x) - g(x)$ is divisible by x_1, so the same inductive proof applies to obtain (5.8) for all $s \in \mathfrak{B}$. This relation then implies (5.9), with the average now over $\mathfrak{B}$.)

(b) Set $p_{\mathcal{H}_{\mathfrak{B}}}(t) = \sum_{j \ge 0} (\dim \mathcal{H}_{\mathfrak{B}}^j) t^j$. Prove that

$$p_{\mathcal{H}_{\mathfrak{B}}}(t) = \prod_{k=1}^{n} (1 + t + \cdots + t^{2k-1}) .$$

Conclude that $\dim \mathcal{H}_\mathfrak{B} = 2^n n! = |\mathfrak{B}|$, that the $\mathfrak{B}$-harmonic polynomials have degrees at most n^2, and that there is a unique (up to a constant multiple) $\mathfrak{B}$-harmonic polynomial of degree n^2. (HINT: Use the calculation of the Hilbert series for $\mathcal{J}_\mathfrak{B}$ in exercise #7 and the proof of Corollary 5.1.7.)

(c) Set $\Delta_\mathfrak{B}(x) = x_1 \cdots x_n \prod_{1 \le i < j \le n} (x_i^2 - x_j^2)$. Prove that $\Delta_\mathfrak{B}(x)$ is $\mathfrak{B}$-harmonic and that $\mathcal{H}_\mathfrak{B}$ is spanned by $\Delta_\mathfrak{B}(x)$ and its partial derivatives of all orders. (HINT: Use the proof of Theorem 5.1.8, with $\mathfrak{S}_n$ replaced by $\mathfrak{B}$ and $\Delta(x)$ replaced by $\Delta_\mathfrak{B}(x)$. Here $\text{sgn}(s)$ is defined for $s \in \mathfrak{B}$ by the relation $\Delta_\mathfrak{B}(s \cdot x) = \text{sgn}(s)\Delta_\mathfrak{B}(x)$. The divisibility properties needed in the proof follow by the same argument as in (a).)

10. Let the group $\mathfrak{D} = \mathfrak{D}_n = \mathfrak{S}_n \ltimes (\mathbb{Z}_2^n)_{\text{even}}$ be as in exercise #6 and let $\mathcal{P} = \mathbb{C}[x_1, \ldots, x_n]$. Set $\mathcal{J}_\mathfrak{D} = \mathcal{P}^\mathfrak{D}$ and let $(\mathcal{J}_\mathfrak{D})_+$ be the $\mathfrak{D}$-invariant polynomials with constant term 0. We say that a polynomial $h(x)$ is $\mathfrak{D}$-harmonic if $\partial(f)h(x) = 0$ for all $f \in (\mathcal{J}_\mathfrak{D})_+$. Let $\mathcal{H}_\mathfrak{D}$ be the space of all $\mathfrak{D}$-harmonic polynomials and $\mathcal{H}_\mathfrak{D}^k$ the subspace of homogeneous polynomials of degree k.

(a) Prove that the multiplication map $f, g \mapsto fg$ extends to a linear isomorphism
$$\mathcal{J}_\mathfrak{D} \otimes \mathcal{H}_\mathfrak{D} \longrightarrow \mathcal{P}.$$

(b) Set $p_{\mathcal{H}_\mathfrak{D}}(t) = \sum_{j \ge 0} (\dim \mathcal{H}_\mathfrak{D}^j) t^j$. Prove that

$$p_{\mathcal{H}_\mathfrak{D}}(t) = (1 + t + \cdots + t^{n-1}) \prod_{k=1}^{n-1} (1 + t + \cdots + t^{2k-1}).$$

Conclude that $\dim \mathcal{H}_\mathfrak{D} = n! 2^{n-1} = |\mathfrak{D}|$, that the $\mathfrak{D}$-harmonic polynomials have degrees at most $n(n-1)$, and that there is a unique (up to a constant multiple) $\mathfrak{D}$-harmonic polynomial of degree $n(n-1)$.

(c) Set $\Delta_\mathfrak{D}(x) = \prod_{1 \le i < j \le n} (x_i^2 - x_j^2)$. Prove that $\Delta_\mathfrak{D}(x)$ is $\mathfrak{D}$-harmonic and that $\mathcal{H}_\mathfrak{D}$ is spanned by $\Delta_\mathfrak{D}(x)$ and its partial derivatives of all orders. (HINT: Follow the same approach as in the previous exercise. In this case use the fact that $\mathfrak{D}$ is generated by transpositions and the transformation τ, where $\tau(x_1) = -x_2$, $\tau(x_2) = -x_1$, and $\tau(x_j) = x_j$ for $j \ge 3$. Note that $\Delta_\mathfrak{D}(x)$ changes sign under the action of these generators.)

5.2 Polynomial Invariants

Let G be a reductive linear algebraic group and (ρ, V) a regular representation of G. For each positive integer k, let $V^k = V \oplus \cdots \oplus V$ (k copies); this space should not be confused with the k-fold tensor product $V^{\otimes k} = \bigotimes^k V$. Likewise, let $(V^*)^k$ be the direct sum of k copies of V^*. Given positive integers k and m, consider the algebra $\mathcal{P}((V^*)^k \oplus V^m)$ of polynomials with k covector arguments (elements of V^*) and m vector arguments (elements of V). The action of $g \in G$ on $f \in \mathcal{P}((V^*)^k \oplus V^m)$ is

$$g \cdot f(v_1^*, \ldots, v_k^*, v_1, \ldots, v_m)$$
$$= f(\rho^*(g^{-1})v_1^*, \ldots, \rho^*(g^{-1})v_k^*, \rho(g^{-1})v_1, \ldots, \rho(g^{-1})v_m),$$

where ρ^* is the representation on V^* dual to ρ. We shall refer to a description of (finite) generating sets for $\mathcal{P}((V^*)^k \oplus V^m)^G$, for all k, m, as a *first fundamental theorem* (FFT) for the pair (G, ρ). Here the emphasis is on *explicit formulas* for generating sets; the existence of a finite generating set of invariants (for each k, m) is a consequence of Theorem 5.1.1. In this section we give the FFT for the general linear groups, orthogonal groups, and symplectic groups in their defining representations. We prove a basic case for the orthogonal and symplectic groups; proofs for the general case are deferred to later sections.

5.2.1 First Fundamental Theorems for Classical Groups

Let G be a reductive linear algebraic group and (ρ, V) a regular representation of G. Since $\mathcal{P}((V^*)^k \oplus V^m)^G \supset \mathcal{P}((V^*)^k \oplus V^m)^{\mathbf{GL}(V)}$, an FFT for $\mathbf{GL}(V)$ gives some information about invariants for the group $\rho(G)$, so we first consider this case. The key observation is that $\mathbf{GL}(V)$-invariant polynomials on $(V^*)^k \oplus V^m$ come from the following geometric construction:

There are natural isomorphisms

$$(V^*)^k \cong \mathrm{Hom}(V, \mathbb{C}^k), \qquad V^m \cong \mathrm{Hom}(\mathbb{C}^m, V).$$

Here the direct sum $v_1^* \oplus \cdots \oplus v_k^*$ of k covectors corresponds to the linear map

$$v \mapsto [\langle v_1^*, v \rangle, \ldots, \langle v_k^*, v \rangle]$$

from V to $\mathbb{C}^k$, whereas the direct sum $v_1 \oplus \cdots \oplus v_m$ of m vectors corresponds to the linear map

$$[c_1, \ldots, c_m] \mapsto c_1 v_1 + \cdots + c_m v_m$$

from $\mathbb{C}^m$ to V. This gives an algebra isomorphism

$$\mathcal{P}((V^*)^k \oplus V^m) \cong \mathcal{P}(\mathrm{Hom}(V, \mathbb{C}^k) \oplus \mathrm{Hom}(\mathbb{C}^m, V)),$$

with the action of $g \in \mathbf{GL}(V)$ on $f \in \mathcal{P}(\mathrm{Hom}(V, \mathbb{C}^k) \oplus \mathrm{Hom}(\mathbb{C}^m, V))$ becoming

$$g \cdot f(x, y) = f(x\rho(g^{-1}), \rho(g)y). \tag{5.13}$$

We denote the vector space of $k \times m$ complex matrices by $M_{k,m}$. Define a map

$$\mu : \mathrm{Hom}(V, \mathbb{C}^k) \oplus \mathrm{Hom}(\mathbb{C}^m, V) \longrightarrow M_{k,m}$$

by $\mu(x, y) = xy$ (composition of linear transformations). Then

$$\mu(x\rho(g^{-1}), \rho(g)y) = x\rho(g)^{-1}\rho(g)y = \mu(x, y)$$

for $g \in \mathbf{GL}(V)$. The induced homomorphism μ^* on $\mathcal{P}(M_{k,m})$ has range in the $\mathbf{GL}(V)$-invariant polynomials:

$$\mu^* : \mathcal{P}(M_{k,m}) \longrightarrow \mathcal{P}(\mathrm{Hom}(V, \mathbb{C}^k) \oplus \mathrm{Hom}(\mathbb{C}^m, V))^{\mathbf{GL}(V)} \,,$$

where, as usual, $\mu^*(f) = f \circ \mu$ for $f \in \mathcal{P}(M_{k,m})$. Thus if we let $z_{ij} = \mu^*(x_{ij})$ be the image of the matrix entry function x_{ij} on $M_{k,m}$, then z_{ij} is the *contraction* of the ith covector position with the jth vector position:

$$z_{ij}(v_1^*, \ldots, v_k^*, v_1, \ldots, v_m) = \langle v_i^*, v_j \rangle \,.$$

Theorem 5.2.1. (Polynomial FFT for $\mathbf{GL}(V)$) *The map*

$$\mu^* : \mathcal{P}(M_{k,m}) \longrightarrow \mathcal{P}((V^*)^k \oplus V^m)^{\mathbf{GL}(V)}$$

is surjective. Hence $\mathcal{P}((V^*)^k \oplus V^m)^{\mathbf{GL}(V)}$ *is generated (as an algebra) by the contractions* $\{\langle v_i^*, v_j \rangle : i = 1, \ldots, k, \, j = 1, \ldots, m\}$.

We shall prove this theorem in Section 5.4.2.

We next consider the case in which $\rho(G)$ leaves invariant a nondegenerate symmetric or skew-symmetric bilinear form (see Section 3.2.4). The basic case is the defining representations of the orthogonal and symplectic groups . Here we obtain the invariant polynomials by the following modification of the geometric construction used for $\mathbf{GL}(V)$.

Let $V = \mathbb{C}^n$ and define the symmetric form

$$(x, y) = \sum_i x_i y_i \quad \text{for } x, y \in \mathbb{C}^n \,. \tag{5.14}$$

Write $\mathbf{O}_n$ for the orthogonal group for this form. Thus $g \in \mathbf{O}_n$ if and only if $g^t g = I_n$. Let SM_k be the vector space of $k \times k$ complex *symmetric* matrices B, and define a map $\tau : M_{n,k} \longrightarrow SM_k$ by $\tau(X) = X^t X$. Then

$$\tau(gX) = X^t g^t g X = \tau(X) \quad \text{for } g \in \mathbf{O}_n \text{ and } X \in M_{n,k} \,.$$

Hence $\tau^*(f)(gX) = \tau^*(f)(X)$ for $f \in \mathcal{P}(SM_k)$, and we obtain an algebra homomorphism

$$\tau^* : \mathcal{P}(SM_k) \longrightarrow \mathcal{P}(V^k)^{\mathbf{O}_n} \,.$$

For example, given $v_1, \ldots, v_k \in \mathbb{C}^n$, let $X = [v_1, \ldots, v_k] \in M_{n,k}$ be the matrix with these vectors as columns. Then $X^t X$ is the $k \times k$ symmetric matrix with entries (v_i, v_j). Hence under the map τ^* the matrix entry function x_{ij} on SM_k pulls back to the $\mathbf{O}_n$-invariant quadratic polynomial

$$\tau^*(x_{ij})(v_1, \ldots, v_k) = (v_i, v_j)$$

on $(\mathbb{C}^n)^k$ (the contraction of the ith and jth vector positions using the symmetric form).

Now assume that $n = 2m$ is even. Let J_n be the $n \times n$ block-diagonal matrix

$$J_n = \text{diag}[\underbrace{\kappa, \ldots, \kappa}_{m}], \qquad \text{where } \kappa = \begin{bmatrix} 0 & 1 \\ -1 & 0 \end{bmatrix}.$$

Define the skew-symmetric form

$$\omega(x, y) = (x, J_n y) \tag{5.15}$$

for $x, y \in \mathbb{C}^n$, and let $\mathbf{Sp}_n$ be the invariance group of this form. Thus $g \in \mathbf{Sp}_n$ if and only if $g^t J_n g = J_n$. Let AM_k be the vector space of $k \times k$ complex *skew-symmetric* matrices A, and define a map $\gamma : M_{n,k} \longrightarrow AM_k$ by $\gamma(X) = X^t J_n X$. Then

$$\gamma(gX) = X^t g^t J_n g X = \gamma(X) \quad \text{for } g \in \mathbf{Sp}_n \text{ and } X \in M_{n,k} .$$

Hence $\gamma^*(f)(gX) = \gamma^*(f)(X)$ for $f \in \mathcal{P}(AM_k)$, and we obtain an algebra homomorphism

$$\gamma^* : \mathcal{P}(AM_k) \longrightarrow \mathcal{P}(V^k)^{\mathbf{Sp}_n} .$$

As in the orthogonal case, given $v_1, \ldots, v_k \in \mathbb{C}^n$, let $X = [v_1, \ldots, v_k] \in M_{n,k}$. Then the skew-symmetric $k \times k$ matrix $X^t J_n X$ has entries $(v_i, J_n v_j)$. Hence the matrix entry function x_{ij} on AM_k pulls back to the $\mathbf{Sp}_n$-invariant quadratic polynomial

$$\gamma^*(x_{ij})(v_1, \ldots, v_k) = \omega(v_i, v_j) \cdot$$

(the contraction of the ith and jth positions, $i < j$, using the skew form).

Theorem 5.2.2. (Polynomial FFT for $\mathbf{O}_n$ and $\mathbf{Sp}_n$) *Let $V = \mathbb{C}^n$.*

1. *The homomorphism $\tau^* : \mathcal{P}(SM_k) \longrightarrow \mathcal{P}(V^k)^{\mathbf{O}_n}$ is surjective. Hence the algebra of $\mathbf{O}_n$-invariant polynomials in k vector arguments is generated by the quadratic polynomials $\{(v_i, v_j) : 1 \leq i \leq j \leq k\}$.*
2. *Suppose n is even. The homomorphism $\gamma^* : \mathcal{P}(AM_k) \longrightarrow \mathcal{P}(V^k)^{\mathbf{Sp}_n}$ is surjective. Hence the algebra of $\mathbf{Sp}_n$-invariant polynomials in k vector arguments is generated by the quadratic polynomials $\{\omega(v_i, v_j) : 1 \leq i < j \leq k\}$.*

We shall prove this theorem in Section 5.4.3. An immediate consequence is the following FFT for invariant polynomials of k covector and m vector variables:

Corollary 5.2.3. *Let $V = \mathbb{C}^n$.*

1. *Let $G = \mathbf{O}_n$. Then $\mathcal{P}((V^*)^k \oplus V^m)^G$ is generated by the quadratic polynomials (v_i, v_j), (v_p^*, v_q^*), and $\langle v_p^*, v_i \rangle$, for $1 \leq i, j \leq m$ and $1 \leq p, q \leq k$.*
2. *Let $G = \mathbf{Sp}_n$ with n even. Then $\mathcal{P}((V^*)^k \oplus V^m)^G$ is generated by the quadratic polynomials $\omega(v_i, v_j)$, $\omega(v_p^*, v_q^*)$, and $\langle v_p^*, v_i \rangle$, for $1 \leq i, j \leq m$ and $1 \leq p, q \leq k$.*

Proof. The G-invariant bilinear form gives an isomorphism $\varphi : (V^k)^* \cong V^k$ as G-modules. In the orthogonal case $(v_p^*, v_q^*) = (\varphi(v_p^*), \varphi(v_q^*))$ and $\langle v_p^*, v_i \rangle = (\varphi(v_p^*), v_i)$, so (1) follows from Theorem 5.2.2 (1). The same argument applies in the symplectic case, with the form $(\cdot, \cdot)$ replaced by the skew form ω. $\square$

We now prove a linear algebra result that will be used to obtain the FFT for the orthogonal and symplectic groups and will be essential for all cases of the *second fundamental theorem.*

Lemma 5.2.4. *Let* $V = \mathbb{C}^n$ *and take the maps* μ, τ, *and* γ *as above.*

1. *The image of* μ *consists of all matrices* Z *with* $\mathrm{rank}(Z) \leq \min(k,m,n)$.
2. *The image of* τ *consists of all symmetric matrices* B *with* $\mathrm{rank}(B) \leq \min(k,n)$.
3. *The image of* γ *consists of all skew-symmetric matrices* A *with* $\mathrm{rank}(A) \leq \min(k,n)$.

Proof. (1): Let $Z \in M_{k,m}$ have rank $r \leq \min(k,m,n)$. We may assume $k \leq m$ (otherwise replace Z by Z^t). Then by row and column reduction of Z we can find $u \in \mathbf{GL}(k)$ and $w \in \mathbf{GL}(m)$ such that

$$uZw = \begin{bmatrix} I_r & 0 \\ 0 & 0 \end{bmatrix},$$

where I_r denotes the $r \times r$ identity matrix and 0 denotes a zero matrix of the appropriate size to fill the matrix. Take

$$X = \begin{bmatrix} I_r & 0 \\ 0 & 0 \end{bmatrix} \in M_{k,n} \quad \text{and} \quad Y = \begin{bmatrix} I_r & 0 \\ 0 & 0 \end{bmatrix} \in M_{n,m}.$$

Then $XY = uZw$, so we have $Z = \mu(u^{-1}X, Yw^{-1})$.

(2): Given $B \in SM_k$, we let $Q(x,y) = (Bx, y)$ be the associated symmetric bilinear form on $\mathbb{C}^k$. Let r be the rank of B. Let $N = \{x \in \mathbb{C}^k : Bx = 0\}$ and choose an r-dimensional subspace $W \subset \mathbb{C}^k$ such that $\mathbb{C}^k = W \oplus N$. Then $Q(N, W) = 0$, so this decomposition is orthogonal relative to the form Q. Furthermore, the restriction of Q to W is nondegenerate. Indeed, if $w \in W$ and $(Bw, w') = 0$ for all $w' \in W$, then $(Bw, w' + z) = (w, Bz) = 0$ for all $w' \in W$ and $z \in N$. Hence $Bw = 0$, which forces $w \in W \cap N = \{0\}$. Assume $W \neq 0$ and let B be the restriction of Q to W. Then by Lemma 1.1.2 there is a B-orthonormal basis $f_1, \ldots, f_r$ for W. Take $f_{r+1}, \ldots, f_k$ to be any basis for N. Then $Q(f_i, f_j) = 0$ for $i \neq j$, $Q(f_i, f_i) = 1$ for $1 \leq i \leq r$, and $Q(f_i, f_i) = 0$ for $r < i \leq k$. Let $\{e_i : i = 1, \ldots, n\}$ be the standard basis for $\mathbb{C}^k$ and define $g \in \mathbf{GL}(k, \mathbb{C})$ by $g f_i = e_i$ for $1 \leq i \leq k$. Then

$$(Bg^{-1}e_i, g^{-1}e_i) = Q(f_i, f_j) = \begin{cases} \delta_{ij} & \text{if } 1 \leq i, j \leq r, \\ 0 & \text{otherwise}. \end{cases}$$

Hence $B = g^t \begin{bmatrix} I_r & 0 \\ 0 & 0 \end{bmatrix} g$. Since $r \leq \min(k,n)$, we can set $X = \begin{bmatrix} I_r & 0 \\ 0 & 0 \end{bmatrix} g \in M_{n,k}$. Then $B = X^t X$ as desired.

(3): Given $A \in AM_k$ of rank $r = 2p$, we let $\Omega(x,y) = (Ax,y)$ be the associated skew-symmetric form on $\mathbb{C}^k$. We follow the same argument as in (2) (now using Lemma 1.1.5) to obtain a basis $\{f_1,\ldots,f_k\}$ for $\mathbb{C}^k$ such that

$$\Omega(f_{2i},f_{2i-1}) = -\Omega(f_{2i-1},f_{2i}) = 1 \quad \text{for } i = 1,\ldots,p,$$

and $\Omega(f_i,f_j) = 0$ otherwise. Define $g \in \mathbf{GL}(k)$ by $gf_i = e_i$ for $i = 1,\ldots,k$. Then $A = g^t \begin{bmatrix} J_r & 0 \\ 0 & 0 \end{bmatrix} g$. Since $r \leq \min(k,n)$, we can set $X = \begin{bmatrix} J_r & 0 \\ 0 & 0 \end{bmatrix} g \in M_{n,k}$. From the relation $\kappa^2 = -I_2$ we see that $A = X^t J_n X$. $\qquad\square$

Combining Lemma 5.2.4 with Theorems 5.2.1 and 5.2.2, we obtain the following special case of the *second fundamental theorem* (SFT) for the classical groups (the complete SFT will be proved in Section 12.2.4):

Corollary 5.2.5. (SFT, Free Case) *Let $V = \mathbb{C}^n$.*

1. *If $n \geq \min(k,m)$ then $\mu^* : \mathcal{P}(M_{k,m}) \longrightarrow \mathcal{P}((V^*)^k \oplus V^m)^{\mathbf{GL}(V)}$ is bijective. Let z_{ij} be the contraction $\langle v_i^*, v_j \rangle$. The polynomials $\{z_{ij} : 1 \leq i \leq k, 1 \leq j \leq m\}$ are algebraically independent and generate $\mathcal{P}((V^*)^k \oplus V^m)^{\mathbf{GL}(V)}$.*

2. *If $n \geq k$ then $\tau^* : \mathcal{P}(SM_k) \longrightarrow \mathcal{P}(V^k)^{\mathbf{O}_n}$ is bijective. Let b_{ij} be the orthogonal contraction (v_i, v_j). The polynomials $\{b_{ij} : 1 \leq i \leq j \leq k\}$ are algebraically independent and generate $\mathcal{P}(V^k)^{\mathbf{O}_n}$.*

3. *If n is even and $n \geq k$ then $\gamma^* : \mathcal{P}(AM_k) \longrightarrow \mathcal{P}(V^k)^{\mathbf{Sp}_n}$ is bijective. Let a_{ij} be the symplectic contraction $\omega(v_i, v_j)$. The polynomials $\{a_{ij} : 1 \leq i < j \leq k\}$ are algebraically independent and generate $\mathcal{P}(V^k)^{\mathbf{Sp}_n}$.*

Proof. From the FFT, the maps μ^*, τ^*, and γ^* are surjective. By Lemma 5.2.4 the maps μ, τ, and γ are surjective when $n \geq \min(m,k)$. This implies that μ^*, τ^*, and γ^* are also injective. The contractions z_{ij} (respectively b_{ij} or a_{ij}) are the images by these maps of the linear coordinate functions on the $k \times m$ (respectively $k \times k$ symmetric or skew-symmetric) matrices. Hence they are algebraically independent and generate the algebra of invariants. $\qquad\square$

5.2.2 Proof of a Basic Case

We begin the proof of the FFT for the orthogonal and symplectic groups by establishing the following special case of Theorem 5.2.2.

Proposition 5.2.6. *Let ω be a symmetric or skew-symmetric nonsingular bilinear form on V. Let $G \subset \mathbf{GL}(V)$ be the group preserving ω. Suppose $f \in \mathcal{P}(V^n)^G$, where $n = \dim V$. Then f is a polynomial function of the quadratic G-invariants $\{\omega(v_i,v_j) : 1 \leq i \leq j \leq n\}$.*

Proof. We may assume that f is a homogeneous polynomial. Since $-I \in G$, a G-invariant homogeneous polynomial of odd degree must be zero. Hence we may assume f homogeneous of degree $2k$.

Orthogonal Case: Fix a basis $\{e_1,\ldots,e_n\}$ for V such that the form ω is given by (5.14) and identify V with $\mathbb{C}^n$ by this basis. We identify V^n with M_n by viewing the columns of $X \in M_n$ as n vectors in V. The action of G on V^n becomes left multiplication on M_n and f becomes a polynomial function on M_n. We must prove that there is a polynomial F on the space SM_n of symmetric matrices so that

$$f(X) = F(X^t X) \tag{5.16}$$

(as in Theorem 5.2.2).

We proceed by induction on n. We embed $\mathbb{C}^{n-1} \subset \mathbb{C}^n$ as $\mathrm{Span}\{e_1,\ldots,e_{n-1}\}$. Write $G = G_n$ and embed G_{n-1} into G_n as the subgroup fixing e_n. Consider the restriction of f to the set

$$U = \{[v_1,\ldots,v_n] : (v_1,v_1) \neq 0\}.$$

Let $[v_1,\ldots,v_n] \in U$. There exist $\lambda \in \mathbb{C}$ with $\lambda^2 = (v_1,v_1)$ and $g_1 \in G$ such that $g_1 v_1 = \lambda e_n$ (Lemma 1.1.2). Set $v'_i = \lambda^{-1} g_1 v_{i+1}$ for $i = 1,\ldots,n-1$. Then we can write

$$f(v_1,\ldots,v_n) = f(g_1 v_1,\ldots,g_1 v_n) = \lambda^{2k} f(e_n, v'_1,\ldots,v'_{n-1})$$
$$= (v_1,v_1)^k \widehat{f}(v'_1,\ldots,v'_{n-1}),$$

where $\widehat{f}(v'_1,\ldots,v'_{n-1}) = f(e_n, v'_1,\ldots,v'_{n-1})$. Note that in the case $n = 1$, $\widehat{f}$ is a constant, and the proof is done.

Now assume $n > 1$. Define $v''_j = v'_j - t_j e_n$, where $t_j = (v'_j, e_n)$ for $j = 1,\ldots,n-1$. Then $v_j = v''_j \oplus t_j e_n$ and we can expand $\widehat{f}$ as a polynomial in the variables $\{t_j\}$ with coefficients that are polynomial functions of $v''_1,\ldots,v''_{n-1}$:

$$\widehat{f}(v'_1,\ldots,v'_{n-1}) = \sum_I \widehat{f}_I(v''_1,\ldots,v''_{n-1}) t^I. \tag{5.17}$$

Observe that for $g \in G_{n-1}$ one has

$$\widehat{f}(g v'_1,\ldots,g v'_{n-1}) = f(g e_n, g v'_1,\ldots,g v'_{n-1}) = \widehat{f}(v'_1,\ldots,v'_{n-1}).$$

Thus $\widehat{f}$ is invariant under G_{n-1}. Furthermore, the variables t_j are unchanged under the substitution $v'_j \mapsto g v'_j$, since $(g v'_j, e_n) = (v'_j, g^{-1} e_n) = (v'_j, e_n)$. It follows by the linear independence of the monomials $\{t^I\}$ that the polynomials $\widehat{f}_I$ are invariant under G_{n-1}.

The induction hypothesis applied to G_{n-1} furnishes polynomials φ_I in the variables $\{(v''_i, v''_j) : 1 \leq i \leq j \leq n-1\}$ such that $\widehat{f}_I = \varphi_I$. Now we have

$$t_i = \lambda^{-1}(g_1 v_{i+1}, e_n) = \lambda^{-1}(v_{i+1}, g_1^{-1} e_n) = (v_{i+1}, v_1)/(v_1, v_1).$$

Also $(v'_i, v'_j) = (v''_i, v''_j) + t_i t_j$. Hence

$$(v_i'', v_j'') = \lambda^{-2}(g_1 v_{i+1}, g_1 v_{j+1}) - \frac{(v_{i+1}, v_1)(v_{j+1}, v_1)}{(v_1, v_1)^2}$$

$$= \frac{(v_{i+1}, v_{j+1})(v_1, v_1) - (v_{i+1}, v_1)(v_{j+1}, v_1)}{(v_1, v_1)^2}.$$

Substituting this formula for (v_i'', v_j'') in the polynomials φ_I and using the expansion (5.17), we conclude that there are a positive integer r_1 and a polynomial function φ_1 on M_n such that

$$f(X) = (v_1, v_1)^{-r_1} \varphi_1(X^t X)$$

for all $X = [v_1, \ldots, v_n] \in M_n$ such that $(v_1, v_1) \neq 0$.

By the same argument, there are a positive integer r_2 and a polynomial function φ_2 on M_n such that

$$f(X) = (v_2, v_2)^{-r_2} \varphi_2(X^t X)$$

for all $X = [v_1, \ldots, v_n] \in M_n$ such that $(v_2, v_2) \neq 0$. Denote the entries of the matrix $u = X^t X$ by u_{ij}. We have shown that

$$(u_{11})^{r_1} \varphi_2(u) = (u_{22})^{r_2} \varphi_1(u) \tag{5.18}$$

on the subset of SM_n where $u_{11} \neq 0$ and $u_{22} \neq 0$. Since both sides of the equation are polynomials, (5.18) holds for all $u \in SM_n$. Since $\{u_{ij} : 1 \leq i \leq j \leq n\}$ are coordinate functions on SM_n, it follows that $\varphi_2(u)$ is divisible by $(u_{22})^{r_2}$. Thus $\varphi_1(u) = (u_{11})^{r_1} F(u)$ with F a polynomial, and hence $f(X) = F(X^t X)$ for all $X \in M_n$, completing the induction.

Symplectic Case: Let $n = 2l$ and fix a basis $\{e_1, \ldots, e_n\}$ for V so that the form ω is given by (5.15). We identify V with $\mathbb{C}^n$ by this basis and view f as a polynomial function on M_n, just as in the orthogonal case. We must prove that there is a polynomial F on the space AM_n of skew-symmetric matrices such that

$$f(X) = F(X^t J_n X) \tag{5.19}$$

(as in Theorem 5.2.2).

We embed $\mathbb{C}^{n-2} \subset \mathbb{C}^n$ as the span of $e_1, \ldots, e_{n-2}$. Write $G = G_l$ and embed G_{l-1} into G_l as the subgroup fixing e_{n-1} and e_n. Consider the restriction of f to the set

$$W = \{[v_1, \ldots, v_n] : (v_1, J_n v_2) \neq 0\}.$$

Let $[v_1, \ldots, v_n] \in W$ and set $\lambda = \omega(v_1, v_2)$. Then by Lemma 1.1.5 there exists $g_1 \in G_l$ such that $g_1 v_1 = \lambda e_{n-1}$ and $g_1 v_2 = e_n$. Set $v_i' = g_1 v_{i+2}$ for $1 \leq i \leq n-2$. Then

$$f(v_1, \ldots, v_n) = f(g_1 v_1, \ldots, g_1 v_n) = f(\lambda e_{n-1}, e_n, v_1', \ldots, v_{n-2}').$$

If $g \in G_{l-1}$ then $f(\lambda e_{n-1}, e_n, g v_1', \ldots, g v_{n-2}') = f(\lambda e_{n-1}, e_n, v_1', \ldots, v_{n-2}')$. Thus in the expansion

$$f(\lambda e_{n-1}, e_n, v_1', \ldots, v_{n-2}') = \sum_{k=0}^{2k} \lambda^k f_k(v_1', \ldots, v_{n-2}') , \qquad (5.20)$$

the coefficients $f_k(v_1', \ldots, v_{n-2}')$ are invariant under G_{l-1}. Note that in the case $l = 1$, f_k is a constant and the proof is done. In any case (5.20), which was derived under the assumption $\omega(v_{n-1}, v_n) \neq 0$, holds for all $\lambda \in \mathbb{C}$ and all $v_1', \ldots, v_{n-2}'$ in $\mathbb{C}^n$.

Assume that $l > 1$ and the result is proved for $l - 1$. Given $v_1', \ldots, v_{n-2}'$ in $\mathbb{C}^n$, define

$$v_j'' = v_j' - s_j e_{n-1} - t_j e_n ,$$

where $t_j = \omega(v_j', e_{n-1})$ and $s_j = -\omega(v_j', e_n)$. Then

$$\omega(v_j'', e_{n-1}) = \omega(v_j'', e_n) = 0 ,$$

so that $v_j'' \in \mathbb{C}^{n-2}$. We expand f_k as a polynomial in the variables $\{t_j, s_j\}$ with coefficients that are polynomial functions of $v_1'', \ldots, v_{n-2}''$:

$$f_k(v_1', \ldots, v_{n-2}') = \sum_{I,J} f_{I,J,k}(v_1'', \ldots, v_{n-2}'') s^I t^J . \qquad (5.21)$$

Observe that for $g \in G_{l-1}$ the variables s_j and t_j are unchanged under the substitution $v_j' \mapsto g v_j'$, since

$$\omega(g v_j', e_p) = \omega(v_j', g^{-1} e_p) = (v_j', e_p)$$

for $p = n - 1, n$. It follows by the linear independence of the monomials $\{s^I t^J\}$ that the polynomials $f_{I,J,k}$ are invariant under G_{l-1}.

The induction hypothesis furnishes polynomials $\varphi_{I,J,k}$ in the variables $\omega(v_i'', v_j'')$, for $1 \leq i \leq j \leq n - 2$, such that $f_{I,J,k} = \varphi_{I,J,k}$. We calculate that

$$\omega(v_i'', v_j'') = \omega(v_{i+2}, v_{j+2}) + 2(s_i t_j - s_j t_i) .$$

In terms of the original variables $v_1, \ldots, v_n$, we have

$$s_i = -\omega(g_1 v_{i+2}, e_n) = -\omega(v_{i+2}, g_1^{-1} e_n) = -\omega(v_{i+2}, v_2)/\omega(v_1, v_2) ,$$

and similarly $t_i = \omega(v_{i+2}, v_1)/\omega(v_1, v_2)$. Substituting these formulas for $\omega(v_i'', v_j'')$, s_i, and t_i in the polynomials $\varphi_{I,J,k}$ and using expansion (5.21), we conclude that there are a positive integer r_1 and a polynomial function φ_1 on AM_n such that

$$f(X) = \omega(v_1, v_2)^{-r_1} \varphi_1(X^t J_n X)$$

for all $X = [v_1, \ldots, v_n] \in M_n(\mathbb{C})$ such that $\omega(v_1, v_2) \neq 0$.

By the same argument, there are a positive integer r_2 and a polynomial function φ_2 on AM_n such that

$$f(X) = \omega(v_3, v_4)^{-r_2} \varphi_2(X^t J_n X)$$

for all $X = [v_1, \ldots, v_n] \in M_n(\mathbb{C})$ such that $\omega(v_3, v_4) \neq 0$. Denote the entries of the matrix $u = X^t J_n X$ by u_{ij}. We have shown that

$$(u_{12})^{r_1} \varphi_2(u) = (u_{34})^{r_2} \varphi_1(u)$$

on the subset of AM_n where $u_{12} \neq 0$ and $u_{34} \neq 0$. Just as in the orthogonal case, this implies that $\varphi_1(u) = (u_{12})^{r_1} F(u)$ with F a polynomial. Hence $f(X) = F(X^t J_n X)$ for all $X \in M_n$. $\square$

5.2.3 Exercises

1. Let $G = \mathbf{SL}(2, \mathbb{C})$. Suppose that $0 \neq T \in \mathrm{Hom}_G(\mathcal{P}^k(\mathbb{C}^2) \otimes \mathcal{P}^l(\mathbb{C}^2), \mathcal{P}^r(\mathbb{C}^2))$.
   (a) Prove that $r = k + l - 2s$ with $s \leq \min(k, l)$.
   (b) Let x, y be the variables for $\mathbb{C}^2$. Prove that

   $$T(f \otimes g)(x, y) = c \sum_{j=0}^{s} \binom{s}{j} (-1)^j \frac{\partial^s f}{\partial x^{s-j} \partial y^j}(x, y) \frac{\partial^s g}{\partial x^j \partial y^{s-j}}(x, y)$$

   for all $f \in \mathcal{P}^k(\mathbb{C}^2)$ and $g \in \mathcal{P}^l(\mathbb{C}^2)$ with c a constant depending on T.
   (HINT: Recall that $\mathbf{SL}(2, \mathbb{C}) = \mathbf{Sp}_2(\mathbb{C})$ and use the FFT for $\mathbf{Sp}_2(\mathbb{C})$.)

5.3 Tensor Invariants

Let G be a reductive group and (ρ, V) a regular representation of G. For all nonnegative integers k, m we have a regular representation $\rho_{k,m}$ of G on the space $V^{\otimes k} \otimes V^{* \otimes m}$ of *mixed tensors of type* (k, m) (see Section B.2.2 for notation). We shall say that we have a *tensor version* of the *first fundamental theorem* (FFT) for the pair (G, ρ) if we have explicit spanning sets for the space

$$\left(V^{\otimes k} \otimes V^{* \otimes m}\right)^G \tag{5.22}$$

of G-invariant tensors for every k, m.

   We pursue the following strategy to obtain the FFT (polynomial form) for the classical groups stated in Section 5.2.1: We will prove the tensor form of the FFT for $\mathbf{GL}(V)$ in the next section. Using this and the special case in Proposition 5.2.6 of the polynomial FFT for $\mathbf{O}(V)$ and $\mathbf{Sp}(V)$, we will then prove the general tensor FFT for these groups. Finally, in Sections 5.4.2 and 5.4.3 we will use the tensor version of the FFT for the classical groups in their defining representations to obtain the polynomial version.

5.3.1 Tensor Invariants for $\mathbf{GL}(V)$

Assume that V is a finite-dimensional complex vector space. Let $\mathbf{GL}(V)$ act on V by the defining representation ρ, and let ρ^* be the dual representation on V^*. For all integers $k, m \geq 0$ we have the representations $\rho_k = \rho^{\otimes k}$ on $V^{\otimes k}$ and $\rho_m^* = \rho^{*\otimes m}$ on $V^{*\otimes m}$. Since there is a natural isomorphism $(V^*)^{\otimes m} \cong (V^{\otimes m})^*$ as $\mathbf{GL}(V)$ modules (see Section B.2.2), we may view ρ_m^* as acting on $(V^{\otimes m})^*$. We denote by $\rho_{k,m} = \rho^{\otimes k} \otimes \rho^{*\otimes m}$ the representation of $\mathbf{GL}(V)$ on $V^{\otimes k} \otimes (V^{\otimes m})^*$.

To obtain the tensor form of the FFT for $\mathbf{GL}(V)$, we must find an explicit spanning set for the space of $\mathbf{GL}(V)$ invariants in $V^{\otimes k} \otimes (V^{\otimes m})^*$. The center λI ($\lambda \in \mathbb{C}^\times$) of $\mathbf{GL}(V)$ acts on $V^{\otimes k} \otimes (V^{\otimes m})^*$ by $\rho_{k,m}(\lambda I)x = \lambda^{k-m}x$. Hence there are no invariants if $k \neq m$, so we only need to consider the representation $\rho_{k,k}$ on $V^{\otimes k} \otimes (V^{\otimes k})^*$.

Recall that when W is a finite-dimensional vector space, then $W \otimes W^* \cong \mathrm{End}(W)$ as a $\mathbf{GL}(W)$-module (Example 5 in Section 1.5.2). We apply this to the case $W = V^{\otimes k}$, where the action of $g \in \mathbf{GL}(V)$ on $\mathrm{End}(V^{\otimes k})$ is given by $T \mapsto \rho_k(g)T\rho_k(g)^{-1}$. Thus

$$(V^{\otimes k} \otimes V^{*\otimes k})^{\mathbf{GL}(V)} \cong \mathrm{End}_{\mathbf{GL}(V)}(V^{\otimes k}) \,. \tag{5.23}$$

This reduces the tensor FFT problem to finding a spanning set for the commutant of $\rho_k(\mathbf{GL}(V))$.

Let $\mathfrak{S}_k$ be the group of permutations of $\{1, 2, \ldots, k\}$. In Section 4.2.4 we defined a representation σ_k of $\mathfrak{S}_k$ on $V^{\otimes k}$ by

$$\sigma_k(s)(v_1 \otimes \cdots \otimes v_k) = v_{s^{-1}(1)} \otimes \cdots \otimes v_{s^{-1}(k)} \,.$$

From Theorem 4.2.10 we know that the algebras $\mathcal{A} = \mathrm{Span}\,\rho_k(\mathbf{GL}(V))$ and $\mathcal{B} = \mathrm{Span}\,\sigma_k(\mathfrak{S}_k)$ are mutual commutants in $\mathrm{End}(V^{\otimes k})$.

We now apply this result to obtain the tensor version of the FFT for $\mathbf{GL}(V)$. Let $e_1, \ldots, e_n$ be a basis for V and let $e_1^*, \ldots, e_n^*$ be the dual basis for V^*. We use the notation of Section 4.2.4: for a multi-index $I = (i_1, \ldots, i_k)$ with $1 \leq i_j \leq n$, set $|I| = k$ and $e_I = e_{i_1} \otimes \cdots \otimes e_{i_k}$ and $e_I^* = e_{i_1}^* \otimes \cdots \otimes e_{i_k}^*$. Let Ξ be the set of all ordered pairs (I, J) of multi-indices with $|I| = |J| = k$. The set $\{e_I \otimes e_J^* : (I, J) \in \Xi\}$ is a basis for $V^{\otimes k} \otimes (V^{\otimes k})^*$.

For $s \in \mathfrak{S}_k$ define a tensor C_s of type (k, k) by

$$C_s = \sum_{|I|=k} e_{s \cdot I} \otimes e_I^* \,. \tag{5.24}$$

Theorem 5.3.1. Let $G = \mathbf{GL}(V)$. The space of G invariants in $V^{\otimes k} \otimes V^{*\otimes k}$ is spanned by the tensors $\{C_s : s \in \mathfrak{S}_k\}$.

Proof. Let $T : V^{\otimes k} \otimes V^{*\otimes k} \longrightarrow \mathrm{End}(\bigotimes^k V)$ be the natural isomorphism (see Appendix B.2.2). For $s \in \mathfrak{S}_k$ we have

$$T(C_s)e_J = \sum_I \langle e_I^*, e_J \rangle e_{s \cdot I} = e_{s \cdot J} \,.$$

Thus $T(C_s) = \sigma_k(s)$. From Theorem 4.2.10 and (5.23) it follows that

$$T : \left[V^{\otimes k} \otimes V^{*\otimes k}\right]^G \longrightarrow \mathrm{Span}\{\sigma_k(s) : s \in \mathfrak{S}_k\}$$

is a linear isomorphism, so the tensors C_s span the G-invariants. $\square$

There is an alternative version of Theorem 5.3.1 coming from the self-dual property of the G-module $V^{\otimes k} \otimes V^{*\otimes k}$ and the natural isomorphism $V^{*\otimes k} \cong (V^{\otimes k})^*$. For $s \in \mathfrak{S}_k$ we define $\lambda_s \in V^{*\otimes k} \otimes V^{\otimes k}$ by

$$\langle \lambda_s, v_1 \otimes \cdots \otimes v_k \otimes v_1^* \otimes \cdots \otimes v_k^* \rangle = \langle v_1^*, v_{s(1)} \rangle \cdots \langle v_k^*, v_{s(k)} \rangle$$

for $v_1, \ldots, v_k \in V$ and $v_1^*, \ldots, v_k^* \in V^*$. We call λ_s the *complete contraction* defined by s.

Corollary 5.3.2. *The space of* $\mathbf{GL}(V)$*-invariant tensors in* $V^{*\otimes k} \otimes V^{\otimes k}$ *is spanned by the complete contractions* $\{\lambda_s : s \in \mathfrak{S}_k\}$.

Proof. Let $B : V^{\otimes k} \otimes V^{*\otimes k} \longrightarrow V^{*\otimes k} \otimes V^{\otimes k}$ be the natural duality map. It suffices to show that $\lambda_s = B(C_s)$. Let $v_1, \ldots, v_k \in V$ and $v_1^*, \ldots, v_k^* \in V^*$. Then for $s \in \mathfrak{S}_k$ we have

$$\langle B(C_s), v_1 \otimes \cdots \otimes v_k \otimes v_1^* \otimes \cdots \otimes v_k^* \rangle = \sum_{|I|=k} \prod_{j=1}^{k} \langle v_{s(j)}^*, e_{i_j} \rangle \prod_{j=1}^{k} \langle e_{i_j}^*, v_j \rangle .$$

By the expansion $v = \sum_i \langle e_i^*, v \rangle e_i$ the right side of the previous equation is

$$\prod_{j=1}^{k} \langle v_1^* \otimes \cdots \otimes v_k^*, v_{s(1)} \otimes \cdots \otimes v_{s(k)} \rangle = \langle \lambda_s, v_1 \otimes \cdots \otimes v_k \otimes v_1^* \otimes \cdots \otimes v_k^* \rangle .$$

This holds for all v_i and v_i^*, so we have $B(C_s) = \lambda_s$ as claimed. $\square$

5.3.2 Tensor Invariants for $\mathbf{O}(V)$ and $\mathbf{Sp}(V)$

We now obtain an FFT (tensor version) when G is $\mathbf{O}(V)$ or $\mathbf{Sp}(V)$ in its defining representation. Since $V \cong V^*$ as a G-module via the invariant form ω, we only need to find a spanning set for $(V^{\otimes m})^G$. The element $-I_V \in G$ acts by $(-1)^m$ on $V^{\otimes m}$, so there are no nonzero invariants if m is odd. Thus we assume that $m = 2k$ is even.

The $\mathbf{GL}(V)$ isomorphism $V^* \otimes V \cong \mathrm{End}(V)$ and the G-module isomorphism $V \cong V^*$ combine to give a G-module isomorphism

$$T : V^{\otimes 2k} \xrightarrow{\ \cong\ } \mathrm{End}(V^{\otimes k}) , \tag{5.25}$$

which we take in the following explicit form: If $u = v_1 \otimes \cdots \otimes v_{2k}$ with $v_i \in V$, then $T(u)$ is the linear transformation

$$T(u)(x_1 \otimes \cdots \otimes x_k) = \omega(x_1, v_2)\omega(x_2, v_4) \cdots \omega(x_k, v_{2k})v_1 \otimes v_3 \cdots \otimes v_{2k-1}$$

for $x_i \in V$. That is, we use the invariant form to change each v_{2i} into a covector, pair it with v_{2i-1} to get a rank-one linear transformation on V, and then take the tensor product of these transformations to get $T(u)$. We extend ω to a nondegenerate bilinear form on $V^{\otimes k}$ for every k by

$$\omega(x_1 \otimes \cdots \otimes x_k, y_1 \otimes \cdots \otimes y_k) = \prod_{i=1}^{k} \omega(x_i, y_i).$$

Then we can write the formula for T as

$$T(v_1 \otimes \cdots \otimes v_{2k})x = \omega(x, v_2 \otimes v_4 \otimes \cdots \otimes v_{2k}) v_1 \otimes v_3 \otimes \cdots \otimes v_{2k-1}$$

for $x \in V^{\otimes k}$.

The identity operator $I_V^{\otimes k}$ on $V^{\otimes k}$ is G-invariant, of course. We can express this operator in tensor form as follows: Fix a basis $\{f_p\}$ for V and let $\{f^p\}$ be the dual basis for V relative to the form ω; thus $\omega(f_p, f^q) = \delta_{pq}$. Set $\theta = \sum_{p=1}^{n} f_p \otimes f^p$ (where $n = \dim V$). Then the $2k$-tensor

$$\theta_k = \underbrace{\theta \otimes \cdots \otimes \theta}_{k} = \sum_{p_1, \ldots, p_k} f_{p_1} \otimes f^{p_1} \otimes \cdots \otimes f_{p_k} \otimes f^{p_k}$$

satisfies $T(\theta_k) = I_V^{\otimes k}$. It follows that θ_k is G-invariant. Since the action of G on $V^{\otimes 2k}$ commutes with the action of $\mathfrak{S}_{2k}$, the tensors $\sigma_{2k}(s)\theta_k$ are also G-invariant, for any $s \in \mathfrak{S}_{2k}$. These tensors suffice to give an FFT for the orthogonal and symplectic groups.

Theorem 5.3.3. *Let G be $\mathbf{O}(V)$ or $\mathbf{Sp}(V)$. Then $\left[V^{\otimes m}\right]^G = 0$ if m is odd, and*

$$\left[V^{\otimes 2k}\right]^G = \mathrm{Span}\{\sigma_{2k}(s)\theta_k : s \in \mathfrak{S}_{2k}\}.$$

Before proving this theorem, we restate it to incorporate the symmetries of the tensor θ_k. View $\mathfrak{S}_{2k}$ as the permutations of the set $\{1, 2, \ldots, 2k-1, 2k\}$. Define $\widetilde{\mathfrak{S}}_k \subset \mathfrak{S}_{2k}$ as the subgroup that permutes the ordered pairs $\{(1,2), \ldots, (2k-1, 2k)\}$:

$$(2i-1, 2i) \rightarrow (2s(i)-1, 2s(i)) \quad \text{for } i = 1, \ldots, k \text{ with } s \in \mathfrak{S}_k.$$

Let $\mathfrak{N}_k \subset \mathfrak{S}_{2k}$ be the subgroup generated by the transpositions $2j-1 \longleftrightarrow 2j$ for $j = 1, \ldots, k$. Then $\mathfrak{N}_k \cong (\mathbb{Z}_2)^k$ is normalized by $\widetilde{\mathfrak{S}}_k$. Thus $\mathfrak{B}_k = \widetilde{\mathfrak{S}}_k \mathfrak{N}_k$ is a subgroup of $\mathfrak{S}_{2k}$. Note that $\mathrm{sgn}(s) = 1$ for $s \in \widetilde{\mathfrak{S}}_k$.

There are $\mathbf{GL}(V)$-module isomorphisms

$$S^k(S^2(V)) \cong \left[V^{\otimes 2k}\right]^{\mathfrak{B}_k} \tag{5.26}$$

(the tensors fixed by $\sigma_{2k}(\mathfrak{B}_k)$) and

$$S^k(\wedge^2(V)) \cong [V^{\otimes 2k}]^{\mathfrak{B}_k,\mathrm{sgn}} \tag{5.27}$$

(the tensors transforming by the character $s \mapsto \mathrm{sgn}(s)$ for $s \in \mathfrak{B}_k$). Now $\sigma_{2k}(s)\theta_k = \theta_k$ for $s \in \widetilde{\mathfrak{S}}_k$. Furthermore, if we set $\varepsilon = 1$ when ω is symmetric and $\varepsilon = -1$ when ω is skew-symmetric, then

$$\sum_{p=1}^{n} f^p \otimes f_p = \varepsilon \sum_{n=1}^{n} f_p \otimes f^p \tag{5.28}$$

(where $n = \dim V$). Thus from (5.28) we conclude that for $t \in \mathfrak{B}_k$,

$$\sigma_{2k}(t)\theta_k = \begin{cases} \theta_k & \text{if } \varepsilon = 1, \\ \mathrm{sgn}(t)\theta_k & \text{if } \varepsilon = -1. \end{cases} \tag{5.29}$$

Thus

$$\theta_k \in \begin{cases} S^k(S^2(V)) & \text{when } \omega \text{ is symmetric}, \\ S^k(\wedge^2(V)) & \text{when } \omega \text{ is skew-symmetric}. \end{cases} \tag{5.30}$$

Since $\sigma_{2k}(st)\theta_k = \pm\sigma_{2k}(s)\theta_k$ for $s \in \mathfrak{S}_{2k}$ and $t \in \mathfrak{B}_k$, we see that Theorem 5.3.3 is equivalent to the following assertion:

Theorem 5.3.4. *Let $\Xi_k \subset \mathfrak{S}_{2k}$ be a collection of representatives for the cosets $\mathfrak{S}_{2k}/\mathfrak{B}_k$. Then $\left[V^{\otimes 2k}\right]^G = \mathrm{Span}\{\sigma_{2k}(s)\theta_k : s \in \Xi_k\}$.*

We next give a dual version of this FFT in terms of tensor contractions, as we did for $\mathbf{GL}(V)$. A *two-partition* of the set $\{1, 2, \ldots, 2k\}$ is a set of k two-element subsets

$$x = \{\{i_1, j_1\}, \ldots, \{i_k, j_k\}\} \tag{5.31}$$

such that $\{i_1, \ldots, i_k, j_1, \ldots, j_k\} = \{1, 2, \ldots, 2k\}$. For example,

$$x_0 = \{\{1, 2\}, \{3, 4\}, \ldots, \{2k-1, 2k\}\}.$$

We label the pairs $\{i_p, j_p\}$ in x so that $i_p < j_p$ for $p = 1, \ldots, k$. The set X_k of all two-partitions of $\{1, 2, \ldots, 2k\}$ is a homogeneous space for $\mathfrak{S}_{2k}$, relative to the natural action of $\mathfrak{S}_{2k}$ as permutations of $\{1, 2, \ldots, 2k\}$. The subgroup $\mathfrak{B}_k$ is the isotropy group of the two-partition x_0 above, so the map $s \mapsto s \cdot x_0$ gives an identification

$$\mathfrak{S}_{2k}/\mathfrak{B}_k = X_k \tag{5.32}$$

as a homogeneous space for $\mathfrak{S}_{2k}$. Given $x \in X_k$ as in (5.31), define the *complete contraction* $\lambda_x \in (V^{\otimes 2k})^*$ by

$$\langle \lambda_x, v_1 \otimes \cdots \otimes v_{2k} \rangle = \prod_{p=1}^{k} \omega(v_{i_p}, v_{j_p}).$$

Observe that for $s \in \mathfrak{S}_{2k}$,

$$\omega(\sigma_{2k}(s)\theta_k, v_1 \otimes \cdots \otimes v_{2k}) = \omega(\theta_k, v_{s(1)} \otimes \cdots \otimes v_{s(2k)})$$

$$= \prod_{p=1}^{k} \omega(v_{s(2p-1)}, v_{s(2p)}) \ .$$

Thus $(\sigma_{2k}(s)\theta_k)^* = \lambda_{s \cdot x_0}$, where for $u \in V^{\otimes 2k}$ we denote by u^* the linear form on $V^{\otimes 2k}$ corresponding to u via ω. Hence Theorem 5.3.4 has the following equivalent dual statement:

Theorem 5.3.5. *The complete contractions $\{\lambda_x : x \in X_k\}$ are a spanning set for the G-invariant 2k-multilinear forms on V.*

We now begin the proof of Theorem 5.3.3. Since $V \cong V^*$ as a G-module, it suffices to consider G-invariant tensors $\lambda \in V^{*\otimes k}$. The key idea is to shift the action of G from $V^{*\otimes 2k}$ to $\mathrm{End}\, V$ by introducing a *polarization variable* $X \in \mathrm{End}\, V$. This transforms λ into a $\mathbf{GL}(V)$-invariant linear map from $V^{\otimes k}$ to the G-invariant polynomials on $\mathrm{End}(V)$ of degree k. Proposition 5.2.6 allows us to express such polynomials as covariant tensors that are automatically G invariant. By this means, λ gives rise to a unique mixed $\mathbf{GL}(V)$-invariant tensor. But we know by the FFT for $\mathbf{GL}(V)$ that all such tensors are linear combinations of complete vector–covector contractions. Finally, setting the polarization variable $X = I_V$, we find that λ is in the span of the complete contractions relative to the form ω.

In more detail, let $G \subset \mathbf{GL}(V)$ be any subgroup for the moment. Given $\lambda \in V^{*\otimes k}$, we can define a polynomial function Φ_λ on $\mathrm{End}(V) \oplus V^{\otimes k}$ by

$$\Phi_\lambda(X, w) = \langle \lambda, X^{\otimes k} w \rangle \quad \text{for } X \in \mathrm{End}\, V \text{ and } w \in V^{\otimes k} \ .$$

Since $X, w \mapsto \Phi_\lambda(X, w)$ is a polynomial of degree k in X and is linear in w, we may also view Φ_λ as an element of $\mathcal{P}^k(\mathrm{End}\, V) \otimes V^{*\otimes k}$. We recover λ from Φ_λ by evaluating at $X = I_V$, so the map $\lambda \mapsto \Phi_\lambda$ is injective.

Let L and R be the representations of G and $\mathbf{GL}(V)$, respectively, on $\mathcal{P}^k(\mathrm{End}\, V)$ given by

$$L(g)f(X) = f(g^{-1}X), \quad R(h)f(X) = f(Xh) \quad \text{for } X \in \mathrm{End}\, V \ .$$

Here $f \in \mathcal{P}^k(\mathrm{End}\, V)$, $g \in G$, and $h \in \mathbf{GL}(V)$. We then have the mutually commuting representations $L \otimes 1$ of G and $\pi = R \otimes \rho_k^*$ of $\mathbf{GL}(V)$ on $\mathcal{P}^k(\mathrm{End}\, V) \otimes V^{*\otimes k}$.

Lemma 5.3.6. *The function Φ_λ has the following transformation properties:*

1. $(L(g) \otimes 1)\Phi_\lambda = \Phi_{g \cdot \lambda}$ for $g \in G$ (where $g \cdot \lambda = \rho_k^*(g)\lambda$).
2. $\pi(h)\Phi_\lambda = \Phi_\lambda$ for $h \in \mathbf{GL}(V)$.

Furthermore, $\lambda \mapsto \Phi_\lambda$ is a bijection from $V^{*\otimes k}$ to the space of $\pi(\mathbf{GL}(V))$ invariants in $\mathcal{P}^k(\mathrm{End}\, V) \otimes V^{*\otimes k}$. Hence the map $\lambda \mapsto \Phi_\lambda$ defines a linear isomorphism

$$\left[V^{*\otimes k}\right]^G \cong \left[\mathcal{P}^k(\mathrm{End}\, V)^{L(G)} \otimes V^{*\otimes k}\right]^{\mathbf{GL}(V)} \ . \tag{5.33}$$

Proof. Let $g \in G$, $X \in \operatorname{End} V$, and $w \in V^{\otimes k}$. Then

$$\Phi_\lambda(g^{-1}X, w) = \langle \lambda, \rho_k(g)^{-1}X^{\otimes k}w \rangle = \langle \rho_k^*(g)\lambda, X^{\otimes k}w \rangle \,,$$

so (1) is clear. Let $h \in \mathbf{GL}(V)$. Then

$$\begin{aligned}(\pi(h)\Phi_\lambda)(X, w) &= \Phi_\lambda(Xh, \rho_k(h)^{-1}w) = \langle \lambda, (Xh)^{\otimes k}\rho_k(h)^{-1}w \rangle \\ &= \Phi_\lambda(X, w)\,,\end{aligned}$$

which gives (2).

We have already observed that the map $\lambda \mapsto \Phi_\lambda$ is injective. It remains only to show that every $F \in \mathcal{P}^k(\operatorname{End} V) \otimes V^{*\otimes k}$ that is invariant under $\pi(\mathbf{GL}(V))$ can be written as Φ_λ for some $\lambda \in V^{*\otimes k}$. We view F as a polynomial map from $\operatorname{End} V$ to $V^{*\otimes k}$. Then the invariance property of F can be expressed as

$$\rho_k^*(h)^{-1}F(X) = F(Xh) \quad \text{for } h \in \mathbf{GL}(V) \text{ and } X \in \operatorname{End} V\,.$$

Set $\lambda = F(I_V)$ and view Φ_λ likewise as a polynomial map from $\operatorname{End} V$ to $V^{*\otimes k}$. Then for $h \in \mathbf{GL}(V)$ we have

$$\Phi_\lambda(h) = \rho_k^*(h)^{-1}F(I_V) = F(h)$$

by the invariance of F. Since F is a polynomial, it follows that $\Phi_\lambda(X) = F(X)$ for all $X \in \operatorname{End} V$. Thus $F = \Phi_\lambda$. Since $\mathbf{GL}(V)$ is a reductive group and $L(G)$ commutes with $\pi(\mathbf{GL}(V))$, we obtain (5.33) from (1), (2), and the fact that the subspace $\mathcal{P}^k(\operatorname{End} V)^{L(G)} \otimes V^{*\otimes k}$ is invariant under $\mathbf{GL}(V)$. $\square$

Completion of proof of Theorem 5.3.3. Let $n = \dim V$ and fix a basis $\{e_1, \dots, e_n\}$ for V. Let $\{e_1^*, \dots, e_n^*\}$ be the dual basis for V^*. We identify V with $\mathbb{C}^n$ and $\mathbf{GL}(V)$ with $\mathbf{GL}(n, \mathbb{C})$ by this basis. We can then define an antiautomorphism $g \mapsto g^t$ (matrix transpose) of $\mathbf{GL}(V)$ via this identification. We may assume that the basis $\{e_1, \dots, e_n\}$ is chosen so that $\mathbf{O}(V)$ is the group that preserves the symmetric form (x, y) on $\mathbb{C}^n$ and $\mathbf{Sp}(V)$ is the group that preserves the skew-symmetric form $(J_n x, y)$, as in Section 5.2.1.

Let $\lambda \in \left[V^{*\otimes 2k}\right]^G$. Then by Lemma 5.3.6 we have

$$\Phi_\lambda \in \left[\mathcal{P}^{2k}(M_n)^{L(G)} \otimes V^{*\otimes 2k}\right]^{\mathbf{GL}(V)}.$$

Furthermore, $\lambda = \Phi_\lambda(I)$. By Proposition 5.2.6 we can express

$$\Phi_\lambda(X, w) = \begin{cases} F_\lambda(X^t X, w) & \text{when } G = \mathbf{O}(V)\,, \\ F_\lambda(X^t J_n X, w) & \text{when } G = \mathbf{Sp}(V)\,, \end{cases}$$

where F_λ is a polynomial on $SM_n \oplus V^{\otimes 2k}$ when $G = \mathbf{O}(V)$, or on $AM_n \oplus V^{\otimes 2k}$ when $G = \mathbf{Sp}(V)$. We view F_λ as an element of $\mathcal{P}^k(SM_n) \otimes V^{*\otimes 2k}$ (respectively of $\mathcal{P}^k(AM_n) \otimes V^{*\otimes 2k}$). Note that

$$\langle \lambda, w \rangle = \Phi_\lambda (I_n, w) = \begin{cases} F_\lambda (I_n, w) & \text{when } G = \mathbf{O}(V) \,, \\ F_\lambda (J_n, w) & \text{when } G = \mathbf{Sp}(V) \,. \end{cases}$$

The next step is to translate the $\mathbf{GL}(V)$-invariance of Φ_λ into an appropriate invariance property of F_λ. We have

$$\pi(g) \Phi_\lambda (X, w) = \begin{cases} F_\lambda (g^t X^t X g, \rho_k(g)^{-1} w) & \text{when } G = \mathbf{O}(V) \,, \\ F_\lambda (g^t X^t J_n X g, \rho_k(g)^{-1} w) & \text{when } G = \mathbf{Sp}(V) \,. \end{cases}$$

Thus the action of $g \in \mathbf{GL}(n, \mathbb{C})$ becomes the action

$$g \cdot f(z, w) = f(g^t z g, \rho_k(g)^{-1} w) \,,$$

where f is a polynomial on $SM_n \oplus V^{\otimes 2k}$ or $AM_n \oplus V^{\otimes 2k}$.

Given a matrix $x = [x_{ij}] \in M_n(\mathbb{C})$, define

$$\Theta(x) = \sum_{i,j} x_{ij} e_i^* \otimes e_j^* \in V^* \otimes V^* \,.$$

Then Θ gives linear isomorphisms

$$AM_n \cong \wedge^2 V^*, \qquad SM_n \cong S^2(V^*) \,. \tag{5.34}$$

Let $g \in \mathbf{GL}(V)$ have matrix $\gamma = [g_{ij}]$ relative to the basis $\{e_i\}$ for V. Then

$$\Theta(\gamma^t x \gamma) = \sum_{p,q} \sum_{i,j} g_{pi} x_{pq} g_{qj} (e_i^* \otimes e_j^*)$$

$$= \sum_{p,q} x_{pq} \rho_2^*(g)^{-1} (e_p^* \otimes e_q^*) = \rho_2^*(g)^{-1} \Theta(x) \,.$$

Thus the isomorphisms in (5.34) are $\mathbf{GL}(V)$-module isomorphisms, so they give rise to $\mathbf{GL}(V)$-module isomorphisms

$$\mathcal{P}^k (AM_n) \cong S^k (\wedge^2 V) \,, \qquad \mathcal{P}^k (SM_n) \cong S^k (S^2(V)) \,.$$

Hence F_λ corresponds to a $\mathbf{GL}(V)$-invariant mixed tensor $C \in V^{* \otimes 2k} \otimes V^{\otimes 2k}$ under these isomorphisms. This means that

$$F_\lambda(A, w) = \langle A^{\otimes k} \otimes w, C \rangle \tag{5.35}$$

for all A in either $S^2(V^*)$ or $\wedge^2 V^*$, as appropriate. Since

$$(S^2(V))^{\otimes k} \subset V^{\otimes 2k} \quad \text{and} \quad (\wedge^2 V^*)^{\otimes k} \subset V^{* \otimes 2k}$$

are $\mathbf{GL}(V)$-invariant subspaces, we may project C onto the space of $\mathbf{GL}(V)$-invariant tensors in $V^{* \otimes 2k} \otimes V^{\otimes 2k}$ without changing (5.35). Thus we may assume that C is $\mathbf{GL}(V)$-invariant. By the FFT for $\mathbf{GL}(V)$ (tensor form), we may also assume that C is a *complete contraction*:

$$C = \sum_{|I|=2k} e^*_{s\cdot I} \otimes e_I$$

for some $s \in \mathfrak{S}_{2k}$. When $G = \mathbf{O}(V)$ we take $A = I_n$ to recover the original G-invariant tensor λ as

$$\langle \lambda, w \rangle = F_\lambda(I_n, w) = \langle \Theta(I_n)^{\otimes k} \otimes w, C \rangle$$
$$= \sum_{|I|=2k} \langle e_I, \Theta(I_n)^{\otimes k} \rangle \langle w, e^*_{s\cdot I} \rangle = \langle \sigma_{2k}(s)\Theta(I_n)^{\otimes k}, w \rangle .$$

When $G = \mathbf{Sp}(V)$, we likewise take $A = J_n$ to get $\langle \lambda, w \rangle = \langle \sigma_{2k}(s)\Theta(J_n)^{\otimes k}, w \rangle$. Since

$$\theta^*_k = \begin{cases} \Theta(I_n)^{\otimes k} & \text{when } G = \mathbf{O}(V) , \\ \Theta(J_n)^{\otimes k} & \text{when } G = \mathbf{Sp}(V) , \end{cases}$$

we conclude that $\lambda = \sigma_{2k}(s)\theta^*_k$. $\square$

5.3.3 Exercises

1. Let $V = \mathbb{C}^n$ and $G = \mathbf{GL}(n, \mathbb{C})$. For $v \in V$ and $v^* \in V^*$ let $T(v \otimes v^*) = vv^* \in M_n$ (so $u \mapsto T(u)$ is the canonical isomorphism between $V \otimes V^*$ and M_n). Let $T_k = T^{\otimes k} : (V \otimes V^*)^{\otimes k} \longrightarrow (M_n)^{\otimes k}$ be the canonical isomorphism, and let $g \in G$ act on $x \in M_n$ by $g \cdot x = gxg^{-1}$.
(a) Show that T_k intertwines the action of G on $(V \otimes V^*)^{\otimes k}$ and $(M_n)^{\otimes k}$.
(b) Let $c \in \mathfrak{S}_k$ be a cyclic permutation $m_1 \to m_2 \to \cdots \to m_k \to m_1$. Let λ_c be the G-invariant contraction $\lambda_c(v_1 \otimes v^*_1 \otimes \cdots \otimes v_k \otimes v^*_k) = \prod_{j=1}^k \langle v^*_{m_j}, v_{m_{j+1}} \rangle$ on $(V \otimes V^*)^{\otimes k}$. Set $X_j = T(v_j \otimes v^*_j)$. Prove that

$$\lambda_c(v_1 \otimes v^*_1 \otimes \cdots \otimes v_k \otimes v^*_k) = \text{tr}(X_{m_1} X_{m_2} \cdots X_{m_k}) .$$

(HINT: Note that $T(Xv \otimes v^*) = XT(v \otimes v^*)$ for $X \in M_n$, and $\text{tr}(T(v \otimes v^*)) = v^*v$.)
(c) Let $s \in \mathfrak{S}_k$ be a product of disjoint cyclic permutations $c_1, \ldots, c_r$, where c_i is the cycle $m_{1,i} \to m_{2,i} \to \cdots \to m_{k_i,i} \to m_{1,i}$. Let λ_s be the G-invariant contraction

$$\lambda_s(v_1 \otimes v^*_1 \otimes \cdots \otimes v_k \otimes v^*_k) = \prod_{i=1}^r \prod_{j=1}^{k_i} \langle v^*_{m_{j,i}}, v_{m_{j+1,i}} \rangle .$$

Set $X_j = T(v_j \otimes v^*_j)$. Prove that

$$\lambda_s(v_1 \otimes v^*_1 \otimes \cdots \otimes v_k \otimes v^*_k) = \prod_{i=1}^r \text{tr}(X_{m_{1,i}} X_{m_{2,i}} \cdots X_{m_{k_i,i}}) .$$

2. Use the previous exercise to show that the algebra of polynomials f on M_n such that $f(gXg^{-1}) = f(X)$ for all $g \in \mathbf{GL}(n,\mathbb{C})$ is generated by the polynomials $u_1, \ldots, u_n$ with $u_i(X) = \mathrm{tr}(X^i)$.

3. Find results analogous to those in Exercises #1 and #2 for $\mathbf{O}(n,\mathbb{C})$ and for $\mathbf{Sp}_{2n}(\mathbb{C})$.

4. Let $G = \mathbf{GL}(n,\mathbb{C})$ with $n \geq 2$.

   (a) Use the result of Exercise #1 to find a basis for the G-invariant linear functionals on $\bigotimes^2 M_n$.

   (b) Prove that there are no nonzero skew-symmetric G-invariant bilinear forms on M_n. (HINT: Use (a) and the projection from $\bigotimes^2 M_n$ onto $\bigwedge^2 M_n$.)

   (c) Use the result of Exercise #1 to find a spanning set for the G-invariant linear functionals on $\bigotimes^3 M_n$.

   (d) Define $\omega(X_1, X_2, X_3) = \mathrm{tr}([X_1, X_2]X_3)$ for $X_i \in M_n$. Prove that ω is the unique G-invariant skew-symmetric linear functional on $\bigotimes^3 M_n$, up to a scalar multiple. (HINT: Use the result in (c) and the projection from $\bigotimes^3 M_n$ onto $\bigwedge^3 M_n$.)

5. Let $G = \mathbf{O}(V,B)$, where B is a symmetric bilinear form on V (assume that $\dim V \geq 3$). Let $\{e_i\}$ be a basis for V such that $B(e_i, e_j) = \delta_{ij}$.

   (a) Let $R \in (V^{\otimes 4})^G$. Show that there are constants $a, b, c \in \mathbb{C}$ such that

   $$R = \sum_{i,j,k,l} \{a\delta_{ij}\delta_{kl} + b\delta_{ik}\delta_{jl} + c\delta_{il}\delta_{jk}\} e_i \otimes e_j \otimes e_k \otimes e_l .$$

   (HINT: Determine all the two-partitions of $\{1,2,3,4\}$.)

   (b) Use (a) to find a basis for the space $[S^2(V) \otimes S^2(V)]^G$. (HINT: Symmetrize relative to tensor positions 1, 2 and positions 3, 4.)

   (c) Use (b) to show that $\dim \mathrm{End}_G(S^2(V)) = 2$ and that $S^2(V)$ decomposes into the sum of two inequivalent irreducible G modules. (HINT: $S^2(V) \cong S^2(V)^*$ as G-modules.)

   (d) Find the dimensions of the irreducible modules in (c). (HINT: There is an obvious irreducible submodule in $S^2(V)$.)

6. Let $G = \mathbf{O}(V,B)$ as in the previous exercise.

   (a) Use part (a) of the previous exercise to find a basis for $[\bigwedge^2 V \otimes \bigwedge^2 V]^G$. (HINT: Skew-symmetrize relative to tensor positions 1, 2 and positions 3, 4.)

   (b) Use (a) to show that $\dim \mathrm{End}_G(\bigwedge^2 V) = 1$, and hence that $\bigwedge^2 V$ is irreducible under G. (HINT: $\bigwedge^2 V \cong \bigwedge^2 V^*$ as G-modules.)

7. Let $G = \mathbf{Sp}(V, \Omega)$, where Ω is a nonsingular skew form on V (assume that $\dim V \geq 4$ is even). Let $\{f_i\}$ and $\{f^j\}$ be bases for V such that $\Omega(f_i, f^j) = \delta_{ij}$.

   (a) Show that $(V^{\otimes 4})^G$ is spanned by the tensors

   $$\sum_{i,j} f_i \otimes f^i \otimes f_j \otimes f^j , \quad \sum_{i,j} f_i \otimes f_j \otimes f^i \otimes f^j , \quad \sum_{i,j} f_i \otimes f_j \otimes f^j \otimes f^i .$$

   (b) Use (a) to find a basis for the space $[\bigwedge^2 V \otimes \bigwedge^2 V]^G$. (HINT: Skew-symmetrize relative to tensor positions 1, 2 and positions 3, 4.)

(c) Use (b) to show that $\dim \operatorname{End}_G(\bigwedge^2 V) = 2$ and that $\bigwedge^2 V$ decomposes into the sum of two inequivalent irreducible G modules. (HINT: $\bigwedge^2 V \cong \bigwedge^2 V^*$ as a G-module.)

(d) Find the dimensions of the irreducible modules in (c). (*Hint*: There is an obvious irreducible submodule in $\bigwedge^2 V$.)

8. Let $G = \mathbf{Sp}(V, \Omega)$ as in the previous exercise.

(a) Use part (a) of the previous exercise to find a basis for $\left[S^2(V) \otimes S^2(V) \right]^G$. (HINT: Symmetrize relative to tensor positions 1, 2 and positions 3, 4.)

(b) Use (a) to show that $\dim \operatorname{End}_G(S^2(V)) = 1$ and hence $S^2(V)$ is irreducible under G. (HINT: $S^2(V) \cong S^2(V)^*$ as a G-module.)

5.4 Polynomial FFT for Classical Groups

Now that we have proved the tensor version of the FFT for the classical groups, we obtain the polynomial version of the FFT by viewing polynomials as tensors with additional symmetries.

5.4.1 Invariant Polynomials as Tensors

Let G be a reductive group and (ρ, V) a regular representation of G. In this section we prove that a tensor FFT for (G, ρ) furnishes a spanning set for $\mathcal{P}(V^k \times V^{*m})^G$ for all k, m.

Consider the torus $\mathbb{T}_{k,m} = (\mathbb{C}^\times)^k \times (\mathbb{C}^\times)^m$. The regular characters of $\mathbb{T}_{k,m}$ are given by

$$t \mapsto t^{[\mathbf{p},\mathbf{q}]} = \prod_{i=1}^k x_i^{p_i} \prod_{j=1}^k y_j^{q_j} \,,$$

where $t = (x_1, \dots x_k, y_1, \dots, y_m) \in \mathbb{T}_{k,m}$ and $[\mathbf{p}, \mathbf{q}] \in \mathbb{Z}^k \times \mathbb{Z}^m$. Let $\mathbb{T}_{k,m}$ act on $V^k \times V^{*m}$ by

$$t \cdot z = (x_1 v_1, \dots, x_k v_k, y_1 v_1^*, \dots, y_m v_m^*)$$

for $z = (v_1, \dots, v_k, v_1^*, \dots, v_m^*) \in V^k \oplus V^{*m}$. This action commutes with the action of G on $V^k \times V^{*m}$, so G leaves invariant the weight spaces of $\mathbb{T}_{k,m}$ in $\mathcal{P}(V^k \times V^{*m})$. These weight spaces are described by the degrees of homogeneity of $f \in \mathcal{P}(V^k \oplus V^{*m})$ in v_i and v_j^* as follows: For $\mathbf{p} \in \mathbb{N}^k$, $\mathbf{q} \in \mathbb{N}^m$ set

$$\mathcal{P}^{[\mathbf{p},\mathbf{q}]}(V^k \oplus V^{*m}) = \{ f \in \mathcal{P}(V^k \oplus V^{*m}) : f(t \cdot z) = t^{[\mathbf{p},\mathbf{q}]} f(z) \} \,.$$

Then

$$\mathcal{P}(V^k \oplus V^{*m}) = \bigoplus_{\mathbf{p} \in \mathbb{N}^k} \bigoplus_{\mathbf{q} \in \mathbb{N}^m} \mathcal{P}^{[\mathbf{p},\mathbf{q}]}(V^k \oplus V^{*m}) \,,$$

and this decomposition is $\mathbf{GL}(V)$-invariant. Thus

$$\mathcal{P}(V^k \oplus V^{*m})^G = \bigoplus_{\mathbf{p} \in \mathbb{N}^k} \bigoplus_{\mathbf{q} \in \mathbb{N}^m} \left[\mathcal{P}^{[\mathbf{p},\mathbf{q}]}(V^k \oplus V^{*m})\right]^G . \tag{5.36}$$

We now give another realization of these weight spaces. Given $\mathbf{p} \in \mathbb{N}^k$ and $\mathbf{q} \in \mathbb{N}^m$ we set

$$V^{* \otimes \mathbf{p}} \otimes V^{\otimes \mathbf{q}} = V^{* \otimes p_1} \otimes \cdots \otimes V^{* \otimes p_k} \otimes V^{\otimes q_1} \otimes \cdots \otimes V^{\otimes q_m} .$$

We view the space $V^{* \otimes \mathbf{p}} \otimes V^{\otimes \mathbf{q}}$ as a G-module with G acting by ρ on each V tensor product and by ρ^* on each V^* tensor product. Let $\mathfrak{S}_\mathbf{p} = \mathfrak{S}_{p_1} \times \cdots \times \mathfrak{S}_{p_k}$, acting as a group of permutations of $\{1, \ldots, |p|\}$ as in Section 4.4.4. Then we have a representation of $\mathfrak{S}_\mathbf{p} \times \mathfrak{S}_\mathbf{q}$ on $V^{* \otimes \mathbf{p}} \otimes V^{\otimes \mathbf{q}}$ (by permutation of the tensor positions) that commutes with the action of G. Define $|p| = \sum p_i$ and $|q| = \sum q_i$.

Lemma 5.4.1. *Let* $\mathbf{p} \in \mathbb{N}^k$ *and* $\mathbf{q} \in \mathbb{N}^m$. *There is a linear isomorphism*

$$\mathcal{P}^{[\mathbf{p},\mathbf{q}]}(V^k \oplus V^{*m})^G \cong \left[(V^{* \otimes |\mathbf{p}|} \otimes V^{\otimes |\mathbf{q}|})^G\right]^{\mathfrak{S}_\mathbf{p} \times \mathfrak{S}_\mathbf{q}} . \tag{5.37}$$

Proof. We have the isomorphisms

$$\mathcal{P}(V^k \oplus V^{*m}) \cong S(V^{*k} \oplus V^m)$$
$$\cong \underbrace{S(V^*) \otimes \cdots \otimes S(V^*)}_{k \text{ factors}} \otimes \underbrace{S(V) \otimes \cdots \otimes S(V)}_{m \text{ factors}}$$

(see Proposition C.1.4) as $\mathbf{GL}(V)$-modules. These give a G-module isomorphism

$$\mathcal{P}^{[\mathbf{p},\mathbf{q}]}(V^k \oplus V^{*m}) \cong S^{[\mathbf{p}]}(V^*) \otimes S^{[\mathbf{q}]}(V) , \tag{5.38}$$

where $S^{[\mathbf{p}]}(V^*) = S^{p_1}(V^*) \otimes \cdots \otimes S^{p_k}(V^*)$ and $S^{[\mathbf{q}]}(V) = S^{q_1}(V) \otimes \cdots \otimes S^{q_m}(V)$. We also have a G-module isomorphism

$$S^r(V) \cong \left[V^{\otimes r}\right]^{\mathfrak{S}_r} \subset V^{\otimes r} ,$$

with $\mathfrak{S}_r$ acting by permuting the tensor positions. Combining this with (5.38) we obtain the linear isomorphisms

$$\mathcal{P}^{[\mathbf{p},\mathbf{q}]}(V^k \oplus V^{*m})^G \cong \left[S^{[\mathbf{p}]}(V^*) \otimes S^{[\mathbf{q}]}(V)\right]^G$$
$$\cong \left[(V^{* \otimes |\mathbf{p}|} \otimes V^{\otimes |\mathbf{q}|})^{\mathfrak{S}_\mathbf{p} \times \mathfrak{S}_\mathbf{q}}\right]^G .$$

This implies (5.37), since the actions of G and $\mathfrak{S}_\mathbf{p} \times \mathfrak{S}_\mathbf{q}$ mutually commute. $\qquad \square$

5.4.2 *Proof of Polynomial FFT for* **GL(V)**

We now prove the polynomial version of the first fundamental theorem for $\mathbf{GL}(V)$, acting in its defining representation on V (Theorem 5.2.1). Using the notation of this theorem and of Section 5.4.1, we must show that for each $\mathbf{p} \in \mathbb{N}^k$ and $\mathbf{q} \in \mathbb{N}^m$, the space $\mathcal{P}^{[\mathbf{p},\mathbf{q}]}(V^k \oplus (V^*)^m)^{\mathbf{GL}(V)}$ is spanned by monomials of the form

$$\prod_{i=1}^{k}\prod_{j=1}^{m}\langle v_i, v_j^* \rangle^{r_{ij}} . \tag{5.39}$$

The subgroup $\mathbb{T} = \{\zeta I_V : \zeta \in \mathbb{C}^\times\}$ in $\mathbf{GL}(V)$ acts on $\mathcal{P}^{[\mathbf{p},\mathbf{q}]}(V^k \oplus (V^*)^m)$ by the character $\zeta \mapsto \zeta^{|\mathbf{q}|-|\mathbf{p}|}$. Hence

$$\mathcal{P}^{[\mathbf{p},\mathbf{q}]}(V^k \oplus V^{*m})^{\mathbf{GL}(V)} = 0 \quad \text{if } |\mathbf{p}| \neq |\mathbf{q}| .$$

Therefore, we may assume that $|\mathbf{p}| = |\mathbf{q}| = n$. By Lemma 5.4.1,

$$\mathcal{P}^{[\mathbf{p},\mathbf{q}]}(V^k \oplus V^{*m})^{\mathbf{GL}(V)} \cong \left[(V^{*\otimes n} \otimes V^{\otimes n})^{\mathbf{GL}(V)} \right]^{\mathfrak{S}_\mathbf{p} \times \mathfrak{S}_\mathbf{q}} . \tag{5.40}$$

From Theorem 5.3.1 we know that the space $(V^{*\otimes n} \otimes V^{\otimes n})^{\mathbf{GL}(V)}$ is spanned by the complete contractions λ_s for $s \in \mathfrak{S}_n$. Hence the right side of (5.40) is spanned by the tensors

$$|\mathfrak{S}_\mathbf{p} \times \mathfrak{S}_\mathbf{q}|^{-1} \sum_{(g,h) \in \mathfrak{S}_\mathbf{p} \times \mathfrak{S}_\mathbf{q}} (\sigma_n^*(g) \otimes \sigma_n(h))\lambda_s ,$$

where s ranges over $\mathfrak{S}_n$. Under the isomorphism (5.40), the action of $\mathfrak{S}_\mathbf{p} \times \mathfrak{S}_\mathbf{q}$ disappears and these tensors correspond to the polynomials

$$\begin{aligned}
F_s(v_1, \ldots, v_k, v_1^*, \ldots, v_m^*) &= \langle \lambda_s, v_1^{\otimes p_1} \otimes \cdots \otimes v_k^{\otimes p_k} \otimes v_1^{*\otimes q_1} \otimes \cdots \otimes v_m^{*\otimes q_m} \rangle \\
&= \langle v_1^{\otimes p_1} \otimes \cdots \otimes v_k^{\otimes p_k}, w_1^* \otimes \cdots \otimes w_n^* \rangle \\
&= \prod_{i=1}^{n} \langle w_i, w_i^* \rangle ,
\end{aligned}$$

where each w_i is v_j for some j and each w_i^* is $v_{j'}^*$ for some j' (depending on s). Obviously F_s is of the form (5.39).

5.4.3 *Proof of Polynomial FFT for* **O(V)** *and* **Sp(V)**

We now obtain the FFT for the action of $G = \mathbf{O}(V)$ or $G = \mathbf{Sp}(V)$ on $\mathcal{P}(V)$ (Theorem 5.2.2). We use the notation of this theorem and of Section 5.4.1, and we will follow the same argument as for $\mathbf{GL}(V)$ to deduce the polynomial version of the FFT from the tensor version. Let $\mathbf{p} \in \mathbb{N}^k$. Since $-I$ is in G and acts by $(-1)^{|\mathbf{p}|}$ on $\mathcal{P}^{[\mathbf{p}]}(V^k)$, we have $\mathcal{P}^{[\mathbf{p}]}(V^k)^G = 0$ if $|\mathbf{p}|$ is odd. Therefore, we may assume that $|\mathbf{p}| = 2m$. We now show that the space $\mathcal{P}^{[\mathbf{p}]}(V^k)^G$ is spanned by the monomials

$$\prod_{i,j=1}^{k} \omega(v_i, v_j)^{r_{ij}} \tag{5.41}$$

of weight $\mathbf{p}$, where ω is the bilinear form fixed by G. This will prove the FFT (polynomial version) for G.

By Lemma 5.4.1,

$$\mathcal{P}^{[\mathbf{p}]}(V^k)^G \cong \left[\left(V^{*\otimes 2m} \right)^G \right]^{\mathfrak{S}_{\mathbf{p}}}. \tag{5.42}$$

The space $(V^{*\otimes 2m})^G$ is spanned by the tensors $\sigma_{2m}^*(s)\theta_m^*$ for $s \in \mathfrak{S}_{2m}$ (Theorem 5.3.3). Hence the right side of (5.42) is spanned by the tensors

$$|\mathfrak{S}_{\mathbf{p}}|^{-1} \sum_{g \in \mathfrak{S}_{\mathbf{p}}} \sigma_{2m}^*(gs)\theta_m^* \,,$$

where s ranges over $\mathfrak{S}_{2m}$. Under the isomorphism (5.42), the action of $\mathfrak{S}_{\mathbf{p}}$ disappears and these tensors correspond to the polynomials

$$F_s(v_1, \ldots, v_k) = \langle \sigma_{2m}^*(s)\theta_m^*, v_1^{\otimes p_1} \otimes \cdots \otimes v_k^{\otimes p_k} \rangle = \prod_{i=1}^{k} \omega(u_i, u_{k+i}) \,,$$

where each u_i is one of the vectors v_j (the choice depends on s). Thus F_s is of the form (5.41).

5.5 Irreducible Representations of Classical Groups

We have already used the defining representation of a classical group G to study the structural features of the group (maximal torus, roots, weights). We now use the FFT to find the commuting algebra of G on the exterior algebra of the defining representation. Using this we obtain explicit realizations of the fundamental irreducible representations (except the spin representations).

5.5.1 Skew Duality for Classical Groups

We begin with some general constructions of operators on the exterior algebra over a finite-dimensional vector space V. Let $\bigwedge^\bullet V$ be the exterior algebra over V (see Section C.1.4). This is a finite-dimensional graded algebra. For $v \in V$ and $v^* \in V^*$ we have the *exterior product operator* $\varepsilon(v)$ and the *interior product operator* $\iota(v^*)$ on $\bigwedge^\bullet V$ that act by $\varepsilon(v)u = v \wedge u$ and

$$\iota(v^*)(v_1 \wedge \cdots \wedge v_k) = \sum_{j=1}^{k} (-1)^{j-1} \langle v^*, v_j \rangle v_1 \wedge \cdots \wedge \widehat{v_j} \wedge \cdots \wedge v_k$$

for $u \in \bigwedge V$ and $v_i \in V$ (here $\widehat{v}_j$ means to omit v_j). These operators change the degree by one: $\varepsilon(v) : \bigwedge^p V \longrightarrow \bigwedge^{p+1} V$ and $\iota(v^*) : \bigwedge^p V \longrightarrow \bigwedge^{p-1} V$. Also, the interior product operator is an *antiderivation*:

$$\iota(v^*)(w \wedge u) = (\iota(v^*)w) \wedge u + (-1)^k w \wedge (\iota(v^*)u)$$

for $w \in \bigwedge^k V$ and $u \in \bigwedge V$.

Define the *anticommutator* $\{a, b\} = ab + ba$ for elements a, b of an associative algebra. As elements of the algebra $\mathrm{End}(\bigwedge V)$, the exterior product and interior product operators satisfy the *canonical anticommutation relations*

$$\{\varepsilon(x), \varepsilon(y)\} = 0, \quad \{\iota(x^*), \iota(y^*)\} = 0, \quad \{\varepsilon(x), \iota(x^*)\} = \langle x^*, x \rangle I \qquad (5.43)$$

for $x, y \in V$ and $x^*, y^* \in V^*$. Indeed, the first two relations follow immediately from skew symmetry of multiplication in $\bigwedge V$, and the third is a straightforward consequence of the formula for $\iota(x^*)$ (the verification is left to the reader).

Interchanging V and V^*, we also have the exterior and interior multiplication operators $\varepsilon(v^*)$ and $\iota(v)$ on $\bigwedge^\bullet V^*$ for $v \in V$ and $v^* \in V^*$. They satisfy

$$\varepsilon(v^*) = \iota(v^*)^t, \qquad \iota(v) = \varepsilon(v)^t. \qquad (5.44)$$

We denote by ρ the representation of $\mathbf{GL}(V)$ on $\bigwedge V$:

$$\rho(g)(v_1 \wedge \cdots \wedge v_p) = gv_1 \wedge \cdots \wedge gv_p$$

for $g \in \mathbf{GL}(V)$ and $v_i \in V$. It is easy to check from the definition of interior and exterior products that

$$\rho(g)\varepsilon(v)\rho(g^{-1}) = \varepsilon(gv), \quad \rho(g)\iota(v^*)\rho(g^{-1}) = \iota((g^t)^{-1}v^*). \qquad (5.45)$$

We define the *skew Euler operator* E on $\bigwedge V$ by

$$E = \sum_{j=1}^d \varepsilon(f_j)\iota(f_j^*),$$

where $d = \dim V$ and $\{f_1, \ldots, f_d\}$ is a basis for V with dual basis $\{f_1^*, \ldots, f_d^*\}$.

Lemma 5.5.1. *The operator E commutes with $\mathbf{GL}(V)$ and acts by the scalar k on $\bigwedge^k V$. Hence E does not depend on the choice of basis for V. If $T \in \mathrm{End}(\bigwedge V)$ and $T : \bigwedge^k V \longrightarrow \bigwedge^{k+p} V$ for all k, then $[E, T] = pT$.*

Proof. Let $g \in \mathbf{GL}(V)$ have matrix $[g_{ij}]$ relative to the basis $\{f_i\}$. Relations (5.45) imply that

$$\rho(g)E\rho(g)^{-1} = \sum_{i,k} \left\{ \sum_j g_{ij}(g^{-1})_{jk} \right\} \varepsilon(f_i)\iota(f_k^*) = E,$$

so E commutes with $\mathbf{GL}(V)$. Clearly, $E(f_1 \wedge \cdots \wedge f_k) = kf_1 \wedge \cdots \wedge f_k$. Given $1 \leq i_1 < \cdots < i_k \leq d$, choose a permutation matrix $g \in \mathbf{GL}(V)$ such that $gf_p = f_{i_p}$

for $p = 1, \ldots, k$. Then

$$
\begin{aligned}
E(f_{i_1} \wedge \cdots \wedge f_{i_k}) &= \rho(g) E \rho(g)^{-1} (f_{i_1} \wedge \cdots \wedge f_{i_k}) = \rho(g) E(f_1 \wedge \cdots \wedge f_k) \\
&= k \rho(g)(f_1 \wedge \cdots \wedge f_k) = k f_{i_1} \wedge \cdots \wedge f_{i_k} \, .
\end{aligned}
$$

Hence $Eu = ku$ for all $u \in \bigwedge^k V$. $\qquad\square$

As a particular case of the commutation relations in Lemma 5.5.1, we have

$$
[E, \varepsilon(v)] = \varepsilon(v), \quad [E, \iota(v^*)] = -\iota(v^*) \quad \text{for } v \in V \text{ and } v^* \in V^* \, . \tag{5.46}
$$

Now suppose $G \subset \mathbf{GL}(V)$ is an algebraic group. The action of G on V extends to regular representations on $V^{\otimes m}$ and on $\bigwedge V$. Denote by Q_k the projection from $\bigwedge V$ onto $\bigwedge^k V$. Then Q_k commutes with G and we may identify $\mathrm{Hom}(\bigwedge^l V, \bigwedge^k V)$ with the subspace of $\mathrm{End}_G(\bigwedge V)$ consisting of the operators $Q_k A Q_l$, where $A \in \mathrm{End}_G(\bigwedge V)$ (these are the G-intertwining operators that map $\bigwedge^l V$ to $\bigwedge^k V$ and are zero on $\bigwedge^r V$ for $r \neq l$). Thus

$$
\mathrm{End}_G(\textstyle\bigwedge V) = \bigoplus_{0 \le l, k \le d} \mathrm{Hom}_G(\textstyle\bigwedge^l V, \textstyle\bigwedge^k V) \, .
$$

Let $\mathcal{T}(V)$ be the tensor algebra over V (see Appendix C.1.2). There is a projection operator $P : \mathcal{T}(V) \longrightarrow \bigwedge V$ given by

$$
Pu = \frac{1}{m!} \sum_{s \in \mathfrak{S}_m} \mathrm{sgn}(s) \sigma_m(s) u \quad \text{for } u \in V^{\otimes m} \, .
$$

Obviously P commutes with the action of G and preserves degree, so we have

$$
\mathrm{Hom}_G(\textstyle\bigwedge^l V, \textstyle\bigwedge^k V) = \{ PRP : R \in \mathrm{Hom}_G(V^{\otimes l}, V^{\otimes k}) \} \, . \tag{5.47}
$$

We now use these results and the FFT to find generators for $\mathrm{End}_G(\bigwedge V)$ when $G \subset \mathbf{GL}(V)$ is a classical group.

General Linear Group

Theorem 5.5.2. *Let $G = \mathbf{GL}(V)$. Then $\mathrm{End}_G(\bigwedge V)$ is generated by the skew Euler operator E.*

Proof. From Theorem 4.2.10 we know that $\mathrm{Hom}_G(V^{\otimes l}, V^{\otimes k})$ is zero if $l \neq k$, and is spanned by the operators $\sigma_k(s)$ with $s \in \mathfrak{S}_k$ when $l = k$. Since $P \sigma_k(s) P = \mathrm{sgn}(s) P$, we see that $\mathrm{End}_G(\bigwedge^p V) = \mathbb{C} I$. Thus if $A \in \mathrm{End}_G(\bigwedge V)$, then A acts on $\bigwedge^p V$ by a scalar a_p. Let $f(x)$ be a polynomial such that $f(p) = a_p$ for $p = 0, 1, \ldots, d$. Then $A = f(E)$. $\qquad\square$

Corollary 5.5.3. *In the decomposition $\bigwedge V = \bigoplus_{p=1}^d \bigwedge^p V$, the summands are irreducible and mutually inequivalent $\mathbf{GL}(V)$-modules.*

Proof. This follows from Theorems 4.2.1 and 5.5.2. $\qquad\square$

Orthogonal and Symplectic Groups

Now let Ω be a nondegenerate bilinear form on V that is either symmetric or skew symmetric. Let G be the subgroup of $\mathbf{GL}(V)$ that preserves Ω. In order to pass from the FFT for G to a description of the commutant of G in $\mathrm{End}(\bigwedge V)$, we need to introduce some operators on the tensor algebra over V.

Define $C : V^{\otimes m} \longrightarrow V^{\otimes(m+2)}$ by

$$Cu = \theta \otimes u \quad \text{for } u \in V^{\otimes m},$$

where $\theta \in (V \otimes V)^G$ is the 2-tensor corresponding to Ω as in Section 5.3.2. Define $C^* : V^{\otimes m} \longrightarrow V^{\otimes(m-2)}$ by

$$C^*(v_1 \otimes \cdots \otimes v_m) = \Omega(v_{m-1}, v_m)v_1 \otimes \cdots \otimes v_{m-2}.$$

Then C and C^* commute with the action of G on the tensor algebra.

For $v^* \in V^*$ define $\kappa(v^*) : V^{\otimes m} \longrightarrow V^{\otimes(m-1)}$ by evaluation on the first tensor place:

$$\kappa(v^*)(v_1 \otimes \cdots \otimes v_m) = \langle v^*, v_1 \rangle v_2 \otimes \cdots \otimes v_m.$$

For $v \in V$ define $\mu(v) : V^{\otimes m} \longrightarrow V^{\otimes(m+1)}$ by left tensor multiplication:

$$\mu(v)(v_1 \otimes \cdots \otimes v_m) = v \otimes v_1 \otimes \cdots \otimes v_m.$$

For $v \in V$ let $v^\sharp \in V^*$ be defined by $\langle v^\sharp, w \rangle = \Omega(v, w)$ for all $w \in V$. The map $v \mapsto v^\sharp$ is a G-module isomorphism. If we extend Ω to a bilinear form on $V^{\otimes k}$ for all k as in Section 5.3.2, then

$$\Omega(Cu, w) = \Omega(u, C^*w) \quad \text{and} \quad \Omega(\mu(v)u, w) = \Omega(u, \kappa(v^\sharp)w)$$

for all $u, w \in V^{\otimes k}$ and $v \in V$.

Lemma 5.5.4. *Let G be $\mathbf{O}(V, \Omega)$ (if Ω is symmetric) or $\mathbf{Sp}(V, \Omega)$ (if Ω is skew symmetric). Then the space $\mathrm{Hom}_G(V^{\otimes l}, V^{\otimes k})$ is zero if $k + l$ is odd. If $k + l$ is even, this space is spanned by the operators $\sigma_k(s)A\sigma_l(t)$, where $s \in \mathfrak{S}_k$, $t \in \mathfrak{S}_l$, and A is one of the following operators:*

1. *CB with $B \in \mathrm{Hom}_G(V^{\otimes l}, V^{\otimes(k-2)})$,*
2. *BC^* with $B \in \mathrm{Hom}_G(V^{\otimes(l-2)}, V^{\otimes k})$, or*
3. *$\sum_{p=1}^d \mu(f_p)B\kappa(f_p^*)$ with $B \in \mathrm{Hom}_G(V^{\otimes(l-1)}, V^{\otimes(k-1)})$. Here $\{f_p\}$ is a basis for V, $\{f_p^*\}$ is the dual basis for V^*, and $d = \dim V$.*

Proof. Recall that there is a canonical $\mathbf{GL}(V)$-module isomorphism

$$V^{\otimes k} \otimes V^{*\otimes l} \cong \mathrm{Hom}(V^{\otimes l}, V^{\otimes k})$$

(see Section B.2.2). Denote by $T_\xi \in \mathrm{Hom}_G(V^{\otimes l}, V^{\otimes k})$ the operator corresponding to $\xi \in (V^{\otimes k} \otimes V^{*\otimes l})^G$. We view ξ as a linear functional on $V^{*\otimes k} \otimes V^{\otimes l}$. From Theorem

5.3.5 and the G-module isomorphism $v \mapsto v^\sharp$ between V and V^* defined by Ω, we may assume that ξ is a complete contraction:

$$\langle \xi, v_1^\sharp \otimes \cdots \otimes v_k^\sharp \otimes v_{k+1} \otimes \cdots \otimes v_{2r} \rangle = \prod_{p=1}^{r} \Omega(v_{i_p}, v_{j_p}) \, .$$

In this formula $2r = k+l$ and $\{(i_1, j_1), \dots, (i_r, j_r)\}$ is a two-partition of $\{1, \dots, 2r\}$ with $i_p < j_p$ for each p.

Case 1: There exists a p such that $j_p \leq k$. Since we can permute the tensor positions in $V^{*\otimes k}$ by $\mathfrak{S}_k$, we may assume that $i_p = 1$ and $j_p = 2$. Let ζ be the complete contraction on $V^{*\otimes(k-2)} \otimes V^{\otimes l}$ obtained from ξ by omitting the contraction corresponding to the pair $(1, 2)$. Then for $u \in V^{\otimes l}$, $u^* \in V^{*\otimes(k-2)}$, and $v, w \in V$ we have

$$\langle T_\xi u, v^\sharp \otimes w^\sharp \otimes u^* \rangle = \langle \xi, v^\sharp \otimes w^\sharp \otimes u^* \otimes u \rangle = \Omega(v, w) \langle \zeta, u^* \otimes u \rangle$$
$$= \langle C T_\zeta u, u^* \rangle \, .$$

Here we have used the relation $\Omega(v, w) = \langle \theta, v^\sharp \otimes v^\sharp \rangle$. Hence $T_\xi = CB$ with $B = T_\zeta \in \mathrm{Hom}_G(V^{\otimes l}, V^{\otimes(k-2)})$.

Case 2: There exists a p such that $i_p > k$. Since we can permute the tensor positions in $V^{\otimes l}$ by $\mathfrak{S}_l$, we may assume that $i_p = k+1$ and $j_p = k+2$. Let ζ be the complete contraction on $V^{*\otimes k} \otimes V^{\otimes(l-2)}$ obtained from ξ by omitting the contraction corresponding to the pair $(k+1, k+2)$. Then for $u \in V^{\otimes(l-2)}$, $u^* \in V^{*\otimes k}$, and $v, w \in V$ we have

$$\langle T_\xi(v \otimes w \otimes u), u^* \rangle = \langle \xi, u^* \otimes v \otimes w \otimes u \rangle = \Omega(v, w) \langle \zeta, u^* \otimes u \rangle$$
$$= \langle T_\zeta C^*(v \otimes w \otimes u), u^* \rangle \, .$$

Hence $T_\xi = BC^*$ with $B = T_\zeta \in \mathrm{Hom}_G(V^{\otimes(l-2)}, V^{\otimes k})$.

Case 3: There exists a p such that $i_p \leq k$ and $j_p > k$. Since we can permute the tensor positions in $V^{*\otimes k}$ by $\mathfrak{S}_k$ and in $V^{\otimes l}$ by $\mathfrak{S}_l$, we may assume that $i_p = 1$ and $j_p = k+1$. Let ζ be the complete contraction on $V^{*\otimes(k-1)} \otimes V^{\otimes(l-1)}$ obtained from ξ by omitting the contraction corresponding to the pair $(1, k+1)$. Then for $v \in V$, $v^* \in V^*$, $u \in V^{\otimes(l-1)}$, and $u^* \in V^{*\otimes(k-1)}$ we have

$$\langle T_\xi(v \otimes u), v^* \otimes u^* \rangle = \langle \xi, v^* \otimes u^* \otimes v \otimes u \rangle = \langle v^*, v \rangle \langle \zeta, u^* \otimes u \rangle$$
$$= \sum_{p=1}^{d} \langle \mu(f_p) T_\zeta \kappa(f_p^*)(v \otimes u), v^* \otimes u^* \rangle \, .$$

Hence T_ξ is given as in (3).

Every two-partition satisfies at least one of these three cases, so the lemma is proved. $\square$

Theorem 5.5.5. (Ω symmetric) *Let* $G = \mathbf{O}(V, \Omega)$. *Then* $\mathrm{End}_G(\bigwedge V)$ *is generated by the skew Euler operator* E.

Proof. In this case the tensor θ is symmetric. Hence $PC = 0$ and $C^*P = 0$. For $v \in V$ and $u \in V^{\otimes m}$ we have

$$P\mu(v)u = P(v \otimes u) = v \wedge Pu = \varepsilon(v)Pu \tag{5.48}$$

(from the definition of multiplication in $\bigwedge V$). Furthermore, for $v^* \in V^*$,

$$\kappa(v^*)Pu = \frac{1}{m}\iota(v^*)Pu . \tag{5.49}$$

This identity follows from the formula

$$P(v_1 \otimes \cdots \otimes v_m) = \frac{1}{m}\sum_{j=1}^{m}(-1)^{j-1}v_j \otimes (v_1 \wedge \cdots \wedge \widehat{v_j} \wedge \cdots \wedge v_m) ,$$

obtained by summing first over $\mathfrak{S}_{m-1}$ and then over the m cosets of $\mathfrak{S}_{m-1}$ in $\mathfrak{S}_m$. From (5.48), (5.49), and Lemma 5.5.4 we conclude that

$$\mathrm{Hom}_G(\textstyle\bigwedge^l V, \bigwedge^k V) = 0 \quad \text{if } l \neq k .$$

Furthermore, $\mathrm{End}_G(\bigwedge^k V)$ is spanned by operators of the form

$$A = \sum_{p=1}^{d} P\mu(f_p)B\kappa(f_p^*)P = \frac{1}{l-1}\sum_{p=1}^{d}\varepsilon(f_p)PBP\iota(f_p^*) , \tag{5.50}$$

where $B \in \mathrm{End}_G(V^{\otimes(k-1)})$ and $d = \dim V$. Since $\varepsilon(f_p)E = (E+1)\varepsilon(f_p)$ by (5.46), it follows by induction on k that A is a polynomial in E. The theorem now follows by (5.47). □

Corollary 5.5.6. (Ω symmetric) *In the decomposition* $\bigwedge V = \bigoplus_{p=1}^{d}\bigwedge^p V$, *the summands are irreducible and mutually inequivalent* $\mathbf{O}(V, \Omega)$-*modules.*

Proof. The proof proceeds by using the same argument as in Corollary 5.5.3, but now using Theorem 5.5.5. □

   Now assume that $\dim V = 2n$ and Ω is skew-symmetric. Let $G = \mathbf{Sp}(V, \Omega)$ and define

$$X = -\frac{1}{2}PC^*P , \quad Y = \frac{1}{2}PCP . \tag{5.51}$$

These operators on $\bigwedge V$ commute with the action of G, since C, C^*, and P commute with G on tensor space.

Lemma 5.5.7. *The operators* X *and* Y *in (5.51) satisfy the commutation relations*

$$[Y, \varepsilon(v)] = 0 , \quad [X, \iota(v^*)] = 0 , \quad [Y, \iota(v^\sharp)] = \varepsilon(v) , \quad [X, \varepsilon(v)] = \iota(v^\sharp)$$

for all $v \in V$ and $v^* \in V^*$. Furthermore, $[E,Y] = 2Y$, $[E,X] = -2X$, and $[Y,X] = E - nI$.

Proof. Fix a basis $\{e_1, \ldots, e_n, e_{-1}, \ldots, e_{-n}\}$ for V such that

$$\Omega(e_i, e_j) = \text{sgn}(i)\, \delta_{i,-j} \quad \text{for } i, j = \pm 1, \ldots, \pm n \,.$$

Then $\{-e_{-1}^{\sharp}, \ldots, -e_{-n}^{\sharp}, e_1^{\sharp}, \ldots, e_n^{\sharp}\}$ is the dual basis for V^*. From (5.48) we see that

$$Y = \frac{1}{2} \sum_{j=1}^{n} \varepsilon(e_j)\varepsilon(e_{-j}) - \varepsilon(e_{-j})\varepsilon(e_j) = \sum_{j=1}^{n} \varepsilon(e_j)\varepsilon(e_{-j}) \,.$$

Likewise, we have

$$X = -\sum_{j=1}^{n} \iota(e_j^{\sharp})\iota(e_{-j}^{\sharp}) \,. \tag{5.52}$$

In this case the Euler operator on $\bigwedge^{\bullet} V$ can be written as

$$E = \sum_{j=1}^{n} \left(\varepsilon(e_{-j})\iota(e_j^{\sharp}) - \varepsilon(e_j)\iota(e_{-j}^{\sharp}) \right) \,.$$

From (5.43) we calculate that Y commutes with $\varepsilon(v)$ and that X commutes with $\iota(v^*)$. We also have

$$[\varepsilon(v), \iota(v^*)] = 2\varepsilon(v)\iota(v^*) - \langle v^*, v \rangle = -2\iota(v^*)\varepsilon(v) + \langle v^*, v \rangle \tag{5.53}$$

for $v \in V$ and $v^* \in V^*$. Using the Leibniz formula $[ab, c] = [a, c]b + a[b, c]$ for elements a, b, and c in an associative algebra and the relations (5.53), we calculate

$$[Y, \iota(v^{\sharp})] = \sum_{j=1}^{n} [\varepsilon(e_j), \iota(v^{\sharp})]\varepsilon(e_{-j}) + \varepsilon(e_j)[\varepsilon(e_{-j}), \iota(v^{\sharp})]$$

$$= \sum_{j=1}^{n} \left(2\varepsilon(e_j)\iota(v^{\sharp})\varepsilon(e_{-j}) - \langle v^{\sharp}, e_j \rangle \varepsilon(e_{-j}) \right.$$

$$\left. - 2\varepsilon(e_j)\iota(v^{\sharp})\varepsilon(e_{-j}) + \langle v^{\sharp}, e_{-j} \rangle \varepsilon(e_j) \right)$$

$$= \sum_{j=1}^{n} \left(-\Omega(v, e_j)\varepsilon(e_{-j}) + \Omega(v, e_{-j})\varepsilon(e_j) \right) \,.$$

Since $v = \sum_{j=1}^{n} \Omega(v, e_{-j})e_j - \Omega(v, e_j)e_{-j}$, we conclude that $[Y, \iota(v^{\sharp})] = \varepsilon(v)$. This implies that $[X, \varepsilon(v^{\sharp})] = \iota(v)$, since X is the operator adjoint to $-Y$ and $\varepsilon(v)$ is the operator adjoint to $\iota(v^{\sharp})$, relative to the bilinear form on $\bigwedge V$ defined by Ω.

The commutation relations involving E follow from Lemma 5.5.1. It remains to calculate

$$[Y,X] = \sum_{j=1}^{n} \left([\varepsilon(e_j), X]\varepsilon(e_{-j}) + \varepsilon(e_j)[\varepsilon(e_{-j}), X] \right)$$

$$= -\sum_{j=1}^{n} \left(\iota(e_j^\#)\varepsilon(e_{-j}) + \varepsilon(e_j)\iota(e_{-j}^\#) \right).$$

Since $\iota(e_j^\#)\varepsilon(e_{-j}) = -\varepsilon(e_{-j})\iota(e_j^\#) + I$, we have

$$[Y,X] = -nI + \sum_{j=1}^{n} \left(\varepsilon(e_{-j})\iota(e_j^\#) - \varepsilon(e_j)\iota(e_j^\#) \right) = E - nI$$

as claimed. $\square$

Define $\mathfrak{g}' = \mathrm{Span}\{X,Y,E-nI\}$. From Lemma 5.5.7 we see that $\mathfrak{g}'$ is a Lie algebra isomorphic to $\mathfrak{sl}(2,\mathbb{C})$.

Theorem 5.5.8. (Ω skew-symmetric) *The commutant of* $\mathbf{Sp}(V,\Omega)$ *in* $\mathrm{End}(\bigwedge V)$ *is generated by* $\mathfrak{g}'$.

Before proving this theorem, we apply it to obtain the fundamental representations of $\mathbf{Sp}(V,\Omega)$.

Corollary 5.5.9. ($G = \mathbf{Sp}(V,\Omega)$) *There is a canonical decomposition*

$$\bigwedge V \cong \bigoplus_{k=0}^{n} F^{(n-k)} \otimes \mathcal{H}^k, \tag{5.54}$$

as a $(G, \mathfrak{g}')$-*module, where* $\dim V = 2n$ *and* $F^{(k)}$ *is the irreducible* $\mathfrak{g}'$-*module of dimension* $k+1$. *Here* $\mathcal{H}^k$ *is an irreducible* G-*module and* $\mathcal{H}^k \not\cong \mathcal{H}^l$ *for* $k \neq l$.

Proof. The eigenvalues of $E - nI$ on $\bigwedge V$ are $n, n-1, \ldots, -n+1, -n$, so the only possible irreducible representations of $\mathfrak{g}'$ that can occur in $\bigwedge V$ are the representations $F^{(k)}$ with $k = 0, 1, \ldots, n$. Now

$$X(e_{-1} \wedge e_{-2} \wedge \cdots \wedge e_{-k}) = 0$$

for $1 \leq k \leq n$, since $\Omega(e_{-i}, e_{-j}) = 0$ for $i, j \geq 0$. Hence the $\mathfrak{g}'$-module generated by $e_{-1} \wedge e_{-2} \wedge \cdots \wedge e_{-k}$ is isomorphic to $F^{(k)}$, by Proposition 2.3.3, so all these $\mathfrak{g}'$-modules actually do occur in $\bigwedge V$. Now we apply Theorems 4.2.1 and 5.5.8 to obtain the decomposition of $\bigwedge V$. $\square$

In Section 5.5.2 we will obtain a more explicit description of the G-modules $\mathcal{H}^k$ that occur in (5.54). We now turn to the proof of Theorem 5.5.8. By (5.47) it suffices to show that the operators PRP, with $R \in \mathrm{Hom}_G(V^{\otimes l}, V^{\otimes k})$, are in the algebra generated by Y, X, and E. The proof will proceed by induction on $k+l$. We may assume that $k+l = 2r$ for some integer r, by Lemma 5.5.4.

Lemma 5.5.10. *Let $k + l$ be even. Then the space $\mathrm{Hom}_G(\bigwedge^l V, \bigwedge^k V)$ is spanned by operators of the following forms:*

1. *YQ with $Q \in \mathrm{Hom}_G(\bigwedge^l V, \bigwedge^{k-2} V)$,*
2. *QX with $Q \in \mathrm{Hom}_G(\bigwedge^{l-2} V, \bigwedge^k V)$,*
3. *$\sum_{p=1}^{2n} \varepsilon(f_p) Q\iota(f_p^*)$ with $Q \in \mathrm{Hom}_G(\bigwedge^{l-1} V, \bigwedge^{k-1} V)$, where $\{f_p\}$ is a basis for V and $\{f_p^*\}$ is the dual basis for V^*.*

Proof. We know that $\mathrm{Hom}_G(\bigwedge^l V, \bigwedge^k V)$ is spanned by operators $P\sigma_k(s)A\sigma_l(t)P$, with $s \in \mathfrak{S}_k$, $t \in \mathfrak{S}_l$, and A given in cases (1), (2), and (3) of Lemma 5.5.4. Since $P\sigma_k(s) = \mathrm{sgn}(s)P$ and $\sigma_l(t)P = \mathrm{sgn}(t)P$, it suffices to consider the operators PAP in the three cases. In case (1) we have $PAP = PCBP = 2YQ$ with $Q = PBP \in \mathrm{Hom}_G(\bigwedge^l V, \bigwedge^{k-2} V)$. In case (2) we have $PAP = PBC^*P = -2QX$ with $Q = PBP \in \mathrm{Hom}_G(\bigwedge^{l-2} V, \bigwedge^k V)$. In case (3) we can take $Q = PBP \in \mathrm{Hom}_G(\bigwedge^{l-1} V, \bigwedge^{k-1} V)$ by (5.50). $\qquad\square$

Completion of proof of Theorem 5.5.8.   We have reduced the proof to the follow-. ing assertion:

($\star$) The space $\mathrm{Hom}_G(V^{\otimes l}, V^{\otimes k})$, for $k + l$ even, is spanned by products of the operators Y, X, and E.

When $k = l = 0$, assertion ($\star$) is true, with the convention that empty products are 1. We assume that ($\star$) is true for $k + l < 2r$ and we take $k + l = 2r$. From Lemma 5.5.10 we see that ($\star$) will be true provided we can prove the following:

($\star\star$) If $Q \in \mathrm{Hom}_G(\bigwedge^{l-1} V, \bigwedge^{k-1} V)$, then $\sum_{p=1}^n \varepsilon(e_p) Q\iota(e_{-p}^\sharp) - \varepsilon(e_{-p}) Q\iota(e_p^\sharp)$ is a linear combination of products of the operators Y, X, and E.

(Here we have taken the basis $\{f_p\}$ for V to be $\{e_p, e_{-p}\}$ in Case (3) of Lemma 5.5.10.) To complete the inductive step, we will also need to prove the following variant of ($\star\star$):

($\star\star\star$) If $Q \in \mathrm{Hom}_G(\bigwedge^{l-1} V, \bigwedge^{k-1} V)$, then $\sum_{p=1}^n \iota(e_p^\sharp) Q\iota(e_{-p}^\sharp) - \iota(e_{-p}^\sharp) Q\iota(e_p^\sharp)$ is a linear combination of products of the operators Y, X, and E.

If $k = l = 1$, then $Q \in \mathbb{C}$. Hence the operator in ($\star\star$) is a constant multiple of E and the operator in ($\star\star\star$) is a constant multiple of X, so both assertions are true in this case. Now assume that ($\star\star$) and ($\star\star\star$) are true when $k + l < 2r$. Take $k + l = 2r$. By the inductive hypothesis for ($\star$), we can write Q as a linear combination of operators RS, where $R \in \{Y, X, E\}$ and $S \in \mathrm{Hom}_G(\bigwedge^a V, \bigwedge^b V)$ with $a + b = 2(r-2)$.

Case 1: $R = Y$. Since $[Y, \varepsilon(v)] = 0$, we see that ($\star\star$) holds for Q by the induction hypothesis applied to S. Since $[Y, \iota(v^\sharp)] = \varepsilon(v)$, the operator in ($\star\star\star$) can be written as $YA + B$, where

$$A = \sum_{p=1}^n \iota(e_p^\sharp) S\iota(e_{-p}^\sharp) - \iota(e_{-p}^\sharp) S\iota(e_p^\sharp), \quad B = -\sum_{p=1}^n \varepsilon(e_p^\sharp) S\iota(e_{-p}^\sharp) - \varepsilon(e_{-p}^\sharp) S\iota(e_p^\sharp).$$

By induction $(\star\star)$ and $(\star\star\star)$ are true with Q replaced by S. Hence $(\star\star\star)$ holds for Q.

Case 2: $R = X$. Since $[Y, \iota(v^\sharp)] = 0$, we see that $(\star\star\star)$ holds for Q by the induction hypothesis applied to S. Since $[X, \varepsilon(v)] = \iota(v^\sharp)$, the operator in $(\star\star)$ can be written as $-XA + B$, where now

$$A = \sum_{p=1}^n \varepsilon(e_p^\sharp) S \iota(e_{-p}^\sharp) - \varepsilon(e_{-p}^\sharp) S \iota(e_p^\sharp) , \quad B = \sum_{p=1}^n \iota(e_p^\sharp) S \iota(e_{-p}^\sharp) - \iota(e_{-p}^\sharp) S \iota(e_p^\sharp) .$$

By induction $(\star\star)$ and $(\star\star\star)$ are true with Q replaced by S. Hence $(\star\star)$ holds for Q.

Case 3: $R = E$. The commutation relations in Lemma 5.5.7 show that the operators in $(\star\star)$ and $(\star\star\star)$ can be written as $(E \pm 1)A$. Here A is obtained by replacing Q by S in $(\star\star)$ or $(\star\star\star)$. By induction we conclude that $(\star\star)$ and $(\star\star\star)$ hold for S, and hence they hold for Q. $\square$

5.5.2 Fundamental Representations

Let G be a connected classical group whose Lie algebra $\mathfrak{g}$ is semisimple. An irreducible rational representation of G is uniquely determined by the corresponding finite-dimensional representation of $\mathfrak{g}$, which is also irreducible (in the following, *representation* will mean finite-dimensional representation). Theorem 3.2.6 establishes a bijection between the irreducible representations of $\mathfrak{g}$ and the set $P_{++}(\mathfrak{g})$ of dominant integral weights (with equivalent representations being identified and a set of positive roots fixed). When $G = \mathbf{SL}(2, \mathbb{C})$, we showed in Section 2.3.2 that every irreducible representation of $\mathfrak{g}$ is the differential of a representation of G by giving an explicit model for the representation and the actions of $\mathfrak{g}$ and G. We would like to do the same thing in the general case.

We begin with the so-called *fundamental representations*. Recall from Section 3.1.4 that the elements of $P_{++}(\mathfrak{g})$ are of the form $n_1 \varpi_1 + \cdots + n_l \varpi_l$ with $n_i \in \mathbb{N}$, where $\varpi_1, \ldots, \varpi_l$ are the fundamental weights. An irreducible representation whose highest weight is ϖ_k for some k is called a *fundamental representation*. We shall give explicit models for these representations and the action of $\mathfrak{g}$ and G (for the orthogonal groups this construction will be completed in Chapter 6 with the construction of the spin representations and spin groups).

Special Linear Group

It is easy to construct the fundamental representations when G is $\mathbf{SL}(n, \mathbb{C})$.

Theorem 5.5.11. *Let $G = \mathbf{SL}(n, \mathbb{C})$. The representation σ_r on the rth exterior power $\bigwedge^r \mathbb{C}^n$ is irreducible and has highest weight ϖ_r for $1 \leq r < n$.*

Proof. From Corollary 5.5.3 we know that $\bigwedge^r \mathbb{C}^n$ is an irreducible $\mathbf{GL}(n,\mathbb{C})$-module. Hence it is also irreducible for $\mathbf{SL}(n,\mathbb{C})$. It remains only to determine its highest weight. Take the positive roots and triangular decomposition $\mathfrak{sl}(n,\mathbb{C}) = \mathfrak{n}^- + \mathfrak{h} + \mathfrak{n}^+$ as in Theorem 2.4.11. Then $\mathfrak{n}^+$ consists of the strictly upper-triangular matrices. Define $u_r = e_1 \wedge \cdots \wedge e_r$. Then u_r is annihilated by $\mathfrak{n}^+$ and has weight ϖ_r for $r = 1, \ldots, n-1$. $\qquad\square$

Remark 5.5.12. For $r = n$ the space $\bigwedge^n \mathbb{C}^n$ is one-dimensional and σ_n is the trivial representation of $\mathbf{SL}(n,\mathbb{C})$.

Special Orthogonal Group

Let B be the symmetric form (2.9) on $\mathbb{C}^n$ and let $G = \mathbf{O}(\mathbb{C}^n, B)$. Let $G^\circ = \mathbf{SO}(\mathbb{C}^n, B)$ (the identity component of G).

Theorem 5.5.13. *Let σ_r denote the representation of G on $\bigwedge^r \mathbb{C}^n$ for $1 \le r \le n$ associated with the defining representation σ_1 on $\mathbb{C}^n$.*

1. *Let $n = 2l + 1 \ge 3$ be odd.*
   If $1 \le r \le l$, then $(\sigma_r, \bigwedge^r \mathbb{C}^n)$ is an irreducible representation of G° with highest weight ϖ_r for $r \le l - 1$ and highest weight $2\varpi_l$ for $r = l$.
2. *Let $n = 2l \ge 4$ be even.*
   (a) *If $1 \le r \le l - 1$, then $(\sigma_r, \bigwedge^r \mathbb{C}^n)$ is an irreducible representation of G° with highest weight ϖ_r for $r \le l - 2$ and highest weight $\varpi_{l-1} + \varpi_l$ for $r = l - 1$.*
   (b) *The space $\bigwedge^l \mathbb{C}^n$ is irreducible under the action of G. As a module for G° it decomposes into the sum of two irreducible representations with highest weights $2\varpi_{l-1}$ and $2\varpi_l$.*

Proof. From Corollary 5.5.6 we know that $(\sigma_r, \bigwedge^r \mathbb{C}^n)$ is an irreducible G-module for $1 \le r \le n$.

(1): If $n = 2l + 1$ is odd, then $G = G^\circ \cup (-I)G^\circ$. Hence $\bigwedge^r \mathbb{C}^n$ is an irreducible G°-module for $1 \le r \le n$. To determine its highest weight when $r \le l$, label the basis for $\mathbb{C}^{2l+1}$ as $e_0 = e_{l+1}$ and $e_{-i} = e_{2l+2-i}$ for $i = 1, \ldots, l$. Take the positive roots and triangular decomposition $\mathfrak{n}^- + \mathfrak{h} + \mathfrak{n}^+$ of $\mathfrak{g} = \mathfrak{so}(\mathbb{C}^n, B)$ as in Theorem 2.4.11. Set $u_r = e_1 \wedge \cdots \wedge e_r$ for $r = 1, \ldots, l$. Then $\mathfrak{n}^+ u_r = 0$, since the matrices in $\mathfrak{n}^+$ are strictly upper triangular. Hence u_r is a highest-weight vector of weight

$$\varepsilon_1 + \cdots + \varepsilon_r = \begin{cases} \varpi_r & \text{for } 1 \le r \le l - 1, \\ 2\varpi_l & \text{for } r = l. \end{cases}$$

(2): Suppose $n = 2l$ is even. Label a B-isotropic basis for $\mathbb{C}^n$ as $\{e_{\pm i} : i = 1, \ldots, l\}$, where $e_{-i} = e_{n+1-i}$. Define $g_0 \in G$ by $g_0 e_l = e_{-l}$, $g_0 e_{-l} = e_l$, and $g_0 e_i = e_i$ for $i \ne \pm l$. Then $G = G^\circ \cup g_0 G^\circ$ and $\mathrm{Ad}(g_0)\mathfrak{h} = \mathfrak{h}$.

Let $\mathfrak{g} = \mathfrak{so}(\mathbb{C}^n, B)$ and take the positive roots and triangular decomposition $\mathfrak{g} = \mathfrak{n}^- + \mathfrak{h} + \mathfrak{n}^+$ as in Theorem 2.4.11. Since

$$\text{Ad}^*(g_0)\varepsilon_i = \begin{cases} \varepsilon_i & \text{for } i = 1,\ldots,l-1, \\ -\varepsilon_i & \text{for } i = l, \end{cases}$$

we have $\text{Ad}^*(g_0)(\Phi^+) = \Phi^+$ and hence $\text{Ad}(g_0)\mathfrak{n}^+ = \mathfrak{n}^+$. Let $u_r = e_1 \wedge \cdots \wedge e_r$ for $r = 1,\ldots,l$. Then u_r has weight

$$\varepsilon_1 + \cdots + \varepsilon_r = \begin{cases} \varpi_r & \text{for } 1 \le r \le l-2, \\ \varpi_{l-1} + \varpi_l & \text{for } r = l-1, \\ 2\varpi_l & \text{for } r = l. \end{cases} \tag{5.55}$$

Because $\mathfrak{n}^+$ consists of strictly upper-triangular matrices, we have $\mathfrak{n}^+ u_r = 0$, and thus u_r is a highest-weight vector. Let T_r be the $\mathfrak{g}$-cyclic subspace of $\bigwedge^r V$ generated by v_r. Then T_r is an irreducible $\mathfrak{g}$-module by Proposition 3.3.9, and hence it is an irreducible G°-module by Theorems 2.2.2 and 2.2.7. If $r < l$, then $g_0 u_r = u_r$, so T_r is invariant under G. Hence $T_r = \bigwedge^r \mathbb{C}^n$ in this case, proving part (a).

Now let $r = l$. Let $v_\pm = e_1 \wedge \cdots \wedge e_{l-1} \wedge e_{\pm l}$. Then v_+ is a highest-weight vector of weight $2\varpi_l$. Since $g_0 v_+ = v_-$ and $\text{Ad}(g_0)\mathfrak{n}^+ = \mathfrak{n}^+$, we have

$$X v_- = g_0 \text{Ad}(g_0)(X) v_+ = 0 \quad \text{for } X \in \mathfrak{n}^+ .$$

Thus v_- is a highest-weight vector of weight $\varepsilon_1 + \cdots + \varepsilon_{l-1} - \varepsilon_l = 2\varpi_{l-1}$. By Proposition 3.3.9 the cyclic spaces $U_\pm$ generated by $v_\pm$ under the action of $\mathfrak{g}$ are irreducible as $\mathfrak{g}$-modules, and hence they are irreducible as G°-modules by Theorems 2.2.2 and 2.2.7. We have $U_+ \cap U_- = \{0\}$, since U_+ is irreducible and inequivalent to U_-. Since $g_0 U_\pm = U_\mp$, the space $U_+ + U_-$ is invariant under G. Hence

$$\bigwedge^l \mathbb{C}^n = U_+ \oplus U_-$$

by the irreducibility of $\bigwedge^l \mathbb{C}^n$ under G. This proves part (b). $\square$

Symplectic Group

Let $G = \mathbf{Sp}(\mathbb{C}^{2l}, \Omega)$, where Ω is a nondegenerate skew-symmetric form. Corollary 5.5.9 gives the decomposition of $\bigwedge \mathbb{C}^{2l}$ under the action of G. Now we will use the theorem of the highest weight to identify the isotypic components in this decomposition. As in Section 5.5.1 we let $\theta \in (\bigwedge^2 V)^G$ be the G-invariant skew 2-tensor corresponding to Ω. Let Y be the operator of exterior multiplication by $(1/2)\theta$, and let $X = -Y^*$ (the adjoint operator relative to the skew-bilinear form on $\bigwedge V$ obtained from Ω). Set $H = lI - E$, where E is the skew Euler operator. Then we have the commutation relations

$$[H,X] = 2X , \quad [H,Y] = -2Y , \quad [X,Y] = H ,$$

by Lemma 5.5.7. Set $\mathfrak{g}' = \text{Span}\{X,Y,H\}$. Then $\mathfrak{g}' \cong \mathfrak{sl}(2,\mathbb{C})$ and $\mathfrak{g}'$ generates the commuting algebra $\text{End}_G(\bigwedge V)$ by Theorem 5.5.8. From formula (5.52) we can view X as a skew-symmetric Laplace operator. This motivates the following terminology:

Definition 5.5.14. An element $u \in \bigwedge \mathbb{C}^{2l}$ is Ω-*harmonic if* $Xu = 0$. *The space of* Ω-*harmonic elements is denoted by* $\mathcal{H}(\bigwedge \mathbb{C}^{2l}, \Omega)$.

Since $X : \bigwedge^p \mathbb{C}^{2l} \longrightarrow \bigwedge^{2p-2} \mathbb{C}^{2l}$ shifts degree by two, an element u is Ω-harmonic if and only if each homogeneous component of u is Ω-harmonic. Thus

$$\mathcal{H}(\bigwedge \mathbb{C}^{2l}, \Omega) = \bigoplus_{p \geq 0} \mathcal{H}(\bigwedge^p \mathbb{C}^{2l}, \Omega) ,$$

where $\mathcal{H}(\bigwedge^p \mathbb{C}^{2l}, \Omega) = \{u \in \bigwedge^p \mathbb{C}^{2l} : Xu = 0\}$. Because X commutes with G, this space is invariant under G.

Theorem 5.5.15. *Let* $G = \mathbf{Sp}(\mathbb{C}^{2l}, \Omega)$ *and* $\mathfrak{g}' = \mathrm{Span}\{X, Y, H\}$ *as above.*

1. If $p > l$ *then* $\mathcal{H}(\bigwedge^p \mathbb{C}^{2l}, \Omega) = 0$.
2. Let $F^{(k)}$ *be the irreducible* $\mathfrak{g}'$-*module of dimension* $k + 1$. *Then*

$$\bigwedge \mathbb{C}^{2l} \cong \bigoplus_{p=0}^{l} \left\{ F^{(l-p)} \otimes \mathcal{H}(\bigwedge^p \mathbb{C}^{2l}, \Omega) \right\} \tag{5.56}$$

as a $(\mathfrak{g}', G)$-*module.*
3. If $1 \leq p \leq l$, *then* $\mathcal{H}(\bigwedge^p \mathbb{C}^{2l}, \Omega)$ *is an irreducible* G-*module with highest weight* ϖ_p.

Proof. We already observed in the proof of Corollary 5.5.9 that the irreducible representations of $\mathfrak{g}'$ that occur in $\bigwedge \mathbb{C}^{2l}$ are $F^{(l-p)}$, where $p = 0, 1, \ldots, l$. By definition of H and X the space $\mathcal{H}(\bigwedge^p \mathbb{C}^{2l}, \Omega)$ consists of all $\mathfrak{g}'$ highest-weight vectors of weight $l - p$. Theorem 4.2.1 implies that G generates the commutant of $\mathfrak{g}'$ in $\mathrm{End}(\bigwedge \mathbb{C}^{2l})$, so Theorem 4.2.12 (applied to $\mathfrak{g}'$) furnishes the decomposition (5.56). For the same reason, we see that $\mathcal{H}(\bigwedge^p \mathbb{C}^{2l}, \Omega)$ is an irreducible G-module. It remains only to find its highest weight.

Fix the triangular decomposition $\mathfrak{g} = \mathfrak{n}^- + \mathfrak{h} + \mathfrak{n}^+$ as in Section 2.4.3. Since the matrices in $\mathfrak{n}^+$ are strictly upper triangular, the p-vector $u_p = e_1 \wedge \cdots \wedge e_p$ is a highest-weight vector of weight ϖ_p. Since $\Omega(e_i, e_j) = 0$ for $i, j = 1, \ldots, l$, it follows from (5.52) that $Xu_p = 0$. Thus u_p is Ω-harmonic and is a joint highest-weight vector for $\mathfrak{g}$ and for $\mathfrak{g}'$. $\qquad \square$

Corollary 5.5.16. *The map* $\mathbb{C}[t] \otimes \mathcal{H}(\bigwedge \mathbb{C}^{2l}, \Omega) \longrightarrow \bigwedge \mathbb{C}^{2l}$ *given by* $f(t) \otimes u \mapsto f(\theta) \wedge u$ *(exterior multiplication) is a* G-*module isomorphism. Thus*

$$\bigwedge^k \mathbb{C}^{2l} = \bigoplus_{p=0}^{[k/2]} \theta^p \wedge \mathcal{H}(\bigwedge^{k-2p} \mathbb{C}^{2l}, \Omega) . \tag{5.57}$$

Hence $\bigwedge^k \mathbb{C}^{2l}$ *is a multiplicity-free* G-*module and has highest weights* ϖ_{k-2p} *for* $0 \leq p \leq [k/2]$ *(where* $\varpi_0 = 0$*).*

Proof. Since $Hu_p = (l-p)u_p$ and $2Y$ is exterior multiplication by θ, we have

$$F^{(l-p)} = \bigoplus_{k=0}^{l-p} \mathbb{C}\theta^k \wedge u_p$$

(notation of Proposition 2.3.3). Now use (5.56). $\qquad \square$

Corollary 5.5.17. *The space* $\mathcal{H}(\bigwedge^p \mathbb{C}^{2l}, \Omega)$ *has dimension* $\binom{2l}{p} - \binom{2l}{p-2}$ *for* $p = 1, \ldots, l$.

Proof. For $p = 1$ we have $\mathcal{H}(\mathbb{C}^{2l}, \Omega) = \mathbb{C}^{2l}$, so the dimension is as stated (with the usual convention that $\binom{m}{q} = 0$ when q is a negative integer). Now use induction on p and (5.57).   □

We can describe the space $\mathcal{H}(\bigwedge^p \mathbb{C}^{2l}, \Omega)$ in another way. Let $v_i \in \mathbb{C}^{2l}$. Call $v_1 \wedge \cdots \wedge v_p$ an *isotropic p-vector* if $\Omega(v_i, v_j) = 0$ for $i, j = 1, \ldots, p$.

Proposition 5.5.18. *The space* $\mathcal{H}(\bigwedge^p \mathbb{C}^{2l}, \Omega)$ *is spanned by the isotropic p-vectors for* $p = 1, \ldots, l$.

Proof. Let F_p be the space spanned by the isotropic p-vectors. Clearly F_p is invariant under G. Any linearly independent set $\{v_1, \ldots, v_p\}$ of isotropic vectors in $\mathbb{C}^{2l}$ can be embedded in a canonical symplectic basis, and G acts transitively on the set of canonical symplectic bases (cf. Lemma 1.1.5). Since $u_p \in F_p \cap \mathcal{H}(\bigwedge^p \mathbb{C}^{2l}, \Omega)$, it follows that $F_p \subset \mathcal{H}(\bigwedge^p \mathbb{C}^{2l}, \Omega)$. Hence we have equality by the irreducibility of the space of Ω-harmonic p-vectors.   □

5.5.3 Cartan Product

Using skew duality we have constructed the fundamental representations of a connected classical group G whose Lie algebra is semisimple (with three exceptions in the case of the orthogonal groups). Now we obtain more irreducible representations by decomposing tensor products of representations already constructed.

Given finite-dimensional representations (ρ, U) and (σ, V) of G we can form the tensor product $(\rho \otimes \sigma, U \otimes V)$ of these representations. The weight spaces of $\rho \otimes \sigma$ are

$$(U \otimes V)(\nu) = \sum_{\lambda + \mu = \nu} U(\lambda) \otimes V(\mu). \tag{5.58}$$

In particular,

$$\dim(U \otimes V)(\nu) = \sum_{\lambda + \mu = \nu} \dim U(\lambda) \dim V(\mu). \tag{5.59}$$

Decomposing $U \otimes V$ into isotypic components for G and determining the multiplicities of each component is a more difficult problem that we shall treat in later chapters with the aid of the Weyl character formula. However, when ρ and σ are irreducible, then by the theorem of the highest weight (Theorem 3.2.5) we can identify a particular irreducible component that occurs with multiplicity one in the tensor product.

Proposition 5.5.19. *Let* $\mathfrak{g}$ *be a semisimple Lie algebra. Let* (π^λ, V^λ) *and* (π^μ, V^μ) *be finite-dimensional irreducible representations of* $\mathfrak{g}$ *with highest weights* $\lambda, \mu \in P_{++}(\mathfrak{g})$.

1. *Fix highest-weight vectors $v_\lambda \in V^\lambda$ and $v_\mu \in V^\mu$. Then the $\mathfrak{g}$-cyclic subspace $U \subset V^\lambda \otimes V^\mu$ generated by $v_\lambda \otimes v_\mu$ is an irreducible $\mathfrak{g}$-module with highest weight $\lambda + \mu$.*
2. *If ν occurs as the highest weight of a $\mathfrak{g}$-submodule of $V^\lambda \otimes V^\mu$ then $\nu \preceq \lambda + \mu$.*
3. *The irreducible representation $(\pi^{\lambda+\mu}, V^{\lambda+\mu})$ occurs with multiplicity one in $V^\lambda \otimes V^\mu$.*

Proof. The vector $v_\lambda \otimes v_\mu$ is $\mathfrak{b}$-extreme of weight $\lambda + \mu$. Hence U is irreducible by Proposition 3.3.9 and has highest weight $\lambda + \mu$, which proves (1).

Set $M = V^\lambda \otimes V^\mu$. By Theorem 3.2.5 and (5.58) the weights of M are in the set $\lambda + \mu - Q_+(\mathfrak{g})$ and the weight space $M(\lambda + \mu)$ is spanned by $v_\lambda \otimes v_\mu$. Thus $\dim M^{\mathfrak{n}}(\lambda + \mu) = 1$, which implies $\text{mult}_M(\pi^{\lambda+\mu}) = 1$ by Corollary 3.3.14. This proves (2) and (3). $\qquad\square$

We call the submodule U in (1) of Proposition 5.5.19 the *Cartan product* of the representations (π^λ, V^λ) and (π^μ, V^μ).

Corollary 5.5.20. *Let G be the group $\mathbf{SL}(V)$ or $\mathbf{Sp}(V)$ with $\dim V \geq 2$, or $\mathbf{SO}(V)$ with $\dim V \geq 3$. If π^λ and π^μ are differentials of irreducible regular representations of G, then the Cartan product of π^λ and π^μ is the differential of an irreducible regular representation of G with highest weight $\lambda + \mu$. Hence the set of highest weights of irreducible regular G-modules is closed under addition.*

Proof. This follows from Proposition 5.5.19 and Theorems 2.2.2 and 2.2.7. $\qquad\square$

Theorem 5.5.21. *Let G be the group $\mathbf{SL}(V)$ or $\mathbf{Sp}(V)$ with $\dim V \geq 2$, or $\mathbf{SO}(V)$ with $\dim V \geq 3$. For every dominant weight $\mu \in P_{++}(G)$ there exists an integer k such that $V^{\otimes k}$ contains an irreducible G-module with highest weight μ. Hence every irreducible regular representation of G occurs in the tensor algebra of V.*

Proof. Suppose that $G = \mathbf{SL}(\mathbb{C}^{l+1})$ or $\mathbf{Sp}(\mathbb{C}^{2l})$ and let $n = l+1$ or $n = 2l$, respectively. From Theorems 5.5.11 and 5.5.15 we know that $\bigwedge \mathbb{C}^n$ contains irreducible representations of G with highest weights $\varpi_1, \ldots, \varpi_l$. These weights generate the semigroup $P_{++}(G) = P_{++}(\mathfrak{g})$ (see Section 3.1.4).

Now let $G = \mathbf{SO}(n, \mathbb{C})$ with $n \geq 3$. From Theorem 5.5.13 we know that $\bigwedge \mathbb{C}^n$ contains irreducible representations of G with highest weights $\varpi_1, \ldots, \varpi_{l-2}, 2\varpi_{l-1}, 2\varpi_l$, and $\varpi_{l-1} + \varpi_l$ when $n = 2l$ is even, and $\bigwedge \mathbb{C}^n$ contains irreducible representations of G with highest weights $\varpi_1, \ldots, \varpi_{l-1}$, and $2\varpi_l$ when $n = 2l+1$ is odd. These weights generate the semigroup $P_{++}(G)$ by Proposition 3.1.19, so we may apply Theorem 3.2.5 and Corollary 5.5.20 to complete the proof. $\qquad\square$

5.5.4 Irreducible Representations of $\mathbf{GL(V)}$

We now extend the theorem of the highest weight to the group $G = \mathbf{GL}(n, \mathbb{C})$. Recall from Section 3.1.4 that $P_{++}(G)$ consists of all weights

$$\mu = m_1 \varepsilon_1 + \cdots + m_n \varepsilon_n, \quad \text{with} \quad m_1 \geq \cdots \geq m_n \quad \text{and} \quad m_i \in \mathbb{Z}. \tag{5.60}$$

Define the dominant weights

$$\lambda_i = \varepsilon_1 + \cdots + \varepsilon_i \quad \text{for} \quad i = 1, \ldots, n. \tag{5.61}$$

Note that the restriction of λ_i to the diagonal matrices of trace zero is the fundamental weight ϖ_i of $\mathfrak{sl}(n, \mathbb{C})$ for $i = 1, \ldots, n-1$. If μ is given by (5.60) then

$$\mu = (m_1 - m_2)\lambda_1 + (m_2 - m_3)\lambda_2 + \cdots + (m_{n-1} - m_n)\lambda_{n-1} + m_n \lambda_n.$$

Hence the elements of $P_{++}(G)$ can also be written uniquely as

$$\mu = k_1 \lambda_1 + \cdots + k_n \lambda_n, \quad \text{with} \quad k_1 \geq 0, \ldots, k_{n-1} \geq 0 \quad \text{and} \quad k_i \in \mathbb{Z}.$$

The restriction of μ to the diagonal matrices of trace zero is the weight

$$\mu_0 = (m_1 - m_2)\varpi_1 + (m_2 - m_3)\varpi_2 + \cdots + (m_{n-1} - m_n)\varpi_{n-1}. \tag{5.62}$$

Theorem 5.5.22. *Let $G = \mathbf{GL}(n, \mathbb{C})$ and let μ be given by (5.60). Then there exists a unique irreducible rational representation (π_n^μ, F_n^μ) of G such that the following hold:*

1. *The restriction of π_n^μ to $\mathbf{SL}(n, \mathbb{C})$ has highest weight μ_0 given by (5.62).*
2. *The element zI_n of G (for $z \in \mathbb{C}^\times$) acts by $z^{m_1 + \cdots + m_n} I$ on F_n^μ.*

Define $\check{\pi}_n^\mu(g) = \pi_n^\mu(g^t)^{-1}$ for $g \in G$. Then the representation $(\check{\pi}_n^\mu, F_n^\mu)$ is equivalent to the dual representation $((\pi_n^\mu)^*, (F_n^\mu)^*)$.

Proof. Let (π_0, V) be the irreducible regular representation of $\mathbf{SL}(n, \mathbb{C})$ with highest weight μ_0 whose existence follows from Theorem 5.5.21. We extend π_0 to a representation π of $\mathbf{GL}(n, \mathbb{C})$ on V by using (2) to define the action of the center zI_n, $z \in \mathbb{C}^\times$, of $\mathbf{GL}(n, \mathbb{C})$. We must show that this definition is consistent. Note that if $h = zI_n \in \mathbf{SL}(n, \mathbb{C})$, then $z^n = 1$. However, $\pi_0(h) = cI$ for some scalar c. By considering the action of h on the highest-weight vector, we see that $\pi_0(h) = z^p I$, where $p = m_1 + \cdots + m_{n-1} - (n-1)m_n$. Hence $z^{m_1 + \cdots + m_n} = z^p$, as needed. Property (1) of the theorem uniquely determines the restriction of π to $\mathbf{SL}(n, \mathbb{C})$ by Theorems 2.2.5 and 3.2.5. Property (2) uniquely determines the extension of π_0 to $\mathbf{GL}(n, \mathbb{C})$, since $\mathbf{GL}(n, \mathbb{C}) = \mathbb{C}^\times \cdot \mathbf{SL}(n, \mathbb{C})$. Thus we may define $\pi_n^\mu = \pi$ and $F_n^\mu = V$.

Let $\mathfrak{g} = \mathfrak{sl}(n, \mathbb{C})$ and let $\mathfrak{g} = \mathfrak{n}^- + \mathfrak{h} + \mathfrak{n}^+$ be the usual triangular decomposition of $\mathfrak{g}$ with $\mathfrak{n}^- = (\mathfrak{n}^+)^t$. Let v_0 be a highest-weight vector for π_0. Then

$$\check{\pi}_0(\mathfrak{n}^-)v_0 = \pi_0(\mathfrak{n}^+)v_0 = 0 \quad \text{and} \quad \check{\pi}_0(Y)v_0 = -\langle \mu_0, Y \rangle v_0 \quad \text{for } Y \in \mathfrak{h}.$$

Thus v_0 is a lowest-weight vector of weight $-\mu_0$ for $\check{\pi}_0$. Since $\check{\pi}_0$ is irreducible, it is isomorphic to π_0^* by Theorems 3.2.5 and 3.2.13. Since $\check{\pi}(zI_n) = z^m I$, where $m = -m_1 - \cdots - m_n$, it follows that $\check{\pi} \cong \pi^*$. $\qquad \square$

5.5.5 Irreducible Representations of $\mathbf{O}(\mathbf{V})$

We now determine the irreducible regular representations of the full orthogonal group in terms of the irreducible representations of the special orthogonal group.

We use the following notation: Let B be the symmetric bilinear form (2.9) on $\mathbb{C}^n$. Let $G = \mathbf{O}(\mathbb{C}^n, B)$, so that $G^\circ = \mathbf{SO}(\mathbb{C}^n, B)$. Let H be the diagonal subgroup of G, $H^\circ = H \cap G^\circ$, and $N^+ = \exp(\mathfrak{n}^+)$ the subgroup of upper-triangular unipotent matrices in G, as in Theorem 2.4.11. Let (π^λ, V^λ) be the irreducible representation of G° with highest weight $\lambda \in P_{++}(G^\circ)$.

When $n = 2l + 1$ is odd, then $\det(-I) = -1$, and we have $G^\circ \times \mathbb{Z}_2 \cong G$ (direct product). In this case $H \cong H^\circ \times \{\pm I\}$. If (ρ, W) is an irreducible representation of G, then $\rho(-I) = \varepsilon I$ with $\varepsilon = \pm 1$ by Schur's lemma, since $-I$ is in the center of G and $\rho(-I)^2 = 1$. Hence the restriction of ρ to G° is still irreducible, so $\dim W^{\mathfrak{n}^+} = 1$ by Corollary 3.3.14. The action of H on $W^{\mathfrak{n}^+}$ is by some character $\chi_{\lambda,\varepsilon}(h, a) = \varepsilon h^\lambda$ for $h \in H^\circ$, where $\varepsilon = \pm$ and

$$\lambda = \lambda_1 \varepsilon_1 + \cdots + \lambda_l \varepsilon_l \; ; \quad \text{with } \lambda_1 \geq \cdots \geq \lambda_l \geq 0 \, ,$$

is the weight for the action $\mathfrak{h}$ on $W^{\mathfrak{n}^+}$. Furthermore, $\rho|_{G^\circ}$ is equivalent to (π^λ, V^λ). We set $V^{\lambda,\varepsilon} = V^\lambda$ and extend π^λ to a representation $\pi^{\lambda,\varepsilon}$ of G by $\pi^{\lambda,\varepsilon}(-I) = \varepsilon I$. Clearly, $\pi^{\lambda,\varepsilon} \cong \rho$. Conversely, we can start with π^λ and extend it in two ways to obtain irreducible representations $\pi^{\lambda,\pm}$ of G in which $-I$ acts by $\pm I$. Thus we have classified the irreducible representations of G in this case as follows:

Theorem 5.5.23. *The irreducible regular representations of $G = \mathbf{O}(n, \mathbb{C})$ for n odd are of the form $(\pi^{\lambda,\varepsilon}, V^{\lambda,\varepsilon})$, where λ is the highest weight for the action of G°, $\varepsilon = \pm$, and $-I \in G$ acts by εI.*

Assume for the rest of this section that $n = 2l$ is even. In this case $H = H^\circ$ and the relation between irreducible representations of G and of G° is more involved. Fix $g_0 \in \mathbf{O}(\mathbb{C}^n, B)$ with $g_0 e_l = e_{l+1}$, $g_0 e_{l+1} = e_l$, and $g_0 e_i = e_i$ for $i \neq l, l+1$. Then $G = G^\circ \cup g_0 G^\circ$ and $g_0^2 = I$. The element g_0 is not in the center of G, and $\{I, g_0\} \ltimes G^\circ \cong G$ (semidirect product). We have $g_0 H g_0^{-1} = H$, $\mathrm{Ad}(g_0)\mathfrak{n}^+ = \mathfrak{n}^+$, and g_0 acting on the weight lattice by

$$g_0 \cdot \varepsilon_l = -\varepsilon_l , \quad g_0 \cdot \varepsilon_i = \varepsilon_i \quad \text{for } i = 1, \ldots, l-1 \, .$$

To obtain the irreducible representations of G, we start with the irreducible representation (π^λ, V^λ) of G°. Let $\mathcal{O}[G; V^\lambda]$ be the vector space of all regular V^λ-valued functions on G, and set

$$I(V^\lambda) = \{f \in \mathcal{O}[G; V^\lambda] : f(xg) = \pi(x)f(g) \quad \text{for } x \in G^\circ \text{ and } g \in G\} \, .$$

We define the *induced representation* $\rho = \mathrm{Ind}_{G^\circ}^G(\pi)$ to be the right translation action of G on $I(V^\lambda)$. Since G° is of index 2 in G, we can decompose $I(V^\lambda)$ into two invariant subspaces under G° as follows: Define

$$I_1(V^\lambda) = \{f \in I(V^\lambda) : f(xg_0) = 0 \quad \text{for all } x \in G^\circ\}$$

(the functions supported on G°), and

$$I_0(V^\lambda) = \{f \in I(V^\lambda) : f(x) = 0 \quad \text{for all } x \in G^\circ\}$$

(the functions supported on $g_0 G^\circ$). These subspaces are invariant under $\rho(G^\circ)$, and $I(V^\lambda) = I_1(V^\lambda) \oplus I_0(V^\lambda)$. The operator $\rho(g_0)$ interchanges the spaces $I_1(V^\lambda)$ and $I_0(V^\lambda)$ and maps $I_1(V^\lambda)^{\mathfrak{n}^+}$ onto $I_0(V^\lambda)^{\mathfrak{n}^+}$.

The map $f \mapsto f(1) \oplus f(g_0)$ is a G° isomorphism between $I(V^\lambda)$ and $V^\lambda \oplus V^\lambda$, and we have $\rho|_{G^\circ} \cong \pi^\lambda \oplus \pi_0^\lambda$, where π_0^λ is the representation $\pi_0^\lambda(x) = \pi^\lambda(g_0^{-1} x g_0)$. Since $\mathrm{Ad}\, g_0(\mathfrak{n}^+) = \mathfrak{n}^+$, the representation π_0^λ has highest weight $g_0 \cdot \lambda$; hence it is equivalent to $\pi^{g_0 \cdot \lambda}$. Thus

$$\dim I(V^\lambda)^{\mathfrak{n}^+} = 2, \tag{5.63}$$

and from Corollary 3.3.14 we have

$$I(V^\lambda) = \mathrm{Span}\, \rho(G^\circ) I(V^\lambda)^{\mathfrak{n}^+}. \tag{5.64}$$

Fix a highest-weight vector $0 \neq v_0 \in (V^\lambda)^{\mathfrak{n}^+}$ and for $x \in G^\circ$ define

$$f_1(x) = \pi^\lambda(x)v_0, \quad f_1(xg_0) = 0, \qquad f_0(x) = 0, \quad f_0(xg_0) = \pi^\lambda(x)v_0.$$

Then $f_1 \in I_1(V^\lambda)$, $f_0 \in I_0(V^\lambda)$, and $\rho(g_0)f_1 = f_0$. For $h \in H$ and $x \in G^\circ$ we have

$$\rho(h)f_1(x) = \pi(xh)v_0 = h^\lambda f_1(x),$$
$$\rho(h)f_0(xg_0) = \pi(xg_0 h g_0^{-1})v_0 = h^{g_0 \cdot \lambda} f_0(xg_0),$$

and $d\rho(X)f_1 = d\rho(X)f_0 = 0$ for $X \in \mathfrak{n}^+$. The functions f_1 and f_0 give a basis for $I(V^\lambda)^{\mathfrak{n}^+}$.

We now determine whether ρ is irreducible. Recall that λ is of the form

$$\lambda = \lambda_1 \varepsilon_1 + \cdots + \lambda_l \varepsilon_l, \qquad \lambda_1 \geq \cdots \geq \lambda_{l-1} \geq |\lambda_l|.$$

Since g_0 changes λ_l to $-\lambda_l$, we may assume $\lambda_l \geq 0$.

Case 1: $g_0 \cdot \lambda \neq \lambda$ (this occurs when $\lambda_l > 0$). Suppose $W \subset I(V^\lambda)$ is a nonzero G-invariant subspace. The space $W^{\mathfrak{n}^+}$ is nonzero and decomposes under the action of H as

$$W^{\mathfrak{n}^+} = W^{\mathfrak{n}^+}(\lambda) \oplus W^{\mathfrak{n}^+}(g_0 \cdot \lambda)$$

(a direct sum because the weights λ and $g_0 \cdot \lambda$ are distinct). Since

$$\rho(g_0)W^{\mathfrak{n}^+}(\lambda) = W^{\mathfrak{n}^+}(g_0 \cdot \lambda),$$

we have $\dim W^{\mathfrak{n}^+} \geq 2$. Hence $W^{\mathfrak{n}^+} = I(V^\lambda)^{\mathfrak{n}^+}$ by (5.63), and so $W = I(V^\lambda)$ by (5.64). Thus $I(V^\lambda)$ is irreducible in this case, and we write $\rho = \rho^\lambda$.

Case 2: $g_0 \cdot \lambda = \lambda$ (this case occurs when $\lambda_l = 0$). Define $\varphi_\pm = f_1 \pm f_0$. Then $\{\varphi_+, \varphi_-\}$ is a basis for $I(V^\lambda)^{n^+}$, and $\rho(g_0)\varphi_\pm = \pm\varphi_\pm$. Since $g_0 \cdot \lambda = \lambda$, we know that φ_+ and φ_- are extreme vectors of weight λ for G°. Define

$$V^{\lambda,\pm} = \mathrm{Span}\,\rho(G^\circ)\varphi_\pm .$$

These spaces are invariant under G° and g_0, and hence they are invariant under G. From (5.64) we have

$$I(V^\lambda) = V^{\lambda,+} \oplus V^{\lambda,-} . \tag{5.65}$$

Let $\pi^{\lambda,\pm}$ be the restriction of ρ to $V^{\lambda,\pm}$. Since

$$\dim \left(V^{\lambda,\pm}\right)^{n^+} = 1 , \tag{5.66}$$

G° acts irreducibly on $V^{\lambda,\pm}$. Thus (5.65) is the decomposition of $I(V^\lambda)$ into irreducible subrepresentations. Notice that $\pi^{\lambda,-} = \det \otimes \pi^{\lambda,+}$.

Now let (σ, W) be an irreducible representation of G. There exist a dominant weight $\lambda \in P_{++}(G^\circ)$ and a subspace $V^\lambda \subset W$ on which $\sigma|_{G^\circ}$ acts by the representation π^λ. Since G° is reductive, there is a G°-invariant projection $P : W \longrightarrow V_n^\lambda$. Using this projection we define

$$S : W \longrightarrow I(V^\lambda) , \qquad S(w)(g) = P(\sigma(g)w) .$$

Since $S(w)(1) = w$, it is clear that S is injective and $S\sigma(g) = \rho(g)S$ for $g \in G$. Thus we may assume $W \subset I(V^\lambda)$. It follows from the analysis of the induced representation that

$$\sigma = \begin{cases} \rho^\lambda & \text{if } \lambda_l \neq 0, \\ \pi^{\lambda,\pm} & \text{if } \lambda_l = 0. \end{cases}$$

Note that when $\lambda_l \neq 0$ then $\dim W^{n^+} = 2$, whereas if $\lambda_l = 0$ then $\dim W^{n^+} = 1$.

We may summarize this classification as follows:

Theorem 5.5.24. *Let $n \geq 4$ be even. The irreducible regular representations (σ, W) of $\mathbf{O}(n, \mathbb{C})$ are of the following two types:*

1. *Suppose $\dim W^{n^+} = 1$ and $\mathfrak{h}$ acts by the weight λ on W^{n^+}. Then g_0 acts on this space by εI ($\varepsilon = \pm$) and one has $(\sigma, W) \cong (\pi^{\lambda,\varepsilon}, V^{\lambda,\varepsilon})$.*
2. *Suppose $\dim W^{n^+} = 2$. Then $\mathfrak{h}$ has two distinct weights λ and $g_0 \cdot \lambda$ on W^{n^+}, and one has $(\sigma, W) \cong (\rho^\lambda, V^\lambda)$.*

5.5.6 Exercises

1. Let $G = \mathbf{Sp}(\mathbb{C}^4, \Omega)$, where the skew form Ω is given by (2.6). Consider the representation ρ of G on $\bigwedge^2 \mathbb{C}^4$.

(a) Find the weights and a basis of weight vectors for ρ. Express the weights in terms of ε_1, ε_2 and verify that the set of weights is invariant under the Weyl group of G.

(b) Set $X = \iota(e_{-1})\iota(e_1) + \iota(e_{-2})\iota(e_2)$, where $\iota(x)$ is the graded derivation of $\bigwedge \mathbb{C}^4$ such that $\iota(x)y = \Omega(x,y)$ for $x,y \in \mathbb{C}^4$, and we label the basis for $\mathbb{C}^4$ by $e_{-1} = e_4$ and $e_{-2} = e_3$. Show that

$$X(u \wedge v) = \sum_{p=1}^{2} \det \begin{bmatrix} \Omega(e_p,u) & \Omega(e_{-p},u) \\ \Omega(e_p,v) & \Omega(e_{-p},v) \end{bmatrix} \quad \text{for } u,v \in \mathbb{C}^4 \ .$$

(c) Let $\mathcal{H}^2 = \mathrm{Ker}(X) \subset \bigwedge^2 \mathbb{C}^4$ (this is an irreducible G-module with highest weight ϖ_2). Use the formula in (b) to find a basis for $\mathcal{H}^2$. (HINT: $\mathcal{H}^2$ is the sum of weight spaces.)

(d) Use Proposition 5.5.18 to find a basis for $\mathcal{H}^2$.

2. Let (π_r, F_r) be the rth fundamental representation of $G = \mathbf{Sp}(l, \mathbb{C})$. Show that the weights of F_r are the transforms under the Weyl group W_G of the set of dominant weights ϖ_r, ϖ_{r-2}, ..., ϖ_1 if r is odd or the set of dominant weights ϖ_r, ϖ_{r-2}, ..., ϖ_2, 0 if r is even.

5.6 Invariant Theory and Duality

We shall now combine the FFT and the general duality theorem from Chapter 4 in the context of polynomial-coefficient differential operators to obtain *Howe duality*.

5.6.1 Duality and the Weyl Algebra

Let V be an n-dimensional vector space over $\mathbb{C}$ and let $x_1,\ldots,x_n$ be coordinates on V relative to a basis $\{e_1,\ldots,e_n\}$. Let $\xi_1,\ldots,\xi_n$ be the coordinates for V^* relative to the dual basis $\{e_1^*,\ldots,e_n^*\}$. We denote by $\mathbb{D}(V)$ the algebra of *polynomial coefficient differential operators* on V. This is the subalgebra of $\mathrm{End}(\mathcal{P}(V))$ generated (as an associative algebra) by the operators

$$D_i = \frac{\partial}{\partial x_i} \quad \text{and} \quad M_i = \text{multiplication by } x_i \quad (i = 1,\ldots,n) \ .$$

Since $(\partial/\partial x_i)(x_j f) = (\partial x_j/\partial x_i)f + x_j(\partial f/\partial x_i)$ for $f \in \mathcal{P}(V)$, these operators satisfy the *canonical commutation relations*

$$[D_i, M_j] = \delta_{ij}I \quad \text{for } i,j = 1,\ldots,n \ . \tag{5.67}$$

The algebra $\mathbb{D}(V)$ is called the *Weyl algebra*. From (5.67) it is easily verified that the set of operators $\{M^\alpha D^\beta : \alpha, \beta \in \mathbb{N}^n\}$ is a (vector-space) basis for $\mathbb{D}(V)$, where we write

$$M^\alpha = M_1^{\alpha_1} \cdots M_n^{\alpha_n} , \qquad D^\beta = D_1^{\beta_1} \cdots D_n^{\beta_n} .$$

Define $\mathbb{D}_0(V) = \mathbb{C}I$, and for $k \geq 1$ let $\mathbb{D}_k(V)$ be the linear span of all products of k or fewer operators from the generating set $\{D_1, \ldots, D_n, M_1, \ldots, M_n\}$. This defines an increasing filtration of the algebra $\mathbb{D}(V)$:

$$\mathbb{D}_k(V) \subset \mathbb{D}_{k+1}(V) , \quad \bigcup_{k \geq 0} \mathbb{D}_k(V) = \mathbb{D}(V) , \text{ and } \mathbb{D}_k(V) \cdot \mathbb{D}_m(V) \subset \mathbb{D}_{k+m}(V) .$$

Recall that $\mathbf{GL}(V)$ acts on $\mathcal{P}(V)$ by the representation ρ, where $\rho(g)f(x) = f(g^{-1}x)$ for $f \in \mathcal{P}(V)$. We view $\mathbb{D}(V)$ as a $\mathbf{GL}(V)$-module relative to the action $g \cdot T = \rho(g)T\rho(g^{-1})$ for $T \in \mathbb{D}(V)$ and $g \in \mathbf{GL}(V)$. For $g \in \mathbf{GL}(V)$ with matrix $[g_{ij}]$ relative to the basis $\{e_1, \ldots, e_n\}$, we calculate that

$$\rho(g)D_j\rho(g^{-1}) = \sum_{i=1}^n g_{ij}D_i , \qquad \rho(g)M_i\rho(g^{-1}) = \sum_{j=1}^n g_{ij}M_j . \tag{5.68}$$

Thus $\mathbf{GL}(V)$ preserves the filtration of $\mathbb{D}(V)$, and the action of $\mathbf{GL}(V)$ on each subspace $\mathbb{D}_k(V)$ is regular.

We can now obtain the general *Weyl algebra duality theorem*.

Theorem 5.6.1. *Let G be a reductive algebraic subgroup of $\mathbf{GL}(V)$. Let G act on $\mathcal{P}(V)$ by $\rho(g)f(x) = f(g^{-1}x)$ for $f \in \mathcal{P}(V)$ and $g \in G$. There is a canonical decomposition*

$$\mathcal{P}(V) \cong \bigoplus_{\lambda \in \mathcal{S}} E^\lambda \otimes F^\lambda , \tag{5.69}$$

as a module for the algebra $\mathbb{D}(V)^G \otimes \mathcal{A}[G]$, where $\mathcal{S} \subset \widehat{G}$, F^λ is an irreducible regular G-module in the class λ, and E^λ is an irreducible module for $\mathbb{D}(V)^G$ that uniquely determines λ.

Proof. Set $\mathcal{R} = \mathbb{D}(V)$ and $L = \mathcal{P}(V)$. Then L has countable dimension as a vector space and ρ is a locally regular representation of G, since $\mathcal{P}^k(V)$ is a regular G-module for each integer k. We shall show that L, $\mathcal{R}$, and G satisfy conditions (i), (ii), and (iii) of Section 4.2.1.

Let $0 \neq f \in \mathcal{P}(V)$ be of degree d. Then there is some $\alpha \in \mathbb{N}^n$ with $|\alpha| = d$ such that $0 \neq D^\alpha f \in \mathbb{C}$. Given any $\varphi \in \mathcal{P}(V)$, let $M_\varphi \in \mathbb{D}(V)$ be the operator of multiplication by φ. Then $\varphi \in \mathbb{C}M_\varphi D^\alpha f$. This proves that $\mathcal{R}$ acts irreducibly on $\mathcal{P}(V)$ (condition (i)). Conditions (ii) and (iii) hold because they are true for the action of $\mathbf{GL}(V)$ on $\mathbb{D}(V)$. $\qquad \square$

To use Theorem 5.6.1 effectively for a particular group G we must describe the algebra $\mathbb{D}(V)^G$ in explicit terms. For this we will use the following results: Let

$$\mathrm{Gr}(\mathbb{D}(V)) = \bigoplus_{k \geq 0} \mathrm{Gr}^k\big(\mathbb{D}(V)\big)$$

be the graded algebra associated with the filtration on $\mathbb{D}(V)$ (see Appendix C.1.1). Let $T \in \mathbb{D}(V)$. We say that T has *filtration degree* k if $T \in \mathbb{D}_k(V)$ but $T \notin \mathbb{D}_{k-1}(V)$, and we write $\deg(T) = k$. We write

$$\mathrm{Gr}(T) = T + \mathbb{D}_{k-1}(V) \in \mathrm{Gr}^k\left(\mathbb{D}(V)\right)$$

when $\deg(T) = k$. The map $T \mapsto \mathrm{Gr}(T)$ is an algebra homomorphism from $\mathbb{D}(V)$ to $\mathrm{Gr}(\mathbb{D}(V))$. From (5.67) it is easily verified that

$$\deg(M^\alpha D^\beta) = |\alpha| + |\beta| \quad \text{for } \alpha, \beta \in \mathbb{N}^n ,$$

and hence the set $\{\mathrm{Gr}(M^\alpha D^\beta) : |\alpha| + |\beta| = k\}$ is a basis for $\mathrm{Gr}^k\left(\mathbb{D}(V)\right)$. Thus the nonzero operators of filtration degree k are those of the form

$$T = \sum_{|\alpha| + |\beta| \leq k} c_{\alpha\beta} M^\alpha D^\beta , \tag{5.70}$$

with $c_{\alpha\beta} \neq 0$ for some pair α, β such that $|\alpha| + |\beta| = k$ (note that the filtration degree of T is generally larger than the order of T as a differential operator). If T in (5.70) has filtration degree k, then we define the *Weyl symbol* of T to be the polynomial $\sigma(T) \in \mathcal{P}^k(V \oplus V^*)$ given by

$$\sigma(T) = \sum_{|\alpha| + |\beta| = k} c_{\alpha\beta} x^\alpha \xi^\beta . \tag{5.71}$$

Using (5.67), one shows by induction on k that any monomial of degree k in the operators $D_1, \ldots, D_n$, $M_1, \ldots, M_n$ is congruent (modulo $\mathbb{D}_{k-1}(V)$) to a unique ordered monomial $M^\alpha D^\beta$ with $|\alpha| + |\beta| = k$. This implies that $\sigma(ST) = \sigma(S)\sigma(T)$ for $S, T \in \mathbb{D}(V)$, and hence $\sigma : \mathbb{D}(V) \longrightarrow \mathcal{P}(V \oplus V^*)$ is an algebra homomorphism. Since $\rho(g)\mathbb{D}_k(V)\rho(g^{-1}) = \mathbb{D}_k(V)$, there is a representation of $\mathbf{GL}(V)$ on $\mathrm{Gr}^k\left(\mathbb{D}(V)\right)$.

Lemma 5.6.2. *The Weyl symbol map gives a linear isomorphism*

$$\mathbb{D}_k(V) \cong \bigoplus_{j=0}^{k} \mathcal{P}^j(V \oplus V^*) \tag{5.72}$$

as $\mathbf{GL}(V)$*-modules, for all* k.

Proof. From (5.67) it is easy to show (by induction on filtration degree) that $\sigma(T) = \sigma(S)$ if $\mathrm{Gr}(T) = \mathrm{Gr}(S)$. Thus σ gives a linear isomorphism

$$\mathrm{Gr}^k\left(\mathbb{D}(V)\right) \overset{\cong}{\longrightarrow} \mathcal{P}^k(V \oplus V^*) . \tag{5.73}$$

From (5.68) we see that D_i transforms as the vector e_i under conjugation by $\mathbf{GL}(V)$, whereas M_i transforms as the dual vector e_i^*. Since $\mathbf{GL}(V)$ acts by algebra automorphisms on $\mathrm{Gr}(\mathbb{D}(V))$ and on $\mathcal{P}(V \oplus V^*)$, this implies that (5.73) is an isomorphism of $\mathbf{GL}(V)$-modules. Set

$$\mathbb{L}_j = \mathrm{Span}\{M^\alpha D^\beta : |\alpha| + |\beta| = j\} \quad \text{for } j = 0, 1, 2, \ldots.$$

Then $\mathbb{L}_j$ is a finite-dimensional subspace of $\mathbb{D}(V)$ that is invariant under $\mathbf{GL}(V)$, and

$$\mathbb{D}_k(V) = \bigoplus_{j=0}^{k} \mathbb{L}_j. \tag{5.74}$$

The canonical quotient map $\mathbb{D}_k(V) \longrightarrow \mathrm{Gr}^k\left(\mathbb{D}_k(V)\right)$ restricts to give an isomorphism $\mathbb{L}_k \cong \mathrm{Gr}^k\left(\mathbb{D}(V)\right)$ of $\mathbf{GL}(V)$-modules. This implies the lemma. $\qquad\square$

Theorem 5.6.3. *Let G be a reductive subgroup of $\mathbf{GL}(V)$. Let $\{\psi_1, \ldots, \psi_r\}$ be a set of polynomials that generates the algebra $\mathcal{P}(V \oplus V^*)^G$. Suppose $T_j \in \mathbb{D}(V)^G$ are such that $\sigma(T_j) = \psi_j$ for $j = 1, \ldots, r$. Then $\{T_1, \ldots, T_r\}$ generates the algebra $\mathbb{D}(V)^G$.*

Proof. Let $\mathcal{J} \subset \mathbb{D}(V)^G$ be the subalgebra generated by $T_1, \ldots, T_r$. Then $\mathbb{D}_0(V)^G = \mathbb{C}I \subset \mathcal{J}$. Let $S \in \mathbb{D}_k(V)^G$ have filtration degree k. We may assume by induction on k that $\mathbb{D}_{k-1}(V)^G \subset \mathcal{J}$. Since $\sigma(S) \in \mathcal{P}^k(V \oplus V^*)^G$ by Lemma 5.6.2, we can write

$$\sigma(S) = \sum_{j_1, \ldots, j_r} c_{j_1 \cdots j_r} \, \psi_1^{j_1} \cdots \psi_r^{j_r},$$

where $c_{j_1 \cdots j_r} \in \mathbb{C}$. Set

$$R = \sum_{j_1, \ldots, j_r} c_{j_1 \cdots j_r} \, T_1^{j_1} \cdots T_r^{j_r}.$$

Although R is not unique (it depends on the enumeration of the T_j), we have $\sigma(R) = \sigma(S)$, since σ is an algebra homomorphism. Hence $R - S \in \mathbb{D}_{k-1}(V)$ by Lemma 5.6.2. By the induction hypothesis, $R - S \in \mathcal{J}$, so we have $S \in \mathcal{J}$. $\qquad\square$

Corollary 5.6.4. (Notation as in Theorem 5.6.3) *Suppose $T_1, \ldots, T_r$ can be chosen so that $\mathfrak{g}' = \mathrm{Span}\{T_1, \ldots, T_r\}$ is a Lie subalgebra of $\mathbb{D}(V)^G$. Then in the duality decomposition (5.69) the space E^λ is an irreducible $\mathfrak{g}'$-module that determines λ uniquely.*

Proof. The action of elements of $\mathfrak{g}'$ as differential operators on $\mathcal{P}(V)$ extends to a representation $\rho' : U(\mathfrak{g}') \longrightarrow \mathrm{End}(\mathcal{P}(V))$ (see Appendix C.2.1). The assumption on $T_1, \ldots, T_r$ implies that $\rho'(U(\mathfrak{g}')) = \mathbb{D}(V)^G$. Hence the irreducible $\mathbb{D}(V)^G$-modules are the same as irreducible $\mathfrak{g}'$-modules. $\qquad\square$

In the following sections we will use the FFT to show that the assumptions of Corollary 5.6.4 are satisfied when V is a multiple of the defining representation of a classical group G. This will give the *Howe duality* between the (finite-dimensional) regular representations of G occurring in $\mathcal{P}(V)$ and certain irreducible representations (generally infinite-dimensional) of the *dual* Lie algebra $\mathfrak{g}'$.

5.6.2 GL(n)–GL(k) *Schur–Weyl Duality*

Let $G = \mathbf{GL}(n, \mathbb{C})$ act on $V = (\mathbb{C}^n)^k$ as usual by $g \cdot (x_1, \ldots, x_k) = (gx_1, \ldots, gx_k)$, where each $x_j \in \mathbb{C}^n$ is a column vector with entries x_{ij}. We view x_{ij} as functions on V. Then $\mathcal{P}(V)$ is the polynomial ring in the variables

$$\{x_{ij} : 1 \leq i \leq n, 1 \leq j \leq k\}.$$

We will identify V with the space $M_{n,k}$ of $n \times k$ complex matrices by the map $(x_1, \ldots, x_k) \mapsto [x_{ij}] \in M_{n,k}$. With this identification the action of G on V becomes left multiplication on $M_{n,k}$. This gives a representation ρ of G on $\mathcal{P}(M_{n,k})$ by $\rho(g)f(v) = f(g^{-1}v)$ for $f \in \mathcal{P}(M_{n,k})$.

Define $G' = \mathbf{GL}(k, \mathbb{C})$ and let G' act on $M_{n,k}$ by matrix multiplication on the right. This action of G' obviously commutes with the action of G and gives rise to a representation ρ' of G' on $\mathcal{P}(M_{n,k})$, where $\rho'(g)f(v) = f(vg)$ for $f \in \mathcal{P}(M_{n,k})$. The elements in the standard basis $\{e_{ij}\}_{1 \leq i,j \leq k}$ for $\mathfrak{g}' = \mathfrak{gl}(k, \mathbb{C})$ act by the operators

$$E_{ij} = \sum_{p=1}^{n} x_{pi} \frac{\partial}{\partial x_{pj}} \tag{5.75}$$

on $\mathcal{P}(M_{n,k})$ (as in Section 1.3.7). Note that the operators E_{ij} preserve the spaces $\mathcal{P}^m(M_{n,k})$ of homogeneous polynomials of degree m.

Theorem 5.6.5. *Let $G = \mathbf{GL}(n, \mathbb{C})$ acting by left multiplication on $V = M_{n,k}$. Set $\mathfrak{g}' = \mathrm{Span}\{E_{ij} : 1 \leq i, j \leq k\}$. Then $\mathfrak{g}'$ is a Lie algebra in $\mathbb{D}(V)$ that is isomorphic to $\mathfrak{gl}(k, \mathbb{C})$, and it generates the algebra $\mathbb{D}(V)^G$.*

Proof. The FFT for G (Theorem 5.2.1) asserts that $\mathcal{P}(V \oplus V^*)^G$ is generated by the quadratic polynomials $z_{ij} = \sum_{p=1}^{n} x_{pi} \xi_{pj}$ for $1 \leq i, j \leq k$, where ξ_{pj} are the coordinates on V^* dual to the coordinates x_{ij} on V. From (5.75) we see that $\sigma(E_{ij}) = z_{ij}$. Hence we may apply Theorem 5.6.3. $\square$

Corollary 5.6.6. *In the canonical duality decomposition*

$$\mathcal{P}(M_{n,k}) \cong \bigoplus_{\lambda \in \mathcal{S}} E^{\lambda} \otimes F^{\lambda} \tag{5.76}$$

(where $\mathcal{S} \subset \widehat{G}$), each summand $E^{\lambda} \otimes F^{\lambda}$ is contained in $\mathcal{P}^m(M_{n,k})$ for some integer m (depending on λ) and is irreducible under $\mathbf{GL}(k, \mathbb{C}) \times \mathbf{GL}(n, \mathbb{C})$. Hence $\mathcal{P}(M_{n,k})$ is multiplicity-free as a $\mathbf{GL}(k, \mathbb{C}) \times \mathbf{GL}(n, \mathbb{C})$-module.

Proof. This follows from Theorem 5.6.5 and Corollary 5.6.4. $\square$

Now we shall use the theorem of the highest weight to find the representations that occur in the decomposition of ρ. In $\mathbf{GL}(n, \mathbb{C})$ we have the subgroups D_n of invertible diagonal matrices, N_n^+ of upper-triangular unipotent matrices, and N_n^- of lower-triangular unipotent matrices. We set $B_n = D_n N_n^+$ and $\overline{B}_n = D_n N_n^-$. We extend a regular character χ of D_n to a character of B_n (resp. $\overline{B}_n$) by $\chi(hu) = \chi(vh) = \chi(h)$ for $h \in D_n$, $u \in N_n^+$, and $v \in N_n^-$. A weight $\mu = \sum_{i=1}^n \mu_i \varepsilon_i$ of D_n is called *nonnegative* if $\mu_i \geq 0$ for all i. Recall that the weight μ is *dominant* if $\mu_1 \geq \mu_2 \geq \cdots \geq \mu_n$.

When μ is dominant, we denote by (π_n^μ, F_n^μ) the irreducible representation of $\mathbf{GL}(n, \mathbb{C})$ with highest weight μ (see Theorem 5.5.22). If μ is dominant and nonnegative, we set

$$|\mu| = \sum_{i=1}^n \mu_i \quad \text{(the } size \text{ of } \mu\text{)} .$$

In this case it is convenient to extend μ to a dominant weight of D_l for all $l > n$ by setting $\mu_i = 0$ for all integers $i > n$. We define

$$\text{depth}(\mu) = \min\{k : \mu_{k+1} = 0\} .$$

Thus we may view μ as a dominant integral weight of $\mathbf{GL}(l, \mathbb{C})$ for any $l \geq \text{depth}(\mu)$. If μ is a nonnegative dominant weight of depth k, then

$$\mu = m_1 \lambda_1 + \cdots + m_k \lambda_k ,$$

with $\lambda_i = \varepsilon_1 + \cdots + \varepsilon_i$ and $m_1, \ldots, m_k$ strictly positive integers.

By Proposition 4.2.5 the irreducible finite-dimensional regular representations of $G = \mathbf{GL}(k, \mathbb{C}) \times \mathbf{GL}(n, \mathbb{C})$ are outer tensor products $(\pi_k^\mu \widehat{\otimes} \pi_n^\nu, F_k^\mu \otimes F_n^\nu)$. For $i = 1, \ldots, \min\{k, n\}$ we denote by Δ_i the ith *principal minor* on $M_{k,n}$ (see Section B.2.5). We denote by $\mathcal{P}(M_{k,n})^{N_k^- \times N_n^+}$ the subspace of polynomials on $M_{k,n}$ that are fixed by left translations by N_k^- and right translations by N_n^+.

Theorem 5.6.7. *The space of homogeneous polynomials on $M_{k,n}$ of degree d decomposes under the representation ρ of $\mathbf{GL}(k, \mathbb{C}) \times \mathbf{GL}(n, \mathbb{C})$ as*

$$\mathcal{P}^d(M_{k,n}) \cong \bigoplus_\nu (F_k^\nu)^* \otimes F_n^\nu , \qquad (5.77)$$

with the sum over all nonnegative dominant weights ν of size d and $\text{depth}(\nu) \leq r$, where $r = \min\{k, n\}$. Furthermore,

$$\mathcal{P}(M_{k,n})^{N_k^- \times N_n^+} = \mathbb{C}[\Delta_1, \ldots, \Delta_r] \qquad (5.78)$$

is a polynomial ring on r algebraically independent generators.

Proof. We may assume $k \geq n$ (otherwise, we use the map $x \mapsto x^t$ to interchange k and n). Let x_{ij} be the ij-entry function on $M_{k,n}$ and let $m \in \mathbb{N}$. Then

$$\rho(a, b) x_{ij}^m = a_i^{-m} b_j^m x_{ij}^m \quad \text{for } a \in D_k \text{ and } b \in D_n .$$

Since the functions x_{ij} generate the algebra $\mathcal{P}(M_{k,n})$, we see that the weights of $D_k \times D_n$ on $\mathcal{P}(M_{k,n})$ are given by pairs $(-\mu, \nu)$, where μ and ν are nonnegative weights. For every such pair with ν dominant, write $\nu = m_1 \lambda_1 + \cdots + m_n \lambda_n$ with $\mathbf{m} = [m_1, \ldots, m_n] \in \mathbb{N}^n$. Define

$$\Delta_{\mathbf{m}}(x) = \prod_{i=1}^{n} \Delta_i(x)^{m_i} \quad \text{for } x \in M_{k,n} .$$

For $i = 1, \ldots, n$ the principal minor Δ_i satisfies

$$\Delta_i(a^{-1}xb) = a^{-\lambda_i} b^{\lambda_i} \Delta_i(x)$$

for $a \in \overline{B}_k$ and $b \in B_n$. Hence $\Delta_{\mathbf{m}}$ is an $(N_k^- \times N_n^+)$-invariant polynomial with weight $\lambda = (-\nu, \nu)$. From Corollary 5.6.6 we conclude that all the representations of $\mathbf{GL}(n, \mathbb{C})$ with nonnegative highest weight occur in $\mathcal{P}(M_{k,n})$, and no others. Since $\Delta_{\mathbf{m}}$ is a lowest-weight vector of weight $-\nu$ for the left action of $\mathbf{GL}(k, \mathbb{C})$, the representation F_n^ν of $\mathbf{GL}(n, \mathbb{C})$ is paired with the representation $(F_k^\nu)^*$ of $\mathbf{GL}(k, \mathbb{C})$ in decomposition (5.76). The degree of the polynomial $\Delta_{\mathbf{m}}$ is $\sum i m_i = |\nu|$. Hence the space spanned by the left and right translates of $\Delta_{\mathbf{m}}$ is an irreducible module for $\mathbf{GL}(k, \mathbb{C}) \times \mathbf{GL}(n, \mathbb{C})$ that is isomorphic to the outer tensor product $(F_k^\mu)^* \otimes F_n^\mu$. This proves (5.77).

The argument just given shows that the set of polynomials $\{\Delta_{\mathbf{m}} : \mathbf{m} \in \mathbb{N}^n\}$ spans $\mathcal{P}(M_{k,n})^{N_k^- \times N_n^+}$. This set is linearly independent, since the weights of its elements are all distinct. Hence the functions $\Delta_1, \ldots, \Delta_n$ are algebraically independent. $\quad\square$

Corollary 5.6.8. *As a module for* $\mathbf{GL}(k, \mathbb{C}) \times \mathbf{GL}(n, \mathbb{C})$,

$$S(\mathbb{C}^k \otimes \mathbb{C}^n) \cong \bigoplus_\mu F_k^\mu \otimes F_n^\mu ,$$

with the sum over all nonnegative dominant weights μ of depth $\leq \min\{k, n\}$.

Proof. Define a representation σ of $\mathbf{GL}(k, \mathbb{C}) \times \mathbf{GL}(n, \mathbb{C})$ on $\mathcal{P}(M_{k,n})$ by

$$\sigma(y, z)f(x) = f(y^t x z) \quad \text{for } y \in \mathbf{GL}(k, \mathbb{C}) \text{ and } z \in \mathbf{GL}(n, \mathbb{C}) .$$

Then σ is the representation on $\mathcal{P}(M_{k,n})$ associated with the twisted action $x \mapsto (y^t)^{-1} x z^{-1}$. Relative to this twisted action, we have $M_{k,n} \cong (\mathbb{C}^k)^* \otimes (\mathbb{C}^n)^*$ as a module for $\mathbf{GL}(k, \mathbb{C}) \times \mathbf{GL}(n, \mathbb{C})$. Hence

$$\mathcal{P}(M_{k,n}) \cong S(\mathbb{C}^k \otimes \mathbb{C}^n)$$

when $\mathbf{GL}(k, \mathbb{C}) \times \mathbf{GL}(n, \mathbb{C})$ acts by the representation σ on $\mathcal{P}(M_{k,n})$ and acts in the standard way on $S(\mathbb{C}^k \otimes \mathbb{C}^n)$. On the other hand, by Theorem 5.5.22, the automorphism $y \mapsto (y^t)^{-1}$ on $\mathbf{GL}(k, \mathbb{C})$ transforms the representation π_k^μ to its dual. Hence the corollary follows from decomposition (5.77). $\quad\square$

5.6.3 O(n)–𝔰𝔭(k) *Howe Duality*

Let $G = \mathbf{O}(n, \mathbb{C})$ be the group preserving the symmetric bilinear form $(x, y) = x^t y$ on $\mathbb{C}^n$. Take $V = M_{n,k}$ with G acting by left multiplication. We also view V as $(\mathbb{C}^n)^k$ and use the notation of the previous section concerning differential operators on $\mathcal{P}(V)$.

For $1 \leq i, j \leq k$ define

$$\Delta_{ij} = \sum_{p=1}^{n} \frac{\partial^2}{\partial x_{pi} \partial x_{pj}} \quad \text{and} \quad M_{ij} = \text{multiplication by } \sum_{p=1}^{n} x_{pi} x_{pj}$$

as operators on $\mathcal{P}(V)$. Note that $\Delta_{ij} = \Delta_{ji}$ and $M_{ij} = M_{ji}$. These operators are in $\mathbb{D}(V)^G$. Indeed, M_{ij} is multiplication by the G-invariant function (x_i, x_j), so it clearly commutes with the action of G on $\mathcal{P}(V)$. We have

$$\rho(g) \frac{\partial}{\partial x_{pi}} \rho(g^{-1}) = \sum_{q=1}^{n} g_{qp} \frac{\partial}{\partial x_{qi}}$$

for $g = [g_{pq}] \in \mathbf{GL}(n, \mathbb{C})$. Hence if $g \in G$ then the equation $gg^t = I$ gives

$$\rho(g) \Delta_{ij} \rho(g^{-1}) = \sum_{q,r=1}^{n} \left\{ \sum_{p=1}^{n} g_{qp} g_{rp} \right\} \frac{\partial^2}{\partial x_{qi} \partial x_{rj}} = \Delta_{ij} .$$

Theorem 5.6.9. *Let $G = \mathbf{O}(n, \mathbb{C})$ acting by left multiplication on $V = M_{n,k}(\mathbb{C})$. Set $\mathfrak{g}' = \text{Span}\{E_{ij} + (n/2)\delta_{ij}, M_{ij}, \Delta_{ij} : 1 \leq i, j \leq k\}$. Then $\mathfrak{g}'$ is a Lie subalgebra of $\mathbb{D}(V)^G$ that is isomorphic to $\mathfrak{sp}(k, \mathbb{C})$. Furthermore, $\mathfrak{g}'$ generates the algebra $\mathbb{D}(V)^G$.*

Proof. The operators $\{\Delta_{ij}\}$ and $\{M_{ij}\}$ satisfy

$$[\Delta_{ij}, \Delta_{rs}] = 0 , \quad [M_{ij}, M_{rs}] = 0 . \tag{5.79}$$

The commutator of a second-order differential operator $\partial^2/\partial x \partial y$ with the operator of multiplication by a function $\varphi(x,y)$ is the first-order differential operator $(\partial\varphi/\partial y)\partial/\partial x + (\partial\varphi/\partial x)\partial/\partial y + \partial^2\varphi/\partial x \partial y$. Thus

$$[\Delta_{ij}, M_{rs}] = \sum_{p=1}^{n} \left(\frac{\partial}{\partial x_{pi}}(x_r, x_s) \right) \frac{\partial}{\partial x_{pj}} + \left(\frac{\partial}{\partial x_{pj}}(x_r, x_s) \right) \frac{\partial}{\partial x_{pi}}$$

$$+ \sum_{p=1}^{n} \frac{\partial^2}{\partial x_{pi} \partial x_{pj}}(x_r, x_s) .$$

This gives

$$[\Delta_{ij}, M_{rs}] = \delta_{ri}\left(E_{sj} + \frac{n}{2}\delta_{sj}\right) + \delta_{si}\left(E_{rj} + \frac{n}{2}\delta_{rj}\right) \tag{5.80}$$
$$+ \delta_{rj}\left(E_{si} + \frac{n}{2}\delta_{si}\right) + \delta_{sj}\left(E_{ri} + \frac{n}{2}\delta_{ri}\right) ,$$

where E_{ij} is the vector field given by (5.75). These vector fields satisfy the commutation relations of the elementary matrices e_{ij} (see Section 1.3.7). Hence we have

$$\left[E_{ij} + \frac{n}{2}\delta_{ij}, E_{rs} + \frac{n}{2}\delta_{rs}\right] = \delta_{jr}\left(E_{is} + \frac{n}{2}\delta_{is}\right) - \delta_{is}\left(E_{rj} + \frac{n}{2}\delta_{rj}\right) . \tag{5.81}$$

The operator $[E_{ij}, M_{rs}]$ is multiplication by the function

$$\sum_{p,q=1}^{n} x_{pi} \frac{\partial}{\partial x_{pj}} (x_{qr} x_{qs}) = \sum_{p=1}^{n} (\delta_{jr} x_{ps} x_{pi} + \delta_{js} x_{pi} x_{pr}) .$$

Thus

$$\left[E_{ij} + \frac{n}{2} \delta_{ij}, M_{rs} \right] = \delta_{jr} M_{is} + \delta_{js} M_{ir} . \tag{5.82}$$

The commutator of a vector field $\varphi \partial/\partial x$ with the differential operator $\partial^2/\partial y \partial z$ is the operator $-(\partial \varphi/\partial y)\partial^2/\partial x \partial z - (\partial \varphi/\partial z)\partial^2/\partial x \partial y - (\partial^2 \varphi/\partial y \partial z)\partial/\partial x$. This gives

$$[E_{ij}, \Delta_{rs}] = -\sum_{p,q} \left\{ \frac{\partial x_{pi}}{\partial x_{qr}} \frac{\partial^2}{\partial x_{pj} \partial x_{qs}} + \frac{\partial x_{pi}}{\partial x_{qs}} \frac{\partial^2}{\partial x_{pj} \partial x_{qr}} \right\} .$$

Hence we have

$$\left[E_{ij} + \frac{n}{2} \delta_{ij}, \Delta_{rs} \right] = -\delta_{ir} \Delta_{js} - \delta_{is} \Delta_{jr} . \tag{5.83}$$

These commutation relations show that $\mathfrak{g}'$ is a Lie subalgebra of $\mathbb{D}(V)$.

Take $\mathfrak{sp}(k, \mathbb{C}) \subset \mathfrak{gl}(2k, \mathbb{C})$ as the matrices X such that $X^t J + J X = 0$, where $J = \begin{bmatrix} 0 & I_k \\ -I_k & 0 \end{bmatrix}$. Then $X \in \mathfrak{sp}(k, \mathbb{C})$ if and only if X has the block form

$$X = \begin{bmatrix} A & B \\ C & -A^t \end{bmatrix} , \quad \text{with } A, B, C \in M_k(\mathbb{C}) \text{ and } B = B^t, C = C^t . \tag{5.84}$$

Define a linear map $\varphi : \mathfrak{g}' \longrightarrow \mathfrak{sp}(k, \mathbb{C})$ by $\varphi(E_{ij} + (n/2)\delta_{ij}) = e_{ij} - e_{k+j,k+i}$, $\varphi(M_{ij}) = e_{i,j+k} + e_{j,i+k}$, and $\varphi(\Delta_{ij}) = -e_{i+k,j} - e_{j+k,i}$ for $1 \leq i, j \leq k$. From (5.84) we see that φ is a linear isomorphism. Using the commutation relations (5.80)–(5.83), it is easy to check that φ is a Lie algebra homomorphism.

To prove the last statement of the theorem, we calculate the operator symbols

$$\sigma(M_{ij}) = \sum_{p=1}^{n} x_{pi} x_{pj} , \quad \sigma(\Delta_{ij}) = \sum_{p=1}^{n} \xi_{pi} \xi_{pj} , \quad \sigma\left(E_{ij} + \frac{n}{2} \delta_{ij} \right) = \sum_{p=1}^{n} x_{pi} \xi_{pj} .$$

The FFT for G (Corollary 5.2.3) asserts that $\mathcal{P}(V \oplus V^*)^G$ is generated by these polynomials. It follows from Theorem 5.6.3 that $\mathfrak{g}'$ generates $\mathbb{D}(V)^G$. $\square$

Corollary 5.6.10. $(G = \mathbf{O}(n, \mathbb{C})$ with $n \geq 3)$ *In the canonical duality decomposition*

$$\mathcal{P}(V) \cong \bigoplus_{\lambda \in \mathcal{S}} E^\lambda \otimes F^\lambda \tag{5.85}$$

under the joint action of $\mathbb{D}(V)^G$ and G (where $\mathcal{S} \subset \widehat{G}$), each $\mathbb{D}(V)^G$-module E^λ is an irreducible infinite-dimensional representation of $\mathfrak{sp}(k, \mathbb{C})$.

Proof. Apply Theorem 5.6.9 and Corollary 5.6.4. $\square$

5.6.4 Spherical Harmonics

We shall examine in more detail the Howe duality for the pair $\mathbf{O}(n,\mathbb{C})\text{--}\mathfrak{sp}(k,\mathbb{C})$ when $k = 1$ (recall that $\mathfrak{sp}(1,\mathbb{C}) \cong \mathfrak{sl}(2,\mathbb{C})$). From the duality decomposition in Corollary 5.6.10 we will obtain the classical expansion of polynomials in spherical harmonics.

Let $G = \mathbf{O}(n,\mathbb{C})$ be the group that preserves the bilinear form $(x,y) = x^t y$ on $\mathbb{C}^n$ (assume that $n \geq 3$), and let $G^\circ = \mathbf{SO}(n,\mathbb{C})$. We denote by ρ the representation of G on $\mathcal{P}(\mathbb{C}^n)$ given by $\rho(g)f(x) = f(g^{-1}x)$ for $g \in G$, $x \in \mathbb{C}^n$, and $f \in \mathcal{P}(\mathbb{C}^n)$. Let $X = -(1/2)\Delta$, $Y = (1/2)r^2$, and $H = -E - (n/2)$, where

$$\Delta = \sum_{p=1}^{n} \left(\frac{\partial}{\partial x_p}\right)^2, \qquad r^2 = \sum_{p=1}^{n} x_p^2, \qquad E = \sum_{p=1}^{n} x_p \frac{\partial}{\partial x_p}.$$

The commutation relations

$$[E,\Delta] = -2\Delta, \quad [E,r^2] = 2r^2, \quad [\Delta,r^2] = 4\big(E + (n/2)\big) \qquad (5.86)$$

from Section 5.6.3 show that $\{X,Y,H\}$ is a TDS triple: $[H,X] = 2X$, $[H,Y] = -2Y$, $[X,Y] = H$. We set $\mathfrak{g}' = \operatorname{Span}\{X,Y,H\} \subset \mathbb{D}(\mathbb{C}^n)^G$.

To determine the explicit form of the decomposition of $\mathcal{P}(\mathbb{C}^n)$ in Corollary 5.6.10 in this case, recall from Theorem 4.2.1 that the $(\mathfrak{g}',G)$-modules appearing in the decomposition are mutually inequivalent and that F^λ is uniquely determined by E^λ. Consequently, our strategy will be to find the modules E^λ using the representation theory of $\mathfrak{sl}(2,\mathbb{C})$, and then to determine the associated G-module F^λ using the classification of the representations of G in Section 5.5.5. The key step will be to find polynomials that are highest-weight vectors both for $\mathfrak{g} = \mathfrak{so}(n,\mathbb{C})$ and for $\mathfrak{g}'$.

We begin with the highest-weight vectors for $\mathfrak{g}'$. We say that a polynomial $f \in \mathcal{P}(\mathbb{C}^n)$ is *G-harmonic* if $\Delta f = 0$. Let

$$\mathcal{H}(\mathbb{C}^n) = \{f \in \mathcal{P}(\mathbb{C}^n) : \Delta f = 0\}$$

be the space of all *G*-harmonic polynomials and let $\mathcal{H}^k(\mathbb{C}^n)$ be the harmonic polynomials that are homogeneous of degree k. Since $\Delta : \mathcal{P}^k(\mathbb{C}^n) \longrightarrow \mathcal{P}^{k-2}(\mathbb{C}^n)$, a polynomial f is harmonic if and only if each homogeneous component of f is harmonic. Thus

$$\mathcal{H}(\mathbb{C}^n) = \bigoplus_{k \geq 0} \mathcal{H}^k(\mathbb{C}^n).$$

Because Δ commutes with the action of G, the spaces $\mathcal{H}^k(\mathbb{C}^n)$ are invariant under G. The FFT for G (Corollary 5.2.3) implies that the constant-coefficient G-invariant differential operators on $\mathbb{C}^n$ are polynomials in Δ. Hence the notion of *G-harmonic polynomial* here is consistent with the use of the term in Section 5.1.2 for the symmetric group.

Let $\{X,Y,H\} \subset \mathbb{D}(\mathbb{C}^n)^G$ be the TDS triple introduced above. If $f \in \mathcal{H}^k(\mathbb{C}^n)$ then

$$Xf = 0 \quad \text{and} \quad Hf = -\left(k + \frac{n}{2}\right)f. \tag{5.87}$$

Hence f is a highest-weight vector for $\mathfrak{g}'$ of weight $-(k + \frac{n}{2})$.

For each integer $k \geq 0$ we can find a harmonic polynomial homogeneous of degree k that is also a highest-weight vector for $\mathfrak{g}$. To do this, we identify $\mathbb{C}^n$ with $(\mathbb{C}^n)^*$ via the form (x,y). For $x \in \mathbb{C}^n$ we write x^k for the polynomial $y \mapsto (x,y)^k$; then

$$\Delta(x^k) = k(k-1)(x,x)x^{k-2}.$$

Hence if x satisfies $(x,x) = 0$ (so that x is an *isotropic vector*), then $x^k \in \mathcal{H}^k(\mathbb{C}^n)$. Let B be the symmetric form (2.9) on $\mathbb{C}^n$ with matrix $S = [\delta_{i,n+1-j}]$. Let

$$\mathfrak{so}(\mathbb{C}^n, B) = \mathfrak{n}^- + \mathfrak{h} + \mathfrak{n}^+$$

be the triangular decomposition in Theorem 2.4.11. The standard basis vector e_1 is isotropic for B and it is $(\mathfrak{h} + \mathfrak{n}^+)$-extreme of weight ε_1. There exists $T \in \mathbf{GL}(n,\mathbb{C})$ such that $T^t T = S$ (the Cholesky decomposition). Then $\mathrm{Ad}(T)\mathfrak{so}(\mathbb{C}^n, B) = \mathfrak{g}$. Set $\varphi = Te_1$ and $\mathfrak{b} = \mathrm{Ad}(T)(\mathfrak{h} + \mathfrak{n}^+)$. Then φ is an isotropic vector relative to the bilinear form (x,y), and it is $\mathfrak{b}$-extreme of weight $T^t \varepsilon_1$. (Note that in terms of fundamental highest weights for $\mathfrak{g}$ one has $T^t \varepsilon_1 = \varpi_1$ when $n \geq 5$, while $T^t \varepsilon_1 = \varpi_1 + \varpi_2$ when $n = 4$.) Hence for any positive integer k the polynomial φ^k is G-harmonic and $\mathfrak{b}$-extreme of weight $k\varpi_1$. Thus it is a highest-weight vector for $\mathfrak{g}$ and also for $\mathfrak{g}'$.

Let $\mathcal{E}^k$ be the cyclic $\mathfrak{g}'$-module generated by φ^k. By (5.87) we have

$$\mathcal{E}^k = \bigoplus_{p \geq 0} \mathbb{C}r^{2p}\varphi^k \tag{5.88}$$

(direct sum, since each summand is homogeneous of a different degree). Thus every function in $\mathcal{E}^k$ can be written uniquely as $\psi(r^2)\varphi^k$ for a polynomial $\psi \in \mathbb{C}[t]$.

Theorem 5.6.11. (Notation as above)

1. $\mathcal{E}^k$ is an irreducible $\mathfrak{g}'$-module.
2. There is an injective $(\mathfrak{g}' \times G)$-intertwining map from $\mathcal{E}^k \otimes \mathcal{H}^k(\mathbb{C}^n)$ to $\mathcal{P}(\mathbb{C}^n)$ such that $\psi(r^2)\varphi^k \otimes f \mapsto \psi(r^2)f$ for $\psi \in \mathbb{C}[t]$ and $f \in \mathcal{H}^k(\mathbb{C}^n)$.
3. $\mathcal{H}^k(\mathbb{C}^n)$ is an irreducible G°-module with highest weight $k\varpi_1$ when $n \neq 4$, and highest weight $k(\varpi_1 + \varpi_2)$ when $n = 4$.
4. $\mathcal{P}(\mathbb{C}^n) \cong \bigoplus_{k \geq 0} \mathcal{E}^k \otimes \mathcal{H}^k(\mathbb{C}^n)$ as a $(\mathfrak{g}' \times G)$-module, where the equivalence is given on each summand by the intertwining map in (2).

Proof. Following the strategy outlined above, we begin by determining the irreducible $\mathfrak{g}'$-modules in $\mathcal{P}(\mathbb{C}^n)$. For $\mu \in \mathbb{C}$ let $\mathcal{M}_\mu$ be the $\mathbb{C}$ vector space with basis $\{v_j : j \in \mathbb{N}\}$. Define linear operators $\pi_\mu(X)$, $\pi_\mu(Y)$, and $\pi_\mu(H)$ on $\mathcal{M}_\mu$ by $\pi_\mu(Y)v_j = v_{j+1}$, $\pi_\mu(H)v_j = (\mu - 2j)v_j$, and

$$\pi_\mu(X)v_j = j(\mu - j + 1)v_{j-1},$$

for $j \in \mathbb{N}$ (where $v_{-1} = 0$). From the commutation relations in the proof of Lemma 2.3.1 it is easy to verify that $(\mathcal{M}_\mu, \pi_\mu)$ is a representation of $\mathfrak{g}'$. Furthermore, if $\mu \neq 0, 1, 2, \ldots$ then $(\mathcal{M}_\mu, \pi_\mu)$ is irreducible (by the same argument as in the proof of Proposition 2.3.3).

($\star$) Each irreducible $\mathfrak{g}'$-module in $\mathcal{P}(\mathbb{C}^n)$ is isomorphic to $\mathcal{M}_{-(k+(n/2))}$ for some nonnegative integer k, and all values of k occur.

To prove ($\star$), let V be an irreducible $\mathfrak{g}'$-submodule in $\mathcal{P}(\mathbb{C}^n)$ and take $0 \neq f \in V$. Since f is a polynomial, there exists an integer p such that $X^p f \neq 0$ and $X^{p+1} f = 0$. Each homogeneous component of f is annihilated by X^{p+1}, so we may assume that f is homogeneous. Replacing f by $X^p f$, we may also assume that $X f = 0$. Let U_f be the cyclic $\mathfrak{g}'$-module generated by f. By (5.87) we have

$$U_f = \bigoplus_{j \geq 0} \mathbb{C} r^{2j} f. \qquad (5.89)$$

The sum (5.89) is direct, since each summand is homogeneous of a different degree. Hence $U_f \cong \mathcal{M}_{-(k+(n/2))}$ as a $\mathfrak{g}'$ module if f is homogeneous of degree k, where the equivalence is given by the map $r^{2j} f \mapsto v_j$ for $j = 0, 1, 2, \ldots$ (this is a special case of Theorem 3.2.5). Since V is irreducible, we must have $V = U_f$. To show that all values of k occur, take $f = \varphi^k$. This proves ($\star$) and statement (1) of the theorem. Note that the $\mathfrak{g}'$-modules are mutually inequivalent for different values of k (this is obvious from the eigenvalues of H; it also follows from Theorem 3.2.5).

We next prove (2), where it is understood that $\mathfrak{g}'$ acts only on the factor $\mathcal{E}^k$ and G acts only on factor $\mathcal{H}^k(\mathbb{C}^n)$ in $\mathcal{E}^k \otimes \mathcal{H}^k(\mathbb{C}^n)$. The intertwining property is easily checked. Suppose $\sum_j \psi_j(r^2) f_j = 0$. We may assume that $\{f_j\}$ is linearly independent and by homogeneity that $\psi_j(r^2) = c_j r^2$. This immediately implies $c_j = 0$, proving injectivity of the map.

Now we prove (3). By Theorem 4.2.1 the multiplicity space for the irreducible $\mathfrak{g}'$-module of type $\mathcal{E}^k$ is an irreducible G-module. But from (2) we know that this multiplicity space contains a G-submodule isomorphic to $\mathcal{H}^k(\mathbb{C}^n)$. Hence $\mathcal{H}^k(\mathbb{C}^n)$ must be an irreducible G-module. To see that it is already irreducible under G°, let $\mathcal{F}^k$ be the cyclic $\mathfrak{g}$-module generated by φ^k. Since φ^k is a $\mathfrak{b}$-extreme vector, we know that $\mathcal{F}^k$ is an irreducible $\mathfrak{g}$-module by Proposition 3.3.9, and hence it is also an irreducible G°-module by Theorem 2.2.7. We can take $g_0 \in G$ such that $G = G^\circ \cup g_0 G^\circ$ and $\rho(g_0)\varphi^k = \varphi^k$ (see Section 5.5.5). Hence $\mathcal{F}^k$ is an irreducible G-invariant subspace of $\mathcal{H}^k(\mathbb{C}^n)$, so the spaces must coincide.

Now that we know all the $\mathfrak{g}'$-isotypic components of $\mathcal{P}(\mathbb{C}^n)$ from ($\star$), the decomposition (4) follows from (2), (3), and Corollary 5.6.10. $\qquad \square$

From parts (2) and (4) of Theorem 5.6.11 we obtain the following analogue of Theorem 5.1.4 (with the difference that now the space of G-harmonic polynomials is not finite-dimensional).

Corollary 5.6.12. *The map* $\mathbb{C}[r^2] \otimes \mathcal{H}(\mathbb{C}^n) \longrightarrow \mathcal{P}(\mathbb{C}^n)$ *given by* $f \otimes u \mapsto fu$ *(pointwise multiplication of functions) is a linear bijection. Thus*

$$\mathcal{P}^k(\mathbb{C}^n) = \bigoplus_{p=0}^{[k/2]} r^{2p} \mathcal{H}^{k-2p}(\mathbb{C}^n). \tag{5.90}$$

The space of harmonic polynomials of degree k is the analogue for the orthogonal group of the space spanned by the isotropic k-vectors for the symplectic group (see Theorem 5.5.15).

Proposition 5.6.13. *The space $\mathcal{H}^k(\mathbb{C}^n)$ is spanned by the polynomials x^k, where x is an isotropic vector in $\mathbb{C}^n$.*

Proof. Let $\mathcal{L}_k = \text{Span}\{x^k : x \text{ isotropic in } \mathbb{C}^n\}$. Then $0 \neq \mathcal{L}_k \subset \mathcal{H}^k(\mathbb{C}^n)$ and $\mathcal{L}_k$ is G-invariant. Hence $\mathcal{L}_k = \mathcal{H}^k(\mathbb{C}^n)$ by the irreducibility of $\mathcal{H}^k(\mathbb{C}^n)$.   □

5.6.5 Sp(n)–$\mathfrak{so}$(2k) *Howe Duality*

Let $G = \mathbf{Sp}(n, \mathbb{C}) \subset \mathbf{GL}(2n, \mathbb{C})$ be the group preserving the skew-symmetric bilinear form

$$\omega(x, y) = x^t J y \quad \text{for } x, y \in \mathbb{C}^{2n}, \quad \text{where } J = \begin{bmatrix} 0 & I_n \\ -I_n & 0 \end{bmatrix}.$$

Take $V = (\mathbb{C}^{2n})^k$ with G acting by $g \cdot (x_1, \dots, x_k) = (gx_1, \dots, gx_k)$, where $x_j \in \mathbb{C}^{2n}$. We use the same notation as in the previous sections.

For $1 \leq i, j \leq k$ define

$$D_{ij} = \sum_{p=1}^{n} \left\{ \frac{\partial^2}{\partial x_{pi} \partial x_{p+n,j}} - \frac{\partial^2}{\partial x_{p+n,i} \partial x_{pj}} \right\}$$

and

$$S_{ij} = \text{multiplication by } \sum_{p=1}^{n} (x_{pi} x_{p+n,j} - x_{p+n,i} x_{pj})$$

as operators on $\mathcal{P}(V)$. Note that $D_{ij} = -D_{ji}$ and $S_{ji} = -S_{ij}$. In particular, $D_{ii} = 0$ and $S_{ii} = 0$. We claim that these operators are in $\mathbb{D}(V)^G$. Indeed, S_{ij} is multiplication by the G-invariant function $\omega(x_i, x_j)$, so it clearly commutes with the action of G on $\mathcal{P}(V)$. If

$$g = \begin{bmatrix} A & B \\ C & D \end{bmatrix} \in \mathbf{Sp}(n, \mathbb{C}) \quad \text{with } A, B, C, D \in M_n,$$

then the equation $g^t J g = J$ is equivalent to the relations $A^t C = C^t A$, $B^t D = D^t B$, and $A^t D = I + C^t B$. Using these relations, a calculation like that done in Section 5.6.3 shows that $\rho(g) D_{ij} \rho(g^{-1}) = D_{ij}$. Thus $D_{ij} \in \mathbb{D}(V)^G$.

Theorem 5.6.14. *Let $G = \mathbf{Sp}(n, \mathbb{C})$ act by left multiplication on $V = M_{2n,k}$. Set $\mathfrak{g}' = \text{Span}\{E_{ij} + n\delta_{ij}, S_{ij}, D_{ij} : 1 \leq i, j \leq k\}$. Then $\mathfrak{g}'$ is a Lie subalgebra of $\mathbb{D}(V)^G$ that is isomorphic to $\mathfrak{so}(2k, \mathbb{C})$. Furthermore, the algebra $\mathbb{D}(V)^G$ is generated by $\mathfrak{g}'$.*

Proof. The operators $\{D_{ij}\}$ and $\{S_{ij}\}$ satisfy

$$[D_{ij}, D_{rs}] = 0 \quad \text{and} \quad [S_{ij}, S_{rs}] = 0. \tag{5.91}$$

As in the case of $\mathbf{O}(n, \mathbb{C})$–$\mathfrak{sp}(k)$ duality, we calculate that

$$[D_{ij}, S_{rs}] = \sum_{p=1}^{n} \left(\frac{\partial}{\partial x_{pi}} \omega(x_r, x_s) \right) \frac{\partial}{\partial x_{p+n,j}}$$
$$- \sum_{p=1}^{n} \left(\frac{\partial}{\partial x_{p+n,i}} \omega(x_r, x_s) \right) \frac{\partial}{\partial x_{pi}} + D_{ij}(\omega(x_r, x_s)).$$

This gives

$$[D_{ij}, S_{rs}] = \delta_{ri}(E_{sj} + n\delta_{sj}) - \delta_{rj}(E_{si} + n\delta_{si}) \tag{5.92}$$
$$+ \delta_{sj}(E_{ri} + n\delta_{ri}) - \delta_{si}(E_{rj} + n\delta_{rj}),$$

where now

$$E_{ij} = \sum_{q=1}^{2n} x_{qi} \frac{\partial}{\partial x_{qj}}.$$

We also calculate that

$$E_{ij}(\omega(x_r, x_s)) = \sum_{p=1}^{n} \left\{ x_{pi} \frac{\partial}{\partial x_{pj}} \omega(x_r, x_s) + x_{p+n,i} \frac{\partial}{\partial x_{p+n,j}} \omega(x_r, x_s) \right\}$$
$$= \delta_{jr}\omega(x_i, x_s) + \delta_{js}\omega(x_r, x_i).$$

Hence

$$[E_{ij} + n\delta_{ij}, S_{rs}] = \delta_{jr}S_{is} + \delta_{js}S_{ri}. \tag{5.93}$$

A calculation similar to that in the previous case gives

$$[E_{ij} + \delta_{ij}, D_{rs}] = -\delta_{ir}D_{js} + \delta_{is}D_{jr}. \tag{5.94}$$

These commutation relations show that $\mathfrak{g}'$ is a Lie subalgebra of $\mathbb{D}(V)^G$.

Let B be the symmetric bilinear form on $\mathbb{C}^{2k}$ with matrix $S = \begin{bmatrix} 0 & I_k \\ I_k & 0 \end{bmatrix}$. Then $X \in \mathfrak{so}(\mathbb{C}^{2k}, B)$ if and only if it has the block form

$$X = \begin{bmatrix} A & B \\ C & -A^t \end{bmatrix}, \quad \text{with } A, B, C \in M_k(\mathbb{C}) \text{ and } B = -B^t, \, C = -C^t. \tag{5.95}$$

Define a linear map $\varphi : \mathfrak{g}' \longrightarrow \mathfrak{so}(\mathbb{C}^{2k}, B)$ by $\varphi(S_{ij}) = e_{i,j+k} - e_{j,i+k}$, $\varphi(D_{ij}) = e_{i+k,j} - e_{j+k,i}$, and $\varphi(E_{ij} + n\delta_{ij}) = e_{ij} - e_{k+j,k+i}$, for $1 \leq i, j \leq k$. From (5.95) we see that φ is a linear isomorphism. It is a straightforward exercise, using (5.81) and (5.91)–(5.94), to verify that φ is a Lie algebra homomorphism.

The operators spanning $\mathfrak{g}'$ have the following symbols:

$$\sigma(S_{ij}) = \sum_{p=1}^{n}(x_{pi}x_{p+n,j} - x_{p+n,i}x_{pj}), \quad \sigma(D_{ij}) = \sum_{p=1}^{n}(\xi_{pi}\xi_{p+n,j} - \xi_{p+n,i}\xi_{pj}),$$

$$\sigma(E_{ij} + \delta_{ij}) = \sum_{p=1}^{2n} x_{pi}\xi_{pj}.$$

The FFT for G (Corollary 5.2.3) asserts that $\mathcal{P}(V \oplus V^*)^G$ is generated by these polynomials. It follows from Theorem 5.6.3 that $\mathfrak{g}'$ generates $\mathbb{D}(V)^G$. $\square$

Corollary 5.6.15. $(G = \mathbf{Sp}(n,\mathbb{C}))$ *In the canonical duality decomposition*

$$\mathcal{P}(V) \cong \bigoplus_{\lambda \in \mathcal{S}} E^\lambda \otimes F^\lambda \qquad (5.96)$$

under the joint action of $\mathbb{D}(V)^G$ *and* G *(where* $\mathcal{S} \subset \widehat{G}$*), each irreducible* $\mathbb{D}(V)^G$-*module* E^λ *is an irreducible infinite-dimensional representation of* $\mathfrak{so}(2k,\mathbb{C})$.

Proof. Apply Corollary 5.6.4. $\square$

5.6.6 Exercises

1. Let $G = \mathbf{GL}(n,\mathbb{C})$ and $V = M_{n,p} \oplus M_{n,q}$. Let $g \in G$ act on V by $g \cdot (x \oplus y) = gx \oplus (g^t)^{-1}y$ for $x \in M_{n,p}$ and $y \in M_{n,q}$. Note that the columns x_i of x transform as vectors in $\mathbb{C}^n$ and the columns y_j of y transform as covectors in $(\mathbb{C}^n)^*$.
   (a) Let $\mathfrak{p}_-$ be the subspace of $\mathbb{D}(V)$ spanned by the operators of multiplication by the functions $(x_i)^t \cdot y_j$ for $1 \leq i \leq p$, $1 \leq j \leq q$. Let $\mathfrak{p}_+$ be the subspace of $\mathbb{D}(V)$ spanned by the operators $\Delta_{ij} = \sum_{r=1}^{n}(\partial/\partial x_{ri})(\partial/\partial y_{rj})$ for $1 \leq i \leq p$, $1 \leq j \leq q$. Prove that $\mathfrak{p}_\pm \subset \mathbb{D}(V)^G$.
   (b) Let $\mathfrak{k}$ be the subspace of $\mathbb{D}(V)$ spanned by the operators $E_{ij}^{(x)} + (k/2)\delta_{ij}$ (with $1 \leq i,j \leq p$) and $E_{ij}^{(y)} + (k/2)\delta_{ij}$ (with $1 \leq i,j \leq q$), where $E_{ij}^{(x)}$ is defined by equation (5.75) and $E_{ij}^{(y)}$ is similarly defined with x_{ij} replaced by y_{ij}. Prove that $\mathfrak{k} \subset \mathbb{D}(V)^G$.
   (c) Prove the commutation relations $[\mathfrak{k},\mathfrak{k}] \subset \mathfrak{k}$, $[\mathfrak{k},\mathfrak{p}_\pm] = \mathfrak{p}_\pm$, $[\mathfrak{p}_-,\mathfrak{p}_+] \subset \mathfrak{k}$.
   (d) Set $\mathfrak{g}' = \mathfrak{p}_- + \mathfrak{k} + \mathfrak{p}_+$. Prove that $\mathfrak{g}'$ is isomorphic to $\mathfrak{gl}(p+q,\mathbb{C})$, and that $\mathfrak{k} \cong \mathfrak{gl}(p,\mathbb{C}) \oplus \mathfrak{gl}(q,\mathbb{C})$.
   (e) Prove that $\mathbb{D}(V)^G$ is generated by $\mathfrak{g}'$. (HINT: Use Theorems 5.2.1 and 5.6.3. Note that there are four possibilities for contractions to obtain G-invariant polynomials on $V \oplus V^*$:

   (i) vector and covector in V;   (iii) vector from V and covector from V^*;
   (ii) vector and covector in V^*;   (iv) covector from V and vector from V^*.

   Show that the contractions of types (i) and (ii) furnish symbols for bases of $\mathfrak{p}_\pm$, and that contractions of type (iii) and (iv) furnish symbols for a basis of $\mathfrak{k}$.)

5.7 Further Applications of Invariant Theory

In this final section we give several more applications of the FFT for $\mathbf{GL}(n, \mathbb{C})$ using the abstract Capelli theorem from Chapter 4.

5.7.1 Capelli Identities

Let V be $S^2(\mathbb{C}^n)$ or $\wedge^2 \mathbb{C}^n$ and let $G = \mathbf{GL}(n, \mathbb{C})$. Then V is a G-module as a G-invariant subspace of $\mathbb{C}^n \otimes \mathbb{C}^n$. Let τ be the corresponding representation of G on $\mathcal{P}(V)$, given by

$$\tau(g)f(x) = f(g^{-1}x) \quad \text{for } f \in \mathcal{P}(V),\ x \in V,\ \text{and } g \in \mathbf{GL}(n, \mathbb{C}) \,.$$

Let $\mathfrak{g} = \mathfrak{gl}(n, \mathbb{C})$ and let $Z(\mathfrak{g})$ be the center of the universal enveloping algebra $U(\mathfrak{g})$.

Theorem 5.7.1. *Suppose* $T \in \mathbb{D}(V)^G$. *Then there exists* $z \in Z(\mathfrak{g})$ *such that* $T = d\tau(z)$. *Hence* $\mathbb{D}(V)^G$ *is commutative and* $\mathcal{P}(V)$ *is a multiplicity-free G-module.*

The theorem will follow from Theorem 4.2.13 once we show that $\mathbb{D}(V)^G \subset d\tau(U(\mathfrak{g}))$. For this, we will use the tensor form of the FFT for G. If $x \in V$ and $\xi \in V^*$ then

$$x = \sum_{i,j=1}^n x_{ij}\, e_i \otimes e_j \quad \text{and} \quad \xi = \sum_{i,j=1}^n \xi_{ij}\, e_i^* \otimes e_j^* \,. \tag{5.97}$$

Here $x_{ij} = x_{ji}$ and $\xi_{ij} = \xi_{ji}$ when $V = S^2(\mathbb{C}^n)$ (resp. $x_{ij} = -x_{ji}$ and $\xi_{ij} = -\xi_{ji}$ when $V = \wedge^2 \mathbb{C}^n$). The function x_{ij} for $1 \le i \le j \le n$ (resp. $1 \le i < j \le n$) give linear coordinates on V. Define

$$\varphi_{pq} = \sum_{i=1}^n x_{qi} \xi_{pi} \,,$$

viewed as a quadratic polynomial on $V \oplus V^*$ via the coordinates (5.97) (p and q are interchanged in this definition because of equation (5.101) below). Obviously φ_{pq} is invariant under $\mathbf{GL}(n, \mathbb{C})$. Following the general pattern of classical invariant theory, we now prove that all invariants are expressible in terms of these quadratic invariants.

Lemma 5.7.2. *If* $f \in \mathcal{P}(V \oplus V^*)^G$ *then f is a polynomial in* $\{\varphi_{pq} : 1 \le p, q \le n\}$.

Proof of Lemma 5.7.2 We may assume that f is homogeneous. By considering the action of $-I \in G$, we see that f is of bidegree (k, k) for some k, and we prove the result by induction on k. If $k = 0$ then $f \in \mathbb{C}$, so the result is true. Now assume the result for $k - 1$. We can consider f to be an element of $(S^k(V) \otimes S^k(V^*))^*$. By the FFT (Corollary 5.3.2) it suffices to take f as a complete contraction and to evaluate f on the tensors $x^{\otimes k} \otimes \xi^{\otimes k}$, where $x \in V$ and $\xi \in V^*$, since tensors of this form span $S^k(V) \otimes S^k(V^*)$ (Lemma B.2.3). With x and ξ given as in (5.97), we have

$$x^{\otimes k} \otimes \xi^{\otimes k} = \sum \left\{ \left(x_{i_1 j_1} \cdots x_{i_k j_k} \xi_{p_1 q_1} \cdots \xi_{p_k q_k} \right) \right.$$
$$\left. \times e_{i_1} \otimes e_{j_1} \otimes \cdots \otimes e_{i_k} \otimes e_{j_k} \otimes e_{p_1}^* \otimes e_{q_1}^* \otimes \cdots \otimes e_{p_k}^* \otimes e_{q_k}^* \right\}.$$

(In this expression and the following ones, the symbol $\sum$ indicates summation from 1 to n of all the indices i_r, j_s, p_u, q_v that occur.) We may assume (by the symmetries of x_{ij} and ξ_{pq}) that f contracts the first position in $(\mathbb{C}^n)^{\otimes 2k}$ (labeled by the index i_1) with the first position in $((\mathbb{C}^n)^*)^{\otimes 2k}$ (labeled by the index p_1). This first contraction gives the tensor

$$\sum \left\{ \left(\sum_i x_{ij_1} \xi_{iq_1} \right) \left(x_{i_2 j_2} \cdots x_{i_k j_k} \xi_{p_2 q_2} \cdots \xi_{p_k q_k} \right) \right. \tag{5.98}$$
$$\left. \times e_{j_1} \otimes e_{i_2} \otimes \cdots \otimes e_{i_k} \otimes e_{j_k} \otimes e_{q_1}^* \otimes e_{p_2}^* \otimes \cdots \otimes e_{p_k}^* \otimes e_{q_k}^* \right\}$$

in $(\mathbb{C}^n)^{\otimes(2k-1)} \otimes ((\mathbb{C}^n)^*)^{\otimes(2k-1)}$. By the same symmetry conditions, we may assume that f performs one of the following contractions on the tensor (5.98):

(i)   Vectors in the first position (index j_1) contract with covectors in the first position (index q_1).
(ii) Vectors in the first position (index j_1) contract with covectors in the second position (index p_2).

In case (i) we have, after the second contraction, the tensor

$$\left(\sum_i \varphi_{ii} \right) \sum \left\{ \left(x_{i_2 j_2} \cdots x_{i_k j_k} \xi_{p_2 q_2} \cdots \xi_{p_k q_k} \right) \right.$$
$$\left. \times e_{i_2} \otimes e_{j_2} \otimes \cdots \otimes e_{i_k} \otimes e_{j_k} \otimes e_{p_2}^* \otimes e_{q_2}^* \otimes \cdots \otimes e_{p_k}^* \otimes e_{q_k}^* \right\}$$

in $S^{(k-1)}(V) \otimes S^{(k-1)}(V^*)$. The remaining $2k - 2$ contractions of f on this tensor yield a function that by the inductive hypothesis is a polynomial in the functions $\{\varphi_{pq}\}$. This completes the inductive step in case (i).

We now look at case (ii). After the second contraction we have the tensor

$$\sum \left\{ \left(\sum_{i,j} x_{ij} \xi_{ip_2} \xi_{jq_2} \right) \left(x_{i_2 j_2} \cdots x_{i_k j_k} \xi_{p_3 q_3} \cdots \xi_{p_k q_k} \right) \right. \tag{5.99}$$
$$\left. \times e_{i_2} \otimes e_{j_2} \otimes \cdots \otimes e_{i_k} \otimes e_{j_k} \otimes e_{p_2}^* \otimes e_{q_2}^* \otimes \cdots \otimes e_{p_k}^* \otimes e_{q_k}^* \right\}$$

(note that after contracting we have relabeled the index q_1 as p_2). Let t be an indeterminate, and set

$$\zeta_{pq} = \xi_{pq} + t \sum_{i,j} x_{ij} \xi_{ip} \xi_{jq}.$$

We observe that ζ_{pq} has the same symmetry properties as ξ_{pq} and that the coefficient of t in the tensor

$$\sum \Big\{ \big(x_{i_2 j_2} \cdots x_{i_k j_k} \zeta_{p_2 q_2} \cdots \zeta_{p_k q_k} \big)$$

$$\times \, e_{i_2} \otimes e_{j_2} \otimes \cdots \otimes e_{i_k} \otimes e_{j_k} \otimes e_{p_2}^* \otimes e_{q_2}^* \otimes \cdots \otimes e_{p_k}^* \otimes e_{q_k}^* \Big\} \tag{5.100}$$

is a constant times the projection of (5.99) onto $S^{k-1}(V) \otimes S^{k-1}(V^*)$. The remaining $2k - 2$ contractions of f on (5.100) yield a function that by the inductive hypothesis is a polynomial in the functions $\sum_j x_{qj} \zeta_{pj}$ for $1 \le p, q \le n$. Since

$$\sum_j x_{qj} \zeta_{pj} = \sum_j x_{qj} \xi_{pj} + t \sum_{j,r,s} x_{qj} x_{rs} \xi_{pr} \xi_{js} = \varphi_{pq} + t \sum_s \varphi_{sq} \varphi_{ps} \,,$$

this completes the inductive step in case (ii). □

Proof of Theorem 5.7.1 On V we use coordinates x_{ij} as in (5.97) and we write ∂_{ij} for the corresponding partial differentiation. Let $\{e_{pq}\}$ be the usual basis of $\mathfrak{gl}(n, \mathbb{C})$. Then a direct calculation yields

$$d\tau(e_{pq}) = -2 \sum_j x_{qj} \partial_{pj} \,. \tag{5.101}$$

Hence, in the notation just introduced, we see that the symbol of $d\tau(e_{pq})$ is $-2\varphi_{pq}$. From Lemma 5.7.2 and the same induction argument as in Theorem 5.6.3, it follows that $\mathbb{D}^G(V) \subset d\tau(U(\mathfrak{g}))$. Thus we can apply the abstract Capelli theorem (Theorem 4.2.13). □

5.7.2 Decomposition of $\mathbf{S(S^2(V))}$ under $\mathbf{GL(V)}$

Let $G = \mathbf{GL}(n, \mathbb{C})$ and let SM_n be the space of symmetric $n \times n$ complex matrices. We let G act on SM_n by $g, x \mapsto (g^t)^{-1} x g^{-1}$. Let ρ be the associated representation of G on $\mathcal{P}(SM_n)$:

$$\rho(g)f(x) = f(g^t x g) \quad \text{for } f \in \mathcal{P}(SM_n) \,.$$

Note that $SM_n \cong S^2(\mathbb{C}^n)^*$ (the symmetric bilinear forms on $\mathbb{C}^n$) as a G-module relative to this action, where a matrix $x \in SM_n$ corresponds to the symmetric bilinear form $\beta_x(u, v) = u^t x v$ for $u, v \in \mathbb{C}^n$. Thus as a G-module,

$$\mathcal{P}(SM_n) \cong \mathcal{P}(S^2(\mathbb{C}^n)^*) \cong S(S^2(\mathbb{C}^n)) \,.$$

From Theorem 5.7.1 we know that $\mathcal{P}(SM_n)$ is multiplicity-free as a G-module. We now obtain the explicit form of the decomposition of $\mathcal{P}(SM_n)$.

Theorem 5.7.3. *The space of homogeneous polynomials on SM_n of degree r decomposes under $\mathbf{GL}(n, \mathbb{C})$ as*

$$\mathcal{P}^r(SM_n) \cong \bigoplus_\mu F_n^\mu \tag{5.102}$$

with the sum over all nonnegative dominant weights $\mu = \sum_i \mu_i \varepsilon_i$ *of size* $2r$ *such that* $\mu_i \in 2\mathbb{N}$ *for all i. Furthermore,*

$$\mathcal{P}(SM_n)^{N_n^+} = \mathbb{C}[\widetilde{\Delta}_1, \ldots, \widetilde{\Delta}_n] \,, \tag{5.103}$$

where $\widetilde{\Delta}_i$ *denotes the restriction of the ith principal minor to the space of symmetric matrices. The functions* $\widetilde{\Delta}_1, \ldots, \widetilde{\Delta}_n$ *are algebraically independent.*

Proof. We follow the same line of argument as in Theorem 5.6.7. In this case the algebra $\mathcal{P}(SM_n)$ is generated by the matrix entry functions x_{ij} with $i \leq j$, and

$$\rho(h)x_{ij} = h_i h_j x_{ij} \quad \text{for } h = \mathrm{diag}[h_1, \ldots, h_n] \ .$$

Thus the weights of D_n on $\mathcal{P}(SM_n)$ are nonnegative. Suppose $f \in \mathcal{P}(SM_n)^{N_n^+}(\mu)$. Then for $u \in N_n^+$ and $h \in D_n$ one has $f(u^t h^2 u) = h^\mu f(I_n)$. By Lemma B.2.7 the orbit of I_n under $D_n N_n^+$ consists of all $x \in SM_n$ for which $\Delta_i(x) \neq 0$ for $i = 1, \ldots, n$. Hence $f(I_n) \neq 0$ and f is uniquely determined by $f(I_n)$. If $h^2 = I_n$, then $f(I_n) = h^\mu f(I_n)$. Taking $h = \mathrm{diag}[\pm 1, \ldots, \pm 1]$, we see that $\mu_i \in 2\mathbb{N}$ for $i = 1, \ldots, n$.

Now we show that all such even dominant integral highest weights occur. For $i = 1, \ldots, n$ the polynomial $\widetilde{\Delta}_i$ on SM_n is nonzero and transforms by

$$\widetilde{\Delta}_i(u^t h x h u) = h^{2\lambda_i} \widetilde{\Delta}_i(x) \quad \text{for } h \in D_n \text{ and } u \in N_n^+ \,,$$

where $\lambda_i = \varepsilon_1 + \cdots + \varepsilon_i$. The function $\widetilde{\Delta}_{\mathbf{m}}$ is thus a highest-weight vector with weight

$$\mu = 2m_1 \lambda_1 + \cdots + 2m_n \lambda_n \ .$$

As a polynomial on SM_n it has degree $|\mu|/2$. The decomposition (5.102) now follows from Theorem 5.7.1 and the theorem of the highest weight. The argument for the algebraic independence of the set of functions $\widetilde{\Delta}_1, \ldots, \widetilde{\Delta}_n$ and (5.103) is the same as in the proof of Theorem 5.6.7.   □

Corollary 5.7.4. *The space* $S(S^2(\mathbb{C}^n))$ *is isomorphic to* $\bigoplus_\mu F_n^\mu$ *as a module for* $\mathbf{GL}(n, \mathbb{C})$, *where the sum is over all weights* $\mu = \sum_{i=1}^n 2m_i \lambda_i$ *with* $m_i \in \mathbb{N}$.

5.7.3 Decomposition of $\mathbf{S}(\bigwedge^2(\mathbf{V}))$ under $\mathbf{GL}(\mathbf{V})$

Let AM_n be the space of skew-symmetric $n \times n$ matrices and let $G = \mathbf{GL}(n, \mathbb{C})$ act on AM_n by $g, x \mapsto (g^t)^{-1} x g^{-1}$. Let $\rho(g)f(x) = f(g^t x g)$ be the associated representation of G on $\mathcal{P}(AM_n)$. Note that $AM_n \cong \bigwedge^2(\mathbb{C}^n)^*$ (the skew-symmetric bilinear forms on $\mathbb{C}^n$) as a G-module relative to this action. Thus as a G-module,

$$\mathcal{P}(AM_n) \cong \mathcal{P}(\bigwedge^2(\mathbb{C}^n)^*) \cong S(\bigwedge^2 \mathbb{C}^n) \ .$$

From Theorem 5.7.1 we know that $\mathcal{P}(AM_n)$ is multiplicity-free. We now obtain the explicit form of its decomposition into irreducible G-modules. Let Pf_i be the ith principal Pfaffian on AM_n for $i = 1, \ldots, k$, where $k = [n/2]$ (see Section B.2.6).

Theorem 5.7.5. *The space of homogeneous polynomials on AM_n of degree r decomposes under $\mathbf{GL}(n, \mathbb{C})$ as*

$$\mathcal{P}^r(AM_n) \cong \bigoplus_\mu F_n^\mu \tag{5.104}$$

with the sum over all nonnegative dominant integral weights $\mu = \sum \mu_i \varepsilon_i$ such that

$$|\mu| = 2r, \quad \mu_{2i-1} = \mu_{2i} \quad for\ i = 1, \ldots, k, \quad and \quad \mu_{2k+1} = 0 \tag{5.105}$$

(the last equation only when $n = 2k + 1$ is odd). The functions $\mathrm{Pf}_1, \ldots, \mathrm{Pf}_k$ are algebraically independent and $\mathcal{P}(AM_n)^{N_n^+} = \mathbb{C}[\mathrm{Pf}_1, \ldots, \mathrm{Pf}_k]$.

Proof. We follow the same line of argument as in Theorem 5.7.3, with the principal minors replaced by the principal Pfaffians. The functions $x_{ij} - x_{ji}$ $(1 \le i < j \le n)$ generate $\mathcal{P}(AM_n)$. Since

$$\rho(h)(x_{ij} - x_{ji}) = h_i h_j (x_{ij} - x_{ji}) \quad \text{for } h = \mathrm{diag}[h_1, \ldots, h_n],$$

we see that all the weights of D_n on $\mathcal{P}(AM_n)$ are nonnegative.

Suppose $0 \ne f \in \mathcal{P}(AM_n)^{N_n^+}(\mu)$. If $n = 2k$ is even, set $x_0 = J \oplus \cdots \oplus J$ (with k summands), where

$$J = \begin{bmatrix} 0 & 1 \\ -1 & 0 \end{bmatrix}.$$

When $n = 2k + 1$ is odd, set $x_0 = J \oplus \cdots \oplus J \oplus 0$. Lemma B.2.8 implies that $f(x_0) \ne 0$ and f is determined by $f(x_0)$. For $h = \mathrm{diag}[h_1, \ldots, h_n] \in D_n$ we have

$$h x_0 h = (h_1 h_2) J \oplus (h_3 h_4) J \oplus \cdots \oplus (h_{n-1} h_n) J$$

when n is even, whereas

$$h x_0 h = (h_1 h_2 J) \oplus (h_3 h_4) J \oplus \cdots \oplus (h_{n-2} h_{n-1}) J \oplus 0$$

when n is odd. Thus the stabilizer H_0 of x_0 in D_n consists of all matrices

$$h = \mathrm{diag}[z_1, z_1^{-1}, \ldots, z_k, z_k^{-1}] \qquad \text{when } n = 2k \text{ is even};$$
$$h = \mathrm{diag}[z_1, z_1^{-1}, \ldots, z_k, z_k^{-1}, z_{k+1}] \qquad \text{when } n = 2k + 1 \text{ is odd},$$

where $z_i \in \mathbb{C}^\times$. For $h \in H_0$,

$$f(x_0) = f(h x_0 h) = h^\mu f(x_0) = f(x_0) z \prod_{i=1}^k z_i^{\mu_{2i-1} - \mu_{2i}},$$

where $z = 1$ when $n = 2k$ and $z = (z_{k+1})^{\mu_{2k+1}}$ when $n = 2k + 1$. Hence μ satisfies (5.105).

Now we construct a highest-weight vector for each nonnegative dominant weight μ that satisfies (5.105). From (B.19) we have

$$\mathrm{Pf}_i(u^t h x h u) = \Delta_{2i}(h)\,\mathrm{Pf}_i(x)$$

for $h \in D_n$ and $u \in N_n^+$. Hence Pf_i is a highest-weight vector of weight $\lambda_{2i} = \varepsilon_1 + \cdots + \varepsilon_{2i}$. For integers $m_i \geq 0$ define

$$\mathrm{Pf}_{\mathbf{m}}(x) = \prod_{i=1}^{[n/2]} \mathrm{Pf}_i(x)^{m_i} \quad \text{for } x \in AM_n .$$

This polynomial is a highest-weight vector with weight $\mu = \sum_{i=1}^{k} m_i \lambda_{2i}$. The weights of this form are precisely the nonnegative dominant weights that satisfy (5.105). The decomposition of $\mathcal{P}(AM_n)$ and the algebraic independence of the principal Pfaffians $\mathrm{Pf}_1, \ldots, \mathrm{Pf}_k$ now follow as in the proof of Theorem 5.7.3. $\square$

Corollary 5.7.6. *The space* $S(\bigwedge^2(\mathbb{C}^n))$ *is isomorphic to* $\bigoplus_\mu F_n^\mu$ *as a module for* $\mathbf{GL}(n, \mathbb{C})$, *where the sum is over all weights* $\mu = \sum_{i=1}^{[n/2]} m_i \lambda_{2i}$ *with* $m_i \in \mathbb{N}$.

5.7.4 Exercises

1. Let $\mathcal{H}^k(\mathbb{C}^n)$ be the space of $\mathbf{O}(n)$-harmonic polynomials on $\mathbb{C}^n$.
   (a) Show that $\mathcal{H}^0(\mathbb{C}^n) = \mathbb{C}$ and $\mathcal{H}^1(\mathbb{C}^n) = \mathbb{C}^n$.
   (b) Prove that $\dim \mathcal{H}^k(\mathbb{C}^n) = (-1)^k\{\binom{-n}{k} - \binom{-n}{k-2}\}$ for $k \geq 2$, where $\binom{-n}{k} = (-n)(-n-1)\cdots(-n-k+1)/k!$ is the negative binomial coefficient. (HINT: Recall that $\dim \mathcal{P}^k(\mathbb{C}^n) = (-1)^k \binom{-n}{k}$ and use Corollary 5.6.12.)
2. Let $\mathbf{O}(n)$ act on $M_{n,k}$ by left matrix multiplication and consider the $\mathbf{O}(n)$–$\mathfrak{sp}(k)$ duality in Section 5.6.3. To each irreducible regular representation of $\mathbf{O}(n)$ that occurs in $\mathcal{P}(M_{n,k})$ there is a corresponding irreducible infinite-dimensional representation of $\mathfrak{sp}(k)$. Can you describe some properties of these Lie algebra representations? (HINT: The case $n = 1$ was done in Section 5.6.4; see also Section 3.2.1.)

5.8 Notes

Section 5.1.1. Theorem 5.1.1 was a major landmark in the history of invariant theory; the proof given is due to Hurwitz. In the case of binary forms, the space $F^{(n)}$ of all binary forms of degree n furnishes an irreducible representation of $G = \mathbf{SL}(2, \mathbb{C})$ (Proposition 2.3.5). Hence the ring $\mathcal{P}(F^{(n)})^G$ is finitely generated by Theorem 5.1.1. For each integer k the character (and hence the multiplicities) of $\mathcal{P}^k(F^{(n)})$ is known by Theorem 4.1.20. In particular, the dimension of the space $\mathcal{P}^k(F^{(n)})^G$ is known.

For small n, sets of basic generators for $\mathcal{P}(F^{(n)})^G$ have been found. For general values of n explicit descriptions of sets of basic invariants are lacking, although various bounds on degrees and asymptotic results are known (see Howe [69], Popov [120], and Wehlau [158]).

Section 5.1.2. For a general treatment of invariant polynomials and harmonic polynomials, see Helgason [67, Chapter III]. The statements and proofs of Theorems 5.1.4 and 5.1.8 are valid for the ring of invariant polynomials and the harmonic polynomials relative to any finite group generated by reflections (see Chevalley [37]); for the Weyl groups of types BC and D see Exercises 5.1.3.

Section 5.2.1. The terminology *first fundamental theorem* (FFT) for invariants is due to Weyl [164]. In most treatments of this theory only one of the two (equivalent) problems is considered. Our approach is novel in that we do a simultaneous analysis of both problems using special cases of each to derive the general theorem. There are very few pairs (G, ρ) for which an FFT is known (see Vust [151], Procesi [122], Schwarz [133] and Howe [71]).

Section 5.3.2. The proof of Theorem 5.3.3 is based on Atiyah, Bott, and Patodi [4]. A key step in the proof (Lemma 5.3.6) is a tensor algebra version of the classical *polarization operators* as used by Weyl [164].

Sections 5.4.2 and 5.4.3. In Weyl [164] the polynomial form of the FFT for the classical groups is proved by induction using the *Cappelli identity* and polarization operators; see also Fulton and Harris [52, Appendix F] for this approach.

Section 5.5.1. The basic result of this section is that the commutant of the action G on the exterior algebra is generated (as an associative algebra) by operators of degree two, and these operators themselves generate a finite-dimensional Lie algebra. This was exploited by Weil [159]. For further developments along this line, see Howe [72, Chapter 4].

Sections 5.5.2 and 5.5.3. The fundamental representations (for any simple Lie algebra) were first constructed by Cartan [27]. The special feature of the classical Lie algebras is the realization of the fundamental representations in $\bigwedge V$ (where V is the defining representation) and the explicit description of the centralizer algebras via the FFT. For $\mathfrak{sl}(n, \mathbb{C})$ and $\mathfrak{so}(n, \mathbb{C})$ it is easy to bypass the FFT in this construction; for $\mathfrak{sp}(n, \mathbb{C})$ it is also possible to avoid the FFT but the details are considerably more lengthy (see Bourbaki [13, Chapitre VIII, §13]). The method of obtaining all irreducible representations by taking the highest component in a tensor product of fundamental representations appears first in Cartan [27].

Section 5.6.2. In Howe [72] there is an extensive discussion of the relations between $\mathbf{GL}(k)$–$\mathbf{GL}(n)$ duality and the FFT.

Sections 5.6.3 and 5.6.5. The use of the FFT to obtain the commutant of a classical group G in the polynomial coefficient differential operators was initiated by Howe [70] (this influential paper was written in 1976). The unifying principle is that the ring of G-invariant differential operators is generated by a set of operators of filtration degree two that span a finite-dimensional Lie algebra.

Section 5.6.4. For a treatment of spherical harmonics by the same method used for the symmetric group in Section 5.1.2 see Helgason [67, Chapter III] or Stein and Weiss [137, Chapter IV]. See Howe–Tan [73] for some applications of these results to non-commutative harmonic analysis.

Section 5.7.1. Let $T \in \mathbb{D}(V)^G$. A particular choice of an element $z \in Z(\mathfrak{g})$ such that $T = d\tau(z)$ gives a so-called *Capelli identity*. The case $V = S^2(\mathbb{C}^n)$ was carried out in Gårding [53] (see also Turnbull [145]). See Howe [70], Howe and Umeda [75], and Kostant and Sahi [91] for the general *Capelli problem*.

Sections 5.7.2 and 5.7.3. For an alternative approach to these results, see Howe and Umeda [75].

Chapter 6
Spinors

Abstract In this chapter we complete the construction of the fundamental representations for the orthogonal Lie algebras using *Clifford algebras* and their irreducible representations on spaces of *spinors*. We show that the orthogonal Lie algebras are isomorphic to Lie algebras of quadratic elements in Clifford algebras. From these isomorphisms and the action of Clifford algebras on spinors we obtain the fundamental representations that were missing from Chapter 5, namely those with highest weights ϖ_l for $\mathfrak{so}(2l+1, \mathbb{C})$ and ϖ_{l-1}, ϖ_l for $\mathfrak{so}(2l, \mathbb{C})$. We then show that these Lie algebra representations correspond to regular representations of the *spin groups*, which are twofold covers of the orthogonal groups. With the introduction of the spin groups and the spin representations we finally achieve the property proved for the special linear groups and symplectic groups in Chapter 5, namely that every finite-dimensional representation of $\mathfrak{so}(n, \mathbb{C})$ is the differential of a unique regular representation of $\mathbf{Spin}(n, \mathbb{C})$. The chapter concludes with a description of the real forms of the spin groups.

6.1 Clifford Algebras

We begin by constructing new associative algebras, the *Clifford algebras*, which are related to exterior algebras of vector spaces as the algebras of polynomial coefficient differential operators (the Weyl algebras) are related to the symmetric algebras of vector spaces.

6.1.1 Construction of $\mathbf{Cliff}(V)$

Let V be a finite-dimensional complex vector space with a symmetric bilinear form β (for the moment we allow β to be possibly degenerate).

R. Goodman, N.R. Wallach, *Symmetry, Representations, and Invariants*,
Graduate Texts in Mathematics 255, DOI 10.1007/978-0-387-79052-3_6,
© Roe Goodman and Nolan R. Wallach 2009

Definition 6.1.1. A *Clifford algebra* for (V, β) is an associative algebra $\mathrm{Cliff}(V, \beta)$ with unit 1 over $\mathbb{C}$ and a linear map $\gamma : V \longrightarrow \mathrm{Cliff}(V, \beta)$ satisfying the following properties:

(C1) $\{\gamma(x), \gamma(y)\} = \beta(x, y)1$ for $x, y \in V$, where $\{a, b\} = ab + ba$ is the *anticommutator* of a, b.

(C2) $\gamma(V)$ generates $\mathrm{Cliff}(V, \beta)$ as an algebra.

(C3) Given any complex associative algebra $\mathcal{A}$ with unit element 1 and a linear map $\varphi : V \longrightarrow \mathcal{A}$ such that $\{\varphi(x), \varphi(y)\} = \beta(x, y)1$, there exists an associative algebra homomorphism $\widetilde{\varphi} : \mathrm{Cliff}(V, \beta) \longrightarrow \mathcal{A}$ such that $\varphi = \widetilde{\varphi} \circ \gamma$:

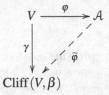

It is easy to see that an algebra satisfying properties (C1), (C2), and (C3) is unique (up to isomorphism). Indeed, if $\mathcal{C}$ and $\mathcal{C}'$ are two such algebras with associated linear maps $\gamma : V \longrightarrow \mathcal{C}$ and $\gamma' : V \longrightarrow \mathcal{C}$, then property (C3) provides algebra homomorphisms $\widetilde{\gamma} : V \longrightarrow \mathcal{C}'$ and $\widetilde{\gamma}' : V \longrightarrow \mathcal{C}$ such that $\gamma' = \widetilde{\gamma}' \circ \gamma$ and $\gamma = \widetilde{\gamma} \circ \gamma'$. It follows that $\widetilde{\gamma} \circ \widetilde{\gamma}'$ is the identity map on $\gamma(V)$ and hence it is the identity map on $\mathcal{C}$ by property (C2). Likewise, $\widetilde{\gamma}' \circ \widetilde{\gamma}$ is the identity map on $\mathcal{C}$. This shows that $\widetilde{\gamma} : \mathcal{C}' \longrightarrow \mathcal{C}$ is an algebra isomorphism.

To prove existence of a Clifford algebra, we start with the tensor algebra $\mathcal{T}(V)$ (see Appendix C.1.2) and let $\mathcal{J}(V, \beta)$ be the two-sided ideal of $\mathcal{T}(V)$ generated by the elements

$$x \otimes y + y \otimes x - \beta(x, y)1, \quad x, y \in V.$$

Define $\mathrm{Cliff}(V, \beta) = \mathcal{T}(V)/\mathcal{J}(V, \beta)$ and let $\gamma : V \longrightarrow \mathrm{Cliff}(V, \beta)$ be the natural quotient map coming from the embedding $V \hookrightarrow \mathcal{T}(V)$. Clearly this pair satisfies (C1) and (C2). To verify (C3), we first factor φ through the map $\widehat{\varphi} : \mathcal{T}(V) \longrightarrow \mathcal{A}$ whose existence is provided by the universal property of $\mathcal{T}(V)$. Then $\widehat{\varphi}(\mathcal{J}(V, \beta)) = 0$ so we obtain a map $\widetilde{\varphi}$ by passing to the quotient.

Let $\mathrm{Cliff}_k(V, \beta)$ be the span of 1 and the operators $\gamma(a_1) \cdots \gamma(a_p)$ for $a_i \in V$ and $p \leq k$. The subspaces $\mathrm{Cliff}_k(V, \beta)$, for $k = 0, 1, \ldots$, give a *filtration* of the Clifford algebra:

$$\mathrm{Cliff}_k(V, \beta) \cdot \mathrm{Cliff}_m(V, \beta) \subset \mathrm{Cliff}_{k+m}(V, \beta).$$

Let $\{v_i : i = 1, \ldots, n\}$ be a basis for V. Since $\{\gamma(v_i), \gamma(v_j)\} = \beta(v_i, v_j)$, we see from (C1) that $\mathrm{Cliff}_k(V, \beta)$ is spanned by 1 and the products

$$\gamma(v_{i_1}) \cdots \gamma(v_{i_p}), \quad 1 \leq i_1 < i_2 < \cdots < i_p \leq n,$$

where $p \leq k$. In particular,

$$\mathrm{Cliff}(V, \beta) = \mathrm{Cliff}_n(V, \beta) \quad \text{and} \quad \dim \mathrm{Cliff}(V, \beta) \leq 2^n.$$

The linear map $v \mapsto -\gamma(v)$ from V to $\mathrm{Cliff}(V,\beta)$ satisfies (C3), so it extends to an algebra homomorphism

$$\alpha : \mathrm{Cliff}(V,\beta) \longrightarrow \mathrm{Cliff}(V,\beta)$$

such that $\alpha(\gamma(v_1)\cdots\gamma(v_k)) = (-1)^k \gamma(v_1)\cdots\gamma(v_k)$. Obviously $\alpha^2(u) = u$ for all $u \in \mathrm{Cliff}(V,\beta)$. Hence α is an automorphism, which we call the *main involution* of $\mathrm{Cliff}(V,\beta)$. There is a decomposition

$$\mathrm{Cliff}(V,\beta) = \mathrm{Cliff}^+(V,\beta) \oplus \mathrm{Cliff}^-(V,\beta) \,,$$

where $\mathrm{Cliff}^+(V,\beta)$ is spanned by products of an even number of elements of V, $\mathrm{Cliff}^-(V,\beta)$ is spanned by products of an odd number of elements of V, and α acts by ± 1 on $\mathrm{Cliff}^{\pm}(V,\beta)$.

6.1.2 Spaces of Spinors

From now on we assume that V is a finite-dimensional complex vector space with nondegenerate symmetric bilinear form β. In the previous section we proved the existence and uniqueness of the Clifford algebra $\mathrm{Cliff}(V,\beta)$ (as an abstract associative algebra). We now study its irreducible representations.

Definition 6.1.2. Let S be a complex vector space and let $\gamma : V \longrightarrow \mathrm{End}(S)$ be a linear map. Then (S,γ) is a *space of spinors* for (V,β) if

(S1) $\{\gamma(x),\gamma(y)\} = \beta(x,y)I$ for all $x,y \in V$.
(S2) The only subspaces of S that are invariant under $\gamma(V)$ are 0 and S.

If (S,γ) is a space of spinors for (V,β), then the map γ extends to an irreducible representation

$$\widetilde{\gamma} : \mathrm{Cliff}(V,\beta) \longrightarrow \mathrm{End}(S)$$

(by axioms (C1), (C2), and (C3) of Section 6.1.1). Conversely, every irreducible representation of $\mathrm{Cliff}(V,\beta)$ arises this way. Since $\mathrm{Cliff}(V,\beta)$ is a finite-dimensional algebra, a space of spinors for (V,β) must also be finite-dimensional.

Let (S,γ) and (S',γ') be spaces of spinors for (V,β). One says that (S,γ) is *isomorphic* to (S',γ') if there exists a linear bijection $T : S \longrightarrow S'$ such that $T\gamma(v) = \gamma'(v)T$ for all $v \in V$.

Theorem 6.1.3. *Assume that β is a nondegenerate bilinear form on V.*

1. *If $\dim V = 2l$ is even, then up to isomorphism there is exactly one space of spinors for (V,β), and it has dimension 2^l.*
2. *If $\dim V = 2l + 1$ is odd, then there are exactly two nonisomorphic spaces of spinors for (V,β), and each space has dimension 2^l.*

Proof. Let $\dim V = n$. We begin by an explicit construction of some spaces of spinors. Fix a pair W, W^* of dual maximal isotropic subspaces of V relative to β, as in Section B.2.1. We identify W^* with the dual space of W via the form β and write $\beta(x^*, x) = \langle x^*, x \rangle$ for $x \in W$ and $x^* \in W^*$. When n is even, then $V = W^* \oplus W$ and

$$\beta(x + x^*, y + y^*) = \langle x^*, y \rangle + \langle y^*, x \rangle \tag{6.1}$$

for $x, y \in W$ and $x^*, y^* \in W^*$. When n is odd, we take a one-dimensional subspace $U = \mathbb{C}e_0$ such that $\beta(e_0, e_0) = 2$ and $\beta(e_0, W) = \beta(e_0, W^*) = 0$. Then $V = W \oplus U \oplus W^*$ and

$$\beta(x + \lambda e_0 + x^*, y + \mu e_0 + y^*) = \langle x^*, y \rangle + 2\lambda\mu + \langle y^*, x \rangle \tag{6.2}$$

for $x, y \in W$, $x^*, y^* \in W^*$, and $\lambda, \mu \in \mathbb{C}$.

We shall identify $\bigwedge^p W^*$ with $C^p(W)$, the space of p-multilinear functions on W that are *skew-symmetric* in their arguments, as follows (see Appendix B.2.4): Given p elements $w_1^*, \ldots, w_p^* \in W^*$, define a skew-symmetric p-linear function ψ on W by

$$\psi(w_1, \ldots, w_p) = \det\left[\langle w_i^*, w_j \rangle\right].$$

Since ψ depends in a skew-symmetric way on $w_1^*, \ldots, w_p^*$, we obtain in this way a bijective linear map from $\bigwedge^p(W^*)$ to $C^p(W)$. Set

$$C^\bullet(W) = \bigoplus_{p=0}^{\dim W} C^p(W),$$

and give $C^\bullet(W)$ the multiplication coming from the isomorphism with $\bigwedge W^*$. If $w^* \in W^*$ then $\varepsilon(w^*) \in \operatorname{End} C^\bullet(W)$ will denote the operator of left exterior multiplication by w^*, as in Section 5.5.1. We have $\varepsilon(w^*): C^p(W) \longrightarrow C^{p+1}(W)$ given by

$$\varepsilon(w^*)\psi(w_0, \ldots, w_p) = \sum_{j=0}^{p} (-1)^j \langle w^*, w_j \rangle \psi(w_0, \ldots, \widehat{w_j}, \ldots, w_p) \tag{6.3}$$

for $\psi \in C^p(W)$, where the notation $\widehat{w_j}$ means to omit w_j. Since $x^* \wedge y^* = -y^* \wedge x^*$ for $x^*, y^* \in W^*$, the exterior multiplication operators satisfy

$$\varepsilon(x^*)\varepsilon(y^*) = -\varepsilon(y^*)\varepsilon(x^*). \tag{6.4}$$

For $w \in W$, we let $\iota(w): \bigwedge^p W^* \longrightarrow \bigwedge^{p-1} W^*$ be the *interior product* with w. Under the isomorphism $\bigwedge W^* \cong C^\bullet(W)$ above, the operator $\iota(w)$ becomes evaluation in the first argument:

$$\iota(w)\psi(w_2, \ldots, w_p) = \psi(w, w_2, \ldots, w_p)$$

for $\psi \in C^p(W)$ and $w, w_2, \ldots, w_p \in W$. By the skew symmetry of ψ we have

$$\iota(x)\iota(y) = -\iota(y)\iota(x) \quad \text{for } x, y \in W. \tag{6.5}$$

By (5.43) the interior and exterior multiplication operators satisfy the *anticommutation relations*

$$\{\varepsilon(x^*), \iota(x)\} = \langle x^*, x \rangle I \quad \text{for } x \in W \text{ and } x^* \in W^* . \tag{6.6}$$

When $\dim V$ is even, we combine these operators to obtain a linear map

$$\gamma : V \longrightarrow \text{End}C^\bullet(W) , \quad \gamma(x + x^*) = \iota(x) + \varepsilon(x^*) \tag{6.7}$$

for $x \in W$ and $x^* \in W^*$. From (6.4), (6.5), and (6.6) we calculate that

$$\{\gamma(a), \gamma(b)\} = \beta(a, b)I \quad \text{for } a, b \in V . \tag{6.8}$$

When $\dim V$ is odd, we define two linear maps $\gamma_\pm : V \longrightarrow \text{End}(C^\bullet(W))$ by

$$\gamma_\pm(w + \lambda e_0 + w^*)u = (\iota(w) + \varepsilon(w^*) \pm (-1)^p \lambda)u \quad \text{for } u \in C^p(W) ,$$

where $w \in W$, $w^* \in W^*$, and $\lambda \in \mathbb{C}$. Thus the restrictions of $\gamma_\pm$ to $W \oplus W^*$ are the maps for the even-dimensional case. Also, since $\varepsilon(w^*)$ increases degree by one, whereas $\iota(w)$ decreases degree by one, it follows that

$$\{\gamma_\pm(w), \gamma_\pm(e_0)\} = 0 , \quad \{\gamma_\pm(w^*), \gamma_\pm(e_0)\} = 0 .$$

From these relations and (6.6) we calculate that

$$\{\gamma(x + x^*), \gamma(y + y^*)\} = (\langle x^*, y \rangle + \langle y^*, x \rangle)I ,$$
$$\{\gamma_\pm(x + \lambda e_0 + x^*), \gamma_\pm(y + \mu e_0 + y^*)\} = (\langle x^*, y \rangle + \langle y^*, x \rangle + 2\lambda\mu)I ,$$

for $x, y \in W$, $x^*, y^* \in W^*$, and $\lambda, \mu \in \mathbb{C}$. By (6.1) and (6.2) we see that the pair $(C^\bullet(W), \gamma)$ and the pairs $(C^\bullet(W), \gamma_\pm)$ satisfy condition (S1) in Definition 6.1.2.

We now show that these pairs satisfy condition (S2). It is enough to do this when n is even. Let $Y \subset C^\bullet(W)$ be a nonzero subspace that is invariant under $\gamma(V)$. If we can show that $1 \in Y$, then it will follow that $Y = C^\bullet(W)$, since the restriction of γ to W^* is the left multiplication representation of the algebra $C^\bullet(W)$. Take $0 \neq y \in Y$ and write

$$y = y_0 + y_1 + \cdots + y_p \quad \text{with } y_j \in C^j(W) \text{ and } y_p \neq 0 .$$

Let $\{e_1, \ldots, e_l\}$ be a basis for W, where $l = n/2$, and let $\{e_{-1}, \ldots, e_{-l}\}$ be the basis for W^* such that $\beta(e_i, e_{-j}) = \delta_{ij}$. Then y_p contains some term

$$ce_{-j_1} \wedge \cdots \wedge e_{-j_p} \quad \text{with } 1 \leq j_1 < \cdots < j_p \leq l \text{ and } c \neq 0 .$$

Since y_p is the term of top degree in y, we have

$$\gamma(e_{j_1}) \cdots \gamma(e_{j_p})y = \iota(e_{j_1}) \cdots \iota(e_{j_p})y = \pm c \in Y .$$

This implies that $Y = C^\bullet(W)$ and proves (S2). $\square$

The uniqueness assertions in Theorem 6.1.3 are consequences of the following lemma, which can be viewed as the spinor analogue of Theorem 3.2.5 and Corollary 3.3.14.

Lemma 6.1.4. *Suppose (S', γ') is a space of spinors for (V, β). Let $n = \dim V$.*

1. *Set $Z = \bigcap_{w \in W} \mathrm{Ker}\, \gamma'(w)$. Then $Z \neq 0$.*
2. *Fix $0 \neq z_0 \in Z$. If n is even, then there is a unique spinor-space isomorphism $T : (C^\bullet(W), \gamma) \longrightarrow (S', \gamma')$ such that $T1 = z_0$. If n is odd, then there is a unique spinor-space isomorphism $T : (C^\bullet(W), \gamma_c) \longrightarrow (S', \gamma')$ such that $T1 = z_0$, where $c = \pm 1$ is determined by $\gamma'(e_0)z_0 = cz_0$. In particular, $\dim Z = 1$ in both cases.*
3. *If n is odd the spin spaces $(C^\bullet(W), \gamma_+)$ and $(C^\bullet(W), \gamma_-)$ are not equivalent.*

Proof. Take a basis $e_{\pm i}$ for V as above. We have

$$Z = \bigcap_{i=1}^{l} \mathrm{Ker}\, \gamma'(e_i)$$

(where $n = 2l$ or $2l + 1$). Now $\gamma'(e_1)^2 = 0$, since $\beta(e_1, e_1) = 0$. Hence $\mathrm{Ker}\, \gamma'(e_1) \neq 0$. If $u \in \mathrm{Ker}\, \gamma'(e_1)$ then

$$\gamma'(e_1)\gamma'(e_j)u = -\gamma'(e_j)\gamma'(e_1)u = 0 \quad \text{for } j = 1, 2, \ldots, l \,.$$

(This equation also holds for $j = 0$ when n is odd.) Hence $\mathrm{Ker}\, \gamma'(e_1)$ is invariant under $\gamma'(e_j)$. By the same argument

$$(\mathrm{Ker}\, \gamma'(e_1)) \cap (\mathrm{Ker}\, \gamma'(e_2)) \neq 0 \,,$$

and this space is invariant under $\gamma'(e_j)$ for $j = 1, \ldots, l$ (and also for $j = 0$ when n is odd). Continuing in this way, we prove (1) by induction on l.

Define a linear map $T : \bigwedge W^* \longrightarrow S'$ by $T1 = z_0$ and

$$T(e_{-j_1} \wedge \cdots \wedge e_{-j_p}) = \gamma'(e_{-j_1}) \cdots \gamma'(e_{-j_p})z_0$$

for $1 \leq j_1 < \cdots < j_p \leq l$ and $p = 1, 2, \ldots, l$. From the relations (S1) we see that the range of T is invariant under $\gamma'(V)$. Hence T is surjective by (S2).

We next prove that T intertwines the representations γ and γ'. By the definition of T we have $T\gamma(w^*) = \gamma'(w^*)T$ for $w^* \in W^*$. Let $w \in W$. Then

$$T(\gamma(w)1) = \gamma'(w)z_0 = 0 \,.$$

Assume that $T(\gamma(w)u) = \gamma'(w)T(u)$ for all $u \in \bigwedge^p(W^*)$. Take $w^* \in W^*$ and $u \in \bigwedge^p(W^*)$. Since $\gamma(w) = \iota(w)$ is an antiderivation of $\bigwedge^\bullet(W^*)$, we have

$$\begin{aligned}
T(\gamma(w)(w^* \wedge u)) &= T(\langle w^*, w\rangle u - \gamma(w^*)\gamma(w)u) \\
&= \langle w^*, w\rangle T(u) - \gamma'(w^*)\gamma'(w)T(u) = \gamma'(w)\gamma'(w^*)T(u) \\
&= \gamma'(w)T(w^* \wedge u) \,,
\end{aligned}$$

where we have used relations (S1) and the induction hypothesis. Hence by induction we conclude that $T\gamma(w) = \gamma'(w)T$ for all $w \in W$. In case n is odd, then for $u \in \bigwedge^p(W^*)$ we have

$$\begin{aligned}
T(\gamma_c(e_0)u) &= c(-1)^p T(u) = c(-1)^p \gamma'(u)z_0 \\
&= (-1)^p \gamma'(u)\gamma'(e_0)z_0 = \gamma'(e_0)\gamma'(u)z_0 \\
&= \gamma'(e_0)T(u) .
\end{aligned}$$

The intertwining property implies that $\operatorname{Ker} T = 0$, since $\gamma(W + W^*)$ acts irreducibly. Hence T is an isomorphism of spaces of spinors. In particular, $\dim Z = 1$. This completes the proof of (2).

It remains to prove that $(S, \gamma_\pm)$ are inequivalent spaces of spinors when n is odd. If there existed a linear isomorphism $R : \bigwedge W^* \longrightarrow \bigwedge W^*$ such that $R\gamma_+(v) = \gamma_-(v)R$ for all $v \in V$, then R would commute with $\iota(w)$ and $\varepsilon(w^*)$ for all $w \in W$ and $w^* \in W^*$. Hence by irreducibility, $R = \lambda I$ for some nonzero $\lambda \in \mathbb{C}$. This would imply that $\gamma_-(e_0) = \gamma_+(e_0)$, which is a contradiction. This proves (3).   □

6.1.3 Structure of Cliff(V)

We now use the spin spaces to determine the structure of the Clifford algebras.

Proposition 6.1.5. *Suppose* $\dim V = n$ *is even. Let* (S, γ) *be a space of spinors for* (V, β). *Then* $(\operatorname{End} S, \gamma)$ *is a Clifford algebra for* (V, β). *Thus* $\operatorname{Cliff}(V, \beta)$ *is a simple algebra of dimension* 2^n. *The map* $\gamma : V \longrightarrow \operatorname{Cliff}(V, \beta)$ *is injective, and for any basis* $\{v_1, \ldots, v_n\}$ *of* V *the set of all ordered products*

$$\gamma(v_{i_1}) \cdots \gamma(v_{i_p}), \quad \text{where } 1 \leq i_1 < \cdots < i_p \leq n \tag{6.9}$$

(empty product = 1), is a basis for $\operatorname{Cliff}(V, \beta)$.

Proof. Let $\widetilde{\gamma} : \operatorname{Cliff}(V, \beta) \longrightarrow \operatorname{End} S$ be the canonical algebra homomorphism extending the map $\gamma : V \longrightarrow \operatorname{End} S$. Since $\widetilde{\gamma}$ is an irreducible representation, we know by Corollary 4.1.7 that $\widetilde{\gamma}(\operatorname{Cliff}(V, \beta)) = \operatorname{End} S$. Since $\dim S = 2^{n/2}$ by Theorem 6.1.3, it follows that $\dim(\operatorname{Cliff}(V, \beta)) \geq 2^n$. But in Section 6.1.1 we saw that the elements in (6.9) are a spanning set for $\operatorname{Cliff}(V, \beta)$, so $\dim(\operatorname{Cliff}(V, \beta)) \leq 2^n$. Thus equality holds, and $\widetilde{\gamma}$ is an isomorphism. In particular, γ is injective.   □

Before considering the Clifford algebra for an odd-dimensional space, we introduce another model for the spin spaces that proves useful for calculations. Assume that $\dim V = 2l$ is even. Take β-isotropic spaces W, W^* and a basis $e_{\pm i}$ for V as in Section 6.1.2. Set

$$U_i = \bigwedge \mathbb{C}e_{-i} = \mathbb{C}1 \oplus \mathbb{C}e_{-i} \quad \text{for } i = 1, \ldots, l .$$

Then each two-dimensional subspace U_i is a graded algebra with ordered basis $\{1, e_{-i}\}$ and relation $e_{-i}^2 = 0$. Since $W^* = \mathbb{C}e_{-1} \oplus \cdots \oplus \mathbb{C}e_{-l}$, Proposition C.1.8 furnishes an isomorphism of graded algebras .

$$\bigwedge{}^{\bullet}(W^*) \cong U_1 \widehat{\otimes} \cdots \widehat{\otimes} U_l, \tag{6.10}$$

where $\widehat{\otimes}$ denotes the skew-commutative tensor product. If we ignore the algebra structure and consider $\bigwedge W^*$ as a vector space, we have an isomorphism $\bigwedge W^* \cong U_1 \otimes \cdots \otimes U_l$. This furnishes an algebra isomorphism

$$\mathrm{End}\left(\bigwedge W^*\right) \cong \mathrm{End}\, U_1 \otimes \cdots \otimes \mathrm{End}\, U_l. \tag{6.11}$$

Notice that in this isomorphism the factors on the right mutually commute. To describe the operators $\gamma(x)$ in this tensor product model, let $J = \{j_1, \ldots, j_p\}$ with $1 \leq j_1 < \cdots < j_p \leq l$. Under the isomorphism (6.10) the element $e_{-j_1} \wedge \cdots \wedge e_{-j_p}$ corresponds to $u_J = u_1 \otimes \cdots \otimes u_l$, where

$$u_i = \begin{cases} e_{-i} & \text{if } i \in J, \\ 1 & \text{if } i \notin J. \end{cases}$$

We have

$$e_{-i} \wedge e_{-j_1} \wedge \cdots \wedge e_{-j_p} = \begin{cases} 0 & \text{if } i \in J, \\ (-1)^r e_{-j_1} \wedge \cdots \wedge e_{-i} \wedge \cdots \wedge e_{-j_p} & \text{if } i \notin J, \end{cases}$$

where r is the number of indices in J that are less than i, and e_{-i} appears in position $r+1$ on the right side. Thus the exterior multiplication operator $\varepsilon(e_{-i})$ acts on the basis $\{u_J\}$ by

$$A_{-i} = H \otimes \cdots \otimes H \otimes \underbrace{\begin{bmatrix} 0 & 0 \\ 1 & 0 \end{bmatrix}}_{i\text{th place}} \otimes I \otimes \cdots \otimes I,$$

where $H = \mathrm{diag}[1, -1]$, I is the 2×2 identity matrix, and we enumerate the basis for U_i in the order $1, e_{-i}$. For the interior product operator we likewise have

$$\iota(e_i)(e_{-j_1} \wedge \cdots \wedge e_{-j_p}) = \begin{cases} (-1)^r e_{-j_1} \wedge \cdots \wedge \widehat{e_{-i}} \wedge \cdots \wedge e_{-j_p} & \text{if } i \in J, \\ 0 & \text{if } i \notin J. \end{cases}$$

Thus $\iota(e_i)$ acts on the basis $\{u_J\}$ by

$$A_i = H \otimes \cdots \otimes H \otimes \underbrace{\begin{bmatrix} 0 & 1 \\ 0 & 0 \end{bmatrix}}_{i\text{th place}} \otimes I \otimes \cdots \otimes I.$$

It is easy to check that the operators $\{A_{\pm i}\}$ satisfy the canonical anticommutation relations (the factors of H in the tensor product ensure that $A_i A_j = -A_j A_i$). This gives a matrix algebra proof that $S = U_1 \otimes \cdots \otimes U_l$ together with the map $e_{\pm i} \mapsto A_{\pm i}$ furnishes a space of spinors for (V, β).

Now assume that $\dim V = 2l + 1$ is odd with $l \geq 1$, and set

$$A_0 = H \otimes \cdots \otimes H \qquad (l \text{ factors}).$$

Then $A_0^2 = 1$ and $A_0 A_{\pm i} = -A_{\pm i} A_0$ for $i = 1, \ldots, l$. Hence we can obtain models for the spinor spaces $(S, \gamma_{\pm})$ by setting $S = U_1 \otimes \cdots \otimes U_l$, with $e_{\pm i}$ acting by $A_{\pm i}$ and e_0 acting by $\pm A_0$.

Proposition 6.1.6. *Suppose* $\dim V = 2l + 1$ *is odd. Let* (S, γ_+) *and* (S, γ_-) *be the two inequivalent spaces of spinors for* (V, β), *and let* $\gamma : V \longrightarrow \operatorname{End} S \oplus \operatorname{End} S$ *be defined by* $\gamma(v) = \gamma_+(v) \oplus \gamma_-(v)$. *Then* $(\operatorname{End} S \oplus \operatorname{End} S, \gamma)$ *is a Clifford algebra for* (V, β). *Thus* $\operatorname{Cliff}(V, \beta)$ *is a semisimple algebra and is the sum of two simple ideals of dimension* 2^{n-1}. *The map* $\gamma : V \longrightarrow \operatorname{Cliff}(V, \beta)$ *is injective. For any basis* $\{v_1, \ldots, v_n\}$ *of* V *the set of all ordered products* $\gamma(v_{i_1}) \cdots \gamma(v_{i_p})$, *where* $1 \leq i_1 < \cdots < i_p \leq n$ *(empty product* $= 1$*), is a basis for* $\operatorname{Cliff}(V, \beta)$.

Proof. Let $l \geq 1$ (the case $\dim V = 1$ is left to the reader) and use the model $S = U_1 \otimes \cdots \otimes U_l$ for spinors, with $\gamma_{\pm}(e_{\pm i}) = A_{\pm i}$ and $\gamma_{\pm}(e_0) = \pm A_0$ (notation as above). Let

$$\widetilde{\gamma} : \operatorname{Cliff}(V, \beta) \longrightarrow \operatorname{End} S \oplus \operatorname{End} S$$

be the canonical extension of the map γ. We calculate that

$$[A_i, A_{-i}] = I \otimes \cdots \otimes I \otimes H \otimes I \otimes \cdots \otimes I$$

(with H in the ith tensor place). Hence

$$[A_1, A_{-1}][A_2, A_{-2}] \cdots [A_l, A_{-l}] = H \otimes \cdots \otimes H = A_0 .$$

Since $\gamma(e_{\pm i}) = A_{\pm i} \oplus A_{\pm i}$, while $\gamma(e_0) = A_0 \oplus (-A_0)$, we have

$$[\gamma(e_1), \gamma(e_{-1})] \cdots [\gamma(e_l), \gamma(e_{-l})] \gamma(e_0) = (A_0 A_0) \oplus (-A_0 A_0) = I \oplus (-I) .$$

Thus the image of $\widetilde{\gamma}$ contains the operator $I \oplus (-I)$. Hence the image contains the operators $I \oplus 0$ and $0 \oplus I$. Using Proposition 6.1.5, we conclude that

$$\widetilde{\gamma}(\operatorname{Cliff}(V, \beta)) = \operatorname{End} S \oplus \operatorname{End} S .$$

Since we already have the upper bound $2^{2l+1} = 2 \dim(\operatorname{End} S)$ for $\dim \operatorname{Cliff}(V, \beta)$ from Section 6.1.1, it follows that $\widetilde{\gamma}$ is an isomorphism. The proof of the rest of the proposition is now the same as that of Proposition 6.1.5. $\qquad\square$

Let V be odd-dimensional. Decompose $V = W \oplus \mathbb{C} e_0 \oplus W^*$ as in Section 6.1.2. Set $V_0 = W \oplus W^*$ and let β_0 be the restriction of β to V_0. Recall that $\operatorname{Cliff}^+(V, \beta)$ is the subalgebra of $\operatorname{Cliff}(V, \beta)$ spanned by the products of an even number of elements of V.

Lemma 6.1.7. *There is an algebra isomorphism* $\mathrm{Cliff}(V_0, \beta_0) \cong \mathrm{Cliff}^+(V, \beta)$. *Hence* $\mathrm{Cliff}^+(V, \beta)$ *is a simple algebra.*

Proof. Let $\gamma : V \longrightarrow \mathrm{Cliff}(V, \beta)$ be the canonical map. For $v \in V_0$ we define $\varphi(v) = i \gamma(e_0) \gamma(v)$. Then $\varphi(v) \in \mathrm{Cliff}^+(V, \beta)$ and

$$\{\varphi(x), \varphi(y)\} = -\gamma(e_0)\gamma(x)\gamma(e_0)\gamma(y) - \gamma(e_0)\gamma(y)\gamma(e_0)\gamma(x)$$
$$= \{\gamma(x), \gamma(y)\}$$

for $x, y \in V_0$. Thus φ extends to an algebra homomorphism

$$\widetilde{\varphi} : \mathrm{Cliff}(V_0, \beta_0) \longrightarrow \mathrm{Cliff}^+(V, \beta) .$$

Let $\gamma_0 : V_0 \longrightarrow \mathrm{Cliff}(V_0, \beta_0)$ be the canonical map. For $v_1, \ldots, v_k \in V$ the anticommutation relation $\{\gamma(e_0), \gamma(v_j)\} = 0$ gives

$$\widetilde{\varphi}(\gamma_0(v_1) \cdots \gamma_0(v_k)) = i^k \gamma(e_0)\gamma(v_1) \cdots \gamma(e_0)\gamma(v_k) = i^p \gamma(e_0)^k \gamma(v_1) \cdots \gamma(v_k) ,$$

where $p = 3k - 2$. Since $\gamma(e_0)^2 = 1$, we see that

$$\widetilde{\varphi}(\gamma_0(v_1) \cdots \gamma_0(v_k)) = \begin{cases} i^p \gamma(e_0)\gamma(v_1) \cdots \gamma(v_k) & \text{for } k \text{ odd}, \\ i^p \gamma(v_1) \cdots \gamma(v_k) & \text{for } k \text{ even}. \end{cases}$$

It follows from Propositions 6.1.5 and 6.1.6 that φ is injective. Since $\mathrm{Cliff}(V_0, \beta_0)$ and $\mathrm{Cliff}^+(V, \beta)$ both have dimension $2^{\dim V_0}$, φ is an isomorphism. $\qquad \square$

6.1.4 Exercises

1. (a) Show that $\mathrm{Cliff}(V, 0) = \bigwedge V$, the *exterior algebra* over V (see Section C.1.4).
   (b) Let β be any symmetric bilinear form β and give $\mathrm{Cliff}(V, \beta)$ the filtration of Section 6.1.1. Show that the graded algebra $\mathrm{Gr}(\mathrm{Cliff}(V, \beta))$ is isomorphic to $\bigwedge V$.

2. Let $V = W \oplus W^*$ be an even-dimensional space, and let β be a bilinear form on V for which W and W^* are β-isotropic and in duality, as in Section 6.1.2.
   (a) Let (S, γ) be a space of spinors for (V, β). Show that $\bigcap_{w^* \in W^*} \mathrm{Ker}(\gamma(w^*))$ is one-dimensional.
   (b) Let $S' = \bigwedge W$. For $w \in W$, $w^* \in W^*$ define $\gamma'(w + w^*) = \varepsilon(w) + \iota(w^*)$ on S', where $\varepsilon(w)$ is the exterior product operator and $\iota(w^*)$ is the interior product operator. Show that (S', γ') is a space of spinors for (V, β).
   (c) Fix $0 \neq u \in \bigwedge^l W$, where $l = \dim W$. Show that there is a unique spinor-space isomorphism T from $(\bigwedge W^*, \gamma)$ to $(\bigwedge W, \gamma')$ such that $T(1) = u$. Here $\gamma(w + w^*) = \iota(w) + \varepsilon(w^*)$ and γ' is the map in (b).

3. (a) Show that when $n \geq 3$, the polynomial $x_1^2 + \cdots + x_n^2$ in the commuting variables $x_1, \ldots, x_n$ cannot be factored into a product of linear factors with coefficients in $\mathbb{C}$.

(b) Show that $x_1^2 + \cdots + x_n^2 = 2(x_1 e_1 + \cdots + x_n e_n)^2$ when the multiplication on the right is done in the Clifford algebra $\text{Cliff}(\mathbb{C}^n, \beta)$ and $e_1, \ldots, e_n$ is a β-orthonormal basis for $\mathbb{C}^n$. Here β is the bilinear form on $\mathbb{C}^n$ with matrix I_n.

(c) Let (S, γ) be a space of spinors for $(\mathbb{C}^n, \beta)$. Consider the Laplace operator $\Delta = (1/2) \sum_{i=1}^n (\partial / \partial x_i)^2$ acting on $C^\infty(\mathbb{R}^n, S)$ (smooth functions with values in S). Show that Δ can be factored as D^2, where

$$D = \gamma(e_1) \frac{\partial}{\partial x_1} + \cdots + \gamma(e_n) \frac{\partial}{\partial x_n} \quad \text{(the \emph{Dirac operator}).}$$

4. Let V be a complex vector space with a symmetric bilinear form β. Take a basis $\{e_1, \ldots, e_n\}$ for V such that $\beta(e_i, e_j) = \delta_{ij}$.

   (a) Show that if i, j, k are distinct, then $e_i e_j e_k = e_j e_k e_i = e_k e_i e_j$, where the product is in the Clifford algebra for (V, β).

   (b) Show that if $A = [a_{ij}]$ is a symmetric $n \times n$ matrix, then $\sum_{i,j=1}^n a_{ij} e_i e_j = (1/2) \text{tr}(A)$ (product in the Clifford algebra for (V, β)).

   (c) Show that if $A = [a_{ij}]$ is a skew-symmetric $n \times n$ matrix, then $\sum_{i,j=1}^n a_{ij} e_i e_j = 2 \sum_{1 \le i < j \le n} a_{ij} e_i e_j$ (product in the Clifford algebra for (V, β)).

5. Let (V, β) and $e_1, \ldots, e_n$ be as in the previous exercise. Let $R_{ijkl} \in \mathbb{C}$ for $1 \le i, j, k, l \le n$ be such that

   (i) $R_{ijkl} = R_{klij}$,

   (ii) $R_{jikl} = -R_{ijkl}$,

   (iii) $R_{ijkl} + R_{kijl} + R_{jkil} = 0$.

   (a) Show that $\sum_{i,j,k,l} R_{ijkl} e_i e_j e_k e_l = (1/2) \sum_{i,j} R_{ijji}$, where the multiplication of the e_i is in the Clifford algebra for (V, β). (HINT: Use part (a) of the previous exercise to show that for each l, the sum over distinct triples i, j, k is zero. Then use the anticommutation relations to show that the sum with $i = j$ is also zero. Finally, use part (b) of the previous exercise to simplify the remaining sum.)

   (b) Let $\mathfrak{g}$ be a Lie algebra and B a symmetric nondegenerate bilinear form on $\mathfrak{g}$ such that $B([x, y], z) = -B(y, [x, z])$. Let $e_1, \ldots, e_n$ be an orthonormal basis of $\mathfrak{g}$ relative to B. Show that $R_{ijkl} = B([e_i, e_j], [e_k, e_l])$ satisfies conditions (i), (ii), and (iii). (See Section 10.3 for more examples of such tensors.)

6. Show that there exists an action of $\mathbf{O}(V, \beta)$ on $\text{Cliff}(V, \beta)$ by automorphisms such that $g \cdot \gamma(v) = \gamma(gv)$ for $g \in \mathbf{O}(V, \beta)$ and $v \in V$.

6.2 Spin Representations of Orthogonal Lie Algebras

We now construct an isomorphism between the orthogonal Lie algebra $\mathfrak{so}(V, \beta)$ and a Lie subalgebra of the associative algebra $\text{Cliff}(V, \beta)$ (an analogous embedding of the orthogonal Lie algebra $\mathfrak{so}(2l, \mathbb{C})$ into a Weyl algebra was constructed in Section 5.6.5).

6.2.1 Embedding $\mathfrak{so}(V)$ in Cliff(V)

Given $a, b \in V$ we define $R_{a,b} \in \mathrm{End}(V)$ by $R_{a,b}v = \beta(b, v)a - \beta(a, v)b$. Since

$$\beta(R_{a,b}x, y) = \beta(b, x)\beta(a, y) - \beta(a, x)\beta(b, y) = -\beta(x, R_{a,b}y),$$

we have $R_{a,b} \in \mathfrak{so}(V, \beta)$.

Lemma 6.2.1. *The linear transformations $R_{a,b}$, with a, b ranging over V, span $\mathfrak{so}(V, \beta)$.*

Proof. First consider the case in which $V = W \oplus W^*$ has dimension $2l$, where W and W^* are maximal isotropic. For $x, y \in W$ and $x^*, y^* \in W^*$, we have

$$R_{x,x^*}(y + y^*) = \langle x^*, y \rangle x - \langle y^*, x \rangle x^*,$$

$$R_{x,y}(x^* + y^*) = \langle x^*, y \rangle x - \langle y^*, x \rangle y, \quad R_{x^*,y^*}(x + y) = \langle y^*, y \rangle x^* - \langle x^*, x \rangle y^*.$$

Fix a basis $\{e_i\}$ for W and $\{e_{-i}\}$ for W^* as in Section 6.1.2, and let $e_{i,j}$ be the elementary transformation on V that carries e_i to e_j and the other basis vectors to 0. Then from the formulas above we see that

$$R_{e_i, e_{-j}} = e_{i,j} - e_{-j,-i},$$

$$R_{e_i, e_j} = e_{i,-j} - e_{j,-i}, \quad R_{e_{-i}, e_{-j}} = e_{-i,j} - e_{-j,i}.$$

From the results of Section 2.4.1 these operators furnish a basis for $\mathfrak{so}(V, \beta)$.

When $\dim V = 2l + 1$, we take $V = W \oplus \mathbb{C}e_0 \oplus W^*$ as in Section 6.1.2, with W, W^* as above and $\beta(e_0, e_0) = 2$. Let $V_0 = W \oplus W^*$ and let β_0 be the restriction of β to V_0. The additional transformations in $\mathfrak{so}(V, \beta)$ besides those in $\mathfrak{so}(V_0, \beta_0)$ are

$$R_{x,e_0}(y + \lambda e_0 + y^*) = \lambda x - \langle y^*, x \rangle e_0, \quad R_{x^*,e_0}(y + \lambda e_0 + y^*) = \lambda x^* - \langle x^*, y \rangle e_0$$

for $x, y \in W$, $x^*, y^* \in W^*$, and $\lambda \in \mathbb{C}$. Thus

$$R_{e_i, e_0} = e_{i,0} - e_{0,-i}, \quad R_{e_{-i}, e_0} = e_{-i,0} - e_{0,i}.$$

From the results of Section 2.4.1 we see that the lemma holds in this case also. □

Since $R_{a,b}$ is a skew-symmetric bilinear function of the vectors a and b, it defines a linear map

$$R : \wedge^2 V \longrightarrow \mathfrak{so}(V, \beta), \qquad a \wedge b \mapsto R_{a,b}.$$

This map is easily seen to be injective, and by Lemma 6.2.1 it is surjective. We calculate that

$$[R_{a,b}, R_{x,y}] = R_{u,y} + R_{x,v} \quad \text{with } u = R_{a,b}x \text{ and } v = R_{a,b}y \tag{6.12}$$

for $a, b, x, y \in V$. Thus R intertwines the representation of $\mathfrak{so}(V, \beta)$ on $\wedge^2 V$ with the adjoint representation of $\mathfrak{so}(V, \beta)$.

The key step for an explicit construction of the spin representations is to embed $\mathfrak{so}(V,\beta)$ into $\text{Cliff}(V,\beta)$ as follows:

Lemma 6.2.2. *Define a linear map* $\varphi : \mathfrak{so}(V,\beta) \longrightarrow \text{Cliff}_2(V,\beta)$ *by* $\varphi(R_{a,b}) = (1/2)[\gamma(a),\gamma(b)]$ *for* $a,b \in V$. *Then* φ *is an injective Lie algebra homomorphism, and*

$$[\varphi(X),\gamma(v)] = \gamma(Xv) \tag{6.13}$$

for $X \in \mathfrak{so}(V,\beta)$ *and* $v \in V$.

Proof. Since $a,b \mapsto (1/2)[\gamma(a),\gamma(b)]$ is bilinear and skew symmetric, the map φ extends uniquely to a linear map on $\mathfrak{so}(V,\beta)$. We first verify (6.13) when $X = R_{a,b}$. We calculate that

$$
\begin{aligned}
[[\gamma(a),\gamma(b)],\gamma(v)] &= \gamma(a)\gamma(b)\gamma(v) - \gamma(b)\gamma(a)\gamma(v) \\
&\quad -\gamma(v)\gamma(a)\gamma(b) + \gamma(v)\gamma(b)\gamma(a) \\
&= -\gamma(a)\gamma(v)\gamma(b) + \beta(b,v)\gamma(a) + \gamma(b)\gamma(v)\gamma(a) \\
&\quad -\beta(a,v)\gamma(b) + \gamma(a)\gamma(v)\gamma(b) - \beta(a,b)\gamma(b) \\
&\quad -\gamma(b)\gamma(v)\gamma(a) + \beta(b,v)\gamma(a) \\
&= 2\gamma(R_{a,b}v) ,
\end{aligned}
$$

where we have used the Clifford relations (C1) to permute $\gamma(a)$ and $\gamma(b)$ with $\gamma(v)$.

Now we check that φ preserves Lie brackets. Let $X = R_{a,b}$ and $Y = R_{x,y}$ with $a,b,x,y \in V$. From the Jacobi identity and (6.13) we calculate that

$$
\begin{aligned}
4[\varphi(X),\varphi(Y)] &= [[\gamma(a),\gamma(b)],[\gamma(x),\gamma(y)]] \\
&= [[[\gamma(a),\gamma(b)],\gamma(x)],\gamma(y)] + [\gamma(x),[[\gamma(a),\gamma(b)],\gamma(y)]] \\
&= 2[\gamma(R_{a,b}x),\gamma(y)] + 2[\gamma(x),\gamma(R_{a,b}y)] = 4\varphi(A+B) ,
\end{aligned}
$$

where $A = R_{u,y}$ and $B = R_{x,v}$ with $u = R_{a,b}x$ and $v = R_{a,b}y$. But by (6.12) we have $A + B = [X,Y]$, so φ is a homomorphism.

Finally, to verify that φ is injective, note that

$$2\varphi(R_{e_i,e_{-i}}) = [\gamma(e_i),\gamma(e_{-i})] = 2\gamma(e_i)\gamma(e_{-i}) - 1 . \tag{6.14}$$

Hence in the tensor product model for the space of spinors from Section 6.1.3, $2\varphi(R_{e_i,e_{-i}})$ acts by

$$2A_i A_{-i} - 1 = I \otimes \cdots \otimes I \otimes \underbrace{H}_{i\text{th place}} \otimes I \otimes \cdots \otimes I .$$

It follows that the restriction of φ to the diagonal algebra $\mathfrak{h}$ of $\mathfrak{so}(V,\beta)$ is injective. If $\dim V = 2$, then $\mathfrak{so}(V) = \mathfrak{h}$ and hence φ is injective. If $\dim V \geq 3$, let $\mathfrak{k} = \text{Ker}(\varphi)$. Then $\mathfrak{k}$ is an ideal that intersects $\mathfrak{h}$ in zero. If $\mathfrak{k} \neq 0$, then by Theorem 2.4.1 it would contain some root vector X_α, and hence it would contain $[X_{-\alpha},X_\alpha] \in \mathfrak{h}$, which is a contradiction. Hence $\text{Ker}(\varphi) = 0$ in all cases. $\qquad\square$

6.2.2 Spin Representations

We now obtain the spin representations of the orthogonal Lie algebra $\mathfrak{so}(V,\beta)$ from the representation of $\mathrm{Cliff}(V,\beta)$ on the spaces of spinors.

Assume that $\dim V$ is even and fix a decomposition $V = W \oplus W^*$ with W and W^* maximal β-isotropic subspaces. Let $(C^\bullet(W), \gamma)$ be the space of spinors defined in the proof of Theorem 6.1.3. Define the *even* and *odd* spin spaces

$$C^+(W) = \bigoplus_{p \text{ even}} C^p(W), \qquad C^-(W) = \bigoplus_{p \text{ odd}} C^p(W).$$

Then

$$\gamma(v) : C^\pm(W) \longrightarrow C^\mp(W) \quad \text{for } v \in V, \tag{6.15}$$

so the action of $\gamma(V)$ interchanges the even and odd spin spaces. Denote by $\tilde{\gamma}$ the extension of γ to a representation of $\mathrm{Cliff}(V,\beta)$ on $C^\bullet(W)$.

Let $\varphi : \mathfrak{so}(V,\beta) \longrightarrow \mathrm{Cliff}(V,\beta)$ be the Lie algebra homomorphism in Lemma 6.2.2. Set $\pi(X) = \tilde{\gamma}(\varphi(X))$ for $X \in \mathfrak{so}(V,\beta)$. Since $\varphi(X)$ is an even element in the Clifford algebra, equation (6.15) implies that $\pi(X)$ preserves the even and odd subspaces $C^\pm(W)$. We define

$$\pi^\pm(X) = \pi(X)|_{C^\pm(W)},$$

and call $\pi^\pm$ the *half-spin representations* of $\mathfrak{so}(V,\beta)$. Notice that the labeling of these representations by $\pm$ depends on a particular choice of the space of spinors. In both cases the representation space has dimension 2^{l-1} when $\dim V = 2l$.

Proposition 6.2.3. ($\dim V = 2l$) *The half-spin representations $\pi^\pm$ of $\mathfrak{so}(V,\beta)$ are irreducible with highest weights $\varpi_\pm = (\varepsilon_1 + \cdots + \varepsilon_{l-1} \pm \varepsilon_l)/2$. The weights are*

$$(\pm\varepsilon_1 \pm \cdots \pm \varepsilon_l)/2 \tag{6.16}$$

(an even number of minus signs for π^+ and an odd number of minus signs for π^-), and each weight has multiplicity one.

Proof. Take a β-isotropic basis $e_{\pm i}$ as in Section 2.4.1, and for each ordered index $I = \{1 \le i_1 < \cdots < i_p \le l\}$ set $u_I = e_{-i_1} \wedge \cdots \wedge e_{-i_p}$ (with $u_\emptyset = 1$). The diagonal subalgebra $\mathfrak{h}$ of $\mathfrak{g} = \mathfrak{so}(V,\beta)$ has basis $\{R_{e_i, e_{-i}} : i = 1, \ldots, l\}$, and from (6.14) we have

$$\pi(R_{e_i, e_{-i}}) = \iota(e_i)\varepsilon(e_{-i}) - (1/2).$$

Hence the action of $\mathfrak{h}$ on u_I is given by

$$\pi(R_{e_i, e_{-i}}) u_I = \begin{cases} -(1/2)u_I & \text{if } i \in I, \\ (1/2)u_I & \text{if } i \notin I. \end{cases}$$

Thus we see that u_I is a weight vector of weight

$$\lambda_I = \frac{1}{2}\left\{\sum_{i\notin I}\varepsilon_i - \sum_{i\in I}\varepsilon_i\right\}.$$

This formula shows that the weight λ_I uniquely determines the index I; hence each weight has multiplicity one. There are precisely two dominant weights, namely $\varpi_+ = \lambda_\emptyset$ and $\varpi_- = \lambda_{\{l\}}$ (note that $\varpi_- = \varpi_{l-1}$ and $\varpi_+ = \varpi_l$ in the notation of Section 3.1.4). Indeed, for any other choice of I there exists $i < l$ such that the co-efficient of ε_i in λ_I is negative, and so λ_I is not dominant. It follows from Theorem 4.2.12 that the $\mathfrak{g}$-cyclic subspaces generated by 1 and by e_{-l} are irreducible, and $C^\bullet(W)$ is the direct sum of these two subspaces. However, $\pi(U(\mathfrak{g}))1 \subset C^+(W)$ and $\pi(U(\mathfrak{g}))e_{-l} \subset C^-(W)$, so in fact $C^+(W)$ and $C^-(W)$ must coincide with these cyclic subspaces and hence are irreducible. $\qquad\square$

Now assume $\dim V = 2l+1$. Fix a decomposition

$$V = W \oplus \mathbb{C}e_0 \oplus W^*$$

with W and W^* maximal β-isotropic subspaces, as in Section 6.1.2. Take the space of spinors $(C^\bullet(W), \gamma_+)$ defined in the proof of Theorem 6.1.3. Define a representation of $\mathfrak{so}(V,\beta)$ on $C^\bullet(W)$ by $\pi = \widetilde{\gamma}_+ \circ \varphi$, where $\varphi : \mathfrak{so}(V,\beta) \longrightarrow \mathrm{Cliff}(V,\beta)$ is the homomorphism in Lemma 6.2.2 and $\widetilde{\gamma}_+$ is the canonical extension of γ_+ to a representation of $\mathrm{Cliff}(V,\beta)$ on $C^\bullet(W)$. We call π the *spin representation* of $\mathfrak{so}(V,\beta)$. The representation space has dimension 2^l when $\dim V = 2l+1$.

Proposition 6.2.4. ($\dim V = 2l+1$) *The spin representation of $\mathfrak{so}(V,\beta)$ is irreducible and has highest weight $\varpi_l = (\varepsilon_1 + \cdots + \varepsilon_{l-1} + \varepsilon_l)/2$. The weights are*

$$(\pm\varepsilon_1 \pm \cdots \pm \varepsilon_l)/2 , \qquad\qquad (6.17)$$

and each weight has multiplicity one.

Proof. From the definition of γ_+, we see that the diagonal subalgebra of $\mathfrak{so}(V,\beta)$ has the same action as in the even-dimensional case treated in Proposition 6.2.3. Thus the weights are given by (6.17) and have multiplicity one. The only dominant weight is ϖ_l, so π is irreducible by Corollary 3.3.14. $\qquad\square$

6.2.3 Exercises

1. Recall the isomorphism $\mathfrak{so}(3,\mathbb{C}) \cong \mathfrak{sl}(2,\mathbb{C})$ from Theorem 2.2.2. Show that the spin representation of $\mathfrak{so}(3,\mathbb{C})$ is the representation of $\mathfrak{sl}(2,\mathbb{C})$ on $\mathbb{C}^2$.
2. Recall the isomorphism $\mathfrak{so}(4,\mathbb{C}) \cong \mathfrak{sl}(2,\mathbb{C})\oplus\mathfrak{sl}(2,\mathbb{C})$ from Theorem 2.2.2. Show that the half-spin representations of $\mathfrak{so}(4,\mathbb{C})$ are the two representations $x\oplus y \mapsto x$ and $x\oplus y \mapsto y$ of $\mathfrak{sl}(2,\mathbb{C})\oplus\mathfrak{sl}(2,\mathbb{C})$ on $\mathbb{C}^2$. (HINT: The irreducible representations of $\mathfrak{sl}(2,\mathbb{C})\oplus\mathfrak{sl}(2,\mathbb{C})$ are outer tensor products of irreducible representations of $\mathfrak{sl}(2,\mathbb{C})$.)

3. Let $\mathfrak{g}$ be a simple Lie algebra and let $\mathfrak{g} = \mathfrak{n}^- + \mathfrak{h} + \mathfrak{n}^+$ as in Corollary 2.5.25. Let $l = \dim \mathfrak{h}$ be the rank of $\mathfrak{g}$ and let $B(X,Y) = \mathrm{tr}(\mathrm{ad}\,X\,\mathrm{ad}\,Y)$ for $X, Y \in \mathfrak{g}$ be the Killing form. Then B is a nondegenerate symmetric form on $\mathfrak{g}$, and $\mathrm{ad} : \mathfrak{g} \longrightarrow \mathfrak{so}(\mathfrak{g}, B)$.

   (a) Set $W = \mathfrak{n}^+ + \mathfrak{u}$, where $\mathfrak{u}$ is a maximal B-isotropic subspace in $\mathfrak{h}$. Show that W is a maximal B-isotropic subspace of $\mathfrak{g}$. (HINT: The weights of $\mathrm{ad}(\mathfrak{h})$ on W are the positive roots with multiplicity one and 0 with multiplicity $[l/2]$.)

   (b) Let π be the spin representation of $\mathfrak{so}(\mathfrak{g}, B)$ if l is odd or either of the half-spin representations of $\mathfrak{so}(\mathfrak{g}, B)$ if l is even. Show that the representation $\pi \circ \mathrm{ad}$ of $\mathfrak{g}$ is $2^{[l/2]}$ copies of the irreducible representation of $\mathfrak{g}$ with highest weight ρ (see Lemma 3.1.21). (HINT: Use (a) and Propositions 6.2.3 and 6.2.4 to show that ρ is the only highest weight of $\pi \circ \mathrm{ad}$ and that it occurs with multiplicity $2^{[l/2]}$. Now apply Theorem 4.2.12.)

6.3 Spin Groups

We now study the action of the orthogonal group $\mathrm{O}(V, \beta)$ as automorphisms of the algebra $\mathrm{Cliff}(V, \beta)$.

6.3.1 Action of O(V) on Cliff(V)

Let $g \in \mathrm{O}(V, \beta)$. The defining relations for the Clifford algebra imply that the map $\gamma(v) \mapsto \gamma(gv)$, for $v \in V$, extends to an automorphism of $\mathrm{Cliff}(V, \beta)$. We shall show that this automorphism is given by an invertible element $\widetilde{g}$ in $\mathrm{Cliff}(V, \beta)$. The proof will depend on the following elementary geometric result:

Lemma 6.3.1. *Suppose $x, y \in V$ and $\beta(x,x) = \beta(y,y) \neq 0$.*

1. *If $x - y$ is nonisotropic, then $s_{x-y}x = y$.*
2. *If $x - y$ is isotropic, then $x + y$ is nonisotropic and $s_y s_{x+y} x = y$.*
3. *$\mathrm{O}(V, \beta)$ is generated by reflections.*

Proof. We may assume that $\beta(x,x) = 1$. Obviously,

$$2\beta(x, x-y) = \beta(x-y, x-y),$$

since $\beta(x,x) = \beta(y,y)$. This implies (1). Now assume that $x - y$ is isotropic. Then

$$2\beta(x,y) = \beta(x,x) + \beta(y,y) = 2\beta(x,x) = 2,$$

so $\beta(x+y, x+y) = 4\beta(x,x) = 4$ and $\beta(x+y, x) = 2$. Thus

$$s_{x+y}x = x - \frac{2\beta(x+y,x)}{\beta(x+y,x+y)}(x+y) = -y \,,$$

which implies (2).

Let $g \in \mathbf{O}(V,\beta)$. Take a β-orthonormal basis $\{v_i : i = 1,\ldots,n\}$ for V and set $y = gv_n$. By (1) and (2) there is an $r \in \mathbf{O}(V,\beta)$ that is a product of at most two reflections, such that $rv_n = y$. Hence $r^{-1}gv_n = v_n$. Since $r^{-1}g$ preserves $(v_n)^\perp$, we may assume by induction on n that $r^{-1}g$ is a product of reflections. Hence g is also a product of reflections, proving (3). $\square$

On $\mathrm{Cliff}(V,\beta)$ there is the *main antiautomorphism* τ (transpose) that acts by

$$\tau(\gamma(v_1)\cdots\gamma(v_p)) = \gamma(v_p)\cdots\gamma(v_1) \quad \text{for } v_i \in V \,.$$

(It follows from Propositions 6.1.5 and 6.1.6 that this formula unambiguously defines τ.) We define the *conjugation* $u \mapsto u^*$ on $\mathrm{Cliff}(V,\beta)$ by $u^* = \tau(\alpha(u))$, where α is the main involution (see Section 6.1.1). For $v_1,\ldots,v_p \in V$ we have

$$(\gamma(v_1)\cdots\gamma(v_p))^* = (-1)^p\gamma(v_p)\cdots\gamma(v_1) \,.$$

In particular,

$$\gamma(v)^* = -\gamma(v), \qquad \gamma(v)\gamma(v)^* = -\frac{1}{2}\beta(v,v) \quad \text{for } v \in V \,. \qquad (6.18)$$

Suppose $v \in V$ is nonisotropic and normalized so that $\beta(v,v) = -2$. Then

$$\gamma(v)\gamma(v)^* = \gamma(v)^*\gamma(v) = 1 \,,$$

so we see that $\gamma(v)$ is an invertible element of $\mathrm{Cliff}(V,\beta)$ with $\gamma(v)^{-1} = \gamma(v)^*$. Furthermore, for $y \in V$ we can use the Clifford relations to write

$$\gamma(v)\gamma(y)\gamma(v) = (\beta(v,y) - \gamma(y)\gamma(v))\gamma(v)$$
$$= \gamma(y) + \beta(v,y)\gamma(v) = \gamma(s_v y) \,,$$

where $s_v y = y + \beta(v,y)v$ is the orthogonal reflection of y through the hyperplane $(v)^\perp$. Note that

$$\gamma(v)\gamma(y)\gamma(v) = \alpha(\gamma(v))\gamma(y)\gamma(v)^* \,.$$

Thus the $*$-twisted conjugation by $\gamma(v)$ on the Clifford algebra corresponds to the reflection s_v on V.

In general, we define

$$\mathbf{Pin}(V,\beta) = \{x \in \mathrm{Cliff}(V,\beta) : x \cdot x^* = 1 \text{ and } \alpha(x)\gamma(V)x^* = \gamma(V)\} \,.$$

Since $\mathrm{Cliff}(V,\beta)$ is finite-dimensional, the condition $x \cdot x^* = 1$ implies that x is invertible, with $x^{-1} = x^*$. Thus $\mathbf{Pin}(V,\beta)$ is a subgroup of the group of invertible elements of $\mathrm{Cliff}(V,\beta)$. The defining conditions are given by polynomial equations in the components of x, so $\mathbf{Pin}(V,\beta)$ is a linear algebraic group. The calculation above shows that $\gamma(v) \in \mathbf{Pin}(V,\beta)$ when $v \in V$ and $\beta(v,v) = -2$.

Theorem 6.3.2. *There is a unique regular homomorphism*

$$\pi : \mathbf{Pin}(V,\beta) \longrightarrow \mathbf{O}(V,\beta)$$

such that $\alpha(x)\gamma(v)x^* = \gamma(\pi(x)v)$ *for* $v \in V$ *and* $x \in \mathbf{Pin}(V,\beta)$. *Furthermore,* π *is surjective and* $\mathrm{Ker}(\pi) = \pm 1$.

Proof. Let $x \in \mathbf{Pin}(V,\beta)$. Since $\gamma : V \longrightarrow \mathrm{Cliff}(V,\beta)$ is injective, there is a unique transformation $\pi(x) \in \mathbf{GL}(V)$ such that $\alpha(x)\gamma(v)x^* = \gamma(\pi(x)v)$ for $v \in V$. Clearly the map $x \mapsto \pi(x)$ is a regular representation of $\mathbf{Pin}(V,\beta)$. Furthermore, for $v \in V$ and $x \in \mathbf{Pin}(V,\beta)$, we can use (6.18) to write

$$\beta(\pi(x)v, \pi(x)v) = -2\alpha(x)\gamma(v)x^*x\gamma(v)^*\alpha(x)^*$$
$$= \beta(v,v)\alpha(xx^*) = \beta(v,v),$$

since $x^*x = x^*x = 1$. This shows that $\pi(x) \in \mathbf{O}(V,\beta)$.

Suppose $x \in \mathrm{Ker}(\pi)$. We shall prove that x is a scalar λ, which by the condition $x^*x = 1$ will imply that $\lambda = \pm 1$ as claimed. Write $x = x_0 + x_1$ with $x_0 \in \mathrm{Cliff}^+(V,\beta)$ and $x_1 \in \mathrm{Cliff}^-(V,\beta)$. Then $\alpha(x) = x_0 - x_1$. But $x^* = x^{-1}$, so $\alpha(x)\gamma(v) = \gamma(v)x$ for all $v \in V$. Thus

$$(x_0 - x_1)\gamma(v) = \gamma(v)(x_0 + x_1). \tag{6.19}$$

Comparing even and odd components in (6.19), we see that x_0 is in the center of $\mathrm{Cliff}(V,\beta)$ and that $x_1\gamma(v) = -\gamma(v)x_1$ for all $v \in V$. In particular, x_1 commutes with the Lie algebra $\mathfrak{g} = \varphi(\mathfrak{so}(V,\beta))$ of Lemma 6.2.2. Suppose $\dim V$ is even. In this case $\mathrm{Cliff}(V,\beta)$ is a simple algebra, so x_0 is a scalar λ. Also, $\widetilde{\gamma}(x_1) = 0$ by Schur's lemma, since $\widetilde{\gamma}(x_1)$ is a $\mathfrak{g}$ intertwining operator between the half-spin spaces, which are inequivalent as $\mathfrak{g}$-modules. Now suppose $\dim V$ is odd. Since x_0 is in the center of $\mathrm{Cliff}^+(V,\beta)$, it must be a scalar by Lemma 6.1.7. On the other hand, by Schur's lemma again $\widetilde{\gamma}_\pm(x_1) = \mu_\pm I$ for some $\mu_\pm \in \mathbb{C}$. But x_1 anticommutes with e_0 and $\widetilde{\gamma}_\pm(e_0)$ is invertible, so $\mu_\pm = 0$. Hence $x = x_0 = \lambda$.

It remains only to show that π is surjective. Set $G = \pi(\mathbf{Pin}(V,\beta))$ and let s_v be the orthogonal reflection on V determined by a nonisotropic vector v. We have already calculated that G contains all such reflections. But $\mathbf{O}(V,\beta)$ is generated by reflections, by Lemma 6.3.1, so this completes the proof of Theorem 6.3.2. □

Since $\mathbf{O}(V,\beta)$ is generated by reflections, the surjectivity of the map π implies the following alternative description of the **Pin** group:

Corollary 6.3.3. *The elements* $-I$ *and* $\gamma(v)$, *with* $v \in V$ *and* $\beta(v,v) = -2$, *generate the group* $\mathbf{Pin}(V,\beta)$.

Finally, we introduce the spin group. Assume $\dim V \geq 3$. Define

$$\mathbf{Spin}(V,\beta) = \mathbf{Pin}(V,\beta) \cap \mathrm{Cliff}^+(V,\beta).$$

When $\dim V = 2l$ we fix a β-isotropic basis $\{e_1, \ldots, e_l, e_{-1}, \ldots, e_{-l}\}$ for V. Thus

$$\beta(e_i, e_j) = \delta_{i+j} \quad \text{for } i, j = \pm 1, \ldots, \pm l. \tag{6.20}$$

When $\dim V = 2l + 1$ we fix a β-isotropic basis $\{e_0, e_1, \ldots, e_l, e_{-1}, \ldots, e_{-l}\}$ for V. Thus (6.20) holds, $\beta(e_0, e_j) = 0$ for $j \neq 0$, and $\beta(e_0, e_0) = 1$. For $j = 1, \ldots, l$ and $z \in \mathbb{C}^\times$, define

$$c_j(z) = z\gamma(e_j)\gamma(e_{-j}) + z^{-1}\gamma(e_{-j})\gamma(e_j).$$

For $z = [z_1, \ldots, z_l] \in (\mathbb{C}^\times)^l$ set $c(z) = c_1(z_1) \cdots c_l(z_l)$.

Lemma 6.3.4. *The map* $z \mapsto c(z)$ *is a regular homomorphism from* $(\mathbb{C}^\times)^l$ *to* **Spin**(V, β). *The kernel of this homomorphism is*

$$K = \{(z_1, \ldots, z_l) : z_i = \pm 1 \text{ and } z_1 \cdots z_l = 1\},$$

and the image is an algebraic torus $\widetilde{H}$ *of rank l.*

Proof. In the tensor product model for the spin spaces, $c(z)$ is the operator

$$\begin{bmatrix} z_1 & 0 \\ 0 & z_1^{-1} \end{bmatrix} \otimes \cdots \otimes \begin{bmatrix} z_l & 0 \\ 0 & z_l^{-1} \end{bmatrix} \tag{6.21}$$

(see Section 6.1.3). Hence the map $z \mapsto c(z)$ is a regular homomorphism from $(\mathbb{C}^\times)^l$ to the group of invertible elements of $\text{Cliff}(V, \beta)$. If $c(z)$ is the identity, then each factor in the tensor product (6.21) is $\pm I$ and the overall product is the identity. This implies that $z \in K$.

We have $c_j(z) \in \text{Cliff}^+(V, \beta)$ by definition. Also,

$$c_j(z)^* = z\gamma(e_{-j})\gamma(e_j) + z^{-1}\gamma(e_j)\gamma(e_{-j}) = c_j(z^{-1}).$$

Hence $c_j(z)^* c_j(z) = 1$. From (6.21) and the tensor product realization from Section 6.1.3 in which $\gamma(e_i) = A_i$, we calculate that

$$c_j(z)\gamma(e_i)c_j(z)^* = \begin{cases} z^2\gamma(e_j) & \text{if } i = j, \\ z^{-2}\gamma(e_{-j}) & \text{if } i = -j, \\ \gamma(e_i) & \text{otherwise,} \end{cases} \tag{6.22}$$

for $j = \pm 1, \ldots, \pm l$ (and $c_j(z)\gamma(e_0)c_j(z)^* = \gamma(e_0)$ when $\dim V$ is odd). Thus $c_j(z) \in$ **Spin**(V, β) for $j = 1, \ldots, l$, which implies $c(z) \in$ **Spin**(V, β).

Let $\widetilde{H} = \{c(z) : z \in (\mathbb{C}^\times)^l\}$. Then $\widetilde{H} \cong (\mathbb{C}^\times)^l/K$ as an abelian group. The subgroup K can be described as the intersection of the kernels of the characters $z \mapsto w_j$ of $(\mathbb{C}^\times)^l$ for $j = 1, \ldots, l$, where we set

$$w_1 = z_1 z_2^{-1} z_3 \cdots z_l, \qquad w_2 = z_1 z_2 z_3^{-1} \cdots z_l, \quad \ldots,$$

$$w_{l-1} = z_1 \cdots z_{l-1} z_l^{-1}, \qquad w_l = z_1 \cdots z_l.$$

Thus we can view each character w_j as a function on $\widetilde{H}$. Using the relations

$$(z_1)^2 = \frac{w_1 w_2 \cdots w_{l-1}}{w_l^{l-3}}, \quad (z_2)^2 = \frac{w_l}{w_1}, \quad \ldots, \quad (z_l)^2 = \frac{w_l}{w_{l-1}}$$

in (6.21), we find that $c(z)$ is the transformation

$$\begin{bmatrix} (w_l)^{-l+2} & 0 \\ 0 & (w_1 \cdots w_l)^{-1} \end{bmatrix} \otimes \begin{bmatrix} w_l & 0 \\ 0 & w_1 \end{bmatrix} \otimes \begin{bmatrix} w_l & 0 \\ 0 & w_2 \end{bmatrix} \otimes \cdots \otimes \begin{bmatrix} w_l & 0 \\ 0 & w_{l-1} \end{bmatrix}$$

in the tensor product model for the spin spaces. Thus $\widetilde{H}$ consists of all the transformations of this form with $w_1, \ldots, w_l \in \mathbb{C}^\times$. This shows that $\widetilde{H}$ is an algebraic torus of rank l in $\mathbf{Spin}(V, \beta)$ with coordinate functions $w_1, \ldots, w_l$. $\qquad \square$

Let $H \subset \mathbf{SO}(V, \beta)$ be the maximal torus that is diagonalized by the β-isotropic basis $\{e_i\}$ for V.

Theorem 6.3.5. *The group* $\mathbf{Spin}(V, \beta)$ *is the identity component of* $\mathbf{Pin}(V, \beta)$, *and the homomorphism* $\pi : \mathbf{Spin}(V, \beta) \longrightarrow \mathbf{SO}(V, \beta)$ *is surjective with* $\mathrm{Ker}(\pi) = \{\pm 1\}$. *The subgroup* $\widetilde{H} = \pi^{-1}(H)$ *and*

$$\pi(c(z)) = \begin{cases} \mathrm{diag}[z_1^2, \ldots, z_l^2, z_l^{-2}, \ldots, z_1^{-2}] & (\dim V = 2l), \\ \mathrm{diag}[z_1^2, \ldots, z_l^2, 1, z_l^{-2}, \ldots, z_1^{-2}] & (\dim V = 2l+1). \end{cases}$$

Hence $\widetilde{H}$ *is a maximal torus in* $\mathbf{Spin}(V, \beta)$. *Each semisimple element of* $\mathbf{Spin}(V, \beta)$ *is conjugate to an element of* $\widetilde{H}$.

Proof. Since a reflection has determinant -1, $\mathbf{SO}(V, \beta)$ is generated by products of an even number of reflections. By Corollary 6.3.3, $\mathbf{Spin}(V, \beta)$ is generated by ± 1 and products of an even number of elements $\gamma(v)$ with $\beta(v, v) = -2$. Thus by Theorem 6.3.2, π is surjective with kernel ± 1. Since $\mathbf{SO}(V, \beta)$ is connected and $c(-1, \ldots, -1) = -I$, we have $-I$ in the identity component of $\mathbf{Spin}(V, \beta)$. Hence $\mathbf{Spin}(V, \beta)$ is connected.

The formula for $\pi(c(z))$ is immediate from (6.22), and it shows that $\pi(\widetilde{H}) = H$. Since $\mathrm{Ker}(\pi) \subset \widetilde{H}$, we have $\pi^{-1}(H) = \widetilde{H}$. The last statements of the theorem now follow from Theorems 2.1.5 and 2.1.7. $\qquad \square$

Theorem 6.3.6. *The Lie algebra of* $\mathbf{Spin}(V, \beta)$ *is* $\varphi(\mathfrak{so}(V, \beta))$, *where* φ *is the isomorphism of Lemma 6.2.2.*

Proof. Since $\mathbf{Spin}(V, \beta)$ is a subgroup of the invertible elements of $\mathrm{Cliff}(V, \beta)$, we may identify $\mathrm{Lie}(\mathbf{Spin}(V, \beta))$ with a Lie subalgebra $\mathfrak{g}$ of $\mathrm{Cliff}(V, \beta)$. We have

$$\dim \mathbf{Spin}(V, \beta) = \dim \mathbf{SO}(V, \beta)$$

by Theorem 6.3.5, so we only need to show that $\mathfrak{g} \supset \varphi(\mathfrak{so}(V, \beta))$.

We know from Theorem 2.4.11 that $\mathfrak{so}(V, \beta)$ is generated by $\mathfrak{n}^+$ and $\mathfrak{n}^-$. By Lemma 6.2.1 these subalgebras are spanned by elements $R_{x,y}$, where $x, y \in V$ satisfy

$$\beta(x,x) = \beta(x,y) = \beta(y,y) = 0$$

(Span$\{x,y\}$ is an isotropic subspace). Therefore, it suffices to show that $\varphi(R_{x,y}) \in \mathfrak{g}$. We have $\gamma(x)^2 = \gamma(y)^2 = 0$ and $\gamma(x)\gamma(y) = -\gamma(y)\gamma(x)$. Hence $\varphi(R_{x,y}) = \gamma(x)\gamma(y)$ and $(\gamma(x)\gamma(y))^2 = -\gamma(x)^2\gamma(y)^2 = 0$. Define

$$u(t) = \exp t\varphi(R_{x,y}) = I + t\gamma(x)\gamma(y)$$

for $t \in \mathbb{C}$. Then the map $t \mapsto u(t)$ is a one-parameter subgroup of unipotent elements in $\mathrm{Cliff}^+(V,\beta)$. It is easy to see that these elements are in $\mathbf{Spin}(V,\beta)$. Indeed, for $z \in V$ we have

$$\gamma(x)\gamma(y)\gamma(z)\gamma(x)\gamma(y) = -\gamma(x)\gamma(y)\gamma(x)\gamma(z)\gamma(y) + \gamma(x)\gamma(y)\beta(x,z)\gamma(y)$$
$$= \frac{1}{2}\beta(y,y)\beta(x,z)\gamma(x) = 0,$$

from which it follows that

$$u(t)\gamma(z)u(-t) = \gamma(z) + t[\gamma(x)\gamma(y),\gamma(z)] = \gamma(z) + t\gamma(R_{x,y}z)$$

is in V. Hence $u(t) \in \mathbf{Spin}(V,\beta)$, so $\varphi(R_{x,y}) \in \mathfrak{g}$, which completes the proof. $\qquad\square$

Corollary 6.3.7. *Let P be the weight lattice of $\mathfrak{so}(V,\beta)$. For $\lambda \in P_{++}$ there is an irreducible regular representation of $\mathbf{Spin}(V,\beta)$ with highest weight λ.*

Proof. The spin representation (when $\dim V$ is odd) or half-spin representations (when $\dim V$ is even) furnish the fundamental representations not obtainable from $\bigwedge V$ (cf. Theorem 5.5.13). Every regular representation of $\mathbf{SO}(V,\beta)$ gives a regular representation of $\mathbf{Spin}(V,\beta)$ via the covering map π of Theorem 6.3.5. Thus λ occurs as the highest weight of a suitable Cartan product of fundamental representations, by Corollary 5.5.20. $\qquad\square$

6.3.2 Algebraically Simply Connected Groups

Let G and H be connected linear algebraic groups, and let $\pi : H \longrightarrow G$ be a surjective regular homomorphism. We call π a *covering homomorphism* if $\mathrm{Ker}(\pi)$ is finite. The group G is said to be *algebraically simply connected* if every covering homomorphism $\pi : H \longrightarrow G$ is an isomorphism of algebraic groups.

Theorem 6.3.8. *Let G be a connected linear algebraic group and let $\mathfrak{g} = \mathrm{Lie}(G)$. Suppose every finite-dimensional representation of $\mathfrak{g}$ is the differential of a regular representation of G. Then G is algebraically simply connected.*

Proof. Let $\pi : H \longrightarrow G$ be a covering homomorphism, where $H \subset \mathbf{GL}(n,\mathbb{C})$ is a connected linear algebraic subgroup. We shall use a general result about linear algebraic groups that will be proved in Chapter 11; namely, that H is a connected

Lie group (Theorems 1.4.10 and 11.2.9). We have $\mathrm{Lie}(H) = \mathfrak{h} \subset \mathfrak{gl}(n,\mathbb{C})$, and since $\mathrm{Ker}(\pi)$ is finite, it has Lie algebra 0. Hence $d\pi : \mathfrak{h} \longrightarrow \mathfrak{g}$ is an isomorphism of Lie algebras. Thus the inverse of $d\pi$ furnishes a representation $\rho : \mathfrak{g} \longrightarrow \mathfrak{gl}(n,\mathbb{C})$. By assumption there is a regular representation $\sigma : G \longrightarrow \mathbf{GL}(n,\mathbb{C})$ such that $d\sigma = \rho$. We claim that

$$\pi \circ \sigma(g) = g \quad \text{for all } g \in G. \tag{6.23}$$

Indeed, if $X \in \mathfrak{g}$ then $\pi(\sigma(\exp X)) = \pi(\exp \rho(X)) = \exp X$ by Proposition 1.3.14. Since G is generated by $\{\exp X : X \in \mathfrak{g}\}$, this proves (6.23). The same argument shows that $\sigma \circ \pi(h) = h$ for all $h \in H$. Hence π is an isomorphism. □

Corollary 6.3.9. *Let G be $\mathbf{SL}(n,\mathbb{C})$ for $n \geq 2$, $\mathbf{Spin}(n,\mathbb{C})$ for $n \geq 3$, or $\mathbf{Sp}(n,\mathbb{C})$ for $n \geq 1$. Then G is algebraically simply connected. Furthermore, for every $\lambda \in P_{++}(\mathfrak{g})$ there is a unique irreducible regular representation of G with highest weight λ.*

Proof. We know that G is connected by Theorems 2.2.5 and 6.3.5. Let $\mathfrak{g} = \mathrm{Lie}(G)$. Then, by Theorem 3.3.12, every finite-dimensional representation of $\mathfrak{g}$ is completely reducible. Hence every finite-dimensional representation of $\mathfrak{g}$ is the differential of a regular representation of G by Theorem 5.5.21 and Corollary 6.3.7. Thus we may apply Theorem 6.3.8. The existence and uniqueness of an irreducible regular representation with highest weight λ follow by taking Cartan products of the fundamental representations, as in Section 5.5.3. □

Remark 6.3.10. If a linear algebraic group G is simply connected as a real Lie group (in the manifold topology), then G is algebraically simply connected. This can be proved as follows: With the notation of the proof of Theorem 6.3.8, one defines

$$\sigma(\exp X) = \exp \rho(X) \quad \text{for } X \in \mathfrak{g}.$$

The Campbell–Hausdorff formula can be used to show that there is a neighborhood U of 0 in $\mathfrak{g}$ such that

$$\sigma(\exp X \exp Y) = \sigma(\exp X)\sigma(\exp Y) \quad \text{for } X,Y \in U.$$

Since $\exp U$ generates G, we can apply the monodromy principle for simply connected Lie groups to conclude that σ extends to a representation of G (cf. Hochschild [68, Chapter IV, Theorem 3.1]). Thus $\mathrm{Ker}(\pi) = \{1\}$, so π is an isomorphism of Lie groups. It follows by Corollary 11.1.16 that π is also an isomorphism of algebraic groups.

6.3.3 Exercises

1. Establish the following isomorphisms of linear algebraic groups:
   (a) $\mathbf{Spin}(3,\mathbb{C}) \cong \mathbf{SL}(2,\mathbb{C})$ (b) $\mathbf{Spin}(4,\mathbb{C}) \cong \mathbf{SL}(2,\mathbb{C}) \times \mathbf{SL}(2,\mathbb{C})$
   (c) $\mathbf{Spin}(5,\mathbb{C}) \cong \mathbf{Sp}(2,\mathbb{C})$ (d) $\mathbf{Spin}(6,\mathbb{C}) \cong \mathbf{SL}(4,\mathbb{C})$
   (HINT: For (a) and (b) see Section 2.2.2; for (c) and (d) see Exercises 2.4.5.)

2. Let $G = H = \mathbb{C}^\times$ and let $p > 1$ be an integer. Show that $z \mapsto z^p$ is a nontrivial covering homomorphism from H to G; hence $\mathbb{C}^\times$ is not algebraically simply connected.

3. Let G be the group of upper-triangular 2×2 unipotent matrices. Prove that G is algebraically simply connected.

4. Let $H = \mathbb{C}^\times \times \mathbf{SL}(n, \mathbb{C})$ with $n \geq 2$. Define $\pi : H \longrightarrow \mathbf{GL}(n, \mathbb{C})$ by $\pi(z, g) = zg$ for $z \in \mathbb{C}^\times$ and $g \in \mathbf{SL}(n, \mathbb{C})$. Prove that π is a nontrivial covering homomorphism; hence $\mathbf{GL}(n, \mathbb{C})$ is not algebraically simply connected.

5. Let $V = \mathbb{C}^n$ with nondegenerate bilinear form β. Let $\mathcal{C} = \mathrm{Cliff}(V, \beta)$ and identify V with $\gamma(V) \subset \mathcal{C}$ by the canonical map γ. Let α be the automorphism of $\mathcal{C}$ such that $\alpha(v) = -v$ for $v \in V$, let τ be the antiautomorphism of $\mathcal{C}$ such that $\tau(v) = v$ for $v \in V$, and let $x \mapsto x^*$ be the antiautomorphism $\alpha \circ \tau$ of $\mathcal{C}$. Define the *norm function* $\Delta : \mathcal{C} \longrightarrow \mathcal{C}$ by $\Delta(x) = x^* x$. Let $\mathcal{L} = \{x \in \mathcal{C} : \Delta(x) \in \mathbb{C}\}$.
   (a) Show that $\lambda + v \in \mathcal{L}$ for all $\lambda \in \mathbb{C}$ and $v \in V$.
   (b) Show that if $x, y \in \mathcal{L}$ and $\lambda \in \mathbb{C}$, then $\lambda x \in \mathcal{L}$ and

$$\Delta(xy) = \Delta(x)\Delta(y) , \quad \Delta(\tau(x)) = \Delta(\alpha(x)) = \Delta(x^*) = \Delta(x) .$$

Hence $xy \in \mathcal{L}$ and $\mathcal{L}$ is invariant under τ and α. Prove that $x \in \mathcal{L}$ is invertible if and only if $\Delta(x) \neq 0$. In this case $x^{-1} = \Delta(x)^{-1} x^*$ and $\Delta(x^{-1}) = 1/\Delta(x)$.
   (c) Let $\Gamma(V, \beta) \subset \mathcal{L}$ be the set of all products $w_1 \cdots w_k$, where $w_j \in \mathbb{C} + V$ and $\Delta(w_j) \neq 0$ for all $1 \leq j \leq k$ (k arbitrary). Prove that $\Gamma(V, \beta)$ is a group (under multiplication) that is stable under α and τ.
   (d) Prove that if $g \in \Gamma(V, \beta)$ then $\alpha(g)(\mathbb{C} + V)g^* = \mathbb{C} + V$. ($\Gamma(V, \beta)$ is called the *Clifford group*; note that it contains $\mathbf{Pin}(V, \beta)$.)

6.4 Real Forms of **Spin**$(n, \mathbb{C})$

Nondegenerate bilinear forms on a real vector space are classified by their signature, which determines the form up to equivalence. In this section we study the associated real Clifford algebras and spin groups as real forms of the complex algebras and groups.

6.4.1 Real Forms of Vector Spaces and Algebras

Let V be an n-dimensional vector space over $\mathbb{C}$. A *real form* of V is an $\mathbb{R}$-subspace V_0 of V (looked upon as a vector space over $\mathbb{R}$) such that

$$V = V_0 \oplus i V_0 \tag{6.24}$$

as a real vector space. We note that (6.24) implies that $V_0 \cap iV_0 = \{0\}$ and $\dim_{\mathbb{R}} V_0 = n$. For example, let $V = \mathbb{C}^n$ and take $V_0 = \mathbb{R}^n$, considered as an $\mathbb{R}$-subspace of $\mathbb{C}^n$ using the inclusion $\mathbb{R} \subset \mathbb{C}$.

Lemma 6.4.1. *Let V_0 be an n-dimensional real subspace of V such that $V_0 \cap iV_0 = \{0\}$. Then V_0 is a real form of V.*

Proof. Since $\dim_{\mathbb{R}} V = 2n$ and $V_0 \cap iV_0 = \{0\}$, one has $\dim_{\mathbb{R}}(V_0 + iV_0) = 2n$. Thus $V = V_0 \oplus iV_0$. $\qquad\qquad\square$

If $\mathcal{A}$ is a finite-dimensional algebra over $\mathbb{C}$ then a *real form* of $\mathcal{A}$ as an algebra is a real form $\mathcal{A}_0$ of $\mathcal{A}$ as a vector space such that $\mathcal{A}_0$ is a real subalgebra of $\mathcal{A}$ (considered as an algebra over $\mathbb{R}$). For example, let $\mathcal{A} = M_n(\mathbb{C})$ and $\mathcal{A}_0 = M_n(\mathbb{R})$.

Lemma 6.4.2. *Let $\mathcal{A}$ be an n-dimensional algebra over $\mathbb{C}$. Suppose that $e_1, \ldots, e_n$ is a basis of $\mathcal{A}$ over $\mathbb{C}$ such that*

$$e_i e_j = \sum_k a_{ij}^k e_k \quad \text{with } a_{ij}^k \in \mathbb{R} \text{ for all } i, j, k. \tag{6.25}$$

Then $\mathcal{A}_0 = \mathrm{Span}_{\mathbb{R}}\{e_1, \ldots, e_n\}$ is a real form of $\mathcal{A}$ as an algebra.

This is just a direct reformulation in terms of bases of the definition of real form of an algebra over $\mathbb{C}$.

If V is a finite-dimensional vector space over $\mathbb{C}$ and if V_0 is a real form then from (6.24) we may define an $\mathbb{R}$-linear isomorphism $\sigma : V \to V$ by

$$\sigma(v_0 + iw_0) = v_0 - iw_0 \quad \text{for } v_0, w_0 \in V_0.$$

We will call σ the *complex conjugation on V corresponding to V_0*.

If V is a vector space over $\mathbb{C}$ and if σ is an $\mathbb{R}$-linear endomorphism of V such that $\sigma^2 = I$ and $\sigma(zv) = \bar{z}\sigma(v)$ for $z \in \mathbb{C}$, $v \in V$, then we will call σ a *complex conjugation* on V. If σ is a complex conjugation on V, set

$$V_0 = \{v \in V : \sigma(v) = v\}.$$

Clearly, V_0 is a real subspace of V. Since $\sigma(iv) = -i\sigma(v)$, we see that $V = V_0 \oplus iV_0$. Thus V_0 is a real form and σ is the conjugation on V corresponding to V_0.

If $\mathcal{A}$ is a finite-dimensional algebra over $\mathbb{C}$ and σ is an automorphism of $\mathcal{A}$ as an algebra over $\mathbb{R}$ such that σ is a conjugation of $\mathcal{A}$ as a vector space, then σ is called a *conjugation* on $\mathcal{A}$ (as an algebra). As in the case of vector spaces, if we set

$$\mathcal{A}_0 = \{a \in \mathcal{A} : \sigma(a) = a\},$$

then $\mathcal{A}_0$ is a real subalgebra of $\mathcal{A}_0$ that is a real form of $\mathcal{A}$ as an algebra.

6.4.2 Real Forms of Clifford Algebras

Let V be a finite-dimensional complex vector space and let β be a nondegenerate symmetric bilinear form on V. We say that the pair (V_0, β_0) is a *real form* of (V, β) if

1. V_0 is a real form of the vector space V, and
2. $\beta_0 = \beta|_{V_0 \times V_0}$ is a real-valued nondegenerate bilinear form on V_0.

Example

Let $V = \mathbb{C}^n$ and $\beta(z, w) = z_1 w_1 + \cdots + z_n w_n$. Fix $1 \le p \le n$. Let $\{e_1, \ldots, e_n\}$ be the standard basis of $\mathbb{C}^n$ and define

$$
f_j = \begin{cases} i e_j & \text{for } j = 1, \ldots, p, \\ e_j & \text{for } j = p+1, \ldots, n. \end{cases}
$$

Let $V_{0,p} = \mathrm{Span}_{\mathbb{R}}\{f_1, \ldots, f_n\}$ and set $\beta_{0,p} = \beta|_{V_{0,p} \times V_{0,p}}$. Lemma 6.4.1 implies that $V_{0,p}$ is a real form of V. Since $\beta(f_j, f_k) = \pm \delta_{j,k}$, we see that $(V_{0,p}, \beta_{0,p})$ is a real form of (V, β). Define a bilinear form h_p on $\mathbb{R}^n$ by

$$
h_p(x, y) = -x_1 y_1 - \cdots - x_p y_p + x_{p+1} y_{p+1} + \cdots + x_n y_n,
$$

and let $T : V_{0,p} \longrightarrow \mathbb{R}^n$ by $T\left(\sum x_j f_j\right) = [x_1, \ldots, x_n]$. Then $\beta(v, w) = h_p(Tv, Tw)$ for $v, w \in V_0$. Thus $\beta_{0,p}$ has signature $(n - p, p)$; in particular, $\beta_{0,n}$ is negative definite.

Let (V_0, β_0) be a real form of (V, β) and let $\{e_1, \ldots, e_n\}$ be a basis of V_0 over $\mathbb{R}$. Then $\{e_1, \ldots, e_n\}$ is a basis of V over $\mathbb{C}$. Thus

$$
\{\gamma(e_{i_1}) \cdots \gamma(e_{i_r}) : 0 \le r \le n, 1 \le i_1 < \cdots < i_r \le n\}
$$

is a basis of $\mathrm{Cliff}(V, \beta)$ (see Propositions 6.1.5 and 6.1.6). We also note that since

$$
\gamma(e_i)\gamma(e_j) + \gamma(e_j)\gamma(e_i) = \beta(e_i, e_j),
$$

this basis satisfies (6.25); hence the $\mathbb{R}$-span of this basis is a real form of $\mathrm{Cliff}(V, \beta)$. We note that this real form is just the subalgebra generated over $\mathbb{R}$ by $\{1, \gamma(V_0)\}$ and so it is independent of the choice of basis. We will denote this subalgebra by $\mathrm{Cliff}(V_0, \beta_0)$.

6.4.3 Real Forms of **Pin**(n) and **Spin**(n)

Let (V_0, β_0) be a real form of (V, β). We will use the same notation σ for the complex conjugation on V with respect to V_0 and the complex conjugation on $\mathrm{Cliff}(V, \beta)$ with respect to $\mathrm{Cliff}(V_0, \beta_0)$. We note that $\sigma(\gamma(v)) = \gamma(\sigma(v))$. We observe that

$$\sigma(\gamma(v)) = \gamma(\sigma(v)) \quad \text{for } v \in V. \tag{6.26}$$

Let $u \mapsto u^*$ be defined as in Section 6.3.1. Then a direct calculation using (6.26) shows that

$$\sigma(u)^* = \sigma(u^*) \quad \text{for } u \in \mathrm{Cliff}(V, \beta). \tag{6.27}$$

We also note that

$$\sigma(1) = 1 \quad \text{and} \quad \sigma(\mathrm{Cliff}^+(V, \beta)) = \mathrm{Cliff}^+(V, \beta). \tag{6.28}$$

Thus (6.27) and (6.28) imply that $\sigma(\mathbf{Pin}(V, \beta)) = \mathbf{Pin}(V, \beta)$ and $\sigma(\mathbf{Spin}(V, \beta)) = \mathbf{Spin}(V, \beta)$ (see Section 6.3.1). Using the description of the linear algebraic group structures on $\mathbf{Pin}(V, \beta)$ and $\mathbf{Spin}(V, \beta)$ in Section 6.3.1, we see that $\sigma|_{\mathbf{Pin}(V,\beta)}$ and $\sigma|_{\mathbf{Spin}(V,\beta)}$ define complex conjugations of $\mathbf{Pin}(V, \beta)$ and $\mathbf{Spin}(V, \beta)$ as algebraic groups (see Section 1.7.1). We will denote the corresponding real forms of $\mathbf{Pin}(V, \beta)$ and $\mathbf{Spin}(V, \beta)$ by $\mathbf{Pin}(V_0, \beta_0)$ and $\mathbf{Spin}(V_0, \beta_0)$, respectively.

Theorem 6.4.3. *If β_0 is negative definite on V_0, then the groups $\mathbf{Pin}(V_0, \beta_0)$ and $\mathbf{Spin}(V_0, \beta_0)$ are compact real forms of $\mathbf{Pin}(V, \beta)$ and $\mathbf{Spin}(V, \beta)$, respectively.*

Proof. Let $\{e_1, \ldots, e_n\}$ be a basis of V_0 such that $\beta(e_i, e_j) = -\delta_{ij}$. In this basis V_0 is identified with $\mathbb{R}^n$ and $\beta_0(x, y) = -(x_1 y_1 + \cdots + x_n y_n)$. Then V becomes $\mathbb{C}^n$ and $\beta(z, w) = -(z_1 w_1 + \cdots + z_n w_n)$. With these conventions we define

$$\mathbf{Pin}(V, \beta) = \mathbf{Pin}(n, \mathbb{C}) \quad \text{and} \quad \mathbf{Spin}(V, \beta) = \mathbf{Spin}(n, \mathbb{C}).$$

If $u \in \mathbf{Pin}(V_0, \beta_0)$ and $v_0, w_0 \in V_0$, then $u\gamma(v_0)u^* = \gamma(\pi(u)v_0)$, $\pi(u)v_0 \in V_0$, and $(\pi(u)v_0, \pi(u)w_0) = (v_0, w_0)$. Thus $\pi : \mathbf{Pin}(V_0, \beta_0) \longrightarrow \mathbf{O}(n)$. The obvious variant of Lemma 6.3.1 implies that $\pi(\mathbf{Pin}(V_0, \beta_0)) = \mathbf{O}(n)$ and that

$$\mathrm{Ker}(\pi|_{\mathbf{Pin}(V_0, \beta_0)}) = \{\pm I\}.$$

Thus $\mathbf{Pin}(V_0, \beta_0)$ is compact. Since $\mathbf{Spin}(V_0, \beta_0)$ is closed in $\mathbf{Pin}(V_0, \beta_0)$, it is also compact. $\square$

Following the proof above we will use the notation $\mathbf{Pin}(n)$ for $\mathbf{Pin}(V_0, \beta_0)$ and $\mathbf{Spin}(n)$ for $\mathbf{Spin}(V_0, \beta_0)$ when $V = \mathbb{C}^n$. More generally, we write

$$\mathbf{Pin}(p, q) = \mathbf{Pin}(V_{0,p}, \beta_{0,p}), \qquad \mathbf{Spin}(p, q) = \mathbf{Spin}(V_{0,p}, \beta_{0,p})$$

in the notation of Section 6.4.2, where $q = n - p$.

6.4.4 Exercises

1. Let $\mathcal{A} = M_2(\mathbb{C})$. Let $J = \left[\begin{smallmatrix} 0 & 1 \\ -1 & 0 \end{smallmatrix}\right] \in M_2(\mathbb{C})$. Let $\sigma(X) = -J\overline{X}J$ with $\overline{X}$ given by complex conjugation of the matrix entries. Show that σ is a complex conjugation of $\mathcal{A}$ as an algebra and that the corresponding real form is isomorphic to the quaternions (see Section 1.1.4).

2. Let $V = \mathbb{C}^2$. Take $\beta_0 = \beta_{0,2}$ and $V_0 = V_{0,2}$ as in the example of Section 6.4.2. Show that $\mathrm{Cliff}(V_0, \beta_0)$ is isomorphic to the quaternions (see Section 1.1.4 and the previous exercise).

3. Let $\mathbb{F} = \mathbb{R}$, $\mathbb{C}$, or $\mathbb{H}$ and let $\mathcal{A} = M_k(\mathbb{F})$ be the $k \times k$ matrices over $\mathbb{F}$, viewed as an algebra over $\mathbb{R}$ relative to matrix addition and multiplication.
   (a) Show that $\mathcal{A}$ is a simple algebra over $\mathbb{R}$.
   (b) Let $n = 2l \leq 6$. Let $V_{0,p}$ be the real form of $\mathbb{C}^n$ as in the example of Section 6.4.2. Show that if $p = 0, 1, \ldots, l$ then there exists $\mathbb{F}$ as above such that $\mathrm{Cliff}(V_{0,p}, \beta_{0,p}) \cong M_k(\mathbb{F})$ for some k.

4. Verify the following isomorphisms of compact Lie groups:
   (a) $\mathbf{Spin}(3) \cong \mathbf{SU}(2)$ (b) $\mathbf{Spin}(4) \cong \mathbf{SU}(2) \times \mathbf{SU}(2)$
   (c) $\mathbf{Spin}(5) \cong \mathbf{Sp}(2)$ (d) $\mathbf{Spin}(6) \cong \mathbf{SU}(4)$

5. Verify the following isomorphisms of Lie groups, where G° denotes the identity component of G as a Lie group:
   (a) $\mathbf{Spin}(1,2)^\circ \cong \mathbf{SL}(2, \mathbb{R})$ (b) $\mathbf{Spin}(2,2)^\circ \cong \mathbf{SL}(2, \mathbb{R}) \times \mathbf{SL}(2, \mathbb{R})$
   (c) $\mathbf{Spin}(1,3)^\circ \cong \mathbf{SL}(2, \mathbb{C})$ (d) $\mathbf{Spin}(3,2)^\circ \cong \mathbf{Sp}_4(\mathbb{R})$
   (e) $\mathbf{Spin}(3,3)^\circ \cong \mathbf{SL}(4, \mathbb{R})$ (f) $\mathbf{Spin}(4,2)^\circ \cong \mathbf{SU}(2,2)$
   (g) $\mathbf{Spin}(5,1)^\circ \cong \mathbf{SU}^*(4) \cong \mathbf{SL}(2, \mathbb{H})$ (see Exercises 1.7.3, #2)

6. Let $C(n,+)$ be the real Clifford algebra of signature $(n,0)$, and $C(n,-)$ the real Clifford algebra of signature $(0,n)$. Show that $C(n,-) \otimes C(2,+) \cong C(n+2,+)$ and $C(n,+) \otimes C(2,-) \cong C(n+2,-)$. (HINT: Take generators e_i for $C(n,-)$ and f_j for $C(2,+)$ so that $\{e_i, e_j\} = -2\delta_{i,j}$ and $\{f_i, f_j\} = 2\delta_{i,j}$. Set $u_i = e_i \otimes f_1 f_2$ for $i = 1, \ldots, n$, and $u_{n+1} = 1 \otimes f_1$, $u_{n+2} = 1 \otimes f_2$. Then $\{u_i, u_j\} = 2\delta_{i,j}$.)

6.5 Notes

Section 6.1.1. Clifford algebras were introduced by Clifford [39] as a unification of Hamilton's quaternions and Grassmann's exterior algebra.

Section 6.1.2. The systematic mathematical development of spaces of spinors began with Brauer and Weyl [21]; Dirac had previously coined the term in connection with his theory of the spinning electron (see Weyl [163]). Since then spinors have come to play a fundamental role in geometry, representation theory, and mathematical physics. Our treatment is from Wallach [154].

Section 6.2.1. Our exposition follows Howe [70].

Section 6.2.2. The spin representations of the orthogonal Lie algebras were discovered by Cartan [27].

Section 6.4.2 and *Exercises 6.4.4.* For a more extensive discussion of real forms of Clifford algebras, see Harvey [62] and Porteous [121]. Exercise #6 is from a lecture of R. Bott.

Chapter 7
Character Formulas

Abstract The central result of this chapter is the celebrated Weyl character formula for irreducible representations of a connected semisimple algebraic group G. We give two (logically independent) proofs of this formula. The first is algebraic and uses the theorem of the highest weight, some arguments involving invariant regular functions, and the Casimir operator. The second is Weyl's original analytic proof based on his integral formula for the compact real form of G.

7.1 Character and Dimension Formulas

We begin with a statement of the character formula and derive some of its immediate consequences: the Weyl dimension formula, formulas for inner and outer multiplicities, and character formulas for the commutant of G in a regular G-module. These character formulas will be used to obtain branching laws in Chapter 8 and to express the characters of the symmetric group in terms of the characters of the general linear group in Chapter 9.

7.1.1 Weyl Character Formula

Let G be a connected reductive linear algebraic group with Lie algebra $\mathfrak{g}$. We assume that $\mathfrak{g}$ is semisimple and that G is algebraically simply connected (see Section 6.3.2 for the classical groups, and Section 11.2.4 for the exceptional groups). This excludes $\mathbf{GL}(n,\mathbb{C})$ and $\mathbf{SO}(2,\mathbb{C}) \cong \mathbf{GL}(1,\mathbb{C})$; we will use the results of Section 5.5.4 to take care of this case at the end of this section.

Fix a maximal algebraic torus H of G with Lie algebra $\mathfrak{h}$. Then $\mathfrak{h}$ is a Cartan subalgebra of $\mathfrak{g}$, and we let $\Phi \subset \mathfrak{h}^*$ be the set of roots of $\mathfrak{h}$ on $\mathfrak{g}$. We fix a set Φ^+ of positive roots. Let $P = P(\mathfrak{g}) \subset \mathfrak{h}^*$ be the weight lattice and let $P_{++} \subset P$ be the dominant weights relative to Φ^+. Let

R. Goodman, N.R. Wallach, *Symmetry, Representations, and Invariants,*
Graduate Texts in Mathematics 255, DOI 10.1007/978-0-387-79852-3_7,
© Roe Goodman and Nolan R. Wallach 2009

$$\rho = \frac{1}{2} \sum_{\alpha \in \Phi^+} \alpha = \varpi_1 + \cdots + \varpi_l \, ,$$

where ϖ_i are the fundamental weights of $\mathfrak{g}$ (see Lemma 3.1.21). The character $h \mapsto h^\rho$ is well defined on H, since the weight lattice of G coincides with the weight lattice of $\mathfrak{g}$. We define the *Weyl function*

$$\Delta_G = \mathrm{e}^\rho \prod_{\alpha \in \Phi^+} \left(1 - \mathrm{e}^{-\alpha}\right) .$$

Here we view the exponentials e^λ, for $\lambda \in P$, as elements of the group algebra $\mathcal{A}[P]$ of the additive group P of weights (see Section 4.1.1; addition of exponents corresponds to the convolution multiplication in the group algebra under this identification). Thus Δ_G is an element of $\mathcal{A}[P]$. We can also view Δ_G as a function on H.

When $G = \mathbf{SL}(n, \mathbb{C})$ we write $\Delta_G = \Delta_n$. Since e^{ϖ_i} is the character

$$h \mapsto x_1 x_2 \cdots x_i \qquad (h = \mathrm{diag}[x_1, \ldots, x_n])$$

of H in this case, we see that e^ρ is the character $h \mapsto x_1^{n-1} x_2^{n-2} \cdots x_{n-1}$. Since the positive roots give the characters $h \mapsto x_i x_j^{-1}$ for $1 \le i < j \le n$, we have

$$\Delta_n(h) = x_1^{n-1} x_2^{n-2} \cdots x_{n-1} \prod_{1 \le i < j \le n} \left(1 - x_i^{-1} x_j\right) = \prod_{1 \le i < j \le n} (x_i - x_j) \, .$$

For the other classical groups, Δ_G is given as follows. Let $n = 2l$ and

$$h = \mathrm{diag}[x_1, \ldots, x_l, x_l^{-1}, \ldots, x_1^{-1}] \, .$$

Then with Φ^+ taken as in Section 2.4.3, we calculate that

$$\Delta_{\mathbf{SO}(2l)}(h) = \prod_{1 \le i < j \le l} \left(x_i + x_i^{-1} - x_j - x_j^{-1}\right)$$

and

$$\Delta_{\mathbf{Sp}(l)}(h) = \prod_{1 \le i < j \le l} \left(x_i + x_i^{-1} - x_j - x_j^{-1}\right) \prod_{k=1}^{l} (x_k - x_k^{-1}) \, .$$

For $n = 2l + 1$ and $h = \mathrm{diag}[x_1, \ldots, x_l, 1, x_l^{-1}, \ldots, x_1^{-1}]$ in $\mathbf{SO}(\mathbb{C}^{2l+1}, B)$ we have

$$\Delta_{\mathbf{SO}(2l+1)}(h) = \prod_{1 \le i < j \le l} \left(x_i + x_i^{-1} - x_j - x_j^{-1}\right) \prod_{k=1}^{l} (x_k^{1/2} - x_k^{-1/2})$$

(see Theorem 6.3.5 for the interpretation of the square roots to make this function single-valued on the simply connected group $\mathbf{Spin}(2l + 1, \mathbb{C})$).

Recall that the Weyl group W is equal to $\mathrm{Norm}_G(H)/H$. The adjoint representation of G gives a faithful representation σ of W on $\mathfrak{h}^*$, and we define

$$\text{sgn}(s) = \det(\sigma(s)).$$

Since W is generated by reflections, we have $\text{sgn}(s) = \pm 1$. The function Δ_G is *skew-symmetric*:

$$\Delta_G(shs^{-1}) = \text{sgn}(s)\Delta_G(h) \quad \text{for } h \in H. \tag{7.1}$$

Indeed, if we write Δ_G (viewed as a formal exponential sum) as

$$\Delta_G = \prod_{\alpha \in \Phi^+} (e^{\alpha/2} - e^{-\alpha/2}),$$

then the reflection given by a simple root α_i changes the sign of the factor involving α_i and permutes the other factors (see Lemma 3.1.21). Since these reflections generate W, this implies (7.1). Of course, this property can also be verified case by case from the formulas above and the description of W from Section 3.1.1.

A finite-dimensional $\mathfrak{g}$-module V decomposes as a direct sum of $\mathfrak{h}$ weight spaces $V(\mu)$, with $\mu \in P(\mathfrak{g})$ (see Theorem 3.1.16). We write

$$\text{ch}(V) = \sum_{\mu \in P} \dim V(\mu) e^{\mu}$$

as an element in the group algebra $\mathcal{A}[P]$. We may also view $\text{ch}(V)$ as a regular function on H because G is algebraically simply connected. There is a regular representation π of G on V whose differential is the given representation of $\mathfrak{g}$, and

$$\text{ch}(V)(h) = \text{tr}(\pi(h)) \quad \text{for } h \in H$$

(see Corollary 6.3.9 and Theorem 11.2.14).

Theorem 7.1.1 (Weyl Character Formula). *Let $\lambda \in P_{++}$ and let V^{λ} be the finite-dimensional irreducible G-module with highest weight λ. Then*

$$\Delta_G \cdot \text{ch}(V^{\lambda}) = \sum_{s \in W} \text{sgn}(s) e^{s \cdot (\lambda + \rho)}. \tag{7.2}$$

In the Weyl character formula, the character, which is invariant under the action of W, is expressed as a ratio of functions that are skew-symmetric under the action of W (for $G = \mathbf{SL}(2, \mathbb{C})$ it is just the formula for the sum of a finite geometric series). Later in this chapter we shall give two proofs of this fundamental result: an algebraic proof that uses the Casimir operator and the theorem of the highest weight, and an analytic proof (Weyl's original proof). Both proofs require rather lengthy developments of preliminary results. At this point we derive some immediate consequences of the character formula.

We first extend the formula to include the case $G = \mathbf{GL}(n, \mathbb{C})$. Let

$$\mu = m_1 \varepsilon_1 + \cdots + m_n \varepsilon_n \quad \text{with } m_1 \geq m_2 \geq \cdots \geq m_n \text{ and } m_i \in \mathbb{Z}.$$

Let (π_n^μ, F_n^μ) be the irreducible representation of $\mathbf{GL}(n,\mathbb{C})$ associated with μ by Theorem 5.5.22. Define

$$\rho_n = (n-1)\varepsilon_1 + (n-2)\varepsilon_2 + \cdots + \varepsilon_{n-1} \tag{7.3}$$

as an element in the weight lattice for $\mathbf{GL}(n,\mathbb{C})$. Define the function $\Delta_n(h)$, for $h = \mathrm{diag}[x_1,\ldots,x_n]$, by the same formula as for $\mathbf{SL}(n,\mathbb{C})$.

Corollary 7.1.2. *Let* $h = \mathrm{diag}[x_1,\ldots,x_n]$. *Then*

$$\Delta_n(h)\,\mathrm{tr}(\pi_n^\mu(h)) = \sum_{s\in\mathfrak{S}_n} \mathrm{sgn}(s)\, h^{s(\mu+\rho_n)}$$

$$= \det \begin{bmatrix} x_1^{m_1+n-1} & x_1^{m_2+n-2} & \cdots & x_1^{m_n} \\ x_2^{m_1+n-1} & x_2^{m_2+n-2} & \cdots & x_2^{m_n} \\ \vdots & \vdots & \ddots & \vdots \\ x_n^{m_1+n-1} & x_n^{m_2+n-2} & \cdots & x_n^{m_n} \end{bmatrix}. \tag{7.4}$$

Proof. Write $h = z h_0$, where $z \in \mathbb{C}^\times$ and $\det(h_0) = 1$. Then

$$\mathrm{tr}(\pi_n^\mu(h)) = z^{m_1+\cdots+m_n}\,\mathrm{tr}(\pi_n^\mu(h_0)) \,.$$

Here the Weyl group is $\mathfrak{S}_n$. Note that the restriction of ρ_n to the trace-zero diagonal matrices is the weight ρ for $\mathbf{SL}(n,\mathbb{C})$. Thus

$$h^{s(\rho_n+\mu)} = z^{m_1+\cdots+m_n+n(n-1)/2} h_0^{s(\rho+\mu)}$$

for $s \in \mathfrak{S}_n$. We also have

$$\prod_{1\le i<j\le n} (x_i - x_j) = z^{n(n-1)/2}\Delta_n(h_0) \,.$$

Now apply the Weyl character formula for $\mathbf{SL}(n,\mathbb{C})$ to obtain the first equation in (7.4). The determinant formula is an immediate consequence, since $\mathrm{sgn}(s)$ is the usual sign of the permutation s and

$$h^{s\lambda} = x_{s(1)}^{\lambda_1} \cdots x_{s(n)}^{\lambda_n} \quad \text{for } \lambda = \lambda_1\varepsilon_1 + \cdots + \lambda_n\varepsilon_n \,. \qquad \square$$

We now draw some consequences of the Weyl character formula in general.

Corollary 7.1.3 (Weyl Denominator Formula). *The Weyl function is the skew-symmetrization of the character* e^ρ *of H:*

$$\Delta_G = \sum_{s\in W} \mathrm{sgn}(s)\, e^{s\cdot\rho}. \tag{7.5}$$

Proof. Take $\lambda = 0$ in the Weyl Character Formula; then $\mathrm{ch}\,V^0 = 1$. $\qquad\square$

Note that when $G = \mathbf{GL}(n, \mathbb{C})$, then the Weyl denominator formula is the product expansion

$$\prod_{1 \leq i < j \leq n} (x_i - x_j) = \det V(h)$$

of the Vandermonde determinant

$$V(h) = \begin{bmatrix} x_1^{n-1} & x_1^{n-2} & \cdots & x_1 & 1 \\ x_2^{n-1} & x_2^{n-2} & \cdots & x_2 & 1 \\ \vdots & \vdots & \ddots & \vdots & \vdots \\ x_n^{n-1} & x_n^{n-2} & \cdots & x_n & 1 \end{bmatrix} .$$

Let $\lambda \in P_{++}$. For $\mu \in P$ we write $m_\lambda(\mu) = \dim V^\lambda(\mu)$ (the multiplicity of the weight μ in V^λ).

Corollary 7.1.4. *Suppose $\mu \in P$. If $\mu + \rho = t \cdot (\lambda + \rho)$ for some $t \in W$, then*

$$\sum_{s \in W} \operatorname{sgn}(s) m_\lambda(\mu + \rho - s \cdot \rho) = \operatorname{sgn}(t) . \tag{7.6}$$

Otherwise, the sum on the left is zero. In particular, if $\mu = \lambda$ the sum is 1, while if $\mu \in P_{++}$ and $\mu \neq \lambda$, then the sum is zero.

Proof. Expressing Δ_G as an alternating sum over W by the Weyl denominator formula, we can write (7.2) as

$$\sum_{s \in W} \left\{ \sum_{\mu \in P} \operatorname{sgn}(s) m_\lambda(\mu) e^{\mu + s \cdot \rho} \right\} = \sum_{t \in W} \operatorname{sgn}(t) e^{t \cdot (\lambda + \rho)} .$$

We replace μ by $\mu + \rho - s \cdot \rho$ in the inner sum on the left, for each $s \in W$. Interchanging the order of summation, we obtain the identity

$$\sum_{\mu \in P} \left\{ \sum_{s \in W} \operatorname{sgn}(s) m_\lambda(\mu + \rho - s \cdot \rho) \right\} e^{\mu + \rho} = \sum_{t \in W} \operatorname{sgn}(t) e^{t \cdot (\lambda + \rho)} .$$

On the right side of this identity the only weights that appear are those in the W orbit of the dominant regular weight $\lambda + \rho$, and these are all distinct (see Proposition 3.1.20). Now compare coefficients of the exponentials on each side. Note that if $\mu \in P_{++}$ and $\mu \neq \lambda$, then $\mu + \rho$ is not in the W orbit of $\lambda + \rho$, so the coefficient of $e^{\mu + \rho}$ on the left side must vanish. $\qquad \square$

Remark 7.1.5. The steps in the proof just given are reversible, so Corollaries 7.1.3 and 7.1.4 imply the Weyl character formula.

Let (σ, F) be a finite-dimensional regular representation of G. Then F decomposes as a direct sum of irreducible representations. The number of times that a particular irreducible module V^λ appears in the decomposition is the *outer multiplicity* $\operatorname{mult}_F(V^\lambda)$.

Corollary 7.1.6. *The outer multiplicity of V^λ is the skew-symmetrization over $s \in W$ of the multiplicities of the weights $\lambda + \rho - s \cdot \rho$:*

$$\text{mult}_F(V^\lambda) = \sum_{s \in W} \text{sgn}(s) \dim F(\lambda + \rho - s \cdot \rho) . \qquad (7.7)$$

Proof. For any weight $v \in P$, the weight space $F(v)$ has dimension

$$\dim F(v) = \sum_{\mu \in P_{++}} \text{mult}_F(V^\mu) m_\mu(v) .$$

Take $v = \lambda + \rho - s \cdot \rho$ with $s \in W$ in this formula, multiply by $\text{sgn}(s)$, and sum over W. This gives

$$\sum_{s \in W} \text{sgn}(s) \dim F(\lambda + \rho - s \cdot \rho) = \sum_{\mu \in P_{++}} \text{mult}_F(V^\mu) \Big\{ \sum_{s \in W} \text{sgn}(s) m_\mu(\lambda + \rho - s \cdot \rho) \Big\} .$$

But by Corollary 7.1.4 the right-hand side reduces to $\text{mult}_F(V^\lambda)$.   □

Corollary 7.1.7. *Let $\mu, v \in P_{++}$. The tensor product $V^\mu \otimes V^v$ decomposes with multiplicities*

$$\text{mult}_{V^\mu \otimes V^v}(V^\lambda) = \sum_{t \in W} \text{sgn}(t) m_\mu(\lambda + \rho - t \cdot (v + \rho)) . \qquad (7.8)$$

Proof. Set $F = V^\mu \otimes V^\lambda$. By (5.59) we have

$$\dim F(\gamma) = \sum_{\alpha \in P} m_\mu(\alpha) m_v(\gamma - \alpha)$$

for all $\gamma \in P$. Substituting this into (7.7), we obtain

$$\text{mult}_F(V^\lambda) = \sum_{s \in W} \sum_{\alpha \in P} \text{sgn}(s) m_\mu(\alpha) m_v(\lambda - \alpha + \rho - s \cdot \rho) .$$

But by (7.6) the sum over W is zero unless $\alpha = \lambda + \rho - t \cdot (v + \rho)$ for some $t \in W$, and in this case it equals $\text{sgn}(t)$. This gives (7.8).   □

7.1.2 Weyl Dimension Formula

We now obtain a formula for the dimension of the irreducible G-module V^λ, which is the value of $\text{ch}(V^\lambda)$ at 1. The Weyl character formula expresses this character as a ratio of two functions on the maximal torus H, each of which vanishes at 1 (by skew-symmetry), so we must apply l'Hospital's rule to obtain $\dim V^\lambda$. This can be carried out algebraically as follows.

We define a linear functional $\varepsilon : \mathcal{A}[\mathfrak{h}^*] \longrightarrow \mathbb{C}$ by

$$\varepsilon\left(\sum_\beta c_\beta e^\beta\right) = \sum_\beta c_\beta .$$

To motivate this definition, we note that if we set $\varphi = \sum_{\beta \in P} c_\beta e^\beta$ and consider φ as a function on H, then $\varepsilon(\varphi) = \varphi(1)$. For $s \in W$ and $f \in \mathcal{A}[\mathfrak{h}^*]$ we have

$$\varepsilon(s \cdot f) = \varepsilon(f) , \tag{7.9}$$

where we define $s \cdot e^\beta = e^{s(\beta)}$ for $\beta \in \mathfrak{h}^*$.

Fix a W-invariant symmetric bilinear form (α, β) on $\mathfrak{h}^*$, as in Section 2.4.2. For $\alpha \in \mathfrak{h}^*$ define a derivation ∂_α on $\mathcal{A}[\mathfrak{h}^*]$ by $\partial_\alpha(e^\beta) = (\alpha, \beta)e^\beta$. Then

$$s \cdot (\partial_\alpha f) = \partial_{s(\alpha)}(s \cdot f) \tag{7.10}$$

for $s \in W$ and $f \in \mathcal{A}[\mathfrak{h}^*]$. Define the differential operator

$$D = \prod_{\alpha \in \Phi^+} \partial_\alpha .$$

We claim that

$$s \cdot (Df) = \operatorname{sgn}(s)D(s \cdot f) . \tag{7.11}$$

Indeed, if s is a reflection for a simple root, then by (7.10) we see that s changes the sign of exactly one factor in D and permutes the other factors (see Lemma 3.1.21). Since W is generated by simple reflections, this property implies (7.11).

Let $\lambda \in P_{++}$ and define

$$A_{\lambda+\rho} = \sum_{s \in W} \operatorname{sgn}(s) e^{s \cdot (\lambda+\rho)}$$

(the numerator in the Weyl character formula). From (7.11) we have

$$D \cdot A_{\lambda+\rho} = \sum_{s \in W} s \cdot \left(D \cdot e^{\lambda+\rho}\right) = \left\{ \prod_{\alpha \in \Phi^+} (\rho+\lambda, \alpha) \right\} \sum_{s \in W} e^{s \cdot (\rho+\lambda)} . \tag{7.12}$$

Now for $\lambda = 0$ we have $A_\rho = \Delta_G$ by the Weyl denominator formula (7.5). Hence

$$D(\Delta_G) = \left\{ \prod_{\alpha \in \Phi^+} (\rho, \alpha) \right\} \sum_{s \in W} e^{s \cdot \rho} . \tag{7.13}$$

Thus from (7.9) we obtain $\varepsilon(D \cdot A_{\lambda+\rho}) = |W| \prod_{\alpha \in \Phi^+} (\lambda+\rho, \alpha)$. Applying the Weyl character formula and using (7.12), we see that

$$\varepsilon\left(D(\operatorname{ch}(V^\lambda)\Delta_G)\right) = \varepsilon\left(DA_{\lambda+\rho}\right) = |W| \prod_{\alpha \in \Phi^+} (\lambda+\rho, \alpha) . \tag{7.14}$$

Lemma 7.1.8. *If $f \in \mathcal{A}[\mathfrak{h}^*]$, then*

$$\varepsilon\left(D(f\Delta_G)\right) = \varepsilon\left(fD(\Delta_G)\right) . \tag{7.15}$$

Proof. For every subset Q of Φ^+ we define

$$F_Q = \prod_{\beta \in Q}(e^{\beta/2} - e^{-\beta/2}), \qquad \partial_Q = \prod_{\alpha \in Q} \partial_\alpha \, .$$

Then $\Delta_G = F_{\Phi^+}$ and $D = \partial_{\Phi^+}$. We have

$$\partial_\alpha F_Q = \sum_{\beta \in Q}(\alpha, \beta)(e^{\beta/2} + e^{-\beta/2})F_{Q \setminus \{\beta\}} \, .$$

From this equation we verify by induction on $|Q|$ that there are elements $f_P \in \mathcal{A}[\mathfrak{h}^*]$ such that

$$\partial_Q(f\Delta_G) = f\partial_Q(\Delta_G) + \sum_{P \subset \Phi^+} f_P F_P \, ,$$

where P ranges over subsets of Φ^+ with $|P| \geq |\Phi^+| - |Q| + 1$. Hence for $Q = \Phi^+$ we can write

$$D(f\Delta_G) = fD(\Delta_G) + \sum_{P \subset \Phi^+} f_P F_P \, ,$$

where now P ranges over all nonempty subsets of Φ^+. But each term $f_P F_P$ contains a factor $e^{\beta/2} - e^{-\beta/2}$, so $\varepsilon(f_P F_P) = 0$. Hence the lemma follows. $\square$

Theorem 7.1.9 (Weyl Dimension Formula). *The dimension of V^λ is a polynomial of degree $|\Phi^+|$ in λ:*

$$\dim V^\lambda = \prod_{\alpha \in \Phi^+} \frac{(\lambda + \rho, \alpha)}{(\rho, \alpha)} \, . \tag{7.16}$$

Proof. From (7.14) and (7.15) with $f = \mathrm{ch}(V^\lambda)$, we have

$$\varepsilon\big(\mathrm{ch}(V^\lambda)D(\Delta_G)\big) = |W| \prod_{\alpha \in \Phi^+}(\lambda + \rho, \alpha) \, . \tag{7.17}$$

But from (7.13) we have the expansion

$$\varepsilon\big(\mathrm{ch}(V^\lambda)D(\Delta_G)\big) = \Big\{ \prod_{\alpha \in \Phi^+}(\rho, \alpha)\Big\} \varepsilon\Big(\sum_{s \in W}\sum_{\mu \in \mathfrak{h}^*} m_\lambda(\mu)\, e^{\mu + s \cdot \rho}\Big)$$

$$= |W| \dim V^\lambda \prod_{\alpha \in \Phi^+}(\rho, \alpha) \, .$$

Now (7.16) follows from this equation and (7.14). $\square$

Examples

Type A: Let $G = \mathbf{SL}(n, \mathbb{C})$. If $\lambda \in P_{++}$ then λ is the restriction to H of the weight $\lambda_1 \varepsilon_1 + \cdots + \lambda_{n-1}\varepsilon_{n-1}$ with $\lambda_1 \geq \cdots \geq \lambda_{n-1} \geq 0$, where the λ_i are integers. Setting $\lambda_n = 0$, we have

$$(\rho, \varepsilon_i - \varepsilon_j) = j - i, \quad (\lambda + \rho, \varepsilon_i - \varepsilon_j) = \lambda_i - \lambda_j + j - i.$$

Thus from the Weyl dimension formula we get

$$\dim V^\lambda = \prod_{1 \le i < j \le n} \frac{\lambda_i - \lambda_j + j - i}{j - i}. \tag{7.18}$$

For example, the representation V^ρ has dimension $2^{n(n-1)/2}$. For $n = 3$ it happens to be the adjoint representation, but for $n \ge 4$ it is much bigger than the adjoint representation.

Types B and C: Let $G = \mathbf{Spin}(2n+1, \mathbb{C})$ or $\mathbf{Sp}(n, \mathbb{C})$. Then $\rho = \rho_1 \varepsilon_1 + \cdots + \rho_n \varepsilon_n$ with

$$\rho_i = \begin{cases} n - i + (1/2) & \text{for } G = \mathbf{Spin}(2n+1, \mathbb{C}), \\ n - i + 1 & \text{for } G = \mathbf{Sp}(n, \mathbb{C}). \end{cases}$$

Let $\lambda = \lambda_1 \varepsilon_1 + \cdots + \lambda_n \varepsilon_n$ with $\lambda_1 \ge \lambda_2 \ge \cdots \ge \lambda_n \ge 0$ be a dominant integral weight for G. Then we calculate from (7.16) that

$$\dim V^\lambda = \prod_{1 \le i < j \le n} \frac{(\lambda_i + \rho_i)^2 - (\lambda_j + \rho_j)^2}{\rho_i^2 - \rho_j^2} \prod_{1 \le i \le n} \frac{\lambda_i + \rho_i}{\rho_i}.$$

For example, for the smallest regular dominant weight ρ we have $\dim V^\rho = 2^{n^2}$. Note that for the orthogonal group (type B) V^ρ is a representation of $\mathbf{Spin}(2n+1, \mathbb{C})$ but not of $\mathbf{SO}(2n+1, \mathbb{C})$, since ρ is only half-integral.

Type D: Let $G = \mathbf{Spin}(2n, \mathbb{C})$. Then $\rho = \rho_1 \varepsilon_1 + \cdots + \rho_n \varepsilon_n$ with $\rho_i = n - i$. Let $\lambda = \lambda_1 \varepsilon_1 + \cdots + \lambda_n \varepsilon_n$ with $\lambda_1 \ge \cdots \ge \lambda_{n-1} \ge |\lambda_n|$ be a dominant integral weight for G. Then we calculate from (7.16) that

$$\dim V^\lambda = \prod_{1 \le i < j \le n} \frac{(\lambda_i + \rho_i)^2 - (\lambda_j + \rho_j)^2}{\rho_i^2 - \rho_j^2}.$$

In this case, for the smallest regular dominant weight ρ we have $\dim V^\rho = 2^{n^2 - n}$. Since ρ is integral, V^ρ is a single-valued representation of $\mathbf{SO}(2n, \mathbb{C})$.

7.1.3 Commutant Character Formulas

Assume that G is semisimple and algebraically simply connected. For $\lambda \in P_{++}$ let (π^λ, V^λ) be the irreducible representation of G with highest weight λ. Let $N^+ \subset G$ be the unipotent subgroup with Lie algebra $\mathfrak{n}^+$. Denote the character of π^λ by

$$\varphi_\lambda(g) = \mathrm{tr}(\pi^\lambda(g)) \quad \text{for } g \in G.$$

Thus φ_λ is a regular *invariant* function on G (it is constant on conjugacy classes).

Let (π, F) be a rational representation of G. Let $\mathcal{B} = \mathrm{End}_G(F)$ be the commuting algebra for the G action on F. Since G is reductive, the duality theorem (Theorem 4.2.1) implies that F decomposes as

$$F \cong \bigoplus_{\lambda \in \mathrm{Spec}(\pi)} V^\lambda \otimes E^\lambda , \tag{7.19}$$

where $\mathrm{Spec}(\pi) \subset P_{++}$ (the *G-spectrum* of π) is the set of highest weights of the irreducible G representations occurring in F, and E^λ is an irreducible representation of $\mathcal{B}$. Here $g \in G$ acts by $\pi^\lambda(g) \otimes 1$ and $b \in \mathcal{B}$ acts by $1 \otimes \sigma^\lambda(b)$ on the summands in (7.19), where $(\sigma^\lambda, E^\lambda)$ is an irreducible representation of B. We denote the character of the $\mathcal{B}$-module E^λ by χ_λ:

$$\chi_\lambda(b) = \mathrm{tr}(\sigma^\lambda(b)) .$$

Let F^{N^+} be the space of N^+-fixed vectors in F (since $N^+ = \exp(\mathrm{n}^+)$, a vector is fixed under N^+ if and only if it is annihilated by n^+). Then F^{N^+} is invariant under $\mathcal{B}$, and the weight-space decomposition

$$F^{N^+} = \bigoplus_{\lambda \in \mathrm{Spec}(\pi)} F^{N^+}(\lambda)$$

is also invariant under $\mathcal{B}$. From Theorem 4.2.12 we have $E^\lambda \cong F^{N^+}(\lambda)$ as a $\mathcal{B}$-module for $\lambda \in \mathrm{Spec}(\pi)$. Hence

$$\chi_\lambda(b) = \mathrm{tr}(b|_{F^{N^+}(\lambda)}) .$$

This formula for the character χ_λ is not very useful, however. Although the full weight space $F(\lambda)$ is often easy to determine, finding a basis for the N^+-fixed vectors of a given weight is generally difficult. We now use the Weyl character formula to obtain two formulas for the character χ_λ that involve only the full H-weight spaces in F.

Theorem 7.1.10. *For $\lambda \in P_{++}$ and $b \in \mathcal{B}$ one has*

$$\chi_\lambda(b) = \text{coefficient of } x^{\lambda+\rho} \text{ in } \Delta_G(x) \mathrm{tr}_F(\pi(x)b) \tag{7.20}$$

(where $x \in H$).

Proof. We note from (7.19) that

$$\mathrm{tr}_F(\pi(g)b) = \sum_{\lambda \in \mathrm{Spec}(\pi)} \varphi_\lambda(g) \chi_\lambda(b) \qquad \text{for } g \in G \text{ and } b \in \mathcal{B} . \tag{7.21}$$

By the Weyl character formula we have

$$\Delta_G(x)\varphi_\lambda(x) = \sum_{s \in W} \mathrm{sgn}(s) x^{s \cdot (\lambda+\rho)} \qquad \text{for } x \in H .$$

Using this in (7.21) we can write

$$\Delta_G(x)\operatorname{tr}_F(\pi(x)b) = \sum_{\lambda\in\operatorname{Spec}(\pi)}\sum_{s\in W}\operatorname{sgn}(s)\,\chi_\lambda(b)\,x^{s\cdot(\lambda+\rho)}. \qquad (7.22)$$

But the map $(s,\lambda)\mapsto s\cdot(\lambda+\rho)$ from $W\times P_{++}$ to P is injective, since $\lambda+\rho$ is regular (see Proposition 3.1.20). Hence the character $x\mapsto x^{\lambda+\rho}$ occurs only once in (7.22), and has coefficient $\chi_\lambda(b)$ as claimed. $\qquad\square$

We now give the second character formula.

Theorem 7.1.11. *For* $\lambda\in P_{++}$ *and* $b\in\mathcal{B}$ *one has*

$$\chi_\lambda(b) = \sum_{s\in W}\operatorname{sgn}(s)\operatorname{tr}_{F(\lambda+\rho-s\cdot\rho)}(b). \qquad (7.23)$$

In particular,

$$\dim E^\lambda = \sum_{s\in W}\operatorname{sgn}(s)\dim F(\lambda+\rho-s\cdot\rho). \qquad (7.24)$$

Proof. For $\zeta\in\mathbb{C}$ the generalized ζ-eigenspace

$$F_\zeta = \{v\in F : (b-\zeta)^k v = 0 \text{ for some } k\}$$

is invariant under G, and we have $F = \bigoplus_{\zeta\in\mathbb{C}}F_\zeta$. Therefore, replacing F by F_ζ, we need to prove (7.23) only when $b = 1$. Hence we need to prove only (7.24). Since $\dim E^\lambda = \operatorname{mult}_F(V^\lambda)$, this follows from Corollary 7.1.6. $\qquad\square$

7.1.4 Exercises

1. Verify the formulas for the Weyl denominators for the orthogonal and symplectic groups in Section 7.1.1. (*Hint:* Use the formulas for ρ from Section 3.1.4.)
2. Let $G = \mathbf{SL}(2,\mathbb{C})$ and let $\varpi = \varepsilon_1$ be the fundamental weight (the highest weight for the defining representation on $\mathbb{C}^2$). Set $F = V^{p\varpi}\otimes V^{q\varpi}$, where $0\le p\le q$ are integers. Use (7.8) to obtain the *Clebsch–Gordan formula*:

$$V^{p\varpi}\otimes V^{q\varpi} \cong V^{(q+p)\varpi}\oplus V^{(q+p-2)\varpi}\oplus\cdots\oplus V^{(q-p+2)\varpi}\oplus V^{(q-p)\varpi}.$$

3. Use (7.8) to show that $V^{\mu+\nu}$ occurs with multiplicity one in $V^\mu\otimes V^\nu$.
4. Let $G = \mathbf{GL}(n,\mathbb{C})$.
   (a) Use (7.8) to decompose the representation $\mathbb{C}^n\otimes\mathbb{C}^n$.
   (b) Show that (a) gives the decomposition $\mathbb{C}^n\otimes\mathbb{C}^n\cong S^2(\mathbb{C}^n)\oplus\bigwedge^2\mathbb{C}^n$.
5. Let $G = \mathbf{SL}(n,\mathbb{C})$. Take the defining representation π^{ϖ_1} on $\mathbb{C}^n$ and consider the decomposition of the tensor product $\mathbb{C}^n\otimes V^\nu$ for $\nu\in P_{++}$.

(a) Use the fact that the inner multiplicities of π^{ϖ_1} are all 1 with weights $\varepsilon_1, \ldots, \varepsilon_n$ to show that if $\lambda \in P_{++}$ and $t \in W$, then $\lambda + \rho - t \cdot (v + \rho)$ is a weight of π^{ϖ_1} if and only if $t \cdot (v + \rho) = \lambda + \rho - \varepsilon_i$ for some $i \in \{1, \ldots, n\}$.

(b) Use the result in (a) and (7.8) to prove that $\mathbb{C}^n \otimes V^v$ decomposes as a sum of inequivalent irreducible representations (each occurring once).

(c) Show that a necessary condition for V^λ to occur in the decomposition of $\mathbb{C}^n \otimes V^v$ is that $\lambda = v + \varepsilon_i$ for some i.

(d) Show that the condition in (c) is also sufficient provided $v + \varepsilon_i \in P_{++}$. Conclude that $\mathbb{C}^n \otimes V^v \cong \bigoplus V^{v+\varepsilon_i}$, with the sum over $i = 1$ and all other $i \in \{2, \ldots, n\}$ such that $v_{i-1} > v_i$ (where we take $v_n = 0$).

6. Verify the formulas given in Section 7.1.2 for $\dim V^\lambda$ for G of types B, C, and D.

7. (a) Calculate the dimensions of the defining representations of the classical groups from the Weyl dimension formula.

(b) Calculate the dimensions of the other fundamental representations of the classical groups from the Weyl dimension formula and confirm the values obtained from the explicit constructions of these representations in Sections 5.5.2 and 6.2.2.

8. Let $G = \mathbf{SL}(n, \mathbb{C})$. Let $\lambda = m_1 \varpi_1 + \cdots + m_{n-1} \varpi_{n-1}$, with $m_i \in \mathbb{N}$, be a dominant weight of G.

(a) Show that

$$\dim V^\lambda = \prod_{1 \le i < j \le n} \left\{ 1 + \frac{m_i + \cdots + m_{j-1}}{j - i} \right\}.$$

(b) Conclude from (a) that $\dim V^\lambda$ is a monotonically increasing function of m_i for $i = 1, \ldots, n-1$, and it has a minimum when $\lambda = \varpi_1$ or $\lambda = \varpi_{n-1}$. Hence the defining representation of G (and its dual) are the unique representations of smallest dimension.

9. Let $G = \mathbf{Spin}(2n + 1, \mathbb{C})$ with $n \ge 2$ (recall that $\mathbf{Spin}(3, \mathbb{C}) \cong \mathbf{SL}(2, \mathbb{C})$). Let $\lambda = m_1 \varpi_1 + \cdots + m_n \varpi_n$, with $m_i \in \mathbb{N}$, be a dominant weight of G.

(a) Show that

$$\dim V^\lambda = \prod_{1 \le i < j \le n} \left\{ 1 + \frac{m_i + \cdots + m_{j-1}}{j - i} \right\}$$
$$\times \prod_{1 \le i < j \le n} \left\{ 1 + \frac{m_i + \cdots + m_{j-1} + 2(m_j + \cdots + m_{n-1}) + m_n}{2n + 1 - i - j} \right\}$$
$$\times \prod_{1 \le i \le n} \left\{ 1 + \frac{2m_i + \cdots + 2m_{n-1} + m_n}{2n + 1 - 2i} \right\}.$$

(b) Conclude from (a) that $\dim V^\lambda$ is a monotonically increasing function of m_i for $i = 1, \ldots, n$.

(c) Show that $\dim V^{\varpi_1} < \dim V^{\varpi_k}$ for $k = 2, \ldots, n-1$. (HINT: Recall that $\dim V^{\varpi_k} = \binom{2n+1}{k}$ by Theorem 5.5.13 for $k = 1, \ldots, n-1$, and that $\dim V^{\varpi_n} = 2^n$ by Proposition 6.2.4.)

(d) Use (b) and (c) to show that $\dim V^\lambda$ has a unique minimum for $\lambda = \varpi_1$, and hence that the covering map $G \longrightarrow \mathbf{SO}(2n+1, \mathbb{C})$ gives the unique representation of G of smallest dimension.

10. Let $G = \mathbf{Sp}(n, \mathbb{C})$. Let $\lambda = m_1 \varpi_1 + \cdots + m_n \varpi_n$, with $m_i \in \mathbb{N}$, be a dominant weight of G.

(a) Show that

$$\dim V^\lambda = \prod_{1 \le i < j \le n} \left\{ 1 + \frac{m_i + \cdots + m_{j-1}}{j-i} \right\}$$

$$\times \prod_{1 \le i < j \le n} \left\{ 1 + \frac{m_i + \cdots + m_{j-1} + 2(m_j + \cdots + m_n)}{2n+2-i-j} \right\}$$

$$\times \prod_{1 \le i \le n} \left\{ 1 + \frac{m_i + \cdots + m_n}{n+1-i} \right\}.$$

(b) Conclude from (a) that $\dim V^\lambda$ is a monotonically increasing function of m_i for $i = 1, \ldots, n$.

(c) Show that $\dim V^{\varpi_1} < \dim V^{\varpi_k}$ for $k = 2, \ldots, n$. (HINT: Recall that $\dim V^{\varpi_k} = \binom{2n}{k} - \binom{2n}{k-2}$ by Corollary 5.5.17.)

(d) Use (b) and (c) to show that $\dim V^\lambda$ has a unique minimum for $\lambda = \varpi_1$, and hence that the defining representation of G is the unique representation of smallest dimension.

11. Let $G = \mathbf{Spin}(2n, \mathbb{C})$ with $n \ge 4$ (recall that $\mathbf{Spin}(3, \mathbb{C}) \cong \mathbf{SL}(4, \mathbb{C})$ and $\mathbf{Spin}(4, \mathbb{C}) \cong \mathbf{SL}(2, \mathbb{C}) \times \mathbf{SL}(2, \mathbb{C})$). Let $\lambda = m_1 \varpi_1 + \cdots + m_n \varpi_n$, with $m_i \in \mathbb{N}$, be a dominant weight of $\mathfrak{g}$.

(a) Show that

$$\dim V^\lambda = \prod_{1 \le i < j \le n} \left\{ 1 + \frac{m_i + \cdots + m_{j-1}}{j-i} \right\}$$

$$\times \prod_{1 \le i < j \le n} \left\{ 1 + \frac{m_i + \cdots + m_{j-1} + 2(m_j + \cdots + m_{n-1}) + m_n}{2n-i-j} \right\}.$$

(b) Conclude from (a) that $\dim V^\lambda$ is a monotonically increasing function of m_i for $i = 1, \ldots, n$.

(c) Show that if $n > 4$ then $\dim V^{\varpi_1} < \dim V^{\varpi_k}$ for $k = 2, \ldots, n$, whereas if $n = 4$ then $\dim V^{\varpi_1} = \dim V^{\varpi_{n-1}} = \dim V^{\varpi_n} < \dim V^{\varpi_k}$ for $k = 2, \ldots, n-2$. (HINT: Recall that $\dim V^{\varpi_k} = \binom{2n}{k}$ from Theorem 5.5.13 for $k = 1, \ldots, n-2$, and $\dim V^{\varpi_{n-1}} = \dim V^{\varpi_n} = 2^{n-1}$ from Proposition 6.2.3.)

(d) Suppose $n > 4$. Use (b) and (c) to show that $\dim V^\lambda$ has a unique minimum for $\lambda = \varpi_1$, and hence that the covering map $G \longrightarrow \mathbf{SO}(2n, \mathbb{C})$ gives the unique representation of smallest dimension in this case.

(e) Suppose $n = 4$. Show that $\mathbf{SO}(8, \mathbb{C})$ has a unique (single-valued) representation of smallest dimension, namely the defining representation on $\mathbb{C}^8$.

7.2 Algebraic Group Approach to the Character Formula

We now develop an algebraic proof of the Weyl character formula, based on the
following observations: The character of an irreducible regular representation of a
semisimple reductive group is an *invariant function*, and so it is determined by its
restriction to a maximal torus. It is also an eigenfunction for the Casimir operator,
with the eigenvalue determined by the highest weight. Using the *radial part* of the
Casimir operator, we show that the function on the torus given by the product of the
Weyl denominator and the character is the numerator in the Weyl character formula.

7.2.1 Symmetric and Skew-Symmetric Functions

Let $\mathfrak{g}$ be a semisimple Lie algebra. Fix a Cartan subalgebra $\mathfrak{h}$ of $\mathfrak{g}$ and a set Φ^+ of
positive roots for $\mathfrak{h}$ on $\mathfrak{g}$. Fix an invariant bilinear form B on $\mathfrak{g}$ that is positive definite
on $\mathfrak{h}_{\mathbb{R}}$ (the real span of the coroots); see Section 2.4.1 for the classical groups or
Section 2.5.3 in general. We denote by $(\cdot, \cdot)$ the inner product on $\mathfrak{h}_{\mathbb{R}}^*$ defined by B.

Write $P = P(\mathfrak{g})$, $P_{++} = P_{++}(\mathfrak{g})$, and $W = W_{\mathfrak{g}}$ for the weight lattice, the Φ^+-
dominant weights, and the Weyl group of $\mathfrak{g}$. Then P is a free abelian group generated
by the fundamental weights $\varpi_1, \ldots, \varpi_l$, and it has a group algebra $\mathcal{A}[P]$ (see Section
4.1.1). It is convenient to use the exponential notation of Section 7.1.1, so that

$$\mathcal{A}[P] = \bigoplus_{\mu \in P} \mathbb{C} e^\mu$$

as a vector space, and multiplication in $\mathcal{A}[P]$ corresponds to addition of exponents.

There is a canonical representation σ of the Weyl group on $\mathcal{A}[P]$ associated with
the linear action of W on $\mathfrak{h}^*$:

$$\sigma(s)\left(\textstyle\sum_{\mu \in P} c_\mu e^\mu\right) = \textstyle\sum_{\mu \in P} c_\mu e^{s \cdot \mu} \quad \text{for } s \in W .$$

Using the one-dimensional trivial and signum representations of W, we define the
subspaces of *symmetric functions*

$$\mathcal{A}[P]_{\text{symm}} = \{ f \in \mathcal{A}[P] : \sigma(s)f = f \quad \text{for all } s \in W \}$$

and *skew-symmetric functions*

$$\mathcal{A}[P]_{\text{skew}} = \{ f \in \mathcal{A}[P] : \sigma(s)f = \text{sgn}(s)f \quad \text{for all } s \in W \} .$$

The *symmetrizer operator* **Sym**, which acts on e^μ by

$$\mathbf{Sym}(e^\mu) = \frac{1}{|W|} \sum_{s \in W} e^{s \cdot \mu} ,$$

projects $\mathcal{A}[P]$ onto $\mathcal{A}[P]_{\mathrm{symm}}$. Likewise, the *skew-symmetrizer operator* **Alt**, which acts on e^μ by

$$\mathbf{Alt}(e^\mu) = \frac{1}{|W|} \sum_{s \in W} \mathrm{sgn}(s)\, e^{s \cdot \mu} \,,$$

projects $\mathcal{A}[P]$ onto $\mathcal{A}[P]_{\mathrm{skew}}$.

For $\mu \in P$ we define the *elementary skew-symmetric function*

$$A_\mu = \sum_{s \in W} \mathrm{sgn}(s)\, e^{s \cdot \mu} = |W|\, \mathbf{Alt}(e^\mu) \,.$$

We note that

$$A_{t \cdot \mu} = \sum_{s \in W} \mathrm{sgn}(s)\, e^{(ts) \cdot \mu} = \mathrm{sgn}(t) A_\mu \quad \text{for } t \in W \,. \tag{7.25}$$

Recall from Section 3.1.4 that $\mu \in P$ is *regular* if $s_\alpha(\mu) \neq \mu$ for all $\alpha \in \Phi$, where s_α is the root reflection associated to α. We write P^{reg} for the regular weights and P_{++}^{reg} for the regular dominant weights. If μ is not regular it is called *singular*. Since $\mathrm{sgn}(s_\alpha) = -1$, we see from (7.25) that $A_\mu = 0$ if μ is singular.

Lemma 7.2.1. *The functions* $\{A_\mu : \mu \in P_{++}^{\mathrm{reg}}\}$ *give a basis for* $\mathcal{A}[P]_{\mathrm{skew}}$.

Proof. By (7.25) each nonzero elementary skew-symmetric function is determined (up to sign) by a Weyl group orbit in P^{reg}, and we know from Proposition 3.1.20 that each such orbit contains a unique dominant weight. Hence if μ and ν are distinct dominant regular weights, then the orbits $W \cdot \mu$ and $W \cdot \nu$ are disjoint. It follows that the set $\{A_\mu : \mu \in P_{++}^{\mathrm{reg}}\}$ is linearly independent. This set spans $\mathcal{A}[P]_{\mathrm{skew}}$, since $f = \mathbf{Alt} f$ for $f \in \mathcal{A}[P]_{\mathrm{skew}}$. $\qquad\square$

Let

$$\rho = \frac{1}{2} \sum_{\alpha \in \Phi^+} \alpha \,.$$

From Lemma 3.1.21 we know that $\rho = \varpi_1 + \cdots + \varpi_l$ is the sum of the fundamental dominant weights. Hence ρ is regular. Furthermore, every dominant regular weight λ can be written as $\lambda = \mu + \rho$ with μ dominant, since $\lambda = m_1 \varpi_1 + \cdots + m_l \varpi_l$ with $m_j \geq 1$ for all j.

Define the *Weyl function* $\Delta_{\mathfrak{g}} \in \mathcal{A}[P]$ by

$$\Delta_{\mathfrak{g}} = e^\rho \prod_{\alpha \in \Phi^+} (1 - e^{-\alpha}) \,.$$

We showed in Section 7.1.1 that $\Delta_{\mathfrak{g}}$ is skew-symmetric. Now we give an *a priori* proof of the *Weyl denominator formula* (we will use this formula in our algebraic proof of the character formula).

Proposition 7.2.2. *The Weyl function* $\Delta_{\mathfrak{g}}$ *equals* A_ρ.

Proof. We can expand the product defining $\Delta_{\mathfrak{g}}$ to obtain the formula

$$\Delta_{\mathfrak{g}} = e^\rho + \sum_{\emptyset \neq M \subset \Phi^+} \varepsilon_M \, e^{\rho - \langle M \rangle} \,, \tag{7.26}$$

where $\varepsilon_M = \pm 1$ and $\langle M \rangle = \sum_{\alpha \in M} \alpha$ for a subset $M \subset \Phi^+$. Since $\Delta_{\mathfrak{g}}$ is skew-symmetric, there is cancellation in the right-hand side of (7.26) of all terms involving singular weights, and by Lemma 7.2.1 we can write

$$\Delta_{\mathfrak{g}} = A_\rho + \sum_{\mu \in P_{++}^{\text{reg}} \setminus \{\rho\}} c_\mu A_\mu \tag{7.27}$$

for some coefficients c_μ. Comparing (7.26) and (7.27), we see that to prove $c_\mu = 0$ it suffices to prove the following:

If $M \subset \Phi^+$ and $\rho - \langle M \rangle \in P_{++}^{\text{reg}}$, then $M = \emptyset$. \tag{7.28}

To prove this assertion, set $\mu = \rho - \langle M \rangle$. For every simple root α_i we have

$$0 < (\mu, \check{\alpha}_i) = (\rho, \check{\alpha}_i) - (\langle M \rangle, \check{\alpha}_i) = 1 - (\langle M \rangle, \check{\alpha}_i)$$

(where $\check{\alpha}$ denotes the coroot to α). But $(\langle M \rangle, \check{\alpha}_i)$ is an integer, so this implies that

$$(\langle M \rangle, \check{\alpha}_i) \leq 0 \quad \text{for } i = 1, \ldots, l \,. \tag{7.29}$$

Since each positive root is a nonnegative linear combination of simple coroots, we have $\langle M \rangle = \sum_i k_i \check{\alpha}_i$ with $k_i \geq 0$. Thus

$$0 \leq (\langle M \rangle, \langle M \rangle) = \sum_i k_i (\langle M \rangle, \check{\alpha}_i) \leq 0$$

by (7.29). Hence $\langle M \rangle = 0$, proving (7.28). $\qquad\qquad\qquad\qquad\qquad\square$

7.2.2 Characters and Skew-Symmetric Functions

Let $\lambda \in P_{++}$ and let V^λ be the irreducible finite-dimensional $\mathfrak{g}$-module with highest weight λ. Define the *formal character*

$$\chi_\lambda = \mathrm{ch}(V^\lambda) = \sum_{\mu \in P} m_\lambda(\mu) e^\mu \,,$$

where $m_\lambda(\mu) = \dim V^\lambda(\mu)$ is the *multiplicity* of the weight μ in V^λ. Since $m_\lambda(\mu) = m_\lambda(s \cdot \mu)$ for all $s \in W$ (Proposition 3.2.7), the formal character is a symmetric element of $\mathcal{A}[P]$. If we multiply it by the Weyl denominator, we obtain a skew-symmetric element in $\mathcal{A}[P]$. As a preliminary to proving the Weyl character formula, we establish the following form of this element. Recall from Section 3.2.1 that for

$\mu, \lambda \in P$ the relation $\mu \prec \lambda$ means that $\lambda - \mu$ is a (nonempty) sum of positive roots, with repetitions allowed.

Lemma 7.2.3. *There exist integers n_γ such that*

$$\Delta_{\mathfrak{g}} \chi_\lambda = A_{\lambda+\rho} + \sum_\gamma n_\gamma A_\gamma, \tag{7.30}$$

with the sum over $\gamma \in P_{++}^{\mathrm{reg}}$ such that $\gamma \prec \lambda + \rho$.

Proof. By Corollary 3.2.3 we have

$$\chi_\lambda = e^\lambda + \sum_{\mu \prec \lambda} m_\lambda(\mu) e^\mu .$$

Hence by Proposition 7.2.2 we can write $\Delta_{\mathfrak{g}} \chi_\lambda = A_{\lambda \dot+ \rho} + B$, where

$$B = \sum_{s \in W} \sum_{\mu \prec \lambda} \mathrm{sgn}(s) \, m_\lambda(\mu) e^{\mu + s \cdot \rho} .$$

The function B is skew-symmetric and has integer coefficients. So by Lemma 7.2.1 we know that B is an integer linear combination of the elementary skew-symmetric functions A_γ with $\gamma \in P_{++}^{\mathrm{reg}}$ and $\gamma = \mu + s \cdot \rho$ for some $\mu \prec \lambda$ and $s \in W$. Thus

$$\gamma \prec \lambda + s \cdot \rho = \lambda + \rho + (s \cdot \rho - \rho) .$$

Hence to complete the proof of the lemma, it suffices to show that

$$s \cdot \rho - \rho \preceq 0 \quad \text{for all } s \in W . \tag{7.31}$$

Given $s \in W$, we set $Q(s) = \{\alpha \in \Phi^+ : s \cdot \alpha \in -\Phi^+\}$. Obviously

$$s \cdot \rho = \frac{1}{2} \sum_{\alpha \in \Phi^+ \setminus Q(s)} s \cdot \alpha + \frac{1}{2} \sum_{\alpha \in Q(s)} s \cdot \alpha .$$

If $\beta \in \Phi^+$, then either $s^{-1}\beta = \alpha \in \Phi^+$, in which case $\alpha \in \Phi^+ \setminus Q(s)$, or else $s^{-1} \cdot \beta = -\alpha \in -\Phi^+$, with $\alpha \in Q(s)$. Thus $\Phi^+ = \{s \cdot (\Phi^+ \setminus Q(s))\} \cup \{-s \cdot Q(s)\}$ is a disjoint union. It follows that

$$\rho = \frac{1}{2} \sum_{\alpha \in \Phi^+ \setminus Q(s)} s \cdot \alpha - \frac{1}{2} \sum_{\alpha \in Q(s)} s \cdot \alpha .$$

Subtracting this formula for ρ from the formula for $s \cdot \rho$, we obtain

$$s \cdot \rho - \rho = \sum_{\alpha \in Q(s)} s \cdot \alpha ,$$

which proves (7.31). $\qquad \square$

Remark 7.2.4. The Weyl character formula is the assertion that the coefficients n_γ in formula (7.30) are all zero.

7.2.3 Characters and Invariant Functions

We continue the assumptions and notation of the previous section for the semisimple Lie algebra $\mathfrak{g}$ and Cartan subalgebra $\mathfrak{h} \subset \mathfrak{g}$. Now let G be an algebraically simply connected linear algebraic group with Lie algebra $\mathfrak{g}$. Let $H \subset G$ be the maximal algebraic torus with Lie algebra $\mathfrak{h}$. (When $\mathfrak{g}$ is a classical semisimple Lie algebra we have proved the existence of G and H in Chapters 2 and 6; for the general case see Section 11.2.4.)

Since G is algebraically simply connected, the character group $\mathfrak{X}(H)$ is isomorphic to the additive group P, where $\mu \in P$ determines the character $h \mapsto h^\mu$ for $h \in H$. Furthermore, for every $\lambda \in P_{++}$ there exists an irreducible regular representation (π^λ, V^λ) of G whose differential is the representation of $\mathfrak{g}$ with highest weight λ given by Theorems 3.2.5 and 3.2.6. A function f on G is called *invariant* if $f(gxg^{-1}) = f(x)$ for all $x, g \in G$. Denote the regular invariant functions on G by $\mathcal{O}[G]^G$. The group character φ_λ of V^λ, defined by

$$\varphi_\lambda(g) = \mathrm{tr}(\pi^\lambda(g)) \quad \text{for } g \in G,$$

is an element of $\mathcal{O}[G]^G$.

We can view the formal character χ_λ as a regular function on H. Thus for $h \in H$ we have

$$\varphi_\lambda(h) = \chi_\lambda(h) = \sum_{\nu \in P} m_\lambda(\nu) h^\nu.$$

Likewise, for $\mu \in P_{++}^{\mathrm{reg}}$ we consider the skew-symmetric elements A_μ and $\Delta_\mathfrak{g}$ as functions on H:

$$A_\mu(h) = \sum_{s \in W} \mathrm{sgn}(s) h^{s \cdot \mu}, \qquad \Delta_G(h) = h^\rho \prod_{\alpha \in \Phi^+} (1 - h^{-\alpha}).$$

Set $H' = \{h \in H : \Delta_G(h) \neq 0\}$ (the *regular elements* in H). We define a rational function S_μ on H by

$$S_\mu(h) = \frac{A_\mu(h)}{\Delta_G(h)} \quad \text{for } h \in H'.$$

Since μ is regular, we can write $\mu = \lambda + \rho$ with $\lambda \in P_{++}$. After we have proved the Weyl character formula, we will know that S_μ is the restriction to H' of φ_λ. Our strategy now is to use the existence of φ_λ and Lemma 7.2.3 to show that S_μ extends to an invariant regular function on G.

Lemma 7.2.5. *If $\mu \in P_{++}^{\mathrm{reg}}$ then there exists a unique invariant regular function on G whose restriction to H' is S_μ.*

Proof. We will use an induction relative to the partial order $\prec$ on P_{++}^{reg} defined by the positive roots. For this we need the following property:

$(\star)$ There is no infinite subset $\{\lambda_k : k \in \mathbb{N}\}$ in P_{++}^{reg} such that $\lambda_{k+1} \prec \lambda_k$ for all $k \in \mathbb{N}$.

To prove $(\star)$, suppose that $\lambda_j \in P_{++}^{\mathrm{reg}}$ and $\lambda_k \prec \lambda_{k-1} \prec \cdots \prec \lambda_0$. Since λ_0 is regular, there exists $c > 0$ such that $(\lambda_0, \alpha) \geq c$ for all $\alpha \in \Phi^+$. The definition of the root order implies that $\lambda_k = \lambda_0 - Q_k$ with Q_k a sum of at least k (not necessarily distinct) elements of Φ^+. Since $(\lambda_k, Q_k) \geq 0$ we thus obtain the inequalities

$$0 < (\lambda_k, \lambda_k) = (\lambda_k, \lambda_0 - Q_k) \leq (\lambda_k, \lambda_0) = (\lambda_0, \lambda_0) - (\lambda_0, Q_k)$$
$$\leq (\lambda_0, \lambda_0) - kc \ .$$

This implies that $k < c^{-1}(\lambda_0, \lambda_0)$, and hence $(\star)$ holds.

Let $\mu \in P_{++}^{\mathrm{reg}}$ and write $\mu = \lambda + \rho$ with $\lambda \in P_{++}$. We call μ *minimal* if there is no $\gamma \in P_{++}^{\mathrm{reg}}$ such that $\gamma \prec \mu$. When μ is minimal, then by (7.30) we have

$$\Delta_G(h)\varphi_\lambda(h) = A_\mu(h) \quad \text{for } h \in H \ .$$

Hence $S_\mu(h) = \varphi_\lambda(h)$ for $h \in H'$, and so S_μ extends to a regular invariant function on G in this case. Now suppose μ is not minimal. Assume that S_γ extends to a regular invariant function f_γ on G for all $\gamma \in P_{++}^{\mathrm{reg}}$ with $\gamma \prec \mu$. Then by (7.30) we have

$$\Delta_G(h)\varphi_\lambda(h) = A_\mu(h) + \sum_{\gamma \prec \mu} n_\gamma A_\gamma(h)$$
$$= A_\mu(h) + \sum_{\gamma \prec \mu} n_\gamma \Delta_G(h) f_\gamma(h)$$

for $h \in H$. Hence

$$S_\mu(h) = \varphi_\lambda(h) - \sum_{\gamma \prec \mu} n_\gamma f_\gamma(h)$$

for $h \in H'$, and the right side of this equation is the restriction to H' of a regular invariant function on G. It now follows from $(\star)$ and the induction hypothesis that S_μ extends to a regular invariant function on G.

The regular invariant extension of S_μ is unique, since the set

$$G' = \{ghg^{-1} \in H : g \in G, \ h \in H'\}$$

of regular semisimple elements is Zariski dense in G by Theorem 11.4.18. $\square$

7.2.4 Casimir Operator and Invariant Functions

We continue the assumptions and notation of the previous section. We choose elements $e_\alpha \in \mathfrak{g}_\alpha$ and $e_{-\alpha} \in \mathfrak{g}_{-\alpha}$ that satisfy $B(e_\alpha, e_{-\alpha}) = 1$, and a B-orthonormal basis $h_1, \ldots, h_l$ for $\mathfrak{h}$. In terms of this basis for $\mathfrak{g}$ the Casimir operator from Section 3.3.2 corresponds to the element

$$C = \sum_{i=1}^{l} h_i^2 + \sum_{\alpha \in \Phi^+} (e_\alpha e_{-\alpha} + e_{-\alpha} e_\alpha)$$

of the enveloping algebra $U(\mathfrak{g})$.

Let R be the right-translation representation of G on $\mathcal{O}[G]$. The differentiated representation dR of $\mathfrak{g}$ acts by

$$dR(A)f = X_A f \quad \text{for } A \in \mathfrak{g} \text{ and } f \in \mathcal{O}[G] ,$$

where X_A is the left-invariant vector field on G corresponding to A defined in Section 1.3.7. The Lie algebra representation dR extends canonically to an associative algebra representation of $U(\mathfrak{g})$ as differential operators on $\mathcal{O}[G]$ that we continue to denote by dR; see Appendix C.2.1. Let $\lambda \in P_{++}$ and let (π^λ, V^λ) be the irreducible representation of G with highest weight λ.

Proposition 7.2.6. *For $T \in \text{End}(V^\lambda)$ define $f_T \in \mathcal{O}[G]$ by $f_T(g) = \text{tr}(\pi(g)T)$. Then*

$$dR(C)f_T = ((\lambda + \rho, \lambda + \rho) - (\rho, \rho))f_T . \tag{7.32}$$

Thus f_T is an eigenfunction for the differential operator $dR(C)$. In particular, the character φ_λ of π^λ satisfies (7.32).

Proof. By (1.45) we have $dR(A)f_T = X_A f_T = f_{d\pi^\lambda(A)T}$ for all $A \in \mathfrak{g}$. Extending dR and $d\pi^\lambda$ to representations of $U(\mathfrak{g})$, we obtain $dR(C)f_T = f_{d\pi^\lambda(C)T}$. But C acts by the scalar $(\lambda + \rho, \lambda + \rho) - (\rho, \rho)$ on V^λ by Lemma 3.3.8. This proves (7.32). $\square$

The key step in our proof of the Weyl character formula is the following formula for the action of the differential operator $dR(C)$ on invariant regular functions, expressed in terms of constant-coefficient differential operators on the maximal algebraic torus $H \subset G$ and multiplication by the Weyl function Δ_G.

Theorem 7.2.7. *Let φ be an invariant regular function on G. Then, for $t \in H$,*

$$\Delta_G(t)(dR(C)\varphi)(t) = \left(\sum_{i=1}^{l} X_{h_i}^2 - (\rho, \rho) \right) \Delta_G(t)\varphi(t) . \tag{7.33}$$

We shall prove Theorem 7.2.7 by reducing the calculation to the case $G = \mathbf{SL}(2, \mathbb{C})$. Let

$$e = \begin{bmatrix} 0 & 1 \\ 0 & 0 \end{bmatrix}, \quad f = \begin{bmatrix} 0 & 0 \\ 1 & 0 \end{bmatrix}, \quad h = \begin{bmatrix} 1 & 0 \\ 0 & -1 \end{bmatrix} \tag{7.34}$$

be the standard TDS triple in $\mathfrak{sl}(2, \mathbb{C})$.

Lemma 7.2.8. $(G = \mathbf{SL}(2, \mathbb{C}))$ *Let* $\varphi \in \mathcal{O}[G]^G$ *and let* $y = \mathrm{diag}[z, z^{-1}]$ *with* $z \neq 1$.
Then

$$(X_e X_f + X_f X_e)\varphi(y) = \frac{z + z^{-1}}{z - z^{-1}} X_h \varphi(y) . \tag{7.35}$$

Proof. Define $\varphi_n(y) = (\mathrm{tr}(y))^n$ for $n \in \mathbb{N}$ and $y \in G$. Then $\varphi_n \in \mathcal{O}[G]^G$, and we claim that the functions $\{\varphi_n : n \in \mathbb{N}\}$ are a linear basis for $\mathcal{O}[G]^G$. Indeed, an invariant regular function f is determined by its restriction to the diagonal matrices, and $f(y)$ is a Laurent polynomial $p(z, z^{-1})$ when $y = \mathrm{diag}[z, z^{-1}]$. Since

$$\begin{bmatrix} 0 & 1 \\ -1 & 0 \end{bmatrix} \begin{bmatrix} z & 0 \\ 0 & z^{-1} \end{bmatrix} \begin{bmatrix} 0 & 1 \\ -1 & 0 \end{bmatrix}^{-1} = \begin{bmatrix} z^{-1} & 0 \\ 0 & z \end{bmatrix} ,$$

this polynomial satisfies $p(z, z^{-1}) = p(z^{-1}, z)$, so it is a linear combination of the polynomials $z^n + z^{-n}$. An induction on n using the binomial expansion shows that $z^n + z^{-n}$ is in the span of $(z + z^{-1})^k$ for $0 \leq k \leq n$, proving the claim. Hence to prove formula (7.35) in general, it suffices to check it on the functions φ_n.

Let $A \in \mathfrak{sl}(2, \mathbb{C})$ and $y \in \mathbf{SL}(2, \mathbb{C})$. Then

$$X_A \varphi_n(y) = \frac{d}{ds}\bigg|_{s=0} \left(\mathrm{tr}(y + syA) \right)^n = n\varphi_{n-1}(y)\,\mathrm{tr}(yA) . \tag{7.36}$$

Hence for $B \in \mathfrak{sl}(2, \mathbb{C})$ we have

$$\begin{aligned} X_B X_A \varphi_n(y) &= nX_B \varphi_{n-1}(y)\,\mathrm{tr}(yA) + n\varphi_{n-1}(y)\,\mathrm{tr}(yBA) \\ &= n(n-1)\varphi_{n-2}(y)\,\mathrm{tr}(yA)\,\mathrm{tr}(yB) + n\varphi_{n-1}(y)\,\mathrm{tr}(yBA) . \end{aligned}$$

Reversing A and B in this formula and adding the two cases, we obtain

$$\begin{aligned} (X_A X_B + X_B X_A)\varphi_n(y) &= n\varphi_{n-1}(y)\,\mathrm{tr}(y(AB + BA)) \\ &\quad + 2n(n-1)\varphi_{n-2}(y)\,\mathrm{tr}(yA)\,\mathrm{tr}(yB) . \end{aligned}$$

Now take $A = e$, $B = f$, and $y = \mathrm{diag}[z, z^{-1}]$. Since $ef + fe = I$ and $\mathrm{tr}(ye) = \mathrm{tr}(yf) = 0$, we obtain

$$(X_e X_f + X_f X_e)\varphi_n(y) = n\varphi_{n-1}(y)\,\mathrm{tr}(y) = n(z + z^{-1})^n .$$

Since $\mathrm{tr}(yh) = z - z^{-1}$, from (7.36) we have

$$X_h \varphi_n(y) = n\varphi_{n-1}(y)(z - z^{-1}) = n(z + z^{-1})^{n-1}(z - z^{-1}) .$$

Thus if $z \neq 1$ we conclude that

$$(X_e X_f + X_f X_e)\varphi_n(y) = \frac{z + z^{-1}}{z - z^{-1}} X_h \varphi_n(y)$$

for all $n \in \mathbb{N}$, which suffices to prove (7.35). $\qquad\square$

We now return to the case of a general group G. For $\alpha \in \mathfrak{h}^*$ we denote by $h_\alpha \in \mathfrak{h}$ the element such that $\alpha(h) = B(h_\alpha, H)$ for $h \in \mathfrak{h}$. Then $[e_\alpha, e_{-\alpha}] = h_\alpha$, so $\{e_\alpha, e_{-\alpha}, h_\alpha\}$ is a TDS triple. Let $\{e, f, h\}$ be the TDS triple (7.34).

Lemma 7.2.9. *For each* $\alpha \in \Phi^+$ *there is an algebraic group homomorphism* $\psi : \mathbf{SL}(2, \mathbb{C}) \longrightarrow G$ *such that* $d\psi(e) = e_\alpha$, $d\psi(f) = e_{-\alpha}$, *and* $d\psi(h) = h_\alpha$.

Proof. We may assume that G is an algebraic subgroup of $\mathbf{GL}(n, \mathbb{C})$, so that we have $\mathfrak{g} \subset \mathfrak{gl}(n, \mathbb{C})$. Hence there is a Lie algebra representation π of $\mathfrak{sl}(2, \mathbb{C})$ on $\mathbb{C}^n$ such that $\pi(e) = e_\alpha$, $\pi(f) = e_{-\alpha}$, and $\pi(h) = h_\alpha$. By Corollary 2.3.8 there is a regular representation ψ of $\mathbf{SL}(2, \mathbb{C})$ on $\mathbb{C}^n$ with $d\psi = \pi$. Theorem 1.6.2 implies that

$$\psi(\exp(te)) = \exp(te_\alpha) \in G \quad \text{and} \quad \psi(\exp(tf)) = \exp(te_{-\alpha}) \in G$$

for all $t \in \mathbb{C}$. Since $\mathbf{SL}(2, \mathbb{C})$ is generated by the one-parameter unipotent subgroups $t \mapsto \exp(te)$ and $t \mapsto \exp(tf)$ (Lemma 2.2.1), it follows that $\psi(\mathbf{SL}(2, \mathbb{C})) \subset G$. $\square$

Proof of Theorem 7.2.7: Let $\varphi \in \mathcal{O}[G]^G$. Fix $\alpha \in \Phi^+$ and let $\psi : \mathbf{SL}(2, \mathbb{C}) \longrightarrow G$ be the homomorphism in Lemma 7.2.9. Then for $y \in \mathbf{SL}(2, \mathbb{C})$ we have

$$X_{e_\alpha} X_{e_{-\alpha}} \varphi(\psi(y)) = X_e X_f (\varphi \circ \psi)(y),$$

and the analogous formula with α and $-\alpha$ interchanged. The regular function $\varphi \circ \psi$ on $\mathbf{SL}(2, \mathbb{C})$ is invariant, so if $y = \operatorname{diag}[z, z^{-1}]$ and $z \neq 1$, then (7.35) gives

$$(X_{e_\alpha} X_{e_{-\alpha}} + X_{e_{-\alpha}} X_{e_\alpha}) \varphi(\psi(y)) = \frac{z + z^{-1}}{z - z^{-1}} X_{h_\alpha} \varphi(\psi(y)).$$

We now write this equation in terms intrinsic to G. Let $t = \psi(y)$. Since the root α is the image under $d\psi^*$ of the root $\varepsilon_1 - \varepsilon_2$ of $\mathfrak{sl}(2, \mathbb{C})$, we have $t^\alpha = y^{\varepsilon_1 - \varepsilon_2} = z^2$. Hence

$$(X_{e_\alpha} X_{e_{-\alpha}} + X_{e_{-\alpha}} X_{e_\alpha}) \varphi(t) = \frac{1 + t^{-\alpha}}{1 - t^{-\alpha}} X_{h_\alpha} \varphi(t). \tag{7.37}$$

We claim that (7.37) is true for all $t \in H'$. To prove this, let $T_\alpha \subset H$ be the image under ψ of the diagonal matrices in $\mathbf{SL}(2, \mathbb{C})$, and let

$$T^\alpha = \{t \in H : t^\alpha = 1\}.$$

If $t_0 \in T^\alpha$, then $\operatorname{Ad}(t_0) e_{\pm\alpha} = e_{\pm\alpha}$. This implies that the right translation operator $R(t_0)$ commutes with the vector fields $X_{e_{\pm\alpha}}$, so that we have

$$(X_{e_\alpha} X_{e_{-\alpha}} + X_{e_{-\alpha}} X_{e_\alpha}) \varphi(t_1 t_0) = R(t_0)(X_{e_\alpha} X_{e_{-\alpha}} + X_{e_{-\alpha}} X_{e_\alpha}) \varphi(t_1) \tag{7.38}$$

for all $t_1 \in H$. Given $t \in H'$, we can write $t = \exp A$ with $A \in \mathfrak{h}$, and then factor $t = t_1 t_0$ with

$$t_0 = \exp\left(A - \frac{1}{2}\alpha(A) h_\alpha\right) \quad \text{and} \quad t_1 = \exp\left(\frac{1}{2}\alpha(A) h_\alpha\right) \in T_\alpha.$$

Since $\alpha(h_\alpha) = 2$, we have $t_0 \in T^\alpha$; furthermore, $(t_1)^\alpha = t^\alpha \neq 1$. Hence from (7.38) we conclude that (7.37) holds, as claimed.

At this point we have established the formula

$$\Delta_G(t)(dR(C)\varphi)(t) = \Delta_G(t)\left(\sum_i X_{h_i}^2 + \sum_{\alpha \in \Phi^+} \frac{1+t^{-\alpha}}{1-t^{-\alpha}} X_{h_\alpha}\right)\varphi(t) \tag{7.39}$$

for $\varphi \in \mathcal{O}[G]^G$ and $t \in H'$. It remains only to show that (7.39) is the same as (7.33). For $t \in H$ we have

$$\left(\sum_i X_{h_i}^2\right)\Delta_G(t)\varphi(t) = \Delta_G(t)\left(\sum_i X_{h_i}^2\right)\varphi(t) + 2\sum_i \left(X_{h_i}\Delta_G(t)\right)\left(X_{h_i}\varphi(t)\right)$$

$$+ \varphi(t)\left(\sum_i X_{h_i}^2\right)\Delta_G(t). \tag{7.40}$$

If $t = \exp Y \in H'$ with $Y \in \mathfrak{h}$, then $\Delta_G(t) = \prod_{\alpha \in \Phi^+}\left(e^{\alpha(Y)/2} - e^{-\alpha(Y)/2}\right)$. For $A \in \mathfrak{h}$ we use this product formula and logarithmic differentiation to calculate

$$2\Delta_G(t)^{-1}X_A\Delta_G(t) = 2\Delta_G(t)^{-1}\left.\frac{d}{dz}\right|_{z=0}\Delta_G(\exp(Y+zA))$$

$$= \alpha(A)\sum_{\alpha \in \Phi^+}\frac{e^{\alpha(Y)/2} + e^{-\alpha(Y)/2}}{e^{\alpha(Y)/2} - e^{-\alpha(Y)/2}}$$

$$= \alpha(A)\sum_{\alpha \in \Phi^+}\frac{1+t^{-\alpha}}{1-t^{-\alpha}}.$$

Taking $A = h_i$ and summing over i, we obtain

$$2\sum_i \left(X_{h_i}\Delta_G(t)\right)\left(X_{h_i}\varphi(t)\right) = \Delta_G(t)\sum_{\alpha \in \Phi^+}\frac{1+t^{-\alpha}}{1-t^{-\alpha}}\left(\sum_i \alpha(h_i)X_{h_i}\right)\varphi(t)$$

$$= \Delta_G(t)\sum_{\alpha \in \Phi^+}\frac{1+t^{-\alpha}}{1-t^{-\alpha}}X_{h_\alpha}\varphi(t).$$

Finally, from the sum formula for $\Delta_G(t)$ (Proposition 7.2.2) we calculate that

$$X_A^2\Delta_G(t) = \sum_{s \in W} \mathrm{sgn}(s)\langle s\cdot\rho, A\rangle^2 t^{s\cdot\rho}$$

for $A \in \mathfrak{h}$. Taking $A = h_i$ and summing over i, we obtain

$$\sum_i X_{h_i}^2\Delta_G(t) = \sum_{s \in W} \mathrm{sgn}(s)\left(\sum_i \langle s\cdot\rho, h_i\rangle^2\right)t^{s\cdot\rho}$$

$$= \sum_{s \in W} \mathrm{sgn}(s)(s\cdot\rho, s\cdot\rho)t^{s\cdot\rho} = (\rho,\rho)\Delta_G(t).$$

From these calculations and (7.40) we conclude that formulas (7.39) and (7.33) are identical, which completes the proof the theorem.   □

Corollary 7.2.10. *For* $\lambda \in P_{++}^{\text{reg}}$ *the invariant function* S_λ *on* G *is an eigenfunction of the differential operator* $dR(C)$:

$$dR(C)S_\lambda = ((\lambda,\lambda) - (\rho,\rho))S_\lambda . \tag{7.41}$$

Proof. Since $dR(C)$ commutes with left and right translations by elements of G, the functions on both sides of equation (7.41) are G-invariant. Hence it is enough to verify this equation on the set H'. By (7.33) we have

$$dR(C)S_\lambda(t) + (\rho,\rho)S_\lambda(t) = \Delta_G(t)^{-1}\sum_{i=1}^{l} X_{h_i}^2 A_\lambda(t)$$
$$= \Delta_G(t)^{-1}\sum_{s\in W} \text{sgn}(s)\,(s\cdot\lambda, s\cdot\lambda)\,t^{s\cdot\lambda}$$
$$= (\lambda,\lambda)S_\lambda(t)$$

for $t \in H'$.   □

Remark 7.2.11. Note that the character $t \mapsto t^\alpha$ of H, for $\alpha \in P$, is an eigenfunction of the vector field X_{h_i} with eigenvalue $\langle\alpha, h_i\rangle$. Thus the right side of (7.33) can be calculated explicitly when $\varphi|_H$ is given as a linear combination of characters of H.

7.2.5 Algebraic Proof of the Weyl Character Formula

We now prove the Weyl character formula (7.2). Let $\lambda \in P_{++}$ be a dominant integral weight. By (7.30) and Lemma 7.2.5 we can write the character φ_λ of the irreducible representation with highest weight λ as a linear combination of the invariant regular functions S_μ:

$$\varphi_\lambda = S_{\lambda+\rho} + \sum_{\mu\prec\lambda+\rho} n_\mu S_\mu .$$

All the functions in this formula are in $\mathcal{O}[G]^G$ and are eigenfunctions of $dR(C)$, by Proposition 7.2.6 and Corollary 7.2.10. The functions φ_λ and $S_{\lambda+\rho}$ have the same eigenvalue $(\lambda+\rho,\lambda+\rho) - (\rho,\rho)$, whereas the function S_μ for $\mu \prec \lambda+\rho$ has the eigenvalue $(\mu,\mu) - (\rho,\rho)$. We may assume $\mu \in P_{++}$. Then by the argument in the proof of Proposition 3.3.9 there is a strict inequality $(\mu,\mu) < (\lambda+\rho,\lambda+\rho)$. Since eigenfunctions corresponding to distinct eigenvalues are linearly independent, all the coefficients n_μ in the formula are zero. Hence $\varphi_\lambda = S_{\lambda+\rho}$, which is Weyl's character formula.

7.2.6 Exercises

In the following exercises $\mathfrak{g}$ is a semisimple Lie algebra, and the notation follows that of Section 7.2.1.

1. Let V be the irreducible representation of $\mathfrak{g}$ with highest weight ρ.
   (a) Show that $\dim V = 2^r$, where $r = |\Phi^+|$.
   (b) Show that $\rho - \langle Q \rangle$ is a weight of V for every $Q \subset \Phi^+$.
   (c) Define $\xi = e^\rho \prod_{\alpha \in \Phi^+} (1 + e^{-\alpha})$. Show that ξ is invariant under W.
   (d) Prove that ξ is the character of V.
2. Let (π, V) be a finite-dimensional representation of $\mathfrak{g}$.
   (a) Suppose $X \in \mathfrak{g}_\alpha$ and $Y \in \mathfrak{g}_{-\alpha}$. Prove that for all $\mu \in P(\mathfrak{g})$,

$$\text{tr}((\pi(X)\pi(Y) - \langle \mu, [X,Y] \rangle I)|_{V(\mu)}) = \text{tr}(\pi(X)\pi(Y)|_{V(\mu+\alpha)}) \, .$$

(HINT: It suffices to take $X = e_\alpha$ and $Y = f_\alpha$, so $[X,Y] = h_\alpha$. By complete reducibility it suffices to verify the formula when V is irreducible for the TDS triple $\{e_\alpha, f_\alpha, h_\alpha\}$. We may assume $V(\mu + p\alpha) \neq 0$ and $V(\mu + (p+1)\alpha) = 0$ for some integer p. Now apply Proposition 2.3.3 with $n = m + 2p$, where $m = \langle \mu, h_\alpha \rangle$, to calculate the matrix of $\pi(e_\alpha)\pi(f_\alpha)$ relative to the basis $\{v_k\}$.)
   (b) Set $m(\mu) = \dim V(\mu)$ for $\mu \in P(\mathfrak{g})$. Show that

$$\text{tr}(\pi(X)\pi(Y)|_{V(\mu)}) = \sum_{k \geq 0} \langle \mu + k\alpha, [X,Y] \rangle m(\mu + k\alpha) \, ,$$

$$\text{tr}(\pi(Y)\pi(X)|_{V(\mu)}) = \sum_{k \geq 1} \langle \mu + k\alpha, [X,Y] \rangle m(\mu + k\alpha) \, .$$

(HINT: The first formula follows by iterating (a). For the second formula, use $\text{tr}(\pi(X)\pi(Y)|_{V(\mu)}) = \text{tr}(\pi(Y)\pi(X)|_{V(\mu)}) + \langle \mu, [X,Y] \rangle m(\mu).$)
   (c) Assume that V is the irreducible $\mathfrak{g}$-module with highest weight λ. Show that the weight multiplicities of V satisfy *Freudenthal's recurrence formula*

$$((\lambda + \rho, \lambda + \rho) - (\mu + \rho, \mu + \rho)) m_\lambda(\mu) = 2 \sum_{\alpha \in \Phi^+} \sum_{k \geq 1} (\mu + k\alpha, \alpha) m_\lambda(\mu + k\alpha) \, .$$

Note that for $\mu \in \mathcal{X}(V)$ and $\mu \neq \lambda$, the coefficient of $m_\lambda(\mu)$ on the left is positive (see the proof of Proposition 3.3.9). Since $m_\lambda(\lambda) = 1$ and the coefficients are nonnegative, this formula gives a recursive algorithm for calculating weight multiplicities that is more efficient than the alternating formula (7.6). (HINT: The Casimir operator C acts by the scalar $(\lambda + \rho, \lambda + \rho) - (\rho, \rho)$ on V and $\sum_i h_i^2 + 2H_\rho$ acts by $(\mu + \rho, \mu + \rho) - (\rho, \rho)$ on $V(\mu)$. Now use (b) and Lemma 3.3.7 (4).)

7.3 Compact Group Approach to the Character Formula

We now give Weyl's original proof of the character formula, based on integration on a compact real form and the Weyl integral formula. We first study the compact real form, the conjugacy classes of regular semisimple elements, and the Weyl group of a connected semisimple algebraic group. This requires some general results about algebraic groups from Chapter 11; however, for the classical groups direct matrix calculations using the defining representations suffice.

7.3.1 Compact Form and Maximal Compact Torus

Let G be a connected algebraic group whose Lie algebra $\mathfrak{g}$ is semisimple. Let H be a maximal (algebraic) torus in G with Lie algebra $\mathfrak{h}$. Then $\mathfrak{h}$ is a Cartan subalgebra of $\mathfrak{g}$. Let $\Phi = \Phi(\mathfrak{g},\mathfrak{h})$ be the roots of $\mathfrak{h}$ on $\mathfrak{g}$ and let B be the Killing form on $\mathfrak{g}$. For $\alpha \in \Phi$ let $H_\alpha \in \mathfrak{h}$ be the element such that $B(H_\alpha,h) = \langle \alpha,h \rangle$ for all $h \in \mathfrak{h}$. A fundamental result of Weyl asserts that one can choose elements $X_\alpha \in \mathfrak{g}_\alpha$ for each $\alpha \in \Phi$ such that the following holds:

- The triple $\{X_\alpha,X_{-\alpha},H_\alpha\}$ satisfies $[X_\alpha,X_{-\alpha}] = H_\alpha$.
- The linear transformation $\theta : \mathfrak{g} \longrightarrow \mathfrak{g}$ given by $\theta(X_\alpha) = -X_{-\alpha}$ for $\alpha \in \Phi$ and $\theta(h) = -h$ for $h \in \mathfrak{h}$ is a Lie algebra automorphism.
- The subspace

$$\mathfrak{u} = \sum_{\alpha \in \Phi} \mathbb{R}iH_\alpha + \sum_{\alpha \in \Phi} \mathbb{R}(X_\alpha - X_{-\alpha}) + \sum_{\alpha \in \Phi} \mathbb{R}i(X_\alpha + X_{-\alpha}) \qquad (7.42)$$

is a real Lie algebra, and the restriction of B to $\mathfrak{u} \times \mathfrak{u}$ is negative definite.
- There is a compact connected subgroup $U \subset G$ with Lie algebra $\mathfrak{u}$.

For details, see Helgason [66, Theorem III.6.3] or Knapp [86, Theorem 6.11 and Section VII.1]. It is clear from (7.42) that $\mathfrak{g} = \mathfrak{u} + i\mathfrak{u}$; hence U is a compact real form of G. One knows that U is unique up to conjugation; see Sections 11.5.1 and 11.5.2. Let τ be the conjugation of $\mathfrak{g}$ relative to $\mathfrak{u}$. Then $\mathfrak{h}$ is invariant under τ.

Remark 7.3.1. The automorphism θ is an *involution* of $\mathfrak{g}$. We see from the definition of θ and (7.42) that $\tau\theta = \theta\tau$. We shall study such pairs (τ,θ) in great detail in Chapters 11 and 12.

When G is a connected classical group we can take the matrix form of $\mathfrak{g}$ invariant under the involution $\theta(X) = -X^t$ so that the conjugation relative to a compact real form is $\tau(X) = -\overline{X}^t$. The resulting compact real form U of G is given in Section 1.7.2 (note that for the orthogonal groups this is *not* the same matrix form for G that we used in Section 2.1.2). We fix a τ-stable maximal torus H in G as follows:

Type A $(G = \mathbf{SL}(n,\mathbb{C}))$: Let H be the group of all matrices (where $n = l+1$)

$$h(x_1,\ldots,x_l) = \mathrm{diag}[x_1,\ldots,x_n] \quad \text{with } x_j \in \mathbb{C}^\times \text{ and } x_1\cdots x_n = 1 \,.$$

Type C ($G = \mathbf{Sp}(l,\mathbb{C})$): Let H be the group of all matrices

$$h(x_1,\ldots,x_l) = \mathrm{diag}[x_1,\ldots,x_l,x_l^{-1},\ldots,x_1^{-1}] \quad \text{with } x_j \in \mathbb{C}^\times \,.$$

Type D ($G = \mathbf{SO}(2l,\mathbb{C})$): Define

$$R(x) = \frac{1}{2}\begin{bmatrix} (x+x^{-1}) & -\mathrm{i}(x-x^{-1}) \\ \mathrm{i}(x-x^{-1}) & (x+x^{-1}) \end{bmatrix} \quad \text{with } x \in \mathbb{C}^\times \,.$$

Then $x \mapsto R(x)$ is an injective regular homomorphism from $\mathbb{C}^\times$ to $\mathbf{GL}(2,\mathbb{C})$, $R(x)^* = R(\bar{x})$, and $R(x)R(x)^t = I$. (Note that $R(\mathrm{e}^{\mathrm{i}\theta}) \in U$ is rotation through the angle $\theta \in \mathbb{R}/2\pi\mathbb{Z}$.) Let H be the group of all block-diagonal matrices

$$h(x_1,\ldots,x_l) = \begin{bmatrix} R(x_1) & 0 & \cdots & 0 \\ 0 & R(x_2) & \cdots & 0 \\ \vdots & \vdots & \ddots & \vdots \\ 0 & 0 & \cdots & R(x_l) \end{bmatrix} \quad \text{with } x_1,\ldots,x_l \in \mathbb{C}^\times \,. \tag{7.43}$$

When $G = \mathbf{Spin}(2l,\mathbb{C})$ we let H be the inverse image of this torus under the covering map from G to $\mathbf{SO}(2l,\mathbb{C})$ (see Theorem 6.3.5).

Type B ($G = \mathbf{SO}(2l+1,\mathbb{C})$): Take H as the group of all matrices

$$\begin{bmatrix} h(x_1,\ldots,x_l) & 0 \\ 0 & 1 \end{bmatrix} \quad \text{with } x_1,\ldots,x_l \in \mathbb{C}^\times \,.$$

Here $h(x_1,\ldots,x_l) \in \mathbf{SO}(2l,\mathbb{C})$ is defined as in (7.43). For $G = \mathbf{Spin}(2l+1,\mathbb{C})$ let H be the inverse image of this torus under the covering map from G to $\mathbf{SO}(2l+1,\mathbb{C})$.

We return to the compact real form U with Lie algebra $\mathfrak{u}$ given by (7.42). Let $\mathbb{T} = \{x \in \mathbb{C} : |x| = 1\} \cong \mathbb{R}/\mathbb{Z}$. We shall call a Lie group isomorphic to $\mathbb{T}^l$ a *compact torus* of *rank l*. Let

$$\mathfrak{h}_\mathbb{R} = \sum_{\alpha \in \Phi} \mathbb{R}H_\alpha$$

and define $\mathfrak{t} = \mathrm{i}\mathfrak{h}_\mathbb{R} \subset \mathfrak{u}$.

Proposition 7.3.2. *Let $T = \exp\mathfrak{t}$. Then the following hold:*

1. $T = H \cap U$ *and T is a closed maximal commutative subgroup of U.*
2. T *is a compact torus in U of rank $l = \dim H$.*

Proof. (1): Since $H = \exp\mathfrak{h}$, the polar decomposition of G (Theorem 11.5.9) shows that $H \cap U = T$. If $k \in U$ commutes with T, then $\mathrm{Ad}(k)h = h$ for all $h \in \mathfrak{t}$. But $\mathfrak{h} = \mathfrak{t} + \mathrm{i}\mathfrak{t}$, so we have $\mathrm{Ad}(k)h = h$ for all $h \in \mathfrak{h}$. Hence k commutes with H. Since k is a semisimple element of G and H is a maximal algebraic torus in G, this implies that $k \in H$ by Theorem 11.4.10, and hence $k \in T$.

(2): Let $P(G) \subset \mathfrak{h}_{\mathbb{R}}^*$ be the weight lattice of G, relative to the fixed choice of maximal torus H (see Section 3.1.3). Let $h \in \mathfrak{h}_{\mathbb{R}}$. Then $\exp(2\pi ih) = I$ if and only if $\sigma(\exp(2\pi ih)) = I$ for all irreducible regular representations σ of G. Thus

$$\exp(2\pi ih) = I \quad \text{if and only if} \quad \langle \alpha, h \rangle \in \mathbb{Z} \text{ for all } \alpha \in P(G) . \qquad (7.44)$$

The subset of $\mathfrak{h}_{\mathbb{R}}$ defined by (7.44) is the lattice $P(G)^{\vee}$ in $\mathfrak{h}_{\mathbb{R}}$ dual to $P(G)$. Hence the map $h \mapsto \exp(2\pi ih)$ induces a Lie group isomorphism $\mathfrak{h}_{\mathbb{R}}/P(G)^{\vee} \cong T$. This proves that T is a torus of dimension l. $\qquad\qquad\qquad\qquad\qquad\qquad\qquad\qquad\qquad\qquad\qquad\qquad\qquad\square$

Remark 7.3.3. For the classical groups with maximal algebraic torus H taken in the matrix form as above, the maximal compact torus T is obtained by taking the coordinates $x_j \in \mathbb{T}^1$.

7.3.2 Weyl Integral Formula

We continue with the assumptions and notation of the previous section. Now we consider the conjugacy classes in U. Define the map

$$\varphi : (U/T) \times T \longrightarrow U, \qquad \varphi(u \cdot o, h) = uhu^{-1} ,$$

where we denote the coset $\{T\} \in U/T$ by o. We shall show that φ is surjective and obtain the celebrated *Weyl integral formula* for $\varphi^* \omega_U$, where ω_U is the invariant volume form on U (see Appendix D.2.4).

Let

$$\mathfrak{p} = \bigoplus_{\alpha \in \Phi} \mathfrak{g}_\alpha .$$

Then $\mathfrak{g} = \mathfrak{h} \oplus \mathfrak{p}$. Since the complex conjugation automorphism τ preserves $\mathfrak{h}$, it leaves the subspace $\mathfrak{p}$ invariant. Hence there exists an $\mathrm{Ad}(T)$-stable real subspace $V \subset \mathfrak{u}$ such that $\mathfrak{p} = V + iV$. Then $\mathfrak{u} = V \oplus \mathfrak{t}$ and $V \cong \mathfrak{u}/\mathfrak{t}$ as a T-module. In particular, since the roots $\alpha \in \Phi$ take purely imaginary values on $\mathfrak{t}$, we have

$$\det(\mathrm{Ad}(h)|_V) = \prod_{\alpha \in \Phi^+} |h^\alpha|^2 = 1$$

for $h \in T$. Hence U/T is an orientable manifold (see Appendix D.1.3).

The group $U \times T$ acts on $(U/T) \times T$ by left translation:

$$L_{(g,h)}(x \cdot o, y) = (gx \cdot o, hy) \quad \text{for } g, x \in U \text{ and } h, y \in T .$$

We identify the smooth functions on $(U/T) \times T$ with the smooth functions f on $U \times T$ such that $f(gk, h) = f(g, h)$ for all $k \in T$. We then have a bijection between $V \oplus \mathfrak{t}$ and the tangent space of $(U/T) \times T$ at (o, I), where $X \in V$ and $Y \in \mathfrak{t}$ give the tangent vector

$$(X \oplus Y)f = \frac{d}{dt} f\big(\exp(tX), \exp(tY)\big)\Big|_{t=0} \quad \text{for } f \in C^\infty(U/T, T)$$

(see Appendix D.2.3). We use this bijection to identify $V \oplus \mathfrak{t}$ with this tangent space (as a real vector space).

Lemma 7.3.4. *Fix $g \in U$ and $h \in T$ and define the map*

$$\psi = L_{gh^{-1}g^{-1}} \circ \varphi \circ L_{(g,h)} : (U/T) \times T \longrightarrow U .$$

Then $\psi(o, I) = I$ and

$$d\psi_{(o,I)}(X \oplus Y) = \mathrm{Ad}(g)(\mathrm{Ad}(h^{-1})X - X + Y) \tag{7.45}$$

for $X \in V$ and $Y \in \mathfrak{t}$. Furthermore,

$$\det(d\psi_{(o,I)}) = |\Delta_G(h)|^2 , \tag{7.46}$$

where Δ_G is the Weyl function on H.

Proof. By definition,

$$\begin{aligned}
d\psi_{(o,I)}(X \oplus Y) &= \frac{d}{dt}\left\{ gh^{-1}\exp(tX)h\exp(tY)\exp(-tX)g^{-1} \right\}\Big|_{t=0} \\
&= \frac{d}{dt}\left\{ g\exp(t\,\mathrm{Ad}(h^{-1})X)\exp(tY)\exp(-tX)g^{-1} \right\}\Big|_{t=0} \\
&= \mathrm{Ad}(g)(\mathrm{Ad}(h^{-1})X - X + Y) ,
\end{aligned}$$

where we are using an embedding $G \subset \mathbf{GL}(n, \mathbb{C})$ to calculate the differential via the matrix exponential and matrix multiplication. To calculate the determinant, we extend $d\psi_{(o,I)}$ to a complex-linear transformation on $\mathfrak{g}$. Let Φ be the roots of H on $\mathfrak{g}$. The eigenvalues of $\mathrm{Ad}(h^{-1})$ on $\mathfrak{p}$ are $\{h^{-\alpha} : \alpha \in \Phi\}$, with multiplicity one. Hence we have the factorization

$$\det(I - \mathrm{Ad}(h^{-1})\big|_{\mathfrak{g}/\mathfrak{h}}) = h^\rho h^{-\rho} \prod_{\alpha \in \Phi}(1 - h^{-\alpha}) = \Delta_G(h)\Delta_G(h^{-1}). \tag{7.47}$$

Now $\det \mathrm{Ad}(g) = 1$ for $g \in G$, and $h^{-\alpha} = \overline{h^\alpha}$ for $h \in T$, so $\Delta_G(h^{-1}) = \overline{\Delta_G(h)}$. Thus we obtain (7.46) from (7.47) and (7.45). $\qquad \square$

Remark 7.3.5. The function $|\Delta_G(h)|$ is always single-valued on T because $|h^\rho| = 1$ for $h \in T$.

Set
$$T'' = \{h \in T : \Delta_G(h) \neq 0 \text{ and } whw^{-1} \neq h \text{ for } 1 \neq w \in W_G\} .$$

Since Δ_G is a nonzero real-analytic function on T, we know that T'' is open and dense in T. Let $\{X_1, \ldots, X_r\}$ be a basis for $\mathfrak{u}$ such that $\{X_1, \ldots, X_l\}$ is a basis for $\mathfrak{t}$ and $\{X_{l+1}, \ldots, X_r\}$ is a basis for V. Let ω_T, $\omega_{U/T}$, and ω_U be left-invariant volume forms on T, U/T, and U, respectively, normalized relative to this basis as in Appendix D.2.4.

Theorem 7.3.6. *The restriction of φ to $(U/T) \times T''$ is a covering map of degree $|W_G|$. Furthermore,*

$$(\varphi^* \omega_U)_{u \cdot o, h} = |\Delta_G(h)|^2 (\omega_T)_h \wedge (\omega_{U/T})_{u \cdot o} \tag{7.48}$$

for $u \in U$ and $h \in T''$.

Proof. From Lemma 7.3.4 we see that $d\varphi_{(u \cdot o, h)}$ is nonsingular for all $u \in U$ and $h \in T''$. Since $\dim(U/T) + \dim(T) = \dim U$, we conclude from the inverse function theorem that the restriction of φ to $(U/T) \times T''$ is a local diffeomorphism. By Lemma 11.4.15 and Corollary 11.4.13 we have

$$|\varphi^{-1}(uhu^{-1})| = |W_G| \quad \text{for } h \in T'' \text{ and } u \in U .$$

Formula (7.48) now follows from (7.46) and Theorem D.1.17.   □

Corollary 7.3.7 (Weyl Integral Formula). *Let $f \in C(U)$. Then*

$$\int_U f(u) \, du = \frac{1}{|W_G|} \int_T |\Delta_G(h)|^2 \left\{ \int_{U/T} f(uhu^{-1}) \, d\dot{u} \right\} dh , \tag{7.49}$$

where $du = \omega_U$, $d\dot{u} = \omega_{U/T}$, and $dh = \omega_T$.

Proof. The complement of T'' has measure zero in T. Hence we may replace the integral over T by the integral over T'' on the right side of (7.49) without changing the value of the integral. With this done, formula (7.49) follows immediately from Theorems 7.3.6 and D.1.17.   □

Corollary 7.3.8. *The map $\varphi : (U/T) \times T \longrightarrow U$ is surjective. Hence every element of U is U-conjugate to an element of T.*

Proof. Since $(U/T) \times T$ is compact, the image of φ is closed in U. If it were not all of U, there would be a nonzero function $f \in C(U)$ such that $f \geq 0$ and $f = 0$ on the image of U. However, then $f(uhu^{-1}) = 0$ for all $u \in U$ and $h \in T$, and thus

$$\int_U f(u) \, du = 0$$

by (7.49). This contradicts Lemma D.1.15.   □

7.3.3 Fourier Expansions of Skew Functions

We continue with the assumptions and notation of the previous section. Now we also assume that G is semisimple and algebraically simply connected, so the weight lattice of G is $P(\mathfrak{g})$. We write $P = P(\mathfrak{g})$, $P_{++} = P_{++}(\mathfrak{g})$, and $W = W_G$, the Weyl group of G. For $\mu \in P$ we define the elementary skew-symmetric function A_μ on H by

$$A_\mu(h) = \sum_{s \in W} \text{sgn}(s) h^{s \cdot \mu} .$$

We noted in Section 7.2.2 that $A_{t \cdot \mu} = \text{sgn}(t) A_\mu$ for $t \in W$. Recall from Section 3.1.4 that $\mu \in P$ is *regular* if $s_\alpha(\mu) \neq \mu$ for all $\alpha \in \Phi$, where s_α is the root reflection associated to α. We write P_{++}^{reg} for the regular dominant weights. If μ is not regular it is called *singular*. Since $\text{sgn}(s_\alpha) = -1$, we have $A_\mu = 0$ if μ is singular.

We normalize the invariant measures on U and T to have total mass 1.

Lemma 7.3.9. *Suppose* $\lambda, \mu \in P_{++}^{\text{reg}}$. *Then*

$$\frac{1}{|W|} \int_T A_\lambda(t) \overline{A_\mu(t)} \, dt = \begin{cases} 0 \ \text{if } \lambda \neq \mu , \\ 1 \ \text{if } \lambda = \mu . \end{cases}$$

Proof. If $\lambda \neq \mu$ then the orbits $\{s \cdot \lambda : s \in W\}$ and $\{s \cdot \mu : s \in W\}$ are disjoint and have $|W|$ elements, since λ, μ are dominant and regular (see Proposition 3.1.20). Hence for $s, s' \in W$ we have

$$\int_T t^{s \cdot \lambda} \overline{t^{s' \cdot \mu}} \, dt = \int_T t^{s \cdot \lambda - s' \cdot \mu} \, dt = 0 ,$$

since $\int_T t^\alpha \, dt = 0$ for any $0 \neq \alpha \in P$. This implies the orthogonality of A_λ and A_μ.

If $\lambda = \mu$, then the same argument gives

$$\int_T |A_\lambda(t)|^2 \, dt = \sum_{s \in W} \int_T |t^{s \cdot \lambda}|^2 \, dt = |W| . \qquad \square$$

Proposition 7.3.10. *Suppose* $\varphi \in \mathcal{O}[H]$ *satisfies* $\varphi(shs^{-1}) = \text{sgn}(s)\varphi(h)$ *for all* $s \in W$. *Then*

$$\varphi = \sum_{\mu \in P_{++}^{\text{reg}}} c(\mu) A_\mu$$

for unique coefficients $c(\mu) \in \mathbb{C}$, *and*

$$\frac{1}{|W|} \int_T |\varphi(t)|^2 \, dt = \sum_{\mu \in P_{++}^{\text{reg}}} |c(\mu)|^2 . \tag{7.50}$$

Proof. The function φ has an expansion

$$\varphi(h) = \sum_{\mu \in P} c(\mu) h^\mu$$

with $c(\mu) \in \mathbb{C}$ and $c(\mu) = 0$ for all but a finite number of μ. Since this expansion is unique, the skew symmetry of φ is equivalent to $c(s \cdot \mu) = \text{sgn}(s) c(\mu)$ for all $s \in W$. In particular, if μ is singular then $c(\mu) = 0$. Thus we may write the expansion of φ as

$$\varphi(h) = \sum_{s \in W} \sum_{\mu \in P_{++}^{\text{reg}}} c(s \cdot \mu) h^{s \cdot \mu} = \sum_{\mu \in P_{++}^{\text{reg}}} c(\mu) \sum_{s \in W} \text{sgn}(s) h^{s \cdot \mu} = \sum_{\mu \in P_{++}^{\text{reg}}} c(\mu) A_\mu(h) .$$

Now (7.50) follows from Lemma 7.3.9. $\square$

7.3.4 Analytic Proof of the Weyl Character Formula

We continue with the assumptions and notation of the previous section. For every $\lambda \in P_{++}$ there is an irreducible regular representation (π^λ, V^λ) with highest weight λ, by Corollary 6.3.9 and Theorem 11.2.14. Let φ_λ be the character of π^λ. We write $\Delta = \Delta_G$ for the Weyl function and $\rho = \rho_G$. Since $\varphi_\lambda|_H$ is invariant under W, the function $h \mapsto \Delta(h)\varphi_\lambda(h)$ on H is skew-symmetric. Hence by Proposition 7.3.10 we can write

$$\Delta(h)\varphi_\lambda(h) = \sum_{\mu \in P_{++}^{\text{reg}}} c(\mu) A_\mu(h) .$$

Note that because μ in this formula is dominant and regular, $c(\mu)$ is the coefficient of h^μ when we write out this expansion in terms of the characters of H (see the proof of Lemma 7.3.9). We claim that $c(\lambda + \rho) = 1$. Indeed, expanding the product formula for Δ gives

$$\Delta(h) = h^\rho + \sum_{\nu \prec \rho} a_\nu t^\nu$$

for some $a_\nu \in \mathbb{Z}$. Furthermore, by Corollary 3.2.3,

$$\varphi_\lambda(h) = h^\lambda + \sum_{\nu \prec \lambda} m_\lambda(\nu) h^\nu ,$$

where $m_\lambda(\nu)$ is the multiplicity of the weight ν in V^λ. Hence the coefficient of $h^{\lambda + \rho}$ in $\Delta \varphi_\lambda|_H$ is 1. However, since λ is dominant and ρ is regular, we have $\lambda + \rho \in P_{++}^{\text{reg}}$. By the observation above, we conclude that $c(\lambda + \rho) = 1$.

By Proposition 7.3.10,

$$\frac{1}{|W|} \int_T |\varphi_\lambda(t)|^2 |\Delta(t)|^2 \, dt = \sum_{\mu \in P_{++}^{\text{reg}}} |c(\mu)|^2 .$$

Hence by the Weyl integral formula (Corollary 7.3.7),

$$\int_U |\varphi_\lambda(k)|^2 \, dk = \sum_{\mu \in P_{++}^{\text{reg}}} |c(\mu)|^2 . \tag{7.51}$$

The Schur orthogonality relations in Lemma 4.3.3 are also valid for U when the averaging over the finite group G in that lemma is replaced by integration relative to the normalized invariant measure on U (with this change, the proof is the same). This implies that the left side of (7.51) is 1. Since $c(\lambda + \rho) = 1$, this forces $c(\mu) = 0$

for $\mu \neq \lambda + \rho$. Hence $\Delta(h)\varphi_\lambda(h) = A_{\lambda+\rho}(h)$ for $h \in H$, which is the Weyl character formula.

7.3.5 Exercises

In these exercises G is $\mathbf{SL}(n,\mathbb{C})$, $\mathbf{SO}(n,\mathbb{C})$, $\mathbf{Spin}(n,\mathbb{C})$, or $\mathbf{Sp}(n,\mathbb{C})$ (for the orthogonal or spin groups assume $n \geq 3$); H is a maximal algebraic torus of G; and U is a compact real form of G with maximal compact torus $T \subset H$.

1. Let $\mathfrak{t}$ be the (real) Lie algebra of T.
   (a) Verify that $\mathfrak{t} = i\mathfrak{h}_\mathbb{R}$, where $\mathfrak{h}_\mathbb{R} = \sum_{\alpha \in \Phi} \mathbb{R} h_\alpha$ (here h_α is the coroot to α).
   (b) Let $\check{Q} \subset \mathfrak{h}_\mathbb{R}$ be the *coroot lattice* $\sum_{\alpha \in \Phi} \mathbb{Z} h_\alpha$. Show that $\exp(2\pi i \check{Q}) = 1$.
   (c) Assume that G is algebraically simply connected (thus take G as $\mathbf{Spin}(n,\mathbb{C})$ rather than $\mathbf{SO}(n,\mathbb{C})$). Show that the exponential map gives a group isomorphism $T \cong \mathfrak{t}/(2\pi i \check{Q})$, where $\mathfrak{t}$ is viewed as a group under addition. (HINT: Recall that $\check{Q}$ is the dual lattice to the weight lattice and that for every fundamental weight λ there is a representation of G with highest weight λ.)

2. Suppose $h = \exp(ix)$ with $x \in \mathfrak{h}_\mathbb{R}$. Show that

$$|\Delta_G(h)|^2 = 2^r \prod_{\alpha \in \Phi^+} \sin^2\left(\frac{\langle \alpha, x \rangle}{2}\right),$$

   where r is the number of roots of G.

3. Take coordinate functions $x_1, \ldots, x_l$ on H as in Section 2.1.2. Let

$$\Gamma = \{[\theta_1, \ldots, \theta_l] : -\pi \leq \theta_j \leq \pi \text{ for } j = 1, \ldots, l\} \subset \mathbb{R}^l.$$

   Let $h(\theta) \in T$ be the element with coordinates $x_j = \exp(i\theta_j)$ for $\theta \in \Gamma$.
   (a) Show that the density function in the Weyl integral formula is given in these coordinates as follows:
   Type A_l:

$$|\Delta(h(\theta))|^2 = 2^{l(l+1)} \prod_{1 \leq i < j \leq l+1} \sin^2\left(\frac{\theta_i - \theta_j}{2}\right), \qquad (\theta_{l+1} = -(\theta_1 + \cdots + \theta_l))$$

   Type B_l:

$$|\Delta(h(\theta))|^2 = 2^{2l^2} \prod_{1 \leq i < j \leq l} \sin^2\left(\frac{\theta_i - \theta_j}{2}\right) \sin^2\left(\frac{\theta_i + \theta_j}{2}\right) \prod_{1 \leq i \leq l} \sin^2\left(\frac{\theta_i}{2}\right)$$

   Type C_l:

$$|\Delta(h(\theta))|^2 = 2^{2l^2} \prod_{1 \leq i < j \leq l} \sin^2\left(\frac{\theta_i - \theta_j}{2}\right) \sin^2\left(\frac{\theta_i + \theta_j}{2}\right) \prod_{1 \leq i \leq l} \sin^2(\theta_i)$$

   Type D_l:

$$|\Delta(h(\theta))|^2 = 2^{2l(l-1)} \prod_{1 \leq i < j \leq l} \sin^2\left(\frac{\theta_i - \theta_j}{2}\right) \sin^2\left(\frac{\theta_i + \theta_j}{2}\right)$$

(HINT: Use the previous exercise.)

(b) Suppose f is a continuous central function on U. Deduce from the Weyl integral formula that

$$\int_U f(u)\,du = \frac{1}{|W|(2\pi)^l} \int_\Gamma f(h(\theta))\,|\Delta(h(\theta))|^2\,d\theta_1 \cdots d\theta_l,$$

where $d\theta_1 \cdots d\theta_l$ is the l-dimensional Euclidean measure on $\mathbb{R}^l$ and r is the number of roots of G.

(c) Use the W-invariance of $f|_T$ to reduce the domain of integration in (b) from a cube to a simplex.

7.4 Notes

Section 7.1.1. For the case $G = \mathbf{GL}(n, \mathbb{C})$ the Weyl character formula (as a formula for the ratio of two determinants, with no notion of group representations) was first proved by Jacobi (see Weyl [164, p. 203, footnote 12]). For $G = \mathbf{SO}(n, \mathbb{C})$ it was proved by Schur [131]; Weyl [162] subsequently generalized Schur's proof to all compact semisimple Lie groups. See Hawkins [63, Chapter 12] and Borel [17, Chapter III] for detailed historical accounts.

Section 7.1.3. The inspiration for the character formulas of this section was a lecture of D.N. Verma at Rutgers University in 1988.

Section 7.2.3. The inductive proof in Lemma 7.2.5 that the ratio of the Weyl numerator and Weyl denominator extends to a regular function on G is new. It exploits the existence of the global character without requiring an explicit formula for the character.

Section 7.2.4. Formula (7.33) for the action of the Casimir operator on invariant functions is the differential operator analogue of the Weyl integral formula.

Section 7.3.2. The Weyl integral formula (for a general semisimple compact Lie group) appears in Weyl [162]. The surjectivity of the map φ in Corollary 7.3.8 can be proved for the classical groups along the same lines as in Theorem 2.1.7. The proof given here, however, shows the power of the analytic methods and also applies to every connected compact Lie group, once one has the Weyl integral formula.

Section 7.3.4. Our analytic proof of the character formula is the same as that in Weyl [162].

Chapter 8
Branching Laws

Abstract Since each classical group G fits into a descending family of classical groups, the irreducible representations of G can be studied inductively. This gives rise to the *branching problem*: Given a pair $G \supset H$ of reductive groups and an irreducible representation π of G, find the decomposition of $\pi|_H$ into irreducible representations. In this chapter we solve this problem for the pairs $\mathbf{GL}(n,\mathbb{C}) \supset \mathbf{GL}(n-1,\mathbb{C})$, $\mathbf{Spin}(n,\mathbb{C}) \supset \mathbf{Spin}(n-1,\mathbb{C})$, and $\mathbf{Sp}(n,\mathbb{C}) \supset \mathbf{Sp}(n-1,\mathbb{C})$. We show that the representations occurring in $\pi|_H$ are characterized by a simple interlacing condition for their highest weights. For the first and second pairs the representation $\pi|_H$ is multiplicity-free; in the symplectic case the multiplicities are given in terms of the highest weights by a product formula. We prove all these results by a general formula due to Kostant that expresses branching multiplicities as an alternating sum over the Weyl group of G of a suitable partition function. In each case we show that this alternating sum can be expressed as a determinant. The explicit evaluation of the determinant then gives the branching law.

8.1 Branching for Classical Groups

Let G be a reductive linear algebraic group and $H \subset G$ a reductive algebraic subgroup. When an irreducible regular representation of G is restricted to H, it is no longer irreducible, in general, but decomposes into a sum of irreducible H modules. A *branching law* from G to H is a description of the H-irreducible representations and their multiplicities that occur in the decomposition of any irreducible representation of G.

Now assume that G and H are connected. We have already seen that the irreducible representations of G and H are parameterized by their highest weights, so a branching law can be stated entirely in terms of these parameters. By the conjugacy of maximal tori (cf. Section 2.1.2 for the classical groups, and Section 11.4.5 for a general reductive group), we may choose maximal tori T_G in G and T_H in H such that $T_H \subset T_G$. Let λ and μ be dominant integral weights for G and H, respectively,

R. Goodman, N.R. Wallach, *Symmetry, Representations, and Invariants*,
Graduate Texts in Mathematics 255, DOI 10.1007/978-0-387-79852-3_8,
© Roe Goodman and Nolan R. Wallach 2009

and let V^λ and V^μ be the corresponding irreducible G-module and H-module. Set

$$m(\lambda,\mu) = \mathrm{mult}_{V^\lambda}(V^\mu) .$$

A $G \to H$ *branching law* is an explicit description of the multiplicity function $m(\lambda,\mu)$, with λ ranging over all dominant weights of T_G. In case the multiplicities are always either 0 or 1 we say that the branching is *multiplicity-free*.

8.1.1 Statement of Branching Laws

We now state some branching laws for the classical groups that we will prove later in the chapter. Denote by $\mathbb{Z}^n_{++}$ the set of all integer n-tuples $\lambda = [\lambda_1, \ldots, \lambda_n]$ such that

$$\lambda_1 \ge \lambda_2 \ge \cdots \ge \lambda_n .$$

Let $\mathbb{N}^n_{++} \subset \mathbb{Z}^n_{++}$ be the subset of all such weakly decreasing n-tuples with $\lambda_n \ge 0$. For $\lambda \in \mathbb{N}^n$ let $|\lambda| = \sum_{i=1}^n \lambda_i$.

Let $G = \mathbf{GL}(n,\mathbb{C})$. Take $H \cong \mathbf{GL}(n-1,\mathbb{C})$ as the subgroup of matrices $\begin{bmatrix} y & 0 \\ 0 & 1 \end{bmatrix}$, where $y \in \mathbf{GL}(n-1,\mathbb{C})$. For $\lambda = [\lambda_1, \ldots, \lambda_n] \in \mathbb{Z}^n_{++}$ let $(\pi_n^\lambda, F_n^\lambda)$ be the irreducible representation of G from Theorem 5.5.22 with highest weight $\sum_{i=1}^n \lambda_i \varepsilon_i$. Let $\mu = [\mu_1, \ldots, \mu_{n-1}] \in \mathbb{Z}^{n-1}_{++}$. We say that μ *interlaces* λ if

$$\lambda_1 \ge \mu_1 \ge \lambda_2 \ge \cdots \ge \lambda_{n-1} \ge \mu_{n-1} \ge \lambda_n .$$

Theorem 8.1.1. *The branching from* $\mathbf{GL}(n,\mathbb{C})$ *to* $\mathbf{GL}(n-1,\mathbb{C})$ *is multiplicity-free. The multiplicity* $m(\lambda,\mu)$ *is* 1 *if and only if* μ *interlaces* λ.

An easy consequence of this result is the following branching law from $G = \mathbf{GL}(n,\mathbb{C})$ to $H = \mathbf{GL}(n-1,\mathbb{C}) \times \mathbf{GL}(1,\mathbb{C})$:

Theorem 8.1.2. *Let* $\lambda \in \mathbb{N}^n_{++}$. *There is a unique decomposition*

$$F_n^\lambda = \bigoplus_\mu M^\mu \tag{8.1}$$

under the action of $\mathbf{GL}(n-1,\mathbb{C}) \times \mathbf{GL}(1,\mathbb{C})$, *where the sum is over all* $\mu \in \mathbb{N}^{n-1}_{++}$ *such that* μ *interlaces* λ. *Here* $M^\mu \cong F_{n-1}^\mu$ *as a module for* $\mathbf{GL}(n-1,\mathbb{C})$, *and* $\mathbf{GL}(1,\mathbb{C})$ *acts on* M^μ *by the character* $z \mapsto z^{|\lambda|-|\mu|}$.

Next we take $G = \mathbf{Spin}(\mathbb{C}^{2n+1}, B)$ with B as in (2.9). Fix a B-isotropic basis $\{e_0, e_{\pm 1}, \ldots, e_{\pm n}\}$ as in Section 2.4.1. Let $\pi : G \longrightarrow \mathbf{SO}(\mathbb{C}^{2n+1}, B)$ be the covering map from Theorem 6.3.5 and set $H = \{g \in G : \pi(g)e_0 = e_0\}$. Then $H \cong \mathbf{Spin}(2n,\mathbb{C})$.

By Theorem 6.3.6 we may identify $\mathfrak{g} = \mathrm{Lie}(G)$ with $\mathfrak{so}(\mathbb{C}^{2n+1}, B)$ in the matrix realization of Section 2.4.1 and $\mathrm{Lie}(H)$ with $\mathfrak{h} = \{X \in \mathfrak{g} : Xe_0 = 0\}$. Let $\varpi_n = [1/2, \ldots, 1/2] \in \mathbb{R}^n$ and let $\lambda \in \mathbb{N}^n_{++} + \varepsilon\varpi_n$, where ε is 0 or 1. We say that λ is *integral* if $\varepsilon = 0$ and *half-integral* if $\varepsilon = 1$. We identify λ with the dominant weight $\sum_{i=1}^n \lambda_i \varepsilon_i$ for $\mathfrak{g}$ as in Proposition 3.1.20. The half-integral weights are highest weights of representations of G that are not representations of $\mathbf{SO}(\mathbb{C}^{2n+1}, B)$. Likewise, the dominant weights for $\mathfrak{h} = \mathrm{Lie}(H)$ are identified with the n-tuples $\mu = [\mu_1, \ldots, \mu_n]$ such that $[\mu_1, \ldots, \mu_{n-1}, |\mu_n|] \in \mathbb{N}^n_{++} + \varepsilon\varpi_n$.

Theorem 8.1.3. *The branching from* $\mathbf{Spin}(2n+1)$ *to* $\mathbf{Spin}(2n)$ *is multiplicity-free. The multiplicity* $m(\lambda, \mu)$ *is 1 if and only if* λ *and* μ *are both integral or both half-integral and*

$$\lambda_1 \geq \mu_1 \geq \lambda_2 \geq \cdots \geq \lambda_{n-1} \geq \mu_{n-1} \geq \lambda_n \geq |\mu_n| . \tag{8.2}$$

Let $G = \mathbf{Spin}(2n, \mathbb{C})$. By Theorem 6.3.6 we may identify $\mathfrak{g} = \mathrm{Lie}(G)$ with $\mathfrak{so}(2n, \mathbb{C})$ in the matrix realization of Section 2.4.1. Let $\pi : G \longrightarrow \mathbf{SO}(\mathbb{C}^{2n}, B)$ as in Theorem 6.3.5 and set $H = \{g \in G : \pi(g)(e_n + e_{n+1}) = e_n + e_{n+1}\}$. Then $H \cong \mathbf{Spin}(2n-1, \mathbb{C})$ and we identify $\mathrm{Lie}(H)$ with

$$\mathfrak{h} = \{X \in \mathfrak{g} : X(e_n + e_{n+1}) = 0\} .$$

Theorem 8.1.4. *The branching from* $\mathbf{Spin}(2n)$ *to* $\mathbf{Spin}(2n-1)$ *is multiplicity-free. The multiplicity* $m(\lambda, \mu)$ *is 1 if and only if* λ *and* μ *are both integral or both half-integral and*

$$\lambda_1 \geq \mu_1 \geq \lambda_2 \geq \cdots \geq \lambda_{n-1} \geq \mu_{n-1} \geq |\lambda_n| . \tag{8.3}$$

We now turn to the branching law from $G = \mathbf{Sp}(n, \mathbb{C})$ to $H = \mathbf{Sp}(n-1, \mathbb{C})$, where $n \geq 2$. In this case the restriction is not multiplicity-free. The highest weights that occur satisfy a *double interlacing condition* and the multiplicities are given by a product formula.

We take G in the matrix form of Section 2.1.2. Let $\{e_{\pm i} : i = 1, \ldots, n\}$ be an isotropic basis for $\mathbb{C}^{2n}$ as in Section 2.4.1 and take $H = \{h \in G : he_{\pm n} = e_{\pm n}\}$. Let $\lambda \in \mathbb{N}^n_{++}$ be identified with a dominant integral weight for G by Proposition 3.1.20 and let $\mu \in \mathbb{N}^{n-1}_{++}$ be a dominant integral weight for H.

Theorem 8.1.5. ($\mathbf{Sp}(n) \to \mathbf{Sp}(n-1)$ **Branching Law**) *The multiplicity* $m(\lambda, \mu)$ *is nonzero if and only if*

$$\lambda_j \geq \mu_j \geq \lambda_{j+2} \quad \text{for } j = 1, \ldots, n-1 \tag{8.4}$$

(here $\lambda_{n+1} = 0$*). When these inequalities are satisfied let*

$$x_1 \geq y_1 \geq x_2 \geq y_2 \geq \cdots \geq x_n \geq y_n$$

be the nonincreasing rearrangement of $\{\lambda_1, \ldots, \lambda_n, \mu_1, \ldots, \mu_{n-1}, 0\}$*. Then*

$$m(\lambda, \mu) = \prod_{j=1}^n (x_j - y_j + 1) . \tag{8.5}$$

8.1.2 Branching Patterns and Weight Multiplicities

We can use the $\mathbf{GL}_n \to \mathbf{GL}_{n-1}$ branching law to obtain a canonical basis of weight vectors for the irreducible representations of $\mathbf{GL}(n,\mathbb{C})$ and a combinatorial algorithm for weight multiplicities.

Let $\lambda = [\lambda_1,\ldots,\lambda_n] \in \mathbb{N}_{++}^n$. We shall identify λ with its *Ferrers diagram* (also called a *Young diagram*). This diagram consists of p left-justified rows of boxes with λ_i boxes in the ith row. Here p is the largest index i such that $\lambda_i > 0$, and we follow the convention of numbering the rows from the top down, so the longest row occurs at the top. For example, $\lambda = [4,3,1]$ is identified with the diagram

We say that a Ferrers diagram with p rows has *depth* p (the term *length* is often used). The total number of boxes in the diagram is $|\lambda| = \sum_i \lambda_i$.

We can describe the branching law in Theorem 8.1.2 in terms of Ferrers diagrams. All diagrams of the highest weights $\mu \in \mathbb{N}_{++}^{n-1}$ that interlace λ are obtained from the diagram of λ as follows:

Box removal rule.   Remove all the boxes in the nth row of λ (if there are any). Then remove at most $\lambda_k - \lambda_{k+1}$ boxes from the end of row k, for $k = 1,\ldots,n-1$.

We shall indicate this process by putting the integer n in each box of the diagram of λ that is removed to obtain the diagram of μ. For example, if $\lambda = [4,3,1] \in \mathbb{N}_{++}^3$, then $\mu = [3,2]$ interlaces λ. The scheme for obtaining the diagram of μ from the diagram of λ is

$$
\begin{array}{|c|c|c|}
\hline
 & & 3 \\
\hline
\end{array}
$$

Note that an element $y = \mathrm{diag}[I_{n-1},z]$ of $\mathbf{GL}(n-1,\mathbb{C}) \times \mathbf{GL}(1,\mathbb{C})$ acts on the space M^μ in (8.1.2) by the scalar z^ν, where $\nu = |\lambda| - |\mu|$ is the number of boxes containing the integer n.

We can iterate the branching law in Theorem 8.1.2. Let $\mu^{(k)} \in \mathbb{N}_{++}^k$. We say that $\gamma = \{\mu^{(n)},\mu^{(n-1)},\ldots,\mu^{(1)}\}$ is an *n-fold branching pattern* if $\mu^{(k-1)}$ interlaces $\mu^{(k)}$ for $k = n, n-1,\ldots,2$. Call the Ferrers diagram of $\mu^{(n)}$ the *shape* of γ. We shall encode a branching pattern by placing integers in the boxes of its shape as follows:

Branching pattern rule.   Start with the Ferrers diagram for $\mu^{(n)}$. Write the number n in each box removed from this diagram to obtain the diagram for $\mu^{(n-1)}$. Then repeat the process, writing the number $n-1$ in each box removed from the diagram of $\mu^{(n-1)}$ to obtain $\mu^{(n-2)}$, and so forth, down to the diagram for $\mu^{(1)}$. Then write 1 in the remaining boxes.

This rule fills the shape of γ with numbers from the set $\{1, 2, \ldots, n\}$ (repetitions can occur, and not all numbers need appear). For example, if $\gamma = \{\mu^{(3)}, \mu^{(2)}, \mu^{(1)}\}$ with $\mu^{(3)} = [4, 3, 1]$, $\mu^{(2)} = [3, 2]$, and $\mu^{(1)} = [2]$, then we encode γ by

1	1	2	3
2	2	3	
3			

Each n-fold branching pattern thus gives rise to a Ferrers diagram of at most n rows, with each box filled with a positive integer $j \leq n$, such that

(i) the numbers in each row are nondecreasing from left to right, and
(ii) the numbers in each column are strictly increasing from top to bottom.

Conversely, any Ferrers diagram of at most n rows with integers $j \leq n$ inserted that satisfy these two conditions comes from a unique n-fold branching pattern with the given diagram as shape. We shall study such numbered Ferrers diagrams (also called *semistandard tableaux*) in more detail in Chapter 9.

Let T_n be the subgroup of diagonal matrices in $\mathbf{GL}(n, \mathbb{C})$. For $0 \leq k \leq n$ we define $L_{n,k}$ to be the subgroup of $\mathbf{GL}(n, \mathbb{C})$ consisting of all block diagonal matrices

$$g = \begin{bmatrix} x & 0 \\ 0 & y \end{bmatrix} \quad \text{with } x \in \mathbf{GL}(k, \mathbb{C}) \text{ and } y = \text{diag}[y_1, \ldots, y_{n-k}] \in T_{n-k} \,.$$

Thus we have a decreasing chain of subgroups

$$\mathbf{GL}(n, \mathbb{C}) = L_{n,n} \supset L_{n,n-1} \supset \cdots \supset L_{n,1} = T_n$$

connecting $\mathbf{GL}(n, \mathbb{C})$ with its maximal torus T_n.

Proposition 8.1.6. *Let $\lambda \in \mathbb{N}_{++}^n$ and let $\gamma = \{\mu^{(n)}, \ldots, \mu^{(1)}\}$ be an n-fold branching pattern of shape λ. There is a unique flag of subspaces $F_n^\lambda \supset M_{n-1}^\gamma \supset \cdots \supset M_1^\gamma$ such that for $1 \leq k \leq n - 1$ the following hold:*

1. M_k^γ is invariant and irreducible under $L_{n,k}$.

2. $M_k^\gamma \cong F_k^{\mu^{(k)}}$ as a module for the subgroup $\mathbf{GL}(k, \mathbb{C}) \times I_{n-k}$ of $L_{n,k}$.

The element $\text{diag}[I_k, x_{k+1}, \ldots, x_n] \in L_{n,k}$ acts by the scalar $x_{k+1}^{b_{k+1}} \cdots x_n^{b_n}$ on M_k^γ, where $b_j = |\mu^{(j)}| - |\mu^{(j-1)}|$ for $j = 1, \ldots, n$ (with the convention $\mu^{(0)} = 0$).

Proof. This follows from Theorem 8.1.2 by induction on n. ☐

The space M_1^γ in Proposition 8.1.6 is irreducible under T_n; hence it is one-dimensional. Fix a nonzero element $u_\gamma \in M_1^\gamma$. Define $b(\gamma) = [b_1, \ldots, b_n] \in \mathbb{N}^n$, where $b_j = |\mu^{(j)}| - |\mu^{(j-1)}|$ as in the proposition. Then u_γ is a weight vector of weight $b(\gamma)$. We call $b(\gamma)$ the *weight* of γ. If we encode γ by inserting numbers in the Ferrers diagram of λ following the branching pattern rule, then b_j is the number of boxes that contain the integer j (this is the number of boxes that are removed in passing from $\mu^{(j)}$ to $\mu^{(j-1)}$).

Corollary 8.1.7 (Gelfand–Cetlin Basis). *Let $\lambda \in \mathbb{N}^n_{++}$. The set $\{u_\gamma\}$, where γ ranges over all n-fold branching patterns of shape λ, is a basis for F_n^λ. Hence the weights of F_n^λ are in $\mathbb{N}^n$ and have multiplicities*

$$\dim F_n^\lambda(\mu) = \#\{\ \text{n-fold branching patterns of shape } \lambda \text{ and weight } \mu\ \}\ .$$

Proof. This follows from Theorem 8.1.2 by induction on n. $\square$

The weight multiplicities $K_{\lambda\mu} = \dim F_n^\lambda(\mu)$ are called *Kostka coefficients*. Note that these are the coefficients in the character of F_n^λ:

$$\mathrm{ch}_{\mathbf{GL}(n)}(F_n^\lambda)(x) = \sum_{\mu \in \mathbb{N}^n} K_{\lambda\mu} x^\mu \qquad \text{for } x \in T_n\ .$$

Example

Let

$$\gamma = \begin{array}{|c|c|c|c|} \hline 1 & 1 & 2 & 3 \\ \hline 2 & 2 & 3 \\ \cline{1-3} 3 \\ \cline{1-1} \end{array}$$

as above. Then γ has shape $[4,3,1]$ and weight $[2,3,3]$. There is one other threefold branching pattern with the same shape and weight, namely

$$\begin{array}{|c|c|c|c|} \hline 1 & 1 & 2 & 2 \\ \hline 2 & 3 & 3 \\ \cline{1-3} 3 \\ \cline{1-1} \end{array}$$

Hence the weight $[2,3,3]$ has multiplicity 2 in the representation $F_3^{[4,3,1]}$ of $\mathbf{GL}(3,\mathbb{C})$.

8.1.3 Exercises

1. Let $\mathfrak{g} = \mathfrak{so}(2n+1,\mathbb{C})$ and $\mathfrak{h} = \mathfrak{so}(2n,\mathbb{C})$. For $\lambda \in \mathbb{N}^n_{++} + \varepsilon\varpi_n$ let $\mathbf{B}_n^\lambda$ denote the irreducible $\mathfrak{g}$-module with highest weight λ. For $\mu = [\mu_1, \ldots, \mu_n]$ with $[\mu_1, \ldots, \mu_{n-1}, |\mu_n|] \in \mathbb{N}^n_{++} + \varepsilon\varpi_n$ let $\mathbf{D}_n^\mu$ denote the irreducible $\mathfrak{h}$-module with highest weight μ. Here $\varpi_n = [1/2, \ldots, 1/2]$ and $\varepsilon = 0, 1$.

   (a) Use the branching law to show that the defining representation, the spin representation, and the adjoint representation of $\mathfrak{g}$ decompose under restriction to $\mathfrak{h}$ as follows:

$$\mathbf{B}_n^{[1]} \longrightarrow \mathbf{D}_n^{[1]} \oplus \mathbf{D}_n^{[0]} \qquad \text{for } n \geq 2,$$

$$\mathbf{B}_n^{[1/2,\dots,1/2]} \longrightarrow \mathbf{D}_n^{[1/2,\dots,1/2,1/2]} \oplus \mathbf{D}_n^{[1/2,\dots,1/2,-1/2]} \qquad \text{for } n \geq 2,$$

$$\mathbf{B}_n^{[1,1]} \longrightarrow \mathbf{D}_n^{[1,1]} \oplus \mathbf{D}_n^{[1]} \qquad \text{for } n \geq 3.$$

Here $[1] = [1,0,\dots,0]$ and $[1,1] = [1,1,0,\dots,0]$ (zeros added as necessary).

(b) Obtain the results in (a) without using the branching law.

(c) Take $n = 3$ and let $\rho = [5/2, 3/2, 1/2]$ be the smallest dominant regular weight for $\mathfrak{g}$. Find the decomposition of $\mathbf{B}_3^\rho$ under restriction to $\mathfrak{h}$, and express the highest weights of the components in terms of the fundamental weights $\varpi_1 = [1,0,0]$, $\varpi_2 = [1/2, 1/2, -1/2]$, $\varpi_3 = [1/2, 1/2, 1/2]$ of $\mathfrak{h}$. Check by calculating the sum of the dimensions of the components (recall that $\dim \mathbf{B}_n^\rho = 2^{n^2}$).

2. (Notation as in previous exercise) Let $\mathfrak{g} = \mathfrak{so}(2n, \mathbb{C})$ and $\mathfrak{h} = \mathfrak{so}(2n-1, \mathbb{C})$ with $n \geq 2$.

(a) Use the branching law to show that the defining representation, the half-spin representations, and the adjoint representation of $\mathfrak{g}$ decompose under restriction to $\mathfrak{h}$ as follows:

$$\mathbf{D}_n^{[1]} \longrightarrow \mathbf{B}_{n-1}^{[1]} \oplus \mathbf{B}_{n-1}^{[0]},$$

$$\mathbf{D}_n^{[1/2,\dots,1/2,\pm 1/2]} \longrightarrow \mathbf{B}_{n-1}^{[1/2,\dots,1/2]},$$

$$\mathbf{D}_n^{[1,1]} \longrightarrow \mathbf{B}_{n-1}^{[1,1]} \oplus \mathbf{B}_{n-1}^{[1]}.$$

(b) Obtain the results in (a) without using the branching law.

(c) Take $n = 3$ and let $\rho = [2,1,0]$ be the smallest dominant regular weight for $\mathfrak{g}$. Find the decomposition of $\mathbf{D}_3^\rho$ under restriction to $\mathfrak{h}$ and express the highest weights of the components in terms of the fundamental weights $\varpi_1 = [1,0]$ and $\varpi_2 = [\frac{1}{2}, \frac{1}{2}]$ of $\mathfrak{h}$. Check by calculating the sum of the dimensions of the components (recall that $\dim \mathbf{D}_n^\rho = 2^{n^2 - n}$).

3. Let $G = \mathbf{Sp}(n, \mathbb{C})$ and $H = \mathbf{Sp}(n-1, \mathbb{C})$. Denote by $\mathbf{C}_n^\lambda$ the irreducible $\mathbf{Sp}(n, \mathbb{C})$ module with highest weight $\lambda \in \mathbb{N}_{++}^n$. Use the branching law to obtain the decompositions

$$\mathbf{C}_n^{[1]} \longrightarrow \mathbf{C}_{n-1}^{[1]} \oplus 2\mathbf{C}_{n-1}^{[0]},$$

$$\mathbf{C}_n^{[2]} \longrightarrow \mathbf{C}_{n-1}^{[2]} \oplus 2\mathbf{C}_{n-1}^{[1]} \oplus 3\mathbf{C}_{n-1}^{[0]},$$

for the restrictions to H of the defining representation and the adjoint representation of G, where we identify $[k]$ with $[k, 0, \dots, 0]$. Check by comparing the dimensions of both sides. For the defining representation, give a geometric interpretation of the summand $2\mathbf{C}_{n-1}^{[0]}$. In the case of the adjoint representation, show that the summand $3\mathbf{C}_{n-1}^{[0]}$ corresponds to another copy of $\mathfrak{sp}(1, \mathbb{C}) \subset \mathrm{Lie}(G)$ that commutes with $\mathrm{Lie}(H)$.

4. Use the branching law to obtain the following decompositions (notation as in the previous exercise):
   (a) Let $G = \mathbf{Sp}(2, \mathbb{C})$ and $H = \mathbf{Sp}(1, \mathbb{C})$. Let $\rho = [2, 1]$. Show that

$$\mathbf{C}_2^\rho \longrightarrow 2\mathbf{C}_1^{[2]} \oplus 4\mathbf{C}_1^{[1]} \oplus 2\mathbf{C}_1^{[0]}$$

under restriction to H. Check by comparing the dimensions of both sides.
   (b) Let $G = \mathbf{Sp}(3, \mathbb{C})$ and $H = \mathbf{Sp}(2, \mathbb{C})$. Let $\rho = [3, 2, 1]$. Show that

$$\mathbf{C}_3^\rho \longrightarrow 2\mathbf{C}_2^{[3,2]} \oplus 4\mathbf{C}_2^{[3,1]} \oplus 2\mathbf{C}_2^{[3,0]} \oplus 4\mathbf{C}_2^{[2,2]} \oplus 8\mathbf{C}_2^{[2,1]} \oplus 4\mathbf{C}_2^{[2,0]} \oplus 4\mathbf{C}_2^{[1,1]} \oplus 2\mathbf{C}_2^{[1,0]}$$

under restriction to H. Check by comparing the dimensions of both sides.
5. Let $\lambda = [\lambda_1, \dots, \lambda_n] \in \mathbb{N}_{++}^n$ and set $\check{\lambda} = [\lambda_n, \dots, \lambda_1]$.
   (a) By the theorem of the highest weight and Corollary 8.1.7, there is a unique n-fold branching pattern of shape λ and weight λ. What is it?
   (b) Find an n-fold branching pattern of shape λ and weight $\check{\lambda}$, and prove that it is unique.
6. Let $\lambda = [3, 1, 0]$ and consider the representation $(\pi_3^\lambda, F_3^\lambda)$ of $\mathbf{GL}(3, \mathbb{C})$.
   (a) Use the Weyl dimension formula to show that $\dim F_3^\lambda = 15$.
   (b) Find all the branching patterns of shape λ and weight $[3, 1, 0]$, $[2, 2, 0]$, or $[2, 1, 1]$. Show that these weights are the only dominant weights that occur in F_3^λ.
   (c) Use (b) to write down the character of π_3^λ. Check by carrying out the division in the Weyl character formula (with the help of a computer algebra system).

8.2 Branching Laws from Weyl Character Formula

We turn to the proof of the branching laws. We will obtain a general multiplicity formula, due to Kostant, from which we will derive the branching laws for the classical groups.

8.2.1 Partition Functions

Let $H \subset G$ be reductive linear algebraic groups with Lie algebras $\mathfrak{h} \subset \mathfrak{g}$. Fix maximal algebraic tori T_G in G and T_H in H with $T_H \subset T_G$. Let $\mathfrak{t}_\mathfrak{g}$ and $\mathfrak{t}_\mathfrak{h}$ be the corresponding Lie algebras. Let $\Phi_\mathfrak{g}$ be the roots of $\mathfrak{t}_\mathfrak{g}$ on $\mathfrak{g}$ and $\Phi_\mathfrak{h}$ the roots of $\mathfrak{t}_\mathfrak{h}$ on $\mathfrak{h}$. Let $\Phi_\mathfrak{g}^+$ be a system of positive roots for $\mathfrak{g}$. We make the following *regularity assumption:*

(R)   There is an element $X_0 \in \mathfrak{t}_\mathfrak{h}$ such that $\langle \alpha, X_0 \rangle > 0$ for all $\alpha \in \Phi_\mathfrak{g}^+$.

This condition is automatic if $\mathfrak{t}_\mathfrak{g} \subset \mathfrak{h}$.

If $\alpha \in \mathfrak{t}_\mathfrak{g}^*$ we write $\overline{\alpha}$ for the restriction of α to $\mathfrak{t}_\mathfrak{h}$. Because of assumption (R), $\overline{\alpha} \neq 0$ for all $\alpha \in \Phi_\mathfrak{g}$. We can take the system of positive roots for $\mathfrak{h}$ as

$$\Phi_{\mathfrak{h}}^+ = \{\gamma \in \Phi_{\mathfrak{h}} : \langle \gamma, X_0 \rangle > 0\}. \tag{8.6}$$

Evidently, if $\alpha \in \Phi_{\mathfrak{g}}^+$ and $\overline{\alpha} \in \Phi_{\mathfrak{h}}$, then $\overline{\alpha} \in \Phi_{\mathfrak{h}}^+$. We set

$$\rho_{\mathfrak{g}} = \frac{1}{2} \sum_{\alpha \in \Phi_{\mathfrak{g}}^+} \alpha, \qquad \rho_{\mathfrak{h}} = \frac{1}{2} \sum_{\beta \in \Phi_{\mathfrak{h}}^+} \beta. \tag{8.7}$$

We write $\overline{\Phi_{\mathfrak{g}}^+} = \{\overline{\alpha} : \alpha \in \Phi_{\mathfrak{g}}^+\}$ for the set of positive restricted roots. Then we have $\Phi_{\mathfrak{h}}^+ \subset \overline{\Phi_{\mathfrak{g}}^+}$. For each $\beta \in \overline{\Phi_{\mathfrak{g}}^+}$ let $R_\beta = \{\alpha \in \Phi_{\mathfrak{g}}^+ : \overline{\alpha} = \beta\}$ and define

$$\Sigma_0 = \{\beta : \beta \in \Phi_{\mathfrak{h}}^+ \text{ and } |R_\beta| > 1\}, \qquad \Sigma_1 = \overline{\Phi_{\mathfrak{g}}^+} \setminus \Phi_{\mathfrak{h}}^+.$$

Set $\Sigma = \Sigma_0 \cup \Sigma_1$. The *multiplicity* m_β of $\beta \in \Sigma$ is defined as

$$m_\beta = \begin{cases} |R_\beta| & \text{if } \beta \notin \Phi_{\mathfrak{h}}^+, \\ |R_\beta| - 1 & \text{if } \beta \in \Phi_{\mathfrak{h}}^+. \end{cases}$$

Define the *partition function* $\wp_\Sigma$ on $\mathfrak{t}_{\mathfrak{h}}^*$ by the formal identity

$$\prod_{\beta \in \Sigma} \left(1 - e^{-\beta}\right)^{-m_\beta} = \sum_\xi \wp_\Sigma(\xi) e^{-\xi} \tag{8.8}$$

(the usual notion of partition function for a set with multiplicity). Since $\langle \beta, X_0 \rangle > 0$ for all $\beta \in \Sigma$, the partition function has finite values and $\wp_\Sigma(\xi)$ is the number of ways of writing

$$\xi = \sum_{\beta \in \Sigma} c_\beta \beta \qquad (c_\beta \in \mathbb{N}),$$

where each β that occurs is counted with multiplicity m_β.

8.2.2 Kostant Multiplicity Formulas

For dominant weights $\lambda \in P_{++}(\mathfrak{g})$ and $\mu \in P_{++}(\mathfrak{h})$ we denote by $m(\lambda, \mu)$ the multiplicity of the irreducible $\mathfrak{h}$-module with highest weight μ in the irreducible $\mathfrak{g}$-module with highest weight λ. We denote by $W_{\mathfrak{g}}$ and $W_{\mathfrak{h}}$ the Weyl groups of $\mathfrak{g}$ and $\mathfrak{h}$. Other notation follows Section 8.2.1.

Theorem 8.2.1 (Branching Multiplicity Formula). *Assume that the pair* $\mathfrak{g}, \mathfrak{h}$ *satisfies condition* (R). *Then the branching multiplicities are*

$$m(\lambda, \mu) = \sum_{s \in W_{\mathfrak{g}}} \text{sgn}(s) \, \wp_\Sigma\big(\overline{s(\lambda + \rho_{\mathfrak{g}})} - \mu - \overline{\rho_{\mathfrak{g}}}\big),$$

where the bar denotes restriction from t_g *to* $t_\mathfrak{h}$.

Proof. Let φ_λ be the character of the irreducible $\mathfrak{g}$-module with highest weight λ and let ψ_μ be the character of the irreducible $\mathfrak{h}$-module with highest weight μ. If $\overline{\varphi_\lambda}$ is the restriction of the $\mathfrak{g}$-character to $t_\mathfrak{h}$ then by definition of the multiplicities $m(\lambda,\mu)$ we have $\overline{\varphi_\lambda} = \sum_\mu m(\lambda,\mu)\,\psi_\mu$, with the sum over all $\mu \in P_{++}(\mathfrak{h})$.
Applying the Weyl character formula for $\mathfrak{h}$ to ψ_μ, we have an expansion

$$\overline{\varphi_\lambda} \prod_{\gamma \in \Phi_\mathfrak{h}^+} (1 - e^{-\gamma}) = \sum_{\mu \in P_{++}(\mathfrak{h})} \sum_{t \in W_\mathfrak{h}} m(\lambda,\mu)\,\mathrm{sgn}(t)\,e^{t\cdot(\mu+\rho_\mathfrak{h})-\rho_\mathfrak{h}} . \qquad (8.9)$$

However, we can also write

$$\prod_{\gamma \in \Phi_\mathfrak{h}^+} (1 - e^{-\gamma}) = \prod_{\alpha \in \Phi_\mathfrak{g}^+} (1 - e^{-\overline{\alpha}}) \prod_{\beta \in \Sigma} (1 - e^{-\beta})^{-m_\beta} . \qquad (8.10)$$

Hence by the Weyl character formula for $\mathfrak{g}$ and (8.8) we have

$$\overline{\varphi_\lambda} \prod_{\gamma \in \Phi_\mathfrak{h}^+} (1 - e^{-\gamma}) = \sum_{s \in W_\mathfrak{g}} \sum_{\xi \in t_\mathfrak{h}^*} \mathrm{sgn}(s)\,\wp_\Sigma(\xi)\,e^{\overline{s\cdot(\lambda+\rho_\mathfrak{g})}-\overline{\rho_\mathfrak{g}}-\xi}$$

$$= \sum_{\mu \in t_\mathfrak{h}^*} \sum_{s \in W_\mathfrak{g}} \mathrm{sgn}(s)\,\wp_\Sigma(\overline{s\cdot(\lambda+\rho_\mathfrak{g})} - \mu - \overline{\rho_\mathfrak{g}})\,e^\mu . \qquad (8.11)$$

Now if $\mu \in P_{++}(\mathfrak{h})$ then $\mu + \rho_\mathfrak{h}$ is regular, so the weights $t \cdot (\mu + \rho_\mathfrak{h}) - \rho_\mathfrak{h}$, for $t \in W_\mathfrak{h}$, are all distinct (see Proposition 3.1.20). Hence the coefficient of e^μ in (8.9) is $m(\lambda,\mu)$. From the alternative expression for this coefficient in (8.11) we thus obtain the branching multiplicity formula. □

As a special case, we take H as the maximal torus in G. Then $m(\lambda,\mu) = m_\lambda(\mu)$ is the multiplicity of the weight μ in V^λ and $\Sigma = \Phi_\mathfrak{g}^+$. In this case we write $\wp_\Sigma = \wp$ for the partition function, $W_\mathfrak{g} = W$ for the Weyl group, and $\rho_\mathfrak{g} = \rho$.

Corollary 8.2.2 (Weight Multiplicity Formula). *The multiplicity of the weight μ is the alternating sum*

$$m_\lambda(\mu) = \sum_{s \in W} \mathrm{sgn}(s)\,\wp(s\cdot(\lambda+\rho) - \mu - \rho) .$$

8.2.3 Exercises

1. Let Σ be the positive roots for the A_2 root system with simple roots α_1 and α_2. Let $m_1, m_2 \in \mathbb{N}$. Show that $\wp_\Sigma(m_1\alpha_1 + m_2\alpha_2) = \min\{m_1, m_2\} + 1$.
2. Let Σ be the positive roots for the B_2 root system with simple roots $\alpha_1 = \varepsilon_1 - \varepsilon_2$ and $\alpha_2 = \varepsilon_2$. Let $m_1, m_2 \in \mathbb{N}$. Show that $\wp_\Sigma(m_1\alpha_1 + m_2\alpha_2)$ is the number of points $(x,y) \in \mathbb{N}^2$ that satisfy $0 \le x + y \le m_1$ and $0 \le x + 2y \le m_2$.

3. Derive Weyl's character formula from Corollary 8.2.2.
4. Use Corollary 8.2.2 to obtain the weight multiplicities for the irreducible representations of $\mathbf{SL}(2,\mathbb{C})$.

8.3 Proofs of Classical Branching Laws

We now work out the consequences of Kostant's multiplicity formula for the pairs of classical groups in Section 8.1.1.

8.3.1 Restriction from GL(n) to GL(n − 1)

Let $G = \mathbf{GL}(n,\mathbb{C})$ and $H = \mathbf{GL}(n-1,\mathbb{C}) = \{g \in G : ge_n = e_n\}$ with $n \geq 2$. We use the notation of Section 8.2.2, expressing roots and weights in terms of the basis $\{\varepsilon_1,\ldots,\varepsilon_n\}$. Let $\lambda \in \mathbb{Z}^n_{++}$ and $\mu \in \mathbb{Z}^{n-1}_{++}$. If $\lambda_n < 0$ we multiply the given representation of G by $(\det)^m$ for some $m \geq -\lambda_n$. This adds m to every component of λ and μ, so it doesn't change the interlacing property, and allows us to assume that $\lambda_n \geq 0$. We set $\mu_n = 0$.

The roots of $\mathfrak{g}$ are $\pm(\varepsilon_i - \varepsilon_j)$ with $1 \leq i < j \leq n$. Hence the matrix

$$X_0 = \mathrm{diag}[n-1, n-2, \ldots, 1, 0] \in \mathfrak{t}_\mathfrak{h}$$

satisfies condition (R) in Section 8.2.1, and via (8.6) defines the usual sets of positive roots

$$\Phi^+_\mathfrak{g} = \{\varepsilon_i - \varepsilon_j : 1 \leq i < j \leq n\}, \qquad \Phi^+_\mathfrak{h} = \{\varepsilon_i - \varepsilon_j : 1 \leq i < j \leq n-1\}.$$

Thus $\Sigma = \{\varepsilon_1,\ldots,\varepsilon_{n-1}\}$, in the notation of Section 8.2.1. Since Σ is linearly independent, the partition function $\wp_\Sigma$ takes only the values 0 and 1. Clearly

$$\wp_\Sigma(\xi) = 1 \text{ if and only if } \xi = m_1\varepsilon_1 + \cdots + m_{n-1}\varepsilon_{n-1} \text{ with } m_i \in \mathbb{N}. \tag{8.12}$$

In Corollary 7.1.2 we showed that the Weyl character formula is valid for G with the ρ-shift taken as

$$\rho = (n-1)\varepsilon_1 + (n-2)\varepsilon_2 + \cdots + \varepsilon_{n-1} \tag{8.13}$$

(this choice of ρ has the advantage of being a positive dominant weight for G). Thus we can use ρ in (8.13) instead of the half-sum of the positive roots (which is not an integral weight of G). We have $W_\mathfrak{g} = \mathfrak{S}_n$, and for $s \in \mathfrak{S}_n$ we calculate that

$$s \cdot \rho - \rho = \sum_{i=1}^{n} (i - s(i))\,\varepsilon_i.$$

Thus Kostant's multiplicity formula for the pair G, H is

$$m(\lambda, \mu) = \sum_{s \in \mathfrak{S}_n} \mathrm{sgn}(s)\, \wp_\Sigma(s \cdot \lambda + \overline{s \cdot \rho - \rho} - \mu). \qquad (8.14)$$

We will express the right-hand side of this formula as a determinant. Let $s \in \mathfrak{S}_n$. Then (8.12) implies that $\wp_\Sigma(s \cdot \lambda + \overline{s \cdot \rho - \rho} - \mu) \neq 0$ if and only if

$$\lambda_{s(i)} + i - s(i) \geq \mu_i \quad \text{for } i = 1, \ldots, n-1. \qquad (8.15)$$

For $a \in \mathbb{R}$ define

$$[a]_+ = \begin{cases} 1 & \text{if } a \geq 0 \text{ and } a \in \mathbb{Z}, \\ 0 & \text{otherwise}. \end{cases}$$

For $x, y \in \mathbb{R}^n$ and $1 \leq i, j \leq n$ set $a_{ij}(x, y) = [x_i + j - i - y_j]_+$, and let $A_n(x, y)$ be the $n \times n$ matrix with entries $a_{ij}(x, y)$.

We now show that the right side of (8.14) is $\det A_n(\lambda, \mu)$. Since $\lambda_n \geq 0$ and $\mu_n = 0$, we have

$$a_{in}(\lambda, \mu) = [\lambda_i + n - i]_+ = 1.$$

Thus we can write the values of the partition function occurring in the multiplicity formula as

$$\wp_\Sigma(\overline{s \cdot (\lambda + \rho)} - \mu - \overline{\rho}) = \prod_{j=1}^n a_{s(j), j}(\lambda, \mu).$$

It follows that $m(\lambda, \mu) = \det A_n(\lambda, \mu)$.

We now analyze this determinant. The matrix $A_n(\lambda, \mu)$ has the form

$$\begin{bmatrix} [\lambda_1 - \mu_1]_+ & [\lambda_1 - \mu_2 + 1]_+ & \cdots & [\lambda_1 - \mu_n + n - 1]_+ \\ [\lambda_2 - \mu_1 - 1]_+ & [\lambda_2 - \mu_2]_+ & \cdots & [\lambda_2 - \mu_n + n - 2]_+ \\ \vdots & \vdots & \ddots & \vdots \\ [\lambda_n - \mu_1 - n + 1]_+ & [\lambda_n - \mu_2 - n + 2]_+ & \cdots & [\lambda_n - \mu_n]_+ \end{bmatrix}.$$

The following lemma completes the proof of Theorem 8.1.1.

Lemma 8.3.1. *Let* $\lambda, \mu \in \mathbb{Z}_{++}^n$. *Then* $\det A_n(\lambda, \mu) = 1$ *if*

$$\lambda_1 \geq \mu_1 \geq \lambda_2 \geq \cdots \geq \lambda_{n-1} \geq \mu_{n-1} \geq \lambda_n \geq \mu_n.$$

Otherwise, $\det A_n(\lambda, \mu) = 0$.

Proof. We proceed by induction on n, the case $n = 1$ being the definition of the function $a \mapsto [a]_+$. Assume that the lemma is true for $n - 1$ and take $\lambda, \mu \in \mathbb{Z}_{++}^n$. If $\mu_1 > \lambda_1$ then $\mu_1 > \lambda_i$ for $i = 2, \ldots, n$. Hence the first column of $A_n(\lambda, \mu)$ is zero, so $\det A_n(\lambda, \mu) = 0$. If $\mu_1 < \lambda_2$ then $\mu_i \leq \mu_1 < \lambda_2 \leq \lambda_1$ for $i = 2, \ldots, n$, so the first and second rows of $A_n(\lambda, \mu)$ are

$$\begin{bmatrix} 1 & 1 & \cdots & 1 \\ 1 & 1 & \cdots & 1 \end{bmatrix}.$$

Hence $\det A_n(\lambda, \mu) = 0$ in this case also.

If $\lambda_1 \geq \mu_1 \geq \lambda_2$ then $\mu_1 \geq \lambda_i$ for $i = 3, \ldots, n$ also, so all entries in the first column of $A_n(\lambda, \mu)$ are zero except the first entry, which is 1. Hence

$$\det A_n(\lambda, \mu) = \det A_{n-1}(\lambda', \mu'),$$

where $\lambda' = (\lambda_2, \ldots, \lambda_n)$ and $\mu' = (\mu_2, \ldots, \mu_n)$. The inductive hypothesis now implies that $\det A_n(\lambda, \mu) = 1$ if μ' interlaces λ'; otherwise it equals zero. $\square$

Proof of Theorem 8.1.2: Theorem 8.1.1 gives the decomposition of F_n^λ under $\mathbf{GL}(n-1, \mathbb{C})$ as indicated. This decomposition is multiplicity-free and the matrix $y = \mathrm{diag}[I_{n-1}, z]$, for $z \in \mathbb{C}^\times$, commutes with $\mathbf{GL}(n-1, \mathbb{C})$. Hence by Schur's lemma y acts by a scalar z^ν on M^μ for some integer ν. To determine ν, we note that zI_n, for $z \in \mathbb{C}^\times$, acts on F_n^λ by the scalar $z^{|\lambda|}$. Likewise, the element $x = \mathrm{diag}[zI_{n-1}, 1]$ acts on M^μ by the scalar $z^{|\mu|}$. Since $xy = zI_n$, it follows that $|\mu| + \nu = |\lambda|$. $\square$

8.3.2 Restriction from $\mathbf{Spin}(2n+1)$ to $\mathbf{Spin}(2n)$

We use the notation introduced before the statement of Theorem 8.1.3 and in Section 8.2.2. In this case $\mathfrak{g}$ and $\mathfrak{h}$ have the same rank, and we have $\mathfrak{t}_\mathfrak{g} = \mathfrak{t}_\mathfrak{h}$, which we write as $\mathfrak{t}$. The positive roots of $\mathfrak{t}$ on $\mathfrak{g}$ are $\{\varepsilon_i \pm \varepsilon_j : 1 \leq i < j \leq n\} \cup \{\varepsilon_i : 1 \leq i \leq n\}$, whereas the positive roots of $\mathfrak{t}$ on $\mathfrak{h}$ are $\{\varepsilon_i \pm \varepsilon_j : 1 \leq i < j \leq n\}$. Hence in this case Σ is the set $\{\varepsilon_i : 1 \leq i \leq n\}$ and is linearly independent. Thus the partition function $\wp_\Sigma$ takes only the values 0 and 1. We will identify $\mathfrak{t}^*$ with $\mathbb{C}^n$ via the basis $\{\varepsilon_1, \ldots, \varepsilon_n\}$. Let $u, v \in \mathbb{C}^n$. Then

$$\wp_\Sigma(u-v) = 1 \text{ if and only if } u_i - v_i \in \mathbb{Z} \text{ and } u_i \geq v_i \text{ for } i = 1, \ldots, n. \quad (8.16)$$

We have $\rho_\mathfrak{g} = \rho_\mathfrak{h} + \rho_\Sigma$, where

$$\rho_\mathfrak{h} = [n-1, n-2, \ldots, 1, 0] \quad \text{and} \quad \rho_\Sigma = [\tfrac{1}{2}, \ldots, \tfrac{1}{2}].$$

In particular, we observe that every coordinate of $\rho_\mathfrak{g}$ is strictly positive.

The Weyl group for $\mathfrak{g}$ is $W = \mathfrak{Z}_n \mathfrak{S}_n$, where

$$\mathfrak{Z}_n = \langle \sigma_1, \ldots, \sigma_n \rangle, \quad \sigma_i \varepsilon_j = (-1)^{\delta_{ij}} \varepsilon_j$$

(see Section 3.1.1). Thus we can write the branching multiplicity formula in this case as

$$m(\lambda, \mu) = \sum_{\sigma \in \mathfrak{Z}_n} \sum_{s \in \mathfrak{S}_n} \mathrm{sgn}(\sigma) \, \mathrm{sgn}(s) \, \wp_\Sigma((\sigma s) \cdot \lambda + (\sigma s) \cdot \rho_\mathfrak{g} - \rho_\mathfrak{g} - \mu)$$

(since the restriction map is the identity, no bars are needed).

Let $s \in \mathfrak{S}_n$. Every coordinate of $s \cdot \rho_{\mathfrak{g}}$ is strictly positive. Every coordinate of $s \cdot \lambda$ is nonnegative, and the same holds for μ, except for possibly μ_n. Hence if $\sigma \in \mathfrak{Z}_n$ and $\sigma \neq \sigma_n$ or 1, then (8.16) implies that

$$\wp_\Sigma((\sigma s) \cdot \lambda + (\sigma s) \cdot \rho_{\mathfrak{g}} - \rho_{\mathfrak{g}} - \mu) = 0.$$

However, for $\sigma = \sigma_n$ we have $\sigma_n \cdot \rho_\Sigma = \rho_\Sigma - \varepsilon_n$ and $\sigma_n \cdot \rho_{\mathfrak{h}} = \rho_{\mathfrak{h}}$. Also, $s \cdot \rho_\Sigma = \rho_\Sigma$. Thus

$$(\sigma_n s) \cdot \rho_{\mathfrak{g}} - \rho_{\mathfrak{g}} = (\sigma_n s) \cdot \rho_{\mathfrak{h}} - \rho_{\mathfrak{h}} + (\sigma_n s) \cdot \rho_\Sigma - \rho_\Sigma = \sigma_n \cdot (s \cdot \rho_{\mathfrak{h}} - \rho_{\mathfrak{h}}) - \varepsilon_n.$$

These observations let us simplify the branching multiplicity formula to

$$\begin{aligned}
m(\lambda, \mu) = &\sum_{s \in \mathfrak{S}_n} \operatorname{sgn}(s)\, \wp_\Sigma \cdot (s \cdot \lambda - \mu + s \cdot \rho_{\mathfrak{h}} - \rho_{\mathfrak{h}}) \\
&- \sum_{s \in \mathfrak{S}_n} \operatorname{sgn}(s)\, \wp_\Sigma((\sigma_n s) \cdot \lambda - \mu + \sigma_n \cdot (s \cdot \rho_{\mathfrak{h}} - \rho_{\mathfrak{h}}) - \varepsilon_n).
\end{aligned} \tag{8.17}$$

We now write each sum on the right as a determinant.

Let $s \in \mathfrak{S}_n$. Since $s \cdot \rho_{\mathfrak{h}} - \rho_{\mathfrak{h}} = [1 - s(1), \ldots, n - 1 - s(n-1), n - s(n)]$ and

$$\sigma_n \cdot (s \cdot \rho_{\mathfrak{h}} - \rho_{\mathfrak{h}}) = [1 - s(1), \ldots, n - 1 - s(n-1), s(n) - n],$$

we see from (8.16) that

(1) $\wp_\Sigma(s \cdot \lambda + s \cdot \rho_{\mathfrak{h}} - \rho_{\mathfrak{h}} - \mu) \neq 0$ if and only if

$$\lambda_{s(i)} - \mu_i \in \mathbb{Z} \quad \text{and} \quad \lambda_{s(i)} - \mu_i + i - s(i) \geq 0 \quad \text{for } i = 1, \ldots, n.$$

(2) $\wp_\Sigma(\sigma_n s \cdot \lambda + \sigma_n(s \cdot \rho_{\mathfrak{h}} - \rho_{\mathfrak{h}}) - \mu - \varepsilon_n) \neq 0$ if and only if

$$\lambda_{s(i)} - \mu_i \in \mathbb{Z} \quad \text{and} \quad \lambda_{s(i)} - \mu_i + i - s(i) \geq 0 \quad \text{for } i = 1, \ldots, n-1,$$
$$\lambda_{s(n)} - \mu_n \in \mathbb{Z} \quad \text{and} \quad -\lambda_{s(n)} - \mu_n + s(n) - n - 1 \geq 0.$$

Let $a_{ij}(\lambda, \mu) = [\lambda_i - \mu_j + j - i]_+$ as in Section 8.3.1. Then (1) and (2) imply

$$\wp_\Sigma(s \cdot (\lambda + \rho_{\mathfrak{h}}) - \mu - \rho_{\mathfrak{h}}) = \prod_{j=1}^n a_{s(j),j}(\lambda, \mu),$$

$$\wp_\Sigma((\sigma_n s) \cdot (\lambda + \rho_{\mathfrak{h}}) - \mu - \rho_{\mathfrak{h}} - \varepsilon_n) = [-\lambda_{sn} + sn - n - 1 - \mu_n]_+$$
$$\times \prod_{j=1}^{n-1} a_{s(j),j}(\lambda, \mu).$$

Hence $m(\lambda, \mu) = \det A_n(\lambda, \mu) - \det B_n(\lambda, \mu)$, where $A_n(\lambda, \mu)$ is the matrix in Section 8.3.1 and $B_n(\lambda, \mu)$ is the matrix

$$
\begin{bmatrix}
[\lambda_1 - \mu_1]_+ & \cdots & [-\lambda_1 - \mu_n - n]_+ \\
[\lambda_2 - \mu_1 - 1]_+ & \cdots & [-\lambda_2 - \mu_n - n + 1]_+ \\
\vdots & \ddots & \vdots \\
[\lambda_{n-1} - \mu_1 - n + 2]_+ & \cdots & [-\lambda_{n-1} - \mu_n - 2]_+ \\
[\lambda_n - \mu_1 - n + 1]_+ & \cdots & [-\lambda_n - \mu_n - 1]_+
\end{bmatrix}
$$

(all the columns of $B_n(\lambda, \mu)$ except the nth are the same as those of $A_n(\lambda, \mu)$). To complete the proof of Theorem 8.1.3, it suffices to prove the following:

Lemma 8.3.2. *If both conditions*

1. $\lambda_i - \mu_i \in \mathbb{Z}$ *for* $i = 1, \ldots, n$ *and*
2. $\lambda_1 \geq \mu_1 \geq \lambda_2 \geq \cdots \geq \lambda_{n-1} \geq \mu_{n-1} \geq \lambda_n \geq |\mu_n|$

hold, then $\det A_n(\lambda, \mu) = \det B_n(\lambda, \mu) + 1$. *Otherwise,* $\det A_n(\lambda, \mu) = \det B_n(\lambda, \mu)$.

Proof. If condition (1) fails for some i then the ith columns of $A_n(\lambda, \mu)$ and $B_n(\lambda, \mu)$ are zero, so $\det A_n(\lambda, \mu) = 0$ and $\det B_n(\lambda, \mu) = 0$. Hence we may assume that condition (1) holds. Then $\lambda_i - \mu_j \in \mathbb{Z}$ for $1 \leq i, j \leq n$.

We proceed by induction on n. When $n = 1$,

$$
\det A_1(\lambda, \mu) - \det B_1(\lambda, \mu) = [\lambda_1 - \mu_1]_+ - [-\lambda_1 - \mu_1 - 1]_+ .
$$

If $\mu_1 > \lambda_1$ then both terms on the right are zero. Suppose $\lambda_1 - \mu_1 \in \mathbb{Z}$. If $\mu_1 < -\lambda_1$ then $0 \leq \lambda_1 < -\mu_1$ and $-\lambda_1 - \mu_1 > 0$, so we have

$$
[\lambda_1 - \mu_1]_+ = [-\lambda_1 - \mu_1 - 1]_+ = 1
$$

and the difference is zero. Finally, if $\lambda_1 \geq \mu_1 \geq -\lambda_1$, then

$$
[\lambda_1 - \mu_1]_+ = 1, \quad [-\lambda_1 - \mu_1 - 1]_+ = 0 ,
$$

and the difference is 1. This proves the case $n = 1$.

Let $n \geq 2$ and assume that the lemma is true for $n - 1$. If $\mu_1 > \lambda_1$ then all the entries in the first column of $A_n(\lambda, \mu)$ and of $B_n(\lambda, \mu)$ are zero, and so the lemma is true for n in this case.

If $\mu_1 < \lambda_2$ then $\mu_i \leq \mu_1 < \lambda_2 \leq \lambda_1$ for $i = 2, \ldots, n$. Hence the first and second rows of $A_n(\lambda, \mu)$ are $[1 \ 1 \ \cdots \ 1]$. But $\mu_n \leq \mu_1 < \lambda_2 \leq \lambda_1$. Thus $-\lambda_1 - \mu_n - n < 0$ and $-\lambda_2 - \mu_n - n + 1 < 0$, and so the first and second rows of $B_n(\lambda, \mu)$ are $[1 \ 1 \ \cdots \ 1 \ 0]$. It follows that $\det A_n(\lambda, \mu) = 0$ and $\det B_n(\lambda, \mu) = 0$, so the lemma is true for n in this case.

Finally, suppose $\lambda_1 \geq \mu_1 \geq \lambda_2$. Then $[\lambda_1 - \mu_1]_+ = 1$ and all the other entries in the first columns of $A_n(\lambda, \mu)$ and $B_n(\lambda, \mu)$ are zero. Hence

$$
\det A_n(\lambda, \mu) = \det A_{n-1}(\lambda', \mu') \quad \text{and} \quad \det B_n(\lambda, \mu) = \det B_{n-1}(\lambda', \mu') ,
$$

where $\lambda' = [\lambda_2, \ldots, \lambda_n]$ and $\mu' = [\mu_2, \ldots, \mu_n]$. Since λ' and μ' satisfy the hypotheses of the lemma for $n - 1$, we may apply the induction hypothesis to prove the lemma for n in this case also. $\qquad\square$

8.3.3 Restriction from Spin(2n) to Spin(2n − 1)

Let $\mathfrak{g}$ and $\mathfrak{h}$ be as in the statement of the branching law in Theorem 8.1.4. To find the branching multiplicity $m(\lambda, \mu)$, it suffices to consider the case $\lambda_n \geq 0$. Indeed, if $\lambda_n < 0$ we can conjugate $\mathfrak{g}$ by the outer automorphism defined by the orthogonal matrix g_0 that interchanges e_n and e_{-n} and fixes e_i for $i \neq \pm n$. Composing the representation π^λ with this automorphism, we obtain the representation with highest weight $\lambda' = [\lambda_1, \ldots, \lambda_{n-1}, -\lambda_n]$. However, g_0 commutes with the action of $\mathfrak{h}$ on $\mathbb{C}^{2n}$, so the representations π^λ and $\pi^{\lambda'}$ have the same restriction to $\mathfrak{h}$.

We follow the notation of Section 8.2.2. The roots of $\mathfrak{t}_\mathfrak{g}$ on $\mathfrak{g}$ are $\pm(\varepsilon_i \pm \varepsilon_j)$ for $1 \leq i < j \leq n$. Hence the matrix

$$X_0 = \operatorname{diag}[n-1, n-2, \ldots, 1, 0, 0, -1, \ldots, -n+2, -n+1] \in \mathfrak{t}_\mathfrak{h}$$

satisfies condition (R) of Section 8.2.2. Thus the general branching multiplicity formula is valid. Using X_0, we obtain via (8.6) the usual sets of positive roots

$$\Phi_\mathfrak{g}^+ = \{\varepsilon_i \pm \varepsilon_j : 1 \leq i < j \leq n\},$$
$$\Phi_\mathfrak{h}^+ = \{\varepsilon_i \pm \varepsilon_j : 1 \leq i < j \leq n-1\} \cup \{\varepsilon_i : 1 \leq i \leq n-1\}.$$

Thus every root of $\mathfrak{t}_\mathfrak{g}$ restricts to a root of $\mathfrak{t}_\mathfrak{h}$. The root ε_i is the restriction of $\varepsilon_i \pm \varepsilon_n$. Hence $\Sigma_1 = \emptyset$ and $\Sigma = \Sigma_0 = \{\varepsilon_1, \ldots, \varepsilon_{n-1}\}$ with all multiplicities one. Since Σ is linearly independent, the partition function $\wp_\Sigma$ takes only the values 0 and 1. We will identify $\mathfrak{t}_\mathfrak{h}^*$ with $\mathbb{C}^{n-1}$ via the basis $\{\varepsilon_1, \ldots, \varepsilon_{n-1}\}$. Let $u \in \mathbb{C}^{n-1}$. Then

$$\wp_\Sigma(u) = 1 \text{ if and only if } u_i \in \mathbb{Z} \text{ and } u_i \geq 0 \text{ for } i = 1, \ldots, n-1]. \quad (8.18)$$

In this case $\rho_\mathfrak{g} = [n-1, n-2, \ldots, 1, 0]$ and $\overline{\rho_\mathfrak{g}} = [n-1, \ldots, 2, 1]$.

The Weyl group for $\mathfrak{g}$ is $W = 3_n^+ \mathfrak{S}_n$, where $3_n^+ \subset 3_n$ consists of the products $\sigma_1 \cdots \sigma_k$ with k even (see Section 3.1.1). Thus the branching multiplicities are

$$m(\lambda, \mu) = \sum_{\sigma \in 3_n^+} \sum_{s \in \mathfrak{S}_n} \operatorname{sgn}(\sigma) \operatorname{sgn}(s) \wp_\Sigma(\overline{(\sigma s) \cdot (\lambda + \rho_\mathfrak{g})} - \overline{\rho_\mathfrak{g}} - \mu).$$

Let $s \in \mathfrak{S}_n$. Every coordinate of $s \cdot \lambda$ and $s \cdot \rho_\mathfrak{g}$ is nonnegative, and every coordinate of $\overline{\rho_\mathfrak{g}}$ is strictly positive. If $\sigma \in 3_n^+$ and $\sigma \neq 1$, then σ changes the sign of the ith coordinate for some $i < n$. Hence the ith coordinate of $\overline{(\sigma s) \cdot (\lambda + \rho_\mathfrak{g})} - \overline{\rho_\mathfrak{g}} - \mu$ is negative, so we have

$$\wp_\Sigma(\overline{(\sigma s) \cdot (\lambda + \rho_\mathfrak{g})} - \overline{\rho_\mathfrak{g}} - \mu) = 0$$

in this case. Thus we can write the branching multiplicity formula as

$$m(\lambda, \mu) = \sum_{s \in \mathfrak{S}_n} \operatorname{sgn}(s) \wp_\Sigma(\overline{s \cdot (\lambda + \rho_\mathfrak{g})} - \overline{\rho_\mathfrak{g}} - \mu). \quad (8.19)$$

We now express the right-hand side of (8.19) as a determinant. As previously noted, we may assume that $\lambda_n \geq 0$. Let $s \in \mathfrak{S}_n$. Then

$$\overline{s \cdot (\lambda + \rho_{\mathfrak{g}}) - \overline{\rho_{\mathfrak{g}}}}$$
$$= [\lambda_{s(1)} + 1 - s(1), \lambda_{s(2)} + 2 - s(2), \ldots, \lambda_{s(n-1)} + n - 1 - s(n-1)] .$$

This formula shows that if we set $\mu_n = 0$, then we can write the values of the partition function occurring in (8.19) as

$$\wp_{\Sigma}(\overline{s \cdot (\lambda + \rho_{\mathfrak{g}})} - \mu - \overline{\rho_{\mathfrak{g}}}) = \prod_{j=1}^{n} [\lambda_{s(j)} + j - s(j) - \mu_j]_+ ,$$

just as in the case of $\mathbf{GL}(n, \mathbb{C})$ (Section 8.3.1). Hence $m(\lambda, \mu) = \det A_n(\lambda, \mu)$. The proof of Theorem 8.1.4 now follows from Lemma 8.3.1.

8.3.4 Restriction from Sp(n) to Sp(n − 1)

Let $G = \mathbf{Sp}(n, \mathbb{C})$ and $H = \mathbf{Sp}(n - 1, \mathbb{C})$. To obtain the branching law in this case we must modify the method of Section 8.2.2, because there are no G-regular elements in H. However, there is a subgroup $K \cong \mathbf{Sp}(1, \mathbb{C})$ in G that commutes with H with the additional property that HK contains the maximal torus T_G. Thus we can use the same basic approach. We consider the restriction to H of

$$\chi_\lambda \prod_{\gamma \in \Phi_{\mathfrak{h}}^+} (1 - \mathrm{e}^{-\gamma}) ,$$

which can be expressed either in terms of the multiplicities $m(\lambda, \mu)$ or in terms of the Weyl character formula for G. In this case the factors given by the roots of K in the Weyl denominator Δ_G vanish on restriction to T_H, as do the terms in the Weyl numerator. The ratio contributes a *residue* upon restriction to T_H (which is just the polynomial d_K given by the Weyl dimension formula for K). We then invert the remaining factors in the Weyl denominator by an expansion in terms of a suitable partition function $\wp_{\Sigma}$. Finally, we obtain the formula for $m(\lambda, \mu)$ by equating coefficients of e^μ for μ a dominant weight of T_H, as in Section 8.2.2. This procedure yields a general branching multiplicity formula involving an alternating sum (over the cosets $W_K \backslash W_G$) of values of the function $\wp_{\Sigma} d_K$.

The approach to branching formulas just sketched applies quite generally to reductive groups G and reductive subgroups H. However, the partition function $\wp_{\Sigma}$ is complicated in general, and from the formula that one obtains it is difficult even to determine when the multiplicities are nonzero. In particular, for the case $G = \mathbf{Sp}(n, \mathbb{C})$ and $H = \mathbf{Sp}(n - 1, \mathbb{C})$ treated here, the set Σ has multiplicity 2, so the values of $\wp_{\Sigma}$ are not just 0 and 1. This complication will require new combinatorial arguments to obtain the interlacing condition for the highest weights and the explicit multiplicity formula stated in Theorem 8.1.5.

We now turn to the details of the proof. In this case it will be more convenient to use a mixed notation involving both the maximal tori and their Lie algebras. Take $G = \mathbf{Sp}(\mathbb{C}^{2n}, \Omega)$ and the basis $\{e_{\pm i} : i = 1, \ldots, n\}$ for $\mathbb{C}^{2n}$ as in Section 2.4.1. Define

$$K = \{k \in G : ke_{\pm i} = e_{\pm i} \text{ for } i = 1, \ldots, n-1\}.$$

Since the skew form Ω is nondegenerate on $\mathbb{C}e_n + \mathbb{C}e_{-n}$, we have $K \cong \mathbf{Sp}(1, \mathbb{C})$. Let D be the $2n \times 2n$ diagonal matrices in $\mathbf{GL}(2n, \mathbb{C})$. We take as maximal tori $T_G = G \cap D$, $T_H = H \cap D$, and $T_K = K \cap D$. Then $T_G = T_H \times T_K$.

We choose the positive roots $\Phi_{\mathfrak{g}}^+$ of T_G on $\mathfrak{g}$ and positive roots $\Phi_{\mathfrak{h}}^+$ of T_H on $\mathfrak{h}$ as in Section 2.4.3. We take $\Phi_{\mathfrak{k}}^+ = \{2\varepsilon_n\}$. Let

$$S^+ = \{\varepsilon_i \pm \varepsilon_n : i = 1, \ldots, n-1\}.$$

We view all of these roots as characters of T_G via the decomposition $T_G = T_H \times T_K$. Then $\Phi_{\mathfrak{g}}^+ = \Phi_{\mathfrak{h}}^+ \cup \Phi_{\mathfrak{k}}^+ \cup S^+$; therefore, we can write

$$\prod_{\gamma \in \Phi_{\mathfrak{h}}^+} (1 - e^{-\gamma}) = \left(1 - e^{-2\varepsilon_n}\right)^{-1} \prod_{\beta \in S^+} (1 - e^{-\beta})^{-1} \prod_{\alpha \in \Phi_{\mathfrak{g}}^+} (1 - e^{-\alpha}). \qquad (8.20)$$

The Weyl group of G is $W_G = \mathfrak{Z}_n \mathfrak{S}_n$ as for $\mathbf{SO}(2n+1, \mathbb{C})$ (see Section 3.1.1). We identify the Weyl group W_K of K with the subgroup $\{1, \sigma_n\}$ of W_G (where σ_n changes the sign of ε_n). This is consistent with the natural actions of these groups on T_G and T_K and with the decomposition $T_G = T_H \times T_K$.

Let χ_λ be the character of the G-module with highest weight λ. From (8.20) and the Weyl character formula we have

$$\chi_\lambda \prod_{\gamma \in \Phi_{\mathfrak{h}}^+} (1 - e^{-\gamma})$$

$$= \prod_{\beta \in S^+} (1 - e^{-\beta})^{-1} \sum_{s \in W_K \backslash W_G} \mathrm{sgn}(s) \left\{ \sum_{r \in W_K} \frac{\mathrm{sgn}(r) e^{(rs) \cdot (\lambda + \rho_{\mathfrak{g}}) - \rho_{\mathfrak{g}}}}{1 - e^{-2\varepsilon_n}} \right\}.$$

For $s \in W_G$ set $\gamma = s \cdot (\lambda + \rho_{\mathfrak{g}}) = \sum_{i=1}^n \gamma_i \varepsilon_i$. Then

$$\sum_{r \in W_K} \frac{\mathrm{sgn}(r) e^{r \cdot \gamma}}{1 - e^{-2\varepsilon_n}} = x_1^{\gamma_1} \cdots x_{n-1}^{\gamma_{n-1}} \left\{ \frac{x_n^{\gamma_n} - x_n^{-\gamma_n}}{1 - x_n^{-2}} \right\}, \qquad (8.21)$$

where $x_i = e^{\varepsilon_i}$ is the ith coordinate function on T_G as usual.

We can now see what happens when we restrict these formulas to T_H by setting $x_n = 1$. In (8.21) we obtain $(\gamma, \varepsilon_n) e^{\bar{\gamma}}$, where (α, β) is the inner product making the set $\{\varepsilon_1, \ldots, \varepsilon_n\}$ orthonormal. Both of the roots $\varepsilon_i \pm \varepsilon_n$ restrict to ε_i on $\mathfrak{t}_{\mathfrak{h}}$ for $i = 1, \ldots, n-1$. We set $\Sigma = \{\varepsilon_i : i = 1, \ldots, n-1\}$ with multiplicities $m_{\varepsilon_i} = 2$, and we define the partition function $\wp_\Sigma$ as usual by

$$\prod_{i=1}^{n-1}\left(1-e^{-\varepsilon_i}\right)^{-2}=\sum_{\xi\in\mathfrak{t}_\mathfrak{h}^*}\wp_\Sigma(\xi)\,e^{-\xi}.$$

Letting the bar denote restriction to T_H, we then can write

$$\overline{\chi_\lambda}\prod_{\gamma\in\Phi_\mathfrak{h}^+}(1-e^{-\gamma})=\sum_{s\in W_K\backslash W_G}\sum_{\xi\in\mathfrak{t}_\mathfrak{h}^*}\operatorname{sgn}(s)\wp_\Sigma(\xi)(s\cdot(\lambda+\rho_\mathfrak{g}),\varepsilon_n)\,e^{\overline{s\cdot(\lambda+\rho_\mathfrak{g})}-\xi-\overline{\rho_\mathfrak{g}}}.$$

From this point we proceed exactly as in the proof of Theorem 8.2.1 to find the multiplicity $m(\lambda,\mu)$ as the coefficient of e^μ in this formula. The result is

$$m(\lambda,\mu)=\sum_{s\in W_K\backslash W_G}\operatorname{sgn}(s)\,(s\cdot(\lambda+\rho_\mathfrak{g}),\varepsilon_n)\,\wp_\Sigma(\overline{s\cdot(\lambda+\rho_\mathfrak{g})}-\mu-\overline{\rho_\mathfrak{g}}).$$

We now simplify this formula. Using the power-series identity

$$\frac{1}{(1-z)^2}=\sum_{m\geq0}(m+1)z^m,$$

we obtain

$$\wp_\Sigma\left(\textstyle\sum_{i=1}^{n-1}m_i\varepsilon_i\right)=\begin{cases}\prod_{i=1}^{n-1}(m_i+1)&\text{if }m_i\geq0\text{ for }i=1,\dots,n-1,\\0&\text{otherwise}.\end{cases}$$

Let $s\in\mathfrak{S}_n$. Then

$$\overline{s\cdot(\lambda+\rho_\mathfrak{g})}_i=\lambda_{s(i)}+n-s(i)+1>0\quad\text{for}\quad i=1,\dots,n-1.$$

Let $\sigma\in\mathfrak{Z}_n$ and suppose $\sigma\neq1$. Since $\overline{\rho_\mathfrak{g}}=[n,n-1,\dots,2]$ and $\mu_i\geq0$, it follows that some coordinate of $\overline{(\sigma s)\cdot(\lambda+\rho_\mathfrak{g})}-\mu-\overline{\rho_\mathfrak{g}}$ is negative. Hence

$$\wp_\Sigma(\overline{(\sigma s)\cdot(\lambda+\rho_\mathfrak{g})}-\mu-\overline{\rho_\mathfrak{g}})=0$$

in this case. Thus we may take $\sigma=1$ and the multiplicity formula becomes

$$m(\lambda,\mu)=\sum_{s\in\mathfrak{S}_n}\operatorname{sgn}(s)\,(s\cdot(\lambda+\rho_\mathfrak{g}),\varepsilon_n)\,\wp_\Sigma(\overline{s\cdot(\lambda+\rho_\mathfrak{g})}-\mu-\overline{\rho_\mathfrak{g}}).\tag{8.22}$$

For $a\in\mathbb{R}$ we define

$$(a)_+=\begin{cases}a&\text{if }a\text{ is a positive integer},\\0&\text{otherwise}.\end{cases}$$

Using this function we can write

$$\wp_{\Sigma}(\overline{s \cdot (\lambda + \rho_{\mathfrak{g}})} - \mu - \overline{\rho_{\mathfrak{g}}}) = \prod_{i=1}^{n-1}(\lambda_{s(i)} - \mu_i + i - s(i) + 1)_+$$

for $s \in \mathfrak{S}_n$. Setting $\mu_n = 0$, we also have

$$(s \cdot (\lambda + \rho_{\mathfrak{g}}), \varepsilon_n) = \lambda_{s(n)} - \mu_n + n - s(n) + 1 = (\lambda_{s(n)} - \mu_n + n - s(n) + 1)_+ .$$

Thus (8.22) simplifies to

$$m(\lambda, \mu) = \sum_{s \in \mathfrak{S}_n} \text{sgn}(s) \prod_{i=1}^{n}(\lambda_{s(i)} - \mu_i + i - s(i) + 1)_+ . \tag{8.23}$$

For any $x, y \in \mathbb{R}^n$ let $C_n(x,y)$ be the $n \times n$ matrix with i, j entry

$$C_n(x,y)_{ij} = (x_i - y_j + j - i + 1)_+ .$$

Then (8.23) can be expressed as

$$m(\lambda, \mu) = \det C_n(\lambda, \mu) . \tag{8.24}$$

We now establish a combinatorial result that describes when $\det C_n(\lambda, \mu) \neq 0$ and gives a product formula for this determinant. This will complete the proof of Theorem 8.1.5.

Let $\lambda, \mu \in \mathbb{N}_{++}^n$. We will say that $p = (p_1, \ldots, p_n) \in \mathbb{N}_{++}^n$ *interlaces the pair* (λ, μ) if

(1) $\lambda_1 \geq p_1 \geq \lambda_2 \geq \cdots \geq \lambda_{n-1} \geq p_{n-1} \geq \lambda_n \geq p_n$,   and
(2) $p_1 \geq \mu_1 \geq p_2 \geq \cdots \geq p_{n-1} \geq \mu_{n-1} \geq p_n \geq \mu_n$.

We define $P_n(\lambda, \mu)$ to be the set of all $p \in \mathbb{N}_{++}^n$ that interlace the pair (λ, μ). Note that $P_n(\lambda, \mu)$ is nonempty if and only if

$$\lambda_j \geq \mu_j \geq \lambda_{j+2} \quad \text{for } j = 1, \ldots, n, \text{ where } \lambda_{n+1} = \lambda_{n+2} = 0 . \tag{8.25}$$

Lemma 8.3.3. *Let* $\lambda, \mu \in \mathbb{N}_{++}^n$ *with* $n \geq 1$. *Then*

$$\det C_n(\lambda, \mu) = \text{Card}\, P_n(\lambda, \mu) . \tag{8.26}$$

Hence $\det C_n(\lambda, \mu) \neq 0$ *if and only if* (8.25) *holds. Assume this is the case and that*

$$x_1 \geq y_1 \geq x_2 \geq y_2 \geq \cdots \geq x_n \geq y_n$$

is the nonincreasing rearrangement of the set $\{\lambda_1, \ldots, \lambda_n, \mu_1, \ldots, \mu_n\}$. *Then*

$$\det C_n(\lambda, \mu) = \prod_{j=1}^{n}(x_j - y_j + 1) . \tag{8.27}$$

Proof. We shall prove (8.26) and (8.27) by induction on n. When $n = 1$ then $C_1(\lambda, \mu) = (\lambda_1 - \mu_1 + 1)_+$, which is nonzero if and only if $\lambda_1 \geq \mu_1$. If this holds, then $p \in P_1(\lambda, \mu)$ must satisfy $\lambda_1 \geq p \geq \mu_1$. There are $\lambda_1 - \mu_1 + 1$ choices for p, so the lemma is true in this case.

Let $n \geq 2$ and assume that the statement of the lemma is true for $n - 1$. Let $\lambda, \mu \in \mathbb{N}_{++}^n$. We shall consider several cases depending on the position of μ_1 relative to $(\lambda_1, \lambda_2, \lambda_3)$.

Case 1: $\mu_1 > \lambda_1$. Since $\mu_1 > \lambda_i$ for $i = 1, \ldots, n$, the first column of $C_n(\lambda, \mu)$ is zero. Hence $\det C_n(\lambda, \mu) = 0$. But $P_n(\lambda, \mu)$ is empty, so the result holds.

Case 2: $\lambda_1 \geq \mu_1 \geq \lambda_2$. We have $(\lambda_1 - \mu_1 + 1)_+ = \lambda_1 - \mu_1 + 1$ and $(\lambda_i - \mu_1)_+ = 0$ for $i = 2, \ldots, n$. Hence

$$\det C_n(\lambda, \mu) = (\lambda_1 - \mu_1 + 1) \det C_{n-1}(\lambda', \mu'),$$

where $\lambda' = (\lambda_2, \ldots, \lambda_n)$ and $\mu' = (\mu_2, \ldots, \mu_{n-1})$. But $\lambda_1 - \mu_1 + 1$ is the number of choices for p_1 in $P_n(\lambda, \mu)$, whereas the constraint on p_2 is $\lambda_2 \geq p_2 \geq \lambda_3$. Hence

$$\mathrm{Card}\, P_n(\lambda, \mu) = (\lambda_1 - \mu_1 + 1)\, \mathrm{Card}\, P_{n-1}(\lambda', \mu')$$

in this case. By induction we have $\det C_{n-1}(\lambda', \mu') = \mathrm{Card}\, P_{n-1}(\lambda', \mu')$, so we get (8.26). Furthermore, in the nonincreasing rearrangement of λ, μ we have $x_1 = \lambda_1$ and $y_1 = \mu_1$ in this case, so (8.27) also holds by induction.

Case 3: $\lambda_2 > \mu_1 \geq \lambda_3$. Now $(\lambda_2 - \mu_1)_+ = \lambda_2 - \mu_1$ and $(\lambda_i - \mu_1 - i)_+ = 0$ for $i = 3, \ldots, n$. Thus

$$C_n(\lambda, \mu) = \begin{bmatrix} \lambda_1 - \mu_1 + 1 & \lambda_1 - \mu_2 + 2 & \cdots & \lambda_1 - \mu_n + n \\ \lambda_2 - \mu_1 & \lambda_2 - \mu_2 + 1 & \cdots & \lambda_2 - \mu_n + n - 1 \\ 0 & & & \\ \vdots & & \mathbf{B} & \\ 0 & & & \end{bmatrix},$$

where $\mathbf{B}$ is the $(n-2) \times (n-1)$ matrix with $b_{ij} = (\lambda_{i+2} - \mu_{j+1} + j - i)_+$. Subtracting the first column of $C_n(\lambda, \mu)$ from each of the other columns, we see that

$$\det C_n(\lambda, \mu) = \det \begin{bmatrix} \lambda_1 - \mu_1 + 1 & \mu_1 - \mu_2 + 2 & \cdots & \mu_1 - \mu_n + n \\ \lambda_2 - \mu_1 & \mu_1 - \mu_2 + 1 & \cdots & \mu_1 - \mu_n + n - 1 \\ 0 & & & \\ \vdots & & \mathbf{B} & \\ 0 & & & \end{bmatrix}.$$

By the cofactor expansion along the first column of the matrix on the right we obtain

$$\det C_n(\lambda, \mu) = (\lambda_1 - \lambda_2 + 1) \det C_{n-1}(\lambda'', \mu'),$$

where $\lambda'' = [\mu_1, \lambda_3, \ldots, \lambda_n]$. Since $\mu_1 \geq \lambda_3$, the induction hypothesis gives

$$\det C_{n-1}(\lambda'', \mu') = \operatorname{Card} P_{n-1}(\lambda'', \mu') \,.$$

Let $p \in P_n(\lambda, \mu)$. Since $\lambda_2 \geq \mu_1$, the only constraint on p_1 is

$$\lambda_1 \geq p_1 \geq \lambda_2 \,; \tag{8.28}$$

thus the number of choices for p_1 is $\lambda_1 - \lambda_2 + 1$. Since $\mu_1 \geq \lambda_3$, the constraints on p_2 are $\mu_1 \geq p_2 \geq \lambda_3$ and $p_2 \geq \mu_2$. Thus $p \in P_n(\lambda, \mu)$ if and only if (8.28) holds and $p' \in P_{n-1}(\lambda'', \mu')$, where $p' = (p_2, \ldots, p_n)$. Hence by the induction hypothesis,

$$\begin{aligned}
\operatorname{Card} P_n(\lambda, \mu) &= (\lambda_1 - \lambda_2 + 1) \operatorname{Card} P_{n-1}(\lambda'', \mu') \\
&= (\lambda_1 - \lambda_2 + 1) \det C_{n-1}(\lambda'', \mu') = \det C_n(\lambda, \mu) \,.
\end{aligned}$$

Furthermore, in the nonincreasing rearrangement of λ, μ we have $x_1 = \lambda_1$ and $y_1 = \lambda_2$ in this case, so formula (8.27) also holds by induction.

Case 4: $\mu_1 < \lambda_3$. In this case there is no p_2 such that $\mu_1 \geq p_2 \geq \lambda_3$, so $P_n(\lambda, \mu)$ is empty. We must show that $\det C_n(\lambda, \mu) = 0$ also. Since we have $\mu_j < \lambda_3$ for $j = 1, \ldots, n-1$, the first three rows of the matrix $C_n(\lambda, \mu)$ are

$$\begin{bmatrix}
\lambda_1 - \mu_1 + 1 & \lambda_1 - \mu_2 + 2 & \cdots & \lambda_1 - \mu_n + n \\
\lambda_2 - \mu_1 & \lambda_2 - \mu_2 + 1 & \cdots & \lambda_2 - \mu_n + n - 1 \\
\lambda_3 - \mu_1 - 1 & \lambda_3 - \mu_2 & \cdots & \lambda_3 - \mu_n + n - 2
\end{bmatrix} \,.$$

We claim that this matrix has rank 2, and hence $\det C_n(\lambda, \mu) = 0$. Indeed, if we subtract the first column from the other columns and then subtract the third row of the new matrix from the other rows, we get

$$\begin{bmatrix}
\lambda_1 - \lambda_3 + 2 & 0 & \cdots & 0 \\
\lambda_2 - \lambda_3 + 1 & 0 & \cdots & 0 \\
\lambda_3 - \mu_1 - 1 & \mu_1 - \mu_2 + 1 & \cdots & \mu_1 - \mu_n + n - 1
\end{bmatrix} \,.$$

This completes the induction.   □

8.4 Notes

Section 8.1.1. The branching laws $\mathbf{GL}(n) \to \mathbf{GL}(n-1)$ and $\mathbf{Spin}(n) \to \mathbf{Spin}(n-1)$ are well known (Boerner [9], Želobenko [171, Chapter XVIII]). We will give another proof of the $\mathbf{GL}(n) \to \mathbf{GL}(n-1)$ branching law (not based on the Weyl character formula) in Section 12.2.3. The branching law for $\mathbf{Sp}(n) \to \mathbf{Sp}(n-1)$ is in Želobenko [171, §130] and Hegerfeldt [64] (without an explicit multiplicity formula); the method of Želobenko was studied in more detail by Lee [100]. A branching law from $\mathbf{Sp}(n) \to \mathbf{Sp}(n-1) \times \mathbf{Sp}(1)$, with an explicit combinatorial description of the multiplicities, is in Lepowsky [101], [102]. The embedding of $\mathbf{Sp}(n-1,\mathbb{C})$ into $\mathbf{Sp}(n,\mathbb{C})$ factors through an embedding of $\mathbf{Sp}(n-1,\mathbb{C}) \times \mathbf{Sp}(1,\mathbb{C})$ into $\mathbf{Sp}(n,\mathbb{C})$. Therefore the $\mathbf{Sp}(n-1,\mathbb{C})$ multiplicity space arising in Theorem 8.1.5 is an $\mathbf{Sp}(1,\mathbb{C}) = \mathbf{SL}(2,\mathbb{C})$-module, and Wallach and Yacobi [156] show that it is isomorphic to $\bigotimes_{i=1}^{n} V_{x_i - y_i}$, where V_k is the $(k+1)$-dimensional irreducible representation of $\mathbf{SL}(2,\mathbb{C})$. There are extensive published tables of branching laws for classical and exceptional simple Lie algebras (e.g., Tits [143], Bremner, Moody, and Patera [22], and McKay and Patera [110]). There are also interactive computer algebra systems that calculate the branching multiplicities, for example the program LiE (available at http://www-math.univ-poitiers.fr/~maavl/LiE).

Section 8.1.2. The basis in Corollary 8.1.7 is called the *Gel'fand–Cetlin* basis; see Želobenko [171, Chapter X] for further details.

Section 8.2.2. Kostant's original proof of the multiplicity formula is in Kostant [87]. Our proof follows Cartier [29] and Lepowsky [101]. Želobenko [171] gives another approach to branching laws. A treatment of branching laws via classical invariant theory and dual pairs is given by Howe, Tan, and Willenbring [74].

Section 8.3.4. The product formula for the symplectic group branching multiplicity is in Whippman [167] for the cases $n=2$ and $n=3$, along with many other examples of branching laws of interest in physics. The explicit product formula for the multiplicity for arbitrary n was obtained by Miller [112].

Chapter 9
Tensor Representations of GL(V)

Abstract In this chapter we bring together the representation theories of the groups $\mathbf{GL}(n, \mathbb{C})$ and $\mathfrak{S}_k$ via their mutually commuting actions on $\bigotimes^k \mathbb{C}^n$. We already exploited this connection in Chapter 5 to obtain the first fundamental theorem of invariant theory for $\mathbf{GL}(n, \mathbb{C})$. In this chapter we obtain the full isotypic decomposition of $\bigotimes^k \mathbb{C}^n$ under the action of $\mathbf{GL}(n, \mathbb{C}) \times \mathfrak{S}_k$. This decomposition gives the *Schur–Weyl duality pairing* between the irreducible representations of $\mathbf{GL}(n, \mathbb{C})$ and those of $\mathfrak{S}_k$. From this pairing we obtain the celebrated Frobenius character formula for the irreducible representations of $\mathfrak{S}_k$. We then reexamine Schur–Weyl duality and $\mathbf{GL}(k, \mathbb{C})$–$\mathbf{GL}(n, \mathbb{C})$ duality from Chapters 4 and 5 in the framework of *dual pairs* of reductive groups. Using the notion of *seesaw pairs* of subgroups, we obtain reciprocity laws for tensor products and induced representations. In particular, we show that every irreducible $\mathfrak{S}_k$-module can be realized as the weight space for the character $x \mapsto \det(x)$ in an irreducible $\mathbf{GL}(k, \mathbb{C})$ representation. Explicit models (the *Weyl modules*) for all the irreducible representations of $\mathbf{GL}(n, \mathbb{C})$ are obtained using *Young symmetrizers*. These elements of the group algebra of $\mathbb{C}[\mathfrak{S}_k]$ act as projection operators onto $\mathbf{GL}(n, \mathbb{C})$-irreducible invariant subspaces. The chapter concludes with the Littlewood–Richardson rule for calculating the multiplicities in tensor products.

9.1 Schur–Weyl Duality

With the Weyl character formula now available, we proceed to examine the Schur–Weyl duality decomposition of tensor space from Chapter 4 in great detail.

R. Goodman, N.R. Wallach, *Symmetry, Representations, and Invariants*,
Graduate Texts in Mathematics 255, DOI 10.1007/978-0-387-79852-3_9,
© Roe Goodman and Nolan R. Wallach 2009

9.1.1 Duality between GL(n) and $\mathfrak{S}_k$

The tensor space $\bigotimes^k \mathbb{C}^n$ carries representations ρ_k of $G_n = \mathbf{GL}(n, \mathbb{C})$ and σ_k of the symmetric group $\mathfrak{S}_k$ that mutually commute. Let $\mathcal{A}$ (respectively, $\mathcal{B}$) be the sub-algebra of $\mathrm{End}(\bigotimes^k \mathbb{C}^n)$ generated by $\rho_k(\mathbb{C}[G_n])$ (respectively, $\sigma_k(\mathfrak{S}_k)$). Since G_n and $\mathfrak{S}_k$ are reductive groups, these algebras are semisimple. By Schur's commu-tant theorem (Theorem 4.2.10) and the duality theorem (Theorem 4.2.1) there are mutually inequivalent irreducible G_n-modules $F_1, \ldots, F_d$ and mutually inequivalent irreducible $\mathfrak{S}_k$-modules $E_1, \ldots, E_d$ such that

$$\bigotimes^k \mathbb{C}^n \cong \bigoplus_{i=1}^d F_i \otimes E_i \tag{9.1}$$

as an $\mathcal{A} \otimes \mathcal{B}$ module. This decomposition sets up a pairing between (certain) repre-sentations of G_n and (certain) representations of $\mathfrak{S}_k$. To complete the analysis of ρ_k and σ_k, we must determine

(a)   the representations of G_n and $\mathfrak{S}_k$ that occur in (9.1);
(b)   the explicit form of pairing between G_n and $\mathfrak{S}_k$ representations;
(c)   the projection operators on tensor space for decomposition (9.1).

We shall approach these problems from the representation theory of $\mathbf{GL}(n, \mathbb{C})$, which we know through the theorem of the highest weight, using the Weyl character formula and the commutant character formulas (Section 7.1.3). It will be convenient to let n vary, so for $m < n$ we view $\mathbb{C}^m \subset \mathbb{C}^n$ as the column vectors $v = [x_1, \ldots, x_n]^t$ with $x_i = 0$ for $i > m$. Let e_j be the vector with 1 in the jth position and 0 elsewhere. We view $G_m = \mathbf{GL}(m, \mathbb{C}) \subset G_n$ as the subgroup fixing e_i for $i > m$. Let H_n be the subgroup of diagonal matrices in G_n, and let N_n^+ be the group of upper-triangular unipotent matrices. Thus we have the inclusions

$$G_n \subset G_{n+1}, \quad H_n \subset H_{n+1}, \quad N_n^+ \subset N_{n+1}^+ .$$

We parameterize the regular characters of H_n by $\mathbb{Z}^n$ as usual: $\lambda = [\lambda_1, \ldots, \lambda_n] \in \mathbb{Z}^n$ gives the character

$$h \mapsto h^\lambda = x_1^{\lambda_1} \cdots x_n^{\lambda_n} \quad \text{for } h = \mathrm{diag}[x_1, \ldots, x_n] .$$

For $m < n$ we embed $\mathbb{Z}^m \subset \mathbb{Z}^n$ as the elements with $\lambda_i = 0$ for $m < i \leq n$. Thus $\lambda \in \mathbb{Z}^m$ defines a regular character of H_n for all $n \geq m$.

We recall from Section 4.4.4 that the weights of H_n that occur in $\bigotimes^k \mathbb{C}^n$ are all $\lambda \in \mathbb{N}^n$ with $|\lambda| = \lambda_1 + \cdots + \lambda_n = k$. The corresponding weight space is

$$(\bigotimes^k \mathbb{C}^n)(\lambda) = \mathrm{Span}\{e_I : \mu_I = \lambda\} . \tag{9.2}$$

Here we write $e_I = e_{i_1} \otimes \cdots \otimes e_{i_k}$ with $1 \leq i_p \leq n$, and the condition $\mu_I = \lambda$ means that the integer i occurs λ_i times in I, for $1 \leq i \leq n$. The embedding $\mathbb{C}^p \subset \mathbb{C}^n$ for $p < n$ gives an embedding $\bigotimes^k \mathbb{C}^p \subset \bigotimes^k \mathbb{C}^n$. From (9.2) we see that the weight spaces have the following *stability property*: If $\lambda \in \mathbb{N}^p$ with $|\lambda| = k$, then

$$\left(\bigotimes^k \mathbb{C}^p\right)(\lambda) = \left(\bigotimes^k \mathbb{C}^n\right)(\lambda) \quad \text{for all } n \geq p. \tag{9.3}$$

We say that $\lambda \in \mathbb{N}^n$ is *dominant* if $\lambda_1 \geq \lambda_2 \geq \cdots \geq \lambda_n$. We denote by $\mathbb{N}_{++}^n$ the set of all dominant weights in $\mathbb{N}^n$. If λ is dominant and $|\lambda| = k$, we identify λ with the *partition* $k = \lambda_1 + \cdots + \lambda_n$ of k. We write $\mathrm{Par}(k)$ for the set of all partitions of k and $\mathrm{Par}(k,n)$ for the set of all partitions of k with at most n parts. Since any partition of k can have at most k parts, we have $\mathrm{Par}(k,n) = \mathrm{Par}(k)$ for $n \geq k$.

We return to decomposition (9.1). We can identify the representations F_i using Theorem 5.5.22. For $\lambda \in \mathbb{N}_{++}^n$ let $(\pi_n^\lambda, F_n^\lambda)$ be the irreducible representation of G_n with highest weight λ. The highest weight for V_i is some $\lambda \in \mathrm{Par}(k,n)$. From the results of Sections 5.5.2 and 5.5.3 we know that all such λ do occur as extreme weights in $\bigotimes^k \mathbb{C}^n$. Thus $V_i \cong F_n^\lambda$ for a unique λ, so we may label the associated representation E_i of $\mathfrak{S}_k$ as $(\sigma_{n,k}^\lambda, G_{n,k}^\lambda)$. We can then rewrite (9.1) as

$$\bigotimes^k \mathbb{C}^n \cong \bigoplus_{\lambda \in \mathrm{Par}(k,n)} F_n^\lambda \otimes G_{n,k}^\lambda. \tag{9.4}$$

Furthermore, by Theorem 4.2.12 we may take the multiplicity space $G_{n,k}^\lambda$ for the action of G_n to be the space of N_n^+-fixed tensors of weight λ:

$$G_{n,k}^\lambda = \left(\bigotimes^k \mathbb{C}^n\right)^{N_n^+}(\lambda). \tag{9.5}$$

Lemma 9.1.1. *Let λ be a partition of k with p parts. Then for all $n \geq p$,*

$$\left(\bigotimes^k \mathbb{C}^n\right)^{N_n^+}(\lambda) = \left(\bigotimes^k \mathbb{C}^p\right)^{N_p^+}(\lambda). \tag{9.6}$$

Proof. Let $i > p$. From (9.3) we have $\rho_k(e_{i,i+1})u = 0$ for all $u \in (\bigotimes^k \mathbb{C}^n)(\lambda)$, where $e_{ij} \in M_n(\mathbb{C})$ are the usual elementary matrices. Since $\mathfrak{n}_n^+$ is generated (as a Lie algebra) by $e_{12}, e_{23}, \ldots, e_{n-1,n}$ and $N_n^+ = \exp(\mathfrak{n}_n^+)$, we conclude that N_p^+ and N_n^+ fix the same elements of $(\bigotimes^k \mathbb{C}^n)(\lambda)$. $\square$

From Lemma 9.1.1 the space $G_{n,k}^\lambda$ depends only on λ, and will be denoted by G^λ. The representation of $\mathfrak{S}_k$ on G^λ will be denoted by σ^λ. We now study these representations of $\mathfrak{S}_k$ in more detail.

Recall that the conjugacy classes in $\mathfrak{S}_k$ are described by partitions of k (see Section 4.4.4); thus $\mathfrak{S}_k$ has $|\mathrm{Par}(k)|$ inequivalent irreducible representations by Proposition 4.3.8. Furthermore, by Theorem 4.2.1 the representations $\{\sigma^\lambda : \lambda \in \mathrm{Par}(k)\}$ are mutually inequivalent. Hence *every* irreducible representation of $\mathfrak{S}_k$ is equivalent to some σ^λ. We can state this duality as follows:

Theorem 9.1.2 (Schur–Weyl Duality). *Under the action of $\mathbf{GL}(n,\mathbb{C}) \times \mathfrak{S}_k$ the space of k-tensors over $\mathbb{C}^n$ decomposes as*

$$\bigotimes^k \mathbb{C}^n \cong \bigoplus_{\lambda \in \mathrm{Par}(k,n)} F_n^\lambda \otimes G^\lambda.$$

In particular, in the stable range $n \geq k$, every irreducible representation of $\mathfrak{S}_k$ occurs in this decomposition.

Let $\prec$ denote the partial order on the weights of H_n defined by the positive roots $\{\varepsilon_i - \varepsilon_j : 1 \le i < j \le n\}$. Thus $\mu \prec \lambda$ if and only if

$$\mu = \lambda - \sum_{i=1}^{n-1} k_i(\varepsilon_i - \varepsilon_{i+1})$$

with k_i nonnegative integers. In terms of components, this means that

$$\mu_1 \le \lambda_1, \quad \mu_1 + \mu_2 \le \lambda_1 + \lambda_2, \quad \dots, \quad \mu_1 + \cdots + \mu_n \le \lambda_1 + \cdots + \lambda_n.$$

Recall that $\mu \in \mathrm{Par}(k,n)$ determines an induced representation I^μ of $\mathfrak{S}_k$ (Proposition 4.4.6). For $\lambda, \mu \in \mathrm{Par}(k,n)$ we have already defined the *Kostka coefficient* $K_{\lambda\mu}$ as the multiplicity of the weight μ in the irreducible $\mathbf{GL}(n,\mathbb{C})$-module with highest weight λ (see Section 8.1.2). We have the following *reciprocity law* for multiplicities of the irreducible representations of $\mathfrak{S}_n$ in I^μ:

Proposition 9.1.3. *Let $\mu \in \mathrm{Par}(k,n)$. Then the induced module I^μ for $\mathfrak{S}_k$ decomposes as*

$$I^\mu \cong \bigoplus_{\lambda \in \mathrm{Par}(k,n)} K_{\lambda\mu} G^\lambda.$$

In particular, G^μ occurs in I^μ with multiplicity one, and G^λ does not occur in I^μ if $\mu \not\prec \lambda$.

Proof. By Proposition 4.4.6 and Theorem 9.1.2 we have

$$I^\mu \cong \bigoplus_{\lambda \in \mathrm{Par}(k,n)} F_n^\lambda(\mu) \otimes G^\lambda$$

as a module for $\mathfrak{S}_k$. Since $K_{\mu\lambda} = \dim F_n^\lambda(\mu)$, the Kostka coefficients give the multiplicities in the decomposition of I^μ. But we know from the theorem of the highest weight (Theorem 3.2.5) that $K_{\mu\mu} = 1$ and $K_{\mu\lambda} = 0$ if $\mu \not\prec \lambda$. This proves the last statement. $\square$

The irreducible module G^λ can be characterized uniquely in terms of the induced modules I^μ as follows: Give the partitions of k the *lexicographic order*: $\mu \overset{\text{lex}}{>} \lambda$ if for some $j \ge 1$ we have $\mu_i = \lambda_i$ for $i = 1, \dots, j-1$ and $\mu_j > \lambda_j$. Thus

$$[k] \overset{\text{lex}}{>} [k-1,1] \overset{\text{lex}}{>} [k-2,2] \overset{\text{lex}}{>} [k-2,1,1] \overset{\text{lex}}{>} \cdots \overset{\text{lex}}{>} [1,\dots,1]$$

(we omit the trailing zeros in the partitions). If $\mu \overset{\text{lex}}{>} \lambda$ then $\mu_1 + \cdots + \mu_j > \lambda_1 + \cdots + \lambda_j$ for some j. Hence $\mu \not\prec \lambda$, so $K_{\lambda\mu} = 0$. Thus Proposition 9.1.3 gives

$$I^\mu = G^\mu \oplus \bigoplus_{\lambda \overset{\text{lex}}{>} \mu} K_{\lambda\mu} G^\lambda.$$

For example, $G^{[k]} = I^{[k]}$ is the trivial representation. From Corollary 4.4.5 we have

$$I^{[k-1,1]} = \mathbb{C}[\mathfrak{S}_k/\mathfrak{S}_{k-1}] = V \oplus \mathbb{C},$$

where V is the $(k-1)$-dimensional standard representation of $\mathfrak{S}_k$. Hence we see that $G^{[k-1,1]} \cong V$.

9.1.2 Characters of $\mathfrak{S}_k$

Let λ be a partition of k with at most n parts. We shall express the character of the irreducible representation $(\sigma^\lambda, G^\lambda)$ of $\mathfrak{S}_k$ via an alternating sum formula involving the much simpler induced representations I^μ for certain other partitions μ. For every $\mu \in \mathbb{N}^n$ with $|\mu| = k$ we have a subgroup $\mathfrak{S}_\mu \cong \mathfrak{S}_{\mu_1} \times \cdots \times \mathfrak{S}_{\mu_n}$ of $\mathfrak{S}_k$ (see Section 4.4.4). An element $s \in \mathfrak{S}_n$ acts on $\mu \in \mathbb{Z}^n$ by permuting the entries of μ. We write this action as $s, \mu \mapsto s \cdot \mu$. Let $\rho = [n-1, n-2, \ldots, 1, 0] \in \mathbb{N}^n_{++}$.

Theorem 9.1.4. *Let $y \in \mathfrak{S}_k$. Then*

$$\mathrm{ch}_{\mathfrak{S}_k}(G^\lambda)(y) = \sum_{s \in \mathfrak{S}_n} \mathrm{sgn}(s) \,\#\{\text{fixed points of } y \text{ on } \mathfrak{S}_k/\mathfrak{S}_{\lambda+\rho-s\cdot\rho}\}, \qquad (9.7)$$

where the sum is over all $s \in \mathfrak{S}_n$ such that $\lambda + \rho - s \cdot \rho \in \mathbb{N}^n$.

Proof. By Theorem 7.1.11 and equation (9.5), we have

$$\mathrm{ch}_{\mathfrak{S}_k}(G^\lambda) = \sum_{s \in \mathfrak{S}_n} \mathrm{sgn}(s)\, \mathrm{ch}_{\mathfrak{S}_k}\left((\textstyle\bigotimes^k \mathbb{C}^n)(\lambda + \rho - s \cdot \rho)\right).$$

In this formula $\mathfrak{S}_n$ is the Weyl group of $\mathbf{GL}(n, \mathbb{C})$, acting on the characters of the diagonal subgroup H_n by permutations of the coordinates. Using Proposition 4.4.6 we conclude that

$$\mathrm{ch}_{\mathfrak{S}_k}(G^\lambda) = \sum_{s \in \mathfrak{S}_n} \mathrm{sgn}(s)\, \mathrm{ch}_{\mathfrak{S}_k}(I^{\lambda+\rho-s\cdot\rho}). \qquad (9.8)$$

Here $I^\mu = 0$ when any coordinate of μ is negative, since $(\bigotimes^k \mathbb{C}^n)(\mu) = 0$ in that case. Now apply Corollary 4.4.7 to each summand in (9.8) to obtain (9.7). $\square$

We obtain a formula for the dimension of G^λ by setting $y = 1$ in (9.7) and simplifying the resulting determinant, as we did for the Weyl dimension formula.

Corollary 9.1.5. *Let $\lambda \in \mathrm{Par}(k,n)$. Set $\mu = \lambda + \rho$. Then*

$$\dim G^\lambda = \frac{k!}{\mu_1! \mu_2! \cdots \mu_n!} \prod_{1 \le i < j \le n} (\mu_i - \mu_j). \qquad (9.9)$$

Proof. For $\gamma \in \mathbb{N}^n$ with $|\gamma| = k$, the cardinality of $\mathfrak{S}_k/\mathfrak{S}_\gamma$ is the multinomial coefficient $\binom{k}{\gamma} = k!/\prod \gamma_i!$. Hence taking $y = 1$ in (9.7) gives

$$\dim G^\lambda = \sum_{s \in \mathfrak{S}_n} \text{sgn}(s) \binom{k}{\lambda + \rho - s \cdot \rho} \tag{9.10}$$

with the usual convention that $\binom{k}{\gamma} = 0$ if any entry in γ is negative.

We now show that this alternating sum (9.10) has a product formula similar to the Weyl dimension formula for $\dim F^\lambda$. Since

$$\lambda + \rho - s \cdot \rho = [\mu_1 + s(1) - n, \mu_2 + s(2) - n, \dots, \mu_n + s(n) - n]$$

for $s \in \mathfrak{S}_n$, the multinomial coefficient in (9.10) can be written as

$$\binom{k}{\mu} \prod_{i=1}^n \frac{\mu_i!}{(\mu_i + s(i) - n)!} .$$

Dividing the factorials occurring in the product, we obtain

$$\binom{k}{\lambda + \rho - s \cdot \rho} = \binom{k}{\mu} \prod_{i=1}^n a_{i,s(i)} ,$$

where $a_{in} = 1$ and $a_{ij} = \mu_i(\mu_i - 1) \cdots (\mu_i - n + j + 1)$ (product with $n - j$ factors). Hence by (9.10) we have

$$\dim G^\lambda = \binom{k}{\lambda + \rho} \det[a_{ij}] .$$

Now $a_{ij} = \mu_i^{n-j} + p_j(\mu_i)$, where $p_j(x)$ is a polynomial of degree $n - j - 1$. Thus $\det[a_{ij}]$ is a polynomial in $\mu_1, \dots, \mu_n$ of degree $n(n-1)/2$ that vanishes when $\mu_i = \mu_j$ for any $i \neq j$. Hence

$$\det[a_{ij}] = c \prod_{1 \leq i < j \leq n} (\mu_i - \mu_j)$$

for some constant c. Since the monomial $\mu_1^{n-1} \mu_2^{n-2} \cdots \mu_{n-1}$ occurs in $\det[a_{ij}]$ with coefficient 1, we see that $c = 1$. This proves (9.9). $\square$

We shall identify the partition $\lambda = \{\lambda_1 \geq \lambda_2 \geq \cdots \geq \lambda_p > 0\}$ with the *Ferrers diagram* or *shape* consisting of p left-justified rows of boxes, with λ_i boxes in the ith row, as in Section 8.1.2. Define the *dual partition* or *transposed shape* $\lambda^t = \{\lambda_1^t \geq \lambda_2^t \geq \cdots\}$ by

$$\lambda_j^t = \text{length of } j\text{th column of the shape } \lambda .$$

We have $|\lambda^t| = |\lambda|$, and the depth of λ^t is the length of the first row of λ. For example, if $\lambda = [4, 3, 1]$ then $\lambda^t = [3, 2, 2, 1]$. The Ferrers diagram for λ^t is obtained by reflecting the diagram for λ about the diagonal:

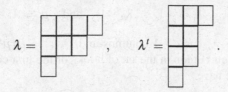

The formula for $\dim G^{\lambda}$ can be described in terms of the Ferrers diagram of λ, using the notion of a *hook*. For each position (i, j) in the diagram, the (i, j) *hook* is the set consisting of the box in position (i, j) together with all boxes in column j below this box and all boxes in row i to the right of this box. For example, if $\lambda = [4, 3, 1]$, then the $(1, 2)$ hook is the set of boxes containing a $\bullet$:

The *length* h_{ij} of the (i, j) hook is the number of boxes in the hook. We can indicate these numbers by inserting h_{ij} into the (i, j) box. For the example above we have

6	4	3	1
4	2	1	
1			

From the following corollary we can calculate $\dim G^{[4,3,1]} = \dfrac{8!}{6 \cdot 4 \cdot 4 \cdot 3 \cdot 2} = 70$.

Corollary 9.1.6 (Hook-Length Formula). *Let $\lambda \in \mathrm{Par}(k)$. Then*

$$\dim G^{\lambda} = \frac{k!}{\prod_{(i,j) \in \lambda} h_{ij}} .$$

Proof. We use induction on the number of columns of the Ferrers diagram of λ. If the diagram has one column, then $\dim G^{\lambda} = 1$, since λ corresponds to the sgn representation. In this case $h_{i1} = k + 1 - i$, so that the product of the hook lengths is $k!$, and hence the formula is true.

Now let λ have n parts and c columns, where $c > 1$. Set $\mu = \lambda + \rho$ and

$$D(\mu) = \prod_{1 \le i < j \le n} (\mu_i - \mu_j)$$

as in Corollary 9.1.5. It suffices to show that

$$D(\mu) \prod_{(i,j) \in \lambda} h_{ij} = \mu_1! \cdots \mu_n! . \tag{9.11}$$

We assume that (9.11) is true for all diagrams with $c - 1$ columns. Set

$$\lambda' = [\lambda_1 - 1, \lambda_2 - 1, \ldots, \lambda_n - 1] \quad \text{and} \quad \mu' = \lambda' + \rho .$$

Then the diagram for λ' has $c - 1$ columns and $D(\mu) = D(\mu')$. Since the set of hooks of λ is the disjoint union of the set of hooks of the first column of λ and the set of hooks of λ', we have

$$\begin{aligned}
D(\mu) \prod_{(i,j) \in \lambda} h_{ij} &= D(\mu') \big(\prod \{\text{hook lengths of } \lambda'\} \big) \\
&\qquad \times (\lambda_1 + n - 1)(\lambda_2 + n - 2) \cdots \lambda_n \\
&= (\mu'_1)! \cdots (\mu'_n)! (\lambda_1 + n - 1)(\lambda_2 + n - 2) \cdots \lambda_n \\
&= \mu_1! \cdots \mu_n!
\end{aligned}$$

(using the induction hypothesis in the next-to-last line). This proves (9.11).   □

From the hook-length formula we see that $\dim G^\lambda = \dim G^{\lambda^t}$. Later in this chapter we shall obtain more information about the representations σ^λ, including an explicit basis for the space G^λ and the fact that σ^{λ^t} is the representation $\sigma^\lambda \otimes \det$.

9.1.3 Frobenius Formula

We now use Schur duality to relate the irreducible characters of $\mathfrak{S}_k$ to the irreducible characters of $\mathbf{GL}(n, \mathbb{C})$, which we know thanks to the Weyl character formula. From this relation we will obtain an algorithm for finding character values that is generally much easier to use than the one from Section 9.1.2.

Suppose $x = \mathrm{diag}[x_1, \ldots, x_n] \in \mathbf{GL}(n, \mathbb{C})$ and $s \in \mathfrak{S}_k$. Let $V = \bigotimes^k \mathbb{C}^n$. The operators $\sigma_k(s)$ and $\rho_k(x)$ on V mutually commute. The function $\mathrm{tr}_V(\sigma_k(s)\rho_k(x))$ is a symmetric polynomial in $x_1, \ldots, x_n$ that depends only on the conjugacy class of s. To calculate this polynomial, let $s \in C(1^{i_1} 2^{i_2} \cdots k^{i_k})$. We may assume that s has the cycle decomposition

$$s = \underbrace{(1)(2) \cdots (i_1)}_{i_1 \ 1\text{-cycles}} \underbrace{(i_1 + 1, i_1 + 2) \cdots (i_1 + 2i_2 - 1, i_1 + 2i_2)}_{i_2 \ 2\text{-cycles}} \cdots .$$

Take the basis $\{e_J : |J| = k\}$ for V. Then for $J = (j_1, \ldots, j_k)$ with $1 \le j_p \le n$ we have $\rho_k(x)\sigma_k(s)e_J = x_J e_{s \cdot J}$, where $x_J = x_{j_1} \cdots x_{j_k}$. Thus

$$\mathrm{tr}_V(\rho_k(x)\sigma_k(s)) = \sum_J x_J \, \delta_{J, s \cdot J} . \tag{9.12}$$

But the indices J such that $s \cdot J = J$ are of the form

$$(\underbrace{a, b, \ldots,}_{i_1 \text{ singles}} \underbrace{a', a', b', b', \ldots,}_{i_2 \text{ pairs}} \underbrace{a'', a'', a'', \ldots,}_{i_3 \text{ triples}} \text{ etc.}) ,$$

where $a, b, \ldots, a', b', \ldots, a'', \ldots$ independently range over $1, 2, \ldots, n$. The sum over the i_1 indices $a, b, \ldots$ in (9.12) contributes a factor

$$(x_1 + x_2 + \cdots + x_n)^{i_1}$$

to $\mathrm{tr}_V(\sigma_k(s)\rho_k(x))$. Likewise, the sum over the i_2 indices $a', b', \ldots$ contributes a factor

$$(x_1^2 + x_2^2 + \cdots + x_n^2)^{i_2}$$

to the trace, and so on. We introduce the *elementary power sums*

$$p_r(x) = x_1^r + x_2^r + \cdots + x_n^r = \mathrm{tr}_{\mathbb{C}^n}(x^r) \quad \text{for } r = 0, 1, 2, \ldots .$$

Then we see from these considerations that the trace in (9.12) is the product

$$\mathrm{tr}_V(\rho_k(x)\sigma_k(s)) = p_1(x)^{i_1} \cdots p_k(x)^{i_k} . \tag{9.13}$$

We now apply the first commutant character formula from Section 7.1.3. Recall that the Weyl denominator for $\mathbf{GL}(n, \mathbb{C})$ is

$$\Delta_n(x) = \prod_{1 \le i < j \le n} (x_i - x_j) .$$

Theorem 9.1.7 (Frobenius Character Formula). *Let $\lambda = [\lambda_1, \ldots, \lambda_n]$ be a partition of k with at most n parts. Set*

$$\mu = \lambda + \rho = [\lambda_1 + n - 1, \lambda_2 + n - 2, \ldots, \lambda_n] .$$

Then $\mathrm{ch}_{\mathfrak{S}_k}(G^\lambda)(C(1^{i_1} \cdots k^{i_k}))$ is the coefficient of x^μ in $\Delta_n(x) p_1(x)^{i_1} \cdots p_k(x)^{i_k}$.

Proof. This follows immediately from (9.13) and Theorem 7.1.10.   □

Examples

1. Let C_m be the conjugacy class of m-cycles in $\mathfrak{S}_k$. Then $i_1 = k - m$ and $i_m = 1$, so $\mathrm{ch}_{\mathfrak{S}_k}(G^\lambda)(C_m)$ is the coefficient of $x^{\lambda + \rho}$ in

$$(x_1 + \cdots + x_n)^{k-m}(x_1^m + \cdots + x_n^m) \prod_{1 \le i < j \le n} (x_i - x_j) . \tag{9.14}$$

We call a monomial $x_1^{a_1} \cdots x_n^{a_n}$ *strictly dominant* if $a_1 > a_2 > \cdots > a_n$. For the simplest case of partitions λ with two parts and cycles of maximum length $m = k$, the strictly dominant terms in (9.14) are $x_1^{k+1} - x_1^k x_2$. Hence

$$\mathrm{ch}_{\mathfrak{S}_k}(G^\lambda)(C_k) = \begin{cases} -1 \text{ for } \lambda = [k-1, 1] , \\ 0 \text{ for } \lambda = [k-j, j] \text{ with } j > 1 . \end{cases}$$

2. Consider the group $\mathfrak{S}_3$, which has three conjugacy classes: $C(1^3) = \{\text{identity}\}$, $C(1^1 2^1) = \{(12), (13), (23)\}$, and $C(3^1) = \{(123), (132)\}$. Its irreducible representations are $\sigma^{[3]}$ (trivial), $\sigma^{[2,1]}$ (standard), and $\sigma^{[1,1,1]}$ (signum). To calculate the character values, we let $x = [x_1, x_2]$ and expand the polynomials

$$\Delta_2(x) p_1(x)^3 = x_1^4 + 2x_1^3 x_2 + \cdots ,$$
$$\Delta_2(x) p_1(x) p_2(x) = x_1^4 + \cdots ,$$
$$\Delta_2(x) p_3(x) = x_1^4 - x_1^3 x_2 + \cdots ,$$

where $\cdots$ indicates nondominant terms. By Theorem 9.1.7 the coefficients of the dominant terms in these formulas furnish the entries in the character table for $\mathfrak{S}_3$. We write χ^λ for the character of the representation σ^λ. For example, when $\lambda = [2, 1]$ we have $\lambda + \rho = [3, 1]$, so the coefficient of $x_1^3 x_2$ in $\Delta_2(x) p_3(x)$ gives the value of $\chi^{[2,1]}$ on the conjugacy class $C(3^1)$. The full table is given in Table 9.1, where the top row indicates the number of elements in each conjugacy class, and the other rows in the table give the character values for each irreducible representation.

Table 9.1 Character table of $\mathfrak{S}_3$.

conj. class:	$C(1^3)$	$C(1^1 2^1)$	$C(3^1)$
# elements:	1	3	2
$\chi^{[3]}$	1	1	1
$\chi^{[2,1]}$	2	0	-1
$\chi^{[1,1,1]}$	1	-1	1

9.1.4 Exercises

1. Show that $I^{[k-2,2]} \cong G^{[k-2,2]} \oplus G^{[k-1,1]} \oplus G^{[k]}$. (HINT: Show that the weight $\mu = [k-2, 2]$ occurs with multiplicity one in the representations $F_2^{[k-1,1]}$ and $F_2^{[k]}$ of **GL**$(2, \mathbb{C})$.)
2. Use the preceding exercise to show that $\dim G^{[k-2,2]} = k(k-3)/2$. Check using the hook-length formula.
3. Show that $\bigotimes^k \mathbb{C}^n$ decomposes under **SL**$(n, \mathbb{C}) \times \mathfrak{S}_k$ as follows:
   (a) Let λ and μ be in $\mathbb{N}_{++}^n$. Prove that $F_n^\lambda \cong F_n^\mu$ as a representation of **SL**$(n, \mathbb{C})$ if and only if $\lambda - \mu = r(\varepsilon_1 + \cdots + \varepsilon_n)$ for some integer r.
   (b) Prove that the decomposition of $\bigotimes^k \mathbb{C}^n$ under **SL**$(n, \mathbb{C}) \times \mathfrak{S}_k$ is the same as the decomposition relative to **GL**$(n, \mathbb{C}) \times \mathfrak{S}_k$.
4. Let $G = $ **SL**$(n, \mathbb{C})$ and let p, q, and r be positive integers. Use the previous exercise to prove the following:
   (a) If $p \not\equiv q \pmod{n}$ then $\text{Hom}_G\left(\bigotimes^p \mathbb{C}^n, \bigotimes^q \mathbb{C}^n\right) = 0$.
   (b) Suppose $q = p + rn$. Let $e_1, \ldots, e_n$ be the standard basis for $\mathbb{C}^n$ and set $\omega = e_1 \wedge \cdots \wedge e_n$. Define $T^+ : \bigotimes^p \mathbb{C}^n \longrightarrow \bigotimes^q \mathbb{C}^n$ by

$$T^+(v_1 \otimes \cdots \otimes v_p) = \omega^{\otimes r} \otimes v_1 \otimes \cdots \otimes v_p ,$$

where $v_i \in \mathbb{C}^n$. Then $\mathrm{Hom}_G(\bigotimes^p \mathbb{C}^n, \bigotimes^q \mathbb{C}^n) = \sigma_q(\mathbb{C}[\mathfrak{S}_q])T^+\sigma_p(\mathbb{C}[\mathfrak{S}_p])$.

(c) Suppose $q = p - rn$. For $v_1, \ldots, v_n$ in $\mathbb{C}^n$ let $\det[v_1 \cdots v_n]$ be the determinant of the matrix with columns $v_1, \ldots, v_n$. Define $T^- : \bigotimes^p \mathbb{C}^n \longrightarrow \bigotimes^q \mathbb{C}^n$ by

$$T^-(v_1 \otimes \cdots \otimes v_p) = \left\{ \prod_{s=0}^{r-1} \det[v_{sn+1}\, v_{sn+2} \cdots v_{(s+1)n}] \right\} v_{rn+1} \otimes \cdots \otimes v_p \,.$$

Then $\mathrm{Hom}_G(\bigotimes^p \mathbb{C}^n, \bigotimes^q \mathbb{C}^n) = \sigma_q(\mathbb{C}[\mathfrak{S}_q])T^-\sigma_p(\mathbb{C}[\mathfrak{S}_p])$.

5. Let $V = (\bigotimes^{n+1} \mathbb{C}^n) \otimes (\mathbb{C}^n)^*$. Show that the space of $\mathbf{SL}(n, \mathbb{C})$-invariant linear functionals on V is spanned by $\varphi_1, \ldots, \varphi_{n+1}$, where for $v_i \in \mathbb{C}^n$ and $v^* \in (\mathbb{C}^n)^*$ we set

$$\varphi_i(v_1 \otimes \cdots \otimes v_{n+1} \otimes v^*) = v^*(v_i) \det[v_1 \cdots \widehat{v_i} \cdots v_{n+1}] \,.$$

(HINT: Use the isomorphism $V^* \cong \mathrm{Hom}(\bigotimes^{n+1} \mathbb{C}^n, \mathbb{C}^n)$ and the previous exercise.)

6. Use the dimension formula to calculate $\dim G^\lambda$ for the following representations of $\mathfrak{S}_k$:

   (a) the standard representation, where $\lambda = [k-1, 1]$.

   (b) the representation G^λ, where $\lambda = [k-r, \underbrace{1, \ldots, 1}_{r}]$ with $1 < r < k$.

   (c) the representation G^λ, where $\lambda = [k-r, r]$ with $2r \le k$.

7. Show that $\mathfrak{S}_4$ has the following irreducible representations: two of dimension 1 (associated with the partition $4 = 4$ and its dual $4 = 1+1+1+1$), two of dimension 3 (associated with the partition $4 = 3+1$ and its dual $4 = 2+1+1$), and one of dimension 2 (associated with the self-dual partition $4 = 2+2$). Check by calculating the sum of the squares of the dimensions.

8. Show that $\mathfrak{S}_5$ has the following irreducible representations: two of dimension 1 (associated with the partition $5 = 5$ and its dual $5 = 1+1+1+1+1$), two of dimension 4 (associated with the partition $5 = 4+1$ and its dual $5 = 2+1+1+1$), two of dimension 5 (associated with the partition $5 = 3+2$ and its dual $5 = 2+2+1$), and one of dimension 6 (associated with the self-dual partition $5 = 3+1+1$). Check by calculating the sum of the squares of the dimensions.

9. Let $\lambda \in \mathrm{Par}(k, n)$ and let h_{ij} be the length of the (i, j) hook of λ. Set $\mu = [\lambda_1 + n - 1, \lambda_2 + n - 2, \ldots, \lambda_n]$. Use the proof of the hook-length formula and the Weyl dimension formula to show that

$$\dim F_n^\lambda = \frac{\mu_1! \cdots \mu_n!}{(n-1)!(n-2)! \cdots 2! \prod_{(i,j) \in \lambda} h_{ij}} \,.$$

10. Let $x = [x_1, x_2, x_3]$. Show (by a computer algebra system or otherwise) that

$$\Delta_3(x)p_1(x)p_3(x) = x_1^6 x_2 - x_1^4 x_2^3 + \cdots,$$
$$\Delta_3(x)p_1(x)^2 p_2(x) = x_1^6 x_2 + x_1^5 x_2^2 - x_1^4 x_2^2 x_3 + \cdots,$$
$$\Delta_3(x)p_2(x)^2 = x_1^6 x_2 - x_1^5 x_2^2 + 2x_1^4 x_2^3 - x_1^4 x_2^2 x_3 + \cdots,$$
$$\Delta_3(x)p_4(x) = x_1^6 x_2 - x_1^5 x_2^2 + x_1^4 x_2^2 x_3 + \cdots,$$

where $\cdots$ indicates nondominant terms.

11. Use Theorem 9.1.7 and the previous exercise to obtain Table 9.2; recall that $\sigma^{[4]}$ is the trivial representation and $\sigma^{[1,1,1,1]}$ is the sgn representation.

Table 9.2 Character table of $\mathfrak{S}_4$.

conj. class:	$C(1^4)$	$C(1^2 2^1)$	$C(1^1 3^1)$	$C(2^2)$	$C(4^1)$
# elements:	1	6	8	3	6
$\chi^{[4]}$	1	1	1	1	1
$\chi^{[3,1]}$	3	1	0	-1	-1
$\chi^{[2,2]}$	2	0	-1	2	0
$\chi^{[2,1,1]}$	3	-1	0	-1	1
$\chi^{[1,1,1,1]}$	1	-1	1	1	-1

12. Let $x = [x_1, x_2, x_3, x_4]$. Show (by a computer algebra system or otherwise) that

$$
\begin{aligned}
\Delta_4(x)p_1(x)^3 p_2(x) &= x_1^8 x_2^2 x_3 + 2x_1^7 x_2^3 x_3 + x_1^6 x_2^4 x_3 - x_1^5 x_2^4 x_3^2 \\
&\quad - 2x_1^5 x_2^3 x_3^2 x_4 + \cdots, \\
\Delta_4(x)p_1(x)p_2(x)^2 &= x_1^8 x_2^2 x_3 + x_1^6 x_2^4 x_3 - 2x_1^6 x_2^3 x_3^2 + x_1^5 x_2^4 x_3^2 + \cdots, \\
\Delta_4(x)p_1(x)^2 p_3(x) &= x_1^8 x_2^2 x_3 + x_1^7 x_2^3 x_3 - x_1^6 x_2^4 x_3 - x_1^5 x_2^4 x_3^2 \\
&\quad + x_1^5 x_2^3 x_3^2 x_4 + \cdots, \\
\Delta_4(x)p_1(x)p_4(x) &= x_1^8 x_2^2 x_3 - x_1^6 x_2^4 x_3 + x_1^5 x_2^4 x_3^2 + \cdots, \\
\Delta_4(x)p_2(x)p_3(x) &= x_1^8 x_2^2 x_3 - x_1^7 x_2^3 x_3 + x_1^6 x_2^4 x_3 - x_1^5 x_2^4 x_3^2 \\
&\quad + x_1^5 x_2^3 x_3^2 x_4 + \cdots, \\
\Delta_4(x)p_5(x) &= x_1^8 x_2^2 x_3 - x_1^7 x_2^3 x_3 + x_1^6 x_2^3 x_3^2 + x_1^5 x_2^3 x_3^2 x_4 + \cdots,
\end{aligned}
$$

where $\cdots$ indicates nondominant terms.

13. Use Theorem 9.1.7 and the previous exercise to obtain Table 9.3; recall that $\sigma^{[5]}$ is the trivial representation and $\sigma^{[1,1,1,1,1]}$ is the sgn representation.

Table 9.3 Character table of $\mathfrak{S}_5$.

conj. class:	$C(1^5)$	$C(1^3 2^1)$	$C(1^1 2^2)$	$C(1^2 3^1)$	$C(1^1 4^1)$	$C(2^1 3^1)$	$C(5^1)$
# elements:	1	10	15	20	30	20	24
$\chi^{[5]}$	1	1	1	1	1	1	1
$\chi^{[4,1]}$	4	2	0	1	0	-1	-1
$\chi^{[3,2]}$	5	1	1	-1	-1	1	0
$\chi^{[3,1,1]}$	6	0	-2	0	0	0	1
$\chi^{[2,2,1]}$	5	-1	1	-1	1	-1	0
$\chi^{[2,1,1,1]}$	4	-2	0	1	0	1	-1
$\chi^{[1,1,1,1,1]}$	1	-1	1	1	-1	-1	1

9.2 Dual Reductive Pairs

We continue developing the duality framework in representation theory and introduce the notion of *dual reductive pairs* of groups, *seesaw pairs*, and the associated *reciprocity laws* for multiplicities.

9.2.1 Seesaw Pairs

Let K and K' be reductive groups and suppose (ρ, Y) is a regular representation of $K \times K'$. The group $K \times K'$ is reductive and its irreducible regular representations are of the form $(\pi \widehat{\otimes} \pi', V^\pi \otimes V^{\pi'})$, where $\pi \in \widehat{K}$ and $\pi' \in \widehat{K'}$ (see Propositions 4.2.5 and 4.2.6). Thus the isotypic decomposition of (ρ, Y) is of the form

$$Y = \bigoplus_{(\pi, \pi') \in \widehat{K} \times \widehat{K'}} m_{\pi, \pi'} V^\pi \otimes V^{\pi'}, \tag{9.15}$$

where $m_{\pi, \pi'}$ is the multiplicity of the irreducible representation $\pi \widehat{\otimes} \pi'$ in ρ.

Proposition 9.2.1. *Let $\mathcal{A}$ (respectively, $\mathcal{A}'$) be the subalgebra of $\mathrm{End}(Y)$ generated by $\rho(K)$ (respectively, $\rho(K')$). The following are equivalent:*

1. *All multiplicities $m_{\pi, \pi'}$ are either 0 or 1, and each $\pi \in \widehat{K}$ (respectively $\pi' \in \widehat{K'}$) occurs at most once in (9.15).*
2. *$\mathcal{A}$ is the commutant of $\mathcal{A}'$ in $\mathrm{End}(Y)$.*

Proof. The implication (2) $\implies$ (1) follows directly from Theorem 4.2.1. Now assume that (1) holds and suppose $m_{\pi, \pi'} = 1$ for some pair (π, π'). The π'-isotypic subspace of Y (viewed as a K'-module) is $Y_{(\pi')} = V^\pi \otimes V^{\pi'}$, since π occurs only paired with π' in Y. Let $T \in \mathrm{End}_{K'}(Y)$. Then T leaves each subspace $Y_{(\pi')}$ invariant, and by the double commutant theorem (Theorem 4.1.13), T acts on $Y_{(\pi')}$ as $T_\pi \otimes I_{\pi'}$, where $T_\pi \in \mathrm{End}(V^\pi)$. Hence $T \in \mathcal{A}$ by Corollary 4.2.4. $\qquad\square$

When the conditions of Proposition 9.2.1 are satisfied, we say that $(\rho(K), \rho(K'))$ is a *dual reductive pair* of subgroups in $\mathbf{GL}(Y)$. Assume that these conditions hold. Then isotypic decomposition of Y is of the form

$$Y = \bigoplus_{(\pi, \pi') \in S} V^\pi \otimes V^{\pi'},$$

where $S \subset \widehat{K} \times \widehat{K'}$ is determined by its projection onto $\widehat{K}$ or its projection onto $\widehat{K'}$. Thus S is the graph of a bijection $\pi \leftrightarrow \pi'$ between a subset $\widehat{K}_Y \subset \widehat{K}$ and a subset $\widehat{K'}_Y \subset \widehat{K'}$. When $K = \mathbf{GL}(n, \mathbb{C})$, $K' = \mathbf{GL}(k, \mathbb{C})$, and $Y = \mathcal{P}^d(M_{n,k})$, this bijection is the duality in Section 5.6.2. When $K = \mathbf{GL}(n, \mathbb{C})$, $K' = \mathfrak{S}_k$, and $Y = \bigotimes^k \mathbb{C}^n$, this bijection is Schur–Weyl duality, where each partition in $\mathrm{Par}(k,n)$ determines an element of $\widehat{K}_Y$ and an element of $\widehat{K}'_Y$.

Suppose we have another pair G, G' of reductive groups such that

$$
\begin{aligned}
K &\subset G, \\
G' &\subset K'.
\end{aligned}
\tag{9.16}
$$

Assume further that ρ extends to a regular representation of the larger group $G \times K'$ on the same space Y (which we continue to denote by ρ) and that $(\rho(G), \rho(G'))$ is a dual reductive pair in $\mathbf{GL}(Y)$. We thus also have a decomposition of Y into $(G \times G')$-isotypic components of the form

$$
Y_{(\sigma, \sigma')} = V^\sigma \otimes V^{\sigma'},
$$

where $\sigma \in \widehat{G}$ and $\sigma' \in \widehat{G}'$. Again, this sets up a bijection $\sigma \leftrightarrow \sigma'$ between a subset $\widehat{G}_Y \subset \widehat{G}$ and a subset $\widehat{G}'_Y \subset \widehat{G}'$. In this situation we will say that the groups $G \times G'$ and $K \times K'$ are a *seesaw pair*, relative to the $(G \times K')$-module Y (the terminology is suggested by the diagonal pairing of the subgroups in (9.16)).

Examples

1. Let $X = M_{n,k+m}$ and $Y = \mathcal{P}^d(X)$. We take the groups $K = \mathbf{GL}(n, \mathbb{C})$, $K' = \mathbf{GL}(k+m, \mathbb{C})$, and the representation ρ of $K \times K'$ on Y as in Section 5.6.2:

$$
\rho(g, g')f(x) = f(g^t x g') \quad \text{for } f \in \mathcal{P}^d(X) \text{ and } (g, g') \in K \times K'.
$$

Set $r = \min\{n, k+m\}$. By Corollary 5.6.8 we have

$$
\rho \cong \bigoplus_{\substack{|\lambda| = d \\ \mathrm{depth}(\lambda) \leq r}} \pi_n^\lambda \,\widehat{\otimes}\, \pi_{k+m}^\lambda
\tag{9.17}
$$

as a representation of $K \times K'$. Let $G = \mathbf{GL}(n, \mathbb{C}) \times \mathbf{GL}(n, \mathbb{C})$ and embed $K \subset G$ as the pairs (k, k), $k \in K$. Let $G' = \mathbf{GL}(k, \mathbb{C}) \times \mathbf{GL}(m, \mathbb{C})$ and embed $G' \subset K'$ as the block-diagonal matrices $a \oplus b$ with $a \in \mathbf{GL}(k, \mathbb{C})$ and $b \in \mathbf{GL}(m, \mathbb{C})$. We have the isomorphism

$$
M_{n,k+m} \cong M_{n,k} \oplus M_{n,m}.
$$

By this isomorphism we view a polynomial function on X as a function $f(u, v)$, where $u \in M_{n,k}$ and $v \in M_{n,m}$. Thus we can extend the representation ρ to $G \times G'$ by

$$
\rho(g, g')f(u, v) = f(a^t u c, b^t v d)
$$

for $g = (a, b) \in G$ and $g' = (c, d) \in G'$. The group $G \times G'$ can be viewed as

$$
(\mathbf{GL}(n, \mathbb{C}) \times \mathbf{GL}(k, \mathbb{C})) \times (\mathbf{GL}(n, \mathbb{C}) \times \mathbf{GL}(m, \mathbb{C})).
$$

Relative to this grouping of the factors we have

$$\mathcal{P}(X) \cong \mathcal{P}(M_{n,k}) \otimes \mathcal{P}(M_{n,m}) .$$

Thus by Theorem 5.6.7 again,

$$\rho \cong \bigoplus_{\mu,\nu} \left(\pi_n^\mu \widehat{\otimes} \pi_n^\nu \right) \otimes \left(\pi_k^\mu \widehat{\otimes} \pi_m^\nu \right) \tag{9.18}$$

as a representation of $G \times G'$, where the sum is over all diagrams μ and ν with $|\mu| + |\nu| = d$, $\mathrm{depth}(\mu) \leq \min\{k,n\}$, and $\mathrm{depth}(\nu) \leq \min\{m,n\}$. This shows that (G,G') acts as a dual pair on $\mathcal{P}^d(X)$.

2. Let $K = \mathbf{GL}(n,\mathbb{C})$ and $K' = \mathfrak{S}_{k+m}$ acting on $Y = \otimes^{k+m}\mathbb{C}^n$ as usual by Schur–Weyl duality. Then

$$Y \cong \bigoplus_{\lambda \in \mathrm{Par}(k+m,n)} F_n^\lambda \otimes G^\lambda \tag{9.19}$$

as a $K \times K'$-module, where G^λ is the irreducible representation of $\mathfrak{S}_{|\lambda|}$ corresponding to the partition λ (Theorem 9.1.2). Thus (K,K') acts as a dual pair on Y. Take $G = \mathbf{GL}(n,\mathbb{C}) \times \mathbf{GL}(n,\mathbb{C})$ and embed K diagonally in G as the pairs (x,x), $x \in K$. Let $G' = \mathfrak{S}_k \times \mathfrak{S}_m$ and embed $G' \subset K'$ as the permutations leaving fixed the sets $\{1,\ldots,k\}$ and $\{k+1,\ldots,k+m\}$. By the isomorphism

$$\otimes^{k+m}\mathbb{C}^n \cong \left(\otimes^k \mathbb{C}^n \right) \otimes \left(\otimes^m \mathbb{C}^n \right) \tag{9.20}$$

we obtain a representation of G on Y that extends the representation of K. We have (G,G') acting as a dual pair on Y, since

$$Y \cong \bigoplus_{\mu \in \mathrm{Par}(k,n)} \bigoplus_{\nu \in \mathrm{Par}(m,n)} (F_n^\mu \otimes F_n^\nu) \otimes (G^\mu \otimes G^\nu) ,$$

by Schur–Weyl duality.

9.2.2 Reciprocity Laws

The interest in seesaw pairs lies in the reciprocity law for multiplicities obtained by decomposing Y into isotypic components relative to $K \times G'$.

Theorem 9.2.2 (Seesaw Reciprocity). *Let $K \times K'$ and $G \times G'$ be a seesaw pair, relative to a $(G \times K')$-module Y. Let $\pi \in \widehat{K}_Y$ and $\pi' \in \widehat{K}'_Y$ determine the same isotypic component of Y as a $(K \times K')$-module, and let $\sigma \in \widehat{G}_Y$ and $\sigma' \in \widehat{G}'_Y$ determine the same isotypic component of Y as a $(G \times G')$-module. Then there is a linear isomorphism*

$$\mathrm{Hom}_K\left(\pi, \mathrm{Res}_K^G(\sigma) \right) \cong \mathrm{Hom}_{G'}\left(\sigma', \mathrm{Res}_{G'}^{K'}(\pi') \right) .$$

Furthermore, if $\tau \in \widehat{K} \setminus \widehat{K}_Y$ *then* $\mathrm{Hom}_K\left(\tau, \mathrm{Res}_K^G(\sigma)\right) = 0$ *for all* $\sigma \in \widehat{G}_Y$. *Likewise, if* $\mu' \in \widehat{G}' \setminus \widehat{G}'_Y$ *then* $\mathrm{Hom}_{G'}\left(\mu', \mathrm{Res}_{G'}^{K'}(\pi')\right) = 0$ *for all* $\pi' \in \widehat{K}'_Y$.

Proof. Let $\sigma \in \widehat{G}$. Then the isotypic decomposition of V_σ as a K-module is

$$V_\sigma \cong \bigoplus_{\tau \in \widehat{K}} \mathrm{Hom}_K(\tau, \mathrm{Res}_K^G(\sigma)) \otimes V_\tau \,,$$

with $k \in K$ acting by $1 \otimes \tau(k)$ on the τ-isotypic component (see Section 4.1.4). Hence the isotypic decomposition of Y as a $(K \times G')$-module is

$$Y \cong \bigoplus_{\tau \in \widehat{K}} \bigoplus_{\sigma' \in \widehat{G}'_Y} \mathrm{Hom}_K(\tau, \mathrm{Res}_K^G(\sigma)) \otimes V_\tau \otimes V_{\sigma'} \,. \tag{9.21}$$

Likewise, for $\pi' \in \widehat{K}'$, the isotypic decomposition of $V_{\pi'}$ as a G'-module is

$$V_{\pi'} \cong \bigoplus_{\mu' \in \widehat{G}'} \mathrm{Hom}_{G'}(\mu', \mathrm{Res}_{G'}^{K'}(\pi')) \otimes V_{\mu'} \,,$$

with $g' \in G'$ acting by $1 \otimes \mu'(g')$ on the μ'-isotypic component. Hence we also obtain the isotypic decomposition of Y as a $(K \times G')$-module in the form

$$Y \cong \bigoplus_{\pi \in \widehat{K}_Y} \bigoplus_{\mu' \in \widehat{G}'} \mathrm{Hom}_{G'}(\mu', \mathrm{Res}_{G'}^{K'}(\pi')) \otimes V_\pi \otimes V_{\mu'} \,. \tag{9.22}$$

Since the decomposition of Y into isotypic components relative to $K \times G'$ is unique by Proposition 4.1.15, the theorem follows by comparing (9.21) and (9.22).   □

We now apply the reciprocity law to Examples 1 and 2 in Section 9.2.1. For diagrams μ and ν of depth at most n, the tensor product representation $\pi_n^\mu \otimes \pi_n^\nu$ is a polynomial representation; it therefore decomposes as

$$\pi_n^\mu \otimes \pi_n^\nu \cong \bigoplus_{\mathrm{depth}(\lambda) \leq n} c_{\mu\nu}^\lambda \, \pi_n^\lambda \,, \tag{9.23}$$

where the multiplicities $c_{\mu\nu}^\lambda$ are nonnegative integers called *Littlewood–Richardson* coefficients (see Corollary 7.1.7). However, if k and m are positive integers and λ is a diagram of depth at most $k + m$, then there is also a decomposition arising from restriction:

$$\mathrm{Res}_{\mathbf{GL}(k,\mathbb{C}) \times \mathbf{GL}(m,\mathbb{C})}^{\mathbf{GL}(k+m,\mathbb{C})} \left(\pi_{k+m}^\lambda\right) = \bigoplus_{\mu,\nu} d_{\mu\nu}^\lambda \, \pi_k^\mu \,\widehat{\otimes}\, \pi_m^\nu \,, \tag{9.24}$$

where the multiplicities $d_{\mu\nu}^\lambda$ are nonnegative integers, μ ranges over diagrams of depth at most k, and ν ranges over diagrams of depth at most m.

Theorem 9.2.3. *Let* λ, μ, ν *be Ferrers diagrams and* k, m, n *positive integers with* $\mathrm{depth}(\lambda) \leq n$, $\mathrm{depth}(\mu) \leq \min\{k, n\}$, *and* $\mathrm{depth}(\nu) \leq \min\{m, n\}$. *Then*

$$c_{\mu\nu}^{\lambda} = \begin{cases} 0 & \text{if } \operatorname{depth}(\lambda) > \min\{n, k+m\}, \\ d_{\mu\nu}^{\lambda} & \text{if } \operatorname{depth}(\lambda) \leq \min\{n, k+m\}. \end{cases} \tag{9.25}$$

In particular, the Littlewood–Richardson coefficients $c_{\mu\nu}^{\lambda}$ in (9.23) do not depend on n, provided $n \geq \max\{\operatorname{depth}(\lambda), \operatorname{depth}(\mu), \operatorname{depth}(\nu)\}$.

Proof. Let $K = \mathbf{GL}(n, \mathbb{C})$, $K' = \mathbf{GL}(k+m, \mathbb{C})$, $G = \mathbf{GL}(n, \mathbb{C}) \times \mathbf{GL}(n, \mathbb{C})$, $G' = \mathbf{GL}(k, \mathbb{C}) \times \mathbf{GL}(m, \mathbb{C})$, and $Y = \mathcal{P}^d(M_{n,k+m})$. By decomposition (9.17) of Y we see that

$$\widehat{K}_Y = \{\lambda : \operatorname{depth}(\lambda) \leq \min\{n, k+m\}\},$$

and the representation $\pi = \pi_n^{\lambda}$ of K is paired with the representation $\pi' = \pi_{k+m}^{\lambda}$ of K'. However, by decomposition (9.18) of Y we have

$$\widehat{G}_Y' = \{(\mu, \nu) : \operatorname{depth}(\mu) \leq \min\{k, n\}, \operatorname{depth}(\nu) \leq \min\{m, n\}\},$$

and the representation $\sigma = \pi_n^{\mu} \widehat{\otimes} \pi_n^{\nu}$ of G is paired with the representation $\sigma' = \pi_k^{\mu} \widehat{\otimes} \pi_m^{\nu}$ of G'. The theorem now follows from Theorem 9.2.2. $\qquad\square$

Let λ be a Ferrers diagram of depth $\leq n$ and let μ be a diagram of depth $\leq n-1$. As in Chapter 8, we say that μ *interlaces* λ if

$$\lambda_1 \geq \mu_1 \geq \lambda_2 \geq \cdots \geq \lambda_{n-1} \geq \mu_{n-1} \geq \lambda_n.$$

Thus λ is obtained from μ by adding a certain number of boxes, with the constraints that the resulting diagram have at most n rows of nonincreasing lengths and that each new box be in a different column.

Corollary 9.2.4 (Pieri's Rule). *Let μ be a diagram of depth $\leq n-1$ and ν a diagram of depth one. Then*

$$\pi_n^{\mu} \otimes \pi_n^{\nu} \cong \bigoplus_{\lambda} \pi_n^{\lambda},$$

where the sum is over all diagrams λ of depth $\leq n$ such that $|\lambda| = |\mu| + |\nu|$ and μ interlaces λ.

Proof. By the branching law from $\mathbf{GL}(n, \mathbb{C})$ to $\mathbf{GL}(n-1, \mathbb{C}) \times \mathbf{GL}(1, \mathbb{C})$ (Theorem 8.1.2), we have

$$d_{\mu\nu}^{\lambda} = \begin{cases} 1 & \text{if } \mu \text{ interlaces } \lambda \text{ and } |\lambda| = |\mu| + |\nu|, \\ 0 & \text{otherwise.} \end{cases}$$

The decomposition of the tensor product now follows from (9.25). $\qquad\square$

For an example of Pieri's rule, let $\mu = \;\begin{array}{c}\boxed{}\boxed{}\\\boxed{}\end{array}\;$ and $\;\nu = \boxed{}\boxed{}$. Then the

diagrams λ occurring in the decomposition of $\pi_n^{\mu} \otimes \pi_n^{\nu}$ for $n \geq 3$ are

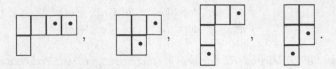

Here $\boxed{\bullet}$ indicates the boxes added to μ (this illustrates the stability property of the tensor decompositions for large n).

We now show that the Littlewood–Richardson coefficients also give multiplicities for the restriction to $\mathfrak{S}_k \times \mathfrak{S}_m$ of an irreducible representation of $\mathfrak{S}_{k+m}$.

Theorem 9.2.5. *Let* $\lambda \in \mathrm{Par}(k+m,n)$. *Then*

$$\mathrm{Res}^{\mathfrak{S}_{k+m}}_{\mathfrak{S}_k \times \mathfrak{S}_m}(\sigma^\lambda) \cong \bigoplus_{\mu,\nu} c^\lambda_{\mu\nu}\, \sigma^\mu \widehat{\otimes} \sigma^\nu \,,$$

where the sum is over all $\mu \in \mathrm{Par}(k,n)$ *and* $\nu \in \mathrm{Par}(m,n)$.

Proof. Assume that λ has depth n. We take the seesaw pairs

$$K = \mathbf{GL}(n,\mathbb{C}) \subset G = \mathbf{GL}(n,\mathbb{C}) \times \mathbf{GL}(n,\mathbb{C})\,,$$
$$G' = \mathfrak{S}_k \times \mathfrak{S}_m \subset K' = \mathfrak{S}_{k+m}\,,$$

acting on $Y = \left(\bigotimes^k \mathbb{C}^n\right) \otimes \left(\bigotimes^m \mathbb{C}^n\right) = \bigotimes^{k+m} \mathbb{C}^n$. By the Schur–Weyl duality decomposition (9.19) of Y as a $(K \times K')$-module we know that $\widehat{K}_Y = \mathrm{Par}(k+m,n)$, and the representation π_n^λ of K is paired with the representation σ^λ of K'. However, from the decomposition (9.20) of Y as a $(G \times G')$-module we have

$$\widehat{G}'_Y = \{(\mu,\nu) : \mu \in \mathrm{Par}(k,n),\, \nu \in \mathrm{Par}(m,n)\}\,,$$

and the representation $\pi_n^\mu \widehat{\otimes} \pi_n^\nu$ of G is paired with the representation $\sigma^\mu \widehat{\otimes} \sigma^\nu$ of G'. The theorem now follows from Theorems 9.2.2 and 9.2.3. $\qquad\square$

Corollary 9.2.6. *Let* $\mu \in \mathrm{Par}(k,n)$ *and* $\nu \in \mathrm{Par}(m,n)$. *Then*

$$\mathrm{Ind}^{\mathfrak{S}_{k+m}}_{\mathfrak{S}_k \times \mathfrak{S}_m}(\sigma^\mu \widehat{\otimes} \sigma^\nu) = \bigoplus_{\lambda \in \mathrm{Par}(k+m,n)} c^\lambda_{\mu\nu}\, \sigma^\lambda\,.$$

Proof. Use Theorem 9.2.5 and Frobenius reciprocity (Theorem 4.4.1). $\qquad\square$

Corollary 9.2.7 (Branching Rule). *Let* $\lambda \in \mathrm{Par}(n)$. *Then*

$$\mathrm{Res}^{\mathfrak{S}_n}_{\mathfrak{S}_{n-1}}(\sigma^\lambda) \cong \bigoplus_\mu \sigma^\mu$$

with the sum over all $\mu \in \mathrm{Par}(n-1)$ *such that* μ *interlaces* λ.

Proof. Take $k = n-1$ and $m = 1$ in Theorem 9.2.5 and use the calculation of the Littlewood–Richardson coefficients in the proof of Corollary 9.2.4. $\qquad\square$

The partitions $\mu \in \mathrm{Par}(n-1)$ that interlace $\lambda \in \mathrm{Par}(n)$ are obtained as follows: Take any p between 1 and $n-1$ such that $\lambda_p > \lambda_{p+1}$, set $\mu_i = \lambda_i$ for all $i \neq p$, and set $\mu_p = \lambda_p - 1$. Thus the Ferrers diagram for μ is obtained by removing one box from some row of the diagram for λ in such a way that the resulting diagram still has rows of nonincreasing length (from top to bottom). For example, when $\lambda = [2,2]$, then the branching law for $\mathfrak{S}_4 \longrightarrow \mathfrak{S}_3$ gives the diagrams

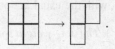

This shows that the restriction of $G^{[2,2]}$ to $\mathfrak{S}_3$ remains irreducible. We can iterate this procedure to branch from $\mathfrak{S}_3$ to $\mathfrak{S}_2$:

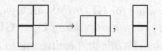

We conclude that the representation $G^{[2,2]}$ of $\mathfrak{S}_4$ has a basis $\{v_+, v_-\}$, where $v_\pm$ are eigenvectors with eigenvalues ± 1 under the action of the nontrivial element of $\mathfrak{S}_2$.

9.2.3 Schur–Weyl Duality and $\mathbf{GL(k)}$–$\mathbf{GL(n)}$ Duality

We now relate Schur–Weyl duality to $\mathbf{GL}(k)$–$\mathbf{GL}(n)$ duality. We embed $\mathfrak{S}_k$ into $\mathbf{GL}(k, \mathbb{C})$ as the permutation matrices and we let H_k be the diagonal subgroup in $\mathbf{GL}(k, \mathbb{C})$. The group $\mathfrak{S}_k$ normalizes H_k, so we have the semidirect product group $\mathfrak{S}_k \ltimes H_k \subset \mathbf{GL}(k, \mathbb{C})$.

Let $X = M_{k,n}$ and $G = \mathbf{GL}(k, \mathbb{C}) \times \mathbf{GL}(n, \mathbb{C})$, as in Section 5.6.2. Consider $\mathcal{P}(X)$ as a module for the subgroup $(\mathfrak{S}_k \ltimes H_k) \times \mathbf{GL}(n, \mathbb{C})$. As a module for $\mathbf{GL}(n, \mathbb{C})$,

$$X \cong \underbrace{(\mathbb{C}^n)^* \oplus \cdots \oplus (\mathbb{C}^n)^*}_{k \text{ summands}},$$

since $g \in \mathbf{GL}(n, \mathbb{C})$ acts by right multiplication on a matrix $x \in X$. Under this isomorphism, $h = \mathrm{diag}[h_1, \ldots, h_k] \in H_k$ acts by $h_j^{-1} I$ on the jth summand and the elements of $\mathfrak{S}_k$ act by permuting the summands. The canonical isomorphism $S(E \oplus F) \cong S(E) \otimes S(F)$ (see Proposition C.1.4) gives an isomorphism

$$\mathcal{P}(X) \cong \underbrace{S(\mathbb{C}^n) \otimes \cdots \otimes S(\mathbb{C}^n)}_{k \text{ factors}} \tag{9.26}$$

as a module for $\mathfrak{S}_k \times \mathbf{GL}(n, \mathbb{C})$, with the usual action of $\mathbf{GL}(n, \mathbb{C})$ on each tensor factor and $\mathfrak{S}_k$ permuting the tensor factors. For the H_k-weight space $\mathcal{P}(X)(\mu)$, where $\mu = \sum_{i=1}^{k} m_i \varepsilon_i$, we thus have the isomorphism

$$\mathcal{P}(X)(\mu) \cong S^{m_1}(\mathbb{C}^n) \otimes \cdots \otimes S^{m_k}(\mathbb{C}^n) .$$

The weight $\varepsilon_1 + \cdots + \varepsilon_k$ of H_k parameterizes the character $x \mapsto \det(x)$ of **GL**$(k, \mathbb{C})$, and we will denote it by $\det_k$. This weight is fixed by $\mathrm{Ad}^*(\mathfrak{S}_k)$, and one has

$$\mathcal{P}(X)(\det_k) \cong S^1(\mathbb{C}^n) \otimes \cdots \otimes S^1(\mathbb{C}^n) = (\mathbb{C}^n)^{\otimes k} \tag{9.27}$$

as a module for $\mathfrak{S}_k \times \mathbf{GL}(n, \mathbb{C})$, with the usual action on $(\mathbb{C}^n)^{\otimes k}$.

Theorem 9.2.8. *Let $X = M_{k,n}(\mathbb{C})$. The $\det_k$-weight space in $\mathcal{P}(X)$ decomposes under the action of $\mathfrak{S}_k \times \mathbf{GL}(n, \mathbb{C})$ as*

$$\mathcal{P}(X)(\det_k) \cong \bigoplus_{\lambda \in \mathrm{Par}(k,n)} \left\{ F_k^\lambda(\det_k) \right\} \otimes F_n^\lambda . \tag{9.28}$$

Here F_n^λ is the irreducible module for $\mathbf{GL}(n, \mathbb{C})$ with highest weight λ, and $\mathfrak{S}_k$ acts on the weight space $F_k^\lambda(\det_k)$ via the embedding $\mathfrak{S}_k \subset \mathbf{GL}(k, \mathbb{C})$. Furthermore, the action of $\mathfrak{S}_k$ on $F_k^\lambda(\det_k)$ is irreducible and equivalent to the representation σ^λ.

Proof. We have $\mathcal{P}(X)(\det_k) \subset \mathcal{P}^k(X)$. Hence Theorem 5.6.7 implies the decomposition (9.28). The irreducibility of $F_k^\lambda(\det_k)$ under $\mathfrak{S}_k$ and its equivalence to σ^λ now follow from Schur–Weyl duality. $\qquad\square$

9.2.4 Exercises

1. Let $k \geq 3$. We say that $\lambda \in \mathrm{Par}(k)$ is a *hook* partition if $\lambda = [k - p, \underbrace{1, \ldots, 1}_{p}]$, where $0 \leq p \leq k - 1$. The Ferrers diagram of λ is thus

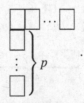

The associated representation $(\sigma^\lambda, G^\lambda)$ of $\mathfrak{S}_k$ is called a *hook representation*. For example, the trivial, signum, and standard representations are hook representations.

(a) Suppose $\lambda \in \mathrm{Par}(k)$ is a hook partition with $1 \leq p \leq k - 1$. Prove that $\sigma^\lambda|_{\mathfrak{S}_{k-1}} \cong \sigma^\mu \oplus \sigma^\nu$, where $\mu, \nu \in \mathrm{Par}(k-1)$ are the two different hook partitions obtained by removing one box from the diagram of λ.

(b) Suppose $\lambda \in \mathrm{Par}(k)$ is *not* a hook partition. Prove that $\sigma^\lambda|_{\mathfrak{S}_{k-1}}$ is the sum of at least three irreducible representations. (HINT: Use the branching law.)

2. For $k \geq 3$ let $\pi_k = \sigma^{[k-1,1]}$ be the standard representation of $\mathfrak{S}_k$. Assume $1 \leq p \leq k - 1$.

(a) Show that $\bigwedge^p \pi_k|_{\mathfrak{S}_{k-1}} \cong (\bigwedge^p \pi_{k-1}) \oplus (\bigwedge^{p-1} \pi_{k-1})$. (HINT: For the case $p = 1$ use the preceding exercise or a direct calculation, and note that $\bigwedge^0 \pi_{k-1}$ is the trivial representation.)

(b) Prove that $\bigwedge^p \pi_k \cong \sigma^{[k-p,1,\ldots,1]}$. (HINT: Use (a) and induction on k to prove that $\bigwedge^p \pi_k|_{\mathfrak{S}_{k-1}} \cong \sigma^{[k-p,1,\ldots,1]} \oplus \sigma^{[k-p-1,1,\ldots,1]}$. Then use part (b) of the preceding exercise to conclude that $\bigwedge^p \pi_k$ must be a single hook representation. Finally, use part (a) of the preceding exercise to identify the hook partition.)

3. Let $\mu \in \mathrm{Par}(n-1)$. Prove that $\mathrm{Ind}_{\mathfrak{S}_{n-1}}^{\mathfrak{S}_n}(\sigma^\mu) \cong \bigoplus_\lambda \sigma^\lambda$, with the sum over all $\lambda \in \mathrm{Par}(n)$ such that μ interlaces λ.

9.3 Young Symmetrizers and Weyl Modules

Now that we have determined the representations and characters of $\mathfrak{S}_k$ through Schur duality and the characters of $\mathbf{GL}(n,\mathbb{C})$, we turn to the problem of finding realizations of the representations of $\mathbf{GL}(n,\mathbb{C})$ in subspaces of tensor space determined by symmetry conditions relative to $\mathfrak{S}_k$.

9.3.1 Tableaux and Symmetrizers

Recall that to a partition λ of k with at most n parts we have associated the following data:

(1) an irreducible representation π_n^λ of $\mathbf{GL}(n,\mathbb{C})$ with highest weight λ;

(2) an irreducible representation σ^λ of $\mathfrak{S}_k$.

In the Schur–Weyl duality pairing between (1) and (2), σ^λ acts on the space

$$G^\lambda = (\bigotimes^k \mathbb{C}^n)^{N_n^+}(\lambda)$$

of N_n^+-fixed k-tensors of weight λ. We now use the symmetry properties of the tensors in G^λ relative to certain subgroups of $\mathfrak{S}_k$ to construct projection operators onto irreducible $\mathbf{GL}(n,\mathbb{C})$ subspaces of type π_n^λ and the projection onto the full $\mathbf{GL}(n,\mathbb{C}) \times \mathfrak{S}_k$ isotypic subspace in $\bigotimes^k \mathbb{C}^n$.

Definition 9.3.1. Let $\lambda \in \mathrm{Par}(k)$ and identify λ with its Ferrers diagram (shape). A *tableau of shape* λ is an assignment of the integers $1, 2, \ldots, k$ to the k boxes of λ, each box receiving a different integer.

If A is a tableau of shape $\lambda = [\lambda_1, \ldots, \lambda_n] \in \mathrm{Par}(k)$, we write A_{ij} for the integer placed in the jth box of the ith row of A, for $i = 1, \ldots, n$ and $j = 1, \ldots, \lambda_i$, and we set $|A| = k$ (we say that A has *size* k). Given a shape λ, we denote by $A(\lambda)$ the tableau

obtained by numbering the boxes consecutively down the columns of the shape from
left to right. For example, if $\lambda = [3,2,1,1]$ then

$$A(\lambda) = \begin{array}{|c|c|c|} \hline 1 & 5 & 7 \\ \hline 2 & 6 \\ \cline{1-2} 3 \\ \cline{1-1} 4 \\ \cline{1-1} \end{array} \quad .$$

We denote by $\mathrm{Tab}(\lambda)$ the set of all tableaux of shape λ. The group $\mathfrak{S}_k$ operates
simply transitively on $\mathrm{Tab}(\lambda)$ by permuting the numbers in the boxes: $(s \cdot A)_{ij} = s(A_{ij})$.

Given a tableau A of size $|A| = k$, we set $i_j = r$ if j occurs in the rth row of A. We
define the k-tensor

$$e_A = e_{i_1} \otimes \cdots \otimes e_{i_k} .$$

Thus the numbers in the first row of A indicate the tensor positions in e_A containing
e_1, the numbers in the second row of A indicate the tensor positions in e_A containing
e_2, and so on. For example, if $\lambda = [3,2,1,1]$ as in the example above, then

$$e_{A(\lambda)} = e_1 \otimes e_2 \otimes e_3 \otimes e_4 \otimes e_1 \otimes e_2 \otimes e_1 .$$

From the definition, we see that e_i occurs λ_i times in e_A, and as A ranges over $\mathrm{Tab}(\lambda)$
the positions of e_i can be arbitrary. It follows that

$$\mathrm{Span}\{e_A : A \in \mathrm{Tab}(\lambda)\} = \left(\textstyle\bigotimes^k \mathbb{C}^n \right)(\lambda) , \qquad (9.29)$$

where $k = |\lambda|$. Thus the tableaux of shape λ label a basis for the tensors of weight
λ, and the action of $\mathfrak{S}_k$ on k-tensors is compatible with the action on tableaux:

$$\sigma_k(s)e_A = e_{s \cdot A} \quad \text{for } s \in \mathfrak{S}_k .$$

A tableau A with r rows gives a partition of the set $\{1,2,\ldots,k\}$ into r disjoint
subsets $R_1,\ldots,R_r$, where R_i is the set of numbers in the ith row of A. We say that
$s \in \mathfrak{S}_k$ *preserves the rows* of A if s preserves each of the subsets R_i. In a similar way,
the c columns of A also give a partition of $\{1,2,\ldots,k\}$ into c disjoint subsets, and
we say that $s \in \mathfrak{S}_k$ *preserves the columns* of A if s preserves each of these subsets.
We define the *row group* and *column group* of A by

$$\mathrm{Row}(A) = \{s \in \mathfrak{S}_k : s \text{ preserves the rows of } A\} ,$$

$$\mathrm{Col}(A) = \{s \in \mathfrak{S}_k : s \text{ preserves the columns of } A\} .$$

Obviously $\mathrm{Row}(A) \cap \mathrm{Col}(A) = \{1\}$. Since e_A is formed by putting e_i in the positions
given by the integers in the ith row of the tableau A, it is clear that $\sigma_k(s)e_A = e_A$ if
and only if $s \in \mathrm{Row}(A)$. Furthermore, if $A, B \in \mathrm{Tab}(\lambda)$ then $e_A = e_B$ if and only if
$B = s \cdot A$ for some $s \in \mathrm{Row}(A)$.

Associated to a tableau A are the *row symmetrizer*

$$\mathbf{r}(A) = \sum_{r \in \text{Row}(A)} r$$

and the *column skew symmetrizer*

$$\mathbf{c}(A) = \sum_{c \in \text{Col}(A)} \text{sgn}(c)\, c$$

in the group algebra $A[\mathfrak{S}_k]$. Notice that

$$\mathbf{r}(A)x = x\mathbf{r}(A) = \mathbf{r}(A) \quad \text{for } x \in \text{Row}(A), \tag{9.30}$$
$$\mathbf{c}(A)y = y\mathbf{c}(A) = \text{sgn}(y)\mathbf{c}(A) \quad \text{for } y \in \text{Col}(A). \tag{9.31}$$

These elements transform by $\mathbf{r}(s \cdot A) = s\mathbf{r}(A)s^{-1}$ and $\mathbf{c}(s \cdot A) = s\mathbf{c}(A)s^{-1}$ for $s \in \mathfrak{S}_k$. If A has at most n rows then the operators $\mathbf{r}(A)$ and $\mathbf{c}(A)$ preserve the space $\left(\bigotimes^k \mathbb{C}^n \right)(\lambda)$, where λ is the shape of A.

Lemma 9.3.2. *Let $\lambda \in \text{Par}(k,n)$. If A has shape λ then $\mathbf{c}(A)e_A$ is nonzero and N_n^+-fixed of weight λ.*

Proof. Suppose first that $A = A(\lambda)$. Let λ^t be the transposed shape (dual partition) to λ. Then

$$e_A = e_1 \otimes e_2 \otimes \cdots \otimes e_{\lambda_1^t} \otimes e_1 \otimes e_2 \otimes \cdots \otimes e_{\lambda_2^t} \otimes \cdots \otimes e_1 \otimes e_2 \otimes \cdots \otimes e_{\lambda_q^t}.$$

The group $\text{Col}(A)$ gives all permutations of positions $1, 2, \ldots, \lambda_1^t$, all permutations of positions $\lambda_1^t + 1, \ldots, \lambda_2^t$, and so on. Hence

$$\mathbf{c}(A)e_A = \kappa\, \omega_{\lambda_1^t} \otimes \omega_{\lambda_2^t} \otimes \cdots \otimes \omega_{\lambda_q^t},$$

where $\omega_i = e_1 \wedge \cdots \wedge e_i$ and κ is a nonzero constant. In particular, $\mathbf{c}(A)e_A \neq 0$. Since each ω_i is fixed by N_n^+, so is $\mathbf{c}(A)e_A$.

Now let A be any tableau of shape λ. There is some $s \in \mathfrak{S}_k$ such that $A = s \cdot A(\lambda)$. Hence $e_A = \sigma_k(s)e_{A(\lambda)}$ and $\mathbf{c}(A) = \sigma_k(s)\mathbf{c}(A(\lambda))\sigma_k(s)^{-1}$. It follows that

$$\mathbf{c}(A)e_A = \sigma_k(s)\mathbf{c}(A(\lambda))e_{A(\lambda)}.$$

But $\sigma_k(\mathfrak{S}_k)$ commutes with $\rho_k(N_n^+)$, so $\mathbf{c}(A)e_A$ is N_n^+-fixed as claimed. $\square$

We now need the following combinatorial result:

Lemma 9.3.3. *Let $\lambda, \mu \in \text{Par}(k)$ with $k \geq 2$. Let A be a tableau of shape λ and B a tableau of shape μ. Suppose either*

(i) $\lambda \overset{\text{lex}}{<} \mu$, *or else*
(ii) $\lambda = \mu$ *and $c \cdot A \neq r \cdot B$ for all $c \in \text{Col}(A)$ and $r \in \text{Row}(B)$.*

Then there exists a pair of distinct integers l, m that are in the same column of A and in the same row of B.

Proof. First suppose $\lambda = \mu$ and argue by contradiction. If the numbers in each row of B appear in distinct columns of A, then there exists $c \in \mathrm{Col}(A)$ such that each row of $c \cdot A$ contains the same set of numbers as the corresponding row of B. Hence there exists $r \in \mathrm{Row}(B)$ such that $r \cdot B = c \cdot A$, contradicting (ii).

Now we suppose $\lambda \overset{\mathrm{lex}}{<} \mu$ and use induction on k. When $k = 2$ then

$$\mu = \boxed{} \quad \text{and} \quad \lambda = \boxed{\begin{array}{c} \\ \end{array}} ,$$

so the lemma is obviously true. Assume the lemma true for all tableaux of size less than k. Let $B = [b_{ij}]$. If $\mu_1 > \lambda_1$, then each of the numbers $b_{11}, \dots, b_{1\mu_1}$ in the first row of B is assigned to one of the λ_1 columns of A. Hence two of these numbers must appear in the same column of A, and we are done.

If $\lambda_1 = \mu_1$ and each of the numbers $b_{11}, \dots, b_{1\mu_1}$ appears in a different column in A, then there exists $c \in \mathrm{Col}(A)$ such that $c \cdot A$ has the same set of numbers in its first row as does B. Thus there exists $r \in \mathrm{Row}(B)$ such that $r \cdot B$ and $c \cdot A$ have the same first row. Now we observe that the lemma is true for the pair A, B if and only if it is true for $c \cdot A, r \cdot B$. We remove the first row of $c \cdot A$ and the first row of $r \cdot B$. The resulting tableaux are of size $k - \lambda_1$ and satisfy condition (i). By induction the lemma holds for them and hence for the original tableaux A, B.   □

Corollary 9.3.4. (Hypotheses of Lemma 9.3.3) *There exists $\gamma \in \mathrm{Col}(A) \cap \mathrm{Row}(B)$ with $\gamma^2 = 1$ and $\mathrm{sgn}(\gamma) = -1$.*

Proof. There exists a pair of numbers l, m in the same column of A and the same row of B, by Lemma 9.3.3. Hence we can take $\gamma \in \mathfrak{S}_k$ to be the transposition of l and m.   □

Proposition 9.3.5. *Let $\lambda \in \mathrm{Par}(k, n)$.*

1. *If A is a tableau of shape λ, then $\mathbf{c}(A)\big(\bigotimes^k \mathbb{C}^n\big)(\lambda) = \mathbb{C}\mathbf{c}(A)e_A$.*
2. *$G^\lambda = \sum_{A \in \mathrm{Tab}(\lambda)} \mathbb{C}\mathbf{c}(A)e_A$.*

Proof. (1): By (9.29) it suffices to consider the action of $\mathbf{c}(A)$ on e_B, for all $B \in \mathrm{Tab}(\lambda)$. If there exist $c \in \mathrm{Col}(A)$ and $r \in \mathrm{Row}(B)$ such that $c \cdot A = r \cdot B$, then

$$\mathbf{c}(A)e_B = \mathbf{c}(A)e_{r \cdot B} = \mathbf{c}(A)e_{c \cdot A} .$$

But $\mathbf{c}(A)e_{c \cdot A} = \mathbf{c}(A)\sigma_k(c)e_A = \mathrm{sgn}(c)\mathbf{c}(A)e_A$, yielding the result. If there does not exist such a pair c, r, we take γ as in Corollary 9.3.4. Then by (9.31) we have

$$\mathbf{c}(A)e_B = \mathbf{c}(A)e_{\gamma \cdot B} = \mathbf{c}(A)\sigma_k(\gamma)e_B = -\mathbf{c}(A)e_B , \qquad (9.32)$$

so $\mathbf{c}(A)e_B = 0$.

(2): When $A \in \mathrm{Tab}(\lambda)$ the tensor $\mathbf{c}(A)e_A$ is N_n^+-fixed of weight λ, by Lemma 9.3.2, and hence it is in G^λ. Also, if $s \in \mathfrak{S}_k$ and $A \in \mathrm{Tab}(\lambda)$ then $\sigma_k(s)\mathbf{c}(A)e_A =$

$\mathbf{c}(s \cdot A)e_{s \cdot A}$, and thus the right side of (2) is nonzero and invariant under $\mathfrak{S}_k$. Since G^λ is irreducible under $\mathfrak{S}_k$, equality must hold in (2). $\qquad \square$

Lemma 9.3.6. *Let* $\lambda, \mu \in \mathrm{Par}(k,n)$.

1. *If* $\mu \overset{\mathrm{lex}}{>} \lambda$ *then* $\mathbf{c}(A)\left(\bigotimes^k \mathbb{C}^n\right)(\mu) = 0$ *for all* $A \in \mathrm{Tab}(\lambda)$.

2. *If* $\mu \overset{\mathrm{lex}}{<} \lambda$ *then* $\mathbf{r}(A)\mathbf{c}(B) = 0$ *for all* $A \in \mathrm{Tab}(\lambda)$ *and* $B \in \mathrm{Tab}(\mu)$.

Proof. (1): Take $B \in \mathrm{Tab}(\mu)$ and let $\gamma \in \mathrm{Col}(A) \cap \mathrm{Row}(B)$ be as in Corollary 9.3.4. Then $\mathbf{c}(A)e_B = 0$ by the same calculation as (9.32). This proves (1).

(2): Take $\gamma \in \mathrm{Col}(B) \cap \mathrm{Row}(A)$ as in Corollary 9.3.4 (with λ and μ interchanged). We have $\mathbf{r}(A)\gamma = \mathbf{r}(A)$ and $\gamma \mathbf{c}(B) = -\mathbf{c}(B)$. But $\gamma^2 = 1$, so we obtain $\mathbf{r}(A)\mathbf{c}(B) = \mathbf{r}(A)\gamma^2\mathbf{c}(B) = -\mathbf{r}(A)\mathbf{c}(B)$. $\qquad \square$

Proposition 9.3.7. *Let* $\lambda \in \mathrm{Par}(k,n)$ *and let* A *be a tableau of shape* λ. *Define* $\mathbf{s}(A) = \mathbf{c}(A)\mathbf{r}(A)$ *as an element of the group algebra of* $\mathfrak{S}_k$.

1. $\mathbf{s}(A)G^\mu = 0$ *for all* $\mu \in \mathrm{Par}(k,n)$ *with* $\mu \neq \lambda$.
2. $\mathbf{s}(A)G^\lambda$ *is spanned by the* N_n^+-*fixed tensor* $\mathbf{c}(A)e_A$ *of weight* λ.

Proof. Suppose $\mu \overset{\mathrm{lex}}{>} \lambda$. Since the weight spaces of H_n are invariant under the group algebra of $\mathfrak{S}_k$, we have

$$\mathbf{s}(A)G^\mu \subset \mathbf{s}(A)\left(\bigotimes^k\mathbb{C}^n\right)(\mu) \subset \mathbf{c}(A)\left(\bigotimes^k\mathbb{C}^n\right)(\mu) = 0$$

by Lemma 9.3.6 (1).

Now suppose $\mu \overset{\mathrm{lex}}{<} \lambda$. By Proposition 9.3.5 it suffices to consider the action of $\mathbf{s}(A)$ on tensors of the form $\mathbf{c}(B)e_B$, for $B \in \mathrm{Tab}(\mu)$. But $\mathbf{r}(A)\mathbf{c}(B) = 0$ by Lemma 9.3.6 (2), so $\mathbf{s}(A)\mathbf{c}(B)e_B = 0$ also. Thus we have proved assertion (1).

Since $G^\lambda \subset \left(\bigotimes^k\mathbb{C}^n\right)(\lambda)$ and $\mathbf{r}(A)e_A = |\mathrm{Row}(A)|e_A$, assertion (2) follows by Proposition 9.3.5 (1) and Lemma 9.3.2. $\qquad \square$

The element $\mathbf{s}(A)$ in Proposition 9.3.7 is called the *Young symmetrizer* corresponding to the tableau A. If A has p rows, then $\mathbf{s}(A)$ operates on the tensor spaces $\bigotimes^k\mathbb{C}^n$ for all $n \geq p$.

Example

Suppose $A = \begin{array}{|c|c|} \hline 1 & 2 \\ \hline 3 \\ \cline{1-1} \end{array}$. Then $\mathbf{r}(A) = 1 + (12)$, $\mathbf{c}(A) = 1 - (13)$. Thus we have

$$\mathbf{s}(A) = (1 - (13))(1 + (12)) = 1 - (13) + (12) - (123).$$

We leave as an exercise to describe the symmetry properties of the 3-tensors in the range of $\mathbf{s}(A)$.

9.3.2 Weyl Modules

Let $\lambda \in \mathrm{Par}(k,n)$. We now show that each Young symmetrizer of shape λ is (up to a normalizing constant) a projection operator onto a $\mathbf{GL}(n,\mathbb{C})$-irreducible subspace of $\bigotimes^k \mathbb{C}^n$. Define

$$\mathcal{E}_k = \left(\bigotimes^k \mathbb{C}^k \right)^{N_k^+}$$

(note that we have taken $n = k$). This space is invariant under $\mathfrak{S}_k$, and from Lemma 9.1.1 and Theorem 9.1.2 we have

$$\mathcal{E}_k = \bigoplus\nolimits'_{\lambda \in \mathrm{Par}(k)} G^\lambda .$$

Thus $\mathcal{E}_k$ contains each irreducible representation of $\mathfrak{S}_k$ exactly once, and we have an algebra isomorphism

$$\mathcal{A}[\mathfrak{S}_k] \cong \bigoplus\nolimits_{\lambda \in \mathrm{Par}(k)} \mathrm{End}(G^\lambda) . \tag{9.33}$$

Let $\mathcal{B}^\lambda$ be the two-sided ideal in $\mathcal{A}[\mathfrak{S}_k]$ that corresponds to $\mathrm{End}(G^\lambda)$ under this isomorphism.

Lemma 9.3.8. *Let A be a tableau of shape $\lambda \in \mathrm{Par}(k)$. Then $\mathbf{s}(A) \in \mathcal{B}^\lambda$ and $\mathbf{s}(A)^2 = \xi_\lambda \mathbf{s}(A)$, where the scalar ξ_λ is nonzero and is the same for all $A \in \mathrm{Tab}(\lambda)$.*

Proof. By Proposition 9.3.7 (2) there exists a linear functional $f_A \in \mathcal{E}_k^*$ such that

$$\mathbf{s}(A)x = f_A(x)\mathbf{c}(A)e_A \quad \text{for } x \in \mathcal{E}_k . \tag{9.34}$$

Hence $\mathbf{s}(A) \in \mathcal{B}^\lambda$ and $\mathbf{s}(A)\mathbf{c}(A)e_A = \xi_A \mathbf{c}(A)e_A$ for some scalar ξ_A. We claim that $\xi_A \neq 0$. Indeed, if $\xi_A = 0$, then we would have $\mathbf{s}(A)G^\lambda = 0$ by Proposition 9.3.5, contradicting Proposition 9.3.7 (2). It follows that

$$\mathbf{s}(A)^2 x = f_A(x)\mathbf{s}(A)\mathbf{c}(A)e_A = \xi_A \mathbf{s}(A)x \quad \text{for } x \in \mathcal{E}_k .$$

Since the representation of $\mathcal{A}[\mathfrak{S}_k]$ on $\mathcal{E}_k$ is faithful, we have $\mathbf{s}(A)^2 = \xi_A \mathbf{s}(A)$ in $\mathcal{A}[\mathfrak{S}_k]$.

It remains to show that ξ_A depends only on λ. If $A, B \in \mathrm{Tab}(\lambda)$ then there is $\gamma \in \mathfrak{S}_k$ such that $B = \gamma \cdot A$. Hence

$$\mathbf{s}(B)^2 = \gamma \mathbf{s}(A)^2 \gamma^{-1} = \xi_A \gamma \mathbf{s}(A) \gamma^{-1} = \xi_A \mathbf{s}(B) ,$$

and so we have $\xi_B = \xi_A$. Thus we may define $\xi_\lambda = \xi_A$ unambiguously, for any $A \in \mathrm{Tab}(\lambda)$. $\qquad\square$

Lemma 9.3.9. *Let $\lambda \in \mathrm{Par}(k)$. Then $\xi_\lambda = k!/\dim G^\lambda$.*

Proof. Let L be the representation of $\mathcal{A}[\mathfrak{S}_k]$ on $\mathcal{A}[\mathfrak{S}_k]$ (viewed as a vector space) given by left convolution. For $A \in \mathrm{Tab}(\lambda)$ define $\mathbf{p}_A = \xi_\lambda^{-1} \mathbf{s}(A)$ in $\mathcal{A}[\mathfrak{S}_k]$. Then $\mathbf{p}_A^2 = \mathbf{p}_A$, so $L(\mathbf{p}_A)$ is idempotent and has range $\mathbf{s}(A)\mathcal{A}[\mathfrak{S}_k]$. Hence

$$\mathrm{tr}(L(\mathbf{p}_A)) = \dim(\mathbf{s}(A)\mathcal{A}[\mathfrak{S}_k]) \, .$$

We can also calculate the trace of $L(\mathbf{p}_A)$ by writing

$$\mathbf{p}_A = \sum_{g \in \mathfrak{S}_k} a_g \, g \, .$$

Then $L(\mathbf{p}_A)$ has matrix $[a_{gh^{-1}}]$ relative to the basis $\{g\}_{g \in \mathfrak{S}_k}$ for $\mathcal{A}[\mathfrak{S}_k]$. Hence

$$\mathrm{tr}(L(\mathbf{p}_A)) = \sum_{g \in \mathfrak{S}_k} a_{gg^{-1}} = a_1 |\mathfrak{S}_k| \, .$$

But $a_1 = \xi_\lambda^{-1}$; therefore, comparing these two trace calculations gives

$$\xi_\lambda = \frac{k!}{\dim(s(A)\mathcal{A}[\mathfrak{S}_k])} \, . \tag{9.35}$$

We know that $\mathbf{s}(A)\mathcal{A}[\mathfrak{S}_k] = \mathbf{s}(A)\mathcal{B}^\lambda$, since $\mathbf{s}(A) \in \mathcal{B}^\lambda$. Take a basis $\{f_i\}$ for G^λ for which $f_1 = c(A)e_A$, and let $\{e_{ij}\}$ be the corresponding elementary matrices. Under the isomorphism (9.33), $\mathbf{s}(A)\mathcal{B}^\lambda$ corresponds to $e_{11} \mathrm{End}(G^\lambda)$ (this follows by (9.34)). Thus $\dim(\mathbf{s}(A)\mathcal{A}[\mathfrak{S}_k]) = \dim G^\lambda$. Now substitute this in (9.35). $\qquad\square$

Theorem 9.3.10. *Let λ be a partition of k with at most n parts. If A is a tableau of shape λ, then the operator $\mathbf{p}_A = (\dim G^\lambda / k!)\mathbf{s}(A)$ projects $\bigotimes^k \mathbb{C}^n$ onto an irreducible $\mathbf{GL}(n, \mathbb{C})$-module with highest weight λ.*

Proof. From Lemma 9.3.9 we know that $\mathbf{p}_A$ is a projection operator, and from Schur–Weyl duality we have

$$\mathbf{p}_A\big(\textstyle\bigotimes^k \mathbb{C}^n\big) \cong \mathbf{p}_A\big(\textstyle\bigotimes_{\mu \in \mathrm{Par}(k,n)} F_n^\mu \otimes G^\mu\big) \cong F_n^\lambda \otimes \mathbf{p}_A G^\lambda$$

as a $\mathbf{GL}(n, \mathbb{C})$-module. By Proposition 9.3.7 we have $\mathbf{p}_A G^\lambda = \mathbb{C}c(A)e_A$. Hence $\mathbf{p}_A\big(\bigotimes^k \mathbb{C}^n\big) \cong F_n^\lambda$ as a $\mathbf{GL}(n, \mathbb{C})$-module. $\qquad\square$

The subspace $\mathbf{p}_A\big(\bigotimes^k \mathbb{C}^n\big)$ of $\bigotimes^k \mathbb{C}^n$ is called the *Weyl module* defined by the tableau A.

Examples

The most familiar examples of Weyl modules are the spaces $S^k(\mathbb{C}^n)$ and $\bigwedge^k \mathbb{C}^n$ corresponding to the symmetrizers and tableaux

$$\mathbf{s}(A) = \sum_{g \in \mathfrak{S}_k} g \quad \text{and} \quad \mathbf{s}(B) = \sum_{g \in \mathfrak{S}_k} \mathrm{sgn}(g)g \, ,$$

where $A = \boxed{1}\,\boxed{2}\cdots\boxed{k}$ and $B = \begin{array}{|c|}\hline 1\\\hline 2\\\hline \vdots\\\hline k\\\hline\end{array}$.

We have already seen that these spaces give irreducible representations of $\mathbf{GL}(n,\mathbb{C})$ with highest weights $k\varepsilon_1$ and $\varepsilon_1 + \cdots + \varepsilon_k$, respectively. For $k = 2$ these are the only possibilities. Now take $k = 3$ and consider the tableau $C = \begin{array}{|c|c|}\hline 1 & 3\\\hline 2\\\cline{1-1}\end{array}$, whose normalized Young symmetrizer is

$$\mathbf{p}_C = \frac{2}{3!}(1-(12))(1+(13)) = \frac{1}{3}(1-(12)+(13)-(321)) \, .$$

The corresponding Weyl module consists of all tensors $u = \sum u_{ijk}\, e_i \otimes e_j \otimes e_k$ such that $u = \mathbf{p}_C u$. Thus the components u_{ijk} of u satisfy the symmetry conditions

$$u_{ijk} = \frac{1}{2}(u_{kji} - u_{jik} - u_{kij}) \quad \text{for} \quad 1 \leq i,j,k \leq n \, . \tag{9.36}$$

By Theorem 9.3.10 the space of all 3-tensors satisfying (9.36) is an irreducible $\mathbf{GL}(n,\mathbb{C})$-module with highest weight $2\varepsilon_1 + \varepsilon_2$. The highest-weight tensor in this space is
$$(1 - \sigma_3(12))e_1 \otimes e_2 \otimes e_1 = e_1 \otimes e_2 \otimes e_1 - e_2 \otimes e_1 \otimes e_1 \, .$$

Note that this space depends on the particular choice of a tableau C of shape $[2,1]$; another choice for C gives a symmetry condition different from (9.36).

9.3.3 Standard Tableaux

Definition 9.3.11. Let $\lambda \in \mathrm{Par}(k)$. The tableau $A \in \mathrm{Tab}(\lambda)$ is *standard* if the entries in each row (resp. each column) of A are increasing from left to right (resp. from top to bottom).

We denote the set of standard tableaux of shape λ by $\mathrm{STab}(\lambda)$. For example, if $\lambda = [2,1] \in \mathrm{Par}(3)$ then $\mathrm{STab}(\lambda)$ consists of the two tableaux

$$\begin{array}{|c|c|}\hline 1 & 3\\\hline 2\\\cline{1-1}\end{array} \qquad \begin{array}{|c|c|}\hline 1 & 2\\\hline 3\\\cline{1-1}\end{array} \, .$$

We know that the corresponding module G^λ is the two-dimensional standard representation of $\mathfrak{S}_3$ in this case. This is a special case of the following general result:

Theorem 9.3.12. *The tensors $\{\mathbf{s}(A)e_A : A \in \mathrm{STab}(\lambda)\}$ give a basis for G^λ.*

The proof will require a combinatorial lemma. We give $\mathrm{STab}(\lambda)$ the lexicographic order $\overset{\text{lex}}{>}$ defined by reading the entries of $A \in \mathrm{STab}(\lambda)$ from left to right along each row from the top row to the bottom row. For example,

$$\begin{array}{|c|c|}\hline 1 & 3 \\\hline 2 \\\cline{1-1}\end{array} \quad \overset{\text{lex}}{>} \quad \begin{array}{|c|c|}\hline 1 & 2 \\\hline 3 \\\cline{1-1}\end{array} .$$

Lemma 9.3.13. *Let* $A, A' \in \mathrm{STab}(\lambda)$ *and suppose* $A \overset{\text{lex}}{>} A'$. *Then* $\mathbf{s}(A')\mathbf{s}(A) = 0$.

Proof. We may assume that

1. the first $p-1$ rows of A and A' are identical;
2. the elements in positions $1, \ldots, q-1$ of the pth rows of A; A' are identical;
3. $m = A_{pq} > m' = A'_{pq}$.

We claim that the number m' must appear in A within the first $q-1$ columns and below the pth row. Indeed, every entry in A to the right and below A_{pq} is greater than m, since A is standard, so m' cannot occur in any of these positions. Furthermore, m' cannot occur in A in a position prior to A_{pq} (in the lexicographic order) because A and A' have the same entries in all of these positions.

Now choose m'' to be the entry in the pth row of A that is in the same column as m'. Then both m' and m'' occur in the pth row of A', so the transposition $\tau = (m', m'')$ is in $\mathrm{Row}(A') \cap \mathrm{Col}(A)$. It follows that

$$\mathbf{s}(A')\mathbf{s}(A) = (\mathbf{s}(A')\tau)(\tau\mathbf{s}(A)) = -\mathbf{s}(A')\mathbf{s}(A) .$$

Hence $\mathbf{s}(A')\mathbf{s}(A) = 0$. $\qquad\square$

Proof of Theorem 9.3.12. We first prove that the set $\{\mathbf{s}(A)e_A : A \in \mathrm{STab}(\lambda)\}$ is linearly independent. Suppose we have a linear relation $\sum_A b_A \mathbf{s}(A)e_A = 0$ with $b_A \in \mathbb{C}$ and the sum over $A \in \mathrm{STab}(\lambda)$. Assume for the sake of contradiction that $b_{A'} \neq 0$ for some A'. Take the smallest such A' (in the lexicographic order). Applying $\mathbf{s}(A')$ to the relation and using Lemmas 9.3.13 and 9.3.8 we conclude that

$$b_{A'}\mathbf{s}(A')^2 e_{A'} = \xi_\lambda b_{A'}\mathbf{s}(A')e_{A'} = 0 .$$

Since $\xi_\lambda \neq 0$, it follows that $b_{A'} = 0$, a contradiction.

Let N_λ be the number of standard tableaux of shape λ and let $f_\lambda = \dim G^\lambda$. From the linear independence just established, it remains only to show that $N_\lambda = f_\lambda$. This is true when $k = 1$. Assume that it is true for all $\lambda \in \mathrm{Par}(k-1)$. Take $\lambda \in \mathrm{Par}(k)$ and denote by $B(\lambda) \subset \mathrm{Par}(k-1)$ the set of all μ that interlace λ. Then by the branching rule from $\mathfrak{S}_k$ to $\mathfrak{S}_{k-1}$ (Corollary 9.2.7) we know that f_λ satisfies the recurrence

$$f_\lambda = \sum_{\mu \in B(\lambda)} f_\mu .$$

We claim that N_λ satisfies the same recurrence. Indeed, the location of the integer k in a standard tableau of shape λ must be in a box at the end of both a row and a column. Removing this box gives a standard tableau of shape $\mu \in B(\lambda)$, and each $\mu \in B(\lambda)$ arises in this way for a unique $A \in \mathrm{STab}(\lambda)$. Conversely, given a standard tableau of shape $\mu \in B(\lambda)$, there is exactly one way to add a box containing k to obtain a standard tableau of shape λ. This proves that

$$N_\lambda = \sum_{\mu \in B(\lambda)} N_\mu \, .$$

Hence by induction, we obtain $f_\lambda = N_\lambda$.   □

9.3.4 Projections onto Isotypic Components

We complete the Schur–Weyl decomposition of tensor space by showing that the sum (over all tableaux of a fixed shape λ) of the Young symmetrizers is (up to a normalizing constant) the projection onto the $\mathbf{GL}(n, \mathbb{C}) \times \mathfrak{S}_k$-isotypic component of $\bigotimes^k \mathbb{C}^n$ labeled by λ. In Section 9.3.2 we obtained the canonical decomposition

$$\mathcal{A}[\mathfrak{S}_k] = \bigoplus_{\lambda \in \mathrm{Par}(k)} \mathcal{B}^\lambda \tag{9.37}$$

of the group algebra of $\mathfrak{S}_k$ as a direct sum of simple algebras, where $\mathcal{B}^\lambda$ is the two-sided ideal in $\mathcal{A}[\mathfrak{S}_k]$ that corresponds to $\mathrm{End}(G^\lambda)$ under the isomorphism in equation (9.33).

Let $\mathbf{P}_\lambda \in \mathcal{B}^\lambda$ be the minimal central idempotent that acts by the identity on G^λ and by zero on G^μ for $\mu \neq \lambda$. Then $\{\mathbf{P}_\lambda : \lambda \in \mathrm{Par}(k)\}$ is the canonical resolution of the identity in $\mathcal{A}[\mathfrak{S}_k]$ (see Section 4.3.4).

Proposition 9.3.14. *The minimal central idempotent in $\mathcal{A}[\mathfrak{S}_k]$ for the shape $\lambda \in$ Par(k) is given by*

$$\mathbf{P}_\lambda = \left(\frac{\dim G^\lambda}{k!} \right)^2 \sum_{A \in \mathrm{Tab}(\lambda)} \mathbf{s}(A) \, ,$$

where $\mathbf{s}(A)$ is the (unnormalized) Young symmetrizer for the tableau A.

Proof. Define

$$\mathbf{q}_\lambda = \sum_{A \in \mathrm{Tab}(\lambda)} \mathbf{s}(A) \, .$$

Since $\mathbf{s}(t \cdot A) = t \mathbf{s}(A) t^{-1}$ for $t \in \mathfrak{S}_k$ and $\mathfrak{S}_k \cdot \mathrm{Tab}(\lambda) = \mathrm{Tab}(\lambda)$, we see that $\mathbf{q}_\lambda t = t \mathbf{q}_\lambda$, so $\mathbf{q}_\lambda$ is a central element in $\mathcal{A}[\mathfrak{S}_k]$. But $\mathbf{s}(A) \in \mathcal{B}^\lambda$ for all $A \in \mathrm{Tab}(\lambda)$, whereas the center of $\mathcal{B}^\lambda$ is one-dimensional and spanned by $\mathbf{P}_\lambda$. Hence $\mathbf{q}_\lambda = \delta_\lambda \mathbf{P}_\lambda$ for some constant δ_λ . Therefore, we need to prove only that $\delta_\lambda = \xi_\lambda^2$, where $\xi_\lambda = k! / \dim G^\lambda$.

The coefficient of 1 in $\mathbf{s}(A)$ is 1, so the coefficient of 1 in $\mathbf{q}_\lambda$ is $k!$. Because $\mathbf{P}_\lambda$ is an idempotent, $\mathbf{q}_\lambda^2 = \delta_\lambda^2 \mathbf{P}_\lambda = \delta_\lambda \mathbf{q}_\lambda$. Hence

$$k! \delta_\lambda = \text{coefficient of 1 in } \mathbf{q}_\lambda^2 \, .$$

To calculate this coefficient, we first observe that for $A \in \text{Tab}(\lambda)$ there is a function $t \mapsto \eta_A(t)$ on $\mathfrak{S}_k$ such that

$$\mathbf{s}(A)\,t\,\mathbf{s}(A) = \eta_A(t)\,\mathbf{s}(A) . \tag{9.38}$$

Indeed, the element $\mathbf{s}(A)\,t\,\mathbf{s}(A)$ is in $\mathcal{B}^\lambda$, so it is determined by its action on the space G^λ. But on this space its range is in $\mathbb{C}e_A$, which implies (9.38).

To determine the function η_A, write $\mathbf{s}(A) = \sum_{r \in \mathfrak{S}_k} \varphi_A(r)r$ as an element of $\mathcal{A}[\mathfrak{S}_k]$. Since $\mathbf{s}(A)^2 = \xi_\lambda\,\mathbf{s}(A)$, the function $r \mapsto \varphi_A(r)$ on $\mathfrak{S}_k$ satisfies

$$\xi_\lambda\,\varphi_A(r) = \sum_{s \in \mathfrak{S}_k} \varphi_A(s)\varphi_A(s^{-1}r) . \tag{9.39}$$

Now for $t \in \mathfrak{S}_k$ we have

$$\mathbf{s}(A)\,t\,\mathbf{s}(A) = \sum_{s,s' \in \mathfrak{S}_k} \varphi_A(s)\varphi_A(s')sts' = \sum_{r \in \mathfrak{S}_k}\left\{ \sum_{s \in \mathfrak{S}_k} \varphi_A(s)\varphi_A(t^{-1}s^{-1}r)\right\}r .$$

From (9.38) we see that $\eta_A(t)$ is the coefficient of 1 in $\mathbf{s}(A)\,t\,\mathbf{s}(A)$. Hence

$$\eta_A(t) = \sum_{s \in \mathfrak{S}_k} \varphi_A(s)\varphi_A(t^{-1}s^{-1}) = \xi_\lambda\,\varphi_A(t^{-1}) .$$

We can now calculate $\mathbf{q}_\lambda^2$ in $\mathcal{A}[\mathfrak{S}_k]$ using (9.38) and the formula just obtained for η_A. Set $A = A(\lambda)$. Then

$$\mathbf{q}_\lambda^2 = \sum_{s,t \in \mathfrak{S}_k} s\mathbf{s}(A)s^{-1}t\mathbf{s}(A)t^{-1} = \xi_\lambda \sum_{s,t \in \mathfrak{S}_k} s\varphi_A(t^{-1}s)\mathbf{s}(A)t^{-1}$$

$$= \xi_\lambda \sum_{r,s,t \in \mathfrak{S}_k} \varphi_A(t^{-1}s)\varphi_A(r)\,srt^{-1} .$$

Calculating the coefficient of 1 in $\mathbf{q}_\lambda^2$ from this last expansion, we obtain

$$k!\delta_\lambda = \xi_\lambda \sum_{r,s \in \mathfrak{S}_k} \varphi_A(r^{-1}s^{-1}s)\varphi_A(r) = k!\xi_\lambda \sum_{r \in \mathfrak{S}_k} \varphi_A(r^{-1})\varphi_A(r)$$

$$= k!\,\xi_\lambda^2 ,$$

where we used (9.39) in the last step. $\square$

We summarize the results we have obtained on the decomposition of tensor space under $\mathbf{GL}(n,\mathbb{C})$ and $\mathfrak{S}_k$ using Young symmetrizers.

Theorem 9.3.15. *Let λ be a partition of k with at most n parts.*

1. *If $U \subset \bigotimes^k \mathbb{C}^n$ is a subspace invariant under $\rho_k(\mathbf{GL}(n,\mathbb{C}))$, then $\mathbf{P}_\lambda U$ is the iso-typic component of U of type F_n^λ for $\mathbf{GL}(n,\mathbb{C})$. In particular, if $\mathbf{s}(A)U = 0$ for all $A \in \text{STab}(\lambda)$, then U does not contain any subrepresentation isomorphic to F_n^λ.*

2. *If* $V \subset \bigotimes^k \mathbb{C}^n$ *is a subspace invariant under* $\sigma_k(\mathfrak{S}_k)$, *then* $\mathbf{P}_\lambda V$ *is the isotypic component of* V *of type* G^λ *for* $\mathfrak{S}_k$. *In particular, if* $\mathbf{s}(A)V = 0$ *for all* $A \in \mathrm{STab}(\lambda)$, *then* V *does not contain any subrepresentation isomorphic to* G^λ.

Proof. (1): Let $\mathcal{A} = \mathrm{Span}\,\rho_k(\mathbf{GL}(n,\mathbb{C}))$ and $\mathcal{B} = \mathrm{Span}\,\sigma_k(\mathfrak{S}_k)$. By Schur's commutant theorem (Theorem 4.2.10) and the double commutant theorem (Theorem 4.1.13) we know that $\mathcal{A} \cap \mathcal{B}$ is the center of $\mathcal{B}$ (and of $\mathcal{A}$). Hence $\mathbf{P}_\lambda \in \mathcal{A} \cap \mathcal{B}$ by Proposition 9.3.14. In particular, $\mathbf{P}_\lambda U \subset U$, so we have

$$\mathbf{P}_\lambda U = U \cap \mathbf{P}_\lambda\big(\textstyle\bigotimes^k \mathbb{C}^n\big) = U \cap \big(F_n^\lambda \otimes G^\lambda\big)\,,\quad,$$

which is the $\mathbf{GL}(n,\mathbb{C})$-isotypic component of type F_n^λ in U.

(2): Use the same argument as (1) with $\mathbf{GL}(n,\mathbb{C})$ replaced by $\mathfrak{S}_k$. $\square$

9.3.5 Littlewood–Richardson Rule

From the examples in Section 9.2.2 it is evident that the Littlewood–Richardson coefficients $c_{\mu\nu}^\lambda$ (first defined by (9.23)) appear in many problems of representation theory of the classical groups. Calculating them by the alternating sum multiplicity formula (7.8) is impractical in general, due to the number of terms and the cancellations in this formula. The *Littlewood–Richardson rule* is a combinatorial algorithm for calculating these integers.

To state the Littlewood–Richardson rule, we need two new notions. Let λ and μ be partitions with at most n parts. We write $\mu \subseteq \lambda$ (μ *is contained in* λ) if the Ferrers diagram of μ fits inside the diagram of λ (equivalently, $\mu_j \leq \lambda_j$ for all j). In this case the *skew Ferrers diagram* λ/μ is the configuration of boxes obtained by removing the boxes of the Ferrers diagram of μ from the boxes of the Ferrers diagram of λ. For example, let $\lambda = [4,3]$ and $\mu = [2,1]$. Mark the boxes in the diagram of λ that are also boxes of μ with $\bullet$; then λ/μ is the set of unmarked boxes:

Definition 9.3.16. A *semistandard skew tableau* of shape λ/μ and weight $\nu \in \mathbb{N}^n$ is an assignment of positive integers to the boxes of the skew Ferrers diagram λ/μ such that

1. if $\nu_j \neq 0$ then the integer j occurs in ν_j boxes for $j = 1,\dots,n$,
2. the integers in each row are nondecreasing from left to right, and
3. the integers in each column are strictly increasing from top to bottom.

By condition (1), the weight ν of a semistandard skew tableau of shape λ/μ satisfies $|\nu| + |\mu| = |\lambda|$. Let $\mathrm{SSTab}(\lambda/\mu, \nu)$ denote the set of semistandard skew tableaux of shape λ/μ and weight ν. There is the following *stability property* for the weights of semistandard skew tableaux: if $p > n$ and we set $v' = [v_1, \ldots, v_n, 0, \ldots, 0]$ (with $p - n$ trailing zeros adjoined), then $\mathrm{SSTab}(\lambda/\mu, v) = \mathrm{SSTab}(\lambda/\mu, v')$ (this is clear from condition (1) in the definition).

For example, there are two semistandard skew tableaux of shape $[3,2]/[2,1]$ and weight $[1,1]$, namely

$$
\begin{array}{|c|c|c|}
\hline
\bullet & \bullet & 1 \\
\hline
\bullet & 2 \\
\cline{1-2}
\end{array}
\quad \text{and} \quad
\begin{array}{|c|c|c|}
\hline
\bullet & \bullet & 2 \\
\hline
\bullet & 1 \\
\cline{1-2}
\end{array}. \tag{9.40}
$$

When $\mu = 0$ we set $\lambda/\mu = \lambda$. A filling of the Ferrers diagram of λ that satisfies conditions (1), (2), and (3) in Definition 9.3.16 is called a *semistandard tableau* of shape λ and weight v. For example, take $\lambda = [2,1] \in \mathrm{Par}(3)$. If $v = [2,1]$, then $\mathrm{SSTab}(\lambda, v)$ consists of the single tableau $\begin{array}{|c|c|}\hline 1 & 1 \\ \hline 2 \\ \cline{1-1}\end{array}$, while if $v = [1,1,1]$, then $\mathrm{SSTab}(\lambda, v)$ consists of the two standard tableaux

$$
\begin{array}{|c|c|}
\hline
1 & 3 \\
\hline
2 \\
\cline{1-1}
\end{array}, \quad
\begin{array}{|c|c|}
\hline
1 & 2 \\
\hline
3 \\
\cline{1-1}
\end{array}.
$$

In general, for any $\lambda \in \mathrm{Par}(k)$, if $v = \lambda$ then the set $\mathrm{SSTab}(\lambda, \lambda)$ consists of a single tableau with the number j in all the boxes of row j, for $1 \le j \le \mathrm{depth}(\lambda)$. At the other extreme, if $v = 1^k \in \mathbb{N}^k$ (the weight $\det_k$) then every $A \in \mathrm{SSTab}(\lambda, v)$ is a standard tableau, since all the entries in A are distinct. Thus in this case we have $\mathrm{SSTab}(\lambda, \det_k) = \mathrm{STab}(\lambda)$.

We already encountered semistandard tableaux in Section 8.1.2, where each n-fold branching pattern was encoded by a semistandard tableau arising from the $\mathbf{GL}_n \to \mathbf{GL}_{n-1}$ branching law. Let $\lambda \in \mathrm{Par}(k, n)$. By Corollary 8.1.7 the irreducible $\mathbf{GL}(n, \mathbb{C})$ module F_n^λ has a basis $\{u_A : v \in \mathbb{N}^n, A \in \mathrm{SSTab}(\lambda, v)\}$, and u_A has weight v for the diagonal torus of $\mathbf{GL}(n, \mathbb{C})$. Thus the Kostka coefficients (weight multiplicities) are $K_{\lambda v} = |\mathrm{SSTab}(\lambda, v)|$.

We now introduce the second new concept needed for the Littlewood–Richardson rule. Call an ordered string $w = x_1 x_2 \cdots x_r$ of positive integers x_j a *word*, and the integers x_j the *letters* in w. If T is a semistandard skew tableau with n rows, then the *row word* of T is the juxtaposition $w_{\mathrm{row}}(T) = R_n \cdots R_1$, where R_j is the word formed by the entries in the jth row of T.

Definition 9.3.17. A word $w = x_1 x_2 \cdots x_r$ is a *reverse lattice word* if when w is read from right to left from the end x_r to any letter x_s, the sequence $x_r, x_{r-1}, \ldots, x_s$ contains at least as many 1's as it does 2's, at least as many 2's as 3's, and so on for all positive integers. A semistandard skew tableau T is an *L–R skew tableau* if $w_{\mathrm{row}}(T)$ is a reverse lattice word.

For example, the two skew tableaux (9.40) have row words 21 and 12, respectively. The first is a reverse lattice word, but the second is not.

Littlewood–Richardson Rule: The L–R coefficient $c_{\mu\nu}^{\lambda}$ is the number of L–R skew tableaux of shape λ/μ and weight ν.

See Macdonald [106], Sagan [128], or Fulton [51] for a proof of the correctness of the L–R rule. The representation-theoretic portion of the proof (based on the branching law) is outlined in the exercises. Note that from their representation-theoretic definition the L–R coefficients have the symmetry $c_{\mu\nu}^{\lambda} = c_{\nu\mu}^{\lambda}$; however, this symmetry is not obvious from the L–R rule. In applying the rule, it is natural to take $|\mu| \geq |\nu|$.

Examples

1. Pieri's rule (Corollary 9.2.4) is a direct consequence of the L–R rule. To see this, take a Ferrers diagram μ of depth at most $n - 1$ and a diagram ν of depth one. Let λ be a diagram that contains μ and has $|\mu| + |\nu|$ boxes. If $T \in \mathrm{SSTab}(\lambda/\mu, \nu)$, then 1 has to occur each of the boxes of T, since $\nu_j = 0$ for $j > 1$. In this case $w_{\mathrm{row}}(T) = 1\cdots 1$ ($|\nu|$ occurrences of 1) is a reverse lattice word, and so T is an L–R skew tableau. Since the entries in the columns of T are strictly increasing, each box of T must be in a different column. In particular, λ has depth at most n. Thus to each skew Ferrers diagram λ/μ with at most one box in each column there is a unique L–R skew tableau T of shape λ/μ and weight ν, and hence $c_{\mu,\nu}^{\lambda} = 1$. If λ/μ has a column with more than one box, then $c_{\mu,\nu}^{\lambda} = 0$. This is Pieri's rule.

2. Let $\mu = [2,1]$ and $\nu = [1,1]$. Consider the decomposition of the tensor product representation $F_n^{\mu} \otimes F_n^{\nu}$ of **GL**$(n,\mathbb{C})$ for $n \geq 4$. The irreducible representations F_n^{λ} that occur have $|\lambda| = |\mu| + |\nu| = 5$ and $\lambda \supset \mu$. The highest (Cartan) component has $\lambda = \mu + \nu = [3,2]$ and occurs with multiplicity one by Proposition 5.5.19. This can also be seen from the L–R rule: The semistandard skew tableaux of shape $[3,2]/[2,1]$ and weight ν are shown in (9.40) and only one of them is an L–R tableau. The other possible highest weights λ are $[3,1,1]$, $[2,2,1]$, and $[2,1,1,1]$ (since we are assuming $n \geq 4$). The corresponding semistandard skew tableaux of shape λ/μ and weight ν are as follows:

In each case exactly one of the skew tableaux satisfies the L–R condition. Hence the L–R rule implies that

$$F_n^{[2,1]} \otimes F_n^{[1,1]} = F_n^{[3,2]} \oplus F_n^{[3,1,1]} \oplus F_n^{[2,2,1]} \oplus F_n^{[2,1,1,1]} . \tag{9.41}$$

In particular, the multiplicities are all one. As a check, we can calculate the dimension of each of these representations by the Weyl dimension formula (7.18). This yields

$$\dim F_n^{[2,1]} \cdot \dim F_n^{[1,1]} = \big((n+1)n(n-1)/3\big) \cdot \big(n(n-1)/2\big) \,,$$

$$\dim F_n^{[3,2]} = (n+2)(n+1)n^2(n-1)/24 \,,$$

$$\dim F_n^{[3,1,1]} = (n+2)(n+1)n(n-1)(n-2)/20 \,,$$

$$\dim F_n^{[2,2,1]} = (n+1)n^2(n-1)(n-2)/24 \,,$$

$$\dim F_n^{[2,1,1,1]} = (n+1)n(n-1)(n-2)(n-3)/30 \,.$$

These formulas imply that both sides of (9.41) have the same dimension.

9.3.6 Exercises

1. Let $\lambda \in \mathrm{Par}(k)$ and let λ^t be the transposed partition. Prove that $\sigma^{\lambda^t} \cong \mathrm{sgn} \otimes \sigma^\lambda$ (HINT: Observe that $\mathrm{Row}(A) = \mathrm{Col}(A^t)$ for $A \in \mathrm{Tab}(\lambda)$; now apply Lemma 9.3.6 and Proposition 9.3.7.)

2. Verify the result of the previous exercise when $k = 3, 4$, or 5 from the character tables of $\mathfrak{S}_3$, $\mathfrak{S}_4$, and $\mathfrak{S}_5$.

3. Let $\lambda \in \mathrm{Par}(k)$ and suppose $\lambda = \lambda^t$. Prove that $\chi^\lambda(s) = 0$ for all odd permutations $s \in \mathfrak{S}_k$, where χ^λ is the character of σ^λ.

4. Verify the result in the previous exercise for the cases $\lambda = [2,1]$ and $\lambda = [2,2]$ directly from the character tables of $\mathfrak{S}_3$ and $\mathfrak{S}_4$.

5. Let $A = \begin{array}{|c|c|}\hline 1 & 2 \\\hline 3 \\\cline{1-1}\end{array}$. Find the symmetry conditions satisfied by the components of the 3-tensors in the range of the normalized Young symmetrizer $\mathbf{p}_A$. What is the dimension of this space?

6. Let $A = \begin{array}{|c|c|}\hline 1 & 3 \\\hline 2 & 4 \\\hline\end{array}$. Find the highest-weight tensor and the symmetry conditions satisfied by the components of the 4-tensors in the range of the normalized Young symmetrizer $\mathbf{p}_A$. What is the dimension of this space?

7. Let $\lambda = [2,2]$. Show that there are two standard tableaux $A_1 \overset{\text{lex}}{<} A_2$ of shape λ, and calculate the corresponding normalized Young symmetrizers.

8. Let $\lambda = [3,1]$. Show that there are three standard tableaux $A_1 \overset{\text{lex}}{<} A_2 \overset{\text{lex}}{<} A_3$ of shape λ, and calculate the corresponding normalized Young symmetrizers.

9. Let $\lambda \in \mathrm{Par}(k)$ and let $A_1 \overset{\text{lex}}{<} \cdots \overset{\text{lex}}{<} A_d$ be an enumeration of the standard tableaux of shape λ (where $d = \dim G^\lambda$). Assume that the following condition is satisfied:

   ($\star$)   For every $1 \le i, j \le d$ with $i \ne j$ there is a pair of numbers that occurs in the same row of A_i and the same column of A_j.

(Note that by Lemma 9.3.13, $(\star)$ is always true for $i < j$.)

(a) Prove that $\mathbf{c}(A_i)\mathbf{r}(A_j) = 0$ for all $1 \le i, j \le d$ with $i \ne j$.

(b) Let $\sigma_{ij} \in \mathfrak{S}_k$ be the element such that $\sigma_{ij}A_i = A_j$. Show that $\mathbf{c}(A_i)\sigma_{ij} = \sigma_{ij}\mathbf{c}(A_j)$, $\mathbf{r}(A_i)\sigma_{ij} = \sigma_{ij}\mathbf{r}(A_j)$, and $\sigma_{ij}\sigma_{jk} = \sigma_{ik}$.

(c) Define $e_{ij} = (d/k!)\mathbf{c}(A_i)\sigma_{ij}\mathbf{r}(A_j)$ for $1 \le i, j \le d$. Show that $e_{ij}e_{pq} = \delta_{jp}e_{iq}$, so that $\{e_{ij}\}$ is a standard elementary matrix basis for the ideal $\mathcal{B}^\lambda$. Conclude that $\mathbf{P}_\lambda = e_{11} + \cdots + e_{dd}$ when $(\star)$ holds.

10. (a) Let $\lambda = [2,1]$. Enumerate $\mathrm{STab}(\lambda)$ and show that condition $(\star)$ of Exercise 9 holds. Conclude that $\mathbf{P}_{[2,1]} = \frac{1}{3}[2 - (123) - (132)]$.

    (b) Obtain the formula for $\mathbf{P}_{[2,1]}$ using the character table of $\mathfrak{S}_3$ and Theorem 4.3.9.

    (c) Repeat (a) for $\lambda = [2,2]$ and conclude that $\mathbf{P}_{[2,2]} = \frac{1}{12}[2 + 2x - y]$, where $x = (13)(24) + \cdots$ is the sum of the four pairs of commuting transpositions in $\mathfrak{S}_4$ and $y = (123) + \cdots$ is the sum of the eight 3-cycles in $\mathfrak{S}_4$.

    (d) Obtain the formula for $\mathbf{P}_{[2,2]}$ using the character table of $\mathfrak{S}_4$ and Theorem 4.3.9.

11. Take $\lambda = [3,2]$ and the tableaux $A_1 = \begin{array}{|c|c|c|}\hline 1 & 2 & 3 \\\hline 4 & 5 \\\cline{1-2}\end{array}$ and $A_5 = \begin{array}{|c|c|c|}\hline 1 & 3 & 4 \\\hline 2 & 5 \\\cline{1-2}\end{array}$ of shape λ. Show that condition $(\star)$ of Exercise #9 is not satisfied and that $\mathbf{r}(A_1)\mathbf{c}(A_5) \ne 0$. In this case the normalized Young symmetrizers $\mathbf{p}_{A_i}$ do not give a system of elementary matrices for the ideal $\mathcal{B}^\lambda$. (See Littlewood [104, §5.4] for more details.)

12. Use the Littlewood–Richardson rule to show that $c_{\mu\nu}^\lambda = 3$ when $\lambda = [4,2,1]$, $\mu = [3,1]$, and $\nu = [2,1]$.

13. Use the Littlewood–Richardson rule to establish the decomposition

$$F_n^{[1,1,1]} \otimes F_n^{[1,1]} = F_n^{[2,2,1]} \oplus F_n^{[2,1,1,1]} \oplus F_n^{[1,1,1,1,1]}$$

of **GL**$(n, \mathbb{C})$ modules when $n \ge 5$. Check the result by calculating the dimensions of each representation from the Weyl dimension formula (7.18). Note that the diagram $[3,1,1]$ does not occur as a highest weight, even though it contains $[1,1,1]$ and is a partition of 5.

The following exercises reduce the proof of the L–R rule to a combinatorial problem.

14. Fix positive integers k, m and embed **GL**$_k \times$ **GL**$_m$ into **GL**$_{k+m}$ as in Section 9.2.2. Let λ be a Ferrers diagram with at most $k + m$ rows and let $\mu \subset \lambda$ be a Ferrers diagram with at most k rows. Define $E_{k,m}^{\lambda/\mu} = \mathrm{Hom}_{\mathbf{GL}_k}(F_k^\mu, F_{k+m}^\lambda)$. Then $E_{k,m}^{\lambda/\mu}$ is a **GL**$_m$ module, since the actions of **GL**$_n$ and **GL**$_k$ on F_{k+m}^λ mutually commute.

    (a) Show that $F_{k+m}^\lambda \cong \bigoplus_{\mu \subset \lambda} F_k^\mu \otimes E_{k,m}^{\lambda/\mu}$ as a **GL**$_k \times$ **GL**$_m$ module.

    (b) Let $n = |\lambda| - |\mu|$. Show that $E_{k,m}^{\lambda/\mu} \cong \bigoplus_{\nu \in \mathrm{Par}(n,m)} c_{\mu\nu}^\lambda F_m^\nu$ as a **GL**$_m$ module.

    (c) Use Proposition 8.1.6 to show that $E_{k,m}^{\lambda/\mu}$ has a **GL**$_m$ weight basis labeled by the set of semistandard skew tableaux of shape λ/μ with entries from the set of integers $\{1, \ldots, m\}$. Hence the character of $E_{k,m}^{\lambda/\mu}$ evaluated on $\mathrm{diag}[x_1, \ldots, x_m]$ is

$$\sum_{\gamma \in \mathbb{N}^m} |\text{SSTab}(\lambda/\mu, \gamma)| x^\gamma .$$

This polynomial is called a *skew Schur function*.

(d) Use part (b) and Corollary 8.1.7 to show that the character of $E_{k,m}^{\lambda/\mu}$ evaluated on $\text{diag}[x_1, \ldots, x_m]$ is also given by

$$\sum_{\nu \in \text{Par}(n,m)} c_{\mu\nu}^\lambda \left\{ \sum_{\gamma \in \mathbb{N}^m} |\text{SSTab}(\nu, \gamma)| x^\gamma \right\}, \quad \text{where } n = |\lambda| - |\mu| .$$

15. *(Continuation of previous exercise)* Let $\nu \in \text{Par}(n, m)$. Define $d_{\mu\nu}^\lambda$ to be the number of L–R skew tableaux of shape λ/μ and weight ν.

(a) Let $\gamma \in \mathbb{N}^m$ with $|\gamma| = n$. The *jeu de taquin* of Schützenberger furnishes a map

$$j : \text{SSTab}(\lambda/\mu, \gamma) \longrightarrow \bigcup_{\nu \in \text{Par}(n,m)} \text{SSTab}(\nu, \gamma)$$

that satisfies $|j^{-1}(A)| = d_{\mu\nu}^\lambda$ for $A \in \text{SSTab}(\nu, \gamma)$ (see Sagan [128, §4.9]). Use this to prove that

$$|\text{SSTab}(\lambda/\mu, \gamma)| = \sum_{\nu \in \text{Par}(n,m)} d_{\mu\nu}^\lambda |\text{SSTab}(\nu, \gamma)| .$$

(b) Use the result of (a) and the previous exercise to show that

$$\sum_{\nu \in \text{Par}(n,m)} d_{\mu\nu}^\lambda \, \text{ch}(F_m^\nu) = \sum_{\nu \in \text{Par}(n,m)} c_{\mu\nu}^\lambda \, \text{ch}(F_m^\nu) .$$

Conclude that $d_{\mu\nu}^\lambda = c_{\mu\nu}^\lambda$ by linear independence of characters.

9.4 Notes

Section 9.1.1. The duality between $\mathbf{GL}(n, \mathbb{C})$ and $\mathfrak{S}_k$ was first presented in Schur's thesis [129] and developed further by Weyl [164]. For a detailed historical account, see Hawkins [63, Chapter 10]. When $n \geq k$, the induced module I^μ is independent of n, by (9.2). Hence the multiplicity of μ as a weight in F_n^λ is also independent of n for $n \geq k$. This is the *stability property* of weight multiplicities for $\mathbf{GL}(n, \mathbb{C})$; see Benkart, Britten, and Lemire [5]. The partial ordering on partitions defined by the positive roots of $\mathbf{GL}(n, \mathbb{C})$ is called the *dominance ordering* in the combinatorial literature (Sagan [128, Definition 2.2.2]).

Section 9.1.2. The characters of $\mathfrak{S}_k$ were first found by Frobenius [47]. The formula for $\dim G^\lambda$ is due to Schur [129, §23]. For the history of the hook-length formula, see Sagan [128, §3.1].

Section 9.1.3. The Frobenius character formula first appeared in Frobenius [49]. For more character calculations and tables see Littlewood [104] and Murnaghan [116].

Sections 9.2.1 and 9.2.2. In these sections we have followed Howe [72]. For the use of dual reductive pairs in representation and invariant theory, see also Howe [70]. The term *seesaw pair* is from Kudla [94].

Section 9.2.3. One can prove directly (without using Schur–Weyl duality) that the representation of $\mathfrak{S}_k$ on $V_k^\lambda(\det)$ is irreducible and then go on to establish the properties of Schur–Weyl duality from **GL**(k)–**GL**(n) duality. For this approach, see Howe [72].

Section 9.3.1. The tableaux and associated symmetrizer operators are due to Young [169]; see also Frobenius [50]. The tensors e_A, for A a tableau, are called *tabloids* in the combinatorial literature, and the extreme tensors $\mathbf{s}(A)e_A$ are called *polytabloids* (see Sagan [128]).

Section 9.3.2. More examples of Weyl modules are in Boerner [9, Chapter V, §5].

Section 9.3.3. The proof of Theorem 9.3.12, which relies on the branching law to give the recursive formula for $\dim G^\lambda$, is standard; see Specht [135]. There are combinatorial proofs based on straightening rules that express the polytabloids $\mathbf{s}(B)e_B$, for $B \in \mathrm{Tab}(\lambda)$, in terms of so-called *standard polytabloids* (Peel [119], Sagan [128]).

Section 9.3.5. The L–R rule was first stated in Littlewood–Richardson [105] with examples but no general proof. The first complete proofs were given in the 1970s. See Sagan [128], Macdonald [106], and Fulton [51] for further history and citations of recent work.

Chapter 10
Tensor Representations of O(V) and Sp(V)

Abstract In this chapter we analyze the action of the orthogonal and symplectic groups on the tensor powers of their defining representations. We show (following ideas of Weyl [164]) that the subspaces of *harmonic tensors* can be decomposed using the theory of Young symmetrizers from Chapter 9. This furnishes models (*Weyl modules*) for all the irreducible representations of the orthogonal and symplectic groups as spaces of harmonic tensors in the image of Young symmetrizers. Our approach involves the interplay of the commuting algebra (a quotient of the *Brauer algebra*) with the representation theory of the orthogonal and symplectic groups. The key observation is that the action of the Brauer algebra on the space of harmonic tensors factors through the action of the symmetric group on tensors.

The Riemannian curvature tensor of a pseudo-Riemannian manifold plays a central role in differential geometry, Lie groups, and physics (through Einstein's theory of general relativity). We use the results of Chapters 9 and the present chapter to analyze the symmetry properties of curvature tensors. We show that the space of all curvature tensors at a fixed point of a manifold is irreducible under the action of the general linear group. Under the orthogonal group, this space decomposes into irreducible subspaces corresponding to *scalar curvature*, *traceless Ricci curvature*, and *Weyl conformal curvature* parts. We determine these subspaces using earlier results in this chapter together with the theorem of the highest weight and the Weyl dimension formula. In the last section of the chapter we apply representation theory to knot theory. We use the invariant theory of the orthogonal group to prove the existence of the *Jones polynomial* (an invariant of oriented links under ambient oriented diffeomorphism).

10.1 Commuting Algebras on Tensor Spaces

Let $G \subset \mathbf{GL}(V)$ be the isometry group of a nondegenerate symmetric or skew-symmetric bilinear form. We determine generators and relations for the algebra of linear transformations on $\bigotimes^k V$ that commute with the action of G.

R. Goodman, N.R. Wallach, *Symmetry, Representations, and Invariants,*
Graduate Texts in Mathematics 255, DOI 10.1007/978-0-387-79852-3_10,

10.1.1 Centralizer Algebra

Let V be a finite-dimensional complex vector space and let G be a reductive algebraic subgroup of $\mathbf{GL}(V)$. Denote by ρ_k the natural representation of G on $\bigotimes^k V$:

$$\rho_k(g)(v_1 \otimes \cdots \otimes v_k) = g v_1 \otimes \cdots \otimes g v_k \qquad \text{for } v_i \in V_. \, .$$

To determine the decomposition of $\bigotimes^k V$ into G-isotypic components, we can use the approach of the previous chapter for $\mathbf{GL}(V)$. Namely, we form the *centralizer algebra*

$$\mathcal{B} = \{ B \in \mathrm{End}(\textstyle\bigotimes^k V) : B\rho_k(g) = \rho_k(g)B \quad \text{for all } g \in G \} \, .$$

Since G is reductive, Theorem 4.2.1 gives a decomposition

$$\textstyle\bigotimes^k V \cong \bigoplus_i E_i \otimes F_i$$

as a module for $\mathcal{A}[G] \otimes \mathcal{B}$, which pairs an irreducible representation E_i of G with an irreducible representation F_i of $\mathcal{B}$ in a unique way. So the problem of decomposing $\bigotimes^k V$ under the action of G is equivalent to the problem of decomposing $\bigotimes^k V$ into isotypic components relative to $\mathcal{B}$.

The next step in this duality program is to determine the structure of the algebra $\mathcal{B}$. Recall the isomorphisms

$$\mathrm{End}\left(\textstyle\bigotimes^k V\right) \cong \left(\textstyle\bigotimes^k V\right) \otimes \left(\textstyle\bigotimes^k V\right)^* \cong \left(\textstyle\bigotimes^k V\right) \otimes \left(\textstyle\bigotimes^k V^*\right)$$

as modules for $\mathbf{GL}(V)$ (see Section B.2.2). Hence

$$\mathcal{B} = \mathrm{End}_G\left(\textstyle\bigotimes^k V\right) \cong \left[\left(\textstyle\bigotimes^k V\right) \otimes \left(\textstyle\bigotimes^k V^*\right)\right]^G \tag{10.1}$$

(a *vector space* isomorphism). Thus if we have an explicit description of a spanning set for the G-invariant mixed tensors of type (k,k), we can use (10.1) to obtain a spanning set for $\mathcal{B}$. Since $\mathcal{B}$ contains the commutant of the $\mathbf{GL}(V)$ action on $\bigotimes^k V$, we have $\sigma_k(\mathfrak{S}_k) \subset \mathcal{B}$, where σ_k is the representation of $\mathfrak{S}_k$ studied in Chapter 9. The complete contractions of k vectors with k covectors correspond to the elements $\sigma_k(s)$ for $s \in \mathfrak{S}_k$, as we saw in Section 5.3.1.

Suppose G leaves invariant a nondegenerate bilinear form ω on V. Then ω defines a G-module isomorphism $V \cong V^*$ and hence an isomorphism

$$\left(\textstyle\bigotimes^k V\right) \otimes \left(\textstyle\bigotimes^k V^*\right) \cong \textstyle\bigotimes^{2k} V$$

of G-modules. Combining this with the $\mathbf{GL}(V)$-module isomorphism above, we obtain a G-module isomorphism $T : \bigotimes^{2k} V \longrightarrow \mathrm{End}(\bigotimes^k V)$. We take T as in Section 5.3.2:

$$T(v_1 \otimes \cdots \otimes v_{2k})u = \omega(u, v_2 \otimes v_4 \otimes \cdots \otimes v_{2k})v_1 \otimes v_3 \otimes \cdots \otimes v_{2k-1} \tag{10.2}$$

for $v_i \in V$ and $u \in \bigotimes^k V$. Here we have extended ω to a bilinear form on $\bigotimes^k V$ by

$$\omega(u_1 \otimes \cdots \otimes u_k, v_1 \otimes \cdots \otimes v_k) = \prod_{i=1}^{k} \omega(u_i, v_i)$$

for $u_i, v_i \in V$. Thus we have a vector-space isomorphism

$$T : \left(\bigotimes^{2k} V\right)^G \xrightarrow{\cong} \mathrm{End}_G\left(\bigotimes^k V\right).$$

Let $\theta_k = T^{-1}(I_{\otimes^k V})$ as in Section 5.3.2. We define an injective homomorphism $\tau : \mathfrak{S}_k \longrightarrow \mathfrak{S}_{2k}$ by

$$\tau(s)(2i-1) = 2s(i)-1, \qquad \tau(s)(2i) = 2i \quad \text{for } i = 1, \ldots, k \quad \text{and } s \in \mathfrak{S}_k$$

($\tau(s)$ permutes $\{1, 3, \ldots, 2k-1\}$ and fixes $\{2, 4, \ldots, 2k\}$ pointwise). Recall the subgroup $\mathfrak{B}_k = \widetilde{\mathfrak{S}}_k \cdot \mathfrak{N}_k$ of $\mathfrak{S}_{2k}$ from Section 5.3.2. Here $\widetilde{\mathfrak{S}}_k$ is the group of permutations of the ordered pairs $(1,2)$, $(3,4)$, $\ldots$, $(2k-1, 2k)$ and $\mathfrak{N}_k \cong (\mathbb{Z}_2)^k$ is the subgroup of $\mathfrak{S}_{2k}$ generated by the transpositions $2j-1 \leftrightarrow 2j$ for $j = 1, \ldots, k$.

Proposition 10.1.1. *Let $G \subset \mathbf{GL}(V)$ be the full group of isometries for a nondegenerate symmetric or skew-symmetric bilinear form ω on V. Let $\Gamma \subset \mathfrak{S}_{2k}$ be a set of representatives for the double cosets $\tau(\mathfrak{S}_k) \backslash \mathfrak{S}_{2k} / \mathfrak{B}_k$. Then*

$$\mathrm{End}_G\left(\bigotimes^k V\right) = \mathrm{Span}\{\sigma_k(s)T(\sigma_{2k}(\gamma)\theta_k) : s \in \mathfrak{S}_k \text{ and } \gamma \in \Gamma\}.$$

Proof. Recall from Theorem 5.3.4 that

$$\mathrm{End}_G\left(\bigotimes^k V\right) = \mathrm{Span}\{T(\sigma_{2k}(s)\theta_k) : s \in \Xi_k\}, \qquad (10.3)$$

where Ξ_k is any set of representatives for the cosets $\mathfrak{S}_{2k} / \mathfrak{B}_k$. From formula (10.2) for T it is clear that

$$\sigma_k(s)T(u) = T(\sigma_{2k}(\tau(s))u) \quad \text{for } s \in \mathfrak{S}_k, \quad u \in \bigotimes^{2k} V. \qquad (10.4)$$

The proposition now follows by (10.3) and (10.4). $\qquad\qquad\square$

Let G and ω be as in Proposition 10.1.1. Set $n = \dim V$ and write

$$\mathcal{B}_k(\varepsilon n) = \mathrm{End}_G\left(\bigotimes^k V\right),$$

where we set

$$\varepsilon = \begin{cases} 1 & \text{if } \omega \text{ is symmetric}, \\ -1 & \text{if } \omega \text{ is skew-symmetric}. \end{cases}$$

The notation is justified, because the group G is determined (up to conjugation in $\mathbf{GL}(V)$) by ε, and hence the algebra $\mathcal{B}_k(\varepsilon n)$ is determined (up to isomorphism as an associative algebra) by $k, \varepsilon n$. We have a homomorphism $\mathbb{C}[\mathfrak{S}_k] \longrightarrow \mathcal{B}_k(\varepsilon n)$ via the representation $s \mapsto \sigma_k(s)$ for $s \in \mathfrak{S}_k$. There is an embedding $\mathcal{B}_k(\varepsilon n) \subset \mathcal{B}_{k+1}(\varepsilon n)$ with $b \in \mathcal{B}_k(\varepsilon n)$ acting on $\bigotimes^{k+1} V$ by $b(u \otimes v) = bu \otimes v$ for $u \in \bigotimes^k V$ and $v \in V$.

Proposition 10.1.1 gives only a spanning set for $\mathcal{B}_k(\varepsilon n)$ as a vector space. To describe the multiplicative structure of $\mathcal{B}_k(\varepsilon n)$ we will choose a specific set of double coset representatives. For this it is convenient to introduce a graphic presentation of the coset space $\mathfrak{S}_{2k}/\mathfrak{B}_k$. We display the set $\{1, 2, \ldots, 2k\}$ as an array of two rows of k labeled dots, with the dots in the top row labeled $1, 3, \ldots, 2k-1$ from left to right, and the dots in the bottom row labeled $2, 4, \ldots, 2k$ from left to right, as shown in Figure 10.1. The group $\mathfrak{S}_{2k}$ acts by permuting the dots according to their labels.

$$
\begin{array}{cccc}
1 & 3 & 5 & 2k-1 \\
\bullet & \bullet & \bullet & \bullet \\
& & & \cdots \\
\bullet & \bullet & \bullet & \bullet \\
2 & 4 & 6 & 2k
\end{array}
$$

Fig. 10.1 A two-row array.

The subgroup $\widetilde{\mathfrak{S}}_k$ of $\mathfrak{S}_{2k}$ permutes the columns of the array, and the subgroup $\mathfrak{N}_k$ interchanges the upper and lower dots in a column. The subgroup $\tau(\mathfrak{S}_k)$ permutes the top row of dots and fixes each dot in the bottom row.

Consider the set X_k of all graphs obtained from the two rows of dots by connecting each dot with exactly one other dot. (A dot in the top row can be connected either with another dot in the top row or with a dot in the bottom row.) An example with $k = 5$ is shown in Figure 10.2.

$$x_1 =$$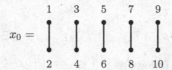

Fig. 10.2 A Brauer diagram.

We call an element of X_k a *Brauer diagram*. There is a natural action of $\mathfrak{S}_{2k}$ on X_k obtained by permuting the labels of the dots. If $x \in X_k$ then $s \cdot x$ is the graph such that dot $s(i)$ is connected to dot $s(j)$ if and only if dot i is connected to dot j in x.

$$
x_0 = \quad
\begin{array}{ccccc}
1 & 3 & 5 & 7 & 9 \\
\bullet & \bullet & \bullet & \bullet & \bullet \\
\mid & \mid & \mid & \mid & \mid \\
\bullet & \bullet & \bullet & \bullet & \bullet \\
2 & 4 & 6 & 8 & 10
\end{array}
$$

Fig. 10.3 Diagram for x_0.

Let x_0 be the graph with each dot in the top row connected with the dot below it; Figure 10.3 shows the case $k = 5$. Then $X_k = \mathfrak{S}_{2k} \cdot x_0$ and $\mathfrak{B}_k$ is the stability subgroup of x_0. Thus we may identify the coset space $\mathfrak{S}_{2k}/\mathfrak{B}_k$ with X_k. For example, the Brauer diagram x_1 in Figure 10.2 is $s \cdot x_0$, where $s \in \mathfrak{S}_{10}$ is the cyclic permutation (2594).

The double coset space $\tau(\mathfrak{S}_k) \backslash \mathfrak{S}_{2k}/\mathfrak{B}_k$ is the set of $\tau(\mathfrak{S}_k)$ orbits on X_k. Let $x \in X_k$ and let r be the number of edges in the diagram of x that connect a dot in the top row with another dot in the top row (call such an edge a *top bar*). The bottom row of x also has r analogous such edges (call them *bottom bars*), and we call x an *r-bar diagram*. All diagrams in the $\tau(\mathfrak{S}_k)$-orbit of x also have r top bars, and there is a unique z in the $\tau(\mathfrak{S}_k)$-orbit of x such that all the edges of z connecting the top and bottom rows are vertical (that is, if z is considered as a 2-partition of $2k$, then every pair $\{2i-1,2j\}$ that occurs in z has $i = j$). We will call such a Brauer diagram (or 2-partition) *normalized*. The normalized diagrams give a set of representatives for the $\tau(\mathfrak{S}_k)$ orbits on X_k. For example, when $k = 3$ and $r = 1$ then there are three orbits of 1-bar diagrams, with normalized representatives indicated in Figure 10.4. These orbits correspond to the 2-partitions $z_1 = \{\{1,2\},\{3,5\},\{4,6\}\}$, $z_2 = \{\{1,5\},\{2,6\},\{3,4\}\}$, and $z_3 = \{\{1,3\},\{2,4\},\{5,6\}\}$.

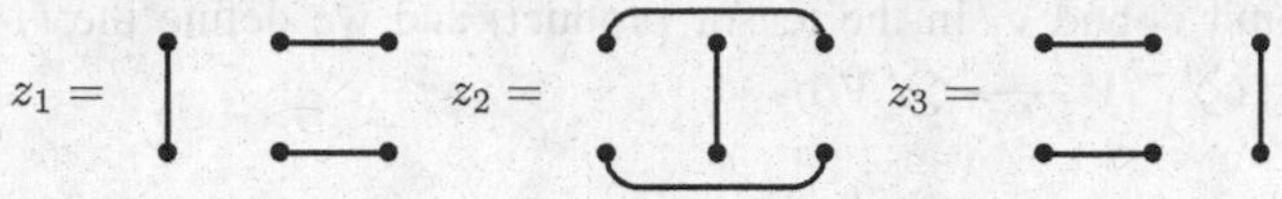

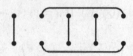

Fig. 10.4 Normalized 1-bar Brauer diagrams.

If z is a normalized Brauer diagram, then for every top bar in z joining the dots numbered $2i-1$ and $2j-1$ there is a corresponding bottom bar joining the dots numbered $2i$ and $2j$. We will say that z contains an (i,j)-*bar* in this case (we take $i < j$). For example, the normalized diagram in the orbit $\tau(\mathfrak{S}_k)x_1$ (where x_1 is as above) is shown in Figure 10.5; it contains a $(2,5)$ bar.

Fig. 10.5 Normalized Brauer diagram with $(2,5)$-bar.

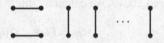

We now determine the operator on tensor space corresponding to a normalized Brauer diagram. For example, the diagram shown in Figure 10.6 contains a single

Fig. 10.6 Diagram for the operator $\tau_{12} = D_{12}C_{12}$.

$(1,2)$-bar corresponding to the tensor $\sigma_{2k}(23)\theta_k$, where (23) is the transposition $2 \leftrightarrow 3$. Since $\sigma_{2k}(23)\theta_k = (\sigma_4(23)\theta_2) \otimes \theta_{2k-2}$, we have

$$T(\sigma_{2k}(23)\theta_k)v_1 \otimes v_2 \otimes u = \left\{ \sum_{p_2} \omega(v_1, f_{p_2})\omega(v_2, f^{p_2}) \right\} \sum_{p_1} f_{p_1} \otimes f^{p_1} \otimes u$$

$$= \omega(v_1, v_2)\theta \otimes u$$

for $v_1, v_2 \in V$ and $u \in V^{\otimes(k-2)}$. Here $\{f_p\}$ and $\{f^p\}$ are bases for V with $\omega(f_p, f^q) = \delta_{pq}$. Thus this 1-bar diagram gives an operator $\tau_{12} = T(\sigma_{2k}(23)\theta_k)$ that is the composition

$$\otimes^k V \xrightarrow{\ C_{12}\ } V^{\otimes(k-2)} \xrightarrow{\ D_{12}\ } \otimes^k V \,,$$

where C_{12} is a *contraction operator*: $C_{12}(v_1 \otimes v_2 \otimes u) = \omega(v_1, v_2)u$ and D_{12} is an *expansion operator*:

$$D_{12}(u) = \sum_p f_p \otimes f^p \otimes u = \theta \otimes u$$

(which is multiplication by the G-invariant tensor θ). These operators obviously intertwine the actions of G on $\bigotimes^k V$ and $\bigotimes^{k-2} V$, showing again that τ_{12} commutes with the action of G on $\bigotimes^k V$.

In general, for any pair $1 \le i < j \le k$ we define the ij-*contraction operator* $C_{ij} : \bigotimes^k V \longrightarrow \bigotimes^{k-2} V$ by

$$C_{ij}(v_1 \otimes \cdots \otimes v_k) = \omega(v_i, v_j)\, v_1 \otimes \cdots \otimes \widehat{v_i} \otimes \cdots \otimes \widehat{v_j} \otimes \cdots \otimes v_k$$

(where we omit v_i and v_j in the tensor product) and we define the ij-*expansion operator* $D_{ij} : \bigotimes^{k-2} V \longrightarrow \bigotimes^k V$ by

$$D_{ij}(v_1 \otimes \cdots \otimes v_{k-2}) = \sum_{p=1}^{n} v_1 \otimes \cdots \otimes \underbrace{f_p}_{i\text{th}} \otimes \cdots \otimes \underbrace{f^p}_{j\text{th}} \otimes \cdots \otimes v_{k-2}.$$

These operators intertwine the action of G and are mutually adjoint, relative to the extension of the invariant form ω to $\bigotimes^k V$:

$$\omega(C_{ij}u, w) = \omega(u, D_{ij}w) \qquad \text{for } u \in \bigotimes^k V, \quad w \in \bigotimes^{k-2} V. \tag{10.5}$$

Set $\tau_{ij} = D_{ij}C_{ij} \in \operatorname{End}_G(\bigotimes^k V)$. If $u = v_1 \otimes \cdots \otimes v_k$ with $v_i \in V$, then

$$\tau_{ij}(u) = \omega(v_i, v_j) \sum_{p=1}^{n} v_1 \otimes \cdots \otimes \underbrace{f_p}_{i\text{th}} \otimes \cdots \otimes \underbrace{f^p}_{j\text{th}} \otimes \cdots \otimes v_k. \tag{10.6}$$

Let $s \in \mathfrak{S}_{2k}$ be the transposition $2i \leftrightarrow 2j - 1$. Then, just as in the example with $i = 1$ and $j = 2$ considered above, we calculate that

$$\tau_{ij} = T(\sigma_{2k}(s)\theta_k). \tag{10.7}$$

The contraction and expansion operators satisfy the symmetry properties

$$C_{ij} = \varepsilon C_{ji} \quad \text{and} \quad D_{ij} = \varepsilon D_{ji}, \tag{10.8}$$

since $\sum_p f_p \otimes f^p = \varepsilon \sum_p f^p \otimes f_p$. Hence $\tau_{ij} = \tau_{ji}$, so the operator τ_{ij} depends only on the set $\{i, j\}$.

Lemma 10.1.2. *Suppose that* $z = \{i_1, j_1\}, \ldots, \{i_r, j_r\} \in X_k$ *is a normalized r-bar Brauer diagram. Then*

$$\tau_{i_p j_p} \tau_{i_q j_q} = \tau_{i_q j_q} \tau_{i_p j_p} \quad \text{for } p \ne q. \tag{10.9}$$

Thus the operator $\tau_z = \prod_{p=1}^{r} \tau_{i_p j_p} \in \mathcal{B}_k(\varepsilon n)$ *is defined independently of the order of the product. There exists* $\gamma \in \mathfrak{S}_{2k}$ *such that* $z = \gamma \cdot x_0$ *and*

$$T(\sigma_{2k}(\gamma)\theta_k) = \tau_z. \tag{10.10}$$

Proof. The commutativity relation (10.9) is clear, since τ_{ij} operates on only the ith and jth tensor positions. We take $\gamma \in \mathfrak{S}_{2k}$ as the product of the transpositions $2i_p \leftrightarrow 2j_p - 1$ for $p = 1, \ldots, r$. Then (10.10) follows directly from (10.6) and the same calculation that gives (10.7). $\qquad\square$

Let $Z_{k,r} \subset X_k$ be the set of normalized r-bar Brauer diagrams, and set

$$Z_k = \bigcup_{r=0}^{[k/2]} Z_{k,r}.$$

Proposition 10.1.3. *Let $n = \dim V$. The algebra $\mathcal{B}_k(\varepsilon n)$ is spanned by the set of operators $\sigma_k(s)\tau_z$ with $s \in \mathfrak{S}_k$ and $z \in Z_k$.*

Proof. Given $z \in Z_k$ take $\gamma \in \mathfrak{S}_{2k}$ as in Lemma 10.1.2. The set Γ of all such γ is then a set of representatives for the double cosets $\tau(\mathfrak{S}_k)\backslash\mathfrak{S}_{2k}/\mathcal{B}_k$. Now apply Proposition 10.1.1. $\qquad\square$

Corollary 10.1.4. *Suppose $n \geq 2k$. Then the set $\{\sigma_k(s)\tau_z : s \in \mathfrak{S}_k, z \in Z_k\}$ is a basis for $\mathcal{B}_k(\varepsilon n)$.*

Proof. As a vector space, $\mathcal{B}_k(\varepsilon n)$ is isomorphic to $\left[\bigotimes^{2k} V^*\right]^G$, with the operator $\sigma_k(s)\tau_z$ corresponding to the complete contraction λ_x for $x = s \cdot z$ (see Theorem 5.3.5). Thus it suffices to show that the set $\{\lambda_x : x \in X_k\}$ is linearly independent. Recall that

$$\lambda_x(v_1 \otimes \cdots \otimes v_{2k}) = \prod \omega(v_i, v_j)$$

(product over all pairs $\{i, j\} \in x$ with $i < j$). Thus for $s \in \mathfrak{S}_{2k}$ we have

$$\lambda_{s \cdot x}(v_1 \otimes \cdots \otimes v_{2k}) = \pm\lambda_x(\sigma_{2k}(s)(v_1 \otimes \cdots \otimes v_{2k})). \tag{10.11}$$

Since $\dim V \geq 2k$, there exist $f_{\pm 1}, \ldots, f_{\pm k} \in V$ such that $\omega(f_i, f_j) = \delta_{i+j}$. Let $u = f_1 \otimes f_{-1} \otimes \cdots \otimes f_k \otimes f_{-k}$. We claim that

$$\lambda_x(u) = \begin{cases} 1 & \text{if } x = x_0, \\ 0 & \text{otherwise}, \end{cases} \tag{10.12}$$

where x_0 is the Brauer diagram with all vertical lines. Indeed, if x contains any bars, then $\lambda_x(u)$ contains a factor $\omega(f_i, f_j)$ with i, j both positive, which is zero. Likewise, if x contains any pairs $\{2i-1, 2j\}$ with $i \neq j$, then $\lambda_x(u)$ contains a factor $\omega(f_i, f_{-j})$ that is also zero.

Suppose now that there is a linear relation

$$\sum_{x \in X_k} c_x \lambda_x = 0.$$

Applying $s \in \mathfrak{S}_{2k}$ and using (10.11) we obtain a relation of the form

$$\sum_{x \in X_k} \pm c_{s \cdot x} \lambda_x = 0.$$

Evaluating this relation on u and using (10.12), we find that $c_{s \cdot x_0} = 0$ for all $s \in \mathfrak{S}_{2k}$. Hence $c_x = 0$ for all $x \in X_k$. $\square$

10.1.2 Generators and Relations

We next study the relations in the algebra $\mathcal{B}_k(\varepsilon n)$.

Lemma 10.1.5. *The operators τ_{ij} satisfy the following relations, where $n = \dim V$ and (il) denotes the transposition of i and l:*

1. *$\tau_{ij}^2 = n\tau_{ij}$.*
2. *$\tau_{ij}\tau_{jl} = \sigma_k(il)\tau_{jl}$ for distinct i, j, l.*
3. *$\sigma_k(s)\tau_{ij}\sigma_k(s)^{-1} = \tau_{s(i),s(j)}$ for all $s \in \mathfrak{S}_k$.*
4. *$\sigma_k(ij)\tau_{ij} = \varepsilon\tau_{ij}$.*

Proof. The contraction and expansion operators satisfy

$$C_{ij}D_{ij} = nI, \qquad\qquad (10.13)$$

which follows from $\sum_{p=1}^n \omega(f_p, f^p) = n$. This implies property (1). To verify (2), note that

$$\tau_{ij}\tau_{jl}(v_1 \otimes \cdots \otimes v_k) = \omega(v_j, v_l)\sum_{p,q} \omega(v_i, f_p)u_{pq},$$

where $u_{pq} = v_1 \otimes \cdots \otimes \underbrace{f_q}_{i\text{th}} \otimes \cdots \otimes \underbrace{f^q}_{j\text{th}} \otimes \cdots \otimes \underbrace{f^p}_{l\text{th}} \otimes \cdots \otimes v_k$. But

$$\sum_p \omega(v_i, f_p)u_{pq} = \varepsilon v_1 \otimes \cdots \otimes \underbrace{f_q}_{i\text{th}} \otimes \cdots \otimes \underbrace{f^q}_{j\text{th}} \otimes \cdots \otimes \underbrace{v_i}_{l\text{th}} \otimes \cdots \otimes v_k,$$

which gives (2). Relations (3) and (4) are simple calculations from the definition of the operators τ_{ij}. $\square$

Fig. 10.7 Diagram for transposition s_r.

We can describe the multiplication in $\mathcal{B}_k(\varepsilon n)$ and the relations in Lemma 10.1.5 in terms of concatenation of Brauer diagrams. Let $s_r \in \mathfrak{S}_k$ be the transposition $r \leftrightarrow r + 1$. It corresponds to the Brauer diagram shown in Figure 10.7. Let z_r be the normalized Brauer diagram with a single $(r, r+1)$ bar corresponding to the operator $\tau_{r,r+1}$, as in Figure 10.8. Since $\mathfrak{S}_k$ is generated by $s_1, \dots, s_{k-1}$, we see from Proposition 10.1.3 and property (3) in Lemma 10.1.5 that the algebra $\mathcal{B}_k(\varepsilon n)$ is generated by the operators $s_1, \dots, s_{k-1}$ and $z_1, \dots, z_{k-1}$.

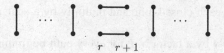

Fig. 10.8 Diagram for operator z_r.

To describe the product xy we place the diagram for x above the diagram for y and join the lower row of dots in x to the upper row of dots in y. When x, y are in $\mathfrak{S}_k$ (so their diagrams have no bars) this procedure obviously gives the multiplication in $\mathfrak{S}_k$. When x or y has bars, we remove the closed loops from the concatenated graph using relation (1) in Lemma 10.1.5.

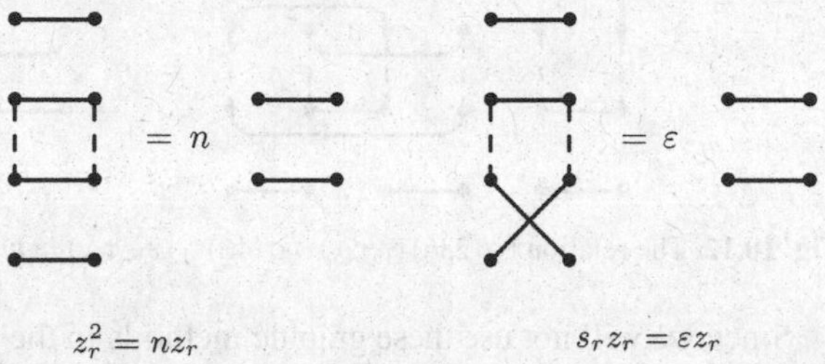

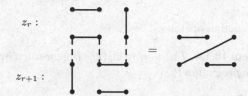

Fig. 10.9 Relations in Brauer algebra.

$$z_r^2 = n z_r \qquad\qquad s_r z_r = \varepsilon z_r$$

We illustrate this procedure with the following examples: Relations (1) and (4) in Lemma 10.1.5 are shown in Figure 10.9. From relation (2) in Lemma 10.1.5 we

z_r :

z_{r+1} :

Fig. 10.10 The relation $z_r z_{r+1} = \sigma_k(r, r+2) z_{r+1}$.

have the relation shown in Figure 10.10. Using this relation and relation (2) from Lemma 10.1.5, we finally get the result shown in Figure 10.11.

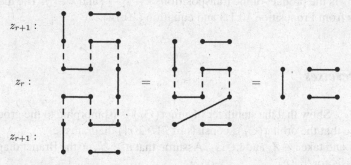

z_{r+1} :

z_r :

z_{r+1} :

Fig. 10.11 The relation $z_{r+1} z_r z_{r+1} = z_{r+1}$.

The general recipe for transforming the concatenated Brauer diagrams of x and y into a scalar multiple of the Brauer diagram for xy is as follows:

1. Delete each closed loop and multiply by a scalar factor of n^r if there are r such loops.
2. Multiply by a factor of ε for every path beginning and ending on the top row of x (or on the bottom row of y).

For example, if $x = \sigma(236)\tau_{35}\tau_{46}$ and $y = \sigma(46)\tau_{12}\tau_{34}\tau_{56}$, then xy is obtained as shown in Figure 10.12.

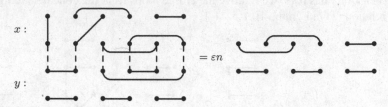

Fig. 10.12 The relation $(\sigma(236)\tau_{35}\tau_{46}) \cdot (\sigma(46)\tau_{12}\tau_{34}\tau_{56}) = \varepsilon n \, \sigma(23)\,\tau_{12}\tau_{34}\tau_{56}$.

Since we will not use these graphic methods in the proofs, we leave the verification of their validity as an exercise.

Define $P_{k-1} = n^{-1}\tau_{k-1,k}$. Then $P_{k-1}^2 = P_{k-1}$ by (1) in Lemma 10.1.5, so

$$P_{k-1} : V^{\otimes k} \longrightarrow V^{\otimes(k-2)} \otimes \mathbb{C}\theta$$

is a projection operator.

Theorem 10.1.6. *The algebra* $\mathcal{B}_k(\varepsilon n)$ *is generated by the operators* $\sigma_k(s)$ *for* $s \in \mathfrak{S}_k$ *and the projection* P_{k-1}.

Proof. From (3) in Lemma 10.1.5 we have

$$\tau_{ij} = n\sigma_k(s)P_{k-1}\sigma_k(s)^{-1} , \tag{10.14}$$

where $s \in \mathfrak{S}_k$ is the product of the transpositions $k-1 \leftrightarrow i$ and $k \leftrightarrow j$. The theorem now follows from Proposition 10.1.3 and equation (10.14). $\qquad\square$

10.1.3 Exercises

1. Let $z \in Z_{k,r}$. Show that the stabilizer of z in $\tau(\mathfrak{S}_k)$ is isomorphic to the group $\mathfrak{B}_r$ and hence that the orbit $\tau(\mathfrak{S}_k)z$ consists of $k!/(2^r r!)$ diagrams.
2. Let $\varepsilon = 1$ and take $z \in Z_k$ and $s \in \mathfrak{S}_k$. Assume that $n \geq 2k$, so the Brauer diagrams label a basis for $\mathcal{B}_k(n)$.
   (a) Show that the Brauer diagram for $\sigma_k(s)\tau_z$ is obtained by applying the permutation s to the top row of z (enumerated as $1, 2, \ldots, k$).
   (b) Show that the Brauer diagram for $\tau_z\sigma_k(s)$ is obtained by applying the permutation s to the bottom row of z (enumerated as $1, 2, \ldots, k$).

3. Verify (10.5).
4. Use the relations from Lemma 10.1.5 to verify the example multiplication via Brauer diagrams given in Figure 10.12.
5. Let $x \in \mathcal{B}_k(\varepsilon n)$ correspond to a Brauer diagram, and suppose that $x \notin \mathcal{B}_{k-1}(\varepsilon n)$. Show that there are elements $a, b \in \mathcal{B}_{k-1}(\varepsilon n)$ and y in the algebra generated by s_{k-1}, z_{k-1} such that $x = ayb$. (HINT: Consider three cases, depending on whether the last two vertices in the diagram for x belong to two, one, or no bars.)
6. Verify the validity of the rules (1) and (2) for the multiplication in $\mathcal{B}_k(\varepsilon n)$ via concatenation of Brauer diagrams. (HINT: Use induction on k and the previous exercise.)

10.2 Decomposition of Harmonic Tensors

Let G be the isometry group of the bilinear form ω. The intersection of the kernels of all the ω-contraction operators on $\bigotimes^k V$ is called the space of *harmonic* (or *completely traceless*) k-tensors. We decompose this space as a module for $G \times \mathfrak{S}_k$.

10.2.1 Harmonic Tensors

We continue the notation of Section 10.1.1. A tensor $u \in \bigotimes^k V$ is called ω-*harmonic* if it is annihilated by all the contraction operators C_{ij} defined in Section 10.1.2. Denote by

$$\mathcal{H}(\textstyle\bigotimes^k V, \omega) = \bigcap_{1 \leq i < j \leq k} \mathrm{Ker}(C_{ij})$$

the space of all ω-harmonic k-tensors. We will simply call these tensors *harmonic* and write $\mathcal{H}(\bigotimes^k V, \omega) = \mathcal{H}(\bigotimes^k V)$ when ω is clear from the context.

Theorem 10.2.1. *The space $\mathcal{H}(\bigotimes^k V)$ is invariant under $\rho_k(G)$ and $\sigma_k(\mathfrak{S}_k)$. Furthermore, the commutant of G on $\mathcal{H}(\bigotimes^k V)$ is $\mathbb{C}[\sigma_k(\mathfrak{S}_k)]$. Hence there is a decomposition*

$$\mathcal{H}(\textstyle\bigotimes^k V) \cong \bigoplus_{\lambda \in \Lambda} E^\lambda \otimes G^\lambda \tag{10.15}$$

as a $(G \times \mathfrak{S}_k)$-module. Here Λ is a subset of $\mathrm{Par}(k)$, G^λ is the irreducible $\mathfrak{S}_k$-module corresponding to the partition λ by Schur–Weyl duality, and E^λ is an irreducible G-module. Furthermore, the modules E^λ that occur are mutually inequivalent.

Proof. Since $C_{ij}\tau_{ij} = C_{ij}D_{ij}C_{ij} = nC_{ij}$, we have

$$\mathrm{Ker}(C_{ij}) = \mathrm{Ker}(\tau_{ij}). \tag{10.16}$$

Hence u is harmonic if and only if $\tau_{ij}u = 0$ for $1 \leq i < j \leq k$. Since τ_{ij} commutes with $\rho_k(G)$, we see that $\mathcal{H}(\bigotimes^k V)$ is invariant under G. Theorem 10.1.6 implies that $\mathcal{H}(\bigotimes^k V)$ is invariant under $\mathcal{B}_k(\varepsilon n)$ and the action of $\mathcal{B}_k(\varepsilon n)$ on $\mathcal{H}(\bigotimes^k V)$ reduces to the action of the group algebra of $\mathfrak{S}_k$. Hence the theorem follows by Theorem 4.2.1. $\qquad\qquad\qquad\qquad\qquad\qquad\qquad\qquad\qquad\qquad\qquad\qquad\qquad\qquad$ $\square$

In the following sections we will determine the set Λ of partitions of k occurring in Theorem 10.2.1 and the corresponding irreducible representations E^λ when G is the symplectic or orthogonal group. Just as in the case of Schur–Weyl duality for $\mathbf{GL}(V)$, the key tool will be the theorem of the highest weight.

Examples

1. When $k = 0$ then $\mathcal{H}(\bigotimes^0 V) = \mathbb{C}$, the trivial G-module.

2. When $k = 1$ there are no contraction operators, so $\mathcal{H}(V) = V$ is an irreducible G-module.

3. Let $k = 2$. By (10.16) we have $\mathcal{H}(\bigotimes^2 V) = \mathrm{Ker}(C_{12}) = \mathrm{Ker}(P)$, where $P = (\dim V)^{-1}\tau_{12}$. Since P is a projection with range $\mathbb{C}\theta$, we have a decomposition

$$\bigotimes^2 V = \mathbb{C}\theta \oplus \mathcal{H}(\bigotimes^2 V).$$

10.2.2 Harmonic Extreme Tensors

We now determine the correspondence between the partitions of k and the representations of G that occur in Theorem 10.2.1. We take $V = \mathbb{C}^n$ and the bilinear form ω on $\mathbb{C}^n$ as in Section 2.1.2 ($\omega = B$ in the symmetric case, and $\omega = \Omega$ in the skew-symmetric case, as given by (2.6) and (2.9)). Let $\widetilde{G} = \mathbf{GL}(n,\mathbb{C})$ and let G be the subgroup of $\widetilde{G}$ preserving ω. Set $\widetilde{\mathfrak{g}} = \mathfrak{gl}(n,\mathbb{C})$ and $\mathfrak{g} = \mathrm{Lie}(G)$. This choice of bilinear form gives the following compatible diagonal and upper-triangular subalgebras of $\mathfrak{g}$ and $\widetilde{\mathfrak{g}}$:

Denote by $\widetilde{\mathfrak{h}}$ the diagonal $n \times n$ matrices and by $\mathfrak{h} = \mathfrak{g} \cap \widetilde{\mathfrak{h}}$ the diagonal matrices in $\mathfrak{g}$. Let $\widetilde{\mathfrak{n}}^+$ be the strictly upper-triangular $n \times n$ matrices and

$$\mathfrak{n}^+ = \mathfrak{g} \cap \widetilde{\mathfrak{n}}^+ \tag{10.17}$$

the strictly upper-triangular matrices in $\mathfrak{g}$. Then by the choice of the bilinear form ω we have

$$\mathfrak{n}^+ = \bigoplus_{\alpha \in \Phi^+} \mathfrak{g}\alpha,$$

where Φ^+ is the system of positive roots for $(\mathfrak{g}, \mathfrak{h})$ from Section 2.4.3. We set $\widetilde{\mathfrak{b}} = \widetilde{\mathfrak{h}} + \widetilde{\mathfrak{n}}^+$ and $\mathfrak{b} = \mathfrak{h} + \mathfrak{n}^+$ (see Theorem 2.4.11). For $\mu \in \widetilde{\mathfrak{h}}^*$ we write $\overline{\mu} = \mu|_{\mathfrak{h}}$.

From Schur–Weyl duality (Theorem 9.1.2) we know that the space of tensors that are $\widetilde{\mathfrak{b}}$-extreme of a fixed dominant integral weight furnishes an irreducible representation of $\mathfrak{S}_k$. Since $\mathfrak{b} \subset \widetilde{\mathfrak{b}}$, a tensor that is $\widetilde{\mathfrak{b}}$-extreme is also $\mathfrak{b}$-extreme. We will show that the converse holds on the space of harmonic tensors. This property, together with the theorem of the highest weight, will be our method for obtaining the explicit form of the G-isotypic decomposition (10.15) of the harmonic tensors.

For a diagonal matrix x with diagonal entries $x_1, \ldots, x_n$, let $\varepsilon_i(x) = x_i$. The weights of $\mathfrak{h}$ are of the form

$$\lambda = \lambda_1 \varepsilon_1 + \cdots + \lambda_l \varepsilon_l \tag{10.18}$$

with λ_i integers and $l = [n/2]$. We shall write $\widetilde{\lambda}$ for the weight of $\widetilde{\mathfrak{h}}$ given by this same formula.

To determine all the weights λ of the $\mathfrak{b}$-extreme tensors in $\bigotimes^k \mathbb{C}^n$, it suffices to consider the case

$$\lambda_1 \geq \lambda_2 \geq \cdots \geq \lambda_p > 0, \tag{10.19}$$

where $p \leq l$ and $\lambda_j = 0$ for $j > p$. Indeed, this is just the condition for dominance when G is $\mathbf{O}(\mathbb{C}^{2l+1}, B)$ or $\mathbf{Sp}(\mathbb{C}^{2l}, \Omega)$ (see Section 3.1.4). For the case $G = \mathbf{O}(\mathbb{C}^{2l}, B)$, let $g_0 \in G$ act by

$$g_0 e_l = e_{l+1}, \quad g_0 e_{l+1} = e_l, \quad \text{and} \quad g_0 e_i = e_i \text{ for } i \neq l, l+1.$$

Then $\mathrm{Ad}(g_0) : \mathfrak{h} \longrightarrow \mathfrak{h}$ induces the transformation $\varepsilon_l \leftrightarrow -\varepsilon_l$, $\varepsilon_i \mapsto \varepsilon_i$ for $i \neq l$ on $\mathfrak{h}^*$. We will denote this transformation by $\lambda \mapsto g_0 \cdot \lambda$ (it gives the outer automorphism of the Dynkin diagram of $\mathfrak{g}$). Note that $g_0 \cdot \Phi^+ = \Phi^+$, so $\mathrm{Ad}(g_0) : \mathfrak{n}^+ \longrightarrow \mathfrak{n}^+$. If $u \in \bigotimes^k \mathbb{C}^n$ is $\mathfrak{b}$-extreme of weight λ, then $\rho_k(g_0)u$ is $\mathfrak{b}$-extreme of weight $g_0 \cdot \lambda$. We can replace λ by $g_0 \cdot \lambda$ if necessary to achieve $\lambda_l \geq 0$. This assumption on λ makes $\widetilde{\lambda}$ a $\widetilde{\mathfrak{b}}$-dominant integral weight.

Let $\mu = \mu_1 \varepsilon_1 + \cdots + \mu_n \varepsilon_n$ with $\mu_1 \geq \cdots \geq \mu_n \geq 0$ be any nonnegative $\widetilde{\mathfrak{b}}$-dominant weight. Write

$$W^k(\lambda) = \left(\bigotimes^k \mathbb{C}^n \right)^{\mathfrak{n}^+}(\lambda), \qquad \widetilde{W}^k(\mu) = \left(\bigotimes^k \mathbb{C}^n \right)^{\widetilde{\mathfrak{n}}^+}(\mu)$$

for the spaces of extreme k-tensors of weights λ and μ relative to $\mathfrak{b}$ and $\widetilde{\mathfrak{b}}$ respectively. The following result will be a basic tool for decomposing the harmonic tensors into $(G \times \mathfrak{S}_k)$-irreducible subspaces.

Proposition 10.2.2. *There are the following dichotomies:*

1. *Assume that λ satisfies (10.19). Then either $W^k(\lambda) \cap \mathcal{H}(\bigotimes^k \mathbb{C}^n) = 0$, or else $W^k(\lambda) \subset \mathcal{H}(\bigotimes^k \mathbb{C}^n)$.*

2. *If μ is a $\widetilde{\mathfrak{b}}$-dominant weight, then either $\widetilde{W}^k(\mu) \cap \mathcal{H}(\bigotimes^k \mathbb{C}^n) = 0$, or else $\widetilde{W}^k(\mu) = W^k(\overline{\mu}) \subset \mathcal{H}(\bigotimes^k \mathbb{C}^n)$.*

Proof. (1): For each G-isotypic subspace E in $\bigotimes^k \mathbb{C}^n$, there is a unique λ satisfying (10.19) that is the weight of a $\mathfrak{b}$-extreme tensor in E. By Theorems 4.2.1 and 4.2.12 (and using the results in Section 5.5.5 relating representations of $\mathbf{SO}(n, \mathbb{C})$ and $\mathbf{O}(n, \mathbb{C})$), we conclude that the algebra $\mathcal{B}_k(\varepsilon n)$ acts irreducibly on $W^k(\lambda)$. Since $W^k(\lambda) \cap \mathcal{H}(\bigotimes^k \mathbb{C}^n)$ is a $\mathcal{B}_k(\varepsilon n)$-invariant subspace of $W^k(\lambda)$, it must be 0 or $W^k(\lambda)$.

(2): Assume $\widetilde{W}^k(\mu) \cap \mathcal{H}(\bigotimes^k \mathbb{C}^n) \neq 0$. Since $\widetilde{W}^k(\mu) \subset W^k(\overline{\mu})$, it follows by (1) that $W^k(\overline{\mu}) \subset \mathcal{H}(\bigotimes^k \mathbb{C}^n)$. Furthermore, $W^k(\overline{\mu})$ is irreducible under $\mathfrak{S}_k$. Indeed, it is irreducible under $\mathcal{B}_k(\varepsilon n)$ by (1), and on the harmonic tensors the action of $\mathcal{B}_k(\varepsilon n)$ is the same as the action of $\mathfrak{S}_k$. By Schur–Weyl duality $\mathfrak{S}_k$ also acts irreducibly on $\widetilde{W}^k(\mu)$. Hence $W^k(\overline{\mu}) = \widetilde{W}^k(\mu)$. $\square$

Corollary 10.2.3. *Let μ be a $\widetilde{\mathfrak{b}}$-dominant weight. Assume that $|\mu| = k - 2r$ for some integer $r \geq 0$, and that $0 \neq W^k(\overline{\mu}) \subset \mathcal{H}(\bigotimes^k \mathbb{C}^n)$. Then $r = 0$ and $W^k(\overline{\mu}) = \widetilde{W}^k(\mu)$.*

Proof. Since μ is $\widetilde{\mathfrak{b}}$-dominant and $|\mu| = k - 2r$, we have $\widetilde{W}^{k-2r}(\mu) \neq 0$. Thus

$$0 \neq (D_{12})^r \widetilde{W}^{k-2r}(\mu) \subset W^k(\overline{\mu}) \,,$$

since the expansion operator D_{12} is injective and commutes with the action of $\mathfrak{g}$. Suppose $r > 0$. Then

$$C_{12}(D_{12})^r \widetilde{W}^{k-2r}(\mu) = (D_{12})^{r-1} \widetilde{W}^{k-2r}(\mu) \neq 0 \,.$$

Therefore $D_{12}^r \widetilde{W}^{k-2r}(\mu)$ contains nonharmonic tensors. This contradicts the assumption $W^k(\overline{\mu}) \subset \mathcal{H}(\bigotimes^k \mathbb{C}^n)$. Thus $r = 0$. It follows by Proposition 10.2.2, (2) that $\widetilde{W}^k(\mu) = W^k(\overline{\mu})$. $\square$

Corollary 10.2.4. *If $p \neq q$ then $\mathrm{Hom}_G\big(\mathcal{H}(\bigotimes^p \mathbb{C}^n), \mathcal{H}(\bigotimes^q \mathbb{C}^n)\big) = 0$. In particular, $[\mathcal{H}(\bigotimes^p \mathbb{C}^n)]^G = 0$ for all $p > 0$.*

Proof. We may assume $p \leq q$. Let $0 \neq T \in \mathrm{Hom}_G(\mathcal{H}(\bigotimes^p \mathbb{C}^n), \mathcal{H}(\bigotimes^q \mathbb{C}^n))$. Since $-I \in G$ acts by $(-1)^p$ on $\bigotimes^p \mathbb{C}^n$, we have $T(-1)^p = (-1)^q T$. Hence $q = p + 2r$ for some integer $r \geq 0$. There exists $\lambda \in \mathfrak{h}^*$ such that $0 \neq W^p(\lambda) \subset \mathcal{H}(\bigotimes^p \mathbb{C}^n)$ and $TW^p(\lambda) \neq 0$. But $TW^p(\lambda) \subset W^q(\lambda)$, so $W^q(\lambda)$ contains nonzero harmonic tensors. Hence by Proposition 10.2.2 we have $W^q(\lambda) \subset \mathcal{H}(\bigotimes^q \mathbb{C}^n)$. This is impossible if $r > 0$, since

$$0 \neq (D_{12})^r W^p(\lambda) \subset W^q(\lambda)$$

and $C_{12}(D_{12})^r = n(D_{12})^{r-1}$ is injective. Hence $p = q$.

For the last statement, take $q = 0$ in the argument above. If $u \in [\mathcal{H}(\bigotimes^p \mathbb{C}^n)]^G$, then $z \mapsto zu$ is a G-intertwining map from $\mathbb{C}$ (as a trivial G-module) to $\mathcal{H}(\bigotimes^p \mathbb{C}^n)$. Hence this map is zero if $p > 0$, so $u = 0$. $\square$

We now establish a simple weight criterion to determine when $\widetilde{\mathfrak{b}}$-extreme tensors are harmonic. If $\mu = \sum_{i=1}^n \mu_i \varepsilon_i$, with

$$\mu_1 \geq \mu_2 \geq \cdots \geq \mu_n \geq 0 \quad \text{and} \quad \sum_{i=1}^n \mu_i = k \,, \tag{10.20}$$

then we identify μ with the corresponding partition of k with at most n parts. Call μ *G-admissible* if it satisfies the following condition:

Type BD $[G = \mathbf{O}(\mathbb{C}^n)]$:   The sum of the lengths of the first two columns of the Ferrers diagram of μ is at most n.

Type C $[G = \mathbf{Sp}(\mathbb{C}^{2l})]$: The first column of the Ferrers diagram of μ has length at most l.

In the symplectic case $(n = 2l)$, the condition for admissibility is $\mu_j = 0$ for $j > l$. In the orthogonal case $(n = 2l$ or $2l + 1)$ there are two types of admissible weights. Either $\mu_j = 0$ when $j > l$, or else

$$\mu = \sum_{i=1}^{p} \mu_i \varepsilon_i + \sum_{i=p+1}^{q} \varepsilon_i \,,$$

where $\mu_1 \geq \cdots \geq \mu_p \geq 2$, $p < q$, and $p + q \leq n$. (When $\mu_i = 1$ for all i, we take $p = 0$ and omit the first summand.)

In terms of the fundamental b-dominant weights $\varpi_i = \varepsilon_1 + \cdots + \varepsilon_i$, the weights

$$\mu = \sum_{i=1}^{l} m_i \varpi_i \,, \quad \text{with } m_i \geq 0 \,, \tag{10.21}$$

are G-admissible. In addition, when G is an orthogonal group, the weights $\mu = \varpi_q$ for $q \leq n$ and

$$\mu = m_1 \varpi_1 + \cdots + m_p \varpi_p + \varpi_q \quad \text{for } m_i \geq 0 \text{ and } m_p \geq 1 \tag{10.22}$$

are also admissible provided $p \leq l$ and $p + q \leq n$.

Theorem 10.2.5. *Let $\mu \in \mathrm{Par}(k,n)$. Then $\widetilde{W}^k(\mu) \subset \mathcal{H}(\bigotimes^k \mathbb{C}^n)$ if and only if μ is G-admissible. In this case $\widetilde{W}^k(\mu) = W^k(\overline{\mu})$.*

Proof. Write $\mu = \sum_i m_i \varpi_i$ and set

$$u = (w_1)^{\otimes m_1} \otimes \cdots \otimes (w_n)^{\otimes m_n} \,, \tag{10.23}$$

where $w_p = e_1 \wedge e_2 \wedge \cdots \wedge e_p$ for $1 \leq p \leq n$. Then u is b̃-extreme of weight μ. We shall show that u is harmonic if and only if μ is G-admissible. This will imply the theorem by Proposition 10.2.2 (2).

Suppose G is orthogonal. In this case w_p is harmonic for all $1 \leq p \leq n$. Indeed, take $1 \leq i < j \leq p$ and let s be the transposition $i \leftrightarrow j$. Then $\sigma_k(s) w_p = -w_p$, whereas from (10.8) we have $C_{ij}\sigma_k(s) = C_{ij}$. Thus

$$C_{ij} w_p = -C_{ij}\sigma_k(s) w_p = -C_{ij} w_p \,,$$

and so $C_{ij} w_p = 0$.

Now suppose G is symplectic. We claim that w_p is harmonic if and only if $p \leq l$. Let $1 \leq i < j \leq p$. If $p \leq l$, then $C_{ij} w_p = 0$, since $\omega(e_i, e_j) = 0$. Conversely, if $p > l$ then

$$C_{l,l+1} w_p = e_1 \wedge \cdots \wedge \widehat{e_l} \wedge \widehat{e_{l+1}} \wedge \cdots \wedge e_p \neq 0 \,,$$

since $\omega(e_l, e_{l+1}) = 1$. Therefore w_p is not harmonic in this case.

It is clear from (10.21), (10.22), and (10.23) that to determine when u is harmonic, it remains only to consider the contractions of $w_p \otimes w_q$. We have thus reduced the proof of the theorem to the following lemma:

Lemma 10.2.6. *Assume* $1 \le p \le q \le n$. *Then* $C_{ij}(w_p \otimes w_q) = 0$ *for all* $1 \le i \le p < j \le p+q$ *if and only if* $p+q \le n$.

Proof. For ease of notation we consider only the case $i = 1$ and $j = p+1$; the argument applies in general. Set $v = w_p \otimes w_q$. In terms of the basis $\{e_I\}$, v is obtained by a double alternation:

$$
v = \frac{1}{p!q!} \sum_{s \in \mathfrak{S}_p} \sum_{t \in \mathfrak{S}_q} \mathrm{sgn}(s)\, \mathrm{sgn}(t)\, e_{s(1)} \otimes \cdots \otimes e_{s(p)} \otimes e_{t(1)} \otimes \cdots \otimes e_{t(q)} .
$$

The contraction operator $C_{1,p+1}$ then replaces $e_{s(1)}$ and $e_{t(1)}$ in each term of the sum by $\omega(e_{s(1)}, e_{t(1)})$, which is zero except when $s(1)+t(1) = n+1$. Since $s(1)+t(1) \le p+q$, it follows that $C_{1,p+1}(v) = 0$ when $p+q \le n$.

Suppose now that $p+q \ge n+1$. Then $C_{1,p+1}(v)$ is given by

$$
\frac{1}{p!q!} \sum_{i=1}^{p} \left\{ \sum_{s} \sum_{t} \mathrm{sgn}(s)\, \mathrm{sgn}(t)\, e_{s(2)} \otimes \cdots \otimes e_{s(p)} \otimes e_{t(2)} \otimes \cdots \otimes e_{t(q)} \right\},
$$

where the inner summation is over $s \in \mathfrak{S}_p$ and $t \in \mathfrak{S}_q$ such that $s(1) = i$ and $t(1) = n+1-i$. For $1 \le i \le p$ we embed $\mathfrak{S}_{p-1}$ in $\mathfrak{S}_p$ as the subgroup fixing i, and we embed $\mathfrak{S}_{q-1}$ in $\mathfrak{S}_q$ as the subgroup fixing $n+1-i$. If $s(1) = i$, then $s = s'\tau_i$, where s' fixes i and τ_i is the transposition $1 \leftrightarrow i$. Making a similar factorization for t, we see that $(-1)^{n-1} pq\, C_{1,p+1}(v)$ can be written as

$$
\sum_{i=n+1-q}^{p} (e_1 \wedge \cdots \wedge \widehat{e_i} \wedge \cdots \wedge e_p) \otimes (e_1 \wedge \cdots \wedge \widehat{e}_{n+1-i} \wedge \cdots \wedge e_q)
$$

(where we omit e_i from the first tensor factor, omit e_{n+1-i} from the second, and take empty wedge products equal to 1). The range of summation is nonempty, since $n+1-q \le p$, and the terms in this sum form a linearly independent set. Hence $C_{1,p+1}(v) \ne 0$, completing the proof of the lemma and Theorem 10.2.5. $\square$

10.2.3 Decomposition of Harmonics for **Sp(V)**

We now apply the results on harmonic extreme tensors to the case of a skew-symmetric form Ω. This yields the decomposition of the space $\mathcal{H}(\bigotimes^k \mathbb{C}^{2l}, \Omega)$ under the joint action of $G = \mathbf{Sp}(\mathbb{C}^{2l}, \Omega)$ and $\mathfrak{S}_k$. Our tools will be Schur–Weyl duality and the characterization of G-admissible **GL**$(2l, \mathbb{C})$-highest weights in Theorem 10.2.5.

Let $\mu = \sum_{i=1}^{l} \mu_i \varepsilon_i$, with $\mu_1 \ge \mu_2 \ge \cdots \ge \mu_l \ge 0$ integers and $|\mu| = k$. We identify μ with a partition of k with at most l parts, as in Section 9.1.1. Let (σ^μ, G^μ) be the irreducible representation of $\mathfrak{S}_k$ associated with μ by Schur–Weyl duality. Write $\overline{\mu}$ for the restriction of μ to the diagonal subalgebra $\mathfrak{h}$ of $\mathfrak{g}$. Let $(\pi^{\overline{\mu}}, V^{\overline{\mu}})$ be the irreducible representation of G with highest weight $\overline{\mu}$.

Theorem 10.2.7. *As a module for* $\mathbf{Sp}(\mathbb{C}^{2l}, \Omega) \times \mathfrak{S}_k$, *the space of* Ω-*harmonic* k-*tensors has isotypic decomposition*

$$\mathcal{H}(\bigotimes^k \mathbb{C}^{2l}, \Omega) \cong \bigoplus_{\mu \in \mathrm{Par}(k,l)} V^{\overline{\mu}} \otimes G^{\mu} . \tag{10.24}$$

Proof. Suppose λ is a $\mathfrak{b}$-dominant weight and $W^k(\lambda) \neq 0$. Then we must have $|\lambda| = k - 2r$ for some integer $r \geq 0$. To see this, take $0 \neq v \in W^k(\lambda)$ and decompose v under $\widetilde{\mathfrak{h}}$ as $v = \sum_{\mu} v_{\mu}$, where μ ranges over the weights of $\widetilde{\mathfrak{h}}$ on $\bigotimes^k \mathbb{C}^{2l}$. Take $\mu = \sum_i \mu_i \varepsilon_i$ with $v_{\mu} \neq 0$. Then $\sum_{i=1}^n \mu_i = k$. Since μ restricts to λ, we also have $\lambda_i = \mu_i - \mu_{n+1-i}$. Thus

$$|\lambda| = \sum_{i=1}^{l} \mu_i - \sum_{i=l+1}^{2l} \mu_i = k - 2 \sum_{i=l+1}^{2l} \mu_i ,$$

which proves the assertion about $|\lambda|$.

Now assume $W^k(\lambda) \subset \mathcal{H}(\bigotimes^k \mathbb{C}^{2l}, \Omega)$. Let $\mu = \widetilde{\lambda}$ (that is, μ is λ viewed as a weight of $\widetilde{\mathfrak{h}}$). Then μ is G-admissible and $|\mu| = k - 2r$ by the argument just given; thus Corollary 10.2.3 implies $|\mu| = k$ and $W^k(\lambda) = \widetilde{W}^k(\mu)$. Furthermore, μ is the unique G-admissible $\widetilde{\mathfrak{b}}$-dominant weight with $\overline{\mu} = \lambda$. Hence the $\mathfrak{h}$ weight-space decomposition of the Ω-harmonic $\mathfrak{b}$-extreme k-tensors is

$$\mathcal{H}(\bigotimes^k \mathbb{C}^{2l}, \Omega)^{\mathfrak{n}^+} = \bigoplus_{\mu \in \mathrm{Par}(k,l)} W^k(\overline{\mu}) .$$

Schur–Weyl duality gives $W^k(\overline{\mu}) = \widetilde{W}^k(\mu) = G^{\mu}$ as a $\mathfrak{S}_k$-module. By Proposition 3.3.9 every nonzero element of $W^k(\overline{\mu})$ generates an irreducible G-module isomorphic to $V^{\overline{\mu}}$. Now apply Theorems 4.2.1 and 4.2.12 to obtain (10.24). $\qquad\square$

We can now obtain the *Weyl modules* for the symplectic group.

Corollary 10.2.8. $(G = \mathbf{Sp}(\mathbb{C}^{2l}, \Omega))$ *Let* λ *be a dominant integral weight on* $\mathfrak{h}$. *Let* $k = |\lambda|$, *so that* λ *determines a partition of* k *with at most* l *parts, and let* A *be a tableau of shape* λ. *Then the irreducible* G-*module with highest weight* λ *is isomorphic to* $\mathbf{s}(A)\mathcal{H}(\bigotimes^k \mathbb{C}^{2l}, \Omega)$, *where* $\mathbf{s}(A)$ *is the Young symmetrizer associated to* A.

Proof. By Proposition 9.3.5 we know that $\mathbf{s}(A)$ projects G^{λ} onto a one-dimensional subspace spanned by a single $\mathfrak{b}$-extreme vector of weight λ and annihilates the spaces G^{μ} for $\mu \neq \lambda$. Hence the corollary follows from (10.24). $\qquad\square$

Examples

1. Assume $l \geq k$. Then the partition $\mu = [1^k]$ is G-admissible. It corresponds to the sgn representation of $\mathfrak{S}_k$ and the highest weight $\varpi = \varepsilon_1 + \cdots + \varepsilon_k$. Hence from

Corollary 10.2.8 we see that the space

$$\mathcal{H}_{\text{skew}}(\otimes^k \mathbb{C}^{2l}, \Omega) = \mathcal{H}(\otimes^k \mathbb{C}^{2l}, \Omega) \cap \bigwedge^k \mathbb{C}^{2l}$$

of harmonic skew-symmetric k-tensors is an irreducible G-module with highest weight ϖ_k. This gives the same realization of the kth fundamental representation of G as in Theorem 5.5.15. Indeed, all the contraction operators C_{ij} are $\pm X$ on the space of skew-symmetric tensors, so we have

$$\mathcal{H}_{\text{skew}}(\otimes^k \mathbb{C}^{2l}, \Omega) = \mathcal{H}(\bigwedge^k \mathbb{C}^{2l}, \Omega)$$

in the notation of Theorem 5.5.15.

2. Let $k = 2$ and assume $l \geq 2$. Then both partitions of 2 are G-admissible and give the representations of G with highest weights $2\varepsilon_1$ and $\varepsilon_1 + \varepsilon_2$. These are paired with the trivial and sgn representations of $\mathfrak{S}_k$, respectively. Because the form Ω is skew-symmetric, every symmetric tensor is harmonic (see Theorem 10.2.5). Hence by Theorem 10.2.7 and Corollary 10.2.8 we have

$$\mathcal{H}(\otimes^2 \mathbb{C}^{2l}, \Omega) = S^2(\mathbb{C}^{2l}) \oplus \mathcal{H}_{\text{skew}}(\otimes^2 \mathbb{C}^{2l}, \Omega) .$$

The first summand on the right is the irreducible G-module with highest weight $2\varepsilon_1$, and the second summand is the irreducible G-module with highest weight $\varepsilon_1 + \varepsilon_2$.

10.2.4 Decomposition of Harmonics for O(2l+1)

Let $G = \mathbf{O}(\mathbb{C}^n, B)$, with $n = 2l + 1$ and the form B given by equation (2.9). We shall decompose the space $\mathcal{H}(\otimes^k \mathbb{C}^n, B)$ of B-harmonic k-tensors under the joint action of G and $\mathfrak{S}_k$. Denote by $\mathbf{A}(k, n)$ the set of all G-admissible $\tilde{\mathfrak{b}}$-dominant weights $\mu = \sum_{i=1}^n \mu_i \varepsilon_i$ with $|\mu| = k$. Recall that this means that the sum of the lengths of the first two columns of the Ferrers diagram for μ does not exceed n. For example, when $k = 4$ and $n = 3$ these weights correspond to the Ferrers diagrams

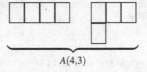

$$A(4,3)$$

Let $\mu \in \mathbf{A}(k, n)$ and let (σ^μ, G^μ) be the irreducible representation of $\mathfrak{S}_k$ associated with μ by Schur duality. Let $(\pi^{\overline{\mu}, \varepsilon}, V^{\overline{\mu}, \varepsilon})$ be the irreducible representation of G with highest weight $\overline{\mu}$ in which $-I$ acts by εI with $\varepsilon = \pm 1$ (see Theorem 5.5.23).

Theorem 10.2.9. ($n = 2l + 1$) *As a module for* $\mathbf{O}(\mathbb{C}^n, B) \times \mathfrak{S}_k$, *the space of B-harmonic k-tensors has isotypic decomposition*

$$\mathcal{H}(\otimes^k \mathbb{C}^n, B) = \bigoplus_{\mu \in \mathbf{A}(k,n)} V^{\overline{\mu}, \varepsilon} \otimes G^\mu , \qquad \text{where } \varepsilon = (-1)^k . \tag{10.25}$$

Proof. Just as in the case of the symplectic group we will use Schur–Weyl duality and the characterization of G-admissible $\mathbf{GL}(n, \mathbb{C})$-highest weights in Theorem 10.2.5. The new feature in this case is that for each $\mathfrak{b}$-dominant weight

$$\lambda = \sum_{i=1}^{p} \lambda_i \varepsilon_i \,,$$

where $p \leq l$, there are two G-admissible $\widetilde{\mathfrak{b}}$-dominant weights that restrict to λ; namely, $\widetilde{\lambda}$ and

$$\lambda^{\natural} = \widetilde{\lambda} + \sum_{i=p+1}^{n-p} \varepsilon_i \,.$$

We shall need the following lemma:

Lemma 10.2.10. *Suppose* $\lambda \in \mathfrak{h}^*$ *and* $W^k(\lambda) \neq 0$. *Then* $|\lambda| \leq k$. *If* $|\lambda| < k$ *then* $|\lambda^{\natural}| = k - 2r$ *for some integer* $r \geq 0$.

Since the proof of this lemma is rather long, we give it at the end of the section.

Completion of proof of Theorem 10.2.9. Suppose λ is a $\mathfrak{b}$-dominant weight and $0 \neq W^k(\lambda) \subset \mathcal{H}(\bigotimes^k \mathbb{C}^n, B)$. Then $|\lambda| \leq k$ by Lemma 10.2.10. If $|\lambda| = k$ let $\mu = \widetilde{\lambda}$, whereas if $|\lambda| < k$, let $\mu = \lambda^{\natural}$. Then μ is G-admissible and $|\mu| = k$ or $|\mu| = k - 2r$ (by Lemma 10.2.10). Now apply Corollary 10.2.3 to conclude that $r = 0$ and $W^k(\lambda) = \widetilde{W}^k(\mu)$. Furthermore, μ is uniquely determined by λ and the condition $|\mu| = k$, since $\widetilde{\lambda}$ and $\lambda^{\natural}$ are the only G-admissible weights that restrict to λ.

Thus the $\mathfrak{h}$ weight-space decomposition of the harmonic $\mathfrak{b}$-extreme k-tensors is

$$\mathcal{H}\left(\bigotimes^k \mathbb{C}^n\right)^{\mathfrak{n}^+} = \bigoplus_{\mu \in \mathbf{A}(k,n)} W^k(\overline{\mu}) \,.$$

Since $W^k(\overline{\mu}) = \widetilde{W}^k(\mu)$, we have $W^k(\overline{\mu}) = G^{\mu}$ as a $\mathfrak{S}_k$-module by Schur–Weyl duality.

The element $-I \in G$ acts by $\varepsilon I = (-1)^k I$ on $\mathcal{H}(\bigotimes^k \mathbb{C}^n)$. By the classification of the representations of G (Section 5.5.5) every nonzero element of $W^k(\overline{\mu})$ generates an irreducible G-module isomorphic to $V^{\overline{\mu},\varepsilon}$. It follows that the G-isotypic components in $\mathcal{H}(\bigotimes^k \mathbb{C}, B)$ are given by (10.25). $\quad\square$

We can now obtain the *Weyl modules* for the orthogonal group in this case.

Corollary 10.2.11. $(G = \mathbf{O}(\mathbb{C}^n, B), n = 2l+1)$ *Let* λ *be a dominant integral weight on* $\mathfrak{h}$. *Let* $k = |\lambda|$ *and* $m = |\lambda^{\natural}|$. *Let* A (*resp.* $A^{\natural}$) *be a tableau of shape* λ (*resp.* $\lambda^{\natural}$). *Let* $\mathbf{s}(A)$ *and* $\mathbf{s}(A^{\natural})$ *be the Young symmetrizers associated to* A *and* $A^{\natural}$ *and set* $\varepsilon = (-1)^k$. *Then* $\mathbf{s}(A)\mathcal{H}(\bigotimes^k \mathbb{C}^n, B)$ *is isomorphic to the irreducible* G-module $V^{\lambda,\varepsilon}$ *and* $\mathbf{s}(A^{\natural})\mathcal{H}(\bigotimes^m \mathbb{C}^n, B)$ *is isomorphic to the irreducible* G-module $V^{\lambda,-\varepsilon}$.

Proof. The weights λ and $\lambda^{\natural}$ are both in $\mathbf{A}(m,n)$. Hence we can use (10.25) and the same argument as in the case of the symplectic group (Corollary 10.2.8), noting that the integers k and m have opposite parity. $\quad\square$

Examples

1. Let $G^\circ = \mathbf{SO}(\mathbb{C}^n, B)$, with n odd. From Corollary 10.2.11 we see that the space

$$\mathcal{H}_{\mathrm{sym}}(\bigotimes^k \mathbb{C}^n, B) = \mathcal{S}^k(\mathbb{C}^n) \cap \mathcal{H}(\bigotimes^k \mathbb{C}^n, B)$$

of harmonic symmetric k-tensors furnishes an irreducible G°-module with highest weight $k\varepsilon_1$ corresponding to the Ferrers diagram

$$\mu = \underbrace{\boxed{}\boxed{}\cdots\boxed{}}_{k}$$

and the trivial representation of $\mathfrak{S}_k$. The space $\mathcal{H}_{\mathrm{sym}}(\bigotimes^k \mathbb{C}^n, B)$ can be described more explicitly. Namely,

$$\mathcal{H}_{\mathrm{sym}}(\bigotimes^k \mathbb{C}^n, B) = \mathrm{Span}\{v^{\otimes k} : v \in \mathbb{C}^n \text{ and } B(v,v) = 0\} \tag{10.26}$$

is the space spanned by the kth powers of *isotropic* vectors in $\mathbb{C}^n$. Indeed, the right side of (10.26) is obviously a G°-invariant subspace, and hence equality holds by irreducibility. This gives another proof of Proposition 5.6.13 when n is odd and shows that $\mathcal{H}_{\mathrm{sym}}(\bigotimes^k \mathbb{C}^n, B)$ is the space of homogeneous G-harmonic polynomials of degree k studied in Section 5.6.4.

2. Let $k = 2$ and assume that $n \geq 5$ is odd. Then both partitions of 2 are G-admissible and give the representations with highest weights $2\varepsilon_1$ and $\varepsilon_1 + \varepsilon_2$. These are paired with the trivial and sgn representations of $\mathfrak{S}_k$, respectively. Because the form B is symmetric, every alternating tensor is harmonic (see Theorem 10.2.5). Hence by Theorem 10.2.9 and Corollary 10.2.11 we have

$$\mathcal{H}(\bigotimes^2 \mathbb{C}^n, B) = \mathcal{H}_{\mathrm{sym}}(\bigotimes^2 \mathbb{C}^n, B) \oplus \bigwedge^2 \mathbb{C}^n ,$$

where the first summand on the right is the irreducible G°-module with highest weight $2\varepsilon_1$ and the second summand is the irreducible G°-module with highest weight $\varepsilon_1 + \varepsilon_2$.

Proof of Lemma 10.2.10.   Write λ as in (10.18), and view λ as a weight of $\widetilde{\mathfrak{h}}$. It is then $\mathfrak{b}$-dominant and G-admissible. We have

$$|\lambda| = k - m \quad \text{for some integer } m \geq 0 . \tag{10.27}$$

To see this, take $0 \neq v \in W^k(\lambda)$ and decompose v under $\widetilde{\mathfrak{h}}$ as $v = \sum_\mu v_\mu$, where μ ranges over the weights of $\widetilde{\mathfrak{h}}$. Take any μ with $v_\mu \neq 0$. Since v_μ is a k-tensor, we have

$$\sum_{i=1}^n \mu_i = k . \tag{10.28}$$

However, μ restricts to λ, so $\lambda_i = \mu_i - \mu_{n+1-i}$. Thus

$$\sum_{i=1}^{l} (\mu_i - \mu_{n+1-i}) = |\lambda| .$$ (10.29)

Subtracting (10.29) from (10.28), we obtain (10.27). If m is even we are done. Suppose $m = 2s + 1$ is odd. We will prove that $|\lambda^\natural| = k - 2r$ for some integer $r \geq 0$. We have

$$|\lambda^\natural| = |\lambda| + n - p = k - 2(s - l + p) .$$

If we can show that

$$s \geq l - p ,$$ (10.30)

then $|\lambda^\natural| = k - 2r$, where $r = s - (l - p) \geq 0$, and the lemma follows.

We now turn to the proof of (10.30). Fix a $\mathfrak{b}$-extreme k-tensor v of weight λ. It decomposes under the action of $\widetilde{\mathfrak{h}}$ as $v = \sum_\mu v_\mu$ with v_μ of weight μ relative to $\widetilde{\mathfrak{h}}$. Set

$$\Sigma(v) = \{\mu \in \widetilde{\mathfrak{h}}^* : v_\mu \neq 0\} .$$

Clearly, $\mu|_{\mathfrak{h}} = \lambda$ for all $\mu \in \Sigma(v)$. Since $|\lambda| = k - 2s - 1$, the components of μ satisfy

$$2s + 1 = 2 \sum_{i=p+1}^{l} \mu_i + \mu_{l+1} + 2 \sum_{i=1}^{p} \mu_{2l+2-i} .$$

It follows that μ_{l+1} is odd and

$$s \geq \sum_{i=p+1}^{l} \mu_i .$$

To obtain the desired inequality (10.30), it thus suffices to prove that there exists $\gamma \in \Sigma(v)$ such that $\gamma_i \neq 0$ for $p + 1 \leq i \leq l$. For this we argue as follows, using the matrix form of $\mathfrak{n}^+$ from Section 2.4.1:

For $p + 1 \leq i \leq l$ let

$$\Sigma_i(v) = \{\mu \in \Sigma(v) : \mu_i = 0\} , \qquad \Sigma^i(v) = \{\mu \in \Sigma(v) : \mu_i \neq 0\} .$$

We can then split $v = v_i + v^i$, where

$$v_i = \sum_{\mu \in \Sigma_i(v)} v_\mu , \qquad v^i = \sum_{\mu \in \Sigma^i(v)} v_\mu .$$

If $\mu \in \Sigma_i(v)$ then $\mu_{2l+2-i} = \mu_i = 0$. Hence if we write v_μ in terms of the basis $\{e_I\}$ of Section 9.1.1, then e_{2l+2-i} cannot occur. Thus $e_{l+1,2l+2-i} v_\mu = 0$. By contrast,

$$e_{i,l+1} v_\mu \neq 0 \quad \text{for} \quad \mu \in \Sigma_i(v) .$$ (10.31)

To prove this, note that the matrices $e = e_{i,l+1}$, $f = e_{l+1,i}$, and $h = e_{ii} - e_{l+1,l+1}$ make up a TDS triple, and

$$h v_\mu = (\mu_i - \mu_{l+1}) v_\mu = -\mu_{l+1} v_\mu .$$

Since $\mu_{l+1} \geq 1$, equation (10.31) follows from the complete reducibility of $\mathfrak{sl}(2, \mathbb{C})$ and Proposition 2.3.3. Set $X = e_{i,l+1} - e_{l+1,2l+2-i} \in \mathfrak{g}_{\varepsilon_i}$. We have

$$X v_\mu = e_{i,l+1} v_\mu \quad \text{for} \quad \mu \in \Sigma_i(v) \,.$$

We now use the fact that v is a $\mathfrak{b}$-extreme tensor. This gives the relation

$$0 = X v = \sum_{\mu \in \Sigma_i(v)} e_{i,l+1} v_\mu + e_{i,l+1} v^i - e_{l+1,2l+2-i} v^i \,. \tag{10.32}$$

Thus the component in each $\widetilde{\mathfrak{h}}$ weight space that occurs in (10.32) must vanish. For $\mu \in \Sigma_i(v)$ the tensor $e_{i,l+1} v_\mu$ is nonzero of weight $\mu + \varepsilon_i - \varepsilon_{l+1}$, whereas the weights of the other terms in (10.32) are of the form $\gamma + \varepsilon_i - \varepsilon_{l+1}$ and $\gamma + \varepsilon_{l+1} - \varepsilon_{2l+2-i}$ with $\gamma \in \Sigma^i(v)$. Hence for every $\mu \in \Sigma_i(v)$ there exists $\gamma \in \Sigma^i(v)$ such that

$$\mu + \varepsilon_i - \varepsilon_{l+1} = \gamma + \varepsilon_{l+1} - \varepsilon_{2l+2-i}$$

(since the coefficient of ε_i on the left side is 1). It follows that

$$\mu_{l+1} = \gamma_{l+1} + 2 \,. \tag{10.33}$$

Choose i and μ such that $\mu \in \Sigma_i(v)$ and μ_{l+l} has the smallest possible value among all weights in $\Sigma_j(v)$, for $p + 1 \leq j \leq l$. Then the weight γ occurring in (10.33) cannot be in $\Sigma_j(v)$ for any $p + 1 \leq j \leq l$, by minimality of μ_{l+1}. Hence γ is a weight in $\Sigma(v)$ with $\gamma_j \neq 0$ for $p + 1 \leq j \leq l$. We already saw that this implies (10.30), so the proof of the lemma is complete. $\square$

10.2.5 Decomposition of Harmonics for **O(2l)**

We now obtain the $G \times \mathfrak{S}_k$ isotypic decomposition of the space of harmonic k-tensors in the most complicated case, namely when $n = 2l$ is even and $G = \mathbf{O}(\mathbb{C}^n, B)$. Denote by $\mathbf{A}(k, n)$ the set of all G-admissible $\widetilde{\mathfrak{b}}$-dominant weights $\mu = \sum_{i=1}^n \mu_i \varepsilon_i$ with $|\mu| = k$. Recall that this means that the sum of the lengths of the first two columns of the Ferrers diagram for μ does not exceed n. We partition $\mathbf{A}(k, n)$ into three disjoint sets as follows:

1. $\mu \in \mathbf{A}_+(k, n)$ if $\mu_i = 0$ for $i \geq l$.
2. $\mu \in \mathbf{A}_-(k, n)$ if there exists $p < l$ such that $\mu_i = 1$ for $p + 1 \leq i \leq n - p$ and $\mu_i = 0$ for $i > n - p$.
3. $\mu \in \mathbf{A}_0(k, n)$ if $\mu_l > 0$ and $\mu_i = 0$ for $i > l$.

For example, when $n = 4$ and $k = 5$ we have the following G-admissible diagrams:

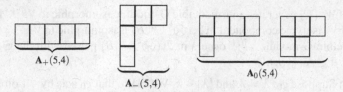

$\mathbf{A}_+(5,4)$

$\mathbf{A}_-(5,4)$

$\mathbf{A}_0(5,4)$

Let $g_0 \in G$ be as in Section 10.2.2; $\mathrm{Ad}^*(g_0)$ fixes $\varepsilon_1,\ldots,\varepsilon_{l-1}$ and sends ε_l to $-\varepsilon_l$. For μ in cases (1) or (2), $\overline{\mu}$ is a $\mathfrak{b}$-dominant weight with $\mu_l = 0$. Hence $g_0 \cdot \overline{\mu} = \overline{\mu}$. Recall from Section 5.5.5 that there are irreducible regular representations $(\pi^{\overline{\mu},\pm}, V^{\overline{\mu},\pm})$ of G that remain irreducible on restriction to $\mathbf{SO}(n,\mathbb{C})$ and have g_0 acting by ± 1. For μ in case (3), $g_0 \cdot \overline{\mu} \neq \overline{\mu}$ and there is a unique irreducible representation $(\rho^{\overline{\mu}}, I(V^{\overline{\mu}}))$ of G determined by $\overline{\mu}$. Let G^μ be the irreducible $\mathfrak{S}_k$-module corresponding to μ by Schur–Weyl duality.

Theorem 10.2.12. *($n = 2l$) As a module for* $\mathbf{O}(\mathbb{C}^n, B) \times \mathfrak{S}_k$*, the space of B-harmonic k-tensors decomposes as*

$$\mathcal{H}(\textstyle\bigotimes^k \mathbb{C}^n, B) = \mathcal{H}_-^{\otimes k} \oplus \mathcal{H}_0^{\otimes k} \oplus \mathcal{H}_+^{\otimes k}, \tag{10.34}$$

with isotypic decompositions

$$\mathcal{H}_\pm^{\otimes k} \cong \bigoplus_{\mu \in \mathbf{A}_\pm(k,n)} V^{\overline{\mu},\pm} \otimes G^\mu, \qquad \mathcal{H}_0^{\otimes k} \cong \bigoplus_{\mu \in \mathbf{A}_0(k,n)} I(V^{\overline{\mu}}) \otimes G^\mu.$$

Proof. Suppose $0 \neq W^k(\lambda) \subset \mathcal{H}(\bigotimes^k \mathbb{C}^n, B)$, where

$$\lambda = \sum_{i=1}^p \lambda_i \varepsilon_i \quad \text{with} \quad \lambda_1 \geq \lambda_2 \geq \cdots \geq \lambda_p > 0. \tag{10.35}$$

Then $|\lambda| = k - 2r$ for some integer $r \geq 0$, by the same argument as in the proof of Theorem 10.2.7.

Case 1: Suppose $g_0 \cdot \lambda = \lambda$ and $|\lambda| = k$. Then $p < l$ in (10.35) because g_0 fixes λ. Let $\mu = \widetilde{\lambda}$. Then $\mu \in \mathbf{A}_\pm(k,n), \overline{\mu} = \lambda$, and from Proposition 10.2.2 and Theorem 10.2.5 we have $W^k(\lambda) = \widetilde{W}^k(\mu)$.

To determine the action of g_0 on $W^k(\lambda)$ in this case, we define

$$u_+ = (w_p)^{m_p} \otimes (w_{p-1})^{m_{p-1}} \otimes \cdots \otimes (w_1)^{m_1},$$

where $w_j = e_1 \wedge e_2 \wedge \cdots \wedge e_j$ and $m_p = \lambda_p, m_{p-1} = \lambda_{p-1} - \lambda_p, \ldots, m_1 = \lambda_1 - \lambda_2$. Then $u_+ \in \widetilde{W}^k(\mu)$ and $g_0 \cdot u_+ = u_+$, since $p < l$. By Schur–Weyl duality

$$\widetilde{W}^k(\mu) = G^\mu = \mathbb{C}[\mathfrak{S}_k] \cdot u_+.$$

Since g_0 commutes with the action of $\mathfrak{S}_k$, it follows that g_0 acts by $+1$ on $W^k(\lambda)$. From the classification of the representations of G in Section 5.5.5, every nonzero element of $W^k(\overline{\mu})$ generates an irreducible G-module isomorphic to $V^{\overline{\mu},+}$. It follows that the $V^{\overline{\mu},+}$-isotypic component in $\mathcal{H}(\bigotimes^k \mathbb{C}^n, B)$ is isomorphic to $V^{\overline{\mu},+} \otimes G^\mu$. Thus the irreducible G-module $V^{\overline{\mu},+}$ occurs in $\mathcal{H}(\bigotimes^k \mathbb{C}^n, B)$ paired with the $\mathfrak{S}_k$-module G^μ for all $\mu \in \mathbf{A}_+(k,n)$.

Case 2: Suppose $g_0 \cdot \lambda = \lambda$ and $|\lambda| < k$. We claim that g_0 acts by -1 on $W^k(\lambda)$ in this case. Indeed, if there were $v_0 \in W^k(\lambda)$ such that $g_0 \cdot v_0 = v_0$, then the G-cyclic subspace V_0 generated by v_0 would be an irreducible G-submodule of $\mathcal{H}(\bigotimes^k \mathbb{C}^n, B)$ isomorphic to $V^{\lambda,+}$ (by the argument in Case 1). But this contradicts Corollary 10.2.4, since we proved in Case 1 that $V^{\lambda,+}$ occurs in $\mathcal{H}^{\otimes q}(\mathbb{C}^n, B)$ with $q = |\lambda| < k$.

Define $\lambda^\natural$ as in Section 10.2.4. We now use the following general result on admissible extensions of weights for $\mathfrak{h}$:

Lemma 10.2.13. *Suppose $\lambda \in \mathfrak{h}^*$ and $W^k(\lambda) \neq 0$. Assume $g_0 \cdot \lambda = \lambda$, $|\lambda| < k$, and that there exists $0 \neq v \in W^k(\lambda)$ such that $g_0 \cdot v = -v$. Then $|\lambda^\natural| = k - 2r$ for some integer $r \geq 0$.*

The proof of this lemma is quite lengthy, although similar to that of Lemma 10.2.10, so we defer it to the end of this section.

We now use Lemma 10.2.13 to complete our analysis of Case 2. Let $\mu = \lambda^\natural$. Then $\overline{\mu} = \lambda$ and $|\mu| = k - 2r$ by Lemma 10.2.13. Since $W^k(\lambda) \subset \mathcal{H}(\bigotimes^k \mathbb{C}^n, B)$, we conclude from Corollary 10.2.3 that $|\mu| = k$ (so $\mu \in \mathbf{A}_-(k,n)$) and $W^k(\lambda) = \widetilde{W}^k(\mu)$. We have already argued that g_0 acts by -1 on $W^k(\lambda)$. From the classification of the representations of G in Section 5.5.5, every nonzero element of $W^k(\overline{\mu})$ generates an irreducible G-module isomorphic to $V^{\overline{\mu},-}$. It follows that the $V^{\overline{\mu},-}$-isotypic component in $\mathcal{H}(\bigotimes^k \mathbb{C}^n, B)$ is isomorphic to $V^{\overline{\mu},-} \otimes G^\mu$. Thus the irreducible G-module $V^{\overline{\mu},-}$ occurs in $\mathcal{H}(\bigotimes^k \mathbb{C}^n, B)$ paired with the $\mathfrak{S}_k$-module G^μ for all $\mu \in \mathbf{A}_-(k,n)$.

Case 3: Suppose $g_0 \cdot \lambda \neq \lambda$. Set $\mu = \widetilde{\lambda}$. Since $|\lambda| = k - 2r$ for some integer $r \geq 0$, we may apply Corollary 10.2.3 to conclude that $|\mu| = k$ (so $\mu \in \mathbf{A}_0(k,n)$) and $W^k(\lambda) = \widetilde{W}^k(\mu)$. Furthermore, for $0 \neq u \in W^k(\overline{\mu})$, one has $\mathbb{C}[G]u \cong I(V^{\overline{\mu}})$, since $g_0 \cdot \lambda \neq \lambda$ (see Section 5.5.5). It follows that the $I(V^{\overline{\mu}})$-isotypic component in $\mathcal{H}(\bigotimes^k \mathbb{C}^n, B)$ is isomorphic to $I(V^{\overline{\mu}}) \otimes G^\mu$. Thus the irreducible G-module $I(V^{\overline{\mu}})$ occurs in $\mathcal{H}(\bigotimes^k \mathbb{C}^n, B)$ paired with the $\mathfrak{S}_k$-module G^μ for all $\mu \in \mathbf{A}_0(k,n)$.

We have now determined all possible G-isotypic components in $\mathcal{H}(\bigotimes^k \mathbb{C}^n, B)$. Combining the results of cases (1), (2), and (3) gives (10.34). $\qquad\square$

We can now obtain the *Weyl modules* for the orthogonal group in this case.

Corollary 10.2.14. *($G = \mathbf{O}(\mathbb{C}^n, B)$, $n = 2l$) Let $\lambda = \sum_{i=1}^l \lambda_i \varepsilon_i$ be a dominant integral weight on $\mathfrak{h}$ ($\lambda_1 \geq \cdots \geq \lambda_{l-1} \geq \lambda_l \geq 0$).*

1. *Suppose $\lambda_l = 0$. Set $k = |\lambda|$ and $m = |\lambda^\natural|$. Let A (resp. $A^\natural$) be a tableau of shape λ (resp. $\lambda^\natural$). Let $\mathbf{s}(A)$ and $\mathbf{s}(A^\natural)$ be the Young symmetrizers associated to A (resp. $A^\natural$). Then $\mathbf{s}(A)\mathcal{H}(\bigotimes^k \mathbb{C}^n, B)$ is isomorphic to the irreducible G-module $V^{\lambda,+}$ and $\mathbf{s}(A^\natural)\mathcal{H}^{\otimes m}(\mathbb{C}^n, B)$ is isomorphic to the irreducible G-module $V^{\lambda,-}$.*
2. *Suppose $\lambda_l > 0$. Set $k = |\lambda|$ and let A be a tableau of shape λ. Let $\mathbf{s}(A)$ be the Young symmetrizer associated to A. Then $\mathbf{s}(A)\mathcal{H}(\bigotimes^k \mathbb{C}^n, B)$ is isomorphic to the irreducible G-module $I(V^\lambda)$.*

Proof. In Case (1) the weight λ is in $\mathbf{A}_+(k,n)$ and the weight $\lambda^{\natural}$ is in $\mathbf{A}_-(m,n)$. In Case (2) the weight λ is in $\mathbf{A}_0(k,n)$. Now use (10.34) and the same argument as in Corollary 10.2.11. $\qquad\qquad\square$

Examples

1. Let $G = \mathbf{SO}(\mathbb{C}^n, B)$, with $n \geq 4$ even. From Corollary 10.2.14 we see that the space

$$\mathcal{H}_{\mathrm{sym}}(\textstyle\bigotimes^k \mathbb{C}^n, B) = S^k(\mathbb{C}^n) \cap \mathcal{H}(\textstyle\bigotimes^k \mathbb{C}^n, B)$$

of harmonic symmetric k-tensors furnishes an irreducible G-module with highest weight $k\varepsilon_1$ corresponding to the Ferrers diagram

$$\mu = \underbrace{\square\,\square\cdots\square}_{k}$$

and the trivial representation of $\mathfrak{S}_k$. Just as in the case in which n is odd we have

$$\mathcal{H}_{\mathrm{sym}}(\textstyle\bigotimes^k \mathbb{C}^n, B) = \mathrm{Span}\{v^{\otimes k} : v \in \mathbb{C}^n \text{ and } B(v,v) = 0\}.$$

This gives another proof of Proposition 5.6.13 when n is even, and shows that $\mathcal{H}_{\mathrm{sym}}(\bigotimes^k \mathbb{C}^n, B)$ is the space of homogeneous G-harmonic polynomials of degree k studied in Section 5.6.4.

2. Let $k = 2$ and assume that $n \geq 6$ is even. Then both partitions of 2 are in $\mathbf{A}_+(2,n)$. Because the form B is symmetric, every alternating tensor is harmonic. Hence by Theorem 10.2.12 and Corollary 10.2.14 we have

$$\mathcal{H}(\textstyle\bigotimes^2 \mathbb{C}^n, B) = \mathcal{H}_{\mathrm{sym}}(\textstyle\bigotimes^2 \mathbb{C}^n, B) \oplus \textstyle\bigwedge^2 \mathbb{C}^n.$$

The first summand on the right is the irreducible G-module with highest weight $2\varepsilon_1$ and the second summand is the irreducible G-module with highest weight $\varepsilon_1 + \varepsilon_2$. The element $g_0 \in G$ acts as the identity in both summands.

Proof of Lemma 10.2.13. By the argument at the beginning of the proof of Theorem 10.2.7 we know that $|\lambda| = k - 2r$ with r a nonnegative integer. Hence $|\lambda^{\natural}| = k - 2r + n - p = k - 2(r - (l - p))$. Thus it suffices to prove

$$r \geq l - p. \tag{10.36}$$

The argument to prove (10.36) is similar to that in Lemma 10.2.10. Fix a $\mathfrak{b}$-extreme k-tensor v of weight λ that satisfies $g_0 v = -v$. It decomposes under the action of $\widetilde{\mathfrak{h}}$ as $v = \sum_{\mu} v_{\mu}$, with v_{μ} a k-tensor of weight μ relative to $\mathfrak{h}$. Set

$$\Sigma(v) = \{\mu : v_{\mu} \neq 0\} \subset \widetilde{\mathfrak{h}}^*.$$

Then $\mu|_{\mathfrak{h}} = \lambda$ for all $\mu \in \Sigma(v)$ and hence $\mu_i = \mu_{2l+1-i}$ for $p < i \leq l$. In particular, $g_0 \cdot \mu = \mu$, since g_0 interchanges ε_l with ε_{l+1} and fixes all other ε_i. Thus $g_0 v_\mu$ also has weight μ, so it must be a multiple of v_μ. But $g_0 \cdot v = -v$, and we conclude that

$$g_0 v_\mu = -v_\mu \quad \text{for all } \mu \in \Sigma(v) . \tag{10.37}$$

This in turn implies that $\mu_l \geq 1$. Indeed, if $\mu_l = 0$ then $\mu_{l+1} = 0$ also, so if v_μ were written in terms of the basis $\{e_I\}$ of Section 9.1.1, e_l and e_{l+1} would not occur. But this would imply $g_0 \cdot v_\mu = v_\mu$, contradicting (10.37). Since $\sum_{i=1}^{2l} \mu_i = k$, it follows that

$$2r = \sum_{i=1}^{2l} \mu_i - \sum_{i=1}^{p} \lambda_i = \mu_{p+1} + \cdots + \mu_{2l} \geq 2 \sum_{i=p+1}^{l} \mu_i .$$

To obtain the desired inequality (10.36), it thus suffices to prove that there exists $\gamma \in \Sigma(v)$ such that $\gamma_i > 0$ for $p+1 \leq i \leq l$.

Define $\Sigma_i(v) = \{\mu \in \Sigma(v) : \mu_i = 0\}$ and $\Sigma^i(v) = \{\mu \in \Sigma(v) : \mu_i > 0\}$ for $p < i < l$. We can then split $v = v_i + v^i$, where

$$v_i = \sum_{\mu \in \Sigma_i(v)} v_\mu , \quad v^i = \sum_{\mu \in \Sigma^i(v)} v_\mu .$$

If $\mu \in \Sigma_i(v)$ then $\mu_{2l+1-i} = \mu_i = 0$. Hence v_μ cannot contain e_{2l+1-i}, and so $e_{l,2l+1-i}v_\mu = 0$. By contrast, $\mu_i = 0$, whereas $\mu_{l+1} = \mu_l \geq 1$, so just as in the proof of (10.31) we have

$$e_{i,l+1}v_\mu \neq 0 \quad \text{for} \quad \mu \in \Sigma_i(v) .$$

Thus

$$X_{\varepsilon_i + \varepsilon_l} v_\mu = (e_{i,l+1} - e_{l,n+1-i})v_\mu = e_{i,l+1}v_\mu \quad \text{for} \quad \mu \in \Sigma_i(v) .$$

We now use the fact that v is a $\mathfrak{b}$-extreme tensor. This gives the relation

$$0 = X_{\varepsilon_i + \varepsilon_l} v = \sum_{\mu \in \Sigma_i(v)} e_{i,l+1}v_\mu + e_{i,l+1}v^i - e_{l,n+1-i}v^i . \tag{10.38}$$

Thus the component in each $\tilde{\mathfrak{h}}$ weight space that occurs in (10.38) must vanish. For $\mu \in \Sigma_i(v)$ we note that $e_{i,l+1}v_\mu$ is a nonzero tensor of weight $\mu + \varepsilon_i - \varepsilon_{l+1}$, whereas the weights of the other terms in (10.38) are of the form $\gamma + \varepsilon_i - \varepsilon_{l+1}$ and $\gamma + \varepsilon_l - \varepsilon_{n+1-i}$ with $\gamma \in \Sigma^i(v)$. Hence for every $\mu \in \Sigma_i(v)$ there exists $\gamma \in \Sigma^i(v)$ such that

$$\mu + \varepsilon_i - \varepsilon_{l+1} = \gamma + \varepsilon_l - \varepsilon_{n+1-i}$$

(since the coefficient of ε_i on the left side is 1). It follows that

$$\mu_l = \gamma_l + 1 . \tag{10.39}$$

Now choose i and μ such that $\mu \in \Sigma_i(v)$ and μ_l has the smallest possible value among all weights in $\Sigma_j(v)$, for $p < j < l$. Then the weight γ occurring in (10.39) cannot be in $\Sigma_j(v)$ for any $p < j < l$, by minimality of μ_l. Hence γ is a weight in

$\Sigma(v)$ with $\gamma_j \neq 0$ for $p < j < l$. We already saw that this implies (10.36), so the proof is complete. $\qquad\qquad\square$

10.2.6 Exercises

1. Let $T : V \otimes V \longrightarrow \mathrm{End}(V)$ be the map (10.2). Show that $u \in V^{\otimes 2}$ is harmonic if and only if $\mathrm{tr}(T(u)) = 0$.
2. Show that $u \in \bigotimes^k V$ is harmonic if and only if $C_{12}\sigma_k(s)u = 0$ for all $s \in \mathfrak{S}_k$. (HINT: Show that the operator D_{12} is injective and use Lemma 10.1.5.)
3. Suppose $V = \mathbb{C}^n$ and $\omega(x, y) = (x, y) = \sum_i x_i y_i$ for $x, y \in \mathbb{C}^n$.
   (a) Show that $C_{ij} = C_{12}$ on $S^k(\mathbb{C}^n)$. (HINT: Consider the action of C_{ij} on the symmetric tensors $x^{\otimes k}$ for $x \in \mathbb{C}^n$ and use Lemma B.2.3.)
   (b) Let $\Delta = \sum_i (\partial/\partial x_i)^2$ be the Laplace operator. Use the form ω to identify the symmetric tensor algebra $S(\mathbb{C}^n)$ with the polynomial algebra $\mathcal{P}(\mathbb{C}^n)$, where the symmetric tensor $x^{\otimes k}$ becomes the polynomial $y \mapsto \omega(x, y)^k$. Thus the contraction operator C_{12} maps $\mathcal{P}^k(\mathbb{C}^n)$ to $\mathcal{P}^{k-2}(\mathbb{C}^n)$. Show that $C_{12} = \Delta$. Hence $f \in \mathcal{P}^k(\mathbb{C}^n)$ is *harmonic* (viewed as a symmetric k-tensor) if and only if $\Delta f = 0$.
4. Let $G = \mathbf{Sp}(\mathbb{C}^{2l}, \Omega)$.
   (a) Show that $S^k(\mathbb{C}^{2l}) \subset \mathcal{H}(\bigotimes^k \mathbb{C}^{2l}, \Omega)$. (HINT: $C_{ij}\sigma_k(ij) = -C_{ij}$ for every transposition $(ij) \in \mathfrak{S}_k$.)
   (b) Show that the natural action of G on $S^k(\mathbb{C}^{2l})$ is irreducible and has highest weight $k\varepsilon_1$. (HINT: Use (a) and Corollary 10.2.8; see also Section 12.2.2.)
5. Let $G = \mathbf{Sp}(\mathbb{C}^{2l}, \Omega)$ with $l \geq 3$. Decompose $\mathcal{H}(\bigotimes^3 \mathbb{C}^{2l}, \Omega)$ under $G \times \mathfrak{S}_3$.
6. Let $G = \mathbf{O}(\mathbb{C}^{2l+1}, B)$. Decompose $\mathcal{H}(\bigotimes^3 \mathbb{C}^{2l+1}, B)$ under $G \times \mathfrak{S}_3$.
7. (Notation as in Section 10.2.4) The space of $\mathfrak{b}$-extreme k-tensors of a given weight does not necessarily contain a $\widetilde{\mathfrak{b}}$-extreme k-tensor. Here is an example.
   (a) Let $\lambda = \varepsilon_1$ and set $u = \theta \otimes e_1$, where θ is the G-invariant 2-tensor corresponding to B. Show that u is a $\mathfrak{b}$-extreme 3-tensor of weight λ.
   (b) Prove that there is no dominant integral weight μ on $\widetilde{\mathfrak{h}}$ such that $\mu|_{\mathfrak{h}} = \varepsilon_1$ and $\sum \mu_i = 3$.
8. Decompose $\mathcal{H}(\bigotimes^2 \mathbb{C}^4, B)$ under $\mathbf{O}(\mathbb{C}^4, B)$ and $\mathbf{SO}(\mathbb{C}^4, B)$.
9. Decompose $\mathcal{H}(\bigotimes^2 \mathbb{C}^{2l}, B)$ under $\mathbf{SO}(\mathbb{C}^{2l}, B)$ for $l \geq 3$.
10. Let $G = \mathbf{O}(\mathbb{C}^{2l}, B)$ with $l \geq 2$. Decompose $\mathcal{H}(\bigotimes^3 \mathbb{C}^{2l}, B)$ under $G \times \mathfrak{S}_3$.
11. Let $G = \mathbf{O}(\mathbb{C}^n, B)$ and $G^\circ = \mathbf{SO}(\mathbb{C}^n, B)$. If $k > 0$ then $\mathcal{H}(\bigotimes^k \mathbb{C}^{2l}, B)$ has no nonzero G-fixed tensors by Corollary 10.2.4. When does it have nonzero G°-fixed elements? Consider the cases n odd and n even separately.

10.3 Riemannian Curvature Tensors

Let (M, g) be a smooth pseudo-Riemannian manifold with $\dim M \geq 2$; at each point $p \in M$ there is a nondegenerate symmetric bilinear form $g_p(\cdot, \cdot)$ on $T_p(M)$ that is a smooth function of p. Let ∇ be the associated *Levi-Civita connection*; this is the unique affine connection on M for which the metric is covariant constant:

$$X(g(Y,Z)) = g(\nabla_X Y, Z) + g(Y, \nabla_X Z) \,, \tag{10.40}$$

and which is torsion-free:

$$\nabla_X Y - \nabla_Y X = [X,Y] \tag{10.41}$$

for all smooth vector fields X, Y, Z on M. Let u, v, w be tangent vectors to M at a point p. The *curvature tensor* $R_p(u,v) \in \operatorname{End} T_p(M)$ is defined by

$$R_p(u,v)w = \left(\nabla_X \nabla_Y Z - \nabla_Y \nabla_X Z - \nabla_{[X,Y]} Z \right)_p \,, \tag{10.42}$$

where X, Y, Z are vector fields on M with $X_p = u$, $Y_p = v$, and $Z_p = w$ (the right side of (10.42) depends only on the values of X, Y, Z at p). For details on pseudo-Riemannian structures and affine connections see Helgason [66, Chapter I].

We now examine the symmetries of the curvature tensor. From its definition, $R_p(v,w)$ is a bilinear function of v, w and satisfies

$$R_p(v,w) = -R_p(w,v) \,. \tag{10.43}$$

Thus there is a unique linear map $R_p : \bigwedge^2 T_p(M) \longrightarrow \operatorname{End}(T_p(M))$ with $R_p(v \wedge w) = R_p(v,w)$.

Using (10.40) one can show that the linear transformation $R_p(v,w)$ determined by $v, w \in T_p(M)$ is *skew-symmetric* relative to g_p:

$$g_p(R_p(v,w)x, y) = -g_p(x, R_p(v,w)y) \quad \text{for } x,y \in T_p(M) \,. \tag{10.44}$$

This equation implies that $R_p(v,w) \in \operatorname{Lie}(\mathbf{O}(g_p))$, where $\mathrm{O}(g_p) \subset \mathbf{GL}(T_p(M))$ is (as usual) the isometry group of the form g_p. If we identify $T_p(M)$ with $T_p(M)^*$ using the form g_p, we have $\operatorname{End}(T_p(M)) \cong \bigotimes^2 T_p(M)$ as an $\mathbf{O}(g_p)$-module, with the skew-symmetric transformations corresponding to $\bigwedge^2 T_p(M)$. Thus we may also view the curvature tensor as a linear map $R_p : \bigwedge^2 T_p(M) \longrightarrow \bigwedge^2 T_p(M)$.

The curvature tensor has additional symmetries: the Jacobi identity for the Lie algebra of vector fields together with (10.41) implies the *first Bianchi identity*

$$R_p(x,y)z + R_p(y,z)x + R_p(z,x)y = 0 \tag{10.45}$$

for all $x, y, z \in T_p(M)$. Furthermore, (10.43), (10.44), and (10.45) give the symmetry

$$g_p(R_p(v,w)x, y) = g_p(R_p(x,y)v, w) \,. \tag{10.46}$$

10.3.1 The Space of Curvature Tensors

Fix a point $p \in M$ and the bilinear form g_p, and consider the possible curvature tensors R_p. Define the *Riemann–Christoffel curvature tensor* $R \in \bigotimes^4 T_p(M)^*$ by

$$R(v,w,x,y) = g_p(R_p(v,w)x, y)$$

for $v, w, x, y \in T_p(M)$. Then from (10.43), (10.44), and (10.46) we see that R has the symmetries

$$\sigma(12)R = -R, \quad \sigma(34)R = -R, \quad \sigma(13)\sigma(24)R = R. \tag{10.47}$$

Here σ is the representation of the symmetric group $\mathfrak{S}_4$ on $\bigotimes^4 T_p(M)^*$ given by permutation of the tensor positions, as in Section 4.2.4. The Bianchi identity (10.45) gives the additional symmetry

$$R + \sigma(123)R + \sigma(321)R = 0. \tag{10.48}$$

For studying the decomposition of curvature tensors under the orthogonal group it is convenient to replace $T_p(M)$ by its complexification $E = T_p(M)_{\mathbb{C}}$. Let Q be the complex-bilinear extension of g_p to E. Let $n = \dim E$ and fix an orthonormal basis $\{e_i\}$ for E relative to Q. We can then identify E with $\mathbb{C}^n$ and identify $\dot{Q}$ with the form $\langle x,y \rangle = \sum_{i=1}^n x_i y_i$ on $\mathbb{C}^n$.

If we use the form $\langle x,y \rangle$ to identify $\mathbb{C}^n$ with $(\mathbb{C}^n)^*$, then we can view an element $R \in S^2(\bigwedge^2 \mathbb{C}^n)$ either as a four-tensor with the symmetries (10.47), or as the skew-symmetric linear transformation $R(v \wedge w) \in \mathrm{End}(\mathbb{C}^n)$ defined by

$$\langle R(v \wedge w)x, y \rangle = R(v,w,x,y)$$

that has the additional symmetry (10.46). The two points of view are equivalent relative to the action of $\mathbf{O}(\mathbb{C}^n, Q)$.

Definition 10.3.1. The space of *curvature tensors* $\mathrm{Curv}(\mathbb{C}^n)$ is the subspace of $S^2(\bigwedge^2 \mathbb{C}^n)$ consisting of all tensors R that also satisfy (10.48).

Notice that whereas the symmetry (10.46) of the tensor R_p follows from the first Bianchi identity (10.45) together with identities (10.43) and (10.44), we are now taking this symmetry as given and imposing the Bianchi identity as an extra condition to define a linear space of curvature tensors. This facilitates the representation-theoretic analysis of $\mathrm{Curv}(\mathbb{C}^n)$.

To understand the role of the Bianchi identity, we define the *Bianchi operator* $b \in \mathrm{End}\left(\bigotimes^4 \mathbb{C}^n\right)$ by

$$b = \frac{1}{3}\left(1 + \sigma(123) + \sigma(321)\right). \tag{10.49}$$

Thus b is the average of the cyclic permutations of the first three tensor positions, and hence b projects onto the subspace of tensors fixed by $\sigma(123)$.

Lemma 10.3.2. $S^2(\bigwedge^2 \mathbb{C}^n)$ *is invariant under the Bianchi operator.*

Proof. Let $R \in S^2(\bigwedge^2 \mathbb{C}^n)$. Since $(123) = (13)(12)$ and $(321) = (23)(12)$, we see that $\sigma(12)bR = b\sigma(12)R = -bR$ and that

$$bR = \frac{1}{3}\big(1 - \sigma(13) - \sigma(23)\big)R = \frac{1}{3}\big(1 - \sigma(24) - \sigma(14)\big)R \qquad (10.50)$$

(note that (10.47) implies $\sigma(23)R = \sigma(14)R$). Since $(34)(13) = (14)(34)$ and $(34)(23) = (14)(34)$, we can use (10.47) and (10.50) to calculate

$$\sigma(34)bR = \frac{1}{3}\big(-R + \sigma(14)R + \sigma(24)R\big) = -bR .$$

Likewise, since $(13)(24)(14) = (14)(34)(12)$, we can use (10.47) and the second formula in (10.50) to calculate

$$\sigma(13)\sigma(24)bR = \frac{1}{3}\big(R - \sigma(13)R - \sigma(14)R\big) = bR .$$

This completes the proof that $bR \in S^2(\bigwedge^2 \mathbb{C}^n)$. $\square$

Let β be the restriction of the Bianchi operator to $S^2(\bigwedge^2 \mathbb{C}^n)$. Then $\mathrm{Curv}(\mathbb{C}^n) = \mathrm{Ker}\,\beta$ and β commutes with the action of **GL**$(n, \mathbb{C})$ on $S^2(\bigwedge^2 \mathbb{C}^n)$. Since multiplication of even-degree elements in the exterior algebra is commutative, we have

$$\bigwedge^4 \mathbb{C}^n = \big(\bigwedge^2 \mathbb{C}^n\big) \wedge \big(\bigwedge^2 \mathbb{C}^n\big) \subset S^2(\bigwedge^2 \mathbb{C}^n) .$$

Conversely, if $u \in \bigwedge^4 \mathbb{C}^n$ then $\sigma(123)u = u$, since (123) is an even permutation. Hence $u = \beta u$. This proves that $\mathrm{Range}(\beta) = \bigwedge^4 \mathbb{C}^n$.

Proposition 10.3.3. *If $n \geq 2$ then* $\mathrm{Curv}(\mathbb{C}^n) \cong F_n^{[2,2]}$ *as a* **GL**$(n, \mathbb{C})$ *module. Hence*

$$\dim \mathrm{Curv}(\mathbb{C}^n) = \frac{1}{12}n^2(n+1)(n-1) . \qquad (10.51)$$

If $n = 2$ or 3 then $\mathrm{Curv}(\mathbb{C}^n) = S^2(\bigwedge^2 \mathbb{C}^n)$. *If $n \geq 4$ then*

$$S^2(\bigwedge^2 \mathbb{C}^n) = \mathrm{Curv}(\mathbb{C}^n) \oplus \big(\bigwedge^4 \mathbb{C}^n\big) \qquad (10.52)$$

is the decomposition into irreducible **GL**$(n, \mathbb{C})$ *modules.*

Proof. Since $\beta^2 = \beta$ is a projection, we have the decomposition

$$S^2(\bigwedge^2 \mathbb{C}^n) = \mathrm{Ker}(\beta) \oplus \mathrm{Range}(\beta) ,$$

which gives (10.52). Thus it remains to prove the first assertion and (10.51).

Set $R = (e_1 \wedge e_2) \otimes (e_1 \wedge e_2)$. Then as an element of $\bigwedge^4 \mathbb{C}^4$,

$$4R = e_1 \otimes e_2 \otimes e_1 \otimes e_2 - e_1 \otimes e_2 \otimes e_2 \otimes e_1$$
$$- e_2 \otimes e_1 \otimes e_1 \otimes e_2 + e_2 \otimes e_1 \otimes e_2 \otimes e_1 .$$

Using the first formula in (10.50), we verify that $\beta R = 0$. It is clear that R is fixed by the upper-triangular unipotent matrices in $\mathbf{GL}(n, \mathbb{C})$ and is an eigenvector for the diagonal matrices $h = \text{diag}[x_1, \ldots, x_n]$ with eigenvalue $2x_1 + 2x_2$. Thus by Proposition 3.3.9 we know that $\text{Curv}(\mathbb{C}^n)$ contains an irreducible $\mathbf{GL}(n, \mathbb{C})$ submodule isomorphic to $F_n^{[2,2]}$. From the Weyl dimension formula we calculate that $\dim F_n^{[2,2]}$ equals the right side of (10.51). On the other hand,

$$\dim S^2(\bigwedge^2 \mathbb{C}^n) = n(n-1)(n^2-n+2)/8 \,.$$

From these formulas we obtain $\dim S^2(\bigwedge^2 \mathbb{C}^n) = \dim F_n^{[2,2]}$ when $n = 2$ or 3. If $n \geq 4$ then $\bigwedge^4 \mathbb{C}^n \cong F_n^{[1,1,1,1]}$ is a nonzero irreducible $\mathbf{GL}(n, \mathbb{C})$ submodule of dimension $n(n-1)(n-2)(n-3)/24$. It follows that

$$\dim S^2(\bigwedge^2 \mathbb{C}^n) = \dim F_n^{[2,2]} + \dim \bigwedge^4 \mathbb{C}^n \,.$$

Hence $\text{Curv}(\mathbb{C}^n) \cong F_n^{[2,2]}$ as $\mathbf{GL}(n, \mathbb{C})$-modules and (10.51) holds. $\qquad \square$

Corollary 10.3.4. *The space* $\text{Curv}(\mathbb{C}^n)$ *is the irreducible Weyl module for* $\mathbf{GL}(n, \mathbb{C})$ *corresponding to the tableau* $A = \begin{array}{|c|c|} \hline 1 & 3 \\ \hline 2 & 4 \\ \hline \end{array}$.

Proof. Since the representation $G^{[2,2]}$ of $\mathfrak{S}_4$ has degree two, the normalized Young symmetrizer is $\mathbf{p}_A = (1/12)\mathbf{c}(A)\mathbf{r}(A)$, where $\mathbf{r}(A) = (1 + \sigma(13))(1 + \sigma(24))$ and $\mathbf{c}(A) = (1 - \sigma(12))(1 - \sigma(34))$. From the definition we see that $\sigma(12)\mathbf{p}_A = -\mathbf{p}_A$ and $\sigma(34)\mathbf{p}_A = -\mathbf{p}_A$. We calculate that $\sigma(13)\sigma(24)$ commutes with $\mathbf{c}(A)$. Hence $\sigma(13)\sigma(24)\mathbf{p}_A = \mathbf{p}_A$, so we have shown that

$$\text{Range}(\mathbf{p}_A) \subset S^2(\bigwedge^2 \mathbb{C}^n) \,.$$

Since $\text{Range}(\mathbf{p}_A) \cong F_n^{[2,2]}$ as a $\mathbf{GL}(n, \mathbb{C})$ module by Theorem 9.3.10, we conclude from (10.52) that $\text{Range}(\mathbf{p}_A) = \text{Curv}(\mathbb{C}^n)$. $\qquad \square$

10.3.2 Orthogonal Decomposition of Curvature Tensors

We now turn to the decomposition of $\text{Curv}(\mathbb{C}^n)$ as an $\mathbf{O}(\mathbb{C}^n, Q)$ module. We have used the action of $\mathfrak{S}_4$ on $\bigotimes^4 \mathbb{C}^n$ to define $\text{Curv}(\mathbb{C}^n)$; now we use contractions relative to the form Q to obtain $\mathbf{O}(\mathbb{C}^n, Q)$ intertwining operators. The *Ricci contraction* $\text{Ric}_Q : S^2(\bigwedge^2 \mathbb{C}^n) \longrightarrow S^2(\mathbb{C}^n)$ defined by

$$\text{Ric}_Q(R)(v, w) = \sum_{i=1}^{n} R(e_i, v, e_i, w) \tag{10.53}$$

is the only nonzero contraction on $S^2(\bigwedge^2 \mathbb{C})$, up to a scalar multiple; note that the right side of (10.53) is symmetric in v and w because of (10.47). If $R \in \text{Curv}(\mathbb{C}^n)$

then $\text{Ric}_Q(R)$ is called the *Ricci curvature tensor* of R. We can perform one more contraction on the Ricci curvature to obtain the *scalar curvature*

$$s_Q(R) = \text{tr}_Q(\text{Ric}_Q(R)) = \sum_{i,j=1}^{n} R(e_i, e_j, e_i, e_j) . \tag{10.54}$$

The map $R \mapsto s_Q(R)$ is an $\mathbf{O}(\mathbb{C}^n, Q)$ intertwining operator from $\text{Curv}(\mathbb{C}^n)$ to the trivial $\mathbf{O}(\mathbb{C}^n, Q)$ module $\mathbb{C}$.

Going in the other direction, we construct elements of $\text{Curv}(\mathbb{C}^n)$ as follows. If $A, B \in \text{End}(\mathbb{C}^n)$ we define $A \bigotimes B \in \text{End}(\bigwedge^2 \mathbb{C}^n)$ by

$$(A \bigotimes B)(v \wedge w) = Av \wedge Bw + Bv \wedge Aw \tag{10.55}$$

(note that the right side of (10.55) is symmetric in A, B and is skew-symmetric in v, w). Now assume that A and B are symmetric transformations relative to the form Q. Define a bilinear form on $\bigwedge^2 \mathbb{C}^n$ by

$$\langle v \wedge w, x \wedge y \rangle = \det \begin{bmatrix} \langle v, x \rangle & \langle v, y \rangle \\ \langle w, x \rangle & \langle w, y \rangle \end{bmatrix} . \tag{10.56}$$

Then one verifies that $A \bigotimes B$ is a symmetric transformation relative to the form (10.56); hence it corresponds to an element of $S^2(\bigwedge^2 \mathbb{C}^n)$. This correspondence depends on the underlying form Q; if $\varphi(x, y) = \langle Ax, y \rangle$ and $\psi(x, y) = \langle Bx, y \rangle$ are the associated symmetric bilinear forms, then $\varphi \bigotimes \psi$ is the four-tensor

$$\begin{aligned} (\varphi \bigotimes \psi)(v, w, x, y) &= \varphi(v, x)\psi(w, y) - \varphi(v, y)\psi(w, x) \\ &\quad + \psi(v, x)\varphi(w, y) - \psi(v, y)\varphi(w, x) , \end{aligned} \tag{10.57}$$

for $v, w, x, y \in \mathbb{C}^n$. From (10.57) it is easy to check that $\varphi \bigotimes \psi$ satisfies the Bianchi identity (10.48). Thus we have a linear map

$$S^2(S^2(\mathbb{C}^n)) \longrightarrow \text{Curv}(\mathbb{C}^n) , \quad A \otimes B + B \otimes A \mapsto A \bigotimes B . \tag{10.58}$$

If $h \in \mathbf{GL}(n, \mathbb{C})$ then $\rho(h)(A \bigotimes B)\rho(h^{-1}) = (hAh^{-1}) \bigotimes (hBh^{-1})$, where ρ is the natural representation of $\mathbf{GL}(n, \mathbb{C})$ on $\bigwedge^2 \mathbb{C}^n$. Since $\mathbf{O}(\mathbb{C}^n, Q)$ preserves the symmetric transformations in $\text{End}(\mathbb{C}^n)$ and $\text{End}(\bigwedge^2 \mathbb{C}^n)$, the map (10.58) intertwines the $\mathbf{O}(\mathbb{C}^n, Q)$ actions. In particular, if $A \in S^2(\mathbb{C}^n)$ then we calculate that

$$\text{Ric}_Q(A \bigotimes Q) = \text{tr}_Q(A)Q + (n-2)A . \tag{10.59}$$

When $n = 2$ then $\dim \text{Curv}(\mathbb{C}^2) = 1$. Since $s_Q(Q \bigotimes Q) = 4$ in this case, it follows from (10.59) that

$$R = \frac{1}{4} s_Q(R) Q \bigotimes Q \quad \text{and} \quad \text{Ric}_Q(R) = \frac{1}{2} s_Q(R) Q$$

for any $R \in \text{Curv}(\mathbb{C}^2)$. From now on we assume $n \geq 3$.

Definition 10.3.5. The space $\mathrm{Weyl}_Q(\mathbb{C}^n)$ of *Weyl conformal curvature tensors* for the form Q is the subspace of $\mathrm{Curv}(\mathbb{C}^n)$ of tensors W satisfying $\mathrm{Ric}_Q(W) = 0$.

We can decompose any $R \in \mathrm{Curv}(\mathbb{C}^n)$ into three parts as follows: Define

$$A = \frac{1}{n-2}\left\{ \mathrm{Ric}_Q(R) - \frac{1}{n}s_Q(R)Q \right\} \quad \text{and} \quad W = R - A \wedge Q - \gamma s_Q(R)Q \wedge Q ,$$

where $\gamma = 1/(2n^2 - n)$. Then $W \in \mathrm{Curv}(\mathbb{C}^n)$ and from (10.59) we calculate that $\mathrm{tr}_Q(A) = 0$ and $\mathrm{Ric}_Q(W) = 0$. Thus $W \in \mathrm{Weyl}_Q(\mathbb{C}^n)$ and we have the decomposition

$$R = \gamma s_Q(R)Q \wedge Q + A \wedge Q + W . \tag{10.60}$$

One calls $\gamma s_Q(R)Q \wedge Q$ the *scalar part*, $A \wedge Q$ the *traceless Ricci part*, and W the *Weyl part* of the curvature tensor R, relative to the form Q.

We now interpret decomposition (10.60) in terms of irreducible representations of $\mathbf{O}(\mathbb{C}^n, Q)$. The map $\alpha \mapsto \alpha Q \wedge Q$ embeds the trivial $\mathbf{O}(\mathbb{C}^n, Q)$-module $\mathbb{C}$ into $\mathrm{Curv}(\mathbb{C}^n)$. Let $\mathcal{H}^2_{\mathrm{sym}}(\mathbb{C}^n, Q)$ be the space of Q-harmonic symmetric two-tensors. By Examples 1 in Sections 10.2.4 and 10.2.5 we know that $\mathcal{H}^2_{\mathrm{sym}}(\mathbb{C}^n, Q)$ is irreducible under $\mathbf{SO}(\mathbb{C}^n, Q)$. Its highest weight (in terms of the fundamental dominant weights) is $2\varpi_1$ when $n \neq 4$ and $2(\varpi_1 + \varpi_2)$ when $n = 4$ (see Theorem 5.6.11). We have

$$\dim \mathcal{H}^2_{\mathrm{sym}}(\mathbb{C}^n, Q) = \frac{1}{2}n(n+1) - 1 = \frac{1}{2}(n+2)(n-1) . \tag{10.61}$$

From (10.59) we see that $A \mapsto A \wedge Q$ is an injective $\mathbf{O}(\mathbb{C}^n, Q)$ intertwining map from $\mathcal{H}^2_{\mathrm{sym}}(\mathbb{C}^n, Q)$ into $\mathrm{Curv}(\mathbb{C}^n)$ and

$$\left\{ \mathbb{C}(Q \wedge Q) \oplus \left(\mathcal{H}^2_{\mathrm{sym}}(\mathbb{C}^n, Q) \wedge Q \right) \right\} \cap \mathrm{Weyl}_Q(\mathbb{C}^n) = 0 .$$

Since Ric_Q is an $\mathbf{O}(\mathbb{C}^n, Q)$ intertwining operator, the space $\mathrm{Weyl}_Q(\mathbb{C}^n)$ is invariant under $\mathbf{O}(\mathbb{C}^n, Q)$. Hence we conclude from (10.60) that there is a decomposition

$$\mathrm{Curv}(\mathbb{C}^n) = \mathbb{C}(Q \wedge Q) \oplus \left(\mathcal{H}^2_{\mathrm{sym}}(\mathbb{C}^n, Q) \wedge Q \right) \oplus \mathrm{Weyl}_Q(\mathbb{C}^n) \tag{10.62}$$

as an $\mathbf{O}(\mathbb{C}^n, Q)$-module. We calculate from (10.62) that

$$\dim \mathrm{Weyl}_Q(\mathbb{C}^n) = \frac{1}{12}n^2(n^2 - 1) - \frac{1}{2}(n+2)(n-1) - 1$$
$$= \frac{1}{12}n(n+1)(n+2)(n-3) . \tag{10.63}$$

In particular, $\mathrm{Weyl}_Q(\mathbb{C}^3) = 0$, so in three dimensions the curvature tensor is completely determined by the Ricci curvature tensor from formula (10.60).

10.3.3 The Space of Weyl Curvature Tensors

We now study the space $\mathrm{Weyl}_Q(\mathbb{C}^n)$ as a representation of the orthogonal group. Using the Q-orthonormal basis e_i we define a Q-isotropic basis for $\mathbb{C}^n$ by

$$f_k = \frac{1}{\sqrt{2}}\left(e_k + i e_{n+1-k}\right), \quad f_{-k} = \frac{1}{\sqrt{2}}\left(e_k - i e_{n+1-k}\right)$$

for $k = 1, \ldots, l$, where $l = [n/2]$; if n is odd then we set $f_0 = e_{l+1}$ (see Section B.2.1). The four-tensor

$$W = (f_1 \wedge f_2) \otimes (f_1 \wedge f_2) \tag{10.64}$$

is in $\mathrm{Curv}(\mathbb{C}^n)$ (see the proof of Proposition 10.3.3); one calculates that $\mathrm{Ric}_Q(W) = 0$. Hence $W \in \mathrm{Weyl}_Q(\mathbb{C}^n)$.

If we use the isotropic basis $\{f_j\}$ then we obtain the matrix form $\mathfrak{g} = \mathfrak{so}(\mathbb{C}^n, B)$ in Section 2.4.1 for the Lie algebra of $\mathbf{SO}(\mathbb{C}^n, Q)$. Let $\mathfrak{g} = \mathfrak{n}^+ + \mathfrak{h} + \mathfrak{n}^-$ be the triangular decomposition as in Section 2.4.3. The matrices in $\mathfrak{n}^+$ are upper triangular relative to the isotropic basis, so $\mathfrak{n}^+ \cdot W = 0$. If $\{\varepsilon_1, \ldots, \varepsilon_l\}$ is the basis for $\mathfrak{h}^*$ as in Section 2.4.1, then W is an eigenvector for $\mathfrak{h}$ with eigenvalue $2\varepsilon_1 + 2\varepsilon_2$.

Assume that $n \geq 5$ (the case $n = 4$ is special and will be considered at the end). Then $\mathfrak{g}$ is a simple Lie algebra; in terms of the fundamental dominant weights we have $2\varepsilon_1 + 2\varepsilon_2 = 2\varpi_2$ (see Section 3.1.3). If $V_n \subset \mathrm{Weyl}_Q(\mathbb{C}^n)$ is the $\mathfrak{g}$-invariant subspace generated by W, then V_n is an irreducible $\mathfrak{g}$-module by Proposition 3.3.9. From the Weyl dimension formula (see Section 7.1.2) and (10.63), we find that $\dim V_n = \dim \mathrm{Weyl}_Q(\mathbb{C}^n)$. Hence $V_n = \mathrm{Weyl}_Q(\mathbb{C}^n)$. We can summarize our results as follows:

Proposition 10.3.6. *When $n \geq 5$ the space of curvature tensors on $\mathbb{C}^n$ has the following structure relative to the orthogonal group:*

1. *The space $\mathcal{H}^2_{\mathrm{sym}}(\mathbb{C}^n, Q) \wedge Q$ corresponding to trace-zero Ricci curvature tensors is irreducible for $\mathbf{SO}(\mathbb{C}^n, Q)$ and has highest weight $2\varpi_1$.*
2. *The space $\mathrm{Weyl}_Q(\mathbb{C}^n)$ of Weyl conformal curvature tensors is irreducible for $\mathbf{SO}(\mathbb{C}^n, Q)$ and has highest weight $2\varpi_2$.*
3. *The space of curvature tensors on $\mathbb{C}^n$ decomposes as*

$$\mathrm{Curv}(\mathbb{C}^n) = \mathbb{C}(Q \wedge Q) \oplus \left(\mathcal{H}^2_{\mathrm{sym}}(\mathbb{C}^n, Q) \wedge Q\right) \oplus \mathrm{Weyl}_Q(\mathbb{C}^n)$$

under $\mathbf{SO}(\mathbb{C}^n, Q)$, with each summand also invariant under $\mathbf{O}(\mathbb{C}^n, Q)$.

Finally, we consider the case $n = 4$ (notation as above). The positive roots of $\mathfrak{g}$ are $\alpha_1 = \varepsilon_1 - \varepsilon_2$ and $\alpha_2 = \varepsilon_1 + \varepsilon_2$. The corresponding root vectors act on the Q-isotropic basis $\{f_{-2}, f_{-1}, f_1, f_2\}$ for $\mathbb{C}^4$ by

$$X_{\alpha_1} f_2 = f_1, \quad X_{\alpha_1} f_{-1} = -f_{-2}, \quad X_{\alpha_2} f_{-2} = f_1, \quad X_{\alpha_2} f_{-1} = -f_2, \tag{10.65}$$

with all other matrix entries zero (see Section 2.4.1). Let $H_{\alpha_1}, H_{\alpha_2}$ be the coroots and define Lie subalgebras

$$\mathfrak{g}_k = \text{Span}\{H_{\alpha_k}, X_{\alpha_k}, X_{-\alpha_k}\} \cong \mathfrak{sl}(2,\mathbb{C})$$

for $k = 1,2$. Then $\mathfrak{g} \cong \mathfrak{g}_1 \oplus \mathfrak{g}_2$ is semisimple. The fundamental highest weights for $\mathfrak{g}$ are $\varpi_1 = \frac{1}{2}(\varepsilon_1 - \varepsilon_2)$ and $\varpi_2 = \frac{1}{2}(\varepsilon_1 + \varepsilon_2)$, and $\mathfrak{n}^+ = \text{Span}\{X_{\alpha_1}, X_{\alpha_2}\}$.

Let W be the four-tensor (10.64). Then $\mathfrak{n}^+ \cdot W = 0$ and

$$H_{\alpha_1} \cdot W = 2\langle \alpha_1, \alpha_2 \rangle W = 0 \,, \quad H_{\alpha_2} \cdot W = 2\langle \alpha_2, \alpha_2 \rangle W = 4W \,. \tag{10.66}$$

Thus W has weight $4\varpi_2$. Let $V \subset \text{Weyl}_Q(\mathbb{C}^4)$ be the $\mathfrak{g}$-invariant subspace generated by W. Then (10.66) implies that $\mathfrak{g}_1 \cdot V = 0$ and V is an irreducible $\mathfrak{g}_2$-module with $\dim V = 5$ (see Section 2.3.1). Thus V is also irreducible under $\mathfrak{g}$.

Let $\tau \in \mathbf{O}(\mathbb{C}^4, Q)$ be the element that fixes f_1 and f_{-1} and interchanges f_2 with f_{-2}. Define $\overline{W} = \tau \cdot W = (f_1 \wedge f_{-2}) \otimes (f_1 \wedge f_{-2})$. Then $\overline{W} \in \text{Weyl}_Q(\mathbb{C}^4)$, since this space is invariant under $\mathbf{O}(\mathbb{C}^n, Q)$. Since $\text{Ad}^* \tau$ interchanges α_1 and α_2 and $\text{Ad}\,\tau$ leaves $\mathfrak{n}^+$ invariant, we have $\mathfrak{n}^+ \cdot \overline{W} = 0$. Since f_{-2} has weight $-\varepsilon_2$, the tensor $f_1 \wedge f_{-2}$ has weight α_1. Hence

$$H_{\alpha_1} \overline{W} = 2\langle \alpha_1, \alpha_1 \rangle W = 4W \,, \quad H_{\alpha_2} \overline{W} = 2\langle \alpha_2, \alpha_1 \rangle W = 0 \,. \tag{10.67}$$

Thus $\overline{W}$ has weight $4\varpi_1$. Let $\overline{V} \subset \text{Weyl}_Q(\mathbb{C}^4)$ be the $\mathfrak{g}$-invariant subspace generated by $\overline{W}$. Then (10.67) implies that $\mathfrak{g}_2 \cdot \overline{V} = 0$ and $\overline{V}$ is an irreducible $\mathfrak{g}_1$-module with $\dim \overline{V} = 5$ (see Section 2.3.1). Thus $\overline{V}$ is also irreducible under $\mathfrak{g}$. Since V and $\overline{V}$ are irreducible with different highest weights (as $\mathfrak{g}$-modules), we know that $V \cap \overline{V} = 0$. Hence $V + \overline{V}$ is a subspace of $\text{Weyl}_Q(\mathbb{C}^4)$ of dimension 10. But we know from (10.63) that $\dim W_Q(\mathbb{C}^4) = 10$, so we conclude that $\text{Weyl}_Q(\mathbb{C}^4) = V \oplus \overline{V}$. On the other hand, since τ interchanges the highest-weight vectors W and $\overline{W}$, the space $\text{Weyl}_Q(\mathbb{C}^4)$ is irreducible under $\mathbf{O}(\mathbb{C}^n, Q)$.

We can summarize this representation-theoretic analysis as follows:

Proposition 10.3.7. *The space of curvature tensors on $\mathbb{C}^4$ has the following structure relative to the orthogonal group:*

1. *The nine-dimensional space $\mathcal{H}^2_{\text{sym}}(\mathbb{C}^4, Q) \wedge Q$ that corresponds to trace-zero Ricci curvature tensors is an irreducible $\mathbf{SO}(\mathbb{C}^4, Q)$-module with highest weight $2(\varpi_1 + \varpi_2)$.*
2. *The ten-dimensional space $\text{Weyl}_Q(\mathbb{C}^4)$ is irreducible under $\mathbf{O}(\mathbb{C}^4, Q)$ and is the sum of two five-dimensional irreducible $\mathbf{SO}(\mathbb{C}^4, Q)$-modules with highest weights $4\varpi_1$ and $4\varpi_2$.*
3. *The space of curvature tensors on $\mathbb{C}^4$ decomposes under $\mathbf{O}(\mathbb{C}^4, Q)$ as*

$$\text{Curv}(\mathbb{C}^4) = \mathbb{C}(Q \wedge Q) \oplus (\mathcal{H}^2_{\text{sym}}(\mathbb{C}^4, Q) \wedge Q) \oplus \text{Weyl}_Q(\mathbb{C}^4) \,.$$

10.3.4 Exercises

1. Use (10.40) to prove (10.44). (HINT: Let V, W, X, Y be smooth vector fields whose values at p are v, w, x, y. Set $\varphi = g(X, Y)$. Then φ is a smooth function on M and $[U, V]\varphi = U(V\varphi) - V(U\varphi)$. Now apply (10.40) to each side of this equation.)

2. Use (10.41) to prove (10.45). (HINT: Let X, Y, Z be smooth vector fields on M whose values at p are x, y, z. Show that (10.41) gives

$$[[X, Y], Z] = \nabla_{[X,Y]} Z + \nabla_Z \nabla_Y X - \nabla_Z \nabla_X Y \, .$$

Deduce that $R_p(x, y)z$ is the value at p of the vector field

$$[\nabla_X, \nabla_Y] Z - \nabla_Z \nabla_X Y + \nabla_Z \nabla_Y X - [[X, Y], Z] \, .$$

Now take the sum of the cyclic permutations of X, Y, Z in this formula and use the Jacobi identity.)

3. Use (10.43), (10.44), and (10.45) to prove (10.46). (HINT: Set $F_{vwxy} = g_p(R_p(v, w)x, y)$. Then $F_{vwxy} + F_{wxvy} + F_{xvwy} = 0$. Now replace y successively by x, w, and v, form the sum of the cyclic permutation of the remaining three vectors, and add the resulting equations. Ten of the twelve terms cancel, and the remaining two terms give (10.46).)

4. (a) Show that $(e_1 \wedge e_2) \otimes (e_1 \wedge e_2)$ satisfies (10.48) and is a **GL**$(n, \mathbb{C})$ highest-weight vector in $\mathrm{Curv}(\mathbb{C}^n)$.

   (b) Use the Weyl dimension formula to show that $\dim F_n^{[2,2]} = n^2(n^2 - 1)/12$ and $\dim F_n^{[1,1,1,1]} = n(n-1)(n-2)(n-3)/24$.

   (c) Use (b) to show that $\dim S^2(\bigwedge^2 \mathbb{C}^n) = \dim F_n^{[2,2]} + \dim F_n^{[1,1,1,1]}$.

5. Prove Proposition 10.3.3 by invoking Theorem 5.7.5. (HINT: Use $\bigwedge^2 \mathbb{C}^n \cong AM_n$ as a **GL**$(n, \mathbb{C})$ module.)

6. Prove Corollary 10.3.4 without using Theorem 9.3.10 by showing directly that $I - \mathbf{p}_A$ coincides with the Bianchi operator on $S^2(\bigwedge^2 \mathbb{C}^n)$.

7. Let $A, B \in \mathrm{End}(\mathbb{C}^n)$ be symmetric. Show that $A \bigcirc\!\!\!\!\wedge B$ satisfies the Bianchi identity (10.48).

8. Prove formula (10.59).

9. Show that the four-tensor (10.64) satisfies the Bianchi identity and has zero Ricci curvature.

10. Let $n \geq 5$. Use the Weyl dimension formula from Section 7.1.2 to show that $\dim V_n = \dim \mathrm{Weyl}_Q(\mathbb{C}^n)$, where V_n is the irreducible **SO**$(\mathbb{C}^n, Q)$ module with highest weight $2\varpi_2$. The cases n even and n odd need separate treatment.

11. Prove part (2) of Proposition 10.3.6 using Corollaries 10.2.8, 10.2.14, and 10.3.4.

10.4 Invariant Theory and Knot Polynomials

In this section we show how the simplest results about invariants for the orthogonal group can be used to give a representation-theoretic proof of the existence and uniqueness of the Jones polynomial in knot theory. This is carried out using an idea of D. Meyer [111] to find nontrivial solutions of the braid relations and work of A. Markov [108] and V. G. Turaev [144] that relates certain functions on the set of braids to invariants of links under orientation-preserving diffeomorphisms of the ambient space. Except for the work of Markov our exposition is self-contained (we use an almost trivial special case of Turaev's method). An exposition of the pertinent material on braids and links and their relationship with the Yang–Baxter equation can be found in Chari–Pressley [32, Chapter 15].

10.4.1 The Braid Relations

Let $\mathfrak{S}_n$ denote the symmetric group on n letters. For $i = 1, \ldots, n-1$ let $s_i \in \mathfrak{S}_n$ be the transposition $(i, i+1)$. If we write T_i for s_i then we have the relations

(B_1) $T_i T_{i+1} T_i = T_{i+1} T_i T_{i+1}$ for $i = 1, \ldots, n-2$,
(B_2) $T_i T_j = T_j T_i$ for $|i - j| > 1$, and
(B_3) $T_i^2 = I$.

One can show that if G is a group with generators T_i, $i = 1, \ldots, n-1$, and if the T_i satisfy (B_1), (B_2), and (B_3), then the correspondence $s_i \mapsto T_i$ extends in a unique manner to a homomorphism of $\mathfrak{S}_n$ onto G (see Coxeter [41] or Grove–Benson [57, Chapter 6]). We will call the pair of relations (B_1) and (B_2) the *braid relations*. We will explain the importance of the braid relations to knot theory later.

We have already seen that if V is a finite-dimensional vector space over $\mathbb{C}$, then we can define a representation ρ of $\mathfrak{S}_n$ on $\bigotimes^n V$ by

$$\rho(s)(v_1 \otimes v_2 \otimes \cdots \otimes v_n) = v_{s^{-1}(1)} \otimes v_{s^{-1}(2)} \otimes \cdots \otimes v_{s^{-1}(n)} \, .$$

This representation can be described in terms of generators as follows: Let

$$\sigma : V \otimes V \longrightarrow V \otimes V \quad with \quad \sigma(v_1 \otimes v_2) = v_2 \otimes v_1$$

be the *flip operator*. Let $n \geq 2$. Given (i, j) with $1 \leq i < j \leq n$ and $x, y \in V$, we define the (i, j) *insertion operator* $\iota_{ij}(x \otimes y) : \bigotimes^{n-2} V \longrightarrow \bigotimes^n V$ by

$$\iota_{ij}(x \otimes y)(v_1 \otimes \cdots \otimes v_{n-2})$$
$$= v_1 \otimes \cdots \otimes v_{i-1} \otimes x \otimes v_i \otimes \cdots \otimes v_{j-2} \otimes y \otimes v_{j-1} \otimes \cdots \otimes v_{n-2} \, .$$

In words, put x in the ith position and y in the jth position. Notice that ι_{ij} extends to a linear map $V \otimes V \to \mathrm{Hom}(\bigotimes^{n-2} V, \bigotimes^n V)$. Given $S \in \mathrm{End}(V \otimes V)$, we define the operator $S_{ij}^{(n)} \in \mathrm{End}(\bigotimes^n V)$ for $1 \le i < j \le n$ by

$$S_{ij}^{(n)}(v_1 \otimes \cdots \otimes v_n)$$
$$= \iota_{ij}(S(v_i \otimes v_j))(v_1 \otimes \cdots \otimes v_{i-1} \otimes v_{i+1} \otimes \cdots \otimes v_{j-1} \otimes v_{j+1} \otimes \cdots \otimes v_n) .$$

In words, extract v_i and v_j from the n-tensor, operate on the 2-tensor $v_i \otimes v_j$ with S, and then reinsert the resulting 2-tensor into positions i and j in a bilinear way.

For example, if we choose S to be the flip operator σ and set $T_i = \sigma_{i,i+1}^{(n)}$ for $i = 1, \ldots, n-1$, then it is clear from the verbal description just given that $T_i = \rho(s_i)$. When $n = 2$ we have $S_{12}^{(2)} = S$, so the first nontrivial case of this construction occurs for $n = 3$. We set $S_{ij} = S_{ij}^{(3)}$ for $1 \le i < j \le 3$.

Suppose that $S \in \mathrm{End}(V \otimes V)$. Set $T_i^{(n)}(S) = S_{i,i+1}^{(n)}$ for $i = 1, \ldots, n-1$. It is clear that the map $S \mapsto T_i^{(n)}(S)$ is linear and multiplicative. Thus we have associative algebra homomorphisms

$$T_i^{(n)} : \mathrm{End}(V \otimes V) \longrightarrow \mathrm{End}(\textstyle\bigotimes^n V) \quad \text{for} \quad i = 1, \ldots, n-1 . \tag{10.68}$$

In particular, if $S \in \mathbf{GL}(V \otimes V)$ then $T_i^{(n)}(S) \in \mathbf{GL}(\bigotimes^n V)$. From the definition we see that the operators $T_1^{(n)}(S), \ldots, T_{n-1}^{(n)}(S)$ satisfy (B_2), for any choice of S. This suggests the following questions:

(1)   When do the operators $T_1^{(n)}(S), \ldots, T_{n-1}^{(n)}(S)$ satisfy (B_1)?
(2)   Does the tensor invariant theory of the classical groups furnish operators S that yield solutions to the braid relations by this procedure?

We know one special case in which both questions have an affirmative answer, namely $S = \sigma$, the flip operator. We now use this case to answer the first question in general.

Lemma 10.4.1. *Let $S \in \mathrm{End}(V \otimes V)$ and set $R = \sigma S$. Then the set of operators $\{T_1^{(n)}(S), \ldots, T_{n-1}^{(n)}(S)\}$ satisfies the braid relations for all $n = 3, 4, \ldots$ if and only if R satisfies the* Yang–Baxter equation

$$R_{12}R_{13}R_{23} = R_{23}R_{13}R_{12} , \tag{10.69}$$

where the operators $R_{ij} = R_{ij}^{(3)}$ act on $V \otimes V \otimes V$.

To prove this result observe that the braid relations are satisfied by the operators $T_1^{(n)}(S), \ldots, T_{n-1}^{(n)}(S)$ for all $n \ge 3$ if and only if the relations are satisfied when $n = 3$. Then do the obvious calculation (which we leave as an exercise).

The Yang–Baxter equation appears in various parts of mathematics and physics, especially in statistical mechanics (lattice models) and in knot theory. The knot-theoretic relationship is basically a consequence of Lemma 10.4.1, as we shall see.

10.4.2 Orthogonal Invariants and the Yang–Baxter Equation

In this section we follow an idea of D. Meyer [111] to generate solutions to the Yang–Baxter equation using operators in $\mathrm{End}_G(V \otimes V)$. Here $V = \mathbb{C}^m$ with $m \geq 2$ and $G = \mathbf{O}(m, \mathbb{C})$ is the group leaving invariant the form $(x, y) = x_1 y_1 + \cdots + x_m y_m$ on $\mathbb{C}^m$. In this case $\mathrm{End}_G(V \otimes V)$ has as a basis $\{I, \sigma, P\}$ with σ the flip operator as in Section 10.4.1 and P the G-invariant projection onto the one-dimensional space $(V \otimes V)^G$. Indeed, these operators span $\mathrm{End}_G(V \otimes V)$ by Theorem 10.1.6, and it is easy to verify that they are linearly independent. If $e_1, \ldots, e_n$ is an orthonormal basis of V, then

$$P(v \otimes w) = \frac{1}{m}(v, w)\theta, \quad \text{where} \quad \theta = \sum_{i=1}^{m} e_i \otimes e_i .$$

Lemma 10.4.2. *Let $R = a\sigma + mbP$ with $a, b \in \mathbb{C}$. Then*

$$(R_{12}R_{13}R_{23} - R_{23}R_{13}R_{12})v_1 \otimes v_2 \otimes v_3$$
$$= b(a^2 + mab + b^2)\{(v_2, v_3)\theta \otimes v_1 - (v_1, v_2)v_3 \otimes \theta\} .$$

This is proved by a rather lengthy but straightforward calculation that we leave as an exercise.

Proposition 10.4.3. *Let $m = \dim V \geq 2$ and take $S \in \mathrm{End}_G(V \otimes V)$. Set $R = \sigma S$.*

1. *Suppose either $S = \lambda I$, or $S = \lambda \sigma$ (for some $\lambda \in \mathbb{C}^{\times}$), or $S = aI + mbP$ with $a, b \in \mathbb{C}$, $b \neq 0$, and $a^2 + mab + b^2 = 0$. Then $S \in \mathrm{End}_G(V \otimes V)$ and R satisfies the Yang–Baxter equation (10.69).*
2. *If $m \geq 3$, then the only invertible operators $S \in \mathrm{End}_G(V \otimes V)$ for which R satisfies the Yang–Baxter equation are those given in (1).*

Proof. For S as in (1), the operator R is of the form λI, $\lambda \sigma$, or $a\sigma + mbP$ with $b \neq 0$ and $a^2 + mab + b^2 = 0$ (note that $\sigma P = P$). In all three cases R satisfies the Yang–Baxter equation (the first case is trivial, the second is satisfied by Lemma 10.4.1, and the third by Lemma 10.4.2). To see that S is invertible in the third case, note that $a \neq 0$ and

$$(aI + mbP)(a^{-1}I + mb^{-1}P) = I + \frac{ma}{b}P + \frac{mb}{a}P + m^2 P$$
$$= I + \frac{m(a^2 + b^2 + mab)}{ab}P = I ,$$

since $P^2 = P$. Hence

$$S^{-1} = a^{-1}I + mb^{-1}P \tag{10.70}$$

in this case. This proves part (1) of the proposition.

We turn to part (2). We may write $S = xI + y\sigma + zP$ with $x, y, z \in \mathbb{C}$. Consider first the case $y \neq 0$. Replacing S by $y^{-1}S$ we may assume that $y = 1$. Thus $R = I + T$ with $T = a\sigma + mbP$. Set $Y = R_{12}R_{13}R_{23} - R_{23}R_{13}R_{12}$. Then

$$Y = [T_{12}, T_{13}] + [T_{12}, T_{23}] + [T_{13}, T_{23}] + T_{12}T_{13}T_{23} - T_{23}T_{13}T_{12} . \tag{10.71}$$

This is proved by observing that $R_{ij} = I + T_{ij}$ and writing out the difference. Consider $Y(e_1 \otimes e_2 \otimes e_3)$ (recall that we are assuming $\dim V \geq 3$). We have

$$[T_{12}, T_{13}](e_1 \otimes e_2 \otimes e_3) = a^2[\sigma_{12}, \sigma_{13}](e_1 \otimes e_2 \otimes e_3) ,$$

because $P_{ij}(v_1 \otimes v_2 \otimes v_3) = 0$ whenever $(v_i, v_j) = 0$ for $i \neq j$. The other commutator terms in (10.71) likewise reduce to the commutators of σ_{ij}. The formula in Lemma 10.4.2 (with R replaced by T) gives

$$(T_{12}T_{13}T_{23} - T_{23}T_{13}T_{12})e_1 \otimes e_2 \otimes e_3 = 0 .$$

Combining these results, we see that

$$Y(e_1 \otimes e_2 \otimes e_3) = a^2([\sigma_{12}, \sigma_{13}] + [\sigma_{12}, \sigma_{23}] + [\sigma_{13}, \sigma_{23}])(e_1 \otimes e_2 \otimes e_3) .$$

Evaluating the right side of this equation, we obtain

$$a^2(e_2 \otimes e_3 \otimes e_1 - e_3 \otimes e_1 \otimes e_2) ,$$

which vanishes only if $a = 0$. Suppose $a = 0$. We calculate $Y(e_1 \otimes e_2 \otimes e_2)$ as follows: We have $[T_{12}, T_{13}](e_1 \otimes e_2 \otimes e_2) = 0$, $[T_{12}, T_{23}](e_1 \otimes e_2 \otimes e_2) = b^2\theta \otimes e_1$, and

$$[T_{13}, T_{23}](e_1 \otimes e_2 \otimes e_2) = b^2 \sum_{i=1}^{m} e_i \otimes e_1 \otimes e_i .$$

Applying Lemma 10.4.2 (with R replaced by T), we obtain

$$(T_{12}T_{13}T_{23} - T_{23}T_{13}T_{12})e_1 \otimes e_2 \otimes e_2 = \sum_{i=1}^{m} e_i \otimes e_1 \otimes e_i .$$

Combining these results, we see that $Y(e_1 \otimes e_2 \otimes e_2) = b^2 u$, where

$$u = (1+b)\sum_{i=1}^{m} e_i \otimes e_i \otimes e_1 + \sum_{i=1}^{m} e_i \otimes e_1 \otimes e_i .$$

Since $u \neq 0$, we have $Y = 0$ in this case only if $b = 0$. This reduces the proof of (2) to the case $S = aI + mbP$, which is covered by Lemma 10.4.2. $\square$

10.4.3 The Braid Group

We begin this section by showing the existence of a *universal group* satisfying the braid relations. The reader who is conversant with free groups should skip to the end of this section, where we give examples and describe the relations between braid groups and knots.

If S is a set then we denote by S^m the set of all m-tuples of elements of S. If $a = (a_1, \ldots, a_p) \in S^p$ and $b = (b_1, \ldots, b_q) \in S^q$ then we will identify $(a, b) \in S^p \times S^q$ with $(a_1, \ldots, a_p, b_1, \ldots, b_q) \in S^{p+q}$. We will also identify S^1 with S.

Let $X = \{x_1, \ldots, x_n\}$ and $Y = \{y_1, \ldots, y_n\}$ be two disjoint n-element sets. Let $S = X \cup Y$ and write

$$H = \{\mathbf{1}\} \cup \bigcup_{m \geq 1} S^m \,.$$

We shall define a map $\Phi : H \to H$ inductively, as follows: We set $\Phi(\mathbf{1}) = \mathbf{1}$ and $\Phi((a)) = ((a))$ for $a \in S$. Define $\Phi((x_i, y_i)) = \mathbf{1}$ for $i = 1, \ldots, n$ and set

$$\Phi((a_1, a_2)) = (a_1, a_2) \quad \text{if } \{a_1, a_2\} \neq \{x_i, y_i\} \text{ for any } i = 1, \ldots, n \,.$$

Assume that we have defined $\Phi((a_1, \ldots, a_m))$ for $2 \leq m \leq r - 1$. Define

$$\Phi((x_i, y_i, a_3, \ldots, a_r)) = \Phi((a_3, \ldots, a_r)) \quad \text{for } i = 1, \ldots, n \,.$$

If $\{a_1, a_2\} \neq \{x_i, y_i\}$ for $i = 1, \ldots, n$ then we define

$$\Phi((a_1, \ldots, a_r)) = (a_1, \Phi((a_2, \ldots, a_r))) \,.$$

Finally, we set $F_n = \Phi(H)$.

We define multiplication on F_n as follows: $\mathbf{1} \cdot f = f \cdot \mathbf{1}$ for all $f \in F_n$, and

$$a \cdot b = \Phi((a, b)) \quad \text{for } a \in S^m \text{ and } b \in S^p \,.$$

We assert that with this multiplication F_n is a group. Given $a = (a_1, \ldots, a_m) \in F_n$, define $b = (b_m, \ldots, b_1)$ by setting $b_i = y_j$ if $a_i = x_j$ and $b_i = x_j$ if $a_i = y_j$. Then $b \in F_n$ and $a \cdot b = b \cdot a = \mathbf{1}$. Thus every element of F_n has an inverse. Given $a = (a_1, \ldots, a_p)$, $b = (b_1, \ldots, b_q)$, and $c = (c_1, \ldots, c_r)$ in F_n, we will show that

$$a \cdot (b \cdot c) = (a \cdot b) \cdot c \tag{10.72}$$

by induction on $p + q + r$. If $p + q + r = 0$ this is clear ($a = b = c = \mathbf{1}$). If one of p, q, r is 0 or 1 then it is also obvious. Now assume that (10.72) holds for $p + q + r < s$ and $p, q, r > 1$. If $p + q + r = s$ and $\{a_p, b_1\}$ and $\{b_q, c_1\}$ are not of the form $\{x_i, y_i\}$ for any i, then from the definition of Φ we see that

$$a \cdot (b \cdot c) = (a \cdot b) \cdot c = (a, (b, c)) = ((a, b), c) \,.$$

Otherwise, either we can replace a, b with $a' = (a_1, \ldots, a_{p-1})$, $b' = (b_2, \ldots, b_q)$ or we can replace b, c with $b'' = (b_1, \ldots, b_{q-1})$, $c'' = (c_2, \ldots, c_r)$. Now apply the inductive hypothesis.

Suppose G is a group. If $S \subset G$ then we say that S *generates* G if no proper subgroup of G contains S.

Lemma 10.4.4. *Suppose* $\{g_1, \ldots, g_n\}$ *generates a group* G. *Then there exists a unique surjective group homomorphism* $\psi : F_n \to G$ *such that* $\psi(x_i) = g_i$ *and* $\psi(y_i) = g_i^{-1}$.

This follows directly from the definition of the multiplication in F_n. The group F_n is called the *free group* on n generators.

If G is a group and if $S \subset G$ then we define $N(S)$ to be the intersection of all subgroups of G containing $\{gsg^{-1} : g \in G, s \in S\}$. Then $N(S)$ is a normal subgroup of G.

We are now ready to define the braid group. Let $S \subset F_{n-1}$ be the subset

$$\{x_i x_j x_i^{-1} x_j^{-1} : |i - j| > 1\} \cup \{x_i x_{i+1} x_i x_{i+1}^{-1} x_i^{-1} x_{i+1}^{-1} : i = 1, \ldots, n-2\}.$$

Define $\mathcal{B}_n = F_{n-1}/N(S)$. The group $\mathcal{B}_n$ is called the (Artin) *braid group*. Let $\tau_i = x_i N(S) \in \mathcal{B}_n$. Then $\{\tau_1, \ldots, \tau_{n-1}\}$ generates $\mathcal{B}_n$ and satisfies the braid relations (B_1) and (B_2) in Section 10.4.1.

Lemma 10.4.5. *If* G *is a group and if* $T_1, \ldots, T_{n-1} \in G$ *satisfy the braid relations* (B_1) *and* (B_2), *then there exists a unique homomorphism* $\psi : \mathcal{B}_n \to G$ *such that* $\psi(\tau_i) = T_i$ *for* $i = 1, \ldots, n-1$.

Examples

1. Let $G = \mathfrak{S}_n$ and let s_i be the transposition $(i, i+1)$ for $i = 1, \ldots, n-1$. Then there is a unique surjective homomorphism $\Psi_n : \mathcal{B}_n \to \mathfrak{S}_n$ such that $\Psi_n(\tau_i) = s_i$. The subgroup $\text{Ker}(\Psi_n)$ is called the *group of pure braids*.

2. Set $G = \mathbb{Z}$ under addition, $\alpha(\tau_i) = 1$, and $\alpha(\tau_i^{-1}) = -1$. Then α extends to a group homomorphism of $\mathcal{B}_n$ onto $\mathbb{Z}$. In particular, we see that $\mathcal{B}_n$ is an infinite group.

3. Let $m \geq 2$ and $a^2 + mab + b^2 = 0$. Let $S \mapsto T_i^{(n)}(S)$ be the map in (10.68) and set

$$\rho_{a,b,m,n}(\tau_i) = T_i^{(n)}(aI + mbP),$$

as an operator on $\bigotimes^n \mathbb{C}^m$, where the projection P is as in Section 10.4.2. Then Proposition 10.4.3 combined with Lemma 10.4.1 implies that $\rho_{a,b,m,n}$ extends to a representation of the braid group $\mathcal{B}_n$ on $\bigotimes^n \mathbb{C}^m$.

We now indicate the connection between the braid group and the topological theory of knots (see Chari and Pressley [32, Chapter 15] for precise definitions and more details). Each element of the braid group $\mathcal{B}_n$ corresponds to a unique equivalence class of *oriented braids* with n strands. Here two braids with n strands are *equivalent* if one can be deformed into the other by an orientation-preserving diffeomorphism of $\mathbb{R}^3$ that fixes the n initial points and n terminal points of the braid. A braid can be pictured by a plane projection with the orientation indicated by arrows and the over–under crossings indicated by gaps. For a braid with n strands, the initial points and final points of the strands can be labeled $1, \ldots, n$. The homomorphism $\Psi : \mathcal{B}_n \longrightarrow \mathfrak{S}_n$ associates to the braid the permutation $j \mapsto j'$, where j' is the label of the final point of the strand beginning at j. The *inverse* of the braid corresponds to the picture with top and bottom rows interchanged. For example, the elements τ_1, τ_1^{-1}, and τ_2 of $\mathcal{B}_3$ correspond to the braids T_1, T_1^{-1}, and T_2 in Figure 10.13. The

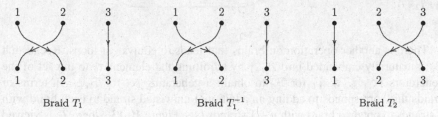

Braid T_1 Braid T_1^{-1} Braid T_2

Fig. 10.13 Some braids with three strands.

multiplication of elements of the braid group corresponds to concatenation of the braids. The braid relation $\tau_1 \tau_2 \tau_1 = \tau_2 \tau_1 \tau_2$ in the braid group is illustrated in Figure 10.14.

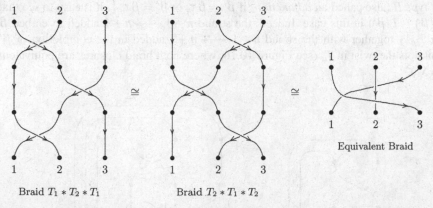

Equivalent Braid

Braid $T_1 * T_2 * T_1$ Braid $T_2 * T_1 * T_2$

Fig. 10.14 Multiplication of braids and the braid relation.

One can show that every braid β can be completed into an *oriented link* $L(\beta)$ (a union of submanifolds of $\mathbb{R}^3$ each diffeomorphic to a circle and each oriented), which is called the *closure* of the braid. For example, the braid $T_1 * T_2 * T_1$ in Figure 10.14 has the closure shown in Figure 10.15. Two links are called *equivalent* ($L_1 \sim L_2$) if one can be mapped to the other by an orientation-preserving diffeomorphism of $\mathbb{R}^3$. Inequivalent braids can give rise to equivalent links. For example, let β be a braid with n strands. If $g \in \mathcal{B}_n$, then $L(\beta) \sim L(g\beta g^{-1})$ (this is obvious geometrically: The action of g twists the n strands added to close β, and then the action of g^{-1} untwists these strands). We say that the braids β and $g\beta g^{-1}$ are related by a *Markov move of type I* (also called a *conjugation*).

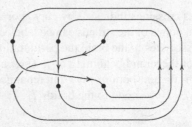

Fig. 10.15 Closure of
$T_1 * T_2 * T_1$.

There is another operation on braids that leads to equivalent loops. Recall that
$\mathcal{B}_n$ is naturally embedded into $\mathcal{B}_{n+1}$ by adjoining the element τ_n to the set of the
generators $\{\tau_1, \ldots, \tau_{n-1}\}$ for $\mathcal{B}_n$ to obtain a generating set for $\mathcal{B}_{n+1}$. In terms of
braids this corresponds to adding an additional untwisted strand to each braid with
n strands to obtain a braid with $n+1$ strands (see Figure 10.13, where T_1 is viewed
as an element of $\mathcal{B}_3$). However, the braid with the extra untwisted strand and its
completion are *not* topologically equivalent to the original braid and its completion,
but for uninteresting reasons (adding a strand introduces an unlinked copy of S^1
to the completion). The topologically relevant embeddings of $\mathcal{B}_n$ into $\mathcal{B}_{n+1}$ are as
follows: We say that a braid $\beta \in \mathcal{B}_n$ is related to a braid $\beta' \in \mathcal{B}_{n+1}$ by a *Markov move
of type II* (also called an *adjunction*) if $\beta' = \beta \tau_n$ or $\beta' = \beta \tau_n^{-1}$. It is easy to see that
$L(\beta) \sim L(\beta')$ in this case. Indeed, the strand $n+1 \longrightarrow n+1$ added to embed β
in $\mathcal{B}_{n+1}$ together with the strand $n+1 \longrightarrow n+1$ added in the completion of β'
unknots the twist in T_n (see Figure 10.16, where both braid closures are equivalent
to S^1).

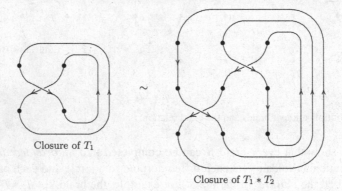

Closure of T_1

$\sim$

Closure of $T_1 * T_2$

Fig. 10.16 Markov move of type II.

The precise relation between equivalence classes of oriented links and equivalence classes of braids is the following:

(a)   Every link is obtained (up to equivalence) as the closure of a braid.
(b)   If β and β' are two braids, then $L(\beta) \sim L(\beta')$ if and only if β and β' are
      related by a sequence of Markov moves of type I and/or type II.

Statement (a) was proved by J. Alexander [2] and statement (b) by A. Markov [108] (see also Morton [113]). These results suggest that knot invariants could be constructed using representation theory, as follows. Suppose (ρ_n, V_n), for $n = 2, 3, \ldots,$ are finite-dimensional representations of $\mathcal{B}_n$ with characters

$$f_n(\beta) = \mathrm{tr}_{V_n}(\rho_n(\beta)).$$

The functions f_n are invariant under conjugation in $\mathcal{B}_n$ for each n, of course. Suppose that this family of representations satisfies the additional *adjunction condition*

$$f_n(\beta) = f_{n+1}(\beta \tau_n) \quad \text{for all } \beta \in \mathcal{B}_n$$

(where on the right side we view β as an element of $\mathcal{B}_{n+1}$). Then we can define a function φ on knots by setting

$$\varphi(L(\beta)) = f_n(\beta) \quad \text{for } \beta \in \mathcal{B}_n.$$

Results (a) and (b) imply that φ is an invariant function defined on all equivalence classes of oriented links (under orientation-preserving diffeomorphisms). In the next section we shall carry out this construction for representations obtained from those in Examples (2) and (3) above.

10.4.4 The Jones Polynomial

Let $\mathcal{B}$ be the *disjoint union* of the braid groups $\mathcal{B}_n$ for all $n \geq 1$. Note that $\mathcal{B}_n$ can be viewed as a *subgroup* of $\mathcal{B}_{n+1}$, since the map $\tau_i \mapsto \tau_i$ for $i = 1, \ldots, n-1$ induces an injective homomorphism $\mathcal{B}_n \longrightarrow \mathcal{B}_{n+1}$. Thus the expression

$$\beta = \tau_{i_1}^{\varepsilon_1} \cdots \tau_{i_l}^{\varepsilon_l}$$

with $1 \leq i_j \leq n-1$ and $\varepsilon_j \in \{1, -1\}$ for $j = 1, \ldots, l$ can be interpreted as an element of $\mathcal{B}_r$ for any $r \geq n$. To be precise as to the group to which β is considered to belong, we will write β_r or $(\tau_{i_1}^{\varepsilon_1} \cdots \tau_{i_l}^{\varepsilon_l})_r$ to indicate membership in $\mathcal{B}_r$. This notation will be important when we consider functions f on $\mathcal{B}$ for which $f((\beta)_n) \neq f((\beta)_r)$ when $r > n$.

We now introduce a family of representations of the braid groups. Define

$$b(m) = \frac{-m - \sqrt{m^2 - 4}}{2} \quad \text{for } m = 2, 3, \ldots .$$

We then have a representation $\rho_{m,n}$ of $\mathcal{B}_n$ on $\bigotimes^n \mathbb{C}^m$ given by

$$\rho_{m,n}(\beta) = \rho_{1,b(m),m,n}(\beta) \quad \text{for } \beta \in \mathcal{B}_n$$

(see Example 3 of Section 10.4.3). Given $\beta \in \mathcal{B}_n$, we set

$$p(b(m),\beta) = (-b(m))^{\alpha(\beta)} \operatorname{tr}(\rho_{m,n}(\beta)) ,$$

where $\beta \mapsto \alpha(\beta)$ is the homomorphism from $\mathcal{B}_n$ to $\mathbb{Z}$ in Example 2 of Section 10.4.3. Thus $\beta \mapsto p(b(m),\beta)$ is the character of the representation $\rho_{m,n}$ twisted by the one-dimensional representation $\beta \mapsto (-b(m))^{\alpha(\beta)}$ of $\mathcal{B}_n$ (this twist is inserted for parts (1) and (3) of the next theorem). For each $\beta \in \mathcal{B}_n$ we thus obtain a sequence $p(b(2),\beta), p(b(3),\beta),\ldots$ of real numbers (it is easy to check that the matrix of $\rho_{m,n}(\beta)$ relative to the standard basis of $\bigotimes^n \mathbb{C}^m$ is real). Part (2) of the following theorem shows that this infinite sequence of character values is determined by a sufficiently large finite subsequence.

Theorem 10.4.6. (Notation as above)

1. If $\beta \in \mathcal{B}_n$ then $p(b(m),(\beta\tau_n)_{n+1}) = p(b(m),(\beta\tau_n^{-1})_{n+1}) = p(b(m),\beta)$.
2. For each $\beta \in \mathcal{B}_n$ there exists a unique polynomial $q(\beta) \in \mathbb{Z}[b,b^{-1}]$ such that

$$p(b(m),\beta) = q(\beta)(b(m),b(m)^{-1}) \quad \text{for all } m = 2,3,\ldots. \tag{10.73}$$

One has $q((\beta\tau_n)_{n+1}) = q((\beta\tau_n^{-1})_{n+1}) = q(\beta)$ and $q(\gamma\beta\gamma^{-1}) = q(\beta)$ for all $\beta,\gamma \in \mathcal{B}_n$.
3. Let β_+, β_0, and β_- be three elements of $\mathcal{B}_n$, where β_+ has the expression

$$\tau_{i_1}^{\varepsilon_1} \tau_{i_2}^{\varepsilon_2} \cdots \tau_{i_k}^{\varepsilon_k} \cdots \tau_{i_l}^{\varepsilon_l}$$

with $\varepsilon_i \in \{1,-1\}$ and $\varepsilon_k = -1$. Assume that β_- is given by the same expression as β_+ except $\varepsilon_k = 1$, and assume that β_0 is given by the same expression as β_+ except $\varepsilon_k = 0$. Then the polynomials $q(\beta_+)$, $q(\beta_-)$, and $q(\beta_0)$ satisfy the relation

$$b^2 q(\beta_+) - b^{-2} q(\beta_-) + (b - b^{-1}) q(\beta_0) = 0 .$$

Proof. The proof of (1) uses an idea of V. G. Turaev [144]. We first recall that if V is a finite-dimensional vector space then $\bigotimes^k \operatorname{End}(V) \cong \operatorname{End}(\bigotimes^k V)$, where

$$(A_1 \otimes A_2 \otimes \cdots \otimes A_k)(v_1 \otimes \cdots \otimes v_k) = A_1 v_1 \otimes \cdots \otimes A_k v_k$$

for $A_i \in \operatorname{End}(V)$ and $v_i \in V$. Notice that this isomorphism is natural (it does not involve a choice of basis of V), and it implies the multiplicative property

$$\operatorname{tr}(A_1 \otimes \cdots \otimes A_k) = \operatorname{tr}(A_1) \cdots \operatorname{tr}(A_k)$$

of the trace.

Set $S = I_m^{\otimes 2} + mbP$ on $\mathbb{C}^m \otimes \mathbb{C}^m$, where $b = b(m)$ and I_m denotes the identity operator on $\mathbb{C}^m$. We shall need the identity

$$\operatorname{tr}((T \otimes I_m) S_{n,n+1}^{(n+1)}) = -b^{-1} \operatorname{tr}(T) \quad \text{for } T \in \operatorname{End}(\bigotimes^n \mathbb{C}^m) . \tag{10.74}$$

It suffices to verify this when $T = A_1 \otimes \cdots \otimes A_n$ with $A_i \in M_m(\mathbb{C})$. If e_{ij} are the usual elementary matrices, then

$$S = I_m \otimes I_m + b\sum_{i,j=1}^{m} e_{ij} \otimes e_{ij}.$$

Hence

$$(T \otimes I_m)S_{n,n+1}^{(n+1)} = T \otimes I + b\sum_{i,j=1}^{m} A_1 \otimes \cdots \otimes A_n e_{ij} \otimes e_{ij}.$$

Thus we have

$$
\begin{aligned}
\operatorname{tr}((T \otimes I_m)S_{n,n+1}^{(n+1)}) &= m\operatorname{tr}(T) + b\sum_{i,j=1}^{m} \operatorname{tr}(A_1)\cdots\operatorname{tr}(A_n e_{ij})\operatorname{tr}(e_{ij}) \\
&= m\operatorname{tr}(T) + b\sum_{i=1}^{m} \operatorname{tr}(A_1)\cdots\operatorname{tr}(A_n e_{ii}) \\
&= m\operatorname{tr}(T) + b\operatorname{tr}(A_1)\cdots\operatorname{tr}(A_n) = (m+b)\operatorname{tr}(T).
\end{aligned}
$$

Since $m = -b - b^{-1}$, this proves (10.74).

We now prove part (1) of the theorem. We have $\alpha(\beta\tau_n) = \alpha(\beta) + 1$. Thus

$$
\begin{aligned}
p(b,(\beta\tau_n)_{n+1}) &= (-b)^{\alpha(\beta)+1}\operatorname{tr}(\rho_{m,n+1}(\beta\tau_n)) \\
&= (-b)^{\alpha(\beta)+1}\operatorname{tr}((\rho_{m,n}(\beta) \otimes I_m)S_{n,n+1}^{(n+1)}) \\
&= (-b)^{\alpha(\beta)}\operatorname{tr}(\rho_{m,n}(\beta)) = p(b,\beta_n),
\end{aligned}
$$

where we have used (10.74) to obtain the last line. The operator $\rho_{m,n+1}(\tau_n^{-1})$ is obtained by replacing b by b^{-1} in the formula for $\rho_{m,n+1}(\tau_n)$ (see (10.70)). Since $\alpha(\beta\tau_n^{-1}) = \alpha(\beta) - 1$, the identity in (1) involving τ_n^{-1} follows from the proof of the identity involving τ_n.

We next prove that

$$b^2 p(b,\beta_+) - b^{-2}p(b,\beta_-) + (b - b^{-1})p(b,\beta_0) = 0, \qquad (10.75)$$

where $\beta_+, \beta_-,$ and β_0 are as in part (3) of the theorem and $b = b(m)$. Given j with $1 \le j \le n-1$, we have

$$\rho_{m,n}(\tau_j) = I_m^{\otimes n} + mbP_{j,j+1}^{(n)}, \qquad \rho_{m,n}(\tau_j^{-1}) = I_m^{\otimes n} + mb^{-1}P_{j,j+1}^{(n)}, \qquad (10.76)$$

by (10.70). Hence

$$b\rho_{m,n}(\tau_j^{-1}) - b^{-1}\rho_{m,n}(\tau_j) = (b - b^{-1})I_m^{\otimes n}. \qquad (10.77)$$

Now $\beta_+, \beta_-,$ and β_0 differ only in the kth tensor position, so from (10.77) with $j = i_k$ we obtain the operator identity

$$b\rho_{m,n}(\beta_+) - b^{-1}\rho_{m,n}(\beta_-) = (b - b^{-1})\rho_{m,n}(\beta_0). \qquad (10.78)$$

Likewise, we have $\alpha(\beta_+) = \alpha(\beta_0) - 1$ and $\alpha(\beta_-) = \alpha(\beta_0) + 1$. Taking the trace of (10.78) and multiplying by $(-b)^{\alpha(\beta_0)}$, we obtain (10.75).

We are left with the proof of part (2) of the theorem (which will imply the entire theorem, by the results already proved). We will prove the existence of a polynomial $q(\beta)$ satisfying (10.73). The uniqueness of $q(\beta)$ then becomes clear, since the values of $b(m)$, for $m = 2, 3, \ldots$, are all different. This implies that $q(\beta)$ has the additional properties in (2), since these properties hold for $p(b(m), \beta)$ for every m.

We prove (10.73) for $\beta \in \mathcal{B}_n$ by induction on n. If $n = 1$ then $\beta = \mathbf{1}_1$ and

$$p(b, \mathbf{1}_1) = m = -b(m) - b(m)^{-1} .$$

Assume that (2) holds for $\mathcal{B}_1, \ldots, \mathcal{B}_{n-1}$ with $n - 1 \geq 1$. We show by induction on l that the following holds:

$(\star)$ Let $\beta = \tau_{i_1}^{\varepsilon_1} \cdots \tau_{i_l}^{\varepsilon_l}$ with $1 \leq i_1, \ldots, i_l \leq n - 1$ and $\varepsilon_i \in \{1, -1\}$. Then there exists $q(\beta) \in \mathbb{Z}[b, b^{-1}]$ such that (10.73) holds.

In the proof m will be fixed, so we write $b = b(m)$. When $l = 0$ then $\beta = \mathbf{1}_n$ (with the usual convention for a product over an empty set of indices). Hence

$$p(b, \beta) = \mathrm{tr}(\rho_{m,n}(\mathbf{1}_n)) = m^n = (-b - b^{-1})^n ,$$

and $(\star)$ is true in this case. Assume that we have proved $(\star)$ for $0, 1, \ldots, l - 1$. Throughout the proof we will use the fact that $p(b, \beta)$ is unchanged if the order of the factors in β undergoes a cyclic permutation. In particular, we may assume that $i_l \geq i_j$ for all j. By (10.75) and the inductive hypothesis it suffices to consider the case $\varepsilon_i = 1$ for all i, which we now assume.

Case 1: Suppose $i_{k-1} = i_k$ for some k with $2 \leq k \leq l$. Take $\beta = \beta_-$ and let β_+ and β_0 be as in part (3) of the theorem. Then β_+ and β_0 are products of $l - 1$ generators, so by the inductive hypothesis and (10.75) we conclude that $(\star)$ holds for β.

Case 2: Assume $i_l < n - 1$. Then we take $\beta_+ = \beta \tau_{n-1}^{-1}$, $\beta_- = \beta \tau_{n-1}$ and $\beta_0 = \beta$ in (10.75). Since we may view β as an element of $\mathcal{B}_{n-1}$, part (1) of the theorem (with n replaced by $n - 1$) gives the identity

$$b^2 p(b, \beta_{n-1}) - b^{-2} p(b, \beta_{n-1}) = -(b - b^{-1}) p(b, \beta_n) .$$

Hence $p(b, \beta_n) = -(b + b^{-1}) p(b, \beta_{n-1})$ and so the inductive hypothesis prevails in this case.

Case 3: Assume $i_l = n - 1$ and that there is no other index that is $n - 1$. Then $(\star)$ follows from part (1) of the theorem (with n replaced by $n - 1$) and the inductive hypothesis.

From now on we assume that no adjacent pairs of indices i_{j-1} and i_j are equal, that $i_l = n - 1$, and that more than one index equals $n - 1$. Let $k < l$ be minimal such that $i_k = n - 1$.

Case 4: Assume $k = 1$. Then $\beta = \tau_{n-1} \beta' \tau_{n-1}$ and so by cyclic permutation we have

$$p(b, \beta) = p(b, \beta' \tau_{n-1} \tau_{n-1}) .$$

Now we are back to Case 1, so the inductive hypothesis implies $(\star)$.

We now assume that $k \geq 2$ and that $(\star)$ holds when fewer than k indices equal $n-1$.

Case 5: Assume that $k = 2$. Then $i_1 = n - 2$ (otherwise we could permute τ_{i_1} and τ_{n-2} and reduce k to 1). Hence

$$\beta = \tau_{n-2}\tau_{n-1}\beta'\tau_{n-1}$$

with β' a product of $l - 3$ generators. We replace β by the cyclically permuted element

$$\beta'\tau_{n-1}\tau_{n-2}\tau_{n-1} = \beta'\tau_{n-2}\tau_{n-1}\tau_{n-2}$$

(here we have used braid relation (B_1)). This element has one fewer index equal to $n-1$, so the inductive hypothesis applies.

Case 6: Assume that $k \geq 3$. Then we claim that with our inductive hypothesis,

$$\beta = \underbrace{\tau_{n-k}\tau_{n-k+1}\cdots\tau_{n-3}\tau_{n-2}\tau_{n-1}}_{k \text{ factors}}\cdots\tau_{n-1}. \tag{10.79}$$

Indeed, we must have $i_{k-1} = n - 2$, since otherwise we could move $\tau_{i_{k-1}}$ to the right of τ_{n-1}, contradicting the choice of k. Assume that

$$\beta = \underbrace{\tau_{i_1}\cdots\tau_{n-p}\tau_{n-p+1}\cdots\tau_{n-2}\tau_{n-1}}_{k \text{ factors}}\cdots\tau_{n-1}$$

with $1 < p < k$. Then the minimality of k, the braid relations, and the fact that no pair of consecutive indices in β are equal imply that this formula holds with p replaced by $p + 1$. Continuing in this way, we obtain (10.79). To evaluate $p(b, \beta)$, we may replace β by the cyclically permuted element

$$\underbrace{\tau_{n-k+1}\cdots\tau_{n-2}\tau_{n-1}}_{k-1 \text{ factors}}\cdots\tau_{n-1}\tau_{n-k} = \underbrace{\tau_{n-k+1}\cdots\tau_{n-2}\tau_{n-1}}_{k-1 \text{ factors}}\cdots\tau_{n-k}\tau_{n-1}.$$

Here we have used the hypothesis $k \geq 3$ to commute τ_{n-k} and τ_{n-1}. Since we have replaced k by $k - 1$, the inductive hypothesis implies that $(\star)$ holds for β. $\square$

Denote by $\mathbb{Q}(x)$ the field of rational functions in one variable with coefficients in the field $\mathbb{Q}$ of rational numbers. From the proof of Theorem 10.4.6, part (2), we obtain the following result:

Proposition 10.4.7. *Let $f, g : \mathcal{B} \to \mathbb{Q}(x)$ be such that the following conditions are satisfied when $h = f$ and $h = g$:*

1. *For all n and all $u, v \in \mathcal{B}_n$, one has $h(uvu^{-1}) = h(v)$.*
2. *If $\beta \in \mathcal{B}_n$ then $h((\beta\tau_n)_{n+1}) = h(\beta_n)$.*
3. *If β_+, β_-, and β_0 are related as in Theorem 10.4.6 (3) then*

$$x^2 h(\beta_+) - x^{-2} h(\beta_-) + (x - x^{-1}) h(\beta_0) = 0.$$

If the additional condition $f(\mathbf{1}_1) = g(\mathbf{1}_1)$ is satisfied, then $f = g$. Furthermore, if $f(\mathbf{1}_1) \in \mathbb{Z}[x, x^{-1}]$ then $f \in \mathbb{Z}[x, x^{-1}]$.

Recall from Section 10.4.3 that $L(\beta)$ is the link in $\mathbb{R}^3$ obtained by closing the braid corresponding to β. We define $q(L(\beta)) = q(\beta)$, where $q(\beta)$ is the polynomial in Theorem 10.4.6. From Theorem 10.4.6 (2) and Markov's results on equivalence of links described at the end of Section 10.4.3, we see that $L \mapsto q(L(\beta))$ is an invariant of links. We have thus shown that there exists a unique invariant q of oriented links with values in $\mathbb{Z}[b, b^{-1}]$ such that $q(\mathbf{1}_1) = -(b + b^{-1})$ and q satisfies (3) in Proposition 10.4.7 with $x = b$. By the uniqueness part of Proposition 10.4.7 we see that q is divisible by $b + b^{-1}$ in $\mathbb{Z}[b, b^{-1}]$. We make the formal change of variable $b = t^{1/2}$ in q and define

$$J(L(\beta))(t, t^{-1}) = -(t^{1/2} + t^{-1/2})^{-1} q(t^{1/2}, t^{-1/2}) \,.$$

This is an element of $\mathbb{Z}[t^{1/2}, t^{-1/2}]$ that satisfies the normalization $J(L(\mathbf{1}_1)) = 1$ and the properties (1) and (2) of Proposition 10.4.7. Relation (3) in that proposition becomes

$$tJ(L(\beta_+)) - t^{-1}J(L(\beta_-)) + (t^{1/2} - t^{-1/2})J(L(\beta_0)) = 0 \,. \tag{10.80}$$

Thus J is the *Jones polynomial* of knot theory (see Chari and Pressley [32, Theorem 15.1.1 and page 498]).

Examples

To calculate the Jones polynomial for the links $L(\tau_j^{\pm 1})$, we can take the trace in (10.76) with $n = j + 1$ to get

$$J(L(\tau_j))(t) = (-1)^{j-1}(t^{1/2} + t^{-1/2})^{j-1} \,. \tag{10.81}$$

Also, since $J(L(\tau_j^{-1}))(t) = J(L(\tau_j))(t^{-1})$, reversing orientation replaces t by t^{-1}. For example, $J(L(\tau_1))(t) = 1$ (recall that $L(\tau_1) \sim S^1$ is an unlinked circle). The same method yields the formulas

$$J(L(\tau_1^k))(t) = (-1)^{k+1} \frac{t^{(k-1)/2}[t^2 + t + 1 + (-1)^k t^{k+1}]}{t+1} \tag{10.82}$$

and $J(L(\tau_1^{-k}))(t) = J(L(\tau_1^k))(t^{-1})$ for $k = 1, 2, 3 \dots$. Notice that the numerator in (10.82) is divisible by $t + 1$ and the result is in $\mathbb{Z}[t^{1/2}]$. Thus the Jones polynomial for the link $L(\tau_1^k)$ has only positive powers of $t^{1/2}$. It follows that this link is not equivalent to the reversed link $L(\tau_1^{-k})$ for $k > 1$.

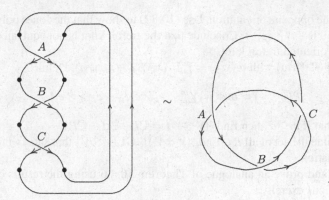

Fig. 10.17 Closure of the braid T_1^3 is a trefoil knot.

10.4.5 Exercises

1. Prove Lemma 10.4.1. (HINT: Write the action of $S \in \text{End}(V \otimes V)$ on the basis $\{e_p \otimes e_q\}$ as $S(e_p \otimes e_q) = \sum_{i,j} c_{pq}^{ij} e_i \otimes e_j$ and show that the braid relation for $T_1(S), T_2(S)$ and the Yang–Baxter equation for $R = \sigma \circ S$ both reduce to the equations

$$\sum_{i,j,k} c_{ur}^{ji} c_{pj}^{ck} c_{ki}^{ba} = \sum_{i,j,k} c_{pq}^{ji} c_{ir}^{ka} c_{jk}^{cb}$$

   for all values of the indices a, b, c and p, q, r.)
2. Prove Lemma 10.4.2.
3. Let $V = \mathbb{C}^m$ and $S = I_m^{\otimes 2} + mbP$ on $V \otimes V$ as in Section 10.4.2. Show that if $b^2 + mb + 1 = 0$ then $\det S = -b^2$. (HINT: Decompose $V \otimes V = \mathbb{C}\theta \oplus W$ with $PW = 0$.)
4. Let $V = \mathbb{C}^2$. For $\lambda \in \mathbb{C}$ let $R \in \text{End}(V \otimes V)$ be the linear transformation that has matrix

$$\begin{bmatrix} 1 & 0 & 0 & 0 \\ 0 & \lambda & 0 & 0 \\ 0 & 1-\lambda^2 & \lambda & 0 \\ 0 & 0 & 0 & 1 \end{bmatrix},$$

   relative to the ordered basis $\{e_1 \otimes e_1, e_2 \otimes e_1, e_1 \otimes e_2, e_2 \otimes e_2\}$ for $V \otimes V$. Show that R satisfies the Yang–Baxter equation.
5. Let $\beta \in \mathcal{B}_{n-1}$. Verify that $\rho_{m,n}(\beta) = \rho_{m,n-1}(\beta) \otimes I_m$ on $\bigotimes^n \mathbb{C}^m = (\bigotimes^{n-1} \mathbb{C}^m) \otimes \mathbb{C}^m$. Use this to obtain the formula $p(b, (\beta)_n) = -(b + b^{-1})p(b, \beta_{n-1})$, which establishes Case 3 in the proof of Theorem 10.4.6.
6. Let $\beta = \tau_2^{-1} \tau_1 \tau_2^{-1} \tau_1 \in \mathcal{B}_3$. Use the method of the proof of Theorem 10.4.6 to calculate $q(\beta)$.
7. Verify the formulas (10.81) and (10.82).
8. The closure of the braid T_1^3 is an oriented trefoil knot (three crossings, alternately over and under) shown in Figure 10.17. The closure of the braid T_1^{-3} is a trefoil

knot with the opposite orientation. Use (10.82) to show that the Jones polynomial of the trefoil is $-t^4 + t^3 + t$. Conclude that the trefoil knot is not equivalent to the oppositely oriented trefoil knot.

9. Let $G = \mathbf{Sp}(\mathbb{C}^{2m}, \omega)$ with $\omega(x, y) = \sum_{i=1}^{m} (x_i y_{i+m} - x_{i+m} y_i)$. Define

$$P(v \otimes w) = \omega(v, w) \sum_{i=1}^{m} (e_i \otimes e_{i+m} - e_{i+m} \otimes e_i).$$

(a) Show that if $m \geq 2$ then $\mathrm{End}_G(V \otimes V) = \mathbb{C}I \oplus \mathbb{C}\sigma \oplus \mathbb{C}P$.

(b) Determine the set of all $R \in \mathrm{End}_G(V \otimes V) \cap \mathbf{GL}(V \otimes V)$ that satisfy the Yang–Baxter equation.

10. Formulate and prove an analogue of Theorem 10.4.6 using the results obtained in the previous exercise.

10.5 Notes

Section 10.1.1. The centralizer algebras and the Brauer diagrams were introduced in Brauer [20]. These algebras were studied in more detail by Brown [24] when G is the orthogonal group. Kerov [85] uses the term *chip* for a Brauer diagram because of the analogy with an integrated circuit chip, where the dots in the top row are the input ports and those in the bottom row are the output ports. For a development of Kerov's approach, see Gavarini and Papi [54] and Gavarini [55].

Section 10.1.2. The relations in Lemma 10.1.5 can be used to define an associative algebra (the *Brauer centralizer algebra*), where the integer εn is replaced by an indeterminate x (or specialized to an arbitrary complex number) and the field $\mathbb{C}$ is replaced by the field of rational functions of x. For $x = n$ a positive integer, this algebra was shown in Brown [25] to be semisimple if and only if $k \leq n+1$, and the decomposition of the algebra into a direct sum of matrix algebras was determined by Brown [24]. The algebra was studied for general x by Hanlon and Wales [59] and Wenzl [160].

Section 10.2.1. In Weyl [163] the contraction operator C_{ij} is called the *ij trace operator*, and harmonic tensors are called *traceless*. We have adopted the term *harmonic* by extension from the terminology for symmetric and skew-symmetric tensors. The decomposition of the full tensor space into a sum of harmonic tensors of *valences* $k, k-2, \ldots$ is given (in broad outline) in Weyl [164, Chapter V, §6 and §7]; see also Brown [24]. In Goodman–Wallach [56, Chapter 10, §3] there is a detailed exposition of this result following the presentation in Benkart–Britten–Lemire [5]. When G is the symplectic group a combinatorial formula for the multiplicities in the decomposition were obtained by Sundaram [140]. For more recent work see Gavarini and Papi [54]. Replacing the defining representation by the adjoint representation of the classical group G, Hanlon [58] has studied the "stable limit" of the decomposition of the tensor algebra over $\mathfrak{g}$ as a module under $\mathrm{Ad}(G)$.

Section 10.2.2. These results are from Weyl [164, Chapter V, §5 and §7], where the introduction of the harmonic (traceless) tensors, on which the action of the Brauer algebra reduces to that of the group algebra of the symmetric group, is described as "some simple prestidigitation." For the symplectic group, this approach works easily, as we show in Section 10.2.3. However, for the orthogonal groups one needs an additional argument (not given by Weyl) that we have stated as Lemmas 10.2.10 and 10.2.13.

Section 10.2.3. When G is $\mathbf{Sp}(l, \mathbb{C})$, then the set of admissible partitions that occurs in the decomposition of the harmonic k-tensors stabilizes as soon as $l \geq k$. Indeed, the first column of a Ferrers diagram $\lambda \in \mathrm{Par}(k)$ has at most k boxes, so if $k \leq l$ then all partitions of k are G-admissible. The G-irreducible subspace of harmonic tensors in Corollary 10.2.8 is called the *Weyl module* corresponding to λ. It furnishes an explicit model for the irreducible representation of G with highest weight λ.

Section 10.2.4. When G is $\mathbf{SO}(n, \mathbb{C})$, then the set of admissible partitions that occurs in the decomposition of the harmonic k-tensors stabilizes as soon as $n \geq k$. Indeed, the first and second columns of a Ferrers diagram $\lambda \in \mathrm{Par}(k)$ have a combined total of at most k boxes, so if $k \leq n$ then all partitions of k are G-admissible. The G-irreducible spaces of harmonic tensors in Corollary 10.2.11 are called *Weyl modules*. See Section 12.2.2 for another realization of the irreducible representation of $\mathbf{SO}(n, \mathbb{C})$ whose highest weight is labeled by a Ferrers diagram with one row of k boxes.

Section 10.3. The Riemannian curvature tensor and the associated Ricci tensor were introduced by Ricci and Levi-Civita. These tensors play a central role in Einstein's theory of general relativity, which was the starting point for Weyl's research in invariant theory, semisimple Lie groups, and representation theory. The decomposition of the space of Riemannian curvature tensors was first obtained by Weyl [161] (see Hawkins [63, Chapter 11] for a detailed historical survey). The main results in this section are stated in Besse [8, Chapter 1, §G]; the product $A \wedge B$ of symmetric bilinear forms was introduced by Kulkarni [95] and Nomizu [117].

Section 10.3.3. The significance of the Weyl curvature tensor in general relativity as the *vacuum curvature* is explained by Dodson and Poston [43, Chapter XII].

Section 10.4.2. The method that has become more traditional for finding solutions to the Yang–Baxter equation is via so-called *quantum groups*. See Chari and Pressley [32] for a guide to this direction, where the term *quantum Yang–Baxter equation* is used.

Section 10.4.4. See Jones [81] for a lucid exposition by the discoverer of the Jones polynomial. See Kassel [84, Chapter X] for an alternative approach to these results.

Chapter 11
Algebraic Groups and Homogeneous Spaces

Abstract We now develop the theory of linear algebraic groups and their homogeneous spaces, as a preparation for the geometric approach to representations and invariant theory in Chapter 12.

11.1 General Properties of Linear Algebraic Groups

We begin by applying some general results from algebraic geometry to study groups and their homomorphisms. We then construct the quotient of an affine algebraic group by a normal algebraic subgroup as a linear algebraic group.

11.1.1 Algebraic Groups as Affine Varieties

Let V be a finite-dimensional complex vector space. We view $\mathbf{GL}(V)$ as the principal open set $\{g \in \mathrm{End}(V) : \det(g) \neq 0\}$ in the vector space $\mathrm{End}(V)$, and we give $\mathbf{GL}(V)$ the Zariski topology (see Sections A.1.1, A.1.2, and A.1.4).

A subgroup $G \subset \mathbf{GL}(V)$ is a *linear algebraic group* if G is a closed subset of $\mathbf{GL}(V)$, relative to the Zariski topology. To see that this agrees with the definition in Section 1.4.1, we observe that the Zariski-closed subsets of $\mathbf{GL}(V)$ are defined by equations of the form

$$f(x_{11}(g), x_{12}(g), \ldots, x_{nn}(g), \det(g)^{-1}) = 0,$$

where f is a polynomial in $n^2 + 1$ variables. Since $\det(g) \neq 0$, we can multiply this equation by $\det(g)^k$ for a suitably large k to obtain a polynomial equation in the matrix coefficients of g.

Recall from Section 1.4.3 that $\mathfrak{gl}(V) = \mathrm{End}(V)$, viewed as a Lie algebra with the bracket $[A, B] = AB - BA$. There is a locally regular representation of $\mathfrak{gl}(V)$ on

R. Goodman, N.R. Wallach, *Symmetry, Representations, and Invariants,*
Graduate Texts in Mathematics 255, DOI 10.1007/978-0-387-79852-3_11,
© Roe Goodman and Nolan R. Wallach 2009

$\mathcal{O}[\mathbf{GL}(V)]$ by left-invariant vector fields, and we have defined $\mathrm{Lie}(G)$ as the Lie subalgebra of $\mathfrak{gl}(V)$ that leaves the ideal $\mathcal{I}_G$ invariant. For $A \in \mathrm{Lie}(G)$ the corresponding left-invariant vector field X_A on $\mathbf{GL}(V)$ induces a derivation of the algebra $\mathcal{O}[G] = \mathcal{O}[\mathbf{GL}(V)]/\mathcal{I}_G$, since it leaves the ideal $\mathcal{I}_G$ invariant. We shall continue to denote this derivation by X_A when we consider its action on $\mathcal{O}[G]$. Thus X_A is a regular vector field on G, and at each point $g \in G$ it determines a tangent vector $(X_A)_g$ in the tangent space $T(G)_g$ (see Section A.3.1).

Theorem 11.1.1. *Let G be a linear algebraic group. For every $g \in G$ the map $A \mapsto (X_A)_g$ is a linear isomorphism from $\mathrm{Lie}(G)$ onto $T(G)_g$. Hence G is a smooth algebraic set and $\dim \mathrm{Lie}(G) = \dim G$.*

Proof. We first show that for fixed $g \in G$, the map $A \mapsto (X_A)_g$ is injective from $\mathrm{Lie}(G)$ to $T(G)_g$. Suppose $(X_A)_g = 0$. Then for $x \in G$ and $f \in \mathcal{O}[\mathbf{GL}(V)]$ we have

$$
\begin{aligned}
(X_A f)(x) &= (X_A f)(x g^{-1} g) = (L(g x^{-1})(X_A f))(g) \\
&= (X_A(L(g x^{-1})f))(g) = 0 \,.
\end{aligned}
$$

Here we have used the left invariance of the vector field X_A on the second line. This shows that $X_A f \in \mathcal{I}_G$ for all $f \in \mathcal{O}[\mathbf{GL}(V)]$. In particular, since $I \in G$, we must have $(X_A f)(I) = 0$ for all regular functions f on $\mathbf{GL}(V)$. Hence $A = 0$ by Lemma 1.4.7.

The dual space $\mathcal{O}[G]^*$ is naturally identified with the subspace of $\mathcal{O}[\mathbf{GL}(V)]$ consisting of the linear functionals that vanish on $\mathcal{I}_G$. In particular, each tangent vector to G at g is also a tangent vector to $\mathbf{GL}(V)$ at g. To show that the map from $\mathrm{Lie}(G)$ to $T(G)_g$ is surjective, it suffices by left invariance to take $g = I$. If $v \in T(G)_I$ then by Lemma 1.4.7 there is a unique $A \in \mathrm{End}(V)$ such that $v = v_A$. We claim that $A \in \mathrm{Lie}(G)$. Take $f \in \mathcal{I}_G$ and $g \in G$. Then

$$
(X_A f)(g) = (L(g^{-1}) X_A f)(I) = X_A(L(g^{-1})f)(I) = v(L(g^{-1})f) \,.
$$

But $L(g^{-1})f \in \mathcal{I}_G$ and v vanishes on $\mathcal{I}_G$. Hence $(X_A f)(g) = 0$. This shows that $X_A \mathcal{I}_G \subset \mathcal{I}_G$, proving that $A \in \mathrm{Lie}(G)$. $\square$

Every affine algebraic set has a unique decomposition into irreducible components (see Section A.1.5). For the case of a linear algebraic group, this decomposition can be described as follows:

Theorem 11.1.2. *Let G be a linear algebraic group. Then G contains a unique subgroup G° that is closed, irreducible, and of finite index in G. Furthermore, G° is a normal subgroup and its cosets in G are both the irreducible components and the connected components of G relative to the Zariski topology.*

Proof. We show the existence of a subgroup G° with the stated properties as follows: Let $G = X_1 \cup \cdots \cup X_r$ be an incontractible decomposition of G into irreducible components (cf. Lemma A.1.12). We label them so that $1 \in X_i$ for $1 \le i \le p$ and $1 \notin X_i$ for $p < i \le r$. We first prove that $p = 1$. Indeed, the set $X_1 \times \cdots \times X_p$ is irreducible, by Lemma A.1.14. Let

$$\mu : X_1 \times \cdots \times X_p \longrightarrow G, \qquad \mu(x_1, \ldots, x_p) = x_1 \cdots x_p.$$

Then μ is a regular map. Set $X_0 = \mu(X_1 \times \cdots \times X_p)$. The Zariski closure $\overline{X_0}$ is irreducible by Lemma A.1.15. Also, $X_i \subset X_0$ for $1 \leq i \leq p$, since $1 \in X_j$ for $1 \leq j \leq p$. Hence $X_i = X_0 = \overline{X_0}$ for $i = 1, \ldots, p$ by the irreducibility of X_i. This is possible only if $p = 1$. Thus

$$1 \in X_1 \quad \text{and} \quad 1 \notin X_i \quad \text{for } i = 2, \ldots, r. \tag{11.1}$$

Let $g \in G$. Since left multiplication by g is a homeomorphism, the decomposition $G = gG = gX_1 \cup \cdots \cup gX_r$ is also incontractible. Hence by the uniqueness of such decompositions (cf. Lemma A.1.12), there is a permutation $\sigma(g) \in \mathfrak{S}_r$ such that $gX_i = X_{\sigma(g)i}$. Clearly the map $\sigma : G \longrightarrow \mathfrak{S}_r$ is a group homomorphism.

If $\sigma(g)1 = i$, then $g = g \cdot 1 \in X_i$. Conversely, if $g \in X_i$, then $1 \in g^{-1}X_i = X_{\sigma(g^{-1})i}$. Hence $\sigma(g^{-1})i = 1$ by (11.1). This shows that

$$X_i = \{g \in G \,|\, \sigma(g)1 = i\}. \tag{11.2}$$

In particular, X_1 is a subgroup of G. For any $g \in G$, the set gX_1g^{-1} is irreducible and contains 1. Hence $gX_1g^{-1} = X_1$, so X_1 is a normal subgroup. It is now clear from (11.2) that each X_i is a coset of X_1. So setting $G^\circ = X_1$, we obtain a subgroup with the properties stated in the theorem.

To prove uniqueness, assume that G_1 is a subgroup of finite index in G that is also a closed, irreducible subset. Let x_iG_1, for $i = 1, \ldots, n$, be the cosets of G_1. Since left multiplication is a homeomorphism, each coset is closed and irreducible. But G is the disjoint union of these cosets, so the cosets must be the irreducible components of G. In particular, G_1 is the unique component of G that contains 1. The complement of G_1 in G is a finite union of components, so it is also closed. Hence G_1 is also open in G, and hence connected, since it is irreducible. $\qquad\square$

In Section 2.2.3 an algebraic group G is defined to be *connected* if the ring $\mathcal{O}[G]$ has no zero divisors. Here are two other equivalent definitions.

Corollary 11.1.3. *Let G be a linear algebraic group. The following are equivalent:*

1. *G is a connected topological space in the Zariski topology.*
2. *G is irreducible as an affine algebraic set.*
3. *The ring $\mathcal{O}[G]$ has no zero divisors.*

Proof. Apply Theorem 11.1.2 and Lemma A.1.10. $\qquad\square$

11.1.2 Subgroups and Homomorphisms

Let $G \subset \mathbf{GL}(n, \mathbb{C})$ be a linear algebraic group. An *algebraic subgroup* of G is a Zariski-closed subset $H \subset G$ that is also a subgroup. The definition of a linear algebraic group in Section 1.4.1 implies that an algebraic subgroup H of G in this sense is also a linear algebraic group as defined there. Furthermore, the inclusion map $\iota : H \subset G$ is regular and $\mathcal{O}[H] \cong \mathcal{O}[G]/\mathcal{I}_H$. Here

$$\mathcal{I}_H = \{f \in \mathcal{O}[G] : f|_H = 0\} \,.$$

In Section 1.4.1 we used this definition of $\mathcal{I}_H$ only when $G = \mathbf{GL}(n,\mathbb{C})$. Since the regular functions on G are the restrictions to G of the regular functions on $\mathbf{GL}(n,\mathbb{C})$, the definition of the ideal $\mathcal{I}_H$ is unambiguous as long as the ambient group G is understood.

Lemma 11.1.4. *Let K be a subgroup of G. Then the closure $\overline{K}$ of K in the Zariski topology is a group, and hence $\overline{K}$ is an algebraic subgroup of G. Furthermore, if K contains a nonempty open subset of $\overline{K}$ then K is closed in the Zariski topology.*

Proof. Let $x \in K$. Then $K = xK \subset x\overline{K}$. Since left multiplication by x is a homeomorphism in the Zariski topology, we know that $x\overline{K}$ is closed. Hence $\overline{K} \subset x\overline{K}$, giving $x^{-1}\overline{K} \subset \overline{K}$. Thus $K \cdot \overline{K} \subset \overline{K}$. Repeating this argument for $x \in \overline{K}$, we conclude that $\overline{K} \cdot \overline{K} \subset \overline{K}$. Since inversion is a homeomorphism, it is clear that $\overline{K}$ is stable under $x \mapsto x^{-1}$. Thus $\overline{K}$ is a subgroup.

Let $U \subset K$ be Zariski open in $\overline{K}$ and nonempty. Take $x \in U$ and set $V = x^{-1}U$. Then $1 \in V \subset K$ and V is Zariski open in $\overline{K}$. Suppose $y \in \overline{K}\backslash K$. Then $yV \subset \overline{K}\backslash K$, since $V \subset K$ and K is a group. Furthermore, yV is an open neighborhood of y in $\overline{K}$. Hence $\overline{K}\backslash K$ is open in $\overline{K}$. Hence K is Zariski closed in G. $\qquad\square$

Regular homomorphisms of algebraic groups always have the following desirable properties:

Theorem 11.1.5. *Let $\varphi : G \longrightarrow H$ be a regular homomorphism of linear algebraic groups. Then $F = \mathrm{Ker}(\varphi)$ is a closed subgroup of G and $\varphi(G)$ is a closed subgroup of H. Hence $\varphi(G)$ is an algebraic group. Furthermore, $\varphi(G^\circ) = \varphi(G)^\circ$.*

Proof. Since φ is continuous in the Zariski topology, it is clear that $\mathrm{Ker}(\varphi)$ is closed. Set $K = \overline{\varphi(G)}$. Then K is an algebraic subgroup of H. By Theorem A.2.8, $\varphi(G)$ contains a nonempty open subset of K. Hence by Lemma 11.1.4 we have $\varphi(G) = K$.

Consider the restriction of φ to G°. The image $\varphi(G^\circ)$ is closed, and hence is irreducible by Lemma A.1.15. The subgroup G° is normal in G and G/G° is finite, so $\varphi(G^\circ)$ has the same properties relative to $\varphi(G)$. Hence $\varphi(G^\circ) = \varphi(G)^\circ$ by Theorem 11.1.2. $\qquad\square$

Remark 11.1.6. See Exercises 11.1.5, #3, for an example of an analytic homomorphism of Lie groups whose range is not closed.

Corollary 11.1.7. *Let $\varphi : G \longrightarrow H$ be a regular homomorphism of linear algebraic groups. Set $K = \varphi(G)$. Let $\iota : K \longrightarrow H$ be the inclusion map and let $\psi : G \longrightarrow K$ be the homomorphism φ, viewed as having image K. Then ι is regular and injective, ψ is regular and surjective, and φ factors as $\varphi = \iota \circ \psi$.*

The last assertion of Corollary 11.1.7 is described by the following diagram:

Corollary 11.1.8. *Let* $\varphi : G \longrightarrow H$ *be a regular homomorphism of linear algebraic groups. Assume that G and H are connected in the Zariski topology and that* $d\varphi : \mathrm{Lie}(G) \longrightarrow \mathrm{Lie}(H)$ *is surjective. Then $\varphi(G) = H$. In particular, if G is a closed subgroup of H and $\dim G = \dim H$, then $G = H$.*

Proof. By Corollary 11.1.3, G and H are irreducible affine algebraic sets. Since the differential of φ maps $T(H)_I$ onto $T(G)_I$, Theorem A.3.4 implies that $\varphi(H)$ is Zariski dense in G. But $\varphi(G)$ is closed by Theorem 11.1.5. Hence $\varphi(G) = H$. The last statement follows by taking φ to be the inclusion map and using Theorem 11.1.1. $\square$

Proposition 11.1.9. *Let G and H be algebraic subgroups of $\mathbf{GL}(n, \mathbb{C})$. Then the algebraic group $G \cap H$ has Lie algebra $\mathrm{Lie}(G) \cap \mathrm{Lie}(H)$.*

Proof. Write $\mathfrak{g} = \mathrm{Lie}(G)$ and $\mathfrak{h} = \mathrm{Lie}(H)$. By Corollary 1.5.5 (1), $\mathrm{Lie}(G \cap H) \subset \mathfrak{g} \cap \mathfrak{h}$. Let $X = G \times H$ (as an affine algebraic set) and define $\varphi : X \longrightarrow \mathbf{GL}(n, \mathbb{C})$ by $\varphi(g, h) = gh^{-1}$. Then φ is a regular map (although it is not a homomorphism of algebraic groups, in general). Set $Y = \overline{\varphi(X)}$ (closure in the Zariski topology) and $F_y = \varphi^{-1}\{y\}$. Then $F_{gh^{-1}} = \{(gz, hz) : z \in G \cap H\}$, and hence $\dim F_{gh^{-1}} = \dim(G \cap H)$ for all $(g, h) \in X$. Since

$$\mathrm{Ker}\, d\varphi_{(1,1)} = \{(A, A) : A \in \mathfrak{g} \cap \mathfrak{h}\} \quad \text{and} \quad d\varphi_{(g,h)} = dL_g \circ dR_{h^{-1}} \circ d\varphi_{(1,1)},$$

we have $\dim \mathrm{Ker}\, d\varphi_{(g,h)} = \dim(\mathfrak{g} \cap \mathfrak{h})$ for all $(g, h) \in X$. Proposition A.3.6 now implies that $\dim(G \cap H) = \dim(\mathfrak{g} \cap \mathfrak{h})$; hence $\mathrm{Lie}(G \cap H) = \mathfrak{g} \cap \mathfrak{h}$. $\square$

Given a regular representation π of an algebraic group G, we will often replace it by the representation $d\pi$ of $\mathfrak{g}$, which we can analyze using techniques of linear algebra. To make this an effective method, we need to relate properties of π with those of $d\pi$. Here is one of the most important results.

Theorem 11.1.10. *Suppose G is a connected algebraic group with Lie algebra $\mathfrak{g}$. Let (π, V) be a regular representation of G. If $W \subset V$ is a linear subspace such that $d\pi(X)W \subset W$ for all $X \in \mathfrak{g}$ then $\pi(g)W \subset W$ for all $g \in G$. Hence if $(d\pi, V)$ is a completely reducible representation of $\mathfrak{g}$, then (π, V) is a completely reducible representation of G.*

Proof. We proved this result in Chapter 2 using the exponential map (Theorem 2.2.7); now we give a purely algebraic proof. Replacing G by $\pi(G)$ and using Theorem 11.1.5, we may take $G \subset \mathbf{GL}(V)$. Set $P = \{h \in \mathbf{GL}(V) : hW \subset W\}$. Then P

is an algebraic subgroup of $\mathbf{GL}(V)$ and $\mathrm{Lie}(P) = \{A \in \mathrm{End}(V) : AW \subset W\}$ (see Example 3 in Section 1.4.3). By assumption, we have $\mathrm{Lie}(G) \subset \mathrm{Lie}(P)$. Hence

$$\mathrm{Lie}(G \cap P) = \mathrm{Lie}(G) \cap \mathrm{Lie}(P) = \mathrm{Lie}(G)$$

by Proposition 11.1.9. Let H be the identity component of $G \cap P$. We just proved that $\dim G = \dim H$. Hence $G = H$ by Corollary 11.1.8. Thus $G \subset P$. $\square$

Proposition 11.1.11. *Let G be a connected linear algebraic group with Lie algebra $\mathfrak{g}$. Suppose $\sigma : G \longrightarrow \mathbf{GL}(n, \mathbb{C})$ is a regular representation and $H \subset \mathbf{GL}(n, \mathbb{C})$ is a linear algebraic subgroup with Lie algebra $\mathfrak{h}$ such that $d\sigma(\mathfrak{g}) \subset \mathfrak{h}$. Then $\sigma(G) \subset H$.*

Proof. By the Hilbert basis theorem (Theorem A.1.2) there is a finite set $f_1, \ldots, f_r$ of regular functions on $\mathbf{GL}(n, \mathbb{C})$ that generate the ideal $\mathcal{I}_H$. Let $V \subset \mathcal{I}_H$ be the subspace spanned by the right translates $R(h)f_i$ for $h \in H$ and $i = 1, \ldots, r$ and let $\rho(h) = R(h)|_V$ for $h \in H$. Then $\dim V < \infty$ and (ρ, V) is a regular representation of H by Proposition 1.4.4. Now if $A \in \mathfrak{g}$, then

$$d(\rho \circ \sigma)(A)V = d\rho(d\sigma(A)V) \subset dR(\mathfrak{h})V \subset V.$$

Hence V is invariant under $\rho(\sigma(G))$ by Theorem 11.1.10. In particular, if $g \in G$ then

$$f_i(\sigma(g)) = (\rho(\sigma(g))f_i)(I) = 0,$$

since $I \in H$ and $\rho(\sigma(g))f \in \mathcal{I}_H$. This proves that $\sigma(G) \subset H$. $\square$

11.1.3 Group Structures on Affine Varieties

In the definition of a linear algebraic group G, we have assumed that the group operations are inherited from an embedding of G in $\mathbf{GL}(n, \mathbb{C})$. We now show that we could take a more abstract point of view, as in the theory of Lie groups (see Appendix D.2.1).

Theorem 11.1.12. *Let X be an affine algebraic set. Assume that X has a group structure such that $x, y \mapsto xy$ and $x \mapsto x^{-1}$ are regular mappings. Then there exist a linear algebraic group G and a group isomorphism $\Phi : X \longrightarrow G$ such that Φ is also an isomorphism of affine algebraic sets.*

Proof. Let $m : X \times X \longrightarrow X$ be the multiplication map $m(x, y) = xy$. Since there is a natural isomorphism $\mathcal{O}[X \times X] \cong \mathcal{O}[X] \otimes \mathcal{O}[X]$, the homomorphism $m^*(f)(x, y) = f(xy)$, for $f \in \mathcal{O}[X]$, can be viewed as a map $\Delta : \mathcal{O}[X] \longrightarrow \mathcal{O}[X] \otimes \mathcal{O}[X]$. Take $f_1, \ldots, f_n$ that generate the algebra $\mathcal{O}[X]$, and write $\Delta(f_i) = \sum_{j=1}^{p} f'_{ij} \otimes f''_{ij}$. Then

$$f_i(xy) = \sum_{j=1}^{p} f'_{ij}(x) f''_{ij}(y) \quad \text{for } x, y \in X, \tag{11.3}$$

as in Proposition 1.4.4. Set $W = \mathrm{Span}\{f'_{ij} : 1 \leq i \leq n, 1 \leq j \leq p\}$ and $V = \mathrm{Span}\{R(y)f_i : y \in X, i = 1, \ldots, n\}$, where $R(y)f(x) = f(xy)$ is the right-translation representation of X on $\mathcal{O}[X]$. Then $V \subset W$ by (11.3), so $\dim V < \infty$. By definition, V is invariant under $R(y)$ for all $y \in X$.

Define $\Phi : X \longrightarrow \mathbf{GL}(V)$ by $\Phi(y) = R(y)|_V$. Then Φ is a group homomorphism. We shall show that it has the following properties:

(1) Φ is injective.

Indeed, if $R(y)f_i = f_i$ for all i, then $f_i(xy) = f_i(x)$ for all $x \in X$, and hence $y = 1$.

(2) Φ is a regular map.

The point evaluations $\{\delta_x\}_{x \in X}$ span V^*. Choose $x_i \in X$ such that $\{\delta_{x_1}, \ldots, \delta_{x_q}\}$ is a basis for V^* and let $\{g_1, \ldots, g_q\}$ be the dual basis for V. Then we can write

$$R(x)g_j = \sum_{i=1}^{q} c_{ij}(x)g_i \quad \text{for } x \in X .$$

Since $c_{ij}(x) = \langle R(x)g_j, \delta_{x_i} \rangle = g_j(x_ix)$, we see that $x \mapsto c_{ij}(x)$ is a regular function on X. This proves (2).

(3) $\Phi(X)$ is closed in $\mathbf{GL}(V)$.

This follows by (2) and the argument in Theorem 11.1.5. Set $G = \Phi(X)$. Then G is an algebraic subgroup of $\mathbf{GL}(V)$ and X is isomorphic to G as an abstract group.

(4) Φ is a biregular map from X to G.

We have $\Phi^*\mathcal{O}[G] = \mathcal{O}[X]$, since the set $\{f_1, \ldots, f_n\} \subset \Phi^*\mathcal{O}[G]$ generates $\mathcal{O}[X]$. Thus for all $f \in \mathcal{O}[X]$ there exists $h \in \mathcal{O}[G]$ with $f = h \circ \Phi$. Hence Φ^{-1} is regular. $\square$

11.1.4 Quotient Groups

Suppose G is a linear algebraic group and $H \subset G$ is an algebraic subgroup. We want to define the quotient space G/H in the context of algebraic groups and spaces. When H is a normal subgroup, then the quotient is an (abstract) group, and we shall show in this section that it has the structure of a linear algebraic group. When H is not normal we must go beyond the setting of affine algebraic spaces to treat G/H, and we defer this until later. In both cases the following representation-theoretic construction is the key tool:

Theorem 11.1.13. *Suppose G is a linear algebraic group and $H \subset G$ is an algebraic subgroup. Let $\mathfrak{h} = \mathrm{Lie}(H)$.*

1. There exist a regular representation (π, V) of G and a one-dimensional subspace $V_0 \subset V$ such that $H = \{g \in G : \pi(g)V_0 = V_0\}$ and $\mathfrak{h} = \{X \in \mathfrak{g} : d\pi(X)V_0 \subset V_0\}$.
2. If H is normal in G then there exists a regular representation (φ, W) of G such that $H = \mathrm{Ker}(\varphi)$ and $\mathfrak{h} = \mathrm{Ker}(d\varphi)$.

Proof. (1): The defining ideal $\mathfrak{I}_H \subset \mathcal{O}[G]$ for H is finitely generated, so there is a finite-dimensional right G-invariant subspace $L \subset \mathcal{O}[G]$ that contains a set of generators for $\mathfrak{I}_H$. Let $\rho(g)$ be the restriction to L of right translation by g, and let $M = L \cap \mathfrak{I}_H$. Then

$$H = \{g \in G : \rho(g)M = M\}. \tag{11.4}$$

Let $A \in \mathfrak{g}$. The left-invariant vector field X_A is a derivation of the algebra $\mathcal{O}[G]$; hence it leaves I_H invariant if and only if it leaves M invariant. Since $d\rho(A)f = X_A f$ for $f \in M$, we conclude that

$$\mathfrak{h} = \{A \in \mathfrak{g} : d\rho(A)M \subset M\}. \tag{11.5}$$

Now take (π, V) to be the dth exterior power of (ρ, L), where $d = \dim M$. Let $\{v_1, \ldots, v_d\}$ be a basis for M and define $V_0 = \mathbb{C}(v_1 \wedge \cdots \wedge v_d) \subset V$. If $g \in G$ and $\rho(g)M = M$, then

$$\pi(g)(v_1 \wedge \cdots \wedge v_d) = \rho(g)v_1 \wedge \cdots \wedge \rho(g)v_d$$

is in V_0. Conversely, if $\pi(g)(v_1 \wedge \cdots \wedge v_d) \in V_0$ then $\rho(g)M = M$ by Lemma 11.1.14 (whose statement and proof we defer to later in this section). Hence by (11.4) we see that the subgroup of G fixing V_0 is H. Likewise, if $X \in \mathfrak{g}$ and $d\rho(X)M \subset M$, then $d\pi(X)V_0 \subset V_0$. The converse also holds by Lemma 11.1.14. Hence from (11.5) we see that the subalgebra of $\mathfrak{g}$ fixing V_0 is $\mathfrak{h}$.

(2): Let (π, V) and V_0 be as in statement (1). Given a regular homomorphism (character) $\chi : H \longrightarrow \mathbb{C}^\times$, set

$$V(\chi) = \{v \in V : \pi(h)v = \chi(h)v \text{ for all } h \in H\}$$

(the χ *weight space* for the action of H). By (1) we know that there exists a character χ_0 such that $V_0 \subset V(\chi_0)$.

Since H is a normal subgroup, there is a natural action of G on the characters and weight spaces for H. Namely, if $g \in G$ and $v \in V(\chi)$, then

$$\pi(h)\pi(g)v = \pi(g)\pi(g^{-1}hg)v = \chi(g^{-1}hg)\pi(g)v.$$

Hence if we write $g \cdot \chi$ for the character $h \mapsto \chi(g^{-1}hg)$, then

$$\pi(g)V(\chi) = V(g \cdot \chi). \tag{11.6}$$

Let $U = \mathrm{Span}\, \pi(G)V(\chi_0)$. Since the weight spaces for distinct characters are linearly independent, it follows from (11.6) that the G orbit of χ_0 is a finite set $\{\chi_0, \ldots, \chi_m\}$ of distinct characters and

$$U = \bigoplus_i V(\chi_i).$$

Set $\sigma = \pi|_U$ and let $W = \mathrm{Comm}(\sigma(H))$ be the commutant of $\sigma(H)$ in $\mathrm{End}(U)$. Since the characters $\{\chi_i\}$ are distinct, we have

$$W = \bigoplus_i \text{End}(V(\chi_i))$$

by Schur's lemma. Furthermore, if $T \in \text{End}(U)$ commutes with $\sigma(H)$, then so does $\sigma(g)T\sigma(g^{-1})$ for $g \in G$, since H is normal in G. Hence W is an invariant subspace for the natural representation of G on $\text{End}(U)$. Let φ be the restriction of this representation to W. Since $h \in H$ acts by a scalar on each space $V(\chi_i)$, we have $H \subset \text{Ker}(\varphi)$. To prove the opposite inclusion, note that if $g \in \text{Ker}(\varphi)$ then $\sigma(g) \in \text{Comm}(W)$. Hence $\sigma(g)$ acts by a scalar in each subspace $V(\chi_i)$ by Theorem 4.2.1. In particular, the subspace V_0 is invariant under $\sigma(g)$, so $g \in H$ by part (1).

Since $\varphi = \text{Ad}|_{\sigma(G)} \circ \sigma$, we have $d\varphi(X)(w) = [d\sigma(X), w]$ for $X \in \mathfrak{g}$ and $w \in W$ by Proposition 1.5.4 and Theorem 1.5.7. Hence $X \in \text{Ker}(d\varphi)$ if and only if $d\sigma(X)$ commutes with all $w \in W$. If $A \in \mathfrak{h}$ then A acts by the scalar $d\chi_i(A)$ on $V(\chi_i)$, so $\mathfrak{h} \subset \text{Ker}(d\sigma)$. Conversely, we see that $\text{Ker}(d\sigma) \subset \mathfrak{h}$ using the same argument just given in the group case and the result of part (1). $\qquad\square$

The following result completes the proof of Theorem 11.1.13:

Lemma 11.1.14. *Let M be a d-dimensional subspace of $\mathbb{C}^n$. Let π be the representation of $\mathbf{GL}(n,\mathbb{C})$ on $\bigwedge^d \mathbb{C}^n$ and let $N = \bigwedge^d M$.*

1. Suppose $g \in \mathbf{GL}(n,\mathbb{C})$ and $\pi(g)N = N$. Then $g \cdot M = M$.
2. Suppose $X \in \mathfrak{gl}(n,\mathbb{C})$ and $d\pi(X)N \subset N$. Then $X \cdot M \subset M$.

Proof. (1): We may assume that $M = \text{Span}\{e_1, \ldots, e_d\}$, where $\{e_j\}$ is the standard basis for $\mathbb{C}^n$. Assume for the sake of contradiction that $g \cdot M \not\subset M$. After performing row and column reductions of g by multiplying on the left and right by block-diagonal matrices $\text{diag}[h, k]$ with $h \in \mathbf{GL}(d,\mathbb{C})$ and $k \in \mathbf{GL}(n-d,\mathbb{C})$, we may assume that g has block matrix form

$$\begin{bmatrix} A & B \\ C & D \end{bmatrix}, \quad \text{where} \quad C = \begin{bmatrix} I_r & 0 \\ 0 & 0 \end{bmatrix}.$$

Here $A \in M_d(\mathbb{C})$ is in reduced row-echelon form, $D \in M_{n-d}(\mathbb{C})$, and I_r is the $r \times r$ identity matrix with $r \geq 1$ (because we have assumed that g does not leave M invariant). Thus there are vectors $f_i \in M$ such that

$$ge_i = \begin{cases} f_i + e_{d+i} & \text{for } 1 \leq i \leq r, \\ f_i & \text{for } r+1 \leq i \leq d. \end{cases}$$

Since $e_1 \wedge \cdots \wedge e_d = \lambda g e_1 \wedge \cdots \wedge g e_d$ for some scalar λ, the set $\{f_1, \ldots, f_d\}$ is a basis for M. Hence the matrix A is invertible, and so $A = I_d$. Thus $f_i = e_i$ and we conclude that

$$ge_1 \wedge \cdots \wedge ge_d = (e_1 + e_{d+1}) \wedge \cdots \wedge (e_r + e_{d+r}) \wedge e_{r+1} \wedge \cdots \wedge e_d.$$

But the right side of this equation is not a multiple of $e_1 \wedge \cdots \wedge e_d$. This contradiction proves part (1).

(2): We proceed by contradiction; as in part (1) we may assume that X has block form

$$\begin{bmatrix} A & B \\ C & D \end{bmatrix}, \quad \text{where} \quad C = \begin{bmatrix} I_r & 0 \\ 0 & 0 \end{bmatrix}.$$

Here $r \geq 1$ because we are assuming that $X \cdot M \not\subset M$. We may replace X by $X - Y$, where $Y = \begin{bmatrix} A & 0 \\ 0 & 0 \end{bmatrix}$, since $Y \cdot M \subset M$. Hence we have $X e_i = e_{d+i}$ for $1 \leq i \leq r$ and $X e_i = 0$ for $r + 1 \leq i \leq d$. Thus

$$d\pi(X)(e_1 \wedge \cdots \wedge e_d) = e_{d+1} \wedge e_2 \wedge \cdots \wedge e_d + \cdots + e_1 \wedge e_2 \wedge \cdots \wedge e_{d+r}.$$

But the right side of this equation is not a multiple of $e_1 \wedge \cdots \wedge e_d$. This contradiction proves (2).   □

Let G be a connected algebraic group and $H \subset G$ a normal algebraic subgroup. We define an algebraic group structure on the abstract group G/H by taking a regular representation (φ, W) of G such that $\mathrm{Ker}(\varphi) = H$, whose existence is provided by Theorem 11.1.13. The group $K = \varphi(G) \subset \mathbf{GL}(W)$ is algebraic, by Theorem 11.1.5. As an abstract group, K is isomorphic to G/H by the map μ such that $\varphi = \mu \circ \pi$, where $\pi : G \longrightarrow G/H$ is the quotient map:

We define $\mathcal{O}[G/H] = \mu^* \mathcal{O}[K]$. This gives G/H the structure of an algebraic group, which a priori might depend on the choice of the representation φ. To show that this structure is unique, we establish the following regularity result for homomorphisms:

Theorem 11.1.15. *Suppose that G, K, and M are algebraic groups and G is connected. Let $\psi : G \longrightarrow K$ and $\varphi : G \longrightarrow M$ be regular homomorphisms. Assume that ψ is surjective and $\mathrm{Ker}(\psi) \subset \mathrm{Ker}(\varphi)$. Let $\mu : K \longrightarrow M$ be the map such that $\varphi = \mu \circ \psi$. Then μ is a regular homomorphism.*

Proof. Because $\mathrm{Ker}(\psi) \subset \mathrm{Ker}(\varphi)$, we can define a homomorphism μ of abstract groups satisfying the commutative diagram

Since ψ is surjective, μ is unique. There is a rational map $\rho : K \longrightarrow M$ such that $\varphi = \rho \circ \psi$ on the domain of $\rho \circ \psi$, by Theorem A.2.9. Hence $\mu = \rho$ on the domain $\mathcal{D}$ of ρ. We shall prove that $\mathcal{D}$ is translation invariant and hence is all of K.

Given $y \in G$ and $f \in \mathcal{O}[M]$, there are $f_i \in \mathcal{O}[M]$ and $c_i \in \mathbb{C}$ such that

$$f(\mu(xy)) = f(\mu(x)\mu(y)) = \sum_i c_i f_i(\mu(x))$$

for all $x \in K$. If $x_0 \in \mathcal{D}$ there exists $h \in \mathcal{O}[K]$ with $h(x_0) \neq 0$ and $(f_i \circ \mu)h \in \mathcal{O}[K]$. Set $g(x) = h(xy^{-1})$ and $g_i = (f_i \circ \mu)h$. Then the function

$$x \mapsto g(x)f(\mu(x)) = \sum_i c_i g_i(xy^{-1})$$

is regular on K, and $g(x_0y) = h(x_0) \neq 0$. Hence $x_0y \in \mathcal{D}$. Thus ρ has domain K, and so by Lemma A.2.1, $\rho = \mu$ is a regular map.   □

Corollary 11.1.16. *Assume that G and K are connected algebraic groups and that $\psi : G \longrightarrow K$ is a bijective regular homomorphism. Then ψ^{-1} is regular, and hence ψ is an isomorphism of algebraic groups.*

Proof. Take $M = G$ and φ as the identity map in Theorem 11.1.15.   □

Theorem 11.1.17. *Let G be a connected linear algebraic group and H a normal algebraic subgroup. Choose a rational representation φ of G with $\mathrm{Ker}(\varphi) = H$, and make G/H into a linear algebraic group by identifying it with $\varphi(G)$.*

1. *The linear algebraic group structure on G/H is independent of the choice of φ, and the quotient map $\pi : G \longrightarrow G/H$ is regular.*
2. *$\pi^*\mathcal{O}[G/H] = \mathcal{O}[G]^{R(H)}$ (the right H-invariant regular functions on G).*

Proof. Assertion (1) is immediate from Theorem 11.1.15 and Corollary 11.1.16. To prove (2), we see from the definition given above of $\mathcal{O}[G/H]$ as $\mu^*\mathcal{O}[K]$ that $\pi^*\mathcal{O}[G/H] \subset \mathcal{O}[G]^{R(H)}$, where R is the right-translation representation on $\mathcal{O}[G]$. For the opposite inclusion, let f be any regular function on G that is right H-invariant. From Proposition 1.4.4 we know that the linear span of the right G translates of f is a finite-dimensional space on which G acts by right translations and H acts trivially. Hence the representation of $\pi(G)$ on this space is regular, by Theorem 11.1.15. Thus the matrix entries of this representation are regular functions on G/H. Since $f(g) = (R(g)f)(1)$, the function f itself is one of these matrix entries; consequently, we conclude that $f \in \pi^*(\mathcal{O}[G/H])$.   □

Corollary 11.1.18. *Let G, H, and K be linear algebraic groups with Lie algebras $\mathfrak{g}$, $\mathfrak{h}$, and $\mathfrak{k}$, respectively. Suppose that G is connected and $H \xrightarrow{\varphi} G \xrightarrow{\psi} K$ is an exact sequence of regular homomorphisms (i.e., φ is injective, $\varphi(H) = \mathrm{Ker}(\psi)$, and ψ is surjective). Then the corresponding sequence of Lie algebra homomorphisms*

$$\mathfrak{h} \xrightarrow{d\varphi} \mathfrak{g} \xrightarrow{d\psi} \mathfrak{k}$$

is also exact: $d\varphi$ is injective, $d\varphi(\mathfrak{h}) = \mathrm{Ker}(d\psi)$, and $d\psi$ is surjective.

Proof. We know that $\varphi(H)$ is closed in G by Theorem 11.1.5; since φ is injective, we also have $H \cong \varphi(H)$ by Corollary 11.1.16, and $\text{Ker}(d\varphi) = 0$. Identifying H with $\varphi(H)$, we may assume from Theorems 11.1.13 and 11.1.17 that $\psi : G \longrightarrow \mathbf{GL}(V)$ is a regular representation, $H = \text{Ker}(\psi)$, $\mathfrak{h} = \text{Ker}(d\psi)$, and $K = \psi(G)$. Hence

$$\dim(d\psi(\mathfrak{g})) = \dim \mathfrak{g} - \dim \mathfrak{h} = \dim G - \dim H .$$

But $\dim G - \dim H = \dim K = \dim \mathfrak{k}$ by Proposition A.3.6. Thus $\dim(d\psi(\mathfrak{g})) = \dim \mathfrak{k}$, so $d\psi$ is surjective. $\qquad\qquad\qquad\qquad\qquad\qquad\qquad\qquad\qquad\qquad\qquad\qquad\qquad\square$

11.1.5 Exercises

1. Let N be the group of matrices $u(z) = \left[\begin{smallmatrix} 1 & z \\ 0 & 1 \end{smallmatrix}\right]$ with $z \in \mathbb{C}$, and let Γ be the subgroup of N consisting of the matrices with $z \in \mathbb{Z}$ an integer. Prove that Γ is Zariski dense in N.
2. Let $G = \mathbf{SL}(2, \mathbb{C})$. Show that every Zariski neighborhood of 1 in G contains unipotent elements, and hence the set of semisimple elements in G is not closed. (HINT: If $f \in \mathcal{O}[G]$ and $f(1) \neq 0$ then $f(u(z))$ is a nonvanishing polynomial in z, where $u(z)$ is the matrix in Exercise #1.)
3. Let $G = \mathbb{R}$ and $H = \mathbb{T}^1 \times \mathbb{T}^1$, where $\mathbb{T}^1 = \{z \in \mathbb{C} : |z| = 1\}$. Define $\varphi(t) = (e^{it}, e^{i\gamma t})$, where γ is an irrational real number. Show that $\varphi : G \to H$ is a Lie group homomorphism such that $\varphi(G)$ is not closed in H.
4. Suppose $\pi : G \longrightarrow H$ is a surjective regular homomorphism of algebraic groups and $\dim G = \dim H$. Prove that $\text{Ker}(\pi)$ is a finite subgroup of the center of G. (HINT: Show that the Lie algebra of $\text{Ker}(\pi)$ is zero.)
5. Let G be a connected linear algebraic group and let $\text{Ad} : G \longrightarrow \mathbf{GL}(\mathfrak{g})$ be the adjoint representation of G. Let $N = \text{Ker}(\text{Ad})$. The group G/N is called the *adjoint group* of G.
   (a) Suppose $\mathfrak{g}$ is a simple Lie algebra. Prove that N is finite.
   (b) Suppose $G = \mathbf{SL}(n, \mathbb{C})$. Find N in this case. The group G/N is denoted by $\mathbf{PSL}(n, \mathbb{C})$ (the *projective linear group*).
6. Let B be a bilinear form on $\mathbb{C}^n$. Define a multiplication $*_B$ on $\mathbb{C}^{n+1}$ by

$$[x, \lambda] *_B [y, \mu] = [x + y, \lambda + \mu + B(x, y)] \quad \text{for } x, y \in \mathbb{C}^n \text{ and } \lambda, \mu \in \mathbb{C} .$$

   (a) Show that $*_B$ defines a group structure on $\mathbb{C}^{n+1}$ with 0 as the identity element, and that multiplication and inversion are regular maps.
   (b) By Theorem 11.1.12 there is a linear algebraic group G_B with $\mathcal{O}[G_B] \cong \mathcal{O}[\mathbb{C}^{n+1}]$ as a $\mathbb{C}$-algebra and $G_B \cong (\mathbb{C}^{n+1}, *_B)$ as a group. Find an explicit matrix realization of G_B. (HINT: Let $f_0(x) = 1$ and $f_i(x) = x_i$ for $x \in \mathbb{C}^{n+1}$. Take the subspace V of $\mathcal{O}[\mathbb{C}^{n+1}]$ spanned by the functions $f_0, f_1, \ldots, f_{n+1}$ and show that it is invariant under right translations relative to the group structure $*_B$. Then show that the matrices for right translation by $y \in \mathbb{C}^{n+1}$ acting on V give an algebraic subgroup of $\mathbf{GL}(n + 2, \mathbb{C})$.)

7. Define a multiplication μ on $\mathbb{C}^\times \times \mathbb{C}$ by

$$\mu([x_1,x_2],[y_1,y_2]) = [x_1y_1, x_2 + x_1y_2] \,.$$

(a) Prove that μ satisfies the group axioms and that the inversion map is regular.
(b) Let $S = (\mathbb{C}^\times \times \mathbb{C}, \mu)$ be the linear algebraic group with regular functions $\mathbb{C}[x_1, x_1^{-1}, x_2]$ and multiplication μ. Let $R(y)f(x) = f(\mu(x,y))$ be the right-translation representation of S on $\mathcal{O}[S]$. Let $V \subset \mathcal{O}[S]$ be the space spanned by the functions x_1 and x_2. Show that V is invariant under $R(y)$, for $y \in S$.
(c) Let $\rho(y) = R(y)|_V$ for $y \in S$. Calculate the matrix of $\rho(y)$ relative to the basis $\{x_1, x_2\}$ of V. Prove that $\rho : S \longrightarrow \mathbf{GL}(2,\mathbb{C})$ is injective, and that $S \cong \rho(S)$ as an algebraic group.

11.2 Structure of Algebraic Groups

We now determine the structure of commutative linear algebraic groups. We then complete the structure theory of complex Lie algebras begun in Chapter 2 by proving the *Levi decomposition*, which splits a Lie algebra into a semisimple subalgebra and a solvable ideal. Returning to linear algebraic groups, we define the *unipotent radical* of an algebraic group and show that the reductive groups are those with trivial unipotent radical. We also prove that connected linear algebraic groups are also connected as Lie groups.

11.2.1 Commutative Algebraic Groups

We begin by extending the multiplicative Jordan decomposition of an invertible matrix to algebraic groups, as follows:

Theorem 11.2.1. *Suppose $G \subset \mathbf{GL}(V)$ is a commutative algebraic group.*

1. *The set G_s of semisimple elements and the set G_u of unipotent elements are subgroups of G.*
2. *There exists a basis for V such that the matrix $[g_{ij}]$ of $g \in G$ is upper triangular and the semisimple component g_s of g is $\mathrm{diag}[g_{11}, \ldots, g_{nn}]$.*
3. *G_s is closed in G and consists of the diagonal matrices in G relative to the basis in (2).*
4. *The map $g \mapsto (g_s, g_u)$ from G to $G_s \times G_u$ is an isomorphism of algebraic groups.*

Proof. To obtain (1), take $x, y \in G$. Since x and y commute, so do x_s and y_s, by Theorem B.1.4. Hence $x_s y_s$ is semisimple and is the semisimple factor of xy. This implies that G_s is a group. The same argument applies to G_u.

For the desired basis in (2), let $\{g_1, \ldots, g_k\}$ be a linear basis for the subalgebra of $\mathrm{End}(V)$ spanned by G. Since these elements commute, we may use the Jordan decomposition and induction on k to find a basis for V such that they are simultaneously upper triangular. Assertion (3) then follows from (2).

To prove (4), use (1) and (3) to see that G_s and G_u are algebraic groups (recall that G_u is always closed in G; see Section 1.6.3). The map $G_s \times G_u \longrightarrow G$ given by multiplication is regular and bijective. Hence it is an isomorphism by Corollary 11.1.16.   □

We now determine the structure of commutative algebraic groups whose elements are semisimple. By Theorem 11.2.1, every such group is isomorphic to an algebraic subgroup of D_n for some n, where D_n is the group of diagonal matrices in $\mathbf{GL}(n, \mathbb{C})$.

Theorem 11.2.2. *Suppose $G \subset D_n$ is a closed subgroup. Then the identity component G° is a torus. Furthermore, there is a finite subgroup $F \subset G$ such that $G^\circ \cap F = \{1\}$ and $G = G^\circ \cdot F$.*

Proof. We first prove that

(1)   the character group $\mathfrak{X}(G)$ is finitely generated and spans $\mathcal{O}[G]$.

Indeed, the characters of D_n are just the monomials $x_1^{p_1} \cdots x_n^{p_n}$ for $p_i \in \mathbb{Z}$, so $x_1, \ldots, x_n$ generate $\mathfrak{X}(D_n)$. Let φ_i be the restriction of x_i to G. Since G is an algebraic subset of D_n, the functions $\varphi_1^{p_1} \cdots \varphi_n^{p_n}$ span $\mathcal{O}[G]$, as p_i ranges over $\mathbb{Z}$. Hence any $\chi \in \mathfrak{X}(G)$ is a linear combination of these functions. But these functions are also characters of G, so by linear independence of characters (Lemma 4.1.18) we have $\chi = \varphi_1^{p_1} \cdots \varphi_n^{p_n}$ for some $p_i \in \mathbb{Z}$. Thus $\mathfrak{X}(G)$ is generated by $\varphi_1, \ldots, \varphi_n$.

Next we observe that

(2)   the group $\mathfrak{X}(G^\circ)$ has no elements of finite order.

Indeed, if $\chi \neq 1$ is a character of G°, then $\chi(G^\circ)$ is a Zariski-connected closed subgroup of $\mathbb{C}^\times$. But $\mathbb{C}^\times$ is irreducible, so $\chi(G^\circ) = \mathbb{C}^\times$. Hence χ cannot have finite order.

By (1) and (2) we see that $\mathfrak{X}(G^\circ) \cong \mathbb{Z}^r$ for some r. We already proved that the restriction map $\rho : \chi \mapsto \chi|_{G^\circ}$ from $\mathfrak{X}(D_n)$ to $\mathfrak{X}(G^\circ)$ is surjective. Since $\mathfrak{X}(D_n)$ and $\mathfrak{X}(G^\circ)$ are free abelian groups, there is a free abelian subgroup $\mathcal{A} \subset \mathfrak{X}(D_n)$ such that $\mathcal{A} \cap \mathrm{Ker}(\rho) = \{1\}$ and

$$\rho : \mathcal{A} \xrightarrow{\cong} \mathfrak{X}(G^\circ)$$

is an isomorphism (see Lang [97, Chapter I, §9]). Then we have

$$\mathfrak{X}(D_n) = \mathcal{A} \cdot \mathrm{Ker}(\rho) \quad \text{(direct product of groups)}.$$

Let $\{\chi_1, \ldots, \chi_r\}$ be a basis of $\mathcal{A}$ and $\{\chi_{r+1}, \ldots, \chi_n\}$ a basis for $\mathrm{Ker}(\rho)$ (as free abelian groups). Then the functions $\{\chi_1, \ldots, \chi_n\}$ generate the algebra $\mathcal{O}[D_n]$ by the argument above, since they generate the group $\mathfrak{X}(D_n)$. Thus the map

$$g \mapsto \Phi(g) = \mathrm{diag}[\chi_1(g), \ldots, \chi_n(g)], \quad \text{for } g \in D_n,$$

is a regular bijection of D_n onto D_n. Hence Φ is an automorphism of algebraic groups by Corollary 11.1.16. Since the functions $\rho(\chi_1), \ldots, \rho(\chi_r)$ generate the algebra $\mathcal{O}[G^\circ]$, the map Φ sends G° bijectively onto the subgroup

$$A = \{\operatorname{diag}[z_1, \ldots, z_r, \underbrace{1, \ldots, 1}_{n-r}] : z_i \in \mathbb{C}^\times\}$$

of D_n. Hence $G^\circ \cong \Phi^{-1}(A)$ is a torus. Let

$$K = \{\operatorname{diag}[\underbrace{1, \ldots, 1}_{r}, z_{r+1}, \ldots, z_n] : z_i \in \mathbb{C}^\times\}.$$

Then $D_n = A \cdot K$ (direct product of algebraic groups). Set $F = \Phi^{-1}(K) \cap G$. It is clear that $F \cap G^\circ = \{1\}$ and $G = F \cdot G^\circ$. Also, F is finite, since $F \cong G/G^\circ$. $\qquad\square$

11.2.2 Unipotent Radical

The complete reducibility (semisimplicity) of an invertible linear transformation is expressed by the triviality of the unipotent factor in its multiplicative Jordan decomposition. Our goal in this section is to obtain an analogous characterization of reductive algebraic groups. We call an algebraic group *unipotent* if all its elements are unipotent. We have the following description of representations of unipotent groups, where V^U denotes the set of vectors fixed by $\rho(U)$:

Theorem 11.2.3 (Engel). *Let G be an algebraic group and $U \subset G$ a normal subgroup consisting of unipotent elements. Suppose (ρ, V) is a regular representation of G. Then there is a G-invariant flag of subspaces*

$$V = V_1 \supset V_2 \supset \cdots \supset V_r \supset V_{r+1} = \{0\} \quad \text{with } V_i \neq V_{i+1}$$

such that $(\rho(u) - I)V_i \subset V_{i+1}$ for $i = 1, \ldots, r$ and all $u \in U$. In particular, $V_r = V^U \neq 0$. Thus if ρ is irreducible, then $V = V^U$ and hence $\rho(U) = \{I\}$.

Proof. We may assume that U is Zariski closed. Indeed, if $G \subset \mathbf{GL}(n, \mathbb{C})$ as an algebraic subgroup, then every unipotent element u in G satisfies $(u - I)^n = 0$. Hence the closure of U is a normal subgroup whose elements are all unipotent.

We proceed by induction on $\dim V$; the case $\dim V = 1$ is obviously true. It suffices to show that $V^U \neq 0$, because then we can apply the induction hypothesis to the G-module V/V^U. We may also assume that V is an irreducible U-module. Since $\rho(x)$ is a unipotent operator for $x \in U$, we have $\operatorname{tr}(\rho(x)) = \dim V$. Hence

$$\operatorname{tr}((\rho(x) - I)\rho(y)) = \operatorname{tr}(\rho(xy)) - \operatorname{tr}(\rho(y)) = 0 \quad \text{for all } x, y \in U.$$

But Corollary 4.1.7 implies that the linear span of the operators $\{\rho(y) : y \in U\}$ is $\mathrm{End}(V)$. Since the trace form is nondegenerate on $\mathrm{End}(V)$, we see that $\rho(x) - I = 0$ for all $x \in U$. $\square$

For any linear algebraic group G we set

$$\mathrm{Rad}_u(G) = \bigcup_{U \subset G} U \qquad (U \text{ normal unipotent subgroup})$$

and call $\mathrm{Rad}_u(G)$ the *unipotent radical* of G.

Lemma 11.2.4. $\mathrm{Rad}_u(G)$ *is a closed normal unipotent subgroup.*

Proof. Let $U_1, U_2 \subset G$ be normal unipotent subgroups of G. Set

$$W = U_1 \cdot U_2 = \{u_1 u_2 : u_i \in U_i\}.$$

Then W is a normal subgroup of G. We will show that the elements of W are unipotent. We may assume $G \subset \mathbf{GL}(n, \mathbb{C})$. Then by Engel's theorem (for the group U_1), there is a G-invariant flag $\{V_i\}$ in $\mathbb{C}^n$ such that

$$(u_1 - 1)V_i \subset V_{i+1} \quad \text{for all } u_1 \in U_1 .$$

Now apply Engel's theorem again (for the group U_2) to the representation of G on V_i for each i. Thus there exists a G-invariant flag $\{V_{ij}\}$ in V_i such that

$$(u_2 - 1)V_{ij} \subset V_{i,j+1} \quad \text{for all } u_2 \in U_2 .$$

Hence for $u_1 \in U_1$ and $u_2 \in U_2$ we have

$$(u_1 u_2 - 1)V_{ij} \subset u_1(u_2 - 1)V_{ij} + (u_1 - 1)V_{ij} \subset u_1 V_{i,j+1} + V_{i+1,j}$$
$$\subset V_{i,j+1} + V_{i+1,j} ;$$

thus $u_1 u_2$ is a unipotent element. This implies that $\mathrm{Rad}_u(G)$ is a normal subgroup. We see that $\mathrm{Rad}_u(G)$ is closed by the argument at the beginning of the proof of Engel's theorem. $\square$

Theorem 11.2.5. *Let G be an (abstract) group and (ρ_i, V_i) a finite-dimensional completely reducible representation of G, for $i = 1, 2$. Then $(\rho_1 \otimes \rho_2, V_1 \otimes V_2)$ is a completely reducible representation of G.*

Proof. It suffices to consider the case $\rho_1 = \rho_2$, since $\rho_1 \oplus \rho_2$ is completely reducible and $\rho_1 \otimes \rho_2$ is a subrepresentation of $(\rho_1 \oplus \rho_2) \otimes (\rho_1 \oplus \rho_2)$. Set $V = V_1 = V_2$ and $\rho = \rho_1 = \rho_2$. Let H be the Zariski closure of $\rho(G)$ in $\mathbf{GL}(V)$. Then the action H on V is completely reducible, since a subspace of V is H-invariant if and only if it is $\rho(G)$-invariant. It thus suffices to show that $V \otimes V$ is completely reducible as an H-module, for the same reason. For this, we may assume that H is connected, by Proposition 3.3.5 and Theorem 11.1.2. But for a connected group, complete reducibility for the group is equivalent to complete reducibility under the action of the Lie algebra (Theorem 11.1.10). Thus we need to prove only the following Lie algebra assertion:

$(\star)$ Suppose $\mathfrak{h} \subset \mathrm{End}(V)$ is a Lie algebra and V is completely reducible under $\mathfrak{h}$. Then $V \otimes V$ is completely reducible under $\mathfrak{h}$.

To prove $(\star)$, we use the decomposition $\mathfrak{h} = \mathfrak{l} \oplus \mathfrak{z}$ with $\mathfrak{l}$ semisimple and $\mathfrak{z}$ the center of $\mathfrak{h}$ (Corollary 2.5.9). We know that $\mathfrak{z}$ acts semisimply on V by Theorem 2.5.3. Thus $V = W_1 \oplus \cdots \oplus W_r$, with W_i irreducible under $\mathfrak{h}$ and $Z \in \mathfrak{z}$ acting by $\lambda_i(Z)I_{W_i}$ on W_i for some homomorphism $\lambda_i : \mathfrak{z} \longrightarrow \mathbb{C}$. Hence W_i is irreducible for $\mathfrak{l}$. From Theorem 3.3.12 we know that $W_i \otimes W_j$ is a completely reducible $\mathfrak{l}$-module. Since $\mathfrak{z}$ acts by $\lambda_i + \lambda_j$ on $W_i \otimes W_j$, it follows that $V \otimes V$ is a completely reducible $\mathfrak{h}$-module. $\square$

Corollary 11.2.6. *Suppose $G \subset \mathbf{GL}(n, \mathbb{C})$ is an algebraic subgroup and the action of G on $\mathbb{C}^n$ is completely reducible. Then G is a reductive group.*

Proof. The right multiplication representation of G on $M_n(\mathbb{C})$ is completely reducible. Hence by Theorem 11.2.5, the k-fold tensor product of this representation on $M_n(\mathbb{C})^{\otimes k}$ is completely reducible for all integers k. Thus the space $\mathcal{P}^k(M_n(\mathbb{C}))$ is completely reducible under the right-translation action of G for all k. Restricting polynomials on $M_n(\mathbb{C})$ to G, we conclude that every finite-dimensional right-invariant subspace of $\mathcal{O}[G]$ is completely reducible as a G-module.

Let (σ, V) be any rational G-module. For $\lambda \in V^*$ define $T_\lambda : V \longrightarrow \mathcal{O}[G]$ by $T_\lambda(v)(g) = \lambda(\sigma(g)v)$ for $v \in V$ and $g \in G$. Then $T_\lambda \circ \sigma(g) = R(g) \circ T_\lambda$, so $W_\lambda = T_\lambda V$ is a finite-dimensional G-submodule of $\mathcal{O}[G]$. Also, if $\{\lambda_1, \ldots, \lambda_n\}$ is a basis for V^* and we set $W = W_{\lambda_1} \oplus \cdots \oplus W_{\lambda_n}$, then the map $T : V \longrightarrow W$ given by $T(v) = T_{\lambda_1}(v) \oplus \cdots \oplus T_{\lambda_n}(v)$ is injective and intertwines the G actions on V and W. By hypothesis each W_{λ_i} is completely reducible under $R(G)$, so W is completely reducible. Since (σ, V) is equivalent to a subrepresentation of (R, W), it is also completely reducible. This proves that G is a reductive group. $\square$

Theorem 11.2.7. *Let G be a linear algebraic group. Then G is reductive if and only if $\mathrm{Rad}_u(G) = \{1\}$.*

Proof. We have $G \subset \mathbf{GL}(V)$ as an algebraic subgroup for some finite-dimensional vector space V. If G is reductive, there is a decomposition $V = \bigoplus_i V_i$, where V_i is an irreducible G-module. Since $\mathrm{Rad}_u(G)$ acts by 1 in every subspace V_i by Theorem 11.2.3, it follows that $\mathrm{Rad}_u(G) = \{1\}$.

Conversely, suppose G is not reductive. Then Corollary 11.2.6 implies that the representation ρ of G on V is not completely reducible. Let

$$\{0\} = V_0 \subset V_1 \subset \cdots \subset V_r = V$$

be a composition series with V_i invariant under G and $W_i = V_i / V_{i-1}$ irreducible. Let π be the representation of G on

$$W = \bigoplus_{i=1}^r W_i$$

(the *semisimplification* of ρ; see Section 4.1.7). Set $H = \pi(G)$. Then H is a closed subgroup of $\mathbf{GL}(W)$ by Theorem 11.1.5 and H is reductive by Corollary 11.2.6. Hence $U = \mathrm{Ker}(\pi) \neq 0$ and $(u-1)V_k \subset V_{k-1}$ for $k = 1, \ldots, r$ and $u \in U$. Thus U is a nontrivial unipotent normal subgroup of G, so we have $\mathrm{Rad}_u(G) \neq \{1\}$. $\square$

Corollary 11.2.8. *Let G be a linear algebraic group. Set $U = \mathrm{Rad}_u(G)$. Then G/U is reductive.*

Proof. Suppose $N/U \subset G/U$ is a unipotent normal subgroup, where N is a normal subgroup of G. The elements of N must be unipotent, by the preservation of the Jordan decomposition under homomorphisms. Hence $N = U$ and $N/U = \{1\}$. Thus G/U is reductive. $\square$

11.2.3 Connected Algebraic Groups and Lie Groups

Recall from Section 1.4.4 that an algebraic group has a unique Lie group structure such that the regular functions are smooth. The purpose of this section is to prove the following result:

Theorem 11.2.9. *If G is a connected linear algebraic group, then G is connected in the Lie group topology.*

Our proof of this theorem will also yield the unipotent generation of a class of connected algebraic groups.

Lemma 11.2.10. *Let G be a linear algebraic group and let N be a Zariski-connected normal subgroup of G. If G/N is Zariski connected, then so is G. If N and G/N are connected in the Lie group topology, then G is connected in the Lie group topology.*

Proof. Fix either of the topologies in the statement of the lemma. Let G° be the identity component of G. Then $N \subset G^\circ$. Let π be the natural projection of G onto G/N. Then $\pi(G^\circ)$ is an open subgroup of G. Hence $\pi(G^\circ)$ is closed, so $\pi(G^\circ) = G/N$. If $g \in G$ then there exists $g_0 \in G^\circ$ such that $\pi(g_0) = \pi(g)$. Hence $\pi(g_0^{-1}g) = 1$. Thus $g_0^{-1}g \in N$, and so $g \in G^\circ N = G^\circ$. $\square$

Let N be the unipotent radical of G. Then Theorem 1.6.2 implies that N is connected in the Zariski topology and in the Lie group topology. Thus to prove Theorem 11.2.9 it is enough to prove it in the case $N = \{1\}$. That is, we may assume that G is reductive, by Theorem 11.2.7 and Corollary 11.2.8.

Let Z be the center of G and let Z° be the identity component of Z in the Zariski topology. Then Z° is an algebraic torus and so is isomorphic to a product of subgroups isomorphic to $\mathbf{GL}(1, \mathbb{C})$. Hence Z° is connected in the Lie group topology. Applying Lemma 11.2.10 again we need only prove Theorem 11.2.9 in the case $Z^\circ = \{1\}$.

The rest of the proof of the theorem goes exactly as the proof of Theorem 2.2.4 once we establish the following result:

Theorem 11.2.11. *Let G be a Zariski-connected reductive linear algebraic group with finite center. Then G is generated by its unipotent elements.*

Proof. Let G' be the subgroup generated by the unipotent elements of G. Then G' is a normal subgroup. If we show that G' has a nonempty Zariski interior then it will be open and closed in the Zariski topology and hence equal to G. We may assume $G \subset \mathbf{GL}(n, \mathbb{C})$ as an algebraic subgroup. Since G is reductive, its Lie algebra $\mathfrak{g}$ acts completely reducibly on $\mathbb{C}^n$ by Theorem 11.1.10. Since the center of G is finite, the center of $\mathfrak{g}$ is 0. Thus $\mathfrak{g}$ is semisimple by Corollary 2.5.9.

Fix a Cartan subalgebra $\mathfrak{h}$ in $\mathfrak{g}$ and let Φ be the root system of $\mathfrak{g}$ with respect to $\mathfrak{h}$. Let $\Phi^+ \subset \Phi$ be a set of positive roots and $\{\alpha_1, \ldots, \alpha_l\} \subset \Phi^+$ the simple roots. Let $h_j \in \mathfrak{h}$ be the coroot to α_j. By Theorem 2.5.20 there exist $e_j \in \mathfrak{g}_{\alpha_j}$ and $f_j \in \mathfrak{g}_{-\alpha_j}$ such that $h_j = [e_j, f_j]$. Set $z_j = h_j + \mathrm{i}(e_j + f_j)$. Then a direct calculation gives

$$[e_j - f_j, z_j] = 2\mathrm{i}z_j.$$

Hence z_j is nilpotent by Lemma 2.5.1. Enumerate $\Phi = \{\alpha_1, \ldots, \alpha_r\}$ and choose $z_{l+j} \in \mathfrak{g}_{\alpha_j} \setminus \{0\}$ for $j = 1, \ldots, r$. Then $\{z_1, \ldots, z_{l+r}\}$ is a basis of $\mathfrak{g}$ consisting of nilpotent elements. Let $n = l + r = \dim G$ and define $\Psi : \mathbb{C}^n \longrightarrow G$ by

$$\Psi(x_1, \ldots, x_n) = \exp(x_1 z_1) \cdots \exp(x_n z_n) \quad \text{for } (x_1, \ldots, x_n) \in \mathbb{C}^n.$$

Then Ψ is a regular map and the image of Ψ is contained in G'. By the product rule,

$$\mathrm{d}\Psi_{(0, \ldots, 0)}(a_1, \ldots, a_n) = \sum_{k=1}^n a_k z_k.$$

Thus $\mathrm{d}\Psi_{(0, \ldots, 0)}$ is bijective. Hence the image of Ψ has nonempty interior in G by Theorem A.3.4. $\qquad\square$

11.2.4 Simply Connected Semisimple Groups

Let $\mathfrak{g}$ be a semisimple Lie algebra. In this section we will prove that there exists a linear algebraic group $\widetilde{G}$ with Lie algebra $\mathfrak{g}$ that is algebraically simply connected. We begin by constructing the *adjoint group*. The group $\mathrm{Aut}(\mathfrak{g})$ of all Lie algebra automorphisms of $\mathfrak{g}$ is a linear algebraic group. We set $G = \mathrm{Aut}(\mathfrak{g})^\circ$.

Proposition 11.2.12. *The Lie algebra of the adjoint group G is $\mathrm{Der}(\mathfrak{g})$, and the adjoint representation* $\mathrm{ad} : \mathfrak{g} \longrightarrow \mathrm{Der}(\mathfrak{g})$ *is a Lie algebra isomorphism.*

Proof. Let $g(t) = \exp(tD)$, for $t \in \mathbb{C}$, be a one-parameter subgroup of $\mathrm{Aut}(\mathfrak{g})$, where $D \in \mathrm{End}(\mathfrak{g})$. Let $X, Y \in \mathfrak{g}$. Differentiating the equation $g(t)[X, Y] = [g(t)X, g(t)Y]$ at $t = 0$, we obtain

$$D[X, Y] = [DX, Y] + [X, DY].$$

Hence $D \in \mathrm{Der}(\mathfrak{g})$. Conversely, if $D \in \mathrm{Der}(\mathfrak{g})$, then for any positive integer n we have

$$D^n[X, Y] = \sum_{k=0}^n \binom{n}{k} [D^k X, D^{n-k} Y]$$

(proof by induction on n). Using this identity in the power series for $\exp(tD)$, we see that $\exp(tD)[X,Y] = [\exp(tD)X, \exp(tD)Y]$. This proves that $\mathrm{Lie}(G) = \mathrm{Der}(\mathfrak{g})$.

The map ad is injective, since the center of $\mathfrak{g}$ is zero. It is surjective by Corollary 2.5.12. $\square$

The adjoint group is not necessarily simply connected. For example, when $\mathfrak{g} = \mathfrak{sl}(2,\mathbb{C})$, then $G \cong \mathbf{SO}(3,\mathbb{C})$. The simply connected group $\widetilde{G}$ in this case is $\mathbf{SL}(2,\mathbb{C})$, and the homomorphism $\mathrm{Ad} : \widetilde{G} \longrightarrow G$ is a twofold covering. In Chapter 6 we constructed $\widetilde{G}$ for $\mathfrak{g}$ of classical type. Now we use the theorem of the highest weight to construct $\widetilde{G}$ in general.

Fix a Cartan subalgebra $\mathfrak{h} \subset \mathfrak{g}$ and a set of positive roots $\Phi^+ \subset \Phi(\mathfrak{g},\mathfrak{h})$. Let $\Delta = \{\alpha_1,\ldots,\alpha_l\}$ $(l = \dim \mathfrak{h})$ be the simple roots in Φ^+, let $\{\varpi_1,\ldots,\varpi_l\}$ be the corresponding set of fundamental weights, and $P_{++} = \sum_{i=1}^l \mathbb{N}\,\varpi_i$ the dominant integral weights relative to Φ^+ (see Section 3.1.4).

For $\lambda \in P_{++}$ let $(\rho^\lambda, V^\lambda)$ be the finite-dimensional irreducible representation of $\mathfrak{g}$ with highest weight λ (it exists by Theorem 3.2.6). For $j = 1,\ldots,l$ let $\mathfrak{s}_j \cong \mathfrak{sl}(2,\mathbb{C})$ be the subalgebra of $\mathfrak{g}$ associated with the simple root α_j (see Section 2.5.3). Then by Corollary 2.3.8 there exist regular representations $(\pi_j^\lambda, V^\lambda)$ of $\mathbf{SL}(2,\mathbb{C})$ such that $d\pi_j^\lambda = \rho^\lambda|_{\mathfrak{s}_j}$. Define $G_j^\lambda = \pi_j^\lambda(\mathbf{SL}(2,\mathbb{C}))$. Then G_j^λ is a connected algebraic subgroup of $\mathbf{SL}(V^\lambda)$ by Theorem 11.1.5.

Lemma 11.2.13. *Let G^λ be the subgroup of $\mathbf{SL}(V^\lambda)$ generated by $G_1^\lambda,\ldots,G_l^\lambda$.*

1. *G^λ is a connected algebraic subgroup of $\mathbf{SL}(V^\lambda)$.*
2. *Assume that $\mathfrak{g}$ is simple. Then ρ^λ is an isomorphism from $\mathfrak{g}$ to $\mathrm{Lie}(G^\lambda)$.*

Proof. To prove (1), define the set $M = G_{j_1} \cdots G_{j_p}$, where $p = l \cdot l!$ and the sequence $j_1,\ldots,j_p$ is the concatenation of the $l!$ sequences $\gamma(1),\ldots,\gamma(l)$ as γ runs over all permutations of $1,\ldots,l$. Then $1 \in M$, so $M^n \subset M^{n+1}$, and we have

$$G^\lambda = \bigcup_{n \geq 1} M^n .$$

Since M is the image of the irreducible set $G_{j_1} \times \cdots \times G_{j_p}$ under the multiplication map, we know from Lemma A.1.15 and Theorem A.2.8 that $\overline{M^n}$ is irreducible and that M^n contains a nonempty open subset of $\overline{M^n}$. By the increasing chain property for irreducible sets (Theorem A.1.19), there is an index r such that $\overline{M^n} \subseteq \overline{M^r}$ for all n, and hence $\overline{G^\lambda} = \overline{M^r}$. Thus G^λ contains a nonempty open subset of $\overline{G^\lambda}$; therefore Lemma 11.1.4 implies that G^λ is closed. Since $G^\lambda = M^r$, it is also irreducible as an affine algebraic set.

To prove (2), note that $\mathrm{Ker}(\rho^\lambda) = 0$, since $\mathfrak{g}$ is simple. Define $\mathrm{Lie}(G^\lambda) = \widetilde{\mathfrak{g}} \subset \mathfrak{sl}(V^\lambda)$. Then $\rho^\lambda(\mathfrak{g}) \subseteq \widetilde{\mathfrak{g}}$, since $G_j^\lambda \subset G^\lambda$ and $\mathfrak{g}$ is generated as a Lie algebra by $\mathfrak{s}_1,\ldots,\mathfrak{s}_l$ (Theorem 2.5.24). Since $\mathfrak{g}$ acts irreducibly on V^λ, so does $\widetilde{\mathfrak{g}}$. By Schur's lemma the center of $\widetilde{\mathfrak{g}}$ acts by scalars, and hence is zero, since $\mathrm{tr}(X) = 0$ for $X \in \widetilde{\mathfrak{g}}$. Thus Corollary 2.5.9 implies that $\widetilde{\mathfrak{g}} = \rho^\lambda(\mathfrak{g}) \oplus \mathfrak{m}$ is semisimple, where $\mathfrak{m}$ is semisimple and commutes with $\rho^\lambda(\mathfrak{g})$. But this forces $\mathfrak{m} = 0$ by irreducibility of the action of $\mathfrak{g}$ on V^λ. $\square$

We now use the groups in Lemma 11.2.13 to construct the desired algebraically simply connected group $\widetilde{G}$ when $\mathfrak{g}$ is simple. For $j = 1, \ldots, l$ write $G_j = G^{\varpi_j}$, and let G be the adjoint group. Since $\mathrm{Lie}(G_j) \cong \mathfrak{g}$, there is a regular homomorphism φ_j from G_j to G arising from the adjoint representation of G_j. Define

$$\Gamma = \{(g_1, \ldots, g_l, g) : \varphi_j(g_j) = g \text{ for } j = 1, \ldots, l\} \subset G_1 \times \cdots \times G_l \times G.$$

Clearly Γ is an algebraic subgroup of $G_1 \times \cdots \times G_l \times G$. Set $\widetilde{G} = \Gamma^\circ$.

Theorem 11.2.14. *Assume that $\mathfrak{g}$ is simple. Then the group $\widetilde{G}$ is algebraically simply connected and its Lie algebra is isomorphic to $\mathfrak{g}$.*

Proof. From the definition of Γ we see that

$$\mathrm{Lie}(\widetilde{G}) = \{(X_1, \ldots, X_l, X) : \mathrm{d}\varphi_j(X_j) = X \text{ for } j = 1, \ldots, l\}$$
$$\subset \mathrm{Lie}(G_1) \oplus \cdots \oplus \mathrm{Lie}(G_l) \oplus \mathrm{Lie}(G).$$

Since $\mathrm{Lie}(G) \cong \mathfrak{g}$ and $\mathrm{d}\varphi_j$ is an isomorphism for each j, it follows that $\mathrm{Lie}(\widetilde{G}) \cong \mathfrak{g}$. Now let $\lambda = \sum_{j=1}^l n_j \varpi_j \in P_{++}$. Fix highest-weight vectors $v_j \in V^{\varpi_j}$ and set

$$v_\lambda = v_1^{\otimes n_1} \otimes \cdots \otimes v_l^{\otimes n_l} \in \left(V^{\varpi_1}\right)^{\otimes n_1} \otimes \cdots \otimes \left(V^{\varpi_l}\right)^{\otimes n_l} = L.$$

The group $G_1 \times \cdots \times G_l \times G$ acts regularly on L, with G_j acting on the jth factor in the tensor product and G acting by the identity.

Let σ be the regular representation of $\widetilde{G}$ on L arising by restriction, and identify $\mathfrak{g}$ with $\mathrm{Lie}(\widetilde{G})$. It is clear from the definition of $\widetilde{G}$ that the representation $\mathrm{d}\sigma$ of $\mathfrak{g}$ is the natural tensor-product action of $\mathfrak{g}$ on L. By Corollary 3.3.14 we can realize the $\mathfrak{g}$-module V^λ as the cyclic submodule generated by v_λ. Then V^λ is invariant under $\sigma(\widetilde{G})$ by Theorem 2.2.7, and the restriction of σ to V^λ is an irreducible representation of $\widetilde{G}$ with differential ρ^λ. Hence $\widetilde{G}$ is algebraically simply connected. $\quad\square$

Corollary 11.2.15. *Assume that $\mathfrak{g} = \mathfrak{g}_1 \oplus \cdots \oplus \mathfrak{g}_r$ is a semisimple Lie algebra, with each $\mathfrak{g}_j$ simple. Let $\widetilde{G}_j$ be the algebraically simply connected group with Lie algebra $\mathfrak{g}_j$ and set $\widetilde{G} = \widetilde{G}_1 \times \cdots \times \widetilde{G}_r$. Then $\widetilde{G}$ has Lie algebra $\mathfrak{g}$ and is algebraically simply connected.*

Remark 11.2.16. When $\mathfrak{g}$ is simple, then the adjoint group has trivial center and is the smallest connected group with Lie algebra $\mathfrak{g}$, while the algebraically simply connected group $\widetilde{G}$ is the largest such group. Any connected algebraic group with Lie algebra $\mathfrak{g}$ is the quotient $\widetilde{G}/D$, where D is a subgroup of the center Z of $\widetilde{G}$. The subgroup Z is finite and isomorphic to the quotient of the weight lattice by the root lattice; hence the order of Z is the determinant of the Cartan matrix for $\mathfrak{g}$. Thus $Z = 1$ if and only if $\mathfrak{g}$ is of type C_n, G_2, F_4 or E_8; see Exercises 3.1.5, #3, for the classical algebras and, e.g., Knapp [86, Appendix C] for the exceptional algebras. Thus the only new examples that come from Theorem 11.2.14 are the groups for the Lie algebras of types E_6 and E_7. However, the arguments we used to prove the existence of $\widetilde{G}$ are also of interest in their own right.

11.2.5 Exercises

1. Let G be a linear algebraic group. Define $H = \bigcap_{\chi \in \mathfrak{X}(G)} \operatorname{Ker} \chi$.
   (a) Prove that H is a closed normal subgroup of G.
   (b) Prove that G/H is isomorphic to a closed subgroup of D_n for some n.
   (c) Prove that $\mathfrak{X}(G) \cong \mathfrak{X}(G/H)$.
   (d) Determine H and G/H when $G = \mathbf{GL}(n, \mathbb{C})$.

11.3 Homogeneous Spaces

The homogeneous spaces for a reductive algebraic group G have a rich geometric structure. The most important examples are the *flag manifolds* G/B, where B is a *Borel subgroup* (the upper-triangular matrices in G in a suitable embedding into $\mathbf{GL}(n, \mathbb{C})$) and the *symmetric spaces* G/K, where K is the fixed-point set of an involution of G. We show how to make G/B into a projective algebraic set. We classify all the symmetric spaces for the classical groups and give explicit models for them as affine algebraic sets.

11.3.1 G-Spaces and Orbits

Let M be a quasiprojective algebraic set (see Appendix A.4.1). An *algebraic action* of a linear algebraic group G on M is a regular map $\alpha : G \times M \longrightarrow M$, written as $(g, m) \mapsto g \cdot m$, such that

$$g \cdot (h \cdot m) = (gh) \cdot m, \quad 1 \cdot m = m,$$

for all $g, h \in G$ and $m \in M$. (Recall from Section A.4.2 that $G \times M$ is a quasiprojective algebraic set.) In general, when the action of G on M is clear from the context, we will often write gm for $g \cdot m$, and we will usually omit the adjective *algebraic*.

Theorem 11.3.1. *For every $x \in M$, the stabilizer G_x of x is an algebraic subgroup of G and the orbit $G \cdot x$ is a smooth quasiprojective subset of M.*

Proof. By the regularity of the map $g \mapsto g \cdot x$, we know that G_x is closed and $G \cdot x$ contains an open subset of its closure (cf. Theorem A.2.8 and the remarks at the end of Section A.4.3). By homogeneity the orbit is thus open in its closure; hence it is quasiprojective. The set of simple points in an orbit is nonempty and invariant under G, so every point in the orbit must be simple. $\square$

Here is one of the most useful consequences of Theorem 11.3.1.

Corollary 11.3.2. *There exists a point $x \in M$ such that $G \cdot x$ is closed in M.*

Proof. Let $y \in M$ and let Y be the closure of $G \cdot y$. Then $G \cdot y$ is open in Y, by the argument in the proof of Theorem 11.3.1, and hence $Z = Y - G \cdot y$ is closed in Y Thus Z is quasiprojective. Furthermore, $\dim Z < \dim Y$ by Theorem A.1.19, and Z is a union of orbits. This implies that an orbit of minimal dimension is closed. $\square$

Here is a converse to Theorem 11.3.1.

Theorem 11.3.3. *Let H be a closed subgroup of a linear algebraic group G.*

1. *There exist a regular action of G on $\mathbb{P}^n$ and a point $x_0 \in \mathbb{P}(V)$ such that H is the stabilizer of x_0. The map $g \mapsto g \cdot x_0$ is a bijection from the coset space G/H to the orbit $G \cdot x_0$. This map endows the set G/H with a structure of a quasi-algebraic variety that (up to regular isomorphism) is independent of the choices made.*
2. *The quotient map from G to G/H is regular.*
3. *If G acts algebraically on a quasiprojective algebraic set M and x is a point of M such that $H \subset G_x$, then the map $gH \mapsto g \cdot x$ from G/H to the orbit $G \cdot x$ is regular.*

Proof. The first assertion in (1) follows from Theorem 11.1.13. The independence of choices and the proofs of (2) and (3) follow by arguments similar to the proof of Theorem 11.1.15, taking into account the validity of Theorem A.2.9 for projective algebraic sets (cf. the remarks at the end of Section A.4.3). $\square$

11.3.2 Flag Manifolds

Let V be a finite-dimensional complex vector space, and let $\bigwedge^p V$ be the pth exterior power of V. We call an element of this space a *p-vector*. Given a p-vector u, we define a linear map

$$T_u : V \longrightarrow \bigwedge^{p+1} V$$

by $T_u v = u \wedge v$ for $v \in V$. Set

$$V(u) = \{v \in V : u \wedge v = 0\} = \mathrm{Ker}(T_u)$$

(the *annihilator* of u in V). The nonzero p-vectors of the form $v_1 \wedge \cdots \wedge v_p$, with $v_i \in V$, are called *decomposable*.

Lemma 11.3.4. *Let $\dim V = n$.*

1. *Let $0 \neq u \in \bigwedge^p V$. Then $\dim V(u) \leq p$ and $\mathrm{Rank}(T_u) \geq n - p$. Furthermore, $\mathrm{Rank}(T_u) = n - p$ if and only if u is decomposable.*
2. *Suppose $u = v_1 \wedge \cdots \wedge v_p$ is decomposable. Then $V(u) = \mathrm{Span}\{v_1, \ldots, v_p\}$. Furthermore, if $V(u) = V(w)$ then $w = cu$ for some $c \in \mathbb{C}^\times$. Hence the subspace $V(u) \subset V$ determines the point $[u] \in \mathbb{P}(\bigwedge^p V)$.*
3. *Let $0 < p < l < n$. Suppose $0 \neq u \in \bigwedge^p V$ and $0 \neq w \in \bigwedge^l V$ are decomposable. Then $V(u) \subset V(w)$ if and only if $\mathrm{Rank}(T_u \oplus T_w)$ is a minimum, namely $n - p$.*

Proof. Let $\{v_1, \ldots, \overset{.}{v}_m\}$ be a basis for $V(u)$. We complete it to a basis for V, and we write

$$u = \sum_J c_J v_J \, ,$$

where $v_J = v_{j_1} \wedge \cdots \wedge v_{j_p}$ for J a p-tuple with $j_1 < \cdots < j_p$ and $c_J \in \mathbb{C}$. When $1 \leq j \leq m$ we have

$$0 = u \wedge v_j = \sum_J c_J v_J \wedge v_j \, .$$

But $v_J \wedge v_j = 0$ if j occurs in J, whereas for fixed j the set

$$\{v_J \wedge v_j \, : \, j \notin J, |J| = p\}$$

is linearly independent. Hence $c_J \neq 0$ implies that J includes all the indices $j = 1, 2, \ldots, m$. In particular, $m \leq p$. If $m = p$ then $c_J \neq 0$ implies that $J = (1, \ldots, p)$, and hence $u = c v_1 \wedge \cdots \wedge v_p$. This proves parts (1) and (2). Part (3) then follows from the fact that $V(u) \subset V(w)$ if and only if $\mathrm{Ker}(T_u \oplus T_w) = \mathrm{Ker}(T_u)$. □

Denote the set of all p-dimensional subspaces of V by $\mathrm{Grass}_p(V)$ (the pth *Grassmannian* manifold). Using part (2) of Lemma 11.3.4, we identify $\mathrm{Grass}_p(V)$ with the subset of the projective space $\mathbb{P}(\bigwedge^k V)$ corresponding to the decomposable p-vectors.

Proposition 11.3.5. $\mathrm{Grass}_p(V)$ *is an irreducible projective algebraic set.*

Proof. We use the notation of Lemma 11.3.4. If u is a p-vector, then $\mathrm{Rank}(T_u) = n - \dim V(u) \geq n - p$. Hence the p-vectors $u \neq 0$ with $\dim V(u) = p$ are those for which all minors of size $n - p + 1$ in T_u vanish. These minors are homogeneous polynomials in the components of u (relative to a fixed basis $\{e_1, \ldots, e_n\}$ for V), so we see that the set of decomposable p-vectors is the zero set of a family of homogeneous polynomials. Hence $\mathrm{Grass}_p(V)$ is a closed subset of $\mathbb{P}(\bigwedge^p V)$. We map

$$\mathbf{GL}(V) \longrightarrow \mathrm{Grass}_p(V) \quad \text{by} \quad g \mapsto [ge_1 \wedge \cdots \wedge ge_p] \, .$$

This is clearly a regular surjective mapping, so the irreducibility of $\mathbf{GL}(V)$ implies that $\mathrm{Grass}_p(V)$ is also irreducible. □

Take $V = \mathbb{C}^n$ and let $X \subset M_{n,p}$ be the open subset of $n \times p$ matrices of maximal rank p. The p-dimensional subspaces of V then correspond to the column spaces of matrices $x \in X$. Since $x, y \in X$ have the same column space if and only if $x = yg$ for some $g \in \mathbf{GL}(k, \mathbb{C})$, we may view $\mathrm{Grass}_p(V)$ as the space of orbits of $\mathbf{GL}(p, \mathbb{C})$ on X. That is, we introduce the equivalence relation $x \sim y$ if $x = yg$; then $\mathrm{Grass}_p(V)$ is the set of equivalence classes.

For $p = 1$ this is the usual model of $\mathrm{Grass}_1(\mathbb{C}^n) = \mathbb{P}^{n-1}$ (see Section A.4.1). For any p it leads to a covering of $\mathrm{Grass}_p(V)$ by affine coordinate patches, just as in the case of projective space, as follows: For $J = (i_1, \ldots, i_p)$ with $1 \leq i_1 < \cdots < i_p \leq n$, let

$$\xi_J(x) = \det \begin{bmatrix} x_{i_1 1} & \cdots & x_{i_1 p} \\ \vdots & \ddots & \vdots \\ x_{i_p 1} & \cdots & x_{i_p p} \end{bmatrix}$$

be the minor determinant formed from rows $i_1, \ldots, i_p$ of $x \in M_{n,p}$. Set

$$X_J = \{x \in M_{n,p} : \xi_J(x) \neq 0\}.$$

As J ranges over all increasing p-tuples the sets X_J cover X. The homogeneous polynomials ξ_J are the so-called *Plücker coordinates* on X (the restriction to X of the homogeneous linear coordinates on $\bigwedge^p \mathbb{C}^n$ relative to the standard basis). Under right multiplication they transform by $\xi_J(xg) = \xi_J(x) \det g$ for $g \in \mathbf{GL}(p, \mathbb{C})$; thus the ratios of the Plücker coordinates are rational functions on $\text{Grass}_p(V)$.

Every matrix in X_J is equivalent (under the right $\mathbf{GL}(p, \mathbb{C})$ action) to a matrix in the affine-linear subspace

$$A_J = \{x \in M_{n,p} : x_{i_r s} = \delta_{rs} \text{ for } r, s = 1, \ldots, p\}.$$

Clearly, if $x, y \in A_J$ and $x \sim y$ then $x = y$. Furthermore, $\xi_J = 1$ on A_J and the $p(n - p)$ matrix coordinates $\{x_{rs} : r \notin J\}$ are the restrictions to A_J of certain Plücker coordinates. For example, let $J = (1, 2, \ldots, p)$. Then $x \in A_J$ is of the form

$$x = \begin{bmatrix} 1 & \cdots & 0 \\ \vdots & \ddots & \vdots \\ 0 & \cdots & 1 \\ x_{p+1,1} & \cdots & x_{p+1,p} \\ \vdots & \ddots & \vdots \\ x_{n1} & \cdots & x_{np} \end{bmatrix}.$$

Given $1 \leq s \leq p$ and $p < r \leq n$, we set $L = (1, \ldots, \widehat{s}, \ldots, p, r)$ (omit s). Then $\xi_L(x) = \pm x_{rs}$ for $x \in A_J$, as we see by column interchanges. In particular,

$$\dim \text{Grass}_p(\mathbb{C}^n) = (n - p)p.$$

Suppose that ω is a bilinear form on V (either symmetric or skew-symmetric). Recall that a subspace $W \subset V$ is *isotropic* relative to ω if $\omega(x, y) = 0$ for all $x, y \in W$. The *quadric Grassmannian* $\mathfrak{I}_p(V)$ is the subset of $\text{Grass}_p(V)$ consisting of all isotropic subspaces. We claim that $\mathfrak{I}_p(V)$ is closed in $\text{Grass}_p(V)$ and hence is a projective algebraic set. To see this, identify V with $\mathbb{C}^n$ by choosing some basis, and let the form ω be represented by the matrix Γ relative to this basis. Then the range of $x \in M_{n,p}$ is ω-isotropic if and only if $x^t \Gamma x = 0$. On each affine chart A_J this equation is quadratic in the matrix coordinates of x. We already observed that these matrix coordinates are the restrictions to A_J of homogeneous coordinates on $\mathbb{P}(\bigwedge^p \mathbb{C}^n)$. Hence $\mathfrak{I}_p(V)$ is closed in $\mathbb{P}(\bigwedge^p \mathbb{C}^n)$ by Lemma A.4.1.

We can now define the *flag manifolds*. Let $0 < p_1 < \cdots < p_k < \dim V$ be integers, and set $\mathbf{p} = (p_1, \ldots, p_k)$. Let $\text{Flag}_{\mathbf{p}}(V)$ consist of all nested chains $V_1 \subset \cdots \subset V_k \subset V$ of subspaces with $\dim V_i = p_i$. We can view $\text{Flag}_{\mathbf{p}}(V)$ as a subset of the product algebraic set

$$\text{Grass}_{\mathbf{p}}(V) = \text{Grass}_{p_1}(V) \times \cdots \times \text{Grass}_{p_k}(V).$$

It follows from Section A.4.2 that $\text{Grass}_{\mathbf{p}}(V)$ is a projective algebraic set. By part (3) of Lemma 11.3.4, $\text{Flag}_{\mathbf{p}}(V)$ is closed in $\text{Grass}_{\mathbf{p}}(V)$, since each inclusion $V(u) \subset V(w)$ between subspaces of V is defined by the vanishing of suitable minors in the matrix for $T_u \oplus T_w$.

The group $\mathbf{GL}(V)$ acts on $\text{Grass}_{\mathbf{p}}(V)$. Fix a basis $\{e_i : i = 1, \ldots, n\}$ for V and set $V_i = \text{Span}\{e_1, \ldots, e_{p_i}\}$. Then it is clear that $\text{Flag}_{\mathbf{p}}(V)$ is the orbit of $x_{\mathbf{p}} = \{V_i\}_{1 \le i \le k}$. The isotropy group $P_{\mathbf{p}}$ of $x_{\mathbf{p}}$ consists of the block upper-triangular matrices

$$\begin{bmatrix} A_1 & \cdots & * \\ \vdots & \ddots & \vdots \\ 0 & \cdots & A_{k+1} \end{bmatrix},$$

where $A_i \in \mathbf{GL}(m_i, \mathbb{C})$, with $m_1 = p_1$, $m_2 = p_2 - p_1, \ldots, m_{k+1} = n - p_k$.

Let $G \subset \mathbf{GL}(n, \mathbb{C})$ be a classical group, in the matrix realization of Section 2.4.1. Let H be the diagonal subgroup of G. Set $\mathfrak{b} = \mathfrak{h} + \mathfrak{n}^+$, where $\mathfrak{h} = \text{Lie}(H)$ and $\mathfrak{n}^+ = \bigoplus_{\alpha \in \Phi^+} \mathfrak{g}_\alpha$ as in Theorem 2.4.11 (recall that $\mathfrak{n}^+$ consists of strictly upper-triangular matrices). Denote by N_n^+ the group of all $n \times n$ upper-triangular unipotent matrices.

Theorem 11.3.6. *Let G be a connected classical group. There is a projective algebraic set X_G on which G acts algebraically and transitively with the following properties:*

1. *There is a point $x_0 \in X_G$ whose stabilizer $B = G_{x_0}$ has Lie algebra $\mathfrak{b}$.*
2. *The group $B = H \cdot N^+$ is a semidirect product, with N^+ connected, unipotent, and normal in B.*
3. *The Lie algebra of N^+ is $\mathfrak{n}^+$, and $N^+ = G \cap N_n^+$.*

Proof. For X_G we take the following flag manifolds:

$G = \mathbf{GL}(n, \mathbb{C})$ or $\mathbf{SL}(n, \mathbb{C})$: Let X_G be the set of all *full flags* $\{V_i\}_{1 \le i \le n}$ with $\dim V_i = i$. Let $x_0 = \{V_i^0\}$ with $V_i^0 = \text{Span}\{e_1, \ldots, e_i\}$, where $\{e_j\}$ is the standard basis for $\mathbb{C}^n$. Then B is the group of all upper-triangular matrices (of determinant 1 in the case of $\mathbf{SL}(n, \mathbb{C})$), and N^+ is the group of all unipotent upper-triangular $n \times n$ matrices. Assertions (1), (2), and (3) in this case are clear from general properties of flag manifolds already established.

$G = \mathbf{Sp}(\mathbb{C}^{2l}, \Omega)$ or $\mathbf{SO}(\mathbb{C}^{2l}, B)$, with Ω and B given by (2.6): Let X be the set of all *isotropic flags* $\{V_i\}_{1 \le i \le l}$, with $\dim V_i = i$ and V_i an isotropic subspace relative to Ω or B. Let x_0 be the flag with $V_i = V_i^0$ for $i = 1, \ldots, l$. The set X is projective, as a closed subset of the projective algebraic set

$$\mathfrak{I}_1(\mathbb{C}^n) \times \mathfrak{I}_2(\mathbb{C}^n) \times \cdots \times \mathfrak{I}_l(\mathbb{C}^n),$$

and the action of G on X is algebraic. To see that the action is transitive, let $x = \{V_i\} \in X_G$ and choose $v_k \in \mathbb{C}^n$ such that $V_i = \text{Span}\{v_k\}_{1 \le k \le i}$. Then there exist $\{v_{-k}\}_{1 \le k \le l}$ such that $\{v_{-k}, v_k\}_{1 \le k \le l}$ is a symplectic (resp. isotropic) basis for $\mathbb{C}^n$. The transformation

$$gv_k = e_k \quad \text{and} \quad gv_{-k} = e_{2l+1-k}, \quad \text{for } k = 1,\ldots,l,$$

is symplectic (resp. orthogonal) by Lemmas 1.1.5 and B.2.2, and it takes x to x_0.

When G is the symplectic group we take $X_G = X$. When $G = \mathbf{SO}(\mathbb{C}^{2l}, B)$ we set $X_G = G \cdot x_0 \subset X$ and fix $g_0 \in \mathbf{O}(\mathbb{C}^{2l}, \omega)$ with $\det g_0 = -1$; for example, $g_0 e_l = e_{l+1}$, $g_0 e_{l+1} = e_l$, and $g_0 e_i = e_i$ for $i \neq l, l+1$. Then $g_0 G g_0^{-1} = G$ and $X = X_G \cup g_0 \cdot X_G$, since $\mathbf{O}(\mathbb{C}^{2l}, B)$ acts transitively on X. Thus G has two orbits on X, each closed, making X_G a projective algebraic set.

Define

$$N^+ = \{g \in G : (g - I)V_i^0 \subset V_{i-1}^0 \text{ for } i = 1,\ldots,l\}.$$

We claim that $N^+ \subset N_n^+$. To prove this, recall that $g \in G$ has a block decomposition

$$g = \begin{bmatrix} A & B \\ C & D \end{bmatrix}$$

with each block $l \times l$. If $g \in N^+$, then $C = 0$ and $A \in N_l^+$ is upper-triangular unipotent. Let J_ε (with $\varepsilon = \pm$) be as in Section 2.1.2. Since g also satisfies $J_\varepsilon g J_\varepsilon^{-1} = (g^t)^{-1}$, a calculation shows that $D = s_l(A^t)^{-1} s_l$, with s_l given by (2.5). Hence $D \in N_l^+$. Thus $g \in N_n^+$, as claimed. Conversely, any element of $G \cap N_n^+$ is in N^+. It follows that

$$\mathrm{Lie}(N^+) = \{T \in \mathrm{Lie}(G) : TV_i^0 \subset V_{i-1}^0 \text{ for } i = 1,\ldots,l\},$$

and the map $T \mapsto I + T$ is a bijection from $\mathfrak{n}^+$ to N^+. Thus N^+ is connected. From the matrix description of $\mathrm{Lie}(G)$ in Section 2.1.2 and from Corollary 1.5.5 we conclude that $\mathrm{Lie}(N^+) = \mathfrak{n}^+$. Clearly H normalizes N^+, so $H \cdot N^+$ is a group that is obviously closed in $\mathbf{GL}(n, \mathbb{C})$. If $g \in B$ there is $h \in H$ such that $(h^{-1}g - I)V_i^0 \subset V_{i-1}^0$ for $i = 1,\ldots,l$. Thus $B = H \cdot N^+$.

$G = \mathbf{SO}(\mathbb{C}^{2l+1}, B)$ with B as in (2.9): Let X_G be the set of flags $\{V_i\}_{1 \leq i \leq l+1}$ such that $\dim V_i = i$ and V_i is isotropic for $i = 1,\ldots,l$. Then X_G is a projective algebraic set by the same argument as in the case $\mathbf{O}(\mathbb{C}^{2l}, B)$. Let x_0 be the flag with $V_i = V_i^0$ for $i = 1,\ldots,l+1$. To see that $G \cdot x_0 = X_G$, let $x = \{V_i\}_{1 \leq i \leq l+1} \in X_G$. There must exist a nonisotropic vector $v_0 \in V_{l+1}$, since V_{l+1} cannot be an isotropic subspace. Choose $v_i \in V_i$ for $i = 1,\ldots,l$ such that $V_i = \mathrm{Span}\{v_1,\ldots,v_i\}$ and let $\{v_{-i}\}_{1 \leq i \leq l}$ be dual to $\{v_i\}_{1 \leq i \leq l}$ and orthogonal to v_0. The map

$$gv_i = e_i, \quad gv_0 = e_l + 1, \quad \text{and} \quad gv_{-i} = e_{2l+2-i}, \quad \text{for } i = 1,\ldots,l,$$

is orthogonal and carries x to x_0. The verification of (2) and (3) is the same as for $\mathbf{SO}(\mathbb{C}^{2l}, B)$. $\qquad \square$

11.3.3 Involutions and Symmetric Spaces

Let G be a connected linear algebraic group, and let θ be an involutive automorphism of G. The differential of θ at 1, which we continue to denote by θ, is then an automorphism of $\mathfrak{g}$ satisfying $\theta^2 = I$. Let $K = G^\theta$ be the fixed-point set of θ. We shall embed the space G/K into G as an affine algebraic subset.

Define

$$g \star y = g y \theta(g)^{-1} \quad \text{for } g, y \in G.$$

We have $(g \star (h \star y)) = (gh) \star y$ for $g, h, y \in G$, so this gives an action of G on itself, which we will call the θ-*twisted conjugation* action. Let

$$Q = \{y \in G : \theta(y) = y^{-1}\}.$$

Then Q is an algebraic subset of G. Since $\theta(g \star y) = \theta(g)y^{-1}g^{-1} = (g \star y)^{-1}$, we have $G \star Q = Q$.

Theorem 11.3.7. *The θ-twisted action of G is transitive on each irreducible component of Q. Hence Q is a finite union of closed θ-twisted G-orbits.*

Proof. Let $y \in Q$. We first show that the tangent space $T(G \star y)_y$ to the orbit $G \star y$ at y coincides with the tangent space $T(Q)_y$ to Q at y. Translating the orbit on the left by y^{-1}, we may identify these tangent spaces with the following subspaces of $\mathfrak{g}$:

$$T(G \star y)_y \cong \{\mathrm{Ad}(y^{-1})A - \theta(A) : A \in \mathfrak{g}\},$$
$$T(Q)_y \cong \{B \in \mathfrak{g} : \mathrm{Ad}(y^{-1})\theta(B) + B = 0\}.$$

To see this, note that

$$\frac{d}{dt}\left(y^{-1}(I + tA)y(I - t\theta(A))\right)\Big|_{t=0} = \mathrm{Ad}(y^{-1})A - \theta(A),$$

whereas the curve $t \mapsto y(I + tB)$ is tangent to Q at y if and only if

$$0 = \frac{d}{dt}\left(y^{-1}\theta(y)(I + t\theta(B))y(I + tB)\right)\Big|_{t=0} = \mathrm{Ad}(y^{-1})\theta(B) + B.$$

Since $y \in Q$, the map $x \mapsto \sigma(x) = \mathrm{Ad}(y^{-1})\theta(x)$ is an involution on $\mathfrak{g}$. Indeed, we have

$$\theta(\mathrm{Ad}(y^{-1})\theta(x)) = \mathrm{Ad}(\theta(y^{-1}))x = \mathrm{Ad}(y)x,$$

so $\sigma(\sigma(x)) = x$. Furthermore, we can describe the tangent spaces introduced above as

$$T(G \star y)_y \cong \{\sigma(A) - A : A \in \mathfrak{g}\}, \qquad T(Q)_y \cong \{B \in \mathfrak{g} : \sigma(B) + B = 0\}.$$

Let $B \in \mathfrak{g}$ be given with $\sigma(B) + B = 0$. Set $A = -(1/2)B$. Then $\sigma(A) - A = B$, so we have proved $T(G \star y)_y = T(Q)_y$.

Now let C be an irreducible component of Q containing y. We have

$$\dim C \le \dim T(Q)_y = \dim T(G \star y)_y = \dim(G \star y) ,$$

with the second equality holding because $G \star y$ is a smooth variety by Theorem 11.3.1. But $G \star y \subset C$, since G is connected and Q is invariant under G. Hence $\dim(G \star y) \le \dim C$, so from the inequality above we have $\dim C = \dim(G \star y)$. Thus $G \star y$ is a Zariski-open neighborhood of y in C. Consequently $G \star y = C$ and each irreducible component of Q is a θ-twisted G-orbit. Since Q has only a finite number of irreducible components, this proves the theorem. $\qquad\square$

Corollary 11.3.8. *Let $P = G \star 1 = \{g\theta(g)^{-1} : g \in G\}$ be the orbit of the identity element under the θ-twisted conjugation action. Then P is a closed irreducible subset of G isomorphic to G/K as a G-space (relative to the θ-twisted conjugation action of G).*

Proof. Since $g \star 1 = 1$ if and only if $g \in K$, the map $\psi : G/K \longrightarrow P$ with $\psi(gK) = g \star 1$ is bijective. This map is regular by Theorem 11.3.3, and its differential at the identity coset sends X to $X - \theta(X)$, for $X \in \mathfrak{g}$. We have the decomposition $\mathfrak{g} = \mathfrak{k} \oplus \mathfrak{p}$, where $\mathfrak{k} = \{A \in \mathfrak{g} : \theta(A) = A\}$ is the Lie algebra of K and

$$\mathfrak{p} = \{B \in \mathfrak{g} : \theta(B) = -B\} .$$

In the proof of Theorem 11.3.7 we showed that $T(P)_1 = \mathfrak{p}$. Since $d\psi_1(B) = 2B$ for $B \in \mathfrak{p}$, it follows that $d\psi_1$ is surjective. By homogeneity, $d\psi_x$ is surjective for all $x \in G/K$. Hence ψ is an isomorphism of varieties. $\qquad\square$

11.3.4 Involutions of Classical Groups

Let $G \subset \mathbf{GL}(n, \mathbb{C})$ be a connected classical group with $\mathrm{Lie}(G)$ a simple Lie algebra. We now prove that the involutions θ and associated symmetric spaces G/K for G can be described in terms of three kinds of geometric structures on $\mathbb{C}^n$:

(i) nondegenerate bilinear forms (symmetric or skew-symmetric);

(ii) polarizations $\mathbb{C}^n = V_+ \oplus V_-$ with $V_\pm$ totally isotropic subspaces relative to a bilinear form (either zero or nondegenerate symmetric or skew-symmetric);

(iii) orthogonal decompositions $\mathbb{C}^n = V_+ \oplus V_-$ relative to a nondegenerate bilinear form (symmetric or skew-symmetric).

In case (i) G is $\mathbf{SL}(n, \mathbb{C})$ and K is the subgroup preserving the bilinear form (two types). For case (ii) G is the group preserving the bilinear form on $\mathbb{C}^n$ (if the form is nondegenerate) or $\mathbf{SL}(n, \mathbb{C})$ (if the form is identically zero) and K is the subgroup preserving the given decomposition of $\mathbb{C}^n$ (three types in all). For case (iii) G is the group preserving the bilinear form and K is the subgroup preserving the given decomposition of $\mathbb{C}^n$ (two types). Thus there are seven types of symmetric spaces that arise in this way.

We will prove that these seven types give all the possible involutive automorphisms of the classical groups (up to inner automorphisms). Our proof will depend on the following characterization of automorphisms of the classical groups:

Proposition 11.3.9. *Let σ be a regular automorphism of the classical group G.*

1. *If $G = \mathbf{SL}(n,\mathbb{C})$ then there exists $s \in G$ such that σ is either $\sigma(g) = sgs^{-1}$ or $\sigma(g) = s(g^t)^{-1}s^{-1}$.*
2. *If $G = \mathbf{Sp}(n,\mathbb{C})$ then there exists $s \in G$ such that $\sigma(g) = sgs^{-1}$.*
3. *If $G = \mathbf{SO}(n,\mathbb{C})$ $(n \neq 2, 4)$ then there exists $s \in \mathbf{O}(n,\mathbb{C})$ such that $\sigma(g) = sgs^{-1}$.*

Proof. Let π be the defining representation of G on $\mathbb{C}^m$ (where $m = n$ in cases (1) and (3), and $m = 2n$ in case (2)). The representation $\pi^\sigma(g) = \pi(\sigma(g))$ also acts irreducibly on $\mathbb{C}^m$. The Weyl dimension formula implies that the defining representation (and its dual, in the case $G = \mathbf{SL}(n,\mathbb{C})$) is the unique representation of dimension m (see Exercises 7.1.4). Note that for $\mathbf{SO}(8,\mathbb{C})$, the two half-spin representations are eight-dimensional, but they are not single-valued on $\mathbf{SO}(8,\mathbb{C})$.

In case (1), we know that $\pi^* \cong \check{\pi}$, where $\check{\pi}(g) = \pi((g^t)^{-1})$ (see Theorem 5.5.22). Hence either $\pi^\sigma \cong \pi$ or $\pi^\sigma \cong \check{\pi}$. Since all these representations act on the same space $\mathbb{C}^n$, this means that there exists $s \in \mathbf{GL}(n,\mathbb{C})$ such that either $\sigma(g) = sgs^{-1}$ or, in the case $G = \mathbf{SL}(n,\mathbb{C})$, $\sigma(g) = s(g^t)^{-1}s^{-1}$ for all $g \in G$. Since λs induces the same automorphism as s, for any $\lambda \in \mathbb{C}^\times$, we can replace s by λs, where $\lambda^m = \det(s)^{-1}$, to obtain $\det(s) = 1$. This proves case (1).

For cases (2) and (3) we have $\sigma(g) = sgs^{-1}$ with $s \in \mathbf{GL}(n,\mathbb{C})$. Let Γ be the matrix of the bilinear form on $\mathbb{C}^m$ associated with G. Then $g^t \Gamma g = \Gamma$ for all $g \in G$. Hence

$$g^t s^t \Gamma s g = s^t (sgs^{-1})^t \Gamma (sgs^{-1}) s = s^t \Gamma s \quad \text{for all } g \in G.$$

This shows that G leaves invariant the bilinear form with matrix $s^t \Gamma s$. Since the representation of G on $\mathbb{C}^m$ is irreducible, the space of G-invariant bilinear forms on $\mathbb{C}^m$ has dimension one by Schur's lemma. Hence $s^t \Gamma s = \mu \Gamma$ for some $\mu \in \mathbb{C}^\times$. Let $\lambda \in \mathbb{C}^\times$ with $\lambda^2 = \mu^{-1}$. Then $(\lambda s)^t \Gamma (\lambda s) = \Gamma$ and $\sigma(g) = (\lambda s)g(\lambda s)^{-1}$ for $g \in G$. Thus λs preserves the form with matrix Γ, and we may replace s by λs to achieve (2) and (3). $\square$

We can now determine all the involutions of the classical groups.

Theorem 11.3.10. *Let θ be an involution of the classical group G. Assume that $\mathrm{Lie}(G)$ is simple. Then θ is given as follows, up to conjugation by an element of G.*

1. *If $G = \mathbf{SL}(n,\mathbb{C})$, then there are three possibilities:*

    a. *$\theta(x) = T(x^t)^{-1}T^t$ for $x \in G$, where $T \in G$ satisfies $T^t = T$. The property $T^t = T$ determines θ uniquely up to conjugation in G. The corresponding bilinear form $B(u,v) = u^t T v$, for $u, v \in \mathbb{C}^n$, is symmetric and nondegenerate.*

b. $\theta(x) = T(x^t)^{-1}T^t$ for $x \in G$, where $T \in G$ satisfies $T^t = -T$. The property $T^t = -T$ determines θ uniquely up to conjugation in G. The corresponding bilinear form $B(u,v) = u^t T v$, for $u, v \in \mathbb{C}^n$, is skew-symmetric and nondegenerate.

c. $\theta(x) = JxJ^{-1}$ for $x \in G$, where $J \in \mathbf{GL}(n, \mathbb{C})$ and $J^2 = I_n$. Let

$$V_\pm = \{v \in \mathbb{C}^n : Jv = \pm v\}.$$

Then $V = V_+ \oplus V_-$ and θ is determined (up to conjugation in G) by $\dim V_+$.

2. If G is $\mathbf{SO}(V, \omega)$ or $\mathbf{Sp}(V, \omega)$, then there are two possibilities:

a. $\theta(x) = JxJ^{-1}$ for $x \in G$, where J preserves the form ω and $J^2 = I$. Let

$$V_\pm = \{v \in V : Jv = \pm v\}.$$

Then $V = V_+ \oplus V_-$, the restriction of ω to $V_\pm$ is nondegenerate, and θ is determined (up to conjugation in G) by $\dim V_+$.

b. $\theta(x) = JxJ^{-1}$ for $x \in G$, where J preserves the form ω and $J^2 = -I$. Let

$$V_\pm = \{v \in V : Jv = \pm iv\}.$$

Then $V = V_+ \oplus V_-$, the restriction of ω to $V_\pm$ is zero, and V_+ is dual to V_- via the form ω. The automorphism θ is uniquely determined (up to conjugation in G).

Proof. We use Proposition 11.3.9. Suppose $G = \mathbf{SL}(n, \mathbb{C})$ and $\theta(x) = J(x^t)^{-1}J^{-1}$. Since

$$x = \theta^2(x) = J(J^t)^{-1}xJ^tJ^{-1} \quad \text{for all } x \in G,$$

we have J^tJ^{-1} commuting with G. Hence $J^t = \lambda J$ for some $\lambda \in \mathbb{C}$ by Schur's lemma. But this implies that $\lambda = \pm 1$, since $J = (J^t)^t = \lambda J^t$. This gives cases (a) and (b), with the uniqueness of θ a consequence of Propositions 1.1.4 and 1.1.6. According to Proposition 11.3.9, the only other possibility for θ is $\theta(x) = JxJ^{-1}$, where $J \in \mathbf{GL}(n, \mathbb{C})$. Since

$$x = \theta^2(x) = J^2xJ^{-2} \quad \text{for all } x \in G,$$

we have J^2 commuting with G. Hence $J^2 = \lambda I$ by Schur's lemma for some $\lambda \in \mathbb{C}$. Replacing J by $\lambda^{-1/2}J$, we may assume that $J^2 = I$. Hence J is semisimple with eigenvalues ± 1 and is determined (up to an inner automorphism of G) by the dimension of its $+1$ eigenspace. This proves (c).

Now suppose G is the orthogonal or symplectic group. Then $\theta(x) = JxJ^{-1}$ for some $J \in G$. By the argument above, $J^2 = \lambda I$ for some $\lambda \in \mathbb{C}$. But J^2 preserves the form ω, so $\lambda^2 = 1$. Thus we have two possibilities: $J^2 = I$ or $J^2 = -I$. In the first case V_+ is orthogonal to V_- relative to ω, since

$$\omega(v_+, v_-) = \omega(Jv_+, Jv_-) = -\omega(v_+, v_-)$$

for $v_\pm \in V_\pm$. Hence the restriction of ω to $V_\pm$ must be nondegenerate. In the second case the restriction of ω to V_+ or to V_- must be zero, since

$$\omega(u,v) = \omega(Ju, Jv) = (\mathrm{i})^2 \omega(u,v) = -\omega(u,v)$$

for $u, v \in V_+$ (and likewise for $u, v \in V_-$). Hence ω gives a nonsingular pairing of V_+ with V_-. The uniqueness of θ (up to an inner automorphism of G) follows from Lemmas B.2.2 and 1.1.5. $\qquad\square$

11.3.5 Classical Symmetric Spaces

We proceed to describe the symmetric spaces for the classical groups in more detail.

Notation. Given the group G and involution θ of G, we set

$$P = \{g\theta(g)^{-1} : g \in G\}, \qquad Q = \{y \in G : \theta(y) = y^{-1}\}.$$

We write s_p for the $p \times p$ matrix (2.5) with 1 on the antidiagonal and 0 elsewhere. Let $\tau(g) = (\bar{g}^t)^{-1}$, the bar denoting complex conjugation relative to the embedding $G \subset \mathbf{GL}(n, \mathbb{C})$. In all cases we will take the matrix form of G and the involution θ to satisfy the following conditions:

(i) $\tau(G) = G$ and G^τ is a compact real form of G (see Section 1.7.2).
(ii) The diagonal subgroup H in G is a maximal torus and $\theta(H) = H$.
(iii) $\tau\theta = \theta\tau$.

It follows from (iii) that $\sigma = \theta\tau$ is also a conjugation on G. This conjugation will play an important role when we study the representation of G on $\mathbb{O}[G/K]$ in Chapter 12.

Involutions Associated with Bilinear Forms

Symmetric Bilinear Form – Type AI. Let $G = \mathbf{SL}(n, \mathbb{C})$ and define the involution $\theta(g) = (g^t)^{-1}$. Then $\theta(g) = g$ if and only if g preserves the symmetric bilinear form $B(u,v) = u^t v$ on $\mathbb{C}^n$. Thus $K = G^\theta = \mathbf{SO}(\mathbb{C}^n, B)$. The θ-twisted action is $g \star y = g y g^t$, and $Q = \{y \in G : y^t = y\}$. A matrix $y \in Q$ defines a symmetric bilinear form $B_y(u,v) = u^t y v$ on $\mathbb{C}^n$. The θ-twisted G-orbit of y corresponds to all the bilinear forms G-equivalent to B_y. Since B_y is nonsingular, there exists $g \in \mathbf{GL}(n, \mathbb{C})$ such that $g \star y = I_n$. Since $\det y = 1$, we have $\det g = \pm 1$; multiplying g by $\mathrm{diag}[-1, 1, \ldots, 1]$ if necessary, we may take $\det g = 1$. Thus Q is a single G-orbit in this case, and hence $Q = P$. By Corollary 11.3.8 we conclude that

$$\mathbf{SL}(n, \mathbb{C})/\mathbf{SO}(\mathbb{C}^n, B) \cong \{y \in M_n(\mathbb{C}) : y = y^t, \det y = 1\}$$

as a G-variety, under the map $gK \mapsto gg^t$. In this case the conjugation $\sigma = \theta\tau$ is given by $\sigma(g) = \bar{g}$.

Skew-Symmetric Bilinear Form – Type AII. Let $G = \mathbf{SL}(2n,\mathbb{C})$. Take

$$\mu = \begin{bmatrix} 0 & 1 \\ -1 & 0 \end{bmatrix},$$

and let $T_n = \mathrm{diag}[\mu,\ldots,\mu]$ be the $2n \times 2n$ skew-symmetric block-diagonal matrix (n blocks). Then $T_n^2 = -I_{2n}$ and $T_n^{-1} = T_n^t$. Define the involution θ by

$$\theta(g) = T_n(g^t)^{-1}T_n^t .$$

Since $\tau(T_n) = T_n$, we have $\theta\tau = \tau\theta$. For $g \in G$, $\theta(g) = g$ if and only if $g^t T_n g = T_n$. This means that g preserves the nondegenerate skew-symmetric bilinear form $\omega(u,v) = u^t T_n v$ on $\mathbb{C}^{2n}$. Thus $K = G^\theta = \mathbf{Sp}(\mathbb{C}^{2n},\omega)$.

The θ-twisted action is $g \star y = gyT_n g^t T_n^t$, and $Q = \{y \in G : (yT_n)^t = -yT_n\}$. A matrix $y \in Q$ defines a nonsingular skew-symmetric bilinear form

$$\omega_y(u,v) = u^t y T_n v , \quad \text{for } u,v \in \mathbb{C}^{2n} ,$$

and the θ-twisted G-orbit of y corresponds to all the bilinear forms equivalent to ω_y. Arguing as in Type AI, we see that Q is a single G orbit and hence $Q = P$. By Corollary 11.3.8 we conclude that

$$\mathbf{SL}(2n,\mathbb{C})/\mathbf{Sp}(\mathbb{C}^{2n},\omega) \cong \{y \in M_n(\mathbb{C}) : yT_n = -(yT_n)^t, \det y = 1\}$$

under the map $gK \mapsto gT_n g^t T_n^t$. In this case the conjugation $\sigma = \theta\tau$ is given by $\sigma(g) = T_n \bar{g} T_n^t$.

Involutions Associated with Polarizations

Zero Bilinear Form – Type AIII. Let $G = \mathbf{SL}(p+q,\mathbb{C})$. For integers $p \leq q$ with $p + q = n$ define

$$J_{p,q} = \begin{bmatrix} 0 & 0 & s_p \\ 0 & I_{q-p} & 0 \\ s_p & 0 & 0 \end{bmatrix} .$$

Then $J_{p,q}^2 = I_n$, so we can define an involution θ of G by

$$\theta(g) = J_{p,q} g J_{p,q} .$$

Since $\tau(J_{p,q}) = J_{p,q}$, we have $\theta\tau = \tau\theta$. The maps $P_\pm = (1/2)(I_n \mp J_{p,q})$ are the projections onto the ± 1 eigenspaces $V_\pm$ of $J_{p,q}$, and

$$\mathbb{C}^n = V_+ \oplus V_- . \tag{11.7}$$

We have $\dim V_+ = \mathrm{tr}(P_+) = (1/2)(n - (q - p)) = p$. The subgroup $K = G^\theta$ consists of all $g \in G$ that commute with $J_{p,q}$. This means that g leaves invariant the decomposition (11.7), so we have

$$K \cong \mathbf{S}(\mathbf{GL}(p,\mathbb{C}) \times \mathbf{GL}(q,\mathbb{C})) \,,$$

the group of all block diagonal matrices $g = \mathrm{diag}[g_1, g_2]$ with $g_1 \in \mathbf{GL}(p,\mathbb{C})$, $g_2 \in \mathbf{GL}(q,\mathbb{C})$, and $\det g_1 \det g_2 = 1$.

In this case $Q = \{y \in G : (yJ_{p,q})^2 = I_n\}$ and the θ-twisted action is

$$g \star y = gyJ_{p,q}g^{-1}J_{p,q} \,.$$

For $y \in Q$ the matrix $z = yJ_{p,q}$ is a nonsingular idempotent. Thus it defines a decomposition

$$\mathbb{C}^n = V_+(y) \oplus V_-(y) \,,$$

where z acts by ± 1 on $V_\pm(y)$. The θ-twisted G orbit of y corresponds to the G-conjugacy class of z, under the map $g \star y \mapsto (g \star y)J_{p,q}$. Hence $G \star y$ is determined by $\dim V_+(y)$, which can be any integer between 0 and n. In particular, the θ-twisted G orbit of I is

$$P = \{y \in \mathbf{GL}(p+q,\mathbb{C}) : (yJ_{p,q})^2 = I_n, \ \mathrm{tr}(yJ_{p,q}) = q - p\} \,.$$

By Corollary 11.3.8 we conclude that

$$\mathbf{SL}(p+q,\mathbb{C})/\mathbf{S}(\mathbf{GL}(p,\mathbb{C}) \times \mathbf{GL}(q,\mathbb{C})) \cong P$$

under the map $gK \mapsto gJ_{p,q}g^{-1}J_{p,q}$. The conjugation $\sigma = \theta\tau$ is given by $\sigma(g) = J_{p,q}(\bar{g}^t)^{-1}J_{p,q}$.

Skew-Symmetric Bilinear Form – Type CI. Let $G = \mathbf{Sp}(\mathbb{C}^{2n}, \Omega)$, where Ω is the skew-symmetric form $\Omega(u,v) = u^t J_n v$ with

$$J_n = \begin{bmatrix} 0 & s_n \\ -s_n & 0 \end{bmatrix} \,.$$

We have $J_n^t = J_n^{-1}$ and $J_n^2 = -I_{2n}$. Thus $J_n \in G$ and the map $\theta(g) = -J_n g J_n$ is an involution on G. Since $\tau(J_n) = J_n$, we see that θ commutes with τ. We can decompose $\mathbb{C}^{2n} = V_+ \oplus V_-$, where J_n acts by $\pm i$ on $V_\pm$. The form Ω vanishes on the subspaces $V_\pm$. Indeed, the projections onto $V_\pm$ are $P_\pm = (1/2)(1 \mp iJ_n)$, and we have $P_+^t = P_-$, since $J_n^t = -J_n$. Thus $P_+^t J_n P_+ = J_n P_- P_+ = 0$, and so $\Omega(P_+u, P_+v) = 0$ (the same holds for P_-). Thus Ω gives a nonsingular pairing between V_- and V_+. In particular, $\dim V_\pm = n$.

The subgroup $K = G^\theta$ consists of all $g \in G$ that commute with J_n. Thus g leaves invariant $V_\pm$. Since g preserves Ω, the action of g on V_- is dual to its action on V_+. Thus

$$K \cong \mathbf{GL}(V_+) \cong \mathbf{GL}(n,\mathbb{C}) \,.$$

The θ-twisted action is $g \star y = g y J_n g^{-1} J_n^{-1}$, and

$$Q = \{y \in G : (yJ_n)^2 = -I_{2n}\} .$$

Let $y \in Q$ and set $z = yJ_n$. Then $z^2 = -I_{2n}$, so we can decompose

$$\mathbb{C}^{2n} = V_+(y) \oplus V_-(y) , \tag{11.8}$$

where z acts by $\pm i$ on $V_\pm(y)$. We claim that $\Omega = 0$ on $V_\pm(y)$. Indeed, the projections onto $V_\pm(y)$ are $P_\pm = (1/2)(I \mp iz)$, and from the relation $y^t J_n = J_n y^{-1}$ we calculate that $z^t J_n = -J_n z$, so this follows just as in the case $y = I_n$. The subspaces $V_\pm(y)$ are thus maximal isotropic for the form Ω, and Ω gives a nonsingular pairing between $V_+(y)$ and $V_-(y)$. Since y is determined by the decomposition (11.8), we see from Lemma 1.1.5 that Q is a single θ-twisted G orbit. Thus

$$P = \{y \in \mathbf{Sp}(n, \mathbb{C}) : (yJ_n)^2 = -I_{2n}\} .$$

By Corollary 11.3.8 we conclude that

$$\mathbf{Sp}(\mathbb{C}^{2n}, \Omega) / \mathbf{GL}(n, \mathbb{C}) \cong P$$

under the map $gK \mapsto gJ_n g^{-1} J_n^t$. The conjugation $\sigma = \theta\tau$ is given by $\sigma(g) = -J_n(\bar{g}^t)^{-1} J_n$.

Symmetric Bilinear Form – Type DIII. Let $G = \mathbf{SO}(\mathbb{C}^n, B)$ with $n = 2l$ even, where $B(u, v) = u^t s_n v$. We define $\Gamma \in \mathbf{GL}(n, \mathbb{C})$ as follows: Let

$$\gamma = \begin{bmatrix} 0 & 1 \\ 1 & 0 \end{bmatrix} .$$

Define the $n \times n$ matrix $\Gamma_n = \mathrm{i}\, \mathrm{diag}[\underbrace{\gamma, \dots, \gamma}_{r}, \underbrace{-\gamma, \dots, -\gamma}_{r}]$ when $l = 2r$ is even, and

$\Gamma_n = \mathrm{i}\, \mathrm{diag}[\underbrace{\gamma, \dots, \gamma}_{r}, 1, -1, \underbrace{-\gamma, \dots, -\gamma}_{r}]$ when $l = 2r+1$ is odd. Then $\Gamma_n s_n = -s_n \Gamma_n$,

$\Gamma_n^t = \Gamma_n$, and $\Gamma_n^2 = -I_n$. Thus $\Gamma_n \in \mathbf{O}(\mathbb{C}, B)$ and the map

$$\theta(g) = -\Gamma_n g \Gamma_n$$

is an involution on G. Since $\tau(\Gamma_n) = -\Gamma_n$, we see that θ commutes with τ.

We can decompose

$$\mathbb{C}^n = V_+ \oplus V_- ,$$

where $V_\pm$ are the $\pm i$ eigenspaces of Γ_n. Since $\Gamma_n s_n = -s_n \Gamma_n$, the form B vanishes on the subspaces $V_\pm$, by the same calculation as in Type CI. As in that case, we have

$$K \cong \mathbf{GL}(V_+) \cong \mathbf{GL}(l, \mathbb{C}) .$$

For this case $Q = \{y \in G : (y\Gamma_n)^2 = -I_n\}$ and the θ-twisted action is $g \star y = gy\Gamma_n g^{-1}\Gamma_n$. For $y \in Q$ the matrix $z = y\Gamma_n$ satisfies $z^2 = -I_n$, so we can decompose

$$\mathbb{C}^n = V_+(y) \oplus V_-(y) \, ,$$

where z acts by $\pm i$ on $V_\pm(y)$. We have

$$z^t s_n = \Gamma_n y^t s_n = \Gamma_n s_n y^{-1} = -s_n \Gamma_n y^{-1} = -s_n z \, .$$

This implies that the subspaces $V_\pm$ are totally isotropic for the form B (by the same calculation as in Type CI). It follows by Lemma B.2.2 that Q is a single θ-twisted G orbit. Thus

$$P = \{y \in \mathbf{SO}(\mathbb{C}^n, B) : (y\Gamma_n)^2 = -I_n\} \, .$$

By Corollary 11.3.8 we conclude that

$$\mathbf{SO}(\mathbb{C}^n, B)/\mathbf{GL}(l, \mathbb{C}) \cong P$$

under the map $gK \mapsto g\Gamma_n g^{-1}\Gamma_n$. The conjugation $\sigma = \theta\tau$ is given by $\sigma(g) = -\Gamma_n(\bar{g}^t)^{-1}\Gamma_n$.

Involutions Associated with Orthogonal Decompositions

Symmetric Bilinear Form – Type BDI. Let $G = \mathbf{SO}(\mathbb{C}^n, B)$, where B is the symmetric bilinear form $B(u, v) = u^t s_n v$ on $\mathbb{C}^n$. For integers $p \leq q$ with $p + q = n$ define $J_{p,q}$ as in Type AIII. We have $J_{p,q}^t = J_{p,q}^{-1} = J_{p,q}$ and $J_{p,q} s_n = s_n J_{p,q}$. Since $g \in G$ if and only if $s_n g^t s_n = g$, we see that $J_{p,q} \in \mathbf{O}(\mathbb{C}^n, B)$. Thus the map

$$\theta(g) = J_{p,q} g J_{p,q}$$

is an involution on G. Clearly, θ commutes with τ. The projections $P_\pm$ onto the ± 1 eigenspaces $V_\pm$ of $J_{p,q}$ commute with s_n. Hence $V_+ \perp V_-$ (relative to the form B), since $P_+^t s_n P_- = s_n P_+ P_- = 0$. We have $\dim V_+ = \mathrm{tr}(P_+) = (1/2)(n - (q - p)) = p$ and $\dim V_- = q$. The subgroup $K = G^\theta$ consists of all $g \in G$ that commute with $J_{p,q}$. This means that g leaves invariant the decomposition (11.7). The restrictions $B_\pm$ of B to $V_\pm$ are nondegenerate, since $V_- \perp V_+$, so we have

$$K \cong \mathbf{S}(\mathbf{O}(V_+, B_+) \times \mathbf{O}(V_-, B_-)) \cong \mathbf{S}(\mathbf{O}(p, \mathbb{C}) \times \mathbf{O}(q, \mathbb{C})),$$

the group of all block diagonal matrices $g = \mathrm{diag}[g_1, g_2]$ with $g_1 \in \mathbf{O}(p, \mathbb{C})$, $g_2 \in \mathbf{O}(q, \mathbb{C})$, and $\det g_1 \det g_2 = 1$.

We have $Q = \{y \in G : (yJ_{p,q})^2 = I_n\}$, and the θ-twisted action is

$$g \star y = gyJ_{p,q} g^{-1} J_{p,q} \, .$$

The G orbits in Q for this action correspond to the G-similarity classes of idempotent matrices $yJ_{p,q}$, with $y \in Q$.

For $y \in Q$ the matrix $z = yJ_{p,q}$ satisfies $z^2 = I_n$, so it gives a decomposition

$$\mathbb{C}^n = V_+(y) \oplus V_-(y), \tag{11.9}$$

where z acts by ± 1 on $V_\pm(y)$. Since $y^t s_n = s_n y^{-1}$, $J_{p,q} y^{-1} = y J_{p,q}$, and $J_{p,q} s_n = s_n J_{p,q}$, we have

$$z^t s_n = J_{p,q} y^t s_n = J_{p,q} s_n y^{-1} = s_n J_{p,q} y^{-1} = s_n z.$$

Hence the same argument that we used when $y = I_n$ shows that the subspaces $V_\pm(y)$ are mutually orthogonal (relative to the form B). This implies that the restrictions of B to $V_\pm$ are nonsingular. Since y is determined by the decomposition (11.9), we see from Lemma B.2.2 that the θ-twisted G orbit of y is determined by the integer

$$\dim V_+(y) = (1/2)(n - \text{tr}(yJ_{p,q})).$$

In particular, $P = \{y \in \mathbf{SO}(\mathbb{C}^n, B) : (yJ_{p,q})^2 = I_n, \ \text{tr}(yJ_{p,q}) = q - p\}$. Corollary 11.3.8 now implies

$$\mathbf{SO}(\mathbb{C}^n, B)/\mathbf{S}(\mathbf{O}(p, \mathbb{C}) \times \mathbf{O}(q, \mathbb{C})) \cong P$$

under the map $gK \mapsto gJ_{p,q}g^{-1}J_{p,q}$. The conjugation $\sigma = \theta\tau$ is given by $\sigma(g) = J_{p,q}(\bar{g}^t)^{-1}J_{p,q}$.

Skew-Symmetric Bilinear Form – Type CII. Let $G = \mathbf{Sp}(\mathbb{C}^{2n}, \Omega)$, where Ω is the skew-symmetric bilinear form $\Omega(u, v) = u^t J_n v$ as in Type CI. For $0 < p \le q$ with $p + q = n$, let $J_{p,q} \in \mathbf{GL}(n, \mathbb{C})$ be as in Type AIII and define

$$K_{p,q} = \begin{bmatrix} J_{p,q} & 0 \\ 0 & J_{p,q} \end{bmatrix}.$$

Since $J_{p,q}^t = J_{p,q}^{-1} = J_{p,q}$ and $s_n J_{p,q} s_n = J_{p,q}$, we have $K_{p,q} \in G$ and $K_{p,q}^2 = I_{2n}$. Thus the map $\theta(g) = K_{p,q}gK_{p,q}$ is an involution on G. Clearly, θ commutes with τ. As in Type BDI, the ± 1 eigenspaces of $K_{p,q}$ give a decomposition

$$\mathbb{C}^{2n} = V_+ \oplus V_- \tag{11.10}$$

that is orthogonal relative to the form Ω. The subgroup $K = G^\theta$ consists of all $g \in G$ that commute with $K_{p,q}$. Since the restrictions of Ω to $V_\pm$ are nondegenerate and $\dim V_+ = \text{tr}(P_+) = (1/2)(2n - \text{tr}(K_{p,q})) = 2p$, we have

$$K \cong \mathbf{Sp}(p, \mathbb{C}) \times \mathbf{Sp}(q, \mathbb{C}),$$

in complete analogy with Type BDI.

Here $Q = \{y \in G : (yK_{p,q})^2 = I_{2n}\}$ and the θ-twisted action is

$$g \star y = gyK_{p,q}g^{-1}K_{p,q}.$$

Let $y \in Q$ and set $z = yK_{p,q}$. Since $y^t J_n = J_n y^{-1}$ and $K_{p,q} J_n = J_n K_{p,q}$, we have $z^t J_n = J_n z$. Thus the ± 1 eigenspaces of z are mutually orthogonal (relative to Ω) and give a decomposition

$$\mathbb{C}^{2n} = V_+(y) \oplus V_-(y). \tag{11.11}$$

The same proof as in Type BDI (using Lemma 1.1.5) shows that the θ-twisted G orbit of y is determined by the integer $\mathrm{tr}(yK_{p,q})$. In particular,

$$P = \{y \in \mathbf{Sp}(\mathbb{C}^{2n}, \Omega) : (yK_{p,q})^2 = I_{2n}, \mathrm{tr}(yK_{p,q}) = 2(q - p)\}.$$

By Corollary 11.3.8 we conclude that

$$\mathbf{Sp}(\mathbb{C}^{2n}, \Omega)/(\mathbf{Sp}(p, \mathbb{C}) \times \mathbf{Sp}(q, \mathbb{C})) \cong P$$

under the map $gK \mapsto gJ_n g^{-1} J_n^{-1}$. The conjugation $\sigma = \theta\tau$ is given by $\sigma(g) = K_{p,q}(\bar{g}^t)^{-1} K_{p,q}$.

11.3.6 Exercises

1. Let G be an algebraic group acting on an affine algebraic variety X. Assume that $\mathcal{J} = \mathcal{O}[X]^G$ is finitely generated as an algebra over $\mathbb{C}$ (if G is reductive this is always true, by Theorem 5.1.1). This action partitions X into G orbits, and every G-invariant function on X is constant on each orbit. An affine variety Y is called the *algebraic quotient* of X by G if there is a regular map $\pi : X \longrightarrow Y$ that is constant on each G-orbit in X, with the following *universal property*: Given any algebraic variety Z and regular map $f : X \longrightarrow Z$ that is constant on G orbits, there exists a unique regular map $\tilde{f}$ such that $f = \tilde{f} \circ \pi$.
   (a) Let Y be the set of maximal ideals of $\mathcal{J}$. Identify the points of Y with the algebra homomorphisms $\mathcal{J} \longrightarrow \mathbb{C}$ by Theorem A.1.3, and define $\pi(x)(f) = f(x)$ for $f \in \mathcal{J}$. This gives a map $\pi : X \longrightarrow Y$. Show that (Y, π) is an algebraic quotient of X by G. (HINT: If Z is an affine variety and $f : X \longrightarrow Z$ is regular and constant on G orbits, then $f^*(\mathcal{O}[Z]) \subset \mathcal{J}$. Hence every homomorphism $\varphi : \mathcal{J} \longrightarrow \mathbb{C}$ determines a homomorphism $\tilde{f}(\varphi) : \mathcal{O}[Z] \longrightarrow \mathbb{C}$, where $\tilde{f}(\varphi)(h) = \varphi(h \circ f)$ for $h \in \mathcal{O}[Z]$. This defines a regular map $\tilde{f}$ such that $f = \tilde{f} \circ \pi$.)
   (b) Show that the universal property of a quotient variety uniquely determines it, up to isomorphism. Write $Y = X//G$ and call π the *canonical map*.
   (c) Suppose G is reductive. Prove that the canonical map is surjective. (HINT: Let $\mathfrak{m} \subset \mathcal{J}$ be a maximal ideal. Then $\mathfrak{m}$ generates a proper ideal in $\mathcal{O}[X]$, since any relation $\sum_i f_i g_i = 1$ with $f_i \in \mathfrak{m}$ and $g_i \in \mathcal{O}[X]$ would give a relation $\sum f_i g_i^{\natural} \in \mathfrak{m} = 1$, where $g \mapsto g^{\natural}$ is the projection onto the G-invariants. By the Hilbert Nullstellensatz there exists $x \in X$ such that all the functions in $\mathfrak{m}$ vanish at x. Hence $\pi(x) = \mathfrak{m}$.)

2. Assume that G is reductive and that X is an affine G-space. Let $Y = X//G$ and let $\pi : X \longrightarrow Y$ be the canonical map.

(a) Let $Z \subset X$ be closed and G-invariant. Prove that $\pi(Z)$ is closed in Y and that the pair $(\pi|_Z, \pi(Z))$ is the algebraic quotient of Z by G. (HINT: Set $\mathcal{R} = \mathcal{O}[X]$, $\mathcal{J} = \mathcal{R}^G$, and let $\mathcal{I} \subset \mathcal{R}$ be the ideal of functions vanishing on Z. The affine ring of Z is $\mathcal{R}/\mathcal{I}$, so the affine ring of $Z//G$ is $(\mathcal{R}/\mathcal{I})^G$. Since G is reductive and $\mathcal{I}$ is G-invariant, $(\mathcal{R}/\mathcal{I})^G \cong \mathcal{J}/\mathcal{I} \cap \mathcal{J}$. But $\mathcal{J}/\mathcal{I} \cap \mathcal{J}$ is the affine ring of $\overline{\pi(Z)}$. Hence it is a model for $Z//G$. Now use the previous exercise to conclude that $\pi(Z) = \overline{\pi(Z)}$.)

(b) Let $\{Z_i\}_{i \in I}$ be any collection of closed, G-invariant subsets of X. Prove that

$$\pi\left(\bigcap_{i \in I} Z_i\right) = \bigcap_{i \in I} \pi(Z_i).$$

Conclude that if U and V are disjoint closed and G-invariant subsets of X, then there exists $f \in \mathcal{O}[X]^G$ such that $f(U) \neq f(V)$. In particular, each fiber of the map $\pi : X \longrightarrow X//G$ contains exactly one closed orbit. (HINT: Let $\mathcal{I}_i$ be the ideal of regular functions vanishing on Z_i. Then the ideal of functions vanishing on $Z = \bigcap_{i \in I} Z_i$ is $\mathcal{I} = \sum_{i \in I} \mathcal{I}_i$, and the affine ring of Z is $\mathcal{R}/\mathcal{I}$. By (a) the affine ring of $\pi(Z)$ is $\mathcal{J}/\sum_{i \in I} \mathcal{J} \cap \mathcal{I}_i$, and $\bigcap_{i \in I} \pi(Z_i)$ is closed with $\sum_{i \in I} \mathcal{J} \cap \mathcal{I}_i$ the ideal of functions vanishing on it.)

3. Let $G = \mathbf{SL}(n, \mathbb{C})$ and let V be the vector space of all symmetric quadratic forms $Q(x) = \sum a_{ij} x_i x_j$ in n variables $x_1, \ldots, x_n$, with $n \geq 2$. The group G acts on V via its linear action on $x = [x_1, \ldots, x_n]^t \in \mathbb{C}^n$. In terms of the symmetric matrix $A = [a_{ij}]$, the action is $g \cdot A = (g^t)^{-1} A g^{-1}$.

(a) Show that the function $D(A) = \det A$ (the *discriminant* of the form) is invariant under G.

(b) Show that every G orbit in V contains exactly one of the forms

$$Q_{n,c}(x) = c x_1^2 + x_2^2 + \cdots + x_n^2, \quad \text{with } c \neq 0,$$

$$Q_r(x) = x_1^2 + \cdots + x_r^2, \quad \text{with } 0 \leq r < n.$$

(c) Show that $\mathcal{P}(V)^G = \mathcal{O}[D]$. (HINT: Define $s : \mathbb{C} \longrightarrow V$ by $s(c) = Q_{n,c}$. Given $f \in \mathcal{P}(V)^G$, let φ be the polynomial $f \circ s$. Show that $f(A) = \varphi(D(A))$ when A is nonsingular, and hence this holds for all A.)

(d) Show that $V//G \cong \mathbb{C}$, with the quotient map $\pi(x) = D(x)$. Show that the closed G orbits are those on which $D \neq 0$ (nonsingular forms) and the point $\{0\}$, and show that the quotient map takes all the nonclosed orbits (the forms of rank $r < n$) to 0.

(e) Show that the G-invariant polynomials can separate the G orbits of the nonsingular forms but cannot separate the orbits of the singular forms. (HINT: Consider the sets $D^{-1}(c)$ for $c \in \mathbb{C}$.)

4. Let H be a connected reductive group and set $G = H \times H$. Let θ be the involution $\theta(x, y) = (y, x)$.

(a) Show that $K = G^\theta = \{(x, x) : x \in H\}$ is isomorphic to H.

(b) Let $P = \{(x, x^{-1}) : x \in H\}$. Show that $P = \{(xy^{-1}, yx^{-1}) : x, y \in H\}$. (HINT: Use Corollary 11.3.8.)

(c) Show that the map $P \longrightarrow H$ given by $(x, x^{-1}) \mapsto x$ is an isomorphism of affine varieties that transforms the θ-twisted action of G on P into the two-sided action of $H \times H$ on H.

5. Let $G = \mathbf{SL}(2, \mathbb{C})$ act on $\mathbb{C}^2$ by left multiplication as usual. This gives an action on $\mathbb{P}^1(\mathbb{C})$. Let $H = \{\operatorname{diag}[z, z^{-1}] : z \in \mathbb{C}^\times\}$ be the diagonal subgroup, let N be the subgroup of upper-triangular unipotent matrices $\left[\begin{smallmatrix} 1 & z \\ 0 & 1 \end{smallmatrix}\right]$, $z \in \mathbb{C}$, and let $B = HN$ be the upper triangular subgroup.

(a) Show that G acts transitively on $\mathbb{P}(\mathbb{C})$. Find a point whose stabilizer is B.

(b) Show that H has one open dense orbit and two closed orbits on $\mathbb{P}(\mathbb{C})$. Show that N has one open dense orbit and one closed orbit on $\mathbb{P}(\mathbb{C})$.

(c) Identify $\mathbb{P}(\mathbb{C})$ with the two-sphere $\mathbf{S}^2$ by stereographic projection and give geometric descriptions of the orbits in (b).

6. (*Same notation as previous exercise*) Let G act on $\mathfrak{g}$ by the adjoint representation $\operatorname{Ad}(g)x = gxg^{-1}$. For $\mu \in \mathbb{C}$ define $X_\mu = \{x \in \mathfrak{g} : \operatorname{tr}(x^2) = 2\mu\}$. Use the Jordan canonical form to prove the following:

(a) If $\mu \neq 0$ then X_μ is a G orbit and $X_\mu \cong G/H$ as a G-space.

(b) If $\mu = 0$ then $X_0 = \{0\} \cup Y$ is the union of two G orbits, where Y is the orbit of $\left[\begin{smallmatrix} 0 & 1 \\ 0 & 0 \end{smallmatrix}\right]$. Show that $Y \cong G/\{\pm 1\}N$ and that Y is not closed in $\mathfrak{g}$.

7. (*Same notation as previous exercise*) Let $Z = \mathbb{P}(\mathfrak{g}) \cong \mathbb{P}^2(\mathbb{C})$ be the projective space of $\mathfrak{g}$, and let $\pi : \mathfrak{g} \longrightarrow Z$ be the canonical mapping.

(a) Show that G has two orbits on Z, namely $Z_1 = \pi(X_1)$ and $Z_0 = \pi(Y)$.

(b) Find subgroups L_1 and L_0 of G such that $Z_i \cong G/L_i$ as a G space. (HINT: Be careful; from the previous problem you know that $H \subset L_1$ and $N \subset L_0$, but these inclusions are proper.)

(c) Prove (without calculation) that one orbit must be closed in Z and one orbit must be dense in Z. Then calculate $\dim Z_i$ and identify the closed orbit. Find equations defining the closed orbit.

8. Let $X = \mathbb{C}^2 \setminus \{0\}$ with its structure as a quasiprojective algebraic set. Then $X = X_1 \cup X_2$, where $X_1 = \mathbb{C}^\times \times \mathbb{C}$ and $X_2 = \mathbb{C} \times \mathbb{C}^\times$ are affine open subsets. Also $f \in \mathcal{O}[X]$ if and only if $f|_{X_i} \in \mathcal{O}[X_i]$ for $i = 1, 2$.

(a) Prove that $\mathcal{O}[X] = \mathbb{C}[x_1, x_2]$, where x_i are the coordinate functions on $\mathbb{C}^2$. (HINT: Let $f \in \mathcal{O}[X]$. Write $f|_{X_1}$ as a polynomial in x_1, x_1^{-1}, x_2 and write $f|_{X_2}$ as a polynomial in x_1, x_2, x_2^{-1}. Then compare these expressions on $X_1 \cap X_2$.)

(b) Prove that X is not a projective algebraic set. (HINT: Consider $\mathcal{O}[X]$.)

(c) Prove that X is not an affine algebraic set. (HINT: By (a) there is a homomorphism $f \mapsto f(0)$ of $\mathcal{O}[X]$.)

(d) Let $G = \mathbf{SL}(2, \mathbb{C})$ and N the upper-triangular unipotent matrices in G. Prove that $G/N \cong \mathbb{C}^2 \setminus \{0\}$, with G acting as usual on $\mathbb{C}^2$. (HINT: Find a vector in $\mathbb{C}^2$ whose stabilizer is N.)

9. Let $G = \mathbf{SL}(2, \mathbb{C}) \times \mathbf{SL}(2, \mathbb{C})$. Let ρ be the representation of G on M_2 given by $\rho(g, h)z = gzh^t$. Let $\pi : \mathbb{C}^2 \times \mathbb{C}^2 \longrightarrow M_2$ by $\pi(x, y) = xy^t$. Identify $\mathbb{P}^3$ with $\mathbb{P}(M_2)$ and let $\tilde{\pi} : \mathbb{P}^1 \times \mathbb{P}^1 \longrightarrow \mathbb{P}^3$ be the map induced by π (the standard embedding of $\mathbb{P}^m \times \mathbb{P}^n$ in $\mathbb{P}^{mn+m+n}$).

(a) Show that the image of $\tilde{\pi}$ is $\{[z] : z \in M_2 \setminus \{0\} \text{ and } \det(z) = 0\}$.

(b) Let G act on $\mathbb{P}^1 \times \mathbb{P}^1$ by the natural action on $\mathbb{C}^2 \times \mathbb{C}^2$ and let G act on $\mathbb{P}^3$ by the representation ρ on M_2. Show that $\tilde{\pi}$ intertwines the G actions.

(c) Show that G has two orbits on $\mathbb{P}^3$ and describe the closed orbit.

10. Let $X = \{x \in M_{4,2} : \text{rank}(x) = 2\}$. For $J = (i_1, i_2)$ with $1 \le i_1 < i_2 \le 4$ let

$$X_J = \{x \in X : \xi_J(x) \ne 0\}, \quad \text{where} \quad \xi_J(x) = \det \begin{bmatrix} x_{i_1 1} & x_{i_1 2} \\ x_{i_2 1} & x_{i_2 2} \end{bmatrix}$$

is the Plücker coordinate corresponding to J.

(a) Let $A_{\{1,2\}} = \{x \in X : x_{ij} = \delta_{ij} \text{ for } 1 \le i, j \le 2\}$. Calculate the restrictions of the Plücker coordinates to $A_{\{1,2\}}$.

(b) Let $\mathbf{GL}(2, \mathbb{C})$ act by right multiplication on X. Show that $X_{\{1,2\}}$ is invariant under $\mathbf{GL}(2, \mathbb{C})$ and $A_{\{1,2\}}$ is a cross-section for the $\mathbf{GL}(2, \mathbb{C})$ orbits.

(c) Let $\pi : X \longrightarrow \text{Grass}_2(\mathbb{C}^4)$ map x to its orbit under $\mathbf{GL}(2, \mathbb{C})$. Let $\mathbf{GL}(4, \mathbb{C})$ act by left multiplication on X and hence also on $\text{Grass}_2(\mathbb{C}^4)$. Show that this action is transitive and calculate the stabilizer of $\pi([e_1 \ e_2])$, where e_i are the standard basis vectors for $\mathbb{C}^4$.

11.4 Borel Subgroups

With the flag manifolds available as a tool, we return to the structure theory of affine algebraic groups. We show that a Borel subgroup B is the unique maximal connected solvable subgroup (up to conjugacy) and that the conjugates of B cover G. An important consequence is that the centralizer of any torus in G is connected.

11.4.1 Solvable Groups

Let G be an (abstract) group. We say that G is *solvable* if there exists a nested chain of subgroups

$$G = G_0 \supset G_1 \supset \cdots \supset G_d \supset G_{d+1} = \{1\}$$

with G_{i+1} a normal subgroup of G_i and G_i/G_{i+1} commutative, for $i = 0, 1, \ldots, d$.

The *commutator subgroup* $\mathcal{D}(G)$ of G is the group generated by the set of commutators $\{xyx^{-1}y^{-1} : x, y \in G\}$. If G_1 is a normal subgroup of G, then G/G_1 is commutative if and only if $G_1 \supset \mathcal{D}(G)$. It follows that G *is solvable if and only if* $G \ne \mathcal{D}(G)$ *and* $\mathcal{D}(G)$ *is solvable.* Define the *derived series* $\{\mathcal{D}^n(G)\}$ of G inductively by

$$\mathcal{D}^0(G) = G, \qquad \mathcal{D}^{n+1}(G) = \mathcal{D}(\mathcal{D}^n(G)).$$

Then G is solvable if and only if $\mathcal{D}^{n+1}(G) = \{1\}$ for some n. In this case, the smallest such n is called the *solvable length* of G.

The archetypical example of a solvable group is the group B_n of $n \times n$ upper-triangular invertible matrices. To see this we observe that the group of unipotent upper-triangular matrices N_n^+ (with ones on the main diagonal) is a normal subgroup of B_n such that B_n/N_n^+ is isomorphic to the group of diagonal matrices. We set $N_{n,r}^+$ equal to the subgroup of N_n^+ consisting of elements such that the second through the rth diagonals are zero. Then $N_{n,r}^+$ is normal in B_n for $r \geq 2$ and $N_{n,r}^+/N_{n,r+1}^+$ is abelian. Note that the isotropy group of any full flag in $\mathbb{C}^n$ is conjugate in $\mathbf{GL}(n,\mathbb{C})$ to B_n and hence is solvable.

We also observe that if S is solvable and if $H \subset S$ is a subgroup then H is solvable. For example, let $G \subset \mathbf{GL}(n,\mathbb{C})$ be a connected classical group. Then the subgroup B in Theorem 11.3.6 is contained in the isotropy group of a full flag and hence is solvable.

The key fact needed to study connected solvable linear algebraic groups is the following result about commutator subgroups (where *closed* refers to the Zariski topology):

Proposition 11.4.1. *Assume that G is a connected linear algebraic group. Then $\mathcal{D}(G)$ is closed and connected.*

Proof. Set $C = \{xyx^{-1}y^{-1} : x, y \in G\}$. Then $C = C^{-1}$, $1 \in C$, and by definition

$$\mathcal{D}(G) = \bigcup_{n \geq 1} C^n \,,$$

where C^n is all products of n commutators. Because C^n is the image of $G \times \cdots \times G$ ($2n$ factors) under the regular map

$$(x_1, y_1, \ldots, x_n, y_n) \mapsto x_1 y_1 x_1^{-1} y_1^{-1} \cdots x_n y_n x_n^{-1} y_n^{-1} \,,$$

we know from Lemma A.1.15 and Theorem A.2.8 that $\overline{C^n}$ is irreducible and that C^n contains a nonempty open subset of $\overline{C^n}$. Now use the same argument as in Lemma 11.2.13 (1) to conclude that $\mathcal{D}(G)$ is closed and irreducible as an affine algebraic set. $\qquad\square$

11.4.2 Lie–Kolchin Theorem

A single linear transformation on $\mathbb{C}^n$ can always be put into upper-triangular form by a suitable choice of basis. The same is true for a connected solvable algebraic group.

Theorem 11.4.2. *Let G be a connected solvable linear algebraic group, and let (π, V) be a regular representation of G. Then there exist characters $\chi_i \in \mathcal{X}(G)$ and a flag*

$$V = V_1 \supset V_2 \supset \cdots \supset V_n \supset V_{n+1} = \{0\}$$

such that $(\pi(g) - \chi_i(g)I)V_i \subset V_{i+1}$ for $i = 1, \ldots n$ and all $g \in G$.

Proof. If $\dim V = 1$ then the theorem holds. Furthermore, if $U \subset V$ is a G-invariant subspace, and the theorem holds for the representations of G on U and on V/U, then it holds for V. Thus we may assume (by induction on $\dim V$) that

(1) G acts irreducibly on V.

The derived group $\mathcal{D}(G)$ is connected and solvable and has smaller dimension than G. Thus we may also assume (by induction on $\dim G$) that

(2)   the theorem holds for the restriction of π to $\mathcal{D}(G)$.

Now assume that (1) and (2) hold. There are regular characters $\theta_i : \mathcal{D}(G) \longrightarrow \mathbb{C}^\times$ and a flag $V = V_1 \supset \cdots \supset V_r \supset V_{r+1} = \{0\}$ with $\pi(x)v \equiv \theta_i(x)v \pmod{V_{i+1}}$ for $x \in \mathcal{D}(G)$ and $v \in V_i$. We first prove that G (acting by conjugation on $\mathcal{D}(G)$) fixes each character θ_i. Indeed, given $x \in \mathcal{D}(G)$, consider the map

$$ g \mapsto [\theta_1(gxg^{-1}), \ldots, \theta_r(gxg^{-1})] $$

from G to $\mathbb{C}^r$. The image consists of the eigenvalues of the operators $\pi(gxg^{-1})$ arranged in some order and possibly with repetitions. But π is a representation of G, so $\pi(gxg^{-1}) = \pi(g)\pi(x)\pi(g)^{-1}$, and hence all these operators have the same set of eigenvalues. Thus the image is a finite subset of $\mathbb{C}^r$. Since G is connected and the map is regular, the image is both connected and finite. This is possible only if the image is one point. This means that $\theta_i(gxg^{-1}) = \theta_i(x)$ for all $g \in G$ and $i = 1, \ldots, r$. If $v \in V$ and $\pi(x)v = \theta_r(x)v$ for all $x \in \mathcal{D}(G)$, then

$$ \pi(x)\pi(g)v = \pi(g)\pi(g^{-1}xg)v = \theta_r(g^{-1}xg)\pi(g)v = \theta_r(x)\pi(g)v . $$

Thus $\pi(x)v = \theta_r(x)v$ for all $v \in V$ and $x \in \mathcal{D}(G)$, since the space of vectors with this property contains the nonzero subspace V_r and is G-invariant. Write $\theta_r = \theta$.

Next we show that $\theta(x) = 1$ for all $x \in \mathcal{D}(G)$. Indeed, if $x = ghg^{-1}h^{-1}$ is a commutator of elements $g, h \in G$, then

$$ \det(\pi(x)) = \det(\pi(g)\pi(h)\pi(g)^{-1}\pi(h)^{-1}) = 1 . $$

Since $\mathcal{D}(G)$ is generated by commutators, it follows that $\det(\pi(x)) = 1$ for all $x \in \mathcal{D}(G)$. But $\det(\pi(x)) = \theta(x)^n$, so the range of θ is contained in the nth roots of unity. Hence $\theta = 1$, since $\mathcal{D}(G)$ is connected.

We have now shown that $\pi(\mathcal{D}(G)) = 1$. This means that the operators $\pi(g)$, for $g \in G$, mutually commute. Hence $\dim V = 1$ by irreducibility of π. Thus the operators $\pi(g)$ act by some character χ of G. This completes the induction. $\square$

Corollary 11.4.3. *Assume that G is a connected and solvable algebraic subgroup of $\mathbf{GL}(V)$. There exists a basis for V such that the elements of G are represented by upper-triangular matrices and the elements of $\mathcal{D}(G)$ have ones along the main diagonal. In particular, $\mathcal{D}(G)$ is unipotent.*

11.4.3 Structure of Connected Solvable Groups

The next theorem should be viewed as a generalization of the multiplicative Jordan decomposition of a single matrix.

Theorem 11.4.4. *Let G be a connected solvable linear algebraic group and let* $N = \text{Rad}_u(G)$. *Then there exists a torus* $T \subset G$ *such that* $G = T \cdot N$. *Furthermore,* $G \cong T \ltimes N$ *(semidirect product) as a group, and* $G \cong T \times N$ *as an affine variety. If* $x \in G$ *is semisimple and commutes with* T, *then* $x \in T$. *In particular,* T *is a maximal torus in* G.

Proof. If $G = N$ there is nothing to prove, so we may assume that N is a proper subgroup of G. We have $\mathcal{D}(G) \subset N$ by Corollary 11.4.3, so $S = G/N$ is a connected commutative algebraic group. From Corollary 11.2.8, S is reductive, and hence S is a torus, by Theorem 11.2.2. Let $\pi : G \longrightarrow S$ be the canonical quotient map. There exists $g \in G$ such that the subgroup generated by $\pi(g)$ is Zariski dense in S (Lemma 2.1.4). Taking the Jordan decomposition of g, we have $g_u \in N$. Hence we may assume that g is semisimple.

Let T be the Zariski closure of the subgroup generated by g. It has the following properties:

(1) $\pi(T) = S$.

This is clear, since $\pi(T)$ is a Zariski-closed subgroup of S (by Theorem 11.1.5) that contains $\pi(g)$.

(2) T is commutative and consists of semisimple elements.

Indeed, we may take g in diagonal form. Then the matrix entry functions x_{ij}, for $i \neq j$, vanish on g and hence vanish on elements of T. Thus T also consists of diagonal matrices.

It follows from (2) that $T \cap N = \{1\}$, so we have an isomorphism $\pi : T \cong S$ by (1). Thus T is a torus and $G = T \cdot N$. We define a group structure on the affine algebraic set $T \times N$ by

$$(t_1, n_1) \cdot (t_2, n_2) = (t_1 t_2, t_2^{-1} n_1 t_2 n_2) \quad \text{for } t_i \in T \text{ and } n_i \in N.$$

This makes $T \times N$ into a linear algebraic group (by Theorem 11.1.12), which we denote by $T \ltimes N$. The regular map $(t, n) \mapsto tn$ is an abstract group isomorphism from $T \ltimes N$ to G. Hence it is an isomorphism of algebraic groups by Corollary 11.1.16.

Let $x \in G$ commute with T. We can write $x = tn$ with $t \in T$ and $n \in N$. Thus $t^{-1}x = n$. Since t and x commute, the element $t^{-1}x$ is semisimple. Hence $n = 1$ and $x \in T$, proving that T is a maximal torus. $\square$

Theorem 11.4.5. *Let* $G = T \cdot N$ *be a connected solvable linear algebraic group, where* T *is a torus and* $N = \text{Rad}_u(G)$. *Let* $g \in G$ *be semisimple.*

1. g is conjugate under the action of N to an element of T.

2. $\mathrm{Cent}_G(g)$ *is connected.*

3. *Let* $b = tn \in G$, *with* $t \in T$ *and* $n \in N$. *Then there exist* $u, v \in N$ *such that* $vbv^{-1} = tu$ *and* $tu = ut$.

Proof. (1): To conjugate g into T, we use induction on $\dim N$. If $\dim N = 0$ there is nothing to prove. Suppose $\dim N = 1$. Let $\mathrm{Lie}(N) = \mathbb{C}X_0$. Then there is a character $\alpha \in \mathfrak{X}(T)$ such that $\mathrm{Ad}(t)X_0 = t^\alpha X_0$ for all $t \in T$. Let $g \in G$ be semisimple and write $g = t\exp(zX_0)$ for some $t \in T$ and $z \in \mathbb{C}$. We may assume $z \neq 0$. We claim that $t^\alpha \neq 1$. Indeed, if $t^\alpha = 1$ then t would commute with $\exp(zX_0)$. Hence by the uniqueness of the Jordan decomposition we would have $g_u = \exp(zX_0) \neq 1$, so g would not be semisimple. For any $y \in \mathbb{C}$ we have

$$\exp(yX_0)g\exp(-yX_0) = t\exp\left((t^{-\alpha} - 1)y + zX_0\right).$$

Hence taking $y = -(t^{-\alpha} - 1)^{-1}z$, we can conjugate g into T in this case using an element of N. Now assume $\dim N > 1$. Let Z be the center of N and $\mathfrak{z} = \mathrm{Lie}(Z)$. Then $Z = \exp\mathfrak{z}$ is a connected abelian algebraic group, $\dim Z \geq 1$, and $\mathrm{Ad}(T)(\mathfrak{z}) = \mathfrak{z}$. Since T is reductive, there is a one-dimensional $\mathrm{Ad}(T)$-invariant subspace $\mathfrak{z}_1 \subset \mathfrak{z}$. Set $Z_1 = \exp\mathfrak{z}_1$. Then Z_1 is a closed normal unipotent subgroup of G. Set $G_1 = T \ltimes (N/Z_1)$. Then $G_1 \cong G/Z_1$; furthermore, the induction hypothesis applies to G_1, since $\dim(N/Z_1) = \dim N - 1$. Let g_1 be the image of g in G_1. Then by induction, we may assume that $g_1 \in T$. Thus $g = tz$ for some $z \in Z_1$. The argument at the beginning of the proof now implies that g is conjugate to an element of T and the conjugation is implemented by an element of N.

(2): By (1), we may assume that $g \in T$. Let $\mathfrak{n} = \mathrm{Lie}(N)$ and take any $s \in G$. Then $s = t\exp X$, where $t \in T$ and $X \in \mathfrak{n}$, so we have

$$gsg^{-1} = t\exp(\mathrm{Ad}(g)X).$$

Hence $s \in \mathrm{Cent}_G(g)$ if and only if $X \in \mathfrak{n}_1$, where $\mathfrak{n}_1 = \mathrm{Ker}(\mathrm{Ad}(g)|_\mathfrak{n} - I)$. Thus $\mathrm{Cent}_G(g) = T \cdot \exp(\mathfrak{n}_1)$. Since $\exp : \mathfrak{n} \longrightarrow N$ is an isomorphism of affine algebraic sets, the product is connected.

(3): Let $b = b_s b_u$ be the Jordan–Chevalley decomposition of b. By (1) there exists $v \in N$ such that $vb_s v^{-1} \in T$. Since $vb_u v^{-1}$ is unipotent, it is in N, and we have the factorization

$$vbv^{-1} = (vb_s v^{-1})(vb_u v^{-1}).$$

But we can also write $vbv^{-1} = tu$, where $u = (t^{-1}vt)nv^{-1} \in N$. Comparing these two decompositions of vbv^{-1}, we see that

$$vb_s v^{-1} = t \quad \text{and} \quad vb_u v^{-1} = u.$$

Hence $tu = ut$ (since b_s and b_u commute), and $vbv^{-1} = tu$. $\square$

Corollary 11.4.6. *Suppose G is a connected solvable linear algebraic group. Let $A \subset G$ be a torus.*

1. *There exists $s \in G$ such that $sAs^{-1} \subset T$. In particular, if A is a maximal torus in G, then $sAs^{-1} = T$. Thus all maximal tori in G are conjugate.*
2. $\mathrm{Cent}_G(A)$ *is connected.*
3. *Let $x \in \mathrm{Cent}_G(A)$ be semisimple. Then there exists a torus $S \subset \mathrm{Cent}_G(A)$ with $A \cup \{x\} \subset S$.*

Proof. Take $g \in A$ such that the subgroup generated by g is Zariski dense in A (Lemma 2.1.4). There exists $s \in G$ such that $sgs^{-1} \in T$. This implies that $sAs^{-1} \subset T$, so (1) holds. Since $\mathrm{Cent}_G(s) = \mathrm{Cent}_G(A)$, we obtain (2) from Theorem 11.4.5.

To prove (3), we use (2) to apply Theorem 11.4.4 to the group $G_1 = \mathrm{Cent}_G(A)$. Thus $G_1 = T_1 \cdot N_1$, where $T_1 \supset A$ is a maximal torus in G_1 and $N_1 = \mathrm{Rad}_u(G_1)$. Now apply Theorem 11.4.5 to x and G_1 to obtain $s_1 \in G_1$ such that $s_1 x s_1^{-1} \in T_1$. Then $S = s_1^{-1} T_1 s_1$ is a torus containing x. Since $s_1^{-1} A s_1 = A$, we also have $A \subset S$. $\square$

11.4.4 Conjugacy of Borel Subgroups

A *Borel subgroup* of an algebraic group G is a maximal connected solvable subgroup.

Theorem 11.4.7. *Let G be a connected linear algebraic group. Then G contains a Borel subgroup B, and all other Borel subgroups of G are conjugate to B. The homogeneous space G/B is a projective variety. Furthermore, if S is any connected solvable subgroup of G such that G/S is a projective variety, then S is a Borel subgroup.*

To prove this theorem we shall use the following geometric generalization of the Lie–Kolchin theorem.

Theorem 11.4.8 (Borel Fixed Point). *Let S be a connected solvable group that acts algebraically on a projective variety X. Then there exists a point $x_0 \in X$ such that $s \cdot x_0 = x_0$ for all $s \in S$.*

Proof. We proceed by induction on $\dim S$, as in the proof of the Lie–Kolchin theorem. The theorem is true when $\dim S = 0$, since $S = \{1\}$ in this case by connectedness. We may assume that the theorem is true for the derived group $\mathcal{D}(S)$. Thus we know that

$$Y = \{x \in X : s \cdot x = x \text{ for all } s \in \mathcal{D}(S)\}$$

is a nonempty closed subset of X (the set of fixed points of a regular map is closed; see Section A.4.3). It is invariant under S, since $\mathcal{D}(S)$ is a normal subgroup. Let $O \subset Y$ be a closed S orbit (it exists by Corollary 11.3.2). Then O is a projective variety, and it is irreducible, since S is connected. On the other hand, if we fix $y \in O$, then $O \cong S/S_y$ as a quasiprojective variety. But $S_y \supset \mathcal{D}(S)$, so S_y is a closed normal subgroup of S. Hence S/S_y is an affine variety, by Theorem 11.1.17. Being both projective and affine, O is a single point by Corollary A.4.9. $\square$

Proof of Theorem 11.4.7. Let B be a Borel subgroup of G of maximum dimension. By Theorem 11.1.13 there exist a representation (π, V) of G and a point $y \in \mathbb{P}(V)$ such that B is the stabilizer of y. Set $X = \mathbb{P}(V)$ and $O = G \cdot y \subset X$. Then $G/B \cong O$ as a quasiprojective set. Set $Z = \overline{O}$ (Zariski closure in X). Then O is open in Z and if $z \in Z - O$ then $\dim G \cdot z < \dim O$. This implies that $\dim G_z > \dim B$. But the identity component of G_z is a connected solvable subgroup. This contradicts our choice of B. We therefore conclude that $Z = O$ is closed in X. If S is another Borel subgroup of G then S has a fixed point gB in $Z \cong G/B$. Thus $S \subset gBg^{-1}$ and hence $S = gBg^{-1}$ by maximality. If S is any connected solvable group such that G/S is a projective variety, then B has a fixed point hS on G/S. Thus $B \subset hSh^{-1}$ and hence $B = hSh^{-1}$ by maximality. $\qquad\square$

Example

Let G be a connected classical group and let B be the connected solvable subgroup in Theorem 11.3.6. The quotient space $X = G/B$ is a projective variety, and hence B is a Borel subgroup.

11.4.5 Centralizer of a Torus

When G is a connected classical group and H is a maximal torus in G we have proved that $\mathrm{Cent}_G(H) = H$ is connected (Theorem 2.1.5). We shall show that the same property holds for the centralizer of any torus in a connected linear algebraic group. This is a powerful technical result, since it will allow us to study centralizers by means of their Lie algebras. Its proof will require all the properties of connected solvable groups and Borel subgroups that we have established together with the following result:

Theorem 11.4.9. *Let G be a connected linear algebraic group and B a fixed Borel subgroup of G. Then*
$$G = \bigcup_{x \in G} xBx^{-1}.$$
Thus every element of G is contained in a Borel subgroup.

Proof. Let $Y = \bigcup_{x \in G} xBx^{-1}$. We first show that Y is closed in G (all topological assertions in the proof will refer to the Zariski topology). To see this we define
$$Z = \{(x, y) : x \in G/B, y \in G, \text{ and } y \cdot x = x\}.$$

Here $y \cdot (gB) = ygB$ denotes the action of G on G/B. We know by Corollary A.4.7 that Z is closed in $(G/B) \times G$, and
$$Y = \{y \in G : (x, y) \in Z \text{ for some } x \in G/B\}.$$

Since G/B is projective, Theorem A.4.10 then implies that Y is closed.

For m a positive integer let $G(m) = \{g \in G : g^m \in Y\}$ and let $\Phi_m : G \longrightarrow G$ be the regular map $\Phi_m(g) = g^m$. Since Y is closed in G and $G(m) = \Phi_m^{-1}(Y)$, it follows that $G(m)$ is closed in G. We claim that

$$G = \bigcup_{m>0} G(m).\tag{11.12}$$

Indeed, given $g \in G$ we let U be the closure of the group $\{g^k : k \in \mathbb{Z}\}$. The identity component U° of U is connected and commutative; hence it is contained in a maximal connected solvable subgroup. Thus Theorem 11.4.7 implies that $U^\circ \subset Y$. Since U/U° is finite, there exists m such that $g^m \in U^\circ$, and hence $g \in G(m)$. This proves (11.12).

Since each set $G(m)$ is closed, we conclude from (11.12) and Proposition A.4.12 that there exists an integer m_0 such that $G = G(m_0)$. Let $\Psi = \Phi_{m_0}$. Then $d\Psi_1(v) = m_0 v$ for $v \in T(G)_1$ (this is most easily seen by embedding G in $\mathbf{GL}(n, \mathbb{C})$ for some n). Thus $d\Psi_1$ is surjective, and hence Theorem A.3.4 implies that $\Psi(G)$ is dense in G. Since $\Psi(G) \subset Y$, we have shown that Y is both dense in G and closed in G. Thus $Y = G$. $\qquad\square$

Theorem 11.4.10. *Let G be a connected linear algebraic group. Suppose $A \subset G$ is a torus. Then $\mathrm{Cent}_G(A)$ is connected. Furthermore, if $x \in \mathrm{Cent}_G(A)$ is semisimple, then there exists a torus $S \subset \mathrm{Cent}_G(A)$ such that $A \cup \{x\} \subset S$.*

Proof. Let $x \in \mathrm{Cent}_G(A)$. By Theorem 11.4.9 there exists a Borel subgroup B containing x. Let $Y \subset G/B$ be the fixed-point set for the action of x on G/B. Then Y is nonempty (since $B \in Y$) and is closed in the projective variety G/B. Hence Y is a projective algebraic set. If $a \in A$ and $gB \in Y$, then $xagB = axgB = agB$. Hence Y is invariant under A, so by Theorem 11.4.8 there exists $g \in G$ such that $xgB = gB$ and $agB = gB$ for all $a \in A$. This means that x and A are contained in the Borel subgroup $B_1 = gBg^{-1}$. Thus $x \in \mathrm{Cent}_{B_1}(A)$. But we know by Corollary 11.4.6 that $\mathrm{Cent}_{B_1}(A)$ is connected. This implies that $M = \mathrm{Cent}_G(A)$ is connected. Indeed, suppose U_1 and U_2 are open subsets of M such that $U_1 \cap U_2 = \emptyset$. Since $A \subset M$ is connected, we may assume that $A \subset U_1$. But we just showed that if $x \in M$ then $x \in \mathrm{Cent}_{B_1}(A)$ for some Borel subgroup $B_1 \supset A$. Since $U_2 \cap A = \emptyset$, this implies that $\mathrm{Cent}_{B_1}(A) \subset U_1$. Hence $U_2 = \emptyset$. The last statement of the theorem now follows by part (3) of Corollary 11.4.6 (since A and x are in the Borel subgroup B_1). $\qquad\square$

11.4.6 Weyl Group and Regular Semisimple Conjugacy Classes

Let G be a connected linear algebraic group whose Lie algebra $\mathfrak{g}$ is semisimple. Fix a maximal algebraic torus in G with Lie algebra $\mathfrak{h}$. Let U be the compact real form of G constructed in Section 7.3.1 relative to $\mathfrak{h}$.

We define the *Weyl group* $W_G = \mathrm{Norm}_G(H)/H$, just as we did in Section 3.1.1 for the classical groups. The adjoint representation of G on $\mathfrak{g}$ restricts to a representation of W_G on $\mathfrak{h}$. Also W_G acts on the character group $\mathfrak{X}(H)$ and acts faithfully as a permutation of the set of roots $\Phi(\mathfrak{g}, \mathfrak{h})$, by the same proof as for the classical groups. In particular, W_G is finite (see Theorem 3.1.1). Let $W(\mathfrak{g}, \mathfrak{h}) \subset \mathbf{GL}(\mathfrak{h})$ be the *algebraic Weyl group* generated by root reflections; see Definition 3.1.8.

Theorem 11.4.11. *The action of W_G on $\mathfrak{h}$ coincides with the action of $W(\mathfrak{g}, \mathfrak{h})$. Furthermore, every coset in W_G has a representative from U.*

Proof. For $\alpha \in \Phi$ and X_α as in (7.42), set

$$u_\alpha = \frac{1}{2}(X_\alpha - X_{-\alpha}) \quad \text{and} \quad v_\alpha = \frac{1}{2i}(X_\alpha + X_{-\alpha}). \tag{11.13}$$

Then $X_\alpha = u_\alpha + iv_\alpha$, and $u_\alpha, v_\alpha \in \mathfrak{u}$. We calculate the action of $\mathrm{ad}\,u_\alpha$ on $h \in \mathfrak{h}$ as follows:

$$[u_\alpha, h] = -\frac{1}{2}\langle \alpha, h \rangle (X_\alpha + X_{-\alpha}) = -i\langle \alpha, h \rangle v_\alpha, \tag{11.14}$$

$$[u_\alpha, v_\alpha] = \frac{1}{4i}[X_\alpha - X_{-\alpha}, X_\alpha + X_{-\alpha}] = \frac{1}{2i}[X_\alpha, X_{-\alpha}] = \frac{1}{2i}H_\alpha. \tag{11.15}$$

From (11.14) we see that for all $s \in \mathbb{C}$,

$$\exp(s\,\mathrm{ad}\,u_\alpha)h = h \quad \text{if } \langle \alpha, h \rangle = 0. \tag{11.16}$$

Taking $h = H_\alpha$ in (11.14), we obtain $\mathrm{ad}(u_\alpha)H_\alpha = i\|\alpha\|^2 v_\alpha$, and hence by (11.15) we have

$$(\mathrm{ad}\,u_\alpha)^2 H_\alpha = -r^2 H_\alpha,$$

where $r = \|\alpha\|/\sqrt{2}$. Continuing in this way, we calculate that

$$\exp(s\,\mathrm{ad}\,u_\alpha)H_\alpha = \cos(rs)H_\alpha - (2i/r)\sin(rs)v_\alpha$$

for all $s \in \mathbb{C}$. In particular, when $s = \pi/r$ the second term vanishes and we have

$$\exp((\pi/r)\,\mathrm{ad}\,u_\alpha)H_\alpha = -H_\alpha. \tag{11.17}$$

Thus by (11.16) and (11.17) we see that the element $g_\alpha = \exp((\pi/r)u_\alpha) \in U$ acts on $\mathfrak{h}$ by the reflection s_α. This proves that every element in the algebraic Weyl group $W(\mathfrak{g}, \mathfrak{h})$ can be implemented by the adjoint action of an element $k \in \mathrm{Norm}_G(H) \cap U$.

To prove the converse, fix a set Φ^+ of positive roots and let

$$\rho = \frac{1}{2}\sum_{\alpha \in \Phi^+}\alpha.$$

If $g \in \mathrm{Norm}_G(H)$, then $\mathrm{Ad}(g)^t \Phi^+ \subset \Phi$ is another set of positive roots, so by Theorem 3.1.9 there is an element $s \in W(\mathfrak{g}, \mathfrak{h})$ such that $s\,\mathrm{Ad}(g)^t \Phi^+ = \Phi^+$. Hence by what has just been proved, there exists $k \in \mathrm{Norm}_G(H) \cap U$ such that $\mathrm{Ad}(k)|_{\mathfrak{h}} = s$, so $\mathrm{Ad}(kg)^t \Phi^+ = \Phi^+$. Thus kg fixes ρ, which means that $\mathrm{Ad}(kg)H_\rho = H_\rho$. The element

H_ρ satisfies $\langle \alpha, H_\rho \rangle \neq 0$ for all $\alpha \in \Phi$ by Lemma 3.1.21. Hence

$$\mathfrak{h} = \{X \in \mathfrak{g} : [H_\rho, X] = 0\} . \tag{11.18}$$

We claim that the one-parameter subgroup

$$\Gamma = \{\exp(tH_\rho) : t \in \mathbb{C}\} \subset H$$

is dense in H (in the Lie group topology). Indeed, the closure (in the Lie group topology) of Γ in G is a closed Lie subgroup of H that has Lie algebra $\mathfrak{h}$ by (11.18) and Theorem 1.3.8, hence coincides with H, since H is connected. But kg commutes with the elements of Γ, so the semisimple and unipotent factors of kg in its Jordan–Chevalley decomposition also commute with Γ, and hence with H. Since the unipotent factor is of the form $\exp X$ with X nilpotent, it follows from (11.18) that $X \in \mathfrak{h}$, and hence $X = 0$. Thus kg is semisimple and commutes with the elements of H, so $kg \in H$, since H is a maximal algebraic torus. This completes the proof of the theorem. $\square$

Remark 11.4.12. For the classical groups, Theorem 11.4.11 can be easily proved on a case-by-case basis using the descriptions of W_G in Section 3.1.1; see Goodman–Wallach [56, Lemma 7.4.3].

Corollary 11.4.13. *The natural inclusion map* $\mathrm{Norm}_U(T)/T \longrightarrow \mathrm{Norm}_G(H)/H$ *is an isomorphism.*

Proof. This follows from Theorem 11.4.11 and Proposition 7.3.2. $\square$

Define the *regular elements* in the maximal torus H as

$$H' = \{h \in H : h^\alpha \neq 1 \text{ for all } \alpha \in \Phi\} .$$

We note that

$(\star)$ $h \in H$ is regular if and only if $(\mathrm{Cent}_G(h))^\circ = H$.

Indeed, let $M = \mathrm{Cent}_G(h)$. We have $X \in \mathrm{Lie}(M)$ if and only if

$$I = \exp(-tX)h\exp(tX)h^{-1} = \exp(-tX)\exp(t\,\mathrm{Ad}(h)X) \quad \text{for all } t \in \mathbb{C} .$$

Differentiating this equation at $t = 0$, we find that $X \in \mathrm{Ker}(\mathrm{Ad}(h) - I)$. If $h^\alpha = 1$ for some root α, then the one-parameter unipotent group $\exp(\mathfrak{g}_\alpha)$ is contained in M. Thus if h is not regular, then M° is strictly larger than H. Conversely, suppose that $h \in H'$. Then $\mathrm{Ker}(\mathrm{Ad}(h) - I) = \mathfrak{h}$. Hence we see that $\mathrm{Lie}(M) = \mathfrak{h}$. Hence $M^\circ = H$. This proves $(\star)$.

Remark 11.4.14. The group $\mathrm{Cent}_G(h)$ is not necessarily H when $h \in H'$. For example, take $G = \mathbf{PSL}(2, \mathbb{C}) = \mathbf{SL}(2, \mathbb{C})/\{\pm I\}$, H the diagonal matrices modulo $\pm I$, and $h = \pm \mathrm{diag}[\mathrm{i}, -\mathrm{i}]$. Then $w = \pm \begin{bmatrix} 0 & 1 \\ -1 & 0 \end{bmatrix}$ commutes with h but is not in H.

If $h \in H'$, then by $(\star)$ we see that $\text{Cent}_G(h) \subset \text{Norm}_G(H)$. Hence if $\text{Cent}_G(h) \neq H$, then there exists $1 \neq w \in W_G$ such that $whw^{-1} = h$. This leads us to define the *strongly regular* elements in H as

$$H'' = \{h \in H : h^\alpha \neq 1 \text{ and } whw^{-1} \neq h \text{ for all } \alpha \in \Phi \text{ and } 1 \neq w \in W_G\}. \quad (11.19)$$

Clearly H'' is an open, dense subset of H. From the remarks preceeding (11.19) we conclude that

$(\star\star)$ $\quad \text{Cent}_G(h) = H$ for every $h \in H''$.

Lemma 11.4.15. *Define a map* $\Psi : (G/H) \times H \longrightarrow G$ *by* $\Psi(gH, h) = ghg^{-1}$. *If* $g \in G$ *and* $h \in H''$, *then*

$$\Psi^{-1}(ghg^{-1}) = \{(gwH, w^{-1}hw) : w \in W_G\},$$

and this set has cardinality $|W_G|$.

Proof. Let $h \in H''$. Suppose $g_1 \in G$ and $h_1 \in H$ satisfy $g_1 h_1 g_1^{-1} = ghg^{-1}$. Set $w = g^{-1}g_1$. Then $wh_1 = hw$. Given any $h_2 \in H$, we have

$$hwh_2w^{-1} = wh_1h_2w^{-1} = wh_2h_1w^{-1} = wh_2w^{-1}h.$$

Hence $wh_2w^{-1} \in \text{Cent}_G(h) = H$ by $(\star\star)$. This shows that $w \in \text{Norm}_G(H)$ and $(g_1, h_1) = (wg, w^{-1}hw)$. Furthermore, $g_1 h g_1^{-1} = ghg^{-1}$ if and only if $g^{-1}g_1 \in H$. Hence w is uniquely determined as an element of W_G. $\qquad \square$

Define $G' = \{ghg^{-1} \in H : g \in G, \ h \in H'\}$. We call the elements of G' the *regular semisimple elements* in G. Fix a set Φ^+ of positive roots, and let $B = HN^+$ be the corresponding Borel subgroup. Given $\alpha \in \mathcal{X}(H)$, we extend α to a character of B by setting $b^\alpha = h^\alpha$ for $b = hn$ with $h \in H$ and $n \in N^+$. Then the regular semisimple elements in B have the following explicit characterization:

Lemma 11.4.16. *An element* b *is in* $B \cap G'$ *if and only if* $b^\alpha \neq 1$ *for all* $\alpha \in \Phi$. *Thus* $B \cap G' = H'N^+$ *is open and Zariski dense in* B.

Proof. Write $b = hn$ with $h \in H$ and $n \in N^+$. By Theorem 11.4.5 (3), b is N^+-conjugate to $b' = hu$, where $u \in N^+$ and $hu = uh$. Hence $b^\alpha = (b')^\alpha$.

By definition, $b \in G'$ if and only $b' \in G'$. If $b^\alpha \neq 1$ for all $\alpha \in \Phi$, then $h \in H'$. Since $u = \exp(X)$ for some $X \in \mathfrak{n}^+$, we have $u \in (\text{Cent}_G(h))^\circ$. Hence $(\star)$ implies that $u = I$, and so $b' = h \in H'$. Conversely, if $b' \in G'$, then b' is semisimple, and so $u = I$. Hence $b' \in H'$ and $b^\alpha \neq 1$ for all $\alpha \in \Phi$. $\qquad \square$

Remark 11.4.17. For $G = \mathbf{SL}(n, \mathbb{C})$, Lemma 11.4.16 asserts that the regular semisimple upper-triangular matrices are exactly those with distinct diagonal entries.

Theorem 11.4.18. *The set* G' *of regular semisimple elements is Zariski dense and open in* G.

Proof. If $f \in \mathcal{O}[G]$ and $f(G') = 0$, then by (11.4.16) we have $f(gH'N^+g^{-1}) = 0$ and hence $f(gBg^{-1}) = 0$ for all $g \in G$. Thus Theorem 11.4.9 implies that $f = 0$, proving the density of G'.

To prove that G' is open, consider the set of *singular elements*

$$S = \{b \in B : b^\alpha = 1 \text{ for some } \alpha \in \Phi\}$$

in B. Clearly S is closed. Note that if $x, y \in G$ and $x^{-1}yx \in S$, then for any $b \in B$ we also have $(xb)^{-1}y(xb) \in S$. Let

$$C = \{(xB, y) : x^{-1}yx \in S\} \subset (G/B) \times G.$$

Then C is closed. Since every element of G is conjugate to an element of B, the complement of G' is the image of C under the projection $(xB, y) \mapsto y$. Since G/B is projective, this image is closed, by Theorem A.4.10. Hence G' is open. $\square$

Remark 11.4.19. Let $G'' = \{xH''x^{-1} : x \in G\}$ be the *strongly regular semisimple elements* in G. Then G'' is open and Zariski dense in G, by the same proof as in Theorem 11.4.18.

11.4.7 Exercises

1. Let (π, V) be a regular representation of the algebraic group G. Prove that the action of G on $\mathbb{P}(V)$ is algebraic.
2. Let G be a connected algebraic group. Show that G is solvable if and only if there is a normal, connected, and solvable algebraic subgroup $H \subset G$ such that G/H is solvable.
3. Let $G \subset \mathbf{GL}(n, \mathbb{C})$ be a solvable group. Show that the Zariski closure of G is solvable.
4. Let $G = \mathbf{GL}(n, \mathbb{C})$, H the diagonal matrices in G, N the upper-triangular unipotent matrices, and $B = HN$. Let X be the space of all flags in $\mathbb{C}^n$.
   (a) Suppose that $x = \{V_1 \subset V_2 \subset \cdots \subset V_n\}$ is a flag that is invariant under H. Prove that there is a permutation $\sigma \in \mathfrak{S}_n$ such that

   $$V_i = \mathrm{Span}\{e_{\sigma(1)}, \ldots, e_{\sigma(i)}\} \quad \text{for } i = 1, \ldots, n.$$

   (HINT: H is reductive and its action on $\mathbb{C}^n$ is multiplicity-free.)
   (b) Suppose that the flag x in (a) is also invariant under N. Prove that $\sigma(i) = i$ for $i = 1, \ldots, n$.
   (c) Prove that if $g \in G$ and $gBg^{-1} = B$, then $g \in B$. (HINT: By (a) and (b), B has exactly one fixed point on $X = G/B$.)
5. Let G be a connected algebraic group and $B \subset G$ a Borel subgroup. Let $P \subset G$ be a closed subgroup.

(a) P is called a *parabolic subgroup* of G if G/P is projective. If P is parabolic, prove that there exists $g \in G$ such that $gBg^{-1} \subset P$. (HINT: B has a fixed point on G/P.)

(b) Suppose that $B \subset P$. Prove that G/P is a projective algebraic set. (HINT: Consider the natural map $G/B \longrightarrow G/P$.)

6. Let G be a connected semisimple group. Let B be a Borel subgroup of G, and $H \subset B$ a maximal torus of G. Suppose that $P \subset G$ is a closed subgroup such that $B \subset P$.

(a) Let $\mathfrak{b} = \mathrm{Lie}(B)$ and let Φ^+ be the positive roots of $\mathfrak{g}$ relative to $\mathfrak{b}$. Prove that $\mathrm{Lie}(P)$ is of the form

$$\mathfrak{b} + \sum_{\alpha \in S} \mathfrak{g}_{-\alpha} \qquad (\star)$$

for some subset S of Φ^+. (HINT: $\mathrm{Lie}(P)$ is invariant under $\mathrm{Ad}(H)$.)

(b) Let $S \subset \Phi^+$ be any subset and let $\{\alpha_1, \dots, \alpha_l\}$ be the simple roots in Φ^+. Prove that the subspace defined by $(\star)$ is a Lie algebra if and only if S satisfies the following properties:

(P1)   If $\alpha, \beta \in S$ and $\alpha + \beta \in \Phi^+$, then $\alpha + \beta \in S$.
(P2)   If $\beta \in S$ and $\beta - \alpha_i \in \Phi^+$ then $\beta - \alpha_i \in S$.

(HINT: $\mathfrak{b}$ is generated by $\mathfrak{h}$ and $\{\mathfrak{g}_{\alpha_i} : i = 1, \dots, l\}$.)

(c) Let R be any subset of the simple roots, and define S_R to be all the positive roots β such that no elements of R occur when β is written as a linear combination of the simple roots. Show that S_R satisfies (P1) and (P2). Conversely, if S satisfies (P1) and (P2), let R be the set of simple roots that do not occur in any $\beta \in S$. Prove that $S = S_R$.

(d) Let $G = \mathbf{GL}(n, \mathbb{C})$. Use (c) to determine all subsets S of Φ^+ that satisfy (P1) and (P2). (HINT: Use Exercise 2.4.5 #2(a).)

(e) For each subset S found in (d), show that there is a closed subgroup $P \supset B$ with $\mathrm{Lie}(P)$ given by $(\star)$. (HINT: Show that S corresponds to a partition of n and consider the corresponding block decomposition of G.)

11.5 Further Properties of Real Forms

We now turn to the structure of a reductive algebraic group G as a real Lie group. We study conjugations and involutive automorphisms of G, and we obtain the *polar decomposition* of G relative to a compact real form U.

11.5.1 Groups with a Compact Real Form

Theorem 11.5.1. *A connected linear algebraic group is reductive if and only if it has a compact real form.*

Proof. A connected linear algebraic group G is also connected in the Lie group topology, by Theorem 11.2.9. If it has a compact real form, then it is reductive, by Theorem 3.3.15. Conversely, if G is reductive and has Lie algebra $\mathfrak{g}$, then $\mathfrak{g} = \mathfrak{z} \oplus [\mathfrak{g}, \mathfrak{g}]$ with $\mathfrak{z}$ the center of $\mathfrak{g}$ and $[\mathfrak{g}, \mathfrak{g}]$ semisimple, by Corollary 2.5.9. Thus G has a compact real form (see Section 7.3.1). $\qquad\square$

Remark 11.5.2. The classical groups are reductive, and we gave explicit constructions of their compact forms in Section 1.7.2.

Assume that G is connected and reductive, and that τ_0 is a conjugation on G (see Section 1.7) such that the corresponding real form is compact. The main result of this section is the following:

Theorem 11.5.3. *Let σ denote either a complex conjugation on G or an involutive automorphism of G such that σ acts by the identity on the identity component of the center of G. Then there exists $g \in G$ such that the automorphism $\tau(x) = g\tau_0(x)g^{-1}$, for $x \in G$, satisfies $\tau\sigma = \sigma\tau$.*

This theorem will be proved at the end of the section. We first prove a lemma needed for an important corollary of the theorem. Throughout this section for notational convenience we shall write τ_0 instead of $d\tau_0$ (it will be clear from the context that the action is on $\mathfrak{g}$); likewise, we shall write σ and τ instead of $d\sigma$ and $d\tau$.

Lemma 11.5.4. *Let G and τ_0 be as above and assume that G has finite center. Set $\mathfrak{g} = \mathrm{Lie}(G)$. Let $U = \{g \in G : \tau_0(g) = g\}$ and identify $\mathfrak{u} = \mathrm{Lie}(U)$ with the space of all $X \in \mathfrak{g}$ such that $\tau_0(X) = X$. Then $\mathrm{tr}(\mathrm{ad}(X)^2) < 0$ for all $0 \neq X \in \mathfrak{u}$.*

Proof. We apply the unitary trick (Section 3.3.4) to the representation $(\mathrm{Ad}|_U, \mathfrak{g})$. Thus there is a positive definite Hermitian inner product $(\cdot, \cdot)$ on $\mathfrak{g}$ such that

$$(\mathrm{Ad}(u)X, \mathrm{Ad}(u)Y) = (X, Y) \quad \text{for } u \in U.$$

This implies that if $Z \in \mathfrak{u}$ and $X, Y \in \mathfrak{g}$ then $([Z, X], Y) = -(X, [Z, Y])$. Thus $\mathrm{ad}\,Z$ is skew-adjoint with respect to $(\cdot, \cdot)$, so it is diagonalizable with purely imaginary eigenvalues. Thus $\mathrm{tr}(\mathrm{ad}(Z)^2) \leq 0$, with equality if and only if $\mathrm{ad}\,Z = 0$. If $\mathrm{ad}\,Z = 0$ then Z is in the Lie algebra of the center of G, which is $\{0\}$. $\qquad\square$

From Theorem 11.5.3 and Lemma 11.5.4 we obtain the uniqueness of compact real forms modulo inner automorphisms.

Corollary 11.5.5. *If G is reductive with finite center and if U_1 and U_2 are compact real forms of G, then there exists $g \in G$ such that $gU_1g^{-1} = U_2$.*

Proof. Let τ_1 and τ_2 be the conjugations of G corresponding to U_1 and U_2. Then Theorem 11.5.3 implies that there exists $g \in G$ such that the automorphism $\tau_3(x) = g\tau_1(x)g^{-1}$, for $x \in G$, satisfies $\tau_3\tau_2 = \tau_2\tau_3$. Let $\mathfrak{u}_3 = \mathrm{Lie}(U_3) = \mathrm{Ad}(g)\mathfrak{u}_1$. Then $\tau_2(\mathfrak{u}_3) = \mathfrak{u}_3$. So

$$\mathfrak{u}_3 = (\mathfrak{u}_3 \cap \mathfrak{u}_2) \oplus (\mathfrak{u}_3 \cap i\mathfrak{u}_2).$$

Now Lemma 11.5.4 implies that if $X \in \mathfrak{u}_3 \cap i\mathfrak{u}_2$ then $0 \leq \mathrm{tr}(\mathrm{ad}(X)^2) \leq 0$. Thus $X = 0$, so we have $\mathfrak{u}_3 \subset \mathfrak{u}_2$. By symmetry $\mathfrak{u}_2 \subset \mathfrak{u}_3$. Thus $\mathfrak{u}_2 = \mathfrak{u}_3$. But this implies that if $X, Y \in \mathfrak{u}_2$ then

$$\tau_2(\exp(X + iY)) = \exp(X - iY) = \tau_3(\exp(X + iY)). \qquad (11.20)$$

Since G is connected, it is generated by $\exp(\mathfrak{g})$. Thus from (11.20) we see that $\tau_2 = \tau_3$, and hence $U_2 = U_3$.   □

We now develop some additional results needed for the proof of Theorem 11.5.3. For $A \in M_n(\mathbb{C})$ we define the *Hermitian adjoint* A^* to be $A^* = \bar{A}^t$, where the bar denotes complex conjugation.

Lemma 11.5.6. *Let $A \in M_n(\mathbb{C})$ be such that $A^* = A$. Let f be a polynomial on $M_n(\mathbb{C})$ such that $f(\exp(mA)) = 0$ for all positive integers m. Then $f(\exp(tA)) = 0$ for all $t \in \mathbb{R}$.*

Proof. There is a basis $\{e_1, \ldots, e_n\}$ of $\mathbb{C}^n$ and $\lambda_i \in \mathbb{R}$ such that $Ae_i = \lambda_i e_i$. Rewriting f in terms of matrix entries with respect to this basis, we see that the lemma reduces to the following assertion:

$(\star)$ Let $f \in \mathbb{C}[x_1, \ldots, x_n]$ and assume that $f(e^{m\lambda_1}, \ldots, e^{m\lambda_n}) = 0$ for all positive integers m. Then $f(e^{t\lambda_1}, \ldots, e^{t\lambda_n}) = 0$ for all $t \in \mathbb{R}$.

To prove $(\star)$, write $f = \sum_I a_I x^I$ and consider the set

$$\left\{ \textstyle\sum_{j=1}^n i_j \lambda_j : I = [i_1, \ldots, i_n] \quad \text{such that } a_I \neq 0 \right\}.$$

We enumerate this set of real numbers as $a_1 > a_2 > \cdots > a_p$. Then there exist complex numbers c_j such that

$$\varphi(t) = f(e^{t\lambda_1}, \ldots, e^{t\lambda_n}) = \sum_{j=1}^p c_j e^{ta_j}.$$

Now $\varphi(m) = 0$ for all positive integers m. Thus

$$0 = \lim_{m \to +\infty} e^{-ma_1} \varphi(m) = c_1.$$

Hence by induction on p we conclude that all the coefficients c_j are zero. Thus $\varphi = 0$ as asserted.   □

Let $(z, w) = z^t \bar{w}$ denote the usual Hermitian inner product on $\mathbb{C}^n$. If $A \in M_n(\mathbb{C})$ satisfies

$$A^* = A \quad \text{and} \quad (Az, z) > 0 \quad \text{for all } z \in \mathbb{C}^n \setminus \{0\},$$

then we say that A is *positive definite*. If A is positive definite then there is an orthonormal basis $\{f_1, \ldots, f_n\}$ of $\mathbb{C}^n$ such that $Af_i = \lambda_i f_i$ and $\lambda_i > 0$. Define $\log A \in M_n(\mathbb{C})$ to be the element X such that $Xf_i = \log(\lambda_i)f_i$ for $i = 1, \ldots, n$. Then $X^* = X$ and $A = \exp X$.

Set $\mathrm{Herm}_n = \{X \in M_n(\mathbb{C}) : X^* = X\}$ and set

$$\Omega_n = \{A \in M_n(\mathbb{C}) : A \text{ is positive definite}\}.$$

Notice that Ω_n is open in the real vector space Herm_n.

Lemma 11.5.7. *The map* $\Psi : \mathrm{Herm}_n \to \Omega_n$ *given by* $\Psi(X) = \exp X$ *is a diffeomorphism of* Herm_n *onto* Ω_n.

Proof. Suppose $A \in \Omega_n$ and $\exp X = \exp Y = A$ with $X, Y \in \mathrm{Herm}_n$. Then

$$\exp(mX) = \exp(mY) = A^m \quad \text{for } m = 1, 2, \dots.$$

Thus Lemma 11.5.6 implies that $\exp(tX) = \exp(tY)$ for all $t \in \mathbb{R}$. Differentiating this equation at $t = 0$ yields $X = Y$. Thus Ψ is one-to-one. The discussion preceding this lemma implies that Ψ is surjective. Thus by the inverse function theorem we need only to show that $\mathrm{d}\Psi_X$ is injective for all $X \in \mathrm{Herm}_n$.

Suppose $X, Y \in \mathrm{Herm}_n$. Let $v, w \in \mathbb{C}^n$. Then

$$(\exp(X + tY)v, w) = (\exp(X)v, w)$$

$$+ t \left\{ \sum_{m \geq 1} \frac{1}{m!} \sum_{i=0}^{m-1} (X^{m-i-1}YX^i v, w) \right\} + \mathrm{O}(t^2).$$

Differentiating at $t = 0$ and using the fact that $X^* = X$, we obtain

$$(\mathrm{d}\Psi_X(Y)v, w) = \sum_{m \geq 1} \frac{1}{m!} \sum_{i=0}^{m-1} (YX^i v, X^{m-i-1}w).$$

Now assume that $Xv = \lambda v$ and $Xw = \mu w$ for some $\lambda, \mu \in \mathbb{R}$. Then

$$(\mathrm{d}\Psi_X(Y)v, w) = \left\{ \sum_{m \geq 1} \frac{1}{m!} \sum_{i=0}^{m-1} \lambda^{m-i-1}\mu^i \right\} (Yv, w). \tag{11.21}$$

Assume now that $\mathrm{d}\Psi_X(Y) = 0$. If $\lambda \neq \mu$ then (11.21) implies that

$$0 = \left\{ \frac{1}{(\lambda - \mu)} \sum_{m \geq 1} \frac{\lambda^m - \mu^m}{m!} \right\} (Yv, w) = \frac{e^\lambda - e^\mu}{(\lambda - \mu)} (Yv, w).$$

Thus $(Yv, w) = 0$, since $e^\lambda \neq e^\mu$. If $\lambda = \mu$ then (11.21) implies that

$$0 = \sum_{m \geq 1} \frac{1}{m!} m\lambda^{m-1} (Yv, w) = e^\lambda (Yv, w).$$

Thus in all cases we see that $(Yv, w) = 0$. Since X is diagonalizable this implies that there is a basis $\{f_1, \dots, f_n\}$ of $\mathbb{C}^n$ such that $(Yf_i, f_j) = 0$ for all i, j. Hence $Y = 0$, proving that $\mathrm{d}\Psi_X$ is injective. $\qquad \square$

In light of Lemma 11.5.7, the inverse map to the map $X \mapsto \exp X$ from Herm_n to Ω_n is given by the matrix-valued function $\log : \Omega_n \longrightarrow \mathrm{Herm}_n$ defined above in terms of eigenspace decompositions. We define

$$A^s = \exp(s \log A) \quad \text{for } A \in \Omega_n \text{ and } s \in \mathbb{R}.$$

Proof of Theorem 11.5.3. Let U be the compact real form of G corresponding to τ_0. We first note that since σ and τ_0 are automorphisms of G as a real Lie group, they map the identity component Z° of the center of G into itself. But by assumption σ and τ_0 commute on Z°, since σ acts by the identity there. Thus it is enough to prove the theorem with G replaced by G/Z°.

Let $\mathfrak{g} = \mathrm{Lie}(G)$ and set

$$B(X,Y) = \mathrm{tr}(\mathrm{ad}(X)\,\mathrm{ad}(Y)) \quad \text{for } X,Y \in \mathfrak{g}.$$

Let U be the compact form of G corresponding to τ_0. Then we have seen that $B(X,X) < 0$ for $X \in \mathfrak{u} = \mathrm{Lie}(U)$, since we are assuming that the center of $\mathfrak{g}$ is 0. Thus if we set

$$(X,Y) = -B(X, \tau_0 Y) \quad \text{for } X,Y \in \mathfrak{g},$$

then $(\cdot, \cdot)$ is a positive definite Hermitian inner product on $\mathfrak{g}$.

We note that if γ is a (complex linear) automorphism of $\mathfrak{g}$ then $\mathrm{ad}(\gamma X) = \gamma \,\mathrm{ad}(X) \gamma^{-1}$ for all $X \in \mathfrak{g}$. Thus

$$B(\gamma X, Y) = B(X, \gamma Y) \quad \text{for all } X,Y \in \mathfrak{g}. \tag{11.22}$$

Set $P = (\tau_0 \sigma)^2$. Then P is an automorphism of $\mathfrak{g}$ (if τ_0 and σ actually commute, then $P = I$). By (11.22) we have

$$(PX,Y) = -B(PX, \tau_0 Y) = -B(X, \sigma \tau_0 \sigma Y) = (X, PY),$$

since $\tau_0^2 = I$. Thus P is self-adjoint with respect to the form $(\cdot, \cdot)$. We assert that P is positive definite. For this we forget the complex structure and observe that if $X \in \mathfrak{g}$ then the endomorphism $\mathrm{ad}(X)\,\mathrm{ad}(Y)$ of $\mathfrak{g}$ (viewed as a real vector space) has trace equal to

$$2\,\mathrm{Re}\,B(X,Y) = \nu(X,Y).$$

By the argument given above, the bilinear form $\langle X,Y \rangle = -\nu(X, \tau_0 Y)$ defines an inner product on $\mathfrak{g}$ as a real vector space, and one has

$$\langle \tau_0 \sigma X, Y \rangle = -\nu(X, \sigma Y) = \langle X, \tau_0 \sigma X \rangle.$$

Thus $\tau_0 \sigma$ is real-linear, symmetric, and invertible, so $P = (\tau_0 \sigma)^2$ has all positive eigenvalues. Hence P is positive definite as asserted. Since $\sigma^2 = \tau_0^2 = I$, we have

$$\sigma P = \sigma \tau_0 \sigma \tau_0 \sigma = P^{-1} \sigma.$$

Similarly, $\tau_0 P = P^{-1}\tau_0$. We now apply Lemma 11.5.7 to see that $\sigma P^s = P^{-s}\sigma$ and $\tau_0 P^s = P^{-s}\tau_0$ for all $s \in \mathbb{R}$. Set $\tau_s = P^{-s}\tau_0 P^s$. Then

$$\sigma\tau_s = \sigma P^{-s}\tau_0 P^s = P^s\sigma\tau_0 P^s = P^s\sigma P^{-s}\tau_0 = P^{2s}\sigma\tau_0 .$$

However, $\tau_s\sigma = P^{-s}\tau_0 P^s\sigma = P^{-2s}\tau_0\sigma$. If s can be chosen such that $P^{4s} = P$, then $\tau_s\sigma = \sigma\tau_s$ by definition of P. Obviously, $s = 1/4$ does the trick. We claim that $P^{1/4} = \mathrm{Ad}\,g$ for some $g \in G$. Indeed, we will prove that there exists $Z \in \mathfrak{g}$ such that

$$P^s = \exp(s\,\mathrm{ad}Z) = \mathrm{Ad}(\exp(sZ)) \quad \text{for all } s \in \mathbb{R} . \tag{11.23}$$

This will complete the proof of the theorem, since $\exp(sZ) \in G$ by Theorem 11.2.9.

We first observe that Lemma 11.5.7 implies that P^s is a Lie algebra automorphism of $\mathfrak{g}$, since $[P^m X, P^m Y] = P^m[X, Y]$ for $m \in \mathbb{Z}$. Now $P^s = \exp(sA)$, where $A = \log P$ is a self-adjoint endomorphism of $\mathfrak{g}$. If we differentiate the equation $[P^s X, P^s Y] = P^s[X, Y]$ at $s = 0$, then we find that

$$A[X, Y] = [AX, Y] + [X, AY] \quad \text{for all } X, Y \in \mathfrak{g} . \tag{11.24}$$

Thus (11.23) is a consequence of the following *inner derivation* property:

$(\star\star)$ If A satisfies (11.24), then there exists $Z \in \mathfrak{g}$ such that $A = \mathrm{ad}Z$.

To prove $(\star\star)$ we note that B is a nondegenerate and $\mathrm{ad}(\mathfrak{g})$-invariant bilinear form on $\mathfrak{g}$. Hence the argument of Corollary 2.5.12 applies verbatim. $\square$

11.5.2 Polar Decomposition by a Compact Form

Let G be a connected reductive linear algebraic group. Let τ be a complex conjugation on G corresponding to a compact real form U (see Theorem 11.5.1). Thus

$$U = \{g \in G : \tau(g) = g\} .$$

Let $\mathfrak{g} = \mathrm{Lie}(G)$ and $\mathfrak{u} = \mathrm{Lie}(U)$. We use the notation $A^* = \bar{A}^t$ for $A \in M_n(\mathbb{C})$.

Lemma 11.5.8. *There exists a regular homomorphism* $\Psi : G \to \mathbf{GL}(n, \mathbb{C})$ *such that Ψ is an isomorphism of G onto its image and such that* $\Psi(\tau(g)) = (\Psi(g)^*)^{-1}$.

Proof. We may assume that $G \subset \mathbf{GL}(V)$ as a Zariski-closed subgroup, where V is an n-dimensional complex vector space. The unitary trick (Section 3.3.4) implies that there exists an inner product $(\cdot, \cdot)$ on V such that $(uv, w) = (v, u^{-1}w)$ for all $u \in U$ and $v, w \in V$. We write the elements of $\mathbf{GL}(V)$ as matrices with respect to a $(\cdot, \cdot)$-orthonormal basis for V. This defines an isomorphism Ψ from G to a subgroup of $\mathbf{GL}(n, \mathbb{C})$ such that

$$\Psi(U) = \{g \in G : \Psi(g)^* = \Psi(g)^{-1}\} .$$

We identify $\mathfrak{g}$ and $\mathfrak{u}$ with Lie subalgebras of $M_n(\mathbb{C})$ using Ψ. Then $X^* = -X$ for $X \in \mathfrak{u}$. Since U is a real form of G, we have $\mathfrak{g} = \mathfrak{u} + i\mathfrak{u}$. It follows that $X^* \in \mathfrak{g}$ for all $X \in \mathfrak{g}$. Since G is connected in the Lie group topology (Theorem 11.2.9), this implies that $\Psi(g)^* \in \Psi(G)$ for each $g \in G$. Furthermore, we have $\Psi(\tau(g)) = (\Psi(g)^*)^{-1}$.

□

Theorem 11.5.9. *The map* $\Phi : U \times \mathfrak{u} \longrightarrow G$ *defined by* $\Phi(u,X) = u\exp(iX)$, *for* $u \in U$ *and* $X \in \mathfrak{u}$, *is a diffeomorphism onto G. In particular, U is connected.*

Proof. By Lemma 11.5.8 we may assume that $G \subset \mathbf{GL}(n, \mathbb{C})$ and $\tau(g) = (g^*)^{-1}$. If $g \in G$ then g^*g is positive definite. Since $(g^*g)^m \in G$ for all $m \in \mathbb{Z}$, Lemma 11.5.6 implies that $(g^*g)^s \in G$ for all $s \in \mathbb{R}$. Also, $s \mapsto (g^*g)^s$ defines a one-parameter group of G as a real Lie group. Thus $(g^*g)^s = \exp(sX)$ for some $X \in \mathfrak{g}$ by Theorem D.2.6. Clearly, $d\tau(X) = -X$. Thus $X \in i\mathfrak{u}$.

For $g \in G$ define $k(g) = g(g^*g)^{-1/2}$. Then

$$k(g)^* = (g^*g)^{-1/2}g^* = (g^*g)^{-1/2}g^*gg^{-1} = (g^*g)^{1/2}g^{-1} = k(g)^{-1}.$$

It is also evident that $k(g) \in U$. Thus the map Φ in the theorem is surjective. If $u\exp(iX) = v\exp(iY)$ with $u,v \in U$ and $X,Y \in \mathfrak{u}$, then

$$\exp(2iX) = (u\exp(iX))^* u\exp(iX) = (v\exp(iY))^* v\exp(iY) = \exp(2iY).$$

Applying Lemma 11.5.6 yields $\exp(itX) = \exp(itY)$ for all $t \in \mathbb{R}$. Thus $X = Y$, and hence $u = v$. This proves that Φ is injective.

If $u \in U$ and $X, Z, W \in \mathfrak{u}$, then

$$d\Phi_{(u,X)}(Z,W) = uZ\exp(iX) + u\,d\Psi_{iX}(iW)$$

(notation as in Lemma 11.5.7). Now $Z^* = -Z$ and $d\Psi_{iX}(iW)^* = d\Psi_{iX}(iW)$. Thus $d\Phi_{(u,X)}$ is injective (see the proof of Lemma 11.5.7). The theorem now follows from the inverse function theorem. □

Theorem 11.5.10. *Let G be a connected reductive linear algebraic group. Let τ be a conjugation on G corresponding to a compact real form U. Let θ be an involutive automorphism of G such that $\tau\theta = \theta\tau$. Set $K = \{g \in G : \theta(g) = g\}$ and $K_0 = K \cap U$. Then K is reductive and K_0 is a compact real form of K that is Zariski dense in K.*

Proof. Since $\theta\tau = \tau\theta$, the restriction of τ to K is a conjugation of K. Hence K_0 is a real form of K that is compact, since U is compact. Thus K is reductive by Theorem 11.5.1. It remains to show that K_0 is Zariski dense in K.

Let $\mathfrak{k} = \mathrm{Lie}(K) = \{X \in \mathfrak{g} : d\theta(X) = X\}$. We first show that if K° is the identity component of K, then $K_0 K^\circ = K$. Indeed, if $g \in K$, then in the notation of Theorem 11.5.9 we have $g = u\exp(iX)$ with $u \in U$ and $X \in \mathfrak{u}$. Now $\theta(U) = U$; hence $d\theta(\mathfrak{u}) = \mathfrak{u}$. Since $\theta(g) = g$, we have $\theta(u) = u$ and $d\theta(X) = X$ by uniqueness of the polar decomposition, so $u \in U \cap K = K_0$ and $X \in \mathfrak{u} \cap \mathfrak{k}$. Since $d\theta$ is complex linear, we have $iX \in \mathfrak{k}$ and hence $\exp(iX) \in K^\circ$. Thus the assertion follows.

Let $B = \{f \in \mathcal{O}[K] : f(K_0) = 0\}$. Set $\mathfrak{k}_0 = \mathrm{Lie}(K_0) \subset \mathfrak{k} = \mathrm{Lie}(K)$. For simplicity of notation, it is convenient here to identify the elements of $\mathfrak{k}$ with the corresponding left-invariant vector fields on K, as in Appendix D.2.2. If $X \in \mathfrak{k}_0$ then $X_g \in T(K_0)_g$ for all $g \in K_0$. Hence $Xf(g) = 0$ for all $f \in B$. Since $\mathfrak{k} = \mathfrak{k}_0 + i\mathfrak{k}_0$, this implies that $XB \subset B$ for all $X \in \mathfrak{k}$. Thus

$$X^m f|_{K_0} = 0 \quad \text{for } X \in \mathfrak{k}, \ f \in B, \ \text{and } m = 1, 2, \ldots.$$

This implies that $R(\exp X)B = B$ for all $X \in \mathfrak{k}$, where $R(g)$ denotes right translation by $g \in G$. We know that $\exp(\mathfrak{k})$ generates the identity component of K in the Lie group topology, by Corollary D.2.3. Hence Theorem 11.2.9 implies that $R(k)B = B$ for $k \in K^\circ$. Thus if $f \in B$, then $f(K_0 K^\circ) = R(K^\circ)f(K_0) = 0$. We have already observed that $K = K_0 K^\circ$, hence $f = 0$. Thus K_0 is Zariski dense in K. $\square$

11.6 Gauss Decomposition

The final structural result that we need for Chapter 12 is the *Gauss decomposition* of G relative to a diagonal torus $A \subset G$ (the factorization of a matrix as a product of a block upper-triangular unipotent matrix, a block-diagonal matrix, and a block lower-triangular unipotent matrix, with the block sizes determined by the multiplicities of the weights of A). The set of elements in G that admit such a decomposition is shown to be Zariski dense. We also obtain a Gauss decomposition for real forms of G when the torus A is *split* relative to the real form.

11.6.1 Gauss Decomposition of $\mathbf{GL}(\mathbf{n}, \mathbb{C})$

Let V be a finite-dimensional complex vector space. Let T be an algebraic torus in $\mathbf{GL}(V)$ and let $\mathfrak{X}(T)$ denote the group of all rational characters of T. By Proposition 2.1.3 there is a finite set $\Sigma \subset \mathfrak{X}(T)$ such that

$$V = \bigoplus_{\chi \in \Sigma} V(\chi),$$

where $V(\chi) = \{v \in V : tv = \chi(t)v \text{ for all } t \in T\}$ is the χ weight space for T. There is a subset $S = \{\chi_1, \ldots, \chi_m\} \subset \Sigma$ such that the map $\Psi : T \to (\mathbb{C}^\times)^m$ given by $\Psi(t) = [\chi_1(t), \ldots, \chi_m(t)]$, for $t \in T$, is a regular isomorphism (see the proof of Theorem 11.2.2). Hence every element $\chi \in \mathfrak{X}(T)$ is uniquely expressed as

$$\chi = \chi_1^{p_1} \cdots \chi_m^{p_m} \quad \text{with } p_i \in \mathbb{Z}. \tag{11.25}$$

Let $\Phi = \{\chi \nu^{-1} : \chi, \nu \in \Sigma, \chi \neq \nu\}$. Then $\mathrm{End}(V)$ decomposes under the adjoint action of T as

$$\mathrm{End}(V) = \mathrm{End}_T(V) \oplus \bigoplus_{\lambda \in \Phi} \mathrm{End}(V)(\lambda), \qquad (11.26)$$

where $\mathrm{End}(V)(\lambda) = \{A \in \mathrm{End}(V) : tAt^{-1} = \lambda(t)A \text{ for all } t \in T\}$ and $\mathrm{End}_T(V)$ is the commutant of T in $\mathrm{End}(V)$.

We order $\mathfrak{X}(T)$ lexicographically relative to the decomposition (11.25). That is,

$$\chi_1^{p_1} \cdots \chi_m^{p_m} > \chi_1^{q_1} \cdots \chi_m^{q_m}$$

if $p_j > q_j$ and $p_i = q_i$ for all $i < j$. We enumerate Σ as $\{v_1, \ldots, v_r\}$ so that $v_i > v_j$ if $i < j$. Set $\dim V = n$ and $m_i = \dim V(v_i)$. Choose a basis $\{e_1, \ldots, e_n\}$ for V such that

$$\{e_1, \ldots, e_{m_1}\} \subset V(v_1), \quad \{e_{m_1+1}, \ldots, e_{m_1+m_2}\} \subset V(v_2), \quad \ldots .$$

Using this basis, we identify $\mathrm{End}(V)$ with $M_n(\mathbb{C})$ and $\mathbf{GL}(V)$ with $\mathbf{GL}(n, \mathbb{C})$. Define

$$L = \{g \in \mathbf{GL}(n, \mathbb{C}) : gtg^{-1} = t \text{ for all } t \in T\}.$$

Then L is a linear algebraic subgroup of $\mathbf{GL}(n, \mathbb{C})$. Let $\mathfrak{l} = \mathrm{Lie}(L)$. Then $\mathfrak{l} = \mathrm{End}_T(V)$.

With the ordering of the characters and basis as above, the following assertions about block forms of matrices are easily deduced from (11.26), where 0_i denotes the $m_i \times m_i$ zero matrix and $*$ denotes a matrix block whose size is determined by the diagonal blocks:

1. L consists of all block-diagonal matrices $\mathrm{diag}[g_1, \ldots, g_r]$ with $g_i \in M_{m_i}(\mathbb{C})$.

2. The T weight space $M_n(\mathbb{C})(\chi)$ for $\chi > 1$ is contained in the space of block upper-triangular matrices of the form $\begin{bmatrix} 0_1 & * & \cdots & * \\ 0 & 0_2 & \cdots & * \\ \vdots & \vdots & \ddots & \vdots \\ 0 & 0 & \cdots & 0_r \end{bmatrix}$.

3. The T weight space $M_n(\mathbb{C})(\chi)$ for $\chi < 1$ is contained in the space of block lower-triangular matrices of the form $\begin{bmatrix} 0_1 & 0 & \cdots & 0 \\ * & 0_2 & \cdots & 0 \\ \vdots & \vdots & \ddots & \vdots \\ * & * & \cdots & 0_r \end{bmatrix}$.

Let V^+ be the unipotent group of all matrices $\begin{bmatrix} I_1 & * & \cdots & * \\ 0 & I_2 & \cdots & * \\ \vdots & \vdots & \ddots & \vdots \\ 0 & 0 & \cdots & I_r \end{bmatrix}$, and let V^- be the unipotent group of all matrices $\begin{bmatrix} I_1 & 0 & \cdots & 0 \\ * & I_2 & \cdots & 0 \\ \vdots & \vdots & \ddots & \vdots \\ * & * & \cdots & I_r \end{bmatrix}$, where I_j is the $m_j \times m_j$ identity matrix.

Lemma 11.6.1 (Gauss Decomposition). *Set $\Omega = V^- L V^+$. Let $\Delta_i(g)$ denote the upper left-hand corner minor of g of size $m_1 + \cdots + m_i$. Then*

$$\Omega = \{g \in \mathbf{GL}(n,\mathbb{C}) : \Delta_i(g) \neq 0 \ \text{for} \ i = 1,\ldots,r-1\} \ .$$

There exist rational maps γ_0, γ_+, and γ_- from $\mathbf{GL}(n,\mathbb{C})$ to L, V^+, and V^- respectively such that if $g \in \Omega$ then g can be written uniquely in the form $g = \gamma_-(g)\gamma_0(g)\gamma_+(g)$.

Proof. This follows by induction on r using the proof of Lemma B.2.6, with the scalar matrix entries x_{ij} in that proof replaced by $m_i \times m_j$ matrices.   □

11.6.2 Gauss Decomposition of an Algebraic Group

Let G be a connected linear algebraic group and let T be an algebraic torus in G. We may assume that G is a Zariski-closed subgroup of $\mathbf{GL}(n,\mathbb{C})$. Thus we can use the notation of Section 11.6.1. Let

$$M = L \cap G = \{g \in G : gtg^{-1} = t \quad \text{for all } t \in T\}$$

be the centralizer of T in G and set $N^\pm = G \cap V^\pm$. Clearly, M and $N^\pm$ are closed subgroups of $\mathbf{GL}(n,\mathbb{C})$. The group M is connected by Theorem 11.4.10 and the groups $N^\pm$ are unipotent, hence connected.

Let $\Omega = V^- L V^+$. Since $I \in \Omega$, Lemma 11.6.1 implies that $\Omega \cap G$ is open and Zariski dense in G.

Theorem 11.6.2. *If $g \in G \cap \Omega$, then $\gamma_0(g)$, $\gamma_+(g)$, and $\gamma_-(g)$ are in G.*

Proof. The key to the theorem is the following assertion:

$(\star)$ $N^- M N^+$ contains a neighborhood of I in G relative to the Zariski topology.

Let us show how $(\star)$ implies the theorem. Suppose $f \in \mathcal{O}[\mathbf{GL}(n,\mathbb{C})]$ vanishes on G. We must show that $f(\gamma_0(g)) = f(\gamma_-(g)) = f(\gamma_+(g)) = 0$ for $g \in \Omega \cap G$. We have $N^- M N^+ \subset \Omega \cap G$. From Lemma 11.6.1 we know that $f \circ \gamma_0$ and $f \circ \gamma_\pm$ are rational functions on G whose domains include $\Omega \cap G$, and each vanishes on the open nonempty subset $N^- M N^+$. Thus each function is identically 0, since G is connected. Hence we only need to prove $(\star)$.

Consider the map $\Psi : N^- \times M \times N^+ \to G$ given by

$$\Psi(v,m,u) = vmu \quad \text{for } v \in N^-, m \in M, n \in U \ .$$

A direct calculation shows that if $V \in \mathrm{Lie}(N^-)$, $Y \in \mathrm{Lie}(M)$, and $U \in \mathrm{Lie}(N^+)$, then $d\Psi_{(1,1,1)}(V,Y,U) = V + Y + U$. Using the adjoint action of T, we decompose $\mathrm{Lie}(G)$ into weight spaces with $\mathrm{Lie}(M) = \mathrm{Lie}(G)(1)$ and

$$\mathrm{Lie}(N^+) = \bigoplus_{\chi > 1} \mathrm{Lie}(G)(\chi) \ , \qquad \mathrm{Lie}(N^-) = \bigoplus_{\chi < 1} \mathrm{Lie}(G)(\chi) \ .$$

Hence $\mathrm{Lie}(G) = \mathrm{Lie}(N^-) \oplus \mathrm{Lie}(M) \oplus \mathrm{Lie}(N^+)$, so $d\Psi$ is bijective at $(1,1,1)$. Since $N^- \times M \times N^+$ and G are smooth affine algebraic sets, this implies that the image of Ψ contains a neighborhood of the identity, by Theorem A.3.4. □

11.6.3 Gauss Decomposition for Real Forms

Let G be a linear algebraic group and let σ be a conjugation on G. Let G_0 be the corresponding real form of G. Let $A \subset G$ be an algebraic torus and assume that $\sigma(A) = A$. Thus $\sigma|_A$ defines a conjugation on A and we denote by A_0 the corresponding real form of A. Let $\mathcal{X}(A)$ denote the set of all regular homomorphisms of A to $\mathbb{C}^\times$. If $\chi \in \mathcal{X}(A)$ then we set $\chi^\sigma(a) = \overline{\chi(\sigma(a))}$ for $a \in A$, with the bar denoting complex conjugation.

Definition 11.6.3. The algebraic torus A is σ-*split* if $\chi^\sigma = \chi$ for all $\chi \in \mathcal{X}(A)$.

Note that when A is σ-split then every $\chi \in \mathcal{X}(A)$ is real-valued on A_0. We now prove that every σ-split torus can be diagonalized so that the action of σ becomes complex conjugation.

Lemma 11.6.4. *Suppose that A is σ-split. There exists a regular homomorphism $\varphi : G \to \mathbf{GL}(n, \mathbb{C})$ such that $\varphi : G \to \varphi(G)$ is a regular isomorphism and such that*

1. $\varphi(\sigma(g)) = \overline{\varphi(g)}$ for all $g \in G$;
2. $\varphi(A) \subset D_n$.

Proof. We use the notation in the proof of Theorem 1.7.5. Let V and ρ be as in that proof. Since (ρ, V) is a regular representation of G, Proposition 2.1.3 furnishes a weight-space decomposition

$$V = \bigoplus_{\chi \in F} V(\chi)$$

with $F \subset \mathcal{X}(A)$ and

$$V(\chi) = \{v \in V : \rho(a) = \chi(a)v \quad \text{for all } a \in A\} .$$

We set $C(f)(g) = \overline{f(\sigma(g))}$ as in the proof of Theorem 1.7.5. Then $C(V) = V$ and $C(\rho(g)v) = \rho(\sigma(g))C(v)$. We have

$$C \cdot V(\chi) = V(\chi^\sigma) = V(\chi) ,$$

since A is σ-split. Thus if we define $V_+ = \{f \in V : C(f) = f\}$, then

$$V_+ = \bigoplus_{\chi \in F} V_+ \cap V(\chi) .$$

Now take a basis $\{v_1, \ldots, v_n\}$ of V_+ over $\mathbb{R}$ with $v_i \in V(\chi_i) \cap V_+$ for some $\chi_i \in F$. Let $\varphi(g)$ be the matrix representation of $\rho(g)$ relative to this basis. Then φ satisfies properties (1) and (2) (see the proof of Theorem 1.7.5). □

Fix a connected linear algebraic group G, a conjugation σ on G, and a σ-split torus A in G. Set $G_0 = \{g \in G : \sigma(g) = g\}$ and $A_0 = \{a \in A : \sigma(a) = a\}$. Let

$$M = \{g \in G : gag^{-1} = a \text{ for all } a \in A\}$$

be the centralizer of A in G. Then $A \subset M$ and $\sigma(M) = M$. Set

$$M_0 = M \cap G_0 = \{m \in M : \sigma(m) = m\} \,.$$

Lemma 11.6.4 implies that we may assume that $G \subset \mathbf{GL}(n, \mathbb{C})$, that $\sigma(g) = \bar{g}$, and that $A \subset D_n$. Fix an order on $\mathfrak{X}(A)$ as in Section 11.6.1. Let $N^{\pm}$ be as in Section 11.6.2. Since $\chi^{\sigma} = \chi$ for all $\chi \in \mathfrak{X}(A)$, we see that $\sigma(N^{\pm}) = N^{\pm}$. Set $N_0^{\pm} = N^{\pm} \cap G_0$.

Theorem 11.6.5. *The subgroups N_0^+ and N_0^- are connected Lie groups. The map $\varphi : N_0^- \times M_0 \times N_0^+ \to G_0$ given by $\varphi(n^-, m, n^+) = n^- m n^+$ is a diffeomorphism onto an open dense subset Ω_0 of G_0 (in the Lie group topology) that is Zariski dense in G.*

Proof. To see that N_0^+ is connected we observe that if $u \in N_0^+$ then u is unipotent (see the matrix form in Section 11.6.1). Let $X = \log u$. Then $\sigma(X) = \sigma(\log u) = \log \sigma(u)$. Thus

$$\log X \in \operatorname{Lie}(G_0) \cap N^+ = \operatorname{Lie}(N_0^+) \,.$$

The curve $t \mapsto \exp(tX)$ joins I to u. The argument for N_0^- is the same.

If $X \in \operatorname{Lie}(N_0^-)$, $Y \in \operatorname{Lie}(M_0)$, and $Z \in \operatorname{Lie}(N_0^+)$ then

$$\begin{aligned} d\varphi_{(n^-, m, n^+)}(X, Y, Z) &= n^- X m n^+ + n^- m Y n^+ + n^- m n^+ Z \\ &= n^-(X + \operatorname{Ad}(m)Y + \operatorname{Ad}(mn^+)Z)mn^+ \,. \end{aligned}$$

Since $X \in \operatorname{Lie}(N^-)$, $\operatorname{Ad}(m)Y \in \operatorname{Lie}(M_0)$, and $\operatorname{Ad}(mn^+)Z \in \operatorname{Lie}(N^+)$, we see that $d\varphi_{(n^-, m, n^+)}(X, Y, Z) = 0$ implies $X = Y = Z = 0$. Thus the image Ω_0 of φ is open in G_0 by the open mapping theorem (Lang [98]). We have already seen that φ is injective; therefore, the inverse function theorem implies that φ is a diffeomorphism onto Ω_0.

It remains to prove that Ω_0 is dense in G_0 in the Lie group topology. We first show that Ω_0 is Zariski dense in G. Let $f \in \mathcal{O}[G]$ vanish on Ω_0. Then using the calculation above we see that if $X \in \mathfrak{g} = \operatorname{Lie}(G)$, then $Xf(\Omega_0) = 0$ (here for notational convenience we are identifying elements of $\mathfrak{g}$ with the corresponding left-invariant vector fields on G). Iterating this argument with f replaced by Xf, we see that $X^k f(\Omega_0) = 0$ for all positive integers k. Let U be an open connected neighborhood of I in G in the Lie group topology such that $U \subset \exp(\mathfrak{g})$. Then since f is analytic, we have

$$R(u)f(\Omega_0) = 0 \quad \text{for } u \in U \,,$$

where $R(u)$ is right translation by u acting on $\mathcal{O}[G]$. Hence if g is in the subgroup of G generated by U, then $R(g)f(\Omega_0) = 0$. Now U generates the identity component of G in the Lie group topology (Proposition 1.3.1). Thus Theorem 11.2.9 implies that $R(g)f(\Omega_0) = 0$ for all $g \in G$. So $f(g) = R(g)f(I) = 0$ for all $g \in G$.

Let Δ_i be the principal minors, as in Section 11.6.1. Then $\Delta_i^\sigma = \Delta_i$ and $\Delta_i|_{G_0}$ is real-valued. We now show that if W is a nonempty open subset of G_0, then Δ_i is not identically 0 on W. Indeed, suppose the contrary. Then, as in the previous paragraph, we would have $X^k \Delta_i(W) = 0$ for all $X \in \mathfrak{g}$ and $k = 1, 2, \ldots$. Hence there would be an open neighborhood U of I in G (in the Lie group topology) such that $R(g)\Delta_i(W) = 0$ for all $g \in U$. Applying the argument above to $R(g)\Delta_i$, we find that $R(g)\Delta_i(W) = 0$ for all $g \in G$. Thus if $w \in W$ then

$$0 = R(w^{-1})\Delta_i(w) = \Delta_i(I) .$$

But $\Delta_i(I) = 1$, so this contradiction proves the assertion. This implies that the set of all $g \in G_0$ such that $\Delta_i(g) \neq 0$ for all i is dense in G_0. This set is easily seen to be Ω_0, which completes the proof. $\square$

11.6.4 Exercises

1. Let $G = \mathbf{GL}(n, \mathbb{C})$, $\sigma(g) = \bar{g}$, and $A = D_n$. Show that A is σ-split.
2. Let $G = \mathbf{GL}(2, \mathbb{C})$, $\sigma(g) = (\bar{g}^t)^{-1}$, and $A = D_2$. Show that A is not σ-split.

11.7 Notes

Section 11.1.1. For general affine algebraic sets, the notion of *irreducible component* is more useful than the weaker topological notion of *connected component* (cf. Appendix A.1.5). An algebraic set can be connected without being irreducible (cf. Exercise #4 in Appendix A). However, in the case of algebraic groups, Theorem 11.1.2 shows that the two notions coincide.

Section 11.1.3. If the condition that X be affine is dropped in Theorem 11.1.12 then the theorem becomes false. There are irreducible *projective* algebraic groups. These groups are the subject of the theory of *abelian varieties* (see Lang [96] and Mumford [115]).

Section 11.1.4. The results of this section are due to Chevalley and Borel (see Borel [16, Chapter II]). Theorem 11.1.17 is from Onishchik and Vinberg [118].

Section 11.2.1. The proof of Theorem 11.2.2 is from Humphreys [77, §16.2].

Section 11.2.2. The proof of Engel's theorem, using Burnside's theorem, is from Borel [16, §4.8].

Section 11.2.4. The construction of the algebraically simply connected group $\widetilde{G}$ associated to a simple Lie algebra $\mathfrak{g}$ is adapted from Chevalley's general treatment of isogenies in [38, Exposé 23]. The existence of $\widetilde{G}$ as a Lie group follows immediately from the construction of the adjoint group and Theorem D.2.9; this topological argument does not show that $\widetilde{G}$ is a linear algebraic group, however.

Section 11.3.3. The results in this section are due to Richardson [124].

Section 11.3.5. The symmetric spaces are labeled according to Cartan's classification (see Helgason [66, Chapter X] or Knapp [86, Chapter VI §10]).

Exercises 11.3.6. The exercises on algebraic quotients are from Kraft [92].

Sections 11.4.1–11.4.5. For most of the results in these sections we have followed the expositions of Borel [16] and Humphreys [77], based on Borel [15].

Section 11.4.6. The treatment of regular and strongly regular semisimple elements is based on Steinberg [138, §2].

Section 11.5.1. Theorem 11.5.3 and its proof are due to Mostow [114]. Lemma 11.5.6 is from Chevalley [33, Chapter I, §IV, Proposition 5].

Chapter 12
Representations on Spaces of Regular Functions

Abstract If G is a reductive linear algebraic group acting on an affine variety X, then G acts linearly on the function space $\mathcal{O}[X]$. In this chapter we will give several of the high points in the study of this representation. We will first analyze cases in which the representation decomposes into distinct irreducible representations (one calls X *multiplicity-free* in this case), give the most important class of such spaces (symmetric spaces), and determine the decomposition of $\mathcal{O}[X]$ as a G-module in this case. We also obtain the *second fundamental theorem* of invariants for the classical groups from this approach. The philosophy in this chapter is that the geometry of the orbits of G in X gives important information about the structure of the corresponding representation of G on $\mathcal{O}[X]$. This philosophy is most apparent in the last part of this chapter, in which we give a new proof of a celebrated theorem of Kostant and Rallis concerning the *isotropy representation* of a symmetric space. This chapter is also less self-contained than the earlier ones. For example, the basic results of Chevalley on invariants corresponding to symmetric pairs are only quoted (although for the pairs of classical type these facts are verified on a case-by-case basis). We also mix algebraic and analytic techniques by viewing G both as an algebraic group and as a Lie group with a compact real form.

12.1 Some General Results

We obtain the isotypic decomposition of $\mathcal{O}[X]$ associated with the action of a reductive group G on the affine variety X. For suitable choices of X this decomposition is multiplicity free and furnishes function-space models for the irreducible regular representations of G.

R. Goodman, N.R. Wallach, *Symmetry, Representations, and Invariants*,   545
Graduate Texts in Mathematics 255, DOI 10.1007/978-0-387-79852-3_12,
© Roe Goodman and Nolan R. Wallach 2009

12.1.1 Isotypic Decomposition of $\mathcal{O}[\mathbf{X}]$

Let X be an affine algebraic set on which the algebraic group G acts regularly. We denote by ρ_X the representation of G on $\mathcal{O}[X]$ given by $\rho_X(g)f(x) = f(g^{-1}x)$ for $f \in \mathcal{O}[X]$. This representation extends to a representation of the group algebra $\mathcal{A}[G]$ that is locally regular: For any finite-dimensional subspace $U \subset \mathcal{O}[X]$, the G-invariant space

$$\mathcal{A}[G]U = \sum_{g \in G} \rho_X(g)U$$

that it generates is finite-dimensional, and the representation of G on $\mathcal{A}[G]U$ is regular (this follows by the same argument used in Section 1.5.1 for the left translation representation on $\mathcal{O}[G]$).

From now on we assume that G is reductive. Let $\widehat{G}$ denote the set of equivalence classes of irreducible regular finite-dimensional representations of G. For $\omega \in \widehat{G}$ let (π_ω, V_ω) be a representation in the class ω. Let (ρ, E) be a locally regular representation of G, for example, the representation $(\rho_X, \mathcal{O}[X])$ as above. Denote by $E_{(\omega)}$ the sum of all the G-irreducible subspaces V of E such that $\rho|_V$ is in the class ω, and call $E_{(\omega)}$ the *isotypic subspace* of type ω. A linear map $T : V_\omega \longrightarrow E$ that intertwines the G actions is called a *covariant of type* ω for the representation (ρ, E). We denote the space of all covariants of type ω by $\mathrm{Hom}_G(\omega, \rho)$.

Proposition 12.1.1. *There is an isotypic decomposition $E = \bigoplus_{\omega \in \widehat{G}} E_{(\omega)}$. Furthermore, for each $\omega \in \widehat{G}$ the map $T \otimes v \mapsto T(v)$ for $T \in \mathrm{Hom}_G(\omega, \rho)$ and $v \in V_\omega$ gives a G-module isomorphism*

$$\mathrm{Hom}_G(\omega, \rho) \otimes V_\omega \cong E_{(\omega)} \tag{12.1}$$

(with trivial G-action on the first factor).

Proof. This follows from Proposition 4.1.15 with $\mathcal{A} = \mathcal{A}[G]$. $\square$

By (12.1) we see that $E_{(\omega)}$ is equivalent to a direct sum of irreducible representations in the class ω. Although this vector space decomposition is not unique if more than one summand occurs, the number of summands (which can be finite or infinite) is uniquely determined. We call this number the *multiplicity* of ω in E and denote it by $\mathrm{mult}_\rho(\omega)$. From (12.1) we have

$$\mathrm{mult}_\rho(\omega) = \dim \mathrm{Hom}_G(\omega, \rho) . \tag{12.2}$$

We say that (ρ, E) is *multiplicity-free* if $\mathrm{mult}_\rho(\omega) \leq 1$ for all $\omega \in \widehat{G}$. When $(\rho_X, \mathcal{O}[X])$ is multiplicity-free, where X is an affine G-space, we also say that X is a *multiplicity-free G-space*.

Now suppose that G is a connected reductive group. Fix a Borel subgroup $B = HN^+$ of G, with H a maximal torus in G and N^+ the unipotent radical of B (see Theorem 11.4.4). Recall that $P(G) \subset \mathfrak{h}^*$ denotes the *weight lattice* of G and $P_{++}(G)$ the *dominant weights*, relative to the system of positive roots determined by N^+

(since the Borel subgroups in G are all conjugate by Theorem 11.4.7, the notations $P(G)$ and $P_{++}(G)$ are unambiguous once B is fixed). For $\lambda \in P(G)$ we denote by $h \mapsto h^\lambda$ the corresponding character of H. We extend this to a character of B by setting $(hn)^\lambda = h^\lambda$ for $h \in H$ and $n \in N^+$.

Since G is connected, an irreducible regular representation (π, V) of G is completely determined (up to equivalence) by the Lie algebra representation $(d\pi, V)$, and hence by its highest weight relative to the subgroup B (Theorem 3.2.5). The subspace V^{N^+} of N^+-fixed vectors in V is one-dimensional, and H acts on it by a character $h \mapsto h^\lambda$, where $\lambda \in P_{++}(G)$. For each such λ we fix a model (π^λ, V^λ) for the irreducible representation with highest weight λ, and we fix a nonzero highest-weight vector $v_\lambda \in (V^\lambda)^{N^+}$. Let $\mathcal{O}[X]^{N^+}$ be the space of N^+-fixed regular functions on X. For every regular character $b \mapsto b^\lambda$ of B, let $\mathcal{O}[X]^{N^+}(\lambda)$ be the N^+-fixed regular functions f of weight λ:

$$\rho_X(b)f = b^\lambda f \quad \text{for } b \in B . \tag{12.3}$$

We can then describe the G-isotypic decomposition of $\mathcal{O}[X]$ as follows:

Theorem 12.1.2. *For $\lambda \in P_{++}(G)$, the isotypic subspace of type π^λ in $\mathcal{O}[X]$ is the span of $\rho_X(G)\mathcal{O}[X]^{N^+}(\lambda)$. This subspace is isomorphic to $V^\lambda \otimes \mathcal{O}[X]^{N^+}(\lambda)$ as a G-module, with action $\pi^\lambda(g) \otimes 1$. Thus*

$$\mathcal{O}[X] \cong \bigoplus_{\lambda \in P_{++}(G)} V^\lambda \otimes \mathcal{O}[X]^{N^+}(\lambda) .$$

Proof. We define a bijection between covariants and highest-weight vectors for the representation ρ_X as follows: Given $T : V^\lambda \longrightarrow \mathcal{O}[X]$, a covariant of type π^λ, we set $\psi(T) = Tv_\lambda$. Then $\psi(T) \in \mathcal{O}[X]^{N^+}(\lambda)$, and T is uniquely determined by $\psi(T)$, since v_λ is a cyclic vector for π^λ. Conversely, given $f \in \mathcal{O}[X]^{N^+}(\lambda)$, there is a unique G-intertwining map $T : V^\lambda \longrightarrow \mathcal{O}[X]$ such that $T(\pi^\lambda(g)v_\lambda) = \rho_X(g)f$, by Theorem 3.2.5. Clearly $\psi(T) = f$, so $\psi : \mathrm{Hom}_G(V^\lambda, \mathcal{O}[X]) \longrightarrow \mathcal{O}[X]^{N^+}(\lambda)$ is bijective. The theorem now follows by the isomorphism (12.1). $\square$

This theorem shows that the G-multiplicities in $\mathcal{O}[X]$ are the dimensions of the spaces $\mathcal{O}[X]^{N^+}(\lambda)$. We have $\mathcal{O}[X]^{N^+}(\lambda) \cdot \mathcal{O}[X]^{N^+}(\mu) \subset \mathcal{O}[X]^{N^+}(\lambda + \mu)$ under pointwise multiplication. Hence the set

$$\mathcal{S}_G(X) = \{\lambda \in P_{++}(G) : \mathcal{O}[X]^{N^+}(\lambda) \neq 0\} \quad \text{(the G-spectrum of X)}$$

is an additive semigroup that completely determines the G-isotypic decomposition of $\mathcal{O}[X]$.

12.1.2 Frobenius Reciprocity

Let G be an affine algebraic group and K a closed subgroup. Given any regular representation (π, V_π) of G, we obtain a regular representation $\mathrm{Res}_K^G(\pi)$ of K by restriction. Conversely, a regular representation (μ, V_μ) of K *induces* a representation of G by the following construction, which we introduced in Section 4.4.2 in the case of a finite group. For any affine algebraic set X and finite-dimensional complex vector space V, let $\mathcal{R}(X;V)$ be the space of all regular functions $f : X \longrightarrow V$. Taking $X = G$ and $V = V_\mu$, we define $I_\mu \subset \mathcal{R}(G; V_\mu)$ to be the subspace of functions such that

$$f(gk) = \mu(k)^{-1} f(g) \quad \text{for all } k \in K \text{ and } g \in G \ .$$

Then I_μ is a vector space that is invariant under left translation by G. The *induced representation* $\tau = \mathrm{Ind}_K^G(\mu)$ of G acts on the space I_μ by $\tau(g)f(x) = f(g^{-1}x)$. The representation τ is locally regular, since it is a subrepresentation of the left-translation representation of G on $\mathcal{R}(G; V_\mu)$.

Theorem 12.1.3. *Let (π, V_π) be a regular representation of G. Then there is a vector-space isomorphism*

$$\mathrm{Hom}_G(\pi, \mathrm{Ind}_K^G(\mu)) \cong \mathrm{Hom}_K(\mathrm{Res}_K^G(\pi), \mu) \ . \tag{12.4}$$

In particular, if G and K are reductive groups and π and μ are irreducible, then the multiplicity of π in $\mathrm{Ind}_K^G(\mu)$ equals the multiplicity of μ in $\mathrm{Res}_K^G(\pi)$.

Proof. The proof of Theorem 4.4.1 applies without change, because the maps defined in that proof are regular when G and K are linear algebraic groups. When G and K are reductive, the reciprocity statement about multiplicities follows from (12.2) and (12.4). $\square$

Now assume that G is reductive and take μ to be the one-dimensional trivial representation of K. In this case the induced representation of G acts on the space

$$\mathcal{R}(G/K) = \{ f \in \mathbb{C}[G] : f(gk) = f(g) \quad \text{for all } k \in K, g \in G \}$$

by left translations. Here we do *not* assume that K is a reductive group, so the (quasiprojective) variety G/K is not necessarily affine. Thus $\mathcal{R}(G/K)$, although it is a ring of functions, is not necessarily the ring of regular functions on an affine variety. Nonetheless, we can give a precise description of its decomposition as a G-module, as follows:

Theorem 12.1.4. *As a G-module under left translation,*

$$\mathcal{R}(G/K) \cong \bigoplus_{\omega \in \widehat{G}} V_\omega \otimes (V_\omega^*)^K \ , \tag{12.5}$$

where $g \in G$ acts by $\pi_\omega(g) \otimes 1$ in each summand and $(V_\omega^*)^K$ is the subspace of K-fixed vectors in V_ω^*.

Proof. By part (2) of Theorem 4.2.7, a function $f \in \mathcal{O}[G]$ is fixed under right translations by K if and only if its components in the decomposition 4.18 are in $\phi_\omega((V_\omega^*) \otimes V_\omega^K)$ for all $\omega \in \hat{G}$. Now replace ω by ω^* to obtain (12.5). $\qquad\square$

12.1.3 Function Models for Irreducible Representations

Suppose G is a connected reductive group. Fix a maximal torus H in G and Borel subgroup $B = HN^+$ with unipotent radical N^+, and define $P_{++}(G)$ relative to the set of positive roots Φ^+ giving N^+. Let N^- be the unipotent group *opposite* to N^+ corresponding to the roots $-\Phi^+$. If we take a matrix form of G such that H is diagonalized and N^+ is upper triangular, then N^- is lower triangular (see Theorem 11.6.2). We shall obtain all the irreducible representations of G as representations induced from characters of the Borel subgroup $B^- = HN^-$.

We begin with the representation of G on the function space

$$\mathcal{R}(N^-\backslash G) = \{ f \in \mathcal{O}[G] : f(\bar{n}g) = f(g) \text{ for } \bar{n} \in N^- \},$$

on which G acts by right translation. We decompose this space into irreducible subspaces as follows: For $\lambda \in P_{++}(G)$ let

$$\varphi_\lambda : V^{\lambda^*} \otimes V^\lambda \longrightarrow \mathcal{O}[G]$$

be the map in Theorem 4.2.7. Choose an N^+-fixed vector $v_\lambda \in V^\lambda$ and an N^--fixed vector $v_\lambda^* \in V^{\lambda^*}$, normalized so that $\langle v_\lambda^*, v_\lambda \rangle = 1$. This can be done, since v_λ^* has weight $-\lambda$ and so is orthogonal to all weight spaces in V^λ except $\mathbb{C}v_\lambda$.

Theorem 12.1.5. *The space $\mathcal{R}(N^-\backslash G)$ decomposes under G as*

$$\mathcal{R}(N^-\backslash G) = \bigoplus_{\lambda \in P_{++}(G)} \varphi_\lambda(v_\lambda^* \otimes V^\lambda) \cong \bigoplus_{\lambda \in P_{++}(G)} V^\lambda . \tag{12.6}$$

Hence it contains every irreducible regular representation of G with multiplicity one.

Proof. By Theorem 3.2.13 the space of N^--fixed vectors in $(V^\lambda)^*$ is spanned by v_λ^*. We use the map $f \mapsto \check{f}$, where $\check{f}(x) = f(x^{-1})$, to change $\mathcal{R}(N^-\backslash G)$ into $\mathcal{R}(G/N^-)$ and right actions into left actions. Then (12.6) follows from Theorem 12.1.4. $\qquad\square$

From the decomposition of $\mathcal{R}(N^-\backslash G)$ we obtain the following function models for the irreducible representations of G:

Theorem 12.1.6. *Let $\lambda \in P_{++}(G)$. Let $\mathcal{R}_\lambda \subset \mathcal{O}[G]$ be the subspace of functions such that*

$$f(\bar{n}hg) = h^\lambda f(g) \quad \text{for } \bar{n} \in N^-, \ h \in H, \text{ and } g \in G . \tag{12.7}$$

Then $\mathcal{R}_\lambda = \varphi_\lambda(v_\lambda^* \otimes V^\lambda)$. Hence $\mathcal{R}_\lambda$ is spanned by the right translates of the function $f_\lambda(g) = \langle v_\lambda^*, \pi^\lambda(g)v_\lambda \rangle$, and the restriction of the right regular representation R of G to $\mathcal{R}_\lambda$ is an irreducible representation with highest weight λ. The function f_λ is uniquely determined by $f(\bar{n}hn) = h^\lambda$ for $\bar{n} \in N^-$, $h \in H$, and $n \in N^+$.

Proof. The space $\mathcal{R}(N^- \backslash G)$ is invariant under left translations by the subgroup H (since H normalizes N^-). The vector v_λ^* has weight $-\lambda$. Hence by (12.6) we see that the subspace $\varphi_\lambda(v_\lambda^* \otimes V^\lambda)$ is the $-\lambda$ weight space for the left H action on $\mathcal{R}(N^- \backslash G)$. Thus this space coincides with $\mathcal{R}_\lambda$.

The subset $N^- H N^+$ is Zariski dense in G and contains a neighborhood of 1, by Theorem 11.6.2. This implies the uniqueness of the function f_λ.   □

We call the function $f_\lambda(g)$ in Theorem 12.1.6 the *generating function* for the representation with highest weight λ. It can be calculated from the fundamental representations of G as follows:

Corollary 12.1.7. *Let $\lambda_1, \ldots, \lambda_r$ be generators for the additive semigroup $P_{++}(G)$. Set $f_i(g) = f_{\lambda_i}(g)$. Let $\lambda \in P_{++}(G)$ and write $\lambda = m_1\lambda_1 + \cdots + m_r\lambda_r$ with $m_i \in \mathbb{N}$. Then*

$$f_\lambda(g) = f_1(g)^{m_1} \cdots f_r(g)^{m_r} \quad \text{for } g \in G. \tag{12.8}$$

Proof. Set $V_i = V^{\lambda_i}$, $v_i = v_{\lambda_i}$, and $v_i^* = v_{\lambda_i}^*$. Then V^λ can be realized as the G-cyclic space generated by the highest-weight vector

$$v_\lambda = v_1^{\otimes m_1} \otimes \cdots \otimes v_r^{\otimes m_r} \in V_1^{\otimes m_1} \otimes \cdots \otimes V_r^{\otimes m_r}$$

(see Proposition 5.5.19). For the dual lowest-weight vector we can take

$$v_\lambda^* = (v_1^*)^{\otimes m_1} \otimes \cdots \otimes (v_r^*)^{\otimes m_r}.$$

From this model for V^λ we see that the function f_λ is given by (12.8).   □

12.1.4 Exercises

1. Suppose the reductive group G acts linearly on a vector space V. The group $\mathbb{C}^\times$ acts on $\mathcal{P}(V)$ via scalar multiplication on V and commutes with G. Hence one has a representation of the group $G \times \mathbb{C}^\times$ on $\mathcal{P}(V)$. Prove that the isotypic decomposition of $\mathcal{P}(V)$ under $G \times \mathbb{C}^\times$ is

$$\mathcal{P}(V) = \bigoplus_{k \geq 0} \bigoplus_{\omega \in \widehat{G}} \mathcal{P}^k(V)_{(\omega)},$$

where $\mathcal{P}^k(V)_{(\omega)}$ is the ω-isotypic component in the homogeneous polynomials of degree k.

2. Suppose G is a reductive group. Let X be an affine algebraic G-space and $\omega \in \widehat{G}$. The space $\left(\mathcal{O}[X] \otimes \mathcal{P}^k(V_\omega) \right)^G$ is called the *covariants of k arguments of type ω*. These are the symmetric, multilinear G-maps from the k-fold product of V_ω to $\mathcal{O}[X]$. Show that this space is finitely generated as a module for the algebra $\mathcal{O}[X]^G$ of G-invariant functions.

3. Let G be a reductive algebraic group. Let ρ be the conjugation representation of G on $\mathcal{O}[G]$ given by $\rho(g)f(x) = f(g^{-1}xg)$ for $f \in \mathcal{O}[G]$. Prove that $\mathcal{O}[G]^{\rho(G)}$ is spanned by the characters $x \mapsto \mathrm{tr}(\pi_\omega(x))$ with $\omega \in \widehat{G}$.

4. Let μ and ν be Ferrers diagrams with $\mathrm{depth}(\mu) \leq k$ and $\mathrm{depth}(\nu) \leq m$. Show that

$$\mathrm{Ind}_{\mathbf{GL}(k,\mathbb{C}) \times \mathbf{GL}(m,\mathbb{C})}^{\mathbf{GL}(k+m,\mathbb{C})} (\pi_k^\mu \,\widehat{\otimes}\, \pi_m^\nu) = \bigoplus_{\mathrm{depth}(\lambda) \leq k+m} c_{\mu\nu}^\lambda \, \pi_{k+m}^\lambda ,$$

where $c_{\mu\nu}^\lambda$ are the Littlewood–Richardson coefficients. (HINT: Use Frobenius reciprocity and (9.23), (9.24), and (9.25).)

5. Show that the generating functions in Corollary 12.1.7 are given as follows, where $\Delta_i(g)$ denotes the ith principal minor of the matrix g and ϖ_i are the fundamental weights of $\mathfrak{g}$:

   (a) Let $G = \mathbf{SL}(n,\mathbb{C})$. Take $r = n - 1$ and $\lambda_i = \varpi_i$ for $i = 1,\dots,n-1$. Then $f_i(g) = \Delta_i(g)$. (HINT: Use Theorem 5.5.11.)

   (b) Let $G = \mathbf{SO}(\mathbb{C}^{2l+1}, B)$, where B is the symmetric form (2.9). Take $r = l$ and $\lambda_i = \varpi_i$ for $i = 1,\dots,l-1$, $\lambda_l = 2\varpi_l$. Then $f_i(g) = \Delta_i(g)$. (HINT: Use Proposition 3.1.19 and Theorem 5.5.13.)

   (c) Let $G = \mathbf{Sp}(\mathbb{C}^{2l}, \Omega)$, where Ω is the skew form (2.6). Take $r = l$ and $\lambda_i = \varpi_i$ for $i = 1,\dots,l$. Then $f_i(g) = \Delta_i(g)$. (HINT: Use Theorem 5.5.15.)

   (d) Let $G = \mathbf{SO}(\mathbb{C}^{2l}, B)$, where B is the symmetric form (2.6). Take $r = l + 1$ and $\lambda_i = \varpi_i$ for $i = 1,\dots,l-2$, $\lambda_{l-1} = \varpi_{l-1} + \varpi_l$, $\lambda_l = 2\varpi_l$, and $\lambda_{l+1} = 2\varpi_{l-1}$. Then $f_i(g) = \Delta_i(g)$ for $i = 1,\dots,l$ and $f_{l+1}(g) = \Delta_l(g_0 g g_0)$, where g_0 is the orthogonal transformation that interchanges e_l and e_{l+1}, as in Section 5.5.5. (HINT: Use Proposition 3.1.19 and Theorem 5.5.13.)

12.2 Multiplicity-Free Spaces

Let G be a reductive algebraic group and X an affine algebraic G-space. Recall that X is called *multiplicity-free* as a G-space if all the irreducible representations of G that occur in $\mathcal{O}[X]$ have multiplicity one. For example, if $X = K$, with K a reductive group and $G = K \times K$ acting on X by left and right multiplication, then X is multiplicity-free relative to G by Theorem 4.2.7. In this section we obtain a geometric criterion for a space to be multiplicity-free, and examine important examples of multiplicity-free spaces.

12.2.1 Multiplicities and B-Orbits

Assume that G is a connected reductive group. Fix a Borel subgroup $B = HN^+ \subset G$ and define $P_{++}(G)$ relative to the set of positive roots giving N^+. Let $\mathfrak{g} = \mathrm{Lie}(G)$ and let $d\rho$ denote the differential of the (locally regular) representation ρ of G on $\mathcal{O}[X]$. For $Y \in \mathfrak{g}$ the operator $d\rho(Y)$ is a vector field on X, and $d\rho(Y)_x$ is the corresponding tangent vector at $x \in X$. When X is a finite-dimensional vector space and the G action is linear, then $d\rho(Y)_x = d\rho(Y)x$ (where now $d\rho(Y) \in \mathrm{End}(X)$ denotes the linear transformation defined in Section 1.5.2).

For a subgroup $M \subset G$ and $x \in X$ we write $M_x = \{m \in M : m \cdot x = x\}$ for the isotropy group at x. Note that if $\mathfrak{m} = \mathrm{Lie}(M)$, then the Lie algebra of M_x is $\mathfrak{m}_x = \{Y \in \mathfrak{m} : d\rho(Y)_x = 0\}$.

Theorem 12.2.1. *Let X be an irreducible affine G-space. Suppose $B \cdot x_0$ is open in X for some point $x_0 \in X$ (equivalently, $\dim \mathfrak{b} = \dim X + \dim \mathfrak{b}_{x_0}$). Then*

1. *X is multiplicity-free as a G-space, and*
2. *if $\lambda \in \mathcal{S}(X)$ then $h^\lambda = 1$ for all $h \in H_{x_0}$.*

Proof. (1): It suffices by Theorem 12.1.2 to show that $\dim \mathcal{O}[X]^N(\lambda) \leq 1$ for all $\lambda \in P_{++}(G)$. Suppose $B \cdot x_0$ is open in X (and hence dense in X, by the irreducibility of X). Then $f \in \mathcal{O}[X]^{N^+}(\lambda)$ is determined by $f(x_0)$, since on the dense set $B \cdot x_0$ it satisfies $f(b \cdot x_0) = b^{-\lambda} f(x_0)$.

(2): Let $\lambda \in \mathcal{S}(X)$ and let $0 \neq f \in \mathcal{O}[X]^N(\lambda)$. If $h \in H_{x_0}$, then $h^\lambda f(x_0) = f(x_0)$. Since $f(x_0) \neq 0$ by the proof of (1), it follows that $h^\lambda = 1$. $\qquad\square$

Remark 12.2.2. The open B-orbit condition is also necessary for X to be multiplicity-free. This is the starting point for the classification of multiplicity-free actions.

Remark 12.2.3. The criterion of Theorem 12.2.1 does not depend on the choice of B, since any other Borel subgroup is of the form $B_1 = gBg^{-1}$ for some $g \in G$. If the B_1 orbit of x_1 is open then the B orbit of $x_0 = g^{-1} \cdot x_1$ is $g^{-1}B_1 \cdot x_1$, which is also open.

Let G be a reductive algebraic group and $K \subset G$ a reductive algebraic subgroup. The pair (G, K) will be called *spherical* if $\dim V_\omega^K \leq 1$ for every irreducible regular representation (ω, V_ω) of G.

Proposition 12.2.4. *The pair (G, K) is spherical if and only if the space $\mathcal{R}(G/K)$ is multiplicity-free as a G-module.*

Proof. This follows immediately from Theorem 12.1.4. $\qquad\square$

Theorem 12.2.5. *Suppose G is a connected reductive group. If there exists a connected solvable algebraic subgroup S of G such that $\mathrm{Lie}(S) + \mathrm{Lie}(K) = \mathfrak{g}$, then (G, K) is spherical.*

Proof. The Lie algebra assumption implies that the map $S \times K \longrightarrow G$ given by multiplication has surjective differential at $(1,1)$. Hence SK is Zariski dense in G by Theorem A.3.4. By Theorem 11.4.7 there exists $x_0 \in G$ such that $x_0 S x_0^{-1} \subset B$. Thus the double coset $Bx_0 K \supset x_0 SK$ is Zariski dense in G. Theorem 12.1.6 now implies that $\dim V^K \leq 1$ for every irreducible G-module V. $\qquad \square$

When (G, K) is a spherical pair an irreducible representation V^λ of G will be called *K-spherical* if $(\dim V^\lambda)^K = 1$. These are precisely the representations that occur in the decomposition of $\mathcal{R}(G/K)$ into G-irreducible subspaces.

12.2.2 B-Eigenfunctions for Linear Actions

We consider linear multiplicity-free actions in more detail. Let G be a connected reductive group with Borel subgroup $B = HN^+$. Let (σ, X) be a regular representation of G. Let $\rho(g)f(x) = f(\sigma(g^{-1})x)$ be the corresponding representation of G on $\mathcal{P}(X)$.

Theorem 12.2.6. *Assume that there exists $x_0 \in X$ such that $\sigma(B)x_0$ is open in X. Let $H_0 = \{h \in H : h \cdot x_0 = x_0\}$. Let $\mathcal{E}(X)$ be the set of all irreducible polynomials $f \in \mathcal{P}(X)$ such that f is a B-eigenfunction and $f(x_0) = 1$. Then the following hold:*

1. *The set $\mathcal{E}(X) = \{f_1, \ldots, f_k\}$ is finite and $k \leq \dim(H/H_0)$, where the polynomial f_i has B-weight λ_i and is homogeneous of degree d_i. Furthermore, the set of weights $\{\lambda_1, \ldots, \lambda_k\}$ is linearly independent over $\mathbb{Q}$ and $h^{\lambda_i} = 1$ for all $h \in H_0$.*
2. *The B-eigenfunctions $f \in \mathcal{P}(X)$, normalized by $f(x_0) = 1$, are the functions*

$$f_{\mathbf{m}} = \prod_{i=1}^{k} f_i^{m_i} \quad \text{with } \mathbf{m} = (m_1, \ldots, m_k) \in \mathbb{N}^k . \tag{12.9}$$

3. *For $r \geq 0$ the homogeneous polynomials of degree r decompose under G as $\mathcal{P}^r(X) = \bigoplus_\lambda V^\lambda$, where the sum is over all $\lambda = \sum_i m_i \lambda_i$ with $\sum_i d_i m_i = r$, and V^λ is the irreducible G-module generated by $f_{\mathbf{m}}$.*

Proof. Let $\{f_1, \ldots, f_k\}$ be any finite subset of $\mathcal{E}(X)$. Since $B \cdot x_0$ is open, each f_i is uniquely determined by its weight λ_i and the normalization $f(x_0) = 1$. Also f_i must be a homogeneous polynomial, since its translates span an irreducible G-module and G leaves invariant the spaces of homogeneous polynomials. The set $\{\lambda_1, \ldots, \lambda_k\}$ must be linearly independent over $\mathbb{Q}$. For if this set satisfied a nontrivial $\mathbb{Q}$-linear relation, then by clearing denominators we would obtain $\mathbf{m} \in \mathbb{N}^k$ and a subset $L \subset \{1, \ldots, k\}$ such that

$$\sum_{i \in L} m_i \lambda_i = \sum_{j \notin L} m_j \lambda_j . \tag{12.10}$$

This would imply that

$$\prod_{i \in L} f_i^{m_i} = \prod_{j \notin L} f_j^{m_j},$$

since the functions on the left and right are B-eigenfunctions with the same weight by (12.10). But this contradicts the assumption that each f_i is irreducible ($\mathcal{P}(X)$ is a unique factorization domain). From linear independence and the fact that $h^{\lambda_i} = 1$ for $h \in H_0$ (by Theorem 12.2.1), we have $k \leq \dim(H/H_0)$.

Let $f \in \mathcal{P}(X)$ be a B-eigenfunction. Then $f(x_0) \neq 0$, so we may assume $f(x_0) = 1$. We can factor f uniquely in $\mathcal{P}(X)$ as

$$f = \prod_{i=1}^{r} p_i^{m_i},$$

with $m_i \in \mathbb{N}$, $p_1, \ldots, p_r$ distinct irreducible polynomials, and $p_i(x_0) = 1$. We shall show that each p_i is a B-eigenfunction and hence $p_i \in \mathcal{E}(X)$.

Let λ be the weight of f. For $b \in B$ we have $\rho(b)f = b^\lambda f$. Thus the factorization of f gives the relation

$$\prod_{i=1}^{r} (\rho(b)p_i)^{m_i} = b^\lambda \prod_{i=1}^{r} p_i^{m_i}. \tag{12.11}$$

Since $\mathcal{P}(X)$ is a unique factorization domain, (12.11) implies that there are scalars $\psi_i(b)$ and a permutation $s(b) \in \mathfrak{S}_r$ such that $\rho(b)p_i = \psi_i(b)p_{s(b)i}$. From the definition we see that $\psi_i(1) = 1$ and $\psi(b_1 b_2) = \psi(b_1)\psi(b_2)$. Evaluating $\rho(b)f_i$ at x_0, we obtain $\psi_i(b) = p_i(b^{-1} \cdot x_0)$, so ψ_i is a regular character of B. Let $x \in X$ and consider the map

$$b \mapsto [p_{s(b)1}(x), \ldots, p_{s(b)r}(x)]$$

from B to $\mathbb{C}^r$. Since $p_{s(b)i}(x) = \psi_i(b)^{-1} p_i(b^{-1}x)$, this map is regular. Since B is connected and the image is a finite set, the map must be constant. Hence $s(b)i = i$ for all b. Thus each p_i is a B-eigenfunction. The assertions about the decomposition of $\mathcal{P}^r(X)$ now follow from Proposition 12.1.1 and Theorem 12.1.2. $\square$

Corollary 12.2.7. *The algebra $\mathcal{P}(X)^{N^+} \cong \mathbb{C}[f_1, \ldots, f_k]$ is a polynomial ring with generators $\mathcal{E}(X)$.*

12.2.3 Branching from $\mathbf{GL(n)}$ to $\mathbf{GL(n-1)}$

We can use the open orbit condition relative to a Borel subgroup to obtain the branching law for the pair $\mathbf{GL}(n-1,\mathbb{C}) \subset \mathbf{GL}(n,\mathbb{C})$ (which was proved in Section 8.3.1 using the Kostant multiplicity formula). Let $G = \mathbf{GL}(n-1,\mathbb{C}) \times \mathbf{GL}(n-1,\mathbb{C})$. View G as a subgroup of $\mathbf{GL}(n,\mathbb{C}) \times \mathbf{GL}(n-1,\mathbb{C})$ by the embedding

$$g \mapsto \begin{bmatrix} g & 0 \\ 0 & 1 \end{bmatrix}, \quad g \in \mathbf{GL}(n-1,\mathbb{C})$$

of $\mathbf{GL}(n-1,\mathbb{C})$ into $\mathbf{GL}(n,\mathbb{C})$. Then G acts on $M_{n,n-1}$ by the restriction of the action of $\mathbf{GL}(n,\mathbb{C}) \times \mathbf{GL}(n-1,\mathbb{C})$. Let $B \subset G$ be the Borel subgroup of all elements

$$\left(\begin{bmatrix} a & 0 \\ 0 & 1 \end{bmatrix}, b \right) \quad \text{with } a,b \in B_{n-1} .$$

Let $x = \begin{bmatrix} y \\ z \end{bmatrix} \in M_{n,n-1}$ with $y \in M_{n-1,n-1}$ and $z \in M_{1,n-1}$. Let $\Delta_i(x) = \Delta_i(y)$ be the ith principal minor of x, for $i = 1,\ldots,n-1$. Define $\Gamma_1(x) = z_1$, and for $i = 2,\ldots,n-1$ let

$$\Gamma_i(x) = \det \begin{bmatrix} y_{1,1} & \cdots & y_{1,i} \\ \vdots & \ddots & \vdots \\ y_{i-1,1} & \cdots & y_{i-1,i} \\ z_1 & \cdots & z_i \end{bmatrix}, \quad x_0 = \begin{bmatrix} 1 & \cdots & 0 \\ \vdots & \ddots & \vdots \\ 0 & \cdots & 1 \\ 1 & \cdots & 1 \end{bmatrix} . \tag{12.12}$$

Note that $\Delta_i(x_0) = \Gamma_i(x_0) = 1$ for $i = 1,\ldots,n-1$.

Lemma 12.2.8. *The B orbit of x_0 is open in $M_{n,n-1}$ and consists of all x such that*

$$\Delta_i(x) \neq 0, \quad \Gamma_i(x) \neq 0 \quad \text{for } i = 1,\ldots,n-1 . \tag{12.13}$$

Proof. Let $g = \left(\begin{bmatrix} a & 0 \\ 0 & 1 \end{bmatrix}, b \right)$, where $a,b \in B_{n-1}$. For $x = \begin{bmatrix} y \\ z \end{bmatrix} \in M_{n,n-1}$ we have

$$g^{-1} \cdot x = \begin{bmatrix} a^t & 0 \\ 0 & 1 \end{bmatrix} \begin{bmatrix} y \\ z \end{bmatrix} b = \begin{bmatrix} a^t y b \\ z b \end{bmatrix} .$$

Thus for $i = 1,\ldots,n-1$ we have

$$\Delta_i(g^{-1} \cdot x) = \Delta_i(a^t y b) = \Delta_i(a)\Delta_i(y)\Delta_i(b) , \tag{12.14}$$

$$\Gamma_i(g^{-1} \cdot x) = \Delta_{i-1}(a)\Gamma_i(x)\Delta_i(b) \tag{12.15}$$

(recall that $\Delta_0 = 1$). Suppose $\Delta_i(x) \neq 0$ for $i = 1,\ldots,n-1$. Choose $a,b \in B_{n-1}$ such that $a^t y b = I$ (this is possible by Lemma B.2.6). Then

$$g^{-1} \cdot x = \begin{bmatrix} 1 & 0 & \cdots & 0 \\ 0 & 1 & \cdots & 0 \\ \vdots & \vdots & \ddots & \vdots \\ 0 & 0 & \cdots & 1 \\ w_1 & w_2 & \cdots & w_{n-1} \end{bmatrix} .$$

Suppose also that $\Gamma_i(x) \neq 0$ for $i = 1,\ldots,n-1$. Then $w_i = \Gamma_i(g^{-1} \cdot x) \neq 0$ by (12.15). Set $h = \mathrm{diag}[w_1^{-1},\ldots,w_{n-1}^{-1}] \in D_{n-1}$ and

$$g_1 = \left(\begin{bmatrix} h & 0 \\ 0 & 1 \end{bmatrix}, h^{-1} \right) \in B .$$

Then

$$(gg_1)^{-1} \cdot x = \begin{bmatrix} h^{-1} & 0 \\ 0 & 1 \end{bmatrix} \begin{bmatrix} 1 & 0 & \cdots & 0 \\ 0 & 1 & \cdots & 0 \\ \vdots & \vdots & \ddots & \vdots \\ 0 & 0 & \cdots & 1 \\ w_1 & w_2 & \cdots & w_{n-1} \end{bmatrix} \quad h = x_0 \,.$$

Thus $x \in B \cdot x_0$. $\square$

The maximal torus H of B is $D_{n-1} \times D_{n-1}$, so the dominant weights of B are given by pairs (μ, λ), where μ and λ are dominant weights for B_{n-1}.

Theorem 12.2.9. *Let $X = M_{n,n-1}(\mathbb{C})$ and let x_0 be given by* (12.12). *The space $\mathcal{P}(X)$ is multiplicity-free under the action of $G = \mathbf{GL}(n-1, \mathbb{C}) \times \mathbf{GL}(n-1, \mathbb{C})$. For $\mathbf{k}, \mathbf{m} \in \mathbb{N}^{n-1}$ and $x \in X$ define*

$$f_{\mathbf{k},\mathbf{m}}(x) = \prod_{i=1}^{n-1} \Delta_i(x)^{k_i} \Gamma_i(x)^{m_i} \,. \tag{12.16}$$

Then the function $f_{\mathbf{k},\mathbf{m}}$ is a normalized B-eigenfunction with weight

$$(\mu, \lambda) = \left(\sum_{i=1}^{n-1} (k_i + m_{i+1}) \varpi_i, \; \sum_{i=1}^{n-1} (k_i + m_i) \varpi_i \right) \tag{12.17}$$

(where $m_n = 0$), and all normalized B-eigenfunctions in $\mathcal{P}(X)$ are of this form. The G-cyclic subspace of $\mathcal{P}(X)$ generated by $f_{\mathbf{k},\mathbf{m}}$ is equivalent to $F_{n-1}^{\mu} \otimes F_{n-1}^{\lambda}$. The G-module decomposition of $\mathcal{P}(X)$ is

$$\mathcal{P}(X) \cong \bigoplus_{\lambda, \mu} F_{n-1}^{\mu} \otimes F_{n-1}^{\lambda} \,, \tag{12.18}$$

with the sum over all pairs $\lambda = p_1 \varepsilon_1 + \cdots + p_{n-1} \varepsilon_{n-1}$ and $\mu = q_1 \varepsilon_1 + \cdots + q_{n-1} \varepsilon_{n-1}$ of dominant weights such that

$$p_1 \geq q_1 \geq p_2 \geq q_2 \geq \cdots \geq p_{n-1} \geq q_{n-1} \geq 0 \,. \tag{12.19}$$

Proof. The action of G is multiplicity-free by Lemma 12.2.8. To find which representations of G occur, we observe that $\Delta_1, \ldots, \Delta_{n-1}$ and $\Gamma_1, \ldots, \Gamma_{n-1}$ are B-eigenfunctions, by (12.14) and (12.15). They are also irreducible polynomials by Lemma B.2.10. Since H has dimension $2(n-1)$, it follows from Theorem 12.2.6 that these polynomials are the complete set $\mathcal{E}(X)$ of normalized irreducible B-eigenfunctions, and that the functions $f_{\mathbf{k},\mathbf{m}}$ are all the normalized B-eigenfunctions in $\mathcal{P}(X)$.

The G-cyclic space generated by $f_{\mathbf{k},\mathbf{m}}$ is irreducible, since this function is a highest-weight vector. The representation of G on this space is equivalent to the (outer) tensor product $F_{n-1}^{\mu} \widehat{\otimes} F_{n-1}^{\lambda}$, where the highest weights μ and λ are given by (12.17) (see Proposition 4.2.5). Writing $\lambda = p_1\varepsilon_1 + \cdots + p_{n-1}\varepsilon_{n-1}$ and $\mu = q_1\varepsilon_1 + \cdots + q_{n-1}\varepsilon_{n-1}$, we have $p_i = q_i + m_i$ and $q_i = k_i + p_{i+1}$. This implies that the pair (λ, μ) satisfies (12.19), since $m_i \geq 0$ and $k_i \geq 0$. Conversely, any pair (λ, μ) that satisfies (12.19) can be written as (12.17) with $k_i \geq 0$ and $m_i \geq 0$. $\qquad\square$

12.2.4 Second Fundamental Theorems

Other examples of multiplicity-free linear actions arise in connection with the *second fundamental theorem* (SFT) for the polynomial invariants of an arbitrary number of vectors and covectors. For this application we begin with a general result for representations on polynomial rings, following the notation of Section 12.2.1.

Suppose (σ, V) is a regular representation of the connected reductive group G. Fix a Borel subgroup $B = HN^+$ of G. We view $\mathcal{P}(V)$ as a G-module relative to the action $\rho(g)f(v) = f(\sigma(g)^{-1}v)$ for $f \in \mathcal{P}(V)$, $v \in V$, and $g \in G$.

Proposition 12.2.10. *Let $\mathcal{I} \subset \mathcal{P}(V)$ be a G-invariant ideal. Then*

$$\mathcal{I} = \mathrm{Span}\{\rho(G)\mathcal{I}^{N^+}\} \quad and \quad \mathcal{P}(V)/\mathcal{I} = \mathrm{Span}\{\bar{\rho}(G)(\mathcal{P}(V)/\mathcal{I})^{N^+}\}, \qquad (12.20)$$

where $\bar{\rho}$ is the representation of G on $\mathcal{P}(V)/\mathcal{I}$. Furthermore, the map $f \mapsto f + \mathcal{I}$ gives a surjection of $\mathcal{P}(V)^{N^+}$ onto $(\mathcal{P}(V)/\mathcal{I})^{N^+}$ with kernel $\mathcal{I}^{N^+}$.

Proof. We first prove (12.20). Since G is reductive and the action of G on $\mathcal{I}$ and on $\mathcal{P}(V)/\mathcal{I}$ is locally regular, there is a decomposition

$$\mathcal{I} \cong \bigoplus_i U_i, \qquad \mathcal{P}(V)/\mathcal{I} \cong \bigoplus_j W_j,$$

with U_i and W_j irreducible regular G-modules. Hence (12.20) follows from Theorems 2.2.7 and 4.2.12, since G is connected.

We write $\pi(f) = f + \mathcal{I}$ for $f \in \mathcal{P}(V)$. Suppose $0 \neq \pi(f) \in (\mathcal{P}(V)/\mathcal{I})^{N^+}$. We can decompose

$$\pi(f) = \sum_{\mu \in \mathfrak{X}(H)} \pi(g_\mu)$$

relative to the maximal torus H, where $g_\mu \in \mathcal{P}(V)$ and $\rho(h)g_\mu = h^\mu g_\mu$ for $h \in H$. We claim that $\bar{\rho}(n)\pi(g_\mu) = \pi(g_\mu)$ for all μ and all $n \in N^+$. Indeed, if $X \in \mathfrak{n}^+ = \mathrm{Lie}(N^+)$ has weight α then $d\bar{\rho}(X)\pi(g_\mu)$ has weight $\mu + \alpha$. But $d\bar{\rho}(X)\pi(f) = 0$, so

$$\sum_{\mu \in \mathfrak{X}(H)} d\bar{\rho}(X)\pi(g_\mu) = 0.$$

Distinct terms in this sum have distinct weights, so $d\bar{\rho}(X)\pi(g_\mu) = 0$ for all μ. Hence $d\bar{\rho}(n)\pi(g_\mu) = 0$, proving the claim.

Take any weight μ such that $\pi(g_\mu) \neq 0$. Then μ is dominant, since $\bar{\rho}(N^+)$ preserves the space $\pi(g_\mu)$. Set $F = \mathrm{Span}\{\bar{\rho}(G)\pi(g_\mu)\} \subset \mathcal{P}(V)/\mathcal{I}$. Then F is an irreducible G-module with highest weight μ, by Proposition 3.3.9. Set

$$E = \text{Span}\{\rho(G)g_\mu\} \subset \mathcal{P}(V) .$$

Then $\pi(E) = F$. Let $E_{(\lambda)}$ be the G-isotypic component of E corresponding to the irreducible G-module with highest weight λ. By Schur's lemma we have $\pi(E_{(\lambda)}) = 0$ if $\lambda \neq \mu$, and $\pi(E_{(\mu)}) = F$. Let f_μ be the projection of g_μ into $E_{(\mu)}$. Then $\pi(f_\mu) = \pi(g_\mu)$, so in particular, $f_\mu \neq 0$. Since f_μ has weight μ, the theorem of the highest weight implies that $f_\mu \in \mathcal{P}(V)^{N^+}$. Hence $\sum_{\mu \in \mathcal{X}(H)} f_\mu \in \mathcal{P}(V)^{N^+}$ and

$$\sum_{\mu \in \mathcal{X}(H)} \pi(f_\mu) = \pi(f) .$$

This proves surjectivity of the map $\mathcal{P}(V)^{N^+} \longrightarrow (\mathcal{P}(V)/\mathcal{J})^{N^+}$. The kernel of this map is $\mathcal{J}^{N^+}$ by definition. $\square$

Corollary 12.2.11. *Suppose*

$$\mathcal{P}(V) \cong \bigoplus_{\lambda \in \mathcal{S}} V^\lambda$$

is multiplicity-free as a G-module. Let $\mathcal{S}_0$ be the set of weights occurring in $\mathcal{J}^{N^+}$. Then

$$\mathcal{J} \cong \bigoplus_{\lambda \in \mathcal{S}_0} V^\lambda \quad and \quad \mathcal{P}(V)/\mathcal{J} \cong \bigoplus_{\lambda \in \mathcal{S} \setminus \mathcal{S}_0} V^\lambda$$

as G-modules.

We now apply these results to obtain the second fundamental theorem (SFT) for each family of classical groups.

General Linear Group

We write $\mathbf{GL}(p, \mathbb{C}) = \mathbf{GL}_p$. Let $\pi_{k,m}$ be the representation of $\mathbf{GL}_k \times \mathbf{GL}_n$ on $\mathcal{P}(M_{k,n})$ given by

$$\pi_{k,m}(g,h)f(x) = f(g^{-1}xh) \quad \text{for } g \in \mathbf{GL}_k \text{ and } h \in \mathbf{GL}_m .$$

Recall from Section 5.2.1 that if we let $g \in \mathbf{GL}_n$ act on $f \in \mathcal{P}(M_{k,n} \times M_{n,m})$ by

$$\pi(g)f(x,y) = f(xg, g^{-1}y) ,$$

then the multiplication map $\mu : M_{k,n} \times M_{n,m} \longrightarrow M_{k,m}$ induces a surjective algebra homomorphism

$$\mu^* : \mathcal{P}(M_{k,m}) \longrightarrow \mathcal{P}(\underbrace{(\mathbb{C}^n)^* \times \cdots \times (\mathbb{C}^n)^*}_{k} \times \underbrace{\mathbb{C}^n \times \cdots \times \mathbb{C}^n}_{m})^{\mathbf{GL}_n}$$

(Theorem 5.2.1). By Corollary 5.2.5 we have $\text{Ker}(\mu^*) = 0$ when $n \geq \min(k,m)$. The SFT describes the ideal $\text{Ker}(\mu^*)$ when $n < \min(k,m)$. If $n < \min(k,m)$ then

$\mu(M_{k,n} \times M_{n,m})$ is the space of all matrices in $M_{k,m}$ of rank at most n, by Lemma 5.2.4 (1). We denote this subset by $\mathcal{DV}_{k,m,n}$ (called a *determinantal variety*). Let $\mathfrak{I}_{k,m,n} \subset \mathcal{P}(M_{k,m})$ be the ideal of polynomials vanishing on $\mathcal{DV}_{k,m,n}$. Thus $\mathrm{Ker}(\mu^*) = \mathfrak{I}_{k,m,n}$, and $\mathcal{P}(M_{k,m})/\mathfrak{I}_{k,m,n}$ is the algebra of regular functions on $\mathcal{DV}_{k,m,n}$. The SFT gives generators for the ideal $\mathfrak{I}_{k,m,n}$ and the decompositions of $\mathfrak{I}_{k,m,n}$ and $\mathcal{O}(\mathcal{DV}_{k,m,n})$ as modules for $\mathbf{GL}_k \times \mathbf{GL}_m$.

Theorem 12.2.12. (SFT for $\mathbf{GL}_n$)   *Assume $n < \min(k,m)$.*

1. *The set of all $(n+1) \times (n+1)$ minors is a minimal generating set for $\mathfrak{I}_{k,m,n}$.*
2. *As a module for $\mathbf{GL}_k \times \mathbf{GL}_m$, the determinantal ideal $\mathfrak{I}_{k,m,n}$ decomposes as*

$$\mathfrak{I}_{k,m,n} \cong \bigoplus_\lambda (F_k^\lambda)^* \otimes F_m^\lambda \,,$$

   where λ runs over all nonnegative dominant weights with $n < \mathrm{depth}(\lambda) \leq \min(k,m)$.
3. *As a module for $\mathbf{GL}_k \times \mathbf{GL}_m$, the space of regular functions on the determinantal variety $\mathcal{DV}_{k,m,n}$ decomposes as*

$$\mathcal{O}[\mathcal{DV}_{k,m,n}] \cong \bigoplus_\lambda (F_k^\lambda)^* \otimes F_m^\lambda \,,$$

   where λ runs over all nonnegative dominant weights with $\mathrm{depth}(\lambda) \leq n$.

Proof. Set $G = \mathbf{GL}_k \times \mathbf{GL}_m$ and take $N^+ = N_k^- \times N_m^+$. Since $\mathfrak{I}_{k,m,n}$ is a G-invariant ideal in $\mathcal{P}(M_{k,m})$, Proposition 12.2.10 implies that

$$\mathfrak{I}_{k,m,n} = \mathrm{Span}\{\pi_{m,k}(G)\mathfrak{I}_{k,m,n}^{N^+}\} \,. \tag{12.21}$$

By Theorem 5.6.7 we have $\mathcal{P}(M_{k,m})^{N^+} = \mathbb{C}[\Delta_1, \ldots, \Delta_r]$, where $r = \min(k,m)$. Since $\Delta_p \in \mathfrak{I}_{k,m,n}$ if and only if $p > n$, we see that

$$\mathfrak{I}_{k,m,n}^{N^+} = \sum_{p=n+1}^r \mathbb{C}[\Delta_1, \ldots, \Delta_r]\Delta_p \,. \tag{12.22}$$

Since G acts by automorphisms of the algebra $\mathcal{P}(M_{k,m})$, we conclude from (12.21) and (12.22) that

$$\mathfrak{I}_{k,m,n} = \sum_{p=n+1}^r \mathcal{P}(M_{k,m})\,\mathrm{Span}\{\pi_{m,k}(G)\Delta_p\} \,. \tag{12.23}$$

Lemma 12.2.13. $\mathrm{Span}\{\pi_{m,k}(G)\Delta_p\}$ *is the space spanned by the set of all $p \times p$ minors. It is isomorphic to $\left(F_k^{\lambda_p}\right)^* \otimes F_m^{\lambda_p}$ as a G-module.*

Proof. We embed $\mathfrak{S}_k$ into $\mathbf{GL}_k$ as the permutation matrices as usual. The space spanned by the set of all $p \times p$ minors is then

$$\mathrm{Span}\{\pi_{k,m}(s,t)\Delta_p : s \in \mathfrak{S}_k, t \in \mathfrak{S}_m\} \,.$$

The function $\Delta_p(x)$ is a lowest-weight vector of weight $-\lambda_p$ for the left action of $\mathbf{GL}_k$ and a highest-weight vector of weight λ_p for the right action of $\mathbf{GL}_m$ (see the proof of Theorem 5.6.7), where $\lambda_p = \varepsilon_1 + \cdots + \varepsilon_p$. Using the explicit model $F_k^{\lambda_p} \cong \bigwedge^p \mathbb{C}^k$ from Theorem 5.5.11, we see that

$$\mathrm{Span}\{\pi_{k,m}(s,t)\Delta_p : s \in \mathfrak{S}_k, t \in \mathfrak{S}_m\} = \mathrm{Span}\{\pi_{k,m}(g)\Delta_p : g \in G\}\,,$$

and that this space is isomorphic to $\left(F_k^{\lambda_p}\right)^* \otimes F_m^{\lambda_p}$. $\qquad\qquad\square$

Completion of Proof of Theorem 12.2.12. By the cofactor expansion of a determinant, we see that the minors of size greater than $n+1$ are in the ideal generated by the minors of size $n+1$. The set of minors of size $n+1$ is linearly independent. Also, each minor is an irreducible polynomial (see Lemma B.2.10). Hence by (12.23) and Lemma 12.2.13 these minors give a minimal generating set for $\mathfrak{I}_{k,m,n}$. This completes the proof of part (1) of the theorem.

To prove parts (2) and (3) of the theorem, recall that the nonnegative dominant weights λ of depth p are of the form

$$\lambda = m_1\lambda_1 + \cdots + m_p\lambda_p \quad \text{with } m_p > 0\,.$$

From Theorem 5.6.7 we know that if $p \leq \min(k,m)$ then the function $\Delta_1^{m_1} \cdots \Delta_p^{m_p}$ is the only lowest-weight vector of weight $-\lambda$ for $\mathbf{GL}_k$ and highest-weight vector of weight λ for $\mathbf{GL}_m$ in $\mathcal{P}(M_{k,m})$ (up to normalization). Since $m_p > 0$ this function is in $\mathfrak{I}_{k,m,n}^{N^+}$ if and only if $p > n$. Hence (2) and (3) now follow from (12.21) and Corollary 12.2.11. $\qquad\qquad\square$

Orthogonal Group

Let SM_k be the space of symmetric $k \times k$ complex matrices. Recall from Section 5.2.1 the map

$$\tau : M_{n,k} \longrightarrow SM_k \quad \text{with } \tau(x) = x^t x\,.$$

By the FFT, the associated homomorphism

$$\tau^* : \mathcal{P}(SM_k) \longrightarrow \mathcal{P}(M_{n,k})^{\mathbf{O}(n)}$$

is surjective (Theorem 5.2.2). By Corollary 5.2.5, τ^* is injective when $n \geq k$. We will find generators for the ideal $\mathrm{Ker}(\tau^*)$ when $n < k$.

Assume $n < k$. By Lemma 5.2.4 (2) the range of τ consists of all symmetric matrices of rank $\leq n$. We denote this subset by $S\mathcal{V}_{k,n}$ (the *symmetric determinantal variety*). Let $S\mathfrak{I}_{k,n}$ be the ideal of polynomials vanishing on $S\mathcal{V}_{k,n}$. Then $\mathrm{Ker}(\tau^*) = S\mathfrak{I}_{k,n}$, and $\mathcal{P}(SM_k)/S\mathfrak{I}_{k,n}$ is the algebra of regular functions on $S\mathcal{V}_{k,n}$.

Let π be the representation of $\mathbf{GL}_k$ on $\mathcal{P}(SM_k)$ given by

$$\pi(g)f(x) = f(g^t x g), \quad \text{for } f \in \mathcal{P}(SM_k)\,.$$

Theorem 12.2.14. (SFT for $\mathbf{O}(n)$) *Assume $n < k$.*

1. *The restrictions to SM_k of the $(n+1) \times (n+1)$ minors is a minimal generating set for the ideal $\mathcal{SI}_{k,n}$.*
2. *As a module for $\mathbf{GL}_k$, the symmetric determinantal ideal $\mathcal{SI}_{k,n}$ decomposes as*

$$\mathcal{SI}_{k,n} \cong \bigoplus_\lambda F_k^\lambda \,,$$

where λ runs over all nonnegative even dominant weights that satisfy $n < \text{depth}(\lambda) \leq k$.
3. *As a module for $\mathbf{GL}_k$, the space of regular functions on the symmetric determinantal variety $\mathcal{SV}_{k,n}$ decomposes as*

$$\mathcal{O}[\mathcal{SV}_{k,n}] \cong \bigoplus_\lambda F_k^\lambda \,,$$

where λ runs over all nonnegative even dominant weights with $\text{depth}(\lambda) \leq n$.

Proof. We follow the same general line of argument as in Theorem 12.2.12. Let $G = \mathbf{GL}_k$ and $N^+ = N_k^+$. Since $\mathcal{SI}_{k,m,n}$ is a G-invariant ideal in $\mathcal{P}(SM_k)$, Proposition 12.2.10 implies that

$$\mathcal{SI}_{k,n} = \text{Span}\left\{\pi(G)\mathcal{SI}_{k,n}^{N^+}\right\}.$$

By Theorem 5.7.3 we have $\mathcal{P}(SM_k)^{N^+} = \mathbb{C}[\widetilde{\Delta}_1,\ldots,\widetilde{\Delta}_k]$. Since $\widetilde{\Delta}_p \in \mathcal{SI}_{k,n}$ if and only if $p > n$, we see that

$$\mathcal{SI}_{k,n}^{N^+} = \sum_{p=n+1}^k \mathbb{C}[\widetilde{\Delta}_1,\ldots,\widetilde{\Delta}_k]\widetilde{\Delta}_p \,.$$

Hence

$$\mathcal{SI}_{k,n} = \sum_{p=n+1}^k \mathcal{P}(SM_k)\,\text{Span}\{\pi(G)\widetilde{\Delta}_p\} \,. \tag{12.24}$$

Let $U_p \subset \mathcal{P}(M_k)$ be the space spanned by the $p \times p$ minors. Then U_p is invariant under the two-sided action of $\mathbf{GL}_k \times \mathbf{GL}_k$, by Lemma 12.2.13. Let V_p be the restrictions to SM_k of the functions in U_p. Then it follows that $\text{Span}\{\pi(G)\widetilde{\Delta}_p\} \subset V_p$. Since $V_p \subset \mathcal{SI}_{k,n}$ if $p > n$, we conclude from (12.24) and the cofactor expansion of a determinant that the ideal $\mathcal{SI}_{k,n}$ is generated by V_{n+1}. The set of $(n+1) \times (n+1)$ minors restricted to SM_k is linearly independent and consists of irreducible polynomials (see Lemma B.2.10). Hence it is a minimal generating set for $\mathcal{SI}_{k,n}$. This proves part (1) of the theorem.

To prove parts (2) and (3), recall from Theorem 5.7.3 that there is a multiplicity-free decomposition

$$\mathcal{P}(SM_k) = \bigoplus_\mu F_k^\mu \,, \tag{12.25}$$

with the sum over all nonnegative dominant weights μ such that μ is even and $\text{depth}(\mu) \leq k$. The even dominant weights λ of depth p are of the form

$$\lambda = 2m_1\lambda_1 + \cdots + 2m_p\lambda_p \quad \text{with } m_p > 0 \,.$$

From Theorem 5.7.3 the highest-weight vector of weight μ for π is the restriction to SM_k of the function $\Delta_{\mathbf{m}}$, where $\mu = 2m_1\lambda_1 + \cdots + 2m_p\lambda_p$. Since $\Delta_{\mathbf{m}}|_{SM_k} \in \mathcal{SJ}_{k,n}$ if and only if μ has depth $\geq n+1$, parts (2) and (3) of the theorem now follow from (12.25) and Corollary 12.2.11. □

Symplectic Group

Let AM_k be the space of skew-symmetric $k \times k$ complex matrices. Assume that n is even and set $K = \mathbf{Sp}(\mathbb{C}^n)$. Recall from Section 5.2.1 the map

$$\gamma : M_{n,k} \longrightarrow AM_k \quad \text{with} \quad \gamma(x) = x^t J_n x.$$

The associated homomorphism

$$\gamma^* : \mathcal{P}(AM_k) \longrightarrow \mathcal{P}(M_{n,k})^K$$

is surjective (Theorem 5.2.2). By Corollary 5.2.5, γ^* is injective when $n \geq k$. We will now find generators for $\mathrm{Ker}(\gamma^*)$ when $n < k$.

Assume $n < k$. By Lemma 5.2.4 (3) the range of γ consists of all skew-symmetric matrices of rank $\leq n$. We denote this subset by $A\mathcal{V}_{k,n}$ (the *alternating determinantal variety*). Let $A\mathcal{J}_{k,n}$ be the ideal of polynomials vanishing on $A\mathcal{V}_{k,n}$. Then we have $\mathrm{Ker}(\gamma^*) = A\mathcal{J}_{k,n}$, and $\mathcal{P}(AM_k)/A\mathcal{J}_{k,n}$ is the algebra of regular functions on $A\mathcal{V}_{k,n}$.

Let π be the representation of $\mathbf{GL}_k$ on $\mathcal{P}(AM_k)$ given by

$$\pi(g)f(x) = f(g^t x g) \quad \text{for} \ f \in \mathcal{P}(AM_k).$$

For $1 \leq p \leq [k/2]$ let Pf_p be the pth principal Pfaffian (see Section B.2.6). Recall that $\mathrm{Pf}_p(x)^2 = \Delta_{2p}(x)$ by Corollary B.2.9, so $\mathrm{Pf}_p \in A\mathcal{J}_{k,n}$ for $p > 2n$. From Theorem 5.7.5 we have the decomposition

$$\mathcal{P}(AM_k) = \bigoplus_{\mu} F_k^{\mu} \tag{12.26}$$

as a $\mathbf{GL}_k$-module, with the sum over all dominant weights

$$\mu = \sum_{i=1}^{[k/2]} m_i \lambda_{2i}. \tag{12.27}$$

Note that if $\mu \neq 0$ then $\mathrm{depth}(\mu) = 2p$, where $p = \max\{i : m_i \neq 0\}$.

Theorem 12.2.15. (SFT for $\mathbf{Sp}(\mathbb{C}^n)$) *Assume $n < k$.*

1. *The set $\{\pi(s)\,\mathrm{Pf}_{(n/2)+1} : s \in \mathfrak{S}_k\}$ is a minimal generating set for $A\mathcal{J}_{k,n}$.*
2. *As a module for $\mathbf{GL}_k$, the alternating determinantal ideal $A\mathcal{J}_{k,n}$ decomposes as*

$$A\mathcal{J}_{k,n} \cong \bigoplus_{\mu} F_k^{\mu},$$

where μ satisfies (12.27) and $n < \mathrm{depth}(\mu) \leq k$.

3. *As a module for $\mathbf{GL}_k$, the space of regular functions on the alternating determinantal variety $\mathcal{AV}_{k,n}$ decomposes as*

$$\mathbb{C}[\mathcal{AV}_{k,n}] \cong \bigoplus_\mu F_k^\mu \,,$$

where μ satisfies (12.27) and $\mathrm{depth}(\mu) \leq n$.

Proof. We follow the same general line of argument as in Theorem 12.2.14. Let $G = \mathbf{GL}_k$ and $N^+ = N_k^+$. Since $\mathcal{AI}_{k,m,n}$ is a G-invariant ideal in $\mathcal{P}(AM_k)$, Proposition 12.2.10 implies that

$$\mathcal{AI}_{k,n} = \mathrm{Span}\{\pi(G)\mathcal{AI}_{k,n}^{N^+}\} \,.$$

By Theorem 5.7.5 we have $\mathcal{P}(AM_k)^{N^+} = \mathbb{C}[\mathrm{Pf}_1,\ldots,\mathrm{Pf}_k]$. Since $\mathrm{Pf}_p \in \mathcal{AI}_{k,n}$ if and only if $p > (n/2)$, we see that

$$\mathcal{AI}_{k,n}^{N^+} = \sum_{p=(n/2)+1}^{k} \mathbb{C}[\mathrm{Pf}_1,\ldots,\mathrm{Pf}_k]\,\mathrm{Pf}_p \,.$$

Hence

$$\mathcal{AI}_{k,n} = \sum_{p=(n/2)+1}^{k} \mathcal{P}(AM_k)\,\mathrm{Span}\{\pi(G)\,\mathrm{Pf}_p\} \,. \tag{12.28}$$

We know from Theorem 5.7.5 that Pf_p is a highest-weight vector of weight λ_{2p} for $\mathbf{GL}_k$. Hence by Theorem 5.5.11, $\pi(g)\,\mathrm{Pf}_p \in \mathrm{Span}\{\pi(s)\,\mathrm{Pf}_p : s \in \mathfrak{S}_k\}$ for all $g \in \mathbf{GL}_k$. The set of functions $\{\pi(s)\,\mathrm{Pf}_p : s \in \mathfrak{S}_k\}$ is linearly independent, since it is a basis for the irreducible $\mathbf{GL}_k$-module with highest weight λ_{2p}. Each function in this set is an irreducible polynomial (Lemma B.2.10). From the Pfaffian expansion (B.18) it is clear that Pf_p is in the ideal generated by these functions when $p > (n/2) + 1$. Hence part (1) of the theorem follows from (12.28).

To prove parts (2) and (3), let μ be given by (12.27). From Theorem 5.7.5 we know that the highest-weight vector of weight μ for π is the function

$$f(x) = \mathrm{Pf}_1(x)^{m_1} \cdots \mathrm{Pf}_p(x)^{m_p} \,,$$

where $2p = \mathrm{depth}(\mu)$ and $m_p \neq 0$. Thus $f \in \mathcal{AI}_{k,n}$ if and only if $p \geq (n/2) + 1$. Hence parts (2) and (3) of the theorem now follow from (12.26) and Corollary 12.2.11. $\qquad\square$

12.2.5 Exercises

1. Use Theorem 12.2.1 to show that the following spaces are multiplicity-free:
   (a) $G = \mathbf{GL}_n \times \mathbf{GL}_k$, $X = M_{n,k}(\mathbb{C})$; action $(g,h) \cdot x = gxh^{-1}$.
   (HINT: Lemma B.2.6.)

(b) $G = \mathbf{GL}_n$, $X = SM_n(\mathbb{C})$; action $g \cdot x = gxg^t$. (HINT: Lemma B.2.7.)

(c) $G = \mathbf{GL}_n$, $X = AM_n(\mathbb{C})$; action $g \cdot x = gxg^t$. (HINT: Lemma B.2.8.)

2. Let G be a connected reductive group with Borel subgroup $B = HN^+ \subset G$. For $0 \neq \lambda \in P_{++}(G)$ take a B-eigenvector $v_\lambda \in V^\lambda$ and let X be the Zariski closure of the orbit $G \cdot v_\lambda$. Then X is a G-invariant affine variety in V^λ, called a *highest vector* variety by Vinberg–Popov [148].

(a) Show that X is a multiplicity-free G-space. (HINT: Let $B^- = HN^-$ be the Borel subgroup opposite to B. Show that B^- has an open orbit on X.)

(b) Let $\mathbb{P}V^\lambda$ be the projective space of lines in V^λ, and let $[v] \in \mathbb{P}V^\lambda$ be the line through a nonzero vector $v \in V^\lambda$. Show that $[G \cdot v_\lambda]$ is closed in $\mathbb{P}V^\lambda$. (HINT: Set $P = \{g \in G : [g \cdot v_\lambda] = [v_\lambda]\}$. Then $[G \cdot v_\lambda] \cong G/P$ as a G variety and G/P is projective, since $B \subset P$.)

(c) Show that $X = G \cdot v_\lambda \cup \{0\}$ and that X is invariant under multiplication by $\mathbb{C}^\times$. (HINT: Use (b) to determine the action of B on X.)

(d) Let $\mathcal{O}[X]^{(n)}$ be the restrictions to X of the homogeneous polynomials of degree n on V^λ. Show that the isotypic decomposition of $\mathcal{O}[X]$ as a G-module is

$$\mathcal{O}[X] = \bigoplus_{n \in \mathbb{N}} \mathcal{O}[X]^{(n)}$$

and that $\mathcal{O}[X]^{(n)}$ is an irreducible G-module isomorphic to $(V^{n\lambda})^*$. (HINT: Let $f_\lambda(x) = \langle v_\lambda^*, x \rangle$ for $x \in X$, where v_λ^* is the lowest-weight vector in $(V^\lambda)^*$. Show that f_λ^n is a B^- eigenfunction of weight $-n\lambda$ for the representation ρ_X, and hence $(V^{n\lambda})^* \subseteq \mathcal{O}[X]^{(n)}$ for all positive integers n. Now use Theorem 12.2.1 to show that if μ occurs as a B^- weight in $\mathcal{O}[X]$, then μ is proportional to λ.)

3. Show that the following pairs (G,X) are multiplicity-free, and find the decomposition of $\mathcal{O}[X]$ as a G-module.

(a) Take $G = \mathbf{GL}(n,\mathbb{C})$ or $G = \mathbf{SL}(n,\mathbb{C})$ acting on $X = \mathbb{C}^n$ by the defining representation.

(b) Let $G = \mathbf{SO}(\mathbb{C}^n, B)$ acting on the *nullcone* $X = \{x \in \mathbb{C}^n : B(x,x) = 0\}$ via its action on $\mathbb{C}^n$.

(HINT: Use the previous exercise.)

4. Let $G = \mathbf{Spin}(7,\mathbb{C})$.

(a) Show that the spin representation (π, V) of G is of dimension 8 and carries a G-invariant nondegenerate form. Thus $\pi : G \to \mathbf{SO}(8,\mathbb{C})$.

(b) Let $\overline{M} = \pi(G)$. Let M be the pullback of $\overline{M}$ to $\mathbf{Spin}(8,\mathbb{C})$ via the covering map $\mathbf{Spin}(8,\mathbb{C}) \longrightarrow \mathbf{SO}(8,\mathbb{C})$. Let M_1 be the usual embedding of $\mathbf{Spin}(7,\mathbb{C})$ in $\mathbf{Spin}(8,\mathbb{C})$ (i.e., the pullback of the group of all matrices in $\mathbf{SO}(8,\mathbb{C})$ of the form

$$\begin{bmatrix} 1 & 0 \\ 0 & g \end{bmatrix} \tag{12.29}$$

with $g \in \mathbf{SO}(7,\mathbb{C})$). Regard $\mathbf{Spin}(8,\mathbb{C})$ as a subgroup of $\mathbf{Spin}(9,\mathbb{C})$ as the pullback of matrices of the form (12.29) with $g \in \mathbf{SO}(8,\mathbb{C})$, so that M and M_1 become subgroups of $\mathbf{Spin}(9,\mathbb{C})$. Show that $\mathbf{Spin}(9,\mathbb{C})/M$ is a multiplicity-free space for $\mathbf{Spin}(9, \mathbb{C})$ but $\mathbf{Spin}(9, \mathbb{C})/M_1$ is not. (HINT: Use the branching formulas $\mathbf{Spin}(9, \mathbb{C}) \to \mathbf{Spin}(8, \mathbb{C}) \to \mathbf{Spin}(7, \mathbb{C})$ for the case of M_1. For M use the branching formula but think of the weights of a half-spin representation of $\mathbf{Spin}(8, \mathbb{C})$ as $\pm\varepsilon_1, \ldots, \pm\varepsilon_4$.)

5. Let $G = \mathbf{SO}(\mathbb{C}^n, B)$ with $n \geq 3$, where B is the symmetric form in equations (2.6) and (2.9). Let $Q(x) = B(x, x)$. Take as Borel subgroup B the upper-triangular matrices in G with maximal torus H the diagonal matrices in G.

   (a) Show that the action of $\mathbb{C}^\times \times G$ on $\mathbb{C}^n$ is multiplicity-free, where $\mathbb{C}^\times$ acts by scalar multiplication. (HINT: Consider the $\mathbb{C}^\times \times B$ orbit of $x_0 = e_1 + e_n$ when n is even and $x_0 = e_1 + e_{l+1} + e_n$ when $n = 2l + 1$ is odd.)

   (b) Show that the irreducible $\mathbb{C}^\times \times B$ eigenfunctions are x_n and Q. (HINT: Calculate the stabilizer in $\mathbb{C}^\times \times H$ of the vector x_0 in (a).)

   (c) Show that for $r \geq 1$,

$$\mathcal{P}^r(\mathbb{C}^n) = \bigoplus_{k+2m=r} Q^m V^{k\varepsilon_1} \qquad (k \geq 0, m \geq 0),$$

   where $V^{k\varepsilon_1}$ is the $\mathbf{SO}(\mathbb{C}^n, B)$-cyclic subspace generated by x_n^k and is an irreducible representation of highest weight $k\varepsilon_1$. (This is the spherical harmonic decomposition of Corollary 5.6.12.)

6. Let $G = \mathbf{Sp}(\mathbb{C}^{2n}, \Omega)$, where Ω is the skew form in (2.6). Take as Borel subgroup B the upper-triangular matrices in G with maximal torus H the diagonal matrices in G.

   (a) Show that the action of G on $\mathbb{C}^{2n}$ is multiplicity-free. (HINT: Consider the B orbit of $e_1 + e_{2n}$.)

   (b) Show that there is one irreducible B eigenfunction, namely x_{2n}. (HINT: Calculate the stabilizer of $e_1 + e_{2n}$ in H.)

   (c) Show that for $k \geq 1$ the space of polynomials homogeneous of degree k is irreducible under G, with highest weight $k\varepsilon_1$ and highest-weight eigenfunction x_{2n}^k.

7. Let V be an n-dimensional vector space. Let $X = \{(x, y) \in V \times V : x \wedge y = 0\}$.

   (a) Show that X is isomorphic as an algebraic variety with the determinantal variety $\mathcal{DV}_{n,2,1}$ (see Section 12.2.4).

   (b) Let $\Phi : V \times \mathbb{C} \mapsto X$ be defined by $\Phi(v, t) = (v, tv)$. Show that the Zariski closure of $\Phi(V \times \mathbb{C})$ is X. (HINT: Consider the pullback of the functions on X as a representation of $\mathbf{GL}(V)$. Show that $\mathcal{DV}_{n,2,1}$ is irreducible and of dimension $n + 1$.)

8. Let $G = \mathbf{GL}_k \times \mathbf{GL}_m$ act on $M_{k,m}$ as usual. Let $p = \min(k, m)$.

   (a) Show that $M_{k,m} = \bigcup_{0 \leq n < p} G \cdot P_n$, where $P_n = \sum_{i \leq n} E_{ii}$.

   (b) Calculate $\dim G \cdot P_n$ and $\dim(\mathcal{DV}_{k,m,n})$.

   (c) Show that the Zariski closure of $G \cdot P_n$ is $\mathcal{DV}_{k,m,n}$. Use this to show that $\mathcal{DV}_{k,m,n}$ is irreducible.

9. Formulate and prove results analogous to those of the previous exercise for the symmetric determinantal variety and the alternating determinantal variety (see Section 12.2.4).

12.3 Regular Functions on Symmetric Spaces

We now make a detailed study of the most important examples of multiplicity-free affine G-spaces with nonlinear G action; namely, the symmetric spaces $X = G/K$, where K is the fixed-point subgroup of an involution on G.

12.3.1 Iwasawa Decomposition

Let G be a connected reductive group with center Z and let θ be an involutive regular automorphism of G. We assume that θ acts by the identity on Z° (this is automatic if G is semisimple). Let $K = \{g \in G : \theta(g) = g\}$, and assume that $K \neq G$. In this section we shall construct a solvable subgroup $AN^+ \subset G$ that is a semidirect product of a torus A with a unipotent group N^+, and prove that $K \cap (AN^+)$ is finite and KAN^+ is Zariski dense in G; this gives the (complexified) *Iwasawa decomposition* of G. Since AN^+ is contained in some Borel subgroup B of G, this will imply that KB is also Zariski dense in G, and hence (G, K) is a spherical pair. Obtaining these results in this general context will require a large number of structural results about G and K that we now develop. Readers interested in only the classical groups can see these results in explicit form in Section 12.3.2.

Let τ be a conjugation of G such that the corresponding real form $U = G^\tau$ is compact (see Sections 1.7.2 and 11.5.1). Theorem 11.5.3 implies that we may assume that $\tau\theta = \theta\tau$. Furthermore, by Lemma 11.5.8 we may assume $G \subset \mathbf{GL}(n, \mathbb{C})$ with $\tau(g) = (\bar{g}^t)^{-1}$. Let $K_0 = U \cap K$. Then K_0 is a compact real form of the reductive group K that is Zariski dense in K by Theorem 11.5.10.

Remark 12.3.1. When G is a classical group, we obtained matrix versions of θ and K in Section 11.3.5 such that the diagonal subgroup $H \subset G$ is a τ-stable maximal torus and $\theta(H) = H$. For each of the seven types of classical symmetric spaces K is a classical group or a homomorphic image of a product of two classical groups. We also gave explicit embeddings of G/K into G as an affine algebraic set.

Let $\mathfrak{g}$ be the Lie algebra of G. For notational convenience we write θ for $d\theta$ and τ for $d\tau$ when it is clear from the context that the action is on $\mathfrak{g}$. The Lie algebra of K is then $\mathfrak{k} = \{X \in \mathfrak{g} : \theta X = X\}$. Set

$$V = \{X \in \mathfrak{g} : \theta X = -X\}.$$

Then $\mathfrak{g} = \mathfrak{k} \oplus V$ as a K-module under $\mathrm{Ad}|_K$. Since θ is a Lie algebra automorphism, the ± 1 eigenspaces of θ satisfy the commutation relations

$$[V, V] \subset \mathfrak{k} \quad \text{and} \quad [\mathfrak{k}, V] \subset V.$$

Hence if $X \in V$, then $(\mathrm{ad}X)^2 : V \longrightarrow V$. For $t \in \mathbb{C}$ we can write

$$\det(tI_V - (\mathrm{ad}X)^2|_V) = t^p\, \delta(X) + \text{terms of higher order in } t, \tag{12.30}$$

with δ a nonzero polynomial on V. We note that $\delta \in \mathcal{P}(V)^K$.

Let $\mathfrak{u} = \mathrm{Lie}(U)$ and $V_0 = V \cap i\mathfrak{u}$. Then V_0 is a real subspace of V, and since V is stable under τ, we have $V = V_0 + iV_0$. We say that a real subspace $\mathfrak{a}_0 \subset V_0$ is *abelian* if $[\mathfrak{a}_0, \mathfrak{a}_0] = 0$. It is *maximal abelian* if it is not contained in any larger abelian subspace of V_0. Such subspaces clearly exist, for dimensional reasons.

Proposition 12.3.2. *Let $\mathfrak{a}_0 \subset V_0$ be a real subspace that is maximal abelian. Set $l = \dim_{\mathbb{R}} \mathfrak{a}_0$. Let $\mathfrak{a}$ be the complexification of $\mathfrak{a}_0$. Let $V' = \{X \in V : \delta(X) \neq 0\}$.*

1. *The lowest-degree term in (12.30) has $p = l$.*
2. *If $h \in \mathfrak{a}_0 \cap V'$ then $\mathfrak{a} = \{X \in V : \mathrm{ad}(h)X = 0\}$.*
3. *If $X \in V'$ then X is semisimple and $\mathrm{Ker}(\mathrm{ad}X|_V)$ is an l-dimensional abelian Lie algebra consisting of semisimple elements.*

Proof. There is a positive definite Hermitian inner product $\langle \cdot \mid \cdot \rangle$ on $\mathfrak{g}$ such that

$$\langle \mathrm{Ad}(u)Y \mid \mathrm{Ad}(u)Z \rangle = \langle Y \mid Z \rangle \quad \text{for } u \in U \text{ and } Y, Z \in \mathfrak{g}, \tag{12.31}$$

and this inner product is real-valued on $\mathfrak{u}$ (see the proof of Theorem 11.5.1). Setting $u = \exp(tX)$ in (12.31) with $X \in \mathfrak{u}$ and differentiating in t at $t = 0$, we obtain

$$\langle \mathrm{ad}X(Y) \mid Z \rangle = -\langle Y \mid \mathrm{ad}X(Z) \rangle \quad \text{for } X \in \mathfrak{u} \text{ and } Y, Z \in \mathfrak{g}. \tag{12.32}$$

Since $V_0 \subset i\mathfrak{u}$, this implies that

$$\langle \mathrm{ad}X(Y) \mid Z \rangle = \langle Y \mid \mathrm{ad}X(Z) \rangle \quad \text{for } X \in V_0 \text{ and } Y, Z \in \mathfrak{g}. \tag{12.33}$$

In particular, for $X \in V_0$ the operator $\mathrm{ad}X$ is self-adjoint on $\mathfrak{g}$ with respect to this inner product. Hence X is semisimple, the eigenvalues of $\mathrm{ad}X$ are real, and

$$\mathrm{Ker}(\mathrm{ad}X|_V) = \mathrm{Ker}((\mathrm{ad}X)^2|_V). \tag{12.34}$$

For $\lambda \in \mathfrak{a}_0^*$ we set

$$\mathfrak{g}^\lambda = \{X \in \mathfrak{g} : \mathrm{ad}H(X) = \lambda(H)X \text{ for all } H \in \mathfrak{a}_0\}.$$

Since $\mathrm{ad}(\mathfrak{a}_0)$ is a commutative set of self-adjoint linear transformations, there is an orthogonal decomposition

$$\mathfrak{g} = \mathfrak{g}^0 \oplus \bigoplus_{\lambda \in \Sigma} \mathfrak{g}^\lambda, \tag{12.35}$$

where $\Sigma = \{\lambda \in \mathfrak{a}_0^* : \lambda \neq 0 \text{ and } \mathfrak{g}^\lambda \neq 0\}$. We call Σ the set of *restricted roots* of $\mathfrak{a}_0$ on $\mathfrak{g}$. We note that since $\theta X = -X$ for $X \in V$ and $\tau X = -X$ for $X \in V_0$, we also have

$$\theta \mathfrak{g}^\lambda = \mathfrak{g}^{-\lambda}, \quad \tau \mathfrak{g}^\lambda = \mathfrak{g}^{-\lambda} \quad \text{for } \lambda \in \Sigma. \tag{12.36}$$

Fix $h \in \mathfrak{a}_0$ with $\lambda(h) \neq 0$ for all $\lambda \in \Sigma$, and set $\Sigma^+ = \{\lambda \in \Sigma : \lambda(h) > 0\}$. Then $\Sigma^+ \cap (-\Sigma^+) = \emptyset$ and we have

$$V_0 = \mathfrak{a}_0 \oplus \bigoplus_{\lambda \in \Sigma^+} \left((\mathfrak{g}^\lambda + \mathfrak{g}^{-\lambda}) \cap V_0 \right).$$

From this direct-sum decomposition the following property is clear:

($\star$)   If $h \in \mathfrak{a}_0$ and $\lambda(h) \neq 0$ for all $\lambda \in \Sigma$, then $\mathfrak{a}_0 = \{X \in V_0 : (\mathrm{ad}\,h)X = 0\}$.

To continue the proof of the proposition, we use the following result:

Lemma 12.3.3. *Let $X, Y \in V_0$. There exists $k_0 \in K_0^\circ$ such that $[\mathrm{Ad}(k_0)X, Y] = 0$.*

Proof. Let $f(k) = \langle \mathrm{Ad}(k)X, Y \rangle$ for $k \in K_0^\circ$. Since K_0° is compact and f is real-valued, f has a critical point, say k_0. Set $Z = \mathrm{Ad}(k_0)X$. If $T \in \mathfrak{k}_0$, then $\exp(tT) \in K_0^\circ$, and hence

$$0 = \frac{d}{dt} f(\exp(tT)k_0) \Big|_{t=0} = \langle [T, Z] \mid Y \rangle = \langle T \mid [Y, Z] \rangle.$$

Taking $T = [Y, Z]$, we find that $[\mathrm{Ad}(k_0)X, Y] = 0$. $\square$

Corollary 12.3.4. *There is a polar decomposition $V_0 = \mathrm{Ad}(K_0^\circ)\mathfrak{a}_0$. Furthermore, if $\mathfrak{a}_1$ is another maximal abelian subspace in V_0, then there exists $k \in K_0^\circ$ such that $\mathrm{Ad}(k)\mathfrak{a}_0 = \mathfrak{a}_1$.*

Proof. This is clear from Lemma 12.3.3 and statement ($\star$). $\square$

If $X \in V_0$, then $\mathrm{Ker}(\mathrm{ad}\,X|_V)$ is the complexification of $\mathrm{Ker}(\mathrm{ad}\,X|_{V_0})$. Hence (12.34), Corollary 12.3.4, and statement ($\star$) imply that

$$l = \min_{X \in V_0} \left\{ \dim \mathrm{Ker}(\mathrm{ad}\,X|_V) \right\} = \min_{X \in V_0} \left\{ \dim \mathrm{Ker}\left((\mathrm{ad}\,X)^2|_V \right) \right\}. \tag{12.37}$$

Set $m = \dim V - l$. Fix $X_1, \ldots, X_{m+1} \in V$ for the moment. For $X \in V$ define

$$\Phi(X) = \mathrm{ad}\,X(X_1) \wedge \cdots \wedge \mathrm{ad}\,X(X_{m+1}).$$

Then $\Phi : V \longrightarrow \bigwedge^{m+1} \mathfrak{g}$ is a regular map and $\Phi(X) = 0$ for $X \in V_0$, since

$$\dim \mathrm{ad}(X)(V) = \dim V - \dim \mathrm{Ker}(\mathrm{ad}(X)|_V) \leq m$$

by (12.37). But V_0 is a real form of V and hence is Zariski dense in V, so we have $\Phi(X) = 0$ for all $X \in V$. Since $X_1, \ldots, X_{m+1}$ were arbitrary in V, this shows that $\dim \mathrm{ad}(X)(V) \leq m$, and hence $\dim \mathrm{Ker}(\mathrm{ad}\,X|_V) \geq l$. From (12.37) we now see that

$$l = \min_{X \in V} \left\{ \dim \mathrm{Ker}(\mathrm{ad}\,X|_V) \right\} = \min_{X \in V} \left\{ \dim \mathrm{Ker}((\mathrm{ad}\,X)^2|_V) \right\}. \tag{12.38}$$

If $X \in V$ let $X = X_s + X_n$ be the Jordan decomposition of X in $\mathfrak{g}$ (Section 1.6.3). It is clear that a Lie algebra automorphism preserves the Jordan decomposition. Hence $\theta(X_s)$ is semisimple and $\theta(X_n)$ is nilpotent. Since $\theta X = -X$ we have $\theta(X_s) = -X_s$ and $\theta(X_n) = -X_n$, so X_s and X_n are in V. Since X_s and X_n commute, we can express

$$(\mathrm{ad}\,X)^2 = (\mathrm{ad}\,X_s + \mathrm{ad}\,X_n)^2 = (\mathrm{ad}\,X_s)^2 + (2\,\mathrm{ad}\,X_s + \mathrm{ad}\,X_n)\,\mathrm{ad}\,X_n.$$

Since the last term on the right is nilpotent, this decomposition shows that the semisimple part of $(\operatorname{ad} X)^2$ is $(\operatorname{ad} X_s)^2$. Because $(\operatorname{ad} X)^2|_V$ and $(\operatorname{ad} X_s)^2|_V$ have the same characteristic polynomial, we see from the definition of $\delta(X)$ that

$$\delta(X) = \delta(X_s) \quad \text{for all } X \in V. \tag{12.39}$$

This implies that

$(\star\star)$ $\delta(X) \neq 0$ if and only if $\dim\left(\operatorname{Ker}(\operatorname{ad} X_s)^2|_V\right) = p$.

Parts (1) and (2) of Proposition 12.3.2 now follow from $(\star)$, (12.38), and $(\star\star)$, since the elements of $\mathfrak{a}_0$ are semisimple.

We now turn to the proof of part (3) of the proposition. Fix $X \in V'$ and set $h = X_s$. Define $\mathfrak{c} = \{Y \in V : [h, Y] = 0\}$. Since $\operatorname{ad}(h)$ is semisimple,

$$\dim \mathfrak{c} = \dim \operatorname{Ker}((\operatorname{ad} h)^2|_V) = l \tag{12.40}$$

by $(\star\star)$ and part (1) of Proposition 12.3.2. Furthermore, we can decompose $\mathfrak{g} = \operatorname{Ker}(\operatorname{ad} h) \oplus [h, \mathfrak{g}]$. We note that $\theta[h, \mathfrak{g}] = [-h, \theta\mathfrak{g}] = [h, \mathfrak{g}]$. Thus

$$V = \mathfrak{c} \oplus ([h, \mathfrak{g}] \cap V).$$

Set $L = [h, \mathfrak{g}] \cap V$. We note that $(\operatorname{ad} h)^2$ defines a bijection of L. If $Y \in \mathfrak{c}$ then $(\operatorname{ad} Y)^2 L \subset L$. Let

$$\mathfrak{c}'' = \{Y \in \mathfrak{c} : (\operatorname{ad} Y)^2|_L \text{ is bijective}\}.$$

Since $h \in \mathfrak{c}''$, we see that $\mathfrak{c}''$ is Zariski open and nonempty, hence Zariski dense in $\mathfrak{c}$. It follows that $\mathfrak{c}''$ spans $\mathfrak{c}$. If $Y \in \mathfrak{c}''$ then $\operatorname{Ker}(\operatorname{ad} Y|_V) \subset \mathfrak{c}$. Hence (12.38) and (12.40) imply that $\operatorname{Ker}(\operatorname{ad} Y|_V) = \mathfrak{c}$. Letting Y run over a basis for $\mathfrak{c}$, we conclude that $\mathfrak{c}$ is abelian.

Set

$$\mathfrak{c}_1 = \operatorname{Span}_{\mathbb{C}}\{Y \in \mathfrak{c}'' : Y \text{ semisimple}\}.$$

Then $\mathfrak{c}_1$ is an abelian subalgebra of V consisting of semisimple elements. Given $\lambda \in \mathfrak{c}_1^*$, set

$$\mathfrak{g}^\lambda = \{Z \in \mathfrak{g} : [Y, Z] = \lambda(Y)Z \quad \text{for all } Y \in \mathfrak{c}_1\}.$$

Then $\mathfrak{g} = \bigoplus_\lambda \mathfrak{g}^\lambda$ and $\mathfrak{g}^0 \cap V = \mathfrak{c}$.

For $Y, Z \in \mathfrak{g}$ set $B(Y, Z) = \operatorname{tr}(\operatorname{ad} Y \operatorname{ad} Z)$. Since G is reductive, $\mathfrak{g}$ decomposes as

$$\mathfrak{g} = \mathfrak{z}(\mathfrak{g}) \oplus [\mathfrak{g}, \mathfrak{g}],$$

where $\mathfrak{z}(\mathfrak{g})$ is the center of $\mathfrak{g}$ (Corollary 2.5.9), and we have $\tau[\mathfrak{g}, \mathfrak{g}] = [\mathfrak{g}, \mathfrak{g}]$. Hence the restriction of B to $[\mathfrak{g}, \mathfrak{g}]$ is nondegenerate. We also have $B(\mathfrak{g}^\lambda, \mathfrak{g}^\mu) = 0$ if $\lambda, \mu \in \mathfrak{c}_1^*$ and $\lambda + \mu \neq 0$. Since $B(\theta Y, \theta Z) = B(Y, Z)$ for all $Y, Z \in \mathfrak{g}$, it follows that $B(V, \mathfrak{k}) = 0$.

We now prove that $\mathfrak{c}_1 = \mathfrak{c}$. The semisimple and nilpotent components of $Y \in \mathfrak{c}$ are also in $\mathfrak{c}$, since $[Y, h] = 0$. Assume that $Y \in \mathfrak{c}$ is nilpotent. Then $\operatorname{ad} Y \operatorname{ad} Z$ is nilpotent for all $Z \in \mathfrak{c}$, since $\mathfrak{c}$ is commutative; therefore $B(Y, Z) = 0$. Since $B(Y, \mathfrak{g}^\lambda) = 0$ for all $\lambda \neq 0$ and $B(Y, \mathfrak{k}) = 0$, we conclude that $B(Y, \mathfrak{g}) = 0$. This implies that $Y \in \mathfrak{z}(\mathfrak{g}) \cap \mathfrak{c}$.

Hence Y is semisimple by Theorem 2.5.3. Thus Y is both nilpotent and semisimple, and so $Y = 0$. This completes the proof of part (3) of Proposition 12.3.2.   □

We now apply these results to obtain the Iwasawa decomposition of G; they will also be used in connection with the Kostant–Rallis theorem. Fix a maximal abelian subspace $\mathfrak{a}_0 \subset V_0$ with complexification $\mathfrak{a}$ as in Proposition 12.3.2. We write $\sigma = \tau\theta$; then σ is another conjugation of G and $G_0 = G^\sigma$ is a (noncompact) real form of G. Recall that a σ-stable algebraic torus A is σ-split if $\chi^\sigma = \chi$ for all characters $\chi \in \mathfrak{X}(A)$, where $\chi^\sigma(a) = \overline{\chi(\sigma(a))}$ for $a \in A$ (the bar denoting complex conjugation). As in Section 11.3.3 we let $Q = \{g \in G : \theta(g) = g^{-1}\}$.

Proposition 12.3.5. *Let $A = \exp(\mathfrak{a})$. Then $A \subset Q$ and the following hold:*

1. A is a τ-stable and σ-split algebraic torus with Lie algebra $\mathfrak{a}$.
2. If A' is any τ-stable algebraic torus in Q, then there exists $k_0 \in K_0^\circ$ such that $k_0 A' k_0^{-1} \subset A$.
3. A is contained in a maximal algebraic torus H in G that is stable under τ and θ.

Proof. Since $\tau(\mathfrak{a}) = \mathfrak{a}$ and $\mathfrak{a}$ is commutative, the elements of $\mathfrak{a}$ are semisimple. Also, $\tau(\exp sZ) = \exp(\bar{s}\tau Z)$ and $\theta(\exp sZ) = \exp(-sZ)$ for $Z \in \mathfrak{a}$ and $s \in \mathbb{C}$, so we see that A is τ-stable and contained in Q. Let A_1 be the identity component of the Zariski closure of A. Then A_1 is a τ-stable algebraic torus in Q by Theorem 11.2.2. It follows that $\mathrm{Lie}(A_1)$ is τ-stable and $\mathfrak{a} \subseteq \mathrm{Lie}(A_1) \subset V$. Hence by maximality of $\mathfrak{a}_0$ we have $\mathrm{Lie}(A_1) = \mathfrak{a}$ and $A = A_1$.

Let $Z = X + iY \in \mathfrak{a}$, where $X, Y \in \mathfrak{a}_0$. Since $\mathfrak{a}_0 \subset V \cap i\mathfrak{u}$, we have

$$\sigma(Z) = \theta\tau(X + iY) = X - iY .$$

Hence if $\chi(\exp Z) = e^{\lambda(X) + i\lambda(Y)}$ is a character of A, where $\lambda \in \mathfrak{a}^*$, then

$$\chi^\sigma(\exp Z) = \overline{\chi(\exp \sigma(Z))} = e^{\lambda(X) + i\lambda(Y)} = \chi(\exp Z) .$$

Hence A is σ-split, proving (1).

If A' is a τ-stable algebraic torus in Q, then $\mathrm{Lie}(A') = \mathfrak{a}_0' + i\mathfrak{a}_0'$, where $\mathfrak{a}_0'$ is an abelian subspace of V_0. By Corollary 12.3.4 there exists $k_0 \in K_0^\circ$ such that

$$\mathrm{Ad}(k_0)\mathfrak{a}_0' \subset \mathfrak{a}_0 .$$

Hence $k_0 A' k_0^{-1} \subset A$, giving (2).

Let H be a τ-stable and θ-stable algebraic torus in G containing A. Let $\mathfrak{h} = \mathrm{Lie}(H)$. If H is not a maximal algebraic torus in G, then there exists a nonzero element $X \in \mathrm{Cent}_\mathfrak{g}(\mathfrak{h})$ such that $\tau(X) = X$ and either $X \in V$ or $X \in \mathfrak{k}$. By Theorem 11.2.2 the identity component of the Zariski closure of the commutative group

$$\{h\exp(sX) : h \in H \text{ and } s \in \mathbb{C}\}$$

is a τ-stable and θ-stable algebraic torus that has larger dimension than H. Thus we may continue this construction until we obtain a maximal algebraic torus, which proves (3).   □

An algebraic torus $A' \subset Q$ is called θ-*anisotropic*. Thus $\theta(a) = a^{-1}$ for all $a \in A'$. Proposition 12.3.5 asserts that every θ-anisotropic τ-stable algebraic torus in G is conjugate under K_0 to a subtorus of the fixed maximal anisotropic τ-stable algebraic torus A. The (complex) dimension of A is called the *rank* of the symmetric space G/K. Note that since K contains the identity component of the center Z° of G and all compact real forms of G/Z° are G-conjugate (by Corollary 11.5.5), it follows that the rank of G/K does not depend on the choice of the conjugation τ.

Definition 12.3.6. Let H be a maximal algebraic torus in G that is τ-stable and θ-stable. Then H is *maximally θ anisotropic* if $H \cap Q$ is maximal as a τ-stable algebraic torus in Q.

Fix a maximal algebraic torus H in G satisfying the conditions of Definition 12.3.6; it exists by Proposition 12.3.5. Set $\mathfrak{h} = \mathrm{Lie}(H)$ and let

$$\mathfrak{g} = \mathfrak{h} + \sum_{\alpha \in \Phi} \mathfrak{g}_\alpha$$

be the rootspace decomposition of $\mathfrak{g}$ relative to H. Since H is θ-stable, we have $\theta(\mathfrak{g}_\alpha) = \mathfrak{g}_{\alpha^\theta}$, where we write $\alpha^\theta = \alpha \circ \theta$. Furthermore,

$$\tau(\mathfrak{g}_\alpha) = \mathfrak{g}_{-\alpha} . \tag{12.41}$$

Indeed, $X \in \mathfrak{h}$ can be written as $X = X_1 + iX_2$, where $X_j \in \mathfrak{h} \cap \mathfrak{u}$. Let $Y \in \mathfrak{g}_\alpha$. Since $\alpha(X_j) \in i\mathbb{R}$, we have

$$[X, \tau(Y)] = \tau([\tau(X), Y]) = \tau([X_1 - iX_2, Y])$$
$$= \left(\overline{\alpha(X_1)} + i\overline{\alpha(X_2)} \right) \tau(Y) = -\langle \alpha, X \rangle \tau(Y) ,$$

proving (12.41).

We now obtain a simple criterion for finding such an algebraic torus that we will apply to the classical groups in Section 12.3.2. A subspace of $\mathfrak{g}$ is a *toral subalgebra* if it is a commutative subalgebra and consists of semisimple elements.

Lemma 12.3.7. *Let H be a θ-stable and τ-stable maximal algebraic torus in G with Lie algebra $\mathfrak{h}$ and root system Φ. Set $\mathfrak{a} = \{X \in \mathfrak{h} : \theta(X) = -X\}$ and*

$$\Phi_0 = \{\alpha \in \Phi : \langle \alpha, X \rangle = 0 \quad \text{for all } X \in \mathfrak{a}\} .$$

The following are equivalent:

(i)   H is maximally θ anisotropic.
(ii) $\mathfrak{a}$ is maximal among all τ-stable toral subalgebras in V.
(iii) θ acts as the identity on $\mathfrak{g}_\alpha$ for all $\alpha \in \Phi_0$.

Assume that H is maximally θ anisotropic. Let $\mathfrak{l}$ be the centralizer of $\mathfrak{a}$ in $\mathfrak{g}$ and let $\mathfrak{m}$ be the centralizer of $\mathfrak{a}$ in $\mathfrak{k}$. Then $\mathfrak{l} = \mathfrak{a} \oplus \mathfrak{m}$.

Proof. The equivalence of conditions (i) and (ii) follows immediately from part (2) of Proposition 12.3.5.

Now we prove the equivalence of conditions (ii) and (iii). Let $\mathfrak{l}$ and $\mathfrak{m}$ be as in the last statement in the lemma. Since $\theta(\mathfrak{l}) = \mathfrak{l}$, we have $\mathfrak{l} = \mathfrak{m} \oplus (\mathfrak{l} \cap V)$. Set $\mathfrak{t} = \{X \in \mathfrak{h} : \theta(X) = X\}$. Then $\mathfrak{h} = \mathfrak{t} + \mathfrak{a}$. If $X \in \mathfrak{l} \cap V$ then we can write

$$X = X_0 + X_1 + \sum_{\alpha \in \Phi} X_\alpha \, ,$$

with $X_0 \in \mathfrak{t}$ and $X_1 \in \mathfrak{a}$. Since $[X, \mathfrak{a}] = 0$ and $[\mathfrak{a}, \mathfrak{t}] = 0$, we have

$$0 = [Y, X] = \sum_{\alpha \in \Phi} \alpha(Y) X_\alpha \quad \text{for all } Y \in \mathfrak{a} \, .$$

Hence $X_\alpha = 0$ for all $\alpha \in \Phi \setminus \Phi_0$. Thus if condition (iii) holds, we can write

$$-X = \theta(X) = X_0 - X_1 + \sum_{\alpha \in \Phi_0} X_\alpha \, .$$

It follows that $X_0 = X_\alpha = 0$ and hence $X = X_1 \in \mathfrak{a}$. This implies that $\mathfrak{a}$ is a maximal τ-stable toral subalgebra of V.

Now suppose that (iii) does not hold. Then there exists a root $\alpha \in \Phi$ such that $\langle \alpha, \mathfrak{a} \rangle = 0$ and θ acts by -1 on $\mathfrak{g}_{\pm \alpha}$. Take a TDS triple $\{e_\alpha, f_\alpha, h_\alpha\}$ with $e_\alpha \in \mathfrak{g}_\alpha$, $f_\alpha \in \mathfrak{g}_{-\alpha}$, and set $X = e_\alpha + f_\alpha$. Then $X \in V$, but $X \notin \mathfrak{a}$, while $[X, \mathfrak{a}] = 0$. Hence $\mathfrak{a} + \mathbb{C}X$ is a toral subalgebra of V properly containing $\mathfrak{a}$, and this subalgebra is τ-stable by (12.41).

The final assertion in the lemma follows from the calculations above. $\square$

Fix a θ-stable and τ-stable maximal algebraic torus H in G that is maximally θ anisotropic, and let $A = H \cap Q$. Define $M = \mathrm{Cent}_K(A)$ (the centralizer of A in K). It is clear from condition (iii) in Lemma 12.3.7 that

$$\mathfrak{m} = \mathfrak{t} + \sum_{\alpha \in \Phi_0} \mathfrak{g}_\alpha \, .$$

Let $L = \mathrm{Cent}_G(A)$. Then L is connected, by Theorem 11.4.10, and $\mathrm{Lie}(L) = \mathfrak{l}$.

Lemma 12.3.8. *The group $L = AM^\circ$ and $M = TM^\circ$. Hence M is connected if and only if T is connected.*

Proof. Let $x \in L$. Then the semisimple and unipotent components x_s and x_u are in L, since they commute with A. We can write $x_u = \exp Y$, where Y is a nilpotent element of $\mathfrak{l}$. Since $\mathfrak{l} = \mathfrak{a} \oplus \mathfrak{m}$, $[\mathfrak{a}, \mathfrak{m}] = 0$, and the elements of $\mathfrak{a}$ are semisimple, it follows that $Y \in \mathfrak{m}$. Hence $x_u \in M^\circ$.

By Theorem 11.4.10 there is an algebraic torus $S \subset L$ such that $A \cup \{x_s\} \subset S$. Let $\mathfrak{s} = \mathrm{Lie}(S) \supset \mathfrak{a}$. If $Z \in \mathfrak{s}$ and $X \in \mathfrak{a}$ then

$$0 = \theta([Z, X]) = [\theta(Z), -X] \, .$$

Hence $Z - \theta(Z)$ is in V and commutes with $\mathfrak{a}$. Thus $Z - \theta(Z) \in \mathfrak{a}$ by Lemma 12.3.7. Hence $\theta(Z) \in \mathfrak{s}$, so we have $\theta(\mathfrak{s}) = \mathfrak{s}$. Thus $\mathfrak{s} = (\mathfrak{s} \cap \mathfrak{k}) \oplus \mathfrak{a}$. Since $S = \exp(\mathfrak{s})$, this shows that $S = A \cdot S_0$, where

$$S_0 = S \cap K = \exp(\mathfrak{s} \cap \mathfrak{k}) \subset M^\circ .$$

Thus we can factor $x_s = ab$ with $a \in A$ and $b \in S_0$. This proves that $x = abx_u$ and hence $L = AM^\circ$. This implies that $M = (A \cap T)M^\circ = TM$. $\qquad\square$

Set $T = H \cap K$. Then T° is an algebraic torus and by Theorem 11.2.2 there is a finite group C such that $T = T^\circ \times C$.

Lemma 12.3.9. *The maximal torus* $H = AT^\circ$ *and* $A \cap T = \{a \in A : a^2 = 1\} \cong (\mathbb{Z}/2\mathbb{Z})^m$, *where* $m = \mathrm{rank}(A)$. *Thus* $C \cong (\mathbb{Z}/2\mathbb{Z})^r$ *for some* r *with* $0 \leq r \leq m$.

Proof. There is a decomposition $\mathfrak{h} = \mathfrak{t} \oplus \mathfrak{a}$, where $\theta = 1$ on $\mathfrak{t}$ and $\theta = -1$ on $\mathfrak{a}$. Clearly we have $\mathrm{Lie}(A) = \mathfrak{a}$ and $\mathrm{Lie}(T) = \mathfrak{t}$. Since

$$H = \exp \mathfrak{h} = \exp(\mathfrak{a}) \exp(\mathfrak{t}) ,$$

we have $H = AT^\circ$. Also, $a \in A \cap T = A \cap K$ if and only if $a = \theta(a) = a^{-1}$. $\qquad\square$

Remark 12.3.10. We shall determine A and T for the seven classical types in Section 12.3.2; for four of the types $T = T^\circ$ is connected, whereas in the remaining three types (AI, CI, and BDI) we have $C \cong (\mathbb{Z}/2\mathbb{Z})^p$.

Since A is σ-split, Lemma 11.6.4 provides an embedding $G \subset \mathbf{GL}(n, \mathbb{C})$ such that $A \subset D_n$ and $\sigma(g) = \bar{g}$. As in the proof of Theorem 11.2.2, we can find a subset $\{i_1, \ldots, i_m\} \subset \{1, \ldots, n\}$ such that the characters

$$a \mapsto \chi_j(a) = x_{i_j} \quad \text{for} \quad a = \mathrm{diag}[x_1, \ldots, x_n]$$

freely generate $\mathfrak{X}(A)$. We fix such a set of characters and we give $\mathfrak{X}(A)$ the corresponding lexicographic order, as in Section 11.6.1. Let the unipotent subgroups $N^\pm$ of G be defined relative to this order, as in Section 11.6.2. Then we claim that

$$\theta(N^+) = N^- . \tag{12.42}$$

Indeed, for $\chi \in \mathfrak{X}(A)$ and $a \in A$, we have

$$\chi^\theta(a) = \chi(\theta(a)) = \chi(a)^{-1} ,$$

since A is θ-anisotropic. Thus θ gives an order-reversing automorphism of $\mathfrak{X}(A)$. This implies (12.42).

A total ordering of $\mathfrak{X}(H)$ will be called *compatible* with the chosen order on $\mathfrak{X}(A)$ if $\mu|_A > \nu|_A$ implies that $\mu > \nu$ in $\mathfrak{X}(H)$. We construct a compatible order on $\mathfrak{X}(H)$ as follows: Let $\Sigma = \{\nu_1, \ldots, \nu_r\} \subset \mathfrak{X}(A)$ be the weights of A on $\mathbb{C}^n$, enumerated so that

$$v_1 > v_2 > \cdots > v_r$$

relative to the order that we have fixed on $\mathcal{X}(A)$. Enumerate the standard basis for $\mathbb{C}^n$ as $e_{j_1}, \ldots, e_{j_n}$ so that e_{j_i} has weight v_1 for $1 \leq i \leq m_1$, weight v_2 for $m_1 + 1 \leq i \leq m_2$, and so forth, as in Section 11.6.1. Each vector e_{j_i} transforms according to a weight μ_{j_i} of H. We give $\mathcal{X}(H)$ the lexicographic order in which $\mu_{j_1} > \mu_{j_2} > \cdots$. This order is clearly compatible with the order on $\mathcal{X}(A)$.

Let Φ^+ be the roots for H that are positive relative to the order just defined. Let B be the Borel subgroup of G defined by the positive system Φ^+. Thus $AN^+ \subset B$. Let $\mathfrak{n}^\pm = \mathrm{Lie}(N^\pm)$ and set $\Phi_1^+ = \Phi^+ \setminus \Phi_0$. Then

$$\mathfrak{n}^+ = \sum_{\alpha \in \Phi_1^+} \mathfrak{g}_\alpha, \qquad \mathfrak{n}^- = \sum_{\alpha \in \Phi_1^+} \mathfrak{g}_{-\alpha}.$$

Lemma 12.3.11. *There are vector-space direct sum decompositions*

$$\mathfrak{g} = \mathfrak{n}^- \oplus \mathfrak{m} \oplus \mathfrak{a} \oplus \mathfrak{n}^+ = \mathfrak{k} \oplus \mathfrak{a} \oplus \mathfrak{n}^+.$$

Hence $N^- M A N^+$ and KAN^+ are open Zariski-dense subsets of G and $K \cap (AN^+)$ is finite.

Proof. Since Φ is the disjoint union $\Phi_0 \cup \Phi_1^+ \cup (-\Phi_1^+)$, we can use the rootspace decomposition of $\mathfrak{g}$ to write $X \in \mathfrak{g}$ as

$$X = \sum_{\beta \in \Phi_1^+} X_{-\beta} + \left\{ H_0 + \sum_{\alpha \in \Phi_0} X_\alpha \right\} + H_1 + \sum_{\beta \in \Phi_1^+} X_\beta,$$

where $H_0 \in \mathfrak{t}$, $H_1 \in \mathfrak{a}$, and $X_\alpha \in \mathfrak{g}_\alpha$. This gives the first decomposition of $\mathfrak{g}$. For the second decomposition, we write

$$X_{-\beta} = X_{-\beta} + \theta(X_{-\beta}) - \theta(X_{-\beta})$$

for $\beta \in \Phi_1^+$, and we note that $X_{-\beta} + \theta(X_{-\beta}) \in \mathfrak{k}$ and $\theta(X_{-\beta}) \in \mathfrak{n}^+$ by (12.42).

Consider the maps $N^- \times M \times A \times N^+ \longrightarrow G$ and $K \times A \times N^+ \longrightarrow G$ given by multiplication in G of the elements from each factor. From the decompositions of $\mathfrak{g}$ we see that differentials of these maps are surjective at 1. Since G is connected, it follows by Theorem A.3.4 that the images are Zariski dense. $\qquad\square$

Theorem 12.3.12. *K is a spherical subgroup of G. If λ is the B-highest weight of an irreducible K-spherical representation of G, then*

$$t^\lambda = 1 \quad \text{for all } t \in T. \tag{12.43}$$

Proof. Lemma 12.3.11 and Theorem 12.2.5 imply that K is spherical. Since $T = K \cap B$, condition (12.43) is satisfied by the highest weight of a K-spherical representation by Theorem 12.2.1. $\qquad\square$

A weight $\lambda \in P_{++}(G)$ satisfying condition (12.43) will be called θ-*admissible*.

12.3.2 Examples of Iwasawa Decompositions

We now work out explicit Iwasawa decompositions for the seven types of classical symmetric spaces G/K associated with an involution θ, following the notation of Section 11.3.4. In all cases the θ-stable and τ-stable maximally θ-anisotropic torus H is diagonal, and $\tau(g) = (\bar{g}^t)^{-1}$.

For each classical type we verify condition (iii) of Lemma 12.3.7, and we describe the maximal θ-anisotropic torus $A = H \cap Q$ and the subgroup $T = H \cap K$. We give a total order on $\mathcal{X}(A)$ and a compatible order on $\mathcal{X}(H)$, following the procedure in Section 12.3.1. We describe the weight decomposition

$$\mathbb{C}^n = V_1 \oplus \cdots \oplus V_r, \tag{12.44}$$

where $G \subset \mathbf{GL}(n, \mathbb{C})$ and A acts on V_i by the character μ_i. The enumeration is chosen such that $\mu_1 > \cdots > \mu_r$. The group M consists of the elements of K that preserve the decomposition (12.44), and N^+ consists of the elements $g \in G$ such that $I - g$ is strictly upper block-triangular relative to the decomposition (12.44). We give the system of positive roots Φ^+ for the compatible order on $\mathcal{X}(H)$, and we find the explicit form of the θ-admissibility condition (12.43) for the Φ^+-dominant weights. The information is summarized in the *Satake diagram*, which is obtained from the Dynkin diagram of $\mathfrak{g}$ by the following procedure:

(S1) If a simple root vanishes on $\mathfrak{a}$, then the corresponding node in the Dynkin diagram is marked by $\bullet$.

(S2) The simple roots that have nonzero restriction to $\mathfrak{a}$ correspond to nodes that are marked by $\circ$; if two simple roots have the same restriction to $\mathfrak{a}$, then the corresponding nodes are joined by a double-pointed arrow.

(S3) The nodes $\circ$ are labeled by the coefficients of the corresponding fundamental weights in the θ-admissible Φ^+-dominant weight (see Section 3.1.4). Nodes joined by a double-pointed arrow have the same coefficient, and nodes marked by $\bullet$ have coefficient zero.

Notation. D_p is the group of invertible diagonal $p \times p$ matrices and $s_p = [\delta_{p+1-i-j}]$ is the $p \times p$ matrix with ones on the antidiagonal and zeros elsewhere. For $a = \mathrm{diag}[a_1, \ldots, a_p] \in D_p$ let $\varepsilon_i(a) = a_i$ and $\check{a} = s_p a s_p = \mathrm{diag}[a_p, \ldots, a_1]$.

Bilinear Forms – Type AI

Here $G = \mathbf{SL}(n, \mathbb{C})$, θ is the involution $\theta(g) = (g^t)^{-1}$, and $K \cong \mathbf{SO}(n, \mathbb{C})$. The maximal torus H of diagonal matrices in G is θ-anisotropic. Hence $A = H$ and $T \cong (\mathbb{Z}/2\mathbb{Z})^{n-1}$ consists of all matrices

$$t = \mathrm{diag}[\delta_1, \ldots, \delta_n], \quad \text{with } \delta_i = \pm 1 \text{ and } \det(t) = 1 .$$

There are no roots that vanish on $\mathfrak{a}$, so condition (iii) of Lemma 12.3.7 is vacuously satisfied. Hence A is a maximal θ-anisotropic torus and $M = T$. Take the characters $\varepsilon_1 > \varepsilon_2 > \cdots > \varepsilon_{n-1}$ as an ordered basis for $\mathcal{X}(A)$. The eigenspace decomposition (12.44) in this case is

$$\mathbb{C}^n = \mathbb{C}e_1 \oplus \cdots \oplus \mathbb{C}e_n$$

and the associated system of positive roots is $\Phi^+ = \{\varepsilon_i - \varepsilon_j : 1 \le i < j \le n\}$.

Let $\lambda = \sum \lambda_i \varepsilon_i$, with $\lambda_i \in \mathbb{N}$, be a weight of H. Suppose $\lambda_1 \ge \cdots \ge \lambda_{n-1} \ge 0$ is Φ^+-dominant. Then $t^\lambda = 1$ for all $t \in T$ if and only if λ_i is even for all i. Thus λ is θ-admissible if and only if

$$\lambda = 2m_1 \varpi_1 + \cdots + 2m_l \varpi_l, \quad m_i \in \mathbb{N},$$

where $\varpi_1, \ldots, \varpi_l$ (with $l = n - 1$) are the fundamental weights.

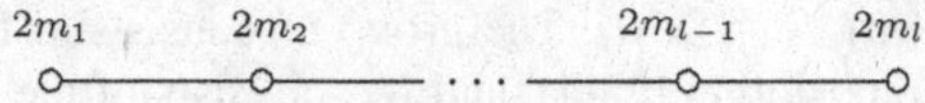

Fig. 12.1 Satake diagram of Type AI.

Bilinear Forms – Type AII

In this case $G = \mathbf{SL}(2n, \mathbb{C})$ and $\theta(g) = T_n(g^t)^{-1} T_n^{-1}$, where $T_n = \mathrm{diag}[\mu, \ldots, \mu]$ (n copies) and $\mu = \begin{bmatrix} 0 & 1 \\ -1 & 0 \end{bmatrix}$. We have $K \cong \mathbf{Sp}(n, \mathbb{C})$. The maximal torus $H = D_{2n} \cap G$ in G is θ-invariant, with $\theta(h) = \mathrm{diag}[x_2^{-1}, x_1^{-1}, \ldots, x_{2n}^{-1}, x_{2n-1}^{-1}]$ for $h = \mathrm{diag}[x_1, \ldots, x_{2n}]$. Thus A consists of all matrices

$$a = \mathrm{diag}[x_1, x_1, \ldots, x_n, x_n] \quad \text{with } x_1 x_2 \cdots x_n = 1,$$

and has rank $n - 1$. We take generators $\chi_1, \ldots, \chi_{n-1}$ for $\mathcal{X}(A)$ as $\chi_i(a) = \varepsilon_{2i-1}(a)$ and we give $\mathcal{X}(A)$ the corresponding lexicographic order. The group T consists of all matrices

$$t = \mathrm{diag}[x_1, x_1^{-1}, \ldots, x_n, x_n^{-1}] \quad \text{with } x_i \in \mathbb{C}^\times,$$

and is a torus of rank n. The roots vanishing on $\mathfrak{a}$ are

$$\Phi_0 = \{\pm(\varepsilon_1 - \varepsilon_2), \pm(\varepsilon_3 - \varepsilon_4), \ldots, \pm(\varepsilon_{2n-1} - \varepsilon_{2n})\},$$

and a calculation shows that θ acts by 1 on $\mathfrak{g}_\alpha$ for $\alpha \in \Phi_0$. Thus condition (iii) of Lemma 12.3.7 is satisfied, and hence A is a maximal θ-anisotropic torus.

The decomposition (12.44) in this case is

$$\mathbb{C}^{2n} = V_1 \oplus \cdots \oplus V_n, \quad \text{where } V_i = \mathbb{C}e_{2i-1} + \mathbb{C}e_{2i}.$$

Note that V_i is nonisotropic for the skew form defined by T_n. One calculates that M consists of the block diagonal matrices

$$m = \mathrm{diag}[g_1, \ldots, g_n], \quad \text{where } g_i \in \mathbf{Sp}(V_i). \tag{12.45}$$

Thus $M \cong \times^n \mathbf{Sp}(1, \mathbb{C})$.

The ordered basis $\varepsilon_1 > \varepsilon_2 > \cdots > \varepsilon_{2n-1}$ for $\mathfrak{X}(H)$ is compatible with the order we have given to $\mathfrak{X}(A)$. Let Φ^+ be the corresponding system of positive roots. Let $\lambda = \sum_{i=1}^{2n-1} \lambda_i \varepsilon_i$, with $\lambda_i \in \mathbb{N}$ and $\lambda_1 \geq \cdots \geq \lambda_{2n-1} \geq 0$, be a Φ^+-dominant weight. Then $t^\lambda = 1$ for all $t \in T$ if and only if $\lambda_{2i-1} = \lambda_{2i}$ for $i = 1, \dots, n-1$ and $\lambda_{2n-1} = 0$. Thus λ is θ-admissible if and only if

$$\lambda = m_2 \varpi_2 + \cdots + m_{l-1} \varpi_{l-1} \quad \text{with } m_i \in \mathbb{N}.$$

Fig. 12.2 Satake diagram of Type AII.

Polarizations – Type AIII

We have $G = \mathbf{SL}(n, \mathbb{C})$ and $\theta(g) = J_{p,q} g J_{p,q}$ with $0 < p \leq q$ and $p + q = n$. Here

$$J_{p,q} = \begin{bmatrix} 0 & 0 & s_p \\ 0 & I_{q-p} & 0 \\ s_p & 0 & 0 \end{bmatrix}.$$

We have $K \cong \mathbf{S}(\mathbf{GL}(p, \mathbb{C}) \times \mathbf{GL}(q, \mathbb{C}))$. The maximal torus $H = D_n \cap G$ is θ-invariant. For $h \in H$ write $h = \mathrm{diag}[a, b, c]$, with $a, c \in D_p$ and $b \in D_{q-p}$. Then $\theta(h) = \mathrm{diag}[\check{c}, b, \check{a}]$. Thus $A \cong D_p$ consists of all matrices

$$h = \mathrm{diag}[a, I_{q-p}, \check{a}^{-1}] \quad \text{with } a \in D_p.$$

We take generators $\chi_1 > \cdots > \chi_p$ for $\mathfrak{X}(A)$ as $\chi_i(h) = \varepsilon_i(a)$, and we give $\mathfrak{X}(A)$ the corresponding lexicographic order. We have $h \in T$ provided

$$h = \mathrm{diag}[a, b_{q-p}, \check{a}] \quad \text{with } a \in D_p, \ b \in D_{q-p}, \text{ and } \det(h) = 1.$$

Thus $T \cong D_{q-1}$ is connected. The roots vanishing on $\mathfrak{a}$ are

$$\Phi_0 = \{\pm(\varepsilon_i - \varepsilon_j) : p+1 \leq i < j \leq q\},$$

and it is obvious that θ acts by 1 on $\mathfrak{g}_\alpha$ for $\alpha \in \Phi_0$. Thus condition (iii) of Lemma 12.3.7 is satisfied, and hence A is a maximal θ-anisotropic torus.

The decomposition (12.44) in this case is

$$\mathbb{C}^{2n} = V_1 \oplus \cdots \oplus V_p \oplus V_0 \oplus V_{-p} \oplus \cdots \oplus V_{-1},$$

where $V_i = \mathbb{C}e_i$, $V_0 = \mathbb{C}e_{p+1} + \cdots + \mathbb{C}e_q$, and $V_{-i} = \mathbb{C}e_{n+1-i}$ for $i = 1, \dots, p$. Here A acts on $V_{\pm i}$ by the character $\chi_i^{\pm 1}$ and acts on V_0 by 1. Hence M consists of the block diagonal matrices

$$x = \begin{bmatrix} a & 0 & 0 \\ 0 & b & 0 \\ 0 & 0 & \check{a} \end{bmatrix}, \quad \text{with } a \in D_p, \ b \in \mathbf{GL}(q-p, \mathbb{C}), \text{ and } \det x = 1.$$

The ordered basis $\varepsilon_1 > \varepsilon_2 > \cdots > \varepsilon_{n-1}$ for $\mathcal{X}(H)$ is compatible with the order we have given to $\mathcal{X}(A)$. Let Φ^+ be the corresponding system of positive roots. Let $\lambda = \sum \lambda_i \varepsilon_i$, with $\lambda_i \in \mathbb{N}$ and $\lambda_1 \geq \cdots \geq \lambda_n$, be a Φ^+-dominant weight. Then $t^\lambda = 1$ for all $t \in T$ if and only if $\lambda_1 = -\lambda_n, \lambda_2 = -\lambda_{n-1}, \ldots, \lambda_p = -\lambda_{q+1}$, and $\lambda_j = 0$ for $p+1 \leq j \leq q$. Thus λ is θ-admissible if and only if

$$\lambda = [\lambda_1, \ldots, \lambda_p, \underbrace{0, \ldots, 0}_{q-p}, -\lambda_p, \ldots, -\lambda_1],$$

where $\lambda_1 \geq \cdots \geq \lambda_p \geq 0$ are arbitrary integers. Thus λ is θ-admissible if and only if

$$\lambda = m_1(\varpi_1 + \varpi_l) + m_2(\varpi_2 + \varpi_{l-1}) + \cdots + m_p(\varpi_p + \varpi_q) \quad \text{with } m_i \in \mathbb{N}.$$

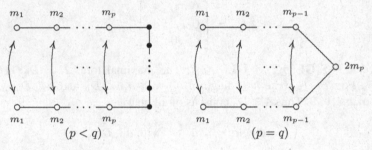

Fig. 12.3 Satake diagram of Type AIII.

Polarizations – Type CI

Let $G = \mathbf{Sp}(\mathbb{C}^{2n}, \Omega)$, where Ω is the bilinear form with matrix $J = \begin{bmatrix} 0 & I_n \\ -I_n & 0 \end{bmatrix}$. We take the involution $\theta(g) = -JgJ$. Here $K \cong \mathbf{GL}(n, \mathbb{C})$ and the maximal torus $H = D_{2n} \cap G$ is θ-anisotropic. Hence $A = H$ and $M = T \cong (\mathbb{Z}/2\mathbb{Z})^n$ consists of all matrices

$$t = \mathrm{diag}[\delta_1, \ldots, \delta_n, \delta_1, \ldots, \delta_n], \quad \text{with } \delta_i = \pm 1.$$

Since $\Phi_0 = \emptyset$, condition (iii) of Lemma 12.3.7 is vacuously satisfied, and hence A is a maximal θ-anisotropic torus. We define an order on $\mathcal{X}(A)$ using the characters

$$\chi_i(h) = \varepsilon_i(a) \quad \text{when } h = \mathrm{diag}[a, a^{-1}],$$

for $i = 1, \ldots, n$. Let Φ^+ be the corresponding system of positive roots. Let $\lambda = \sum_{i=1}^n \lambda \varepsilon_i$ with $\lambda_1 \geq \cdots \geq \lambda_n \geq 0$ be a dominant weight. Then $t^\lambda = 1$ for all $t \in T$ if and only if λ_i is even for all i. Thus λ is θ-admissible if and only if

$$\lambda = 2m_1\varpi_1 + \cdots + 2m_l\varpi_l, \quad \text{with } m_i \in \mathbb{N},$$

where $\varpi_1, \ldots, \varpi_l$ (with $l = n - 1$) are the fundamental weights.

Fig. 12.4 Satake diagram of Type CI.

$$\overset{2m_1}{\circ} \underline{\hspace{1.5cm}} \overset{2m_2}{\circ} \underline{\hspace{1cm}} \cdots \underline{\hspace{1cm}} \overset{2m_{l-1}}{\circ} \!\!\!\Longleftarrow\!\!\! \overset{2m_l}{\circ}$$

Polarizations – Type DIII

Take $G = \mathbf{SO}(\mathbb{C}^n, B)$, with $n = 2l$ even and B the form with matrix s_n. We take the involution $\theta(g) = -\Gamma_n g \Gamma_n$ with Γ_n defined as in Section 11.3.5. As in Type CII, we have $K \cong \mathbf{GL}(l, \mathbb{C})$. The maximal torus $H = D_n \cap G$ is θ-stable. Write elements of H as $h = \mathrm{diag}[a, \check{a}^{-1}]$, where $a = [a_1, \ldots, a_l]$. Then $\theta(h) = \mathrm{diag}[b, \check{b}^{-1}]$, where

$$b = \begin{cases} \mathrm{diag}[a_2, a_1, \ldots, a_{2p}, a_{2p-1}] & \text{when } l = 2p, \\ \mathrm{diag}[a_2, a_1, \ldots, a_{2p}, a_{2p-1}, a_{2p+1}] & \text{when } l = 2p+1. \end{cases}$$

We have $A \cong D_p$ consisting of all $h = \mathrm{diag}[a, \check{a}^{-1}]$ with

$$a = \begin{cases} \mathrm{diag}[x_1, x_1^{-1}, \ldots, x_p, x_p^{-1}] & \text{when } l = 2p, \\ \mathrm{diag}[x_1, x_1^{-1}, \ldots, x_p, x_p^{-1}, 1] & \text{when } l = 2p+1. \end{cases}$$

We take generators $\chi_1, \ldots, \chi_p$ for $\mathcal{X}(A)$ as $\chi_i(a) = \varepsilon_{2i-1}(a)$, where $p = [l/2]$, and we put the corresponding lexicographic order on $\mathcal{X}(A)$. The group T consists of all matrices $h = \mathrm{diag}[a, \check{a}^{-1}]$ with

$$a = \begin{cases} \mathrm{diag}[x_1, x_1, \ldots, x_p, x_p] & \text{when } l = 2p, \\ \mathrm{diag}[x_1, x_1, \ldots, x_p, x_p, x_{p+1}] & \text{when } l = 2p+1. \end{cases}$$

Thus $T = T^\circ$ is a torus of rank p (when l is even) or rank $p+1$ (when l is odd). The roots vanishing on $\mathfrak{a}$ are

$$\pm(\varepsilon_1 + \varepsilon_2), \pm(\varepsilon_3 + \varepsilon_4), \ldots, \pm(\varepsilon_{2p-1} + \varepsilon_{2p}).$$

We leave it as an exercise to check that θ acts by 1 on $\mathfrak{g}_\alpha$ for $\alpha \in \Phi_0$. Thus condition (iii) of Lemma 12.3.7 is satisfied, and hence A is a maximal θ-anisotropic torus.

For $i = 1, \ldots, p$ the χ_i eigenspace for A on $\mathbb{C}^n$ is $V_i = \mathbb{C}e_{2i-1} + \mathbb{C}e_{n-2i+1}$ and the χ_i^{-1} eigenspace is $V_{-i} = \mathbb{C}e_{2i} + \mathbb{C}e_{n-2i+2}$. When $l = 2p+1$ there is also the space $V_0 = \mathbb{C}e_l + \mathbb{C}e_{l+1}$, where A acts by 1. The subspaces $V_{\pm 1}, \ldots, V_{\pm p}$ are B-isotropic, whereas B is nondegenerate on V_0 (when $l = 2p+1$). The space V_{-i} is dual to V_i relative to B, and we have

$$\mathbb{C}^n = V_1 \oplus \cdots \oplus V_p \oplus V_0 \oplus V_{-1} \oplus \cdots \oplus V_{-p}$$

(where we set $V_0 = 0$ when $l = 2p$ is even). From this decomposition we calculate that

$$M \cong \begin{cases} \mathbf{GL}(V_1) \times \cdots \times \mathbf{GL}(V_p) & \text{when } l = 2p, \\ \mathbf{GL}(V_1) \times \cdots \times \mathbf{GL}(V_p) \times \mathbf{SO}(V_0) & \text{when } l = 2p+1. \end{cases} \quad (12.46)$$

Note that $\mathbf{SO}(V_0) \cong \mathbf{GL}(1, \mathbb{C})$, since $\dim V_0 = 2$. We take as ordered basis for $\mathfrak{X}(H)$

$$\varepsilon_1 > -\varepsilon_2 > \varepsilon_3 > -\varepsilon_4 > \cdots > \varepsilon_{2p-1} > -\varepsilon_{2p} \qquad \text{when } l = 2p, \text{ and}$$
$$\varepsilon_1 > -\varepsilon_2 > \varepsilon_3 > -\varepsilon_4 > \cdots > \varepsilon_{2p-1} > -\varepsilon_{2p} > \varepsilon_{2p+1} \quad \text{when } l = 2p+1.$$

This ordering is compatible with the given order on $\mathfrak{X}(A)$. Let Φ^+ be the corresponding system of positive roots for H. We see that Φ^+ is obtained from the positive roots used in Section 2.4.3 by the action of the Weyl group element that transforms the ordered basis $\varepsilon_1 > \varepsilon_2 > \cdots > \varepsilon_{l-1} > \pm\varepsilon_l$ into the ordered basis above (the choice of $\pm$ depending on whether l is even or odd). It follows from the calculations of Section 2.4.3 that, when l is even, the simple roots in Φ^+ are

$$\alpha_1 = \varepsilon_1 + \varepsilon_2, \quad \alpha_2 = -\varepsilon_2 - \varepsilon_3, \ldots, \alpha_{l-2} = -\varepsilon_{l-2} - \varepsilon_{l-1},$$
$$\alpha_{l-1} = \varepsilon_{l-1} + \varepsilon_l, \quad \alpha_l = \varepsilon_{l-1} - \varepsilon_l.$$

If p is odd, then the simple roots in Φ^+ are

$$\alpha_1 = \varepsilon_1 + \varepsilon_2, \quad \alpha_2 = -\varepsilon_2 - \varepsilon_3, \ldots, \alpha_{l-2} = \varepsilon_{l-2} + \varepsilon_{l-1},$$
$$\alpha_{l-1} = -\varepsilon_{l-1} - \varepsilon_l, \quad \alpha_l = -\varepsilon_{l-1} + \varepsilon_l.$$

The simple roots vanishing on $\mathfrak{a}$ in this case are $\alpha_1, \alpha_3, \ldots, \alpha_{2p-1}$. The roots α_{l-1} and α_l have the same restriction to $\mathfrak{a}$.

A weight $\lambda = \sum_{i=1}^l \lambda_i \varepsilon_i$ is Φ^+-dominant if and only if

$$\lambda_1 \geq -\lambda_2 \geq \lambda_3 \geq -\lambda_4 \geq \cdots \geq |\lambda_l|.$$

Let λ be a Φ^+-dominant weight. Then $t^\lambda = 1$ for all $t \in T$ if and only if

$$\lambda_1 = -\lambda_2, \quad \lambda_3 = -\lambda_4, \ldots, \lambda_{2p-1} = -\lambda_{2p}$$

in the case $l = 2p$. When $l = 2p+1$ there is the additional condition $\lambda_{2p+1} = 0$. Writing λ in terms of the fundamental weights, we find that it is θ-admissible if and only if

$$\lambda = \begin{cases} m_2 \varpi_2 + \cdots + m_{l-2} \varpi_{l-2} + 2m_l \varpi_l & l \text{ even}, \\ m_2 \varpi_2 + \cdots + m_{l-3} \varpi_{l-3} + m_{l-1}(\varpi_{l-1} + \varpi_l) & l \text{ odd}. \end{cases}$$

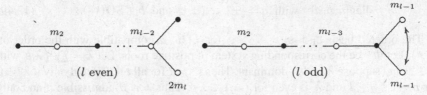

Fig. 12.5 Satake diagram of Type DIII.

Orthogonal Decompositions – Type BDI

Let $G = \mathbf{SO}(\mathbb{C}^n, B)$ and let involution act by $\theta(g) = J_{p,q}gJ_{p,q}$ with $1 \le p \le q$ and $p + q = n$ as in Type AIII. Then $K \cong \mathbf{S}(\mathbf{O}(p,\mathbb{C}) \times \mathbf{O}(q,\mathbb{C}))$. This is the only case in which K is not connected; we have $K^\circ \cong \mathbf{SO}(p,\mathbb{C}) \times \mathbf{SO}(q,\mathbb{C})$. The maximal torus $H = D_n \cap G$ is θ-stable. We can write $h \in H$ as $\mathrm{diag}[a, b, \check{a}^{-1}]$, with $a \in D_p$ arbitrary and b of the form

$$b = \begin{cases} \mathrm{diag}[c, \check{c}^{-1}] & \text{when } n = 2l\,, \\ \mathrm{diag}[c, 1, \check{c}^{-1}] & \text{when } n = 2l+1\,, \end{cases} \tag{12.47}$$

with $c \in D_{l-p}$. We calculate that $\theta(h) = \mathrm{diag}[a^{-1}, b, \check{a}]$. Thus $A \cong D_p$ consists of all diagonal matrices

$$h = \mathrm{diag}[a, I_{q-p}, \check{a}^{-1}] \quad \text{with} \quad a = \mathrm{diag}[a_1, \ldots, a_p] \in D_p\,.$$

We take generators $\chi_1 > \cdots > \chi_p$ for $\mathcal{X}(A)$ as $\chi_i(h) = \varepsilon_i(a)$ and we give $\mathcal{X}(A)$ the corresponding lexicographic order. The group T consists of all diagonal matrices $h = \mathrm{diag}[a, b, \check{a}^{-1}]$, where b is given by (12.47) and $a_i = \pm 1$ for $i = 1, \ldots, p$. Thus $T \cong (\mathbb{Z}/2\mathbb{Z})^p \times D_{l-p}$. The subgroup $T_0 = H \cap K^\circ$ consists of all such diagonal matrices that satisfy the additional condition $a_1 \cdots a_p = 1$.

The roots vanishing on $\mathfrak{a}$ are

$$\{\pm\varepsilon_i \pm \varepsilon_j : p+1 \le i < j \le l\} \quad \text{when } n = 2l\,,$$

$$\{\pm\varepsilon_i \pm \varepsilon_j : p+1 \le i < j \le l\} \cup \{\varepsilon_i : p+1 \le i \le l\} \quad \text{when } n = 2l+1\,.$$

It is clear that θ acts by 1 on $\mathfrak{g}_\alpha$ for $\alpha \in \Phi_0$, since the matrices in $\mathfrak{g}_\alpha$ are of block-diagonal form $\mathrm{diag}[0, x, 0]$ with $x \in M_{q-p}$. Thus condition (iii) of Lemma 12.3.7 is satisfied, and hence A is a maximal θ-anisotropic torus of rank p.

For $i = 1, \ldots, p$ the χ_i eigenspace for A on $\mathbb{C}^n$ is $V_i = \mathbb{C}e_i$ and the χ_i^{-1} eigenspace is $V_{-i} = \mathbb{C}e_{n+1-i}$. The space $V_0 = \mathbb{C}e_{p+1} \oplus \cdots \oplus \mathbb{C}e_q$ is the 1-eigenspace for A. The subspaces $V_{\pm 1}, \ldots, V_{\pm p}$ are B-isotropic, whereas B is nondegenerate on V_0. The space V_{-i} is dual to V_i relative to B. We have

$$\mathbb{C}^n = V_1 \oplus \cdots \oplus V_p \oplus V_0 \oplus V_{-1} \oplus \cdots \oplus V_{-p} \quad (\dim V_0 = q - p)\,. \tag{12.48}$$

From this decomposition we see that M consists of all matrices in block-diagonal form

$$m = \mathrm{diag}[a,b,a] \quad \text{with } a = [\pm 1, \ldots, \pm 1] \text{ and } b \in \mathbf{SO}(V_0) . \tag{12.49}$$

The ordered basis $\varepsilon_1 > \varepsilon_2 > \cdots > \varepsilon_l$ for $\mathcal{X}(H)$ is compatible with the order on $\mathcal{X}(A)$. Let Φ^+ be the corresponding system of positive roots. Let $\lambda = \sum_{i=1}^{l} \lambda_i \varepsilon_i$ with $\lambda_i \in \mathbb{N}$ and suppose λ is Φ^+-dominant. Then $t^\lambda = 1$ for all $t \in T$ if and only if $\lambda_j = 0$ for $p+1 \le j \le l$ and λ_i is even for $i = 1, \ldots, p$. Thus λ is θ-admissible if and only if

$$\lambda = [\lambda_1, \ldots, \lambda_p, \underbrace{0, \ldots, 0}_{l-p}] ,$$

where $\lambda_1 \ge \cdots \lambda_p \ge 0$ are arbitrary even integers. If we require only that $t^\lambda = 1$ for all $t \in H \cap K^\circ$, then the parity condition becomes

$$\lambda_i - \lambda_j \in 2\mathbb{Z} \quad \text{for all } 1 \le i < j \le p .$$

We shall say that λ is K°-*admissible* when this last condition is satisfied. When we write λ in terms of the fundamental dominant weights the admissibility conditions become the following: First assume that $n = 2l + 1$ is odd (Type BI). Then λ is θ-admissible if and only if

$$\lambda = \begin{cases} 2m_1\varpi_1 + \cdots + 2m_{p-1}\varpi_{p-1} + 2m_p\varpi_p & (p < l) , \\ 2m_1\varpi_1 + \cdots + 2m_{l-1}\varpi_{l-1} + 4m_l\varpi_l & (p = l) , \end{cases}$$

where $m_i \in \mathbb{N}$. For λ to be K°-admissible, however, the coefficient of ϖ_p has only to be an integer (not necessarily even) when $p < l$, and the coefficient of ϖ_l has only to be even (not necessarily a multiple of 4) when $p = l$. The Satake diagram is shown in Figure 12.6, where the coefficients shown in parentheses apply to K°-admissible weights.

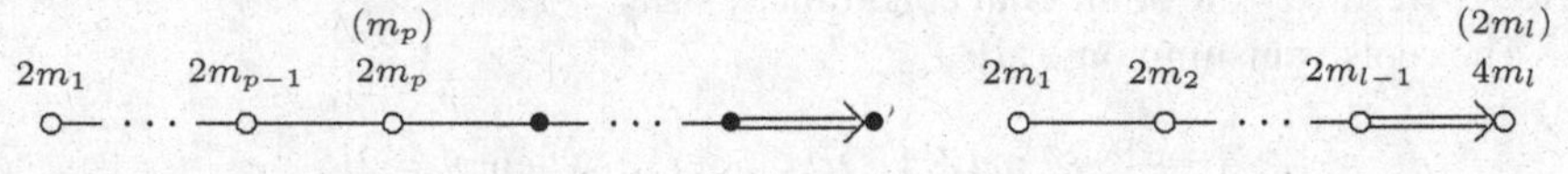

Fig. 12.6 Satake diagrams of Type BI.

Now assume that $n = 2l$ is even (Type DI). If $p < l - 1$ then λ is θ-admissible if and only if

$$\lambda = 2m_1\varpi_1 + 2m_2\varpi_2 + \cdots + 2m_p\varpi_p .$$

If $p = l - 1$ then λ is θ-admissible if and only if

$$\lambda = 2m_1\varpi_1 + 2m_2\varpi_2 + \cdots + 2m_{l-2}\varpi_{l-2} + 2m_{l-1}(\varpi_{l-1} + \varpi_l) .$$

If $p = l$ then λ is θ-admissible if and only if

$$\lambda = 2m_1\varpi_1 + 2m_2\varpi_2 + \cdots + 2m_{l-2}\varpi_{l-2} + 2m_{l-1}\varpi_{l-1} + 2m_l\varpi_l .$$

In all cases the coefficients m_i are nonnegative integers. For λ to be K°-admissible, however, the coefficient of ϖ_p has to be an integer (not necessarily even) only for all $p < l - 1$, whereas the coefficients of ϖ_{l-1} and ϖ_l have to be equal integers (not necessarily even) when $p = l - 1$. When $p = l$ then K°-admissibility is the same as θ-admissibility. The Satake diagram is shown in Figure 12.7, where the coefficients shown in parentheses apply to K°-admissible weights.

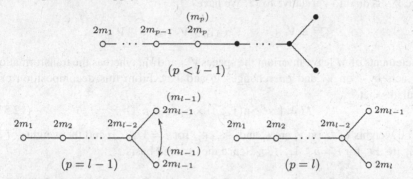

Fig. 12.7 Satake diagrams of Type DI.

Orthogonal Decompositions – Type CII

In this case $G = \mathbf{Sp}(\mathbb{C}^{2l}, \Omega)$, where Ω is the bilinear form with matrix $J = \begin{bmatrix} 0 & s_l \\ -s_l & 0 \end{bmatrix}$. We take the involution $\theta(g) = K_{p,q} g K_{p,q}$, for $1 \le p \le q$ and $p + q = l$, with $K_{p,q}$ as in Section 11.3.5. We have $K \cong \mathbf{Sp}(p, \mathbb{C}) \times \mathbf{Sp}(q, \mathbb{C})$. The maximal torus $H = D_l \cap G$ is θ-stable. We write $h \in H$ as

$$h = \mathrm{diag}[x, \check{x}^{-1}] \text{ with } x = \mathrm{diag}[a, b, c] \quad (a, c \in D_p \text{ and } b \in D_{q-p}). \qquad (12.50)$$

Then $\theta(h) = \mathrm{diag}[y, \check{y}^{-1}]$ with $y = \mathrm{diag}[\check{c}, b, \check{a}]$. Thus $A \cong D_p$ consists of all $h = \mathrm{diag}[x, x]$ with $x = \mathrm{diag}[a, I_{q-p}, \check{a}^{-1}]$, where $a \in D_p$. We take generators $\chi_1, \dots, \chi_p$ for $\mathfrak{X}(A)$ as $\chi_i(h) = \varepsilon_i(a)$ and we give $\mathfrak{X}(A)$ the corresponding lexicographic order. The group T consists of all

$$h = \mathrm{diag}[x, \check{x}^{-1}] \text{ with } x = \mathrm{diag}[a, b, \check{a}] \quad (a \in D_p \text{ and } b \in D_{q-p}).$$

Thus $T \cong D_q$ is connected.

The roots vanishing on $\mathfrak{a}$ are $\pm \varepsilon_i \pm \varepsilon_j$ for $p < i \le j \le q$ and

$$\pm(\varepsilon_1 + \varepsilon_l), \quad \pm(\varepsilon_2 + \varepsilon_{l-1}), \quad \dots, \pm(\varepsilon_p + \varepsilon_{q+1}).$$

A calculation similar to that done above in Type AIII shows that $\theta = 1$ on the corresponding root spaces. Thus condition (iii) of Lemma 12.3.7 is satisfied, and hence A is maximal θ-anisotropic.

For $i = 1, \ldots, p$ the χ_i eigenspace for A on $\mathbb{C}^{2l}$ is $V_i = \mathbb{C}e_i + \mathbb{C}e_{l+i}$ and the χ_i^{-1} eigenspace is $V_{-i} = \mathbb{C}e_{l+1-i} + \mathbb{C}e_{2l+1-i}$. The 1-eigenspace of A is

$$V_0 = \mathbb{C}e_{p+1} + \cdots + \mathbb{C}e_q + \mathbb{C}e_{l+p+1} + \cdots + \mathbb{C}e_{l+q}.$$

The subspaces $V_{\pm 1}, \ldots, V_{\pm r}$ are Ω-isotropic, whereas Ω is nondegenerate on V_0. The space V_{-i} is dual to V_i relative to Ω. We have

$$\mathbb{C}^{2l} = V_1 \oplus \cdots \oplus V_p \oplus V_0 \oplus V_{-1} \oplus \cdots \oplus V_{-p}.$$

The elements of M leave invariant the spaces $V_{\pm i}$ and V_0, whereas the transformation $K_{p,q}$ acts by I on V_0 and interchanges V_i and V_{-i}. From this decomposition one calculates that

$$M \cong \left(\times^p \mathbf{Sp}(1, \mathbb{C}) \right) \times \mathbf{Sp}(q - p, \mathbb{C}). \tag{12.51}$$

The weights of H on V_i are ε_i and $-\varepsilon_{l+1-i}$ for $i = 1, \ldots, p$ and the weights of H on V_0 are $\pm \varepsilon_i$ for $i = p+1, \ldots, q$. Hence the ordered basis

$$\varepsilon_1 > -\varepsilon_l > \varepsilon_2 > -\varepsilon_{l-1} > \cdots > \varepsilon_p > -\varepsilon_{q+1} > \varepsilon_{p+1} > \varepsilon_{p+2} > \cdots > \varepsilon_q$$

for $\mathfrak{X}(H)$ is compatible with the order we have given to $\mathfrak{X}(A)$. Let Φ^+ be the corresponding system of positive roots. Since Φ^+ is obtained from the positive roots used in Section 2.4.3 by the action of the Weyl group element that transforms the ordered basis $\varepsilon_1 > \varepsilon_2 > \cdots > \varepsilon_l$ into the ordered basis above, it follows from the calculations of Section 2.4.3 that the simple roots in Φ^+ are

$$\alpha_1 = \varepsilon_1 + \varepsilon_l, \ \alpha_2 = -\varepsilon_l - \varepsilon_2, \ \ldots, \alpha_{2p-1} = \varepsilon_p + \varepsilon_{q+1}, \ \alpha_{2p} = -\varepsilon_{q+1} - \varepsilon_{p+1},$$
$$\alpha_{2p+1} = \varepsilon_{p+1} - \varepsilon_{p+2}, \ \ldots, \alpha_{l-1} = \varepsilon_{q-1} - \varepsilon_q, \ \alpha_l = 2\varepsilon_q.$$

The simple roots vanishing on $\mathfrak{a}$ are thus $\alpha_1, \alpha_3, \ldots, \alpha_{2p-1}$.

From the description of Φ^+ just given we see that $\lambda = \sum_{i=1}^l \lambda_i \varepsilon_i$ is Φ^+-dominant if and only if

$$\lambda_1 \geq -\lambda_l \geq \lambda_2 \geq \cdots \geq \lambda_p \geq -\lambda_{q+1} \geq \lambda_{p+1} \geq \cdots \geq \lambda_q \geq 0.$$

One has $t^\lambda = 1$ for all $t \in T$ if and only if $\lambda_1 = -\lambda_l$, $\lambda_2 = -\lambda_{l-1}$, $\ldots$, $\lambda_p = -\lambda_{q+1}$, and $\lambda_j = 0$ for $j = p+1, \ldots, q$. Thus λ is θ-admissible if and only if

$$\lambda = [\lambda_1, \ldots, \lambda_p, \underbrace{0, \ldots, 0}_{q-p}, -\lambda_p, \ldots, -\lambda_1],$$

where $\lambda_1 \geq \cdots \geq \lambda_p \geq 0$ are arbitrary integers. When we write λ in terms of the fundamental weights, then it is θ-admissible when there are nonnegative integers m_{2i} such that

$$\lambda = \begin{cases} m_2 \varpi_2 + m_4 \varpi_4 + \cdots + m_{2p} \varpi_{2p} & (2p < l), \\ m_2 \varpi_2 + m_4 \varpi_4 + \cdots + m_{l-2} \varpi_{l-2} + m_l \varpi_l & (2p = l). \end{cases}$$

$(2p < l)$ $(2p = l)$

Fig. 12.8 Satake diagrams of Type CII.

12.3.3 Spherical Representations

We now determine the irreducible representations having a K-fixed vector. We follow the notation of Section 12.3.1: θ is an involution of the connected reductive group G, τ is a conjugation on G whose fixed-point group is a compact real form U, and $\theta\tau = \tau\theta$. We fix a τ-stable and θ-stable maximal torus H in G such that $A = H \cap Q$ is a maximal torus in Q. We set $M = \text{Cent}_K(A)$. Let B be the Borel subgroup of G containing H and N^+ the unipotent subgroup of B that were defined before Lemma 12.3.11.

Theorem 12.3.13. *Let* (π^λ, V^λ) *be an irreducible regular representation of G with highest weight λ (relative to B). The following are equivalent:*

1. *V^λ contains a nonzero vector fixed by K.*
2. *V^λ contains a nonzero vector fixed by MN^+.*
3. *$t^\lambda = 1$ for all $t \in T = H \cap K$.*

Proof. We have already shown (in Theorem 12.3.12) that (1) implies (3) . We observe that condition (2) is equivalent to

$(\star\star)$ $\quad \pi^\lambda(M)v_\lambda = v_\lambda$, where v_λ is a nonzero B eigenvector in V^λ.

This is clear because MN^+ contains the unipotent radical of B, and so $V^{MN^+} \subset \mathbb{C}v_\lambda$. Since $T \subset M$, clearly $(\star\star)$ implies (3).

Suppose (3) holds. We shall prove that $(\star\star)$ holds. We have $M = F \cdot M^\circ$ by Lemma 12.3.8 with $F \subset T$. Since T fixes v^λ, we need only to show that

$$d\pi^\lambda(\mathfrak{g}_\alpha)v_\lambda = 0 \quad \text{for all } \alpha \in \Phi_0. \tag{12.52}$$

Suppose $\alpha \in \Phi_0^+$. Then (12.52) is true, since $\lambda + \alpha$ is not a weight of V^λ. Now by (3) we have $\lambda(\mathfrak{t}) = 0$. But for $\alpha \in \Phi_0$, the coroot h_α is in $\mathfrak{t}$. Hence the reflection s_α fixes λ, since

$$s_\alpha(\lambda) = \lambda - \langle \lambda, h_\alpha \rangle \alpha = \lambda \ .$$

Thus $s_\alpha(\lambda - \alpha) = \lambda - s_\alpha(\alpha) = \lambda + \alpha$. If $\lambda - \alpha$ were a weight of V^λ, then $\lambda + \alpha$ would be a weight also (Proposition 3.2.7), which would be a contradiction. Hence $d\pi^\lambda(\mathfrak{g}_{-\alpha})v_\lambda = 0$.

Thus it remains only to prove that (3) implies (1). For this we will need an additional lemma. We take the conjugations τ and $\sigma = \tau\theta$ on G as in Section 12.3.1 and write $G_0 = G^\sigma$, $K_0 = K^\sigma = K^\tau$, and $A_0 = A^\sigma$ for the corresponding real forms. The group K_0 is compact, whereas G_0 and A_0 are noncompact.

Lemma 12.3.14. *Let $\lambda \in \mathfrak{X}(H)$. Suppose $t^\lambda = 1$ for all $t \in T$. Then $a^\lambda > 0$ for all $a \in A_0$.*

Proof. From Proposition 12.3.5 we know that A is σ-split. Hence by Lemma 11.6.4 there is an embedding $\varphi : G \longrightarrow \mathbf{GL}(n, \mathbb{C})$ such that $\varphi : A \cong D_p$, with A_0 corresponding to the real matrices in D_p (as done for the classical examples in Section 12.3.2). Let $\varphi(a) = \mathrm{diag}[x_1, \ldots, x_p]$ for $a \in A$. Then $a^\lambda = x_1^{m_1} \cdots x_p^{m_p}$ with $m_i \in \mathbb{Z}$. By Lemma 12.3.8 we have $T \cap A = F = \{a \in A : a^2 = 1\}$. Under the isomorphism φ,

$$F \cong \{[\varepsilon_1, \ldots, \varepsilon_p] : \varepsilon_i = \pm 1\} .$$

Thus $a^\lambda = 1$ for all $a \in F$ if and only if $m_i \in 2\mathbb{Z}$ for $i = 1, \ldots, p$. Obviously this implies that $a^\lambda > 0$ when $x_i \in \mathbb{R} \setminus \{0\}$ for $i = 1, \ldots, p$. $\qquad\square$

Completion of proof of Theorem 12.3.13. We assume (3) and hence ($\star\star$). Define

$$v_0 = \int_{K_0} \pi^\lambda(k) v_\lambda \, \mathrm{d}k .$$

Then v_0 is invariant under K_0. Since K_0 is a compact real form of K, we also have v_0 invariant under K (Theorem 11.5.10). To complete the proof, we only need to show that $v_0 \neq 0$ when condition (3) is satisfied.

Let v_λ^* be the lowest-weight vector for the dual representation $(V^\lambda)^*$, normalized so that $\langle v_\lambda^*, v_\lambda \rangle = 1$. Let $f_\lambda(g) = \langle v_\lambda^*, \pi^\lambda(g) v_\lambda \rangle$ be the generating function for π^λ, as in Section 12.1.3. Then

$$\langle v_\lambda^*, v_0 \rangle = \int_{K_0} f_\lambda(k) \, \mathrm{d}k . \tag{12.53}$$

We shall show that

$$f_\lambda(g) \geq 0 \quad \text{for } g \in G_0 . \tag{12.54}$$

Since $f_\lambda(1) = 1$, this will imply that the integral (12.53) is positive, and hence $v_0 \neq 0$.

For $m \in M$, $a \in A$, and $n^\pm \in N^\pm$ we have

$$\begin{aligned}
f_\lambda(n^- m a n^+) &= \langle \pi^{\lambda^*}(n^-)^{-1} v_\lambda^*, \, \pi^\lambda(a) \pi^\lambda(m n^+) v_\lambda \rangle \\
&= \langle v_\lambda^*, \, \pi^\lambda(a) v_\lambda \rangle = a^\lambda ,
\end{aligned}$$

since N^- fixes v_λ^* and MN^+ fixes v_λ by ($\star\star$). Hence $f_\lambda(n^- m a n^+) > 0$ for $a \in A_0$ by Lemma 12.3.14. Since $N_0^- A_0 M_0 N_0^+$ is dense in G_0 (in the Lie group topology) by Theorem 11.6.5, this proves (12.54). $\qquad\square$

Corollary 12.3.15. *The space of regular functions on G/K is isomorphic to $\bigoplus_\lambda V^\lambda$ as a G-module, where λ runs over all θ-admissible dominant weights of H.*

Proof. This follows by Theorems 12.3.12 and 12.3.13. $\qquad\square$

From Corollary 12.3.15 and the Satake diagrams in Section 12.3.2, we conclude that the semigroups of highest weights for spherical representations of the classical groups have the following generators (where l is the rank of G, $K = G^\theta$, and $p = \dim \mathfrak{a}$ is the rank of G/K):

Type AI: $\{2\varpi_1, 2\varpi_2, \ldots, 2\varpi_l\}$ $\quad (p = l)$,

Type AII: $\{\varpi_2, \varpi_4, \ldots, \varpi_{l-1}\}$ $\quad (p = (l-1)/2)$,

Type AIII: $\{\varpi_1 + \varpi_l, \varpi_2 + \varpi_{l-1}, \ldots, \varpi_p + \varpi_{l+1-p}\}$ $\quad (2p \leq l+1)$,

Type CI: $\{2\varpi_1, 2\varpi_2, \ldots, 2\varpi_l\}$ $\quad (p = l)$,

Type DIII: $\{\varpi_2, \varpi_4, \ldots, \varpi_{l-2}, 2\varpi_l\}$ $\quad$ (l even, $p = l/2$),
$\qquad\qquad \{\varpi_2, \varpi_4, \ldots, \varpi_{l-3}, \varpi_{l-1} + \varpi_l\}$ $\quad$ (l odd, $p = (l-1)/2$),

Type BI: $\{2\varpi_1, 2\varpi_2, \ldots, 2\varpi_p\}$ $\quad (p < l)$,
$\qquad\qquad \{2\varpi_1, 2\varpi_2, \ldots, 2\varpi_{l-1}, 4\varpi_l\}$ $\quad (p = l)$,

Type DI: $\{2\varpi_1, 2\varpi_2, \ldots, 2\varpi_p\}$ $\quad (p < l-1)$,
$\qquad\qquad \{2\varpi_1, 2\varpi_2, \ldots, 2\varpi_{l-2}, 2\varpi_{l-1} + 2\varpi_l\}$ $\quad (p = l-1)$,
$\qquad\qquad \{2\varpi_1, 2\varpi_2, \ldots, 2\varpi_{l-2}, 2\varpi_{l-1}, 2\varpi_l\}$ $\quad (p = l)$,

Type CII: $\{\varpi_2, \varpi_4, \ldots, \varpi_{2p}\}$ $\quad (2p \leq l)$.

If we take $K = (G^\theta)^\circ$ in Type *BDI* (the only case in which G^θ is not connected), then the semigroup of highest weights of K-spherical representations has the following generators:

Type BI: $\{2\varpi_1, \ldots, 2\varpi_{p-1}, \varpi_p\}$ $\quad (p < l)$,
$\qquad\qquad \{2\varpi_1, 2\varpi_2, \ldots, 2\varpi_{l-1}, 2\varpi_l\}$ $\quad (p = l)$,

Type DI: $\{2\varpi_1, \ldots, 2\varpi_{p-1}, \varpi_p\}$ $\quad (p < l-1)$,
$\qquad\qquad \{2\varpi_1, 2\varpi_2, \ldots, 2\varpi_{l-2}, \varpi_{l-1} + \varpi_l\}$ $\quad (p = l-1)$,
$\qquad\qquad \{2\varpi_1, 2\varpi_2, \ldots, 2\varpi_{l-2}, 2\varpi_{l-1}, 2\varpi_l\}$ $\quad (p = l)$.

12.3.4 Exercises

1. Let $\gamma(g) = g\theta(g)^{-1}$ be the map from Section 11.3.3 that gives the isomorphism $G/K \cong P$.
   (a) Show that $\gamma(g)^{-1} = \theta(\gamma(g))$ and $\gamma(a) = a^2$ for $a \in A$.
   (b) Prove that if $\gamma(g)$ is in the dense open set N^+MAN^-, then g is in the dense open set N^+AK. (HINT: Let $\gamma(g)$ have the factorization $\gamma(g) = n^+man^-$. Show that $\theta(n) = (n^-)^{-1}$ and $m = m^{-1}$. Conclude that $ma \in A$, so $ma = \tilde{a}^2$ for some $\tilde{a} \in A$ and $g = n\tilde{a}k$, where $k = \tilde{a}\theta(n)^{-1}\theta(g)$. Now show that $k \in K$.)
2. For the symmetric space of type AII:
   (a) Verify that M is given by (12.45). (HINT: $\mathbf{SL}(2, \mathbb{C}) \cong \mathbf{Sp}(1, \mathbb{C})$.)
   (b) Verify that the involution θ acts by 1 on the root space $\mathfrak{g}_\alpha$ for all $\alpha \in \Phi_0$.
   (c) Use Φ_0 and Lemma 12.3.8 to show that $M \cong \times^n \mathbf{Sp}(1, \mathbb{C})$.
3. For the symmetric space of type DIII:
   (a) Verify that the involution θ acts by 1 on $\mathfrak{g}_\alpha$ for all $\alpha \in \Phi_0$.
   (b) Verify that M satisfies (12.46).
   (b) Use Φ_0 and Lemma 12.3.8 to determine M.

4. For the symmetric space of type BDI:
   (a) Verify that M is given by (12.49).
   (b) Verify that Φ_0 is a root system on t of type B or D.
   (c) Use (b) and Lemma 12.3.8 to give another proof of (12.49).

5. For the symmetric space of type CII:
   (a) Verify that the involution θ acts by 1 on $\mathfrak{g}_\alpha$ for all $\alpha \in \Phi_0$.
   (b) Verify that M is given by (12.51) (HINT: $\mathbf{SL}(2,\mathbb{C}) \cong \mathbf{Sp}(1,\mathbb{C})$.)
   (c) Use Φ_0 and Lemma 12.3.8 to give another proof of (12.51).

6. Consider the symmetric space of type AI.
   (a) $2\varpi_1$ is the highest weight of the representation of G on $S^2(\mathbb{C}^n)$. Show that the K-fixed vector is $q = \sum e_i^2$.
   (b) Show that the orbit $G \cdot q$ is an affine open set in $S^2(\mathbb{C}^n)$. (HINT: Identify $S^2(\mathbb{C}^n)$ with the symmetric $n \times n$ matrices, and show that $G \cdot q$ corresponds to the nonsingular matrices.)

7. Consider the symmetric space of type AII.
   (a) ϖ_2 is the highest weight of the representation of G on $\bigwedge^2(\mathbb{C}^{2n})$. Show that the K-fixed vector is the skew-symmetric tensor $q = \sum_{i=1}^{n} e_{2i-1} \wedge e_{2i}$.
   (b) Show that the orbit $G \cdot q$ of q is an affine open set in $\bigwedge^2(\mathbb{C}^{2n})$. (HINT: Identify $\bigwedge^2(\mathbb{C}^{2n})$ with the skew-symmetric $2n \times 2n$ matrices and show that $G \cdot q$ corresponds to the nonsingular matrices.)

8. Consider the symmetric space of type AIII.
   (a) Show that $\varpi_1 + \varpi_l$ (where $l = n - 1$) is the highest weight of the adjoint representation of $G = \mathbf{SL}(n,\mathbb{C})$ on $V = \mathfrak{sl}(n,\mathbb{C})$.
   (c) Find the K-fixed vector for this representation.

9. Consider the symmetric space of type CI.
   (a) Show that $2\varpi_1$ is the highest weight of the adjoint representation of G.
   (b) Find the K-fixed vector for this representation.

10. Consider the symmetric space of type DIII. Find the K-fixed vector for the representation of G with highest weight ϖ_2. (HINT: Recall that this is the representation of G on $\bigwedge^2 \mathbb{C}^n$.)

11. Consider the symmetric space of type BDI. Find the K-fixed vector for the representation of G with highest weight $2\varpi_1$. (HINT: Exercise 12.2.5 #5.)

12. Consider the symmetric space of type CII. Find the K-fixed vector for the representation of G with highest weight ϖ_2. (HINT: See Section 10.2.3, Example #2 for a model of this representation.)

12.4 Isotropy Representations of Symmetric Spaces

We now study the isotropy representation for a symmetric space. The main result of this section (due to Kostant and Rallis [90]) is a generalization of the decomposition of the polynomials on $\mathbb{C}^n$ as a tensor product of the invariant polynomials and the harmonic polynomials (see Section 5.6.4) under the action of the orthogonal group. To see the connection with symmetric spaces, identify $\mathbb{C}^n$ with the tangent space to the symmetric space $\mathbf{SO}(n+1,\mathbb{C})/\mathbf{SO}(n,\mathbb{C})$ at the coset $o = \mathbf{SO}(n,\mathbb{C})$; the action of $\mathbf{SO}(n,\mathbb{C})$ on $\mathbb{C}^n$ is then the *isotropy representation* for this symmetric space.

12.4.1 A Theorem of Kostant and Rallis

Let G be a connected reductive linear algebraic group with center Z. Let θ be an involutive regular automorphism of G. We make the same assumptions and use the same notation as in Section 12.3.1: θ acts by the identity on Z° (this is automatic if G is semisimple). There exists a conjugation τ of G such that the corresponding real form U is compact, by Theorem 11.5.1, and Theorem 11.5.3 implies that we may assume that $\tau\theta = \theta\tau$. Let $\mathfrak{g}$ be the Lie algebra of G and $\mathfrak{u}$ the Lie algebra of U. We will write θ and τ for $d\theta$ and $d\tau$ when it is clear that the action is on $\mathfrak{g}$. Let

$$K = \{g \in G : \theta(g) = g\}, \quad \mathfrak{k} = \{X \in \mathfrak{g} : \theta X = X\}.$$

Then K is reductive, by Theorem 11.5.10.

Set $V = \{X \in \mathfrak{g} : \theta X = -X\}$. Then $\mathfrak{g} = \mathfrak{k} \oplus V$ as a K-module under $\mathrm{Ad}|_K$. Set $\sigma(k) = \mathrm{Ad}(k)|_V$ for $k \in K$. Then (σ, V) is a regular representation of K. Note that we may identify V with the tangent space to G/K at the coset K, with the action of K on V being the natural isotropy representation on the tangent space at a K-fixed point.

Let $\mathcal{P}(V)$ denote the polynomial functions on V and let $\mathcal{P}^j(V)$ denote the space of homogeneous polynomials on V of degree j. As usual, we have a representation μ of K on $\mathcal{P}(V)$ given by

$$\mu(k)f(v) = f(\sigma(k)^{-1}v) \quad \text{for } f \in \mathcal{P}(V), k \in K, \text{ and } v \in V.$$

Let $\mathcal{P}(V)^K = \{f \in \mathcal{P}(V) : \mu(k)f = f \text{ for all } k \in K\}$. Then $\mathcal{P}(V)^K$ is graded by degree. Set

$$\mathcal{P}_+(V)^K = \{f \in \mathcal{P}(V)^K : f(0) = 0\}.$$

Fix a maximal abelian subspace $\mathfrak{a}_0 \subset V_0$. Let $\mathfrak{a} = \mathfrak{a}_0 + i\mathfrak{a}_0$ and

$$M = \{k \in K : \mathrm{Ad}(k)|_\mathfrak{a} = I\}.$$

Then clearly $\tau(M) = M$. Thus M has a compact real form, so M is reductive (for G a classical group we determined M in Section 12.3.2).

To state the Kostant–Rallis theorem we need one more ingredient. We note that the subspace $\mathcal{P}^j(V) \cap (\mathcal{P}(V)\mathcal{P}_+(V)^K)$ is K-invariant. Thus there is a K-invariant subspace $\mathcal{H}^j$ in $\mathcal{P}^j(V)$ such that

$$\mathcal{P}^j(V) = \mathcal{H}^j \oplus \{\mathcal{P}^j(V) \cap (\mathcal{P}(V)\mathcal{P}_+(V)^K)\}. \tag{12.55}$$

Set $\mathcal{H} = \bigoplus_{j\geq 0} \mathcal{H}^j$. The space $\mathcal{H}$ is the analogue of the *spherical harmonics* in Section 5.6.4 (however, it is not associated with a dual reductive pair and is not uniquely determined by (12.55)). We then have the following *separation of variables* theorem for the linear isotropy representation of G/K:

Theorem 12.4.1 (Kostant–Rallis). *The map $h \otimes f \mapsto hf$ (pointwise multiplication) from $\mathcal{H} \otimes \mathcal{P}(V)^K$ to $\mathcal{P}(V)$ is a linear bijection. Furthermore, $\mathcal{H}$ is equivalent to $\mathcal{O}[K/M] = \mathrm{Ind}_M^K(1)$ as a K-module. In particular, if (ρ, F) is an irreducible regular representation of K, then $\mathrm{Hom}_K(F, \mathcal{P}(V))$ is a free $\mathcal{P}(V)^K$-module on $\dim F^M$ generators.*

The last statement of the theorem follows by Frobenius reciprocity (Theorem 12.1.3). The proof of the rest of the theorem will be given later.

The Kostant–Rallis theorem generalizes a celebrated theorem of Kostant concerning the adjoint representation.

Theorem 12.4.2 (Kostant). *Let G be a connected, semisimple, linear algebraic group. Let T be a maximal torus in G. Let $\mathfrak{g}$ be the Lie algebra of G and let $\mu(g)f(X) = f(\mathrm{Ad}(g)^{-1}X)$ for $g \in G$, $f \in \mathcal{P}(\mathfrak{g})$, and $X \in \mathfrak{g}$. Let*

$$\mathcal{P}(\mathfrak{g})^G = \{ f \in \mathcal{P}(\mathfrak{g}) : \mu(g)f = f \text{ for all } g \in G \} .$$

Let $\mathcal{H}$ be a graded $\mu(G)$-invariant subspace of $\mathcal{P}(\mathfrak{g})$ such that

$$\mathcal{P}(\mathfrak{g}) = \mathcal{H} \oplus \{ \mathcal{P}(\mathfrak{g})\mathcal{P}_+(\mathfrak{g})^G \} .$$

Then the map $\mathcal{H} \otimes \mathcal{P}(\mathfrak{g})^G \longrightarrow \mathcal{P}(\mathfrak{g})$ given by $h \otimes f \mapsto hf$ is a linear bijection, and $(\mu, \mathcal{H})$ is equivalent to $\mathrm{Ind}_T^G(1)$ as a representation of G. In particular, if (ρ, F) is an irreducible regular representation of G, then the space $\mathrm{Hom}_G(F, \mathcal{P}(\mathfrak{g}))$ of covariants of type ρ is a free $\mathcal{P}(\mathfrak{g})^G$-module on $\dim F^T$ generators, where F^T is the zero-weight space in F.

Proof. Take $G_1 = G \times G$ in place of G in Theorem 12.4.1 and let $\theta(g, h) = (h, g)$ for $(g, h) \in G_1$. Then G_1 is semisimple and $K = G_1^\theta \cong G$ (embedded diagonally). Let $\mathfrak{g}$ be the Lie algebra of G. Then (σ, V) is equivalent to $(\mathrm{Ad}, \mathfrak{g})$ as a representation of G. The complexification of $\mathfrak{a}_0$ is the Lie algebra of a maximal torus of G. Hence M is a maximal torus in G (see Section 2.1.2). Now apply Theorem 12.4.1.   □

In Section 12.4.3 we will describe the pairs $(K, (\sigma, V))$ covered by Theorem 12.4.1 that are not cases of Theorem 12.4.2. We will give all such pairs that arise from an involution on a simple Lie algebra $\mathfrak{g}$ such that K is a product of classical groups.

12.4.2 Invariant Theory of Reflection Groups

In this section we will discuss some basic results of Chevalley that will be used in our proof of the Kostant–Rallis theorem. Let $K_0 = K \cap U$, $V_0 = \mathrm{i}(V \cap \mathfrak{u})$, and let $\mathfrak{a}_0$ be a fixed maximal abelian subspace of V_0, as in Section 12.4.1. Define

$$N(\mathfrak{a}_0) = \{k \in K_0 : \mathrm{Ad}(k)\mathfrak{a}_0 = \mathfrak{a}_0\},$$

and let $M_0 = \{k \in K_0 : \mathrm{Ad}(k)|_{\mathfrak{a}_0} = I\}$. We look upon $N(\mathfrak{a}_0)/M_0$ as a group of linear transformations of $\mathfrak{a}_0$ and denote it by $W(\mathfrak{a})$. Let the set of restricted roots Σ be as in (12.35). If $\lambda \in \mathfrak{a}_0^*$ then define $H_\lambda \in \mathfrak{a}_0$ by $\langle H_\lambda \mid H \rangle = \lambda(H)$ for $H \in \mathfrak{a}_0$, where $\langle \cdot \mid \cdot \rangle$ is the K_0-invariant inner product on V_0 as in Section 12.3.1. Let $(\lambda, \mu) = \langle H_\lambda \mid H_\mu \rangle$ for $\lambda, \mu \in \mathfrak{a}_0^*$. If $\lambda \in \Sigma$ then we define the reflection s_λ on $\mathfrak{a}_0$ by

$$s_\lambda(H) = H - \frac{2\lambda(H)}{(\lambda, \lambda)} H_\lambda \quad \text{for } H \in \mathfrak{a}_0.$$

Theorem 12.4.3. *If $\lambda \in \Sigma$ then $s_\lambda \in W(\mathfrak{a})$. Furthermore,*

1. $W(\mathfrak{a})$ is a finite group generated by $\{s_\lambda : \lambda \in \Sigma\}$.
2. If $s \in W(\mathfrak{a})$ then there exists $k \in K_0^\circ$ such that $\mathrm{Ad}(k)|_{\mathfrak{a}_0} = s$.
3. If $h \in \mathfrak{a}_0'$ then $|W(\mathfrak{a})h| = |W(\mathfrak{a})|$.

For the adjoint representation (as in Theorem 12.4.2), this was proved in Section 11.4.6. For the case of general symmetric spaces we refer the reader to Helgason [66, Chapter VIII, §2] for a proof.

Theorem 12.4.3 implies that $W(\mathfrak{a})$ is a finite group generated by reflections. The invariant theory of these groups is due to Chevalley. We put the high points of this theory into a theorem that we will need in Section 12.4.5.

Theorem 12.4.4. *Set $\mathcal{P}(\mathfrak{a})^{W(\mathfrak{a})} = \{f \in \mathcal{P}(\mathfrak{a}) : f(sH) = f(H) \text{ for all } s \in W(\mathfrak{a})\}$.*

1. The algebra $\mathcal{P}(\mathfrak{a})^{W(\mathfrak{a})}$ is generated by l homogeneous, algebraically independent elements $g_1, \ldots, g_l$.
2. There exists a $W(\mathfrak{a})$-invariant graded subspace $\mathcal{A} \subset \mathcal{P}(\mathfrak{a})$ such that the map $\mathcal{A} \otimes \mathcal{P}(\mathfrak{a})^{W(\mathfrak{a})} \longrightarrow \mathcal{P}(\mathfrak{a})$, with $a \otimes u \mapsto au$, defines a linear bijection.
3. Let $w = |W(\mathfrak{a})|$. If $\mathcal{A}$ is as in (2), then $\dim \mathcal{A} = w$.

For the classical cases of this result see Section 5.1.2 and Exercises 5.1.3. Proofs for the general case can be found in any of the standard references (Helgason [67, Chapter III, §3 and §5], Bourbaki [12, Chapitre V, §5], Humphreys [78, Chapter 3], Kane [83]).

Examples

We now describe the restricted root systems and a set of algebraically independent generators for the $W(\mathfrak{a})$-invariants for the classical symmetric spaces, following the notation of Section 12.3.2.

1. (Type AI) Since $A = H$, the restricted root system Σ is of type A_{n-1}. Hence $W(\mathfrak{a}) \cong \mathfrak{S}_n$ and $\mathcal{P}(\mathfrak{a})^{W(\mathfrak{a})}$ has algebraically independent generators $h \mapsto \mathrm{tr}(h^{r+1})$ for $r = 1, 2, \ldots, n-1$ (the *elementary power sums*; see Section 9.1.3).

2. (Type AII) The functionals ε_{2i} restrict to ε_{2i-1} on $\mathfrak{a}$ for $i = 1, \ldots, n$. It follows that the restricted root system Σ is of type A_{n-1}. The group $W(\mathfrak{a})$ and the generators for $\mathcal{P}(\mathfrak{a})^{W(\mathfrak{a})}$ are the same as for Type AI.

3. (Type AIII) The functionals ε_i restrict to 0 on $\mathfrak{a}$ for $p + 1 \le i \le q$, and the functionals ε_{n+1-i} restrict to $-\varepsilon_i$ for $1 \le i \le p$. Hence the restricted root system Σ is

$$\{\pm(\varepsilon_i \pm \varepsilon_j) : 1 \le i < j \le p\} \cup \{\varepsilon_i : 1 \le i \le p\} \cup \{2\varepsilon_i : 1 \le i \le p\}$$

when $p < q$. This is the union of the type B_p and type C_p root systems, which is called the BC_p *root system* (it is a *nonreduced* root system: both α and 2α are roots for $\alpha = \varepsilon_i$ with $i = 1, \ldots, p$). When $p = q$, the restricted roots ε_i do not occur, and Σ is of type C_p. In both cases, $W(\mathfrak{a})$ is the group of *signed permutations* (see Section 3.1.1). The polynomials $h \mapsto \mathrm{tr}(h^{2r})$, for $r = 1, \ldots, p$, are algebraically independent and $W(\mathfrak{a})$-invariant. Hence they generate $\mathcal{P}(\mathfrak{a})^{W(\mathfrak{a})}$.

4. (Type CI) Since $A = H$, the restricted root system Σ is of type C_n. Hence $W(\mathfrak{a})$ is the group of signed permutations and $\mathcal{P}(\mathfrak{a})^{W(\mathfrak{a})}$ has algebraically independent generators $h \mapsto \mathrm{tr}(h^{2r})$ for $r = 1, 2, \ldots, n$.

5. (Type DIII) The functionals ε_{2i} restrict on $\mathfrak{a}$ to $-\varepsilon_{2i-1}$ for $1 \le i \le p$. From this it follows that when $l = 2p$ the restricted root system Σ is of type C_p. When $l = 2p + 1$, the functional ε_l restricts on $\mathfrak{a}$ to 0. In this case $\pm\varepsilon_1, \pm\varepsilon_3, \ldots, \pm\varepsilon_{2p-1}$ are also restricted roots, and Σ is of type BC_p. In both cases the group $W(\mathfrak{a})$ and a set of generators for $\mathcal{P}(\mathfrak{a})^{W(\mathfrak{a})}$ are the same as in Type AIII.

6. (Type BDI) The functionals ε_i restrict on $\mathfrak{a}$ to 0 for $p < i \le q$ and the functionals ε_{n+1-i} restrict on $\mathfrak{a}$ to $-\varepsilon_i$ for $1 \le i \le p$. From this it follows that the restricted root system is of type B_p when $p < q$ and of type D_p when $p = q$. For $p < q$ the group $W(\mathfrak{a})$ and a set of generators for $\mathcal{P}(\mathfrak{a})^{W(\mathfrak{a})}$ are the same as in Type AIII. For $p = q$ the group $W(\mathfrak{a})$ consists of all permutations and sign changes of an *even* number of coordinates. In this case a set of generators for $\mathcal{P}(\mathfrak{a})^{W(\mathfrak{a})}$ is given by the polynomials $h \mapsto \mathrm{tr}(h^{2r})$ for $r = 1, \ldots, p - 1$ together with the polynomial $h \mapsto x_1 \cdots x_p$ (where $h = \mathrm{diag}[x_1, \ldots, x_p, -x_p, \ldots, -x_1]$).

7. (Type CII) The functionals ε_i restrict on $\mathfrak{a}$ to 0 for $p < i \le q$, and ε_{n+1-i} restricts on $\mathfrak{a}$ to $-\varepsilon_i$ for $1 \le i \le p$. From this it follows that when $p < q$ the restricted root system Σ is of type BC_p, whereas when $p = q$ the system Σ is of type C_p. In both cases the group $W(\mathfrak{a})$ and a set of generators for $\mathcal{P}(\mathfrak{a})^{W(\mathfrak{a})}$ are the same as in Type AIII.

We conclude this section by recording one more theorem of Chevalley. We note that Corollary 12.3.4 implies that $\mathrm{Ad}(K)\mathfrak{a}$ is Zariski-dense in V. Thus if $f \in \mathcal{P}(V)^K$ and $f|_{\mathfrak{a}} = 0$ then $f = 0$. We also note that if $f \in \mathcal{P}(V)^K$ then $f|_{\mathfrak{a}} \in \mathcal{P}(\mathfrak{a})^{W(\mathfrak{a})}$. The following result is the celebrated Chevalley restriction theorem:

Theorem 12.4.5. *The map* $\mathrm{res} : \mathcal{P}(V)^K \longrightarrow \mathcal{P}(\mathfrak{a})^{W(\mathfrak{a})}$ *defined by* $\mathrm{res}(f) = f|_{\mathfrak{a}}$ *for* $f \in \mathcal{P}(V)^K$ *is an algebra isomorphism.*

We have already shown that the restriction map is injective. For a proof of surjectivity see Helgason [67, Chapter II, §5.2]. We will verify this in the next section by direct calculations when K is a classical group.

12.4.3 Classical Examples

There are 16 pairs $(K, (\sigma, V))$ covered by the Kostant–Rallis theorem in which $\mathfrak{g}$ is simple and K is a product of classical groups (7 pairs with $\mathfrak{g}$ classical and 9 with $\mathfrak{g}$ exceptional); see Helgason [66, Chapter X, §6, Table V]. For the cases in which G is also a classical group, K and θ were determined in Sections 11.3.4 and 11.3.5, and M in Section 12.3.2. For that purpose the matrix forms of G and θ were chosen so that the diagonal subgroup H in G was a maximal torus and $A = H \cap Q$ was a maximal θ-anisotropic torus (where $\theta(g) = g^{-1}$ for $g \in Q$). In the following examples we have chosen the matrix form of G and the involution θ to facilitate the description of V as a K-module. The algebraically independent generating set for $\mathcal{P}(V)^K$ is obtained from the results cited in Section 12.4.2 (Chevalley restriction theorem and the classification of the invariants for finite reflection groups). For the cases with G classical we give generators for $\mathcal{P}(V)^K$ whose restrictions to $\mathfrak{a}$ are the generators of $\mathcal{P}(\mathfrak{a})^{W(\mathfrak{a})}$ given in Section 12.4.2. Note that when M is a finite group the restricted root system coincides with the root system of $\mathfrak{h}$ on $\mathfrak{g}$.

In the following examples, s_n and $\check{x}$ have the same meaning as in Section 12.3.2.

1. (Type AI) Let $G = \mathbf{SL}(n, \mathbb{C})$ and $\theta(g) = (g^t)^{-1}$. Then $K = \mathbf{SO}(n, \mathbb{C})$ and V is the space of symmetric $n \times n$ matrices of trace 0. The action of K on V is $\sigma(k)X = kXk^{-1}$. Here we take $\mathfrak{a}$ to be the diagonal matrices in $\mathfrak{g}$. We have $W(\mathfrak{a}) = W_G = \mathfrak{S}_n$. The polynomials $u_i(X) = \mathrm{tr}(X^{i+1})$, for $i = 1, \ldots, n-1$, restrict on $\mathfrak{a}$ to generators for $\mathcal{P}(\mathfrak{a})^{W(\mathfrak{a})}$. Hence $\mathcal{P}(V)^K$ is the polynomial algebra with generators $u_1, \ldots, u_{n-1}$.

2. (Type AII) Let $G = \mathbf{SL}(2n, \mathbb{C})$. Take $J = \begin{bmatrix} 0 & I_n \\ -I_n & 0 \end{bmatrix}$ and define $\theta(g) = -J(g^t)^{-1}J$. Then $K = \mathbf{Sp}(\mathbb{C}^{2n}, \Omega)$, where Ω is the bilinear form with matrix J. The space V consists of all matrices ($n \times n$ blocks)

$$X = \begin{bmatrix} A & B \\ C & A^t \end{bmatrix} \quad \text{with } \mathrm{tr}(A) = 0, \ B^t = -B, \text{ and } C^t = -C. \tag{12.56}$$

We take $\mathfrak{a} \subset V$ as the matrices

$$X = \begin{bmatrix} Z & 0 \\ 0 & Z \end{bmatrix} \quad \text{with } Z = \mathrm{diag}[z_1, \ldots, z_n] \text{ and } \mathrm{tr}(Z) = 0.$$

From Section 12.4.2 we know that the restricted root system is of type A_{n-1}. The polynomial $u_i(X) = \mathrm{tr}(X^{i+1})$ restricts on $\mathfrak{a}$ to $2\,\mathrm{tr}(Z^{i+1})$. Hence $u_1, \ldots, u_{l-1}$ generate $\mathcal{P}(V)^K$, since their restrictions generate $\mathcal{P}(\mathfrak{a})^{W(\mathfrak{a})}$.

3. (Type AIII) Let $G = \mathbf{SL}(n, \mathbb{C})$. Take $q \geq p > 0$ with $p + q = n$ and the involution $\theta_{q,p}(g) = I_{q,p} g I_{q,p}$, where $I_{q,p} = \begin{bmatrix} I_q & 0 \\ 0 & -I_p \end{bmatrix}$. Then $K = \mathbf{S}(\mathbf{GL}(q, \mathbb{C}) \times \mathbf{GL}(p, \mathbb{C}))$ embedded diagonally and V consists of all matrices in block form

$$v = \begin{bmatrix} 0 & X \\ Y & 0 \end{bmatrix} \quad \text{with } X \in M_{q,p} \text{ and } Y \in M_{p,q} \,. \tag{12.57}$$

As a K-module $V \cong F \oplus F^*$, where $F = M_{q,p}$ with action $\rho(g_1, g_2)X = g_1 X g_2^{-1}$ for $g_1 \in \mathbf{GL}(q, \mathbb{C})$ and $g_2 \in \mathbf{GL}(p, \mathbb{C})$ with $\det(g_1)\det(g_2) = 1$ (we can identify F^* with $M_{p,q}$ with the action $\rho^*(g_1, g_2)Y = g_2 Y g_1^{-1}$). The restriction of ρ to the subgroup $\mathbf{SL}(q, \mathbb{C}) \times \mathbf{SL}(p, \mathbb{C})$ is irreducible and equivalent to the outer tensor product $\mathbb{C}^q \widehat{\otimes} \mathbb{C}^p$ of the defining representations.

In this matrix realization we take $\mathfrak{a} \subset V$ as the matrices v in (12.57) with

$$X = \begin{bmatrix} Z s_p \\ 0_{q-p} \end{bmatrix} \quad \text{and } Y = \begin{bmatrix} s_p Z & 0_{q-p} \end{bmatrix}, \quad \text{with } Z = \mathrm{diag}[z_1, \ldots, z_p] \,. \tag{12.58}$$

The polynomials $u_i(v) = \mathrm{tr}((XY)^i)$ with v as in (12.57) are K-invariant. Since $(Z s_p)(s_p Z) = Z^2$ for Z as in (12.58), the restriction of u_i to $\mathfrak{a}$ is the $W(\mathfrak{a})$-invariant polynomial $Z \mapsto \mathrm{tr}(Z^{2i})$. These polynomials, for $i = 1, \ldots, p$, generate $\mathcal{P}(\mathfrak{a})^{W(\mathfrak{a})}$. Hence $\mathcal{P}(V)^K$ is the polynomial algebra generated by $u_1, \ldots, u_p$.

4. (Type CI) Let $G = \mathbf{Sp}(\mathbb{C}^{2n}, \Omega)$, where Ω is the bilinear form with matrix J as in Example 2, and take $\theta = \theta_{n,n}$ as for type AIII. Then $K \cong \mathbf{GL}(n, \mathbb{C})$ consists of the matrices

$$k = \begin{bmatrix} g & 0 \\ 0 & (g^t)^{-1} \end{bmatrix}, \quad \text{with } g \in \mathbf{GL}(n, \mathbb{C}) \,,$$

whereas V consists of the matrices ($n \times n$ blocks)

$$v = \begin{bmatrix} 0 & X \\ Y & 0 \end{bmatrix} \quad \text{with } X^t = X \text{ and } Y^t = Y \,. \tag{12.59}$$

Let F be the space of $n \times n$ symmetric matrices, and let ρ be the representation of $\mathbf{GL}(n, \mathbb{C})$ on F given by $\rho(g)X = g X g^t$. Then $(\sigma, V) \cong (\rho \oplus \rho^*, F \oplus F^*)$. Here we can identify F^* with F as a vector space, with $g \in \mathbf{GL}(n, \mathbb{C})$ acting by $X \mapsto (g^t)^{-1} X g^{-1}$. Note that (ρ, F) is the irreducible $\mathbf{SL}(n, \mathbb{C})$-module with highest weight $2\varepsilon_1$ (Theorem 5.7.3).

In this realization we take $\mathfrak{a} \subset V$ as the matrices

$$\begin{bmatrix} 0 & X \\ X & 0 \end{bmatrix} \quad \text{with } X = \mathrm{diag}[x_1, \ldots, x_n] \,. \tag{12.60}$$

This is a toral subalgebra of $\mathfrak{g}$ that is conjugate in G to the Lie algebra of the maximal anisotropic torus used in Section 12.3.2. The polynomials $u_i(v) = \mathrm{tr}((XY)^i)$, with v as in (12.59), are K-invariant. The restriction of u_i to $\mathfrak{a}$ is the polynomial $X \mapsto \mathrm{tr}(X^{2i})$. Since the restricted root system is of type C_n, these polynomials, for $i =$

$1,\ldots,n$, generate $\mathcal{P}(\mathfrak{a})^{W(\mathfrak{a})}$. It follows that $u_1,\ldots,u_n$ are algebraically independent generators of $\mathcal{P}(V)^K$.

5. (Type DIII) Let $G = \mathbf{SO}(\mathbb{C}^{2n}, B)$, where B is the bilinear form with matrix $\begin{bmatrix} 0 & I_n \\ I_n & 0 \end{bmatrix}$, and take $\theta(g) = JgJ^{-1}$ with J as in Example 2 (Type AII). Then K is the same as in Example 4 (Type CI), whereas V consists of the matrices ($n \times n$ blocks)

$$v = \begin{bmatrix} 0 & X \\ Y & 0 \end{bmatrix} \quad \text{with } X^t = -X \text{ and } Y^t = -Y. \tag{12.61}$$

Let F be the space of $n \times n$ skew-symmetric matrices, and let ρ be the representation of $\mathbf{GL}(n,\mathbb{C})$ on F given by $\rho(g)X = gXg^t$. Then $(\sigma, V) \cong (\rho \oplus \rho^*, F \oplus F^*)$. Here we can identify F^* with F as a vector space, with $g \in \mathbf{GL}(n,\mathbb{C})$ acting by $X \mapsto (g^t)^{-1}Xg^{-1}$. Note that (ρ, F) is the irreducible $\mathbf{SL}(n,\mathbb{C})$-module with highest weight $\varepsilon_1 + \varepsilon_2$ (Theorem 5.7.5).

In this realization we take $\mathfrak{a} \subset V$ as the matrices

$$v = \begin{bmatrix} 0 & Xs_n \\ s_nX & 0 \end{bmatrix}, \quad X = \begin{cases} \operatorname{diag}[Z, -\check{Z}] & \text{when } n = 2p, \\ \operatorname{diag}[Z, 0, -\check{Z}] & \text{when } n = 2p+1. \end{cases} \tag{12.62}$$

Here $Z = \operatorname{diag}[z_1,\ldots,z_p]$. This is a toral subalgebra of $\mathfrak{g}$ that is conjugate in G to the Lie algebra of the maximal anisotropic torus used in Section 12.3.2. The polynomials $u_i(v) = \operatorname{tr}((XY)^i)$, with v as in (12.61), are K-invariant. The restriction of u_i to $\mathfrak{a}$ is the polynomial $Z \mapsto \operatorname{tr}(Z^{2i})$ (note that $Xs_n = -s_nX$ for X as in (12.62)). Since the restricted root system is of type C_p or BC_p, these polynomials, for $i = 1,\ldots,p$, generate $\mathcal{P}(\mathfrak{a})^{W(\mathfrak{a})}$. It follows that $u_1,\ldots,u_p$ are algebraically independent generators for $\mathcal{P}(V)^K$.

6. (Type BDI) Let $G = \mathbf{SO}(n,\mathbb{C})$ ($g^tg = I_n$ for $g \in G$). Take $p + q = n$, $q \ge p \ge 1$, $\theta = \theta_{q,p}$ as in Example 3 (Type AIII). Then $K = \mathbf{S}(\mathbf{O}(q,\mathbb{C}) \times \mathbf{O}(p,\mathbb{C}))$, embedded diagonally into G, whereas V consists of the matrices

$$v = \begin{bmatrix} 0 & X \\ -X^t & 0 \end{bmatrix} \quad \text{with } X \in M_{q,p}. \tag{12.63}$$

Here (σ, V) is the representation of K on $M_{q,p}$ given by $\sigma(g_1, g_2)X = g_1Xg_2^{-1}$. Restricted to $K^\circ = \mathbf{SO}(q,\mathbb{C}) \times \mathbf{SO}(p,\mathbb{C})$ it is the irreducible representation $\mathbb{C}^q \widehat{\otimes} \mathbb{C}^p$ (outer tensor product of the defining representations) when $p \ne 2$ and $q \ne 2$. For $p = 2$ and $q > 2$ it is the sum of two irreducible representations (recall that $\mathbf{SO}(2,\mathbb{C}) \cong \mathbf{GL}(1,\mathbb{C})$).

In this realization we take $\mathfrak{a} \subset V$ as the matrices v in (12.63) with

$$X = \begin{bmatrix} Zs_p \\ 0 \end{bmatrix}, \quad Z = \operatorname{diag}[z_1,\ldots,z_p].$$

This is a toral subalgebra of $\mathfrak{g}$ that is conjugate in G to the Lie algebra of the maximal anisotropic torus used in Section 12.3.2. The polynomials $u_i(v) = \operatorname{tr}((XX^t)^i)$, with v as in (12.63), are K-invariant. The restriction of u_i to $\mathfrak{a}$ is the polynomial $Z \mapsto \mathrm{tr}(Z^{2i})$. Suppose $p < q$. Then the restricted root system is of type B_p, so it follows that $\{u_1, \ldots, u_p\}$ gives algebraically independent generators for $\mathcal{P}(V)^K$. Now suppose $p = q$. Then the restricted root system is of type D_p. In this case the *Pfaffian* polynomial $\mathrm{Pfaff}(v)$ is K-invariant and restricts to the $W(\mathfrak{a})$-invariant polynomial $Z \mapsto z_1 \cdots z_p$ on $\mathfrak{a}$ (see Section B.2.6). It follows that $\{u_1, \ldots, u_{p-1}, \mathrm{Pfaff}\}$ is a set of algebraically independent generators for $\mathcal{P}(V)^K$ when $p = q$.

7. (Type CII) Let $G = \mathbf{Sp}(\mathbb{C}^{2n}, \omega_n)$, where ω_n is the bilinear form with matrix $T_n = \mathrm{diag}[\mu, \ldots, \mu]$ (n copies) in block-diagonal form with $\mu = \begin{bmatrix} 0 & 1 \\ -1 & 0 \end{bmatrix}$. Take $q \geq p > 0$ with $p + q = n$ and let $\theta = \theta_{2p,2q}$, as in Example 3 (Type AIII). Then $K = \mathbf{Sp}(\mathbb{C}^{2q}, \omega_q) \times \mathbf{Sp}(\mathbb{C}^{2p}, \omega_p)$, embedded diagonally, and V consists of all matrices

$$v = \begin{bmatrix} 0 & X \\ T_p X^t T_q & 0 \end{bmatrix}, \quad \text{with } X \in M_{2q,2p}. \tag{12.64}$$

Here $(k_1, k_2) \in K$ acts on $v \in V$ by $X \mapsto k_1 X k_2^{-1}$ for $k_1 \in K_1 = \mathbf{Sp}(\mathbb{C}^{2q}, \omega_q)$, $k_2 \in K_2 = \mathbf{Sp}(\mathbb{C}^{2p}, \omega_p)$, and $X \in M_{2q,2p}$. Hence the representation (σ, V) is irreducible and equivalent to the outer tensor product $\mathbb{C}^{2q} \widehat{\otimes} \mathbb{C}^{2p}$ of the defining representations of K_1 and K_2.

We take $\mathfrak{a}$ to consist of all matrices (12.64) with

$$X = \begin{bmatrix} Z \\ 0_{q-p} \end{bmatrix}, \quad \text{where } Z = \mathrm{diag}[z_1, z_1, \ldots, z_p, z_p] \in M_{2p}. \tag{12.65}$$

This is a toral subalgebra of $\mathfrak{g}$ that is conjugate in G to the Lie algebra of the maximal anisotropic torus used in Section 12.3.2. The polynomials $u_i(v) = \mathrm{tr}((XX^t)^i)$, with v as in (12.64), are K-invariant. The restriction of u_i to $\mathfrak{a}$ is the polynomial $Z \mapsto \mathrm{tr}(Z^{2i})$. Since the restricted root system is of type BC_p (when $p < q$) or C_p (when $p = q$), it follows that $\{u_1, \ldots, u_p\}$ gives algebraically independent generators for $\mathcal{P}(V)^K$.

8. (Type G) Let $K = (\mathbf{SL}(2, \mathbb{C}) \times \mathbf{SL}(2, \mathbb{C})) / \{(I, I), (-I, -I)\}$ and let (σ, V) be the representation of K on $V = \mathbb{C}^2 \widehat{\otimes} S^3(\mathbb{C}^2)$ (outer tensor product). Here M is isomorphic to $\times^2(\mathbb{Z}/2\mathbb{Z})$. One has $\mathcal{P}(V)^K = \mathbb{C}[u_1, u_2]$ with $\deg u_1 = 2$ and $\deg u_2 = 6$. This example comes from the exceptional group G_2.

9. (Type FI) Let $K = (\mathbf{SL}(2, \mathbb{C}) \times \mathbf{Sp}(3, \mathbb{C})) / \{(I, I), (-I, I)\}$ and let (σ, V) be the representation of K on $\mathbb{C}^2 \widehat{\otimes} F$ (outer tensor product), with F the irreducible representation of $\mathbf{Sp}(3, \mathbb{C})$ having highest weight $\varepsilon_1 + \varepsilon_2 + \varepsilon_3$. Then M is isomorphic with $\times^4(\mathbb{Z}/2\mathbb{Z})$. One has $\mathcal{P}(V)^K = \mathbb{C}[u_1, u_2, u_2, u_4]$ with $\deg u_1 = 2$, $\deg u_2 = 6$, $\deg u_3 = 8$, and $\deg u_4 = 12$. This example comes from the exceptional group F_4.

10. (Type FII) Let $K = \mathbf{Spin}(9, \mathbb{C})$ and let (σ, V) be the spin representation of K. In this case $M \cong \mathbf{Spin}(7, \mathbb{C})$. The restricted root system is of type A_1 and hence $\mathcal{P}(V)^K = \mathbb{C}[u]$ with $\deg u = 2$. This example also comes from the exceptional group F_4.

11. (Type EI) Let $K = \mathbf{Sp}(4, \mathbb{C})$ and take the representation (σ, V) of K on $\bigwedge^4 \mathbb{C}^8 / (\omega \wedge \bigwedge^2 \mathbb{C}^8)$, where ω is a nonzero element of $(\bigwedge^2 \mathbb{C}^8)^K$ (this is the irreducible representation with highest weight $\varepsilon_1 + \varepsilon_2 + \varepsilon_3 + \varepsilon_4$). In this case M is isomorphic to $\times^6(\mathbb{Z}/2\mathbb{Z})$ and $\mathcal{P}(V)^K$ is a polynomial algebra in six generators whose degrees are 2, 5, 6, 8, 9, and 12. This example comes from the exceptional group E_6.

12. (Type EII) Let $K = (\mathbf{SL}(2) \times \mathbf{SL}(6,\mathbb{C}))/\{(I,I),(-I,-I)\}$ and take (σ,V) to be the representation $\mathbb{C}^2 \widehat{\otimes} \bigwedge^3 \mathbb{C}^6$ (outer tensor product). Here M is locally isomorphic to $\mathbf{GL}(1,\mathbb{C}) \times \mathbf{GL}(1,\mathbb{C})$, and the restricted root system is of type F_4. Hence $\mathcal{P}(V)^K$ is a polynomial algebra in four generators with degrees as in Example 9. This example also comes from the exceptional group E_6.

13. (Type EIII) Take $K = (\mathbf{GL}(1,\mathbb{C}) \times \mathbf{Spin}(10,\mathbb{C}))/\{(I,I),(-I,-I)\}$ (here the second $-I$ is the kernel of the covering $\mathbf{Spin}(10,\mathbb{C}) \to \mathbf{SO}(10,\mathbb{C})$). Let (σ,V) be the sum $(\rho_+,F^+) \oplus (\rho_-,F^-)$ of the two half-spin representations of $\mathbf{Spin}(10,\mathbb{C})$ (see Proposition 6.2.3) with $\rho_+(z,I) = zI$ and $\rho_-(z,I) = z^{-1}I$ for $z \in \mathbf{GL}(1,\mathbb{C})$. Here M is isomorphic to $\mathbf{GL}(4,\mathbb{C})$. The restricted root system is of type BC_2. Hence $\mathcal{P}(V)^K$ is a polynomial algebra in two generators, one of degree 2 and the other of degree 4. This example also comes from E_6.

14. (Type EV) Let $K = \mathbf{SL}(8,\mathbb{C})$ and take (σ,V) to be the representation of K on $\bigwedge^4 \mathbb{C}^8$. If we replace K by $\sigma(K)$ then M is isomorphic to $\times^7(\mathbb{Z}/2\mathbb{Z})$. Here $\mathcal{P}(V)^K$ is a polynomial algebra in seven generators whose degrees are 2, 6, 8, 10, 12, 14, and 18. This example comes from the exceptional group E_7.

15. (Type EVI) Let $K = (\mathbf{SL}(2,\mathbb{C}) \times \mathbf{Spin}(12,\mathbb{C}))/\{(I,I),(-I,-I)\}$ (the second $-I$ as in Example 13). In this case (σ,V) is given by $\mathbb{C}^2 \widehat{\otimes} S$ (exterior tensor product) with S a half-spin representation. Then M is locally isomorphic to $\times^3 \mathbf{SL}(2,\mathbb{C})$. The restricted root system is of type F_4. Hence $\mathcal{P}(V)^K$ is a polynomial algebra in four generators whose degrees are as in Example 9. This example also comes from the exceptional group E_7.

16. (Type EVIII) Let $K = \mathbf{Spin}(16,\mathbb{C})$ and take (σ,V) to be a half-spin representation. If we replace K by $\sigma(K)$ then M is isomorphic to $\times^8(\mathbb{Z}/2\mathbb{Z})$. Here $\mathcal{P}(V)^K$ is a polynomial algebra in eight generators whose degrees are 2, 8, 12, 14, 18, 20, 24, and 30. This example comes from the exceptional group E_8.

12.4.4 Some Results from Algebraic Geometry

In this section we will collect a few simple results in algebraic geometry that will be used in our proof of the Kostant–Rallis theorem. Let $\mathcal{P} = \mathbb{C}[x_1,\ldots,x_n]$ and let $\mathcal{P}^j \subset \mathcal{P}$ be the space of polynomials homogeneous of degree j. An ideal $\mathfrak{I}$ in $\mathcal{P}$ will be called *homogeneous* if $\mathfrak{I} = \bigoplus_j \mathfrak{I} \cap \mathcal{P}^j$. Set

$$\mathcal{V}(\mathfrak{I}) = \{x \in \mathbb{C}^n : f(x) = 0 \text{ for all } f \in \mathfrak{I}\}.$$

Lemma 12.4.6. *Let* $\mathfrak{I}$ *be a homogeneous ideal in* $\mathcal{P}$. *Then the following are equivalent:*

1. $\mathcal{V}(\mathfrak{I}) = \{0\}$.
2. There exists $j > 0$ *such that* $\mathfrak{I} \supset \mathcal{P}^j$.

Proof. We first prove that (2) implies (1). If $\mathcal{P}^j \subset \mathfrak{I}$ then $x_i^j \in \mathfrak{I}$ for $i = 1, \ldots, n$. This implies that if $x \in \mathcal{V}(\mathfrak{I})$ then $x_i^j = 0$ for $i = 1, \ldots, n$; thus $x = 0$. We now prove that (1) implies (2). By Corollary A.1.5 for each i there exists $d_i > 0$ such that $x_i^{d_i} \in \mathfrak{I}$. Let $k = \max\{d_i : 1 \le i \le n\}$. If $I \in \mathbb{N}^n$ and $|I| \ge nk$, then some $i_p \ge k$. Thus if $j = nk$ then $x^I \in \mathfrak{I}$ for all I such that $|I| = j$. Hence $\mathcal{P}^j \subset \mathfrak{I}$. $\qquad\square$

Let $d_1, \ldots, d_m$ be strictly positive integers. Set $d = (d_1, \ldots, d_m)$ and

$$V(d) = \mathcal{P}^{d_1} \oplus \cdots \oplus \mathcal{P}^{d_m}.$$

Given $f = [f_1, \ldots, f_m] \in V(d)$, we set $\mathfrak{I}(f) = \sum_i \mathcal{P} f_i$. Since $f_i(0) = 0$ for all i, the zero set of the ideal $\mathfrak{I}(f)$ includes 0.

Lemma 12.4.7. *The set of all* $f \in V(d)$ *such that* $\mathcal{V}(\mathfrak{I}(f)) = \{0\}$ *is Zariski open in* $V(d)$. *It is nonempty if* $m \ge n$.

Proof. Let W be a vector space over $\mathbb{C}$ with basis $w_1, \ldots, w_n$. We grade W by setting

$$W^j = \sum_{d_i = j} \mathbb{C} w_i.$$

We put a grading on $\mathcal{P} \otimes W$ by setting $(\mathcal{P} \otimes W)^k = \sum_{i+j=k} \mathcal{P}^i \otimes W^j$ and we define $T_{j,f} : (\mathcal{P} \otimes W)^j \longrightarrow \mathcal{P}^j$ by

$$T_{j,f}\left(\sum_i \varphi_i \otimes w_i\right) = \sum_i \varphi_i f_i.$$

Then $f \mapsto T_{j,f}$ is a linear map from the vector space $V(d)$ to the vector space of all linear maps from $(\mathcal{P} \otimes W)^j$ to $\mathcal{P}^j$. Set $r_j = \dim(\mathcal{P} \otimes W)^j$ and $s_j = \dim \mathcal{P}^j$. For each j fix a basis $\{w_{ij}\}$ of $(\mathcal{P} \otimes W)^j$ and a basis $\{z_{ij}\}$ of $\mathcal{P}^j$. Let $\{\Delta_{pj}(f)\}$ be the collection of all $s_j \times s_j$ minors of $T_{j,f}$ with respect to these given bases. We note that

$$T_{j,f}(\mathcal{P} \otimes W)^j = \mathfrak{I}(f) \cap \mathcal{P}^j.$$

Applying Lemma 12.4.6, we see that $\mathcal{V}(\mathfrak{I}(f)) \ne \{0\}$ if and only if the maps $T_{j,f}$ are not surjective for $j = 1, 2, \ldots$. This is equivalent to the condition that $\Delta_{pj}(f) = 0$ for all p, j. Thus the set of all $f \in V(d)$ such that $\mathcal{V}(\mathfrak{I}(f)) \ne \{0\}$ is Zariski closed in $V(d)$. The complement of this set is Zariski open and nonempty because for $m \ge n$ the function $f = [x_1^{d_1}, \ldots, x_n^{d_n}, 0, \ldots, 0]$ is in $V(d)$ and $\mathcal{V}(\mathfrak{I}(f)) = \{0\}$. $\qquad\square$

The next result is of a slightly different nature. If $0 \ne f \in \mathcal{P}$ and $\deg f = j$ then $f = f_j + f_{j-1} + \cdots + f_0$ with $f_j \ne 0$ and $f_k \in \mathcal{P}^k$. We define $f_{\text{top}} = f_j$. If $f = 0$ we set $f_{\text{top}} = 0$. If $\mathfrak{I}$ is an ideal in $\mathcal{P}$ we set

$$(\mathcal{P}/\mathcal{I})_j = \sum_{i \leq j} \mathcal{P}^i + \mathcal{I} .$$

Then $(\mathcal{P}/\mathcal{I})_j \subset (\mathcal{P}/\mathcal{I})_{j+1}$ and $\bigcup_j (\mathcal{P}/\mathcal{I})_j = \mathcal{P}/\mathcal{I}$. This filtration gives $\mathcal{P}/\mathcal{I}$ the structure of a filtered algebra over $\mathbb{C}$. Let

$$\mathrm{Gr}^j(\mathcal{P}/\mathcal{I}) = (\mathcal{P}/\mathcal{I})_j / (\mathcal{P}/\mathcal{I})_{j-1} .$$

Then $\mathrm{Gr}(\mathcal{P}/\mathcal{I}) = \bigoplus_{j \geq 0} \mathrm{Gr}^j(\mathcal{P}/\mathcal{I})$ has a natural structure of a graded algebra (see Section C.1.1). If $\mathcal{I}$ happens to be homogeneous then $\mathcal{P}/\mathcal{I}$ inherits a grading and $\mathrm{Gr}(\mathcal{P}/\mathcal{I})$ is isomorphic to $\mathcal{P}/\mathcal{I}$.

Lemma 12.4.8. *Let $\mathcal{I}$ be an ideal in $\mathcal{P}$ and set $\mathcal{J} = \mathrm{Span}\{f_{\mathrm{top}} : f \in \mathcal{I}\}$. Then $\mathcal{J}$ is a homogeneous ideal in $\mathcal{P}$ and $\mathrm{Gr}(\mathcal{P}/\mathcal{I})$ is isomorphic to $\mathcal{P}/\mathcal{J}$ as a graded algebra.*

Proof. If $u \in \mathcal{P}$ then $u = \sum u_i$ with u_i homogeneous of degree i. Thus to show that $u\varphi \in \mathcal{J}$ for all $\varphi \in \mathcal{J}$, we may assume that u and φ are homogeneous. In this case take any $f \in \mathcal{I}$ such that $f_{\mathrm{top}} = \varphi$. Then $(uf)_{\mathrm{top}} = u(f_{\mathrm{top}})$. By definition $\mathcal{J}$ is closed under addition. It is also clear that $\mathcal{J}$ is homogeneous.

Set $\mathcal{P}_j = \sum_{i \leq j} \mathcal{P}^i$. We note that

$$\mathrm{Gr}^j(\mathcal{P}/\mathcal{I}) = (\mathcal{P}_j + \mathcal{I}) / (\mathcal{P}_{j-1} + \mathcal{I}) .$$

We now prove the main assertion of the lemma by giving an explicit isomorphism (which will be used later). Let $f \in \mathcal{P}_j$. Write $f = f_j + u$ with $\deg u \leq j - 1$ and f_j homogeneous of degree j. We define

$$\psi_j(f) = f_j + \mathcal{J} .$$

If $f, g \in \mathcal{P}_j$ and $f - g + \mathcal{I} \subset \mathcal{P}_{j-1} + \mathcal{I}$ then $(f-g)_j \in \mathcal{J}$. Thus ψ_j induces a linear map

$$\overline{\psi}_j : \mathrm{Gr}^j(\mathcal{P}/\mathcal{I}) \longrightarrow (\mathcal{P}/\mathcal{J})^j .$$

It is easy to see that $\overline{\psi}_j$ is bijective. The proof that $\bigoplus_{j \geq 0} \overline{\psi}_j$ is an algebra homomorphism follows from the definition of the multiplication on $\mathrm{Gr}(\mathcal{P}/\mathcal{I})$. $\square$

We will now use the notation $\mathcal{P}$ for $\mathbb{C}[x_1, \ldots, x_m, y_1, \ldots, y_n]$, and we consider $\mathbb{C}[x_1, \ldots, x_m]$ and $\mathbb{C}[y_1, \ldots, y_n]$ as subalgebras of $\mathcal{P}$. We fix an algebraically independent set $\{u_1, \ldots, u_n\}$ of homogeneous elements of $\mathcal{P}$, and for $c = [c_1, \ldots, c_n] \in \mathbb{C}^n$ we define the ideal

$$\mathcal{I}_c = \mathcal{P}(u_1 - c_1) + \cdots + \mathcal{P}(u_n - c_n) .$$

We note that $\mathcal{I}_c$ is a homogeneous ideal if and only if $c = 0$.

Lemma 12.4.9. *Assume that $\mathcal{A} \subset \mathbb{C}[y_1, \ldots, y_n]$ is a finite-dimensional graded subspace such that the map $\mathbb{C}[x_1, \ldots, x_m] \otimes \mathcal{A} \otimes \mathbb{C}[u_1, \ldots, u_n] \longrightarrow \mathcal{P}$ given by $f \otimes a \otimes g \mapsto fag$ is bijective. Then the following hold for all $c \in \mathbb{C}^n$:*

1. *The elements $\{f_{\text{top}} : f \in \mathcal{I}_c\}$ span $\mathcal{I}_0$. In particular, there is a natural isomorphism of $\text{Gr}(\mathcal{P}/\mathcal{I}_c)$ with $\mathcal{P}/\mathcal{I}_0$.*
2. *Every irreducible component of $\mathcal{V}(\mathcal{I}_c)$ has dimension at most m.*

Proof. We first note that $\mathcal{I} \supset \sum_i \mathcal{P}u_i$. Indeed, $(u_i - c_i)_{\text{top}} = u_i$. Hence $\mathcal{I} \supset \mathcal{I}_0$. Let

$$T_c : \mathbb{C}[x_1,\ldots,x_m] \otimes \mathcal{A} \otimes \mathbb{C}[u_1,\ldots,u_m] \longrightarrow \mathcal{P}$$

be defined by $T_c(f \otimes a \otimes \varphi(u_1,\ldots,u_m)) = fa\varphi(u_1 - c_1,\ldots,u_m - c_m)$. Then T_c is a linear bijection. Using this map one sees that

$$\dim(\mathcal{P}/\mathcal{I}_c)_k = \sum_{i+j \le k} \dim(\mathcal{Q}^i \otimes \mathcal{A}^j) \quad \text{with } \mathcal{Q} = \mathbb{C}[x_1,\ldots,x_m] .$$

The right-hand side of this equation is independent of c. Applying Lemma 12.4.8 and these observations we get

$$\dim(\mathcal{P}/\mathcal{I})^j = \dim \text{Gr}^j(\mathcal{P}/\mathcal{I}_c) = \dim \text{Gr}^j(\mathcal{P}/\mathcal{I}_0) = \dim(\mathcal{P}/\mathcal{I}_0)^j .$$

Since $\mathcal{I} \supset \mathcal{I}_0$, this implies that $\mathcal{I} = \mathcal{I}_0$.

Let $\bar{f} = f|_{\mathcal{V}(\mathcal{I}_c)}$ for $f \in \mathcal{P}$. Then using the map T_c we see that $\mathcal{O}[\mathcal{V}(\mathcal{I}_c)]$ is a free $\mathbb{C}[\bar{x}_1,\ldots,\bar{x}_m]$-module with generator set any basis of $\bar{A}$. This implies that $\mathcal{O}[\mathcal{V}(I_c)]$ is integral over the ring $\mathbb{C}[\bar{x}_1,\ldots,\bar{x}_m]$. From this we see that if X is any irreducible component of $\mathcal{V}(I_c)$ and if $y_i = x_i|_X$, then $\mathcal{O}[X]$ is integral over the ring $\mathbb{C}[y_1,\ldots,y_m]$. Thus $\dim X \le m$. $\qquad \square$

We will also need the following ring-theoretic result:

Lemma 12.4.10 (Nakayama). *Let $\mathcal{R}$ be a commutative ring with unit 1. Let $\mathcal{S}$ be a subring of $\mathcal{R}$ with unit 1 and let $\mathcal{I} \subset \mathcal{S}$ be a proper ideal of $\mathcal{S}$. If $\mathcal{R}$ is finitely generated as an $\mathcal{S}$-module, then $\mathcal{I}\mathcal{R} \ne \mathcal{R}$.*

Proof. By the finite generation hypothesis, there exist elements $m_1,\ldots,m_d$ in $\mathcal{R}$ such that $\mathcal{R} = \sum_i \mathcal{S}m_i$. We assume that $\mathcal{I}\mathcal{R} = \mathcal{R}$ and derive a contradiction. Under this assumption there exist $a_{ij} \in \mathcal{I}$ such that $\sum_j a_{ij}m_j = m_i$. Hence

$$\sum_j (\delta_{ij} - a_{ij})m_j = 0 .$$

Let $\mathcal{A} = [b_{ij}]$ be the cofactor matrix of the matrix $[\delta_{ij} - a_{ij}]$. Then

$$0 = \sum_{i,j} b_{ki}(\delta_{ij} - a_{ij})m_j = \det[\delta_{ij} - a_{ij}]m_k .$$

Now $\det[\delta_{ij} - a_{ij}] = 1 + a$ with $a \in \mathcal{I}$ and $(1+a)\mathcal{R} = 0$. But $1 \in \mathcal{R}$, so $1 + a = 0$. Thus $1 = -a \in \mathcal{I}$ and hence $\mathcal{I} = \mathcal{S}$, which is a contradiction. $\qquad \square$

12.4.5 Proof of the Kostant–Rallis Theorem

We now begin the proof of Theorem 12.4.1. We return to the notation of Section 12.4.1. Let $\mathfrak{a}$ be the complexification of $\mathfrak{a}_0$. Using the unitary trick (Section 3.3.4) we obtain a real-valued positive definite inner product $\langle \cdot \mid \cdot \rangle$ on $\mathfrak{u}$ such that

$$\langle \mathrm{Ad}(u)X \mid \mathrm{Ad}(u)Y \rangle = \langle X \mid Y \rangle \quad \text{for } u \in U \text{ and } X, Y \in \mathfrak{u}.$$

Let B denote the complex bilinear extension of $-\langle \cdot \mid \cdot \rangle$ to $\mathfrak{g}$. Then the Hermitian positive definite extension of $\langle \cdot \mid \cdot \rangle$ to $\mathfrak{g}$ is given by

$$\langle X \mid Y \rangle = -B(X, \tau Y) \quad \text{for } X, Y \in \mathfrak{g}.$$

We note that $\tau|_{V_0} = -I$. Thus the restriction of B to V_0 is real-valued and positive definite. Set $V_1 = \{X \in V : B(X, \mathfrak{a}) = 0\}$. Then

$$V = V_1 \oplus \mathfrak{a}. \tag{12.66}$$

If $f \in \mathcal{P}(V_1)$ then we extend f to V by $f(v+h) = f(v)$ for $v \in V_1$ and $h \in \mathfrak{a}$. If $f \in \mathcal{P}(\mathfrak{a})$ then we extend f to V by $f(v+h) = f(h)$ for $v \in V_1$ and $h \in \mathfrak{a}$. In other words, if $p_1 : V \to V_1$ and $p_2 : V \to \mathfrak{a}$ are the natural projections corresponding to (12.66) then we identify $f \in \mathcal{P}(V_1)$ with $p_1^* f$ and $g \in \mathcal{P}(\mathfrak{a})$ with $p_2^* g$. It follows that

(a) the map $f \otimes g \mapsto fg$ (pointwise multiplication) from $\mathcal{P}(V_1) \otimes \mathcal{P}(\mathfrak{a})$ to $\mathcal{P}(V)$ is a graded linear bijection.

Recall from Theorem 12.4.4 that there is a graded subspace $\mathcal{A}$ of $\mathcal{P}(\mathfrak{a})$ with $\dim \mathcal{A} = |W(\mathfrak{a})|$ such that $\mathcal{A}^0 = \mathbb{C}1$. Hence

(b) the map $a \otimes h \mapsto ah$ (pointwise multiplication) from $\mathcal{A} \otimes \mathcal{P}(\mathfrak{a})^{W(\mathfrak{a})}$ to $\mathcal{P}(\mathfrak{a})$ is a graded linear bijection.

We now come to the key lemma.

Lemma 12.4.11. *The map* $\mathcal{P}(V_1) \otimes \mathcal{A} \otimes \mathcal{P}(V)^K \longrightarrow \mathcal{P}(V)$ *given by* $f \otimes a \otimes g \mapsto fag$ *(pointwise multiplication) is a graded linear bijection.*

Proof. We first observe that the Chevalley restriction theorem (Theorem 12.4.5) implies that $\dim \mathcal{P}^j(V)^K = \dim \mathcal{P}^j(\mathfrak{a})^{W(\mathfrak{a})}$. Thus statements (a) and (b) imply that the dimension of the jth graded component of $\mathcal{P}(V_1) \otimes \mathcal{A} \otimes \mathcal{P}(V)^K$ is equal to $\dim \mathcal{P}^j(V)$. Since the map, say ψ, in the lemma respects the gradation, it is enough to show that it is surjective. Clearly, $\psi(1 \otimes 1 \otimes 1) = 1$. Assume that we have shown that $\mathcal{P}^j(V)$ is contained in the image of ψ for $0 \leq j \leq i$. Let $f \in \mathcal{P}^{i+1}(V)$. From statements (a) and (b) we see that

$$f = \sum_{\alpha, \beta} g_\alpha a_{\alpha\beta} h_\beta,$$

where the sum is over suitable homogeneous polynomials $g_\alpha \in \mathcal{P}(V_1)$, $a_{\alpha\beta} \in \mathcal{A}$, and $h_\beta \in \mathcal{P}(\mathfrak{a})^{W(\mathfrak{a})}$ with $\deg g_\alpha + \deg a_{\alpha\beta} + \deg h_\beta = i+1$. Suppose $\deg g_\alpha > 0$. Then

$\deg a_{\alpha\beta} + \deg h_\beta \leq i$, so the inductive hypothesis implies that $a_{\alpha\beta}h_\beta$ is in the image of ψ. Since the image of ψ is invariant under multiplication by $\mathcal{P}(V_1)$, it follows that $g_\alpha a_{\alpha\beta}h_\beta$ is in the image of ψ. We may thus assume that

$$f = \sum_\alpha a_\alpha h_\alpha \,,$$

with a_α a homogeneous element of $\mathcal{A}$, h_α a homogeneous element of $\mathcal{P}(\mathfrak{a})^{W(\mathfrak{a})}$, and $\deg a_\alpha + \deg h_\alpha = i + 1$. By the Chevalley restriction theorem, if $h \in \mathcal{P}^k(\mathfrak{a})^{W(\mathfrak{a})}$, then there exists $\varphi \in \mathcal{P}^k(V)^K$ such that $\varphi|_\mathfrak{a} = h$. Thus if $v_1,\ldots,v_d$ is a basis of V_1 and if $x_1,\ldots,x_d$ are the corresponding linear coordinates on V_1, then

$$\varphi = h + \sum_j x_j u_j \quad \text{with } u_j \in \mathcal{P}^{k-1}(V) \,.$$

We thus can write $h = \varphi - \sum x_j u_j$. Doing this for each h_α, we obtain an expansion $h_\alpha = \varphi_\alpha - \sum_j x_j u_{\alpha,j}$ with φ_α a homogeneous element of $\mathcal{P}(V)^K$ and $\deg u_{\alpha,i} = \deg h_\alpha - 1$. Hence

$$f = \sum_\alpha a_\alpha \varphi_\alpha - \sum_{j,\alpha} x_j a_\alpha u_{\alpha,j} \,. \tag{12.67}$$

Since $\deg \varphi_\alpha = \deg h_\alpha$, we have $\deg a_\alpha + \deg u_{\alpha,j} = i$. The invariance of the image of ψ under multiplication by $\mathcal{P}(V_1)$ implies that the second term in the right-hand side of (12.67) is in the image of ψ. The first term is obviously in the image. This completes the induction. $\square$

Let $g_1,\ldots,g_l$ be homogeneous, algebraically independent elements of $\mathcal{P}(\mathfrak{a})^{W(\mathfrak{a})}$ such that $\mathcal{P}(\mathfrak{a})^{W(\mathfrak{a})}$ is the polynomial algebra $\mathbb{C}[g_1,\ldots,g_l]$. Let $u_i \in \mathcal{P}(V)^K$ be such that each u_i is homogeneous and $u_i|_\mathfrak{a} = g_i$. Then $\mathcal{P}(V)^K$ is the polynomial algebra $\mathbb{C}[u_1,\ldots,u_l]$. From Proposition 12.3.2 we know that there exist elements $h \in \mathfrak{a}_0$ such that

$$\mathfrak{a} = \{X \in V : [h,X] = 0\} \,. \tag{12.68}$$

For any $X \in V$ define $\mathfrak{I}_X = \sum_i \mathcal{P}(V)(u_i - u_i(X))$.

Proposition 12.4.12. *Let $h \in \mathfrak{a}_0$ satisfy (12.68). Then $\mathfrak{I}_h$ is a prime ideal and $\mathcal{V}(\mathfrak{I}_h) = \mathrm{Ad}(K)h = \mathrm{Ad}(K^\circ)h$, where K° is the identity component of K.*

We will show that this proposition implies Theorem 12.4.1. Then we will prove the proposition using the results in Section 12.4.4.

We first show that if $\mathcal{H}$ is defined as in the statement of the Kostant–Rallis theorem, then the map $\mathcal{H} \otimes \mathcal{P}(V)^K \longrightarrow \mathcal{P}(V)$ is a linear graded bijection. To see this we note that if $\mathcal{A} \subset \mathcal{P}(\mathfrak{a})$ is the subspace in Theorem 12.4.4 and we set

$$(\mathcal{P}(V_1)\mathcal{A})^j = \sum_{i=0}^{j} \mathcal{P}^{j-i}(V_1)\mathcal{A}^i = \mathcal{P}^j(V) \cap (\mathcal{P}(V_1)\mathcal{A}) \,,$$

then $(\mathcal{P}(V_1)\mathcal{A})^j \oplus (\mathfrak{I}_0 \cap \mathcal{P}^j(V)) = \mathcal{P}^j(V)$. Thus $\dim \mathcal{H}^j = \dim(\mathcal{P}(V_1)\mathcal{A})^j$. Consequently, as in the proof of Lemma 12.4.11, it is enough to show that

$$\mathcal{H}\mathcal{P}(V)^K = \mathcal{P}(V) \,. \tag{12.69}$$

Since $1 \in \mathcal{H}$, it is clear that $\mathcal{P}^0(V) \subset \mathcal{HP}(V)^K$. Suppose that we have shown that $\mathcal{P}^i(V) \subset \mathcal{HP}(V)^K$ for $0 \leq i < j$. If $f \in \mathcal{P}^j(V)$ then $f = u + \sum f_i u_i$ with $u \in \mathcal{H}^j$ and $\deg f_i < j$. Thus the inductive hypothesis implies that $f_i u_i \in \mathcal{HP}(V)^K$. Since $u \in \mathcal{H}$, we conclude that $f \in \mathcal{HP}(V)^K$. This implies (12.69).

Part (1) of Lemma 12.4.9 implies that as a representation of K, the space $\mathrm{Gr}(\mathcal{P}(V)/\mathcal{I}_h)$ is equivalent to $\mathcal{P}(V)/\mathcal{I}_0$. This representation of K is equivalent to $\mathcal{H}$. Now Proposition 12.4.12 implies that $\mathrm{Ad}(K)h$ is an irreducible closed subset of V and $\mathcal{O}[\mathrm{Ad}(K)h] = \mathcal{P}(V)/\mathcal{I}_h$. Since $M = \{k \in K : \mathrm{Ad}(k)h = h\}$, we conclude that

$$\mathcal{O}[\mathrm{Ad}(K)h] \cong \mathrm{Ind}_M^K(1)$$

as a representation of K, proving the last assertion in Theorem 12.4.1.

We are now left with the proof of Proposition 12.4.12. We first show that $\mathcal{I}_h$ is a radical ideal. To prove this, we claim that it suffices to show the following:

$(\star)$ If $f \in \mathcal{P}(V)/\mathcal{I}_h$ and $f^2 = 0$, then $f = 0$.

Indeed, assume that $(\star)$ holds. Then given $f \in \mathcal{P}(V)/\mathcal{I}_h$ such that $f^r = 0$ for some $r > 2$, we take an integer $p > 1$ with $2^{p-1} < r \leq 2^p$. Since

$$0 = f^{2^p} = \left(f^{2^{p-1}}\right)^2,$$

it then follows from $(\star)$ that $f^{2^{p-1}} = 0$. Hence by induction on r it follows that $f = 0$, which proves that $\mathcal{I}_h$ is a radical ideal.

Now we prove $(\star)$. Recall that $\mathcal{P}(V)/\mathcal{I}_h$ is free $\mathcal{P}(V_1)$-module with generators $a_1 = 1, \ldots, a_w$, where $w = |W(\mathfrak{a})| = \dim \mathcal{A}$. Here we choose a_i to be cosets of a basis of $\mathcal{A}$ modulo $\mathcal{I}_h$. We write

$$a_i a_j = \sum_k f_{ij}^k a_k$$

with $f_{ij}^k \in \mathcal{P}(V_1)$. Since $\mathcal{P}(V)/\mathcal{I}_h$ is commutative, we have $f_{ij}^k = f_{ji}^k$. If $f \in \mathcal{P}(V)/\mathcal{I}_h$, then $f = \sum_i g_i a_i$ with $g_i \in \mathcal{P}(V_1)$. Thus

$$f^2 = \sum_k a_k \left\{ \sum_{i,j} f_{ij}^k g_i g_j \right\}.$$

For $v \in V_1$ and $k = 1, \ldots, w$ let $\varphi_i(v) \in \mathbb{C}[z_1, \ldots, z_w]$ be the quadratic polynomial given by

$$\varphi_k(v)(z_1, \ldots, z_w) = \sum_{i,j} f_{ij}^k(v) z_i z_j.$$

Set $\Phi(v) = [\varphi_1(v), \ldots, \varphi_w(v)]$. Now letting v vary, we obtain a regular map

$$\Phi : V_1 \longrightarrow V(\underbrace{2, \ldots, 2}_{w}) = \mathcal{V}$$

(using the notation of Section 12.4.4). Let Ω be the subset of all $m \in \mathcal{V}$ such that

$$\{x \in \mathbb{C}^w : m_i(x) = 0 \quad \text{for} \quad i = 1, \ldots, w\} = (0).$$

Then Ω is a Zariski-open and dense subset of $\mathcal{V}$ by Lemma 12.4.7. We assert that $\Phi(0) \in \Omega$. Indeed, let $x_1, \ldots, x_d$ be the linear coordinates on V_1 used above. Set

$$\mathcal{R} = \mathcal{P}(V)/\left(\mathfrak{I}_h + \sum_i \mathcal{P}(V)x_i\right),$$

and let $\overline{a}_i$ be the projection of a_i onto $\mathcal{R}$. Then

$$\overline{a}_i \overline{a}_j = \sum_k f_{ij}^k(0)\overline{a}_k . \tag{12.70}$$

Write $\mathcal{X}_h = \{X \in V : u_i(X) = u_i(h) \text{ for } i = 1, \ldots, w\}$ for the zero set of the ideal $\mathfrak{I}_h$. It is clear that $\mathcal{X}_h \cap \mathfrak{a} \supset W(\mathfrak{a})h$, and we know from Theorem 12.4.3 that $|W(\mathfrak{a})h| = w$. By definition, $\mathcal{O}[\mathcal{X}_h]|_{\mathcal{X}_h \cap \mathfrak{a}}$ is the algebra of regular functions on $\mathcal{X}_h \cap \mathfrak{a}$. Now $\mathcal{O}[\mathcal{X}_h]$ is a quotient algebra of $\mathcal{P}(V)/\mathfrak{I}_h$, so $\mathcal{O}[\mathcal{X}_h]|_{\mathcal{X}_h \cap \mathfrak{a}}$ is a quotient algebra of $\mathcal{R}$. But $\dim \mathcal{R} \leq w$. This implies that $\mathcal{X}_h \cap \mathfrak{a}$ is finite. Since $|\mathcal{X}_h \cap \mathfrak{a}| \geq w$, we conclude that $\dim \mathcal{R} = w$ and $\mathcal{R}$ is isomorphic to the algebra of complex-valued functions on the set $\mathcal{X}_h \cap \mathfrak{a} = W(\mathfrak{a})h$. In particular, if $m \in \mathcal{R}$ and $m^2 = 0$, then $m = 0$. We apply this with $m = \sum_i z_i \overline{a}_i$. Then

$$m^2 = \sum_i \varphi_i(0)(z)\overline{a}_i .$$

Suppose $\varphi_i(0)(z) = 0$ for all $i = 1, \ldots, w$. Then $m^2 = 0$ and hence $m = 0$. But $\{\overline{a}_1, \ldots, \overline{a}_w\}$ is linearly independent, so we conclude that $z = 0$. This proves the assertion that $\Phi(0) \in \Omega$.

We therefore see that the set of all $v \in V_1$ such that $\Phi(v) \in \Omega$ is a Zariski-open and nonempty subset Ω_1 of V_1 containing 0. Now suppose that $f \in \mathcal{P}(V)/\mathfrak{I}_h$ and $f^2 = 0$. Then $f = \sum_i g_i a_i$ as above, and

$$\varphi_k(v)(g_1(v), \ldots, g_w(v)) = 0 \quad \text{for } k = 1, \ldots, w \text{ and all } v \in V_1 .$$

Thus if $v \in \Omega_1$, then $g_i(v) = 0$ for $i = 1, \ldots, w$. Since Ω_1 is Zariski dense in V_1, we have $g_i = 0$ for $i = 1, \ldots, w$. Hence $f = 0$.

We have thus proved that $\mathfrak{I}_h$ is a radical ideal. We will now complete the proof of the proposition. For $x \in V$ write

$$\det(tI - \operatorname{ad}x) = t^p D(x) + \text{terms of higher order in } t ,$$

where D is a nonzero polynomial on V. Then clearly $D \in \mathcal{P}(V)^K$. Now $\operatorname{Ad}(K)\mathfrak{a}_0'$ is Zariski open in V_0, so $D|_{\mathfrak{a}_0'} \neq 0$. Hence Proposition 12.3.2 implies that $p = \dim \mathfrak{a} + \dim M$. Thus $D(h) \neq 0$, and so

$$D(x) = D(h) \neq 0 \quad \text{for all } x \in \mathcal{X}_h .$$

We know from Section 12.3.1 that

$$\det(tI - (\operatorname{ad}x)^2|_V) = t^l \delta(x) + \text{terms of higher order in } t .$$

Thus $\delta \in \mathcal{P}(V)^K$ and $\delta(h) \neq 0$. Hence $\delta(x) \neq 0$ for $x \in \mathcal{X}_h$. Proposition 12.3.2 implies that if $x \in \mathcal{X}_h$, then x is semisimple. Thus $\dim \operatorname{Ker}(\operatorname{ad}x|_V) = l$. Hence

$$\dim \mathrm{Ker}(\mathrm{ad}\,x|_{\mathfrak{k}}) = \dim M .$$

Let $K_x^{\circ} = \{k \in K^{\circ} : \mathrm{Ad}(k)x = x\}$. Then $\mathrm{Lie}(K_x^{\circ}) = \mathrm{Ker}(\mathrm{ad}\,x|_{\mathfrak{k}})$. Thus

$$\dim K^{\circ}x = \dim K - \dim M = \dim V_1 .$$

This implies that $K^{\circ}x$ has a nonempty interior in $\mathfrak{X}_h$ relative to the Zariski topology (see Theorem A.1.19). Thus $K^{\circ}x$ is Zariski open in $\mathfrak{X}_h$, since $K^{\circ}x$ is homogeneous. Since K° is connected, we see that $K^{\circ}x$ is irreducible and open. The Zariski closure of $K^{\circ}x$ in $\mathfrak{X}_h$ must be a union of sets of the form $K^{\circ}y$ with $y \in \mathfrak{X}_h$. Furthermore, each of these sets is open and irreducible, and distinct orbits are disjoint.

Assume for the sake of contradiction that $\mathfrak{X}_h \neq K^{\circ}h$. Then from the remarks just made we would have

$$\mathfrak{X}_h = K^{\circ}h \cup K^{\circ}y_1 \cup \cdots \cup K^{\circ}y_q , \tag{12.71}$$

a disjoint union of Zariski-open and Zariski-closed subsets of $\mathfrak{X}_h$ with $q > 0$. We will now show that this leads to a contradiction. The decomposition (12.71) implies that

$$\mathcal{O}[\mathfrak{X}_h] \cong \mathcal{O}[K^{\circ}h] \oplus \mathcal{O}[K^{\circ}y_1] \oplus \cdots \oplus \mathcal{O}[K^{\circ}y_q] \tag{12.72}$$

as a $\mathbb{C}[x_1,\ldots,x_d] = \mathcal{P}(V_1)$-module. Now

$$\mathcal{O}[K^{\circ}h] / \big(\textstyle\sum_{i=1}^d x_i \mathcal{O}[K^{\circ}h] \big) = \mathcal{O}[K^{\circ}h \cap \mathfrak{a}]$$

(as above). Also, $K_0^{\circ}h \supset W(\mathfrak{a})h$ (Theorem 12.4.3). Thus

$$\dim \big(\mathcal{O}[K^{\circ}h] / \textstyle\sum_{i=1}^d x_i \mathcal{O}[K^{\circ}h] \big) \geq w .$$

We have already shown that $\dim \big(\mathcal{O}[\mathfrak{X}_h] / \sum_{i=1}^d x_i \mathcal{O}[\mathfrak{X}_h] \big) = w$. Hence (12.72) implies that

$$\mathcal{O}[K^{\circ}y_j] / \big(\textstyle\sum_{i=1}^d x_i \mathcal{O}[K^{\circ}y_j] \big) = 0 \quad \text{for } j = 1,\ldots,q .$$

We leave it to the reader to check that if $\mathcal{S} = \mathcal{P}(V_1)|_{K^{\circ}y_j}$, then

$$\textstyle\sum_{i=1}^d x_i \mathcal{P}(V_1)|_{K^{\circ}y_j} \neq \mathcal{S} .$$

(HINT: Since $\dim K^{\circ}y_j = \dim V_1$, the restrictions of $x_1,\ldots,x_d$ to $K^{\circ}y_j$ are an algebraically independent set of functions.) Now Lemma 12.4.10 implies that $\mathcal{O}[K^{\circ}y_j] = 0$ for $j = 1,\ldots,q$. This contradiction implies that $\mathfrak{X}_h = K^{\circ}h$. The proof of Proposition 12.4.12, and hence of the Kostant–Rallis theorem, is now complete. $\qquad \square$

12.4.6 Some Remarks on the Proof

Let K be a connected reductive linear algebraic group and let (σ, V) be a regular representation of K. We will now isolate the actual properties of the representations that came into our proof of the Kostant–Rallis theorem. When (σ, V) is the linear isotropy representation for a symmetric space, then there is a subspace $\mathfrak{a}$ in V satisfying the following conditions:

1. The restriction $f \mapsto f|_{\mathfrak{a}}$ defines an isomorphism of $\mathcal{P}(V)^K$ onto a subalgebra $\mathcal{R}$ of $\mathcal{P}(\mathfrak{a})$.
2. The subalgebra $\mathcal{R}$ of $\mathcal{P}(\mathfrak{a})$ is generated by algebraically independent homogeneous elements $u_1, \ldots, u_l$ with $l = \dim \mathfrak{a}$. Furthermore, there exists a graded subspace $\mathcal{A}$ of $\mathcal{P}(\mathfrak{a})$ such that the map $\mathcal{A} \otimes \mathcal{R} \to \mathcal{P}(\mathfrak{a})$ given by $a \otimes r \mapsto ar$ is a linear bijection.
3. There exists $h \in \mathfrak{a}$ such that $|\sigma(K)h \cap \mathfrak{a}| \geq \dim \mathcal{A}$.
4. Let h be as in (3) and set $\mathcal{X}_h = \{v \in V : f(v) = f(h) \text{ for all } f \in \mathcal{P}(V)^K\}$. If $v \in \mathcal{X}_h$ then $\dim Kv = \dim V - \dim \mathfrak{a}$.

Now assume that (σ, V) is any representation of K that satisfies (1)–(4). Set $M = \{k \in K : \sigma(k)h = h\}$. Our proof of Theorem 12.4.1 then applies and we obtain the following result:

Theorem 12.4.13. *Let F be any regular representation of K. Then $\mathrm{Hom}_K(F, \mathcal{P}(V))$ is a free $\mathcal{P}(V)^K$-module with $\dim F^M$ generators.*

Here are some examples that are not linear isotropy representations for symmetric spaces but that nevertheless satisfy conditions (1)–(4). The details will be left as exercises.

1. Let $K = \mathbf{SL}(2, \mathbb{C})$ and let (σ, V) be the representation of K on $S^3(\mathbb{C}^2)$ (i.e., the irreducible four-dimensional representation). Then $\mathcal{P}(V)^K = \mathbb{C}[f]$ with f homogeneous of degree 4 (see Exercises 12.4.7 #17). Let e_1, e_2 be the usual basis of $\mathbb{C}^2$ and let $h = e_1^3 + e_2^3$. If $u = \begin{bmatrix} 0 & i \\ i & 0 \end{bmatrix}$ then $\sigma(u)h = ih$. Set $\mathfrak{a} = \mathbb{C}h$. Thus $\sigma(K)h \cap \mathfrak{a} \supset \{h, -h, ih, -ih\}$. One has $f(h) \neq 0$ and

$$
M = \left\{ \begin{bmatrix} \xi & 0 \\ 0 & \xi^{-1} \end{bmatrix} : \xi^3 = 1 \right\} \cong \mathbb{Z}/3\mathbb{Z}.
$$

We look upon $\mathcal{P}(\mathfrak{a})$ as $\mathbb{C}[t]$. Assuming that $f(h) = 1$, we then have $\mathrm{res}(\mathcal{P}(V)^K) = \mathbb{C}[t^4]$. Take

$$
\mathcal{A} = \mathbb{C}1 \oplus \mathbb{C}t \oplus \mathbb{C}t^2 \oplus \mathbb{C}t^3.
$$

Thus all conditions but (4) have been verified. Condition (4) follows, since f is irreducible and hence $\mathcal{X}_h$ is irreducible. We can thus apply Theorem 12.4.13: if $F^{(k)}$ is the irreducible $(k+1)$-dimensional regular representation of K, then $\mathrm{Hom}_K(F^{(k)}, \mathcal{P}(V))$ is a free $\mathbb{C}[f]$-module on d_k generators, where d_k is the dimension of the space of M-fixed vectors in $F^{(k)}$. These dimensions are

$$
d_{6i+2j} = 2i+1 \quad \text{for } i = 0,1,2,3,\ldots \text{ and } j = 0,1,2,
$$
$$
d_{6i+3+2j} = 2i+2 \quad \text{for } i = 0,1,2,3,\ldots \text{ and } j = 0,1,2.
$$

2. Let $K = \mathbf{Sp}(3, \mathbb{C})$ and let $V \subset \bigwedge^3 \mathbb{C}^6$ be the irreducible K-submodule with highest weight $\varepsilon_1 + \varepsilon_2 + \varepsilon_3$ (see Section 10.2.3). Then $\mathcal{P}(V)^K = \mathbb{C}[f]$ with f an irreducible homogeneous polynomial of degree 4. Let $h = e_1 \wedge e_2 \wedge e_3 + e_4 \wedge e_5 \wedge e_6$. Then $f(h) \neq 0$, so we may normalize f by $f(h) = 1$. Let $\mathfrak{a} = \mathbb{C}h$. Set $u = \begin{bmatrix} 0 & iI_3 \\ iI_3 & 0 \end{bmatrix}$. Then $\sigma(u)h = -ih$. Thus the conditions are satisfied as in Example 1. In this case M is the group of all matrices

$$k = \begin{bmatrix} b & 0 \\ 0 & (b^t)^{-1} \end{bmatrix}, \quad b \in \mathbf{SL}(3, \mathbb{C}).$$

3. Let $K = \mathbf{SL}(6, \mathbb{C})$ and let $V = \bigwedge^3 \mathbb{C}^6$. As in Examples 1 and 2, one has $\mathcal{P}(V)^K = \mathbb{C}[f]$ with f homogeneous of degree 4. We take h and u as in Example 2. Then the conditions (1)–(4) are satisfied and M is the group of all matrices $\begin{bmatrix} b_1 & 0 \\ 0 & b_2 \end{bmatrix}$ with $b_1, b_2 \in \mathbf{SL}(3, \mathbb{C})$.

12.4.7 Exercises

In Exercises 1–3 the notation follows that of Section 12.3.2.

1. For the symmetric space of type DIII:
   (a) Verify that the restricted root system Σ is of type C_p or BC_p (when $l = 2p$ or $l = 2p + 1$, respectively).
   (b) Calculate the multiplicities of the restricted roots.
   (c) Prove that the polynomials $h \mapsto \mathrm{tr}(h^{2r})$ for $r = 1, \ldots, p$ are algebraically independent and generate $\mathcal{P}(\mathfrak{a})^{W(\mathfrak{a})}$.
2. For the symmetric space of type BDI:
   (a) Verify that the restricted root system Σ is of type B_p or D_p (when $p < q$ or $p = q$, respectively).
   (b) Calculate the multiplicities of the restricted roots.
   (c) For the case $p = q$, prove that the polynomials $h \mapsto \mathrm{tr}(h^{2r})$ for $r = 1, \ldots, p - 1$ together with the polynomial $h \mapsto x_1 \cdots x_p$ are algebraically independent and generate $\mathcal{P}(\mathfrak{a})^{W(\mathfrak{a})}$ (see Exercises 5.1.3).
3. For the symmetric space of type CII:
   (a) Verify that the restricted root system Σ is of type BC_p or C_p (when $p < q$ or $p = q$, respectively).
   (b) Calculate the multiplicities of the restricted roots.

In Exercises 4–11 the notation follows that of Section 12.4.3.

4. For the symmetric space of type AI, prove that (σ, V) is the irreducible representation of highest weight $2\varpi_1$ (see Sections 10.2.4 and 10.2.5).
5. For the symmetric space of type AII:
   (a) Verify the block matrix description (12.56) of V.
   (b) Prove that the representation σ of K on V is irreducible with highest weight $\varpi_2 = \varepsilon_1 + \varepsilon_2$ using the results in Section 10.2.3. (HINT: Note that for $X \in V$

one has $XJ = \begin{bmatrix} -B & A \\ -A^t & C \end{bmatrix}$. Hence the map $X \mapsto XJ = \widetilde{X}$ carries V to a subspace $\widetilde{V}$ of the space $A_{2n}(\mathbb{C})$ of skew-symmetric $2n \times 2n$ matrices. Show that $A_{2n}(\mathbb{C}) = \widetilde{V} \oplus \mathbb{C}J$ as a K-module and that the space $\widetilde{V}$ corresponds to the J-*harmonic, skew-symmetric* 2-tensors.)

6. Consider the symmetric space of type AIII:
   (a) Verify that V consists of matrices (12.57).
   (b) Verify that the algebra $\mathfrak{a}$ given by (12.58) is conjugate in G to the algebra used in Section 12.3.2.

7. Consider the symmetric space of type CI:
   (a) Verify that V consists of matrices (12.59).
   (b) Verify that the algebra $\mathfrak{a}$ given by (12.60) is conjugate in G to the algebra used in Section 12.3.2.

8. Consider the symmetric space of type DIII:
   (a) Verify that V consists of matrices (12.61).
   (b) Verify that the subalgebra $\mathfrak{a}$ defined in (12.62) is conjugate in G to the subalgebra used in Section 12.3.2.

9. Consider the symmetric space of type BDI:
   (a) Verify that V consists of matrices (12.63).
   (b) Verify that the subalgebra $\mathfrak{a}$ defined in (12.48) is conjugate in G to the subalgebra used in Section 12.3.2.

10. For the symmetric space of type CII:
    (a) Verify that V is given by (12.64).
    (b) Verify that the subalgebra $\mathfrak{a}$ defined in (12.65) is conjugate in G to the subalgebra used in Section 12.3.2.

11. Verify that all the examples in Section 12.4.3 fit one of these descriptions:
    (a) $\mathfrak{k}$ is semisimple and (σ, V) is irreducible.
    (b) $\mathfrak{k} = [\mathfrak{k}, \mathfrak{k}] \oplus \mathfrak{z}$, where the center $\mathfrak{z}$ of $\mathfrak{k}$ is one-dimensional, and $V \cong F \oplus F^*$ with F an irreducible representation of $[\mathfrak{k}, \mathfrak{k}]$.

12. Let $G = \mathbf{SL}(2, \mathbb{C})$ and let $V = S^2(\mathbb{C}^2)$.
    (a) Show that $\mathcal{P}(V)^G$ is a polynomial algebra in one generator u of degree 2.
    (b) Show that $\mathcal{P}(V)/\mathcal{P}(V)u$ is isomorphic to $\mathrm{Ind}_T^G(1) \cong \bigoplus_{n \geq 0} F^{2k}$, where F^k is the irreducible $(k+1)$-dimensional representation of G and T is the group of diagonal matrices of G.

13. Let $G = \mathbf{SL}(2, \mathbb{C})$ and let $V = S^4(\mathbb{C}^2)$.
    (a) Show that $\mathcal{P}(V)^G$ is a polynomial algebra in two generators of degrees 2 and 3.
    (b) Show that as a representation of G, $\mathcal{P}(V)/\mathcal{P}(V)\mathcal{P}(V)_+^G$ is equivalent to $\mathrm{Ind}_F^G(1)$, where F is the finite subgroup of G defined as follows: Let $\pi : G \to \mathbf{SO}(3, \mathbb{C})$ be defined as in Section 2.2.2. Then $F = \pi^{-1}(F_1)$, where F_1 is the group of diagonal matrices in $\mathbf{SO}(3, \mathbb{C})$. Describe F.

14. Prove Lemma 12.4.6. (HINT: For the implication $(1) \implies (2)$, use Corollary A.1.5 relative to the generators of $\mathcal{J}$ and the coordinate functions on $\mathbb{C}^n$. Proving the opposite implication is easy.)

15. Prove Lemma 12.4.8.

16. Prove Lemma 12.4.9. (HINT: For part (1), use Lemma 12.4.8. Part (2) is most easily proved using the *Hilbert polynomial* of an ideal; see Cox, Little, and O'Shea [40, Chapter 9].)

17. Show that the invariant of degree 4 in Example 1 of Section 12.4.6 can be given as follows: If $u = ae_1^3 + be_1^2 e_2 + ce_1 e_2^2 + de_2^3$ then

$$f(u) = \det \begin{bmatrix} a & b & c & d \\ b & 2c & 3d & 0 \\ 0 & 3a & 2b & c \\ 3a & b & -c & -3d \end{bmatrix}.$$

Use this to show that condition (4) of Section 12.4.6 is satisfied in this case.

18. Describe the degree-4 invariant in Examples 2 and 3 of Section 12.4.6 and carry out the decomposition of $\mathcal{P}(V)$.

12.5 Notes

Section 12.1.3. Theorem 12.1.6 is an algebraic version of the *Borel–Weil theorem*; for the analytic version in terms of holomorphic vector bundles, see Wallach [153, Chapter 6, §6.3]. The functions in $\mathcal{R}_\lambda$ are uniquely determined by their restrictions to the unipotent group N^+. The finite-dimensional space $\mathcal{V}_\lambda \subset \mathcal{O}[N^+]$ of restrictions of functions in $\mathcal{R}_\lambda$ is characterized as the solution space of a system of differential equations (called an *indicator system*) in Želobenko [171, Chapter XVI].

Section 12.2.1. The open B-orbit condition in Theorem 12.2.1 is also a necessary condition for X to be multiplicity-free. This follows easily from the result of Rosenlicht [126] that if B does not have an open orbit on X then there exists a nonconstant B-invariant rational function on X (see Vinberg and Kimelfeld [147]). Likewise, the condition in Theorem 12.2.5 that K have an open orbit on G/B is also a necessary condition for (G, K) to be a spherical pair when K is reductive. For this, one also uses the fact that the quotient $X = G/K$ is an affine algebraic set and $\mathcal{R}(G/K)$ is naturally identified with $\mathcal{O}[X]$ (see Matsushima [109] and Borel and Harish-Chandra [18, Theorem 3.5]). Thus by Proposition 12.2.4, if (G, K) is a spherical pair then G/K is multiplicity-free as a G-space. By the converse to Theorem 12.2.1 this implies that B has an open orbit on G/K, which means that $Kx_0 B$ is open in G for some $x_0 \in G$.

The spherical pairs (G, K) with G connected and K reductive have been classified in Krämer [93]. The term *spherical subgroup* is also applied to any algebraic subgroup L (not necessarily reductive) such that L has an open orbit on G/B. Vinberg and Kimelfeld [147, Theorem 1] show that this orbit condition is necessary and sufficient for the representation $\mathrm{Ind}_L^G(\chi)$ to be multiplicity-free for all regular characters χ of L. For example, any subgroup containing the nilradical of a Borel subgroup is spherical in this sense. (Such subgroups are called *horospherical*.) See Brion [23] for a survey of results in this more general context.

Sections 12.2.2 and 12.2.3. Irreducible linear multiplicity-free actions were classified by Kac [82]. The classification of general linear multiplicity-free actions was done (independently) by Benson and Ratcliff [6] and Leahy [99]. Theorems 12.2.6 and 12.2.9 are from Howe [72]. For further examples see Howe and Umeda [75] and the survey of Benson and Ratcliff [7].

Section 12.2.4. The term *second fundamental theorem* comes from Weyl [164]. See Vust [151], Schwarz [133], Howe [72], and Tan–Zhu [141] for more recent developments.

Section 12.3.1. The Iwasawa decomposition for a general complex reductive group is obtained in Vust [150]. Lemma 12.3.7 is adapted from DeConcini and Procesi [42].

Section 12.3.3. The decomposition into G-irreducible subspaces of the regular functions on the space G/K was first treated in Cartan [28]. Cartan considered a compact real form of G/K; since representative functions on the compact form extend holomorphically to K-invariant functions on G, this is an equivalent formulation of the decomposition problem. Theorem 12.3.13 was proved by Helgason ([65]; see [67, Chapter V, §4.1]). For a proof using algebraic geometry instead of integration over the compact real form of K, see Vust [150].

Section 12.4.1. Theorem 12.4.1 appears in Kostant and Rallis [90]. It has many important applications to the representation theory of real reductive groups (see Kostant [89]). Theorem 12.4.2 on the adjoint representation appears in Kostant [88]. Similar results for the representation of K on $\mathcal{O}[G/K]$ have been obtained by Richardson [125].

Section 12.4.2. Vinberg [146] has extended these results to the setting of a graded Lie algebra such that the corresponding Weyl group is a complex reflection group.

Section 12.4.3. The examples are labeled according to Cartan's classification of symmetric spaces (see Helgason [66, Chapter X, §6, Table V]. For all the cases, in particular when G is exceptional, the Lie algebra $\mathfrak{m}$ of M and the restricted root system can be read off from the Satake diagram; see Araki [3] and Helgason [66, Chapter X, Exercises, Table VI]. The isotropy representations of K on V are obtained in all cases in Wolf [168, §8.11].

Section 12.4.5. For the last part of the proof of Proposition 12.4.12 we could also use the argument that each of the orbits $\mathrm{Ad}(K^{\circ})y_j$ is closed and that in the set $\mathfrak{X}_h$ there is exactly one closed orbit.

Section 12.4.6. Theorem 12.4.13 is related to results of Vinberg [146] and Schwarz [132]. The classification of representations with free modules of covariants is treated in Popov [120, Chapter 5]. Example 1 and Exercises 12.4.7 #12 and #13 give a complete analysis of $\mathcal{P}(V)$ as an $\mathbf{SL}(2,\mathbb{C})$-module for $V \cong S^j(\mathbb{C}^2)$ and $j = 2, 3, 4$. The decomposition for $V \cong \mathbb{C}^2$ has been done in Section 2.3.2. The reader should try to see what can be said about the case of $V = S^5(\mathbb{C}^2)$.

Appendix A
Algebraic Geometry

Abstract We develop the aspects of algebraic geometry needed for the study of algebraic groups over $\mathbb{C}$ in this book. Although we give self-contained proofs of almost all of the results stated, we do not attempt to give an introduction to the field of algebraic geometry or to give motivating examples. We refer the interested reader to Cox, Little, and O'Shea [40], Harris [61], Shafarevich [134], and Zariski and Samuel [170] for more details.

A.1 Affine Algebraic Sets

The basic object in algebraic geometry over $\mathbb{C}$ is a subset of $\mathbb{C}^n$ defined by a finite number of polynomial equations. The geometry of such an *affine algebraic set* is reflected in the algebraic structure of its ring of *regular functions*.

A.1.1 Basic Properties

Let V be a finite-dimensional vector space over $\mathbb{C}$. A complex-valued function f on V is a polynomial of degree $\leq k$ if for some basis $\{e_1, \ldots, e_n\}$ of V one has

$$f\left(\sum_{i=1}^n x_i e_i\right) = \sum_{|I| \leq k} a_I x^I .$$

Here for a multi-index $I = (i_1, \ldots, i_n) \in \mathbb{N}^n$ we write $x^I = x_1^{i_1} \cdots x_n^{i_n}$. This definition is obviously independent of the choice of basis for V. If there exists a multi-index I with $|I| = k$ and $a_I \neq 0$, then we say that f has degree k. If $a_I = 0$ when $|I| \neq k$, then we say that f is *homogeneous* of degree k. Let $\mathcal{P}(V)$ be the set of all polynomials on V, $\mathcal{P}_k(V)$ the polynomials of degree $\leq k$, and $\mathcal{P}^k(V)$ the homogeneous polynomials of degree k. Then $\mathcal{P}(V)$ is a commutative algebra, relative to pointwise multiplication of functions. It is freely generated as an algebra by the linear coordinate functions $x_1,\ldots,x_n$. A choice of a basis for V thus gives rise to an algebra isomorphism $\mathcal{P}(V) \cong \mathbb{C}[x_1,\ldots,x_n]$, the polynomial ring in n variables.

Definition A.1.1. A subset $X \subset V$ is an *affine algebraic set* if there exist functions $f_j \in \mathcal{P}(V)$ such that $X = \{v \in V : f_j(v) = 0 \text{ for } j = 1,\ldots,m\}$.

When X is an affine algebraic set, we define the *affine ring* $\mathcal{O}[X]$ of X to be the functions on X that are restrictions of polynomials on V:

$$\mathcal{O}[X] = \{f|_X : f \in \mathcal{P}(V)\} \,.$$

We call these functions the *regular functions* on X. Define

$$\mathcal{I}_X = \{f \in \mathcal{P}(V) : f|_X = 0\} \,.$$

Then $\mathcal{I}_X$ is an ideal in $\mathcal{P}(V)$, and the map $f + \mathcal{I}_X \mapsto f|_X$ defines an algebra isomorphism $\mathcal{P}(V)/\mathcal{I}_X \cong \mathcal{O}[X]$.

Theorem A.1.2 (Hilbert basis theorem). *Let $\mathcal{I} \subset \mathcal{P}(V)$ be an ideal. Then there is a finite set of polynomials $\{f_1,\ldots,f_d\} \subset \mathcal{I}$ such that every $g \in \mathcal{I}$ can be written as $g = g_1 f_1 + \cdots + g_d f_d$ with $g_i \in \mathcal{P}(V)$.*

Proof. Let $\dim V = n$. Then $\mathcal{P}(V) \cong \mathbb{C}[x_1,\ldots,x_n]$. Since a polynomial in $x_1,\ldots,x_n$ with coefficients in $\mathbb{C}$ can be written uniquely as a polynomial in x_n with coefficients that are polynomials in $x_1,\ldots,x_{n-1}$, there is a ring isomorphism

$$\mathcal{P}(V) \cong \mathcal{R}[x_n], \qquad \text{with } \mathcal{R} = \mathbb{C}[x_1,\ldots,x_{n-1}] \,. \tag{A.1}$$

We call an arbitrary commutative ring $\mathcal{R}$ *Noetherian* if every ideal in $\mathcal{R}$ is finitely generated. For example, the field $\mathbb{C}$ is Noetherian, since its only ideals are $\{0\}$ and $\mathbb{C}$. To prove the theorem, we see from (A.1) that it suffices to prove the following:

(N)   If a ring $\mathcal{R}$ is Noetherian, then the polynomial ring $\mathcal{R}[x]$ is Noetherian

We first show that the Noetherian property for a ring $\mathcal{R}$ is equivalent to the following *ascending chain condition* for ideals in $\mathcal{R}$:

(ACC)   If $\mathcal{I}_1 \subset \mathcal{I}_2 \subset \cdots \subset \mathcal{R}$ is an ascending chain of ideals in $\mathcal{R}$, then there exists an index p such that $\mathcal{I}_j = \mathcal{I}_p$ for all $j \geq p$.

To prove the equivalence of these two conditions, suppose first that $\mathcal{R}$ satisfies (ACC). Given an ideal $\mathcal{I} \subset \mathcal{R}$, we take $f_1 \in \mathcal{I}$ and set $I_1 = \mathcal{R}f_1$. If $\mathcal{I}_1 \neq \mathcal{I}$, we take $f_2 \in \mathcal{I}$ with $f_2 \notin \mathcal{I}_1$ and set $I_2 = \mathcal{R}f_1 + \mathcal{R}f_2$. Then $\mathcal{I}_1 \subset \mathcal{I}_2 \subset \mathcal{I}$. Continuing in this way, we obtain an ascending chain $\mathcal{I}_1, \mathcal{I}_2,\ldots$ of finitely generated ideals in $\mathcal{I}$. By (ACC) there is an index p with $\mathcal{I}_p = \mathcal{I}$, and hence $\mathcal{I}$ is finitely generated. Conversely, suppose $\mathcal{R}$ is Noetherian. Given an ascending chain $\mathcal{I}_1 \subset \mathcal{I}_2 \subset \cdots$ of ideals in $\mathcal{R}$, set $\mathcal{I} = \bigcup_i \mathcal{I}_i$. Then $\mathcal{I}$ is an ideal in $\mathcal{R}$ and hence has a finite set of generators $f_1,\ldots,f_d$ by the Noetherian property. But there exists an index p such that $f_j \in \mathcal{I}_p$ for all j. Thus $\mathcal{I} = \mathcal{I}_p$, so $\mathcal{I}_j = \mathcal{I}_p$ for all $j \geq p$. Hence $\mathcal{R}$ satisfies (ACC).

We now prove (N). Assume that $\mathcal{R}$ is a Noetherian ring and let $\mathcal{I} \subset \mathcal{R}[x]$ be an ideal. Choose a nonzero polynomial $f_1(x) \in \mathcal{I}$ of minimum degree and form the ideal $\mathcal{I}_1 = \mathcal{R}[x] f_1(x) \subset \mathcal{I}$. If $\mathcal{I}_1 = \mathcal{I}$, we are done. If $\mathcal{I}_1 \neq \mathcal{I}$, take $f_2(x)$ as a nonzero polynomial of minimum degree among all elements of $\mathcal{I} \setminus \mathcal{I}_1$ and set $\mathcal{I}_2 = \mathcal{R}[x] f_1(x) + \mathcal{R}[x] f_2(x) \subset \mathcal{I}$. If $\mathcal{I} = \mathcal{I}_2$ we are done. Otherwise, we continue this process of choosing $f_j(x)$ and forming the ideals $\mathcal{I}_j = \mathcal{R}[x] f_1(x) + \cdots + \mathcal{R}[x] f_j(x)$. As long as $\mathcal{I}_j \neq \mathcal{I}$, we can choose $f_{j+1}(x)$ as a nonzero polynomial of minimum degree among all elements of $\mathcal{I} \setminus \mathcal{I}_j$. If $\mathcal{I}_d = \mathcal{I}$ for some d, then $f_1(x), \ldots, f_d(x)$ is a finite set of generators for $\mathcal{I}$ and we are done. We assume, for the sake of obtaining a contradiction, that $\mathcal{I}_j \neq \mathcal{I}$ for all j.

If $f(x) = c_m x^m + \cdots + c_1 x + c_0 \in \mathcal{R}[x]$ and $c_m \neq 0$, we call $c_m x^m$ the *initial term* and c_m the *initial coefficient* of $f(x)$. Let $a_j \in \mathcal{R}$ be the initial coefficient of $f_j(x)$ and let $\mathcal{J} \subset \mathcal{R}$ be the ideal generated by the set $\{a_1, a_2, \ldots\}$ of all initial coefficients. Since $\mathcal{R}$ is Noetherian, there is an integer m so that $\mathcal{J}$ is generated by $a_1, \ldots, a_m$. In particular, there are elements $u_j \in \mathcal{R}$ such that $a_{m+1} = \sum_{j=1}^{m} u_j a_j$. Let d_j be the degree of $f_j(x)$. By the choice of $f_j(x)$ we have $d_j \leq d_k$ for all $j < k$, since $f_k(x) \in \mathcal{I} \setminus \mathcal{I}_{j-1}$ and $f_j(x)$ has minimum degree among all nonzero elements in $\mathcal{I} \setminus \mathcal{I}_{j-1}$. Consider

$$g(x) = \sum_{j=1}^{m} u_j f_j(x) x^{d_{m+1} - d_j} .$$

The initial coefficient of $g(x)$ is a_{m+1}, and $g(x) \in \mathcal{I}_m$. Hence the polynomial $h(x) = f_{m+1}(x) - g(x)$ has strictly smaller degree than $f_{m+1}(x)$. Also, $h(x) \neq 0$, since $f_{m+1}(x) \notin \mathcal{I}_m$. But such a polynomial $h(x)$ cannot exist, by the choice of $f_{m+1}(x)$. This contradiction proves (N). $\square$

Suppose $\mathcal{A} \subset \mathbb{C}[x_1, \ldots, x_n]$ is any collection of polynomials. Let

$$X = \{x \in \mathbb{C}^n : f(x) = 0 \quad \text{for all } f \in \mathcal{A}\} ,$$

and let $\mathcal{I}_X \subset \mathbb{C}[x_1, \ldots, x_n]$ be the set of all polynomials that vanish on X. Then $\mathcal{A} \subset \mathcal{I}_X$ and $\mathcal{I}_X$ is an ideal. By the Hilbert basis theorem there are polynomials $f_1, \ldots, f_d$ that generate $\mathcal{I}_X$. Hence

$$X = \{x \in \mathbb{C}^n : f_j(x) = 0 \quad \text{for } j = 1, \ldots, d\} .$$

Thus it is no loss of generality to require that algebraic sets be defined by a finite number of polynomial equations.

Let $a \in X$. Then

$$\mathfrak{m}_a = \{f \in \mathcal{O}[X] : f(a) = 0\}$$

is a maximal ideal in $\mathcal{O}[X]$, since $f - f(a) \in \mathfrak{m}_a$ for all $f \in \mathcal{O}[X]$ and hence $\dim \mathcal{O}[X]/\mathfrak{m}_a = 1$. We will show that every maximal ideal is of this form. This basic result links algebraic properties of the ring $\mathcal{O}[X]$ to geometric properties of the algebraic set X.

Theorem A.1.3 (Hilbert Nullstellensatz). *Let X be an affine algebraic set. If $\mathfrak{m}$ is a maximal ideal in $\mathcal{O}[X]$, then there is a unique point $a \in X$ such that $\mathfrak{m} = \mathfrak{m}_a$.*

Proof. Consider first the case $X = \mathbb{C}^n$. Define a representation ρ of the algebra $\mathcal{A} = \mathbb{C}[z_1, \ldots, z_n]$ on the vector space $V = \mathbb{C}[z_1, \ldots, z_n]/\mathfrak{m}$ by multiplication:

$$\rho(f)(g + \mathfrak{m}) = fg + \mathfrak{m} \quad \text{for } f, g \in \mathcal{A}.$$

The representation (ρ, V) is irreducible, since $\mathfrak{m}$ is a maximal ideal. Since $\mathcal{A}$ is commutative, we have $\mathcal{A} \subset \mathrm{End}_{\mathcal{A}}(V)$. Hence Schur's lemma (Lemma 4.1.4) implies that $\rho(f)$ is a scalar multiple of the identity. In particular, there exist $a_i \in \mathbb{C}$ such that $z_i - a_i \in \mathfrak{m}$ for $i = 1, \ldots, n$. Set $a = (a_1, \ldots, a_n) \in \mathbb{C}^n$. Then it follows that $f - f(a) \in \mathfrak{m}$ for all $f \in \mathbb{C}[z_1, \ldots, z_n]$. Hence $\mathfrak{m} = \mathfrak{m}_a$.

Now consider an arbitrary algebraic set $X \subset \mathbb{C}^n$ and maximal ideal $\mathfrak{m} \subset \mathcal{O}[X]$. Let $\mathfrak{m}'$ be the inverse image of $\mathfrak{m}$ under the canonical restriction map $f \mapsto f|_X$ from $\mathbb{C}[z_1, \ldots, z_n]$ onto $\mathcal{O}[X]$. Then the quotient rings $\mathcal{O}[X]/\mathfrak{m}$ and $\mathbb{C}[z_1, \ldots, z_n]/\mathfrak{m}'$ are naturally isomorphic. Since $\mathfrak{m}$ is a maximal ideal, these rings are fields, and hence $\mathfrak{m}'$ is a maximal ideal in $\mathbb{C}[z_1, \ldots, z_n]$. By the result just proved there exists $a \in \mathbb{C}^n$ such that $f - f(a) \in \mathfrak{m}'$ for all $f \in \mathbb{C}[z_1, \ldots, z_n]$. Hence $f|_X - f(a) \in \mathfrak{m}$ for all f. If $f \in \mathfrak{I}_X$, then $f|_X = 0$, and so we have $f(a) \in \mathfrak{m}$. But $\mathfrak{m}$ is a proper ideal; thus we conclude that $f(a) = 0$ for all $f \in \mathfrak{I}_X$. Therefore $a \in X$ and $\mathfrak{m} = \mathfrak{m}_a$. $\qquad\square$

If A is an algebra with 1 over $\mathbb{C}$, then $\mathrm{Hom}(A, \mathbb{C})$ is the set of all linear maps $\varphi : A \longrightarrow \mathbb{C}$ such that $\varphi(1) = 1$ and $\varphi(a'a'') = \varphi(a')\varphi(a'')$ for all $a', a'' \in A$ (the *multiplicative linear functionals* on A). When X is an affine algebraic set and $A = \mathcal{O}[X]$, then every $x \in X$ defines a homomorphism φ_x by evaluation:

$$\varphi_x(f) = f(x) \quad \text{for } f \in \mathcal{O}[X].$$

The coordinate functions from an ambient affine space separate the points of X, so the map $x \mapsto \varphi_x$ is injective.

Corollary A.1.4. *Let X be an affine algebraic set, and let $A = \mathcal{O}[X]$. The map $x \mapsto \varphi_x$ is a bijection between X and $\mathrm{Hom}(A, \mathbb{C})$.*

Proof. Let $\varphi \in \mathrm{Hom}(A, \mathbb{C})$. Then $\mathrm{Ker}(\varphi)$ is a maximal ideal of A, so by Theorem A.1.3 there exists $x \in X$ with $\mathrm{Ker}(\varphi) = \mathrm{Ker}(\varphi_x)$. Hence

$$0 = \varphi(f - f(x)) = \varphi(f) - f(x)$$

for all $f \in A$, since $\varphi(1) = 1$. Thus $\varphi = \varphi_x$. $\qquad\square$

Corollary A.1.5. *Let $f_1(x), \ldots, f_s(x)$ and $f(x)$ be polynomials on $\mathbb{C}^n$. Assume that $f(x) = 0$ whenever $f_i(x) = 0$ for $i = 1, \ldots, s$. Then there exist polynomials $g_i(x)$ and an integer r such that*

$$f(x)^r = \sum_{i=1}^{s} g_i(x) f_i(x). \tag{A.2}$$

Proof. Let $x = (x_1, \ldots, x_n) \in \mathbb{C}^n$. Introduce a new variable x_0 and the polynomial $f_0(x_0, x) = 1 - x_0 f(x)$. Take (x_0, x) as coordinates on $\mathbb{C}^{n+1}$, and view $f_1, \ldots, f_s$ as polynomials on $\mathbb{C}^{n+1}$ depending only on x. The assumption on $\{f_i\}$ and f implies that $\{f_0, f_1, \ldots, f_s\}$ have no common zeros on $\mathbb{C}^{n+1}$. Hence by Theorem A.1.3 the ideal generated by these functions contains 1. Thus there are polynomials $h_i(x_0, x)$ on $\mathbb{C}^{n+1}$ such that

$$1 = h_0(x_0, x)(1 - x_0 f(x)) + \sum_{i=1}^{s} h_i(x_0, x) f_i(x) \, . \tag{A.3}$$

Let r be the maximum degree of x_0 in the polynomials h_i. Substituting $x_0 = 1/f(x)$ in (A.3) and multiplying by $f(x)^r$, we obtain (A.2), with $g_i(x) = f(x)^r h_i(1/f(x), x)$. $\qquad \square$

A.1.2 Zariski Topology

Let V be a finite-dimensional vector space over $\mathbb{C}$ and $X \subset V$ an algebraic subset.

Definition A.1.6. A subset $Y \subset X$ is *Zariski closed* in X if Y is an algebraic subset of V. If $0 \neq f \in \mathcal{O}[X]$, then the *principal open subset* of X defined by f is $X^f = \{x \in X : f(x) \neq 0\}$.

Lemma A.1.7. *The Zariski-closed sets of X give X the structure of a topological space. The finite unions of principal open sets X^f, for $0 \neq f \in \mathcal{O}[X]$, are the nonempty open sets in this topology, called the* Zariski *topology on X.*

Proof. We must check that finite unions and arbitrary intersections of algebraic sets are algebraic. Suppose Y_1 is the zero set of polynomials $f_1, \ldots, f_r$ and Y_2 is the zero set of polynomials $g_1, \ldots, g_s$. Then $Y_1 \cup Y_2$ is the zero set of the family of functions $\{f_i g_j : 1 \leq i \leq r, 1 \leq j \leq s\}$ and is thus algebraic.

Given an arbitrary family $\{Y_\alpha : \alpha \in I\}$ of algebraic sets, their intersection is the zero set of a (possibly infinite) collection of polynomials $\{f_1, f_2, \ldots\}$. This intersection is still an algebraic set, however, by the Hilbert basis theorem. By definition, the complement of a proper algebraic subset of X is the union of finitely many sets of the form X^f. $\qquad \square$

Unless otherwise stated, we will use the term *closed set* in this appendix to refer to a *Zariski-closed set*. Let $x_1, \ldots, x_n$ be linear coordinate functions on V determined by a basis for V. Notice that a point $a = (a_1, \ldots, a_n) \in V$ is a closed set, since it is the zero set of the translated coordinate functions $\{x_i - a_i : 1 \leq i \leq n\}$. If $X \subset V$ is any set then $\overline{X}$ will denote the *closure* of X in the Zariski topology (the smallest closed set containing X).

Let V and W be finite-dimensional complex vector spaces. Suppose $X \subset V$ and $Y \subset W$ are algebraic sets and $f : X \longrightarrow Y$. If g is a complex-valued function on Y define $f^*(g)$ to be the function $f^*(g)(x) = g(f(x))$ for $x \in X$. We say that f is a *regular map* if $f^*(g)$ is in $\mathcal{O}[X]$ for all $g \in \mathcal{O}[Y]$. In terms of linear coordinates $x_1, \ldots, x_m$ for V and $y_1, \ldots, y_n$ for W, a regular map is given by the restriction to X of n polynomial functions $y_i = f_i(x_1, \ldots, x_m)$, for $i = 1, \ldots, n$. In particular, when

$Y = \mathbb{C}$ this notion of regular map is consistent with our previous definition. It is clear from the definition that the composition of regular maps is regular.

Lemma A.1.8. *A regular map f between algebraic sets is continuous in the Zariski topology.*

Proof. Let $Z \subset Y$ be a closed set, defined by a set of polynomials $\{g_j\}$, say. Then $f^{-1}(Z)$ is the zero set of the polynomials $\{f^*g_j\}$, and hence is closed. $\qquad\square$

A.1.3 Products of Affine Sets

Let V, W be vector spaces and let $X \subset V$, $Y \subset W$ be affine algebraic sets. Then $X \times Y$ is an affine algebraic set in the vector space $V \oplus W$. To see this, let $f_1, \ldots, f_m \in \mathcal{P}(V)$ be defining functions for X, and let $g_1, \ldots, g_n \in \mathcal{P}(W)$ be defining functions for Y. Extend f_i and g_j to polynomials on $V \times W$ by setting $f_i(v, w) = f_i(v)$ and $g_j(v, w) = g_j(w)$. Then $\{f_1, \ldots, f_m, g_1, \ldots, g_n\}$ is a set of defining functions for $X \times Y$.

By the universal property of tensor products relative to bilinear maps, there is a unique linear map $\mu : \mathcal{P}(V) \otimes \mathcal{P}(W) \longrightarrow \mathcal{P}(V \oplus W)$ such that $\mu(f' \otimes f'')(v, w) = f'(v)f''(w)$. This map clearly preserves multiplication of functions.

Lemma A.1.9. *The map μ induces an isomorphism of commutative algebras*

$$\nu : \mathcal{O}[X] \otimes \mathcal{O}[Y] \longrightarrow \mathcal{O}[X \times Y] .$$

Proof. Since μ is a vector-space isomorphism (see Proposition C.1.4) and the functions in $\mathcal{O}[X \times Y]$ are the restrictions of polynomials on $V \oplus W$, it is clear that ν is surjective and preserves multiplication. We will show that ν is injective. Suppose

$$0 \neq f = \sum_i f_i' \otimes f_i'' \in \mathcal{O}[X] \otimes \mathcal{O}[Y] .$$

Here we may assume that $\{f_i''\}$ is linearly independent and $f_1' \neq 0$. Choose $g_i' \in \mathcal{P}(V)$ and $g_i'' \in \mathcal{P}(W)$ with $f_i' = g_i'|_X$ and $f_i'' = g_i''|_Y$ and set

$$g(v, w) = \sum_i g_i'(v)g_i''(w) .$$

Then $\nu(f) = g|_{X \times Y}$. Choose $x_0 \in X$ such that $f_1'(x_0) \neq 0$. Then by the linear independence of the functions $\{f_i''\}$, we have

$$\sum_i f_i'(x_0) f_i'' \neq 0 .$$

Hence the function $y \mapsto g(x_0, y)$ on Y is nonzero, proving that $\nu(f) \neq 0$. $\qquad\square$

A.1.4 Principal Open Sets

Assume that X is a Zariski-closed subset of a vector space V. Let $f \in \mathcal{O}[X]$ with $f \neq 0$. Define $\psi : X^f \longrightarrow V \times \mathbb{C}$ by $\psi(x) = (x, 1/f(x))$. This map is injective, and we use it to define the structure of an affine algebraic set on the principal open set X^f as follows: Assume that X is defined by $f_1, \ldots, f_n \in \mathcal{P}(V)$. Choose $\tilde{f} \in \mathcal{P}(V)$ such that $\tilde{f}|_X = f$. Then

$$\psi(X^f) = \{(v,t) \in V \times \mathbb{C} : f_i(v) = 0 \text{ for all } i \text{ and } \tilde{f}(v)t - 1 = 0\}.$$

Thus $\psi(X^f)$ is an algebraic set. We define the ring of regular functions on X^f by pulling back the regular functions on $\psi(X^f)$:

$$\mathcal{O}[X^f] = \{\psi^*(g) : g \in \mathcal{P}(V \times \mathbb{C})\}.$$

On $\psi(X^f)$ the coordinate t has the same restriction as $1/\tilde{f}$. Hence we see that the regular functions on X^f are all of the form $g(x_1, \ldots, x_n, 1/\tilde{f})$ with $g(x_1, \ldots, x_n, t)$ a polynomial in $n+1$ variables. In particular, $1/f$ is a regular function on X^f. Note that $1/f$ is not the restriction to X^f of a polynomial on V unless f is constant.

A.1.5 Irreducible Components

Let $X \subset V$ be a nonempty closed set. We say that X is *reducible* if there are nonempty closed subsets $X_i \neq X$, $i = 1, 2$, such that $X = X_1 \cup X_2$. We say that X is *irreducible* if it is not reducible.

Lemma A.1.10. *An algebraic set X is irreducible if and only if $\mathfrak{I}_X$ is a prime ideal ($\mathcal{O}[X]$ has no zero divisors).*

Proof. Suppose X is reducible. There are polynomials $f_i \in \mathfrak{I}_{X_i}$ such that f_1 does not vanish on X_2 and f_2 does not vanish on X_1. Hence $f_i \notin \mathfrak{I}_X$ but $f_1 f_2 \in \mathfrak{I}_X$. Thus $\mathfrak{I}_X$ is not a prime ideal.

Conversely, if $\mathfrak{I}_X$ is not prime, then there exist f_1 and f_2 in $\mathcal{O}[X]$ with $f_1 f_2$ vanishing on X but f_i not vanishing on X. Set $X_i = \{x \in X : f_i(x) = 0\}$. Then $X = X_1 \cup X_2$ and $X \neq X_i$. Hence X is reducible. $\square$

We shall have frequent use for the following density property of irreducible algebraic sets:

Lemma A.1.11. *Let X be an irreducible algebraic set. Every nonempty open subset Y of X is dense in X. Furthermore, if $Y \subset X$ and $Z \subset X$ are nonempty open subsets, then $Y \cap Z$ is nonempty.*

Proof. By assumption, $X \setminus Y$ is a proper closed subset of X. Since $X = (X \setminus Y) \cup \overline{Y}$, the irreducibility of X implies that $\overline{Y} = X$, where $\overline{Y}$ is the closure of Y (in X) in the Zariski topology.

Let Y and Z be nonempty open subsets. If $Y \cap Z = \emptyset$, then $X = Y^c \cup Z^c$ and $Y^c \neq X$, $Z^c \neq X$. This contradicts the irreducibility of X. $\qquad\square$

Lemma A.1.12. *If X is any algebraic set, then there exists a finite collection of irreducible closed sets X_i such that*

$$X = X_1 \cup \cdots \cup X_r \text{ and } X_i \not\subset X_j \text{ for } i \neq j. \qquad (A.4)$$

Furthermore, such a decomposition (A.4) is unique up to a permutation of the indices and is called an incontractible decomposition *of X. The sets X_i are called the* irreducible components *of X.*

Proof. Suppose the lemma is false for some X. Then there must be a decomposition $X = X_1 \cup X_1'$ into closed sets with $X \supset X_1$ properly and the lemma false for X_1. Continuing, we get an infinite strictly decreasing chain $X \supset X_1 \supset X_2 \supset \cdots$ of closed subsets. But this gives an infinite strictly increasing chain $\mathcal{I}_X \subset \mathcal{I}_{X_1} \subset \cdots$ of ideals, which contradicts the ascending chain condition for ideals in $\mathcal{P}(V)$.

By a deletion process, any decomposition of X as a finite union of closed subsets can be written in the form (A.4) with no proper containments among the X_i. Suppose $X = X_1' \cup \cdots \cup X_s'$ is another incontractible decomposition. Then for each index i,

$$X_i' = X \cap X_i' = \bigcup_{j=1}^{r} X_j \cap X_i'.$$

Since X_i' is irreducible, this decomposition has only one nonempty intersection. Thus there exists an index j such that $X_i' \subset X_j$. Similarly, for each index j there exists an index k such that $X_j \subset X_k'$. Hence $i = k$ and $X_i' = X_j$. So $r = s$ and there is a permutation σ such that $X_i' = X_{\sigma(i)}$. $\qquad\square$

Lemma A.1.13. *If X is an irreducible algebraic set then so is X^f for any nonzero function $f \in \mathcal{O}[X]$.*

Proof. Let X be a Zariski-closed subset of $\mathbb{C}^n$. Suppose $u, v \in \mathcal{O}[X^f]$ with $u \neq 0$ and $uv = 0$. There are polynomials g, h in $n + 1$ variables such that $u(x) = g(x, 1/f(x))$ and $v(x) = h(x, 1/f(x))$, for $x = (x_1, \ldots, x_n) \in X^f$. Hence there is an integer k sufficiently large such that $f^k u$ and $f^k v$ are the restrictions to X^f of polynomials $\tilde{u}$ and $\tilde{v}$, respectively. But

$$f(x)\tilde{u}(x)\tilde{v}(x) = f(x)^{2k+1}u(x)v(x) = 0 \quad \text{for } x \in X^f,$$

and obviously $f(x)\tilde{u}(x)\tilde{v}(x) = 0$ if $f(x) = 0$. Since $\mathcal{O}[X]$ is an integral domain and $\tilde{u} \neq 0$, we conclude that $\tilde{v} = 0$. Thus if $x \in X^f$, then $v(x) = \tilde{v}(x)/f(x)^k = 0$. This proves that $\mathcal{O}[X^f]$ is an integral domain. $\qquad\square$

Lemma A.1.14. *Let V and W be finite-dimensional vector spaces. Suppose $X \subset V$ and $Y \subset W$ are irreducible algebraic sets. Then $X \times Y$ is an irreducible algebraic set in $V \oplus W$.*

Proof. We already verified that $X \times Y$ is algebraic. Suppose there are closed subsets Z_1, Z_2 in $V \oplus W$ such that $X \times Y = Z_1 \cup Z_2$. For each $x \in X$ the set $\{x\} \times Y$ is irreducible, since any decomposition of it into proper closed sets would give a decomposition of Y into proper closed sets. Hence for each $x \in X$ either $\{x\} \times Y \subset Z_1$ or else $\{x\} \times Y \subset Z_2$. This induces a decomposition $X = X_1 \cup X_2$, where

$$X_i = \{x \in X : \{x\} \times Y \subset Z_i\}.$$

We claim that each subset X_i is closed. Indeed, let Z_i be the zero set of the functions $\{f_\alpha\} \subset \mathcal{P}(V \oplus W)$ and define

$$X_y^{(i)} = \{x \in X : f_\alpha(x,y) = 0 \text{ for all } \alpha\}$$

for $y \in Y$. Then $X_y^{(i)}$ is closed in X and $X_i = \bigcap_{y \in Y} X_y^{(i)}$, which shows that X_i is closed. From the irreducibility of X it now follows that either $X = X_1$ or else $X = X_2$. Hence either $X \times Y = Z_1$ or else $X \times Y = Z_2$. $\qquad \square$

Lemma A.1.15. *Suppose* $f : X \longrightarrow Y$ *is a regular map between affine algebraic sets. Suppose* X *is irreducible. Then* $\overline{f(X)}$ *is irreducible.*

Proof. Suppose $g, h \in \mathcal{O}[Y]$ and $g(f(x))h(f(x)) = 0$ for all $x \in X$. Then since $\mathcal{O}[X]$ is an integral domain, either $f^*g = 0$ or $f^*h = 0$. Hence either g or h vanishes on $\overline{f(X)}$. This proves that $\mathcal{O}[\overline{f(X)}]$ is an integral domain, and hence $\overline{f(X)}$ is irreducible. $\qquad \square$

A.1.6 Transcendence Degree and Dimension

A regular function f on an affine algebraic set $X \subset \mathbb{C}^n$ is a polynomial in the linear coordinates on $\mathbb{C}^n$. Since these coordinates satisfy relations on X, we should be able to express some of the coordinates (in an implicit algebraic manner) in terms of the remaining coordinates, which should be algebraically independent. In this way, f would become a function of a smaller number of variables. The *dimension* of X would then be the number of algebraically independent variables that remain after this elimination process. To make this notion precise, we need some algebraic preliminaries.

Let $A \subset B$ be commutative rings with 1, and assume that B has no zero divisors. An element $b \in B$ is said to be *integral over* A if b satisfies a monic polynomial

$$b^n + a_1 b^{n-1} + \cdots + a_{n-1} b + a_n = 0 \tag{A.5}$$

with coefficients $a_i \in A$.

Lemma A.1.16. *An element* $b \in B$ *is integral over* A *if and only if there exists a finitely generated* A-submodule $C \subset B$ *such that* $b \cdot C \subset C$.

Proof. Let b satisfy (A.5). Then $A[b] = A \cdot 1 + A \cdot b + \cdots + A \cdot b^{n-1}$ is a finitely generated A-submodule, so we may take $C = A[b]$. Conversely, suppose C exists as stated and is generated by nonzero elements $\{x_1,\ldots,x_n\}$ as an A-module. Since $bx_i \in C$, there are elements $a_{ij} \in A$ such that

$$bx_i - \sum_{j=1}^{n} a_{ij} x_j = 0 \quad \text{for } i = 1,\ldots,n.$$

Since B has no zero divisors, this system of equations implies that $\det[b\delta_{ij} - a_{ij}] = 0$. This determinant is a monic polynomial in b, with coefficients in A. Hence b is integral over A.   □

We now return to the problem of defining the dimension of an irreducible affine algebraic set X using the algebra $\mathcal{O}[X]$. This algebra is finitely generated over $\mathbb{C}$ and has no zero divisors. The following result (the *Noether normalization lemma*) describes the structure of such algebras:

Lemma A.1.17. *Let $\mathbb{F}$ be a field and $B = \mathbb{F}[x_1,\ldots,x_n]$ a finitely generated commutative algebra over $\mathbb{F}$ without zero divisors. Then there exist $y_1,\ldots,y_r \in B$ such that*

1. the set $\{y_1,\ldots,y_r\}$ is algebraically independent over $\mathbb{F}$;
2. every $b \in B$ is integral over the subring $\mathbb{F}[y_1,\ldots,y_r]$.

The integer r is uniquely determined by properties (1) and (2) and is called the transcendence degree *of B over $\mathbb{F}$. A set $\{y_1,\ldots,y_r\}$ with properties (1) and (2) is called a* transcendence basis *for B over $\mathbb{F}$.*

Proof. If $\{x_1,\ldots,x_n\}$ is algebraically independent over $\mathbb{F}$, then we can set $y_i = x_i$. Otherwise, there is a nontrivial relation

$$\sum_J a_J x^J = 0 \tag{A.6}$$

with coefficients $a_J \in \mathbb{F}$. By relabeling the generators x_i if necessary, we may assume that x_1 occurs in (A.6) with a nonzero coefficient. Fix an integer d with $d > j_k$ for all $J = (j_1,\ldots,j_n)$ such that $a_J \neq 0$. Set $M = (1, d, d^2, \ldots, d^{n-1}) \in \mathbb{N}^n$ and introduce the variables

$$y_2 = x_2 - x_1^d, \quad y_3 = x_3 - x_1^{d^2}, \ldots, \quad y_n = x_n - x_1^{d^{n-1}}. \tag{A.7}$$

Then the monomials x^J become

$$x^J = x_1^{J \cdot M} + \sum_{k < J \cdot M} f_k(y_2,\ldots,y_n) x_1^k,$$

where $J \cdot M = j_1 + j_2 d + \cdots + j_n d^{n-1}$ and f_k is a polynomial with integer coefficients. By the choice of d we see that distinct indices J with $a_J \neq 0$ give distinct values of $J \cdot M$. Thus relation (A.6) can be written as

$$\sum_J a_J x_1^{J \cdot M} + f(x_1, y_2,\ldots,y_n) = 0, \tag{A.8}$$

with f a polynomial with coefficients in $\mathbb{F}$ whose degree in x_1 is less than the degree of the first summation in (A.8). Dividing (A.8) by the coefficient $a_J \neq 0$ for which

$J \cdot M$ is the largest, we see that x_1 satisfies a monic polynomial with coefficients in $\mathbb{F}[y_2, \ldots, y_n]$. Hence by (A.7) the elements $x_2, \ldots, x_n$ also satisfy monic polynomials with coefficients in $\mathbb{F}[y_2, \ldots, y_n]$. Thus B is integral over $\mathbb{F}[y_2, \ldots, y_n]$.

If $\{y_2, \ldots, y_n\}$ is algebraically independent, then we are done. Otherwise, we repeat the procedure above until we arrive at an algebraically independent set. The number r is the cardinality of any maximal algebraically independent set in B (see Lang [97, Chapter X, §1, Theorem 1]); hence it is uniquely determined by B. □

Definition A.1.18. Let X be an affine algebraic set. When X is irreducible, $\dim X$ is the transcendence degree of the algebra $\mathcal{O}[X]$. If X is reducible, then $\dim X$ is the maximum of the dimensions of the irreducible components of X.

With this notion of dimension available, we can obtain the following very useful *ascending chain property* for algebraic sets:

Theorem A.1.19.

1. Let M, N be irreducible affine algebraic sets such that $M \subseteq N$. Then $\dim M \leq \dim N$. Furthermore, if $\dim M = \dim N$ then $M = N$.
2. Let $X_1 \subset X_2 \subset \cdots$ be an increasing chain of irreducible affine algebraic subsets of an algebraic set X. Then there exists an index p such that $X_j = X_p$ for $j \geq p$.

Proof. (1): Lemma A.1.17 implies that $\dim M \leq \dim N$. Suppose that the dimensions are equal. To show that $M = N$, it suffices to show that the restriction homomorphism $\sigma : \mathcal{O}[N] \longrightarrow \mathcal{O}[M]$ is injective, by Corollary A.1.4.

Suppose $u \in \mathrm{Ker}(\sigma)$. Take a transcendence basis $S = \{f_1, \ldots, f_r\}$ for $\mathcal{O}[M]$, where $r = \dim M$. Since $M \subseteq N$, the map σ is surjective. Thus there exist $\bar{f}_i \in \mathcal{O}[N]$ such that $\sigma(\bar{f}_i) = f_i$ for $i = 1, \ldots, r$. Clearly, $\bar{S} = \{\bar{f}_1, \ldots, \bar{f}_r\}$ is algebraically independent. Since $\dim N = \dim M = r$, it follows that $\bar{S}$ is a transcendence basis for $\mathcal{O}[N]$. Thus by the Noether normalization lemma there are $b_i \in \mathbb{C}[\bar{f}_1, \ldots, \bar{f}_r]$ such that

$$u^n + b_{n-1}u^{n-1} + \cdots + b_1 u + b_0 = 0.$$

Choose n minimal. Applying the homomorphism σ, we get $\sigma(b_0) = 0$. But this implies that $b_0 = 0$, by the algebraic independence of S. Thus $n = 1$ and $u = 0$.

(2): From part (1) we know that $\dim X_j \leq \dim X_{j+1} \leq \dim X$. Hence there exists an index p such that $\dim X_j = \dim X_p$ for $j \geq p$. Then $X_j = X_p$ for $j \geq p$ by the last part of (1). □

A.1.7 Exercises

1. Let $f : \mathbb{C} \longrightarrow \mathbb{C}$ be a bijective map that is not a polynomial. Show that f is continuous in the Zariski topology but that it is not a regular map.
2. For any subset $X \subset \mathbb{C}^n$ let $\mathcal{I}_X$ be the ideal of all polynomials vanishing on X. For any ideal $\mathcal{J} \subset \mathcal{P}(\mathbb{C}^n)$ let $V(\mathcal{J}) = \{x \in \mathbb{C}^n : f(x) = 0 \text{ for all } f \in \mathcal{J}\}$ be the zero set of $\mathcal{J}$.

(a) Show that the Zariski closure of X is $V(\mathfrak{I}_X)$.

(b) If $\mathfrak{I}, \mathfrak{J}$ are ideals in $\mathcal{P}(\mathbb{C}^n)$, show that $V(\mathfrak{I}\mathfrak{J}) = V(\mathfrak{I} \cap \mathfrak{J})$.

3. Suppose f is an irreducible polynomial (this means that f cannot be factored as a product of polynomials of strictly smaller degree). Show that the zero set of f is irreducible.

4. Let $f_i(t) = g_i(t)/h_i(t)$ be rational functions on $\mathbb{C}^n$ for $i = 1, \dots, r$. Let $X \subset \mathbb{C}^r$ be the Zariski closure of the set $\{[f_1(t), \dots, f_r(t)] : t \in \mathbb{C}^n$ and $h_1(t) \cdots h_r(t) \neq 0\}$. Prove that V is irreducible. (HINT: Show that the ideal $\mathfrak{I}_X$ is prime.)

5. Let $X = \{(x_1, x_2) \in \mathbb{C}^2 : x_1 x_2 = 0\}$.

(a) Show that the irreducible components of X are $X_1 = \{(z, 0) : z \in \mathbb{C}\}$ and $X_2 = \{(0, z) : z \in \mathbb{C}^2\}$ and hence that X is reducible.

(b) Show that X is connected (in the Zariski topology). (HINT: Suppose $X = U \cup V$ with U and V Zariski open, $U \cap V = \emptyset$, and $U \neq \emptyset$. Use (a) to argue that $U \cap X_i = X_i$ for $i = 1, 2$ and hence $V = \emptyset$.)

6. (a) Show that the Zariski-closed subsets of $\mathbb{C}$ are precisely the finite sets.

(b) Show that the Zariski topology on $\mathbb{C} \times \mathbb{C}$ is *not* the product of the Zariski topologies on each factor.

A.2 Maps of Algebraic Sets

The points of an affine algebraic set X correspond to the homomorphisms of the algebra $\mathcal{O}[X]$ into the base field $\mathbb{C}$. We will use this correspondence to obtain the key results concerning regular maps. The translation from geometric to algebraic language leads us to two main problems: When do homomorphisms of an algebra extend to homomorphisms of a larger algebra? If an extension exists, when is it unique? We shall solve these algebraic problems and apply them to obtain geometric properties of maps between affine algebraic sets. We begin with the notion of *rational map* between affine algebraic sets.

A.2.1 Rational Maps

Let A be a commutative ring with 1 and without zero divisors. Then A is embedded in its *quotient field* $\mathrm{Quot}(A)$. The elements of this field are the formal expressions $f = g/h$, where $g, h \in A$ and $h \neq 0$, with the usual algebraic operations on fractions. When X is an irreducible algebraic set, then the algebra $A = \mathcal{O}[X]$ has no zero divisors, so it has a quotient field. We denote this field by $\mathrm{Rat}(X)$ and call it the field of *rational functions* on X.

We may view the elements of $\mathrm{Rat}(X)$ as functions, as follows: If $f \in \mathrm{Rat}(X)$, then we say that f is *defined at a point* $x \in X$ if there exist $g, h \in \mathcal{O}[X]$ with $f = g/h$ and $h(x) \neq 0$. In this case we set $f(x) = g(x)/h(x)$. The *domain* $\mathcal{D}_f$ of f is the subset of

X at which f is defined. It is a dense open subset of X, since it contains the principal open set X^h.

A map f from X to an algebraic set Y is called *rational* if $f^*(\varphi)$ is a rational function on X for all $\varphi \in \mathcal{O}[Y]$. Suppose $Y \subset \mathbb{C}^n$ and y_i is the restriction to Y of the linear coordinate function x_i on $\mathbb{C}^n$. Set $f_i = f^*(y_i)$. Then f is rational if and only if $f_i \in \mathrm{Rat}(X)$ for $i = 1, \ldots, n$. The domain of a rational map f is defined as

$$\mathcal{D}_f = \bigcap_{\varphi \in \mathcal{O}[Y]} \mathcal{D}_{f^*(\varphi)} \, .$$

By Lemma A.1.11 we know that $\mathcal{D}_f = \bigcap_{i=1}^n \mathcal{D}_{f^*(y_i)}$ is a dense open subset of X.

Lemma A.2.1. *Suppose X is irreducible and $f : X \longrightarrow Y$ is a rational map. If $\mathcal{D}_f = X$ then f is a regular map.*

Proof. Let $\varphi \in \mathcal{O}[Y]$. Let $\mathcal{J} \subset \mathcal{O}[X]$ be the set of all functions h such that the function $x \mapsto h(x)\varphi(f(x))$ is regular on X. Then $\mathcal{J}$ is a nonzero ideal in $\mathcal{O}[X]$, since f is a rational map. Suppose $\mathcal{J}$ is a proper ideal. Then it is contained in some maximal ideal. In this case Theorem A.1.3 implies that all $h \in \mathcal{J}$ vanish at some point $x_0 \in X$. Thus $x_0 \notin \mathcal{D}_f$, a contradiction. Hence $1 \in \mathcal{J}$ and $f^*(\varphi) \in \mathcal{O}[X]$ for all $\varphi \in \mathcal{O}[Y]$. $\square$

A.2.2 Extensions of Homomorphisms

Let A be an algebra over $\mathbb{C}$ with unit 1. Given $0 \neq a \in A$, we set

$$\mathrm{Hom}(A, \mathbb{C})^a = \{\varphi \in \mathrm{Hom}(A, \mathbb{C}) \, : \, \varphi(a) \neq 0\} \, .$$

For example, if $A = \mathcal{O}[X]$ for some affine algebraic set X, then $\mathrm{Hom}(A, \mathbb{C})$ is naturally identified with X, by Corollary A.1.4, and $\mathrm{Hom}(A, \mathbb{C})^a$ corresponds to the principal open set X^a. For $a = 1$ we have $\mathrm{Hom}(A, \mathbb{C})^1 = \mathrm{Hom}(A, \mathbb{C})$.

Definition A.2.2. Let A be a subring of a ring B. Then B is *integral over A* if every $b \in B$ is integral over A. Equivalently, the A-module $A[b]$ is finitely generated over A for each $b \in B$.

Theorem A.2.3. *Let B be a commutative algebra over $\mathbb{C}$ with unit 1 and no zero divisors. Suppose that $A \subset B$ is a subalgebra such that $B = A[b_1, \ldots, b_n]$ for some elements $b_i \in B$. Then given $0 \neq b \in B$, there exists $0 \neq a \in A$ such that every $\varphi \in \mathrm{Hom}(A, \mathbb{C})^a$ extends to $\psi \in \mathrm{Hom}(B, \mathbb{C})^b$. In particular, if B is integral over A, then every $\varphi \in \mathrm{Hom}(A, \mathbb{C})$ extends to $\psi \in \mathrm{Hom}(B, \mathbb{C})$.*

Proof. We start with the case $B = A[u]$ for some element $u \in B$. Let $b = f(u)$, where $f(X) = a_n X^n + \cdots + a_0$ with $a_i \in A$. For $g(X) \in A[X]$, denote by $g^\varphi(X) \in \mathcal{O}[X]$ the polynomial obtained by applying φ to the coefficients of $g(X)$. If u is transcendental over A, then for any $\lambda \in \mathbb{C}$ we can define an extension ψ of φ by

$$\psi(g(u)) = g^\varphi(\lambda) \quad \text{for } g \in A[X] .$$ \hfill (A.9)

This extension satisfies $\psi(b) = f^\varphi(\lambda)$. Since $b \neq 0$, the polynomial $f(X)$ is nonzero. Take for $a \in A$ any nonzero coefficient of $f(X)$. If we assume that $\varphi(a) \neq 0$, then $f^\varphi(X) \neq 0$ and any $\lambda \in \mathbb{C}$ with $f^\varphi(\lambda) \neq 0$ will serve to define the desired extension of φ.

Now assume that u is algebraic over A. Then b is also algebraic over A, so there are nonzero polynomials $p(X) = a_m X^m + \cdots + a_0$ and $q(X) = c_n X^n + \cdots + c_0$ (with a_i and c_i in A), whose degrees are minimal and that satisfy $p(u) = 0$ and $q(b) = 0$. We have $a_m \neq 0$ and $c_0 \neq 0$ (otherwise, the polynomial $q(X)/X$ would annihilate b). We will prove that the element $a = a_m c_0$ has the desired property. Note that if u is integral over A and $b = 1$, then $a = 1$, which proves the last assertion of the theorem.

We first observe that for any $g(X) \in A[X]$ the Euclidean division algorithm furnishes polynomials $h(X), r(X) \in A[X]$ with $\deg r(X) < m$, and an integer $d \geq 0$ such that

$$(a_m)^d g(X) = p(X)h(X) + r(X) .$$

In particular, if $g(u) = 0$ then $r(X) = 0$, since $r(u) = 0$ and $\deg r(X) < m$. Thus $(a_m)^d g(X)$ is divisible by $p(X)$ in $A[X]$. Suppose $\varphi \in \text{Hom}(A, \mathbb{C})^a$ and λ is a root of $p^\varphi(X)$. We have $\varphi(a_m)^d g^\varphi(\lambda) = 0$. But $\varphi(a_m) = \varphi(a)/\varphi(c_0) \neq 0$. Hence $g^\varphi(\lambda) = 0$, and so formula (A.9) determines an extension $\psi \in \text{Hom}(B, \mathbb{C})$ of φ. For this extension,

$$0 = \psi(q(b)) = q^\varphi(\psi(b)) .$$

However, 0 is not a root of $q^\varphi(X)$, since $\varphi(c_0) \neq 0$. Thus $\psi(b) \neq 0$, and the theorem is proved when B has a single generator over A.

Let $n \geq 1$. We assume that the theorem is true for all algebras A and all algebras B with $n-1$ generators over A. We shall show that it is true for the case of n generators. Let $B = A[b_1, \ldots, b_n]$ and let $0 \neq b \in B$ be given. Set $\tilde{A} = A[b_1, \ldots, b_{n-1}]$ and $u = b_n$. Then $B = \tilde{A}[u]$. By the proof just given there exists $\tilde{a} \in \tilde{A}$ such that every complex homomorphism $\tilde{\varphi}$ of $\tilde{A}$ satisfying $\tilde{\varphi}(\tilde{a}) \neq 0$ extends to a complex homomorphism ψ of B with $\psi(b) \neq 0$. Now we invoke the induction hypothesis, with B replaced by $\tilde{A}$ and b replaced by $\tilde{a}$. This gives a nonzero element $a \in A$ such that every complex homomorphism φ of A satisfying $\varphi(a) \neq 0$ extends to a complex homomorphism $\tilde{\varphi}$ of $\tilde{A}$ with $\tilde{\varphi}(\tilde{a}) \neq 0$. We complete the induction step by combining these two extension processes. $\qquad\square$

Corollary A.2.4. *Let B be a finitely generated commutative algebra over $\mathbb{C}$ having no zero divisors. Given $0 \neq b \in B$, there exists $\psi \in \text{Hom}(B, \mathbb{C})$ such that $\psi(b) \neq 0$.*

Proof. In Theorem A.2.3 take $A = \mathbb{C}$ and $\varphi(\lambda) = \lambda$ for $\lambda \in \mathbb{C}$. $\qquad\square$

Next we consider the uniqueness of the extensions in Theorem A.2.3. Let $A \subset B$ be a subalgebra, and identify $\text{Quot}(A)$ with the subfield of $\text{Quot}(B)$ generated by A.

For example, if $A = \mathcal{O}[X]$ for an irreducible variety X, and $B = \mathcal{O}[X^f]$ for some nonzero $f \in A$, then $B = A[b] \subset \text{Quot}(A)$, where $b = 1/f$. In this example, every $\psi \in \text{Hom}(B, \mathbb{C})$ such that $\psi(b) \neq 0$ is given by evaluation at a point $x \in X^f$, and hence

ψ is uniquely determined by its restriction to A. The unique restriction property in this example characterizes the general case in which $B \subset \text{Quot}(A)$, as follows:

Theorem A.2.5. *Let B be a finitely generated algebra over $\mathbb{C}$ with no zero divisors. Let $A \subset B$ be a finitely generated subalgebra. Assume that there exists a nonzero element $b \in B$ such that every element of $\text{Hom}(B, \mathbb{C})^b$ is uniquely determined by its restriction to A. Then $B \subset \text{Quot}(A)$.*

Proof. As in Theorem A.2.3, it suffices to consider the case $B = A[u]$ for some element $u \in B$. Let $b = f(u)$, where $f(X) = a_n X^n + \cdots + a_0$ with $a_i \in A$. We first show that if u is not algebraic over A, then the unique extension property does not hold. Indeed, in this case every $\lambda \in \mathbb{C}$ determines an extension $\psi \in \text{Hom}(B, \mathbb{C})$ of $\varphi \in \text{Hom}(A, \mathbb{C})$ by (A.9). The condition $\psi(b) \neq 0$ implies that $f^{\varphi}(X) \neq 0$, so there exist infinitely many choices of λ such that $\psi(b) \neq 0$, as in the proof of Theorem A.2.3. Hence ψ is not uniquely determined by φ.

Now assume that u is algebraic over A. Let $p(X)$ and $q(X)$ be as in the proof of Theorem A.2.3. We shall prove that $m = \deg(p) = 1$. This will imply the theorem, since then $u = -a_0/a_1$. Suppose to the contrary that $m > 1$. Set $K = \text{Quot}(A)$ and define

$$r(X) = a_m \left(X + \frac{a_{m-1}}{m a_m} \right)^m = a_m X^m + a_{m-1} X^{m-1} + \sum_{j=0}^{m-2} d_j X^j$$

in $K[X]$. Then $p(X) \neq r(X)$, since $p(X)$ is irreducible over A. Hence there exists some $j_0 \leq m - 2$ such that $a_{j_0} \neq d_{j_0}$. Set

$$a = (a_m)^{m-1}(a_{j_0} - d_{j_0})c_0 .$$

Then $0 \neq a \in A$, and so by Corollary A.2.4 there exists $\varphi \in \text{Hom}(A, \mathbb{C})$ such that $\varphi(a) \neq 0$. Set $\alpha = \varphi(a_m) \neq 0$ and $\beta = \varphi(a_{m-1})/(m\alpha)$. Then $r^{\varphi}(X) = \alpha(X + \beta)^m$, while the condition $\varphi(a_{j_0}) \neq \varphi(d_{j_0})$ implies that

$$p^{\varphi}(X) \neq \alpha(X + \beta)^m . \tag{A.10}$$

We claim that if $m \geq 2$, then $p^{\varphi}(X)$ must have at least two distinct roots. For if not, then $p^{\varphi}(X)$ would be the mth power of a linear polynomial and hence would have to be $\alpha(X + \beta)^m$, contradicting (A.10).

As in the proof of Theorem A.2.3, each root λ of $p^{\varphi}(X)$ determines an extension $\psi \in \text{Hom}(B, \mathbb{C})$ of φ by (A.9). In particular, $\psi(u) = \lambda$, so distinct roots of $p^{\varphi}(X)$ determine distinct extensions of φ. Also, $0 = \psi(q(b)) = q^{\varphi}(\psi(b))$. But 0 is not a root of $q^{\varphi}(X)$, since $\varphi(c_0) \neq 0$. Thus $\psi(b) \neq 0$. So we conclude that if $m \geq 2$, then there exist $\psi_1, \psi_2 \in \text{Hom}(B, \mathbb{C})$ with $\psi_1(u) \neq \psi_2(u)$, $\psi_i(b) \neq 0$, and $\psi_1 = \psi_2$ on A. This contradicts the unique restriction assumption. Hence $m = 1$. $\qquad\square$

A.2.3 Image of a Dominant Map

Let X and Y be affine algebraic sets and let $f : X \longrightarrow Y$ be a regular map.

Definition A.2.6. The map f is *dominant* if $f(X)$ is dense in Y.

The definition of dominance is equivalent to the property that $f^* : \mathcal{O}[Y] \longrightarrow \mathcal{O}[X]$ is injective. Dominant maps have the following remarkable property (which does not hold for smooth maps of differentiable manifolds):

Theorem A.2.7. *Assume that X and Y are irreducible affine algebraic sets and $f : X \longrightarrow Y$ is a dominant map. Let $M \subset X$ be a nonempty open set. Then $f(M)$ contains a nonempty open subset of Y.*

Proof. Set $B = \mathcal{O}[X]$ and $A = f^*(\mathcal{O}[Y])$. Since X is irreducible, it follows that B has no zero divisors. We may assume that $M = X^b$ for some $0 \neq b \in B$. Let $a \in A$ be as in Theorem A.2.3, and let $a = f^*(\bar{a})$, where $\bar{a} \in \mathcal{O}[Y]$. We claim that

$$f(X^b) \supset Y^{\bar{a}} . \tag{A.11}$$

To prove this, we view the points of Y as the homomorphisms from A to $\mathbb{C}$. For $y \in Y^{\bar{a}}$, the corresponding homomorphism φ satisfies $\varphi(a) = \bar{a}(y) \neq 0$. Hence by Theorem A.2.3, there is an extension ψ of φ to B such that $\psi(b) \neq 0$. In geometric language, this means that ψ is given by evaluation at a point $x \in X^b$. The extension property means that $g(f(x)) = g(y)$ for all $g \in \mathcal{O}[Y]$. Since f is dominant, this implies that $f(x) = y$, proving (A.11). $\square$

Theorem A.2.8. *Let $f : X \longrightarrow Y$ be a regular map between affine algebraic sets. Then $f(X)$ contains an open subset of $\overline{f(X)}$.*

Proof. Let $X_1, \dots, X_r$ be the irreducible components of X. Then

$$\overline{f(X)} = \overline{f(X_1)} \cup \dots \cup \overline{f(X_r)} .$$

The theorem now follows from Theorem A.2.7. $\square$

A.2.4 Factorization of a Regular Map

Let M, N, and P be irreducible affine varieties. Suppose we have regular maps $f : M \longrightarrow N$ and $h : M \longrightarrow P$. If there is a map g that satisfies the commutative diagram

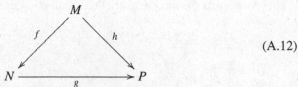

$$\tag{A.12}$$

then h is constant on the fibers of f. Indeed, $f(m) = f(m')$ implies that $h(m) = g(f(m)) = h(m')$. Furthermore, if f is surjective, then any such map g is uniquely determined by f and h.

We now weaken the fiber and surjectivity conditions with the aim of obtaining a rational map g that satisfies the diagram (A.12) in the sense of rational maps.

Theorem A.2.9. *Assume that f and h are dominant and that there is a nonempty open subset U of M such that $f(m) = f(m')$ implies $h(m) = h(m')$ for $m, m' \in U$. Then there exists a rational map $g : N \longrightarrow P$ such that $h = g \circ f$.*

Proof. Consider first the case in which $M = P$ and h is the identity map. We may assume that $U = M^b$ for some $0 \neq b \in \mathcal{O}[M]$. We claim that the conditions of Theorem A.2.5 are satisfied by $A = f^*(\mathcal{O}[N])$ and $B = \mathcal{O}[M]$. Indeed, every homomorphism $\psi : B \longrightarrow \mathbb{C}$ such that $\psi(b) \neq 0$ is given by evaluation at some $x \in M^b$ (cf. Corollary A.1.4). Hence $\psi(f^*p) = p(f(x))$ for $p \in \mathcal{O}[N]$. Since $f(M)$ is dense in N and f is injective on M^b, it follows that x is uniquely determined by the homomorphism $p \mapsto p(f(x))$.

Let x_i, for $i = 1, \ldots, k$, be the coordinate functions defined by some embedding $M \subset \mathbb{C}^k$. Applying Theorem A.2.5, we obtain $q_i \in \mathrm{Rat}(N)$ such that $f^*(q_i) = x_i$ where defined. The rational map g defined by $g(n) = (q_1(n), \ldots, q_k(n))$ for $n \in N$ then satisfies $g \circ f(m) = m$ for m in a dense subset of M, as required.

We now reduce the general case to the one just treated. Let $F : M \longrightarrow N \times P$ with $F(m) = (f(m), h(m))$, and take the projection maps

$$\pi_1 : N \times P \longrightarrow N \quad \text{and} \quad \pi_2 : N \times P \longrightarrow P.$$

Let $L = \overline{F(M)}$ and let $p_i = \pi_i|_L$ for $i = 1, 2$. Then we have the commutative diagram

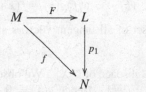

The set L is irreducible and p_1 is a dominant map that is injective on an open set. Hence there is a rational map $k : N \longrightarrow L$ such that $k \circ p_1$ is the identity map on a dense open set. We take $g = p_2 \circ k$ to obtain the desired map. $\square$

A.2.5 Exercises

1. Let X and Y be affine algebraic sets and $\varphi : X \longrightarrow Y$ a regular map.
   (a) Show that the graph $\Gamma_\varphi = \{(x, \varphi(x)) : x \in X\}$ of φ is closed in $X \times Y$.
   (b) Show that the projection map $\pi : \Gamma_\varphi \longrightarrow X$ onto the first coordinate is an isomorphism of affine algebraic sets.

2. Let $\varphi : \mathbb{C}^2 \longrightarrow \mathbb{C}^2$ be the map $\varphi(x,y) = (xy, y)$.
   (a) Show that φ has dense image but is not surjective.
   (b) Show that x is not integral over $\mathbb{C}[xy, y]$.
3. Let $X = \{(x,y) \in \mathbb{C}^2 : y^2 = x^3\}$.
   (a) Show that every element of $\mathcal{O}[X]$ can be written uniquely in the form $P(x) + Q(x)y$, where P and Q are polynomials.
   (b) Let $\varphi : \mathbb{C} \longrightarrow X$ be the map $\varphi(t) = (t^2, t^3)$. Show that φ is bijective and regular, but φ^{-1} is not regular.

A.3 Tangent Spaces

An algebraic set has a *tangent space* at each point, which furnishes a powerful tool for studying regular maps via their differentials. As in calculus, tangent vectors have a geometric description in terms of derivatives, and an algebraic description in terms of linearization of regular functions.

A.3.1 Tangent Space and Differentials of Maps

Let $X \subset \mathbb{C}^n$ be an algebraic set, and let $\mathcal{O}[X]^*$ denote the linear maps from $\mathcal{O}[X]$ (viewed as a $\mathbb{C}$ vector space) to $\mathbb{C}$.

Definition A.3.1. A *tangent vector* at a point $x \in X$ is an element $v \in \mathcal{O}[X]^*$ such that

$$v(fg) = v(f)g(x) + f(x)v(g) \qquad (A.13)$$

for all $f, g \in \mathcal{O}[X]$. The set of all tangent vectors at x is called the *tangent space* $T(X)_x$ of X at x.

Clearly $T(X)_x$ is a linear subspace of $\mathcal{O}[X]^*$. We observe that if v is a tangent vector at x, then $v(1) = v(1 \cdot 1) = 2v(1)$. Hence $v(1) = 0$, so v vanishes on the space of constant functions on X.

We now give an alternative description of the tangent space in terms of maximal ideals. Let $\mathfrak{m}_x \subset \mathcal{O}[X]$ be the maximal ideal of all functions that vanish at x. Then $f - f(x) \in \mathfrak{m}_x$ for any $f \in \mathcal{O}[X]$, and $v(f) = v(f - f(x))$. Hence v is determined by its restriction to $\mathfrak{m}_x$. However, by (A.13) we see that $v(\mathfrak{m}_x^2) = 0$, so v naturally defines an element $\tilde{v} \in (\mathfrak{m}_x/\mathfrak{m}_x^2)^*$. Conversely, given any $\tilde{v} \in (\mathfrak{m}_x/\mathfrak{m}_x^2)^*$, we define a linear functional v on $\mathcal{O}[X]$ by $v(f) = \tilde{v}(f - f(x))$. To verify that v satisfies (A.13), we observe that if $f, g \in \mathcal{O}[X]$, then $(f - f(x))(g - g(x)) \in \mathfrak{m}_x^2$. Hence

$$0 = v(fg) - f(x)v(g) - g(x)v(f) \,,$$

since $v(c) = 0$ for $c \in \mathbb{C}$. This shows that there is a natural isomorphism

$$T(X)_x \cong (\mathfrak{m}_x/\mathfrak{m}_x^2)^* . \tag{A.14}$$

Let X and Y be algebraic sets and $\varphi : X \longrightarrow Y$ a regular map. Then the induced map $\varphi^* : \mathcal{O}[Y] \longrightarrow \mathcal{O}[X]$ is an algebra homomorphism. If $v \in T(X)_x$ then the linear functional $f \mapsto v(\varphi^*(f))$, for $f \in \mathcal{O}[Y]$, is a tangent vector at $y = \varphi(x)$. Indeed, for $f, g \in \mathcal{O}[Y]$ we have

$$v(\varphi^*(fg)) = v(\varphi^*(f)\varphi^*(g)) = v(\varphi^*(f))g(y) + f(y)v(\varphi^*(g)) .$$

We denote this element of $T(Y)_y$ by $d\varphi_x(v)$. At each point $x \in X$ we thus have a linear map $d\varphi_x : T(X)_x \longrightarrow T(Y)_{\varphi(x)}$ that we call the *differential* of φ at x.

Examples

1. If $X = \mathbb{C}^n$, then $\mathcal{O}[X] = \mathcal{P}(\mathbb{C}^n) = \mathbb{C}[x_1, \dots, x_n]$. For $u \in \mathbb{C}^n$ and $f \in \mathcal{P}(\mathbb{C}^n)$ define $D_u f(x) = (\partial/\partial t)f(x + tu)|_{t=0}$ (the *directional derivative* of f along the line $t \mapsto x + tu, t \in \mathbb{C}$). The linear functional $v(f) = D_u f(a)$ is a tangent vector at $a \in \mathbb{C}^n$. If $u = (u_1, \dots, u_n)$, then we can express v in terms of the partial derivative operators $\partial/\partial x_i$:

$$v = \sum_{i=1}^n u_i (\partial/\partial x_i)_a ,$$

where we write $(\partial/\partial x_i)_a$ for the tangent vector $f \mapsto (\partial f/\partial x_i)(a)$ with $a \in \mathbb{C}^n$.

We claim that $T(\mathbb{C}^n)_a$ has basis $\{(\partial/\partial x_1)_a, \dots, (\partial/\partial x_n)_a\}$ and hence has dimension n. Indeed, let a have coordinates $(a_1, \dots, a_n)$. Then it is clear by the Taylor expansion of a polynomial that the ideal $\mathfrak{m}_a$ is generated by the independent linear functions $x_i - a_i$, for $1 \le i \le n$. Thus $v \in T(\mathbb{C}^n)_a$ can be uniquely expressed as $v(f) = D_u f(a)$, where $u = (u_1, \dots, u_n) \in \mathbb{C}^n$ and $u_i = v(x_i - a_i)$. When convenient, we will identify $T(\mathbb{C}^n)_a$ with $\mathbb{C}^n$ by the map $v \mapsto u$.

2. Suppose $X \subset \mathbb{C}^n$ is an algebraic set and $a \in X$. Since $\mathcal{O}[X] = \mathcal{P}(\mathbb{C}^n)/\mathcal{I}_X$, for every $v \in T(X)_a$ there exists $\tilde{v} \in T(\mathbb{C}^n)_a$ with $\tilde{v}(\mathcal{I}_X) = 0$ and $\tilde{v}(f) = v(f + \mathcal{I}_X)$. Conversely, any $\tilde{v}$ with this property induces a tangent vector v to X at a, with $v(f + \mathcal{I}_X) = \tilde{v}(f)$. Hence

$$T(X)_a = \{\tilde{v} \in T(\mathbb{C}^n)_a : \tilde{v}(\mathcal{I}_X) = 0\} .$$

Let $\{f_1, \dots, f_r\}$ be a generating set of polynomials for the ideal $\mathcal{I}_X$. The defining equation for a derivation shows that $\tilde{v}(\mathcal{I}_X) = 0$ if and only if $\tilde{v}(f_i) = 0$ for $i = 1, \dots, r$. Hence if we set $u_j = \tilde{v}(x_j - a_j)$ then $\tilde{v} \in T(X)_a$ if and only if

$$\sum_{j=1}^n u_j \frac{\partial f_i(a)}{\partial x_j} = 0 \quad \text{for } i = 1, \dots, r . \tag{A.15}$$

This is a set of r linear equations for $u = (u_1, \dots, u_n) \in \mathbb{C}^n$. In particular, we see that $\dim T(X)_a = n - \operatorname{rank}(J(a))$, where $J(a)$ is the $r \times n$ matrix $[\partial f_i(a)/\partial x_j]$ of partial derivatives.

3. Let X be an irreducible affine algebraic set. Define

$$m(X) = \min_{a \in X} \dim T(X)_a \, .$$

Let $X_0 = \{a \in X : \dim T(X)_a = m(X)\}$. The points of X_0 are called *smooth*. These are the points at which the matrix $J(a)$ defined above has maximum rank $d = n - m(X)$. Because this condition can be described by the nonvanishing of some $d \times d$ minor of $J(a)$, it follows that X_0 is Zariski dense in X when X is irreducible. If $X_0 = X$ then X is said to be *smooth*.

4. If X is a reducible algebraic set with irreducible components $\{X_i\}$, then we say that X is smooth if each X_i is smooth. We define $m(X) = \max_i \{m(X_i)\}$ in this case.

A.3.2 Vector Fields

Recall that a *derivation* of an algebra A is a linear map $D : A \longrightarrow A$ such that $D(ab) = D(a)b + aD(b)$. If A is commutative and D and D' are derivations of A, then any linear combination of D and D' with coefficients in A is a derivation, and the *commutator* $[D, D'] = DD' - D'D$ is a derivation, as we check by the obvious calculation. Thus the derivations of a commutative algebra A form a Lie algebra $\mathrm{Der}(A)$, which is also an A-module (see Section 4.1.1). In the case $A = \mathcal{O}[X]$, where X is an affine algebraic set, a derivation of A is usually called a *vector field* on X. We denote by $\mathrm{Vect}(X)$ the Lie algebra of all vector fields on X.

Given $L \in \mathrm{Vect}(X)$ and $x \in X$, we define $L_x f = (Lf)(x)$ for $f \in \mathcal{O}[X]$. Then $L_x \in T(X)_x$, by the definition of tangent vector. Conversely, if we have a correspondence $x \mapsto L_x \in T(X)_x$ such that the functions $x \mapsto L_x(f)$ are regular for every $f \in \mathcal{O}[X]$, then L is a vector field on X.

Example

Let $0 \neq f \in \mathcal{O}[X]$ and consider a vector field L on the principal open set X^f. Recall that $\mathcal{O}[X^f]$ is generated by the restrictions to X^f of functions in $\mathcal{O}[X]$ together with f^{-1}. However, since L is a derivation, we have $L(f)f^{-1} + fL(f^{-1}) = L(ff^{-1}) = L(1) = 0$, so that $L(f^{-1}) = -f^{-2}L(f)$. Hence L is completely determined by its action on $\mathcal{O}[X]$. For example, if $X = \mathbb{C}^n$, then

$$L = \sum_{i=1}^{n} \varphi_i \frac{\partial}{\partial x_i} \quad \text{with} \quad \varphi_i = L(x_i) \in \mathbb{C}[x_1, \ldots, x_n, 1/f] \, . \tag{A.16}$$

A.3.3 Dimension

We now show that the notion of dimension defined geometrically via the tangent space coincides with the algebraic definition in terms of transcendence degree.

Theorem A.3.2. *Let X be an algebraic set. Then $m(X) = \dim X$.*

Proof. If X has irreducible components $\{X_i\}$, then $m(X) = \max(m(X_i))$, so we may assume that $X \subset \mathbb{C}^n$ is irreducible. Set $\mathbb{K} = \mathrm{Rat}(X)$, the field of rational functions on X. We prove first that

$$\dim X = \dim_{\mathbb{K}} \mathrm{Der}(\mathbb{K}) . \tag{A.17}$$

For this, we apply the Noether normalization lemma to obtain a transcendence basis $\{u_1, \ldots, u_d\}$ for $B = \mathcal{O}[X]$, where $d = \dim X$. Let $A = \mathbb{C}[u_1, \ldots, u_d]$ and $\mathbb{F} = \mathrm{Quot}(A) \subset \mathbb{K}$. A derivation D of k is determined by an arbitrary choice of d rational functions $Du_i \in \mathbb{F}$, since $\{u_1, \ldots, u_d\}$ is algebraically independent. On the other hand, we claim that every derivation D of $\mathbb{F}$ extends uniquely to a derivation of $\mathbb{K}$. To establish this, we use the fact that the field $\mathbb{K}$ is a finite extension of $\mathbb{F}$, since B is integral over A. Hence the theorem of the primitive element (Lang [97, Chapter VII, §6, Theorem 14]) furnishes an element $b \in \mathbb{K}$ such that $\mathbb{K} = \mathbb{F}(b)$. Let $f(X) \in \mathbb{F}[X]$ be the minimal polynomial for b. Then $f'(b) \neq 0$, so we may define $Db = -f^D(b)/f'(b) \in \mathbb{K}$, where for any polynomial $g(X) = \sum_i a_i X^i \in \mathbb{F}[X]$ we let

$$g^D(X) = \sum_i D(a_i) X^i .$$

If D can be extended to a derivation of $\mathbb{K}$, it must act by

$$D(g(b)) = g^D(b) + g'(b)Db \quad \text{for all } g(X) \in \mathbb{F}[X] . \tag{A.18}$$

To prove that such an extension exists, we must verify that the right side of (A.18) does not depend on the representation of an element of $\mathbb{K}$ as $g(b)$. Indeed, if $g(b) = h(b)$ for some $h(X) \in \mathbb{F}[X]$, then the polynomial $\varphi(X) = g(X) - h(X)$ is of the form $\psi(X)f(X)$ for some $\psi(X) \in \mathbb{F}[X]$. Hence $\varphi'(b) = \psi(b)f'(b)$ and $\varphi^D(b) = \psi(b)f^D(b)$, since $f(b) = 0$. It follows that

$$g'(b)Db - h'(b)Db = \psi(b)f'(b)Db = -\psi(b)f^D(b)$$
$$= -g^D(b) + h^D(b)$$

by the definition of Db. Thus (A.18) is unambiguous and defines the desired extension. This completes the proof of (A.17).

Let $\{f_1, \ldots, f_r\}$ be a set of generators for $\mathcal{I}_X$. A derivation D of $\mathbb{K}$ is uniquely determined by the n functions $\xi_j = Dx_j \in \mathbb{K}$, where x_j are the linear coordinate functions from $\mathbb{C}^n$. It must also satisfy $D(f_i) = 0$ for $i = 1, \ldots, r$. Thus by the chain rule we have

$$\sum_{j=1}^n \frac{\partial f_i}{\partial x_j} \xi_j = 0 \quad \text{for } i = 1, \ldots, r . \tag{A.19}$$

By definition of $m(X)$, system (A.19) has rank $n - m(X)$ over $\mathbb{K}$. Hence the solution space to (A.19) has dimension $m(X)$ over $\mathbb{K}$. Combining this fact with (A.17) completes the proof. □

A.3.4 Differential Criterion for Dominance

We begin with a general criterion for a map to be dominant.

Proposition A.3.3. *Let X and Y be affine algebraic sets and $\psi : X \longrightarrow Y$ a regular map. Assume that Y is irreducible and $\dim Y = m$. Suppose there exists an algebraically independent set $\{u_1, \dots, u_m\} \subset \mathcal{O}[Y]$ such that the set*

$$\{\psi^* u_1, \dots, \psi^* u_m\} \subset \mathcal{O}[X]$$

is also algebraically independent. Then $\psi(X)$ is dense in Y.

Proof. Suppose there exists $0 \neq f \in \mathcal{O}[Y]$ such that $\psi^*(f) = 0$. Since $\dim Y = m$, it follows that f is algebraic over the field $\mathbb{C}(u_1, \dots, u_m)$. Thus there are rational functions $a_j \in \mathbb{C}(x_1, \dots, x_m)$ such that

$$\sum_{j=0}^d a_j(u_1, \dots, u_m) f^j = 0 .$$

We choose the set of functions a_j with d as small as possible ($d \geq 1$, since $f \neq 0$). Take $0 \neq \gamma \in \mathbb{C}[u_1, \dots, u_m]$ such that $\gamma a_j(u_1, \dots, u_m) \in \mathbb{C}[u_1, \dots, u_m]$. Then

$$\sum_{j=0}^d \psi^*(\gamma) a_j(\psi^* u_1, \dots, \psi^* u_m) \psi^*(f^j) = 0 .$$

Since $\psi^*(f^j) = \psi^*(f)^j = 0$ for $j \geq 1$, we obtain the relation

$$\psi^*(\gamma) a_0(\psi^* u_1, \dots, \psi^* u_m) = 0 . \tag{A.20}$$

But we are given that $\psi^* : \mathbb{C}[u_1, \dots, u_m] \longrightarrow \mathbb{C}[\psi^* u_1, \dots, \psi^* u_m]$ is an isomorphism, so we have $\psi^*(\gamma) \neq 0$. Thus (A.20) implies that $a_0(\psi^* u_1, \dots, \psi^* u_m) = 0$, and hence $a_0(u_1, \dots, u_m) = 0$. This contradicts the definition of d. We conclude that $\psi^* : \mathcal{O}[Y] \longrightarrow \mathcal{O}[X]$ is injective. Since Y is irreducible, this implies that $\psi(X)$ is dense in Y. $\qquad\qquad\square$

Let X and Y be irreducible affine algebraic sets and let $\psi : X \longrightarrow Y$ be a regular map. We have the following criterion for ψ to be dominant:

Theorem A.3.4. *Suppose there exists a smooth point $p \in X$ such that $\psi(p)$ is a smooth point of Y and $d\psi_p : T(X)_p \longrightarrow T(Y)_{\psi(p)}$ is bijective. Then $\psi(X)$ is dense in Y.*

Before proving the theorem we establish the following lemma, whose statement and proof are similar to the corresponding result for C^∞ manifolds (the local triviality of the tangent bundle). In the statement of the lemma we identify $T(X)_q$ with a subspace of $\mathbb{C}^n$ as in Example 2 of Section A.3.1.

Lemma A.3.5. *Let $X \subset \mathbb{C}^n$ be closed and irreducible and let $p \in X$ be a smooth point of X. Then there exists an open subset $U \subset X$ with $p \in U$, and regular maps $w_j : U \longrightarrow \mathbb{C}^n$ for $j = 1, \dots, m = \dim X$, such that*

$$T(X)_q = \bigoplus_{j=1}^{m} \mathbb{C}w_j(q) \quad \text{for all } q \in U .$$

Proof. Since p is a smooth point, we have

$$T(X)_p = \{v \in \mathbb{C}^n : (d\varphi)_p (v) = 0 \quad \text{for all } \varphi \in \mathcal{I}_X\}$$

by (A.15). Hence there exist $\varphi_1, \ldots, \varphi_{n-m} \in \mathcal{I}_X$ such that

$$T(X)_p = \{v \in \mathbb{C}^n : (d\varphi_i)_p (v) = 0 \quad \text{for } i = 1, \ldots, n-m\} .$$

This implies that $(d\varphi_1)_p \wedge \cdots \wedge (d\varphi_{(n-m)})_p \neq 0$, so there is an open subset U_1 of $\mathbb{C}^n$ such that $p \in U_1$ and

$$(d\varphi_1)_q \wedge \cdots \wedge (d\varphi_{(n-m)})_q \neq 0 \quad \text{for } q \in U_1 . \tag{A.21}$$

For any point $q \in U_1 \cap X$ we have

$$T(X)_q \subset W_q = \{v \in \mathbb{C}^n : (d\varphi_i)_q (v) = 0 \quad \text{for } i = 1, \ldots, n-m\} .$$

Now $\dim T(X)_q \geq m$, since X is irreducible. Since $\dim W_q = m$ by (A.21), we conclude that

$$T(X)_q = \{v \in \mathbb{C}^n : (d\varphi_i)_q (v) = 0, \quad \text{for } i = 1, \ldots, n-m\} \tag{A.22}$$

at all points $q \in U_1 \cap X$.

Fix a basis $\{e_1 \ldots, e_n\}$ for $\mathbb{C}^n$ such that

$$\det[(d\varphi_i)_p (e_j)]_{1 \leq i,j \leq n-m} \neq 0 .$$

Let $V = \bigoplus_{i=1}^{n-m} \mathbb{C}e_i$ and $W = \bigoplus_{i=n-m+1}^{n} \mathbb{C}e_i$. We then write $x \in \mathbb{C}^n$ as

$$x = \begin{bmatrix} y \\ z \end{bmatrix} \quad \text{with } y \in V \text{ and } z \in W .$$

For $q \in \mathbb{C}^n$ we write the $n \times n$ matrix $J_q = [(d\varphi_i)_q (e_j)]_{1 \leq i,j \leq n}$ in block form as

$$J_q = \begin{bmatrix} A_q & B_q \\ C_q & D_q \end{bmatrix}$$

with A_q of size $(n-m) \times (n-m)$. In terms of the decomposition $\mathbb{C}^n = V \oplus W$ we have $A_q : V \longrightarrow V$ and $B_q : W \longrightarrow V$.

Since $\det A_p \neq 0$, there exists a Zariski-open subset $U_2 \subset U_1$ with $p \in U_2$ and $\det A_q \neq 0$ for all $q \in U_2$. We define regular maps $w_j : U_2 \longrightarrow \mathbb{C}^n = V \oplus W$ by

$$w_j(q) = \begin{bmatrix} -A_q^{-1} B_q e_{n-m+j} \\ e_{n-m+j} \end{bmatrix} \quad \text{for } j = 1, \ldots, m .$$

Clearly $\{w_1(q),\ldots,w_m(q)\}$ is linearly independent. Also,

$$J_q w_j(q) = \begin{bmatrix} (-A_q A_q^{-1} B_q e_{n-m+j} + B_q e_{n-m+j}) \\ * \end{bmatrix} = \begin{bmatrix} 0 \\ * \end{bmatrix}. \qquad (A.23)$$

Set $U = U_2 \cap X$ and let $q \in U$. Then (A.22) and (A.23) imply that

$$(d\varphi_i)_q (w_j(q)) = 0 \quad \text{for } i = 1,\ldots,n-m.$$

Thus $\{w_1(q),\ldots,w_m(q)\}$ is a basis for $T(X)_q$. $\qquad\qquad\qquad\qquad\qquad\qquad$ □

Proof of Theorem A.3.4. Let $p \in X$ satisfy the conditions of the theorem. Take U and $w_1,\ldots,w_m$ as in Lemma A.3.5. Since $\{d\psi_p(w_1(p)),\ldots,d\psi_p(w_m(p))\}$ is a basis for $T(Y)_{\psi(p)}$, there exist functions $u_1,\ldots,u_m$ in $\mathcal{O}[Y]$ such that

$$d\psi_p(w_j(p))u_i = \delta_{ij} \quad \text{for } i,j = 1,\ldots,m.$$

Now for $q \in X$ we can write $d\psi_q(w_j(q))u_i = d(\psi^* u_i)_q(w_j(q))$. Thus there is an open subset $V \subset U$ containing p such that

$$\det[d(\psi^* u_i)_q(w_j(q))]_{1 \le i,j \le m} \ne 0 \quad \text{for all } q \in V.$$

This implies that $\{\psi^* u_1,\ldots,\psi^* u_m\}$ is algebraically independent, so $\{u_1,\ldots,u_m\}$ is algebraically independent. The theorem now follows from Proposition A.3.3. □

Proposition A.3.6. *Let $\varphi : X \longrightarrow Y$ be a dominant regular map of irreducible affine algebraic sets. For $y \in Y$ let $F_y = \varphi^{-1}\{y\}$. Then there is a nonempty open set $U \subset X$ such that $\dim X = \dim Y + \dim F_{\varphi(x)}$ and $\dim F_{\varphi(x)} = \dim \mathrm{Ker}(d\varphi_x)$ for all $x \in U$.*

Proof. Let $d = \dim X - \dim Y$, $S = \varphi^* \mathcal{O}[Y]$, and $R = \mathcal{O}[X]$. Set $\mathbb{F} = \mathrm{Quot}(S)$ and let $B \subset \mathrm{Quot}(R)$ be the subalgebra generated by $\mathbb{F}$ and R (the rational functions on X with denominators in $S \setminus \{0\}$). Since B has transcendence degree d over $\mathbb{F}$, Lemma A.1.17 furnishes an algebraically independent set $\{f_1,\ldots,f_d\} \subset R$ such that B is integral over $k[f_1,\ldots,f_d]$. Taking the common denominator of a set of generators of the algebra B, we obtain $f = \varphi^* g \in S$ such that R_f is integral over $S_f[f_1,\ldots,f_d]$, where $R_f = \mathcal{O}[X^f]$ and $S_f = \varphi^* \mathcal{O}[Y^g]$. By Theorem A.2.7 we can take g such that $\varphi(Y^g) = X^f$.

Define $\psi : X^f \longrightarrow Y^g \times \mathbb{C}^d$ by $\psi(x) = (\varphi(x), f_1(x),\ldots,f_d(x))$. Then

$$\psi^* \mathcal{O}[Y^g \times \mathbb{C}^d] = S_f[f_1,\ldots,f_d],$$

and hence $\mathcal{O}[X^f]$ is integral over $\psi^* \mathcal{O}[Y^g \times \mathbb{C}^d]$. By Theorem A.2.3 every homomorphism from $S_f[f_1,\ldots,f_d]$ to $\mathbb{C}$ extends to a homomorphism from R_f to $\mathbb{C}$. Hence ψ is surjective. Let $\pi : Y^g \times \mathbb{C}^d \longrightarrow Y^g$ by $\pi(y,z) = y$. Then $\varphi = \psi^* \pi$ and $F_y = \psi^{-1}(\{y\} \times \mathbb{C}^d)$. If W is any irreducible component of F_y, then $\mathcal{O}[W]$ is integral over $\psi^* \mathcal{O}[\{y\} \times \mathbb{C}^d]$, and hence $\dim W = d$.

We have $d\varphi_x = d\pi_{\psi(x)} \circ d\psi_x$. The integrality property implies that every derivation of $\mathrm{Quot}(\psi^*(Y^g \times \mathbb{C}^d))$ extends uniquely to a derivation of $\mathrm{Rat}(X^f)$, as in the proof of Theorem A.3.2. Hence $d\psi_x$ is bijective for x in a nonempty dense open set U by Lemma A.3.5. For such x, $\mathrm{Ker}(d\varphi_x) = \mathrm{Ker}(d\pi_{\psi(x)})$ has dimension d. $\square$

A.4 Projective and Quasiprojective Sets

We now turn to the study of projective algebraic sets, which are subsets of projective space defined by homogeneous polynomials. Although projective sets do not have globally defined regular functions, they behave well under mappings because of their compactness properties.

A.4.1 Basic Definitions

Let V be a complex vector space. The *projective space* $\mathbb{P}(V)$ associated with V is the set of lines through 0 (one-dimensional subspaces) in V. For $x \in V \setminus \{0\}$, we let $[x] \in \mathbb{P}(V)$ denote the line through x. The map $p : V \setminus \{0\} \longrightarrow \mathbb{P}(V)$ given by $p(x) = [x]$ is surjective, and $p(x) = p(y)$ if and only if $x = \lambda y$ for some $\lambda \in \mathbb{C}^{\times}$. We denote $\mathbb{P}(\mathbb{C}^{n+1})$ by $\mathbb{P}^n$, and for $x = (x_0, \ldots, x_n) \in \mathbb{C}^{n+1}$ we call $\{x_i\}$ the *homogeneous coordinates* of $[x]$.

If $f(x_0, \ldots, x_n)$ is a homogeneous polynomial in $n+1$ variables and $0 \neq x \in \mathbb{C}^{n+1}$, then $f(x) = 0$ if and only if f vanishes on the line $[x]$. Hence f defines a subset

$$A_f = \{[x] \in \mathbb{P}^n : f(x) = 0\}$$

in projective space. The *Zariski topology* on $\mathbb{P}^n$ is obtained by taking as closed sets the intersections

$$X = \bigcap_{f \in S} A_f \,,$$

where S is any set of homogeneous polynomials on $\mathbb{C}^{n+1}$. Any such set X will be called a *projective algebraic set*. The set

$$p^{-1}(X) \cup \{0\} = \{x \in \mathbb{C}^{n+1} : f(x) = 0 \text{ for all } f \in S\}$$

is closed in $\mathbb{C}^{n+1}$ and is called the *cone over* X.

Every closed set in projective space is definable as the zero locus of a finite collection of homogeneous polynomials, and the descending chain condition for closed sets is satisfied (this is proved by passing to the cones over the sets and using the corresponding results for affine space). Just as in the affine case, this implies that every closed set is a finite union of irreducible closed sets, and any nonempty open subset of an irreducible closed set M is dense in M.

We define a covering of projective space by affine spaces, as follows: For $i = 0, \ldots, n$ let $\mathbb{U}_i^n = \{[x] \in \mathbb{P}^n : x_i \neq 0\}$. Each $\mathbb{U}_i^n$ is an open set in $\mathbb{P}^n$, and every point of $\mathbb{P}^n$ lies in $\mathbb{U}_i^n$ for some i. For $[x] \in \mathbb{U}_i^n$ define the *inhomogeneous coordinates* of $[x]$ to be $y_j = x_j/x_i$ for $j \neq i$. The map

$$\pi_i([x]) = (y_0, \ldots, \widehat{y_i}, \ldots, y_n) \quad (\text{omit } y_i)$$

is a bijection between $\mathbb{U}_i^n$ and $\mathbb{C}^n$. It is also a topological isomorphism (where $\mathbb{U}_i^n$ has the relative Zariski topology from $\mathbb{P}^n$ and $\mathbb{C}^n$ carries the Zariski topology). To see this, let $U = \{y \in \mathbb{C}^n : f(y) \neq 0\}$, with f a polynomial of degree k, and set $g = \left(x_i(\pi_i^* f)\right)^k$. Then g is a homogeneous polynomial of degree k on $\mathbb{C}^{n+1}$, and

$$\pi_i^{-1}(U) \cap \mathbb{U}_i^n = \{[x] : g(x) \neq 0 \text{ and } x_i \neq 0\}.$$

Thus we have a covering by $\mathbb{P}^n$ by the $n+1$ open sets $\mathbb{U}_i^n$, each homeomorphic to the affine space $\mathbb{C}^n$.

Lemma A.4.1. *Let* $X \subset \mathbb{P}^n$. *Suppose that for each* $i = 0, 1, \ldots, n$ *the set* $X \cap \mathbb{U}_i$ *is the zero set of homogeneous polynomials* $f_{ij}(y_0, \ldots, \widehat{y_i}, \ldots, y_n)$ *(where* $\{y_k\}$ *are the inhomogeneous coordinates on* $\mathbb{U}_i$). *Then* X *is closed in* $\mathbb{P}^n$.

Proof. Let d_{ij} be the degree of f_{ij}. Define $g_{ij}(x) = x_i f_{ij}(x_0, \ldots, \widehat{x_i}, \ldots, x_n)$ for $x \in \mathbb{C}^{n+1}$. Then g_{ij} is a homogeneous polynomial of degree $d_{ij} + 1$. For $x_i \neq 0$ we have

$$g_{ij}(x_0, \ldots, \widehat{x_i}, \ldots, x_n) = x_i^{d_{ij}+1} f_{ij}(y_0, \ldots, \widehat{y_i}, \ldots, y_n).$$

Hence $X \cap \mathbb{U}_i = \{[x] : g_{ij}(x) = 0 \quad \text{for all} \quad j\}$. Since $g_{ij}(x) = 0$ when $x_i = 0$ and the sets $X \cap \mathbb{U}_i$ cover X, it follows that $X = \{[x] : g_{ij}(x) = 0 \quad \text{for all} \quad i, j\}$. Thus X is closed in $\mathbb{P}^n$. $\square$

A *quasiprojective algebraic set* is a subset $M \subset \mathbb{P}^n$ defined by a finite set of equalities and inequalities on the homogeneous coordinates of the form

$$f_i(x) = 0 \quad \text{for all } i = 1, \ldots, k \text{ and } g_j(x) \neq 0 \quad \text{for some } j = 1, \ldots, l,$$

where f_i and g_j are homogeneous polynomials on $\mathbb{C}^{n+1}$. In topological terms, M is the intersection of the closed set $Y = \{[x] \in \mathbb{P}^n : f_i(x) = 0 \text{ for all } i = 1, \ldots, k\}$ and the open set $Z = \{[x] \in \mathbb{P}^n : g_j(x) \neq 0 \text{ for some } j\}$.

Examples

1. Any projective algebraic set is quasiprojective, since a closed subset of $\mathbb{P}^n$ always has a finite set of defining equations.

2. The set $\mathbb{U}_i^n$ introduced above is the quasiprojective algebraic set in $\mathbb{P}^n$ defined by $\{x_i \neq 0\}$.

3. Suppose $M \subset \mathbb{U}_i^n$ for some i and $X = \pi_i(M) \subset \mathbb{C}^n$ is an affine algebraic set. Since π_i is a homeomorphism, M is a (relatively) closed subset of the open set $\mathbb{U}_i^n$ and hence is quasiprojective.

A.4.2 Products of Projective Sets

Consider now the problem of putting the structure of a quasiprojective algebraic set on the product of two such sets. We begin with the basic case of projective spaces. Let x and y be homogeneous coordinates on $\mathbb{P}^m$ and $\mathbb{P}^n$. Denote the space of complex matrices of size $r \times s$ by $M_{r,s}$ and view $\mathbb{C}^r = M_{1,r}$ as row vectors. We map $\mathbb{C}^{m+1} \times \mathbb{C}^{n+1} \longrightarrow M_{m+1,n+1}$ by $(x,y) \mapsto xy^t$, where y^t is the transpose of y. The image of $(\mathbb{C}^{m+1} \setminus \{0\}) \times (\mathbb{C}^{n+1} \setminus \{0\})$ consists of all rank-one matrices; hence it is defined by the vanishing of all minors of size greater than 1. These minors are homogeneous polynomials in the matrix coordinates z_{ij} of $z \in M_{m+1,n+1}$. Passing to projective space, we have thus obtained an embedding

$$\mathbb{P}^m \times \mathbb{P}^n \hookrightarrow \mathbb{P}(M_{m+1,n+1}) = \mathbb{P}^{mn+m+n} \qquad (A.24)$$

with closed image. We take this as the structure of a projective algebraic set on $\mathbb{P}^m \times \mathbb{P}^n$.

Now let $X \subset \mathbb{P}^m$ and $Y \subset \mathbb{P}^n$ be projective algebraic sets. The image of $X \times Y$ under the map (A.24) conists of all points $[xy^t]$ such that

$$0 \neq x \in \mathbb{C}^{m+1}, \ \ 0 \neq y \in \mathbb{C}^{n+1}, \ \text{ and } \ f_i(x) = 0, g_i(y) = 0 \ \text{ for } \ i = 1, \ldots k,$$

where f_i and g_i are homogeneous polynomials on $\mathbb{C}^m$ and $\mathbb{C}^n$ that define X and Y, respectively. To see that such sets are projective, consider a homogeneous polynomial $f(x)$ on $\mathbb{C}^{m+1}$ of degree r and a point $[y] \in \mathbb{P}^n$. We shall describe the rank-one matrices $z \in M_{m+1,n+1}$ of the form $z = xy^t$ that satisfy $f(x) = 0$ by giving homogeneous equations in the entries z_{ij} of z. Since $y \neq 0$, the equation $f(x) = 0$ is equivalent to the system of equations

$$y_i^r f(x) = 0 \ \ \text{ for } \ i = 0, \ldots, n+1 .$$

But given that $z_{ij} = x_i y_j$, we can write $y_i^r f(x) = f(y_i x) = \varphi_i(z)$, where $\varphi_i(z) = f(z_{1i}, \ldots, z_{n+1,i})$ is a homogeneous polynomial of degree r. Thus the desired equations are

$$\varphi_i(z) = 0 \ \ \text{ for } \ i = 0, \ldots, n+1 .$$

Interchanging the roles of x and y, we can likewise express homogeneous equalities in y in terms of z, for fixed $[x] \in \mathbb{P}^m$. It follows that the image of $X \times Y$ in $\mathbb{P}^{m+n+mn}$ is closed.

A similar argument for equations of the form $f(x) \neq 0$ shows that the image of $X \times Y$ under the map (A.24) is quasiprojective if X and Y are quasiprojective.

Lemma A.4.2. *Let X be a quasiprojective algebraic set and let*

$$\Delta = \{(x,x) : x \in X\} \subset X \times X$$

be the diagonal. Then Δ is closed.

Proof. Since $X \times X$ is again quasiprojective, it suffices to prove that the diagonal is closed in $\mathbb{P}^n \times \mathbb{P}^n$. In terms of the embedding $\mathbb{P}^n \times \mathbb{P}^n \hookrightarrow \mathbb{P}(M_{n+1,n+1})$, the diagonal consists of the lines $[z]$ where $z = xx^t$ and $x \in \mathbb{C}^{n+1}$. We claim that the diagonal can be described by the homogeneous equation $z^t = z$ and hence is closed. Indeed, if $z = xy^t$, with $x, y \in \mathbb{C}^{n+1} \setminus \{0\}$ and $z^t = z$, then $x_i y_j = x_j y_i$ for all i, j. Pick j such that $y_j \neq 0$. Then $x_i = (x_j y_j^{-1}) y_i$ for all i, so we have $[x] = [y]$ in $\mathbb{P}^n$. $\square$

A.4.3 Regular Functions and Maps

We encounter the basic difference between affine and projective algebraic sets when we consider the notion of a *regular function*. In the affine case we have global functions (polynomials) on an ambient vector space whose restrictions define the regular functions. By contrast, a polynomial $f(x)$ on $\mathbb{C}^{n+1}$, even if homogeneous, does not define a function on $\mathbb{P}^n$ except in the trivial case that f is homogeneous of degree zero (constant). A way around this difficulty is to replace polynomials by rational functions and to *localize* the notion of regularity, as follows:

Let M be an irreducible affine set. Suppose $U \subset M$ is an open subset. Define the *regular functions on* U to be the restrictions to U of rational functions $f \in \mathrm{Rat}(M)$ such that $\mathcal{D}_f \supset U$. The set $\mathcal{O}_M(U)$ of all such functions obviously is a commutative algebra over $\mathbb{C}$. For $U = M$ we have $\mathcal{O}_M(M) = \mathcal{O}[M]$, by Lemma A.2.1, so the terminology is consistent. Replacing U by a point $x \in M$, we define the *local ring* $\mathcal{O}_x$ at x to consist of all rational functions on M that are defined at x. Clearly $\mathcal{O}_x$ is a subalgebra of $\mathrm{Rat}(M)$, and $\mathcal{O}_x = \bigcup_{x \in V} \mathcal{O}_M(V)$, where V runs over all open sets containing x.

This notion of regular function has two key properties:

(restriction)   If $U \subset V$ are open subsets of M and $f \in \mathcal{O}_M(V)$, then $f|_U \in \mathcal{O}_M(U)$.

(locality)   Suppose $f : U \longrightarrow \mathbb{C}$ and for every $x \in U$ there exists $\varphi \in \mathcal{O}_x$ with $\varphi = f$ on some open neighborhood of x. Then $f \in \mathcal{O}_M(U)$.

(We leave the proof of these properties as an exercise.) Thus all regularity properties of functions can be expressed in terms of the local rings. To carry over these constructions to an arbitrary quasiprojective algebraic set, we need the following covering lemma:

Lemma A.4.3. *Suppose X is a quasiprojective algebraic set. There is a finite open covering $X = \bigcup_{\alpha \in A} U_\alpha$ with the following properties:*

1. *For $\alpha \in A$ there are an irreducible affine algebraic set M_α and a homeomorphism $\varphi_\alpha : U_\alpha \longrightarrow M_\alpha$.*
2. *For all $\alpha, \beta \in A$ the maps $\varphi_\beta \circ \varphi_\alpha^{-1} : \varphi_\alpha(U_\alpha \cap U_\beta) \longrightarrow \varphi_\beta(U_\alpha \cap U_\beta)$ are regular.*

Proof. We have $X = U \cap Y$, where U is open and Y is closed in $\mathbb{P}^n$. Set $V_i = \pi_i(\mathbb{U}_i^n \cap X)$ and $M_i = \pi_i(Y \cap \mathbb{U}_i^n)$, where $\pi_i : \mathbb{U}_i^n \longrightarrow \mathbb{C}^n$ is the map defined by inhomogeneous coordinates. Then V_i is an open subset of the affine algebraic set M_i, and so it is a finite union of principal open sets V_{ij}, by Lemma A.1.7. Decomposing the sets M_i into irreducible components and intersecting them with V_{ij}, we obtain a collection of irreducible affine sets M_α (cf. Lemma A.1.13). We take the maps φ_α to be restrictions of the maps π_i^{-1}. The regularity of $\varphi_\beta \circ \varphi_\alpha^{-1}$ follows from the formula

$$(\pi_i \circ \pi_j^{-1})(y) = (y_1/y_i, \ldots, \underbrace{1}_{i\text{th}}, \ldots, \underbrace{1/y_i}_{j\text{th}}, \ldots, y_n/y_i) \,,$$

for $y = (y_1, \ldots, y_n) \in \mathbb{C}^n$. $\qquad\square$

Let X be a quasiprojective algebraic set. We define the *local ring* $\mathcal{O}_x$ at $x \in X$ by carrying over the local rings of the affine open sets U_α via the maps φ_α:

$$\mathcal{O}_x = \varphi_\alpha^*(\mathcal{O}_{\varphi_\alpha(x)}) \quad \text{for } x \in U_\alpha \,.$$

If $x \in U_\alpha \cap U_\beta$, then by the last statement in Lemma A.4.3 we see that $\mathcal{O}_x$ is the same, whether we use φ_α or φ_β. For any open set $U \subset X$ we can now define the ring $\mathcal{O}_X(U)$ of *regular functions* on U using the local rings, just as in the affine case: a continuous function $f : U \longrightarrow \mathbb{C}$ is *regular* if for each $x \in U$ there exists $g \in \mathcal{O}_x$ such that $f = g$ on an open neighborhood of x. One then verifies that the restriction and locality properties hold for the rings $\mathcal{O}_X(U)$.

Definition A.4.4. Let X and Y be quasiprojective algebraic sets. A map $\varphi : X \longrightarrow Y$ is *regular* if φ is continuous and $\varphi^*(\mathcal{O}(U)) \subset \mathcal{O}(\varphi^{-1}(U))$ for all open sets $U \subset Y$.

When X and Y are affine, this notion of regularity agrees with the earlier definition.

Lemma A.4.5. *Let X, Y, and Z be quasiprojective algebraic sets. Then a map $z \mapsto (f(z), g(z))$ from Z to $X \times Y$ is regular if and only if the component maps $f : Z \longrightarrow X$ and $g : Z \longrightarrow Y$ are regular.*

Proof. By Lemma A.4.3, it is enough to check the assertion of the lemma when X, Y, and Z are affine. In this case, it follows immediately from the property $\mathcal{O}[X \times Y] = \mathcal{O}[X] \otimes \mathcal{O}[Y]$. $\qquad\square$

Proposition A.4.6. *Suppose X and Y are quasiprojective algebraic sets and that $\varphi : X \longrightarrow Y$ is a regular map. Then $\Gamma_\varphi = \{(x, \varphi(x)) : x \in X\}$ (the graph of φ) is closed in $X \times Y$.*

Proof. It is enough to consider the case that X and Y are affine, by Lemma A.4.1. A point $(x, y) \in X \times Y$ is in Γ_φ if and only if

$$g(\varphi(x)) - g(y) = 0 \quad \text{for all } g \in \mathcal{O}[Y] \,.$$

Now for $g \in \mathcal{O}[Y]$, the function $f(x, y) = g(\varphi(x)) - g(y)$ is a regular function on $X \times Y$, so Γ_φ is the zero set of a family of regular functions on $X \times Y$. $\qquad\square$

Corollary A.4.7. *Let X, Y be quasiprojective algebraic sets and $\varphi : X \times Y \longrightarrow X$ a regular map. Then $\{(x, y) \in X \times Y : \varphi(x, y) = x\}$ is closed in $X \times Y$.*

Proof. Use the same argument as for Proposition A.4.6.   □

If X is quasiprojective, we denote by $\mathcal{O}[X] = \mathcal{O}_X(X)$ the ring of functions that are everywhere regular on X (the notation is consistent by Lemma A.2.1). If X is affine, then $\mathcal{O}[X]$ separates the points of X. When X is not an affine algebraic set, however, $\mathcal{O}[X]$ may not contain many functions. Here is the extreme case.

Theorem A.4.8. *Let X be an irreducible projective algebraic set. Then $\mathcal{O}[X] = \mathbb{C}$.*

Proof. Assume $X \subset \mathbb{P}^n$ and let $Y \subset \mathbb{C}^{n+1}$ be the cone over X. Then the irreducibility of X implies that Y is irreducible. The ideal $\mathfrak{I}_Y$ is generated by homogeneous polynomials, so the algebra $\mathcal{O}[Y]$ is graded.

Let $f \in \mathcal{O}[X]$. We can consider f as a rational function on Y that is homogeneous of degree 0. The assumption that f is regular everywhere on X means that for all $0 \neq y \in Y$ there exist $p, q \in \mathcal{O}[Y]$, each homogeneous of the same degree, with $q(y) \neq 0$ and $f = p/q$ on a neighborhood of y. Let $\mathfrak{I}$ be the ideal in $\mathcal{P}(\mathbb{C}^{n+1})$ generated by $f_1, \ldots, f_k$ together with all denominators q that occur in this way, and let $f_1, \ldots, f_s$ be homogeneous polynomials that generate $\mathfrak{I}$. The only common zero of $\{f_i\}$ on $\mathbb{C}^{n+1}$ is $y = 0$. Thus by Corollary A.1.5 there is an integer r such that $x_i^r \in \mathfrak{I}$ for $i = 0, 1, \ldots, n$, where $\{x_i\}$ are coordinates on $\mathbb{C}^{n+1}$. Hence $\mathfrak{I}$ contains the space $\mathcal{P}^m(\mathbb{C}^{n+1})$, where $m = r(n+1)$. This implies that if g is a homogeneous polynomial of degree m, then fg is a homogeneous rational function of degree m, and $fg|_Y = h|_Y$, where h is a homogeneous polynomial of degree m.

Let $V \subset \mathcal{O}[Y]$ be the restrictions to Y of the homogeneous polynomials of degree m. It is a finite-dimensional space, and we have just shown that the operator T of multiplication by f maps V into V. Let λ be an eigenvalue of T, and let $0 \neq h \in \mathcal{O}[Y]$ be the corresponding eigenvector. Then $(f - \lambda)h = 0$ in the field $\mathrm{Rat}(Y)$. Hence $f = \lambda$ is a constant.   □

Corollary A.4.9. *If X is an irreducible projective algebraic set that is also isomorphic to an affine algebraic set, then X is a single point.*

Proof. By Theorem A.4.8 we have $\mathcal{O}[X] = \mathbb{C}$. But for every affine algebraic set, the functions in $\mathcal{O}[X]$ separate the points of X. Hence X must consist of one point.   □

Theorems A.2.7 and A.2.8 are also valid when X and Y are quasiprojective algebraic sets (this follows by covering X and Y by irreducible open affine algebraic sets and restricting the map to the sets of the covering). Furthermore, if f is a rational map between affine algebraic sets, then the open set $\mathcal{D}_f$ is a quasiprojective algebraic set, and $f : \mathcal{D}_f \longrightarrow Y$ is a regular map in this new sense. Thus Theorem A.2.9 is also valid for quasiprojective algebraic sets.

If X is quasiprojective and $x \in X$, we define $\dim_x(X) = \dim T(U_\alpha)_x$, where $x \in U_\alpha$ as in Lemma A.4.3. It is easy to see that $\dim_x(X)$ depends only on the local ring $\mathcal{O}_x$ (cf. Theorem A.3.2). We set

$$\dim X = \min_{x \in X} \left\{ \dim_x(X) \right\}.$$

It is clear from this definition of dimension that Theorem A.1.19 holds for quasiprojective algebraic sets. Just as in the affine case, a point $x \in X$ is called *smooth* if $\dim_x(X) = \dim X$. When X is irreducible, the smooth points form a dense open set. If every point of X is smooth then X is said to be *smooth* or *nonsingular*.

An extension of method of proof of Theorem A.4.8 can be used to prove an important theorem that gives a characterization of projective algebraic sets among quasiprojective sets. We will use this result only twice in the text (Theorems 11.4.9 and 11.4.18), and we will leave it to the interested reader to read the proof in, say, Shafarevich [134, Chapter I §5.2 Theorem 3]. The argument is at precisely the same level as the rest of this appendix.

Theorem A.4.10. *Let X, Y be quasiprojective sets with X projective. Let $p(x, y) = y$ for $(x, y) \in X \times Y$. If $C \subset X \times Y$ is closed then $p(C)$ is closed in Y.*

Here is an example of the power of this result.

Corollary A.4.11. *Let X be projective and $f : X \to Y$ a regular map with Y quasiprojective. Then $f(X)$ is closed in Y.*

Proof. Proposition A.4.6 implies that Γ_f is closed in $X \times Y$. Since $f(X)$ is the projection in the second factor of Γ_f, it is closed by Theorem A.4.10. ☐

The final result that we need, which will be used only once in the book (in the proof of Theorem 11.4.9), is the following (the connectedness assumption can be shown to be redundant):

Proposition A.4.12. *Suppose $X \subset \mathbb{C}^m$ is a smooth irreducible affine variety that is also connected as a C^∞ manifold. If $X = \bigcup_{n=1}^\infty X_n$, where each X_n is a Zariski-closed subset, then $X = X_n$ for some n.*

Proof. Assume to the contrary that $X \neq X_n$ for every n. Then X_n is a proper subset of X that is defined by the vanishing of a finite number of nonconstant regular functions on X. The assumption $X = \bigcup_{n=1}^\infty X_n$ thus implies that there is a countable set of nonconstant regular functions $f_1, f_2, \ldots$ such that X is the union of the zero sets of these functions. But since f_k is holomorphic and X is connected, the zero set of f_k is nowhere dense in X in the relative metric topology from $\mathbb{C}^n$. We conclude that X is a countable union of nowhere-dense sets, which contradicts the Baire category theorem. ☐

Appendix B
Linear and Multilinear Algebra

Abstract We recall some basic facts from linear and multilinear algebra in the form needed in the book. For most results we give proofs to make the exposition self-contained, but the reader is assumed to have already encountered linear algebra at this level of abstraction. Standard references are Lang [97] and Jacobson [80].

B.1 Jordan Decomposition

We obtain the Jordan decomposition of a square matrix in the forms needed for algebraic groups, Lie algebras, and representation theory.

B.1.1 Primary Projections

Let M_n denote the algebra of $n \times n$ complex matrices. If $A \in M_n$, then the set $\{I, A, A^2, A^3, \ldots, A^{n^2}\}$ must be linearly dependent, since $\dim M_n = n^2$. Hence there exists a monic polynomial $f(x)$ with $f(A) = 0$. Take any such f and factor it into powers of distinct linear terms:

$$f(x) = \prod_{i=1}^{d} (x - \lambda_i)^{m_i}, \quad \text{where } \lambda_i \neq \lambda_j \text{ for } i \neq j.$$

The function $(x - \lambda_i)^{m_i} / f(x)$ is analytic at $x = \lambda_i$. Let $g_i(x)$ be its Taylor polynomial of degree $m_i - 1$ centered at λ_i. We then have the *partial fraction decomposition*

$$\frac{1}{f(x)} = \sum_{i=1}^{d} \frac{g_i(x)}{(x - \lambda_i)^{m_i}}. \tag{B.1}$$

To verify this, note that the difference between the left and right sides of (B.1) is a rational function with no finite singularities that vanishes at infinity, hence is identically zero. Set $p_i(x) = f(x)g_i(x)(x-\lambda_i)^{-m_i} = g_i(x)\prod_{j\neq i}(x-\lambda_j)^{m_j}$.

Lemma B.1.1. *Define* $P_i = p_i(A) \in M_n$. *Then* $(A-\lambda_i)^{m_i}P_i = 0$ *for* $i = 1,\dots,d$, *and* $\{P_i\}$ *is a resolution of the identity:*

$$\sum_{i=1}^d P_i = I \quad and \quad P_iP_j = \begin{cases} P_i \text{ for } i = j, \\ 0, \text{ otherwise.} \end{cases}$$

Proof. The polynomial $(x-\lambda_i)^{m_i}p_i(x)$ is divisible by $f(x)$ and hence it annihilates A. By (B.1) we see that $\sum_i p_i(x) = 1$, so $\sum_i P_i = I$. Since the polynomial $p_i(x)p_j(x)$ is divisible by $f(x)$ when $i \neq j$, we have $P_iP_j = 0$ for $i \neq j$. This implies that $P_i = P_i(\sum_i P_j) = P_i^2$.   □

Corollary B.1.2. *Let* $S \in M_n(\mathbb{C})$. *The following are equivalent:*

1. *There exists* $g \in \mathrm{GL}(n,\mathbb{C})$ *such that* gSg^{-1} *is a diagonal matrix.*
2. *There exists a basis* $\{v_i\}$ *for* $\mathbb{C}^n$ *consisting of eigenvectors for* S.
3. *There exists a polynomial* $\varphi(x) = \prod_i(x-\lambda_i)$ *with* $\lambda_i \neq \lambda_j$ *for* $i \neq j$ *such that* $\varphi(A) = 0$.

Proof. The implication (1) ⇔ (2) is clear; to see that (2) ⇒ (3), take for $\{\lambda_i\}$ the distinct eigenvalues of S. Now assume that (3) holds and define the projections $\{P_i\}$ as in Lemma B.1.1 relative to the polynomial $\varphi(x)$ and the matrix S. By part (2) of that lemma the range of P_i consists of eigenvectors of S for the eigenvalue λ_i. Hence S has a complete set of eigenvectors.   □

We say that a matrix S is *semisimple* if it satisfies the conditions of Corollary B.1.2. Note that if S is semisimple and $W \subset \mathbb{C}^n$ is a subspace such that $SW \subset W$, then $S|_W$ is semisimple by part (3) of the corollary.

B.1.2 Additive Jordan Decomposition

Theorem B.1.3. *Let* $A \in M_n$. *Then there exist unique matrices* $S,N \in M_n$ *such that*

1. $A = S+N$,
2. S *is semisimple and* N *is nilpotent*,
3. $NS = SN$.

Furthermore, there is a polynomial $\varphi(x)$ *such that* $S = \varphi(A)$.

Proof. Construct the projections $\{P_i\}$ for A as in Lemma B.1.1 and set $S = \sum_i \lambda_i P_i$ and $N = A - S$. By (1) and (2) of that lemma we see that S is semisimple and N is nilpotent. Since $P_i = p_i(A)$, it follows that S is a polynomial in A, and hence $NS = SN$.

To prove uniqueness, suppose the pair of matrices S', N' satisfy (1), (2), and (3). Then $S'A = AS'$ and $N'A = AN'$. Since the projections P_i are polynomials in A, they commute with S' and N'. Hence S' and N' leave invariant the subspaces $V_i = \text{Range}(P_i)$. Since $S'N' = N'S'$, we can pick a basis for V_i such that the matrix for S' in this basis has 0 below the main diagonal, while the matrix for N' has 0 below and on the main diagonal (by the nilpotence of N'). Hence

$$\det(xI - S'|_{V_i}) = \det(xI - A|_{V_i}) = (x - \lambda_i)^{d_i} \quad (d_i = \dim V_i) .$$

Thus the only eigenvalue for $S'|_{V_i}$ is λ_i. Since $S'|_{V_i}$ is semisimple, it follows that $S'|_{V_i} = \lambda_i$ for all i. Hence $S' = S$ and $N' = A - S = N$. $\qquad\square$

We write $A_s = S$ and $A_n = N$ for the semisimple and nilpotent parts of A in the additive Jordan decomposition.

B.1.3 Multiplicative Jordan Decomposition

A matrix $u \in M_n$ is called *unipotent* if $u = I + N$, where N is nilpotent.

Theorem B.1.4. *Let $g \in \mathbf{GL}(n, \mathbb{C})$. There exist unique $s, u \in \mathbf{GL}(n, \mathbb{C})$ such that*

1. $g = su$,
2. s is semisimple and u is unipotent,
3. $us = su$.

Furthermore, there is a polynomial $\varphi(x)$ such that $s = \varphi(g)$.

Proof. Let $g = s + N$ be the additive Jordan decomposition of g. Then $\det(g) = \det(s)$, so $s \in \mathbf{GL}(n, \mathbb{C})$. Set $u = I + s^{-1}N$. Since $sN = Ns$, the operator $s^{-1}N$ is nilpotent, so u is unipotent and $g = su$. Uniqueness follows from the uniqueness of the additive Jordan decomposition. $\qquad\square$

We write $s = g_s$ and $u = g_u$ for the semisimple and unipotent factors in the multiplicative Jordan decomposition of g.

B.2 Multilinear Algebra

In this section we review standard properties of bilinear forms and tensor products of vector spaces, together with some more specialized topics such as matrix factorizations and Pfaffians.

B.2.1 Bilinear Forms

The classical groups associated with bilinear forms were introduced in Section 1.1.2. Here we obtain some additional results about bilinear forms, following the notation of Section 1.1.2. If V is a finite-dimensional vector space over $\mathbb{C}$, then $V^* = \mathrm{Hom}_{\mathbb{C}}(V, \mathbb{C})$ denotes the dual space of V, and we write $\langle v^*, v \rangle = v^*(v)$ for $v^* \in V^*$ and $v \in V$.

For any bilinear form B on $\mathbb{C}^n$, a vector $v \in \mathbb{C}^n$ is *B-isotropic* if $B(v,v) = 0$. A subspace $W \subset \mathbb{C}^n$ is *B-isotropic* if $B(u,v) = 0$ for all $u,v \in W$, and it is *maximal isotropic* if there is no larger B-isotropic subspace containing it.

Lemma B.2.1. *Let B be a nondegenerate bilinear form on $\mathbb{C}^n$. Suppose $W \subset \mathbb{C}^n$ is B-isotropic. Then* $\dim W \leq n/2$.

Proof. For any subspace $U \subset \mathbb{C}^n$ let

$$U^{\perp} = \{x \in \mathbb{C}^n : B(x,u) = 0 \text{ for all } u \in U\}.$$

Since B is nondegenerate, $\dim U + \dim U^{\perp} = n$. When $U = W$ is isotropic, we have $W \subset W^{\perp}$. Hence $2 \dim W \leq \dim W + \dim W^{\perp} = n$. $\square$

Suppose B is a nondegenerate and symmetric bilinear form on $\mathbb{C}^n$ and that $n = 2l$ is even. A basis $\{v_1, \ldots, v_l, v_{-1}, \ldots, v_{-l}\}$ for $\mathbb{C}^n$ that satisfies $B(v_i, v_j) = \delta_{i,-j}$ for $i, j = \pm 1, \ldots, \pm l$ will be called a *B-isotropic basis*. There always exist such bases. For example, let $u_1, \ldots, u_n$ be a B-orthonormal basis, and set

$$v_i = (u_i + i u_{n+1-i})/\sqrt{2}, \qquad v_{-i} = (u_i - i u_{n+1-i})/\sqrt{2}$$

for $i = 1, 2, \ldots, l$. The form B vanishes on the subspaces

$$W = \mathrm{Span}\{v_1, v_2, \ldots, v_l\}, \qquad W^* = \mathrm{Span}\{v_{-1}, v_{-2}, \ldots, v_{-l}\}. \tag{B.2}$$

Since these subspaces have dimension $n/2$, they are *maximal isotropic* relative to B, by Lemma B.2.1. Furthermore, the restriction of B to $W \times W^*$ is nondegenerate and sets up a duality between W and W^*. We have $\mathbb{C}^n = W \oplus W^*$.

When $n = 2l + 1$, a basis $\{v_0, v_1, \ldots, v_l, v_{-1}, \ldots, v_{-l}\}$ for $\mathbb{C}^n$ that satisfies $B(v_i, v_j) = \delta_{i,-j}$ for $i, j = 0, \pm 1, \ldots, \pm l$ will be called a *B-isotropic basis* (even though the vector v_0 is not B-isotropic, since $B(v_0, v_0) = 1$). There always exist such bases, by the same construction as in the even-dimensional case. The subspaces defined in (B.2) are maximal isotropic relative to B, and the restriction of B to $W \times W^*$ is nondegenerate. We have $\mathbb{C}^n = W \oplus \mathbb{C}v_0 \oplus W^*$.

Lemma B.2.2. *Let E be an n-dimensional vector space with a nondegenerate symmetric bilinear form B. Suppose $\{0\} \neq F \subset E$ is a B-isotropic subspace. Then any basis $\{v_1, \ldots, v_k\}$ for F can be extended to a B-isotropic basis for E.*

Proof. We proceed by induction on n. We may assume that $n \geq 2$, since for $n = 1$ there are no isotropic vectors. Set $W = \mathrm{Span}\{v_2, \ldots, v_k\}$. Since $v_1 \notin W$, there exists

$u \in W^\perp$ such that $B(u,v_1) = 1$. Set $v_{-1} = u - (1/2)B(u,u)v_1$. Then $v_{-1} \in W^\perp$, since F is isotropic and $u \in W^\perp$. Also,

$$B(v_{-1}, v_{-1}) = B(u,u) - B(u,u)B(u,v_1) = 0 \,,$$

so v_{-1} is an isotropic vector. If $n = 2$ then $k = 1$ by Lemma B.2.1 and we are done. If $n = 3$ then $k = 1$ for the same reason and $\{v_1, v_{-1}\}^\perp = \mathbb{C}v_0$ for some nonisotropic vector v_0.

Now take $n \geq 4$ and assume that the lemma is true for spaces E of dimension less than n. Set $U = \mathrm{Span}\{v_1, v_{-1}\}$. Then the restriction of B to U is nondegenerate; therefore, $E = U \oplus U^\perp$ and the restriction of B to $U^\perp$ is nondegenerate. Since $W \subset U^\perp$ and $\dim U^\perp = \dim E - 2$, we may apply the induction hypothesis with E replaced by $U^\perp$ and V replaced by W. This gives a B-isotropic basis for $U^\perp$:

$$\{v_2, \ldots, v_n, v_{-2}, \ldots, v_{-n}\} \qquad \text{when } n \text{ is even,}$$
$$\{v_2, \ldots, v_n, v_0, v_{-2}, \ldots, v_{-n}\} \qquad \text{when } n \text{ is odd.}$$

Adjoining $\{v_1, v_{-1}\}$ to this basis, we obtain the desired B-isotropic basis for E. $\quad\square$

B.2.2 Tensor Products

In the following all vector spaces will be over $\mathbb{C}$ (or any field of characteristic zero). Let U and V be vector spaces (not necessarily finite-dimensional). The *tensor product* of U and V is a vector space $U \otimes V$ together with a bilinear mapping $\tau : u, v \mapsto u \otimes v$ from $U \times V \longrightarrow U \otimes V$ satisfying the following *universal mapping property:* Given any vector space W and bilinear map $\beta : U \times V \longrightarrow W$, there exists a unique *linear* map $B : U \otimes V \longrightarrow W$ such that $\beta = B \circ \tau$:

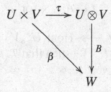

To show that the tensor product exists, let $\{u_i : i \in I\}$ and $\{v_j : j \in J\}$ be bases for U and V, respectively. Define $U \otimes V$ to be the vector space with basis $\{u_i \otimes v_j : (i,j) \in I \times J\}$. Define the map τ by

$$\tau(u,v) = \sum_{i,j} \langle u_i^*, u \rangle \langle v_j^*, v \rangle\, u_i \otimes v_j \,,$$

where $\{u_i^*\}$ is the dual basis for U^* and $\{v_j^*\}$ is the dual basis for V^*. Clearly τ is bilinear. Given β as above, define B on the basis for $U \otimes V$ by $B(u_i \otimes v_j) = \beta(u_i, v_j)$ and extend B to a linear map from $U \otimes V$. This satisfies the universal mapping property and is clearly unique. From the mapping property it follows that the tensor product of two vector spaces is uniquely determined up to isomorphism. In particular, the bilinear map $(u, v) \mapsto v \otimes u$ from $U \times V$ to $V \otimes U$ extends to an isomorphism $U \otimes V \cong V \otimes U$.

The construction of tensor products is *functorial*: Given vector spaces U, V, X, and Y and linear maps $f : U \longrightarrow X$ and $g : V \longrightarrow Y$, there is a unique linear map $f \otimes g : U \otimes V \longrightarrow X \otimes Y$ such that $(f \otimes g)(u \otimes v) = f(u) \otimes g(v)$. Since $f, g \mapsto f \otimes g$ is a bilinear map from $\mathrm{Hom}(U, X) \times \mathrm{Hom}(V, Y)$ to the vector space $\mathrm{Hom}(U \otimes V, X \otimes Y)$, it extends to a linear map

$$\mathrm{Hom}(U, X) \otimes \mathrm{Hom}(V, Y) \longrightarrow \mathrm{Hom}(U \otimes V, X \otimes Y) . \tag{B.3}$$

It is easy to check that this map is injective. In particular, taking $X = Y = \mathbb{C}$, we have a natural embedding $U^* \otimes V^* \hookrightarrow (U \otimes V)^*$ such that $u^* \in U^*$ and $v \in V^*$ give the linear functional

$$\langle u^* \otimes v^*, u \otimes v \rangle = \langle u^*, u \rangle \langle v^*, v \rangle, \quad \text{for } u \in U \text{ and } v \in V .$$

Given a basis $\{u_i\}$ for U and $\{v_j\}$ for V, we can write $x \in U \otimes V$ as

$$x = \sum_{i,j} x^{ij} u_i \otimes v_j ,$$

where x^{ij} are the *components* of x relative to the basis $\{u_i \otimes v_j\}$ for $U \otimes V$. We can express these components in terms of the corresponding dual bases $\{u_i^*\}$ and $\{v_j^*\}$ as $x^{ij} = \langle u_i^* \otimes v_j^*, x \rangle$.

Assume now that U and V are finite-dimensional. From the construction of the basis for $U \otimes V$ we see that $\dim(U \otimes V) = \dim U \dim V$. There is a natural isomorphism between the vector space $\mathrm{Hom}(U^*, V)$ and $U \otimes V$ obtained as follows: Given $u, v \in U \times V$, let $T_{u,v}$ be the rank-one linear transformation

$$T_{u,v}(u^*) = \langle u^*, u \rangle v, \quad \text{for } u^* \in U^* .$$

The map $u, v \mapsto T_{u,v}$ from $U \times V \longrightarrow \mathrm{Hom}(U^*, V)$ is bilinear, so there is a unique linear map $T : U \otimes V \longrightarrow \mathrm{Hom}(U^*, V)$ such that $T(u \otimes v) = T_{u,v}$. In terms of a basis $\{u_i\}$ for U and $\{v_j\}$ for V we have $T(u_i \otimes v_j)u_k^* = \delta_{ik} v_j$, from which we see that T is injective. Since

$$\dim \mathrm{Hom}(U^*, V) = \dim U \dim V = \dim U \otimes V ,$$

it follows that T gives a natural isomorphism $U \otimes V \cong \mathrm{Hom}(U^*, V)$. The trace bilinear form $\langle A, B \rangle = \mathrm{tr}(AB)$, for $A \in \mathrm{Hom}(V, U^*)$ and $B \in \mathrm{Hom}(U^*, V)$, is nondegenerate and gives a natural isomorphism $(\mathrm{Hom}(U^*, V))^* \cong \mathrm{Hom}(V, U^*)$. Thus we have natural isomorphisms

$$(U \otimes V)^* \cong \mathrm{Hom}(V, U^*) \cong U^* \otimes V^* . \tag{B.4}$$

Furthermore, by the universal mapping property, there is a bijective correspondence between bilinear maps $f : U \times V \longrightarrow \mathbb{C}$ and linear functionals $F \in (U \otimes V)^*$ given by

$$\langle F, u \otimes v \rangle = f(u,v), \quad \text{for } u \in U, v \in V .$$

Hence by (B.4) we also have a natural isomorphism of $U^* \otimes V^*$ with the space of all bilinear functions $f : U \times V \longrightarrow \mathbb{C}$. We can write $x^* \in U^* \otimes V^*$ as

$$x^* = \sum_{i,j} x^*_{ij} u^*_i \otimes v^*_j ,$$

where $x^*_{ij} = \langle x^*, u_i \otimes v_j \rangle$ are the components of x^* relative to the basis $\{u^*_i \otimes v^*_j\}$ for $U \otimes V$. The natural pairing between $U \otimes V$ and $U^* \otimes V^*$ is expressed in tensor components as

$$\langle x^*, x \rangle = \sum_{i,j} x^*_{ij} x^{ij} .$$

The naturality property implies that the expression on the right is independent of the choice of bases.

For any other pair X, Y of finite-dimensional vector spaces, the mapping (B.3) is bijective, by dimensional considerations, and it gives a natural isomorphism

$$\mathrm{Hom}(U,X) \otimes \mathrm{Hom}(V,Y) \cong \mathrm{Hom}(U \otimes V, X \otimes Y) .$$

Iterated Tensor Products

Let U, V, and W be three vector spaces. Let τ be the bilinear map

$$\tau : (U \otimes V) \times W \longrightarrow U \otimes (V \otimes W) , \qquad \tau(u \otimes v, w) = u \otimes (v \otimes w) .$$

This map extends to a linear isomorphism $(U \otimes V) \otimes W \cong U \otimes (V \otimes W)$. We note that the space $U \otimes (V \otimes W)$ satisfies the universal mapping property for the pair of vector spaces $(U \otimes V, W)$. Indeed, for every vector space X and bilinear map $\beta : (U \otimes V) \times W \longrightarrow X$, there is a unique bilinear map

$$\tilde{\beta} : U \times (V \otimes W) \longrightarrow X , \quad \text{where } \tilde{\beta}(u, v \otimes w) = \beta(u \otimes v, w) .$$

Hence there is a unique linear map $\tilde{B} : U \otimes (V \otimes W) \longrightarrow X$ such that $\beta = \tilde{B} \circ \tau$. We may thus write $U \otimes V \otimes W$ without ambiguity.

In general, given vector spaces $V_1, \dots, V_p$, we have the p-fold tensor product $V_1 \otimes V_2 \otimes \cdots \otimes V_p$ that satisfies the universal mapping property relative to p-linear maps: If X is a vector space and $f : V_1 \times V_2 \times \cdots \times V_p \longrightarrow X$ is a map that is linear in each argument, then there is a unique linear map

$$F : V_1 \otimes V_2 \otimes \cdots \otimes V_p \longrightarrow X$$

such that $F(v_1 \otimes v_2 \otimes \cdots \otimes v_p) = f(v_1, v_2, \dots, v_p)$. When $V_i = V$ for all i, then we write $V^{\otimes p}$ or $\bigotimes^p V$ for the p-fold tensor product of V with itself, and

$$v^{\otimes p} = \underbrace{v \otimes \cdots \otimes v}_{p \text{ factors}} \quad \text{for } v \in V .$$

Likewise, for $T \in \text{Hom}(V, W)$ we write $T^{\otimes p}$ for the map

$$\underbrace{T \otimes \cdots \otimes T}_{p \text{ factors}} : V^{\otimes p} \longrightarrow W .$$

We have a natural isomorphism $\text{Hom}(V^{\otimes p}, X) \cong L^p(V, X)$, where $L^p(V, X)$ is the space of p-multilinear maps from $V \times \cdots \times V$ (p factors) to X. We call

$$V^{\otimes p} \otimes (V^*)^{\otimes q}$$

the space of *mixed tensors of type* (p, q), relative to V.

Suppose the spaces V_i are finite-dimensional. Just as in the case $p = 2$ there is a natural isomorphism

$$(V_1 \otimes V_2 \otimes \cdots \otimes V_p)^* \cong V_1^* \otimes V_2^* \otimes \cdots \otimes V_p^* .$$

Combining this with the results above, we see that for finite-dimensional vector spaces U and V there are natural isomorphisms

$$\text{End}(U) \otimes \text{End}(V) \cong U^* \otimes U \otimes V^* \otimes V$$
$$\cong U^* \otimes V^* \otimes U \otimes V \cong \text{End}(U \otimes V) .$$

In particular, when $V_i = V$ for all i then $(V^*)^{\otimes p} \cong (V^{\otimes p})^*$.

Contractions

Consider the space $U^* \otimes U$ of mixed tensors of type $(1, 1)$. There is a unique linear functional C (called *tensor contraction*) on $U^* \otimes U$ such that $C(u^* \otimes u) = \langle u^*, u \rangle$. We denote the components of $z \in U^* \otimes U$ relative to a basis $\{u_i^* \otimes u_j\}$ for $U^* \otimes U$ by $z_i^j = \langle u_i \otimes u_j^*, z \rangle$. Then

$$C(z) = \sum_i z_i^i .$$

The canonical isomorphism $U^* \otimes U \cong \text{End}(U)$ transforms C into the linear functional $x \mapsto \text{tr}(x)$ for $x \in \text{End}(U)$. Because of this, a tensor $z \in U^* \otimes U$ is called *traceless* if $C(z) = 0$.

B.2.3 Symmetric Tensors

Let V be a vector space and k a positive integer. The symmetric group $\mathfrak{S}_k$ acts on $V^{\otimes k}$ by permuting the positions of the factors in the tensor product:

$$\sigma_k(s)(v_1 \otimes \cdots \otimes v_k) = v_{s^{-1}(1)} \otimes \cdots \otimes v_{s^{-1}(k)}$$

for $s \in \mathfrak{S}_k$ and $v_1, \ldots, v_k \in V$. Note that $\sigma_k(s)$ moves the vector in the ith position to the vector in the position $s(i)$. Hence $\sigma_k(st) = \sigma_k(s)\sigma_k(t)$ for $s, t \in \mathfrak{S}_k$ and $\sigma_k(1) = 1$. Thus $\sigma_k : \mathfrak{S}_k \longrightarrow \mathbf{GL}(V^{\otimes k})$ is a group homomorphism. Define

$$\mathbf{Sym}(v_1 \otimes \cdots \otimes v_k) = \frac{1}{k!} \sum_{s \in \mathfrak{S}_k} \sigma_k(s)(v_1 \otimes \cdots \otimes v_k) .$$

Then the operator $\mathbf{Sym}$ is the projection onto the space of $\mathfrak{S}_k$-fixed tensors in $V^{\otimes k}$. Its range $S^k(V) = \mathbf{Sym}(V^{\otimes k})$ is the space of *symmetric k-tensors* over V. For example, when $k = 2$, then $S^2(V)$ is spanned by the tensors $x \otimes y + y \otimes x$ for $x, y \in V$.

We can also characterize $S^k(V)$ in terms of a universal mapping property: Given any k-multilinear map

$$f : \underbrace{V \times \cdots \times V}_{k \text{ factors}} \longrightarrow W$$

that is *symmetric* in its arguments (that is, $f \circ \sigma_k(s) = f$ for all $s \in \mathfrak{S}_k$), there is a unique linear map $F : S^k(V) \longrightarrow W$ such that $F(\mathbf{Sym}(v_1 \otimes \cdots \otimes v_k)) = f(v_1, \ldots, v_k)$.

Suppose V is finite-dimensional. Let $\mathcal{P}^k(V^*)$ be the space of homogeneous polynomials of degree k on V^*. This is the linear span of the functions

$$v^* \mapsto \prod_{i=1}^{k} \langle v^*, v_i \rangle \quad \text{with } v_l \in V . \tag{B.5}$$

From (B.5) we get a symmetric k-multilinear map

$$f : \underbrace{V \times \cdots \times V}_{k \text{ factors}} \longrightarrow \mathcal{P}^k(V^*) .$$

By the universal property, there is a unique map $F : S^k(V) \longrightarrow \mathcal{P}^k(V^*)$ such that

$$F(\mathbf{Sym}(v_1 \otimes \cdots \otimes v_k))(v^*) = \prod_{i=1}^{k} \langle v^*, v_i \rangle .$$

In particular, $F(v^{\otimes k})(v^*) = \langle v^*, v \rangle^k$, and this equation uniquely determines F, because of the following result:

Lemma B.2.3 (Polarization). *$S^k(V)$ is spanned by the elements $v^{\otimes k}$ for $v \in V$.*

Proof. Fix $v_1, \ldots, v_k \in V$. It suffices to prove the *polarization identity*:

$$\mathbf{Sym}(v_1 \otimes \cdots \otimes v_k)$$
$$= \frac{1}{k! 2^{k-1}} \sum_{\varepsilon_j = \pm 1} \left(\prod_{j=2}^{k} \varepsilon_j \right) (v_1 + \varepsilon_2 v_2 + \cdots + \varepsilon_k v_k)^{\otimes k} . \tag{B.6}$$

For example, when $k = 2$ this is the identity

$$v_1 \otimes v_2 + v_2 \otimes v_1 = \frac{1}{2} \{ (v_1 + v_2) \otimes (v_1 + v_2) - (v_1 - v_2) \otimes (v_1 - v_2) \} .$$

We can prove (B.6) by Fourier analysis on the abelian group

$$\Gamma = \underbrace{\{\pm 1\} \times \cdots \times \{\pm 1\}}_{k-1 \text{ factors}},$$

as follows: Given any linear functional $\varphi \in S^k(V)^*$, define a function f on Γ by $f(\gamma) = \langle \varphi, (v_1 + \varepsilon_2 v_2 + \cdots + \varepsilon_k v_k)^{\otimes k} \rangle$, where $\gamma = (\varepsilon_2, \dots, \varepsilon_k)$. The characters of Γ are of the form

$$\chi_J(\gamma) = \prod_{j \in J} \varepsilon_j$$

with J ranging over the subsets of $\{2, \dots, n\}$. Hence by Theorem 4.3.9,

$$f(\gamma) = \sum_J \chi_J(\gamma) \widehat{f}(\chi_J), \tag{B.7}$$

where $\widehat{f}(\chi_J) = 1/|\Gamma| \sum_{\gamma \in \Gamma} f(\gamma) \chi_J(\gamma)$. Let $\chi_{\max}(\gamma) = \prod_{j=2}^k \varepsilon_j$ be the character of Γ corresponding to the maximal subset $J = \{2, \dots, k\}$. By expanding the k-fold tensor power, we see directly from the definition of f that the coefficient of $\chi_{\max}$ in (B.7) is $k! \langle \varphi, \mathbf{Sym}(v_1 \otimes \cdots \otimes v_k) \rangle$. Since the characters are linearly independent, we conclude that

$$k! \langle \varphi, \mathbf{Sym}(v_1 \otimes \cdots \otimes v_k) \rangle = \widehat{f}(\chi_{\max}) = \frac{1}{2^{k-1}} \sum_{\gamma \in \Gamma} f(\gamma) \chi_{\max}(\gamma).$$

This holds for all $\varphi \in S^k(V)^*$, and so (B.6) follows.   □

Proposition B.2.4. *Suppose* $\dim V < \infty$. *Then the map carrying* $v^{\otimes k}$ *to the function* $v^* \mapsto \langle v^*, v \rangle^k$ *extends uniquely to a linear isomorphism* $F : S^k(V) \longrightarrow \mathcal{P}^k(V^*)$.

Proof. The existence and uniqueness of F have already been proved, so it remains only to show that F is bijective. Assume that $\dim V = d$ and fix a basis $\{v_1, \dots, v_d\}$ for V. Let $\mathcal{I}_k$ be the set of all finite monotonic sequences

$$M = \{i_1 \leq i_2 \leq \cdots \leq i_k\} \quad \text{with} \quad 1 \leq i_p \leq d.$$

For each such sequence M, we define $v_M = v_{i_1} \otimes v_{i_2} \otimes \cdots \otimes v_{i_k} \in V^{\otimes k}$ and $\varphi_M = F(\mathbf{Sym}(v_M)) \in S^k(V^*)$. Since

$$\mathbf{Sym}(x_1 \otimes \cdots \otimes x_p \otimes x_{p+1} \otimes \cdots \otimes x_k)$$
$$= \mathbf{Sym}(x_1 \otimes \cdots \otimes x_{p+1} \otimes x_p \otimes \cdots \otimes x_k)$$

for all $x_i \in V$, it is clear that the set $\{\mathbf{Sym}(v_M) : M \in \mathcal{I}_k\}$ spans $S^k(V)$. On the other hand,

$$\varphi_M(v^*) = \prod_{p=1}^k \langle v^*, v_{i_p} \rangle.$$

Thus the monomials $\{\varphi_M : M \in \mathcal{I}_k\}$ are a basis for $\mathcal{P}^k(V^*)$, so we conclude that F is an isomorphism.   □

B.2.4 Alternating Tensors

Let V be a vector space. We define an operator **Alt** on $V^{\otimes k}$ by

$$\mathbf{Alt}(v_1 \otimes \cdots \otimes v_k) = \frac{1}{k!} \sum_{s \in \mathfrak{S}_k} \det(s)\, v_{s(1)} \otimes \cdots \otimes v_{s(k)} \,.$$

Then **Alt** is the projection operator onto the space of tensors in $V^{\otimes k}$ that transform by the character $s \mapsto \det(s)$ of $\mathfrak{S}_k$. We define $\bigwedge^k V = \mathbf{Alt}(V^{\otimes k})$, the space of *alternating k-tensors* over V, and we write

$$v_1 \wedge \cdots \wedge v_k = \mathbf{Alt}(v_1 \otimes \cdots \otimes v_k) \quad \text{for } v_i \in V \,.$$

For $s \in \mathfrak{S}_k$ we have $v_{s(1)} \wedge \cdots \wedge v_{s(k)} = \varepsilon\, v_1 \wedge \cdots \wedge v_k$, where $\varepsilon = \det(s) = \pm 1$.

We can also characterize $\bigwedge^k V$ in terms of the following universal mapping property: Given any k-multilinear map

$$f : \underbrace{V \times \cdots \times V}_{k \text{ factors}} \longrightarrow W$$

that is *skew-symmetric* in its arguments (that is, $f \circ \sigma_k(s) = \det(s) f$ for all $s \in \mathfrak{S}_k$), there is a unique linear map $F : \bigwedge^k V \longrightarrow W$ such that

$$F(\mathbf{Alt}(v_1 \otimes \cdots \otimes v_k)) = f(v_1, \ldots, v_k) \,.$$

In particular, when $W = C^k(V^*)$ is the space of all skew-symmetric k-multilinear complex-valued functions on V^*, then $F : \bigwedge^k V \longrightarrow C^k(V^*)$ satisfies

$$F(v_1 \wedge \ldots \wedge v_k)(v_1^*, \ldots, v_k^*) = \det[\langle v_i^*, v_j \rangle] \tag{B.8}$$

for $v_i \in V$ and $v_i^* \in V^*$.

Proposition B.2.5. *Assume* $\dim V = d$. *Then* $\dim \bigwedge^k V = \binom{d}{k}$ *for* $0 \le k \le d$, *and* $\bigwedge^k V = 0$ *for* $k > d$. *Furthermore, the map* F *in* (B.8) *gives a linear isomorphism between* $\bigwedge^k V$ *and* $C^k(V^*)$.

Proof. Fix a basis $\{v_1, \ldots, v_d\}$ for V. Then the set $\{v_{i_1} \wedge \cdots \wedge v_{i_k} : 1 \le i_p \le d\}$ spans $\bigwedge^k V$. Hence by skew symmetry $\bigwedge^k V = 0$ when $k > d$, since at least two of the indices i_p must be the same in this case. For $0 \le k \le d$ we can take

$$\{v_{i_1} \wedge \cdots \wedge v_{i_k} : 1 \le i_1 < \cdots < i_k \le d\} \tag{B.9}$$

as a spanning set for $\bigwedge^k V$. Let $\{v_i^*\}$ be the dual basis for V^*. Then

$$F(v_{i_1} \wedge \cdots \wedge v_{i_k})(v_{j_1}^*, \ldots, v_{j_k}^*) = \begin{cases} 1 & \text{if } i_p = j_p \text{ for all } p, \\ 0 & \text{otherwise.} \end{cases}$$

This implies that the set (B.9) is linearly independent, and so it is a basis for $\bigwedge^k V$. This calculation also shows that F is bijective, since any k-linear function on V^* is determined by its values on the basis $\{v_j^*\}$. $\square$

•

B.2.5 Determinants and Gauss Decomposition

Let M_k be the space of $k \times k$ complex matrices and $M_{k,n}$ the space of $k \times n$ complex matrices. Let N_n^+ denote the group of upper-triangular $n \times n$ matrices with diagonal entries 1, N_n^- the group of lower-triangular $n \times n$ matrices with diagonal entries 1, and $D_{k,n}$ the $k \times n$ matrices $x = [x_{ij}]$ with $x_{ij} = 0$ for $i \neq j$.

For $x \in M_{k,n}$ define the *principal minors*

$$\Delta_i(x) = \det \begin{bmatrix} x_{11} & \cdots & x_{1i} \\ \vdots & \ddots & \vdots \\ x_{i1} & \cdots & x_{ii} \end{bmatrix}$$

for $i = 1, \ldots, \min\{k,n\}$. It is also convenient to define $\Delta_0(x) = 1$.

Lemma B.2.6 (Gauss Decomposition). *Suppose* $x \in M_{k,n}$ *satisfies* $\Delta_i(x) \neq 0$ *for* $i = 1, \ldots, \min\{k,n\}$. *Then there are matrices* $v \in N_k^-$, $u \in N_n^+$, *and* $h \in D_{k,n}$ *such that*

$$x = vhu \,. \tag{B.10}$$

The matrix h is uniquely determined by x, and

$$h_{ii} = \Delta_i(x)/\Delta_{i-1}(x) \quad for \ i = 1, \ldots, \min\{k,n\} \,.$$

If $k = n$ *then* v *and* u *are also uniquely determined by* x.

Proof. We have $x_{11} = \Delta_1(x) \neq 0$. Set $z = x_{11}^{-1}$ and let

$$v_1 = \begin{bmatrix} 1 & 0 & \cdots & 0 \\ -x_{21}z & 1 & \cdots & 0 \\ \vdots & \vdots & \ddots & \vdots \\ -x_{k1}z & 0 & \cdots & 1 \end{bmatrix}, \quad u_1 = \begin{bmatrix} 1 & -zx_{12} & \cdots & -zx_{1n} \\ 0 & 1 & \cdots & 0 \\ \vdots & \vdots & \ddots & \vdots \\ 0 & 0 & \cdots & 1 \end{bmatrix}.$$

The matrices $v_1 \in N_k^-$ and $u_1 \in N_n^+$ are chosen so that

$$v_1 x u_1 = \begin{bmatrix} x_{11} & 0 \\ 0 & y \end{bmatrix} \quad \text{with} \ y \in M_{k-1,n-1} \,.$$

Let $m = \min\{k,n\}$. If $m = 1$ then we are done. Assume $m > 1$. Left multiplication of x by an element of N_k^- adds a multiple of the ith row of x to the jth row, where $i < j$. Likewise, right multiplication of x by an element of N_n^+ adds a multiple of the ith column of x to the jth column, where $i < j$. Thus we have

$$\Delta_i(x) = \Delta_i(v_1 x u_1) = x_{11}\Delta_{i-1}(y) \quad \text{for } i = 1,\ldots,m.$$

Hence $\Delta_i(y) \neq 0$ for $i = 1,\ldots,m-1$, so by induction there exist $v_2 \in N_{k-1}^-$ and $u_2 \in N_{n-1}$ such that $v_2 y u_2 \in D_{k-1,n-1}$. Define $v \in N_k^-$ and $u \in N_n^+$ by

$$v^{-1} = \begin{bmatrix} 1 & 0 \\ 0 & v_2 \end{bmatrix} v_1, \qquad u^{-1} = u_1 \begin{bmatrix} 1 & 0 \\ 0 & u_2 \end{bmatrix}.$$

Then $h = v^{-1}xu^{-1} \in D_{k,n}$ and $\Delta_i(h) = \Delta_i(x)$ for $i = 1,\ldots,\min\{k,n\}$. Since $\Delta_i(h) = h_{11}\cdots h_{ii}$, we have $h_{ii} = \Delta_i(x)/\Delta_{i-1}(x)$.

Let $k = n$. Suppose $x \in M_n$ has two factorizations $x = v_1 h_1 u_1 = v_2 h_2 u_2$ with $v_i \in N_n^-$, $u_i \in N_n^+$, and $h_i \in D_n$. Then

$$v_2^{-1}v_1 = h_2 u_2 u_1^{-1} h_1^{-1}.$$

But the matrix on the left side is lower triangular with 1 on the diagonal, whereas the matrix on the right side is upper triangular. Hence both matrices are the identity; thus $v_1 = v_2$. The same argument applied to x^t shows that $u_1 = u_2$. Thus $h_1 = h_2$ also. $\qquad\square$

When a symmetric matrix has a Gauss factorization, then it also has the following factorization:

Lemma B.2.7 (Cholesky Decomposition). *Suppose $x \in M_n$ is a symmetric matrix and $\Delta_i(x) \neq 0$ for $i = 1,\ldots,n$. Then there exists an upper-triangular matrix $b \in M_n$ such that $x = b^t b$. The matrix b is uniquely determined by x up to left multiplication by a diagonal matrix with entries ± 1.*

Proof. We first find $v \in N_n^+$ such that $v^t x v$ is diagonal. Let u_1 and v_1 be defined as in the proof of Lemma B.2.6. Since x is symmetric, we have $v_1 = u_1^t$. Hence

$$u_1^t x u_1 = \begin{bmatrix} x_{11} & 0 \\ 0 & y \end{bmatrix},$$

where $y \in M_{n-1}$ is symmetric. Since $\Delta_i(y) \neq 0$ for $i = 1,\ldots,n-1$, we may assume by induction that there exists $v_1 \in N_{n-1}$ such that $v_1^t y v_1$ is diagonal. Set

$$v = u_1 \begin{bmatrix} 1 & 0 \\ 0 & v_1 \end{bmatrix} \in N_n^+.$$

Then $v^t x v = \text{diag}[x_{11}, v_1^t y v_1]$ is in diagonal form. Let h be a diagonal matrix such that $h^2 = v^t x v$ and set $b = hv^{-1}$. Then b is upper triangular and $b^t b = (v^t)^{-1}h^2 v^{-1} = x$ as desired.

Suppose we have another factorization $x = c^t c$ with c upper triangular. Since $\Delta_n(x) = (\det c)^2 = (\det b)^2$ and $\Delta_n(x) \neq 0$, the matrices b and c are invertible, and $(c^t)^{-1}b^t = cb^{-1}$. The matrix on the left in this equation is lower triangular, while that on the right is upper triangular. Hence $cb^{-1} = d$ is a diagonal matrix, and $b^t d^2 b = c^t c = b^t b$. Thus $d^2 = 1$, and so the diagonal entries of d are ± 1. $\qquad\square$

B.2.6 Pfaffians and Skew-Symmetric Matrices

Let $A = [a_{ij}]$ be a skew-symmetric $2n \times 2n$ matrix. Given vectors $\mathbf{x}_1, \ldots, \mathbf{x}_{2n}$ in $\mathbb{C}^{2n}$, we define

$$F_A(\mathbf{x}_1, \ldots, \mathbf{x}_{2n}) = \frac{1}{n!2^n} \sum_{s \in \mathfrak{S}_{2n}} \mathrm{sgn}(s) \prod_{i=1}^{n} \omega_A\left(\mathbf{x}_{s(2i-1)}, \mathbf{x}_{s(2i)}\right), \qquad (B.11)$$

where $\omega_A(\mathbf{x}, \mathbf{y}) = \mathbf{x}^t A \mathbf{y}$ is the skew-symmetric bilinear form on $\mathbb{C}^{2n}$ associated to A. Clearly F_A is a multilinear function of $\mathbf{x}_1, \ldots, \mathbf{x}_{2n}$. We claim that it is skew-symmetric. Indeed, given $t \in \mathfrak{S}_{2n}$, set $\mathbf{y}_i = \mathbf{x}_{t(i)}$. Then

$$F_A(\mathbf{x}_1, \ldots, \mathbf{x}_{2n}) = \frac{1}{n!2^n} \sum_{s \in \mathfrak{S}_{2n}} \mathrm{sgn}(s) \prod_{i=1}^{n} \omega_A\left(\mathbf{y}_{t^{-1}s(2i-1)}, \mathbf{y}_{t^{-1}s(2i)}\right)$$

$$= \mathrm{sgn}(t) F_A(\mathbf{y}_1, \ldots, \mathbf{y}_{2n})$$

(for the second equality, replace s by ts in the summation over $\mathfrak{S}_{2n}$ and use the property $\mathrm{sgn}(ts) = \mathrm{sgn}(t)\mathrm{sgn}(s)$). However, up to a scalar multiple, there is only one skew-symmetric multilinear function of $2n$ vectors in $\mathbb{C}^{2n}$, namely the determinant (where we arrange the $2n$ vectors into a $2n \times 2n$ matrix). Hence there is a complex number $\mathrm{Pfaff}(A)$, called the *Pfaffian* of A (or of the skew form ω_A), such that

$$F_A(\mathbf{x}_1, \ldots, \mathbf{x}_{2n}) = \mathrm{Pfaff}(A) \det[\mathbf{x}_1, \ldots, \mathbf{x}_{2n}]. \qquad (B.12)$$

In particular, taking $\mathbf{x}_i = e_i$ (the standard basis for $\mathbb{C}^{2n}$) in (B.11), we obtain the formula

$$\mathrm{Pfaff}(A) = \frac{1}{n!2^n} \sum_{s \in \mathfrak{S}_{2n}} \mathrm{sgn}(s) \prod_{i=1}^{n} a_{s(2i-1),s(2i)}, \qquad (B.13)$$

since $\det[e_1, \ldots, e_{2n}] = 1$.

Let $\mathfrak{B}_n \subset \mathfrak{S}_{2n}$ be the subgroup defined in Section 5.3.2 (the Weyl group of type B_n). Then $|\mathfrak{B}_n| = n!2^n$, and the function

$$s \mapsto \mathrm{sgn}(s) \prod_{i=1}^{n} a_{s(2i-1),s(2i)}$$

is constant on left cosets of $\mathfrak{B}_n$ in $\mathfrak{S}_{2n}$. Hence the formula for the Pfaffian can be written as

$$\mathrm{Pfaff}(A) = \sum_{s \in \mathfrak{S}_{2n}/\mathfrak{B}_n} \mathrm{sgn}(s) \prod_{i=1}^{n} a_{s(2i-1),s(2i)}. \qquad (B.14)$$

We now obtain some properties of the Pfaffian. Let $g \in \mathbf{GL}(2n, \mathbb{C})$. Then $g^t A g$ is skew-symmetric and

$$F_{g^t A g}(\mathbf{x}_1, \ldots, \mathbf{x}_{2n}) = F_A(g\mathbf{x}_1, \ldots, g\mathbf{x}_{2n}).$$

Since $\det[g\mathbf{x}_1, \ldots, g\mathbf{x}_{2n}] = (\det g)\det[\mathbf{x}_1, \ldots, \mathbf{x}_{2n}]$, it follows from (B.12) that

$$\text{Pfaff}(g^t A g) = (\det g)\,\text{Pfaff}(A)\,. \tag{B.15}$$

Let A and B be skew-symmetric matrices of sizes $2k \times 2k$ and $2m \times 2m$, respectively, where $k + m = n$. Form the block-diagonal matrix $A \oplus B = \text{diag}[A, B]$ and calculate (B.14) with A replaced by $A \oplus B$. The only permutations in $\mathfrak{S}_{2k+2m}$ that contribute nonzero terms on the right side of (B.14) are those that permute the first $2k$ indices among themselves and the last $2m$ indices among themselves. Hence the sum in (B.14) in this case can be taken over pairs $(s, t) \in (\mathfrak{S}_{2k}/\mathfrak{B}_k) \times (\mathfrak{S}_{2m}/\mathfrak{B}_m)$. This yields the formula

$$\text{Pfaff}(A \oplus B) = \text{Pfaff}(A)\,\text{Pfaff}(B)\,. \tag{B.16}$$

Let $A = [a_{ij}]$ be a skew-symmetric $2n \times 2n$ matrix. For $k = 1, \ldots, n$ define the truncated matrix $A_{(k)}$ to be the $2k \times 2k$ matrix $[a_{ij}]_{1 \le i,j \le 2k}$ and set

$$\text{Pf}_k(A) = \text{Pfaff}(A_{(k)})\,. \tag{B.17}$$

Then Pf_k is a homogeneous polynomial of degree k in the variables a_{ij}, for $1 \le i < j \le 2k$, which we will call the kth *principal Pfaffian* of A. From (B.14) we calculate, for example, that

$$\text{Pf}_1(A) = a_{12} \quad \text{and} \quad \text{Pf}_2(A) = a_{12}a_{34} - a_{13}a_{24} + a_{14}a_{23}\,.$$

In general, we can express $\text{Pf}(A)$ in terms of Pfaffians of order $n - 1$ formed from a subset of the rows and columns of A, as follows: For $s \in \mathfrak{S}_{2n}$ we define the $2n \times 2n$ skew-symmetric matrix $s \cdot A$ by

$$(s \cdot A)_{ij} = A_{s^{-1}(i), s^{-1}(j)}\,.$$

We embed $\mathfrak{S}_{2n-2} \subset \mathfrak{S}_{2n}$ as the subgroup fixing $2n - 1$ and $2n$. Then from (B.13) we have the following analogue of the cofactor expansion of a determinant:

$$\text{Pf}(A) = \frac{1}{2n} \sum_{s \in \mathfrak{S}_{2n-2} \backslash \mathfrak{S}_{2n}} \text{sgn}(s)\,(s \cdot A)_{2n-1, 2n}\,\text{Pf}_{n-1}(s \cdot A)\,. \tag{B.18}$$

Here the sum is over the $2n(2n - 1)$ right cosets of $\mathfrak{S}_{2n-2}$ in $\mathfrak{S}_{2n}$.

Let $B_{2n} \subset \mathbf{GL}(2n, \mathbb{C})$ be the subgroup of upper-triangular matrices. For $b \in B_{2n}$ and $A \in M_{2n}$ and $1 \le k \le n$ we have the transformation property

$$(b^t A b)_{(k)} = b_{(k)}^t A_{(k)} b_{(k)}\,,$$

where $b_{(k)} = [b_{ij}]_{1 \le i,j \le 2k}$. Hence if A is skew-symmetric, (B.15) gives

$$\text{Pf}_k(b^t A b) = \Delta_{2k}(b)\,\text{Pf}_k(A)\,, \tag{B.19}$$

where $\Delta_{2k}(b) = \det(b_{(k)})$ is the principal minor of b of order $2k$.

Let $J = \begin{bmatrix} 0 & 1 \\ -1 & 0 \end{bmatrix}$. Define the $p \times p$ skew-symmetric matrix

$$J_p = \begin{cases} J \oplus \cdots \oplus J & (n \text{ summands}) & \text{if } p = 2n, \\ J \oplus \cdots \oplus J \oplus 0 & (n+1 \text{ summands}) & \text{if } p = 2n+1. \end{cases}$$

We then have the following analogue of Lemma B.2.7 for skew-symmetric matrices:

Lemma B.2.8 (Skew Cholesky Decomposition). *Let A be a skew-symmetric $p \times p$ matrix. Assume that $\mathrm{Pf}_k(A) \neq 0$ for $k = 1, \ldots, [p/2]$. Then there exists $b \in B_p$ such that $A = b^t J_p b$.*

Proof. Let $N_p^+ \subset B_p$ be the group of all $p \times p$ upper-triangular matrices with diagonal entries 1. Let $n = [p/2]$. We first prove by induction on n that there exist $u \in N_p^+$ and complex numbers $z_1, \ldots, z_n$ such that

$$u^t A u = \begin{cases} z_1 J \oplus z_2 J \oplus \cdots \oplus z_n J & \text{if } p = 2n \text{ is even}, \\ z_1 J \oplus z_2 J \oplus \cdots \oplus z_n J \oplus 0 & \text{if } p = 2n+1 \text{ is odd}. \end{cases} \tag{B.20}$$

For $n = 0$ or 1 there is nothing to prove. Assume that this is true for all skew-symmetric matrices B of size $(p-1) \times (p-1)$ that satisfy $\mathrm{Pf}_k(B) \neq 0$ for $k = 1, \ldots, n-1$. Given a skew-symmetric $p \times p$ matrix A with $p \geq 3$ and $\mathrm{Pf}_1(A) = a_{12} \neq 0$, set $z_1 = a_{12}$ and

$$u_1 = \begin{bmatrix} 1 & 0 & z_1^{-1} a_{23} & \cdots & z_1^{-1} a_{2p} \\ 0 & 1 & -z_1^{-1} a_{13} & \cdots & -z_1^{-1} a_{1p} \\ 0 & 0 & 1 & \cdots & 0 \\ \vdots & \vdots & \vdots & \ddots & \vdots \\ 0 & 0 & 0 & \cdots & 1 \end{bmatrix}.$$

The entries of u_1 are chosen such that

$$u_1^t A = \begin{bmatrix} 0 & z_1 & * & \cdots & * \\ -z_1 & 0 & * & \cdots & * \\ 0 & 0 & * & \cdots & * \\ \vdots & \vdots & \vdots & \ddots & \vdots \\ 0 & 0 & * & \cdots & * \end{bmatrix}.$$

The matrix $u_1^t A u_1$ is then skew-symmetric and has the same first and second columns as $u_1^t A$. Hence $u_1^t A u_1 = z_1 J \oplus B$, where B is a skew-symmetric $(p-2) \times (p-2)$ matrix. We have $\mathrm{Pf}_k(A) = \mathrm{Pf}_k(u_1^t A u_1)$ for $k = 1, \ldots, n$ by (B.19), since the principal minors of u_1 are all 1. Hence

$$\mathrm{Pf}_k(A) = \mathrm{Pf}_1(z_1 J) \, \mathrm{Pf}_{k-1}(B)$$

by (B.16). Since $\mathrm{Pf}_1(z_1 J) = z_1 \neq 0$, we can apply the induction hypothesis to B. Thus there exist $u_2 \in N_{p-2}^+$ and complex numbers $z_2, \ldots, z_n$ such that

$$u_2^t B u_2 = \begin{cases} z_2 J \oplus \cdots \oplus z_n J & \text{if } p \text{ is even,} \\ z_2 J \oplus \cdots \oplus z_n J \oplus 0 & \text{if } p \text{ is odd.} \end{cases}$$

Set $u = u_1(I_2 \oplus u_2)$, where I_2 is the 2×2 identity matrix. Then $u^t A u$ satisfies (B.20). Let $h_i \in \mathbb{C}^\times$ with $h_i^2 = z_i^{-1}$. Set $h = \text{diag}[h_1, h_1, \ldots, h_n, h_n]$ for $p = 2n$ and $h = \text{diag}[h_1, h_1, \ldots, h_n, h_n, 1]$ for $p = 2n + 1$. Then $uh \in B_p$ and $(uh)^t A u h = J_p$. $\square$

Corollary B.2.9. *If $A \in M_{2n}$ is skew-symmetric then $(\text{Pfaff}(A))^2 = \det A$.*

Proof. Since $(\text{Pfaff}(A))^2$ and $\det A$ are polynomial functions of the entries of A, it suffices to prove that they are equal on the dense open set of all A with $\text{Pf}_k(A) \neq 0$ for $k = 1, \ldots, n$. Write $A = b^t J_{2n} b$ with $b \in B_{2n}$. Then

$$\text{Pfaff}(A) = (\det b) \, \text{Pfaff}(J_{2n}) = \det b \, .$$

Since $\det A = (\det b^t)(\det J_{2n})(\det b) = (\det b)^2$, the result follows. $\square$

B.2.7 Irreducibility of Determinants and Pfaffians

Recall that a polynomial $f \in \mathbb{C}[x_1, \ldots, x_n]$ is *irreducible* if it cannot be factored as $f = gh$ with g and h nonconstant polynomials.

Lemma B.2.10. *The following polynomials are irreducible:*

1. *$\det(x)$ on the space of all $n \times n$ matrices, as a polynomial in the variables $\{x_{ij} : 1 \leq i, j \leq n\}$).*
2. *$\det(x)$ on the space of all $n \times n$ symmetric matrices, as a polynomial in the variables $\{x_{ij} : 1 \leq i \leq j \leq n\}$.*
3. *$\text{Pfaff}(x)$ on the space of all $2n \times 2n$ skew-symmetric matrices, as a polynomial in the variables $\{x_{ij} : 1 \leq i < j \leq 2n\}$.*

Proof. Cases (1) and (2): Let $f(x) = \det(x)$, and suppose there is a factorization $f(x) = g(x)h(x)$. In $f(x)$ each variable x_{ii} appears linearly and only in monomials not containing x_{ij} or x_{ji} with $j \neq i$. We may assume that x_{11} occurs in $g(x)$. Then x_{1j} does not occur in $h(x)$ for $1 \leq j \leq n$. Suppose that x_{jj} occurred in $h(x)$ for some $j > 1$. Then x_{1j} would not occur in $g(x)$ and hence would not occur in $f(x)$, a contradiction. Hence none of the variables x_{jj} can occur in $h(x)$. Since $f(x)$ contains the monomial $x_{11} \cdots x_{nn}$, this monomial must occur in $g(x)$. But this forces $h(x)$ to be constant, since $f(x)$ is of degree n.

Case (3): Let $f(x) = \text{Pfaff}(x)$. When $n = 1$, $f(x) = x_{12}$ is irreducible; hence we may assume $n \geq 2$. Suppose there is a factorization $f(x) = g(x)h(x)$. In $f(x)$ each variable x_{ij}, with $i < j$, appears linearly and only in monomials not containing x_{ik} or x_{lj} for $k \neq i$ or $l \neq j$ (see formula (B.14)). We may assume that x_{12} occurs in $g(x)$. Then x_{1j} does not occur in $h(x)$ for $1 \leq j \leq 2n$. Suppose, for the sake of contradiction, that $h(x)$ were not constant; then x_{ij} would occur in $h(x)$ for some

$1 < i < j \leq 2n$. Consequently, x_{1j} could not occur in $g(x)$ and hence it could not occur in $f(x)$, which is a contradiction. Hence $h(x)$ must be constant.   □

Appendix C
Associative Algebras and Lie Algebras

Abstract The theory of group representations and invariants uses several types of associative algebras, which are introduced in this appendix. Further details can be found in standard references such as Lang [97] and Jacobson [80]; for the enveloping algebra of a Lie algebra see Bourbaki [11], Humphreys [76], or Jacobson [79].

C.1 Some Associative Algebras

We recall the basic aspects of filtered and graded associative algebras, and we examine tensor algebras, symmetric algebras, and exterior algebras in more detail.

C.1.1 Filtered and Graded Algebras

Let $\mathcal{A}$ be an associative algebra over $\mathbb{C}$ with unit 1. A *filtration* $\mathcal{F}$ on $\mathcal{A}$ is a family of linear subspaces $\mathcal{F} : \mathbb{C}1 = \mathcal{A}_0 \subseteq \mathcal{A}_1 \subseteq \cdots \subseteq \mathcal{A}_n \subseteq \cdots$ such that

$$\mathcal{A}_j \cdot \mathcal{A}_k \subseteq \mathcal{A}_{j+k}, \qquad \bigcup_{j \geq 0} \mathcal{A}_j = \mathcal{A}.$$

For example, if $\mathcal{A}$ is generated (as an algebra over $\mathbb{C}$) by a linear subspace V, then we obtain a filtration $\mathcal{F} = \mathcal{F}_V$ on $\mathcal{A}$ by setting $\mathcal{A}_0 = \mathbb{C}1$ and

$$\mathcal{A}_n = \mathrm{Span}\{v_1 \cdots v_k : v_j \in V, 0 \leq k \leq n\}.$$

When $\dim \mathcal{A}_n < \infty$ for all n we define the *Hilbert series* for the filtration to be the formal series

$$\Phi_{\mathcal{F}}(t) = \sum_{n=0}^{\infty} t^n \dim \mathcal{A}_n.$$

Let $\mathcal{B}$ be an associative algebra over $\mathbb{C}$ with unit 1. A *gradation* $\mathcal{G}$ on $\mathcal{B}$ is a family of linear subspaces $\mathcal{G} : \mathbb{C}1 = \mathcal{B}^{(0)}, \mathcal{B}^{(1)}, \ldots, \mathcal{B}^{(n)}, \ldots$ such that

$$\mathcal{B}^{(j)} \cdot \mathcal{B}^{(k)} \subseteq \mathcal{B}^{(j+k)}, \qquad \bigoplus_{j \geq 0} \mathcal{B}^{(j)} = \mathcal{B}.$$

When $\dim \mathcal{B}^{(n)} < \infty$ for all n we define the *Hilbert series* for the gradation $\mathcal{G}$ to be the formal series

$$\varphi_{\mathcal{G}}(t) = \sum_{n=0}^{\infty} t^n \dim \mathcal{B}^{(n)}.$$

An ideal $\mathcal{C} \subset \mathcal{B}$ is a *graded ideal* (relative to the given gradation on $\mathcal{B}$) if the subspaces $\mathcal{C}^{(k)} = \mathcal{C} \cap \mathcal{B}^{(k)}$ define a gradation on $\mathcal{C}$. This will be true if and only if

$$\mathcal{C} = \bigoplus_{k \geq 0} \mathcal{C}^{(k)}.$$

Given a filtration $\mathcal{F}$ on an algebra $\mathcal{A}$, we construct the *associated graded algebra* $\mathcal{B} = \mathrm{Gr}(\mathcal{A})$ by setting $\mathcal{B}^{(0)} = \mathbb{C}1$, $\mathcal{B}^{(j)} = \mathcal{A}_j / \mathcal{A}_{j-1}$, and $\mathcal{B} = \bigoplus_{j \geq 0} \mathcal{B}^{(j)}$. We define the multiplication on $\mathcal{B}$ as

$$(x + \mathcal{A}_{j-1}) \cdot (y + \mathcal{A}_{k-1}) = xy + \mathcal{A}_{j+k-1} \quad \text{for } x \in \mathcal{A}_j \text{ and } y \in \mathcal{A}_k.$$

Suppose $\dim \mathcal{A}_n < \infty$ for all n. Since $\dim \mathcal{A}_n - \dim \mathcal{A}_{n-1} = \dim \mathcal{B}^{(n)}$ in this case, the Hilbert series for the filtration $\mathcal{F}$ and associated gradation $\mathcal{G}$ are related by

$$(1-t)\,\Phi_{\mathcal{F}}(t) = \varphi_{\mathcal{G}}(t).$$

Example

Let V be a vector space, and $\mathcal{A} = \mathcal{P}(V)$ the polynomial functions on V. Let $\mathcal{P}_k(V)$ be the subspace of polynomials of degree $\leq k$. This gives a filtration $\mathcal{F}$ on $\mathcal{P}(V)$ that is generated by V^*. Let the multiplicative group $\mathbb{C}^\times$ act on $\mathcal{P}(V)$ by $\rho(t)f(v) = f(tv)$ for $t \in \mathbb{C}^\times$ and $v \in V$. A polynomial f is homogeneous of degree k if $\rho(t)f = t^k f$. Denote by $\mathcal{P}^k(V)$ the space of all polynomials homogeneous of degree k. Then $\mathcal{G} = \{\mathcal{P}^k(V)\}$ is a gradation on $\mathcal{P}(V)$, and $\mathrm{Gr}_{\mathcal{F}}(\mathcal{P}(V)) \cong (\mathcal{P}(V), \mathcal{G})$ via the decomposition

$$\mathcal{P}_k(V) = \mathcal{P}^k(V) \oplus \mathcal{P}_{k-1}(V).$$

When $\dim V = n$, then as a graded algebra $\mathcal{P}(V) \cong \mathbb{C}[x_1, \ldots, x_n]$ by the map $f_i \mapsto x_i$, where $\{f_1, \ldots, f_n\}$ is a basis for V^*.

C.1.2 Tensor Algebra

Representation theory uses several associative algebras, each of which solves a certain *universal mapping problem*. We begin with the *tensor algebra*. Let V be a vector space over $\mathbb{C}$. We construct a *universal* associative algebra $\mathcal{T}(V)$ generated by V, as follows: As in section B.2.2, we let $V^{\otimes k}$ be the k-fold tensor product of V with itself, for $k = 0, 1, 2, \ldots$ (with $V^{\otimes 0} = \mathbb{C}$). We form the algebraic direct sum

$$\mathcal{T}(V) = \bigoplus_{k \geq 0} V^{\otimes k}$$

and define multiplication $\mu : V^{\otimes k} \times V^{\otimes m} \longrightarrow V^{\otimes(k+m)}$ by juxtaposition:

$$\mu(x_1 \otimes \cdots \otimes x_k, y_1 \otimes \cdots \otimes y_m) = x_1 \otimes \cdots \otimes x_k \otimes y_1 \otimes \cdots \otimes y_m$$

for $x_i, y_j \in V$. The multiplication is well defined because the right side is linear in each vector x_i, y_j, and it is clearly associative. We have $V = V^{\otimes 1} \subset \mathcal{T}(V)$, and $\mathcal{T}(V)$ is generated as an algebra by V. The spaces $\{V^{\otimes k}\}_{k \geq 0}$ define a grading on $\mathcal{T}(V)$ that we call the *standard grading*.

The algebra $\mathcal{T}(V)$ is the solution to the following universal mapping problem: Suppose $\mathcal{A}$ is any associative algebra over $\mathbb{C}$, and $\varphi : V \longrightarrow \mathcal{A}$ is any linear map. Then φ extends uniquely to a linear map $\widetilde{\varphi} : \mathcal{T}(V) \longrightarrow \mathcal{A}$ by the formula

$$\widetilde{\varphi}(x_1 \otimes \cdots \otimes x_k) = \varphi(x_1) \cdots \varphi(x_k)$$

for $x_i \in V$ (this is well defined, since the right side is a linear function of each x_i). Since $\mathcal{A}$ is assumed associative, it is clear from the definition of multiplication in $\mathcal{T}(V)$ that $\widetilde{\varphi}$ is an algebra homomorphism. This gives the commutative diagram

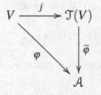

where j is the inclusion map.

C.1.3 Symmetric Algebra

The *symmetric algebra* for a vector space V is the associative algebra generated by V that is universal relative to linear maps $\varphi : V \longrightarrow \mathcal{A}$ (for $\mathcal{A}$ any associative algebra) that satisfy

$$\varphi(x)\varphi(y) = \varphi(y)\varphi(x) \quad \text{for } x, y \in V. \tag{C.1}$$

Thus we require an associative algebra $S(V)$ and a map $\gamma : V \longrightarrow S(V)$ with the following property: Given any linear map $\varphi : V \longrightarrow \mathcal{A}$ that satisfies (C.1), there should be a unique algebra homomorphism $\widehat{\varphi} : S(V) \longrightarrow \mathcal{A}$ such that the diagram

is commutative. We construct $S(V)$ as the quotient of $\mathcal{T}(V)$ modulo the two-sided ideal $\mathcal{C}$ generated by $x \otimes y - y \otimes x$ for $x, y \in V$. We let

$$\gamma : \mathcal{T}(V) \longrightarrow \mathcal{T}(V)/\mathcal{C}$$

be the quotient map. If $\varphi : V \longrightarrow \mathcal{A}$ satisfies (C.1), then the homomorphism $\widetilde{\varphi} : \mathcal{T}(V) \longrightarrow \mathcal{A}$ vanishes on $\mathcal{C}$. Hence $\widetilde{\varphi}$ induces the required homomorphism $\widehat{\varphi} : S(V) \longrightarrow \mathcal{A}$.

Lemma C.1.1.

1. $\mathcal{C}$ is a graded ideal in $\mathcal{T}(V)$, and $\mathcal{C}^{(k)} \overset{\text{def}}{=} \mathcal{C} \cap \bigotimes^k V$ is the subspace spanned by tensors of the form $u - \sigma_k(\tau)u$, where $\tau \in \mathfrak{S}_k$ is a transposition and $u \in \bigotimes^k V$.
2. $\mathcal{C}^{(k)} = \mathrm{Ker}(\mathbf{Sym})$, where $\mathbf{Sym} : \bigotimes^k V \longrightarrow S^k(V)$ is the symmetrization operator. Hence $\bigotimes^k V = S^k(V) \oplus \mathcal{C}^{(k)}$.

Proof. (1): The generators for $\mathcal{C}$ are homogeneous of degree 2, so $\mathcal{C}$ is graded. The second statement is clear from the form of the generators.

(2): For every $s \in \mathfrak{S}_k$ and $u \in \bigotimes^k V$ we have

$$u - \sigma_k(s)u \in \mathcal{C}^{(k)} . \tag{C.2}$$

Indeed, if $s = \tau s_1$ for some transposition τ, then

$$u - \sigma_k(s)u = u - \sigma_k(s_1)u + \sigma_k(s_1)u - \sigma_k(\tau)\sigma_k(s_1)u .$$

But the group $\mathfrak{S}_k$ is generated by transpositions, so we obtain (C.2) by (1) and induction on the number of transpositions in s.

Since $\mathbf{Sym}(\sigma_k(s)u) = \mathbf{Sym}(u)$ for $s \in \mathfrak{S}_k$, we see by (C.2) that $\mathcal{C}^{(k)} \subseteq \mathrm{Ker}(\mathbf{Sym})$. Conversely, if $\mathbf{Sym}(u) = 0$ then $u \in \mathcal{C}^{(k)}$, since

$$u = u - \mathbf{Sym}(u) = \frac{1}{k!} \sum_{s \in \mathfrak{S}_k} (u - \sigma_k(s)u) . \qquad \square$$

Corollary C.1.2. *The quotient map* $\gamma : \mathcal{T}(V) \longrightarrow \mathcal{T}(V)/\mathcal{C}$ *restricts to a vector-space isomorphism* $\bigoplus_{k \geq 0} S^k(V) \cong S(V)$.

Assume now that V is finite-dimensional. Recall that $\mathcal{P}(V^*)$ denotes the polynomial functions on V^*. For $x \in V$ let $\mu(x)$ be the operator on $\mathcal{P}(V^*)$ given by

$$\mu(x)f(v^*) = \langle v^*, x \rangle f(v^*) \quad \text{for } f \in \mathcal{P}(V^*) .$$

Thus $\mu(x)$ is multiplication by the linear function $v^* \mapsto \langle v^*, x \rangle$, and so $\mu(x)\mu(y) = \mu(y)\mu(x)$ for $x, y \in V$. By the universal property, μ extends to an algebra homomorphism $\hat{\mu} : S(V) \longrightarrow \operatorname{End}\mathcal{P}(V^*)$.

Proposition C.1.3. *The map $g \mapsto \hat{\mu}(g)1$ from $S(V)$ to $\mathcal{P}(V^*)$ is an algebra isomorphism.*

Proof. Since $\mathcal{P}(V^*)$ is generated by the functions $\{\mu(x)1 : x \in V\}$, the map is surjective. It is injective by Corollary C.1.2 and Proposition B.2.4. $\qquad\square$

Let $\mathcal{A}$ and $\mathcal{B}$ be associative algebras. We make the vector space $\mathcal{A} \otimes \mathcal{B}$ into an associative algebra relative to the product

$$(a \otimes b)(a' \otimes b') = (aa') \otimes (bb') \quad \text{for } a, a' \in \mathcal{A} \text{ and } b, b' \in \mathcal{B} ,$$

and we call it the *tensor product* of the algebras $\mathcal{A}$ and $\mathcal{B}$. Note that if $\mathcal{A}$ and $\mathcal{B}$ are commutative, then so is $\mathcal{A} \otimes \mathcal{B}$. If $\mathcal{A} = \bigoplus_{i \geq 0} \mathcal{A}^{(i)}$ and $\mathcal{B} = \bigoplus_{j \geq 0} \mathcal{B}^{(j)}$ are graded algebras, then $\mathcal{A} \otimes \mathcal{B}$ is a graded algebra with

$$(\mathcal{A} \otimes \mathcal{B})^{(k)} = \bigoplus_{i+j=k} \mathcal{A}^{(i)} \otimes \mathcal{B}^{(j)} .$$

In the case of symmetric algebras, this construction yields the following natural isomorphism:

Proposition C.1.4. *Let V and W be finite-dimensional vector spaces. The map $v \oplus w \mapsto v \otimes 1 + 1 \otimes w$ from $V \oplus W$ to $V \otimes W$ extends uniquely to an isomorphism of graded algebras $S(V) \otimes S(W) \cong S(V \oplus W)$.*

Proof. By Proposition C.1.3 we can identify $S(V) \otimes S(W)$ with $\mathcal{P}(V^*) \otimes \mathcal{P}(W^*)$ and $S(V \oplus W)$ with $\mathcal{P}(V^* \oplus W^*)$. Given $f \in \mathcal{P}(V^*)$ and $g \in \mathcal{P}(W^*)$, we define a polynomial function fg on $V^* \oplus W^*$ by $fg(v^* \oplus w^*) = f(v^*)g(w^*)$. The map $f \otimes g \mapsto fg$ is bilinear, so it extends uniquely to a linear map

$$\tau : \mathcal{P}(V^*) \otimes \mathcal{P}(W^*) \longrightarrow \mathcal{P}(V^* \oplus W^*) ,$$

which is clearly an algebra homomorphism. Hence τ is uniquely determined by the images of elements of degree one. When f and g are homogeneous of degree 1 (corresponding to elements of V and W), then $\tau(f \otimes 1 + 1 \otimes g)$ is the linear function $(v^*, w^*) \mapsto f(v^*) + g(w^*)$ on $V^* \oplus W^*$, as desired.

We must show that τ is an isomorphism. Let $x_1, \ldots, x_m$ be linear coordinate functions on V^*, and let $y_1, \ldots, y_n$ be linear coordinate functions on W^*. The set of monomials

$$x^{\mathbf{a}} y^{\mathbf{b}} = x_1^{a_1} \cdots x_m^{a_m} y_1^{b_1} \cdots y_n^{b_n} ,$$

for $\mathbf{a} \in \mathbb{N}^m$ and $\mathbf{b} \in \mathbb{N}^n$, is a basis for $\mathcal{P}(V^* \oplus W^*)$. Since τ is an algebra homomorphism, we have $\tau(x^{\mathbf{a}} \otimes y^{\mathbf{b}}) = x^{\mathbf{a}} y^{\mathbf{b}}$. Hence τ is an isomorphism.   □

C.1.4 Exterior Algebra

The *exterior algebra* for a vector space V is the associative algebra generated by V that is universal relative to linear maps ψ from V to associative algebras $\mathcal{A}$ such that

$$\psi(x)\psi(y) = -\psi(y)\psi(x), \quad \text{for } x, y \in V. \tag{C.3}$$

In this case we require an associative algebra $\bigwedge(V)$ and a map $\delta : V \longrightarrow \bigwedge(V)$ with the following property: Given any linear map $\psi : V \longrightarrow \mathcal{A}$ that satisfies (C.3), there should be a unique algebra homomorphism $\check{\psi} : S(V) \longrightarrow \mathcal{A}$ such that the diagram

is commutative. We may construct $\bigwedge(V)$ as the quotient of $\mathcal{T}(V)$ modulo the two-sided ideal $\mathcal{E}$ generated by the elements $x \otimes y + y \otimes x$, for $x, y \in V$, with δ the quotient map. If $\psi : V \longrightarrow \mathcal{A}$ satisfies (C.3), then the universal mapping property of the tensor algebra furnishes a homomorphism $\widetilde{\psi} : \mathcal{T}(V) \longrightarrow \mathcal{A}$ that vanishes on $\mathcal{E}$. Hence $\widetilde{\psi}$ induces the required homomorphism $\check{\psi} : \bigwedge(V) \longrightarrow \mathcal{A}$.

Lemma C.1.5.

1. $\mathcal{E}$ *is a graded ideal in* $\mathcal{T}(V)$, *and* $\mathcal{E}^{(k)} \overset{\text{def}}{=} \mathcal{E} \cap \bigotimes^k V$ *is the subspace spanned by tensors of the form* $u + \sigma_k(\tau)u$, *where* $\tau \in \mathfrak{S}_k$ *is a transposition and* $u \in \bigotimes^k V$.
2. $\mathcal{E}^{(k)} = \mathrm{Ker}(\mathbf{Alt})$, *where* $\mathbf{Alt} : \bigotimes^k V \longrightarrow S^k(V)$ *is the alternation operator. Hence* $\bigotimes^k V = \bigwedge^k V \oplus \mathcal{E}^{(k)}$.

Proof. (1): Since the generators for $\mathcal{E}$ are homogeneous of degree 2, $\mathcal{E}$ is graded. The second statement is clear from the form of the generators.

    (2): For every $s \in \mathfrak{S}_k$ and $u \in \bigotimes^k V$ we have

$$u + \mathrm{sgn}(s)\sigma_k(s)u \in \mathcal{E}^{(k)}. \tag{C.4}$$

Indeed, if $s = \tau s_1$ for some transposition τ, then $\mathrm{sgn}(s) = -\mathrm{sgn}(s_1)$ and

$$u + \mathrm{sgn}(s)\sigma_k(s)u = u + \mathrm{sgn}(s_1)\sigma_k(s_1)u + \mathrm{sgn}(s)\big(\sigma_k(s_1)u + \sigma_k(\tau)\sigma_k(s_1)u\big).$$

But the group $\mathfrak{S}_k$ is generated by transpositions, so we obtain (C.4) by (1) and induction on the number of transpositions in s. Furthermore, since

$$\mathbf{Alt}(\sigma_k(s)u) = \mathrm{sgn}(s)\,\mathbf{Alt}(u) \quad \text{for } s \in \mathfrak{S}_k\,,$$

we see from (C.4) that $\mathcal{E}^{(k)} \subseteq \mathrm{Ker}(\mathbf{Alt})$. Conversely, if $\mathbf{Alt}(u) = 0$ then

$$u = u + \mathbf{Alt}(u) = \frac{1}{k!} \sum_{s \in \mathfrak{S}_k} \big(u + \mathrm{sgn}(s)\sigma_k(s)u\big)\,.$$

Hence $u \in \mathcal{E}^{(k)}$ by (C.4). $\qquad\qquad\qquad\qquad\qquad\qquad\qquad\qquad\qquad\qquad\quad$ $\square$

Corollary C.1.6. *The quotient map* $\mathcal{T}(V) \longrightarrow \mathcal{T}(V)/\mathcal{E}$ *restricts to a vector-space isomorphism* $\bigoplus_{k\geq 0}\bigwedge^k V \cong \bigwedge(V)$.

We use the isomorphism in Corollary C.1.6 to identify $\bigwedge(V)$ with a subalgebra of $\mathcal{T}(V)$. We shall write the product in $\bigwedge(V)$ as $x \wedge y$. If we view x and y as elements of $\mathcal{T}(V)$, then

$$x \wedge y = \mathbf{Alt}(x \otimes y)\,.$$

If $x, y \in V$ then $x \wedge y = -y \wedge x$. In general, for $a \in \bigwedge^p V$ and $b \in \bigwedge^q V$ we have

$$a \wedge b = (-1)^{pq}\, b \wedge a\,.$$

This property of the wedge multiplication is called *skew-commutativity*. If $\{e_i\}$ is a basis for V, then the set of all ordered monomials $e_{i_1} \wedge \cdots \wedge e_{i_k}$, where $i_1 < i_2 < \cdots < i_k$, is a basis for $\bigwedge(V)$.

Definition C.1.7. Let $\mathcal{A} = \bigoplus_{k\geq 0}\mathcal{A}^{(k)}$ and $\mathcal{B} = \bigoplus_{k\geq 0}\mathcal{B}^{(k)}$ be graded associative algebras. The *skew-commutative tensor product* $\mathcal{A}\widehat{\otimes}\mathcal{B}$ is the graded algebra whose kth homogeneous component is $\bigoplus_{i+j=k}\mathcal{A}^{(i)} \otimes \mathcal{B}^{(j)}$, with multiplication of homogeneous elements $a, a' \in \mathcal{A}$ and $b, b' \in \mathcal{B}$ given by

$$(a\widehat{\otimes}b)(a'\widehat{\otimes}b') = (-1)^{\deg a' \deg b}(aa')\widehat{\otimes}(bb')\,. \tag{C.5}$$

We identify $\mathcal{A}$ with the subspace $\mathcal{A}\widehat{\otimes}1$ and $\mathcal{B}$ with the subspace $1\widehat{\otimes}\mathcal{B}$ of $\mathcal{A}\widehat{\otimes}\mathcal{B}$.

In the case of exterior algebras, this construction yields the following skew-symmetric version of Proposition C.1.4:

Proposition C.1.8. *Let* V *and* W *be finite-dimensional vector spaces. The map* $v \oplus w \mapsto v \otimes 1 + 1 \otimes w$ *from* $V \oplus W$ *to* $V \otimes W$ *extends uniquely to an isomorphism of graded algebras* $\bigwedge(V)\widehat{\otimes}\bigwedge(W) \cong \bigwedge(V \oplus W)$.

Proof. Let $\{e_1,\ldots,e_m\}$ be a basis for V and let $\{f_1,\ldots,f_n\}$ be a basis for W. For $1 \leq i_1 < \cdots < i_p \leq m$ and $1 \leq j_1 < \cdots < j_q \leq n$, set

$$\varphi\big(e_{i_1} \wedge \cdots \wedge e_{i_p}, f_{j_1} \wedge \cdots \wedge f_{j_q}\big) = e_{i_1} \wedge \cdots \wedge e_{i_p} \wedge f_{j_1} \wedge \cdots \wedge f_{j_q}\,,$$

where we identify V with the subspace $V \oplus 0$ in $V \oplus W$ (and likewise for W), as we did for the symmetric algebra. The map φ is bilinear, so it extends uniquely to a linear map $\bigwedge(V) \otimes \bigwedge(W) \longrightarrow \bigwedge(V \oplus W)$ that is obviously a vector-space isomorphism. The extended map preserves degree and satisfies (C.5). Hence it is an algebra isomorphism. □

C.1.5 Exercises

1. Show that the Hilbert series for the graded algebra $\mathbb{C}[x_1,\ldots,x_n]$ (with the usual grading by degree) is $\varphi(t) = (1-t)^{-n}$ (expanded in powers of t). (HINT: Use Proposition C.1.4.)
2. Show that the Hilbert series for the graded tensor algebra $\mathcal{T}(\mathbb{C}^n)$ (with the standard grading) is $\varphi(t) = (1-nt)^{-1}$ (expanded in powers of t).
3. Show that the Hilbert series for the graded algebra $\bigwedge \mathbb{C}^n$ is $\varphi(t) = (1+t)^n$. (HINT: Use Proposition C.1.8.)

C.2 Universal Enveloping Algebras

We recall the definition of a Lie algebra, and construct the universal enveloping algebra of a Lie algebra. The main result is the Poincaré–Birkhoff–Witt theorem.

C.2.1 Lie Algebras

If $\mathcal{A}$ is an associative algebra, then from the multiplication in $\mathcal{A}$ we can define the *commutator* $[x,y] = xy - yx$, also called the *Lie bracket*. This product is *skew-symmetric*,

$$[x,y] = -[y,x] \,, \tag{C.6}$$

and a calculation (using associativity in $\mathcal{A}$) shows that it satisfies the *Jacobi identity*

$$[[x,y],z] + [[z,x],y] + [[y,z],x] = 0 \,. \tag{C.7}$$

A vector space $\mathfrak{g}$ with a bilinear product satisfying properties (C.6) and (C.7) is called a *Lie algebra*. If $\mathfrak{g}$ is a Lie algebra and $\mathfrak{h} \subset \mathfrak{g}$, then $\mathfrak{h}$ is a *Lie subalgebra* of $\mathfrak{g}$ if $\mathfrak{h}$ is a linear subspace and $[x,y] \in \mathfrak{h}$ for all $x,y \in \mathfrak{h}$.

When $\mathcal{A}$ is an associative algebra, we denote by $\mathcal{A}_{\mathrm{Lie}}$ the Lie algebra whose underlying vector space is $\mathcal{A}$ and whose Lie bracket is $[x,y]$. If $\mathcal{B}$ is another associative algebra over $\mathbb{C}$ and $\varphi : \mathcal{A} \longrightarrow \mathcal{B}$ is an associative algebra homomorphism, then

$$[\varphi(x), \varphi(y)] = \varphi([x,y]) \quad \text{for } x,y \in \mathcal{A} \,.$$

Thus φ is also a Lie algebra homomorphism from $\mathcal{A}_{\mathrm{Lie}}$ to $\mathcal{B}_{\mathrm{Lie}}$.

Let $\mathfrak{g}$ be any Lie algebra over $\mathbb{C}$. One of the basic results in Lie theory is that $\mathfrak{g}$ can be embedded into an associative algebra such that the Lie bracket on $\mathfrak{g}$ comes from the associative multiplication in the larger algebra. Furthermore, this associative algebra can be constructed such that it has no additional relations beyond those implied by associativity and the Lie bracket relations from $\mathfrak{g}$. To obtain these results, we begin by constructing an algebra satisfying the second property.

We define a *universal enveloping algebra* for $\mathfrak{g}$ to be an associative algebra $\mathfrak{G}$ with 1 over $\mathbb{C}$, together with a Lie algebra homomorphism $j : \mathfrak{g} \longrightarrow \mathfrak{G}$ whose image generates $\mathfrak{G}$ (as an associative algebra). The pair $(\mathfrak{G}, j)$ must satisfy the following *universal mapping property*: Given any associative algebra $\mathcal{A}$ over $\mathbb{C}$ and a Lie algebra homomorphism $\psi : \mathfrak{g} \longrightarrow \mathcal{A}_{\mathrm{Lie}}$, there exists an associative algebra homomorphism $\Psi : \mathfrak{G} \longrightarrow \mathcal{A}$ such that

$$\psi(x) = \Psi(j(x)) \quad \text{for } x \in \mathfrak{g}. \tag{C.8}$$

Thus we have the following commutative diagram:

We can construct a pair $(\mathfrak{G}, j)$ satisfying these requirements as follows: Take the tensor algebra $\mathcal{T}(\mathfrak{g})$ over $\mathfrak{g}$, and let $\mathcal{K}(\mathfrak{g})$ be the two-sided ideal in $\mathcal{T}(\mathfrak{g})$ generated by the elements

$$x \otimes y - y \otimes x - [x, y] \quad \text{for } x, y \in \mathfrak{g}.$$

Define the associative algebra $\mathfrak{G} = \mathcal{T}(\mathfrak{g}) / \mathcal{K}(\mathfrak{g})$. We view $\mathfrak{g} \subset \mathcal{T}(\mathfrak{g})$ as usual, and we define $j : \mathfrak{g} \longrightarrow \mathfrak{G}$ by $j(x) = \pi(x)$, where $\pi : \mathcal{T}(\mathfrak{g}) \longrightarrow \mathfrak{G}$ is the canonical quotient map. Then $j(\mathfrak{g})$ generates $\mathfrak{G}$ as an associative algebra, since $\mathfrak{g}$ generates $\mathcal{T}(\mathfrak{g})$.

Proposition C.2.1. *The pair $(\mathfrak{G}, j)$ satisfies the universal mapping property for an enveloping algebra. If $(\mathfrak{G}', j')$ is another solution to this universal mapping problem for $\mathfrak{g}$, then there is a unique isomorphism of associative algebras $\Psi : \mathfrak{G} \longrightarrow \mathfrak{G}'$ such that $j' = \Psi \circ j$.*

Proof. Given an associative algebra $\mathcal{A}$ and a Lie homomorphism $\psi : \mathfrak{g} \longrightarrow \mathcal{A}$, we can use the universal property of $\mathcal{T}(\mathfrak{g})$ (relative to linear maps) to extend ψ to a unique associative algebra homomorphism, say $\widetilde{\psi}$, from $\mathcal{T}(\mathfrak{g})$ to $\mathcal{A}$. Since

$$\psi([x, y]) = \psi(x)\psi(y) - \psi(y)\psi(x) = \widetilde{\psi}(x \otimes y - y \otimes x),$$

for $x, y \in \mathfrak{g}$, the kernel of $\widetilde{\psi}$ contains the generators of the ideal $\mathcal{K}(\mathfrak{g})$. Hence $\widetilde{\psi}$ passes to the quotient $\mathfrak{G}$ and gives an associative algebra homomorphism Ψ satisfying (C.8). Thus $(\mathfrak{G}, j)$ satisfies the universal mapping property.

Given another such pair $(\mathfrak{G}', j')$, we apply the universal mapping property to j and j' to obtain unique homomorphisms $\Psi : \mathfrak{G} \longrightarrow \mathfrak{G}'$ and $\Psi' : \mathfrak{G}' \longrightarrow \mathfrak{G}$ such that $\Psi \circ j = j'$ and $\Psi' \circ j' = j$. From these relations we see that

$$\Psi \circ \Psi' \circ j' = j' \quad \text{and} \quad \Psi' \circ \Psi \circ j = j.$$

Since $j(\mathfrak{g})$ generates $\mathfrak{G}$ and $j'(\mathfrak{g})$ generates $\mathfrak{G}'$, it follows that Ψ is an isomorphism, with inverse Ψ'. $\qquad\qquad\square$

We shall denote the universal enveloping algebra of $\mathfrak{g}$ by $U(\mathfrak{g})$. The result just proved shows that $U(\mathfrak{g})$ is uniquely determined by $\mathfrak{g}$. Given a Lie algebra $\mathfrak{h}$ and a Lie algebra homomorphism $\varphi : \mathfrak{g} \longrightarrow \mathfrak{h}$, we obtain a unique associative algebra homomorphism $\Phi : U(\mathfrak{g}) \longrightarrow U(\mathfrak{h})$ such that $\Phi \circ j = j \circ \varphi$. Furthermore, if $\mathfrak{k}$ is another Lie algebra and $\psi : \mathfrak{h} \longrightarrow \mathfrak{k}$ is a Lie homomorphism, then the associative algebra homomorphism from $U(\mathfrak{g})$ to $U(\mathfrak{k})$ corresponding to $\psi \circ \varphi$ is the composition $\Psi \circ \Phi$, where $\Psi : U(\mathfrak{h}) \longrightarrow U(\mathfrak{k})$ is the associative algebra homomorphism determined by ψ.

C.2.2 Poincaré–Birkhoff–Witt Theorem

Let $\mathfrak{g}$ be a finite-dimensional Lie algebra over $\mathbb{C}$. The universal enveloping algebra $U(\mathfrak{g})$ is generated by $j(\mathfrak{g})$, so it has a natural increasing filtration

$$\mathbb{C} = U_0(\mathfrak{g}) \subset U_1(\mathfrak{g}) \subset U_2(\mathfrak{g}) \subset \cdots,$$

where $U_n(\mathfrak{g})$ is the subspace spanned by products of n or fewer elements $j(X)$, $X \in \mathfrak{g}$. Our goal now is to show that the canonical map $j : \mathfrak{g} \longrightarrow U(\mathfrak{g})$ is injective and to find a basis for $U_n(\mathfrak{g})$ for all $n \geq 0$.

For $X \in \mathfrak{g}$ we write $j(X) = \widetilde{X}$. Let $d = \dim \mathfrak{g}$ and let $\{A_1, \ldots, A_d\}$ be any ordered basis for $\mathfrak{g}$. For an integer $n \geq 1$ let $\mathfrak{J}_n$ be the set of multi-indices

$$J = (i_1, \ldots, i_k) \quad \text{such that} \quad 1 \leq i_p \leq d, \quad 1 \leq k \leq n, \text{ and } i_1 \leq \cdots \leq i_k.$$

For $J \in \mathfrak{J}_n$ define $\widetilde{A}_J = \widetilde{A}_{i_1} \cdots \widetilde{A}_{i_k}$ (product in $U(\mathfrak{g})$).

Theorem C.2.2 (Poincaré–Birkhoff–Witt). *For every integer $n \geq 1$ the set of ordered monomials*

$$\{\widetilde{A}_J : J \in \mathfrak{J}_n\} \cup \{1\} \tag{C.9}$$

is a basis for $U_n(\mathfrak{g})$. In particular, j is injective, so $X \in \mathfrak{g}$ may be identified with $\widetilde{X} \in U(\mathfrak{g})$.

Proof. We first prove by induction on n that the monomials (C.9) span $U_n(\mathfrak{g})$. This is clear when $n = 0$, since $U_0(\mathfrak{g}) = \mathbb{C}$. For $n \geq 1$, the set consisting of the element 1 together with all ordered monomials $\widetilde{A}_{i_1} \cdots \widetilde{A}_{i_k}$ ($k \leq n$) for all choices of $1 \leq i_j \leq d$

clearly spans $U_n(\mathfrak{g})$. But if $i_p > i_{p+1}$ for some pair of adjacent indices, then we can rearrange the monomial to put this pair of indices in increasing order, since

$$\widetilde{A}_{i_p}\widetilde{A}_{i_{p+1}} = \widetilde{A}_{i_{p+1}}\widetilde{A}_{i_p} + [\widetilde{A}_{i_p}, \widetilde{A}_{i_{p+1}}] .$$

The commutator term contributes an element of $U_{n-1}(\mathfrak{g})$. Continuing this rearranging process, we can rewrite every ordered monomial with at most n factors as a monomial in the form (C.9), modulo $U_{n-1}(\mathfrak{g})$. By induction, this proves the spanning property.

We next prove the linear independence of the set (C.9) when $\mathfrak{g} = \mathfrak{gl}(m,\mathbb{C})$. For $A \in \mathfrak{gl}(m,\mathbb{C})$ let X_A be the vector field on $M_m(\mathbb{C})$ defined by (1.23). The map $A \mapsto X_A$ is a Lie algebra homomorphism, by (1.25). Hence it extends to an associative algebra homomorphism from $U(\mathfrak{g})$ to the polynomial coefficient differential operators on $M_m(\mathbb{C})$ such that

$$\widetilde{A}_{i_1} \cdots \widetilde{A}_{i_k} \mapsto X_{A_{i_1}} \cdots X_{A_{i_k}} .$$

Let $\{f_1,\ldots,f_d\}$ be the basis for $\mathfrak{g}^*$ dual to $\{A_1,\ldots,A_d\}$, where now $d = m^2$. For $J = (i_1,\ldots,i_k) \in \mathcal{J}_n$ we define a polynomial function $f^J(x) = f_{i_1}(x) \cdots f_{i_k}(x)$ and a polynomial coefficient differential operator $X^J = X_{A_{i_1}} \cdots X_{A_{i_k}}$ on $M_n(\mathbb{C})$. Then we claim that

$$X^J f^{J'}(x)\Big|_{x=I} = \begin{cases} 1 & \text{if } J = J', \\ 0 & \text{otherwise.} \end{cases} \tag{C.10}$$

This follows immediately from (1.24) if we use the basis $\{E_{ij}\}$ for $\mathfrak{g}$, since the matrix entry functions x_{ij} give the dual basis. Hence relation (C.10) holds for any basis by the transformation property of a basis and dual basis (see Section 1.1.1). From (C.10) it follows that $\{X^J : J \in \mathcal{J}_n\}$ is linearly independent, and hence the corresponding set of elements $\widetilde{A}_{j_1} \cdots \widetilde{A}_{j_k}$ in $U(\mathfrak{g})$ is also linearly independent.

Finally, we prove the linear independence of the set (C.9) when $\mathfrak{g}$ is any Lie subalgebra of $\mathfrak{gl}(m,\mathbb{C})$. This suffices for all the Lie algebras studied in this book; in fact, by Ado's theorem (see Jacobson [79, Chapter VI, §2]) this assumption holds in general. Choose a basis $\{A_i\}$ for $\mathfrak{gl}(m,\mathbb{C})$ such that $\{A_1,\ldots,A_d\}$ is a basis for $\mathfrak{g}$. The inclusion map $\mathfrak{g} \hookrightarrow \mathfrak{gl}(m,\mathbb{C})$ extends to an associative algebra homomorphism $U(\mathfrak{g}) \longrightarrow U(\mathfrak{gl}(m,\mathbb{C}))$. The ordered monomials for $U(\mathfrak{g})$ in Theorem C.2.2 map to the corresponding ordered monomials in $U(\mathfrak{gl}(m,\mathbb{C}))$, which we have just proved to be part of a linearly independent set. Hence linear independence also holds in $U(\mathfrak{g})$. $\qquad\square$

Corollary C.2.3. *Suppose $\mathfrak{h}$ is a Lie subalgebra of $\mathfrak{g}$.*

1. *The inclusion map $\mathfrak{h} \hookrightarrow \mathfrak{g}$ extends to an injection $U(\mathfrak{h}) \hookrightarrow U(\mathfrak{g})$. Hence $U(\mathfrak{h})$ may be identified with the associative subalgebra of $U(\mathfrak{g})$ generated by $\mathfrak{h}$.*
2. *Suppose $\mathfrak{k}$ is another Lie subalgebra of $\mathfrak{g}$ such that $\mathfrak{g} = \mathfrak{h} \oplus \mathfrak{k}$ (vector space direct sum). Then the bilinear map $U(\mathfrak{h}) \otimes U(\mathfrak{k}) \longrightarrow U(\mathfrak{g})$ given by $a \otimes b \mapsto ab$ is a linear isomorphism.*

Proof. To prove (1), use the argument at the end of the proof of Theorem C.2.2. To prove (2), choose the basis for $\mathfrak{g}$ as in (1) with the additional property that

$\{X_{r+1},\ldots,X_d\}$ is a basis for $\mathfrak{k}$. Each ordered monomial occurring in the basis for $U(\mathfrak{g})$ is then a product of an ordered monomial for $U(\mathfrak{h})$ and an ordered monomial for $U(\mathfrak{k})$. $\qquad\qquad\qquad\qquad\qquad\qquad\qquad\qquad\qquad\qquad\qquad\square$

C.2.3 Adjoint Representation of Enveloping Algebra

Let $\mathfrak{g}$ be a finite-dimensional Lie algebra over $\mathbb{C}$. For $X \in \mathfrak{g}$ the operator $\mathrm{ad}\,X$ on $\mathfrak{g}$ extends uniquely to a derivation of $U(\mathfrak{g})$. Indeed, there is a unique derivation $D(X)$ of $\mathcal{T}(\mathfrak{g})$ such that $D(X)Y = [X,Y]$ for $Y \in \mathfrak{g}$. If $Y,Z \in \mathfrak{g}$ then

$$D(X)\big(Y \otimes Z - Z \otimes Y - [Y,Z]\big) = [X,Y] \otimes Z - Z \otimes [X,Y] - [[X,Y],Z]$$
$$+ Y \otimes [X,Z] - [X,Z] \otimes Y - [Y,[X,Z]]$$

by the Jacobi identity. Hence $D(X)$ preserves the ideal $\mathcal{K}(\mathfrak{g})$ generated by the commutation relations in $\mathfrak{g}$; therefore D_X induces a derivation on the enveloping algebra $U(\mathfrak{g}) = \mathcal{T}(\mathfrak{g})/\mathcal{K}(\mathfrak{g})$, which we continue to denote by $\mathrm{ad}\,X$. The equation

$$[\mathrm{ad}\,X, \mathrm{ad}\,Y] = \mathrm{ad}[X,Y] \quad \text{for } X,Y \in \mathfrak{g},$$

which holds when we view $\mathrm{ad}\,X$, $\mathrm{ad}\,Y$, and $\mathrm{ad}[X,Y]$ as derivations of $\mathfrak{g}$, continues to hold on $U(\mathfrak{g})$, since $\mathfrak{g}$ generates $U(\mathfrak{g})$ as an algebra and both sides of the equation are derivations of $U(\mathfrak{g})$. Thus the map $\mathrm{ad} : \mathfrak{g} \longrightarrow \mathrm{Der}(U(\mathfrak{g}))$ is a representation of $\mathfrak{g}$.

The standard filtration on $U(\mathfrak{g})$ satisfies $[U_k(\mathfrak{g}), U_n(\mathfrak{g})] \subset U_{k+n-1}(\mathfrak{g})$, as one verifies by induction on $k+n$ using the Jacobi identity. Hence the associated graded algebra

$$\mathrm{Gr}(U(\mathfrak{g})) = \bigoplus_{n \geq 0} U_n(\mathfrak{g})/U_{n-1}(\mathfrak{g})$$

is commutative. Since subspaces $U_n(\mathfrak{g})$ are invariant under $\mathrm{ad}\,\mathfrak{g}$ for all $n \geq 0$, there are representations

$$\varphi_n : \mathfrak{g} \longrightarrow \mathrm{End}\big(U_n(\mathfrak{g})/U_{n-1}(\mathfrak{g})\big) \quad \text{for } n = 0, 1, 2, \ldots.$$

(For $n = 1$ we obtain the adjoint representation on $\mathfrak{g}$.) We define the representation $\varphi = \bigoplus_{n \geq 0} \varphi_n$. Then $\varphi(X)$ is a derivation of $\mathrm{Gr}(U(\mathfrak{g}))$ for $X \in \mathfrak{g}$.

For the symmetric algebra $S(\mathfrak{g}) = \mathcal{T}(\mathfrak{g})/\mathcal{C}$ there is a similar construction. There is a unique extension of the operator $\mathrm{ad}\,X$ to a derivation $\delta(X)$ of $S(\mathfrak{g})$, and the subspace $S^n(\mathfrak{g})$ is invariant under $\delta(X)$ for all $n \in \mathbb{N}$. This gives a representation $(\delta, S(\mathfrak{g}))$ of $\mathfrak{g}$.

Theorem C.2.4. *There is a unique algebra isomorphism γ from $S(\mathfrak{g})$ to $\mathrm{Gr}(U(\mathfrak{g}))$ such that $\gamma(Y^n) = Y^n + U_{n-1}(\mathfrak{g})$ for $Y \in \mathfrak{g}$ and $n \in \mathbb{N}$. This isomorphism intertwines the representations δ and φ of $\mathfrak{g}$.*

Proof. Fix a basis $\{X_i : 1 \leq i \leq d\}$ for $\mathfrak{g}$. By the Poincaré–Birkhoff–Witt theorem the cosets

$$\{X_1^{a_1} \cdots X_d^{a_d} + U_{n-1}(\mathfrak{g}) : a_1 + \cdots + a_d = n\}$$

give a basis for $U_n(\mathfrak{g})/U_{n-1}(\mathfrak{g})$. Thus by Proposition B.2.4 and Corollary C.1.2, there is an algebra isomorphism $\gamma : S(\mathfrak{g}) \longrightarrow \mathrm{Gr}(U(\mathfrak{g}))$ that carries the product $X_1^{a_1} \cdots X_d^{a_d}$ in $S(\mathfrak{g})$ to the coset $X_1^{a_1} \cdots X_d^{a_d} + U_{n-1}(\mathfrak{g})$. By the multinomial expansion we have

$$\gamma(Y^n) = Y^n + U_{n-1}(\mathfrak{g}) \quad \text{for all } Y \in \mathfrak{g},$$

and this property uniquely determines γ (Lemma B.2.3).

Let $X \in \mathfrak{g}$. The derivation $\delta(X)$ acts on $Y^n \in S^n(\mathfrak{g})$ by

$$\delta(X)Y^n = n[X,Y]Y^{n-1} \quad \text{(product in } S(\mathfrak{g})).$$

On the other hand, in $U(\mathfrak{g})$ we have

$$\mathrm{ad}(X)Y^n \in n[X,Y]Y^{n-1} + U_{n-1}(\mathfrak{g}).$$

It follows again by Lemma B.2.3 that γ intertwines the two $\mathfrak{g}$ actions. $\qquad\square$

Corollary C.2.5. *There is a unique linear isomorphism* $\omega : S(\mathfrak{g}) \longrightarrow U(\mathfrak{g})$ *such that* $\omega(Y^n) = Y^n$ *for* $Y \in \mathfrak{g}$ *and* $n \in \mathbb{N}$. *This isomorphism intertwines the representation* φ *of* $\mathfrak{g}$ *on* $S(\mathfrak{g})$ *and the adjoint representation* ad *of* $\mathfrak{g}$ *on* $U(\mathfrak{g})$.

Proof. Let $\mathcal{K}_n(\mathfrak{g}) = \mathcal{K}(\mathfrak{g}) \cap U_n(\mathfrak{g})$. Then $\{\mathcal{K}_n(\mathfrak{g})\}$ is a filtration of the ideal $\mathcal{K}(\mathfrak{g})$, and by Theorem C.2.2 we see that

$$\mathcal{T}_n(\mathfrak{g}) = S^n(\mathfrak{g}) \oplus \mathcal{K}_n(\mathfrak{g}). \tag{C.11}$$

We may identify $S(\mathfrak{g})$ with the subspace $\bigoplus_{n \geq 0} S^n(\mathfrak{g})$ of $\mathcal{T}(\mathfrak{g})$, by Corollary C.1.2, and this is an isomorphism of $\mathfrak{g}$-modules. For $z \in S(\mathfrak{g})$ we define

$$\omega(z) = z + \mathcal{K}(\mathfrak{g}) \in U(\mathfrak{g}).$$

Then (C.11) shows that ω is bijective. Clearly, $\omega(Y^n) = Y^{\otimes n} + \mathcal{K}(\mathfrak{g})$ for $Y \in \mathfrak{g}$, and this property uniquely determines ω by Lemma B.2.3. $\qquad\square$

Appendix D
Manifolds and Lie Groups

Abstract The purpose of this appendix is to collect the essential parts of manifold and Lie group theory in a convenient form for the body of the book. The philosophy is to give the main definitions and to prove many of the basic theorems. Some of the more difficult results are stated with appropriate references that the careful reader who is unfamiliar with differential geometry can study.

D.1 C^∞ Manifolds

We introduce the notion of smooth manifold and the associated concepts of tangent space, differential forms, and integration relative to a top-degree form.

D.1.1 Basic Definitions

Let X be a Hausdorff topological space with a countable basis for its topology. Then an *n-chart* for X is a pair (U, Φ) of an open subset U of X and a continuous map Φ of U into $\mathbb{R}^n$ such that $\Phi(U)$ is open in $\mathbb{R}^n$ and Φ is a homeomorphism of U onto $\Phi(U)$. A C^∞ *n-atlas* for X is a collection $\{(U_\alpha, \Phi_\alpha)\}_{\alpha \in I}$ of n-charts for X such that

- the collection of sets $\{U_\alpha\}_{\alpha \in I}$ is an open covering of X,
- the maps $\Phi_\beta \circ \Phi_\alpha^{-1} : \Phi_\alpha(U_\alpha \cap U_\beta) \longrightarrow \Phi_\beta(U_\alpha \cap U_\beta)$ are of class C^∞ for all $\alpha, \beta \in I$.

Examples

1. Let $X = \mathbb{R}^n$ and take $U = \mathbb{R}^n$ and Φ the identity map. Then (U, Φ) is an n-chart for X and $\{(U, \Phi)\}$ is a C^∞ atlas.

2. Let $X = S^n = \{(x_1, \ldots, x_{n+1}) \in \mathbb{R}^{n+1} : x_1^2 + \cdots + x_{n+1}^2 = 1\}$ with the topology as a closed subset of $\mathbb{R}^{n+1}$. Let $S_{i,+}^n = \{(x_1, \ldots, x_{n+1}) \in S^n : x_i > 0\}$ and $S_{i,-} = \{(x_1, \ldots, x_{n+1}) \in S^n : x_i < 0\}$ for $i = 1, \ldots, n+1$. Define

$$\Phi_{i,\pm}(x) = (x_1, \ldots, x_{i-1}, x_{i+1}, \ldots, x_{n+1})$$

for $x \in S_{i,\pm}^n$ (the projection of $S_{i,\pm}$ onto the hyperplane $\{x_i = 0\}$). Then $\Phi_{i,\pm}(S_{i,\pm}^n) = B_n = \{x \in \mathbb{R}^n : x_1^2 + \cdots + x_n^2 < 1\}$ and

$$\Phi_{i,\pm}^{-1}(x_1, \ldots, x_n) = (x_1, \ldots, x_{i-1}, \pm(1 - x_1^2 - \cdots - x_n^2)^{1/2}, x_i, \ldots, x_n). \tag{D.1}$$

Thus each $(U, \Phi_{i,\pm})$ is a chart. The sets $\{S_{i,\pm}^n\}$ cover S^n. From (D.1) it is clear that

$$\{(S_{i,\varepsilon}^n, \Phi_{n,\varepsilon}) : 1 \leq i \leq n+1, \varepsilon = \pm\}$$

is a C^∞ n-atlas for X.

3. Let $f_1, \ldots, f_k$ be real-valued C^∞ functions on $\mathbb{R}^n$ with $k \leq n$. Let X be the set of points $x \in \mathbb{R}^n$ such that $f_i(x) = 0$ for all $i = 1, \ldots, k$ and some $k \times k$ minor of the $k \times n$ matrix

$$D(x) = \left[\frac{\partial f_i}{\partial x_j}(x) \right]$$

is nonzero (note that X might be empty). Give X the subspace topology in $\mathbb{R}^n$. For $1 \leq i_1 < \cdots < i_k \leq n$ let $D_{i_1, i_2, \ldots, i_k}(x)$ be the $k \times k$ matrix formed by rows $i_1, \ldots, i_k$ of $D(x)$. Define

$$U_{i_1, i_2, \ldots, i_k} = \{x \in X : \det D_{i_1, i_2, \ldots, i_k}(x) \neq 0\}.$$

Then these sets constitute an open covering of X.

We now construct an $(n-k)$-atlas for X as follows: Given $x \in X$, choose indices $1 \leq i_1 < \cdots < i_k \leq n$ such that $x \in U_{i_1, i_2, \ldots, i_k}$. Let $1 \leq p_1 < \cdots < p_{n-k} \leq n$ be the complementary set of indices:

$$\{i_1, \ldots, i_k\} \cup \{p_1, \ldots, p_{n-k}\} = \{1, \ldots, n\}.$$

For $y \in \mathbb{R}^n$ we define $u_q(y) = f_{i_q}(y)$ for $1 \leq q \leq k$ and $u_{k+q}(y) = y_{p_q}$ for $1 \leq q \leq n-k$. Then

$$\det \left[\frac{\partial u_i}{\partial x_j}(y) \right] \neq 0 \quad \text{for } y \in U_{i_1, i_2, \ldots, i_k}.$$

Set $\Psi(y) = (u_1(y), \ldots, u_n(y))$. The inverse function theorem (see Lang [98]) implies that then there exists an open subset $V_x \subset \mathbb{R}^n$ containing x such that $W_x = \Psi(V_x)$ is open in $\mathbb{R}^n$ and Ψ is a bijection from V_x onto W_x with C^∞ inverse map. From this result we see that if we define

$$\Phi_x(y) = (y_{p_1}, \ldots, y_{p_{n-k}}) \quad \text{for } y \in U_{i_1, i_2, \ldots, i_k} \cap V_x,$$

then Φ_x is a homeomorphism onto its image, which is open in $\mathbb{R}^{n-k}$. Set $U_x = U_{i_1,i_2,\dots,i_k} \cap V_x$. Then $\{(U_x, \Phi_x)\}_{x \in X}$ is a C^∞ $(n-k)$-atlas for X.

4. Let $X \subset \mathbb{C}^n$ be an irreducible affine variety of dimension m (Appendix A.1.5). Let X_0 be the set of smooth points of X (Appendix A.3.1, Example 3). Endow X_0 with the subspace topology as a subset of $\mathbb{C}^n$ (which we look upon as $\mathbb{R}^{2n}$). We show how to find a $2m$-dimensional C^∞ structure on X_0. Let $f_1,\dots,f_p$ generate the ideal of X. Each function f_i is a polynomial in the complex linear coordinates $z_1,\dots,z_n$ on $\mathbb{C}^n$. We define the $p \times n$ matrix

$$F(x) = \left[\frac{\partial f_i}{\partial z_j}(x) \right].$$

Fix $x \in X_0$. The tangent space $T_x(X)$ (in the sense of algebraic varieties) has dimension m over $\mathbb{C}$. This implies that there exist indices $1 \leq i_1 < \cdots < i_{n-m} \leq p$ and $j_1 < \cdots < j_{n-m}$ such that the minor $N(x)$ formed using rows $i_1,\dots,i_{n-m}$ and columns $j_1,\dots,j_{n-m}$ of $F(x)$ is nonzero. Hence there is a Zariski-open subset U of $\mathbb{C}^n$ containing x such that $N(y) \neq 0$ for all $y \in U$. We identify $\mathbb{C}^n$ with $\mathbb{R}^{2n}$ by $z_j = x_j + \mathrm{i}x_{n+j}$, where $\mathrm{i} = \sqrt{-1}$ and $x_1,\dots,x_{2n}$ are real coordinates on $\mathbb{R}^{2n}$. We write $f_i = g_i + \mathrm{i}g_{p+i}$ for $i = 1,\dots,p$, with g_k a real-valued polynomial in $x_1,\dots,x_{2n}$. Define the $2p \times 2n$ matrix

$$D(y) = \left[\frac{\partial g_i}{\partial x_j}(y) \right].$$

Let $M(y)$ be the minor formed from rows $i_1, \dots, i_{n-m}, p+i_1, \dots, p+i_{m-n}$ and columns $j_1, \dots, j_{n-m}, n+j_1, \dots, n+j_{n-m}$ of $D(y)$. The Cauchy–Riemann equations imply that

$$M(y) = |N(y)|^2 \tag{D.2}$$

(see Helgason [67, Chapter VIII, §2 (7)]). Hence $D(y)$ has rank $2(n-m)$ when $y \in U$.

Let

$$Y = \{y \in \mathbb{C}^n : f_{i_k}(y) = 0 \text{ for } k = 1,\dots,n-m\}.$$

Let V be a Zariski-open subset of $\mathbb{C}^n$ such that $x \in V$ and $V \cap Y$ is irreducible. Then $\dim V \cap Y = m$ and $V \cap X \subset V \cap Y$. Since $\dim V \cap X = m$, this implies that $V \cap X = V \cap Y$ (Theorem A.1.19). Now use (D.2) to construct a C^∞ $2m$-atlas for $V \cap X$ by the method of Example 3. Given $x \in X_0$ choose an element (U, Φ) in this atlas with $x \in U$ and denote it by (U_x, Φ_x). Then $\{(U_x, \Phi_x) : x \in X_0\}$ defines a C^∞ atlas on X_0 that gives rise to a $2m$-dimensional C^∞ structure.

Returning to the general concepts associated with manifolds, suppose X has a C^∞ n-atlas $\mathcal{A} = \{(U_\alpha, \Phi_\alpha)\}_{\alpha \in I}$. We say that an n-chart (U, Φ) is *compatible* with $\mathcal{A}$ if whenever $U \cap U_\alpha \neq \emptyset$ then the maps $\Phi \circ \Phi_\alpha^{-1}$ and $\Phi_\alpha \circ \Phi^{-1}$ between the open subsets $\Phi_\alpha(U \cap U_\alpha)$ and $\Phi(U \cap U_\alpha)$ of $\mathbb{R}^n$ are of class C^∞.

If X is a Hausdorff space with a countable basis for its topology then an n-*dimensional C^∞ structure* on X is an n-atlas $\mathcal{A}$ for X such that every n-chart of X that is compatible with $\mathcal{A}$ is contained in $\mathcal{A}$.

Lemma D.1.1. *If $\mathcal{A}$ is a C^∞ n-atlas for X then $\mathcal{A}$ is contained in a unique n-dimensional C^∞ structure for X.*

Proof. If $\mathcal{B}$ is the collection of all n-charts compatible with $\mathcal{A}$, then the definition of atlas implies that $\mathcal{B} \supset \mathcal{A}$. The chain rule for differentiation implies that a chart compatible with $\mathcal{A}$ is compatible with $\mathcal{B}$. Thus $\mathcal{B}$ is a C^∞ structure on X. The chain rule also implies the uniqueness. $\qquad\qquad\qquad\qquad\qquad\qquad\qquad\qquad \Box$

Definition D.1.2. A pair $(X, \mathcal{A})$ of a Hausdorff topological space with a countable basis for its topology and an n-dimensional C^∞ structure on X will be called an n-dimensional C^∞ *manifold*.

If $M = (X, \mathcal{A})$ we will write $x \in M$ for $x \in X$, and we will say that a chart for X is a chart for M if it is in $\mathcal{A}$. Each example above is a C^∞ manifold with C^∞ structure corresponding to the atlas constructed there. We now give some more important examples.

5. (Products) Let $M = (X, \mathcal{A})$ and $N = (Y, \mathcal{B})$ be manifolds of dimensions m and n respectively. Given $(U, \Phi) \in \mathcal{A}$ and $(V, \Psi) \in \mathcal{B}$, we define $(\Phi \times \Psi)(x, y) = (\Phi(x), \Psi(y))$ for $(x, y) \in U \times V$. Then the set

$$\{(U \times V, \Phi \times \Psi) : (U, \Phi) \in \mathcal{A}, (V, \Psi) \in \mathcal{B}\}$$

is a C^∞ $(m+n)$-atlas for $X \times Y$ (with the product topology). This defines the structure of a C^∞ $(m+n)$-dimensional manifold on $X \times Y$. We use the notation $M \times N$ for the corresponding C^∞ manifold and call it the *product manifold*.

6. (Submanifolds) Let $M = (X, \mathcal{A})$ be a C^∞ manifold and let U be an open subset of X. Then the set

$$\{(V \cap U, \Phi|_{U \cap V}) : (V, \Phi) \in \mathcal{A}\}$$

is a C^∞ atlas for U. Thus U is a C^∞ manifold called an *open submanifold* of M.

7. (Covering Manifolds) Let $M = (X, \mathcal{A})$ be a C^∞ manifold. Let $\pi : Y \longrightarrow X$ be a covering space that is a Hausdorff space with a countable basis for its topology. Recall that this means that

(a) π is a continuous surjective mapping;
(b) for all $x \in X$ there exists an open neighborhood U of x in X such that $\pi^{-1}(U) = \bigcup_\alpha V_\alpha$ is a disjoint union of open sets, and $\pi : V_\alpha \longrightarrow U$ is a homeomorphism.

We say that the neighborhood U in (b) is *evenly covered* by π. We will now show how to find a C^∞ atlas on Y. If $x \in M$ let $(W_x, \Psi_x) \in \mathcal{A}$ with $x \in W_x$. Let U_x be an evenly covered neighborhood of x. Set $V_x = U_x \cap W_x$ and $\Phi_x = \Psi_x|_{V_x}$. Then $(V_x, \Phi_x) \in \mathcal{A}$ and V_x is evenly covered by π. Let

$$\pi^{-1}(V_x) = \bigcup_\alpha V_{x,\alpha}$$

with π a homeomorphism of $V_{x,\alpha}$ onto V_x. Set $\Phi_{x,\alpha}(y) = \Phi_x(\pi(y))$ for $y \in V_{x,\alpha}$. We note that the collection $\{V_{x,\alpha}\}$, with $x \in X$ and α running through all of the appropriate indices, is an open covering of Y. We leave it to the reader to check that $\{(V_{x,\alpha}, \Phi_{x,\alpha})\}$ is a C^∞ n-atlas for Y ($n = \dim M$) and that $\pi : V_{x,\alpha} \longrightarrow V_x$ is a diffeomorphism for all $x \in X$ and all α.

Definition D.1.3. Let $M = (X, \mathcal{A})$ and $N = (Y, \mathcal{B})$ be C^∞ manifolds. A C^∞ *map* $f : M \longrightarrow N$ is a continuous map $f : X \longrightarrow Y$ such that whenever $(V, \Psi) \in \mathcal{B}$ and $(U, \Phi) \in \mathcal{A}$ satisfy $U \subset f^{-1}(V)$, then the map $\Psi \circ f \circ \Phi^{-1} : \Phi(U) \longrightarrow \Psi(V)$ is of class C^∞ (relative to the atlases $\mathcal{A}$ and $\mathcal{B}$).

We will look upon $\mathbb{C}$ as $\mathbb{R}^2$ with the C^∞ structure as in Example 1 above. We denote by $C^\infty(M; \mathbb{R}^n)$ the space of all C^∞ maps of M into $\mathbb{R}^n$ (with the C^∞ structure as in Example 1 above). We write $C^\infty(M) = C^\infty(M; \mathbb{R})$ and $C^\infty(M; \mathbb{C}) = C^\infty(M, \mathbb{R}^2)$. These latter examples are algebras under pointwise addition and multiplication of functions.

We return to the examples above. In Example 1 the usual advanced calculus notion of C^∞ coincides with our definition. In Example 2 the map $\iota : S^n \longrightarrow \mathbb{R}^{n+1}$ with $\iota(x) = x$ is of class C^∞. The map $\iota : X \longrightarrow \mathbb{R}^n$ with $\iota(x) = x$ in Examples 3 and 4 is of class C^∞. In Example 5 the projections on each of the factors are of class C^∞.

Let M be a C^∞ manifold. If f is a real-valued function on M we denote by $\text{supp}(f)$ (the *support* of f) the closure of the set $\{x \in M : f(x) \neq 0\}$. Let $\{U_\alpha\}_{\alpha \in I}$ be an open covering of M as a topological space. Then a *partition of unity* subordinate to the covering is a countable set $\{\varphi_i : 1 \leq i < N\} \subset C^\infty(M)$ with $N \leq \infty$ (the set might be finite) such that the following hold for all $x \in M$:

1. $0 \leq \varphi_i(x) \leq 1$ for all $1 \leq i < N$.
2. There is an open neighborhood U of x such that $\text{Card}\{i : \varphi_i|_U \neq 0\} < \infty$.
3. $\sum_{i<N} \varphi_i(x) = 1$.
4. If $1 \leq i < N$ then there exists $\alpha \in I$ such that $\text{supp}(\varphi_i) \subset U_\alpha$.

Theorem D.1.4. *For each open covering $\mathcal{U}$ of M there exists a partition of unity subordinate to $\mathcal{U}$.*

For a proof see Warner [157].

We now give an example of how partitions of unity are used. Let A and B be closed subsets of M such that $A \cap B = \emptyset$. Let $U = M \setminus A$ and $V = M \setminus B$. Then $U \cup V = M$. Let $\{\varphi_i\}$ be a partition of unity subordinate to the open covering $\{U, V\}$. Let $S = \{i : \text{supp}(\varphi_i) \subset V\}$ and set

$$f(x) = \sum_{i \in S} \varphi_i(x).$$

Note that condition (2) above implies that the sum is actually finite for each x and that f defines an element of $C^\infty(M)$. We note that if $x \in B$ then $f(x) = 0$. Now consider $x \in A$. If $i \notin S$ then $\text{supp}(\varphi_i) \subset U$ by condition (4) above. Hence if $i \notin S$, $\varphi_i(x) = 0$. Thus

$$1 = \sum_i \varphi_i(x) = \sum_{i \in S} \varphi_i(x) = f(x).$$

We have thus shown that there exists $f \in C^\infty(M)$ such that f is identically 0 on B and identically 1 on A.

D.1.2 Tangent Space

Let M be an n-dimensional C^∞ manifold. If $x \in M$ then a *point derivation* of $C^\infty(M)$ at x is a linear functional L on $C^\infty(M)$ such that

$$L(fg) = f(x)L(g) + g(x)L(f) \quad \text{for all } f, g \in C^\infty(M) .$$

Note that if $f(x) = c$ for all $x \in M$ is a constant function, then $L(f^2) = 2cL(f)$. But $f^2 = cf$, so $cL(f) = 2cL(f)$. Hence $L(f) = 0$. If L_1 and L_2 are point derivations at x then so is $a_1 L_1 + a_2 L_2$ for $a_1, a_2 \in \mathbb{R}$. Thus the point derivations at x form a vector space over $\mathbb{R}$.

Definition D.1.5. A point derivation at x is called a *tangent vector*. The *tangent space* of M at x is the vector space $T(M)_x$ of all point derivations of M at x.

Lemma D.1.6. *Let $x \in M$ and $f \in C^\infty(M)$. If f vanishes in an open neighborhood U of x then $Lf = 0$ for all $L \in T(M)_x$.*

Proof. Let $B = \mathrm{supp}(f)$. Then $B \cap U = \emptyset$. Take $\varphi \in C^\infty(M)$ such that $\varphi(x) = 0$ and φ is identically 1 on B; then $\varphi f = f$. If $L \in T(M)_x$ then

$$Lf = L(\varphi f) = \varphi(x)Lf + f(x)L\varphi = 0 . \qquad \square$$

Let M and N be C^∞ manifolds. If $f : M \longrightarrow N$ is a C^∞ map then we define $df_x : T(M)_x \longrightarrow T(N)_{f(x)}$ by

$$df_x(L)\varphi = L(\varphi \circ f) \quad \text{for } L \in T(M)_x \text{ and } \varphi \in C^\infty(N) .$$

The map df_x is called the *differential* of f at x.

Lemma D.1.7. *Let U be an open submanifold of the C^∞ manifold M, and define $\iota : U \longrightarrow M$ by $\iota(x) = x$. Then ι is a C^∞ map, and the map $d\iota_x : T(U)_x \longrightarrow T(M)_x$ is a linear bijection for each $x \in U$.*

Proof. Let $x \in U$. Let (V, Φ) be a chart for M such that $x \in V$. Then $\Phi(V)$ is open in $\mathbb{R}^n$ and by the definition of chart $\Phi(U \cap V)$ is also open in $\mathbb{R}^n$. Choose open sets W_1, W_2 in $\mathbb{R}^n$ such that

$$\Phi(x) \in W_1 \subset \overline{W_1} \subset W_2 \subset \Phi(U \cap V) .$$

Let $A = \Phi^{-1}(\overline{W_1})$ and let $B = M \setminus \Phi^{-1}(W_2)$. Let $\varphi \in C^\infty(M)$ be identically 1 on A and identically 0 on B. Given $f \in C^\infty(U)$, define

$$(\varphi f)(x) = \begin{cases} \varphi(x)f(x) & \text{for } x \in U, \\ 0 & \text{otherwise.} \end{cases}$$

Then $\varphi f \in C^\infty(M)$. Suppose $L \in T(U)_x$ satisfies $d\iota_x(L) = 0$. Then for all $f \in C^\infty(U)$ we have $0 = d\iota_x(L)(\varphi f) = L((\varphi f)|_U)$. Since φf agrees with f in a neighborhood of x, Lemma D.1.6 implies that $Lf = 0$ for all $f \in C^\infty(U)$, and hence $L = 0$. This proves that $d\iota_x$ is injective. Given $L \in T(M)_x$, define $\widetilde{L}f = L(\varphi f)$ for $f \in C^\infty(U)$. The first part of the argument implies that $d\iota_x(\widetilde{L}) = L$. So $d\iota_x$ is surjective. $\qquad\square$

We now determine $T(\mathbb{R}^n)_p$ for each $p \in \mathbb{R}^n$. Given a point p, we define a linear functional $\left(\frac{\partial}{\partial x_i}\right)_p$ on $C^\infty(\mathbb{R}^n)$ by

$$\left(\frac{\partial}{\partial x_i}\right)_p(f) = \frac{\partial f}{\partial x_i}(p) \quad \text{for } f \in C^\infty(\mathbb{R}^n).$$

The Leibniz rule implies that $\left(\frac{\partial}{\partial x_i}\right)_p$ is a tangent vector at p.

Lemma D.1.8. *The set* $\left\{\left(\frac{\partial}{\partial x_1}\right)_p, \ldots, \left(\frac{\partial}{\partial x_n}\right)_p\right\}$ *is a basis for* $T_p(\mathbb{R}^n)$.

Proof. If $f \in C^\infty(\mathbb{R}^n)$, then by the fundamental theorem of calculus

$$f(x) = f(p) + \int_0^1 \frac{d}{dt} f(p + t(x - p)) \, dt = f(p) + \sum_i (x_i - p_i) g_i(x), \qquad \text{(D.3)}$$

with $g_i(x) = \int_0^1 \frac{\partial f}{\partial x_i}(p + t(x - p)) \, dt$. We note that $g_i(p) = \left(\frac{\partial}{\partial x_i}\right)_p(f)$. Suppose that $L \in T(\mathbb{R}^n)_p$. Then by (D.3) we have

$$L(f) = f(p)L(1) + \sum_i L(x_i - p_i)\left(\frac{\partial}{\partial x_i}\right)_p(f) + \sum_i (p_i - p_i)L(g_i)$$

$$= \sum_i L(x_i - p_i)\left(\frac{\partial}{\partial x_i}\right)_p(f).$$

This formula for $L(f)$ shows that the set $\left\{\left(\frac{\partial}{\partial x_1}\right)_p, \ldots, \left(\frac{\partial}{\partial x_n}\right)_p\right\}$ spans the tangent space at p. The set is linearly independent, since $\left(\frac{\partial}{\partial x_1}\right)_p(x_j) = \delta_{ij}$. $\qquad\square$

Definition D.1.9. Let M be an n-dimensional C^∞ manifold and let U be an open subset of M. A *system of local coordinates* on U is a set $\{u_1, \ldots, u_n\}$ of C^∞ functions on U such that if $\Phi(x) = (u_1(x), \ldots, u_n(x))$ for $x \in U$, then (U, Φ) is a chart for M.

Obviously, if (U, Φ) is a chart for M and $\Phi(x) = (u_1(x), \ldots, u_n(x))$, then $\{u_1, \ldots, u_n\}$ is a system of local coordinates for M on U, and this is the way all systems are obtained.

Let $\{u_1, \ldots, u_n\}$ be a system of local coordinates on U and let (U, Φ) be the corresponding chart. If $x \in U$ then

$$d\Phi_x : T(U)_x \longrightarrow T(\Phi(U))_{\Phi(x)}.$$

We note that $\Phi^{-1} : \Phi(U) \longrightarrow U$ is also C^∞, and $\mathrm{d}\Phi^{-1}_{\Phi(x)}$ is the inverse mapping. Thus $T(U)_x$ is isomorphic as a vector space to $T(\Phi(U))_{\Phi(x)}$, which is isomorphic to $T(\mathbb{R}^n)_{\Phi(x)}$ by Lemma D.1.8. We write

$$\left(\frac{\partial}{\partial u_i}\right)_x = \mathrm{d}\Phi^{-1}_{\Phi(x)}\left(\frac{\partial}{\partial x_i}\right)_{\Phi(x)}.$$

Then $\left\{ \left(\frac{\partial}{\partial u_1}\right)_x, \dots, \left(\frac{\partial}{\partial u_n}\right)_x \right\}$ is a basis of $T(U)_x$, and hence it is a basis for $T(M)_x$ by Lemma D.1.7.

It is convenient to look upon $T(\mathbb{R}^n)_x$ as $\mathbb{R}^n$ as follows: If $v \in \mathbb{R}^n$ define the *directional derivative* v_x at x by

$$v_x \cdot f = \frac{d}{dt} f(x + tv)\Big|_{t=0}.$$

Then Lemma D.1.8 implies that $T(\mathbb{R}^n)_x = \{ v_x : v \in \mathbb{R}^n \}$. Thus the map $v \mapsto v_x$ gives a linear isomorphism between $\mathbb{R}^n$ and $T(\mathbb{R}^n)_x$.

Definition D.1.10. Let M and N be C^∞ manifolds with N a subset of M. Let $\iota(x) = x$ for $x \in N$. We call N a *submanifold* of M if ι is a C^∞ map and $\mathrm{d}\iota_x$ is injective for each $x \in N$.

Examples 2 and 3 of Section D.1.1 are submanifolds of $\mathbb{R}^n$. A submanifold is not necessarily a topological subspace (i.e., the topology on N as a manifold is not necessarily the relative topology coming from M).

Definition D.1.11. Let M be a C^∞ manifold. A *vector field* on M is an assignment $x \mapsto X_x \in T(M)_x$ for $x \in M$ such that for all $f \in C^\infty(M)$, the function $x \mapsto X_x f$ is an element of $C^\infty(M)$.

We write $(Xf)(x) = X_x(f)$. Thus a vector field defines an endomorphism of $C^\infty(M)$ as a vector space over $\mathbb{R}$. If X and Y are vector fields on M and $x \in M$, then $[X,Y]_x f = X_x(Yf) - Y_x(Xf)$. A direct calculation shows that $[X,Y]_x$ is a point derivation at x (see Section 1.3.7). Hence the assignment $x \mapsto [X,Y]_x$ defines a vector field on M.

Example

When $M = \mathbb{R}^n$ and $v \in \mathbb{R}^n$, then $x \mapsto v_x$ is a vector field (a *constant field*). More generally, for a C^∞ mapping $F : \mathbb{R}^n \longrightarrow \mathbb{R}^n$, the assignment $x \mapsto F(x)_x$ is a vector field. Every vector field on $\mathbb{R}^n$ is given in this way (see Exercises D.1.4 #3).

D.1.3 *Differential Forms and Integration*

Definition D.1.12. Let M be a C^∞ manifold and p a nonnegative integer. A *differential p-form* on M is an assignment $x \mapsto \omega_x$ with ω_x an alternating p-multilinear form on $T(M)_x$ (see Section B.2.4) such that if $X_1, \ldots, X_p$ are vector fields on M then $x \mapsto \omega_x((X_1)_x, \ldots, (X_p)_x)$ defines a C^∞ function on M. Let $\Omega^p(M)$ denote the space of all differential p-forms on M.

Given $f \in C^\infty(M)$ and $\omega \in \Omega^p(M)$, we define $f\omega$ by $(f\omega)_x = f(x)\omega_x$. This makes $\Omega^p(M)$ into a $C^\infty(M)$-module. If M and N are C^∞ manifolds and $f : M \longrightarrow N$ is a C^∞ map, then we define $f^* : \Omega^p(N) \longrightarrow \Omega^p(M)$ by

$$(f^*\omega)_x(v_1, \ldots, v_p) = \omega_{f(x)}(df_x(v_1), \ldots, df_x(v_p)) \ .$$

If $U \subset M$ is an open submanifold (Example 5 of Section D.1.1) and ι is the inclusion map of U into M, then we write $\omega|_U = \iota^*\omega$ for $\omega \in \Omega^p(M)$.

Given $f \in C^\infty(M)$ we define $df_x(v) = v(f)$ for $v \in T(M)_x$. This notation is consistent with our earlier definition of differential if we identify $T(\mathbb{R})_x$ with $\mathbb{R}$. Let $\{u_1, \ldots, u_n\}$ be a system of local coordinates on an open subset U of M. Given indices $1 \le i_1 < i_2 < \cdots < i_p \le n$, a point $x \in U$, and $v_1, \ldots, v_p \in T(M)_x$, we set

$$(du_{i_1} \wedge \cdots \wedge du_{i_p})_x(v_1, \ldots, v_p) = (du_{i_1})_x \wedge \cdots \wedge (du_{i_p})_x (v_1, \ldots, v_p) \ .$$

It is easily seen that $du_{i_1} \wedge \cdots \wedge du_{i_p} \in \Omega^p(U)$. In fact,

$$du_{i_1} \wedge \cdots \wedge du_{i_p} = \Phi^*(dx_{i_1} \wedge \cdots \wedge dx_{i_p})$$

with $x_i(x) = x_i$ (the usual coordinates on $\mathbb{R}^n$). Recall that $\{(du_1)_x, \ldots, (du_n)_x\}$ is the dual basis to the basis $\{(\partial/\partial u_1)_x, \ldots, (\partial/\partial u_n)_x\}$ of $T(M)_x$ (see Section D.1.2). Hence for every $\omega \in \Omega^p(U)$ there exist unique functions $a_{i_1 \ldots i_p} \in C^\infty(U)$ such that

$$\omega = \sum_{1 \le i_1 < \cdots < i_p \le n} a_{i_1 \ldots i_p} \, du_{i_1} \wedge \cdots \wedge du_{i_p} \ . \tag{D.4}$$

We will be particularly interested in $\Omega^p(M)$ for $p = \dim M$, so that we can define integration of functions on M in an intrinsic way.

Definition D.1.13. A manifold M is *orientable* if there exists an atlas $\mathcal{A}$ for M with the following property: If $(U, \Phi), (V, \Psi) \in \mathcal{A}$ and if $\{u_1, \ldots, u_n\}$ and $\{v_1, \ldots, v_n\}$ are the corresponding systems of local coordinates on U and V, respectively, then

$$\frac{\partial(v_1, \ldots, v_n)}{\partial(u_1, \ldots, u_n)}(x) = \det\left[\left(\frac{\partial}{\partial u_i}\right)_x v_j\right] > 0 \quad \text{for all } x \in U \cap V \ . \tag{D.5}$$

By the chain rule condition (D.5) is equivalent to the condition

$$(dv_1 \wedge \cdots \wedge dv_n)_x = \alpha(x)(du_1 \wedge \cdots \wedge du_n)_x \quad \text{for } x \in U \cap V \ ,$$

where $\alpha \in C^\infty(U \cap V)$ with $\alpha(x) > 0$ for all $x \in U \cap V$.

Assume that M is orientable. An *orientation* of M is an atlas $\mathcal{A}$ for M such that condition (D.5) is satisfied for every pair of elements of $\mathcal{A}$ and $\mathcal{A}$ is maximal with respect to this property. If $\mathcal{A}$ is an orientation of M and (U, Φ) is a chart for M then (U, Φ) will be said to be *compatible with the orientation* if $(U, \Phi) \in \mathcal{A}$.

Theorem D.1.14. *An n-dimensional manifold M is orientable if and only if there exists $\omega \in \Omega^n(M)$ such that $\omega_x \neq 0$ for all $x \in M$. Assume that this condition holds. The set of all charts (U, Φ) for M such that*

$$\omega|_U = a(\mathrm{d}u_1 \wedge \cdots \wedge \mathrm{d}u_n),$$

with $\{u_1, \ldots, u_n\}$ the corresponding system of local coordinates on U and $a(x) > 0$ for all $x \in U$, forms an orientation of M.

Proof. Let $\mathcal{A}$ be an atlas defining an orientation of M. Take a partition of unity $\{\varphi_i\}$ subordinate to the covering $\{U : (U, \Phi) \in \mathcal{A}\}$ of M. Choose a chart $(U_i, \Phi_i) \in \mathcal{A}$ such that $\mathrm{supp}(\varphi_i) \subset U_i$. Let $\omega_i = \Phi_i^*(\mathrm{d}x_1 \wedge \cdots \wedge \mathrm{d}x_n)$ and define

$$\omega_x = \sum_{\{i \,:\, x \in U_i\}} \varphi_i(x) (\omega_i)_x.$$

Then the positivity condition (D.5) implies that $\omega_x \neq 0$ for all $x \in M$. The formula for ω implies that $x \mapsto \omega_x$ defines an element of $\Omega^n(M)$. Proving the second part of the theorem is easier and is left to the reader. $\square$

If $\omega \in \Omega^n(M)$ and $\omega_x \neq 0$ for all $x \in M$ then we will call ω a *volume form*. Our next task is to define integration with respect to volume forms.

Fix a volume form ω on M and let $\mathcal{A}$ be the orientation corresponding to ω. Let $C_c(M)$ denote the space of all continuous real-valued functions on M with compact support. Suppose that $f \in C_c(M)$ and $\mathrm{supp}(f) \subset U$ with $(U, \Phi) \in \mathcal{A}$. We can write $\omega|_U = a \Phi^*(\mathrm{d}x_1 \wedge \cdots \wedge \mathrm{d}x_n)$ with $a \in C^\infty(U)$ and $a(p) > 0$ for $p \in U$. We set

$$\int_M f\omega = \int_{\Phi(U)} a(\Phi^{-1}(x)) f(\Phi^{-1}(x)) \, \mathrm{d}x_1 \cdots \mathrm{d}x_n. \tag{D.6}$$

Suppose $(V, \Psi) \in \mathcal{A}$ is another chart with $\mathrm{supp}(f) \subset V$. We have

$$\omega|_V = b\Psi^*(\mathrm{d}x_1 \wedge \cdots \wedge \mathrm{d}x_n).$$

If $\{u_1, \ldots, u_n\}$ and $\{v_1, \ldots, v_n\}$ are the local coordinates corresponding to (U, Φ) and (V, Ψ), then

$$b\frac{\partial(v_1, \ldots, v_n)}{\partial(u_1, \ldots, u_n)} = a \quad \text{on } U \cap V.$$

The change of variables theorem of advanced calculus (Lang [98]) implies that

$$\int_{\Phi(U)} a(\Phi^{-1}(x)) f(\Phi^{-1}(x)) \, dx_1 \cdots dx_n$$

$$= \int_{\Phi(U \cap V)} a(\Phi^{-1}(x)) f(\Phi^{-1}(x)) \, dx_1 \cdots dx_n$$

$$= \int_{\Psi(U \cap V)} b(\Psi^{-1}(x)) f(\Psi^{-1}(x)) \, dx_1 \cdots dx_n$$

$$= \int_{\Psi(V)} b(\Psi^{-1}(x)) f(\Psi^{-1}(x)) \, dx_1 \cdots dx_n .$$

Thus the formula (D.6) is justified.

We will now define the integral of a general $f \in C_c(M)$ using a partition of unity. Let $\{\varphi_i\}$ and $\{\psi_j\}$ be partitions of unity subordinate to $\{U : (U, \Phi) \in \mathcal{A}\}$. Fix, for each index i (resp. j), a chart $(U_i, \Phi_i) \in \mathcal{A}$ (resp. a chart $(V_j, \Psi_j) \in \mathcal{A}$) such that $\mathrm{supp}(\varphi_i) \subset U_i$ (resp. $\mathrm{supp}(\psi_j) \subset V_j$). Then we assert that

$$\sum_i \int_M \varphi_i f \, \omega = \sum_j \int_M \psi_j f \, \omega . \tag{D.7}$$

Indeed, $\varphi_i f = \sum_j \psi_j \varphi_i f$ and $\psi_j f = \sum_i \varphi_i \psi_j f$ (note that since f has compact support, each sum has only a finite number of nonzero terms). Thus we have

$$\sum_i \int_M \varphi_i f \, \omega = \sum_{j,i} \int_M \psi_j \varphi_i f \, \omega = \sum_j \int_M \psi_j f \, \omega .$$

This proves the assertion. We use formula (D.7) to define the *integral* of f with respect to ω and denote it by $\int_M f \, \omega$.

The basic properties of the advanced calculus notion of integral (such as additivity) carry over to our case. The following two results will be particularly important:

Lemma D.1.15. *Let $f \in C_c(M)$. If $f(x) \geq 0$ for all $x \in M$ and if $f(x) > 0$ for some $x \in M$ then $\int_M f \, \omega > 0$.*

This is clear from the definition.

Theorem D.1.16. *Let M and N be C^∞ manifolds. Let ω be a volume form on N and let $f \in C_c(N)$ and let Φ be a diffeomorphism of M onto N. Then*

$$\int_N f \, \omega = \int_M (f \circ \Phi) \, \Phi^* \omega .$$

When $M = N$ we can write this in a somewhat more suggestive form. In this case $\Phi^* \omega = \varphi \, \omega$ with $\varphi \in C^\infty(M)$. Then the formula in the theorem reads

$$\int_M f \, \omega = \int_M (f \circ \Phi) |\varphi| \, \omega . \tag{D.8}$$

This is proved by taking a partition of unity subordinate to the orientation determined by ω, pulling back via Φ, and then using the change of variables theorem one chart at a time.

If $f \in C_c(M, \mathbb{C})$ (the complex-valued continuous functions with compact support)
we write $f = f_1 + \mathrm{i}f_2$, with $f_1, f_2 \in C_c(M)$, and we set

$$\int_M f\,\omega = \int_M f_1\,\omega + \mathrm{i}\int_M f_2\,\omega \,.$$

We end this section with one additional result that we will need in our analytic
proof of the Weyl character formula. Let M be a C^∞ manifold with volume form ω.
Let $\pi : Y \longrightarrow M$ be a finite covering space (see Example 7, Section D.1.1). Here
finite means that $\pi^{-1}(x)$ is finite for each $x \in M$. Endow Y with the manifold structure in Example 7, Section D.1.1. We will denote this manifold by N. We assume
that M and N are connected; this implies that $|\pi^{-1}(x)|$ is independent of $x \in M$; we
will denote this number by d (the *degree* of the covering). Also, since π is locally a
diffeomorphism, $\pi^*\omega$ is a volume form for N.

Theorem D.1.17. *Let* $f \in C_c(M)$. *Then* $\int_N (f \circ \pi)\pi^*\omega = d\int_M f\,\omega$.

Proof. This is obvious if $\mathrm{supp}(f)$ is contained in an evenly covered chart. In general
use a partition of unity subordinate to a covering by evenly covered charts.   □

D.1.4 Exercises

1. Let $X = \mathbb{R}$. Set $\Psi(x) = x^3$. Show that $\{(X, \Psi)\}$ is a C^∞ atlas for X that is not
   contained in the C^∞ structure corresponding to Example 1 in Section D.1.1.
2. Let M be the C^∞ manifold corresponding to Example 3 in Section D.1.1. Let
   $\iota(x) = x$ for $x \in M$, so $\iota : M \longrightarrow \mathbb{R}^n$. Show that $\mathrm{d}\iota_x : T(M)_x \longrightarrow T(\mathbb{R}^n)_x$ is
   injective and that $\mathrm{d}\iota_x(T(M)_x) = \{v_x : v \in \mathbb{R}^n, (\mathrm{d}f_i)_x(v_x) = 0\}$. In particular, for
   the example of S^n conclude that $\mathrm{d}\iota_x(T(S^n)_x) = \{v_x : (x,v) = 0\}$, where $(\cdot,\cdot)$ is
   the usual inner product on $\mathbb{R}^n$.
3. A C^∞ curve in a C^∞ manifold M is a C^∞ map $\sigma : (a,b) \longrightarrow M$ (here $(a,b) =$
   $\{t \in \mathbb{R} : a < t < b\}$). Define $\sigma'(t) = \mathrm{d}\sigma_t(d/dt)_t$. Show that $T(M)_p$ is the set of
   all $\sigma'(t)$ with $\sigma(t) = p$.
4. Let $M = S^1 \times S^1$, where we take $S^1 = \{e^{\mathrm{i}\theta} : \theta \in \mathbb{R}\} \subset \mathbb{C}$.
   (a) Show that the map $f : \mathbb{R}^2 \longrightarrow M$ given by $f(x,y) = (e^{\mathrm{i}x}, e^{\mathrm{i}y})$ is C^∞.
   (b) Let $Y = f(\{x, \sqrt{2}x\} : x \in \mathbb{R}\})$ and define $\Phi(f(x, \sqrt{2}x)) = x$. Show that Φ is
   a bijection between Y and $\mathbb{R}$.
   (c) Endow Y with the topology that makes Φ a homeomorphism. Endow Y with
   the C^∞ structure containing the chart (Y, Φ). Let N denote this C^∞ manifold. Show
   that N is a submanifold of M.
   (d) Show that the relative topology of Y in M is not the same as the given topology.
5. Let M be an n-dimensional C^∞ manifold. Prove that a correspondence $x \mapsto$
   $X_x \in T(M)_x$ is a vector field if and only if for each system of local coordinates
   $\{u_1, \ldots, u_n\}$ on U there exist C^∞ functions $a_1, \ldots, a_n$ on U such that

$$X_x = \sum_i a_i(x) \left(\frac{\partial}{\partial u_i}\right)_x \quad \text{for } x \in U .$$

(HINT: It is enough to check this for U that cover M.)

6. Show that the vector fields on a C^∞ manifold form a Lie algebra under the vector field bracket.

7. Define $\omega \in \Omega^n(\mathbb{R}^{n+1})$ by

$$\omega = \sum_{i=1}^{n+1} (-x_i)^{i+1} dx_1 \wedge \cdots \wedge dx_{i-1} \wedge dx_{i+1} \wedge \cdots \wedge dx_n .$$

Let $\iota : S^n \longrightarrow \mathbb{R}^{n+1}$ be the usual injection ($\iota(x) = x$). Show that $\iota^*\omega$ defines a volume form on S^n.

8. Let $n = 2$ in the preceding exercise. Let $\Phi : \mathbb{R} \longrightarrow S^1$ be defined by $\Phi(t) = (\cos t, \sin t)$. Calculate $\Phi^*\omega$.

9. Let M be the submanifold of $\mathbb{R}^n$ corresponding to Example 3 in Section D.1.1. If $f \in C^\infty(\mathbb{R}^n)$ define $\nabla f(x) \in \mathbb{R}^n$ by $(\nabla f(x), v) = v_x(f)$ for $v \in \mathbb{R}^n$, where $(\cdot, \cdot)$ is the usual inner product on $\mathbb{R}^n$. Define $\omega \in \Omega^{n-k}(\mathbb{R}^n)$ by

$$\omega_x\big((v_1)_x, \ldots, (v_{n-k})_x\big)$$
$$= (dx_1 \wedge \cdots \wedge dx_n)_x \big(\nabla f_1(x), \ldots, \nabla f_k(x), (v_1)_x, \ldots, (v_{n-k})_x\big) .$$

Let $\iota : M \longrightarrow \mathbb{R}^n$ be the usual injection ($\iota(x) = x$). Show that $\iota^*\omega$ defines a volume form on M. Relate this exercise to exercise #7.

10. Verify formula (D.8).

D.2 Lie Groups

Lie groups and their homogeneous spaces stand at the intersection of the theory of manifolds and group theory. We present the general properties of Lie groups and their Lie algebras, and the theory of group-invariant integration.

D.2.1 Basic Definitions

Let G be a C^∞ manifold such that the underlying set has the structure of a group. We write $m(x,y) = xy$ (the group multiplication) and $\eta(x) = x^{-1}$ (the group inverse). We say that G is a *Lie group* if $m : G \times G \longrightarrow G$ (see Example 5 in Section D.1.1) and $\eta : G \longrightarrow G$ are of class C^∞.

Examples

1. Let $G = \mathbf{GL}(n, \mathbb{R}) = \{X \in M_n(\mathbb{R}) : \det X \neq 0\}$. Then G is open in $M_n(\mathbb{R})$ (which we look upon as $\mathbb{R}^{n^2}$). The map m is clearly C^∞, and the map η is C^∞ by Cramer's rule.

2. Let $G \subset \mathbf{GL}(n, \mathbb{R})$ be any subgroup that is closed (relative to the topology of $\mathbf{GL}(n, \mathbb{R})$). In Section 1.3.5, G is shown to have a Lie group structure that is compatible with its topology as a closed subspace of $\mathbf{GL}(n, \mathbb{R})$.

3. Let $G \subset \mathbf{GL}(n, \mathbb{C})$ be a linear algebraic group (see Section 1.4.1). Then we can view G as a closed subgroup of $\mathbf{GL}(2n, \mathbb{R})$, and hence give G a Lie group structure compatible with the closed subgroup topology.

If G and H are Lie groups then a *Lie group homomorphism* of G to H is a group homomorphism $\varphi : G \longrightarrow H$ that is C^∞. We say that a Lie group homomorphism φ is a *Lie group isomorphism* if φ is a diffeomorphism.

If G and H are Lie groups with the underlying subset of H a subgroup of G then H is said to be a *Lie subgroup* of G if H is a submanifold of G. As in the case of submanifolds a Lie subgroup does not necessarily have the subspace topology (see Exercises D.1.4 #4).

If G and H are Lie groups then $G \times H$ is easily seen to be a Lie group with the product C^∞ structure and the product group structure.

D.2.2 Lie Algebra of a Lie Group

Let G be a Lie group. Let $L_y : G \longrightarrow G$ be the left translation map defined by $L_y x = yx$ for $y, x \in G$. Then L_y is of class C^∞ and $(L_y)^{-1} = L_{y^{-1}}$ by the associative rule. We look upon a vector field on G as a derivation of the algebra $C^\infty(G)$. That is, if X is a vector field and $f \in C^\infty(G)$ then $Xf \in C^\infty(G)$ is defined by $Xf(y) = X_y f$ for $y \in G$. The definition of tangent vector implies that

$$X(fg)(y) = (Xf)(y)g(y) + f(y)(Xg)(y) \quad \text{for } f, g \in C^\infty(G) \, ,$$

and hence X is a derivation of $C^\infty(G)$. We set $L_y^* f = f \circ L_y$. Then a vector field is said to be *left invariant* if for each $y \in G$ one has $L_y^* \circ X = X \circ L_y^*$. This property can be stated in terms of tangent vectors as

$$d(L_y)_x X_x = X_{yx} \quad \text{for all } x, y \in G \, . \tag{D.9}$$

We set $\mathrm{Lie}(G)$ equal to the vector space of all left-invariant vector fields on G. Denote the identity element of G by 1.

Lemma D.2.1. *The map $X \mapsto X_1$ defines a linear bijection between $\mathrm{Lie}(G)$ and $T(G)_1$. If $X, Y \in \mathrm{Lie}(G)$ then $[X, Y] \in \mathrm{Lie}(G)$. Thus $\mathrm{Lie}(G)$ is a Lie algebra over $\mathbb{R}$ of dimension $n = \dim G$.*

Proof. Equation (D.9) implies that if $X_1 = 0$ then $X_g = 0$. Thus the map is injective. If $v \in T(G)_1$ then let $\sigma : (-\varepsilon, \varepsilon) \longrightarrow G$ be a C^∞ curve with $\sigma(0) = 1$ and $\sigma'(0) = v$ (see Exercises D.1.4 #3). Let $\Psi : G \times (-\varepsilon, \varepsilon) \longrightarrow G$ by $\Psi(g,t) = g\sigma(t)$. Since G is a Lie group, Ψ is a C^∞ map. Set

$$X_g f = \frac{d}{dt} f(g\sigma(t)) \Big|_{t=0}$$

for $f \in C^\infty(G)$. Then $g \mapsto X_g f$ defines a C^∞ function on G, and so $g \mapsto X_g$ defines a vector field. We note that $X_1 = v$ and

$$d(L_g)_x X_x f = X_x(f \circ L_g) = \frac{d}{dt}(f \circ L_g)(x\sigma(t)) \Big|_{t=0}$$

$$= \frac{d}{dt} f(gx\sigma(t)) \Big|_{t=0} = X_{gx} f \, .$$

Thus $X \in \mathrm{Lie}(G)$. Hence the map in the lemma is surjective.

Let $X, Y \in \mathrm{Lie}(G)$ and $g \in G$. Then for $f \in C^\infty(G)$ we have

$$L_g^*([X,Y]f) = L_g^*(XYf - YXf) = L_g^*(XYf) - L_g^*(YXf)$$

$$= X(L_g^*(Yf)) - Y(L_g^*(Xf)) = XYL_g^*f - YXL_g^*f$$

$$= [X,Y]L_g^*f \, .$$

Thus $[X,Y] \in \mathrm{Lie}(G)$. $\qquad\square$

In light of the above lemma we call $\mathrm{Lie}(G)$ the *Lie algebra* of G.

Let G be a Lie group. The following basic theorem is proved in Section 1.3.6 when G is a closed subgroup of $\mathbf{GL}(n, \mathbb{R})$ (see Warner [157] for a proof in general):

Theorem D.2.2. *There exists a unique C^∞ map* $\exp : \mathrm{Lie}(G) \longrightarrow G$ *such that* $\exp(0) = 1$ *and*

$$\frac{d}{dt} f(\exp(tX)) = (Xf)(\exp(tX)) \quad \text{for all } X \in \mathrm{Lie}(G), \ f \in C^\infty(G), \text{ and } t \in \mathbb{R} \, .$$

Furthermore, if $X \in \mathrm{Lie}(G)$ and if $\sigma : \mathbb{R} \longrightarrow G$ is a C^∞ curve with $\sigma(0) = 1$ and $\sigma'(t) = X_{\sigma(t)}$ for all $t \in \mathbb{R}$, then $\sigma(t) = \exp(tX)$.

We note that if $X \in \mathrm{Lie}(G)$ then $\sigma_X(t) = \exp(tX)$ is a C^∞ curve with $\sigma_X(0) = 1$ and $\sigma_X'(0) = X_1$. But $\sigma_X'(0) = d\exp_0((X)_0)$, where $(X)_0$ is the tangent vector to the real vector space $\mathrm{Lie}(G)$ at 0 corresponding to X. Thus we see that $d\exp_0$ is a linear bijection between $T(\mathrm{Lie}(G))_0$ and $T(G)_1$. Using a chart containing 1 and applying the inverse function theorem, we can find an open neighborhood U_0 of 0 in $\mathrm{Lie}(G)$ such that $\exp(U_0) = U$ is open in G and $\exp : U_0 \longrightarrow U$ is a diffeomorphism. Let $\log : U \longrightarrow U_0$ be the inverse map to $\exp|_{U_0}$. Then $(U, \log)$ is a chart for G.

Corollary D.2.3. *Suppose that G is a connected Lie group. Then G is generated (as a group) by $\exp(\mathrm{Lie}\, G)$.*

Proof. Since G is connected, it is generated by any neighborhood U of 1 (see Exercises D.2.5, #1). Take $U = \exp(U_0)$ with U_0 as above. $\square$

Given Lie groups G and H and a Lie group homomorphism $\varphi : G \longrightarrow H$, we define $d\varphi : \mathrm{Lie}(G) \longrightarrow \mathrm{Lie}(H)$ by $d\varphi(X)_1 = d\varphi_1(X_1)$.

Lemma D.2.4. *The map $d\varphi$ is a Lie algebra homomorphism:*

$$d\varphi([X,Y]) = [d\varphi(X), d\varphi(Y)] \quad \text{for } X, Y \in \mathrm{Lie}(G) .$$

Proof. If $f \in C^\infty(H)$ then $X(f \circ \varphi) = (d\varphi(X)f) \circ \varphi$ by the left-invariance of X. Hence $[X,Y](f \circ \varphi) = ([d\varphi(X), d\varphi(Y)]f) \circ \varphi$. This implies that $d\varphi([X,Y])_1 = ([d\varphi(X), d\varphi(Y)])_1$. $\square$

Lemma D.2.5. *Let G and H be Lie groups. Suppose $\varphi : G \longrightarrow H$ is a Lie group homomorphism. Then $\varphi(\exp(X)) = \exp(d\varphi(X))$ for all $X \in \mathrm{Lie}(G)$.*

Proof. Let $\sigma(t) = \exp(tX)$ and $\mu(t) = \varphi(\exp(tX))$ for $t \in \mathbb{R}$. Then

$$\mu'(t) = d\varphi_{\sigma(t)}\sigma'(t) = d\varphi_{\sigma(t)}X_{\sigma(t)} = d\varphi(X)_{\mu(t)} .$$

Thus Theorem D.2.2 implies that $\mu(t) = \exp(t d\varphi(X))$. $\square$

Theorem D.2.6. *Let G be a Lie group and let H be a closed subgroup of G. Then H has a structure of a Lie group such that the inclusion map $\iota : H \longrightarrow G$ is a Lie group homomorphism, and*

$$d\iota(\mathrm{Lie}(H)) = \{X \in \mathrm{Lie}(G) : \exp(tX) \in H \quad \text{for all } t \in \mathbb{R}\} .$$

Proof. In the case $G = \mathbf{GL}(n,\mathbb{R})$ this is Theorem 1.3.11. For a proof for nonlinear groups see Warner [157]. $\square$

For $g \in G$ we define $\mathrm{Inn}(g)(x) = gxg^{-1}$. Then $\mathrm{Inn}(g)$ defines a Lie group automorphism of G, called an *inner automorphism*. We define $\mathrm{Ad}(g) = d\,\mathrm{Inn}(g)$.

Lemma D.2.7. *The map $\mathrm{Ad} : G \longrightarrow \mathbf{GL}(\mathrm{Lie}(G))$ is a Lie group homomorphism.*

This lemma is proved in Section 1.3.3 when G is a closed subgroup of $\mathbf{GL}(n,\mathbb{R})$. For a proof when G is not a linear group see Warner [157].

Theorem D.2.8. *Let G be a topological group. Assume that*

1. *there is an open neighborhood U of the identity element in G such that $u^{-1} \in U$ for all $u \in U$;*
2. *there is a surjective homeomorphism $\Phi : U \longrightarrow B_r(0) \subset \mathbb{R}^m$ for some $r > 0$, and $\Phi(u^{-1}) = -\Phi(u)$ for all $u \in U$ (where $B_r(0) = \{x \in \mathbb{R}^m : ||x|| < r\}$);*
3. *there are an s with $0 < s < r$ and a C^∞ map $F : B_s(0) \times B_s(0) \longrightarrow B_r(0)$ such that $uv \in U$ and $\Phi(uv) = F(\Phi(u), \Phi(v))$ for all $u, v \in \Phi^{-1}(B_s(0))$.*

Then there exists a Lie group structure on G compatible with the topological group structure.

Proof. Use the same argument as in the proof of Theorem 1.3.12, but with the logarithm map replaced by the map Φ.   □

Theorem D.2.8 yields an easy proof of the fact that if $H \subset G$ is a closed normal subgroup of a Lie group G, then G/H has the structure of a Lie group.

Theorem D.2.9. *Suppose G is a connected Lie group and $\pi : H \longrightarrow G$ is a covering space. Let e be the identity element of G and choose $e_0 \in \pi^{-1}(e)$. Then H has a structure of a Lie group with identity e_0 such that π is a Lie group homomorphism.*

Proof. Let $L_g : G \longrightarrow G$ be the left translation map $L_g(x) = gx$. Then for each $h \in H$ there exists a unique homeomorphism $\tilde{L}_h : H \longrightarrow H$ such that

$$\tilde{L}_h(e_0) = h \quad \text{and} \quad \pi(\tilde{L}_h(x)) = L_{\pi(h)}(\pi(x)) \quad \text{for all } x \in H .$$

We assert that the product $m(u,v) = \tilde{L}_u(v)$ makes H a group. The identity map has the same property assumed for $\tilde{L}_{e_0}$; thus $m(e_0,u) = u$. By definition $m(u,e_0) = u$. Hence e_0 is an identity element for the multiplication m. If $u \in H$ then there exists a unique $v \in H$ such that $\tilde{L}_u(v) = e_0$. To prove that H is a group it remains only to show that the associative rule is satisfied.

We note that $L_x \circ L_y = L_{xy}$ and $\pi(m(x,y)) = xy$. Since

$$m(x,y) = \tilde{L}_{m(x,y)}(e_0) = \tilde{L}_x \circ \tilde{L}_y(e_0) ,$$

we have $\tilde{L}_{m(x,y)} = \tilde{L}_x \circ \tilde{L}_y$, which proves associativity of multiplication.

We must now prove that $m : H \times H \longrightarrow H$ is continuous. We note that if we set $\mu(x,y) = xy$ for $x,y \in G$, then there is a unique lift $\tilde{\mu} : H \times H \longrightarrow H$ of μ such that $\tilde{\mu}(e_0,e_0) = e_0$. Since m is another such lift we see that $m = \tilde{\mu}$. The existence of a Lie group structure on H such that π is a C^∞ map now follows from Theorem D.2.8.   □

D.2.3 Homogeneous Spaces

Let G be a Lie group and let H be a closed subgroup of G. Then H is a Lie subgroup of G (Theorem D.2.6). Let $Y = G/H$ with the quotient topology. We now show how to put a C^∞ structure on Y yielding a C^∞ manifold M such that the left multiplication map $G \times M \longrightarrow M$ given by $g, m \mapsto gm$ is of class C^∞.

Let $\mathfrak{g} = \mathrm{Lie}(G)$ and let $\mathfrak{h} = d\iota(\mathrm{Lie}(H))$ (where $\iota(h) = h$ for $h \in H$). Take any subspace V in $\mathfrak{g}$ such that $\mathfrak{g} = \mathfrak{h} \oplus V$ and define $\Phi : V \times \mathfrak{h} \longrightarrow G$ by

$$\Phi(v,X) = \exp(v)\exp(X) .$$

Then $d\Phi_{(0,0)}(v_0,X_0) = v_1 + X_1$. Thus there are open neighborhoods U_0 and W_0 of 0 in V and $\mathfrak{h}$, respectively, such that $\Phi(U_0 \times W_0) = U_1$ is open in G and Φ is a diffeomorphism from $U_0 \times W_0$ onto U_1. We now choose open neighborhoods of 0, $U_0' \subset U_0$ and $W_0' \subset W_0$, such that

(a)   If $U_1' = \Phi(U_0' \times W_0')$ and $x, y \in U_1'$, then xy^{-1} and $y^{-1}x$ are in U_1 .

(b) $U_1' \cap H \subset \exp(W_0)$.

This can be done because the maps $G \times G \longrightarrow G$ given by $x, y \mapsto xy^{-1}$ and $x, y \mapsto y^{-1}x$ are C^∞, and because $\exp(W_0)$ contains a neighborhood of 1 in H.

(c)   If $v_1, v_2 \in U_0'$ and if $h_1, h_2 \in H$, then $\exp(v_1)h_1 = \exp(v_2)h_2$ implies that $h_1 = h_2$ and $v_1 = v_2$.

Indeed, we have $\exp(v_2)^{-1}\exp(v_1) = h_2 h_1^{-1}$. Thus (a) implies that $h_2 h_1^{-1} \in U_1 \cap H$ and (b) implies that $h_2 h_1^{-1} = \exp(X)$ with $X \in W_0$. Thus

$$\Phi(v_1, 0) = \exp(v_1) = \exp(v_2)\exp(X) = \Phi(v_2, X) .$$

We conclude that $v_1 = v_2$ and $X = 0$. Hence $h_2 h_1^{-1} = 1$ and assertion (c) is proved.

Let $\pi : G \longrightarrow G/H$ be defined by $\pi(g) = gH$. The map π is called the *natural projection*. Set $\overline{U} = \pi(U_1')$ and define $\Psi(\pi(\exp(v))) = v$ for $v \in U_0'$. Then $(\overline{U}, \Psi)$ defines a chart for G/H. Given $g \in G$ set $g(xH) = gxH$. This will be called the *natural action* of G on G/H. Define $\Psi_g(gx) = \Psi(x)$ for $x \in G/H$. Then a straightforward argument using (a), (b), and (c) shows that $\{(g\overline{U}, \Psi_g)\}_{g \in G}$ is a C^∞ atlas for G/H. We have thus sketched the proof of the following result:

Theorem D.2.10. *Let G be a Lie group and let H be a closed subgroup of G. Then there exists a C^∞ structure on G/H of dimension $\dim G - \dim H$ such that the natural projection is of class C^∞ and the map $G \times G/H \longrightarrow G/H$ given by the natural action is of class C^∞.*

Examples

1. The standard action of the orthogonal group $\mathbf{O}(n+1, \mathbb{R})$ on $\mathbb{R}^{n+1}$ is transitive on the n-sphere S^n. This gives the homogeneous space $\mathbf{O}(n+1, \mathbb{R})/\mathbf{O}(n, \mathbb{R})$.

2. The Grassmann manifold $\mathrm{Grass}_k(\mathbb{R}^n)$ of k-planes in $\mathbb{R}^n$ is isomorphic to

$$\mathbf{O}(n, \mathbb{R})/\mathbf{O}(k, \mathbb{R}) \times \mathbf{O}(n-k, \mathbb{R})$$

as a homogeneous space for $\mathbf{O}(n, \mathbb{R})$.

D.2.4 Integration on Lie Groups and Homogeneous Spaces

Let G be a Lie group and set $\mathfrak{g} = \mathrm{Lie}(G)$. Let $\{X_1, \ldots, X_n\}$ be a basis of $\mathfrak{g}$. From Lemma D.2.1 it follows that $\{(X_1)_g, \ldots, (X_n)_g\}$ is a basis of $T(G)_g$ for each $g \in G$. There is a unique element $\omega_g \in \bigwedge^n(T(G)_g)^*$ such that $\omega_g((X_1)_g, \ldots, (X_n)_g) = 1$. We claim that $g \mapsto \omega_g$ defines an element of $\Omega^n(G)$. To prove this, let $\{y_1, \ldots, y_n\}$ be a system of local coordinates on $U \subset G$. Then

$$(X_i)_g = \sum_j u_{ij}(g)\left(\frac{\partial}{\partial y_j}\right)_g \quad \text{with} \ u_{ij} \in C^\infty(U) \, .$$

Since $\big\{(X_1)_g, \ldots, (X_n)_g\big\}$ is a basis of $T(G)_g$, we have $\det[u_{ij}(g)] \neq 0$ for all $g \in U$. This implies that every vector field X on G can be written as

$$X = \sum_i a_i X_i \quad \text{with} \ a_i \in C^\infty(G) \, .$$

From this property it is clear that $g \mapsto \omega_g$ is indeed in $\Omega^n(G)$. Since $\omega_g \neq 0$ for all $g \in G$, we conclude that ω is a volume form on G.

If $g \in G$, then

$$\begin{aligned}
(L_g^*\omega)_x\big((X_1)_x, \ldots, (X_n)_x\big) &= \omega_{gx}\big((dL_g)_x(X_1)_x, \ldots, (dL_g)_x(X_n)_x\big) \\
&= \omega_{gx}\big((X_1)_{gx}, \ldots, (X_n)_{gx}\big) \\
&= \omega_x\big((X_1)_x, \ldots, (X_n)_x\big) \, .
\end{aligned}$$

Thus $L_g^*\omega = \omega$. This calculation implies that if $f \in C_c(G)$, then

$$\int_G f\omega = \int_G (f \circ L_g)\,\omega \, . \tag{D.10}$$

Notice that ω is determined up to a nonzero real scalar multiple. Having fixed ω, we write

$$\int_G f(g)\,dg = \int_G f\omega \, . \tag{D.11}$$

With this notation (D.10) becomes

$$\int_G f(xg)\,dg = \int_G f(g)\,dg \quad \text{for} \ x \in G \, . \tag{D.12}$$

For $g, x \in G$ we set $R_x g = gx$. Then R_x defines a diffeomorphism of G.

Lemma D.2.11. *If* $f \in C_c(G)$ *and* $x \in G$ *then*

$$\int_G f(gx)\,dg = |\det \mathrm{Ad}(x)| \int_G f(g)\,dg \, .$$

Proof. By the left invariance of ω we can write

$$\int_G f(gx)\,dg = \int_G f(x^{-1}gx)\,dg = \int_G f \circ \mathrm{Inn}(x^{-1})\,\omega \, .$$

Thus by Theorem D.1.16, with $\Phi = \mathrm{Inn}(x)$, we have

$$\int_G f(gx)\,\mathrm{d}g = \int_G f\,\Phi^*\omega\,.$$

Since $\mathrm{d}\Phi = \mathrm{Ad}(x)$, we see from the definition of ω that

$$
\begin{aligned}
(\Phi^*\omega)_g\big((X_1)_g,\ldots,(X_n)_g\big) &= \omega_1\big((\mathrm{Ad}(x)X_1)_1,\ldots,(\mathrm{Ad}(x)X_n)_1\big) \\
&= \det(\mathrm{Ad}(x))\omega_1\big((X_1)_1,\ldots,(X_n)_1\big) \\
&= \det(\mathrm{Ad}(x))\omega_g\big((X_1)_g,\ldots,(X_n)_g\big)
\end{aligned}
$$

for $g \in G$ and $X_1,\ldots,X_n \in \mathfrak{g}$. Hence $\mathrm{Inn}(x)^*\omega = \det(\mathrm{Ad}(x))\omega$. The lemma now follows from formula (D.8). $\square$

We define the *modular function* δ of G to be $\delta(g) = |\det\mathrm{Ad}(g)|$. Since $\mathrm{Ad}(xy) = \mathrm{Ad}(x)\mathrm{Ad}(y)$, we have $\delta(xy) = \delta(x)\delta(y)$, so δ is a Lie homomorphism of G into the multiplicative group $\mathbb{R}_{>0}$ of positive real numbers. By Lemma D.2.11 we can write

$$\int_G f(gx)\delta(g)\,\mathrm{d}g = \delta(x)\int_G f(g)\delta(gx^{-1})\,\mathrm{d}g\,.$$

Since $\delta(gx^{-1}) = \delta(g)\delta(x)^{-1}$, we obtain the integral formula

$$\int_G f(gx)\delta(g)\,\mathrm{d}g = \int_G f(g)\delta(g)\,\mathrm{d}g \tag{D.13}$$

for $x \in G$ and $f \in C_c(G)$. This shows that $\delta(g)\,\mathrm{d}g$ is a *right-invariant* measure on G.

We say that G is *unimodular* if $\delta(g) = 1$ for all $g \in G$. In this case the left-invariant measure $\mathrm{d}g$ on G is also right invariant (the converse also holds).

Lemma D.2.12. *A compact Lie group is unimodular.*

Proof. Since $\delta(g) > 0$ for all $g \in G$, $\delta(G)$ is a compact subgroup of $\mathbb{R}_{>0}$. The only such subgroup is $\{1\}$, by the Archimedean property of the real numbers. $\square$

If G is compact, then $1 \in C_c(G)$ and $0 < c = \int_G \omega < \infty$. Replacing ω by $c^{-1}\omega$, we can achieve

$$\int_G \mathrm{d}g = 1\,. \tag{D.14}$$

We will call $\mathrm{d}g$ the *normalized invariant measure* on G if it satisfies (D.14).

Assume that G is a Lie group and that H is a closed subgroup of G. Give G/H the C^∞ structure defined in Section D.2.3. We assume, for simplicity, the following conditions (where Ad is the adjoint representation of G on $\mathfrak{g}$):

(i)   If $g \in G$, then $\det\mathrm{Ad}(g) = 1$.
(ii)   If $h \in H$, then $\det(\mathrm{Ad}(h)|_{\mathfrak{h}}) = 1$.

Under these assumptions (which could be weakened), we now show how to put a G-invariant volume form ω on G/H. For $g \in G$ let l_g be the transformation $l_g x = gx$ on G/H. Then the desired form ω must satisfy $l_g^*\omega = \omega$ for all $g \in G$.

We use the notation of Section D.2.3, and write $m = \dim G - \dim H$. We begin by defining a form $\eta \in \Omega^m(G)$ as follows: Choose an m-dimensional subspace $V \subset \mathfrak{g}$ such that $\mathfrak{g} = V \oplus \mathfrak{h}$. Then we can identify

$$V^* = \{\lambda \in \mathfrak{g}^* : \langle \lambda, \mathfrak{h} \rangle = 0\} . \tag{D.15}$$

Thus $\bigwedge^m V^*$ is identified with a subspace of $\bigwedge^m \mathfrak{g}^*$. Fix $0 \neq \nu \in \bigwedge^m V^*$ and define

$$\eta_g\big((X_1)_g, \ldots, (X_m)_g\big) = \nu(X_1, \ldots, X_m) .$$

Then, just as in the case of the invariant volume form on G, we see that $\eta \in \Omega^m(G)$ and $L_g^* \eta = \eta$ for all $g \in G$. We note that if $Z_1, \ldots, Z_m \in \mathfrak{h}$, then by (D.15) we have

$$\nu(X_1 + Z_1, \ldots, X_m + Z_m) = \nu(X_1, \ldots, X_m) .$$

Let $\{X_1, \ldots, X_m\}$ be a basis of V and let $\{X_{m+1}, \ldots, X_n\}$ be a basis of $\mathfrak{h}$. If $h \in H$ then the linear transformation $\mathrm{Ad}(h)$ has a matrix of block form $\left[\begin{smallmatrix} A & 0 \\ B & C \end{smallmatrix}\right]$, relative to the basis $\{X_1, \ldots, X_n\}$ for $\mathfrak{g}$. Here C is the matrix of $\mathrm{Ad}(h)|_{\mathfrak{h}}$ relative to the basis $\{X_{m+1}, \ldots, X_n\}$ for $\mathfrak{h}$. Thus

$$\nu(\mathrm{Ad}(h)X_1, \ldots, \mathrm{Ad}(h)X_m) = \det(A)\, \nu(X_1, \ldots, X_m) .$$

Since $\det A \det C = \det(\mathrm{Ad}(h))$, assumptions (i) and (ii) thus imply that $\det A = 1$. We therefore see that

$$L_g^* \eta = \eta, \quad R_h^* \eta = \eta \quad \text{for } g \in G \text{ and } h \in H . \tag{D.16}$$

We are now ready to define the form ω. If $g \in G$ and $v_1, \ldots, v_m \in T(G/H)_{gH}$, choose $X_1, \ldots, X_m \in \mathfrak{g}$ such that $d\pi_g((X_i)_g) = v_i$. We assert that

$(\star)$ the scalar $\eta_g((X_1)_g, \ldots, (X_m)_g)$ depends only on gH and $v_1, \ldots, v_m$.

Indeed, let $w_i = dl_{g^{-1}} v_i \in T(G/H)_H$. Since $l_g \circ \pi = \pi \circ L_g$, we have $w_i = d\pi_1\big((X_i)_1\big)$. If $X_i' \in \mathfrak{g}$ also satisfies $d\pi_1\big((X_i')_1\big) = w_i$ for $i = 1, \ldots, m$, then $X_i - X_i' \in \mathfrak{h}$. Thus

$$\eta_1\big((X_1)_1, \ldots, (X_n)_1\big) = \eta_1\big((X_1')_1, \ldots, (X_n')_1\big) .$$

If $\pi(g) = \pi(g')$, then $g' = gh$ for some $h \in H$. Hence (D.16) implies $(\star)$.

We also note that if $\{X_1, \ldots, X_m\}$ is a basis for V then

$$\left\{ d\pi_1(X_1)_1, \ldots, d\pi_1(X_m)_1 \right\}$$

is a basis of $T(G/H)_H$. Hence there is an open neighborhood U of 1 in G such that $\left\{ d\pi_g((X_1)_g), \ldots, d\pi_g((X_m)_g) \right\}$ is a basis of $T(G/H)_{gH}$ for $g \in U$. Thus if we set

$$\omega_{gH}\big(d\pi_g(X_1)_g, \ldots, d\pi_g(X_m)_g\big) = \eta_g\big((X_1)_g, \ldots, (X_m)_g\big) ,$$

then we have defined ω in $\Omega^m(\pi(U))$. Now $l_{g^{-1}} : l_g \pi(U) \longrightarrow \pi(U)$. Thus for each $g \in G$ we have a differential form $\omega^g \in \Omega^m(l_g \pi(U))$ given by $\omega^g = l_g^* \omega$. Property $(\star)$ above implies that

$$\omega_x^{g_1} = \omega_x^{g_2} \quad \text{for } x \in l_{g_1}(\pi(U)) \cap l_{g_2}(\pi(U)) .$$

We have thus defined ω on G/H. This proves the following result:

Theorem D.2.13. *Let G be a Lie group and let H be a closed subgroup of G. Assume that $\det \mathrm{Ad}(g) = 1$ and $\det \mathrm{Ad}(h)|_{\mathrm{Lie}(H)} = 1$ for all $g \in G$ and $h \in H$. Set $m = \dim G/H$. Then there exists a volume form $\omega \in \Omega^m(G/H)$ such that $l_g^* \omega = \omega$ for all $g \in G$.*

D.2.5 Exercises

1. We look upon $\mathbb{C}$ as $\mathbb{R}^2$ as usual. Then $S^1 = \{z \in \mathbb{C} : |z| = 1\}$ is a group under complex multiplication.
   (a) Use the C^∞ structure as in Example 2 of Section D.1.1 to show that S^1 is a Lie group.
   (b) Define $\mathbb{T}^1 = S^1$ and inductively define $\mathbb{T}^{n+1} = \mathbb{T}^n \times \mathbb{T}^1$ as product of Lie groups for $n \geq 1$. Show that the submanifold in Exercises D.1.4 #4 is a Lie subgroup of $\mathbb{T}^2$.
2. For $1 \leq k < n$ let $P \subset \mathbf{GL}(n, \mathbb{R})$ be the group of block upper-triangular matrices $g = \begin{bmatrix} A & B \\ 0 & D \end{bmatrix}$ with $A \in \mathbf{GL}(k, \mathbb{R})$, $B \in M_{k,n-k}(\mathbb{R})$, and $D \in \mathbf{GL}(n-k, \mathbb{R})$. Given $g \in \mathbf{GL}(k, \mathbb{R})$, let $\mu(g) \subset \mathbb{R}^n$ be the subspace spanned by the first k columns of g. Show that μ sets up an isomorphism between $\mathbf{GL}(n, \mathbb{R})/P$ and the k-Grassmannian $\mathrm{Grass}_k(\mathbb{R})$ as differentiable manifolds (see Example 2 of Section D.2.3).
3. One has $G_1(\mathbb{R}^n) = \mathbb{P}^1(\mathbb{R}^n)$. Show that the natural map $S^{n-1} \longrightarrow \mathbb{P}^1(\mathbb{R}^n)$ is a C^∞ covering.
4. Show that $\mathbf{GL}(n, \mathbb{R})$ and $\mathbf{O}(n, \mathbb{R})$ are unimodular groups.
5. Let G be the subgroup of upper-triangular matrices in $\mathbf{GL}(n, \mathbb{R})$. Calculate the modular function of G.
6. Show that the volume form on S^{n-1} given as in Exercises D.1.4 #7 is a scalar multiple of the one given in Section D.2.4. (HINT: Use Example 1 in Section D.2.3.)

References

1. Agricola, I., Invariante Differentialoperatoren und die Frobenius-Zerlegung einer G-Varietät, *J. of Lie Th.* **11** (2001), 81–109.
2. Alexander, J., Topological invariants of knots and links, *Trans. Amer. Math. Soc.* **20** (1923), 275–306.
3. Araki, S. I., On root systems and an infinitesimal classification of irreducible symmetric spaces, *J. Math. Osaka City Univ.* **13** (1962), 1–34.
4. Atiyah, M., R. Bott, and V. K. Patodi, On the heat equation and the index theorem, *Invent. Math.* **19** (1973), 279–330.
5. Benkart, G., D. Britten, and F. Lemire, *Stability in Modules for Classical Lie Algebras—A Constructive Approach*, Memoirs of the AMS **430** (1990), American Mathematical Society, Providence, R.I..
6. Benson, C., and G. Ratcliff, A classification of multiplicity-free actions, *J. of Algebra* **181** (1996), 152–186.
7. Benson, C., and G. Ratcliff, On Multiplicity Free Actions, in *Representations of Real and p-ádic Groups* (Lecture Note Series, IMS vol. 2), ed. by E. C. Tan and C. B. Zhu, World Scientific, Singapore, 2004, pp. 221–303.
8. Besse, A. *Einstein Manifolds* (Ergebnisse der Mathematik und ihrer Grenzgebiete; 3. Folge, Bd. 10), Springer-Verlag, Berlin/Heidelberg, 1987.
9. Boerner, H., *Representations of Groups with Special Consideration for the Needs of Modern Physics*, North-Holland, New York, 1970.
10. Bourbaki, N., *Éléments de mathématique, Fascicule XXIII, Algèbre*, Chapitre VIII, *Modules et Anneaux Semi-Simples*, Hermann, Paris, 1958.
11. Bourbaki, N., *Éléments de mathématique, Fascicule XXVI, Groupes et algèbres de Lie*, Chapitre I, *Algèbres de Lie*, Hermann, Paris, 1960.
12. Bourbaki, N., *Éléments de mathématique, Fascicule XXXIV, Groupes et algèbres de Lie*, Chapitres IV, V & VI, Hermann, Paris, 1968.
13. Bourbaki, N., *Éléments de mathématique, Fascicule XXXVIII, Groupes et algèbres de Lie*, Chapitres VII & VIII, Hermann, Paris, 1975.
14. Bourbaki, N., *Éléments de mathématique, Groupes et algèbres de Lie*, Chapitre IX, Masson, Paris, 1982.
15. Borel, A., Groupes linéaires algèbriques, *Annals of Math.* (2) **64** (1956), 20–82.
16. Borel, A., *Linear Algebraic Groups* (Second Enlarged Edition), Springer-Verlag, New York, 1991.
17. Borel, A., *Essays in the History of Lie Groups and Algebraic Groups* (History of Mathematics **21**), American Mathematical Society, Providence, R.I., 2001.
18. Borel, A., and Harish-Chandra, Arithmetic subgroups of algebraic groups, *Annals of Math.* (2) **75** (1962), 485–535.
19. Brauer, R., Eine Bedingung für vollständige Reduzibilität von Darstellungen gewönlicher und infinitesimaler Gruppen, *Math. Zeitschrift* **41** (1936), 330–339.
20. Brauer, R., On algebras which are connected with the semisimple continuous groups, *Annals of Math.* (2) **38** (1937), 857–872.
21. Brauer, R., and H. Weyl, Spinors in n dimensions, *Amer. J. of Math.* **57** (1935), 425–449.
22. Bremner, M. R., R. V. Moody, and J. Patera, *Tables of Dominant Weight Multiplicities for Representations of Simple Lie Algebras* (Monographs and Textbooks in Pure and Appl. Math. **90**), Marcel Dekker, New York, 1985.
23. Brion, M., Spherical varieties: An introduction, in *Topological Methods in Algebraic Transformation Groups* (Prog. Math. **80**), ed. by H. Kraft, T. Petrie, and G. Schwarz, Birkhäuser, Basel, 1989, pp. 11–26.
24. Brown, W. P., An algebra related to the orthogonal group, *Michigan Math. J.* **3** (1955), 1–22.
25. Brown, W. P., The semisimplicity of ω_f^n, *Annals of Math.* (2) **63** (1956), 324–335.
26. Cartan, É., Sur la réduction à sa forme canonique de la structure d'un groupe de transformations fini et continu, *Amer. J. of Math.* **18** (1896), 1–61; reprinted in *Oeuvres Complètes* **1**, Part 1, Gauthier-Villars, Paris, 1952, pp. 293–353.

27. Cartan, É., Les groupes projectifs qui ne laissent invariante aucune multiplicité plane, *Bull. Soc. Math. de France* **41** (1913), 53–96; reprinted in *Oeuvres Complètes* **1**, Part 1, Gauthier-Villars, Paris, 1952, pp. 355–398.

28. Cartan, É., Sur la détermination d'un système orthogonal complet dans un espace de Riemann symétrique clos, *Rend. Circ. Mat. Palermo* **53** (1929), 217–252; reprinted in *Oeuvres Complètes* **1**, Part 2, Gauthier-Villars, Paris, 1952, pp. 1045–1080.

29. Cartier, P., On H. Weyl's character formula, *Bull. Amer. Math. Soc.* **67** (1961), 228–230.

30. Casimir, H., Über die Konstruktion einer zu den irreduziblen Darstellungen halbeinfacher kontinuierlicher Gruppen gehörigen Differentialgleichung, *Proc. Kon. Acad. Amsterdam* **34** (1931), 844–846.

31. Casimir, H., and B. L. van der Waerden, Algebraischer Beweis der vollständigen Reduzibilität der Darstellungen halbeinfacher Liescher Gruppen, *Math. Ann.* **111** (1935), 1–12.

32. Chari, V., and A. Pressley, *A Guide to Quantum Groups*, Cambridge University Press, Cambridge, 1994.

33. Chevalley, C., *Theory of Lie Groups I*, Princeton University Press, Princeton, 1946.

34. Chevalley, C., Sur la classification des algèbres de Lie simples et de leurs représentations, *C.R. Acad. Sci. Paris* **227** (1948), 1136–1138.

35. Chevalley, C., *Théorie des groupes de Lie II*, Groupes algèbriques, Hermann, Paris, 1951.

36. Chevalley, C., *Théorie des groupes de Lie III*, Groupes algèbriques, Hermann, Paris, 1954.

37. Chevalley, C., Invariants of finite groups generated by reflections, *Amer. J. Math.* **77** (1955), 778–782.

38. Chevalley, C., *Séminaire sur la classification des groupes de Lie algèbriques*, Ecole Norm. Sup., Paris, 1956–1958.

39. Clifford, W. K., Applications of Grassmann's extensive algebra, *Amer. J. Math.* **1** (1878), 350–358.

40. Cox, D., J. Little, and D. O'Shea, *Ideals, Varieties, and Algorithms*, Springer-Verlag, New York, 1992.

41. Coxeter, H. S. M., Discrete groups generated by reflections, *Annals of Math.* (2) **35** (1934), 588–621.

42. DeConcini, C., and C. Procesi, Complete symmetric varieties, in *Invariant Theory* (Lecture Notes in Mathematics **996**), Springer-Verlag, Heidelberg, 1983, pp. 1–44.

43. Dodson, C. T. J., and T. Poston, *Tensor Geometry: The Geometric Viewpoint and its Uses, Second Edition* (Graduate Texts in Mathematics **130**), Springer-Verlag, New York, 1991.

44. Dynkin, E. B., The structure of semisimple algebras, *Uspehi Mat. Nauk* **2** (1947), no. 4 (20), 59–127 (Russian); English translation: *Amer. Math. Soc. Translation* **17**, 1950; reissued in *Lie Groups* (Translations Series 1, Volume 9), American Mathematical Society, Providence, R.I., 1962, pp. 328–469.

45. Fässler, A., and E. Stiefel, *Group Theoretical Methods and Their Applications*, Birkhäuser, Boston, 1992.

46. Frobenius, F. G., Über Gruppencharactere, *Sitz. Preuss. Akad. Wiss. Berlin* (1896), 985–1021; reprinted in *Gesammelte Abhandlungen* **3**, Springer-Verlag, Heidelberg, 1968, pp. 1–37.

47. Frobenius, F. G., Über die Darstellung der endlichen Gruppen durch lineare Substitutionen, *Sitz. Preuss. Akad. Wiss. Berlin* (1897), 944–1015; reprinted in *Gesammelte Abhandlungen* **3**, Springer-Verlag, Heidelberg, 1968, pp. 82–103.

48. Frobenius, F. G., Über Relationen zwischen den Charackteren einer Gruppe und denen ihrer Untergruppen, *Sitz. Preuss. Akad. Wiss. Berlin* (1898), 501–515; reprinted in *Gesammelte Abhandlungen* **3**, Springer-Verlag, Heidelberg, 1968, pp. 104–118.

49. Frobenius, F. G., Über die Charactere der symmetrischen Gruppe, *Sitz. Preuss. Akad. Wiss. Berlin* (1900), 516–534; reprinted in *Gesammelte Abhandlungen* **3**, Springer-Verlag, Heidelberg, 1968, pp. 148–166.

50. Frobenius, F. G., Über die charakterischen Einheiten der symmetrischen Gruppe, *Sitz. Preuss. Akad. Wiss. Berlin* (1903), 328–358; reprinted in *Gesammelte Abhandlungen* **3**, Springer-Verlag, Heidelberg, 1968, pp. 244–274.

51. Fulton, W., *Young Tableaux: With Applications to Representation Theory and Geometry* (London Mathematical Society Student Texts **35**), Cambridge University Press, Cambridge, 1997.

52. Fulton, W., and J. Harris, *Representation Theory, a First Course* (Graduate Texts in Mathematics **129**), Springer-Verlag, New York, 1991.

53. Gårding, L., Extension of a formula by Cayley to symmetric determinants, *Proc. Edinburgh Math. Soc.* (2) **8** (1947), 73–75.

54. Gavarini, F., and P. Papi, Representations of the Brauer algebra and Littlewood's restriction rules, *J. of Algebra* **194** (1997), 275–298.

55. Gavarini, F., A Brauer algebra theoretic proof of Littlewood's restriction rules, *J. of Algebra* **212** (1999), 240–271.

56. Goodman, R., and N. R. Wallach, *Representations and Invariants of the Classical Groups*, Cambridge University Press, Cambridge, 1998. Reprinted with corrections 1999, 2003.

57. Grove, L. C., and C. T. Benson, *Finite Reflection Groups*, Springer-Verlag, New York, 1985.

58. Hanlon, P., On the decomposition of the tensor algebra of the classical Lie algebras, *Advances in Math.* **56** (1985), 238–282.

59. Hanlon, P., and D. Wales, On the decomposition of Brauer's centralizer algebras, *J. of Algebra* **121** (1989), 409–445.

60. Harish-Chandra, On some applications of the universal algebra of a semi-simple Lie algebra, *Trans. Amer. Math. Soc.* **70** (1951), 28–96.

61. Harris, J., *Algebraic Geometry, a First Course* (Graduate Texts in Mathematics **133**), Springer-Verlag, New York, 1992.

62. Harvey, F. R., *Spinors and Calibrations* (Perspectives in Mathematics **9**), Academic Press, San Diego, 1990.

63. Hawkins, T., *Emergence of the Theory of Lie Groups: An Essay in the History of Mathematics 1869–1926*, Springer-Verlag, New York, 2000.

64. Hegerfeldt, G. C., Branching theorem for the symplectic groups, *J. Math. Physics* **8** (1967), 1195–1196.

65. Helgason, S., A duality for symmetric spaces with applications to group representations, *Advances in Math.* **5** (1970), 1–154.

66. Helgason, S., *Differential Geometry, Lie Groups, and Symmetric Spaces* (Pure and Applied Mathematics **80**), Academic Press, New York, 1978.

67. Helgason, S., *Groups and Geometric Analysis* (Pure and Applied Mathematics **113**), Academic Press, Orlando, 1984.

68. Hochschild, G., *The Structure of Lie Groups*, Holden-Day, San Francisco, 1965.

69. Howe, R., The classical groups and invariants of binary forms, in *Proceedings of Symposia in Pure Mathematics* **48**, American Mathematical Society, Providence, R.I., 1988, pp. 133–166.

70. Howe, R., Remarks on classical invariant theory, *Trans. Amer. Math. Soc.* **313** (1989), 539–570.

71. Howe, R., The first fundamental theorem of invariant theory and spherical subgroups, in *Proceedings of Symposia in Pure Mathematics* **56**, Part I, American Mathematical Society, Providence, R.I., 1994, pp. 333–346.

72. Howe, R., Perspectives on invariant theory: Schur duality, multiplicity-free actions and beyond, in *The Schur Lectures (1992)* (Israel Mathematical Conference Proceedings **8**), American Mathematical Society, Providence, R.I., 1995, pp. 1–182.

73. Howe, R., and E. C. Tan, *Non-Abelian Harmonic Analysis: Applications of* $\mathbf{SL}(2, \mathbb{R})$ (Universitext), Springer-Verlag, New York, 1992.

74. Howe, R., E. C. Tan, and J. F. Willenbring, Reciprocity algebras and branching, in *Groups and Analysis: The Legacy of Hermann Weyl* (LMS Lecture Note Series **354**), ed. by K. Tent, Cambridge University Press, Cambridge, 2008, pp. 191–231.

75. Howe, R., and T. Umeda, The Capelli identity, the double commutant theorem, and multiplicity-free actions, *Math. Ann.* **290** (1991), 565–619.

76. Humphreys, J. E., *Introduction to Lie Algebras and Representation Theory* (Graduate Texts in Mathematics **9**), Springer-Verlag, New York, 1980.

77. Humphreys, J. E., *Linear Algebraic Groups* (Graduate Texts in Mathematics **21**), Springer-Verlag, New York, 1975.

78. Humphreys, J. E., *Reflection Groups and Coxeter Groups* (Cambridge Studies in Advanced Mathematics **29**), Cambridge University Press, Cambridge, 1990.

79. Jacobson, N., *Lie Algebras*, Wiley-Interscience, New York, 1962; reprinted: Dover, New York, 1979.

80. Jacobson, N., *Basic Algebra II*, W. H. Freeman, San Francisco, 1980.

81. Jones, V. F. R., *Subfactors and Knots* (CBMS Regional Conference Series in Mathematics **80**), American Mathematical Society, Providence, R.I., 1991.

82. Kac, V. G., Some remarks on nilpotent orbits, *J. of Algebra* **64** (1980), 190–213.

83. Kane, R., *Reflection Groups and Invariant Theory* (CMS Books in Mathematics), Springer-Verlag, New York, 2001.

84. Kassel, C., *Quantum Groups* (Graduate Texts in Mathematics **155**), Springer-Verlag, New York, 1995.

85. Kerov, S.V., Realizations of representations of the Brauer semigroup, *J. Soviet Math.* **47**, (1989), 2503–2507.

86. Knapp, A. W., *Lie Groups Beyond an Introduction, Second Edition* (Progress in Mathematics **140**), Birkhäuser, Boston, 2002.

87. Kostant, B., A formula for the multiplicity of a weight, *Trans. Amer. Math. Soc.* **93** (1959), 53–73.

88. Kostant, B., Lie group representations on polynomial rings, *Amer. J. Math.* **85** (1963), 327–404.

89. Kostant, B., On the existence and irreducibility of certain series of representations, in *Lie Groups and their Representations* ed. by I. M. Gelfand, Halsted, New York, 1975, pp. 231–329.

90. Kostant, B., and S. Rallis, Orbits and Lie group representations associated to symmetric spaces, *Amer. J. Math.* **93** (1971), 753–809.

91. Kostant, B., and S. Sahi, The Capelli identity, tube domains and the generalized Laplace transform, *Advances in Mathematics* **87** (1991), 71–92.

92. Kraft, H. P., Geometrische Methoden in der Invariententheorie (Aspekte der Mathematik **D1**), Vieweg, Brauschweig/Wiesbaden, 1985.

93. Krämer, M., Sphärische Untergruppen in kompakten zusammenhängenden Liegruppen, *Compositio Math.* **38** (1979), 129–153.

94. Kudla, S., Seesaw dual reductive pairs, in *Automorphic Forms of Several Variables* (Taniguchi Symposium, Katata, 1983), Birkhäuser, Boston, 1983, pp. 244–268.

95. Kulkarni, R. S., On the Bianchi Identities, *Math. Ann.* **199** (1972), 175–204.

96. Lang, S., *Abelian Varieties*, Interscience-Wiley, New York, 1959.

97. Lang, S., *Algebra*, Addison-Wesley, Reading, MA, 1971.

98. Lang, S., *Real and Functional Analysis*, 3rd ed., (Graduate Texts in Mathematics **142**), Springer-Verlag, New York, 1993.

99. Leahy, A. S., A classification of multiplicity-free representations, *J. Lie Th.* **8** (1998), 367–391.

100. Lee, C. Y., On the branching theorem of the symplectic groups, *Canad. Math. Bull.* **17** (1974), 535–545.

101. Lepowsky, J., Representations of semisimple Lie groups and an enveloping algebra decomposition, Ph.D. Thesis, Massachusetts Institute of Technology, 1970.

102. Lepowsky, J., Multiplicity formulas for certain semisimple Lie groups, *Bull. Amer. Math. Soc.* **77** (1971), 601–605.

103. Littelmann, P., A generalization of the Littlewood-Richardson rule, *J. of Algebra* **30** (1990), 328–368.

104. Littlewood, D. E., *The Theory of Group Characters and Matrix Representations of Groups*, Clarendon Press, Oxford, 1940. Second edition: 1950.

105. Littlewood, D. E., and A. R. Richardson, Group characters and algebra, *Philos. Trans. Roy. Soc. London*, Ser. A **233** (1934), 99–142.

106. Macdonald, I. G., *Symmetric Functions and Hall Polynomials* (Oxford mathematical monographs), Clarendon Press, Oxford, 1979. Second edition: 1995.

107. Mackey, G. W., *The Scope and History of Commutative and Noncommutative Harmonic Analysis* (History of Mathematics **5**), American Mathematical Society, Providence, R.I., 1992.

108. Markov, A., Über die freie Äquivalenz der geschlossenen Zöpfe, *Mat. Sb.* **1** (1935), 73–78.

109. Matsushima, Y., Espaces homogènes de Stein des groupes de Lie complexes, *Nagoya Math. J.* **16** (1960), 205–218.

110. McKay, W. G., and J. Patera, *Tables of Dimensions, Indices, and Branching Rules for Representations of Simple Lie Algebras* (Lecture Notes in Pure and Appl. Math. **69**), Marcel Dekker, New York, 1981.

111. Meyer, D., State models for link invariants from the classical Lie groups, in *Knots 90*, Walter de Gruyter & Co., Berlin, 1992, pp. 559–592.

112. Miller, W., A Branching Law for Symplectic Groups, *Pacific J. Math* **16** (1966), 341–346.

113. Morton, H., Threading knot diagrams, *Math. Proc. Cambridge Philos. Soc.* **99** (1986), 247–260.

114. Mostow, G. D., A new proof of É. Cartan's theorem on the topology of semi-simple groups, *Bull. Amer. Math. Soc.* **55** (1949), 969–980.

115. Mumford, D., *Abelian Varieties* (Tata studies in Math., 2nd ed.), Oxford University Press, Oxford, 1975.

116. Murnaghan, F. D., *The Theory of Group Representations*, John Hopkins Press, Baltimore, 1938; reprinted: Dover, New York, 1963.

117. Nomizu, K., On the decomposition of generalized curvature tensor fields, in *Differential Geometry, in honor of K. Yano*, Kinokuniya, Tokyo, 1972, pp. 335–345.

118. Onishchik, A. L., and E. B. Vinberg, *Lie Groups and Algebraic Groups*, Springer-Verlag, Heidelberg, 1990.

119. Peel, M. H., Specht modules and symmetric groups, *J. of Algebra* **36** (1975), 88–97.

120. Popov, V. L., *Groups, Generators, Syzygies, and Orbits in Invariant Theory* (Translations of Mathematical Monographs **100**), American Mathematical Society, Providence, R.I., 1992.

121. Porteous, I. R., *Topological Geometry* (2nd ed.), Cambridge University Press, Cambridge, 1981.

122. Procesi, C., The invariant theory of $n \times n$ matrices, *Advances in Math.* **19** (1976), 306–381.

123. Procesi, C., *Lie Groups: An Approach through Representations and Invariants* (Universitext), Springer, New York, 2007.

124. Richardson, R. W., On orbits of algebraic groups and Lie groups, *Bull. Austral. Math. Soc.* **25** (1982), 1–28.

125. Richardson, R. W., Orbits, invariants, and representations associated to involutions of reductive groups, *Invent. Math.* **66** (1982), 287–313.

126. Rosenlicht, M., On quotient varieties and the affine embedding of certain homogeneous spaces, *Trans. Amer. Math. Soc.* **101** (1961), 211–223.

127. Rossmann, W., *Lie Groups: An Introduction Through Linear Groups*, Oxford University Press, Oxford, 2002.

128. Sagan, B. E., *The Symmetric Group*, Wadsworth & Brooks/Cole, Pacific Grove, CA, 1991. Second edition: Springer-Verlag, New York, 2001.

129. Schur, I., Über eine Klasse von Matrizen, die sich einer gegebenen Matrix zuordnen lassen, Dissertation, 1901; reprinted in *Gesammelte Abhandlungen* **1**, Springer-Verlag, Heidelberg, 1973, pp. 1–72.

130. Schur, I., Neue Begründung der Theorie der Gruppencharacktere, *Sitz. Preuss. Akad. Wiss. Berlin* (1905), 406–432; reprinted in *Gesammelte Abhandlungen* **3**, Springer-Verlag, Heidelberg, 1973, pp. 143–169.

131. Schur, I., Neue Anwendungen der Integralrechnung auf Probleme der Invariantentheorie I, II, III, *Sitz. Preuss. Akad. Wiss. Berlin* (1924), 189–208, 297–321, 346–355; reprinted in *Gesammelte Abhandlungen* **2**, Springer-Verlag, Heidelberg, 1973, pp. 440–494.

132. Schwarz, G. W., Representations of simple Lie groups with a free module of covariants, *Invent. Math.* **50** (1978), 1–12.

133. Schwarz, G. W., On classical invariant theory and binary cubics, *Ann. Inst. Fourier, Grenoble* **37** (1987), 191–216.

134. Shafarevich, I. R., *Basic Algebraic Geometry 1*, 2nd Edition, Springer-Verlag, Berlin, 1994.

135. Specht, D., Die irreduziblen Darstellungen der symmetrischen Gruppe, *Math. Zeitschrift* **39** (1935), 696–711.

136. Springer, T. A., *Linear Algebraic Groups* (Progress in Mathematics **9**), Birkhäuser, Boston, 1981. Second edition: 1998.

137. Stein, E. M., and G. Weiss, *Introduction to Fourier Analysis on Euclidean Spaces*, Princeton University Press, Princeton, 1971.

138. Steinberg, R., Regular elements of semisimple algebraic groups, *Inst. Hautes Études Sci. Publ. Math.* **25** (1965), 49–80.

139. Sternberg, S., *Group Theory and Physics*, Cambridge University Press, Cambridge, 1994.

140. Sundaram, S., On the combinatorics of representations of $\mathbf{Sp}(2n, \mathbb{C})$, Ph.D. Thesis, Massachusetts Institute of Technology, 1986.

141. Tan, E. C., and C. B. Zhu, Poincaré series of holomorphic representations, *Indag. Mathem.*, N.S., **7** (1995), 111-126.

142. Terras, A., *Fourier Analysis on Finite Groups and Applications* (London Mathematical Society student texts **43**), Cambridge University Press, Cambridge, 1999.

143. Tits, J., *Tabellen zu den einfachen Lie Gruppen und ihren Darstellungen* (Lecture Notes in Math. **40**), Springer-Verlag, Berlin, 1967.

144. Turaev, V. G., The Yang-Baxter equation and invariants of links, *Invent. Math.* **92** (1988), 527–553.

145. Turnbull, H. W., Symmetric determinants and the Cayley and Capelli operators, *Proc. Edinburgh Math. Soc.* (2) **8** (1947), 76–86.

146. Vinberg, E. B., The Weyl group of a graded Lie algebra, *Izv. Akad. Nauk SSSR, Ser. Mat.* **40** (1976), 488–526, 709 (Russian). English translation: *Math. USSR Izvestija* **10** (1976), 463–495.

147. Vinberg, E. B., and B. N. Kimelfeld, Homogeneous domains on flag manifolds and spherical subgroups, *Funktsional Anal. i Prilozhen.* **12** (1978), no. 3, 12–19. English translation: *Func. Anal. Appl.* **12** (1978), 168–174.

148. Vinberg, E. B., and V. L. Popov, On a class of quasihomogeneous affine varieties, *Izv. Akad. Nauk SSR* **36** (1972), 749–763. English translation: *Math. USSR Izvestija* **6** (1972), 743–758.

149. Vogan, D., *Representations of Real Reductive Lie Groups* (Progress in Mathematics **15**), Birkhäuser, Boston, 1981.

150. Vust, Th., Opération de groupes réductifs dans un type de cônes presque homogènes, *Bull. Soc. Math. France* **102** (1974), 317–333.

151. Vust, Th., Sur la théorie des invariants des groupes classiques, *Ann. Inst. Fourier, Grenoble* **26** (1976), 1–31.

152. Wallach, N. R., Induced representations of Lie algebras and a theorem of Borel–Weil, *Trans. Amer. Math. Soc.* **136** (1969), 181–187.

153. Wallach, N. R., *Harmonic Analysis on Homogeneous Spaces*, Marcel Dekker, New York, 1973.

154. Wallach, N. R., *Real Reductive Groups I* (Pure and Applied Mathematics **132**), Academic Press, San Diego, 1988.

155. Wallach, N. R., Invariant differential operators on a reductive Lie algebra and Weyl group representations, *J. Amer. Math. Soc.* **6** (1993), 779–816.

156. Wallach, N. R., and Yacobi, O., A multiplicity formula for tensor products of $\mathbf{SL}_2$ modules and an explicit $\mathbf{Sp}_{2n}$ to $\mathbf{Sp}_{2n-2} \times \mathbf{Sp}_2$ branching formula, in *Symmetry in Mathematics and Physics* (Conference in honor of V. S. Varadarajan), Contemp. Math., American Mathematical Society, Providence, R.I., 2009, (to appear).

157. Warner, F. W., *Foundations of Differentiable Manifolds and Lie Groups*, Scott, Foresman, Glenview, Ill, 1971. Reprinted with corrections: Springer-Verlag, New York, 1983.

158. Wehlau, D. L., Constructive invariant theory, in *Proceedings of Symposia in Pure Mathematics* **56**, Part I, American Mathematical Society, Providence, R.I., 1994, pp. 377–383.

159. Weil, A., *Introduction à l'étude des variétés Kähleriennes*, Hermann, Paris, 1958.
160. Wenzl, H., On the structure of Brauer's centralizer algebras, *Annals of Math.* (2) **128** (1988), 173–193.
161. Weyl, H., Reine Infinitesimalgeometrie, *Math. Zeitschrift* **2** (1918), 384–411; reprinted in *Gesammelte Abhandlungen* **2**, Springer-Verlag, Heidelberg, 1968, pp. 1–28.
162. Weyl, H., Theorie der Darstellung kontinuierlicher halbeinfacher Gruppen durch lineare Transformationen. I, II, III und Nachtrag, *Math. Zeitschrift* **23** (1925), 271–309; **24** (1926), 328–376, 377–395, 789–791; reprinted in *Selecta Hermann Weyl*, Birkhäuser Verlag, Basel, 1956, pp. 262–366.
163. Weyl, H., *The Theory of Groups and Quantum Mechanics* (2nd ed., trans. by H. P. Robertson), Dutton, New York, 1932; reprinted: Dover, New York, 1949.
164. Weyl, H., *The Classical Groups, Their Invariants and Representations*, Princeton University Press, Princeton, 1939. Second edition, with supplement, 1946.
165. Whitehead, J. H. C., On the decomposition of an infinitesimal group, *Proc. Cambridge Phil. Soc.* **32** (1936), 229–237; reprinted in *Mathematical Works*, Vol. I, Macmillan, New York, 1963, pp. 281–289.
166. Whitehead, J. H. C., Certain equations in the algebras of a semi-simple infinitesimal group, *Quart. J. Math.* (2) **8** (1937), 220–237; reprinted in *Mathematical Works*, Vol. I, Macmillan, New York, 1963, pp. 291–308.
167. Whippman, M. L., Branching rules for simple Lie groups, *J. Math. Physics* **6** (1965), 1534–1539.
168. Wolf, J. A., *Spaces of Constant Curvature*, 5th ed., Publish or Perish Press, Wilmington, DE, 1984.
169. Young, A., On quantitative substitutional analysis II, *Proc. London Math. Soc.* (1) **34** (1902), 361–397.
170. Zariski, O., and P. Samuel, *Commutative Algebra I*, D. Van Nostrand, Princeton, 1958.
171. Želobenko, D. P., *Compact Lie Groups and Their Representations*, Izdat. "Nauka." Moscow, 1970 (Russian). English translation: Translations of Mathematical Monographs, Vol. 40, American Mathematical Society, Providence, R.I., 1973.

Index of Symbols

A

$\mathbf{A}(k,n)$ (admissble dominant weights)   446
$\mathcal{A}[G]$ (group algebra)   176
$\mathcal{A}[P]_{\text{skew}}$ 342
$\mathcal{A}[P]_{\text{symm}}$ 342
$[a]_+$ 374
ad   54
$\text{Ad}(g)$ 25, 54
$\mathcal{A}\widehat{\otimes}\mathcal{B}$ (skew-commutative tensor product)   667
$\mathcal{A}_{\text{Lie}}$ 668
$\widetilde{\alpha}$ (highest root)   100
Alt (skew symmetrizer)   653
AM_k (skew-symmetric matrices)   240
A_μ 343
A_n (nilpotent Jordan component)   645
$(a)_+$ 381
A_s (semisimple Jordan component)   645
$\text{Aut}(\mathcal{A})$ 46
$\text{Aut}(\mathfrak{g})$ 54, 122
$\mathcal{A}\mathcal{V}_{k,n}$ 562
$A \otimes\!\!\!\wedge\, B$ 456

B

$\mathfrak{B}_k$ (type BC Weyl group)   235, 427
$\mathfrak{B}_k(\varepsilon n)$ 427
B_n (upper-triangular matrices)   36
$B_r(A)$ 20

C

$C(1^{i_1} 2^{i_2} \cdots k^{i_k})$ (conjugacy class)   222
$\mathbf{c}(A)$ (column skew symmetrizer)   409
$C^+(W)$ (even spin space)   314
$C^-(W)$ (odd spin space)   314

$\mathbb{C}[G]^G$ 211
$C_c(M)$ 684
$\text{Cent}_K(A)$ 572, 585
χ_ρ (character of ρ)   211
ch V (character of V)   187
$\text{Cliff}(V, \beta)$ 302
$c_{\mu\nu}^\lambda$ (Littlewood–Richardson coefficent)   403
$\text{Col}(A)$ 408
C^ω 30
$\text{Comm}(\mathcal{S})$ 184
$\text{Conj}(G)$ (conjugacy classes)   211
$C^p(W)$ (p-multilinear skew functions)   304
C_π 166
C_s (tensor invariant)   247
$\text{Curv}(\mathbb{C}^n)$ (curvature tensors)   453

D

$\mathcal{D}(G)$ (commutator subgroup)   519
$\mathcal{D}(\mathfrak{g})$ (derived algebra)   109
$\mathbb{D}(V)$ 278
$(\partial/\partial x_i)_a$ 629
Δ (set of simple roots)   99
Δ_G (Weyl function)   330
Δ_i (ith principal minor)   283, 654
$\text{Der}(\mathcal{A})$ 40
$\text{Der}(\mathcal{O}[G])$ 40
$\det_k$ 406
$\mathcal{D}_f$ (domain of rational function)   622
d_λ (degree of representation)   199
D_n (diagonal matrices)   36, 492
$\mathfrak{D}_n$ (type D Weyl group)   236
$\mathcal{D}^n(G)$ (derived subgroup)   519
$d\varphi$ 30
$\mathcal{D}\mathcal{V}_{k,m,n}$ (determinantal variety)   559

Subject Index

A

adjoint group   490, 497
adjoint representation   25, 54
   differential of   54
   irreducibility of   105
   of subgroup   54
affine algebraic set   612
   affine ring of   612
   ascending chain property   621
   dimension   621
   graph of regular map   627
   group structure on   484
   incontractible decomposition of   618
   irreducible   617
   irreducible components   618
   principal open subset   615
   products of   616
   rational function   622
   reducible   617
   regular functions on   612
   regular map   615
   smooth point   630
   Zariski-closed subset   615
algebra
   automorphism group of   46
   conjugation on   324
   derivation of   630
   real form of   324
algebra of invariants   227
   finite generation of   227
algebraic quotient   516
algebraic torus   39
   centralizer   526
   generator of   71
   maximal   72
   maximally θ anisotropic   571

   rank   39
   rank of   70
   σ-split   541, 543
   θ anisotropic   571
alternating group   219
alternating tensors   653
anti-symmetric   *see* skew-symmetric
anticommutator   260
antipode   192
associative algebra   176
   as Lie algebra   45
   associated graded algebra   662
   filtration on   661
   gradation on   662
   invertible elements in   45
   Lie bracket on   668
   skew-commutative tensor product   667
   tensor product   665
atlas   675
   real analytic   30

B

basic invariants   227
   degrees of   228
basis
   B-symplectic   7
   B-isotropic   646
   pseudo-orthonormal   5, 8
bialgebra   191
Bianchi operator   453
bilinear form   3
   isometry group of   3
   nondegenerate   3
   positive definite   4
   signature of   5
   skew-symmetric   6